D0811341

TRENDS IN ERGONOMICS/ HUMAN FACTORS V

TRENDS IN ERGONOMICS/ HUMAN FACTORS VOLUMES

General Editors

Anil MITAL

Ergonomics Research Laboratory
Department of Mechanical and Industrial Engineering
University of Cincinnati
Cincinnati, Ohio, U.S.A.

NORTH-HOLLAND
AMSTERDAM • NEW YORK • OXFORD • TOKYO

TRENDS IN ERGONOMICS/ HUMAN FACTORS V

Proceedings of the Annual International
Industrial Ergonomics and Safety Conference
held in New Orleans, Louisiana, 8–10 June, 1988

The Official Conference of the International Foundation
for Industrial Ergonomics and Safety Research

Edited by

Fereydoun AGHAZADEH
Industrial Engineering Department
Louisiana State University
Baton Rouge, Louisiana, U.S.A.

1988

NORTH-HOLLAND
AMSTERDAN • NEW YORK • OXFORD • TOKYO

ISBN: 0 444 70442 6

Published by:

ELSEVIER SCIENCE PUBLISHERS B.V.
P.O. Box 1991
1000 BZ Amsterdam
The Netherlands

Sole distributors for the U.S.A. and Canada:

ELSEVIER SCIENCE PUBLISHING COMPANY, INC.
52 Vanderbilt Avenue
New York, N.Y. 10017
U.S.A.

PRINTED IN THE NETHERLANDS

To Mitra, Sanaz, and Monty

PREFACE

Even though ergonomics as a discipline has been in existence for over a century, it has been applied extensively only in the past few decades. As a result, productivity has increased, injuries have decreased, and jobs and workplaces have become more acceptable to the workers. In this context scientists and practitioners were invited to present and share their findings and experiences at the International Industrial Ergonomics and Safety Conference, which was held in New Orleans, Louisiana on June 8–10, 1988. Members of an International Program Advisory Board reviewed the submitted papers and accepted a limited number for presentation at the conference and for inclusion in this volume.

The volume contains 133 papers authored by 237 international researchers and practitioners. The papers discuss a wide range of subjects from basic research to industrial applications. The volume should be of value to scientists, educators, and practitioners of ergonomics/human factors and safety.

I would like to express my sincere gratitude to the authors for their contributions and to the publishers for their cooperation in publishing this volume.

Many individuals have contributed to the success of this project. Members of the Program Advisory Board offered me valuable suggestions and spent their invaluable time to review the articles. Dr. Edward McLaughlin, Dean of Engineering, and Dr. Lawrence Mann, Chairman of the Industrial Engineering Department at Louisiana State University encouraged me and provided me with the necessary resources. Mrs. Terri Wagner assisted me in the preparation of the final manuscript, I am indebted to all of them.

Special thanks go to my wife, Mitra, for her support, understanding, and patience.

Fereydoun Aghazadeh

Baton Rouge, Louisiana, U.S.A.
June, 1988

CONTENTS

Dedication v

Preface vii

I. METHODOLOGY

A Database Management System for Ergonomics Information Analysis
J.-G. Chen, J.B. Peacock, and R.E. Schlegel 3

Analysis of "Error" Variances
A.C. Bittner, Jr. and S.J. Morrissey 15

Statistical Isolation of Timesharing Components:
Demonstration and Interpretation
A.C. Bittner, Jr. and D.L. Damos 21

Random Sampling of Domain Variance (RSDV)
R.J. Wherry, Jr. 29

Human Reliability: A THERP Application at Automatic Storage and
Retrieval System
B.M. Pulat 37

The Importance of Ergonomics/Human Factors Consideration in the
Development of Expert Systems
A.J. Gonzalez and C.H. Lee 45

Ergon-Expert – A Knowledge-Based Approach to the Design of Workplaces
V. Rombach and W. Laurig 53

An Analysis of Human Motions Using Micro Computers
K.B. Lee, S.H. Oh, S.Y. Choi, S.D. Lee, J.H. Chung, and K.D. Lee 63

Human–Machine Modeling with AutoCAD
O.K. Eyada, J.E. Fernandez, R.J. Marley, and T.B. DeGreve 71

Including Anthropometry into the "AutoCAD"-Microcomputer System for Aiding Engineering Drafting
J. Grobelny 77

The Effects of Input Devices on Task Performance
J.E. Fernandez, M. Cihangirli, D.L. Hommertzheim, and I. Sabuncuoglu 83

II. COGNITIVE ERGONOMICS

Is there a Relationship between Working Memory Capacity and Hemispheric Specialization?
M.L. Turner 93

A Comparison of Three Display Formats for Multi-Dimensional Systems
J.-G. Chen, D. Deal, and V. Jeyakumar 101

Redundant Information as a Means of Aiding Performance in Human–Machine Interface
L.-E. Warg 109

Application of Technical Writing Principles to Questionnaire Design
G.E. Fey 115

III. HUMAN PERFORMANCE

Production Standards and Feedback as Determinants of Worker Productivity and Satisfaction in a Repetitive Manufacturing Task
B. Das and A.A. Shikdar 127

A Graphical Representation of Group Decision Performance
V. Dubrovsky 135

A Probability Model for Group Inspection
A.M. Waikar, F. Aghazadeh, and K.S. Lee 143

Evaluation of Two Work Schedules in a Mining Operation
J.C. Duchon 151

A Methodology for Assigning Variable Relaxation Allowances: Visual Strain, Illumination, and Mental Strain
J.H. Goldberg and A. Freivalds 161

Dual Roles of Visual Feedback on Postural Stability
D. Shuman and K.D. White 169

The Effects of Differing Levels of Demand on the Performance of Executive Work
P.R. McCright 177

Aging and the Performance of Computer-Interactive Tasks: Job Design and Stress Potential
S.J. Czaja and J. Sharit 185

The Effects of Instruction on Finger Strength Measurements: Applicability of the Caldwell Regimen
V.J. Berg, D.J. Clay, F.A. Fathallah, and V.L. Higginbotham 191

IV. HUMAN–MACHINE INTERACTION

Design of Interfaces of Hand-Held, Two-Wheeled Devices
H.-J. Bullinger, W.F. Muntzinger, and R. Eckert 201

Ergonomics of Powered Hand Tools on Assembly Line Work
M. Rauko, S. Herranen, and M. Vuori 211

An Investigation of Quantitative Measures of Joystick Control
K. Behbehani, S.N. Imrhan, and G.V. Kondraske 219

Elbow Flexion Strength of Young Females
L.J. Robinson-Edwin, S. Kumar, and H. Clarkson 227

A Comparison of the Results of Recent Studies in Wrist-Twisting Strength
S.N. Imrhan 235

The Effects of Spatial Ability on Automobile Navigation
J.F. Antin, T.A. Dingus, M.C. Hulse, and W.W. Wierwille 241

An Ergonomic Evaluation of the Farm Tractor
J.M. Usher, F. Aghazadeh, and R. Azimullah 249

Ergonomics of Mechanical Transplanting
M.E. Wright, T.R. Way, and M.D. Miller 257

V. HUMAN–COMPUTER INTERFACE

Evaluation of Ocular and Musculoskeletal Subjective Discomfort Score Responses from VDT Operators
H.T. Zwahlen and C.C. Adams, Jr. 267

Inter-Ocular Correlation: Spatial Factors
C.B. Woods, K.D. White, and J.H. Krantz 275

Changes in Temporal Instability of Lateral and Vertical Phorias of the VDT Operators
T. Marek, C. Noworol, W. Pieczonka-Osikowska, J. Przetacznik, and W. Karwowski 283

Effects of Message Type, Screen Message Location, and Rate of Presentation on VDU Message Legibility
S.J. Morrissey 291

The Effect of CRT Quality on Visual Fatigue
M. Miyao, J.S. Allen, S.S. Hacisalihzade, S.A. Cronin, and L.W. Stark 297

Visual Strain Evaluation of VDT Operators Using a Laser Optometer
K.S. Lee, A.M. Waikar, and O. Ostberg 305

Ergonomic Seats for Computer Workstations
K.H.E. Kroemer 313

A Keyboard to Increase Productivity and Reduce Postural Stress
S.W. Hobday 321

VI. WORKPLACE DESIGN

Ergonomic Applications and Control of Chemical Aerosols
J.D. McGlothlin, P.A. Jensen, M.G. Gressel, and W.A. Heitbrink 333

Ergonomic Evaluation of Cleanroom Environments
J.D. Ramsey, J.L. Smith, and Y.G. Kwon 345

Estimation of an Operator's Body Posture in Sagittal and Frontal Planes
E. Chlebicka 351

Ergonomically Designed Chemical Plant Control Room
R.B. Combs and F. Aghazadeh 357

An Effective Methodology to the Design of Work Environment for a Fast-Food Restaurant
S.E. Martinez, K. Kengskool, and J.C. Valdes 365

Human Issues in Automated (Hybrid) Factories
A. Mital and L.J. George 373

System Reliability with Human Components
C.M. Klein and J.A. Ventura 379

Self-Reports in Ergonomics: Agreement between Workers and Ergonomist
K. Woodcock Webb 387

An Ethnic Anthropometric Survey as an Educational Tool
D.E. Malzahn, J.E. Fernandez and C.-H. Kim 395

The Effects of Floor Types on Standing Tolerance in Industry
M.S. Redfern and D.B. Chaffin 401

Walking Condition and Floor Type Effects on Kinematic Variables of Gait
M.A. El-Nawawi, S.B. Moshref, and A.A. Zeinelabidien 407

Evaluation of Office Chairs: A Validation Study
R.R. Bishu, D.J. Cochran and M.W. Riley 417

Double Curvature Backrest: Results of a Pilot Study
A. Mital 425

VII. ENVIRONMENTAL STRESSES

Comparisons of Air vs. Liquid Microenvironmental Cooling for Persons Performing Work while Wearing Protective Clothing
P.A. Bishop, S.A. Nunneley J.R. Garza, and S.H. Constable 433

Biophysical Evaluation of Handwear for Cold Weather Use by Petroleum (POL) Handlers
W.R. Santee, T.L. Endrusick, and L. Wells 441

The Role of Textile Material in Clothing on Thermoregulatory Responses to Intermittent Exercise
R. Nielsen and T.L. Endrusick 449

A Methodology for Assigning Variable Relaxation Allowances: Manual Work and Environmental Conditions
A. Freivalds and J.H. Goldberg 457

A Subjective Evaluation of a Microclimate Cooling System
F. Tayyari 465

An Efficient Method of Verifying Heat Stress Problems in Industry
D.C. Alexander and L.A. Smith 471

Thermal Characteristics of Surgical Gown Materials
R.E. Schlegel, S.D. Cheatwood, and B. Schmidt 479

Thermal Comfort Simulation Tests for the Improvement of a Hot Environment in a Glasswork
L. Banhidi, L. Fabó, and J. Szerdahelyi 485

VIII. EFFECTS OF NOISE AND VIBRATION ON PERFORMANCE

Aspects of Consensus Standards Development in Biodynamics
J.C. Guignard 495

Combined Effects of Whole Body Vibration and Noise on Performance in a Multiple Reaction-Time-Task
G. Notbohm and E. Gros 505

Effects of Mechanical Vibration on Peripheral Body Temperature and the Heat Emission Pattern of the Hand
E. Abdel-Moty and T.M. Khalil 513

Pilot Subject Evaluation of Whole-Body Vibration from an Underground Mine Haulage Vehicle
T.G. Bobick, S. Gallagher, and R.L. Unger 521

Shipboard Evaluation of Motion Sickness Incidence
A.C. Bittner, Jr. and J.C. Guignard 529

Retrofit Noise Control Modifications for an Underground Mine Haulage Vehicle
T.G. Bobick 541

Vibration and Noise Problems Associated with Gardening Equipment
K.S. Lee, F. Aghazadeh, and A.M. Waikar 549

IX. SAFETY MANAGEMENT AND INJURY CONTROL

Job Load and Hazard Analysis: A Co-operative Approach to Identify and to Prevent Ergonomic and Other Hazards at Work
M. Mattila 559

Effective Group Routines for Improving Accident Prevention Activities and Accident Statistics
N. Carter and E. Menckel 567

Evaluation of Safety Attributes in Access Systems
K. Häkkinen, J. Pesonen, and S.Väyrynen 573

Pitfalls in the Use of Accident Data to Focus Ergonomic Attention
K. Woodcock Webb 583

Shiftwork and Safety: A Review of the Literature and Recent Research Results
J.A. Wagner 591

Prevention Strategies Adopted by Select Countries for Work-Related Musculoskeletal Disorders from Repetitive Trauma
V. Putz-Anderson 601

Robots as an Ergonomic Control to Musculoskeletal Injuries in the Workplace
A. Genaidy, G. Roncini, R. Dawood, and A. Mital 613

Unsafe Behavioral Responses of Agricultural Equipment Operators to Interruption of Machine Function
R.L. Hull 621

Factors Reported in Amputation and Other Injury Cases at Mechanical Power Presses between 1976 and 1984
J.R. Etherton 629

Comparative Hazardousness of Metalworking Machines
R.C. Jensen and J.R. Etherton 639

Safety Perceptions and Use Patterns for ATVs and Other Motorized Recreational Vehicles
E.W. Karnes, S.D. Leonard, T. Schneider, W. Pedigo, D. Krupa, and E. Madigan 647

Improved Safety Directions: Safer Work with Grinding Machines
A. Seppälä 657

Occupational Hazards Involved with Chimney Sweeping
P. Tiitta 663

Scale Values for Warning Symbols and Words
S.D. Leonard, E.W. Karnes, and T. Schneider 669

Safety Issues Relating to Agricultural Machines Modified for Disabled Operators
T.L. Wilkinson and W.E. Field 675

Age Related Differences in Injuries in the Home
S.N. Imrhan 683

Aging and Environmental Safety
P.K.H. Kim 689

Management of Carpal Tunnel Syndrome
B.T. Harter, Jr., K.C. Harter, and F.W. Archer 699

X. CONTROL OF BACK INJURIES

Comprehensive Back Injury Prevention Program: An Ergonomics Approach for Controlling Back Injuries in Health Care Facilities
J.W. Aird, P. Nyran, and G. Roberts 705

Sudden-Movement/Unexpected Loading as a Factor in Back Injuries
T.J. Stobbe and R.W. Plummer 713

Patient Handling Devices: An Ergonomic Approach to Lifting Patients
B.D. Owen 721

Hospital Bed Design and Operation – Effect on Incidence of Low Back Injuries among Nursing Personnel
D. Nestor 729

Which Subpopulations of the Mining Industry are at a Higher- or Lower-than-Average Risk for Back Injury Problems?
S.J. Butani 741

XI. OCCUPATIONAL BIOMECHANICS AND STRENGTH MEASUREMENT

Ergonomics, Human Rights and Placement Tests for Physically Demanding Work
R.D.G. Webb and D.W. Tack 751

A Multivariate Analysis of Directional Movement Time
D.E. Malzahn, J.E. Fernandez, R.J. Marley, and J. Dahalan 759

Prediction of Maximum Dynamic Strength from Multiple Repetitions with a Submaximal Load
T.L. Doolittle and K. Kaiyala 767

Distortion of True Body Weight and Stature from Self Estimation
S.N. Imrhan, V. Imrhan, and R. Pin 775

Repeatability of Static and Isokinetic Maximum Voluntary Back Strength Exertions 779
S. Gallagher

Strength Testing May Be an Effective Placement Tool for the Railroad Industry
P.B. McMahan 787

Changes in Postural Stability, Performance, Perceived Exertion and Discomfort with Manipulative Activity in a Sustained Stooped Posture
R. Wickstrom, A. Bhattacharya, and R. Shukla 795

A Comparison between Isokinetic Trunk Strength and Standard Static Strength Tests
J.W. Yates and W. Karwowski 803

The Effect of Starting Position and Speed of Movement on Maximum Strength in Dynamic Strength Testing
E. Asoudegi, T. Stobbe, and M. Jaraiedi 811

Influence of Workload and Anthropometric Measures on Musculoskeletal Complaints
R. Paluch and A. Łazor 821

Effects of Handle Length and Bolt Orientation on Torque Strength Applied during Simulated Maintenance Tasks
S. Deivanayagam and T. Weaver 827

XII. MANUAL MATERIALS HANDLING

The Psychophysical Approach: The Valid Measure of Lifting Capacity
J.E. Fernandez and M.M. Ayoub 837

Systems Comparison of Two Advanced Methods for Measuring Angular Displacement of Torso
I. Gilad, D.B. Chaffin, M. Redfern, and S.N. Byun 847

Transfer Functions of Trunk Muscle Extensions during Motion
W.S. Marras 857

Subjective Judgement of Load Heaviness and Psychophysical Approach to Manual Lifting
W. Karwowski and A. Burkhardt 865

Psychophysical and Physiological Responses to Asymmetric Lifting
A. Garg and J. Banaag 871

Dynamic Biomechanical Model for Asymmetrical Lifting
H.C. Chen and M.M. Ayoub 879

The Validity of Predetermined Motion Time Systems in Setting Work Standards for Manual Materials Handling
A.M. Genaidy, A. Mital, M. Obeidat, and S. Puppala 887

The Comparison of the Arm Crank and Cycle Exercise Tests for the Prediction of the Aerobic and Anaerobic Work Performance in Manual Material Handling Tasks
V. Louhevaara and P. Teråslinna 895

Does the Advantage of Lordotic Posture over Straight Back Posture Translate into Extra Lifting Capacity?
A. Mital 903

XIII. WORK PHYSIOLOGY

On the Development of an Index of Physical Fitness
S. Nanthavanij and H. Gage 911

Isometric–Isotonic Effort and Transcutaneous O_2, CO_2, Heat
C.A. Cacha 919

Effects of Posture on the Metabolic Expenditure Required to Lift a 50-Pound Box
S. Gallagher and T.G. Bobick 927

Oxygen Consumption, Heart Rate, and Perceived Exertion during Walking in Snow with Boots of Differing Weights
J. Smolander, V. Louhevaara, T. Hakola, E. Ahonen, and T. Klen 935

Evaluating the Cardiovascular Fitness of Downs Syndrome Individuals
K.H. Pitetti, J.E. Fernandez, J.A. Stafford, and N.B. Stubbs 941

Physiological Responses while Playing a Video Game
J.E. Fernandez, A.D. Akin, C.L. Collins, and J.F. Virgilio 949

XIV. ERGONOMICS IN REHABILITATION

Ergonomics Considerations for the Reduction of Physical Task Demands of Low Back Pain Patients
E. Abdel-Moty, T.M. Khalil, S.S. Asfour, R.S. Rosomoff, and H.L. Rosomoff 959

Psychological Indicators of Recovery from Back Pain
M.J. Colligan, E.F. Krieg, S.E. Besing, and T. Bennett 969

Effectiveness of Agressive Treatment of Back Pain
T.M. Khalil, S.S. Asfour, S.M. Waly, R.S. Rosomoff, and H.L. Rosomoff 977

Returning to Work: The Need to Measure Physical Abilities
W.K. Stoeffler, A.M. Genaidy, and D.M. Lyth 985

Industrial Safety Programs in Sheltered Workshops
P.C. Witbeck and G.R. Simons 993

The Cardiovascular Fitness of Non-Downs Syndrome, Moderately Mentally Retarded Individuals as an Additional Indice for Job Placement
K.H. Pitetti, J.E. Fernandez, D.C. Pizarro, N.B. Stubbs, and J.A. Stafford 999

XV. INDUSTRIAL APPLICATIONS

Committee Approach to Ergonomic Cases
F.T. Doxie 1009

Implementing Ergonomics Projects
J.L. Wick 1017

Practical Ergonomic Application
F.L. Mc Atee, Jr. 1023

The Validity of Using Existing Ergonomic Norms in Industrial Work Design – Some Case Studies
S.P. Dutta 1027

Practical Applications in Ergonomics
M.P. Klym 1037

An Ergonomic Evaluation of a Steel Tube Furniture Manufacturer: Assembly Area Assessment and Recommendations
L.J.H. Schulze and J.J. Congleton 1045

An Ergonomic Evaluation of a Steel Tube Furniture Manufacturer: Warehouse Assessment and Recommendations
L.J.H. Schulze and J.J. Congleton 1053

An Ergonomic Investigation of Railroad Yard Worker Tasks
S.R. Kuciemba, G.B. Page, and C.J. Kerk 1061

Workload Reduction for Drivers of Heavy Shovel Loaders in Underground Mining
D. Lorenz, W.F. Muntzinger, and M. Dangelmaier 1069

Financial Trader Workstation
J. Holenstein 1077

XVI. APPENDIX

Constitution of International Foundation for Industrial Ergonomics and Safety Research 1085

XVII. AUTHOR INDEX

Author Index 1093

I

METHODOLOGY

Trends in Ergonomics/Human Factors V
F. Aghazadeh (Editor)
© Elsevier Science Publishers B.V. (North-Holland), 1988

A DATABASE MANAGEMENT SYSTEM FOR ERGONOMICS INFORMATION ANALYSIS

Jen-Gwo Chen,[1] J. Brian Peacock,[2] Robert E. Schlegel[3]

[1] Department of Industrial Engineering
University of Houston
Houston, Texas 77004

[2] CPC Group, Engineering North
General Motors Corporation
Pontiac, Michigan 48058-1493

[3] School of Industrial Engineering
University of Oklahoma
Norman, Oklahoma 73019

The database management system for ergonomics information analysis using dBASE III Plus has been developed to provide self-evaluation for the worker, job monitoring and analysis for the supervisor, and necessary information for the ergonomist. The implementation of relational database management system in this paper avoids the present expert system-database interface difficulties. The system includes the screen, guideline and analysis modules. A 5-point scale is adapted to allow the user to screen the existing job with the Ergonomics Information Analysis System (EIAS) checklist in the screen module. The general ergonomics design guideline module can be used as an on-the-job training tool. Finally, a job analysis comparison can also be performed between several different profiles to develop a better job design.

1. INTRODUCTION

Like many professions, ergonomics has now reached a stage where its techniques are being applied by practitioners who have not had the benefit of substantial education and training in the scientific basis. Clearly, the danger lies in the mis-application of the techniques through inadequate analysis of problems and improper selection and implementation of solutions.

There are many levels of problem analysis ranging from simple observation to indepth investigation involving instrumental and statistical measurements. The principal advantage of the casual observation is that it is quick and often results in a concensus. However unsystematic observations may produce non representative results. The checklist approach has the objective of guiding the observer to the scope of an investigation, the ranges of expected observations and the spectrum of possible remedies. The checklist can also form the basis for routine data capture. This approach is particularly pertient in situations where individual observers may converge to a limited set of possibilities due to their training and experience, through an expansion of the scope of both the observations and the remedies.Many ergonomics checklists have been developed for both general and specific purposes [1,2,3]. The one described in this paper has been developed for application in automobile assembly plant [4].

One difficulty with checklists is that ergonomics knowledge may be too broad to apply for practitioners without access to advise from trained experts. Furthermore, the ergonomics information might come from different departments and could be used and updated by different users. Good data protection and easy access to the system are two major considerations. The application of an existing commercial database management system is one answer to this problem. However, the application of an expert system and database management system is expected to become one of the most important application in the next decade [5]. This paper presents a methodology for the expert database system development in ergonomics.

2. CONCEPTUAL OUTLINE OF ERGONOMICS INFORMATION ANALYSIS SYSTEM

The Ergonomics Information Analysis System (EIAS) takes two forms. The short form is largely qualitative in that the ergonomist is asked to rate, on a 5-point category-ratio scale, the seriousness of the each aspect of the problem and to record some reference data. This information should be readily generated by the ergonomist at a problem site, with the aid of his/her checklist or through self-evaluation by an operator. The system also provides for the collection of reference, followup and remedy evaluation data. This qualitative information, when accumulated, will help to identify the predominant problem areas and probable remedy effectiveness.

Where a serious deficiency is identified in a particular aspect of a case, the investigator is directed to the quantitative details that should be measured. This long form of data collection provides the opportunity for both routine and discretionary quantitative item acquisition.

The analysis of this data will include the selection and enumeration of records that have prescribed characteristics, such as all records involving powered hand tool in the trim shop that were associated with wrist problems. A second level of report will include the creation of cross tabulations from which associations between predictor and response variables are identified. For example, the association between back injuries and working posture in the sealer line could be assessed. Finally, quantitative variables could be related through regression methods, as in the case of time on the job and repetitive motion trauma incidence. A general design guideline will help the user to retrieve and to learn ergonomics principles leading to continuous improvements. Possible causes and suggestions related to the contributing factors may be stored in the database to broaden the scope of the practitioner's armory.

The theoretical basis of the short screening checklist is the "range-theory" proposed by Borg [6]. In brief, this theory states that "all individuals perceive roughly the same degree of exertion while performing dynamic work at their respective maximal physical capacity." Oborne [7], using this scale, found that all individuals determined almost the same comfort zone in whole body vibration. Ljunggren [8] stated that the category-ratio is a reliable tool for making self-ratings. Other research results [9,10] have shown that observer self-rating has a linear relationship with heart rate via the category-ratio scale.

The EIAS adapted a 5-point category-ratio scale to rate task components since the scale provided reliable information in physical work stress self-evaluation. The EIAS provides general design guidelines to tell the user how to avoid unnecessary problems and to improve performance. Possible factors contributing to the problem are listed. Possible causes and suggestions are provided based through interactive communications and production rules generated from well established ergonomics sources.

3. ERGONOMICS INFORMATION ANALYSIS SYSTEM COMPONENTS

The EIAS checklist include five sections: case identification, problem description, job description, operator/operation interaction and remedy. The checklist is shown in the appendix.

A. Case identification section

This section deals with the identification of a case with regard to its date and time of occurence and a sequential index number with any calendar year. General personal and anthropometric data such as age, gender, height and weight, and case reference are recorded in this section. Referral information is recorded for keeping track of a particular case and for analyzing the interaction between the problem indentifiers and problem solvers. The other records from different departments or other record systems within the same department, such as photograph and video, are also indicated in this section.

B. Problem description section

The problem in the work area may relate to an injury, poor product quality, absenteeism, high turnover or complaints. The last three problems are indicators of poor quality of work life. Injury severity, body parts involved, injury types and causes provide the basic information for accident investigation. The human causes of low product quality may involve a lack of understanding of the required quality level, inappropriate feedback, lack of skill or lack of motivation. The result of these human deficiencies may be the performance of an incorrect or inaccurate operation or the omission of an operation. Commonly such errors may be precipitated or compounded by a problem with a previous operation, the incoming quality of a part, an unsharpened or poorly adjusted tool or the nature of the workplace.

C. Job description section

The job description section is designed for data regarding the job organization, environment, workplace, parts and tools. The job organization refers to the level of complexity, repetition rate of the job, social interaction between workers, and satisfaction from the job itself or the environment.The job environment subsection records possible physical factors that contribute to the problem, such as noise, light, temperature and ventilation. Workplace characteristics are described in terms of reach, fit and obstacles. An example of an easy reach and fit is a small part assembled on a large horizontal surface at about elbow height. A difficult reach is found where the operator must bend and work under the work surface. Fit problems are identified where the operator must reach through narrow openings, work inside a vehicle compartment or in a confined space such as a pit. The presence of obstacles in the workplace such as steps, tools, parts or carriers should be identified and rated. Part size, weight and material are rated for further ergonomics investigation. Under tool classification, handling, vibration and weight may be the essential contributory factors to a problem.

D. Operator/operation interaction section

A coarse indication of postural factors can be obtained by ranking the static positions of the base, legs, trunk, shoulders, neck, arm, wrists and hands. For example, a wide stance with a flexed trunk and the head bent to one side as is seen in some sealer and welding operations would impose considerable postural stress. High static loads from holding a heavy part or tool overhead or at arm's length would add to this stress. High dynamic loads involving large ranges of movement may produce both biomechanical and physiological stresses. The heart rate should be recorded in this case as a complement to the physical stress evaluation. The information load is related to the complexity of the job in terms of sensory, attention, perceptual, decision making, problem solving or motor skills required. An example of high information load may occur where there is large product differentiation, narrow tolerances and relatively inexperienced operators.

E. Remedy section

It is appropriate in any analysis and intervention system of this nature to document the changes that are recommended or implemented. Thus, an objective evaluation can be made of their effectiveness and efficiency. The changes may involve engineering modifications, placement, job organization,
training/methods improvement, or the use of personal protective equipment. An extention of the placement process is the selection and assignment of operators for jobs based on their non-pathological characteristics. For example, people may all be assigned to jobs that suit their inherent or acquired characteristics such as size, strength, dexterity, vision or experience.

4. INDEPTH EIAS ANALYSIS

More detailed information should be recorded through the measurement of those potential contributory factors or those information already available. For example, quantitative descriptions of the workplace (Figure 1) involve measurement of the vertical and horizontal distance of the job, parts stack and tool using the feet of the operator at the time of the actual operation as a reference. Where the operator moves with the product then his/her foot position at the final part of the cycle could be used as a reference. The job location is determined as the position of the operator's hands in contact with the tool or part as the operation is performed. Also the orientation of the part of the product on which the job is being performed should be identified using the operator's station as he/she waits for the car as a reference. A more detailed quantitative description of the part would include its spatial dimensions, its weight and its material.

Another example of indepth analysis is the remedy section, it could be appropriate to classify the details of engineering intervention as shown in Figure 1 for engineering modification. A set of production rules is implemented to provide the user with possible causes, suggestions and comments through the indepth quantitative assessment and interactive dialogue. The production rules for the EIAS are based on a checklist to investigate potential problems. This is a prototype knowledge base for ergonomics analysis. As an example, rule 13 in the job description section is given below:

•
•

3-3 Job Workplace
Job Vertical Height
Job Horizontal Distance
Parts Vertical Height
Parts Horizontal Distance
Tool Verticla Height
Tool Horizontal Distance
Walking Distance (steps)
Vertical Distance Moved
Job Orientation -upper surface, lower surface, left side, right side, front surface, rear surface

3-4 Parts
Length
Depth
Weight
Sharp Edges
Material - sheet metal, sealer, paint, cast/machined metal, fiberglass or plastic

•
•

5-1-1 Environment
Thermal - lower temperature, higher temperature, air movement increase, air movement decrease, radiant heat shield
Noise - process change, process isolation, baffels
Light - higher intensity, lower intensity, location, color characteristics, local lighting

5-1-2 Workplace - vehicle, parts stock, platcform, pit, obstruction, tool location, information location, materials handling equipment

5-1-3 Tool - torque rating, handle length, handle diameter, weight, handle angle and trigger type

5-1-4 Parts - size, material and weight

•
•

Figure 1. An Example of EIAS Long Form.

Rule 13 - to identify possible causes for tool handling problem
IF (1) rating of tool handle exceeds safety threshold
IF (1) it is difficult to hold the handle
(2) rating of part material exceeds safety threshold
THEN (a) Handle size is inadequate
(b) Redesign tool handle based on hand size
ELSE (a) Handle shape is inadequate
(b) A greater force-bearing area can reduce the hazard
ELSE (a) Lack of handle
(b) Provide appropriate handle
ELSE (a) Tool handle is appropriate

5. GENERAL ERGONOMICS DESIGN GUIDELINE MODULE

General ergonomics design guidelines are provided in this module to teach the user about the basic concept of ergonomics during the job design stage. The guidelines for job environment is an example:

General Design Guideline for Job Environment

For Noise
1. Reduce noise by controlling and isolating noise source
2. Improve hearing by providing appropriate hearing aids
3. Keep noise level equal or below 80 db

For Lighting
1. Eliminate bright light source in the visual field
2. Provide shades or diffusers for all lights
3. The angle between the horizontal and the line from the eye to an overhead lamp should be more than 30 degrees
4. Several low power lamps are better than a high power lamp
5. Avoid reflective colors or materials on machine table tops, switch panels, etc.

For Thermal
1. Recommended room temperatures are:
Sedentary/mental job - 20 ^{o}C
Sedentary light manual job - 19 ^{o}C
Standing light manual job - 18 ^{o}C
Standing heavy manual job - 17 ^{o}C
Severe work - 15 or 16 ^{o}C
2. Relative humidity should not fall below 30% in winter and 40 - 60% in summer

For Ventilaiton
1. Open window in summer if no artificial ventilation
2. Avoid smoking in work area
3. Provide air-conditioning if temperature exceeds 32 ^{o}C

Clearly these rules would be modified or expanded as the knowledge base is developed.

6. SOFTWARE STRUCTURE OF ERGONOMICS INFORMATION ANALYSIS SYSTEM

The system includes three modules: short checklist, long form analysis and design guideline (Figure 2). There may be various interrelationships between the ergonomics information sources. One advantage of a relational database management system is that the relationship between the data is not stored with the data. Thus, the system allows the user to access the data in its natural form without constraints of a preconceived structure. Three different variables are used in this system: numeric, character and data. The unique filename through user's last name initial and last four digits for worker's social security number is the key for each separate database. The qualitative data is controlled by the specific dataname in the checklist. The system allows the user to edit and to update the data through that specific filename and dataname. The hierarchial program structure allows the user to access the desirable module through a control menu. Depending on the user's choice, it will branch to either CLIST.PRG to manage the EIAS checklist, DLIST.PRG to long form quantitative data collection, GLINE.PRG to access the general design guideline and EIASA.PRG to EIAS analysis (Figure 3). E Each module also has independent database structures for each section. The filename and dataname are the connection between these modules. For example, five database structures are created in CLIST.PRG corresponding with the five sections mentioned previously.

7. DISCUSSION

The EIAS checklist serves multiple purposes; i.e., guiding and documenting ergonomics data aquisition problem identificaiton, job screening and monitoring, job design etc. The indepth analysis and production rules help the user to look into the problem in more detail and find out the possible causes. The general design guidelines provide an educational tool for the user for basic ergonomics training.

The combination of the commercial database management system and ergonomics principles provide a basis for an ergonomics expert system development to help the practitioner to perform their job more effectively. Another advantage of this system is to allow different authorized users to share a common database and to provide information for further analysis.

When fully developed through modification and expansion of the modification rules and through refinement of the control structure this system will have the characteristics of a widely usable expert
system. As such it will aid in routine task analysis, problem identification and treatment, and reserach.

8. REFERENCES

[1] Rohmert, W., AET - A New Job-Analysis Method, **Ergonomics,** 28(1) (1985) 245-254.

[2] Calisto, G.W., Jiang, B.C. and Cheng, S.H., A Checklist for Carpal Tunnel Syndrome, **Proceedings of the 30th Human Factors Society Meeting**, 1438-1442 (1986).

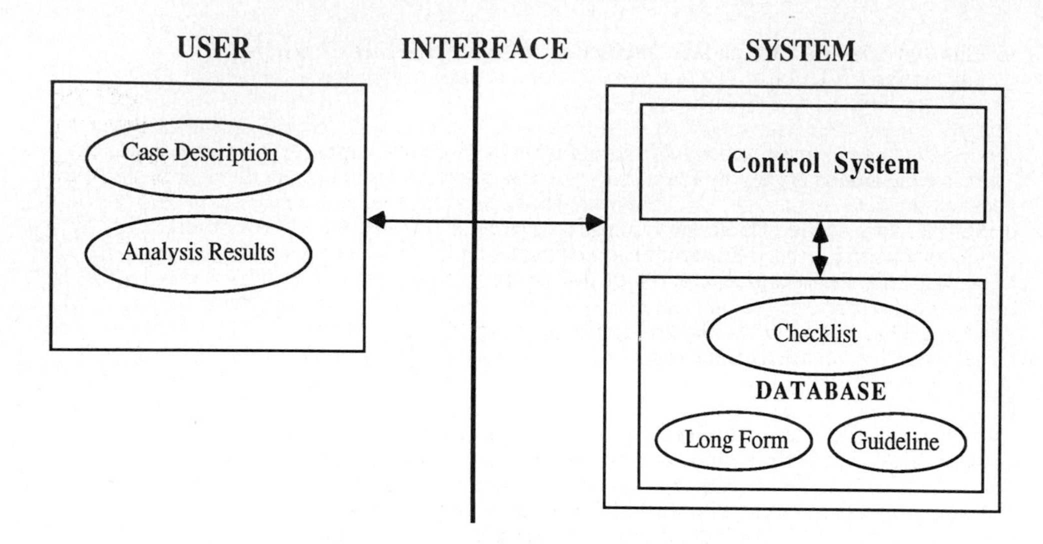

Figure 2. Diagram of System Structure.

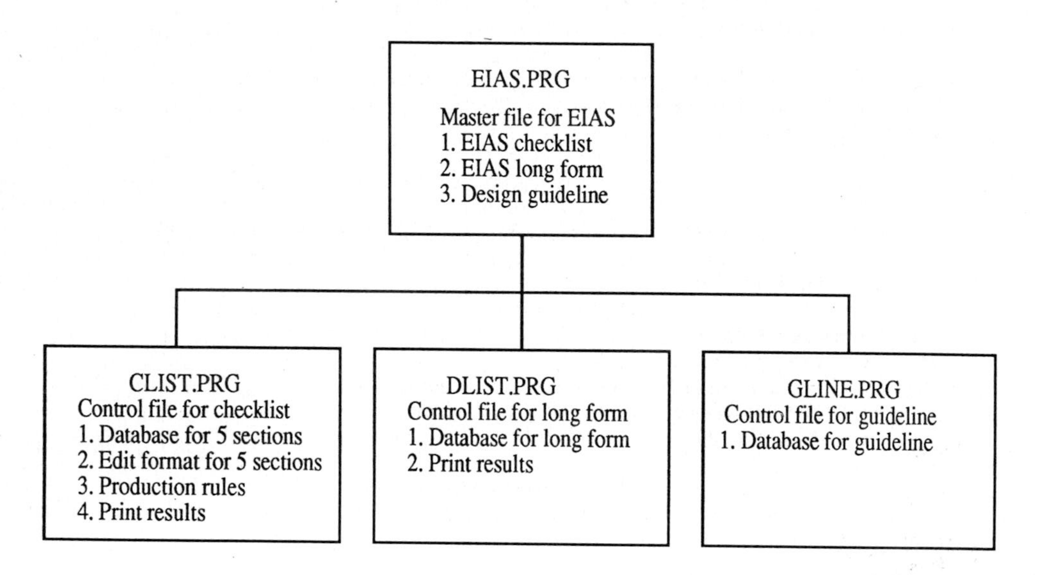

Figure 3. Basic Software Structure for EIAS.

[3] Edosomwan, J.A., Ergonomic Issue in Computer-Aided Manufacturing, in: Asfour, S.S., (ed.), **Trends in Ergonomics/Human Factors IV** (North-Hollan, Amsterdam, 1987) 17-25.

[4] Peacock, J.B., Ergonomics Information Analysis System, Report to General Motors Assembly Plant (Oklahoma City), (1985).

[5] Smith, J.M., Expert Database Systems: A Database Perspective,"in: Kerschberg, L., (ed.), **Expert Database Ssytem** (Benjamin/Cummings, Menlo Park, 1986), pp. 3-15.

[6] Borg, G., Interindividual Scaling and Perception of Muscular Force, **Kungl Fysioqrafiska Sallskapets i Lund Forhanlingar,** 12(31) (1961) 117-125.

[7] Oborne, D.J., Examples of the Use of Rating Scales in Ergonomics Research, **Applied Ergonomics,** 7(4) (1976) 201-204.

[8] Ljunggren, G., Observer Ratings of Perceived Exertion in Relation to Self Ratings and Heart Rate, **Applied Ergonomics,**17(2) (1986) 117-125.

[9] Borg, G., Physical Performance and Perceived Exertion, **Studia Psychological et Paedagogica, Series Altera, Investiqationes 11,** Gleerup, Lund, (1962).

[10]Borg, G. and Noble, B.I., Perceived Exertion, in: Willmore, I.H., (ed.), **Exercise and Sport Science Reviews,** (Academic Press, New York, 1974).

APPENDIX

ERGONOMICS INFORAMTION ANALYSIS SYSTEM (EIAS) CHECKLIST

Section1. Case Identification

1-1 Date - Month/Day/Year
1-2 Time - Hours:Minutes in 24-hour foramt
1-3 Job Code - four digits code representing investigated job
1-4 Case Number - four digits code representing case number
1-5 Department - four characters abbreviation of investigated department
1-6 Section - four digits code representing investigated section
1-7 Worker ID - last four digits code representing worker's social security number
1-8 Shift - 1, 2 or 3
1-9 Age
1-10 Gender - male or female
1-11 Height - in inches

1-12 Weight - in pounds
1-13 Referred by - worker himself, medical personnel, supervisor, safety/ ergonomics personnel, industrial engineer, quality assurance department and other
1-14 Referred to - safety/ergonomics personnel, tooling department, maintenance department, industrial engineering department, placement department, supervisor and other
1-15 Seniority - time to begin the present job in Month/Day/Year
1-16 Other record sources - medical, placement, tooling, maintenance, industrial engineering department, placement department, supervisors and other
1-17 Other record systems - photograph, video, other

Section 2. Problem Description

2-1 Injury
2-1-1 Injury Severity

No Injury	Very Mild	Mild	Moderate	Severe	Very Severe
0	1	2	3	4	5

2-1-2 Injury Parts - foot, ankle, lower leg, knee, thigh, hip, pelvis, sacrum, lumbar region, thoracic region, neck, head, abdomen, chest, shoulder girdle, shoulder, upper arm, elbow, forearm, wrist, palm, finger, thumb, eyes and ears
2-1-2 Injury Types - abrasion, burn, open wound, contusion, fracture, dislocation, foreign body, strain/sprain, other
2-1-3 Injury Symptoms - numb, pain, swollen, hot, tingling, stiff, red, white, other
2-1-4 Injury Causes - trip, fall, falling object, sharp part, sharp tool, blow, pinch, repetitive motions, high torque, other
2-2 Product/Operation Quality
2-2-1Poor Product Quality

No Problem	Very Mild	Mild	Moderate	Severe	Very Severe
0	1	2	3	4	5

2-2-2 Causes of Poor Quality - poorly incoming materials, poorly adjusted tool, awkward workplace, inappropriate feedback, lack of visual feedback, lack of strength, lack of dexterity, inadequate selection/assignment, inadequate/lack of training, inadequate/ lack of supervision, inadequate/lack of motivation, other
2-3 Quality of Work Life
2-3-1 Poor Quality Types - abseenteeism, turnover, complaints, medical visits
2-3-2 Causes of Poor Life Quality - Physical, informational, environmentaal, social, economic stress, otherr

Section 3. Job Description

3-1 Job Organization
3-1-1 Complexity
3-1-2 Repetition Rate (Pace)

- 3-1-3 Social Interaction
- 3-1-4 Job Satisfaction

3-2 Physical Environment
- 3-2-1 Noise
- 3-2-2 Lighting
- 3-2-3 Thermal Environment
- 3-2-4 Ventilation

3-3 Job Workplace
- 3-3-1 Reach
- 3-3-2 Fit
- 3-3-3 Obstacles

3-4 Parts
- 3-4-1 Size
- 3-4-2 Weight
- 3-4-3 Material

3-5 Tool
- 3-5-1 Handling
- 3-5-2 Vibration
- 3-5-3 Weight
- 3-5-4 Trigger

Section 4. Operator/Operation Interaction

4-1 Posural Discomfort - rank 1 or 2 can be tolerated
Mild <==== 1 2 3 4 5 ====> Severe

4-2 Posture Discomfort Unit - base, leg, trunk, shoulder, neck, arm, wrist, hand

4-3 Biomechanical Description
- 4-3-1 Static Load
- 4-3-2 Dynamic Load

4-4 Physical Description
- 4-4-1 Heart Rate

4-5 Information Load
- 4-5-1 Sensory Load
- 4-5-2 Cognitive Load
- 4-5-3 Motor Load

Section 5. Remedy

5-1 Engineering Modifications
- 5-1-1 Environment
- 5-1-2 Workplace
- 5-1-3 Part
- 5-1-4 Tool

5-2 Placement
- 5-2-1 Noise
- 5-2-2 Open Pits
- 5-2-3 Ground Work
- 5-2-4 Power Tools

- 5-2-5 Repetitive Gripping
- 5-2-6 Repetitive Bending
- 5-2-7 Power Tools
- 5-2-8 Lifting Limit
- 5-2-9 Marked Wrist Deviation
- 5-2-10 Other

5-3 Job Organization

- 5-3-1 Rotation
- 5-3-2 Enlargement
- 5-3-3 Rest Pauses
- 5-3-4 Autonomy

5-4 Training and Methods Improvement

- 5-4-1 Supervisor
- 5-4-2 Ergonomics Staff
- 5-4-3 Medical Staff
- 5-4-4 Special Course
- 5-4-5 Industrial Engineer

5-5 Personal Protective Equipment - glove, pad, splint, other

Trends in Ergonomics/Human Factors V
F. Aghazadeh (Editor)

ANALYSIS OF "ERROR" VARIANCES

Alvah C. Bittner, Jr.

Analytics Incorporation
Willow Grove, PA 19090

and

Stephen J. Morrissey

Auburn, AL 36830

Repeated Measures Analysis of Variance (RANOVA) traditionally uses"error terms" based on between subject and within subject effects to derive F-ratios. It is shown that these error effects may be used to derive additional information on the effects of the treatment or independent variables. This technique of error analysis in RANOVA can facilitate examination of differential subject responses to treatments. The theory and the applications of expanded error term analysis for use in human factors and ergonomic research is discussed. Examples show how these techniques can be used in the analysis of individual differences and for study of overall subject-treatment effects. It is concluded that the application of the modified error analysis technique enhances the overall understanding of human factors-ergonomics research results.

1. INTRODUCTION

A popular experimental design for human factors and ergonomics research is the multifactor repeated measures or split-plot design. In this design, subjects are blocked on one variable and receive all combinations of the other variables, which results in subjects being nested within one variable and crossed with all levels of the others. The traditional analysis of this class of designs involves developing F-ratios with the appropriate mean square (MS) terms for the main effects as a numerator, and the subjects-by-condition-within-group MS term as the denominator or "error" term. Table 1 illustrates a typical expected mean square (EMS) table for a three factor split plot type design with one within factor and two between factors with the traditional F-ratios used in analysis of each effect.

It is not widely recognized that the "error" term components used as the denominators contain information relevant to the concerns of the human factors and ergonomics practitioner. These "error" terms can be used to develop valuable statistical and explanatory information which is not examined in traditional RANOVA tests (e.g., Table 1). This report discusses the analysis of error variance for multifactor RANOVA's.

Table 1: Analysis of variance source table for a typical three factor split plot design with one between and two within variables. Adapted from Winer, (1971).

Source of variation	EMS	Traditional F-Ratio	Recommended SCWG Ratios
<u>Between subjects</u>			
(1) A	$\sigma^2_\epsilon + qr\sigma^2_\pi + nqr\sigma^2_\alpha$	(1)/(2)	(1)/(2)
(2) Subj w. groups	$\sigma^2_\epsilon + qr\sigma^2_\pi$	---	(2)/(11)
<u>Within subjects</u>			
(3) B	$\sigma^2_\epsilon + r\sigma^2_{\beta\pi} + npr\sigma^2_\beta$	(3)/(5)	(3)/(5)
(4) AB	$\sigma^2_\epsilon + r\sigma_{\beta\pi} + npr\sigma^2_{\alpha\beta}$	(4)/(5)	(4)/(5)
(5) B x subj w. groups	$\sigma^2_\epsilon + r\sigma^2_{\beta\pi}$	---	(5)/(11)
(6) C	$\sigma^2_\epsilon + q\sigma^2_{\gamma\pi} + npq\sigma^2_\gamma$	(6)/(8)	(6)/(8)
(7) AC	$\sigma^2_\epsilon + q\sigma^2_{\gamma\pi} + nq\sigma^2_{\alpha\gamma}$	(7)/(8)	(7)/(8)
(8) C x subj w. groups	$\sigma^2_\epsilon + q\sigma^2_{\gamma\pi}$	---	(8)/(11)
(9) BC	$\sigma^2_\epsilon + \sigma^2_{\beta\gamma\pi} + np\alpha^2_{\alpha\beta\gamma}$	(9)/(11)	(9)/(11)
(10) ABC	$\sigma^2_\epsilon + \sigma^2_{\beta\gamma\pi} + n\sigma^2{}_{\alpha\beta\gamma}$	(10)/(11)	(10)/(11)
(11) BC x subj w. groups	$\sigma^2_\epsilon + \sigma^2_{\beta\gamma\pi}$	---	---

2. ANALYSIS OF ERROR TERMS

The subjects-by-condition within-groups component (SCWG) for any multifactor RANOVA is well known to reflect individual differences in performance. Winer (1971), in particular, shows how one may estimate reliability based on this sum-of-squares (e.g., from the ratios of terms (2) and the sum of terms (2) plus (11) of Table 1). However it is not well known that, the SCWG interactions (e.g. terms (5) and (8) of Table 1) also contain individual differences components because not all subjects are affected the same way by the treatments. Traditionally these individual difference terms are not examined.

Tests of the significance for a particular effect are traditionally made by forming a ratio between the effect MS and a SCWG term whose EMS contains all of the components present in the main effect MS less the effect of interest. Table 1 illustrates this with, for example, the numerator of the test ratio for main effect A being (1), and the denominator being term (2). As compared to term (1), term (2) lacks only $nqr\sigma_{\alpha}^{2}$, the effect of interest. If it were possible to separate the individual difference components that are contained in the SCWG terms, then tests of their statistical significance could be developed. Unfortunately in Table 1, there are no EMS terms that are exactly appropriate for analysis of the SCWG effects. Thus, it is necessary to use an approximation technique to develop the needed F-ratios from the terms of Table 1. The remainder of this paper discusses how this can be done and provides illustrative examples from the human factors-ergonomics literature.

Table 1 contains four error terms each of which contains information of potential interest to the human factors-ergonomics researcher; Subjects Within Groups, term (2); B x Subjects Within Groups, term (5); C x Subjects Within Groups, term (8); and BC x Subjects Within Groups, term (11). Examining Table 1 it can be seen that there is no one EMS term present that contains only the simple error effect, σ_{ε}^{2}, which could be used as a denominator to form F-ratios. However, if one of the SCWG effects consisted of only σ_{ε}^{2}, then this term could be used to evaluate the other SCWG effects. Analytic considerations suggest that the higher order interactions are less likely to occur, or be substantial, than are the lower order interactions. Thus, one can assume that the highest order SCWG interaction term (11) could serve as the denominator for evaluating the lower order SCWG terms (2), (5), and (8). Importantly, this is a conservative procedure as the presence of the higher order interaction term would bias the resulting F-test towards non-significance. This approximation may be used to identify significant subject and interaction effects.

Realization that there are individual differences in treatment effects may have both practical and theoretical implications for the human factors-ergonomics practitioner (c.f., Bittner, et al., 1983; Guignard, et al., 1982; Lentz, et al., 1982; Woldstad, et al., 1982). In analysis of results from a vibration exposure study, Woldstad et al. (1982) used this technique and reported a marginally significant subject-by-treatment within group interaction ($p<0.06$). Following up on this, additional detailed analysis revealed that one subject did not show nearly as large a performance decrement during severe vibration as did the other subjects. As revealed in experimental notes, this subject had developed a idiosyncratic strategy for fixating his hand to the equipment in a manner that reduced biomechanical interference. This led the authors to develop a hand fixation methodology to provide similar stability for other subjects (with consequent reduction in the vibration performance decrement). The examination of the subject interaction "error" components consequently led to a method for reducing vibration effects and improving operator performance.

A second illustration of within effect error analysis is derived from recent investigation of motion sickness (Lentz, 1981). This study required each subject (pilots in training) be tested for motion sickness sensitivity on two different days. On one day, the subject was given an acupressure treatment and on the other day the subject was given no treatment. The order of treatment can thus be considered to be a between-subject factor with two levels. The design may be viewed as multifactor repeated measures design, with one between-subjects factor (Order) and two within-subjects-factors; Method (Treatment) and Repetitions (Days). Table 2 gives the summary for this study using term (11) of Table 1 as the denominator for the SCWG effects.

Table 2 reveals, not unexpectedly, a very significant subject effect ($F(22,22) = 12.59$; $p<0.001$). Also, a significant days by subjects within groups interaction effect may be seen ($F(22,22) = 2.19$; $p<0.05$). This should alert the analyst that there are individual differences in habituation or adaptation to the treatment. In this case, some subjects did show habituation to motion while other subjects tended toward sensitization. This result naturally led to the researchers posing two questions of practical interest: (1) How do motion- sensitive subjects who habituate to motion differ from motion sensitive subjects who do not habituate, and (2) Are these differences related to success in flight training? These differential effects to treatment would probably not been found and a valuable source of insight into human performance lost without the "effect error analysis" techniques.

3. CONCLUSIONS

Human factor researchers often assume that differences across subjects in their response to a treatment merely reflect the random perturbations of measurement error. This paper demonstrates that the subject-by-treatment-within-groups interaction terms contains valuable information regarding individual differences in treatment effect which may be examined via an analyis of the "error variances". A generalizable procedure for analyzing error variance is also demonstrated. Illustrations of practical and theoretical results of such analyses are provided by study of several research reports in the human factors-ergonomics literature. It is concluded that "error variance" analysis can facilitate the study of questions that are of practical and theoretical concern to the human factors-ergonomics researcher.

Table 2: Summary table for re-analysis of study of Lentz, et al. (1982).

SOURCE	df	MSE	F	P
Order (O)	1	4.68	.03	---
Subj w. Groups	22	159.88	12.59	<0.0001
Repetition (R)	1	13.65	.49	---
R X O	1	20.72	.75	---
R X Subj w. Groups	22	27.72	2.19	<0.05
Method (M)	1	52.21	1.45	---
M X O	1	105.84	2.95	---
M X Subj w. Groups	22	35.89	2.83	<0.01
R X M	1	2.10	.17	---
R X M X O	1	2.22	.17	---
R X M X Subj	22	12.70		

4. REFERENCES

Bittner, A.C., Jr., Guignard, J.C., Woldstad, J.C., and Carter, R.C. (1983). Vibration affects on digit-symbol coding. Journal of Low Frequency Noise and Vibration, 2, p. 169-175.

Guignard, J.C., Bittner, A.C., Jr., and Carter, R.C. (1982). Methodological investigations of vibration affects on performance of three tasks. Journal of Low Frequency Noise and Vibration, 1, p. 169-175.

Lentz, M.J. (1982). Two experiments on laboratory-induced motion sickness: I. Acupressure II. Repeated Exposure. Research Report NAMRL-1288. Penscola, FL: Naval Aerospace Medical Research Laboratory.

Winer, B.J. (1971). Statistical Principles in Experimental Design. New York: McGraw-Hill Company.

Woldstad J., Bittner, A.C., Jr., and Guignard, J.C. (1982). A methodological investigation of subject input/output related error during vibration. Proceedings, 26th Annual Meeting of the Human Factor Society.

Trends in Ergonomics/Human Factors V
F. Aghazadeh (Editor)
Elsevier Science Publishers B.V. (North-Holland), 1988

STATISTICAL ISOLATION OF TIMESHARING COMPONENTS: DEMONSTRATION AND INTERPRETATION

Alvah C. BITTNER Jr.
Analytics Inc.
Willow Grove, PA 19090

Diane L. DAMOS
USC, ISSM
Los Angeles, CA 90089

An objective statistical method for isolating timesharing components of performance is described and demonstrated. During the first stage of this method, partial correlation analysis is used to remove the variance attributable to the single-task measures. During the second stage, the resulting dual-task partial correlations are subjected to a form of factor analysis. Data from both Sverko (1977) and Wickens, Mountford, and Schreiner (1981) were reanalyzed and the results compared with their earlier analyses, as well as with the reanalysis of the latter study by Ackerman, Schneider and Wickens (1984). The characterization of timesharing factors and the question of a general timesharing factor are discussed. The timesharing components are interpreted in terms of individual differences in capabilities for *switching between tasks*.

1. INTRODUCTION

1.1. Background

During the last decade, there have been several attempts to isolate a general timesharing ability that affects only multiple-task performance (Hawkins, Rodriquez, and Reicher, 1979; Jennings and Chiles, 1977; Sverko, 1977; Wickens, Mountford, and Schreiner, 1981). All of these attempts have met with little success.

Ackerman, Schneider, and Wickens (1984) have recently described a number of methodological problems that may have obscured the existence of a general timesharing ability in these studies. These authors discuss solutions to these problems and present a method for identifying such an ability using Procrustean Rotation. Although we generally agree with the solutions to the methodological problems presented by Ackerman et al. (e.g., Damos, Bittner, Kennedy, and Harbeson, 1981), we feel that their proposed analytical method suffers from two major problems. First, their method is based upon a relatively sophisticated tool which, in addition to a number of limitations, relies heavily upon subjective judgements by its practitioners (Harman, 1976, pp. 336-360). Second, Ackerman, et al. have applied this tool to mixtures of single-task and dual-task measures, thereby clouding the identification of unique dual-task abilities. These problems have led us to consider alternate approaches for isolating one or more timesharing components of performance (Bittner and Damos, 1986). (Throughout this paper we have used the statistical term "components" rather than "skills" or "abilities" to avoid the controversy surrounding these latter terms.)

1.2. Purpose

This paper describes a method for isolating timesharing components that is less subjective than the Procrustean Method. The described method, in addition, is intended to be easily applied and uses computer routines that are readily available. Although several packages would have served as well, all calculations were performed using the the *BMDP Statistical Software* (Dixon, 1981). This statistical package was selected for illustrative purposes as it is widely available in industry, government and academia. To demonstrate this approach, data from both Sverko (1977) and Wickens, Mountford, and Schreiner (1981) were reanalyzed. The results reported in this paper for the Wickens, Mountford, and Schreiner data are also compared to results obtained by Ackerman et al. (1984). Although the reports by Sverko (1977), Wickens et al. (1981), and Ackerman et al. (1984) showed little, if any, evidence for multiple timesharing components, substantial evidence is reported in this paper. This evidence will be shown to support the interpretation that timesharing components in the Sverko (1977) and Wickens et al. studies largely reflect individual differences in capabilities for *switching between tasks.*

2. METHOD AND RESULTS

2.1. Statistical Approach

The statistical analysis is conducted in two stages. During the first stage, partial correlation analysis is used to remove the variance attributable to the single-task measures from the dual-task measures. Specifically, all of the single-task measures are partialed out of all of the dual-task measures using the BMDP6R Program (Dixon 1981, pp. 509-518). (This program also provides for easy examination of individual dual-task scores with the single-task variance removed.) Although it may seem more appropriate to partial out only the single-task variance of the tasks that compose each dual-task combination, there are two reasons for taking this approach. First, in most cases, all of the single-task measures are correlated with all of the dual-task measures (both of the examples given in this report show this pattern of correlation). If dual-task performance is determined both by one or more unique timesharing components, as well as by several single-task components, it is necessary to account for the contribution from all of the single tasks to identify the unique timesharing components. Second, by partialing out all variance that is associated with single-task scores, the maximum amount of variance that may be attributed to timesharing components becomes apparent. Thus, the first stage of analysis is aimed at calculation of the dual-task intercorrelations free of all single-task variation.

During the second stage, the resulting dual-task partial correlations are subjected to a form of factor analysis. The BMDP4M principal factor analysis with a minimum (1.1) eigenvalue cutoff and varimax rotation was used (Fame, Jennrich, and Sampson, 1981). This option was selected primarily because it provides for an easy and straightforward analysis.

In summary, the proposed statistical approach involves factor analysis of the dual-task measures from which all the single-task variance has been removed.

2.2. Two Applications

For illustrative purposes, the statistical approach described above will be applied to the data of Wickens et al. (1981) and Sverko (1977). In both examples, scores are included from each task of a dual-task combination. Thus, X(Y) refers to the score of Task X performed with Task Y and Y(X) refers to the score of Task Y when performed with Task X.

Wickens, Mountford, and Schreiner (1981). In this experiment, subjects performed a critical tracking task (T), a number classification task (C), an auditory running memory task (A), and a line judgement task (L). The subject performed each task alone and then all dual combinations, with the exception of the auditory running memory task which was not performed with itself. Bittner and Damos (1986) show the intercorrelations of the four single and 15 dual-task measures from the Wickens et al. experiment (derived from Ackerman, et al., 1982). Bittner and Damos also show the partial intercorrelations between the 15 dual-task measures with all the single-task variance removed. In Bittner and Damos, it was noted that the partialed intercorrelation matrix contained significant positive correlations between the C, L, and A measures, although significant and substantial ($p<.00001$) portions of the variance of each measure had been extracted. The specific percentages removed were: T(T), 46%; T(C), 72%; T(L), 53%; T(A), 84%; C(T), 68%; C(C), 73%; C(L), 59%; L(T), 70%; L(C), 53%; L(L), 40%; L(A), 63%; A(T), 53%; A(C), 45%; and, A(L), 34%. Of interest in the partialed matrix, the T dual-task measures showed no substantial correlation with the other dual-task measures although highly correlated among themselves. This pattern of partial correlations suggested one or more factors accounting for the relation between the C, L, and A dual measures with another factor accounting for the T dual measures.

Principal Factor Analysis of the partial correlation matrix resulted in a four-factor solution that explained 72.4% of the variation. Table 1 shows the rotated factor solution. Examining this table, Factor 1 may be seen to have substantial ($\geq .5$) loadings of .831 on A(C), .799 on L(A), .741 on C(A), .712 on A(L) and .630 on A(T). This pattern of loadings involving "A" suggests labeling this a "dual auditory running memory task" factor. Similarly, Factor 2 has substantial loadings of .831 on T(C), .820 on T(A), .812 on T(L), and .680 on T(T), which suggests this a "dual tracking task" factor. Factors 3 and 4 appear to be mixes of the L and C tasks in dual combinations. Factor 3, tentatively labeled a "dual L/C task" factor, has substantial loadings of .821 on L(C), .785 on L(L), .727 on C(C), .552 on C(L) and .537 on A(T). In contrast, Factor 4 has its largest loadings of .835 on C(T) and.771 on L(T). This factor could be labelled "dual L/C (T)" or perhaps "dual discrete task with tracking". Table 1 altogether reflects dual components that are independent of the single-task measures, but are associated with the tasks or their characteristics.

TABLE 1. ROTATED FACTOR LOADINGS FOR DUAL-TASKS WITH SINGLE-TASK VARIANCES PARTIALED-OUT FOR WICKENS ET AL. (1981)

	FACTOR 1	FACTOR 2	FACTOR 3	FACTOR 4
T(T) 1	.180	.680	-.007	.355
T(C) 2	-.130	.831	.015	-.138
T(L) 3	.145	.812	.050	-.406
T(A) 4	-.061	.820	.312	-.186
C(T) 5	.045	-.037	.135	.835
C(C) 6	.275	.155	.727	.205
C(L) 7	.128	-.171	.552	.523
C(A) 8	.741	.234	.035	.176
L(T) 9	.270	-.268	.210	.771
L(C) 10	.227	.080	.821	-.002
L(L) 11	.130	.265	.785	.293
L(A) 12	.799	.185	.151	.037
A(T) 13	.630	-.206	.537	-.049
A(C) 14	.831	-.088	.222	.089
A(L) 15	.712	-.289	.307	.166

Sverko (1977). In this experiment, the subjects performed four tasks and their six combinations: Pursuit rotor (PR), digit processing (DP), mental arithmetic (MA), and auditory discrimination (AD). Bittner and Damos (1986), it is noteworthy, contains the correlations between the four single-task measures as well as the 12 measures obtained from the six dual-task combinations. Bittner and Damos also show the partial correlations of the dual measure where all the single-task variance has been removed. As in the earlier analysis, significant ($p<.0001$) proportions of the dual-tasks measures were partialed out. Specifically, these were: PR(DP), 47%; PR(MA), 56%; PR(AD), 72%; DP(PR), 26%; DP(MA), 29%; DP(AD), 66%; MA(PR), 82%; MA(DP), 64%; MA(AD), 74%; AD(PR), 60%; AD(DP), 50%; and, AD(MA), 48%. Contrasted with the partial correlation table for Wickens et al. (1981), the structure of the resulting partial correlation matrix for Sverko (1977) showed some added complexity. A correlation of -.42 between DP(PR) and PR(DP), for example, indicated a performance trade-off not seen in the earlier analysis. However, there were prominent clusters of modest positive correlations between scores of the same task performed in different combinations (e.g., PR(DP), PR(MA), and PR(AD)). This pattern of partial correlations suggests multiple factors with each factor associated with the timesharing component of a specific task.

Principal Factor Analysis resulted in a four-factor solution that accounted for 59.8% of the variance in the partial correlation matrix. Table 2 shows the rotated factor solution. Examining this table, it may be noted that the factors are each identified with specific tasks in dual combination: Factor 1 with AD; Factor 2 with PR; Factor 3 with MA; and Factor 4 with DP. Factor 1, for example, has substantial loadings of .824 on AD(DP), .773 on AD(MA), and .710 on AD(PR). Thus, Table 2 presents dual-task capabilities that are defined by specific tasks in dual combinations.

TABLE 2. ROTATED FACTOR LOADINGS FOR DUAL-TASKS WITH SINGLE-TASK VARIANCES PARTIALED OUT FOR SVERKO (1977)

	FACTOR 1	FACTOR 2	FACTOR 3	FACTOR 4
PR(DP) 1	-.004	.719	.130	-.241
PR(MA) 2	.069	.613	.091	.032
PR(AD) 3	-.023	.760	-.063	.129
DP(PR) 4	-.085	-.414	-.023	.728
DP(MA) 5	.189	.311	-.092	.642
DP(AD) 6	.277	-.025	.209	.690
MA(PR) 7	-.010	.049	.764	-.020
MA(DP) 8	.008	.019	.854	-.087
MA(AD) 9	.153	.107	.631	.334
AD(PR) 10	.710	.011	.035	.127
AD(DP) 11	.824	-.109	.033	.038
AD(MA) 12	.773	.198	.036	.136

3. DISCUSSION

The major purposes of this report were: to demonstrate an objective technique for identifying one or more timesharing abilities; and consider the implications of the results of its application. For purposes of this demonstration, dual-task data from Wickens et al. (1981) and Sverko (1977) were reanalyzed using the proposed technique. In the following sections, the results of these reanalysis will be contrasted with earlier analyses and implications for a general timesharing factor considered.

3.1. Contrasts with Previous Analyses

The statistical approach described in this report has two features that distinguish it from earlier approaches. First, the demonstrated method is based on "factor analysis" of dual-task partial correlations from which all single-task variance has been removed. These dual-task relations are not obscured by the presence of single-task components, which according to Ackerman et al. (1984) have contaminated earlier analyses of the same data (i.e., Wickens et al., 1981; Sverko, 1977). The second feature of the demonstrated approach, which differs from that of Ackerman et al. (1984), is that it does not require human intervention. The demonstrated technique involves successive applications of statistical options drawn from the *BMDP Statistical Software* (Dixon,1981). In contrast, the Procrustean Method, advocated by Ackerman et al., requires the user to specify the target structure (cf., Harman, 1976, pp. 336-360). Potential problems with such specification are numerous, and include: *A posteriori* specifications of structure; as well as the possibility, noted by Hurley and Cattell (1962), for making almost any data fit almost any hypothesis. The present method is characterized by its objective analysis of dual-task relations after removal of single-task components.

The above distinctions must be considered when comparing our results with those obtained previously. For example, the present four-factor solution and the four-factor solution of Sverko (1977) initially appear to be similar. Sverko found four rotated factors identified by loadings on each task performed in combination with other tasks but also with equally substantial loadings on the tasks performed alone. For example, his Factor III (which is analogous to the first factor found in this report) had loadings of .89 for AD(DP), .87 for AD(PR), and .85 for AD(MA), but also .92 for AD. Sverko (1977, p.14), therefore, could not clearly ascribe his results to timesharing factors. In contrast, the present factor solution contains four dual factors which are uncorrelated with the single-task measures.

Comparison of the Ackerman et al. (1984) and our reanalyses of the Wickens et al. (1981) data also requires attention to differences between methods. Ackerman et al. report a four-factor solution. The first factor is identified by single- and dual-task measures of T, the second factor is identified by single- and dual-task measures of both C and L, and the third factor is identified by single- and dual-task measures of A. Because both single- and dual-task measures have substantial loadings on these factors, none was identified as a timesharing factor by Ackerman et al. However, the fourth factor was identified by Ackerman et al. as a timesharing factor. Identification of this as a timesharing factor is questionable as *single-task measures* had moderate loadings on it.

The present solution for the Wickens et al. data also yields four factors. However, one factor is clearly identified with Task A, another with Task T, and the remaining two with both Tasks C and L. Since all of the single-task variance was extracted from the data before the PFA was performed, these four factors can clearly be identified as task-specific timesharing factors. The differences between the present results and those of Ackerman et al. are not surprising, given the differences in analytic approaches.

Although initially appearing similar, the results in this report differ substantially from those previously obtained. The primary difference lies in the form of the dual-task relations that result when single-task variation is removed.

3.2. The Question of a General Timesharing Factor

The major question of much of the previous dual-tasks research has concerned the existence of a single general timesharing ability (e.g., Wickens et al., 1981; Sverko, 1977). Our rotated solutions for earlier data show little direct evidence for such an ability or factor (cf.,

Tables 1 and 2). However, some indirect evidence for a general factor was noted during the factor analyses. For example, the first unrotated factor resulting from separate analyses of Wickens et al. and Sverko accounted for 32.7% and 20.7% of their respective data. This indirect evidence, it is noteworthy, is not inconsistent with a general factor related to individual differences in task switching capabilities as posited by Spearman (1930). However, the presence of substantial numbers of dual factors in Tables 1 and 2 provide evidence against the concept of *only* a single general timesharing factor or ability. Interestingly, Wickens (1984) has recently argued for multiple timesharing capabilities, based on his review of previous research. This report as well as previous research consequently support the concept of multiple timesharing factors.

3.3. Characterization of Timesharing Factors

Our solution of the Sverko (1977) data resulted in four factors, each of which was identified with one of the experimental tasks. Similarly, the first and second factors from our solution of the Wickens et al. (1981) data were clearly identified with the A and T tasks, respectively, while the third and fourth factors were identified with the C and L tasks. Initially this pattern of results appeared uninterpretable; indeed, we earlier stated that there was no *a priori* basis for timesharing factors being identified with specific experimental tasks (Bittner and Damos, 1986). However, review of our own earlier results (Damos, Smist and Bittner, 1983) as well as other related findings (e.g., Boettcher, 1986; Spearman, 1930) has revealed such a basis: the task-related factors reflect individual differences in capabilities for switching between specific tasks. This basis is discussed in the following.

Spearman (1930) early posited a general factor (*p*) related to switching between tasks as indicated in the previous section (3.2). Similar to the present study, evidence for this factor came in part from studies of times-to-complete single tasks and pairs of tasks, when done alternately. Though able to establish the relative independence of task and switching capabilities, the analytic approach used at the time could not address the possibility of multiple task-related factors. However, these early results suggest the possibility of task-related switching factors such as seen in Tables 1 and 2. Our own findings further indicate the importance of individual differences in switching capabilities (Damos, Smist and Bittner, 1983). In our study, we found substantial intertask switching capability differences between two groups who were initially classified as using "massed" and "alternating" strategies when accomplishing a dual-task (with a serial-memory and a digit-classification subtask). (This initial classification, it is noteworthy, revealed that only a relatively small percentage (22%) of subjects exhibited a "simultaneous" strategy, most adopted either an alternating or massed strategy.) Group differences in switching capabilities were subsequently assessed over 25 trials which required switching (alternating) between the memory and the classification tasks. One-half of the difference between the mean memory-classification switching task RTC (correct response time) and the sum of the average RTCs for the single classification and memory tasks was used to estimate switching times. The "massed" subjects were found to average about *250ms longer* than the "alternating" subjects, although the two groups did not differ in their average single-task performances. This result pointed out the presence of significant individual differences in switching capabilities, not related to single-task performance. This result first led to the observation that similar individual differences in the times-to-switch could yield task-related factors, if they were task-specific (i.e., relatively constant for switching either to, or from, a specific task but relatively uncorrelated across tasks). Pertinently, Boettcher (1986), using a paradigm requiring switching between various combinations of tasks, has recently reported just such differences (in his tabulated results). His results indicate consistent individual differences in the times to switch to a specific task from any of a number of alternatives. Boettcher's (1986) results as well as that of Spearman (1930) and Damos et al. (1983) altogether point toward interpretation of individual differences in switching capabilities as the source of the task-related timesharing components isolated in the present investigation.

3.4. Summary

An objective statistical method for isolating timesharing components of performance has been demonstrated using historical data. Tables 1 and 2 contain task-related timesharing components isolated in the present demonstrations which are respectively based on Sverko (1977) and Wickens, et al.(1981). The task-related timesharing components isolated in the present demonstration may be interpreted in terms of individual differences in capabilities for *switching between tasks*.

4. REFERENCES

Ackerman, P.L., Schneider, W., and Wickens, C.D. (1982). *Individual differences and time-sharing ability: A critical review and analysis* (Tech. Report HARL-ONR-8102). Champaign, IL; University of Illinois, Human Attention Research Laboratory.

Ackerman, P.L., Schneider, W., and Wickens, C.D. (1984). Deciding the existence of a time-sharing ability: A combined methodological and theoretical approach. *Human Factors, 26*, 71-82.

Bittner, A.C., Jr. and Damos, D.L. (1986). *Demonstration of a statistical method for isolating timesharing components* (NBDL-86R001). New Orleans, LA: Naval Biodynamics Laboratory.

Boettcher, K.L. (1986). A descriptive model for time to switch among subtasks. *Proceedings of the Tenth Psychology in the DoD Symposium*.. Colorado Springs, CO: USAF Academy.

Damos, D.L., Bittner, A.C., Jr., Kennedy, R.S., and Harbeson, M.M. (1981). The effects of extended practice on dual-task tracking performance. *Human Factors, 23*, 627-631.

Damos, D.L., Smist, T.E., and Bittner, A.C., Jr. (1983). Individual differences in multiple task performance as a function of response strategy. *Human Factors, 25*, 215-226.

Dixon, W.J. (Ed.). (1981) *BMDP statistical software (1981 edition)*. Los Angeles: University of California Press.

Fame, J.W., Jennrich, R.I., and Sampson, P.F. (1981). P4M factor analysis. In W.J. Dixon (Ed.), *BMDP statistical software*. (pp. 480-499). Los Angeles: University of California Press.

Harman, H.H. (1976). *Modern factor analysis (3rd ed.)*. Chicago: University of Chicago 1976.

Hawkins, H.L., Rodriguez, E., and Reicher, G.M. (1979). *Is timesharing a general ability?* (Tech. Report No. 3). Eugene, OR; University of Oregon, Center for Cognitive and Perceptual Research.

Hurley, J.R., and Cattell, R.B. (1962). The Procrustes Program: Producing direct rotation to test a hypothesized factor structure. *Behavioral Science, 1*, 258-262.

Jennings, A.E., and Chiles, W.D. (1977). An investigation of timesharing ability as a factor in complex performance. *Human Factors, 19*, 535-547.

Spearman, C. "G" and after - a school to end schools. In C. Murchison (Ed.), *Psychologies of 1930*.(pp. 339-366).Worcester, MA: Clark University Press.

Sverko, B. (1977). *Individual differences in time-sharing performance* (Tech. Report No. ARL-77-4/AFOSR-77-4). Savoy, IL; University of Illinois, Aviation Research Laboratory.

Wickens, C.D.,Mountford, S.J., and Schreiner, W. (1981). Multiple resources, task-hemispheric integrity, and individual differences in time-sharing. *Human Factors, 23*, 211-229.

Wickens, C.D. (1984). *Engineering psychology and human performance*. Columbus, Ohio: Charles E. Merrill.

Trends in Ergonomics/Human Factors V
F. Aghazadeh (Editor)

RANDOM SAMPLING OF DOMAIN VARIANCE (RSDV)

Robert J. WHERRY, Jr.

The Robert J. Wherry, Jr. Company
562 Mallard Drive
Chalfont, Pennsylvania 18914

This paper discusses the Random Sampling of Domain Variance (RSDV) approach to experimental design. This new approach avoids the inherent problems and limitations of ANOVA designs when investigating complex real-world situations. The rationale and procedures for accomplishing laboratory studies using the RSDV approach is presented. Several alternative methods for analyzing the data from RSDV studies are also discussed.

1. INTRODUCTION

Results from human factors laboratory studies do not generalize well to real-world problems. This frequently heard criticism has a great deal of merit. Adams [1], in discussing approximately 40 years of vigilance research and the outcomes of more than a thousand studies, disagrees with those who contend vigilance researchers wasted their time. He defends a researcher's right to investigate anything he wants to in the laboratory, but believes that human factors may have been foolish to spend that amount of time on what may have been a relatively unimportant problem. However, the lack of generalization of research results to solving real-world problems is not limited to vigilance studies; it is pervasive to almost all aspects of psychological and human factors research. One cannot seriously accept a suggestion that all human factors problems investigated in the last fifty years were relatively unimportant ones. We must look elsewhere for an explanation! It appears that something may be basically wrong with the way applied research is being carried out in laboratory settings. Adams, and others as well, have suggested that laboratory studies may be the wrong way to go about solving applied problems, and that field and simulation studies may be more appropriate. This paper contends that the real difficulty lies, not in trying to investigate real problems in the laboratory, but in the experimental designs that have been used.

The following section reviews the advantages and disadvantages of field, laboratory, and simulation studies. Experimental designs traditionally used in human factors studies are then examined. Why they have lead researchers to collect data that cannot generalize to applied situations is explained. A new approach to experimental design known as random sampling of domain variance is then presented. The procedures for using the new approach to plan an experiment and analyze its collected data are described.

2. TYPES OF APPLIED STUDIES

Three traditional types of applied studies have been used to scientifically investigate human behavior: field, laboratory, and simulator studies. Each type has its separate advantages and disadvantages. These advantages and disadvantages should be fully understood by potential researchers before they decide which is the most appropriate way for them to proceed when investigating an applied problem. The pertinent question for each type of study is, "What is being sampled?"

2.1. Field Studies

With typical field studies, researchers go to a field location and collect *naturally occurring* data. The major advantage of true field studies is that researchers get to observe a sample of the real people of interest performing a sample of real tasks under a sample of real conditions. Unfortunately, the samples of persons, tasks, and conditions observed during field trials may not happen to be statistically representative of all of the people, tasks, and situations of interest. Further, because researchers have little or no control over what will occur, they may be unprepared to record the specific data that might be of most interest. Finally, because of incomplete data collection, precise conditions that brought about unique and inexplicable subject behavior may not be able to be replicated to ascertain if that same behavior would occur again under similar or identical circumstances. It is, of course, possible to exercise some control over some field situations. For example, military units typically undergo a variety of field (or "fleet") tests in which the unit being evaluated is confronted with prearranged conditions. Controlled field situations are usually expensive to carry out and difficult to arrange.

2.2. Laboratory Studies

The major advantages of laboratory studies of phenomena of interest stem from the experimenter's ability to exercise control over the events that do occur during the laboratory situations. The three major advantages are that: (a) *the experimenter can be fully prepared to observe and/or record the behavior being studied*, (b) *the experimenter or others can, if desired, repeat the sequence of events that occurred to validate obtained results*, and (c) *the conditions can be systematically varied to determine concomitant variation with selected criteria*. The disadvantages with typical laboratory studies are that, unless carefully planned, they also may be unrepresentative of the real tasks and situations of concern, and, consequently, may not yield results that generalize to the real-world. Historically, results from laboratory studies have generalized poorly to real-world situations.

2.3. Simulator Studies

An example of a simulator study might be one in which subjects perform "man-in-the-loop" tasks in a complex weapon system simulator. While field and laboratory studies have a relatively long history, use of simulators is relatively new. Simulators can be thought of as representing a middle ground between laboratory and field studies. Man-in-the-loop simulation studies have the advantages of being accomplished under controlled conditions, permitting needed data to be collected relatively easily, and being able to represent far more of the complexity of the situations of interest. The major disadvantages are that such simulators are costly and time-consuming to develop, training of subjects to use the simulators can also be costly and time-consuming, and simulators may fail to capture the desired realism of actual systems and operational missions. With simulators, researchers may choose to either observe samples of behavior that happen to occur (as they might in a field situation), or they may attempt to exert influence over the test conditions that the subjects will confront.

3. TRADITIONAL EXPERIMENTAL DESIGNS

3.1 The Analysis of Variance Designs

For the past fifty years, experimental designs used most frequently by psychologists and human factors researchers for both laboratory and controlled field and simulator studies have been derived from the Analysis of Variance (ANOVA) approach. ANOVA designs permit the total variance of the criterion variable (i.e., the subjects' performance scores) to be partitioned into separate, independent components for each experimentally controlled variable (and combination of those variables). To accomplish this requires that all of the main effects (i.e., the experimental variables being investigated) be made *independent* of one another. ANOVA does this by requiring each level of any experimental variable to be used a proportional number of times (and at least once) with each level of every other main effect. Despite their general popularity among psychologists, ANOVA designs suffer from several serious flaws that make them inappropriate for investigating many applied problems.

3.2. Disadvantages of ANOVA Designs

A major disadvantage of ANOVA designs derives from the requirement for independence of main effects. The disadvantage is that, as the number of both main effects and levels within main effects become large, ANOVA designs tend to require massive amounts of data cases. For example, to investigate ten main effects with ten levels in each would require (for a full factorial design) a total of 10^{10} data cases. To illustrate the difficulty of conducting such a study, it would take a research team 2000 years to collect that amount of data -- even if they could collect 500 cases per hour and were willing to work 24 hours per day! The unfortunate result of this constraint has been that few researchers collect data on many variables with many levels in each. This fact has been documented by Simon [2] in 1975. He found that over 92 percent of 240 ANOVA studies published in the *Human Factors* journal from 1958 to 1972 investigated three or less main effects and explained, on the average, only about 45 percent of the criterion variance. Over 98 percent of all of the published studies investigated four or fewer variables and, on the average, accounted for only 61 percent of the variance. These disappointing results are not limited to human factors studies alone; similar results had been found by Dunnett [3] in 1966 for four major American Psychological Association journals. It is obvious that most applied problems have a host of possible experimental variables, each of which usually has many levels of interest. For this reason alone, standard ANOVA designs cannot be expected to handle the inherent complexity of most applied tasks.

There is yet another major reason for avoiding the ANOVA approach for many applied projects. The reason is that ANOVA studies have historically tended to produce results that appear to have little practical value in solving real-world problems. It is, as mentioned earlier, generally acknowledged that results of the vast majority of laboratory experiments have not generalized to (i.e., cannot be successfully applied to) complex, real-world problems. Actually, this result may also stem from ANOVA's requirement for independent main effects, and should not be particularly surprising if one considers what ANOVA designs require researchers to do. In the real-world, it is not unusual for several important variables to be correlated. In tactical military situations, for example, missions and scenarios are, to a large extent, dictated by the relative capabilities of the own, friendly, and hostile systems. Thus, the frequency of certain types of missions (and, consequently, certain types of tasks) is related to the frequency of the occurrence of those situations. But ANOVA forces investigators to make situation and task variables independent. When variables that are related in the real-world are arbitrarily forced to be independent in research studies, the samples of performance data collected cannot be representative of the real-world.

3.3 Modifications to ANOVA Designs

The inherent weakness of the ANOVA approach for investigating highly complex (i.e., multivariate) real-world situations have been recognized by human factors researchers. Meister [4] in 1985 summarizes many of the variations to ANOVA's replicated factorial designs advocated by various researchers. These include single-observation factorial designs, hierarchical (nested) designs, blocking designs, fractional factorial designs, and central-composite designs. What the approaches have in common is that each attempts to reduce the total number of data cases that would have been required by a replicated complete factorial design. Unfortunately, they accomplish their objective by finding ways to reduce the number of main effects and/or the number of levels within an effect, to confound main effects with interactions of other main effects, and/or to eliminate replications that might have been desirable. Because they are all variations on the ANOVA approach, they still require all main effects to be independent of one another. Thus, they all still require the researcher to collect data that may be unrepresentative of the real-world situations that the researchers profess to be interested in. For this reason, they are all likely to produce results that will not generalize to the real-world.

4. THE RANDOM SAMPLING OF DOMAIN VARIANCE APPROACH

4.1. Purpose of RSDV Designs

Recently, Wherry, Jr. [5] developed and tested a new experimental design methodology that abandons the ANOVA requirement for independence of main effects. The new methodology is referred to as the Random Sampling of Domain Variance (RSDV) approach, and is advocated for conducting virtually all applied experiments. RSDV's major purpose is to enhance the likelihood that experimental results derived from laboratory studies will generalize to the real-world situations of interest. It offers a potential way to explain discrepencies between laboratory, field, and simulator studies of the same problem area. RSDV maintains all the advantages of controlled experiments, but does not constrain the number of experimental variables or levels within variables that can be investigated.

4.2. Rationale for the RSDV Approach; Sampling from Specified Domains

The RSDV approach, as will be seen, is a natural extension of random sampling theory, on which all inferential statistics are founded. For example, ANOVA designs and virtually all statistical tests of significance are based on the presumption that the sample of data to be analyzed is a random one. The basic argument is that if one desires *unbiased* results from any sample of some particular population, one must randomly sample from that population (or domain) of interest. The domains of interest for applied human factors problems, are usually particular human tasks in the context in which they occur in the real-world. RSDV procedures require that these task and situational domains of interest be carefully specified prior to conducting laboratory studies. *Domain specification* consists of gathering data on all of the important variables (task and situation) that are likely to occur in real-world tasks and situations of interest. It also requires estimating the frequencies with which various levels of all variables are likely to occur as well as the extent to which the domain variables are related. Once the data for the domain specifications have been established in a mathematical format, they can be stored for subsequent retrieval by a computer. A special random-sampling computer program can then used to repeatedly select as many random samples from the specified task and situation domain as desired. Each domain example selected by the computer represents a true, single, random example from the specified domain. The particular task and situation for a selected sample can then be created (or simulated) for subsequent laboratory testing of personnel by noting the randomly selected level of each domain variable.

Because the *problem situations* have been randomly drawn from the estimated specifications of the real-world tasks and situations, human performance results from using them should be similar to having gone into the field and collected a true random sample of actual real-world tasks. Thus, by using the RSDV approach, experimenters know that results obtained should, in the long run, provide unbiased estimates of the means, standard deviations, and correlations, not only for the domain variables, but also for the human performance measures being collected.

5. ANALYSIS OF RSDV DATA

Having randomly sampled the domains of interest and then collected performance data on the sampled tasks and situations, the performance data must be analyzed. Because experimental variables in RSDV designs are not necessarily independent of one another, a somewhat different multivariate approach to the analysis of RSDV experimental data must be taken than that used with ANOVA data. Three approaches that are applicable for the analyses of various types of RSDV studies are multiple correlation, factor analysis, and canonical correlation. Actually, as is now widely recognized, ANOVA analysis is merely a special case of multiple correlation or multiple regression that applies when all of the predictor variables are statistically independent of one another. While recognition of this correspondence was not always the case, many of the computer programs now used for accomplishing data analyses for ANOVA studies are standard (or slightly modified) multiple regression or multiple correlation programs.

5.1. Multiple Correlation Methods

Multiple correlation/regression approaches, for RSDV data, represent tests of the null hypothesis that state that each of the predictor variables (i.e., the task and situation domain descriptor variables) is unrelated to the criterion variable(s) (i.e., the measure(s) of task performance). Actually, there are three methods that can be used to accomplish multiple correlation: accretion, matrix inversion, and deletion. With the accretion method, the first variable selected is the one that explains the largest amount of the criterion's variance. If the *best* predictor variable can explain a significant portion of the criterion's unexplained variance, then its relationship to all of the remaining variables and the criterion is removed from the (correlation or regression) matrix and the procedure is continued until none of the nonselected variables is found to explain a significant portion of the remaining unexplained criterion variance. In the matrix inversion method, all variables are selected, regardless of whether they are significant or not. The unexplained criterion variance becomes what in ANOVA corresponds to the *error* variance. The deletion method actually starts by using the matrix inversion approach and, having first selected all of the predictor variables, the significance of the raw score predictor weights for each variable are computed. If any are found to be *non-significant*, the least significant one is dropped, and the procedure is begun again without including the rejected variables. This process is repeated until the beta weights for all of the remaining variables are found to be significant. It is not surprising that the accretion and deletion methods usually obtain identical solutions for the multiple correlation (**R**) which is the square root of the explainable criterion variance. All three methods are identical when all predictor variables are found to be significant. Further, all three methods will derive identical beta weights for all selected variables, providing the same variables are selected. That is, the order in which variables are selected is unimportant as long as the same variables are ultimately selected. The tests of significance used in multiple correlation/regression are also identical to those used to test the significance of the experimental variables in ANOVA. Thus, while the concept of using the RSDV approach for the design of applied experiments is new, the appropriate RSDV data analyses techniques have an extremely long history and are widely accepted as appropriate.

Since one may use multiple correlation to determine the percent of variance of the criterion that is explained by the experimental variables, the significance of the overall **R** is tested as it would be in any multiple correlation analysis by the equation

$$F = (R^2 / (1 - R^2)) \cdot ((N-m-1) / m) \quad , \quad df = m, \; N-m-1 ,$$

where **N**is the number of data cases and **m** is the number of predictors. It is obvious that the ratio of R^2 to $1-R^2$ (i.e., the ratio of the explained variance to the unexplained variance) must by larger to still be considered significant as the ratio $(N-m-1) / m$ get smaller. One of the advantages of the RSDV approach is that each experimental variable can usually be treated as a linear variable (that accounts for only one degree of freedom) rather than having to use $k - 1$ dichotomous variables to represent the **k** mutually exclusive levels of that variable. In this way, the ratio of **N** to **m** can be kept fairly high with far fewer subjects. It is also possible in RSDV studies to use nonlinear terms (e.g., quadratic (X^2), cubic (X^3), etc.) of experimental variables as predictor variables to determine if their relationship with the criterion variable is curvilinear. In the same way, possible *interactions* among the experimental variables and the criterion can be tested by using the products of two experimental variables (e.g., X_1X_2)

5.2. Factor Analytic Methods

Factor analysis is another technique that can be used to analyze data from RSDV studies. With this approach, the matrix of correlations is computed for all of the experimental variables (i.e., task and situation domain variables) and the criterion variable(s). Using standard factor analytic approaches, one can then determine (a) how many independent dimensions exist in the task and situation domain and (b) the relationship of each variable (including the criterion) to the independent factors. If all subjects have been tested on each of the random samples of tasks and situations, then a slightly different factor analytic approach can be used. Since each subject has a criterion score (e.g., time to perform) on each task, the matrix of correlations of those scores across subjects can be obtained and factor analyzed. Wherry, Jr. [6, 7] used this latter approach to determine the number of underlying internal processes (i.e., perceptual, cognitive, motor, etc.) needed to explain individual differences in times to perform various tactical situation assessment tasks.

5.3. Canonical Correlation Methods

Finally, when multiple criteria (e.g., time and accuracy) are important to the investigator, canonical correlation techniques can be used to determine the overlap between the predictor set of variables and the criterion set. This approach might be especially useful in determining *time and accuracy tradeoffs*.

Regardless of which data analysis technique is used for the analysis of the RSDV data, the results obtained should correspond closely to those that would have been found if the same technique had been used on a truly random sample of real-world data. Thus, the results obtained from the RSDV study should generalize to and be applicable to the applied domain of interest to the researcher.

6. THE AMOUNT OF DATA NEEDED IN RSDV STUDIES

6.1. The Number of Samples to Use in an RSDV Study

An issue in using the RSDV approach is how to determine an appropriate number of samples from the task and situation domains. It is basically the same question that one must ask in any experimental design. The answer lies in one's willingness to make type I errors (i.e., concluding that variables were significant when they really were not), type II errors (concluding that variables were not significant when they really were), and the cost of collecting data cases. The objective should be to obtain a sample size that permits each variable in the domain to be adequately represented in the sample.

It should be kept in mind that use of the RSDV approach requires researchers to become familiar with the domains they desire to investigate. That is, they cannot randomly sample a domain until it has been specified. This additional task seemed unnecessary with ANOVA studies. Actually, this is not quite true; levels of *random* variables in ANOVA studies should be selected randomly after having specified the probable frequencies of those levels in the population. Few ANOVA researchers attend to this requirement, even though they analyze their data and draw conclusions as if their variables were random ones. Even if ANOVA researchers did follow the prescribed procedures to determine the levels of a random variable to use, the independence of variables requirement constrains ANOVA designs to a relatively few levels for each variable. Even though the mean and variance of each variable's levels may yield unbiased estimates of its population mean and variance, those estimates will be far more discrepant than the corresponding values obtained in an RSDV study. Consider an ANOVA design with four main effects that utilize two, three, four, and five levels respectively. The total number of different tasks/situations will be 120 (i.e., 2x3x4x5). The expected variance of the means of the four experimental variables will be those variables' actual (i.e., population) variance divided respectively by two, three, four, and five. In RSDV studies that use 120 different randomly sampled tasks and situations, the expected variance of the means of all experimental variables will be their respective population variances divided by 120. Thus for similar sample sizes, RSDV studies permit all variables to be represented far more adequately than do ANOVA studies.

6.2. The Number of Subjects to Use in RSDV Studies

This issue is identical to the above if each subject performs only one simulated task in the study. In such a study one obviously needs more subjects than experimental variables (i.e., N must be greater than **m**). However, RSDV designs can be highly useful for determining the contribution of individual differences to overall task performance when each subject performs each of the randomly selected tasks. In these repeated measures designs, dichotomous (i.e., "0" or "1") *between subjects* variables can be created for each subject and used as potential predictors of the criterion variable. Raw score predictor weights obtained for these variables indicate the subjects' relative criterion performance above and beyond that which can be predicted from task and situation variables. When repeated measures RSDV designs are used, the order in which the tasks are presented to subjects can be recorded and various functions of the order (e.g., linear, quadratic, cubic, etc.) can also be used as predictor variables for identifying possible learning and fatigue effects. If a sufficient number of tasks and situations are used, it would even be possible to derive individual learning/fatigue curves for each subject.

7. SUMMARY OF STRENGTHS AND WEAKNESSES OF RSDV

The RSDV approach is particularly applicable when investigators desire to do laboratory research in complex (multivariate) domains, and when they desire to be able to generalize there findings to particular real-world situations. The RSDV approach permits the researcher to include all of the suspected important variables in a single study. This frees the investigator from having to arbitrarily curtail investigation of many potentially important variables (as is often necessary in ANOVA designs). RSDV also randomly samples cases from a theoretical population that contains all levels of each variable in the proportion to which they are expected to be found in the real-world situations of interest. Because of this, the means and variances of each predictor variable can be expected to be highly representative of the true means and variances of those variables in the real-world of interest. It follows that the obtained correlations of each variable with the criterion should, under those conditions, yield unbiased estimates of the true correlation between each predictor and the criterion.variable(s). Also, because of the way cases are randomly sampled from the theoretical domain, the proportions of data collected for various parts of the domain will yield a cost-effective sample of that domain (i.e., each portion of the domain is expected to be represented in the overall sample in proportion to its occurrence in the real-world).

The RSDV approach does require the investigator to specify the domains of interest. This takes time and effort and regardless of how careful the RSDV researcher is in doing this, there is no assurance that the domains have not been erroneously specified. When that is the case, the researcher will unknowing claim that RSDV results generalize to the real-world of interest when, in fact, they do not; they only generalize to the specified domain. While it may always be the case that no domain specifications will be entirely accurate, it should not discourage researchers from attempting to do their best at specifying the domains. Further, the specified domains can be published so other investigators know what the specifications were.

Finally, some researchers may object to the RSDV approach because it will not provide the *independent contributions* of each experimental variable to the criterion. This objection will be raised especially in the case where two domain descriptor variables are specified as being related in the real-world. Several comments seem appropriate to refute this objection. First, many of the domain variables may be actually and appropriately specified as independent from one another. When this is the case, the RSDV sample intercorrelations of those variables will be very close to zero and the proportion of criterion variance attributable to them would be highly similar to that found if ANOVA procedures for *random* variables had been followed. Certainly the standard tests of significance for whether a correlation coefficient differs significantly from zero for data sampled at random from a population are well established. Second, many of the specified domain variables may actually be unrelated. In the case where both related variables are unrelated to the criterion, RSDV studies should conclude that neither was related to the criterion variable. Third, even when two variables were believed to be (and were specified as being) somewhat related, the multiple correlation analysis may find that having first selected the experimental variable with the highest correlation with the criterion, the remaining variable will be unable to account for any of the criterion's remaining unexplained variance. In such a case, it would seem reasonable to assume that a causal relationship existed between the first variable and the criterion, but not the second variable. Given a choice of having to accomplish a number of different ANOVA studies or a single RSDV study, it would seem advantageous to do the RSDV study first.

REFERENCES

[1] Adams, J. A. Criticisms of vigilance research: a discussion. Human Factors, 29 (1988) 737-740.
[2] Simon, C. W. Evaluation of basic and applied research: (1) pragmatic criteria. Paper presented at 83rd annual American Psychological Association convention, Chicago (1975)
[3] Dunnette, M. D. Fads, fashions, and folderol in psychology. American Psychologist 21 (1966) 343-352.
[4] Meister, D. Behavioral analysis and measurement techniques. (John Wiley & Sons, New York 1985)
[5] Wherry, Jr., R. J. Theoretical developments for identifying underlying internal processes: Volume 3; Random sampling of domain variance: a new experimental methodology. Naval Air Development Center Report 86105-60 VOL 3 (1986)
[6] Wherry, Jr., R. J. Internal processes for using displayed tactical information: Study #1.Technical Report 1800.31-TR-06, Analytics, Willow Grove, PA (1986)
[7] Wherry, Jr., R. J. Internal processes for using displayed tactical information: Study #2.Technical Report 2100.03-TR-01, Analytics, Inc., Willow Grove, PA (1986)

Trends in Ergonomics/Human Factors V
F. Aghazadeh (Editor)

HUMAN RELIABILITY: A THERP APPLICATION AT AUTOMATIC STORAGE AND RETRIEVAL SYSTEM

B. Mustafa PULAT

AT&T Network Systems
Oklahoma City Works
7725 West Reno Avenue
Oklahoma City, OK 73126-0060

Measuring and predicting the effect of the operator(s) on equipment/system performance is a topic receiving increasing attention. This paper presents the application of one of the available methods, namely THERP to estimate AS/RS functional as well as total system reliability in terms of inventory balance accuracy.

1. INTRODUCTION

The topic of estimating system reliability has attracted much attention in the literature since early sixties. Ankerbrandt [1], Goldman and Stattery [2], Sandler [3], and Green and Bourne [4] give thorough treatment of the subject matter.

Identifying the hardware components of a system and carrying out reliability analysis is usually a trivial task for an engineer. However, if a system is composed of equipment and human elements, then one has to identify all human tasks (units of behavior), estimate reliability of each, and combine these with equipment reliabilities in order to estimate total system reliability.

Several differences exist between humans and hardware components of a system in terms of characteristics that lead to reliability analysis. First, most equipment operate in an open-loop system. This means equipment errors ordinarily go uncorrected until an attendant detects the error and takes corrective action. However, humans often function in a closed-loop system, where they detect and correct their own errors. Another confounding factor is that machines mostly assume binary states (success or failure) in terms of operating characteristics. Human error does not possess this logic. Furthermore, one cannot often assume independence of tasks when humans are the predominant actors. It is at least for these reasons that one needs to approach the problem of estimating man-machine system reliability with caution.

2. PROBLEM STATEMENT

Oklahoma City Works of AT&T Technologies has recently operationalized an AS/RS (Automatic Storage and Retrieval System) for automating warehouse related activities. This is a mini-load AS/RS which accepts materials from the receiving dock, determines shortage requirements, routes material to either stocking or shortage fill, performs directed stocking, responds to material pull requests, accumulates selects for shop delivery, and performs cycle counts. A bar code system is used to track materials on the conveyor systems, as well as to assure proper stocking and selecting. On-line software systems monitor and control the operations of cranes. Furthermore, receiving, stocking, pick, and audit transactions are communicated with the MRP system on real time basis.

The study reported in this paper was triggered by management request to review the reliability of AS/RS. For Class A user status of MRP-II, an SBA performance - Storeroom Balance Accuracy of a minimum 95 percent is necessary. Thus, implicit in this charter were the steps of:
a) Reviewing the procedural steps of material handling via AS/RS.
b) Calculating a "percent process capability" (expected performance) in terms of material count accuracy for the system.
c) Matching existing SBA performance with expected performance.
d) Making improvement recommendations if SBA and/or expected performance fall below the 95 percent mark.

This paper reports on the first two steps.

3. SYSTEM RELIABILITY MODELS

Reliability (R) is the probability that a system will perform satisfactorily. Its complement is the probability that a system will fail, and is expressed as 1-R.

Reliability mathematics aim at calculating system reliability where components are connected in a functional relationship such as series connections, active redundancy, standby redundancy, and active redundancy with a dependency. Ireson [5] and Calabro [6] may provide information on mathematical reliability to an interested reader.

The point of interest in this paper is on a technique which aims at combining human and equipment reliability to yield system reliability. Although one may run into methodological risks in search of such a procedure according to Adams [7], all have not been lost. A number of procedures have been developed during the last 25 years that deal with systems safety, which one can borrow from, since they focus on evaluating all elements of a system (humans as well as hardware components) in order to predict system failure probability. The Accident Prevention Manual for Industrial Operations [8] classifies these into inductive and deductive categories. The inductive method of analysis considers components of a system and how they contribute to success or failure of the system as a whole. The FMEA - Failure Modes and Effects Analysis and CHA - Construction Hazard Analysis may be given as examples to the inductive methods. Conversely, the deductive methods tell us how a system may fail. Several examples of such methods are the Fault Tree analysis, MORT, THERP, and other decision tree approaches. Nertney [9] and Johnson [10] give more details of these techniques. A deductive analytical procedure was selected for the purposes of this study, namely HRA (Human Reliability Analysis) event tree approach with significant inputs from THERP and SHERB.

3. THERP - TECHNIQUE FOR HUMAN ERROR RATE PREDICTION

The Sandia Corporation [11] - now Sandia National Laboratories - developed a procedure called THERP in order to quantify the likelihood of human error in manufacturing operations. It uses rigorous data gathering and statistical procedures to calculate basic error rates (BER) for industrial tasks. SHERB (Sandia Human Error Rate Bank) (Rigby, [12]) is a compilation of error-rate data based on THERP (Swain, [13]).

4. HRA EVENT TREE

Swain and Guttman [14] propose the use of HRA Event Tree approach in estimating the reliability of a system composed of human, hardware, and software components. This technique considers a man-machine system to possess various functionalities to attain a common goal. For each functionality, one performs detailed task analysis in order to pinpoint equipment and human tasks and precedence relationships between those tasks. Then, flowcharts are developed in network form spanning the binary states - success and failure (errors) - of each task with BERs and equipment failure rates assigned to human and machine tasks respectively. For those tasks that require segments of a behavioral sequence, joint probabilities are calculated. These probabilities may have to be adjusted by a "recovery factor" if the conditions necessitate.

Recovery factors are assigned to tasks where the operator takes extra care for correct performance due to an extraneous factor such as a process checker or a supervisor monitoring performance (looking over the shoulder). Dependencies between tasks are handled by calculating conditional probabilities according to a "dependence level" classification scheme as summarized in Table 1.

Table 1. Equations for conditional probabilities of success on Task "N" given success probability (n) on Task "N-1"

DEPENDENCE LEVEL	SUCCESS EQUATION
ZD	$Pr[S_N \mid S_{N-1} \mid ZD] = n$
LD	$Pr[S_N \mid S_{N-1} \mid LD] = \frac{1 + 19\,n}{20}$
MD	$Pr[S_N \mid S_{N-1} \mid MD] = \frac{1 + 6\,n}{7}$
HD	$Pr[S_N \mid S_{N-1} \mid HD] = \frac{1 + n}{2}$
CD	$Pr[S_N \mid S_{N-1} \mid CD] = 1.0$

Dependence levels between successive tasks are assigned based on several guidelines including spatial, temporal, and actor relationships among events. The possibilities are five-fold: ZD: zero dependence, LD: low dependence, MD: medium dependence, HD: high dependence, and CD: complete dependence.

Next, an event tree diagram is drawn with the complete success path(s) delineated. Each task on this path has a success probability associated with it. Thus, it is an elementary task to calculate the Functional Success Probability (FSP). FFP - Functional Failure Probability is then complement of FSP:

$$FFP = 1 - FSP$$

Assuming independence between the various functionalities, the System Failure Probability (SFP) is calculated as follows:

$$SFP = \sum_i FFP_i \qquad i = 1,\ldots,m$$

where,

m = # of functionalities considered.

Then, system success probability is 1 - SFP = SSP.

5. METHODOLOGY

The HRA Event Tree Technique was applied to obtain AS/RS mini-load success probability. Six major functionalities were assessed to include almost all system error potential:

1. Storeroom receiving
2. Stocking
3. Pick and delivery
4. Accumulation
5. Bin link
6. Bin unlink

For each functionality, detailed flowcharts were drawn to display the sequence of events. This resulted in several conditional branches for which branching probabilities were calculated after field observations. Equipment tasks were assigned success probabilities based on experience with those equipment. Human tasks were evaluated through either judgemental task data (Woodson, [15]) or empirical evidence such as those available from THERP, SHERB, and AIR - American Institute for Research Data Store (Payne and Altman, [16]). Recovery factors and task dependency levels were assigned through observations.

6. RESULTS

Calculated success probabilities for the above-mentioned functionalities ranged between 95.14 percent for pick and delivery, and 99.8 percent for bin linking. These results are in line with the relative complexities of the two functions. The System Success Probability (SSP) or process capability was calculated as 87.4 percent. However, this score refers to the condition where there are no inventory discrepancies between computer records for a particular part in a specific bin and actual count in that bin. At ∓5 percent allowed discrepancy, and our most recent experiences with SBA, expected process capability performance goes up to 94.7 percent.

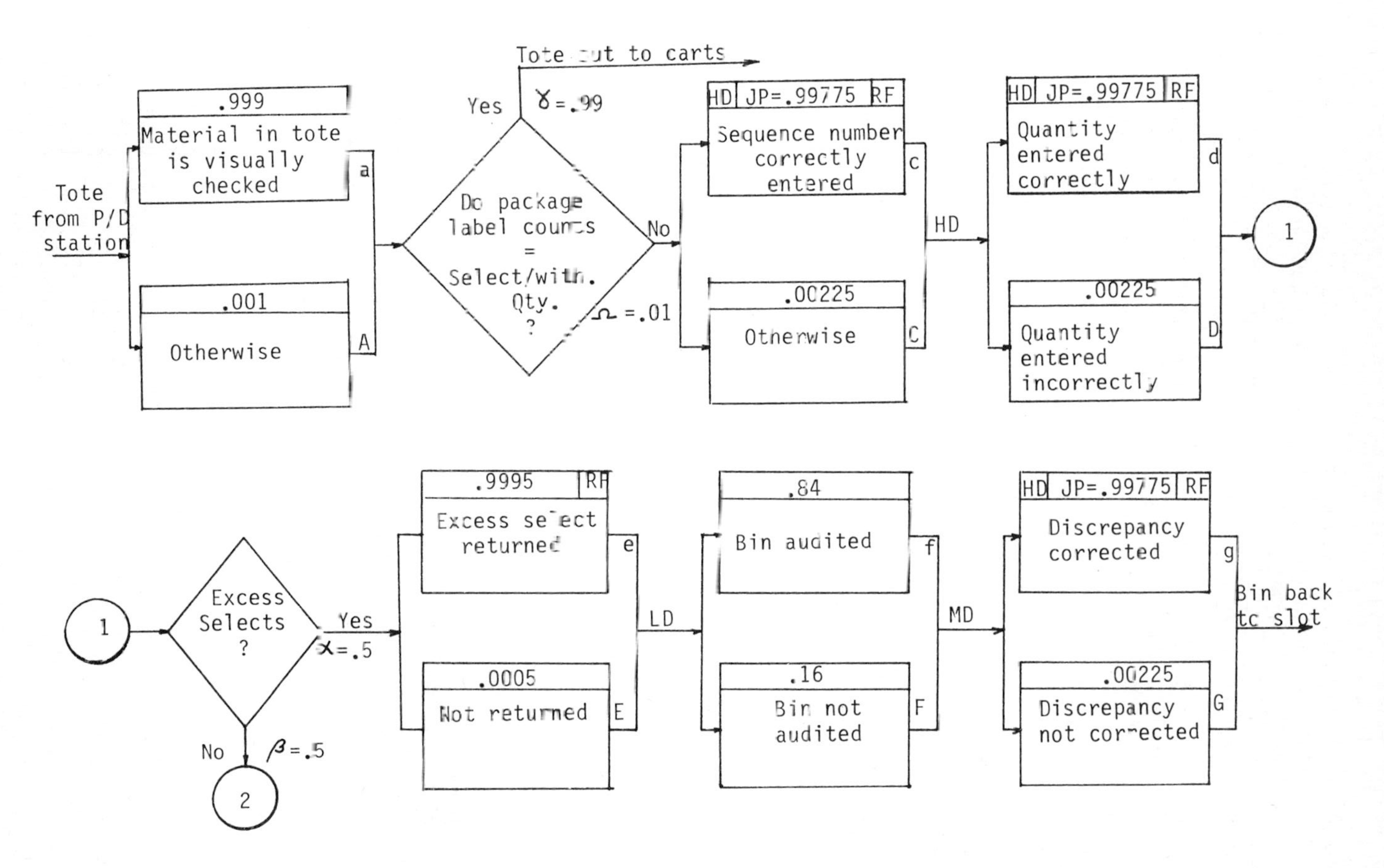

Figure 1. A portion of the operation flowchart of accumulator function.

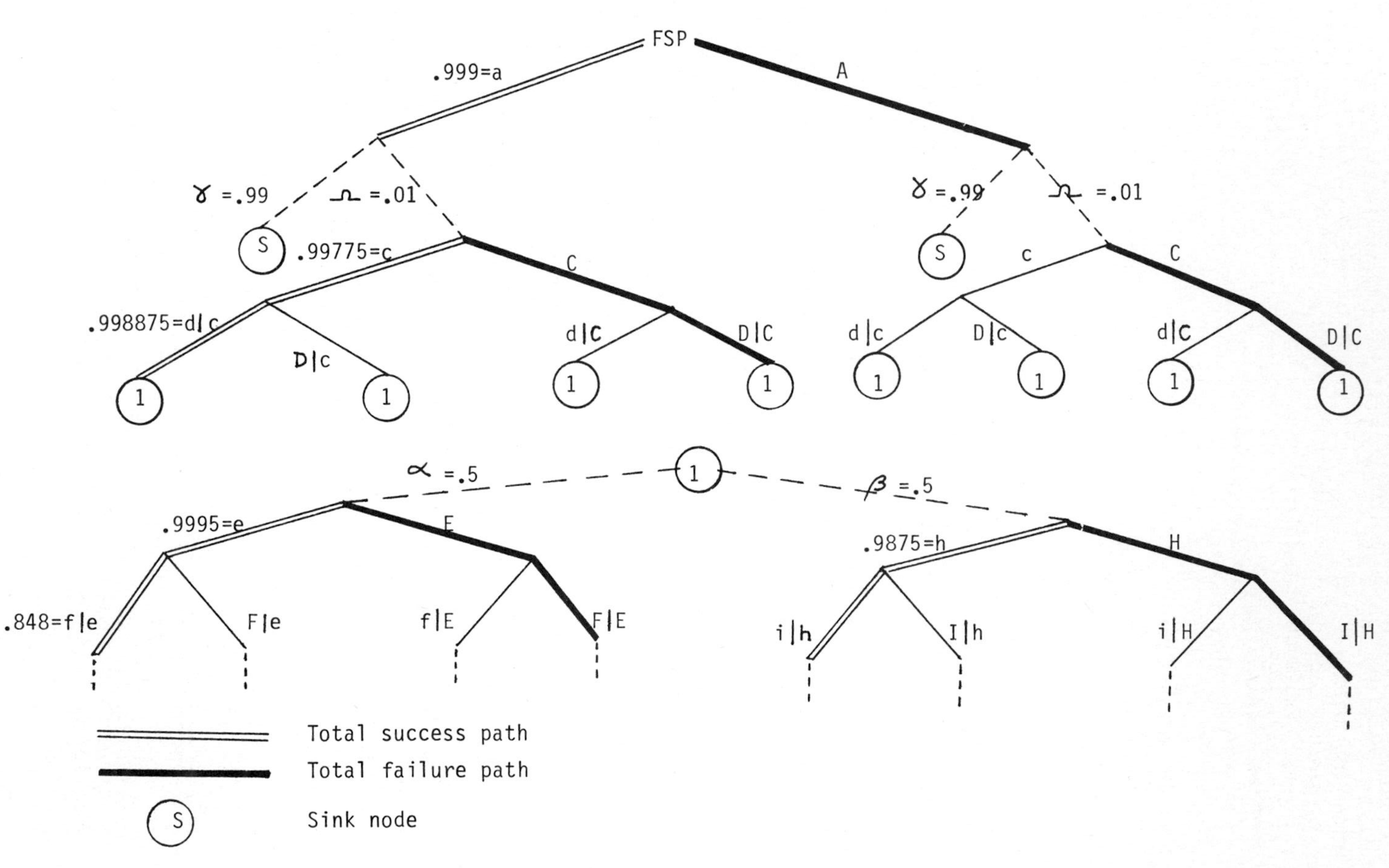

Figure 2. A portion of the HRA Event Tree Diagram for accumulator function.

This is very close to the 95 percent minimum requirement level, therefore, our conclusion was that the process is capable of meeting the minimum requirement. Several improvement ideas were tested and implemented, however, these are beyond the scope of this study. Actual experiences with the modified system suggest that we are achieving higher levels of accuracy.

Figures 1 and 2 provide portions of the accumulator function operation flowchart and HRA Event Tree Diagram respectively.

7. CONCLUSIONS

Estimating the reliability of a system composed of human and equipment elements is not an easy task. There are problems in calculating human reliability, as well as combining these with equipment reliability to yield system reliability as stated by Adams [7]. However, one may be able to use combinations of well accepted procedures to reach this goal with fairly high level of confidence until exact techniques are developed.

REFERENCES

[1] Ankerbrandt, F. L., (Ed.) Maintainability Design, Elizabeth, NJ, Engineering Publishers, 1963.

[2] Goldman, A. S. and Slattery, T. B., Maintainability: A Major Element of System Effectiveness, New York, John Wiley, 1964.

[3] Sandler, G. H., System Reliability Engineering, Englewood Cliffs, NJ, Prentice Hall, 1963.

[4] Green, A. E. and Bourne, A. J., Reliability Technology, New York, Wiley Interscience, 1972.

[5] Ireson, W. G., Reliability Handbook, McGraw Hill Book Co., New York, 1966.

[6] Calabro, S. R., Reliability Principles and Practices, McGraw Hill Book Co., New York, 1962.

[7] Adams, J. A., Issues in Human Reliability, Human Factors, Vol. 24, No. 1, 1982, pp. 1-10.

[8] Accident Prevention Manual for Industrial Operations: Administration and Programs, National Safety Council, Chicago, 1981, pp. 95-98.

[9] Nertney, R. J., Practical Applications of System Safety Concepts, Professional Safety, Vol. 22, No. 2, February 1977.

[10] Johnson, W. G., MORT Safety Assurance Systems, Marcel Dekker, Inc., New York, 1980.

[11] Sandia Corporation, Reduction of Human Error in Industrial Production, SCIM 93-62. (14)1, June 1962.

[12] Rigby, L. V., Sandia Human Error Rate Bank (SHERB), Presented at Symposium on Man-Machine Effectiveness Analysis: Techniques and Requirements, Santa Monica, CA, June 15, 1967.

[13] Swain, A. D., THERP (Technique for Human Error Rate Prediction), Proceedings of the Symposium on Quantification of Human Performance, Subcommittee on Human Factors, Electronic Industries Association, Albuquerque, New Mexico, Aug. 17-19, 1964.

[14] Swain, A. D. and Guttman, H. E., Handbook of Human Reliability Analysis with Emphasis on Nuclear Power Plant Applications, U.S. Nuclear Regulatory Commission, Washington, D.C., NUREG/CR - 1278, 1983.
[15] Woodson, W. E., Human Factors Design Handbook, McGraw Hill Book Co., 1981, p. 989.
[16] Payne, D. and Altman, J. W., An Index of Electronic Equipment Operability: Data Store, Pittsburgh: The American Institute for Research, 1962.

BIOGRAPHICAL SKETCH

B. Mustafa Pulat received his B.Sc. and M.Sc. degrees in industrial engineering from the Middle East Technical University in Ankara, Turkey. His Ph.D. in the same field is from North Carolina State University, Raleigh, NC. Pulat has been conducting basic research and industrial application studies in the area of ergonomics for the last ten years. He is currently a planning engineer at the AT&T Oklahoma City Works responsible for material flows.

Trends in Ergonomics/Human Factors V
F. Aghazadeh (Editor)

THE IMPORTANCE OF ERGONOMICS/HUMAN FACTORS CONSIDERATION IN THE DEVELOPMENT OF EXPERT SYSTEMS

Avelino J. GONZALEZ and Chin H. LEE

Department of Computer Engineering and Department of Industrial Engineering and Management Systems
University of Central Florida
Orlando, Florida

The objective of this paper is to suggest ergonomics design criteria for the knowledge engineer to assist him/her in the performance of this function. Its main thrust is to describe sound ergonomics design of the knowledge extraction process. Such a process represents the interactions between the knowledge engineer and the expert or experts, and includes the interview process as well as the knowledge base verification process. Special attention will be paid to the design of the physical environment in which these processes will take place. The paper will serve as a guide to institutions interested in developing in-house knowledge engineering expertise for a medium to large application project. This guideline can also be used for the evaluation of multiple design alternatives for specific user groups.

1. INTRODUCTION

Ergonomics/human factors considerations in the design of man-machine interfaces has, in the past decade or so, made the transition from an art to a science. Human factors engineers have determined design criteria in the layout of such important systems such as aircraft cockpits and nuclear power plant control panels.

However, little work has been done in applications where there is no direct man-machine interface, but rather a man-man interface with a machine being of auxiliary importance. Such is the case in the development of expert systems, where the main interaction is between the system developer and the domain expert. The machine plays an auxiliary role, yet its importance must not be understated because, ultimately, its interface with the end-user may spell the difference between an otherwise successful system and failure.

This paper will attempt to shed some light on the need for ergonomic design criteria in the development of expert systems.

2. EXPERT SYSTEMS BACKGROUND

Expert systems is a branch of the larger field of artificial intelligence (AI). It is concerned with building computer systems which use heuristic processes to solve complex problems. Such problems can generally only be solved by individuals highly knowledgable in the domain of the problem (experts). Experts often use techniques which are not easily described by algorithms since they use their vast experience which is manifested as hunches, rules of thumb, etc. This problem solving methodology has not been easily represented by computers using traditional means because thay could not easily handle symbolic computing.

Expert systems research has been ongoing since the birth of AI research in the mid-50's. However, the name was not universally accepted until the emergence of the system MYCIN [1] in the late seventies and early 80's. MYCIN uses the knowledge of a hematologist to diagnose blood disorders. Numerous other systems have been developed in the last ten years [2, 3, 4], thus giving the field expert systems the dubious distinction of being "mature" technology in less than a decade after its emergence into the world of real life applications.

One of the key features of expert systems is the separation of the "inferencing mechanism" from its "knowledge base." The inferencing mechanism represents the logic-based chain of reasoning from which deductions are made from statements. For the purpose of illustration, simplifications are being made here. Nevertheless, the means of reasoning can be the same, regardless of whether the statements are in the area of medicine or in weather forecasting. This common inferencing mechanism, however, still requires knowledge in order to solve a problem, and that is what is referred to as a knowledge base.

This concept has resulted in the development of so called "expert systems shells" which use a specific inferencing mechanism and some auxiliary features to assist the developer in building the expert system. The developer only has to construct the knowledge base in order to have an expert system. While they have their disadvantages as well as their advantages, shells have proved to be very popular and they represent the preferred means of building small expert systems. Some medium and large systems have also been successfully built using shells.

The process of developing the knowledge base is commonly referred to as knowledge engineering. The objective is to take the knowledge existing in a domain and represent it in a way which the inference mechanism can use in order to arrive at solution to the problem. The knowledge can reside in documents, but in most cases it is found only in the expert's mind in terms of rules. Thus, it must be extracted before it can be represented in the computer. This process is called "knowledge extraction" and it can be a tedious one, consisting of numerous interview sessions between the knowledge engineer and the expert.

It is exactly this process which can benefit from ergonomics considerations in order to optimize it.

3. KNOWLEDGE EXTRACTION

Like the larger heuristic process of which it is a part, there is no model for knowledge extraction. Different people do it in different ways, and certain techniques work well in some instances and not in others. However, there are some general concepts that apply in a large majority of cases. One of these is the concept of incremental development. Incremental development (ID), entails an iterative process where the KE extensively extracts knowledge from an expert until a certain limit is reached where the knowledge gathered can be self-sufficient, yet it does not necessarily represent the entire knowledge base. Such groups of knowledge are commonly referred to as "chunks" of knowledge. For example, if an expert system is developed to diagnose an automobile, then one such chunk can be the knowledge to diagnose the electrical system of the car. Another chunk would be the corresponding knowledge for the engine, etc.

Once one chunk has been identified and represented in the shell, then the ID process calls for a checkout of the knowledge collected in the chunk of interest. When the checkout is done, then knowledge extraction would again begin on another chunk, etc., until the knowledge base is completed.

In most cases, the knowledge engineer (KE) knows little about the domain. Therefore, most of the interviews early in the extraction process deal with simply educating the KE. This is also true if the KE is knowledgeable about the domain (but not an expert!) except it takes correspondingly less time. Once the KE feels comfortable enough with the domain then he begins the extraction of the expert knowledge. This will be referred to in this paper as Phase I of the extraction process.

Experts can sometimes initially look upon expert systems and the KE as either a threat to their status or as a nuisance. For the ultimate success of the expert system, a cooperative effort between the KE and the expert is a necessity. Anything less than that threatens the viability of the process. Many failed systems can point to this as a major contributor to failure. It is thus important for the KE to win the trust and cooperation of the expert as early as possible.

Onething that can help accomplish this is to hold all Phase I interviews at the expert's home workspace if at all possible. For one thing, this is a good choice from a logistical standpoint because the expert has quick access to all supporting documentation and other "props" which he may need to illustrate a point. More importantly, however, it gives the expert the feeling that "Mohammed is going to the Mountain" rather than vice-versa. Thirdly, people generally prefer to go through stressful situations on their home turf. Although the above is not strictly ergonomic design criteria, it is a human factors consideration which should be considered during the knowledge extraction process.

In the process of incremental development, the time comes when the KE has captured a more-or-less self-contained chunk of knowledge from the expert, and is now ready to represent it electronically on the computer. For this purpose, the expert is usually brought to where the expert system resides. ID calls for constant and consistent evaluation and feedback by the expert. After all, it is the expert who must determine whether the expert system is adequately solving the problem. This review process is referred to in this paper as Phase II. It may involve uncovering of new knowledge, but only as a by-product of the review. No new knowledge is actively being sought at this time.

Phase II requires the use of an interface of some kind to the computer where the expert system resides. This phase requires the use of visual display terminal (VDT), a CRT in an overwhelming majority of cases, and a keyboard to communicate with computers. This is the area where ergonomics consideration will play a large part.

4. ERGONOMIC/HUMAN FACTORS CRITERIA

A number of recommendations for VDT workplace design is available. However, the following list covers a wide range of ergonomics considerations related to the display side of the man/machine interface:

Workspace: One of the key problems in the implementation of Phase II is that during the system checkout, the KE as well as the expert (or experts must be able to see the screen through which the I/O with the expert system must be carried out. Most computer workplaces are intended for operation by one person. When two (or more) people work at the same time, it can become quite cumbersome, especially if these people also require the extensive use of documents (manuals, blueprints, etc.)

One answer for this is to design a workplace which can comfortably seat three or four people, allow easy viewing of the CRT (or CRT's), and provide ample and accessible working area to spread out manuals or documents, for the experts as well as for the KE. The application of anthropometric data is required in the design of work space in order to accommodate the most of user population.

Safety: VDT generally will not expose users to such immediate hazards as radiation or explosion, however, the stress, both physical and mental, of prolonged use should be considered.

Viewing Angle: The optimum viewing angle is usually perpendicular to the surface of the VDT. For this purpose the VDT screen should be tilted back slightly and the angle should be adjustable by the user. The recommended viewing angle for a CRT is between 19 and 38 degrees off center. [5]

Viewing Distance: The recommended viewing distance is generally given as 24 inches. This figure can be as low as 16 inches or as high as several feet as long as the display remains legible to the viewer. Viewing distance differences between the display and other work surfaces should be kept to a minimum to avoid eye fatigue caused by frequent refocusing.[6] When users are required to touch the screen the view- ing distance should be less than 28 inches.[5]

VDT: The VDT screen should be reflection free. If the reflection occurs due to tilting the screen, an antireflective filter should be provided.

Color: The use of color depends on the nature of the task. In general, background and foreground colors should not be the cause of visual problems. However, too much or inappropriate use of color can actually cause eye fatigue and degrade the performance of users.

Controls: When possible the viewer should be able to adjust the display's visual characteristics (luminance and contrast) to suit ambient light conditions. The type of cursor control should also fit the application requirements.

As mentioned above, Phase II requires the attention of ergonomics principles in the design of the workplace.

Once the cycle of Phases I and II has been executed the number of times necessary in the course of developing the larger system composed of various chunks of knowledge, Phase III will finally arrive. Up to that point, the various chunks have been reviewed independently, but no system checkout has been done. This is of what Phase III will consist. At this point, it is highly desirable to make the expert system accessible to the expert so that he can exercise it at his leisure without the influencing presence of the KE. Since in most cases the expert is not a computer-oriented individual (otherwise they probably would not need a KE), it is important that the prototype be easily accessible and its interface be "friendly." At this point, the design of this interface is very independent of the application and the expert involved, so it is necessary to define the human factors criteria to be followed. The following factors are needed to be considered in the human factors criteria:

Coding: The way in which information is presented for visual interpretation on displays is called coding. Different types of coding include color, shape, size, position, intensity, and so on. These modalities have different coding levels that represent effectiveness in representing information.

Selectivity: This is the ability of the user to get the exac information he needs from the expert system.

Feedback: This factor relates to user friendliness. The user will feel more comfortable with the system if he is fully prompted.

Redundancy: Too much redundancy can be counter productive. An example would be mixed coding such as numbers, colors and shapes all representing the same things.

Display Response: It is important that the system response rate and the user response rate are closely matched. If the information is presented too briefly or delayed too long, the user's performance will suffer.

Character recognition: This factor is related to legibility. Characteristic size and font style affect recognition.

Interactivity: This is a measure of the frequency with which the operator must respond to the displayed information.

Compatibility: The displayed information should be compatible with the user's knowledge and skill.[5]

Readability: The degree to which the user understands the information is a measure of readability.[5]

Format: The amount of information displayed and its density is a function of the display format. The layout can be simple or complex, although a complex layout may reduce performance.[7]

Grouping: The way in which information is organized is called grouping. There are four fundamental grouping techniques; sequence, frequency, function and importance. Data can be organized by a combination of these methods.[5]

Some additional considerations are as follows:

- Except for those in the high-end of the spectrum in terms of sophistication, most shells do not have a very flexible user interface. This means that often times, a certain amount of user (not developer!) education is required. Such may tend to discourage experts. The KE may see fit to develop a sub-set of the instructions specifically for use in the expert system in question.

- If the system is being developed in a sophisticated LISP environment commonly found in single-user LISP machines, the expert may have difficulty in even accessing such machines. Although no real solution exists for this, it may be one of the considerations made when the hardware platform is being chosen.

- If system errors occur during a consultation by an expert (as opposed to a knowledge base error which results when an incorrect solution is chosen), the expert will need assistance in resolving the error. Therefore, either the KE or someone with system knowledge must be readily available for the expert to contact when this happens.

Although it was the authors' intention to include as many ergonomics/human factors considerations as possible in the development of expert systems, still the above list is not complete. However, a useful ergonomics/human factors data base exists and should be used in expert system's development. Furthermore, information processing abilities and expectations of the user should not be overlooked.

5. CONCLUSION

Development of an expert system knowledge base can be a difficult task. A large part of the difficulty is in the knowledge extraction effort. Human factors/ergonomic considerations can make the process more palatable and thus increase productivity. Such considerations should be made part of an expert system's functional specifications.

REFERENCES

[1] Shortliffe, E. H., (1976), Computer-Based Medical Consultation: MYCIN, New York: American Elsevier.

[2] Gershman, A., "Building a Geological Expert for Dymeter Interpretation." Proceedings of the European Conference on Artificial Intelligence, pp. 139-140, (1982).

[3] Garchnia, J., "PROSPECTOR: An Expert System for Minera Exploration." In Machine Intelligence, Infotech State of the Art Report 9, No. 3, (1981).

[4] Kahn, G. and McDermott, J., "The MUD System", Proceedings of the First Conference on AI Applications, IEEE Computer Society, (December 1984).

[5] Bailey, R. W., Human Performance Engineering, Prentice-Hall, Englewood Cliffs, NJ (1982).

[6] NCR Corp., Design Guidelines, Ergonomics. (Dayton, OH 1986).

[7] Tullis, T. S.,"The Formatting of Alphanumeric Displays: A Review and Analysis". Human Factors, Vol. 25, (December 1983).

Trends in Ergonomics/Human Factors V
F. Aghazadeh (Editor)

ERGON-EXPERT - A KNOWLEDGE-BASED APPROACH TO THE DESIGN OF WORKPLACES

Volker Rombach and Wolfgang Laurig

Institut für Arbeitsphysiologie an der Universität Dortmund
Abteilung Ergonomie
Ardeystraße 67
D-4600 Dortmund 1
Federal Republic of Germany

Current results from the development of a knowledge-based system for the ergonomic design of workplaces are presented. The system follows the concept of prospective ergonomics as proposed at the Louisville conference in 1986. A prototype system is devoted to the widely known and explored problem of back pain. Special attention was paid to the user interface in order to improve the acceptance of the system and to strengthen its educational aspect. The knowledge base of the system contains basic knowledge for dealing with ergonomic problems relating to manual materials handling, as well as data and models for the forces acting on the various parts of the human body.

1. INTRODUCTION

A considerable number of health hazards, often resulting in illness, are to be found in many of today's workplaces. Increased costs are incurred as a consequence of losing the sick workers' productivity. Industrial hygiene research has produced numerous guidelines, rules and limitations, in addition to theoretical models, to describe and overcome this problem. However, progress has hitherto been slow and the number of workers forced into early retirement (as a result of back pain, for example) continues to rise.

The form of ergonomics predominantly supported by plant physicians corresponds more to traditional industrial hygiene, with its aim of health-hazard prevention, than to industrial ergonomics. The approach advocated by industrial ergonomics, promising both health-hazard prevention and profit maximization, is accordingly largely unknown in Germany. Unfortunately, in Germany as elsewhere, it is only possible to assume even a basic scientific background with regard to industrial ergonomic design and workplace organization for a few, mainly large, companies.

In the Federal Republic of Germany attempts have been made since about the early 70's to improve occupational safety and to introduce new concepts of work organization and participation. This is reflected in the proclamation of a political program for the "humanization of work". In the USA attention seems to center primarily on improving productivity. Interest in the improvement of occupational health as `compelled' by legislation in the form of the "Occupational Safety and Health Act" (OSHAct, 1970) and the foundation of the "Occupational Safety and Health Administration" (OSHA) is also expressly oriented towards the interests of the national economy.

A concept of prospective ergonomics was proposed at the conference in Louisville in 1986 in an attempt to overcome the seemingly conflicting interests of health-hazard prevention and profitability maximization. The Louisville paper [1] suggested the utilization of an expert system containing knowledge about the worker and the workplace which could provide a tool to minimize health hazards without reducing productivity. The knowledge-based approach to the analysis of manual materials handling tasks as presented by Karwowski et al. [2] provided a strong impetus for the development of the system. This paper aims to present the results achieved to date in the development of the aforementioned expert system.

2. THE APPROACH

According to Taylor and Corlett [3], an important object of a knowledge-based design is not only "to inform the user of unacceptable factor constraints", but also "to constrain ergonomic factors by invoking appropriate models". The usefulness of paying attention to ergonomic factors can therefore be demonstrated by selecting a problem with known limit values for unacceptable factors.

Rogers [4] states that "most industrial safety or loss control managers will acknowledge that low back pain is the single most important factor in workers' compensation costs". Nevertheless, practitioners in general express a variety of reservations with regard to the recommendations for the safe design of lifting and carrying activities contained in guidelines (e.g., NIOSH [5]). One plausible explanation for such reservations is that the complexity of a particular case of materials handling in an individual company does not always permit the application of simple guidebook recommendations. An expert-support system is therefore being developed in an attempt to overcome the remaining resistance in industry to ergonomic proposals.

The complexity of manual materials handling tasks necessitates the fundamental design decision of dividing the knowledge-based system into modules. The modules are then linked in a problem-solving hierarchy. The first module always starts with an ergonomic analysis of the specific operation.

In the analysis step all the "case facts" for the work sequence under examination are collected. The first input is a step by step description of the work sequence according to the process chart principle. A series of standardized operation descriptions (e.g., "standing: without or with lifting a load" - "walking: without or with carrying a load") is provided in advance for this purpose. Supplementary quantitative details are demanded for each of these operations (e.g., for "lifting or carrying load": load mass in kg, load dimensions in cm, grasp height, release height...). The determination of the "case facts" ends with questions about the persons planned for the activity (sex, age, weight, height, body dimensions and statements about the persons´ physical condition using terms such as poor, medium and good).

An "assessment module" is used in the next step. The guiding hypothesis in the development of this module is the necessity of eliminating as many health risks as possible from the sequence.

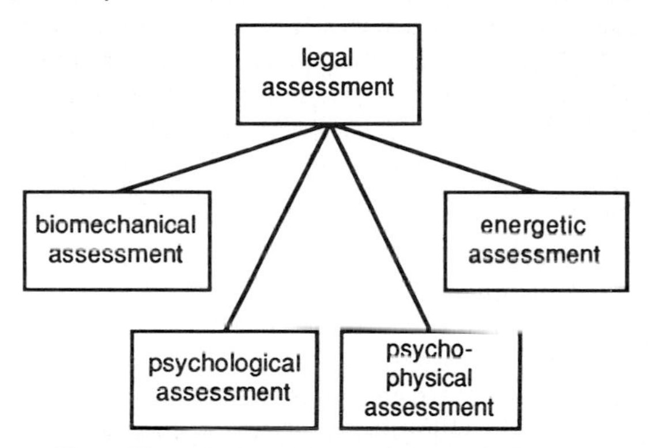

Fig. 1: The assessment module of Ergon-Expert

According to the results of biomechanical studies, the risk of musculo-skeletal injuries and illness due to overexertion can be estimated by means of biomechanical model calculations, as suggested for example by Morris et al. [6], Chaffin [7], Ayoub and El-Bassoussi [8], and Jäger et al. [9]. Ergon-Expert will include a three-dimensional dynamic biomechanical model for the prediction of compressive forces on the lumbar discs [10]. The output of the "assessment module" allows interactive testing of the various design possibilities at workstations as shown in Fig. 2.

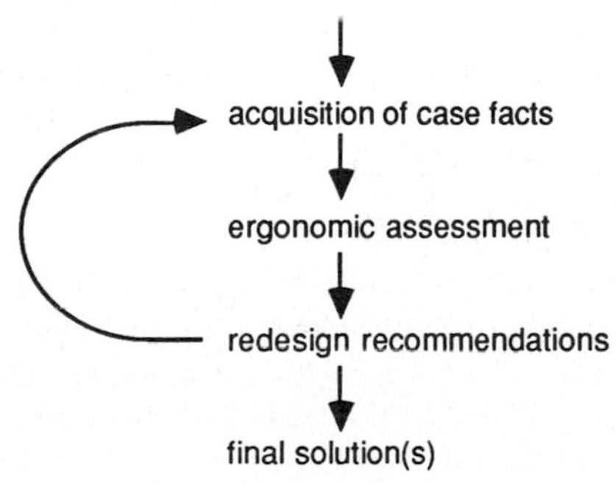

Fig 2: "generate and test" cycle in Ergon-Expert

The main problem presented by the concept of a knowledge-based system is how to model the knowledge of an industrial ergonomist through the combination of such complex types of modules. The development tool for the knowledge-acquisition and linking processes therefore becomes the principal key in solving the problem of the knowledge-based approach in industrial ergonomics.

3. THE DEVELOPMENT TOOL

The knowledge structures found in the knowledge-acquisition process are varied. They include rules (e.g., design guidelines), taxonomies, and a lot of detailed information about the various aims of a work sequence. The utilization of an expert system shell which is capable of processing knowledge in as many forms as possible minimizes the necessity of having to transform the knowledge structures. In this way the knowledge base remains transparent to the user and easy to expand.

Given the above requirements, an expert system shell had to be chosen which is a member of the class of hybrid tools. The tool which seemed to best meet the demands made by the approach is NEXPERT OBJECT [11]. Besides being able to process various knowledge representations, it is also a tool with a very flexible user interface. Most importantly, however, NEXPERT OBJECT runs on quite a number of different computer systems, for instance on IBM-AT and compatible microcomputers under MS-Windows, the Apple Macintosh line, VAXstations, Unix-machines and the HP-9000 series. This feature meets the need for a system which can be used not only in a research laboratory, but at almost any workplace. The knowledge-base kernel containing the domain knowledge of an expert system is easily transferred between these machines with only minor modifications.

NEXPERT OBJECT is a window-based expert system shell. This means that the information presented to the user is displayed in windows which can overlap, be resized or hidden during a session. The display can either be manipulated by the user or controlled by the system. In addition, attention has been paid to integrating the shell into the "typical" operating system environment. Thus the amount of time required to get used to the system is reduced and errors due to a false mental model of how the system works are minimized.

The inference engine of the system is capable of both backward and forward chaining and can be controlled by a "strategy" operator during run time. This option enables control of the run-time behaviour of the system from within the knowledge base by switching between a hypothesis-driven and a data-driven evaluation of a problem. Uncertainty factors can be used to express uncertainty regarding conclusions. External data can be utilized during a session due to an interface to database and spreadsheet files, as well as to the reading of external sensor data. The information collected in a session can also be exported into database or spreadsheet files. At any time during an inference process the current status of the evaluation of a given hypothesis can be displayed. In addition to the display of the state of all the variables, a semantic network of hypotheses and data can help the user to understand the interdependencies between hypotheses and data (Fig.3). These features, along with the ability to ask questions such as "why is hypothesis 'x' true?", contribute to the transparency of the inference process.

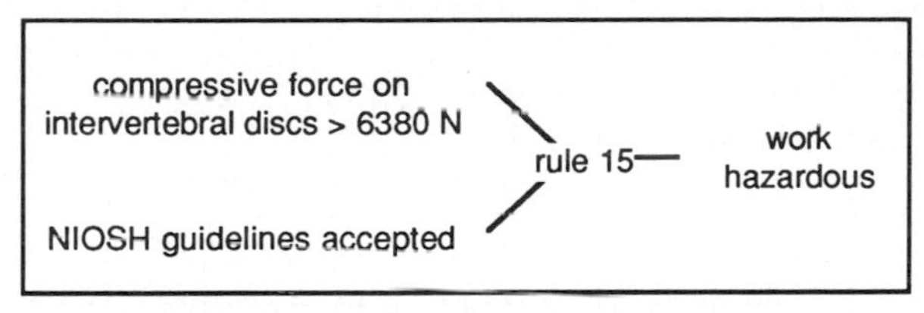

Fig.3: Example of a simple rule network in NEXPERT OBJECT

Three different types of user interfaces are available. First, there is the basic interface where the developer is at all times able to show, for information purposes, texts and graphics as actions triggered by rules. Next, there is an interface to a programming language like "C" or "PASCAL" which can be used to program a user interface, communicate with the expert system, or use external routines for calculations. The third possibility is to use a graphic object-oriented interface as a replacement for the basic interface. The programming interface can also be used to control any device connected to the computer system. This feature could, for instance, be used to control a video or audio cassette recorder in order to elucidate certain conclusions reached by the system. The use of some of these features in the development of the system is revealed in the following chapter.

4. THE USER INTERFACE OF ERGON-EXPERT

The simple and provable rule that the user interface of any software should be adapted to the users of the software has led to long discussion and research relating to the necessary look (and feel) of the user interface of a software system. Although "the" user interface has not been realized (and probably never will be), at least many guidelines have been developed (e.g., [12] and [13]). Since acceptance of a piece of software and user productivity depend very much on the degree of adaptation of the user interface to the users, research into user characteristics has become an important aspect in modern software development. One method employed recently to make systems more useable and acceptable for different classes of users, for instance in the design of ALFIE [3], is the utilization of adaptable or self-adapting user interfaces (as described by Palme [14], for example).

Ergon-Expert is designed for the following users:

- industrial engineers concerned with the design of workplaces
- human factors experts
- OSHA-inspectors

- plant physicians
- students of workplace design

Since the users´ knowledge about ergonomics, computer systems, and Ergon-Expert varies, the user interface of the expert system has to fulfill quite a number of demands. These include among other things

- a high degree of self-explainability
- robustness against incorrect input
- adaptability of the user interface to the needs of experienced and lay users

Other more general demands are derived from [15] and specific requirements from [12].

These demands led to the decision to use a highly interactive user interface with special attention being paid to the explanation facilities. Although the prototype currently under development does not meet all of these demands, the main development guidelines are to

- allow, wherever possible, the user to choose between possible answers to a question
- give the user access to detailed background information on the problems under investigation
- inform the user about incorrect or illogical input as soon as possible

The provision of some of these features in the expert system shell further justifies its choice. The current prototype employs the technique of displaying information, whenever useful or necessary, in the form of graphics and text. An example is provided in Fig. 4. The ability to display at all times the status of the inference process by investigating the current values of both hypotheses and variables, together with information about reasons for the confirmation or rejection of any hypothesis, gives valuable help when analysing a workplace. These facilities are complemented by the tracing facility of the shell and the aforementioned network display which provide an elegant means of investigating the conclusions in their contexts.

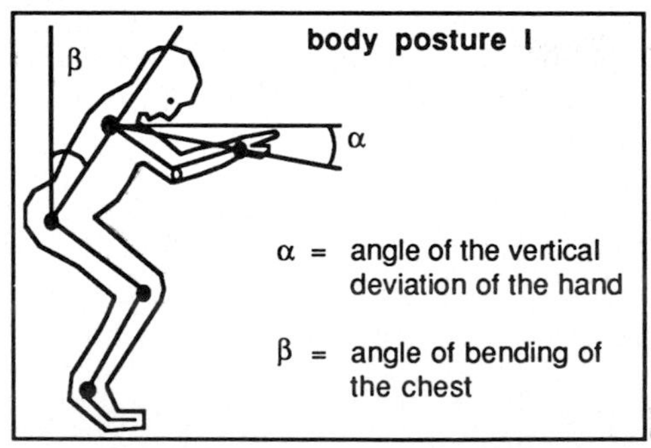

Fig. 4: Example of an explanation screen as used in the prototype of Ergon-Expert

In the final expert system a user interface with an enhanced degree of adaptation to both the user and the operating system environment will be employed. This interface will follow the concept of an object-oriented interactive user interface as used in operating systems which have their roots in the research laboratories of XEROX-PARC [16]. The user will interact with the system by means of a mouse device and a keyboard and the interface will be completely graphical. Fig. 5 provides an example of how the interface might look.

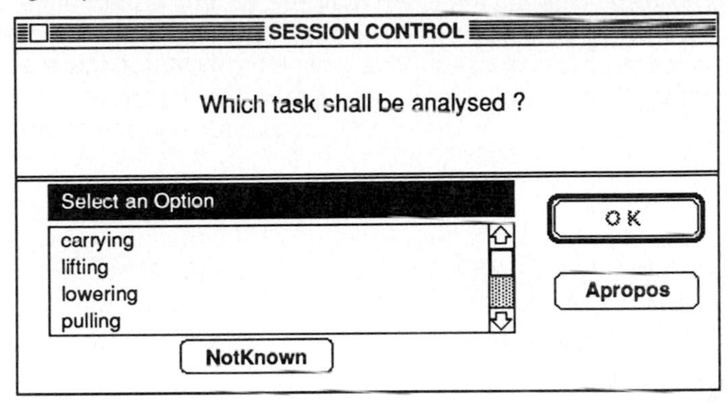

Fig. 5: Example screen of Ergon-Expert's user interface

5. CURRENT STATE OF DEVELOPMENT

The prototype of the expert system under development at the moment will be used to answer questions such as

- which work sequences are hard to describe and where are improvements necessary?
- how should the user interface be adapted to the user?
- which parameters have been found with regard to the run-time behaviour of the system?
- which insights have been gained about the structure of the domain knowledge?

In the course of the project lay and expert users will evaluate the user interface so that improvements can be made. The concept of user participation will be employed throughout the project.

The development is currently being carried out on an Apple Macintosh II. Portation to an IBM-AT compatible microcomputer is also planned towards the end of the project.

6. CONCLUSIONS

The development of an expert-support system for workplace design has shown itself to be a worthwhile approach. Industrial engineers and human factors experts both in industry and research laboratories have few reservations about the development of such a system. Consequently, expert knowledge is readily available from them. One of the main barriers to the acceptance of such a system has proven to be the user interface and the degree of its adaptation to the user. The development of the system presented here follows the concept of rapid prototyping by keeping close contact to the intended usership. The application of this technique ensures that development errors are kept to a minimum.

7. ACKNOWLEDGEMENTS

The research project Ergon-Expert is funded in part by the German Ministry for Research and Technology (BMFT) under 01HK8263 within the framework of the research program "humanization of work".

REFERENCES

[1] Laurig, W., Prospective Ergonomics - a new approach to industrial ergonomics. In: Karwowski, W. (ed.), Trends in Ergonomics/Human Factors III (Elsevier, Amsterdam, 1986) pp. 41-50

[2] Karwowski, W., Mulholland, N.O., Ward, T.L., Jagannathan, V., Kirchner, R.L. jr., LIFTAN: An experimental expert system for analysis of manual lifting tasks. Ergonomics 10 (1986) pp. 1213-1234

[3] Taylor, N.K. and Corlett, E.N., ALFIE - auxiliary logistics for industrial engineers. International Journal of Industrial Ergonomics, 2 (1987) pp. 15-25

[4] Rodgers, S.H., Working with backache (Perinton Press, New York, 1985)

[5] NIOSH (National Institute for Occupational Safety and Health), Work practices guide for manual lifting. Publication No. 81-122 (DHHS (NIOSH), Cincinnati/Oh, 1981)

[6] Morris, J.M., Lucas, D.B., Bresler, B., Role of the trunk in stability of the spine. J. Bone Joint Surg. (Am) 43A (1961) pp. 327-351

[7] Chaffin, D.B., A computerized biomechanical model - development of and use in studying gross body actions. J. Biomech. 2 (1969) pp. 429–441

[8] Ayoub, M.M. and El-Bassoussi, M.M., Dynamic biomechanical model for sagittal plane lifting activities. In: International Ergonomics Association (ed.), Proceedings of the 6th Congress of the Ergonomics Association "Old world, new world, one world" (Univ. of Maryland, College Park/Md, 1976) pp. 355-361

[9] Jäger, M., Luttmann, A., Laurig, W., The load on the spine during the transport of dustbins. Applied Ergonomics 15 (1984) pp. 91-98

[10] Jäger, M., Biomechanisches Modell des Menschen zur Analyse und Beurteilung der Belastung der Wirbelsäule bei der Handhabung von Lasten. Fortschritt-Berichte VDI, Reihe 17, Nr. 33 (VDI, Düsseldorf, 1987)

[11] Neuron Data Inc., Nexpert Object - Fundamentals. (Neuron Data Inc., Palo Alto/Ca, 1987)

[12] Smith,S.L. and Mosier, J.N. (ed.), Design guidelines for user-system interface software (The MITRE Corp., Bedford/Mass., 1984)

[13] Watts, R.A., A friendly interface for the lay user. In: International conference on man/machine systems. IEE conference publications, no. 212 (University of Manchester, Manchester/UK, 1982) pp. 64-67

[14] Palme, J., A human computer interface encouraging user growth In: Sime, M.E. and Coombs, M.J. (eds.), Designing for human-computer communication. (Academic Press, London, 1983) pp. 139-156

[15] Gaines, B.R. and Shaw, M.L.G., Dialog shell design. In: Proceedings of the IFIP conference INTERACT' 84 (North Holland, Amsterdam, 1984) pp. 344-349

[16] Purvy, R., Farrell, J. and Klose, P., The design of Star's Records Processing: Data Processing for the noncomputer professional. In: ACM Transactions on Office Information Systems, 1 (1983) pp. 3-24

Trends in Ergonomics/Human Factors V
F. Aghazadeh (Editor)

AN ANALYSIS OF HUMAN MOTIONS USING MICRO COMPUTERS

Prof. Keun Boo Lee , S.H. Oh and S.Y. Choi
Dept. of Industrial Engineering
Chongju University
Chungbook , KOREA 310

Drs. S. D. Lee and J.H. chung
Dept. of Industrial Engineering
Unversity of Dong-A

Dr. K. D. Lee
Dept. of Neuropsychiatry
Kang Nam General Hospital

The object of this reascarch is to develop an interactive computerized graphic program for graphic output of velocity, acceleration and motion range of body task reference point. Human motions can be reproduced by scanning(rate=1/60) the vidicon image,at same time,C.-O.G. of body segment group, and the results are stored in the micro computer memory.
The results of this study can be extended to simulation and reproduction of human motions for optimal task design.

1. INTRODUCTION

In mordern times the relationshop between human and machines is not always smoothing although human is the beneficiary of mechanical civilization.This assures us that the complete understanding of human who uses machine is necessary in designing tools and machine.
The tendency that human, machine and tools are regarded as a system is prevalent in developing the principle and the sets of data which is applicable in designing man-machine system including enviromental factors.[1]
This research about the human motion which is important element in developing the system is exploitation of the data obtained from analyzing the motions directly or indirectly in designing advanced motion pattern.The enhancement of computer technology is helpful in human motion study.
Computer technology overcomes the batch processing, which is the limit of that of the past, by providing interactive and graphical facilities.[7] Traditional studies focused on mathematical modeling measured the motion variables on the basis of computational and geometrical characteristics. The

recent tendency is that measured data is entried in computer and represented by graphic form,which is understood more easily then numericai listing,through man-computer interaction. From this view point our research is necessary and important in the motion study.The focus of this research is the demonstration of the path of motion by analyzing human motion through the computerized model and reproduction of human motion in the graphic form. The result of this research can provide the basic concept applicable in the society of industry, sports and biology and attibute to the real time analysis with computer and human factor equipment for experiment.

2.BACKGROUND

2.1. Human Kinematics

Human Kinematics, the science that examines human motions and analyzes them, treats skeletons, joint connecting them,muscle activities, their origin and insection, and variables which characterize human motion as the center of gravity. Generally there are two methods in measuring them i.e., direct method and indirect method.The former measures directly human body and human motion by goniometer.But the latter records them as a video image using cinema camera or T.V. camera and takes required data such as coordinate from it.[2,3,8,9]

2.2. Video Image Processing

Video image processing which means a processing that is the basis of pattern recognition is the system which transforms artificial eye recognizing two dimension objects directly and inputed screen into a digitized image suitable to computer process.

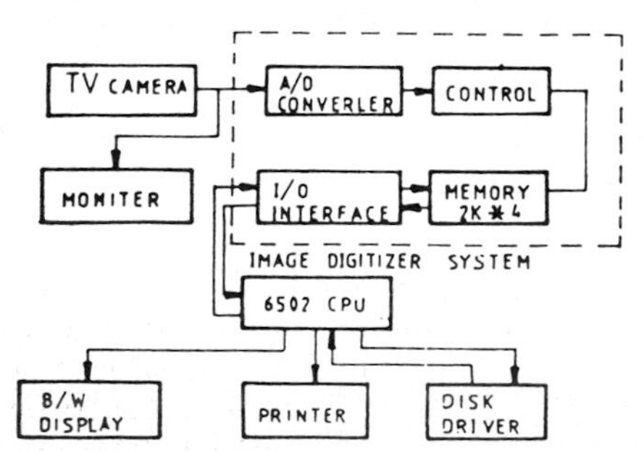

Figure 1. Video image processing system block diagram.

2.2.1. Vidicon Signal

T.V. camera is a unit which transforms optical signal into electrical signal using vidicon. And real time data can be obtained from it. Vidicon scans the image at 30 f.p.s. But in order to remove flicker it divides one frame into two and scans them one by one. Therefore the rate is 1/60.

2.2.2. Extraction of image

Computer memory is arranged by M x N matrix 280 x 192 bit is divided into 3x3 bit and this unit area is taken as a pixel. The value of brightness of this pixel becomes 0-9 levels which 9 bits can take.The interpretation of consecutive video image bases on regarding each frame as a consecution of stop video image frame. When objects moves,frame round it is sampled.Then, since the sampled frame becomes a stop video image, the brightness constant.[4]

2.2.3. Constraction System

In this paper,as seen in Fig.1, image signal is inputed into computer memory through digitizer by DMA so that it is quantized. The gray level is designed in binary. The quantized video image data is stored into secondary memory device and prcessed in the sequence as seen in Fig.2.

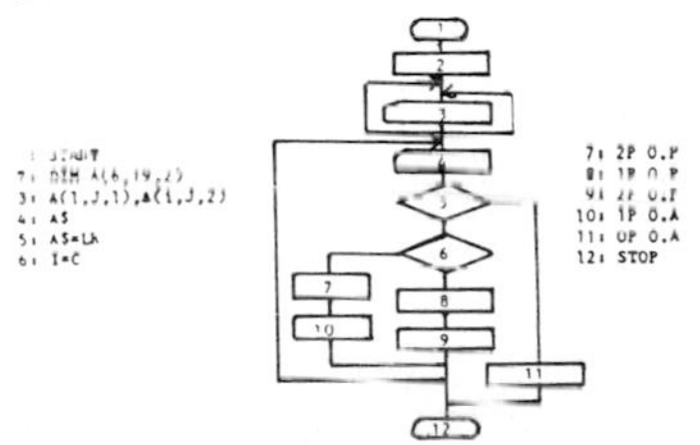

Fig.2. Processing sequence.

2.3. Measurement of the Center of Gravity

In order to locate the center of gravity and measure forces in X,Y, and Z directions force platform system is constructed as Fig.3.

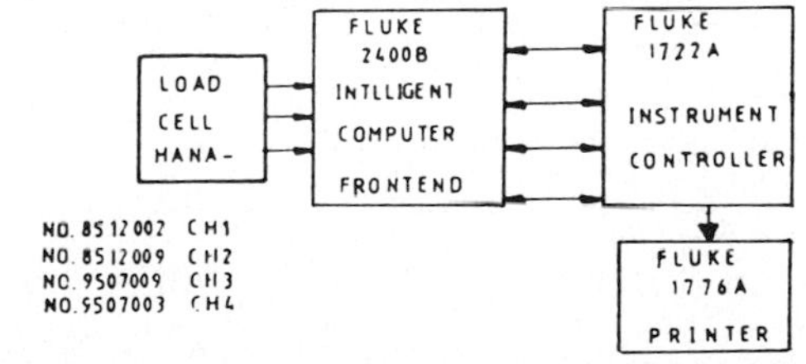

Fig.3. Force plateform block diagram

The data processing method for the storage of obtained data and further research is shown in Fig.4.

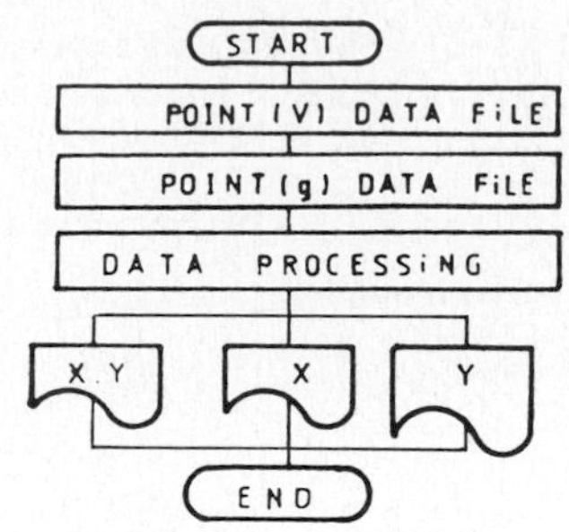

Fig.4. Data Processing sequence

3. EXPERIMENT AND INVESTIGATION

The experimental procedure is as follows;
1) Make the subject perform three types of actions on the force platform. (carring brick in one hand , carring brick in two hand , jumping)
2) After projecting the image by vidicon image processing and measuring the voltage , store obtained data in the disk.
3) Transform the inputed image data into three dimensional coordinate using graphic program and store in the file.
4) Check the image error using monitor.
5) Draw the stick picture using the revised coordinate data.
6) If error between measured values is less than given tolerance , stop otherwise go to 2). [6]

3.1. Image Processing for the Motion Study

In order to analyze the motion , body segments are constructed as seen in table 1.

Table 1. Body Segment

Number	Mark	Joint names
1	R W	Right Wrist Joint
2	R E	Right Elbow Joint
3	R S	Right Shoulder Joint
4	L S	Left Shoulder Joint
5	SST	Superior Sternum
6	G M	Mandible
7	L E	Left Elbow Joint
8	L W	Left Wrist Joint
9	R A	Right Ankle Joint
10	R K	Right Knee Joint
11	HP1	Right Heep Pelbic
12	HP2	Left Heep Pelbic
13	L K	Left Knee Joint
14	L A	Left Ankle Joint

The 9 frame of carring and jumping motion (Picture 1,2,3) is repersented into stick picture using image processing as seen in picturel.(See A,B,C) Fig.5 (See A,B,C)shows moving animation on the CRT which overlaps nine motions and reproduces them in one scene,which makes it possible to analyze order,direction,angle,and velocity of the motion. Fig.6 (See A,B,C) is the stick picture which indicates time on the horizontal axis.

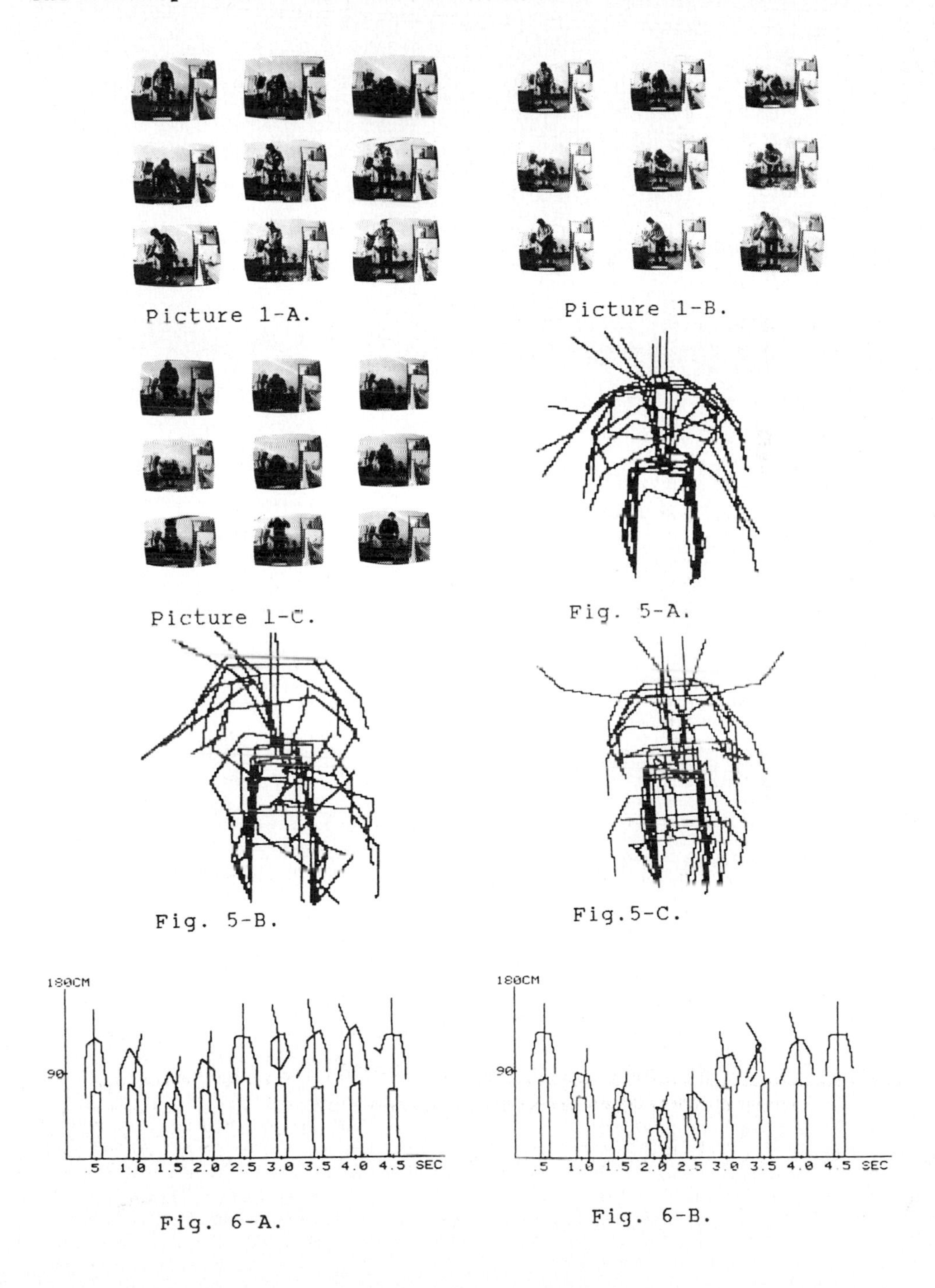

Picture 1-A.

Picture 1-B.

Picture 1-C.

Fig. 5-A.

Fig. 5-B.

Fig.5-C.

Fig. 6-A.

Fig. 6-B.

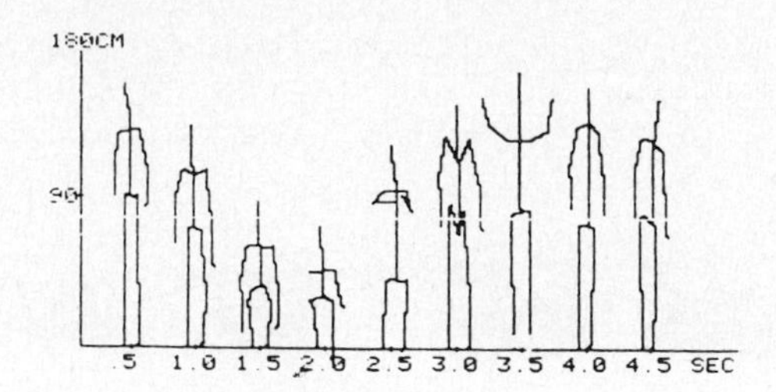

Fig. 6-C.

3.2 Measurement and Analysis of C.O.G.

The changing process of C.O.G. of the working subject is converted into voltage which is generated at load cells installed 4 angles of force platform. Suppose Foo,Foy,Fxo,Fxy are the forces of 4 angles respectively, Then the verticial force (Fz) is the coordinate of C.O.G. is (3.1) and (3.2);

$$X=X/2*[1+(Fxo+Fxy)-(Foo+Foy)/Fz] \quad ..... \quad (3.1)$$

$$Y=Y/2*[1+(Foy+Fxy)-(Foo+Fxo)/Fz] \quad ..... \quad (3.2)$$

X : the width of force platform

Y : the height of force platform

(3.1) and (3.2) are represented at X-Y coodinated system. (Sec Fig.7-A,B,C)

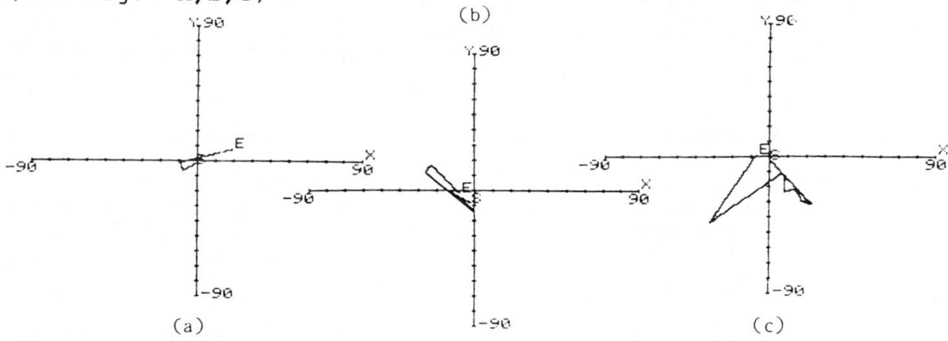

Fig.7-Λ. The changing process of C.O.G. for one-hand brick transportation motion.

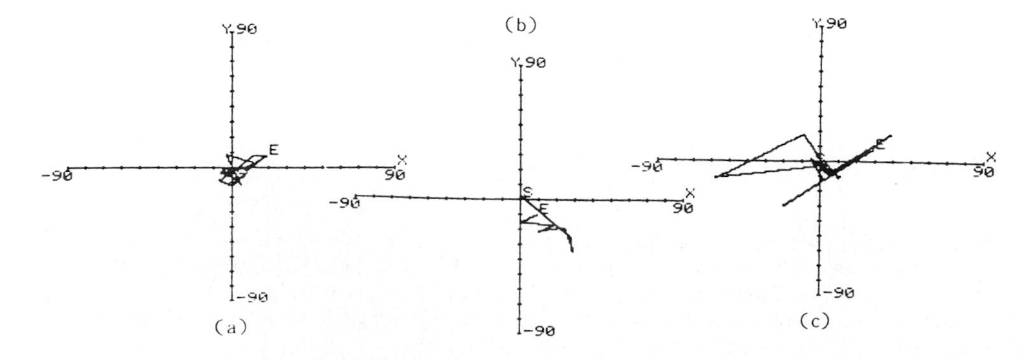

Fig.7-B The changing process of C.O.G. for two-hand brick transportation motion.

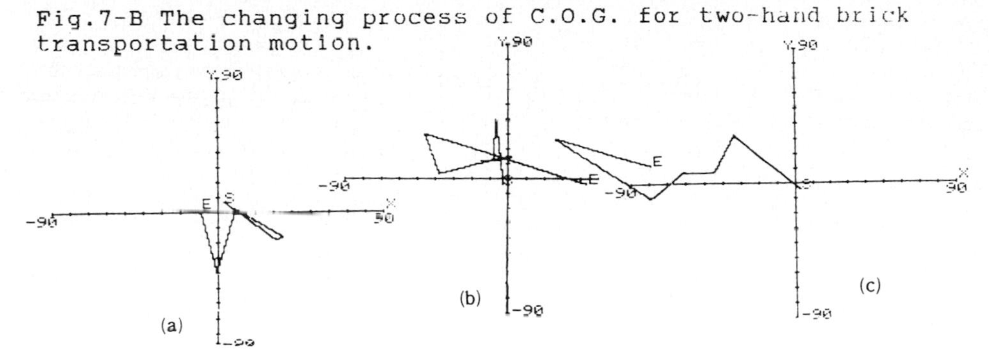

Fig. 7-C. The changing process of C.O.G. for jumping motion.

3.3. Investigation and Analysis

The investigation discussed in this section is as follows;

A. Result of analysis through image processing method.

1) The optimal work space required to do work can be found.
2) Each transformation of the main joint part through stick picture can be observed in detail.
3) The sequence of motion can be specified.
4) Torque analysis can be made easily because of the availability of time series data concerning the displacement for each part of the body.

B. Result of analysis through the center of gravity.

1) The changing mechanism of the C.O.G. at each work in Fig. 7 can be observed from the change of the C.O.G. according to the work speed.
2) Moreover, high degree of the C.O.G. change is found when the landing posture is not good at the start and the end point of jumping motion.
3) A large amount of the C.O.G. change is also observed in the case of brick carriage motion with one hand according to the degree of upper body bending and torque.
4) In the case of brick carriage motion with two hands , the C.O.G. is moved around the working point according to the degree of the knee joint bending.

4. CONCLUSION

Human body motion analysis plays an important role in the application of human engineering by giving fundamental data for the human factors engineering and work study. But there is many problems to be solved. To deal with them,our research performs an analysis using two methods. Through above research we can detect the change of body and make an efficient application because motion path can be reproduced. If we use the above mathod and the computer graphic technique,we can expect high efficiency. During our study,we find many problems as follows and we suggest further research in the following;

1) Noise problem against surrounding (background) when taking a picture of human body motion.
2) Appropriate illumination of photographing human body motion (especially background light of the subject) that is , it is difficult to treat brightness.
3) It is difficult to determine luminescence and the size of the land mark. (loss of reliability in setting x-y coordinate)
4) Appropriate sampling interval determination problem of photographing continuous movement motion. (If we use disk as a secondary memory to store image data , there may be a missing for the lateness of speed)

For further research , 3-dimensional analysis method which can deal with force and torque more easily is expected to get better data for analysis.

REFERENCES

[1] K.S. Park, Human Factors Engineering (Yongji press,KOREA ,1983)

[2] Y.K. Kim , An Analysis of Human Motion Using Computers, Seoul National Univ. , M.S thesis , 1983.

[3] K.B. Lee , Video Image Processing on Apple II P.C and Its Application to Anthopometry and Motion Analysis,J.of H.E.S.K. 4(1985)11-16.

[4] J.S. Lee and K.S. Choi , A study on the Moving Distance and Velocity Measurement of 2-D Moving Object Using a Microcomputer , J.of K.I.E.E 23 (1986) 76-86.

[5] M.W.Lee and K.H.Jung, A Biomechanical Study on Kinetic Posture, Center-of-Gravity, Acceleration and their Effects on the Maximum Capacity of Weight-lifting , J.of K.I.I.E 11(1985) 87-100.

[6] I.Kaoyaki, The Analysis of Human Motions Using Computers , J.of PIXEL 13(1983) 130-131.

[7] Boysen,J.P.Interactive Computer Graphics in the Study of Human Body Planner Motion under Free-Fall Condition, J. Biomech, 10(1977)783-787.

[8] Cappozo, A. Leo, A General Computing Method for the Analysis of Human Location, J. Biomech, 8(1975)307-320.

[9] Winter, D.A., Biomechanics of Human Movement (John Willy and Sons.,1979)

Trends in Ergonomics/Human Factors V
F. Aghazadeh (Editor)

HUMAN-MACHINE MODELING WITH AUTOCAD

Osama K. EYADA, Jeffrey E. FERNANDEZ, Robert J. MARLEY
The Wichita State University
Wichita, KS 67208

Thomas B. DeGREVE
AT&T Data Systems
Little Rock, AR 72209

AutoCAD, an inexpensive yet effective computer-aided drafting software, is proposed as an alternative system for ergonomic design for those who are unable to utilize larger and more expensive systems. As an example, the design of a general office chair was developed using ergonomic principles and standard anthropometric data bases.

1. INTRODUCTION

Human-machine modeling is a valuable tool for human engineering and ergonomic design. Sophisticated technology exists that allows the engineer to combine kinematic and dynamic modeling utilizing the largest and most updated data bases available. Systems such as SAMMIE, GRASP, CADAM, and ADAM have been used successfully [1,2].

These systems can be very costly, however. Richard Davids [1] notes that expenses for installing such systems can be measured in terms of operator training time, a host computer system (including startup and maintenance costs), software costs (which can easily reach up to 50% of the total cost), and the costs associated with space requirements. Thus, an entire system can cost the user between $200,000 and $500,000 to install.

Obviously, such systems are beyond the fiscal realities of many organizations which have the need for human-machine modeling in design work. Therefore, this paper proposes an alternative system of performing such modeling at considerably less expense with the use of AutoCAD.

2. HUMAN-MACHINE MODEL SYSTEMS

The development and expanded usage of computer-aided drafting (CAD) systems in recent years has benefited those involved in ergonomic design. Many CAD systems have been developed specifically for human-machine modeling. Some of the well known systems include SAMMIE, GRASP, REACH, ADAM-EVE, CYBERMAN, COMBIMAN, and BOEMAN. Leppanen and Mattila [3] compiled a sizable list of such working systems and their primary functions.

New systems are being developed continually to aid the ergonomist in decision making. Some of these newer systems, such as SAFEWORK [4] and ERGONOMIST [5], are designed specifically for micro-computer application. The wide variety of systems and their associated hardware requirements pose important considerations for those in need of these capabilities. As mentioned earlier, the financial constraints on many smaller businesses and industries are such that they simply cannot afford the more powerful systems. These users may benefit from inexpensive yet versatile systems for ergonomic applications.

2.2. AutoCAD

AutoCAD is one of many computer-aided drafting and design software package that can be configured to operate on most PC compatibles as well as larger systems. Written by Autodesk, Inc., AutoCAD is readily available in most markets for about $3,000. The version referred to in this paper is the 2.62 release. Most CAD software packages have the capability of being tailored to a specific application. In this project AutoCAD was tailored by creating a data base and rule structure using the AUTOLISP language.

3. EXAMPLE

This paper provides an example of how ergonomic design principles can be applied to a particular situation with AutoCAD. The example given here is the design of a general office chair. The chair is drawn to scale with information provided by the user.

Figure 1 shows the first menu selection in this example. The user may select from the "Standard" or "Custom" option. The "Standard" option utilizes selected U.S. adult anthropometric data [6] needed for the the design of this chair.

If the "Standard" option is selected, the user may choose which segment of the population to be used for design purposes (Figure 2). These include the 5th, 50th, 95th percentiles for males, females and combined. After selection of the appropriate population, AutoCAD then produces the completed engineering design of the chair (Figure 3). Some dimensions of the chair design were taken from Ayoub, et al., [7]. The "Custom" option allows the user to input the critical anthropometric measures (of a special population or an individual) for this chair design.

The final design output provides the engineer data concerning the recommended seat pan width and pan depth, popliteal height, and lumbar support as illustrated in Figure 3.

4. CONCLUSION

This paper has provided an example of human-machine modeling with the use of AutoCAD. AutoCAD is one of the relatively inexpensive, yet versatile CAD systems that are well suited for PC use making it very popular with industries unable to afford larger and more powerful systems. It is hoped that these users would consider utilizing a system such as AutoCAD to meet their need for human-machine modeling.

A majority of educational facilities also have some form of a CAD system. Thus versatile systems, such as AutoCAD, can provide a good opportunity for students to apply the principles of ergonomics to their engineering designs.

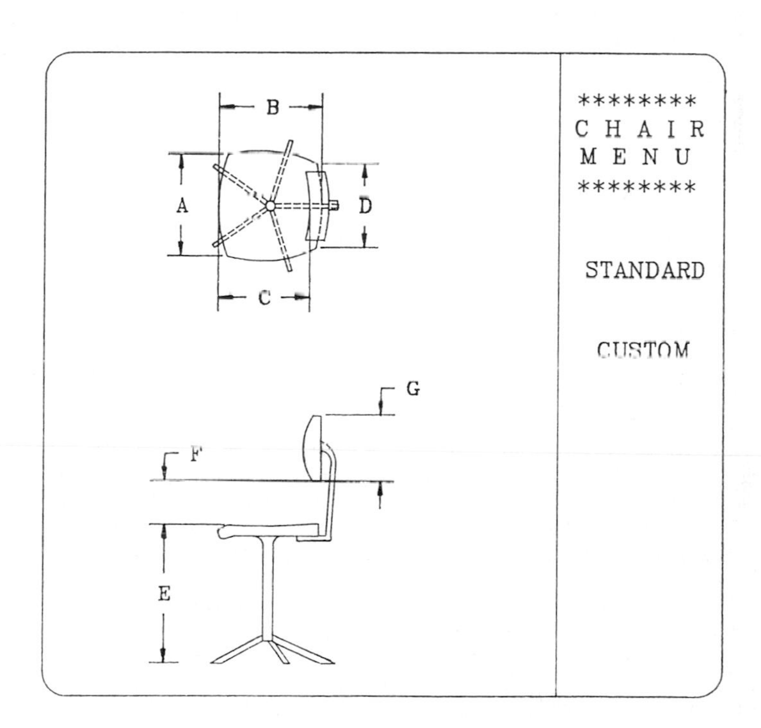

FIGURE 1
Beginning menu for example of chair design with AutoCAD.

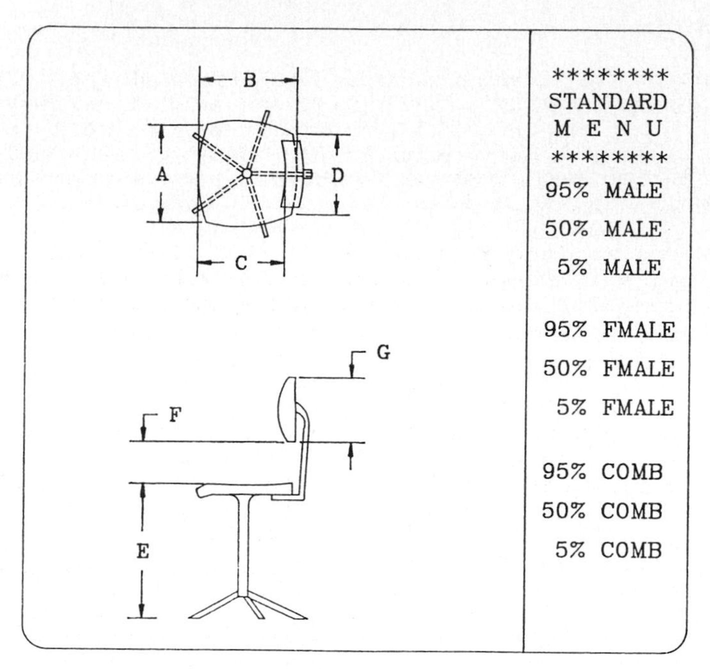

FIGURE 2
Selection menu for particular population.

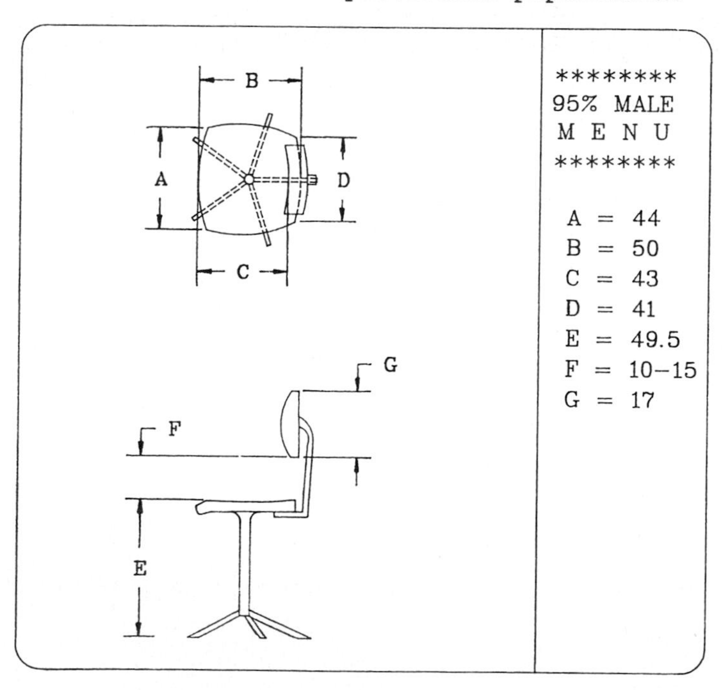

FIGURE 3
Completed drawing with dimensions of the office chair.

REFERENCES

[1] Davids, R.C., (1986). Human factors design tool for factory application. SME Technical Paper No. MM86-1073.

[2] Rahimi, M., (1987). Human factors engineering and safety in robotics and automation. Human Factors Society Bulletin. 30(7), 3-5.

[3] Leppanen, M., and Mattila, M. (1987). Including ergonomics in computer aided design with a 3-D man-model. In Asfour, S.S. (Ed.) Trends in Ergonomics/Human Factors IV. North Holland: Elsevier.

[4] Fortin, C., Gilbert, R., Schiettekatte, J., Carrier, R., Belanger, A., Dechamplain, B., and Lachapelle, M. (1987). SAFEWORK: A micro computer-aided workstation design and analysis. In Asfour, S.S. (Ed.) Trends in Ergonomics/Human Factors IV. North Holland: Elsevier.

[5] DeGreve, T.B. (1985). "A Workplace Design Expert System". Unpublished Masters Thesis. Texas Tech University. Lubbock, TX.

[6] Pheasant, S. (1986). Bodyspace: Anthropometry, Ergonomics and Design. London: Taylor and Francis.

[7] Ayoub, M.M., Fernandez, J.E., and Smith, J.L. (1984). Design of Workplace. Unpublished Departmental Report: Institute for Ergonomics Research, Texas Tech University, Lubbock, TX.

Trends in Ergonomics/Human Factors V
F. Aghazadeh (Editor)

INCLUDING ANTHROPOMETRY INTO THE "AUTOCAD"-MICROCOMPUTER SYSTEM FOR AIDING ENGINEERING DRAFTING

Jerzy Grobelny

Institute of Industrial
Engineering and Management
Technical University of Wrocław
Poland

AutoCAD is one of the best professional microcomputer CAD (Computer Aided Drafting) systems. In the paper a way to construct a graphical anthropometry database inside AutoCAD is presented. The database contains basic anthropometric informations in the form of two dimensional body segments and three dimensional models of functional reach envelopes. Dimensions of these elements are adequate to appropriate population percentile characteristics. All "body" segments are stored as AutoCAD files and can be inserted into a current drawing.

1. INTRODUCTION.

New products designing is such a stage of human activity in which the ergonomical knowledge may be used in the most effective way. Supplying the required knowledge to designers as well as special methods and techniques elaborated on the basis of ergonomics or the related fields is the indispesable condition which allows to apply ergonomical knowledge on the stage of designing. Computer models and methods play a particular part in this area.

Computer aided design models within a range of relation between the man and a designed object were developed at the beginning of the seventies (Roebuck et al. [8], Leppanen and Mattila [5]). The models of anthropometric data form the most numerous group among the first attempts of implementation of the computer ergonomical problems. High costs of equipment, programming and processing caused that these were mostly the specific programs using for military purposes.

Development of microelectronics created new prospects in this field. Cheap microcomputers of good use parameters more and more often become an instrument of the designer's work. Microcomputer implementation of the existing models or construction of the new ones which aid the design from the ergonomical point of view is one of the tasks of modern ergonomics. It is also a chance for dissemination of the methods and ways of thinking of ergonomists during realization of designing tasks. Recently Fortin et al. [3] presented a conception of a system SAFEWORK which is to realize many functions connected withthe ergonomical design on the computers of the IBM PC type. Abdel-Moty and Khalil [1] proposed a special system to aid the designing of VDT sitting work-places. These both proposals belong to the traditional trend of constructing the specialized systems of "computer aided ergonomics" (Leppanen and Mattila, [5]).

A second possible direction of activity, which is postulated in this work and may intensify a practical influence of the anthropometric data on a shape and spatial relations of the recently designed products, is an installation of proper data "inside" the professional CAD systems. Such a procedure is proposed by Leppanen and Mattila [5].

2. DESIGNING IN THE AUTOCAD SYSTEM.

AutoCAD belongs to small systems which aid the design (Voisinet, [9], Raker and Rice, [7]). At the same time it is the entirely professional system. In 1986 it was found decidedly the best in the class of scientific and technical systems in the inquiry of the journal "Chip".

Through full utilization of possibilities of the system AutoCAD entails applications of many special (and expensive) peripherial units (e.g. light pen, plotter, digitizer, etc.), still it is also possible to use many potentialities of the system in "minimal" configuration of the microcomputer IBM PC-XT with graphical printer.

Basic possibilities of the AutoCAD correspond to a general idea of such a type systems (Voisinet, [9]) and may be defined by the funcions accessible for a user. These functions may be arranged in the following groups:

1) geometrical generators,

2) geometrical modifiers,

3) text editor,

4) dimensioning,

5) moduli of drawings' storage and edition.

The access to particular instructions from each above group is organized in a very interesting and convenient way. They may be introduced in the mode of "menu" (for non-advanced users) as well as in the mode of the commands written from the keyboard.

Group 1) includes the commands wich allow to draw on the monitor basic geometrical objects of any size and at any locations such as: POINT, LINE, CIRCLE and ARC.

Group 2) includes the following commands: ERASE, REDRAW, ZOOM, MOVE objects and INSERT - introduction of the previously defined "blocks" (any graphical elements) into the created drawings - in any scale and at any angle.

In group 3) there are the commands of writing the text and defining its form and dimensions (TEXT, STYLE).

Group 4) includes the instructions which allow to dimension automatically lenghts and angles (DIM) as well as areas of the closed figures (AREA).

The commands from the group 5) allow to store drawings on the peripherial units (floppy or hard disks). The command SAVE allow to store a drawing formed to the moment of this command recall. The drawing stored owing to this instruction may be introduced to the computer memory (and processed further on) with the help of the command EDIT. The instruction WBLOCK enables to store a given drawing or its part in a special format of the BLOCK. A block stored on the disk may be introduced to the created drawing with the help of the instruction INSERT (group 2). It is interesting that there is also a possibility of "programming" in a language of the AutoCAD commands. Such a possibility appears when one's own

"MENU" is being constructed. A choice of any option from the MENU causes performing a sequence of commands attributed to this option. Such a way of programming was used during implementation of exemplary anthropometric data to the AutoCAD system. A detailed description of the procedure will be given in the next section.

3. ANTHROPOMETRIC DATA IN THE AUTOCAD SYSTEM.

A basic idea of systems designed to aid the ergonomical design, both the latest version of SAMMIE, (Kingsley, [4]) and the previous systems, such as for example BOEMAN (Roebuck et al., [8]) is storage in the computer of some ergonomical data and the algorithms of their processing for the needs of rational designing. With reference to spatial conditions of the designed products and work-places the idea of two- and three-dimensional manikins is being mainly used. The data concerning dimensions of human body stored in the computer are accesible in the form of manikins graphical picture. One may control it by setting it up on the monitor in proper positions simulating the phases of interaction with a designed object. These basic possibilities may be realized by utilization of the AutoCAD system. For that purpose it is necessary:

1°) to place in the system "a base of dimensional data" concerning the elements of a human body interesting for a designer,

2°) to work out the way of use of this base.

Since models of particular parts of the body will be "recalled" during the designing process it is most convenient to store particular elements as BLOCKS.

The detailed procedure which enables to form a proper base of dimensional data in the AutoCAD system (stage 1°) is presented below:

1. Start the system and choose appropriate dimensions of the drawing (LIMITS, ZOOM) which enables to obtain a proper precision.
2. Introduce a choosen element of the manikin with the help of suitable drawing procedures according to the population data. Mark the joint points of this element with the others.
3. Recall the WBLOCK procedure. This procedure will enable to store the drawn element as a BLOCK of an appriopriate name and to determine the point according to which the element will be "inserted" to another drawings and round which it will be able to be rotated.
4. Repeat the points 2 and 3 until all elements are stored.

In the exemplary implementation the elements of a flat manikin of a size of a 95 percentile man (for Polish population) were stored in such a way. Only the function LINE was used to draw all the elements (processing of straight lines is most effective).

Storage of all constructional elements of a manikin in the form of blocks in the AutoCAD system enables to use each of them for building a manikin in a proper orientation on the monitor. It happens by recalling every element through the name of a block in which it was stored. This "recall" is realized by the instruction "INSERT".

However such a procedure is not too effective because the user must remember the names of particular blocks and introduce them from the keyboard. These inconveniencies may be still ommited by construction and use of one's own MENU offered by the AutoCAD. An arbitrary MENU in the AutoCAD system is shown in the right part of the screen in the form of (maximally) 20 labels, i.e. the expressions defining particular options of the menu. Each menu may be introduced to the system through the command "MENU". After receiving this command the system asks about the name of a set in which the required menu-set is stored. This name is being written in the DOS format with an extension "MNU". Any menu-set is an ordinary ASCII set. Fig. 1 presents exemplary manikin's positions of the 95c

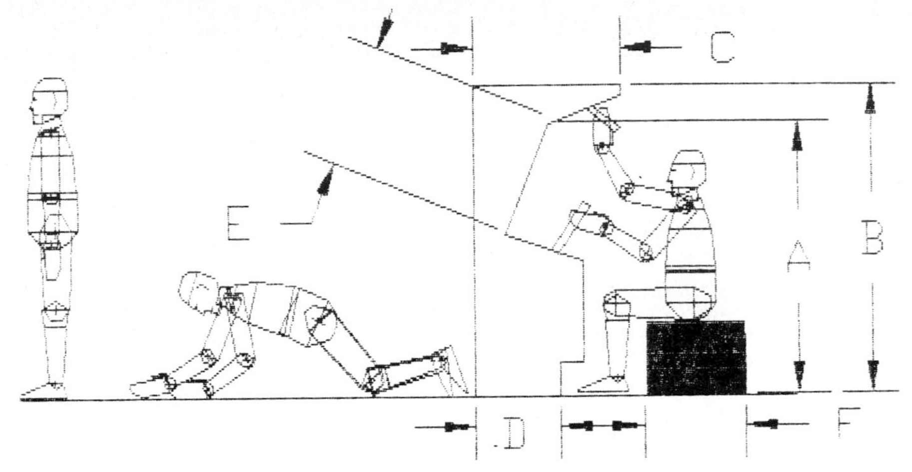

[AUTO]insert;fo95;drag;\ 1;drag;\ insert;sh95;drag;+
\ 1;drag;\ insert;th95;\ ... etc.

FIGURE 1

Exemplary manikin positions and menu option [AUTO] with which presented positions were generated.

individual (for Polish population). The manikin was built of stored elements with the help of AUTO menu shown in the figure. Realization of this option causes succesive dragging the elements of a manikin into the screen with the help of cursor-keys and setting them up in any orientation. It is possible owing to the command DRAG. The codes such as: fo, sh, th define the names of the respective elements of a manikin (foot, shank, thigh).

Drawings in the AutoCAD system are generally made on a plane. Each drawing plane may be defined by a user as a layer. Since it is also possible to "lift" one layer in relation to another one (perpendicularly to the screen plane) and to attribute any "thickness" to drawings on every layer, then there is a chance to obtain simple three-dimensional drawings (together with an option of observation at any angle - VPOINT). It is not indeed the modelling of a "solid" type (with a possibility of using solids in the space) but the discussed possibilities considerably extend the traditional two-dimensional engineering design. However realization of such a research necessitates elaboration of appriopriate data in the way adequate to the possibilities of AutoCAD. Although there is a possibility of storing the elements built of the layers as the BLOCKS, nevertheless it is not possible to transfer a conception of manikin into the three-dimensional space. It results from the fact that manipulating the blocks in three dimensions (moving and rotating) is only possible in the planes and in relation to the axes perpendicular to the layers of the drawing. Therefore one may store the data concerning a 3D manikins through the layers but in practice manipulating such a manikin will be limited to moving and turning once drawn position. Besides the elements of which one may built particular blocks are limited to cuboidal figures according to the presented features of the system (attribution of "thickness" to the layers).

The purposes which a manikin is to serve may be still attained in three dimensions in a little another way. Many data concerning the work space analysis is used in studies of models and prototypes in the form of solids. In particular, it concerns envelopes of functional reaches or complex functional movements (Roebuck, [8]). These are the solids approximating the results of real anthropometric research works. The solids of such a type may be easily drawn with the help of a set of their sections (layers) in the AutoCAD system. However even more reasonably it is to draw and store the exact results of the appropriate research all the more since many anthropometric research works directly structurally correspond to the conception of layers in the drawings of the AutoCAD.

For example the results of studies on human functional reaches presented by Kennedy and reported by McCormick [2] have just such a structure. Figure 2 presents an examplary picture of a structure

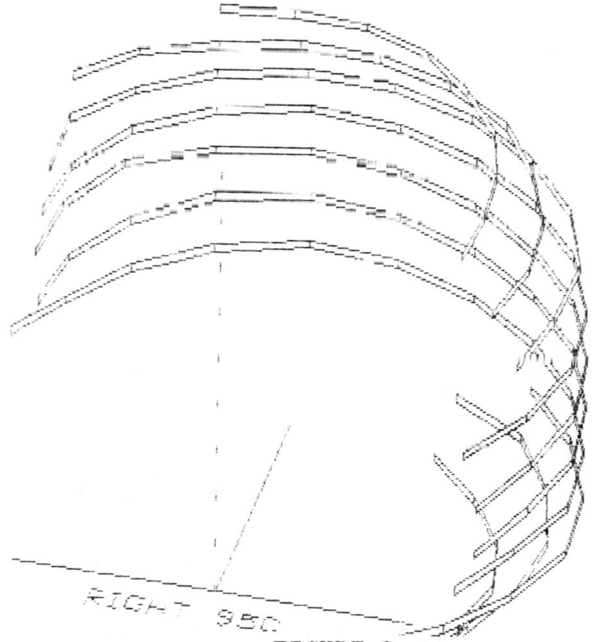

FIGURE 2
Right hand functional reach model stored as a block in AutoCAD.

of functional reach for the right hand of 95 percentile indyvidual from the Polish male population. The quantities and a structure of results were assumed exactly according to the data presented in the work of Nowak [6]. To obtain a greater clarity every layer was assumed to be 2 cm thick and the results on each of the examined layers were joined by segments.

5. SUMMARY AND CONCLUSIONS.

The presented proposals enable to enrich in a relatively easy way the widely disseminated computer aided designing system (AutoCAD) by the basic anthropometric data. Implementation of these data inside the system allow to treat them like other drawings made in the system. Therefore one may use the anthropometric data according to the system's possibilities, i.e. - recall the proper drawings, erase them, move, change their scale, location, etc. One may also build an appropriate menu - which leads to the higher designers work effectiveness through decrease of a number of necessary operations on the keyboard. In this way one also needn't remember the names of the stored data sets.

Wide possibilities of the AutoCAD system within the range of drafts will enable to generate documentation for performers which is verified in anthropometric respect. The use of specialistic systems will not be needed in simple cases. Such a procedure will decrease costs of the anthropometric research and speed up the performance time of many projects.

6. REFERENCES.

[1] Abdel-Moty, E.and Khalil T.,M., The Use of Personal Microcomputers in the Design and Analysis of the VDT Sitting Workplace, in: Asfour, S.S., (ed), Trends in Ergonomics/Human Factors IV (North-Holland,1987).

[2] McCormick, E., Human Factors in Engineering and Design (McGraw-Hill Book Company, 1976).

[3] Fortin, C. et al., SAFEWORK: A Microcomputer-Aided Workstation Design and Analysis, in: Asfour, S.S., (ed.), Trends in Ergonomics/Human Factors IV (North-Holland, 1987).

[4] Kingsley, E.C., Sammie 3-D Graphics for Human Factors Applications, in: Eurographics'82 - Proceedings of the International Conference and Exhibition,Manchester, U.K., (North-Holland, 1987).

[5] Leppanen, M. and Mattila, M., Including Ergonomics in Computer Aided Design with 3-D Man Model, in: Asfour, S.S., (ed.), Trends in Ergonomics/Human Factors IV (North-Holland, 1987).

[6] Nowak, E., Określenie przestrzeni pracy kończyn górnych dla potrzeb projektowania stanowisk roboczych, Prace i Materiały, IWP, Zeszyt 30, (1975).

[7] Raker, D. and Rice, H., Inside AutoCAD, New Riders Publishing, Thousands Oaks, California, (1985).

[8] Roebuck, J. et al., Engineering Anthropometry Methods, Wiley and Sons, (1975).

[9] Voisinet, D., Introduction to Computer Aided Drafting, McGraw-Hill Book Company, (1986).

Trends in Ergonomics/Human Factors V
F. Aghazadeh (Editor)

THE EFFECTS OF INPUT DEVICES ON TASK PERFORMANCE

Jeffrey E. FERNANDEZ, Mihriban CIHANGIRLI,
Donald L. HOMMERTZHEIM, and Ihsan SABUNCUOGLU

Department of Industrial Engineering,
The Wichita State University
Wichita, KS 67208 (USA)

The effects of four input devices: mouse, joystick, trackball and touchscreen on three different tasks were studied. The tasks were a keyboard simulation, tracking an object and pointing to a target box. The results showed that the devices had a significant effect on the performance. Furthermore, the most appropriate device was different for each task. In general, the touchscreen provided the fastest but the least accurate input whereas the trackball was the most accurate but the slowest. The joystick and the mouse were reasonably fast and accurate. Finally, the implications in the rehabilitation field were also discussed.

1. INTRODUCTION

Using computer-based technology, a human operator interfaces with a computer through input devices to gain information and issue commands. In a large number of these applications, an operator is required to direct the computer to carry out actions, the results of which are then displayed to the operator for further consideration [1]. This requires a "good" or sometimes "accurate" method of pointing that enables users to indicate to the computer their selection of a specific element on the computer display, or to move the cursor or pointer from one position on the screen to another. Applications may include: drawing a picture, selecting a menu item among several presented on the screen by positioning the cursor over the target, etc. [2].

Previous researchers found that it was necessary to select the best device for the task at hand considering the design requirements. Therefore, when selecting the best device, the relative importance of speed and accuracy of the task had to be taken into account [3].

Whitfield, Ball and Bird [4] compared on-display (touchpanel) and off-display (touchpad) devices using three tasks. The first task was menu selection. The second task was the selection of an item from a 16x16 tabular display representing an information retrivial operation. The third task was the selection of a target from an array of randomly distributed items on the display, with varying spacing between the target and the non-targets. In this third task, a trackball was also tested. It was concluded that the performance and the subjective reactions of the off-display devices were not significantly different that the performance and the subjective reactions of the on-display devices.

Albert [5] compared the touchscreen, the light pen, the data tablet with puck, the trackball, the position joystick, the force joystick and the keyboard. The task consisted of moving the cursor to a location and "entering" the position. He showed that the most accurate device was the trackball while the fastest and the least accurate device was the touchscreen in this positioning task.

In a study, Card and Burr [6] compared the mouse, the rate controlled isometric joystick, the step keys and the text keys in a text selection on a CRT. It was concluded that the mouse was the best device in terms of speed and accuracy.

Goodwin [7] compared the light pen, the light gun and the keyboard for three different tasks: arbitrary cursor positioning, sequencial cursor positioning, and check-reading. For arbitrary cursor positioning, the target characters were 10 numbers randomly positioned on the display which had to be replaced by X's in numerical order, from 0 to 9. In the sequencial cursor task, the display contained 10 letter "M"'s randomly positioned on the display which had to be replaced with an X, moving from top to bottom of the display. The check-reading task required the subject to read a paragraph of text, find 10 letter substitution errors randomly positioned, and replace the errors with the letter X. He found the light pen and the light gun to be superior to the keyboard for all of these basic tasks.

Finally, Ritchie and Turner [8] studied graphic data entry. The first task they studied was the selection of a particular item from a displayed array, a numerical component value for input data or a program control instruction. The second task was the freehand sketching of pictorial data in the form of line drawings. The third task was the digitization of original hard copy such as maps by tracing. They compared the joystick, trackball, mouse, light pen and electronic data tablets. The trackball and the joystick were the best for accurate cursor positioning. A transparent data tablet was suitable for sketching and tracing purposes.

The purpose of this study was to determine the most suitable input device(s) for tasks with different requirements such as a task requiring accuracy, another requiring speed and yet another one requiring both accuracy and speed.

2. METHODS

2.1. Apparatus

Four different types of computer input devices were considered. Three of these devices were mechanical: a Microsoft mouse, a Premium II joystick and a trackball. The fourth device was a Carrol touchscreen. The tests were performed on a Zenith IBM compatible personal computer.

2.2. Subjects

Twenty male subjects participated in this study. Their ages were between 22 and 37. They were senior or graduate level students, and were either intermediate or frequent computer users. Therefore, they were familiar with the use of computers and input devices.

2.3. Procedure

A computer program was developed to test subjects for three different tasks and four input devices. The program also recorded the performance measures of the subjects into a file for later analysis.

The first task involved entering five-letter words into the computer letter-by-letter. The performance measure of this test was speed (duration to complete the task in seconds) since the test continued until the word was correctly entered. The second task required the subjects to track a moving object on the screen at a moderate rate. The accuracy, i.e. the closeness (in terms of radial distance) of the user's cursor to the object (in inches), was the performance measure. The third task tested the ability of the subject to position a cursor as fast as possible inside four boxes that were presented on the screen in a square pattern. Succesively smaller and smaller boxes were presented as the test advanced, requiring more accuracy to position the cursor correctly. The performance measure was the speed of the subject to complete the task (duration in seconds).

The test consisted of 12 subtests as shown in Table 1. The order of the subtests was randomized by the PROC PLAN random test order procedure of the SAS package.

TABLE 1
List of Subtests

Test no.	Input Dev./ Task	Test no.	Input Dev./ Task
1	Mouse/ Keyboard	7	Trackball/ Keyboard
2	Mouse/ Tracking	8	Trackball/ Tracking
3	Mouse/ Target box	9	Trackball/ Target box
4	Joystick/ Keyboard	10	Touchscreen/ Keyboard
5	Joystick/ Tracking	11	Touchscreen/ Tracking
6	Joystick/ Target box	12	Touchscreen/ Target box

2.4. Statistical Design

The data collected were analyzed by the SAS (version 5.16) package on the IBM 3081 mainframe computer. A completely randomized one-way Analysis of Variance was conducted to determine whether the devices had a significant effect on the subjects' performances.

Duncan's Multiple Range Test was performed as a post hoc analysis to investigate the differences between the devices at the $\underline{p} < 0.05$.

3. RESULTS AND DISCUSSIONS

Devices had a significant effect on the performance speed in keyboard ($\underline{p} < 0.006$) and target box ($\underline{p} < 0.006$), and a significant effect on the accuracy of the tracking task ($\underline{p} < 0.006$). This result was consistent with the previous studies.

The best device varied for each task. The device with the best performance for the keyboard task was the touchscreen. The mouse was rated second followed by the joystick and the trackball as shown in Figure 1.

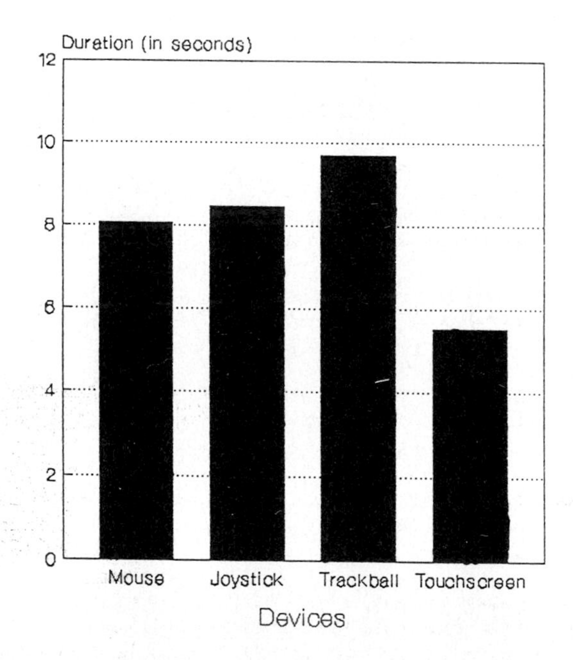

FIGURE 1
The mean performance scores for the keyboard task.

There was no significant difference in the performance of the trackball, the mouse and the joystick in terms of accuracy for the tracking of an object. The touchscreen was the least accurate as shown Figure 2 and was significantly different than the other three devices.

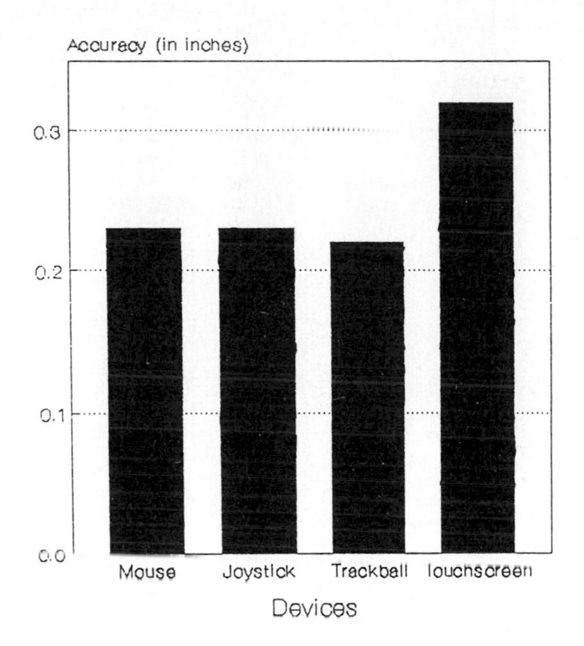

FIGURE 2
The mean performance scores for the tracking task.

The touchscreen and the joystick were the two preferred devices in terms of speed for the target box task as shown in Figure 3. The mouse was rated the second and the trackball was the slowest.

Therefore, when designing computer systems for the human interface, the characteristics and requirements of the task should be analyzed and established prior to the selection of the input devices. This implies that further research in this field may be required as the need arises.

Since most of the interactive systems involve several tasks that require speed and accuracy in varying degrees, the data provided here, can make the system designer more aware of the trade-off involved.

Currently, these same tests are being administered to a group of disabled subjects that have severe physical impairments, especially in the arms and the hands. Since many household appliances and tools or equipment in the office and at the work place can be controlled by computers using actuators, it is important to determine which input devices can be controlled best by this class of users.

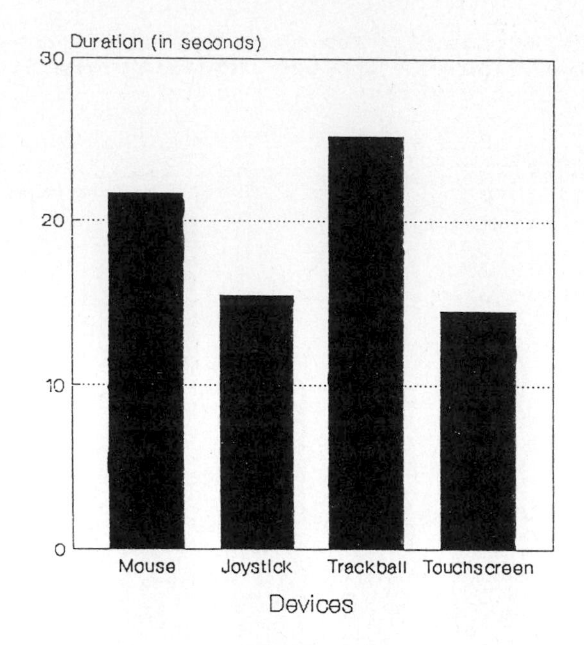

FIGURE 3
The mean performance scores of the target box task.

4. CONCLUSIONS

The conclusions drawn from this study are that the devices had a significant effect on the performance measures, and the best input device varied with the task. The touchscreen was superior for the keyboard task. The trackball was the best for the tracking of an object. Both the touchscreen and the joystick were the preferred devices for the target box. In general, the touchscreen was the fastest, but the least accurate. The trackball was the most accurate but the slowest. Finally, the joystick and mouse were reasonably fast and accurate. Since the best device is different depending on the requirements and characterictics of the tasks, detailed analysis should be done for the selection of the input devices.

REFERENCES

[1] Maher P. K. C., and Bell, H. V. (1977). The man-machine interface - a new approach. Proceedings of International Conference on Displays for Man-Machine Systems, 122-125.

[2] Sheridan, T. B. (1984). Supervisory control of remote manipulation, vehicles, and dynamic processes: experiments in command and display aiding. In Rouse, W. B. (Ed.), Advances in Man-Machine Systems Research (Vol.1). London: JAI Press Inc.

[3] Sanders, M. S., and McCormick, E. J. (1987). Human Factors in Engineering and Design. New York: McGraw Hill Company.

[4] Whitfield, D., Ball, R. G., and Bird, J. M. (1983). Some comparisons of on-display and off-display touch input devices for interaction with computer generated displays. Ergonomics, 26(11), 1033-1053.

[5] Albert, A. (1982). The effect of graphic input devices on performance in a cursor positioning task. Proceedings of the Human Factors Society 26th Annual Meeting. Santa Monica: Human Factors Society.

[6] Card, S. K. and Burr, B. J. (1978). Evaluation of mouse, rate-controlled isometric joystick, step keys, and text keys for task selection on a CRT. Ergonomics, 21(8), 601-613.

[7] Goodwin, N. C. (1975). Cursor positioning on an electronic display using lightpen, lightgun, or keyboard for three basic tasks. Human Factors, 17(3), 289-295.

[8] Ritchie, G. J., and Turner, J. A. (1975). Input devices for interactive graphics. International Journal of Man-Machine Studies, 7, 639-660.

II

COGNITIVE ERGONOMICS

Trends in Ergonomics/Human Factors V
F. Aghazadeh (Editor)

IS THERE A RELATIONSHIP BETWEEN WORKING MEMORY CAPACITY AND HEMISPHERIC SPECIALIZATION?

Marilyn L. TURNER

The Wichita State University
Department of Psychology
Wichita, Kansas 67208

The experiment investigated whether working memory (WM) span is influenced by hemispheric specialization. Correlations were found between cognitive abilities (reading comprehension and SAT scores) and verbal WM spans that were significantly similar to correlations found between spatial WM spans and cognitive abilities. The WM span/ability relationship was still present for subjects with high verbal and low quantitative abilities. And, in addition, when quantitative skills were factored out of the span/ability correlations, the relationship remained significant. On the other hand, the baseline word spans did NOT correlate with the two ability measures). Thus, the findings reported suggest that WM span may predict reading comprehension when the span is measured against a verbal reading- or a spatial arithmetic-related background task However, it is pointed out that these implications are based on the extent that the arithmetic transformation background task can be considered spatially related. That is, only to the extent that processing arithmetic transformations is spatial processing, WM spans measured while performing an arithmetic transformation, and WM spans measured while performing a verbal reading task, can be considered as reflecting the same central WM capacity, rather than reflecting a verbal and spatial WM capacities. Therefore, although these findings indicate that the WM span may not be influenced by hemispheric specialization, further investigations of WM spans measured while processing different kinds of spatial information were suggested.

1. INTRODUCTION

It is generally accepted that higher level cognitive functioning makes demands on a limited capacity system that affects the available workload, and most memory theorists assign this function to a working memory (WM) structure. Baddeley & Hitch (1974) proposed a model of WM suggesting this limited memory system plays a crucial role in higher level functioning. Their model includes processing and structural components, i.e., a limited-in-capacity central executive that is in control of at least two maintenance systems, an articulatory loop, and a visual-spatial scratch pad. They suggest the central executive can be thought of as a flexible work space that uses part of its capacity for processing incoming information, and stores products of that processing in any remaining capacity. A good analogy of WM is a cabinet maker's work bench, i.e., a work space that must be used to hold necessary tools and the partially finished cabinets being built. Although other models of WM have been developed (e.g., Kintsch & van Dijk, 1978; Schneider & Detweller, 1987), virtually every conceptualization of WM assumes that there is a limitation in the amount of information that can be kept active at any given time, and that this limitation affects consequent processing required in cognitive tasks. Empirical support for the importance of the WM limitation in reading comprehension was found by Daneman & Carpenter (1980, 1983) and Daneman & Green (1986). They hypothesized that WM is used to represent reading skills and, in addition, the products of comprehension, e.g., pronoun referents, facts and propositions. Daneman & colleagues developed a dual task in which subjects read increasingly larger sets of unrelated sentences, and at the end of each set recalled the last word of each sentence in that set. The maximum set-size in which the last words were correctly recalled was considered each subject's reading span, i.e., a measure reflecting WM storage capacity. In several experiments, Daneman & colleagues found high correlations between reading span and comprehension, ranging from 0.72 to 0.90, and between reading span and Verbal Scholastic Achievement Test (VSAT) scores, ranging from 0.49 to 0.59. From these remarkable findings, they argued that the reading span reflects a WM capacity NOT allocated to processing sentences, and further, that more capacity would remain for "storage" of to-be-remembered information for good than poor comprehenders, because good comprehenders have more efficient reading strategies than poor comprehenders. The implication is that the WM span measure is dependent on the type of background task used while measuring the span, and that it must include reading to correlate with individual variation in reading comprehension. On the other hand, Turner & Engle (1986) have shown that a complex WM span measured in a dual task is highly related to comprehension, even when the skills used in a verbal background task are clearly unrelated to reading. These latter results suggest that the background task in this complex WM measure does not need to be "reading" related to cause a correlation between the WM span and reading comprehension. One possible explanation of these findings may be that people are good readers because they have a large central WM capacity independent of the task being performed. Another possible explanation of these findings may be that people are good reading comprehenders because they have a large VERBAL WM capacity, independent of the VERBAL task being performed. The implication is that WM span may be influenced by hemispheric specialization.

Research investigating the brain-behavior relationship has shown that the two hemispheres clearly have different functions. The general consensus is that the left hemisphere is specialized for language and serial order, and spatial information is processed in the right hemisphere. Thus, it may be that WM spans measured while processing different types of verbal tasks correlate with comprehension because the memory and processing components of the task are tapping a WM located in the left hemisphere. This possibility suggests that WM spans measured while processing spatial tasks may NOT correlate with reading comprehension, a verbal cognitive ability. Therefore, a major purpose of this study was to test whether WM spans measured while processing verbal and spatial information correlate with comprehension.

2. METHOD

2.1. Subjects and Design

There were 243 USC students that participated as subjects in the experiment. Each subject completed two complex WM span tasks, one traditional span task, and the Nelson-Denny Standardized Reading Comprehension test (ND). Task order was counter-balanced to prevent practice and fatigue effects. The three WM span measures were: (1) a complex span task wherein subjects read sets of unrelated sentences that gradually increased in number, verified whether each sentence made sense, and then recalled the last word in each sentence (sentence-word, SW task), (2) a similar complex span task in which the subjects verified the answers to strings of arithmetic operations, each followed by a to-be-remembered word (operation-word, OW task), and, also (3) a baseline memory span task wherein subjects were asked to remember increasingly larger sets of words.

2.2. Materials

The to-be-recalled words in the SW, OW and simple word tasks were selected from the most common, 4-6 letters in length, one-syllable concrete nouns published in the Francis & Kucera (1982) frequency norms. Sentences for the SW task were generated to make sense using these words as the designated last-word. The unrelated sentences were each from 11 to 16 words long, and were either "correct" or "incorrect". "Correct" sentences made semantic and syntactic sense, e.g., 'The grades for our finals will be posted outside the classroom door.'. "Incorrect" sentences were made non-sensical by reversing the order of the last four to six pre-terminal words, e.g., 'The grades for our finals will classroom the outside posted be door.'. The arithmetic operation strings in the OW task consisted of two arithmetic operations and a stated final answer, e.g., [(9 / 3) + 2 = 5], or [(3 x 4) - 3 = 9]. Approximately half of the operation strings in each trial listed an answer after the equal sign that was correct, and half listed an answer that was incorrect by a least + or - 4, e.g., [(9 / 3) + 2 = 1].

2.3. General Procedure

Form F of the Nelson-Denny was administered in which subjects were given 20 minutes to silently read each of 8 passages and answer the 36 multiple choice questions. In all span tasks subjects heard, saw and read aloud series of stimuli projected one at a time from transparencies onto a large screen. The items were prerecorded on cassette tape, and played through a Sony recorder with presentation rate determined by the length of the items (423 ms for the SW, and 541 ms for the OW stimuli). The experimenter initiated the first trial by projecting the first item on the screen, and at the same time, playing the recording of that item. As soon as subjects finished reading the item aloud, paced by the recording, the next item was presented. While reading each item, subjects decided whether an item was correct or incorrect, checking their answer sheets if the item was correct. After a series of items were read and verified, subjects were cued to write the to-be-remembered words in any order on their answer sheets. There were three trials at each set size, and the set size was gradually increased from 2 to 5 in the two complex SW and OW tasks. In the simple word span task subjects saw, heard and read aloud words presented at a rate of one per second. The number of words presented prior to the "recall" cue was gradually increased from 2 to 7. When the recall cue was heard, subjects wrote the words in any order. Separate spans were defined as the maximum number of items recalled for each of the three tasks. Comprehension scores were defined as the number of correct answers in the Nelson-Denny standardized comprehension test.

3. RESULTS

Table 1 shows the Pearson Product Moment correlation coefficients calculated between span and comprehension measures. The major goal of this experiment was to test whether WM spans measured while processing spatial information also correlate with comprehension.

TABLE 1

SPAN MEASURES	COMPREHENSION MEASURES		
	Nelson-Denny	Verbal SAT	Quant SAT
Sentence-Word (SW) span	0.37*	0.28*	0.26*
Operation-Word (OW) span	0.40*	0.34*	0.33*
Simple word span	0.07	0.08	0.09

* $r(241)$, $p < 0.0005$. N= 243.

Table 1 shows the correlational coefficients central to this question, i.e., the correlations between Nelson-Denny accuracy scores and SW, OW and simple word spans, 0.37, 0.40 and 0.07 respectively. The correlations suggest that comprehension is related to a central WM span, whether measured against a verbal reading-related, or a spatial arithmetic-related background task. As expected, the simple word spans did not correlate with reading comprehension, VSAT or QSAT scores.

These results imply that the background task does NOT need to be "verbal" to produce a correlation between the span and reading comprehension. This suggests individuals may be good or poor reading comprehenders because of a large or small central WM capacity, not because of more or less efficient verbal reading skills. However, these similar correlations may be simply due to good readers also having good spatial, quantitative skills and using both verbal and spatial skills efficiently. Then, the SW/ND correlation could occur because the SW task causes verbal reading skills to be invoked and the residual central WM capacity is reflected by the number of words recalled. The OW task would lead to spatial arithmetic skills being invoked and, since verbal and quantitative skills tend to be correlated (0.54 between VSAT & QSAT in the current sample), the central WM capacity reflected by the OW task would tend to be similar to that reflected by the SW task.

This possible confound was approached in two statistical analyses of the data, a group analysis and a partial correlation analysis. In the group analysis subjects were divided into four groups based on their VSAT and QSAT scores: (1) HH, those subjects achieving high scores on both VSAT and QSAT, (2) HL, those achieving high scores on VSAT and low scores on QSAT, (3) LH, those achieving low scores on VSAT and high scores on QSAT, and (4) LL, those achieving low scores on VSAT and QSAT. The high scores were 1/2 standard deviation (SD) above the mean, and low scores were 1/2 SD below the mean of each sample distribution. The critical OW/ND correlation (0.47) for Group HL, consisting of individuals with poor quantitative skills, was similar to the OW/ND correlation (0.38) for Group LH, consisting of those individuals with good quantitative skills. When comparing these two correlations ($r = 0.47$ and 0.38 standardized to $z = 0.60$ and 0.48) no significant difference was found, $z = 0.45$, $p > 0.33$. The important point is that the individuals in the HL and LH groups did not have similar quantitative skills to use when performing the OW task, yet their performance in the OW task DID equally and significantly correlate with reading comprehension, suggesting the WM span/comprehension correlations may be reflecting differences in a central WM capacity, independent of specific verbal and spatial skills required in the complex span task. Also, when quantitative skills (QSAT) were factored out of the correlations, the OW/ND ($r=0.25$) and SW/ND ($r=0.25$) relationships were still present. These significant partial correlations further suggest the complex span measures of WM capacity are independent of the particular spatial, quantitative skills used in the processing component of the span task, i.e., the arithmetic operations.

4. GENERAL DISCUSSION

The primary question was whether the complex span, reflecting WM limitations, may be influenced by hemispheric specialization. If the complex span correlates with comprehension when measured against verbal reading-related or spatial arithmetic-related background tasks, then the measure can be considered independent of hemispheric specialization. That is, the complex span may be reflecting a central WM capacity rather than a verbal WM located in the left hemisphere, and a spatial WM located in the right hemisphere. The experiment clearly demonstrated good readers remembered more words than poor readers regardless of whether the background task required verbal reading or spatial arithmetic skills. The possibility that good readers also have good quantitative skills, leading to a spurious OW/ND correlation, was also discounted. OW spans for individuals having good reading and poor math skills (HL), and the reverse (LH), equally correlated with reading comprehension. In addition, the relationship between OW (and SW) with comprehension was still present when quantitative skills (QSAT) were factored out of the association. These findings imply that a complex span reflecting a central WM capacity does not need to be "verbal or reading" related to generate a significant correlation with reading comprehension. Importantly, however, it should be pointed out that these implications can only be concluded to the extent that the arithmetic transformation background task can be considered spatially related. WM spans measured while performing another type of background spatial task may NOT reflect the same underlying processing limitation of higher level cognitive abilities (i.e., a central working memory), as those WM spans measured while performing a verbal task. Therefore, although these findings indicate that the WM span may not be influenced by hemispheric specialization, further investigations of WM spans measured while processing different kinds of spatial information are necessary to further define the working memory limitation in the amount of information that affects the available workload when performing any higher level cognitive task.

REFERENCES

Baddeley, A. D. & Hitch, G. (1974). Working Memory. In G.H. Bower (Ed.), _The Psychology of learning and Motivation_, _8_, NY: Academic Press.

Daneman, M. & Carpenter, P. A. (1980). Individual differences in working memory and reading. _Journal of Verbal Learning and Verbal Behavior_, _19_, 450-466.

Daneman, M. & Carpenter, P. A. (1983). Individual differences in integrating information between and within sentences. _Journal of experimental Psychology: Learning, Memory, and Cognition_, _9_, 561-583.

Daneman, M. & Green, I. (1986). Individual differences in comprehending and producing words in context. _Journal of Memory & Language_, _25_, 1 - 18.

Francis, W. N. & Kucera, H. (1982). _Frequency Analysis of English Language._ Boston, Mass.: Houghton Mifflin.

Kintsch, W. & van Dijk, T. A. (1978). Toward a model of text comprehension and production. Psychological Review, 85, 363-394.

Nelson, M. S., & Denny, E. D. (1973). The Nelson-Denny Reading Test. Boston: Houghton Mifflin.

Schneider, W. & Detweller, M. (1987). A connectionist/control architecture for working memory. In G. H. Bower (Ed.), The Psychology of Learning and Motivation, 21, NY: Academic Press.

Turner, M. L. (1987). Cognitive abilities as a function of working memory. Proceedings of the Human Factors Society: Urbana-Champaign. (in press).

Turner, M. L. & Engle, R. W. (1986). Working memory capacity. Proceedings of the Human Factors Society, 2,1273-1277.

Trends in Ergonomics/Human Factors V
F. Aghazadeh (Editor)

A COMPARISON OF THREE DISPLAY FORMATS FOR MULTI-DIMENSIONAL SYSTEMS

Jen-Gwo Chen, Don Deal, V. Jeyakumar

Department of Industrial Engineering
University of Houston
4800 Calhoun Blvd.
Houston, Texas 77004

The objective of this study is to evaluate an enhanced integrated information system display for use in system monitoring, diagnosis, and problem detection. The research concentrates on the comparison of three different color-coded display formats (bar chart, pie chart, and polar graph), with and without digital scale. Three, five and seven component representations in a single display, constituting low, medium and high task complexity, were investigated. The information displays were used in component and trend comparisons for data monitoring in a common environment. In the experiment for component comparison of static data, results based on reaction time suggested that the bar chart display was superior. In the trending task bar charts and polar graphs with digital scaling produced better reaction times and greater accuracy in interpreting data. Interviews with subjects suggested a preference for bar charts without digital scaling and polar charts with or without digital scaling. Discussion of new findings and confirmation of previous survey results is included.

1. INTRODUCTION

As the complexity of technological systems has increased, the problem of monitoring and evaluating information has emerged as a major concern. This problem is particularly severe in sophisticated, complex systems which require monitoring of numerous status displays. The primary objective has become one of designing a system to present large amounts of information in such a manner that the worker can monitor, evaluate and react in a timely manner. Thus, a concise, integrated display has become a key factor in successful system design.

Warning annunciators (visual or auditory) have been widely used to indicate system status abnormality. A complex system, such as represented by the system console in a nuclear power plant, may require thousands of these. A major problem which emerges is that annunciator confusion can lead to human error which may actually aggravate a situation (e.g., the Three Mile Island incident). Traditionally, analog or digital displays, such as mechanical gauges and CRT displays, have been combined with annunciators to provide more detailed information with the intent of reducing confusion. However, this can lead to the necessity of scanning several displays in order to identify the malfunctioning component. This, in turn, can increase workload, stress and reaction time.

In the past simple approaches, such as displaying face figures or polygons, have been used to combine the representations of the status of several systems into one display. As an example, iconic displays have been examined by Wood et. al. [1] in the consideration of nuclear power plant safety. While this display appeared useful as a decision-making aid, the recognition of out-of-tolerance figures required considerable training. In addition, recognition time was increased in status monitoring.

The objective of this study is to evaluate an enhanced integrated information system display for use in system monitoring, diagnosis and problem detection. The research concentrates on the comparison of three different color-coded display formats (histogram, pie chart and polar graph), with and without digital scaling. Three, five and seven component representations in a single display, constituting low, medium and high task complexity, were investigated.

2. METHOD

2.1 SUBJECTS/APPARATUS

Fourteen male subjects with personal computer experience volunteered for participation in the study.

An IBM PC-AT microcomputer was used to display graphical information and queries on an EGA color graphics monitor.

2.2 INDEPENDENT VARIABLES

Three levels of task complexity were investigated in this study. Three, five and seven component representations in a single display corresponded, respectively, to low, medium and high task complexity levels.

Display format consisted of three types: 2-D histogram, pie chart and polar graph (see Figure 1). A weighted polar graph had been employed by Chen and Peacock [2] in a previous study to display physical work stress profiles. Here, radial sectors were used to represent physical components wherein the sector angle indicated the weight or importance of the component. Both area and shape were used to indicate overall system status and particular deviation patterns. In the current study polar graphs with equal angles was employed.

In addition to presentation format, the option of including a digital scale along with the graphic display represented another control variable whose effects were to be investigated.

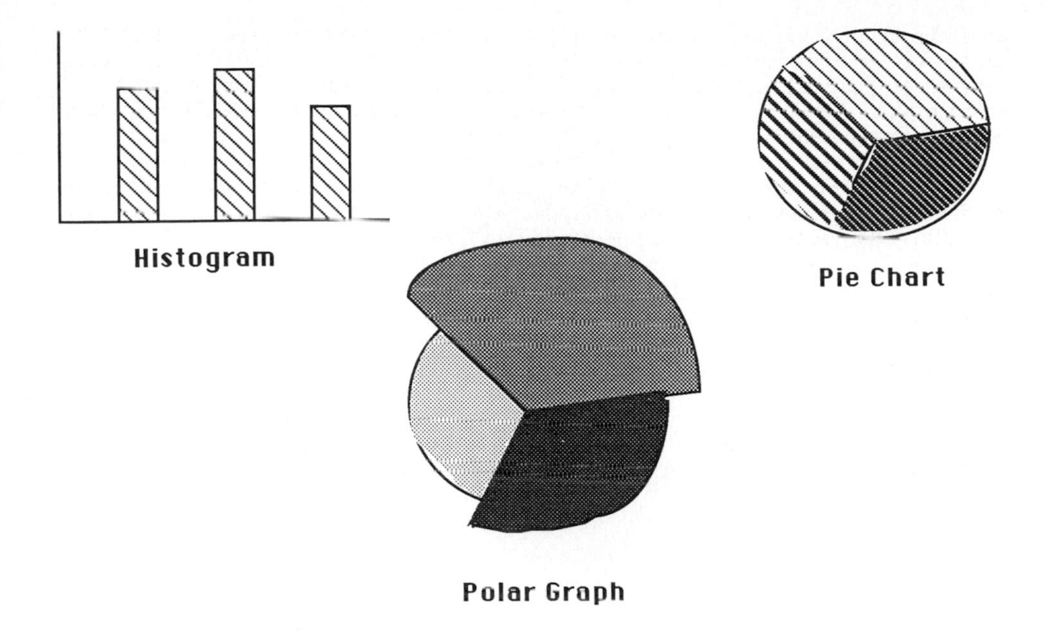

Figure 1. Example of Three Display Formats.

In the component comparison experiment, static data displayed in the various formats was presented to the subject in a test of magnitude comparison accuracy and reaction time. In the trend comparison task, the displays were updated every second over a period of 20 seconds with individual components either increasing, decreasing or remaining stable, independently (thus, direction of component movement was a factor only in the trend comparison task). Again, reaction time and accuracy of responses were to be recorded in this experiment.

2.3 DEPENDENT VARIABLES

Three dependent measures were recorded in the investigation of the various display formats: reaction time, accuracy and subjective preference.

2.4 CONTROL VARIABLES

At the various task complexity levels, the format of the display and the presence or absence of a digital scale represented controls. The same color pattern was utilized in all cases. Questions were presented in positive tone.

2.5 PROCEDURE

Subjects were trained using a set of 12 displays representing different formats, with and without digital scale at the three complexity levels. All subjects completed three tests: comparison between static components, component trending comparison and a declaration of subjective preference. In the component comparison experiment, a set of questions were posed for response based on the displayed data --- e.g., "Is component A greater than B ?" The subjects were instructed to answer either Yes or No by depressing the "Y" or "N" keys. Comparisons between all components were investigated with 3, 10 and 21 comparisons over the three complexity levels for each display format. In the trending experiment the display was updated every second, with 0.5 scale unit, for 20 seconds. Each individual component in the display might be increasing, decreasing or unchanging during the duration of the display; one component would be selected for trend testing. All displays in each experiment were presented in random order. Tasks were self-paced, but the subject was instructed to respond as quickly and as accurately as possible.

3. RESULTS

Mean reaction times and accuracy figures for the various task complexity levels, display formats and scale display options are presented in Tables 1 & 2. All data was subjected to analysis of variance testing.

Table 1. Results of Component Comparison Experiment.

Format	Task Complexity	Scaling	Reaction Time (Second)	Accuracy (%)
Bar Chart	Low	With	3.7	97.8
		Without	3.5	97.8
	Medium	With	3.7	100.0
		Without	3.8	99.3
	High	With	4.2	98.8
		Without	4.1	98.1
Pie Chart	Low	With	4.1	97.8
		Without	4.3	97.8
	Medium	With	4.5	98.7
		Without	4.4	98.7
	High	With	4.8	99.0
		Without	4.6	96.8
Polar Graph	Low	With	3.9	97.8
		Without	3.9	95.6
	Medium	With	4.5	98.7
		Without	4.4	98.0
	High	With	4.8	99.0
		Without	4.8	98.4

Table 2. Results of Trend Reading Experiment.

Format	Task Complexity	Scaling	Reaction Time (Second)	Accuracy (%)
Bar Chart	Low	With	2.6	95.2
		Without	2.9	97.6
	Medium	With	2.9	94.3
		Without	2.8	94.3
	High	With	3.0	100.0
		Without	2.9	95.9
Pie Chart	Low	With	3.3	97.6
		Without	4.2	92.9
	Medium	With	3.2	97.0
		Without	4.6	72.9
	High	With	3.4	99.0
		Without	5.0	71.4
Polar Graph	Low	With	2.7	97.6
		Without	2.7	95.2
	Medium	With	2.8	94.3
		Without	3.1	98.6
	High	With	3.1	95.9
		Without	3.0	100.0

In the component comparison experiment, the inclusion or exclusion of a digital scale was not a significant factor in either reaction time or accuracy of responses. With respect to reaction time, the format of the display was found to be a significant factor at all complexity levels, with the bar chart format being superior to both the polar graph and pie chart. The formatting factor showed no significant effects on accuracy values. Table 3 summarizes statistical testing results for this experiment.

In the trend reading experiment, the reaction times for the bar chart and polar graph formats were statistically indistinguishable, but both exhibited significantly better times than those for the pie chart format. This observation is valid at all task complexity levels. The inclusion of a digital scale also proved to reduce response times at all complexity levels. At the low task complexity level the format and scale factors showed no major effects on the accuracy of responses. However, at the medium and high ranges, the bar and polar graphs exhibited better accuracy values; and the inclusion of a digital scale produced greater accuracy, on average,

Table 3. Analysis Results of Component Comparison Experiment.

Task Complexity Level	Significant Treatment Factor	Performance Ranking	
		Reaction Time	Accuracy
Low	Format (α < 0.01)	Bar < pie (α = 0.01) Bar < polar (α = 0.05)	No effects
Medium	Format (α < 0.01)	Bar < polar, pie (α= 0.01)	No effects
High	Format (α< 0.01)	Bar < polar, pie (α= 0.01)	No effects

than its omission. In most test cases for this experiment, the pie chart without digital scale stood out as the poorest design. Results for this experiment, based on the analysis of data, point towards the recommendation of a bar chart or polar graph with digital scale. Testing results are in given in Table 4.

Participants interviewed subsequent to testing expressed a preference for the following display types: bar chart without digital scale and polar graph with or without digital scale.

4. DISCUSSION

As noted above, in a simple component comparison of static data the bar chart presentation format is to be recommended. Subjects found that the interpretation of length scale from oriented angle (as in the polar graph format) required more time than the one with an upright angle (as in the bar chart format). Bar chart formats have been suggested for the representation of multivariate graphical information in previous studies [3, 4], especially when the task requires the treatment of components separately rather integrated as a single unit. Results in this study confirm this viewpoint.

In the dynamic situation (the more prevalent real-world case), wherein information is updated periodically so that monintored levels are continuously changing, we find that the bar chart and polar graph formats allow for quicker reaction times with higher accuracy in the interpretation of data. This is true, as well, for the inclusion of a digital scale. On this basis, we can recommend a bar or polar graph with digital scale for representation of data in common information monitoring functions.

Table 4. Analysis Results of Trend Reading Experiment.

Task Complexity Level	Significant Treatment Factor	Performance Ranking	
		Reaction Time	Accuracy
Low	Format ($\alpha < 0.01$)	Polar, bar < pie ($\alpha = 0.01$)	No effects
	Scale ($\alpha \simeq 0.45$)	W < W/0 ($\alpha = 0.05$)	
Medium	Format ($\alpha < 0.01$)	Bar, polar < pie ($\alpha = 0.01$)	Bar, polar > pie ($\alpha = 0.01$)
	Scale ($\alpha < 0.01$)	W < W/0 ($\alpha = 0.01$)	W > W/0 ($\alpha = 0.05$)
	Interaction ($\alpha < 0.01$)	Pie W/0 > others ($\alpha = 0.01$)	Pie W/0 < others ($\alpha = 0.01$)
High	Format ($\alpha < 0.01$)	Bar, polar < pie ($\alpha = 0.01$)	Bar, polar > pie ($\alpha = 0.01$)
	Scale ($\alpha < 0.01$)	W < W/0 ($\alpha = 0.05$)	W > W/0 ($\alpha = 0.01$)
	Interaction ($\alpha < 0.01$)	Pie W/0 > others ($\alpha = 0.01$)	Pie W/0 < others ($\alpha = 0.01$)

Further studies are planned for the evaluation of various formats in three dimensions and for cases in which monitored parameters are correlated.

5. REFERENCES

[1] Wood. D.D., Wise, J.A., and Hanes, L.F., An Evaluation of Nuclear Power Plant Safety Parameter Display Systems, **Proceedings of the 25th Human Factors Society Meeting** (1981), 110-114.

[2] Chen, J., Schlegel, R., and Peacock, B., A Prototype Expert System for Physical Work Stress Analysis, In: Asfour, S., (Ed.), **Trends in Ergonomics/Human Factors IV** (North-Holland, Amsterdam 1987), 541-546.

[3] Wickens, C.D., et.al., Display/Cognitive Interface: the Effect Information Integration Requirements of Display Formatting for C3 Displays, Technical Report EPL-85-3/AFHRL-RADC-85-1, University of Illinois, Engineering Psychological Laboratory.

[4] Gehlen, J.R. and Schwartz, D., Display Formatting: An Expert System Application, **Proceedings of the 31st Human Factors Society Meeting** (1987), 1320.

Trends in Ergonomics/Human Factors V
F. Aghazadeh (Editor)

REDUNDANT INFORMATION AS A MEANS OF AIDING PERFORMANCE IN HUMAN-MACHINE INTERFACE

Lars-Erik Warg

Department of Occupational Medicine
Örebro Medical Center Hospital
S-701 85 ÖREBRO
Sweden

INTRODUCTION

Today, there is an impressive body of literature focusing on different aspects of new technology. This technology typically involves changes in the human-machine interface. However several investigations have pointed out that these changes do not necessarily benefit the operator (see e.g. Bainbridge,1987; Brehmer, 1986). Designers are in position to enrich the human-machine interface, but so far they often have taken little consideration of some important knowledge generated by the behavioural sciences, notably psychology.

The purpose of the present contribution, is to assert that designers should build in overlapping stimuli/information in technical systems. Today, technical systems are typically constructed so that every form of overlapping information is removed. Another word for this kind of information is redundancy, a notion often recognized in psychological studies of concept attainment and verbal learning. Here redundancy refers to the perfect covariance of two or more dimensions. That is, values on one dimension are perfectly predictable from knowledge of values on another dimension.

There are at least two ways in which redundant information could aid human performance in situations where new technology is involved: by making it easier to detect the target information, and by aiding man to decide if the correct steps have been taken as a response to the detected stimuli.

To detect the information of importance and to be able to use it as performance feedback, is of course consequential for all humans working with any technical apparatus. In fact, these two aspects have a great impact on how well humans cope with their task.

Before I continue my description of redundancy and its role in new technology, I would like to give an example of the concept in a setting that most of us are familiar with.

MODERN CARS AND INFORMATION OF VELOCITY

Consider yourself sitting in a Volvo or any other modern, comfortable car. You are driving along a straight and broad countryroad. It is a warm day so you have turned on the airconditioner and you are listening to your favorite music on the radio. In short, you are having quite a nice time there in the car.

After driving half an hour, you are requested to estimate the velocity of the car without looking at the speedometer. I guess that many of you would underestimate the speed. The main reason for this is I think, that you are not receiving any redundant information about how fast you are actually driving. In modern cars, this type of information is almost entirely provided by the speedometer; all other redundant information has been eliminated by design. You can not hear any roar from the engine or the wind and at least in a Volvo, you will hardly feel the bumps in the road. In addition information about the status of the car is primarily perceived via the visual system from a single source. At the same time as designers are "improving" ergonomics, they are expanding the amount of information available to the driver to such an extent that some car instrument panels will soon rival those found in airplanes.

I hope that this example has demonstrated that modern technological advances often eliminate redundant information. Those of us who have experience with older cars, know that they really can provide redundant information about the velocity. In fact, that might have been a reason why we bought a new car: we perceived this information as something unwanted and uncomfortable.

Of course I do not mean that designers should build in redundancy by making new cars behave and sound like old cars! Hopefully designers can combine new technology with psychological knowledge about human functioning to produce even safer and more comfortable cars.

DETECTION OF RELEVANT INFORMATION

By definition, redundant information is somehow unnecessary since it is predictable from other types of information. However, it probably has a facilitating effect on detectability of relevant information. To continue with an example from the traffic environment, consider the automatic traffic signals. Here, colour and the position of the light are completely redundant in informational content. The stop signal always appears at the top of the row and is always red. In a similar way, the "go"signal is always at the bottom and green (the example is borrowed from Bourne, Ekstrand & Dominowski, 1971).

Usually when we are approaching the traffic signal, we respond to the colour of the signal. In other situations, however, we may respond more to the position of the signal, e.g. when facing the sun which makes it difficult to see the actual colours. In addition, one may be colourblind and still detect the valid information by checking the positon of the signal.

Automatic traffic signals are a good example of how built in redundancy might be utilized when designing new technology. In the example above focus was on detectability. We can easily apply redundancy and detectability to situations where complicated technology is involved. Consider the person monitoring a computer control panel for normal plant operations. If there is a problem in the process or control system, the demand on the operator rises quickly. However, he/she could not take any measures to correct the problem without actually having detected the "problem" signal. If this signal is provided both visually and audibly (redundant) compared to either one alone, the time between signal and detection should be shortened. In other words, the human system is provided with redundant information, which facilitates the detectability of the signal. No doubt, when it comes to signals for an emergency designers have learned their lesson: they often appeal to two of our senses, vision and hearing.

Although there surely is much more that may be said about redundant information and detectability. I now turn to another aspect of redundancy which is of great importance for humans from a cognitive psychology point of view. Namely, redundant information may serve an important role as performance feedback.

REDUNDANCY AND FEEDBACK

When things are running smoothly in the interaction with new technology, humans can most of the time rely on what Rasmussen (1985) labels skill-based and rule-based behaviour. The former behaviour represents sensory-motor performance which takes place without direct conscious control. It is more like a smooth, automated and highly integrated pattern of behaviour. In rule-based behaviour, on the other hand, the operator perceives information as signs and activates stored rules or prior experience.

However, when faced with an unfamiliar situation for which no rules for control are available, the operator has to rely on knowledge-based behaviour (Rasmussen 1985). Here the control of performance is at such a high level that the internal structure of the system is explicitly represented by a mental model. In fact, in order to control a system, a person must develop a model of the system he wishes to control (Conant & Ashby, 1970). However, modern technology makes it very hard for man to develop such a model. This is mainly due to the fact that these systems tend to make the relationship between the operator and the task indirect (Brehmer, 1987). The operator can never receive direct information about what goes on in the system; only representations are provided. A similar situation exists for output. An operator may observe the current status of the system but nevertheless receive little or no feedback indicating how that output came about.

Often the need to understand the system and to control it (and thereby developing a model of it), forces the operator to conduct experiments with the system. The system is seldom designed however to allow for this activity and responds dogmatically with "invalid command", "error due to ..." or some such message.

When the operator has to rely on the formerly mentioned knowledge-based behaviour, he/she has to formulate hypotheses about how the system is working. Then he has to test these by a trial and error strategy. In this testing process, the feedback which the operator receives from the system is, of course, of great importance for any steps later taken.

The position taken here is that redundant information can form a base for a better understanding of the technical system. Although this type of information by definition is unnecessary, it is far from clear that people have a conception of it as useless information. More likely, they will consider the redundant information as independent evidence about some event or criterion they are having hypotheses about. At least to some extent, this could have a facilitating effect on performance (Armelius, 1976).

MAN AS A PERCEPTUAL SYSTEM

The ideas expressed in this paper, suggest that designers of new technology should take the human perceptual system for what it is: a distinct mode of functioning. In many man-machine systems, perception has mainly been regarded as a means of providing the analytical system with data to process. Much of the research on ergonomics, for instance, has been focussed on speed and accuracy in reading figures from different types of instruments, e.g. comparing dials and counters in these respects.

To consider man as a perceptual system implies that human-machine problems also includes more fundamental problems: how to provide an overview of the process, how to give detailed information about subsystems and components, what type of information should be presented etc (for a more extensive reading, see Osborne, 1987).

In this context, I think it is relevant to stress two findings in research on visual information processing.

Two noticeable aspects of visual memory are its enormous capacity and long duration. No doubt, all of us can vividly remember some scenes from childhood. If these scenes include people, we very often are able to give a detailed description of how these people were standing, what clothes (colours included) they were wearing and perhaps even their facial expressions. This suggests that memory is quite good for visual material. If we want to make this manifest experimentally, we can measure the memory be recognition. There are estimates saying that a person can recognize close to one million pictures. Similary, a typical finding in visual memory experiments, is that pictures seem to afford a great deal of information to the observer. What should be noted especially, is that this is done in a short time and that it seems very easy to retain the information (Brehmer, 1986; Spoer & Lehmkuhle, 1982).

I think that few would disagree with me when I say that so far, this enormous capacity humans have to recognize visual information, has only partially been taken advantage of in the design of many technical systems. Presenting pictures or providing redundant information are only variations on the same theme: the information provided by technical systems should fit man´s perceptual system.

It is quite clear that the organization of a stimulus (information) affects how that stimulus is processed. It will influence both detectability and interpretations of it. For an operator facing displays or even panels in an entire control room, it certainly is of importance how he/she perceives a changed pattern in these stimuli.

In a series of experiments, Prinzmetal and Banks (1977) explored the principle of good continuation, that is, one of the general "laws" of perceptual organization that the Gestalt psychologists in the 1920´s discovered.
In these new experiments, results suggested that pattern recognition occurs in a global-to-locus fashion. In other words, when subjects were shown an entire array of objects, they first organized elements into larger patterns. This is an indication that people (at least in some settings), tend to process patterns of items before analyzing the component elements. Modern technology prescribes operators to process and react to absolute values rather than to patterns. Again we have an example of a technical design that might put some extra strain on the operator coping with his task in a man-machine interface.

CONCLUSIONS

In this paper, I have pointed to a means that I think could aid human performance when interacting with modern technology. Specifically, I have suggested the construction of technical systems that provide the operator with rich and redundant information. Until now, many such systems seem to require that man is a robust and error-free information processor with a great capability to react to absolute values of stimuli.

Findings in experiments conducted in the behavioural sciences, notably cognitive psychology and visual information processing, strongly suggest that this assumption is not correct. Faced with the problem of changing the human or the technical system, the choice ought to be easy.

ACKNOWLEDGEMENTS

I am indebted to Steven Linton for his comments on the ideas expressed in this paper as well as on my English and to Kerstin Åkerstedt for her prompt and skilful typing of the manuscript.

REFERENCES

Armelius, K. (1976). Cue intercorrelation and redundancy in probabilistic inference tasks. *Umeå Psychological Reports*, Supplement No. 1.
Bainbridge, L. (1987). Ironies of Automation. In: J. Rasmussen, K. Duncan and J. Leplat (Eds), *New Technology and Human Error*. Chichester: John Wiley & Sons Ltd, pp 271-283.
Brehmer, B. (1986). Man as an Operator of Systems. In: H.E. Peterson and W. Schneider (Eds), *Human Computer Communications in Health Care*. Elsevier Science Publishers B.V. (North-Holland), pp 7-15.
Brehmer, B. (1987). Development of Mental Models for Decision in Techno-

logical Systems. In: J. Rasmussen, K. Duncan and J. Leplat (Eds), *New Technology and Human Error*. Chichester: John Wiley & Sons Ltd, pp 111-120.
Bourne, L.E., Ekstrand, B.R., and Dominowski, R.L. (1971). *The Psychology of Thinking*. New Jersey: Prentice-Hall, inc.
Conant, R.R., and Ashby, W.R. (1970). Every good regulator of a system must be a model of that system. *International Journal of System Science*, 1, 59-74.
Osborne, D.J. (1987). *Ergonomics at work*. Chichester: John Wiley & Sons Ltd.
Prinzmetal, W., and Banks, W.P. (1977). Good continuation affects visual detection. *Perception & Psychophysics*, 21, 389-395.
Rasmussen, J. (1985). Human Error Data. Facts or Fiction? *Risø National Laboratory*, Roskilde, Denmark, Risø-M-2499.
Spoehr, K.T., and Lehmkuhle, S.W. (1982). *Visual Information Processing*. San Francisco: W.H. Freeman and Company.

Trends in Ergonomics/Human Factors V
F. Aghazadeh (Editor)

APPLICATION OF TECHNICAL WRITING PRINCIPLES TO QUESTIONNAIRE DESIGN

Gail Emily Fey

Human Factors and Usability Design
IBM
Department 54Y/685
Neighborhood Road
Kingston, NY 12401

In gathering data to measure criteria, researchers frequently find themselves compiling questionnaires to be completed by or administered to respondents. The results from the administration of a questionnaire ultimately depend on the questions asked. Drawing on a variety of textbooks, periodicals, and technical reports, this paper presents summary information on questionnaire design. Topics addressed in this paper include pre-questionnaire considerations and wording of the questions. It is prescriptive as well as descriptive in its coverage of questionnaires in general and questions in particular. The author, a former technical writer, points out the similarities between good writing and good questionnaires.

1. INTRODUCTION

Given that the subject of designing questionnaires has been explored by many sources, this author has opted to forego generating one more vanilla here's-how-to-develop-a-good-questionnaire paper. Although the basics of questionnaire design bear repeating, this paper does so with a new flavor -- that of applying technical writing principles to questionnaire design.

Effective technical writers practice the following: (1)Considering the Audience, (2)Defining the Topic, (3)Clarity, (4)Simplicity, (5)Brevity, (6)Specificity, (7)Shunning Wordiness, (8)Avoiding Pomposity, (9)Careful Use of Abstraction, (10)Eliminating Unnecessary Jargon, (11)Appropriate Point of View, (12)Limited Use of Passive Voice, (13)Exemplification, (14)Logical Organization, (15)Cultivating Interest, (16)Remembering Instructions, and (17)Attending to Format. Questionnaire designers can employ these with equal success.

2. PRE-QUESTIONNAIRE ACTIVITIES

CONSIDERING THE AUDIENCE: "We can write prose [a questionnaire] that fails for reasons more important than an unclear style. If we ignore what our readers [respondants] have to know if they're to understand our ideas, then what we write [ask] will surely bewilder them."[1] In the above quote, Williams was speaking of prose, but his point is equally applicable to questionnaires. We should anticipate where respondants will have difficulties, need coaching or reassuring, want visual relief, get tired, misunderstand, become impatient, and so on. The following list presents several points pertinent to analyzing the audience for a questionnaire:

- Will respondants have <u>background knowledge</u> of the subject we are asking about?
- Are respondants privy to the <u>vocabulary</u> of the subject? For example, do they know we mean "printed circuit board" when we ask about PCB's? Or, are polychlorinated biphenyls the only PCB's they've heard of?

- Are there any sensitivities we should be aware of? For example ethnic or social issues.
- Are there any major linguistic differences? Be it due to race, literacy, occupation, geography, religion, or other reasons, people attach different meanings to words. For instance, to some rural southerners, "ill" means angry whereas "sick" refers to medical and mental diseases.[2]
- Is the range in the educational level of the respondants wide? Those with lower educational levels respond radically different from those with moderate, high, and very high educational levels.[3]
- What has personal experience shown with questionnaires of this type? Technical writers use personal experience as a valuable source of information in predicting where their readers may have difficulties. As a result of being confounded by an explanation, mislead by instructions, frustrated by unnecessary detail or a lack of necessary detail, left in the dark by unfamiliar terminology, and so forth, writers learn where they can go wrong. By the same token, question designers can recall their own experiences with questionnaires. Which ones did they complete and why?

Additionally, questionnaire designers will want to be considerate of their respondants by remembering the following:

- Respondants have time constraints. If we want them to spend their time answering questions for no immediate benefit, the process needs to be as painless as possible.
- The care with which respondants read and complete the questionnaire varies. "They [questionnaire authors] may assume that respondents will provide truthful or accurate answers when there is no basis for thinking that they will or when it is completely unrealistic to suppose that they will. If it is necessary for respondents to go to a considerable amount of trouble in order to answer a question, it is highly likely that a large number of individuals will either leave the question blank, answer it erroneously or at random, or simply provide an answer based upon what they think is expected, a guess, or their best estimate."[4]
- No one wants to appear stupid. Give respondants every opportunity to save face by making things clear. Many will not confess they do not know what you mean or that they are confused.
- Respondants do not have to be intellectually dull to misunderstand or become confused. In a study conducted by Lees-Haley[5] a group of graduate students was asked whether or not their car had a dual piston twin engine hydro-carburetor. Two of the respondants were described as being highly intelligent with Master's degrees. They answered affirmatively to his question, even though a dual piston twin engine hydro-carburetor does not exist.

DEFINING THE TOPIC: In addition to audience analysis, identifying what you want to find out is a fundamental pre-questionnaire activity. "If you spend time at the beginning clarifying what you are after, you can make your data simpler to analyze. If you don't clarify your research problem, even flashy statistics may not be able to help."[6]

A solid reason should lie behind each question on a questionnaire. We should not query respondants simply because a similar question appeared on another survey. Unnecessary questions lengthen the survey, tire respondants, and waste time in providing and tracking the answers.

Knowing exactly what you want to find out from the respondant will make wording the questions easier too. For composing questions, your approach (AFTER identifying exactly what you want to find out) might be to devise a question that you think asks what you want to learn. Then, revisit and rework it.

3. WRITING THE QUESTIONS

The technical writing principles outlined here are interrelated. Application of one necessarily affects the others. For instance, use of the passive voice partly determines the question length; the simplicity of writing influences question clarity; and so on.

CLARITY: "Everything that can be thought at all can be thought clearly. Everything that can be said can be said clearly." [7] Communicators of any type, be they speakers, writers, signers, questioners, and others have a message they desire to transmit. The clearer the message and the vehicle by which it is transmitted, the better chance the recipient will interpret the message as intended. And, (this is especially critical to questionnaire design) when the recipient correctly interprets the message, the better the chances for a relevant and accurate reply.

Practically speaking, though, clarity is easier sought than achieved. The other technical communication principles outlined here will help produce clarity in survey questions. As a corollary to the quote above, we can add "Everything that can be written at all can be written clearly. Every question that can be asked can be asked clearly."

SIMPLICITY: Thoreau bemoaned complexity and urged us to "simplify, simplify". Strunk and White [8] encouraged simplicity of writing through avoiding: overwriting, overstating, explaining too much, using awkward constructions, using fancy words, and using the offbeat rather than the standard.

In questionnaires, simplification can ease a respondant's cognitive load. M.A. Collen et al.(1969) [9] reported on a major study of the semantic difficulty of questions and the variability of responses conducted by the Kaiser-Permanente group. In comparing two sets of questions studied, Collen indicated that the set of questions designed to elicit more information from the patient placed multi-level logic demands on patients. That is, the questions presented more than one set of logical alternatives. Whereas the second set of questions had one layer of logic demands. Budd [10] suggests that complex questions requiring multi-level logic, such as those identified by Collen, are difficult for respondants to answer. multi-level logic, are difficult for respondants to answer. Examples from the question sets compared in the Collen study follow:

Question with multi-level logic demands (From Kaiser-Permanente questionnaire):
In the past 6 months have you often
had nausea (sick to stomach) or vomiting Yes No
been troubled by excessive gas or bloating Yes No

Question with single-level logic demands (From Cornell Medical Index):
Do you often vomit (throw up) Yes No

In striving for simplicity in questionnaires, we should also keep in mind that negatively worded questions can turn a simple inquiry into a source of confusion. For example, the question:
Do you agree that the safety officer in your area of the plant should not physically interfere in an effort to protect employees from physical harm?
could be phrased more simply by omitting "not".

Another difficulty with negatively worded questions is that the answer options must not leave the respondant confused. In the example just cited, the answer options of "yes" and "no" might cause respondants to answer the opposite of what they intend. However, the use of negatively worded questions is one way of preventing respondants from falling into a response pattern to positively worded questions.

BREVITY: Brevity of both the questionnaire and the individual questions naturally follow from correct application of the other communication points explained in this paper. Using active voice, having non-convoluted sentence structures, knowing exactly what we want to find out, not asking needless questions, writing clearly, keeping things simple, avoiding wordiness -- all lend themselves to abbreviating questionnaires.

The admonition to be brief does not mean brevity at all costs. For instance, packing several thoughts into one question just to keep a questionnaire short is unwise. Granted, the resulting questionnaire may be brief, but the questions themselves won't be. And, the results may not be interpretable.

SPECIFICITY: Writers know technical documents are not the place for ambiguity. If, for example, the information accompanying a new, expensive piece of medical equipment indicates only that you should turn on the machine, but fails to tell you how or where the power switch is located, then you will most likely spend some time guessing how to turn on the equipment. Or, in a worse case, you will damage the equipment and/or yourself, and experience avoidable frustration.

One especially relevant facet of the questionnaire to which specificity applies is the instructions. Instructions should be quite short and to the point, doing just what their name states: instructing, not explaining or justifying or telling respondants things they don't need to know. Tell the respondants EXACTLY what you want them to do. For instance, if you intend for respondants to circle a number on a scale that appears below a question, you may think that saying nothing or something like *indicate your answer below* is clear enough. It is clearer to precisely steer the respondants by stating something similar to *circle a number on the scale below that best represents your attitude.*

One concern that may arise as a result of specificity is that concrete, specific questions limit respondants. They may to a certain extent, but they also clarify what information you are seeking and ease result analysis.[11]

SHUNNING WORDINESS: Using many words to convey a simple message is wordiness. If we muddy the waters with unnecessary verbiage, respondants may misunderstand and furnish inaccurate information. Sparsity of words is crisper. It keeps the questionnaire short. As a result, respondants are not left with the impression that completing the questionnaire is going to be a long and painful process. Suppose we want to know how many times a respondant uses Product XX in a week. We can simply ask:
How many times per week do you use Product XX?
Or, we could obscure the message with extra words and say:
In your usage of the product being evaluated, what is the frequency on a weekly basis with which you utilize this previously mentioned product?

AVOIDING POMPOSITY: Communication suffers when the communicator hides a message in affectation. Brusaw, et.al., [12] defined affectation as "the use of language that is more technical, or showy, than is necessary to communicate information". Technical writers are warned against it. Questionnaire designers would benefit from the warning as well. Trying to impress with pompous words will likely confuse and possibly intimidate respondants into supplying answers they don't really mean.

CAREFUL USE OF ABSTRACTION: Abstractions do not have to be philosophical, obscure terms. They can be as simple and frequently used as "government", "education", or "safety". In questionnaires, the danger of overusing abstract terms lies in the potential for respondants misunderstanding and therefore supplying erroneous information. S. Chase (1938) [13] stated that we could just as effectively use the word "blab" when we use a high-order abstraction in trying to communicate. An example from Payne [14] of a typical "blab-blab" [overly abstract] question follows:
Should our country (BLAB) be more active (BLAB) in world affairs (BLAB)?

This is not to say we should never use abstract terms. Adept technical writers use both abstract and concrete words in a mutually supportive fashion. However, when it comes to questionnaires, abstractions should be used judiciously.

ELIMINATING UNNECESSARY JARGON: Every field has its terminology. We should, however, be sure any audience we are communicating with concerning that field of knowledge is familiar with the terminology. If we do not, our message will be garbled.

Social scientists may, for example, be totally comfortable with words like "sibling" and "marital status". But some people may not know what you mean if you ask them about their marital status or how many siblings they have. Some could assume you want to know how many members are in the respondant's family. Not wishing to appear ignorant, the respondant would supply you with his or her version of the requested information, albeit incorrect.

In a study by Suchman et al.(1958) [15] even so simple a question as:
Are you a (a) Male (b) Female?
was liable to misinterpretation. The error rate for the above question was 11 percent until it was modified to read:
Are you a (a) Man (b) Woman?
One way to alleviate the problem of inappropriate use of jargon is to have someone outside of your own field of expertise read the questionnaire.

As for acronyms, basic technical writing etiquette dictates that an acronym be spelled out the first time it appears. Subsequently, the acronym may be used alone. The same etiquette applies to questionnaires.

APPROPRIATE POINT OF VIEW: Point of view is the attitude a writer exhibits toward the reader and is demonstrated in the writer's use of personal pronouns (I, you, he, she, we, they). By extending the concept into the realm of questionnaires we can say the questioner's point of view is conveyed to the respondant by the questionnaire's use of personal pronouns.

Williams, [16] pointed out that making the audience the agent of the action involves the reader in the experience. It makes the information the writer is trying to convey immediately relevant. When asking questions, researchers are typically concerned with the respondant's experiences. To alleviate distancing of readers (in our case, respondants) from the information, Williams suggests using the audience's point of view; that is, using the word "you". Rather than use the third person, impersonal point of view (he,she, it):
On the average, in one year, how often does the respondant use the XXX graphics tool?
questions can be reworked to involve the respondant by using the second person personal point of view (you):
On the average, in one year, how often do you use the XXX graphics tool?

When a technical writer avoids using "you", the results can be awkward. When a questioner does it, the results can be misunderstanding, misinformation, or no information at all. Questionnaire designers should consider that using the impersonal point of view may inhibit the respondant's willingness to open up and supply information they are not obliged to give.

LIMITED USE OF PASSIVE VOICE: When improperly applied, a passive construction permits misunderstanding, contorted sentence structures, wordiness, and other undesireable results. Although students of writing have repeatedly heard the warning to avoid the passive, they also realize that like any communication tool, it can be used effectively too. As a general rule-of-thumb, though, prefer the active voice.

According to Bernstein [17] the active voice is the shortest distance between two points. It goes forward. The agent of the action appears first, then the action, then the object acted upon. Passive constructions require the audience to back-pedal. Moreover, usually the passive voice needs more words, directly contradictory to the goal of appropriate brevity in communication. The following are illustrations of the active and passive voices:
Active Voice -- *The nurse incorrectly lifted the patient.*
Passive Voice -- *The patient was lifted incorrectly by the nurse.*

In this example, the damage isn't too dramatic. We get the meaning even though we do have to move backward to find out who did the lifting. The increase in the number of words is minor -- from six to eight. But, consider what occurs when the message is more complex. The following might appear on a questionnaire for the evaluation of software panels:
In the product panels just examined by you, what menu items were used by you to gain the most helpful information?
What we want to know can be stated actively by:
In the product panels you just examined, what menu items gave you the most helpful information?
Questioners and writers also need to be aware of the potential problem of mixing the passive and active voices which can result in confusion. For instance:
The engineers were fired and left.
Were the engineers left behind, or did they leave?

EXEMPLIFICATION: Using the specific to exemplify the general is a technique employed by most if not all effective communicators. Examples give readers (and respondants) something concrete to grasp. A common piece of technical documentation, a computer user's guide, if well written, typically contains examples for: tasks the user might want to accomplish, screens that could appear, and command syntax.

Questionnaire designers can utilize exemplification in both the instructions and questions. For instance, if we want respondants to shade in a vertical scale to indicate their evaluation of some characteristic, we can include an exemplary scale already shaded in to illustrate exactly how the respondant should indicate an answer. Or, several suggestions can be included for respondants to consider when answering a question. To the question:
How often do you experience headaches? We might add:
For example, your answer could be "daily", "twice a week", "never had one", or some other descriptor.

4. ADDITIONAL CONSIDERATIONS

LOGICAL ORGANIZATION: "If we can't find a way to organize our ideas clearly for our specific audience, then what we write [ask] will almost certainly lack that sense of direction and purpose that all coherent prose [a questionnaire] demands."[18] Organization is critical to a technical document. Brusaw, et al., [19] indicated that readers need to mentally shape and structure the information being presented. Writers choose from many different organizational structures e.g., chronological, hierarchical, general-to-specific, specific-to-general, and others. Organization is no less essential to the questionnaire.

One method for organizing the questions is to use the funnel approach which places the general, easy questions before the more specific, difficult questions.[20] Another way to organize a questionnaire is to group the questions by subject. For instance, multiple questions on each of several environmental pollutants such as asbestos, radiation, carbon monoxide, and lead, should not be intermingled. Group the asbestos-related questions together, the radiation-related questions together, and so on.

Regardless of the overall organization, questions appearing early on should not rely on information that comes later. That statement may sound terribly obvious. But, in reworking and reorganizing questions, you can inadvertently mis-sequence questions.

CULTIVATING INTEREST: Communicators quickly learn to "hook" their audiences: speakers relate anecdotes, novelists use various techniques to draw readers past the first few pages, technical writers put the main idea up front so prospective readers know what will be discussed. Questionnaire designers can and should entice their audiences as well. According to D.A. Dillman (1978) [21], the first item in a questionnaire determines whether or not the questionnaire gets discarded or completed.

Warwick and Lininger [22] stated that a good opening question is one that is pleasant, interesting, and easy. They gave an example of a conversational opener that encourages respondants to express themselves in a positive light:

We are interested in how people are getting along financially these days. Would you say that you and your family are better or worse off financially than you were a year ago, or about the same?

Moreover, placing general, easy questions at the beginning builds the respondant's confidence. Kahn and Cannell (1957) [23], feel respondants may be uneasy about whether they can even do what is being asked of them in completing a questionnaire or interview. Most of us like to continue doing things we feel capable of. Fostering that feeling of capability is an incentive to the respondant to enter into the arena of the questionnaire.

REMEMBERING INSTRUCTIONS: Conveying instructions is fundamental to technical communication. Questionnaire instructions should adhere to all the principles of good writing, the most essential one being, in my opinion, specificity. In addition to the other writing principles explained, the following pertain to instructions:

- Use the imperative when possible. The previous sentence is a good example. Rather than:
  The questionnaire designer should use the imperative when possible,
  it is very clear to state it as above.

- If you want respondants to do something sequentially, number the steps. And, if it is important that the respondants do the steps in the sequence in order, you may want to make that clear by telling them. A simple statement like:
  Do steps 1-5 in the order they appear. It is important that you not change the order.

ATTENDING TO FORMAT: Most of us have been subjected to poorly formatted information. For instance, a computer manual with very little white space, a hard-to-read typeface, unbroken text, no color -- in other words, nothing visually pleasing. The same poor formatting can occur in questionnaires.

Format comprises the physical, usability, and aesthetic characteristics of a questionnaire. Some of the formatting items questionnaire designers should consider are: inclusion of heads and running headers/footers; perfect grammer; sufficient white space; readable type size, typeface, and line length; and preference for justified or unjustified text.

5. SUMMARY

The main goal in designing a questionnaire is to convey a message to the respondants, that being "Here is what we want to know." If the message is conveyed clearly enough, the respondants can in turn supply the information desired. The potential for respondant misunderstanding and the collection of erroneous data is insidious in questionnaire design. "The most troublesome errors in questionnaires do not arise from bad judgment after due consideration of doubtful points; they creep in unwittingly, even in 'obviously simple' questions."[24] Writing those obviously simple questions well, through the application of effective technical communication principles, can alleviate at least some degree of questionnaire error.

REFERENCES

[1] Williams, J.M., Style: Ten Lessons in Clarity and Grace (Scott, Foresman, and Co., Glenview, 1981) p.2.
[2] Lees-Haley, P.R., The Questionnaire Design Handbook (Rubicon, Huntsville, 1980) p.37.
[3] Lees-Haley, p.36
[4] Lees Haley, p.31.
[5] Lees Haley, p.31
[6] Krull, R., personal communication.
[7] Williams, J.M., p.1., quoting Ludwig Wittgenstein.
[8] Strunk, W. Jr., and White, E.B., The Elements of Style (Macmillan, New York, 1979) p.viii.
[9] Collen, M.A., et.al., "Reliability of a Self-Administered Medical Questionnaire," Arch. Internal Medicine 123. 664-681. In Budd, M.A., Acquisition of Automated Medical Histories by Questionnaires, Contract #HSM11069264, (The National Center for Health Services Research and Development, 1970) pp.14-15.
[10] Budd, p.15. [11] Krull, R., personal communication.
[12] Brusaw, C.T., et al., Handbook of Technical Writing, 2nd Ed. (St. Martin's Press, 1982) p.26.
[13] Chase, S., The Tyranny of Words, Harcourt, Brace & Co., 1938. In Payne, S., The Art of Asking Questions (Princeton University Press, Princeton, 1951) p.149-150.
[14] Payne, p.150.
[15]
Budd, p.14.
[16] Williams, p.126.
[17] Bernstein, T.M., The Careful Writer (Atheneum, New York, 1981) pp.13-15.
[18] Williams, p.2.
[19] Brusaw, p.424.
[20] Cannell, C.F. and Kahn, R.L. The collection of data by interviewing In L.Festinger and D. Katz (Eds.), Research Methods in the Behavioral Sciences, New York: Dryden Press. In Meister, D., Behavioral Analysis and Measurement Methods (John Wiley & Sons, New York, 1985) p.367,379.

[21] Dillman, D.A. Mail and Telephone Surveys: The Total Design Method. New: John Wiley, 1978. In Lees-Haley, p.40.
[22] Warwick, D.P. and Lininger, C.A., The Sample Survey: Theory and Practice (McGraw Hill, New York, 1975) p.149.
[23] Kahn and Cannell, 1957, The Dynamics of Interviewing. In Lees-Haley, p.39.
[24] Kidder, L.H., Sellitz, Wrightsman, and Cook's Research Methods in Social Relations, 4th Ed. (Holt, Rinehart and Winston, New York, 1981) p.163.

III

HUMAN PERFORMANCE

Trends in Ergonomics/Human Factors V
F. Aghazadeh (Editor)

PRODUCTION STANDARDS AND FEEDBACK AS DETERMINANTS OF WORKER PRODUCTIVITY AND SATISFACTION IN A REPETITIVE MANUFACTURING TASK

Biman Das and Ashraf A. Shikdar
Department of Industrial Engineering
Technical University of Nova Scotia
Halifax, Nova Scotia, Canada B3J 2X4

The main results of two separate studies are presented in which production standards and feedback were provided singly or jointly to determine their effects on worker productivity and satisfaction in a repetitive manufacturing task. Only the combination of an assigned hard production standard in the presence of production quantity and quality feedback had a significant positive effect on worker productivity and satisfaction both at the same time. The increases in quantity and quality output were 13 and 15% respectively, compared to the control group (first study). The provision of an assigned hard standard failed to improve worker productivity but had no adverse effect on worker satisfaction. The incorporation of quantity and quality feedback had no significant effect on worker productivity but had a positive impact on worker satisfaction. A progressive increase in the assigned standards (100, 130, 140 and 150% normal) with feedback (quantity and quality) improved worker productivity significantly up to the provision of 140% of normal and feedback. The increases in quantity and quality output were 14 and 10% respectively, compared to the control group (second study). The provision of a participative standard with feedback was significantly inferior to an assigned 130% standard and feedback. The provision of assigned and participative standards with feedback improved worker satisfaction significantly. No significant difference in worker satisfaction was found among the standards and feedback conditions. Monetary incentive added no incremental performance or satisfaction gain.

1. INTRODUCTION

Worker productivity and satisfaction are of paramount importance in determining the effectiveness of an industrial organization. Workers engaged in repetitive manufacturing task often find the work monotonous, boring, fatiguing and unmotivating. This in turn results in reduced worker productivity and satisfaction and higher absenteeism and causes detrimental effects on worker physical and mental well-being. Goal setting and performance feedback appear to have a positive effect in making the work more challenging, interesting, less boring and creating more attention especially in a repetitive task. Furthermore, if the task is performed on an ergonomically designed workstation, it is likely that the task will be less fatiguing (Das 1987). The potential of goal setting and performance feedback as major components of job design approach has not been fully recognized as yet. These concepts have not been fully utilized in the real-world task situation to improve worker productivity and satisfaction.

Research studies have demonstrated that specific hard goals produce a higher performance level than easy or "do your best" goals (Locke 1968, Kim and Hamner 1976, Locke et al. 1981, Locke and Latham 1984). A hard goal is not a sufficient condition to produce high performance. Performance feedback is necessary for goal setting to be effective. Goal setting with performance feedback was found superior to goal setting or feedback alone in terms of worker productivity and satisfaction (Erez 1977, Becker 1978, Das 1982a, 1982b, Ivancevich 1982). Goals could be either assigned to the operator or set participatively with the operator. Both assigned and participatively set goals produced conflicting results with regard to the superiority of one over the other in terms of worker productivity and satisfaction (Latham and Yulk 1975, Locke et al. 1981, Latham and Steel 1983, Locke and Latham 1984). Monetary incentive is believed to have a positive effect on worker performance through goal acceptance and on worker satisfaction. However, in the laboratory setting, monetary incentive had no impact on worker productivity and satisfaction (Das 1982a, 1982b).

Most of the studies in the past employed simulated or relatively simple tasks and goals were set arbitrarily or based on operator's past performance. Measured standards were not used. Consequently, the question "how hard is hard" cannot be defined precisely. Most studies dealt with worker performance, worker satisfaction was seldom considered. Systematic controlled experiments have seldom been performed with a repetitive manufacturing task employing measured standards under an ergonomically designed manufacturing work system to determine the effects of goal setting and performance feedback, singly or jointly, on worker productivity and satisfaction.

The terms "production standards" and "production feedback" were used in this research instead of "goal setting" and "performance feedback", respectively. The objective of this paper is to present the main results of two separate studies in which production standards and feedback were presented singly or jointly to operators in a repetitive manufacturing task to determine their effects on worker productivity and satisfaction under an ergonomically designed workstation and employing a measured standard.

2. STUDY 1: EFFECTS OF PRODUCTION STANDARDS AND FEEDBACK SINGLY OR JOINTLY ON WORKER PRODUCTIVITY AND SATISFACTION

The main objective of this study was to determine whether production standards and/or production feedback could be provided singly or jointly to operators to improve worker productivity and satisfaction in a repetitive manufacturing task. The details of the experimental method were described elsewhere (Das 1982a, 1982b), only the essentials of the method are highlighted below.

2.1 Method

A drill press operation was chosen for this research to represent a realistic repetitive manufacturing task. The task involved an operator drilling four holes into a previously prepared steel plate while he or she was in a seated position. A power feed drill press with jig and fixture was used to produce connector plates in the Machine Tool Laboratory, North Carolina State University.

The production standards were determined by means of Methods-Time Measurement (MTM). The production/time standard for the operation was 240 holes/hour (100% normal). The percentage of cycle time that was machine controlled was 52%. The hard production standard was determined on the basis that external work elements would be performed at a pace of 130% of normal standard to achieve an overall production standard of 112% normal or 268 holes/hour. An electric totalizing counter was used to provide quantity feedback and the production quality feedback was presented in terms of percentage of good holes drilled in a unit time. Appropriate ergonomic principles and data were employed to facilitate the design of a manufacturing work system (Das 1987).

Experiments were performed upon 56 college students who were paid $3.50 per hour. The operators were trained individually for one hour in the task performance and asked to perform the same task for another hour under a specific experimental condition. There were 7 groups consisting of eight subjects in each group. The seven groups were subjected to the following experimental conditions dealing with production standard (PS) and production feedback (PF): (1) No PS/PF (Control), (2) PS: 100% normal, (3) PS: 130% normal, (4) PF: Quantity, (5) PF: Quantity and quality, (6) PS: 130% normal + PF: Quantity and quality, and (7) PS: 130% normal + PF: Quantity and quality + Monetary Incentive (MI): 1:1, piece rate.

Worker productivity was determined in terms of production quantity (number of holes) and quality (number of good holes) output. Worker satisfaction scores were determined by employing modified Job Diagnostic Survey (JDS) scales (Hackman and Lawler 1971, Hackman and Oldham 1975). The modified JDS scales included the following job or work dimensions: (1) skill variety, (2) task identity, (3) task significance, (4) autonomy, (5) production feedback, (6) production standard, (7) working conditions, and (8) pay. Each subject was asked to answer the questionnaire, which consisted of 18 questions on a seven-point Likert-type scales regarding his or her perception of the various job attributes that were actually present.

During both training and experimental work sessions, quantity and quality output data were collected at the end of each of the production quarters. Worker satisfaction scores were collected at the end of both training and experimental work sessions.

2.2 Results and Discussion

A statistical analysis of the quantity and quality output and worker satisfaction scores was made through the use of the Statistical Analysis System (SAS) computer program (Helwig and Council 1979). The Student's t test was employed to perform a comparative analysis of worker productivity and satisfaction among the groups. The results are presented in Table 1.

Production Standards. The provision of an assigned specific quantitative or 100% normal production standard had no significant effect on worker productivity but had a significant positive effect on worker satisfaction.

The hard production standard of 130% normal had no significant effect on worker productivity but had a highly significant positive effect on worker satisfaction. The assigned hard production standard failed to generate a high level of effort from the operator. The research results contradicted those obtained by Locke (1968). The non-acceptance of the assigned hard standard by 37% of the subjects (Group 3) was probably responsible for such results.

Production Feedback. Quantity feedback had no significant effect on quantity output but had a significant positive effect on quality output. Consequently, definitive conclusions could not be made regarding the overall improvement in worker productivity. This experimental condition had no significant effect on worker satisfaction.

Quantity and quality feedback had no significant effect on worker productivity but had a highly significant positive effect on worker satisfaction.

Combination of Production Standard and Feedback. The provision of a hard production standard in combination with quantity and quality feedback had a highly significant positive effect on worker productivity and satisfaction. The increases in quantity and quality output were 12.84 and 14.46% respectively, compared to the control group. The average increase in worker satisfaction score was 7.66, compared to the control group.

Table 1. Comparative analysis of worker productivity and satisfaction among groups (first study): Student's t test

Main experimental conditions	Comparison between groups (experimental conditions)	Calculated Student's t value: Worker productivity Quantity	Worker productivity Quality	Worker satisfactio
Assigned normal and hard standards	2 (100%) vs. 1 (control)	1.49	1.08	2.39*
	3 (130%) vs. 1	0.82	0.04	3.27**
	3 vs. 2	-0.65	-1.12	0.87
Feedback	4 (quantity) vs. 1	1.46	2.24*	0.57
	5 (quantity & quality) vs. 1	-0.07	0.60	3.44**
	5 vs. 4	-1.56	1.75*	2.78**
Assigned hard standard + feedback	6 (130% + quantity & quality) vs. 1	4.42**	3.97**	6.09**
	6 vs. 3	4.70**	3.73**	2.72**
	6 vs. 5	3.82**	4.50**	2.82**
Assigned hard standard + feedback + monetary incentive	7 (130% + quantity & quality + monetary incentive) vs. 1	3.80**	3.03**	4.20**
	7 vs. 6	-0.47	-0.81	-1.83*

Notes: 1. The tabulated Student's t values for 5% = 1.68 (* $p<0.05$, significant), and for 1% = 2.42 (** $p<0.01$, highly significant).
2. Negative sign = decrease in group means.
3. Adjusted mean for Group 1 (control): quantity = 230.72 holes/hr., quality = 212.27 holes (good)/hr., and worker satisfaction score = 30.31).

This experimental condition was far superior to the hard production standard or the quantity and quality feedback alone, in terms of worker productivity and satisfaction.

Combination of Production Standard and Feedback and Monetary Incentive. The conclusions stated above were found equally true, when the operators were given the same experimental condition as above along with monetary incentive (1:1, piece rate for good holes only with a guaranteed base rate of $3.50 per hour, Group 7). However, a comparison of Groups 7 and 6 showed that the monetary incentive had no significant effect on worker productivity and had a significant negative effect on worker satisfaction. The possible reasons for such results were the perception of the subjects that the monetary incentive was inadequate and the experimenter was enticing them to high performance level with insufficient monetary incentive.

3. STUDY 2: DETERMINATION OF THE SPECIFIC PRODUCTION STANDARD WITH FEEDBACK TO MAXIMIZE WORKER PRODUCTIVITY AND SATISFACTION

The main objective of this study was to determine the effects of various levels of assigned standards and participative standard with feedback to find the specific standard and feedback that would maximize worker productivity and/or satisfaction in a repetitive manufacturing task.

3.1 Method

This investigation was conducted in the Machine Shop, Technical University of Nova Scotia (Das and Shikdar 1986, Das et al. 1987). Basically this study employed a similar experimental task and methodology as the first study.
The subjects in this study were also college students who were paid $5.00 per hour. For the participative standard, the subject was asked by the experimenter to decide upon a production standard above 100% normal that he or she would like to attempt. It was pointed out that workers in the past generally performed between 100 and 150% normal.

Fifty-six subjects participated in this experiment. The seven groups consisting of eight subjects in each group were given the following experimental conditions: (1) No PF/PS (control), (2) PS: 100% normal + PF, (3) PS:130% normal + PF, (4) PS:140% normal + PF, (5) PS:150% normal + PF, (6) PS: Participative + PF, and (7) PS:140% normal + PF + MI. It should be noted that production feedback (PF) included both quantity and quality feedback.

3.2 Results and Discussion

A comparative analysis of quantity and quality output and worker satisfaction (adjusted group means) among the groups was made using the Student-Newman-Kuel's (SNK) range test. The selected results from the SNK range test are presented in Table 2.

Table 2. Comparative analysis of worker productivity and satisfaction among groups (second study): Student-Newman-Kuel's (SNK) range test

Main experimental conditions	Comparison between groups (experimental conditions)	Differences in adjusted means: Worker productivity Quantity	Worker productivity Quality	Worker satisfactio
Assigned normal standard + feedback	2 (100% + feedback) vs. 1 (control)	9.26**	0.98	5.04**
Assigned hard standard + feedback	3 (130% + feedback) vs. 1	26.94**	14.06**	5.42**
	3 vs. 2	17.68**	13.96**	0.38
	4 (140% + feedback) vs. 1	33.77**	19.69**	4.00
	4 vs. 2	24.51**	19.59**	-1.04
	4 vs. 3	6.83**	5.63**	-1.42
	5 (150% + feedback) vs. 1	29.97**	11.84**	4.60**
	5 vs. 4	-3.80**	-7.84**	-0.60
Participative standard + feedback	6 (participative + feedback) vs. 1	26.25**	11.53**	5.06**
	6 vs. 2	16.99**	11.43**	-0.41
	6 vs. 3	-0.69	-2.52**	-0.37
	6 vs. 4	-7.52**	-8.15**	-1.06
Best standard (assigned 140%) + feedback + monetary incentive	7 (140% + feedback + monetary incentive) vs. 1	35.31**	16.50**	5.61**
	7 vs. 4	1.54*	-3.18**	1.61

Notes: 1. Production feedback included both quantity and quality feedback.
2. *<p = 0.05 (significant), **<p = 0.01 (highly significant).
3. Negative sign = decrease in group adjusted means.
4. Adjusted mean for Group 1 (control): Quantity = 234.24 holes/hr., quality = 207.70 holes (good)/hr., and worker satisfaction score = 35.14.

Assigned Normal Standard and Feedback. The incorporation of an assigned normal (100%) standard and feedback (Group 2) had a significant positive effect on quantity output but had no significant effect on quality output. Therefore, no conclusive inference could be made with regard to the overall improvement in worker productivity. This experimental condition had a highly significant positive effect on worker satisfaction.

Assigned Hard Standard and Feedback. The provision of an assigned hard standard of 130% normal and feedback led to a significantly better performance than an assigned 100% standard and feedback but no significant difference in worker satisfaction was found between the two experimental conditions.

The incorporation of a still harder assigned standard of 140% normal with feedback led to a further improvement in worker productivity when compared to the provision of an assigned hard standard of 130% normal and feedback. The increases in quantity and quality output for Group 4 were 14.42 and 9.48%, respectively, compared to the control group. No significant difference in worker satisfaction was found between the two conditions. For Group 4, the average increase in worker satisfaction score was 4.00, compared to the control group.

The further increase in the assigned hard standard (150% normal) with feedback failed to increase worker productivity when a comparison was made with the provision of an assigned hard standard of 140% normal and feedback. In fact a significant deterioration in performance took place. The result was in contradiction with the findings of Locke (1982), who observed no decline in performance as the goal level was increased progressively to a very high (impossible) level. Again, no significant improvement in worker satisfaction was found.

Participative Standard and Feedback. A comparison between Groups 6 (participative standard + feedback) and 4 (assigned 140% normal + feedback) showed that the worker productivity of the latter group was significantly better than the former. Consequently the assigned standard of 140% normal (with feedback) is regarded as the best standard for the present repetitive manufacturing task. The provision of a participative standard and feedback was not superior to an assigned normal or hard standard and feedback, in terms of worker satisfaction.

Combination of Best or Assigned 140% Normal Standard and Feedback and Monetary Incentive. The provision of monetary incentive along with an assigned 140% normal standard and feedback (Group 7), had a significant improvement in quantity output when compared to the provision of an assigned 140% normal standard and feedback (Group 4) but the quality output deteriorated significantly. The probable reason for such an outcome could be the subjects were working faster to reach the (quantity output) standard and did not give adequate attention to maintain the quality level. Considering both quantity and quality output, the assigned hard standard of 140% normal with feedback proved to be the best condition for improving worker productivity in this task.

The difference in worker satisfaction score of 1.61 between Groups 7 and 4 was not significant. Hence, monetary incentive had no impact on worker satisfaction. The possible reason for such result was the lack of opportunity to earn a substantial monetary gain in one hour production time.

4. CONCLUSIONS

The following conclusions are drawn from the results of the two studies performed in the context of a repetitive manufacturing task:

1. Only the combination of an assigned hard production standard (130% normal) in the presence of production feedback on quantity and quality had a significant positive effect on both worker productivity and satisfaction. The increases in quantity and quality output were 13 and 15%, respectively, compared to the control group (first study).
2. The provision of an assigned hard standard alone failed to improve worker productivity, however, it had no adverse impact on worker satisfaction.
3. The incorporation of quantity and quality feedback had no significant effect on worker productivity but had a favorable impact on worker satisfaction.
4. A progressive increase in the assigned standards (100, 130, 140 and 150% normal) with feedback (quantity and quality) improved worker productivity significantly up to the provision of an assigned standard of 140% normal and feedback. The maximum increases in quantity and quality output were 14 and 10%, respectively, compared to the control group (second study). No further improvement in worker productivity resulted as a consequence of an assigned harder standard of 150% normal and feedback.
5. The provision of a participative standard with feedback was significantly inferior to an assigned 130% normal standard and feedback in terms of worker productivity.
6. The provision of assigned and participative standards with feedback improved worker satisfaction significantly. No significant difference in worker satisfaction was found among the standards and feedback conditions.
7. Monetary incentive had no beneficial effect on worker productivity and satisfaction.

ACKNOWLEDGEMENT

The first study was partially funded by the Economic Development Administration, U.S. Department of Commerce. While the second study was funded by the Social Sciences and Humanities Research Council of Canada (Grant 410-84-0579)

REFERENCES

[1] Becker, L.J., Joint effects of feedback and goal setting on performance: A field study of residential energy conservation, Journal of Applied Psychology, 1978, 73(4), 428-433.

[2] Das, B., Effects of production feedback and standards on worker productivity in a repetitive production task, IIE Transactions, 1982a,, 14(1), 27-37.

[3] Das, B., Effects of production feedback and standards on worker satisfaction and job attitudes in a repetitive production task, IIE Transactions, 1982b, 14(3), 193-203.

[4] Das, B., An ergonomic approach to designing a manufacturing work system, International Journal of Industrial Ergonomics, 1987, 1(3), 231-240.

[5] Das, B. and Shikdar, A.A., Optimum application of production standards and feedback to maximize worker productivity in a repetitive manufacturing task. Proceedings of the Human Factors Association of Canada, Vancouver, British Columbia, August 22-23, 1986, 165-168.

[6] Das, B., Worrall, B. and Shikdar, A.A., Production standards and feedback affecting worker satisfaction in a repetitive manufacturing task, Proceedings of the International Industrial Engineering Conference, Washington, D.C., May 17-20, 1987, 240-244.

[7] Erez, M., Feedback: A necessary condition for goal setting-performance relationship, Journal of Applied Psychology, 1977, 62(5), 624-627.

[8] Hackman, J.R. and Lawler, E.E. III, Employee reaction to job characteristics, Journal of Applied Psychology Monograph, 1971, 55(3), 259-289.

[9] Hackman, J.R. and Oldham, G.R., Development of job diagnostic survey, Journal of Applied Psychology, 1975, 60(2), 159-170.

[10] Helwig, J.T. and Council, K.A. (eds.), SAS User's Guide 1979 Edition, SAS Institute Inc., Cary, North Carolina, 1979.

[11] Ivancevich, J.M., Subordinates reactions to performance appraisal interviews: A test of feedback and goal setting techniques, Journal of Applied Psychology, 1982, 67(5), 581-587.

[12] Kim, J.S. and Hamner, W.C., Effects of performance feedback and goal setting on productivity and satisfaction in an organizational setting, Journal of Applied Psychology, 1976, 61(1), 48-57.

[13] Latham, G.P. and Steele, T.P., The motivational effects of participation versus assigned goal setting on performance, Academy of Management Journal, 1983, 26(3), 406-417.

[14] Latham, G.P. and Yulk, G.A., Assigned versus participative goal setting with educated and uneducated woodworkers, Journal of Applied Psychology, 1975, 60(3), 299-302.

[15] Locke, E.A., Toward a theory of task motivation and incentives, Organizational Behavior and Human Performance, 1968, 3, 157-189.

[16] Locke E.A. and Latham, G.P., Goal setting: A motivational technique that works!, Prentice-Hall, Inc., New Jersey, 1984.

[17] Locke, E.A., Shaw, K.N., Saari, L.M. and Latham, G.P., Goal setting and task performance: 1969-1980, Psychological Bulletin, 1981, 90(1), 125-152.

Trends in Ergonomics/Human Factors V
F. Aghazadeh (Editor)

A GRAPHICAL REPRESENTATION OF GROUP DECISION PERFORMANCE

Vitaly Dubrovsky

School of Management
Clarkson University
Potsdam, NY 13676

This paper discusses the problem of representation of group decision performance. It states that the primary goal of group decision making is not a quality decision, but rather the acceptance of a decision by group members. It suggests a new interpretation of group decision making as agreement development with two outcomes - public and private acceptance. It introduces notions of an "agreement situation", "agreement action" and "logical time" of agreement development. On the basis of these notions, the paper suggests a graphical representation of the group decision performance.

1. THE PROBLEM.

Group decision performance is a traditional object of social psychology. As is group performance in general, the group decision performance is viewed as a transformation of an initial situation into a group outcome carried by a process of group interaction, that is interaction among group members [1]. The initial situation is constituted by three categories of factors: (1) individual factors (members skills, knowledge, attitudes, etc.); (2) group factors (group structure, size, etc.); (3) environment factors (type of task; reward structure; time constraints; interaction media, etc.). The group outcome is constituted by: (1) performance outcome (i.e. solution to the problem); (2) impact on the group (i.e. group cohesiveness); and (3) impact on the members (i.e. attitudes, satisfaction, perceptions, etc.).

There are two different paradigms of analysis of group performance. Although both paradigms are based on the above framework and analyze the group interaction by means of content analysis, they treat the interaction in different ways. The first paradigm is centered on the process of verbal interaction, or communication acts of group members, while the initial situation and performance outcome are viewed as the background of the communication process ([2], [3], [4]). The interaction process is described in terms of a distribution of communication acts over time (phases), categories, and group members, as well as by postdiscussion satisfaction and judgments about influence and participation. The deficiency of this approach is its relevance to patterns of group communication in general rather than to performances of specific decision tasks [1],[5].

The second paradigm is centered on the group outcome as determined by the initial situation, while the interaction process is viewed as a mediator between the situation and outcome. Although this paradigm uses content analysis categories which are relevant to the decision task performed by a group, it has its own deficiencies. This paradigm describes the group interaction process in terms of frequencies, distributions, and the rate of various categories of members' acts for the entire process (see [6] for review). It does not take into consideration dynamics of group interaction, changes of the interaction over time, or sequence of relevant events leading to a group outcome. Due to the deficiencies, neither paradigm can provide the conceptual link between group interaction and group performance. In other words, we still do not have an adequate representation of the process of group performance. As a result, thousands of studies of group performance have yielded very little knowledge about why some groups perform better than others or how to improve the performance of the given group on a given task [5].

Recent studies comparing group decision performance in different communication media were inspired by new computerized technologies, such as teleconferencing, office automation, and decision support systems. The differences in group decision performance for the different communication conditions have been usually explained by general social and psychological differences of the electronic media from face-to-face communication such as absence of nonverbal cues [7], reduced perceived social presence [8], or reduced social context information [9]. Although providing important information about group interaction in different communication media, these studies have added little to our understanding of the group decision performance per se.

Hackman and Morris [5] suggest that for future progress in the study of group performance, it is necessary to do the following:

1. Build a number of smaller theories for types of group tasks being performed instead of a single general theory of group performance.

2. Represent group interaction in its dynamics and changes over time, rather than by averages and frequencies of categories of acts generalized over the entire performance period.

3. Describe the interaction on the relatively molar level to represent those aspects of the interaction which are uniquely important in affecting performance outcome for the type of task being performed.

In this paper we attempt to follow these recommendations. As the first step we will analyze a specific group task - group decision making.

2. GROUP DECISION MAKING AS AGREEMENT DEVELOPMENT

2.1. Why People Make Decisions in Groups?

The assumption of early studies was that people usually make important decisions in groups because the group has a greater sum of total knowledge and information, can generate a greater number of approaches to a problem, and have greater chances to detect errors or, in other words, group effort will lead to a higher quality decision. The results of the early experimental studies did not support this assumption but rather implied that individual decisions are superior to group ones (see [10] for review).

For a possible explanation, different authors referred to a number of group "liabilities": Steiner to "process losses", Maier and Solem to "social pressure", Hoffman and Maier to "individual domination", Janis to "group think", and others. Since, in this perspective, group interation was viewed in terms of contributions of group members to the group performance outcome, the "group effectiveness problem" was to establish a group interaction that "capitalizes upon the total pool of information and provides for great interstimulation of ideas without any loss of innovative creativity due to social restraints" [11, p. 268]. In pursuit of the solution, the later studies compared different ways of group organization, leadership styles and structures of group interaction (e.g., "brainstorming"). Summarizing the results in regard to such "structured" decisions, McGrath [1, p. 132] concludes that the superiority of individual performance (larger number and higher quality of ideas produced in a shorter time) is a "very robust empirical finding".

The very fact that people make important decisions in groups, despite the inferior solutions and "liabilities", suggests that: (1) the reason for group decisions is not, or at least not the only higher quality solution; and (2) that the solution to the problem is not the most appropriate, or at least not the only criterium of the effectiveness of group decision performance. As Thibaut and Kelley [11] put it, as long as group members are interdependent in attaining their goals, the decisions regarding the goals and means must be widely accepted and well understood by the group members to ensure their commitment to the implementation. The understanding, acceptance, and commitment are the primary reasons for group decision making and that the group effectiveness must be evaluated appropriately: "A low-quality solution that has good acceptance can be more effective than a higher-quality solution that lacks acceptance" [12, p. 240].

2.2. Two Processes in Group Decision Making.

Laughlin proposed a decision-task continuum anchored by "intellective" and "judgmental" tasks: "An intellective task involves a demonstrably correct solution, whereas a judgmental task involves group concensus on some nondemonstrable behavioral, ethical, aesthetic, or attitudinal judgment" [14, p. 273]. Following McGrath [1] we will call the group judgmental judgmental decisions Group Decision Making (GDM). Below we will discuss only this kind of group decisions.

Since, by definition, there is no correct solution in GDM, the effectiveness of the GDM has at least two components: (1) whether or not the the decision was made; and (2) the degree of acceptance of the decision by group members. Futhermore, a decision made by group is the outcome of GDM performance as it appears externally. Internally, in terms of group interaction, it appears as an agreement of all group members on the solution, that is as "public acceptance" of a decision (e.g., a consensus). Similarly, the degree of acceptance, or "private acceptance" of a decision by group members externally appears as a commitment, while internally it can be measured by the correspondence of a group decision to postagreement attitudes of the group members.

The public and private acceptance are results of two diferent, although interdependent group interaction processes: group "agreement development" and "mutual persuasion" of group members. Not only are the results of these processes (private and public acceptance) different, but also the initial situations are different. In our studies [15], [16] we observed that in many experimental discussions, the person who talked first, before hearing opinions of others, suggested the solutions which were different from their private prediscussion choice. The rate of such shifted initial suggestion can vary substantially. In one case it was as low as 16% of 36 experimental discussions, while in another case the "initial shift" occurred in 82% of 12 pilot discussions. Although we still do not have a satisfactory explanation for the "initial choice shift" phenomenon, it is clear that private pre-decision choices and initial decision suggestions are different things.

Agreement development and mutual persuasion are interdependent processes. For example, on one hand, the use of majority rule can produce a genuine change in minority's preferences [17] while, on the other hand, the information influence of novel arguments exchanged during a group discussion is a good predictor of a group choice shift [18].

2.3. The Process of Agreement Development.

2.3.1. Agreement Situations and Agreement Acts.

A group decision can be viewed as a solution to the problem with which all group members have agreed. Dynamically, or in terms of the agreement development, a group decision can be represented as a final "agreement situation" when all group members have agreed upon one "decision proposal" made by one or more group members. In other agreement situations, each decision proposal is supported only by part of the group members. The agreement development process can be represented as a sequence of changes from one agreement situation into another, until the final situation, group decision, is reached. Each situation can be characterized by a set of proposals with group members supporting each proposal.

A change of an agreement situation can be attributed to one or more of the following acts (or absence of acts) performed by group members in relation to decision proposals: (1) "introducing" a new decision proposal; (2) restating a decision proposal by the member that has introduced or already has agreed with it; (3) "carrying-over" a proposal, or arguing in favor of the proposal without restating it, or arguing against other proposals, or doing nothing by the member previously supporting the proposal; (4) "agreeing" with a proposal by the member previously supporting another proposal; (5) "joining" a proposal, or agreeing or restating the proposal by a member that previously did not introduce or join any proposal, or abandoned a previously supported proposal; and (6) "abandoning" a proposal, or giving it up or denouncing by the member previously supporting it without agreeing with another proposal or introducing a new one. All these acts, except carrying-over, change a situation. Introducing, and abandoning change a set of proposals, joining changes a distribution of support across proposals, while agreeing changes both a set of proposals and a distribution of support across remaining proposals.

2.3.2. Logical Time, Agreement States, and Agreement Steps.

On one extreme, in orderly face-to-face discussions, when only one person speaks at a time, situations are changed by only one act, introducing, agreeing, joining, or abandoning. Typically such an act is accompanied by one or more carrying-overs. On the another extreme, in electronic mail discussions, several messages can be sent approximately at the same time in response to the same discussion situation. Sometimes it happens in face-to-face discussions when more than one person speaks at a time. For example when several members say "I agree" at the same time in response to a proposal. All these actions contribute to a combined change of a situation, and the next act will respond to this new situation. This fact can be represented by assigning all the acts contributing to the combined change of a situation to the same "logical interval", that is "time" of change of one situation into another. Each interval is situated between two "logical moments", corresponding to an initial and resulting situation. Since situations can have only discrete changes in proposals and supporters, "logical time" is also determined as discrete, with each logical interval equal to one "logical-time-unit". By the same token, all other communication acts, made by group members between the two situations must be allocated to the appropriate logical interval.

Initial state of agreement development can be characterized by the agreement situation with no decision proposals, and the logical time equal to 0. All other states can be characterized by two parameters: (1) a current agreement situation and (2) a logical moment equal to the ordinal number of the current situation in the sequence of situations beginning with the initial one. In these terms, the agreement development process can be represented as a sequence of steps, each changing an agreement situation during one logical-time-unit. The logical time of the agreement development is equal to the number of logical steps resulted in a group decision.

2.3.3. Logical and Physical Time of Agreement Development.

Logical time can be characterized as internal to the process time, or time measured in relation to the process. If there are no changes in the process, no logical time elapsed. For example, in a stalemate discussion situation, a lot of physical time can be spent in arguing, while the logical time of agreement development will be frozen at the same logical moment.

The division of group discussion in the logical intervals of agreement development is a radical departure from the traditional division of a discussion in arbitrary 5-minute chunks. Comparison of 5-minute chunks of face-to-face discussion with 5-minute chunks of electronic mail discussion will lead nowhere. Moreover, it is doubtful that the appropriate comparison physical time interval can be determined. An appropriate way of comparison is a comparison of discussion chunks of 1-logical-unit with the following synchronization of logical and physical time.

Synchronization of logical and physical time of agreement development can be achieved through establishing the correspondence between logical and physical time moments. For example, time measurement may show that the 7-th agreement situation in the group discussion occurred at 6:30 p.m. on October 12, 1986. The synchronization can provide useful measures for GDM performance efficiency and can be used for comparison of different group organization patterns, leadership styles, communication media, and other GDM conditions.

2.3.4. A Graphical Representation of Agreement Development.

The agreement development performance of a group as a whole can be represented as a trajectory of agreement situation changes against two orthogonal scales: (1) Potential Decision Scale (vertical); and (2) Logical Time Scale (horizontal) (Figure 1). The Potential Decision Scale can vary from a set of several arbitrary ordered potential decisions to area(s) of multi-dimensional parametric space. The Logical Time Scale is a directed line divided in equal intervals by logical moments with zero moment at the origin. Each agreement state can be shown by ID's of the supporting members for each decision on the intersection of the horizontal proposal line and the vertical logical moment line. Corresponding agreement acts (introducing, joining, carrying-over, etc.) can be shown by different symbols inside the appropriate intervals. All other communication acts of the group can be represented by a special table attached to the Logical Time Scale. In this table numbers are vertically distributed across categories of acts and horizontally across intervals of the Logical Time Scale (see Figure 1).

2.3.4. Group Interaction in Agreement Development.

The graph described above represents group interaction during agreement development in terms of group members acts and with emphasis on results of the acts. Supplementary representation of the group interaction with emphasis on participation, or contributions of individual group members can be done in a similar way. For this purpose, two orthogonal scales can be used: (1) the Group Members Scale (vertical); and (2) the Logical Time Scale. The group members scale is a discrete scale with group members' ID's arbitrarily ordered. All agreement acts made by a group member are shown by a symbol at the intersection of the member's ID horizontal line and the appropriate logical interval vertical column. One symbol should be used for a carrying-over act and another for all other agreement acts. It is convenient to attach the two graphs along the Logical Time Scale.

Also, the same kind of tables can be attached as in the previous graph. However, in this case, one table for each member should be presented in the same vertical order as the ID's on the Group Members Scale. Each table represents a distribution of communication acts of an individual group member across the categories and logical time intervals. All the tables together represent the distribution of the communication acts across group members, categories, and logical time intervals.

CONCLUSION

Following the recommendations of Hackman and Morris (Introduction) this paper:

1. Introduces a theory of group decision making as agreement development.

2. Suggests a two-projection representation of group interaction during agreement development. The first projection is centered on results of interaction. It represents the agreement development as a sequence of changes of an agreement situation under the influence of group interaction. The second projection is centered on contributions of individual group members to the group agreement development effort.

3. Group Decision Making is represented on the relatively molar level and

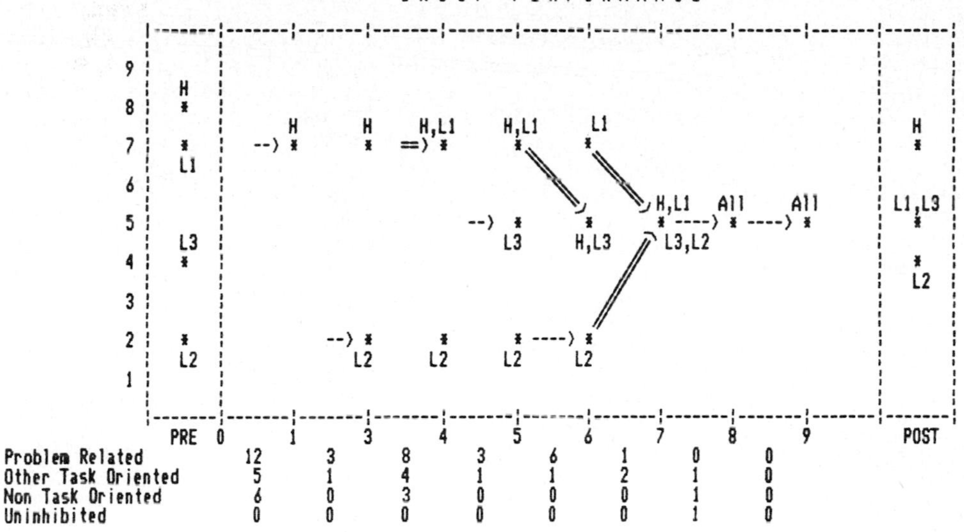

	0–1	1–3	3–4	4–5	5–6	6–7	7–8	8–9
Problem Related	12	3	8	3	6	1	0	0
Other Task Oriented	5	1	4	1	1	2	1	0
Non Task Oriented	6	0	3	0	0	0	1	0
Uninhibited	0	0	0	0	0	0	1	0

MEMBER PARTICIPATION

	PRE	0	1	3	4	5	6	7	8	9	POST
H	8		--> 7				===> 5	----> 5			7
L1	7				==> 7			===> 5		----> 5	5
L2	?			--> 2			----> 2	===> 5		----> 5	4
L3	4					--> 5				----> 5	5

	0–1	1–3	3–4	4–5	5–6	6–7	7–8	8–9
H:								
Problem Related	7	0	2	0	6	1	0	0
Other Task Oriented	2	1	2	0	1	2	0	0
Non Task Oriented	0	0	0	0	0	0	0	0
Uninhibited	0	0	0	0	0	0	0	0
L1:								
Problem Related	5	3	4	0	0	0	0	0
Other Task Oriented	0	0	1	0	0	0	1	0
Non Task Oriented	0	0	0	0	0	0	0	0
Uninhibited	0	0	0	0	0	0	0	0
L2:								
Problem Related	0	0	0	0	0	0	0	0
Other Task Oriented	2	0	0	0	0	0	0	0
Non Task Oriented	0	0	3	0	0	0	1	0
Uninhibited	0	0	0	0	0	0	1	0
L3:								
Problem Related	0	0	2	3	0	0	0	0
Other Task Oriented	1	0	1	1	0	0	0	0
Non Task Oriented	6	0	0	0	0	0	0	0
Uninhibited	0	0	0	0	0	0	0	0

EXPLANATIONS:

1. An experimental group of four (one high status member - H and three low status members - L1,L2,L3) had task to come to concensus on a choice-dilemma problem. A decision had to be odds of 1,2 3,4,5,6,7,8, or 9 chances in 10 - the lowest odds of success acceptable in order to pursue more risky but also more attractive alternative.
2. Symbols used: "PRE" = prediscussion private choices; "POST" = postdiscusion private choices; "*" = decision proposal; "-->" = introducing; "---->" = restating; "==>" = joining; "===>" = agreeing; no symbol is used for carrying-over.

in terms relevant to a group agreement development performance outcome.

There is another, higher molar level of group decision, which we did not discuss here. At this level, group decision making can be described as a sequence of stages constituting a group decision making script. A study of a GDM script, or better scripts is next on our research agenda.

REFERENCES

[1] McGrath, J., E. Groups: Interaction and Performance. (Prentice Hall, New Jersey, 1984)

[2] Bales, R.F., Interaction Process Analysis (Addison-Wesley, Cambridge, Mass., 1950)

[3] Bales, R.F. and Strodtbeck, F.L. Phases in Group Problem Solving, Journal of Abnormal and Social Psychology, 1951, 46 (1951) 485-495.

[4] Fisher B.A., Decision Emergence: Phases in Group Decision-Making, in Backer, S.L., (ed.), Speach Monographs, Vol. 37, 1 (1970) 53-66.

[5] Hackman, J.r. and Morris, C.G., Group Tasks, Group Interaction Process, and Group Performance Effectiveness: A Review and Proposed Intergration., in L. Berkowitz (Ed.) Advances in Experimental Social Psychology, Vol. 8, (Academic Press, New York, 1975)

[6] Myers, D.G. and Lamm, H., The Group Polarization Phenomenon, Psychological Bulletin, Vol. 83, No. 4 (1976) 602-627.

[7] Krueger, G.P. and Chapanis, A., Conferencing and Teleconferencing in Three Communication Modes as A Function of The Number of Conferees, Ergonomics, 23, 2 (1980) 102-122.

[8] Christie, B., Human Factors of Information Technology in The Office (Wiley, New York, 1985)

[9] Kiesler, S., Siegel, J., and McGuire, T., Social Psychological Aspects of Computer-Mediated Communication, American Psychologist, 39, 10 (1984) 1123-1134.

[10] Hoffman, L.R., Group Problem Solving, in L. Berkowitz (Ed.) Advances in Experimental Social Psychology, Vol. 2 (Academic Press, New York, 1965) pp. 99-132.

[11] Thibaut, J.W. and Kelley, H.H. The Social Psychology of Grops (Wiley, New York, 1959)

[12] Maier, N.R.F., Assets and Liabilities in Group Problem Solving: The Need for An Integrative Function, Psychological Review, 74, 4 (1967) 239-249.

[14] Laughlin, P.R. and Early, P.C. Social Combination Models, Persuasive Arguments Theory, Social Comparison Theory, and Choice Shift, Journal of Personality And Social Psychology, 42, 2 (1982) 273-280.

[15] Siegel, J., Dubrovsky, V., Kiesler, S., and McGuire, T. W., Group Processes in Computer-Mediated Communication, Organizational Behavior And Human Decision Processes 37 (1986) 157-187.

[16] Dubrovsky, V., Kiesler, S. and Sethna, B. Effects of Status in Computer-Mediated and Face-to-face Decision Making Groups, (in preparation).

[17] McCauly, C.R., Extremity Shifts, Risky Shifts and Attitude Shifts after Group Decision, European Journal of Social Psychology, 2 (1972) 417-436.

[18] Vinokur, A., Trope, Y., and Burnstein, E., A Decision Making Analysis of Persuasive Argumentation and The Choice-Shift Effect, Journal of Experimental Social Psychology, 11 (1975) 127-148.

Trends in Ergonomics/Human Factors V
F. Aghazadeh (Editor)

A PROBABILITY MODEL FOR GROUP INSPECTION

A. M. WAIKAR, F. AGHAZADEH, and K. S. LEE

Department of Industrial Engineering
Louisiana State University
Baton Rouge, LA 70803 U.S.A.

A probability model was developed and tested to predict performance of different size groups of inspectors in a visual inspection task. The acceptance was assumed to be based on agreement about whether the product is defective by a specific number of individuals in the group (decision rule). The model predicted that certain combinations of group size and decision rule can increase the defectives rejected and the correct inspections. It was concluded that it may be possible to improve quality through the use of group inspection.

1. INTRODUCTION

It is generally acknowledged that inspectors make errors which affect the product cost and quality. Researchers have attempted to determine task conditions which lead to inspection errors and to design inspection systems that minimize these errors [1]. However, inspection errors continue to occur in various industries. Number of recall campaigns by different automobile manufacturers is an example of this situation. Manufacturers of several products are wasting billions of dollars as a result of faulty inspection.

Industrial quality control programs, for critical applications where defects must be avoided at all costs often use reinspection to ensure higher outgoing quality [2]. The goal is to obtain practically defect free outgoing lots. A few studies have been reported that attempt to improve inspection system performance, considering limitations of the inspector, to obtain such high quality [2], [3]. These studies have examined performance of teams of inspectors to achieve higher outgoing quality.

A model of group performance in vigilance studies has been first reported by Wiener [4]. The general conclusion was that group performance is better than individual performance under most conditions on most tasks. He used a visual monitoring task with one, two and three observers and found group performance to be better than individual performance.

However, the actual performance of the group was less than that predicted by the model for group performance.

There is some experimental literature on the performance of multi-person teams in military vigilance tasks which bear some analogy to industrial inspection. Schafer [5] studied auditory vigilance tasks and concluded that a significant increase in detection probability can be realized by adding more than one observer, although little is gained by adding more than two observers. Baker et al. [6] extended Schafer's study to determine the number of observers needed to assure hundred percent detection of signals in a visual task.

Drury et al. [2] reported a study that examined various ways in which two inspectors can be combined to achieve enhanced system performance. They concluded that two inspectors are better than one and that better inspectors give better system performance.

This paper presents a simple probability model for combining responses of inspectors in a group, as a team response for accepting or rejecting an item, to improve quality. The objective was to predict the performance of the groups of inspectors in which the members are working independently and the acceptance or rejection of an item is a groups decision.

2. THE MODEL.

The model assumes that probability of detection of the inspectors for a specific inspection task can be estimated. It uses basic principles of probability theory to determine the expected outcomes where a group of inspectors is detecting defects in a visual inspection task. It takes into account the group size and decision rule for predicting performance of the group.

Group size is defined as the number of inspectors in a group independently detecting the defects. The decision of one inspector to accept or reject the item is assumed not to affect the decision of other inspectors. This may be ensured in real life by physically isolating the members of the group and preventing any communication between them.

The decision rule can vary from a single-response decision rule to an all-response decision rule. The single-response rule requires rejection response from any one member of the group for rejection of the product inspected. The all-response rule requires that all members of the group reject the product for it to be classified as a defective. Thus, in the single-response rule, each successive inspector will examine only the defects missed by the previous inspectors. In real life situation, this can be ensured by removing the defectives from the lot as soon as any one of the inspectors in the group registers a defect. In the all-response decision rule, each inspector will be presented the same number of products during inspection. In this case, the inspected item will be rejected as a defective only when all the members of the group report the item to be defective. It is easily possible to record the responses of all the group members. This will allow the evaluation of the performance of the groups based on the use of intermediate decision rules such as having two or three members to detect a defect before the defective item is rejected.

It is assumed that the decision process for each inspector is a Bernoulli process with acceptance or rejection being the only possible outcomes. Figure 1 shows a probability tree diagram that was constructed based on the assumption of independence for individual inspector responses for the single response decision rule. The model calculates the probabilities for different outcomes of the tree branches resulting in correct rejection, correct acceptance, incorrect acceptance (type I error), and incorrect rejection (type II error) under a given decision rule.

The following describes the model in which

let n = number of defectives in the lot inspected,
N = number of good products in the lot inspected,
Pi = probability of correct rejection for the ith inspector, and
P'i = probability of "false alarm" for the ith inspector.

Consider the case of one response decision rule where the product is rejected after first response from any group member for its rejection. Thus, each successive inspector will detect only the defects missed by the previous inspector. In this case the first inspector is presented with (N+n) targets. He will detect nP1 defects and will have NP'1 false alarms. Thus he will register (nP1+NP'1) defects. The second inspector will be presented (N+n) - (nP1+NP'1) = N(1-P'1) + n(1-P1) products. He will detect [n(1-P1)]P2 defects and will have N(1-P'1)P'2 false alarms. Therefore by induction, the (r+1)th inspector will get

$$N(1-P'1) \ldots (1-P'r) + n(1-P1) \ldots (1-Pr)$$

$$= N \prod_{i=1}^{r} (1-P'i) + n \prod_{i=1}^{r} (1-Pi) \text{ products} \ldots\ldots (1)$$

Therefore,

$$\text{Targets presented to the (r+1)th inspector} = N \prod_{i=1}^{r} (1-P'i) + n \prod_{i=1}^{r} (1-Pi) \ldots\ldots\ldots (2)$$

If there are only r inspectors in the system, then equation 2 represents the outgoing lot with $\prod_{i=1}^{r} (1-Pi)$ defectives in it.

If m is the acceptable percent defective level, then the number of defectives outgoing must be

$$n \prod_{i=1}^{r} (1- Pi) \leq m(N+n) \quad \ldots\ldots\ldots\ldots (3)$$

Therefore, if Pi's and n can be estimated, the number of inspectors necessary to obtain close to 100 percent detection can be estimated using the above model.

With decision rule requiring all member responses, every inspector must register a defect for rejection of the product presented. Therefore if there are r inspectors in the group, the probability that a defective is rejected will be

$$Pr(R) = P1\ P2 \ldots\ldots Pr = \prod_{i=1}^{r} Pi$$

Then, [number of detections expected] = D = n.Pr(R)

and the defects outgoing = d = n−D = $n\,(1- \prod_{i=1}^{r} Pi)$.

Therefore, $d = n\,[1 - \prod_{i=1}^{r} Pi]$ should be $\leq m(N+n)$

The expected number of false alarms will be

$$= N\,[P'1.P'2.P'3 \ldots P'r] = N\,[\prod_{i=1}^{r} P'i]$$

The mathematical model can be modified to predict performance of the groups under any decision rule by determining the appropriate result for each branch of the probability tree. The performance measures predicted by the model were the "Percent Defectives Rejected" (PDR) and "Percent Correct Inspections" (PCI), defined as follows:

$$PDR = \frac{\text{No. of defectives rejected}}{\text{Total no.of defectives}}$$

$$PCI = \frac{\text{No. of good parts accepted + No. of defectives rejected}}{\text{Total No. of parts inspected}}$$

To validate the model, performance data of the groups reported by Waikar et al. [8] was compared with the predictions of the model. The experimental data reported by Waikar et al. [8] consisted of PDR and PCI values for four group sizes (1, 2, 3, 4) for four different incoming quality levels. The probabilities of detection and the probabilities of the false alarms for the individual inspectors in the group were used as input to the model to compute the model predictions.

3. RESULTS

Table 1 shows the comparison of the PDR (Percent Defectives Rejected) values from the experimental data reported by Waikar et al. [8] and the PDR values predicted by the model.

As seen from Table 1, for the single response decision rule, the trends of changes in PDR with changes in group size appeared to be the same for all levels of incoming quality. The model slightly overestimated the values for group sizes of two and three and gave very close values for group sizes of one and four for all incoming quality levels except 2%. At 2% level, the values predicted by the model were quite close to the experimental values for group sizes of one and two. However, for group sizes three and four, the model overestimated the experimental values of PDR by almost 30 to 40%. This may be because of the difficulty in estimating individual probabilities of detection of the group members at such low level of incoming quality.

In the case of all-response decision rule, the model underestimated the values of this variable compared to the experimental results for levels of incoming quality of 26, 18 and 10% with an increase in group size. The deviation between model predictions and experimental values was between 10 and 30%, the maximum deviation being for a group size of four at incoming quality of 18%. At low levels of defectives (2% incoming quality) the model underestimated the values of PDR in comparison to the experimental values. Maximum deviation was about 60% for a group size of three. Again, this may be because of the difficulty in estimating individual probabilities of detection of the group members at such low level of incoming quality. The trends of change in PDR with increase in group size shown by the experimental results were in general agreement with that shown by the model.

Table 2 shows the comparison of the PCI (Percent Correct Inspections) values from the experimental data reported by Waikar et al. [8] and the PCI values predicted by the model.

As seen from Table 2, for the single response decision rule, the trends of changes in PCI with changes in group size appeared to be the same for model predictions and experimental values at all levels of incoming quality. The model overestimated the values of PCI compared to the experimental values at all incoming quality levels except 2%, where it underestimated PCI.

In the case of all-response decision rule, for incoming quality levels of 26 and 18% the model underestimated the values of PCI for all levels of the variable group size. The deviation was about 8% for a group size of four at 26 and 18% incoming quality levels. The amount of deviation increased as the group size increased from one to four. At an incoming quality level of 10% the model and experimental values seemed to agree closely for all group sizes. At an incoming quality level of 2%, the model overestimated the values of PCI, maximum deviation being for a group size of two. In general, the trend of changes (increasing or decreasing) in PCI appeared to be the same for both the model predictions and the experimental values.

4. DISCUSSION AND CONCLUSIONS

As the group size increased, probability of detection and the PDR for the group increased for the single response decision rule. This was also found to be true when the computations of PDR were based on intermediate decision rules. Probability of detection and PDR for the group decreased for the all-response decision rule. This compared favorably with the results reported by Schaffer [5] and Wiener [4]. Percent correct inspections increased with group size primarily because of the significant reduction in the number of false alarms in the case of all-response decision rule.

It was thus concluded that the outgoing quality level may be improved in visual inspection tasks, by designing the inspection system to use multiple inspectors and appropriate decision rule for accepting or rejecting the products. A choice of a single response decision rule will maximize the percent defectives rejected but simultaneously increase the number of false alarms, thus reducing the total correct inspections. A choice of all-response decision rule will improve the total correct inspections but will not eliminate maximum number of defectives. Designing the inspection system to include one of the intermediate decision rule should however provide a good compromise.

4. REFERENCES

[1] Dorris, A.L. and Foote, B.L., Inspection Errors and Statistical Quality Control: A Survey, AIIE Transactions (1978) pp. 184.

[2] Waikar, A., Lee, K., Aghazadeh, F., Quality Improvement Using Multiple Inspectors As A Group, Trends in Ergonomics and Human Factors, IV, Eslevier Publishers, (North Holland) Netherlands (1987) pp 81-88.

[3] Drury, C.G., Karwan, M.H. and Vanderwarker, D.R., The Two -Inspector Problem, IIE Transactions (1986) pp. 174-181.

[4] Wiener, E.L., The Performance of Multiman Monitoring Teams, Human Factors 4, (1964) pp. 179-183.

[5] Schafer, T. H., Detection of a Signal by Several Observers, U.S. Naval Electronics Laboratory, San Diego, California, USNEL Rep. No.1Ø1 (1949)

[6] Baker, R. A., Ware, J. and Sipowicz, R., Signal Detection by Multiple Monitors, Psychology Record, 12, (1962) pp. 113-117.

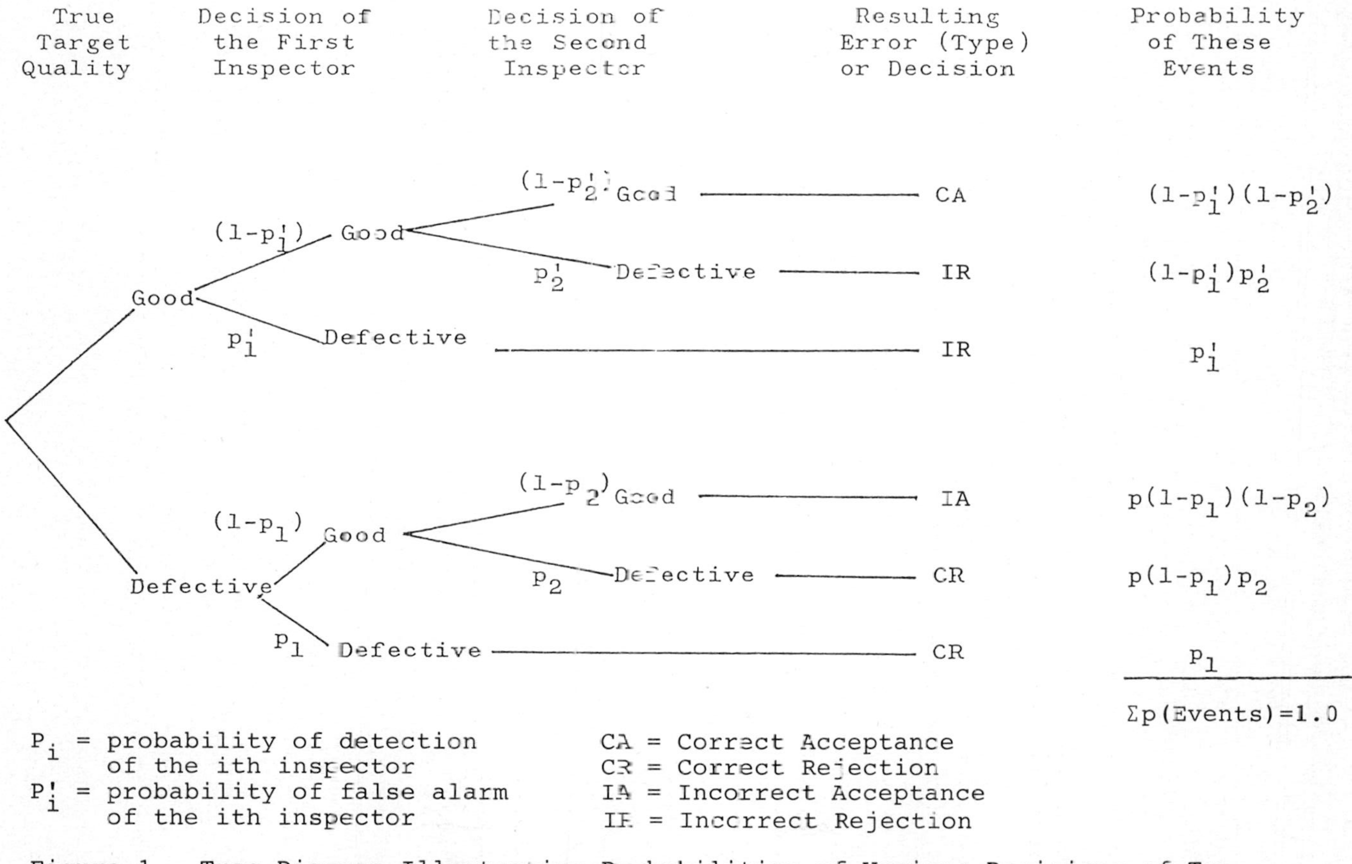

Figure 1. Tree Diagram Illustrating Probabilities of Various Decisions of Two Inspector Group for Single Response Decision Rule.

Table 1. COMPARISON OF PERCENT DEFECTIVES REJECTED
(EXPERIMENTAL VALUES VS. MODEL PREDICTIONS)

(a) Single Response Decision Rule

Incoming Quality	26%		18%		10%		2%	
Group Size	Exp.	Model	Exp.	Model	Exp.	Model	Exp.	Model
1	74.0	74.0	75.0	75.0	75.0	76.0	25.0	25.0
2	88.4	94.0	88.8	94.5	80.0	94.9	50.0	47.5
3	98.0	98.4	97.2	98.6	100.0	98.8	100.0	60.6
4	100.0	99.6	100.0	99.7	100.0	99.7	100.0	72.4

(b) All Response Decision Rule

Incoming Quality	26%		18%		10%		2%	
Group Size	Exp.	Model	Exp.	Model	Exp.	Model	Exp.	Model
1	74.0	74.0	75.0	75.0	75.0	76.0	25.0	25.0
2	74.0	57.0	72.2	58.2	70.0	60.0	50.0	6.2
3	69.2	41.3	66.6	43.0	65.0	44.7	62.5	1.5
4	61.5	31.8	55.5	33.5	60.0	35.3	50.0	0.3

Table 2. COMPARISON OF PERCENT CORRECT INSPECTIONS
(EXPERIMENTAL VALUES VS. MODEL PREDICTIONS)

(a) Single Response Decision Rule

Incoming Quality	26%		18%		10%		2%	
Group Size	Exp.	Model	Exp.	Model	Exp.	Model	Exp.	Model
1	90.5	91.3	92.0	93.0	94.0	94.4	94.4	94.6
2	91.0	95.1	93.0	94.5	91.0	92.8	92.0	91.2
3	91.0	95.2	91.0	92.2	89.5	90.3	89.5	85.9
4	90.0	94.5	88.5	90.9	87.0	88.4	87.0	83.0

(b) All Response Decision Rule

Incoming Quality	26%		18%		10%		2%	
Group Size	Exp.	Model	Exp.	Model	Exp.	Model	Exp.	Model
1	90.5	91.4	92.0	93.0	94.0	94.4	94.5	94.6
2	91.5	88.7	93.0	92.4	94.5	95.9	95.0	98.9
3	91.5	84.8	93.0	89.7	94.5	94.4	96.2	98.9
4	90.0	82.2	93.0	86.0	95.5	93.5	97.0	98.7

Trends in Ergonomics/Human Factors V
F. Aghazadeh (Editor)
Elsevier Science Publishers B.V. (North-Holland), 1988

EVALUATION OF TWO WORK SCHEDULES IN A MINING OPERATION

James C. Duchon

U.S. Bureau of Mines
Twin Cities Research Center
Human Factors
Minneapolis, MN 55417

Research in the area of shift work has found that it produces a variety of negative consequences, specifically, sleep deficiencies, performance decrements, fatigue, gastrointestinal disorders, and social and marital problems. Therefore, management needs to be able to objectively measure the degree of worker satisfaction and the adequacy of particular shift schedules. This paper presents results of a method designed by the Bureau of Mines to assess these problems in a taconite mining company. The site is divided into two groups of plant and pit workers, having two different rotating shift schedules. Informal conversation, a plant-wide vote, and survey items clearly indicate that the continuous 28-day-phase advance shift schedule is less accepted by the workers than is the discontinuous 21-day-phase advance shift schedule. Results indicate differences between the two groups on certain variables such as sleep quantity and quality, eating, and physical and mental exhaustion. It is concluded that the survey represents a valid instrument, in that it is sensitive to variables known to be affected by working irregular hours, and that it discriminates between the workers on the two different work schedules.

1. INTRODUCTION

There has been much research which shows the negative health, safety and performance effects of working irregular hours or shift work. Recently, several popular journals, magazines and newspapers have reported these effects. These reports have been published in "occupational" and "safety" journals, and read by management personnel responsible for the scheduling practices and implementation of shift work schedules. Given that shift work is a fact of life in many industrial settings, it is not clear just what actions, if any, should be taken at particular industrial settings. This paper describes a method used to "diagnose" a shift work schedule in use at a surface mine in northern Minnesota and which, also, might suggest possible changes. A similar methodology used on the rubber and plastic industry was reported elsewhere [1].

Specifically, the goals of this report are to 1) show how a survey may differentiate between a "satisfactory" and an "unsatisfactory" shift schedule, 2) describe the results, comparing both schedules and 3) describe worker preference for factors that should be taken into account when creating new shift schedules.

2. THE INSTRUMENT

Basically, there are two interacting sets of factors that directly contribute to the negative consequences of shift work. The first are the influences of circadian rhythms. It is now known that there are clearly defined circadian rhythms, such as sleep-wake and digestive rhythms, that regulate many bodily functions. As shift workers attempt to work during periods when the body is geared for inactivity or sleep, as during a night shift, fatigue, poor mood states, and lower levels of some types of performance will result. By the same token when sleep is taken during the daytime, sleep is lessened both quantitatively and qualitatively, since the body is warming up for normal daytime activity. Studies show that at least a week may be needed for these bodily rhythms to synchronize with the new schedule [2]. A second set of factors that create problems for shift workers are the socio-psychological influences [3]. These influences disrupt the routine of the shift worker due to family, social pressure, and job stresses. The shortened sleep of the shift worker is then in part due to noise and family responsibilities, interrupting daytime sleep episodes.

Given the above issues related to disturbances experienced by shift workers, a survey was developed to tap into these factors as a basis for investigating the symptoms often associated with shift work. The survey was developed to support three purposes. First, it was designed to document the effects of working irregular hours. Second, it was designed to have "diagnostic" qualities by indicating how well employees are functioning on their shift, according to various factors known to be important. Finally, it was designed to pinpoint changes that could be made to improve the shift work system.

A survey was selected as the primary data collection instrument due to its simplicity and ease of information gathering. A drawback is that the data collected are subjective and, therefore, have inherent problems with reliability and validity. However, as will be shown, data collected from this survey do corroborate previous research and do distinguish between the two groups of shift workers. Also, survey items eliciting self-reported sleep start and end times have been validated with physiological measures [4].

3. THE POPULATION

The site was a taconite producing surface mine, located in northern Minnesota. The mine was divided into the pit and plant workers. The pit workers are on a 21-day major cycle phase advance weekly rotation schedule with all Saturdays and Sundays off. The plant crews are on a continuous 28-day phase advance schedule. The plant workers have seven Day shifts, off one, seven night shifts, off four, and seven afternoon-evening shifts, and off two.

The pit workers are considered to be on a "better" schedule than the plant workers for several reasons. First, the safety director indicated that the plant workers have been complaining for some time about their schedule. There were few complaints from the pit workers.

A vote indicated that the majority of plant workers, but not of the pit workers, were willing to try a new schedule. Finally, from a logical standpoint the pit schedule seems more desirable, since it allocates all weekends off, is more regular, and does not have as many night shifts in a row; five versus seven straight nights.

Demographics are presented in table 1. Response rate was 98%. In general the two groups are about the same. The pit crew are slightly older on the average (2.8 years), which may contribute to some of the differences on the survey.

TABLE 1. DEMOGRAPHICS FOR PIT AND PLANT WORKERS.

VARIABLES	Pit	Plant
Female	0	1
Male	42	60
Single/Divorced/Separated	9	13
Married	33	48
Have Children	29	47
Graduated from High School (pct)	88.1	93.4
Ave. Number of Years at Present Occupation	6.7	7.1

4. RESULTS

4.1 Satisfaction with Current Schedule

As indicated previously, it was assumed a priori that the pit workers were more satisfied than the plant workers with their shift schedule. Figure 1 indicates the plant worker's degree of satisfaction with the current work schedule, as compared to the pit workers. On a scale of 1 (Totally satisfied) to 5 (Not at all satisfied), plant workers averaged 3.8 while the pit crew averaged 2.4. Figure 2 indicates the percentage of employees who feel that a better schedule would improve different aspects of their lives. It is obvious that the plant workers believe that a different schedule would improve the quality of their life in all categories. The results were not statistically analyzed, since the pit and plant were considered populations (not samples), where means are exact parameters, not estimates.

When asked why they would change jobs (Figure 3), the plant workers identified the most critical problem as hours of work, which had a lower priority for the pit workers, who identified job security as most critical.

4.2 Perception of Work Load by Shift

Figures 4, 5, and 6 show how workers feel after working each shift, relative to three variables: feeling sleepy, physically tired and experiencing aches and pains. As expected, working the night shift

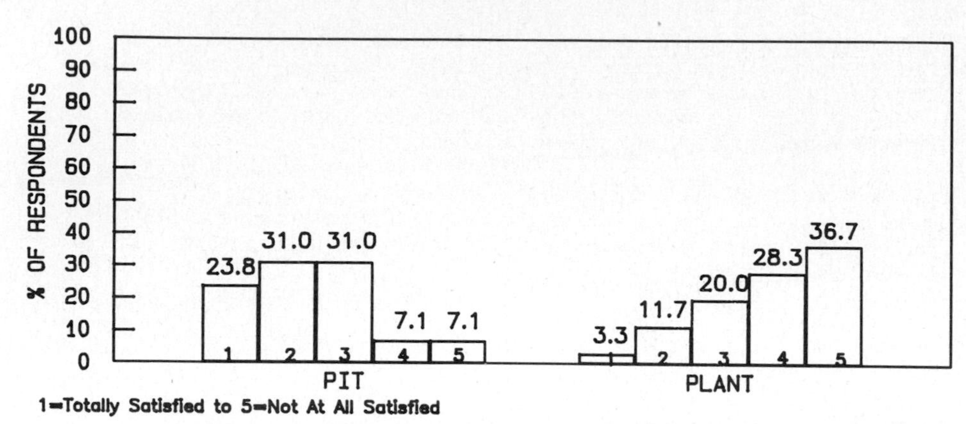

FIGURE 1. – RESPONSE TO SATISFACTION WITH WORK SCHEDULE.

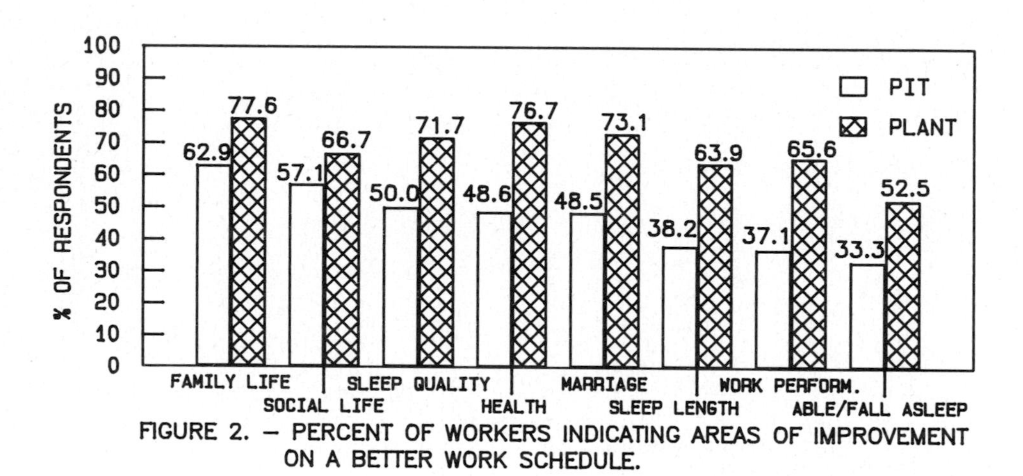

FIGURE 2. – PERCENT OF WORKERS INDICATING AREAS OF IMPROVEMENT ON A BETTER WORK SCHEDULE.

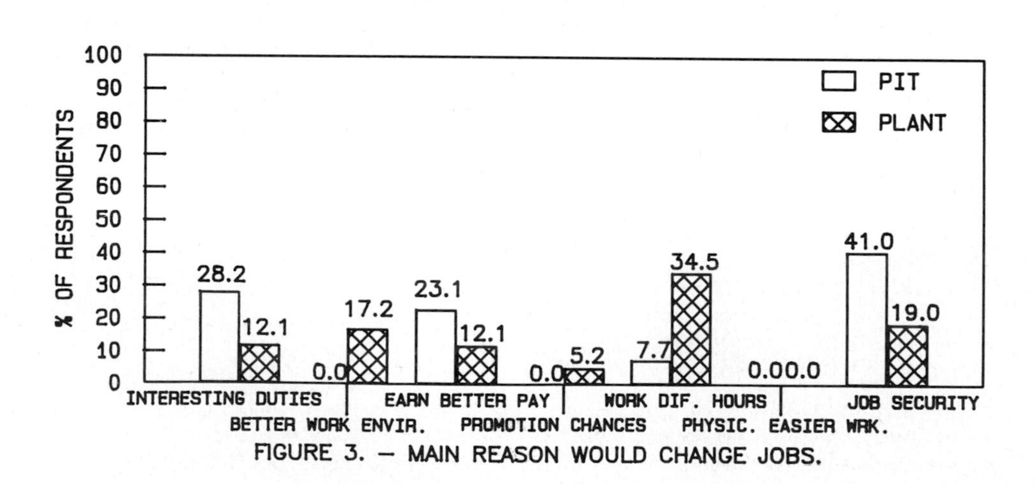

FIGURE 3. – MAIN REASON WOULD CHANGE JOBS.

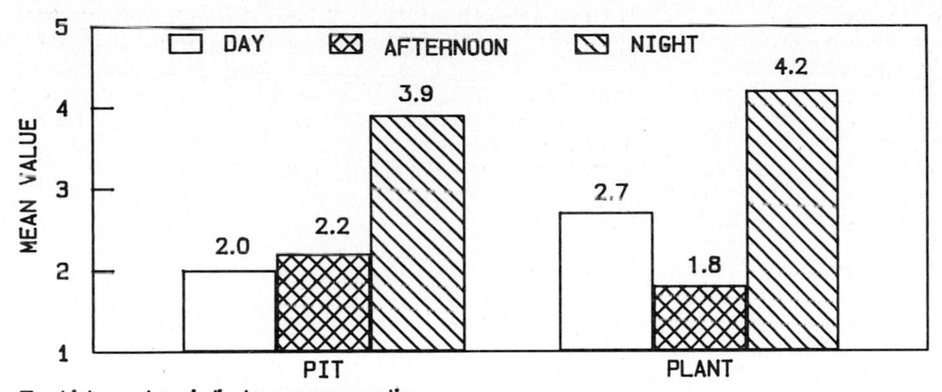

The higher values indicate a more negative condition than the lower values.

FIGURE 4. – HOW SLEEPY AFTER WORKING EACH SHIFT.

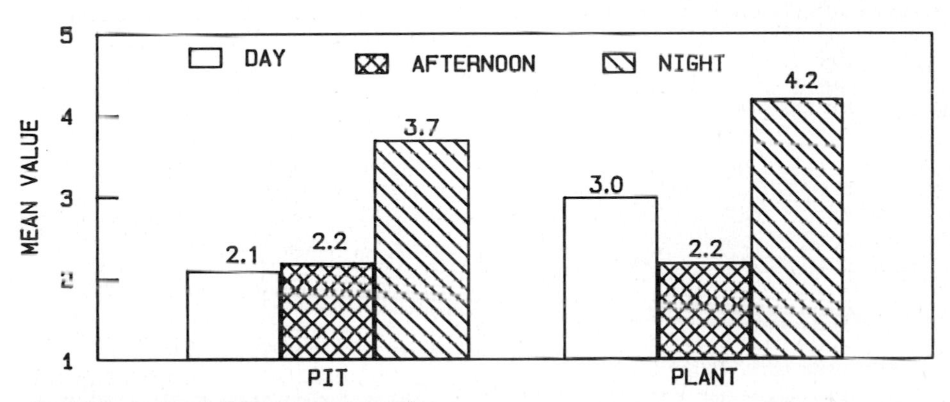

The higher values indicate a more negative condition than the lower values.

FIGURE 5. – HOW PHYSICALLY TIRED AFTER WORKING EACH SHIFT.

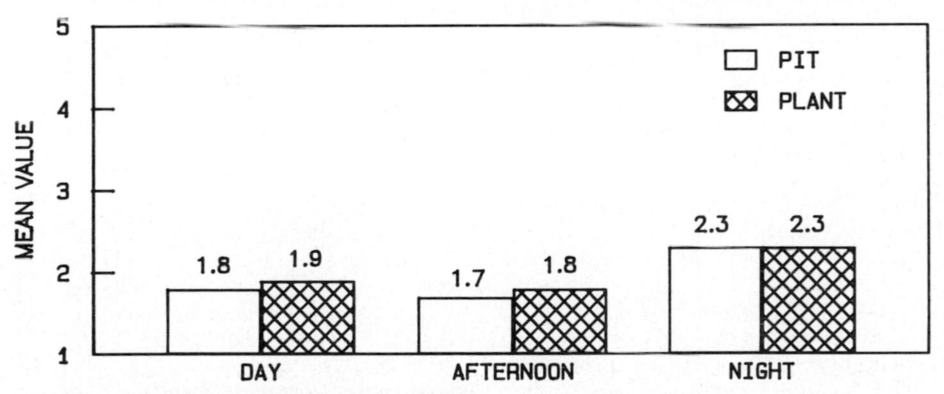

1 = Totally Satisfied to 5 = Not At All Satisfied

FIGURE 6. – AMOUNT OF ACHES/PAINS AFTER WORKING EACH SHIFT.

has a more negative effect than the other shifts. More interesting is the finding that the plant workers reported that they were more sleepy and physically tired on the night shift than the pit workers. The plant workers also feel more sleepy and physically tired after the Day shifts than the pit workers. These negative effects are apparently not due simply to the more physically demanding plant work since these increases are not seen for the afternoon-evening shifts.

4.3 Adaptation to Shift work

Adaptation to shift work occurs if workers begin to feel better or the same as the work week goes on. When workers complain of feeling worse, this signals a problem in adapting to the schedule. Many shift work researchers believe that a "bad" shift schedule can cause negative consequences in the health and performance of shift workers [5].

The plant workers on the Day and afternoon-evening shifts report that on the average they feel about the same as the week goes on in terms of physical exhaustion, mental exhaustion, minor aches and pains and work performance (Figures 7, 8, 9, 10). While on the night shift, however, they report that they feel worse as the week goes on. Their responses are similar to the pit crew. The main difference is that the plant workers, more than the pit workers, report that their performance also gets worse, as opposed to the pit workers who report no change.

The most consistent and probably the most important effect of shift work is on sleep problems. These effects are most dramatic for the plant workers: 41% indicated that they had difficulty falling asleep, 46.7% indicated that they had difficulty staying asleep and 59.3% indicated that they were NOT satisfied with the amount of sleep they were getting. The responses of the pit workers were far more favorable: 21.4%, 19.0% and 23.8%, respectively. Average sleep length on each shift was calculated from sleep start and sleep end times. As expected, workers sleep less when working the night shift (5.9 hours, averaged for both the pit and plant workers) and sleep the most when on the afternoon-evening shift (7.4 hours). When plant workers are compared to the pit workers, their sleep lengths after Day shift work are about the same. Plant workers' sleep lengths after afternoon-evening and night shifts are about 30 or 35 minutes less on the average than those for the pit workers. Since it is well established that working shift work affects sleep quality and length, it could be argued that the plant schedule in some way affects their sleep in a negative way, especially when compared to pit workers. This is supported by the fact that the sleep lengths of pit and plant workers "when free to plan their day" are nearly identical (8.4 hours).

Another common finding in shift work studies is that shift workers often have gastrointestinal disorders [6]. It is reasonable to speculate that this is at least partially due to poor eating habits. For this reason, a series of questions were included to explore eating habits as a function of shift. For these items one of the largest differences between the pit and plant workers is shown in Figure 11. On all three shifts the proportion of workers reporting "eat at different times" each day, is higher for the plant than the pit. This is especially true during the night shift.

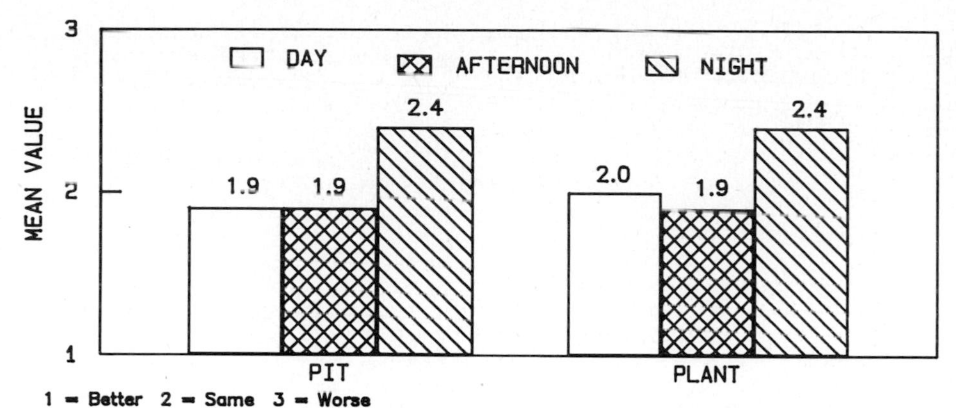

1 = Better 2 = Same 3 = Worse

FIGURE 7. – PHYSICAL EXHAUSTION AS THE WEEK GOES ON FOR EACH SHIFT

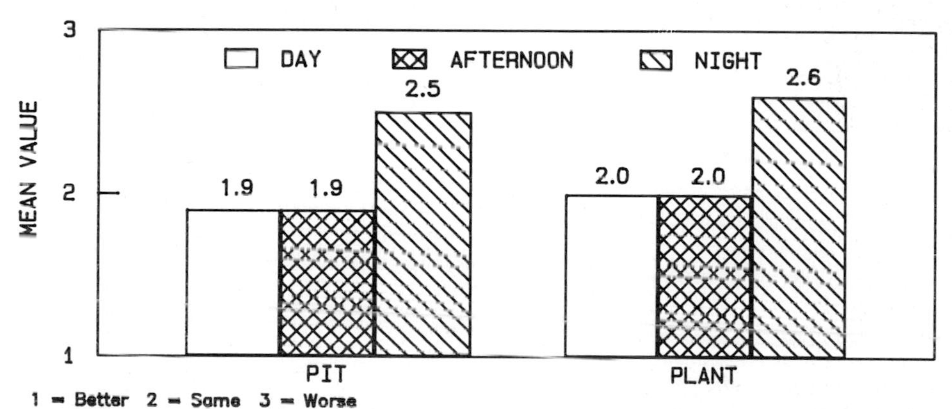

1 = Better 2 = Same 3 = Worse

FIGURE 8. – MENTAL EXHAUSTION AS THE WEEK GOES ON FOR EACH SHIFT.

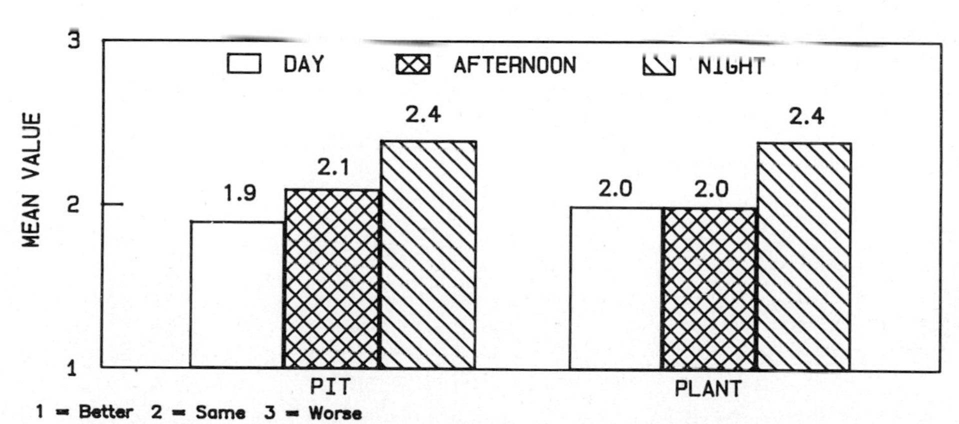

1 = Better 2 = Same 3 = Worse

FIGURE 9. – MINOR ACHES/PAINS AS THE WEEK GOES ON FOR EACH SHIFT

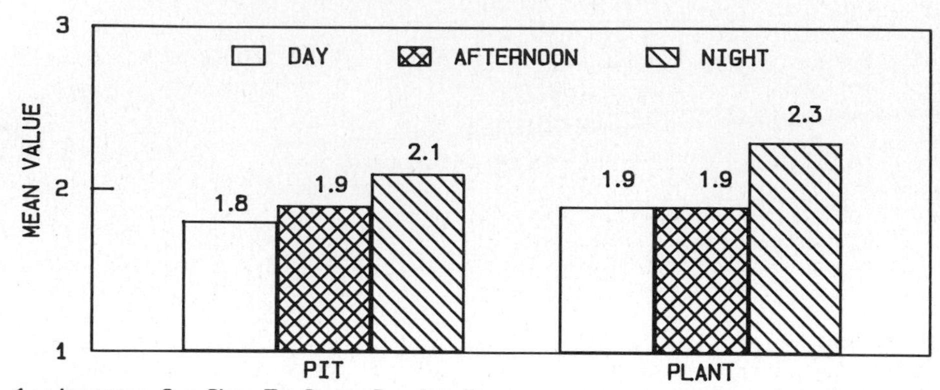

FIGURE 10. – WORK PERFORMANCE AS THE WEEK GOES ON FOR EACH SHIFT.

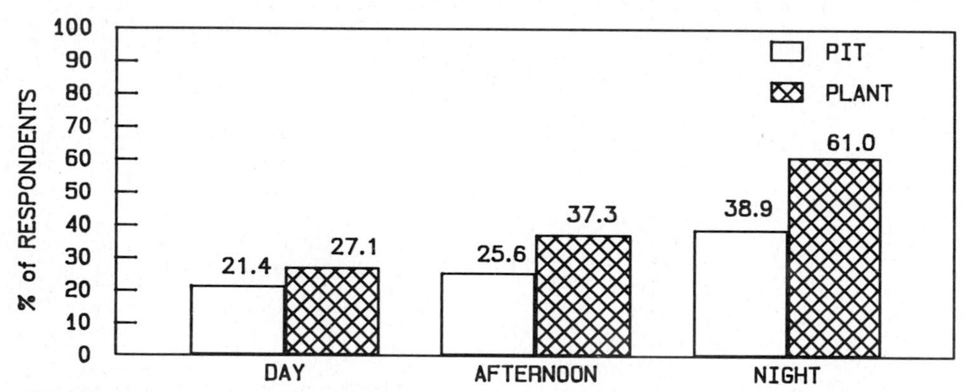

FIGURE 11. – EATING AT DIFFERENT TIMES EACH DAY AS A FUNCTION OF SHIFT

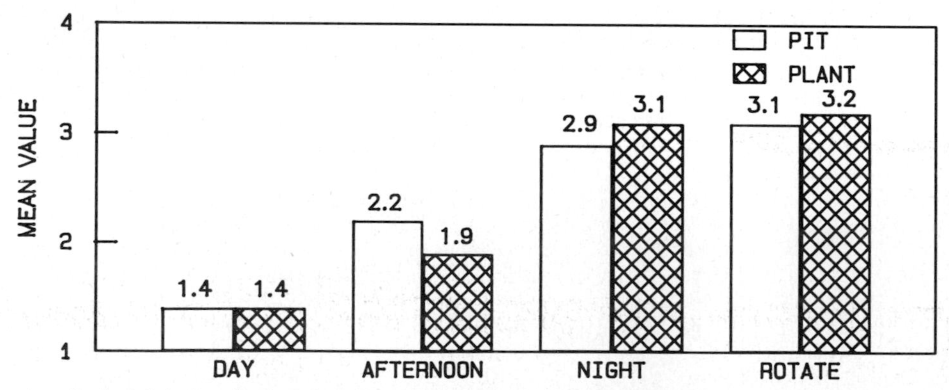

FIGURE 12. – AVERAGE RANKING ON SHIFT PREFERENCE.

4.4 Shift Preference

The following results indicate the preference and acceptability of some aspects of shift rotation. These questions reveal what these workers find desirable in the design of shift schedules.

Workers were asked which shift change-overs were easiest and hardest on their ability to maintain alertness. The hardest shift change for the pit and plant workers was the change from the day to night shift: 66.7% and 75.4%, respectively, indicated this change. The easiest shift change was going from the afternoon-evening to the Day shift: 46.4% and 49.1% of the pit and plant workers, respectively, noted this change.

Both plant and pit workers, in general, favor weekly rotation. However, 40% of the plant crew as opposed to only 21% of the pit crew, favor a rotation longer than every week. A schedule rotating every 4 weeks was the most popular response following one rotating every week.

When asked to rank shift preferences (Figure 12), both plant and pit workers on the average ranked the night shift and rotating shifts about equally as low. This may suggest that permanent night shifts may be equally as acceptable (or unacceptable) as rotating shifts.

Plant and pit workers rate working on the weekends about equally as bad, 4.2 and 4.1, respectively, on a scale of 1 to 6 (1=very acceptable to 6=very unacceptable).

A question designed to ascertain worker opinion on whether "having 4 days off in a row makes it worth working 7 days straight on each shift" was asked of the plant workers. The results showed that the majority (61%) felt that "having 4 days off in a row does not make up for working 7 days straight."

Plant workers (83%) reported that they would favor the long weekend (4 days off) following the night shift as opposed to following the day shift.

5. SUMMARY

The results show that this survey differentiates between workers on the two schedules. In general it would be accurate to say that the plant workers responded in the negative direction in most of the areas tapped in the survey. The question which directly asked if they would be willing to try another schedule showed that 77% of the plant workers, as compared to only 31% of the pit workers, would be willing.

The exact causes of plant worker dissatisfaction are unclear. It is clear that there are social effects i.e., family life (77.6% said "would improve"), marriage (73.1% said "would improve" with a different schedule) and 91.1% said family would prefer a different work schedule. There are also adverse physical effects of shift work, such as poor sleep quality and quantity, feelings of fatigue and sleepiness (especially on the night shift), health (76.7% said would improve on a better shift), not feeling lively, alert or able to work hard. The results of the plant worker's eating habits are disturbing, especially in comparison to the pit workers. Workers responded in a negative way to questions relating to eating habits while working on the night shift and to a lesser extent on the day and afternoon shift.

Plant workers on the day shift report greater fatigue than do pit workers on the day shift. Although sleep lengths for the two groups are nearly identical (6.9 hours), plant workers complain of being more sleepy after working the day shift (yet less sleepy after the Afternoon shift). They also complain of being more physically tired after working the day shift. One possible reason for these problems may be due to the larger perceived physical work load of plant workers, which would have a greater effect on them during the day shift due to their short sleep hours. The lighter load of the pit workers would then have less of an effect.

In conclusion the survey does differentiate between the "good" and "bad" schedule on several variables, including sleep, physical exhaustion and eating. However, further validation is still required. These results indicate that, for the plant workers, a new schedule may improve their conditions. Certain items show the workers have similar preferences and opinions on some aspects of scheduling. For example the result that indicated that the night shift and a rotating shift are about EQUALLY dissatisfying may open the possibility of permanent shifts.

REFERENCES

[1] Tepas, D., Armstrong, D., Carlson, M., Duchon, J., Gersten, A., and Lezotte. (1985) Changing Industry to Continuous Operations: Different Strokes for Different Plants. Behavior Research Methods, Instruments and Computers, 17(6), pp.670-676.

[2] Monk, T. (1986) Advantages and Disadvantages of Rapidly Rotating Shift Schedules - A Circadian Viewpoint. Human Factors, 28(5), pp. 553-557.

[3] Kogi, K. (1982) Sleep Problems in Night and Shift Work, Journal of Human Ergol., 11, Suppl., pp. 217-231.

[4] Tepas, D., Walsh, J., Armstrong, D. (1981) Comprehensive Study of the Sleep of Shift Workers. In: The Twenty-four Workday: Proceedings of a Symposium on Variations in Work-sleep Schedules (NIOSH Publication No. 81-127), Cincinnati, OH: DHHS, pp. 419-433.

[5] Cziesler, C., Moore-Ede, M. and Coleman, R. (1982) Rotating Shift Work Schedules that Disrupt Sleep are Improved by Applying Circadian Principles. Science, 217, pp.460-463.

[6] Tasto, D., Colligan, M., Skjei, E., Polly, S., (1978) Health Consequences of Shift Work. U.S. Department of Health, Education, and Welfare Publication No. NIOSH 78-154 (Washington, D.C.: Government Printing Office).

Trends in Ergonomics/Human Factors V
F. Aghazadeh (Editor)

A METHODOLOGY FOR ASSIGNING VARIABLE RELAXATION ALLOWANCES: VISUAL STRAIN, ILLUMINATION, AND MENTAL STRAIN

Joseph H. GOLDBERG and Andris FREIVALDS

Department of Industrial & Management Systems Engineering
The Pennsylvania State University
207 Hammond Building
University Park, PA 16802

Variable Relaxation Allowances (VRA's) are added to the normal time required by a qualified worker to compensate for special task and/or environmental demands. While tables of VRA's exist, their underlying bases remain suspect. The bases for Visual Strain, Poor Lighting, and Mental Strain VRA's are challenged in this paper, by noting the relationship between reported performance and task parameters. While present VRA's are reasonable in isolated cases, some task allowances were underestimated by at least an order of magnitude. The present system of VRA assignment must be modified to account for a broader range of task-related parameters to accurately compensate for demanding work conditions.

1. INTRODUCTION

The normal time for a job is the time that a qualified operator would need to perform the job, working at a standard performance level. An allowance is added to the normal time to account for interruptions and delays so as to provide a fair and readily maintainable standard. These allowances are classified as personal needs, fixed and variable relaxation, and unavoidable delays. Personal needs allowances are typically assigned 5% of the normal time, and the fixed relaxation allowance 4%. For the variable relaxation allowance (VRA), a variety of schemes have been developed, with the International Labour Office [6] being most typically accepted. These were developed through consensus agreements between management and workers across many industries and have not been directly substantiated. On the other hand, since the 1960's much work has been done in developing specific ergonomic standards for the health and safety of the U.S. worker. The purpose of this paper is therefore, to examine how well these standards compare and to propose a more comprehensive scheme of VRA's to account for current ergonomic standards.

2. DEVELOPMENT OF BASES FOR VISUAL STRAIN AND LIGHTING VRA'S

2.1. Target Visibility Considerations

Jobs causing visual strain due to precise tasks, and those suffering under inadequate illumination require additional rest allowances. The ILO [6] Visual Strain VRA allows no (0%) additional rest allowance for 'Fairly Fine Work', 2% allowance for 'Fine or Exacting Work', and 5% allowance for

'Very Fine or Very Exacting work'. These refer to the required precision of visual task requirements, in terms of required resolution and other factors that might influence task visibility. A newer system for defining Visual Strain VRAs has since been provided [7]. The category 'Eye Strain,' listed example jobs with appropriate points, which are summed to define the VRA. Although this method allows a further breakdown of VRAs, it is not necessarily more accurate, and is also missing important background bases of tabled values. ILO stated that lighting conditions, glare, flicker, illumination, color, working distance, and allowed time are all visual strain-producing factors in this system, but no mention of how they should be measured or applied was given.

An alternative visual strain VRA scheme may be proposed, based on visual search literature. In searching for a target presented on a background, several factors strongly interact in influencing search time and visual strain [2, 4, 8]: (1) The Background luminance (L), measured in Foot-Lamberts (FLs); (2) The Contrast (C), defined by $|L_{target}-L_{background}|/L_{background}$; (3) The Observation Time (T), where the average duration of an eye fixation is about 0.2 second [8]; (4) The Target Visual Angle (A), in arc minutes; (5) Spatial or Temporal Location Uncertainty, where required C is multiplied by 1.5; (6) Moving Targets, where C is multiplied by 2.78; (7) 'Real World' Conditions, with C multiplied by 2.51 outside the laboratory; (8) Required Accuracy of greater than the original 50% thresholds. These visibility curves may serve as a basis for defining VRA's in a task if percentage targets detected (%DET) is modeled as a function of the above task factors. This prediction equation, with R^2=.88, was:

$$\%DET = 81\ C^{.20}\ L^{.045}\ T^{-.003}\ A^{.199} \tag{1}$$

where parameters are defined in the text. The percentage of targets detected may be used to define VRA's by specifying a percentile range for descriptions of population abilities. At least 95% target detection defines a visual task without significant problems, equating to the ILO "Fairly Fine Work" category with associated 0% allowance. At least 50% detection is the "Fine or Exacting Work" category, with 2% allowance, and less than half of targets detected defines "Very Fine or Very Exacting Work" with associated 5% allowance. It must be stressed that Equation 1 model does not directly define VRAs; instead, it defines absolute target detection ability, which in turn can be used to define a relaxation allowance.

2.2. Simulations

In order to compare the VRA definitions derived from Equation 1 with those from the ILO [7] method, three simulations were carried out. The first required inspection of resistor values in a circuit assembly. Estimates of mean time per eye fixation, with no magnification, were made. The second simulation was of reading a newspaper in a moving vehicle, such as a bus. Again, eye fixation estimates were made, and the newspaper was assumed to be moving, due to the jarring in the vehicle. Thirdly, a continuous inspection task of seeking pulled threads on cloth moving at 10 feet/second was simulated. Here, the normal area of vision provided a 4 inch diameter circle for inspection of the moving target. The results of this simulation, in Table 1, showed that the ILO allowances, varying between 0-1% for these tasks, were insensitive to the many task conditions influencing visibility. The computed allowances ranged from 2-5%, with concomitant target detection of 40-85%. Thus, the ILO methodology was not sensitive to task visibility conditions in assignment of visual strain VRA's.

TABLE 1. Visual Task Simulations, Comparing ILO with Computed VRA's

Task	ILO [7] VRA	VRA Computation from Blackwell's Model					
		A (min)	T (sec)	C/ adjustment	L (FL)	Percent Detect	VRA (%)
Resistor Inspection	0%	5.7	0.2	0.5/3.75	10	85.3%	2%
Reading Newspaper	1%	2.9	0.2	0.6/6.98	35	72.1%	2%
Cloth Inspection	1%	2.1	0.03	0.05/6.98	25	40.8%	5%

2.3. Illumination and Performance Considerations

While Equation 1 gives required task luminance and contrast as a function of visibility factors within a task, other methods define required illumination as a function of task and observer characteristics. The ILO [6] system lists three different VRA's for poor lighting: 'Slightly below recommended value' (0% allowance), 'Well below' (2%), and 'Quite Inadequate' (5%). A table with six ranges of recommended illumination, from 5 to 1000 FC, was provided. The Committee on Recommendations for Quality and Quantity of Illumination of the IES [5, 11] broke down visual tasks into categories, each containing three recommended illuminances (in FC). The appropriate illumination within a category is chosen after a structured consideration of task and worker characteristics. For purposes of assigning VRA's, a task that is slightly below recommended guidelines can be considered to be within the same illumination category, whereas a task that is well below adequate illumination is logically one category beneath its recommended illumination. A task with quite inadequate illumination may be defined as being more than one category substandard.

TABLE 2. Recommended Illumination, Adapted from [11]

Cate-gory	Illumination (FC)	Description of Activities (A-C: Areas; D-I: Visual Tasks)
A	2-3-5	Public areas with dark surroundings
B	5-7.5-10	Simple orientation for short visits
C	10-15-20	Workspaces with occasional visual tasks
D	20-30-50	High contrast or large size
E	50-75-100	Medium contrast or small size
F	100-150-200	Low contrast or very small size
G	200-300-500	Low contrast or very small size, prolonged
H	500-750-1000	Very prolonged and exacting visual tasks
I	1000-1500-2000	Very special tasks; low contrast, small size

The literature has repeatedly provided evidence that increasing task illumination results in better performance (given that glare is not a problem), as measured by task completion time. A quantitative estimation of magnitude of percentage performance improvement can provide a check of VRA adequacy, using the above definitions. In the presented models below, Time refers to the mean task performance time (seconds), whereas FC refers to the task illumination (Footcandles). Smith [14] modeled individuals' time to repeatedly insert the tip of one needle into the 4-7 arc min. eye of another needle as:

$$\text{Time} = 25.9 - 7.0 \log(\text{FC}) + 1.45 \log(\text{FC})^2 \quad (2)$$

This model showed that performance times fell with increasing illumination, but to a lesser and lesser degree. Smith and Rea [15] measured the time required to find errors in paragraphs for subjects viewing legible words printed on white paper. Regression of their data indicated:

$$\text{Time} = 37.4 - 2.99 \log(\text{FC}) + .44 \log(\text{FC})^2 \quad (3)$$

In another reading task, Bennett, Chitlangia, and Pangrekar [1] reported the time to read a 450-word, pencil-written article as a cubic function of task illumination:

$$\text{Time} = 251.8 - 33.96 \log(\text{FC}) + 6.15 [\log(\text{FC})]^2 - .37 [\log(\text{FC})]^3 \quad (4)$$

Required VRA's may be computed by determining the percentage performance loss as illumination is reduced, as shown in Table 3. With age and task weighting factors, Smith's [14] task had a recommended illumination of 200 FC. Within the same illumination category [RQQ, 11], the modeled performance time increased by over 7% as the illumination was lowered. This time is already beyond the 5% Quite Inadequate ILO VRA; times increased by up to 25% when only a 5% VRA would be given. Smith and Rea's [15] fairly easy reading task only requires 30 FC, by RQQ [11] standards. Within an illumination category, time only increased by 1%, which was in close agreement with the Slightly Below Recommended Value (0% allowance) ILO VRA category. In the next lower category, times increased by 2 to 3%, agreeing with the ILO 2% allowance. Two illumination categories below the recommended value, modeled performance times increased by up to 4.8%, in close agreement with the 'Quite Inadequate', 5% VRA. Bennett, et al.'s [1] reading task was assigned 75 FC by RQQ [11] standards. Decreasing the illumination to the next lower category increased performance times by 3-5%, a bit greater than the 2% ILO VRA for 'Well Below' recommended lighting. The next lower illumination category produced times that were 6-8% greater than that at recommended levels, somewhat slower than the 5% allowance for 'Inadequate illumination'.

3. DEVELOPMENT OF BASES FOR MENTAL STRAIN VRA'S

The mental strain VRA includes 1% allowance for a 'Fairly Complex' task, 4% for 'Complex or Wide Span of Attention', and 8% for 'Very Complex' tasks [6]. Mental strain was originally defined by "prolonged concentration-for instance, trying to remember a long and complicated process," and can be found when attending to a number of machines, or as the result of anxiety [6, p.257]. Using this system, it is highly likely that there will be much variability between work raters, due to its loose definitions. Investigation of the basis and adequacy of these VRAs necessarily requires objective evidence of changing cognitive productivity with fatigue or time on the job. Performance changes may be measured as changes in accuracy for constant task performance time, or as changes in performance time for a constant accuracy. Both of these will be used below, to assess fatigue effects.

TABLE 3. Comparison of Performance Loss with Assigned VRA's

Study	Illum. (FC)	Modeled Time (Sec.)	Performance Loss (%)	ILO [6] VRA Category	VRA (%)
Smith [14]	200.0	16.46	--	Recommended	--
	100.0	17.70	7.53	Slightly Below	0%
	50.0	18.93	15.01	Well Below	2%
	20.0	20.56	24.91	Inadequate	5%
Smith and Rea [15]	30.0	34.28	--	Recommended	--
	20.0	34.65	1.08	Slightly Below	0%
	15.0	34.92	1.87	Well Below	2%
	10.0	35.29	2.95	Well Below	2%
	7.5	35.55	3.70	Inadequate	5%
Bennett, et al. [1]	75.0	207.31	--	Recommended	--
	50.0	210.04	1.32	Slightly Below	0%
	30.0	213.86	3.16	Well Below	2%
	15.0	219.76	6.00	Inadequate	5%

The effects of mental fatigue are typically investigated by measuring performance over a sufficiently long observation period. Okogbaa [9] had subjects perform reading and mental arithmetic tasks over a five hour period. Controlled-accuracy performance time, as reported by Shell, Okogbaa, and Mital [13] and Okogbaa and Shell [10], was modeled as:

$$\text{Reading Rate (words/min.)} = 70\ e^{-(.0007\ t)} \tag{5}$$

$$\text{Problem Rate (probs./min.)} = 12.2\ e^{(.0003\ t)} \tag{6}$$

where: t = Time on task (minutes)

Computed performance declines thus ranged from 3-4% after one hour, to 7-15% after 4 hours. The reading task was timed and standardized, much like a GRE or SAT test, with a comprehension test given following each passage. Due to a high level of required comprehension, the reading task can be considered to require a "wide span of attention" and a 4% ILO VRA. The problem-solving task required thought before responding, and "represented a high information load." These were selected such that 95% of the population of the chosen skill level could complete the problems, so the task will be considered complex (as opposed to very complex), again receiving a 4% VRA.

Colquhoun [3] had subjects inspect photographic strips for small disks, with strips presented at 50 per minute, and one second viewing time per strip. Half of the 32 subjects received a rest of 5 minutes (an 8.33% VRA) midway in the one hour period. The mean percentage of faults detected, shown in Table 4, degraded by about 3% without rest, but less than 1% with rest. Comparing this fatigue effect between groups, after one hour, rest provided a 3% advantage in performance. Colquhoun's inspection task is complex, by ILO [6] standards, requiring a 'wide span of attention', due to time constraint and visibility considerations. A 4% VRA would be applied to this element, agreeing well with the performance degradation results.

TABLE 4. Percentage Fault Detection, from Colquhoun [3]

Time (min)	NO-REST GROUP Faults Detected (%)	NO-REST GROUP Accuracy Change (%)	REST GROUP Faults Detected (%)	REST GROUP Accuracy Change (%)	DIFFERENCE Accuracy Change (%)
0-10	98.5	-	99.2	-	-
11-20	98.4	0.10	98.9	0.30	-0.20
21-30	97.8	0.71	99.1	0.10	0.61
31-40	96.0	2.54	99.3	-0.10*	2.64
41-50	95.4	3.15	98.9	0.30	2.85
51-60	95.2	3.35	98.7	0.50	2.85

*Detection was slightly improved following rest

Schmidtke [12] measured the recovery time needed for compensation of mental fatigue, as related to duration of work and required level of performance. His task consisted of mental arithmetic; a subject computed the sum/difference of three random numbers, and subtracted successive answers. The duration of forced continuous additions was varied from 5 to 30 minutes, and the length of required rest pauses was modeled such that less than 5 percent of problems were missed. A total working period (with pauses) of 4 hours was maintained. The duration of rest pauses from this task can be modeled as:

$$\text{Pause Time (Min.)} = .0068\ T_w^{1.114}\ L^{2.065} \tag{7}$$

where: Tw = Duration of work periods (Min.)
L = Task difficulty (Prob/min)

These modeled pause times are the required rest time for maintaining performance; to model the required percentage of rest time required for this mental addition task, it is assumed that the total working time is the number of working cycles multiplied by the summation of the cycle work and pause times. Solving for percentage rest allowance:

$$\text{VRA (\%)} = 100\ [\ 1 - (.0068\ T_w^{.114}\ L^{2.065} + 1)^{-1}\ ] \tag{8}$$

These VRA's rapidly increase with time up to about 30 minutes, then stabilize for longer continuous performance times. The allowances are clearly related to task difficulty, with 6 problems/minute demanding rest of over 35% of total time. The ILO allowances of 1-8% are clearly inadequate for many task complexities, underrating required rest by an order of magnitude or more.

4. DISCUSSION

The basis for ILO [6] VRA's for Visual Strain was challenged by considering well-modeled task visibility factors. By using Equation 1 to determine percentage target detection, and applying accepted work design heuristics, the visibility conditions for visual strain were defined. Three examples compared computed VRA's with stated ILO VRA's. ILO [7] VRA's underpredicted computed VRA's (Table 1), in that the former were not able to account for changes in the visibility factors that have a major influence on visual performance. Equation 1 is suggested as an alternative schema for setting Visual Strain VRA's.

Poor Lighting ILO VRA's were challenged by considering recommended IES task illuminations. All values within the same illumination category received no VRA, values in the next lower category received 2% allowance, and those two categories below received the full 5% VRA. While VRA's via this method undercompensated performance losses in the manual performance study, they were in close agreement with performance losses in the latter two reading studies (Table 3). Whether the degree of hand eye coordination, or the lighting requirements is responsible for the divergence is unknown. However, the close agreement in the reading studies is promising.

Accepted 'Mental Strain' VRA's, varying from 0-8% of the normal time, will undercompensate task performance changes for cognitive tasks lasting longer than 2-3 hours. In mental arithmetic and reading, performance losses of 3-4% per hour have been demonstrated (Table 4, Equation 7). Though an objective definition of task complexity is required, the ILO VRA's appeared to compensate performance loss up to 1-2 hours. Schmidtke's study clearly indicated that greater task complexity or difficulty requires greater VRA's, as ILO originally proposed. In fact, increasing the problem rate from 1 to 6 per minute increased required VRA's from 1% to over 30% (Equation 8). Conservative assignment of VRA's should be made on the basis that a task will be performed for at least two hours.

The models presented here may be used to define VRA's for very specific tasks, but the large variability between similar tasks may preclude this as a general technique. The human performance literature can be used as a basis for assigning VRA's, but more across-task models are still required. Better VRA models will insure that performance losses are compensated with adequate rest; productivity can then be increased during work intervals. Further research will be able to provide a generic and accurate methodology for assignment of visual and cognitive work VRA's.

5. REFERENCES

[1] Bennett, C. A., Chitlangia, A., and Pangrekar, A., "Illumination Levels and Performance of Practical Visual Tasks," Proceedings of the 21st Annual Meeting of the Human Factors Society, 1977, pp. 322-325.

[2] Blackwell, H.R., "Brightness Discrimination Data for the Specification of the Quantity and Quality of Illumination," Illuminating Engineering, 1952, p. 602.

[3] Colquhoun, W.P, "The Effect of a Short Rest-Pause on Inspection Efficiency," Ergonomics, 1959, 2:367-372.

[4] Illuminating Engineering Society (IES), IES Lighting Handbook, 3rd Ed., Illuminating Engineering Society, 1959.

[5] IES, IES Lighting Handbook Reference Volume, 1981, Illuminating Engineering Society of North America, New York, NY.

[6] International Labour Office (ILO), Introduction to Work Study, 1st Ed., International Labour Office, 1957.

[7] ILO, Introduction to Work Study, 3rd Ed., International Labour Office, Geneva, 1981.

[8] Murdoch, J.B., Illumination Engineering - From Edison's Lamp to the Laser, New York, MacMillan, 1985.

[9] Okogbaa, O.G., "An Empirical Model for Mental Work Output and Fatigue," Unpublished Doctoral Dissertation, University of Cincinnati, 1983.

[10] Okogbaa, O.G., and Shell, R.L., "The Measurement of Knowledge Worker Fatigue," IIE Transactions, 1986, 12: 335-342.

[11] RQQ, "Selection of Illuminance Values for Interior Lighting Design (RQQ Report No. 6)," Journal of the IES, 1980, 4: 188-190.

[12] Schmidtke, H., "Disturbance of Processing of Information," in Simonson, E., and Weiser, P. (Eds.), Psychological Aspects and Physiological Correlates of Work and Fatigue, Springfield, Il: Thomas, 1976.

[13] Shell, R.L., Okogbaa, O.G., and Mital, A., "Data Transformation Concepts and Empirical Model Development for Mental Work Output and Fatigue," in Mital, A. (Ed.), Trends in Ergonomics/Human Factors I, Amsterdam: North-Holland, 1984, pp. 233-240.

[14] Smith, S. W., "Performance of Complex Tasks under Different Levels of Illumination," Journal of the IES, 1976, 7: 235-242.

[15] Smith, S. W., and Rea, M. S., "Proofreading under Different Levels of Illumination," Journal of the IES, 1978, 10: 47-52.

Trends in Ergonomics/Human Factors V
F. Aghazadeh (Editor)

DUAL ROLES OF VISUAL FEEDBACK ON POSTURAL STABILITY

Dennis SHUMAN and Keith D. WHITE*

Departments of Psychology and Electrical Engineering
University of Florida
Gainesville, Florida 32611

Realtime body sway data was utilized in the control of a surrounding pattern's movement, thus altering the gain of the visual feedback loop. Utilizing stimulus feedback gains ranging from -2 to +2, body sway frequency spectra confirmed true feedback control by vision. The spectra exhibited large peaks and valleys that displayed frequency consistency across stimulus gain magnitudes but reversed with stimulus gain polarity. Both the *direction* of movement of the retinal image and the movement of the retinal image *relative to that expected* seem to be parameters affecting sway performance. Implications for a hierarchical model of equilibrium control are discussed.

1. INTRODUCTION

Although many sensory inputs can provide spatial orientation information, it is well recognized that the visual, vestibular, and proprioceptive senses usually play the dominant roles. The nervous system uses its own program to weight these sensory inputs together with information from past experience. Problems arise when the received information is insufficient, misleading, novel, or conflicting because then disorientation and possible physical disturbances can occur. It is difficult to get a good objective measure of disorientation, but one method has been to measure postural stability or body sway. A greater understanding of the role of *visual* feedback in spatial orientation seems particularly important for the space environment, where vestibular and proprioceptive feedbacks are altered by microgravity whereas visual feedback per se is not. Vision provides important feedback about body sway even on earth. A person overlooking a vista where only distant objects are visible will exhibit increased body sway due to the decrease in retinal image motions accompanying sways. Similarly, a new prescription for corrective lenses can produce disorientation due to the altered relationship between head movement and retinal image motion. The effect of altering the relationship between a subject's head movement and the resulting retinal image motion (visual feedback) is here studied neither by manipulating viewing distance nor by inducing optical distortion, but rather by moving the visible surrounding according to various quantitative relationships with realtime body sway measurements. [1-7]

* Supported by NIH grant #R23-EY03640 and by grants from UF DSR to KDW. This paper is based on a thesis submitted in partial fulfillment of the requirements for the degree of Doctor of Philosophy by DS.

2. METHODS

2.1. Subjects

There were 33 volunteer subjects (17 male and 16 female), all within the age range of 20 to 40 years old. All were naive of the experimental conditions. None had any history of postural problems, unusual dizziness, or afflictions of any of the sensory organs involved with balance. Corrective lenses previously prescribed were worn during the experiment.

2.2. Apparatus

The multi-microcomputer based Position Sensor System (PSS) was developed to sample continuously the position and orientation of a standing subject's head within a finite volume that is large enough to accommodate the extreme possible limits of postural sway in any direction [8,9]. A unique feature of the PSS is that its operation is based on acoustics. The principle involved is that the distance from a point sound source to a point sound receiver can be determined by measuring the propagation time of a transmitted sound front. The PSS samples the distances from two acoustic sources secured to a subject's head, to four fixed microphones positioned within the subject's surroundings. This allows it to resolve rotation as well as translations, and in addition, to use redundant data for error detection and correction. The raw position data are stored for later analysis. The data are also processed online to determine the subject's head position and orientation, some specified function of which is used to generate analog stimulus control signals. Stimulus generation is based upon a vertically oriented cylindrical screen (77 cm radius) within which the subject stands, and a projector positioned along the cylinder's axis of rotation above the subject's head. The projector is a shadowcasting device which projects a vertical grating onto the screen that totally encompasses a properly positioned subject's field of view. The grating moves horizontally across the screen as dictated by the analog control signal. The delay between sampling the subject's position and moving the stimulus was about 115 msec.

2.3. Experimental Environment

We attempted to minimize the influences on postural stability of sensory cues other than those provided by the visual stimulus. The extraneous sensory cues of concern were visual, auditory, and proprioceptive. To prevent unwanted visual cues, the experiment was conducted in a black room and monitor screens were covered. In order to mask auditory cues from equipment and people, the subjects wore headphones transducing white noise. The normal human stance (feet side by side and slightly apart) provides proprioceptive cues from ankle joint rotation (mainly in the sagittal plane) and from differential weight on the two legs. In attempting to reduce or alter proprioceptive cues, alternative stances have sometimes been used, such as one foot or heel to toe stances to enhance lateral sway [10]. For this study, the wobble board was developed to alter proprioceptive cues. It consisted of a circular board resting on a moderately inflated tire inner tube and it performed like a pneumatic spring. The results of a three subject pilot study indicated that visual stimuli exerted the greatest effect on lateral body sway when the feet were slightly apart on the wobble board, followed by feet together on the wobble board, then a one foot stance on a firm surface and finally a heel to toe stance on a firm surface. Therefore, feet apart on the wobble board was chosen for the present study. Also this natural standing position seems to be less of a strain on subjects' leg muscles than the other stances considered. The type of shoe worn by a subject might affect postural stability, such as very thick or spongy soles, so subjects were requested to be barefoot.

2.4. Procedure

The subjects were positioned on the wobble board so that their body was aligned with the axis of rotation of the cylindrical screen, and the proper initial head orientation was indicated by a momentary fixation light centered on the screen. The visual stimulus consisted of a projected pattern of light and dark vertical stripes, each with a visual subtense of 2.5°. The light stripe illuminance was set at 0.03 footcandles along the screen surface at eye level. The subjects were told that their task was to keep as still as they could while looking at the point where they had seen the fixation light. They were told that the vertical stripes might move, and that they should not follow the movement with their eyes but rather let the stripes sweep by.

An experiment consisted of 30 trials, each having a duration of approximately 26 secs (256 head position determinations per trial at a 10 Hz sampling rate). The first 15 trials consisted of 8 feedback-controlled stimulus conditions, 4 simulated stimulus jitter conditions, an eyes closed condition, and a stationary stimulus condition performed twice. These trials were performed in random order (different for each subject), and the second 15 trials used the same trial conditions performed in reverse order to counterbalance for order effects. The stimulus gain in feedback conditions is the relationship between a stimulus movement and its causative subject movement, defined so that a normalized stimulus gain of +1 would result in the stimulus following the subject's head (line of sight) and thus a "stabilized" retinal image. The 8 stimulus gains used were +/-2, +/-1.5, +/-1, and +/-0.5, displaying both an algebraic and a geometric progression of values. During the feedback conditions an unwanted stimulus movement jitter due to noise in the PSS was superimposed on the desired stimulus movement. The magnitude of this jitter was directly proportional to the absolute value of the stimulus gain. In order to ascertain the influence of this jitter artifact on postural stability and thereby to be able to extract its contribution from data obtained during feedback trials, 4 conditions were used in which the stimulus movement was solely jitter, simulating that artifactually produced in the feedback conditions. These control conditions were appropriate for the absolute values of the stimulus gains employed (2, 1.5, 1, and 0.5) and are also referred to as "baseline" conditions.

2.5. Analysis

For each trial with each subject, the stored raw position sample data were used to calculate head location in lateral (X axis) and anteroposterior (Y axis) coordinates and rotational orientation. Individual Fourier power spectra of the processed X axis, Y axis and rotational data for all subjects for each trial were determined using a Fast Fourier Transform (FFT). Within subjects and sway orientations, the power spectra of identical trial conditions were averaged and gains were calculated as the ratios of power spectra for each of the feedback conditions relative to the corresponding jitter simulation condition. The data were further collapsed by averaging these gains across subjects, and by grouping the gains into 0.5 Hz wide frequency bins from 0 to 5 Hz. The resulting output data were the means and standard errors (across subjects) of all the above sway spectral gains within each frequency bin and for each measurement axis (X, Y, and rotation).

3. RESULTS

Sway gains computed within subjects and then averaged across subjects are shown as a function of frequency in Figures 1 and 2. In these graphs, each sway gain plot shows on the ordinate the ratios of spectral power for two trial conditions, where each trial condition is designated in the legend by a number-letter code. In this code the number shows stimulus gain, the letter F indicates a feedback condition, and the letter B indicates a control "baseline" condition which simulates the stimulus jitter artifact present in feedback conditions at that gain. A feedback condition can have either a positive or a negative sign, whereas the corresponding control condition does not require a directional sign.

The standard errors of measurement typically showed values less than 1 dB. Such a narrow range of the measurement uncertainty seems rather surprising given that individual differences in body sway performance are readily

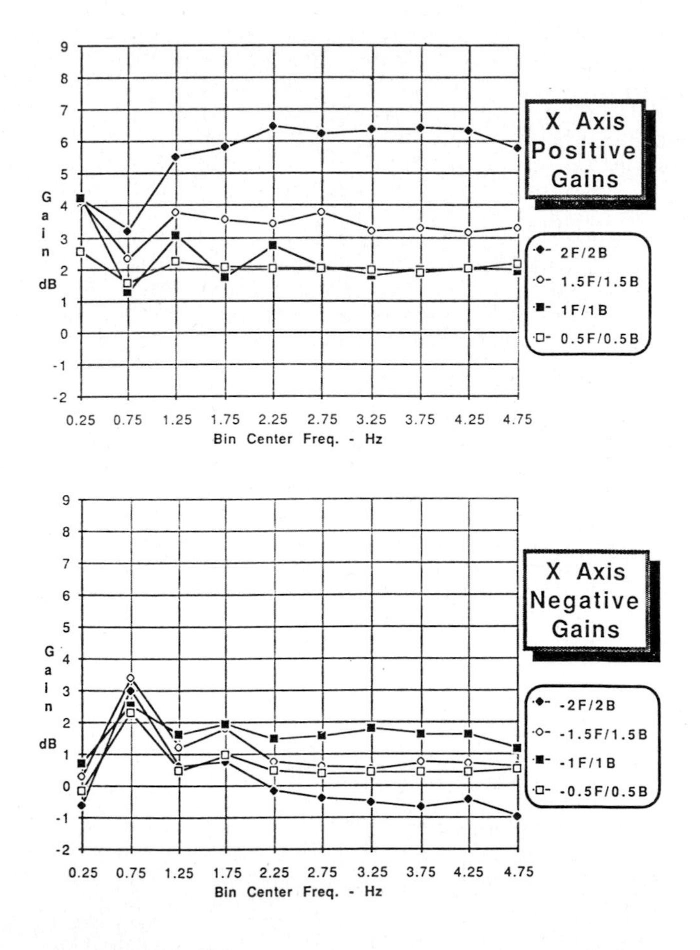

Figure 1. Lateral Sway Gains with (a) Positive and (b) Negative Stimulus Feedback Gains. Each point plots a mean power ratio.

apparent even to casual observation. The present method of calculating sway gains emphasizes *relative* changes in a subject's own body sway performance between trial conditions, and in that regard normalizes every subject with respect to his/her own individual performances [10]. Inasmuch as the individual performances were extremely diverse but the sway gain measures were more nearly comparable across subjects, it appears that sway gain is a particularly suitable index to be averaged across subjects.

Figure 1a shows mean lateral (X axis) sway gains for *positive* stimulus gains relative to their corresponding jitter simulation conditions. For low frequencies the sway gains for all stimulus gains are approximately equal (4 dB) with the exception of +0.5 stimulus gain. At the higher frequencies, the sway gains *increase* with increasing stimulus gains, again with the exception of +0.5 stimulus gain. Note that all the sway gain plots show a 1 to 2 dB *valley* at 0.75 Hz bin center frequency. Figure 1b shows mean

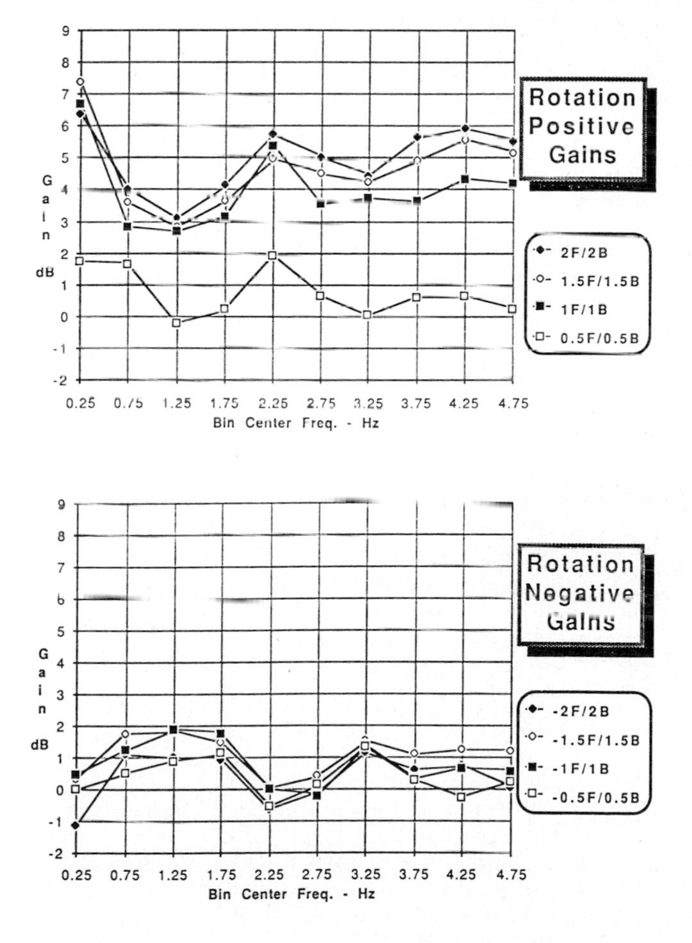

Figure 2. Rotational Sway Gains with (a) Positive and (b) Negative Stimulus Feedback Gains. Each point plots a mean power ratio.

lateral (X axis) sway gains for *negative* stimulus gains relative to their corresponding jitter simulation conditions. For low frequencies the sway gains for all stimulus gains are almost 0 dB. At higher frequencies, the sway gains spread out slightly yet not as much as with, and with lower values than observed using positive stimulus feedback (Figure 1a). Contrastingly, with the exception of -0.5 stimulus gain, the sway gains now *decrease* with increasingly negative stimulus gain. Note that all the sway gain plots show a 1 to 3 dB *peak* in the interval of 0.5 to 1 Hz. Figure 2a shows rotational head movement gains for *positive* stimulus gains. These have similar effects on head rotation, except that the stimulus gain of +0.5 exerts substantially less magnitude of effect although similar frequency shaping. Positive stimulus gains are most influential at low sway frequencies and show a slight positive ordering with stimulus gain on sway gains above 2.75 Hz. The *valleys* first seen in the 0.75 Hz frequency bin with the lateral sway gains (Figure 1a) are again seen here, but they are now much wider and centered about 1.25 Hz. The *valleys* are now followed by *peaks* in the 2.25 Hz frequency bin. Rotational sway gains resulting from *negative* stimulus feedback, as shown in Figure 2b, are generally independent of the feedback magnitude, with the amplitudes averaging about 1 dB. However, the frequency spectra show *peaks* analogous to the peaks in lateral sway gains (Figure 1b) although now much wider and centered about 1.25 Hz. These *peaks* are now followed by *valleys* in the 2.25 Hz frequency bin.

There is a pattern of symmetry in the occurrences of *peaks* and *valleys*. With positive feedback, valleys in lateral sway gains were matched by broader valleys followed by peaks in rotational gains. With negative feedback, peaks in lateral sway gains were matched by broader peaks followed by valleys in rotational gains. These features were absent from the data for anteroposterior sway gains and for jitter sway gains (relative to a stationary stimulus) in all three measurement axes (data not shown). This is reasonable since the stimulus movement was not correlated in these situations with body sway, and thus true feedback was not present.

4. DISCUSSION

In attempting to characterize the stabilizing effect of visual feedback on body sway, standard control system analysis procedures can be employed. The performance of a system can be studied under altered conditions introduced by opening the feedback loop and introducing various feedback gains and phases (delays). Under such conditions, the transient and steady state performances of the system can be ascertained by introducing appropriate stimulus movements (superimposed on the visual feedback stimulus motion) while monitoring the resultant body sway behaviors. The present characterization consists of analyzing the steady state performance for a single input frequency of 0 Hz with various feedback gains and one feedback delay time.

The present results demonstrate that visual input acts as a feedback loop in the body's postural stabilizing system, since altering the relationship between body movement and visual feedback affects postural stability in a manner consistent with control system theory. This is observed most clearly in the lateral (X axis) data where the frequency-dependent signatures of body sway gain spectra tend to be arranged in direct order with positive stimulus gains and in inverse order with negative stimulus gains.

The peaks and valleys observed in sway gain spectra were quite unexpected and might be suspect as experimental artifacts except that they only occur in those subsets of the results where the stimulus was spatially and temporally correlated to body sway [11]. The present findings are also striking in light of the large amplitudes and consistency of the peaks and valleys within groupings, and that the results with positive and negative feedback are such good mirror images of each other. These peaks and valleys are reminiscent of the poles and zeros in the complex domain introduced by control system transfer functions. Their reversals when the sign of feedback changes is in accord with how such a change can affect the closed loop transfer function of a system. Their differences in sharpness and location in the lateral and rotational measurement axes might reflect the movement of poles and zeros in the complex domain as a function of the different physical dynamics required for control of these bodily responses.

Another unexpected finding, most apparent in the rotational data although also discernible in the lateral sway data, is the differentiation of the frequency spectra between +0.5 stimulus gain and other positive stimulus gains. In attempting to explain this finding, it is useful to introduce the concept of "expected visual feedback." This is equivalent to determining the geometric relationship between the movement of a retinal image and its causative subject's head movement given that subject-to-object distance has been correctly determined by the subject [1]. Consider that the magnitude of movement of a retinal image resulting from a given subject movement is inversely proportional to the distance to a seen stationary object. Without an expectation based on this distance, changes in subject-to-object distance would be equivalent to changes in stimulus negative feedback gain, and these would result in changes in postural stability. When subject-to-object distance is very large then body sways do increase, possibly because resulting movements of the retinal images become undetectably small. But over a substantial range of viewing distances visual feedback is almost uniformly effective in controlling body sway, and this tends to confirm our present supposition. The expected visual feedback is stabilizing in that the *direction* of movement of the retinal image results in corrective postural responses. This expected visual feedback also has a *magnitude*. With the use of this concept, it can be seen that the +0.5 stimulus gain is a particularly salient experimental condition. It is the only stimulus gain which provides visual feedback in the expected *direction* but with less than the expected *magnitude*. Therefore, movement *direction* groups this condition with the negative stimulus gains. However, having "less negative" visual feedback *relative to that expected* also groups this condition with the positive stimulus gains.

It may therefore be asked whether the sway results with +0.5 stimulus gain are like those with the other positive stimulus gains or instead like those with the negative stimulus gains. In the rotational orientation, the amplitude of sway gain in the +0.5 stimulus gain condition is substantially lower than with the other positive stimulus gains (which are all quite similar to each other), especially at the low end of the frequency band. In fact, the mean sway gain amplitude across the band in this condition is about the same as with negative stimulus gains. However, the shape (peaks and valleys) of the frequency spectrum in this condition is strongly identified with positive stimulus gains. For lateral sways, grouping of the +0.5 stimulus gain condition is less readily evident but nevertheless seems consistent with the same general descriptions. *Direction* of movements of the retinal image correlates with the amplitudes of the sway gain spectra, especially at the low end of the frequency band. Movements of the retinal image *relative to expected* visual feedback lead to the appearance of peaks and valleys in the sway gain spectra.

We are impressed by the ease with which these findings about *visual* feedback fit into the model of a "hierarchically organized system for equilibrium control" that was originally proposed by Nashner et al to illustrate the lower and higher level roles of *vestibular* feedback [5]. Sensitivity to movement *direction* may be presumed to be largely independent of the external environment and even of the optical characteristics of the visual system. The sensitivity to movement *relative to expected* visual feedback presumably is strongly dependent on the external environment and on the optical characteristics, and therefore needs learning and experience for its calibration. Movement *direction* might then perform the dual roles of providing stabilizing information (low level) and of functioning as a reference (higher level) for calibrating *expected* visual feedback magnitudes. A relevant anecdotal report is that people acclimatized to their corrective lenses are relatively stable with or without them, as compared to with a novel correction. This suggests that experience enables recalculation for the magnitude of expected movements of the retinal images, to match the provisos of changing optical status.

In summary, we have investigated the effects on body sway of manipulating positive and negative visual feedback gain. Both the amplitudes and spectral signatures of responses were found to be unusually reliable between subjects, but they showed some unexpected features. Those features can be understood surprisingly well under a hierarchical model of equilibrium control that was first proposed to explain vestibular feedback.

REFERENCES

[1] Cohn, T.E., Lasley, D.J., Dister, R., and Tong, M. (1985) Static visual stimulus that causes postural instability, Investigative Ophthalmology and Visual Science, 26(supplement 3), p. 142.
[2] Gantchev, G.N., Draganova, N., and Dunev, S. (1981) Role of visual feedback in postural control, Agressologie, 22(A), pp. 59-62.
[3] Gantchev, G.N., and Koitcheva, V. (1981) The role of the visual feedback gain in the control of voluntary body movements, Agressologie, 22(A), pp. 55-57.
[4] Gonshor, A., and Jones, G.M. (1980) Postural adaptation to prolonged optical reversal of vision in man, Brain Research, 192, pp. 239-248.
[5] Nashner, L.M., Black, F.O., and Wall, C. (1982) Adaptation to altered support and visual conditions during stance: patients with vestibular deficits, The Journal of Neuroscience, 2(5), pp 536-544.
[6] Soechting, J.F., and Berthoz, A. (1979) Dynamic role of vision in the control of posture in man, Experimental Brain Research, 36, pp 551-561.
[7] Vidal, P.P., Bethoz, A., and Millanvoye, M. (1982) Difference between eye closure and visual stabilization in the control of posture in man, Aviation, Space, and Environmental Medicine, 53(2), pp. 166-170.
[8] Shuman, D. (1981a) Preliminary design of a computerized acoustic position sensor for body sway measurements, unpublished paper, University of Florida, Gainesville, FL.
[9] Shuman, D. (1981b) An acoustic positions sensor for body sway measurements, unpublished paper, University of Florida, Gainesville, FL.
[10] White, K.D., Post, R.B. and Leibowitz, H.W. (1980) Saccadic eye movements and body sway, Science, 208, pp. 621-623.
[11] Lestienne, F., Soechting, J., and Berthoz, A. (1977) Postural readjustments induced by linear motion of visual scenes, Experimental Brain Research, 28, pp. 363-384.

Trends in Ergonomics/Human Factors V
F. Aghazadeh (Editor)

THE EFFECTS OF DIFFERING LEVELS OF DEMAND ON THE PERFORMANCE OF EXECUTIVE WORK

Paul R. MCCRIGHT

Department of Industrial Engineering
Kansas State University
Manhattan, Kansas USA

An important question facing many managers is how to improve the productivity of office workers. This study examines the effects of variations in job demands on the performance of subjects working on administrative tasks typical of executive jobs. Performance was measured using two outcomes of the work: productivity and quality. The study showed a significant positive relationship between job demand and both productivity and quality. These results indicate that managers and designers of work can positively influence both the quantity and quality of work performed by increasing the demands placed on the worker.

1. INTRODUCTION

As the American economy continues to move from an agrarian and manufacturing-based economy to an information-based economy, the performance of executive and professional workers becomes more crucial to the overall effectiveness of the nation's business community. Measuring the performance of so-called knowledge workers is generally much more difficult than measuring the performance of most production workers.

As American businesses improve the performance of their knowledge workers, the nation will become more competitive. This study investigates the effects of demand on the productivity and quality of the work of knowledge workers. Very few studies have examined the relationship between characteristics of a job (such as demand) and performance. Even fewer have specifically looked at the performance of knowledge workers.

Hackman and Oldham [1] developed an important model which suggests that characteristics of an individual's job can influence the individual's performance on the job and his/her personal responses to the job. This model, known as the job characteristics model, links five job characteristics (skill variety, task identity, task significance, autonomy, and feedback) to satisfaction, motivation, and performance. They developed a standard questionnaire to evaluate both satisfaction and motivation. Performance measures are more job-specific and therefore more difficult to develop.

In a review of thirteen studies which related task characteristics to performance, Griffin, Welsh, and Moorhead [2] found that the performance

---

This study was supported in part by a grant from the National Institute of Mental Health, Washington, D.C.

measures which had been used were "at best only moderately valid and meaningful and at worst potentially invalid and meaningless." (P. 662) No two studies in their sample evaluated employee performance in exactly the same way, making comparisons difficult.

I reviewed 20 studies using the job characteristics model and found that the concepts used as dependent (outcome) variables in these studies are not highly standardized. The most commonly used were general satisfaction, internal work motivation, supervisory satisfaction, growth satisfaction, performance (or productivity), and job satisfaction. My review of the correlations between the independent and dependent variables, where provided in these studies, indicates that growth satisfaction is easiest to predict and performance (productivity) is the most difficult.

Other researchers have proposed alterations or extensions of the basic job characteristics model. Karasek [3] suggested that job demand should be included as a job characteristic. He defined job demand to be an aggregate of the stresses which arise from workload pressures. Related to Karasek's job demand characteristic is Maher's [4] job content. He performed a series of laboratory studies which showed that higher performance is related to higher job content and lower performance is related to lower job content. Very little research has studied the effects of job characteristics on productivity or quality.

Certainly, any student of job design would agree that a valid area for future research is examining how the design of work influences these seldom studied, but vitally important, outcomes to the work experience. While the research presented herein is not the only research needed in this area, this project does attempt to shed some light on these relationships.

2. DEFINITION OF VARIABLES

Karasek [3] defined job demands to be stress sources. Kasl [5] found that students of work stress frequently equate conflict, ambiguity, and overload with role stressors. Rizzo, House, and Lirtzman [6] defined role ambiguity as the "lack of the necessary information available to a given organizational position." They considered role conflict to be incongruence or incompatible requirements of a person's role. According to Karasek [3 p.287], "Overload is usually defined as occurring when the environmental situation poses demands which exceed the individual's capabilities to meet them."

Job demand can be considered to be some combination of the concepts of role ambiguity, role conflict, and role overload. A high level of job demand exists when the worker is expected to do more work than can be reasonably accomplished with the resources (including ability, time, equipment, and co-workers) available. Job demand is a global concept which encompasses a number of specific constraints, deficiencies, and expectations. Therefore, the formal definition used in this research is

> Job demand is the global set of expectations and constraints imposed on the worker within the work context.

Thus, according to this definition, the amount of work to be done, the quality standard, and deadlines would be expectations. Role ambiguities and conflicts, time limits, resource deficiencies, and other problems would be constraints.

As was noted previously, few performance variables have been studied frequently. When they have been studied, these variables have generally been subjectively or coarsely measured. Productivity is a performance outcome of extreme importance to work organizations and is of particular interest to industrial engineers. In this study, productivity is a measure only of the quantity of assigned work completed in a given period of time.

Quality of work output has not been the subject of a controlled study by researchers using the job characteristics model. Therefore, no conclusive evidence can be cited for the effects of various job characteristics on the quality of output produced by a worker. Quality is an aspect of performance which is of vital importance to managers and industrial engineers. In this study, quality is considered to be the thoroughness with which each task is done. This outcome variable has been included in this study because of the lack of understanding of the relationship between job characteristics and quality.

3. EFFECTS OF DEMAND ON PRODUCTIVITY

A number of researchers have examined the relationship between demand and productivity. In a study of owner-managers of hurricane damaged small businesses, Anderson [7] found that perceived stress and organizational performance had a curvilinear relationship similar to an inverted-U. Ivancevich and Mattison [8] imply an inverted-U relationship when they assert that workers perform best when demand is moderate and that performance may suffer from either too much or too little demand. In a 1970 study of little league baseball players, McGrath [9] found "an inverted-U relation between demand and absolute level of performance." (P. 379)

In this study, demand is a function of work assigned to the subject since the assignment of a task implies the expectation that the task will be completed. The need to establish face validity required the low demand working condition to be sufficiently high to appear reasonable for a business situation. The level of the high demand working condition was also somewhat constrained by the need to appear reasonable to subjects. What this study actually tests is the effects of lower and higher amounts of job demand across a normal business range. Therefore, although an inverted-U relationship is expected across the entire range of possible demand levels, this research does not explore either extreme. Over the range of demand created in this experiment, productivity is expected to increase as demand increases.

The productivity hypothesis is based on (1) the idea that most workers will try to accomplish what is expected of them; therefore, they will do more work when more is expected of them, and (2) higher levels of demand create more stimulating jobs leading to higher states of arousal and higher levels of performance. These ideas lead to hypothesis H1 below.

> H1. Increased levels of Job Demand tend to produce increased levels of Productivity within typical levels of demand.

4. EFFECTS OF DEMAND ON QUALITY

I have been unable to find any studies of the effects of demand on quality of work output. At low levels of job demand, increases in demand may result in increases in productivity with little effect on quality because the worker could devote time and attention to producing an acceptable level of quality. However, as the time required to do more work and manage stress increases, the time available to concentrate on quality is reduced. Thus, a point would occur in any job where attempts to increase productivity to meet increasing demand would result in an inability to devote sufficient time to maintain quality. At this point, quality would begin to decrease with increases in demand. What this suggests is a relationship between demand and quality resembling an inverted-J.

Very low levels of demand may lead to boredom on the part of the worker. As the worker becomes bored, he/she may become less attentive to the work and, in fact, quality may suffer. If this effect were strong, the expected relationship between demand and quality would resemble an inverted-U more than an inverted-J.

Since the levels of demand achieved in this study are not expected to be either very low or very high, it is unlikely that a boredom effect will be felt. Because this study tests only two levels of demand, the results cannot show more than a straight line relationship. If the high demand condition is sufficiently high, the degradation effects of high demand may be seen. The following hypothesis was tested in this study:

> H2. Increased levels in Job Demand tend to produce decreased levels of Quality.

5. OVERVIEW OF RESEARCH METHODOLOGY

This research used a laboratory simulation to test the hypothesized relationships between job demand and two outcome variables of interest: productivity and quality. The experiment was a simulation of executive work using the in-basket technique. Subjects were divided into groups working under differing levels of job demand.

5.1 Subjects

The subjects in the experiment were 66 students in a graduate level engineering course in management of organizations. Four subjects were removed from the final sample because they reported difficulty reading and understanding English. Cells were balanced on the basis of gender and nationality. All subjects participated voluntarily in the experiment.

5.2 The Experiment

An executive in-basket simulation was developed for the experiment. It contained a variety of tasks constituting a single "job." This design involved complex tasks of a general managerial nature, provided a high face validity, and allowed the manipulation of job demand while leaving the tasks essentially the same. Measurements for productivity and quality were also developed.

Nine executive-level tasks were developed for the in-basket. These included a budgeting exercise, a hiring exercise, and other tasks of varying importance. In order to frame the tasks composing the in-basket, create an atmosphere of believability, and provide background information to subjects, a scenario was prepared. The scenario describes a state-wide bank, a rural region in the state, and conditions at a particular branch in this region. The simulated job is the manager of this branch. Subjects were told it is their first day as the branch manager and that they must prepare for a meeting with their supervisor. The meeting was scheduled for one hour and fifteen minutes after the starting time of the experimental session. This limits preparation time for the meeting to the allotted time of the work session. The in-basket contains an agenda for the meeting, which lists discussion topics related to some of the materials included in the in-basket.

Job demand conditions were manipulated primarily through the agenda and a message received during the experimental session. Agendas for the low demand condition contain four agenda items and request the subject to prepare "ideas on each agenda item." Agendas for the high demand condition contain these same four items, one additional task, and request the subject to prepare a "proposed course of action" on the agenda items. Agendas for the high demand condition also state that the subject's second-level supervisor has asked to attend this meeting.

After 30 minutes of work, all subjects were interrupted by the experimenter, who gave them a telephone message. In the low demand condition, the meeting was postponed by one-half hour. In the high demand condition, an agenda item was added to the agenda.

5.3 Scoring

Productivity was measured using a weighting scheme. Persons who participated in the experimental pre-test were later asked to rank the nine tasks in order of the amount of time needed to do a complete job on each task. They were then asked to assign a percentage of the total time to each task (with a total of 100 percent). These percentage assignments were then averaged to determine a percentage weighting to be assigned to each task. The reliability coefficient (alpha) was .92 for this measure.

Subjects' output was reviewed to determine which tasks were completed by the subjects during the work period. Each subject was "credited" with the task weight corresponding to each completed task. The sum of these task weight credits became each subject's individual productivity score.

Three trained raters reviewed the questionnaires and all work output from each of the subjects. All of the raters were intimately acquainted with the experiment and the materials in the in-basket. The consensual technique suggested by Amabile [10] for creativity assessment was used for scoring quality. This technique was used because determining the quality of executive work is, like determining creativity, largely judgemental. The raters reviewed the work output independently and recorded quality scores for each subject. The three scores for each outcome were then averaged to determine an overall score which was the final measurement. Interjudge reliability was .84.

6. RESULTS

Items in a post-experiment questionnaire tested the demand condition. T-tests show that the demand manipulation was correctly perceived by the subjects (with a statistical significance of $p \leq .05$).

Subjects in the high demand condition had a higher average productivity (59.31) than subjects in the low demand condition (40.21). Table 1 shows the results of an analysis of variance (ANOVA) performed on the productivity measure. The main effect of job demand was statistically significant. The results for the productivity variable show that increased demand tends to increase subject productivity, which supports H1.

TABLE 1. ANOVA FOR PRODUCTIVITY RESULTS

Source of Variation	Sum of Squares	DF	Mean Square	F
Job Demand	3681.667	1	3681.667	11.879*
Residual	17365.667	58	299.408	
Totals	21047.333	59	356.734	

* Significant at $p \leq .0005$ (1-tailed).

Quality of output ratings for each subject were averaged to form the quality measure. Subjects in the high demand condition received higher quality ratings (4.12) than subjects in the low demand condition (4.62).

TABLE 2. ANOVA FOR QUALITY RESULTS

Source of Variation	Sum of Squares	DF	Mean Square	F
Job Demand	3.852	1	3.852	6.517*
Residual	34.328	60	0.572	
Totals	38.157	61	0.626	

* Significant at $p \leq .01$ (1-tailed).

Table 2 shows the results of an ANOVA performed on the quality measure. These results provide strong support for a relationship between demand and quality; however, the relationship supported by these results is <u>opposite</u> to the relationship of H2 (increased demand leads to decreased quality). These data show that increased job demand leads to increased levels of quality.

7. DISCUSSION OF RESULTS

The results of this study provide strong support for hypothesis H1, which states that increased levels of job demand tend to produce increased levels of productivity within typical levels of demand. This result is compatible with results of previous studies, notably Anderson [7] and Ivancevich and Mattison [8].

Discussions with subjects after completion of their work periods revealed that subjects in both demand conditions found the experiment stimulating. However, subjects in the high demand condition frequently expressed more enthusiasm for the experiment with such comments as "I wanted to keep working on the tasks" and "I'd really love to know what happens in the meeting." These and other remarks provide anecdotal evidence that subjects working in the high demand condition experienced a higher level of arousal during the experiment. Although most subjects appeared somewhat tired after the experiment, many high demand subjects nevertheless appeared quite alert.

My conclusion is that subjects in the high demand condition received more stimulation and experienced a higher level of arousal than subjects working in the low demand condition. This led to more involvement in the scenario and a greater intrinsic motivation to accomplish the tasks set before them. This high level of motivation resulted in more commitment on the part of the subjects as they put forth more effort to master the situation. Because they were more interested and involved, they accomplished more and this higher level of accomplishment was reflected in their productivity scores.

In addition to a greater involvement in the scenario, subjects in the high demand condition may have been responding to the greater expectations implicit in this condition. The agendas for the meeting listed topics for discussion. Subjects were expected to assume these topics were tasks expected to be completed prior to the meeting. Subjects in the high demand condition were given more tasks to accomplish than subjects in the low demand condition. Thus, subjects faced with greater expectations may work harder in order to perform up to the level expected of them. Therefore, the two explanations for the positive effect of demand on productivity are (1) that subjects experiencing higher demand were more aroused and more involved in the exercise and (2) that subjects work harder when they perceive the level of expectations to be higher.

Hypothesis H2 stated that increased levels in job demand tend to produce decreased levels of quality. The results of this study show that increases in job demand do not produce decreases in quality. Instead, increases in job demand produce <u>increases</u> in the quality of output. H2 was based on a hypothesized inverted-J relationship between demand and quality. The results of this study do not support such a relationship.

Since productivity (the quantity of output) is closely related to the quality of output both conceptually and empirically (correlated at .46 in this study), perhaps the same psychological forces are at work in the quality variable as were described for the productivity variable. The similarities between these two variables lead me to suspect the results to be similar for both productivity and quality. According to this explanation, subjects in the high demand condition should be more stimulated by the situation and they should react with more commitment and intrinsic motivation. Higher levels of commitment and intrinsic motivation would be expected to lead to a greater desire to produce high quality output. This desire for high quality would lead to the subject increasing his/her effort to produce high quality. Such a causal chain would explain the results of this study.

8. CONCLUSION

The knowledge gained about productivity and quality of work output make this study one of practical importance since these two outcomes are universally important in work settings and have seldom been studied. The conclusions about the effects of demand on these outcomes may be valuable to managers and practitioners of job design. Thus managers who wish to improve the quantity and quality of work accomplished by their executive (and possibly administrative) employees may wish to increase the demands placed on these employees as long as these demands are perceived as reasonable by the employee.

REFERENCES

[1] Hackman, J. R. and Oldham, G. R., Journal of Applied Psychology 16 (1976) 250.
[2] Griffin, R. W., Welsh, A. and Moorhead, G., Academy of Management Review 6 (1981) 655.
[3] Karasek, R. A., Jr., Administrative Science Quarterly 24 (1979) 285.
[4] Maher, J. R., New Perspectives in Job Enrichment (Van Nostrand, New York, 1971).
[5] Kasl, S. V., Epidemiological Contributions to the Study of Work Stress, in: Cooper, C. L. and Payne, R., (eds.), Stress at Work (Wiley and Sons, New York, 1978).
[6] Rizzo, J., House, R. J. and Lirtzman, S. I., Administrative Science Quarterly 15 (1970).
[7] Anderson, C. R., Journal of Applied Psychology 61 (1976) 30.
[8] Ivancevich, J. M. and Mattison, M. T., Stress and Performance, in: Steers, R. M. and Porter, L. W., (eds.), Motivation and Work Behavior (McGraw-Hill, San Francisco, 1983) pp. 375-384.
[9] McGrath, J. E., Social and Psychological Factors in Stress (Holt, Rinehart, and Winston, New York, 1970).
[10] Amabile, T. M., The Social Psychology of Creativity (Springer-Verlag, New York, 1983).

Trends in Ergonomics/Human Factors V
F. Aghazadeh (Editor)

AGING AND THE PERFORMANCE OF COMPUTER-INTERACTIVE TASKS: JOB DESIGN AND STRESS POTENTIAL

Sara J. Czaja and Joseph Sharit

Department of Industrial Engineering
State University of New York at Buffalo
Amherst, New York, 14260

1. INTRODUCTION

Issues surrounding aging and work performance are receiving increased attention among researchers and policy makers. This interest is partially due to the decline in work activity among older adults which creates a potential for a marked increase in the economic dependency burden. It is also due to the significant growth in the number of middle-aged people in the labor force. These trends create a need to understand how age-related changes in function impact on the performance of jobs and also which type of jobs are most suitable for older adults given that there are functional changes which occur with age.

When addressing these issues, one important question to consider is how the influx of computer technology into work settings impacts on the work life of the elderly. This is important because computer and communication technologies are increasingly being used in most work settings, and these technologies change the nature of work. Overall, there is a shift in emphasis in job demands which is characterized by a reduction in physical demands and an increase in information-processing requirements. In this sense, technology promotes employment opportunities for older people; declines in health and physical capability are a common reason for early retirement [1]. Also, computer technology makes work at home a more likely option. However, technology often increases the pacing, workload, and, in some cases, information-processing demands of tasks. This may generate stress, especially for older workers, as there are functional changes such as declines in response speed which occur with age [2].

Most research has demonstrated that pacing and lack of control over workload are major contributors to job stress [3]. The potential consequences of job stress which include physical as well as psychological outcomes are also well documented. However, to date research examining the impact of work and technology on the health of the other worker has been minimal. Weg postulated [4] that increased age results in an "exhaustion of adaptive capacities" and a decline in physical capacities. Thus the potential for job stress is a critical issue with respect to older workers.

This paper will discuss the potential for the manifestation of stress for older people performing computer-interactive tasks. The approach taken will be based on an analysis of the information-processing demands associated with this type of work in relation to age-related changes in cognitive performance. Finally, a methodological issue surrounding the analysis of stress among older people will be discussed. This is a critical component of any effort directed towards aging and job design as there are age-related differences in responses to traditional stress measures, such as arousal, which need to be understood in order to ensure that the demands associated with tasks are appropriately evaluated.

2. A FRAMEWORK FOR STRESS ANALYSIS

The ability to effectively design computer-interactive tasks for the elderly requires answers to questions concerning how these tasks might adversely affect older persons. Most current models and theories of stress do not assume a strictly response-based approach that regards stress as a dependent variable or a strictly stimulus-based approach which views stress as an independent variable. Instead, the emphasis is on the discrepancy between demand and capability as the basis for the existence of stress, the expression of which is treated as a psychophysiological response [5]. This approach is consistent with McGrath [6] who views stress as resulting from an imbalance between demand and capacity. Given that computer-interactive tasks are largely characterized by their information-processing requirements, assessing the potential for stress requires an understanding of the relationship between the cognitive demands of the task and the information-processing requirements of the human.

The proposed framework for analyzing the existence of stress in the elderly for computer-interactive tasks is consistent with House's [7] paradigm of stress research where individual, situational, and response factors are all considered. Using this paradigm for stress analysis, the existence for stress can be viewed as deriving from a mapping of various sources of stress to various responses. The sources would consist primarily of aging as the individual factor and the computer-interactive task as the situation factor. The latter can be divided into the intrinsic and perceived demands of the task. The salient characteristics of each of these factors must be understood in order to identify an appropriate methodology for analyzing the potential for stress in the task situation.

2.1 Task Characteristics

It is proposed that two important ways computers change the nature of work with respect to potential for inducing stress is in terms of pacing requirements and task complexity. Since computers have the capability of controlling the rate of information flow, they allow tasks which were traditionally unpaced to become machine-paced. For some classes of tasks such as data entry, computers impose tight external control over information processing rate. Rigid pacing of a task by a computer is often cited as a factor contributing to stress, e.g., [8]. Although no formal studies have been performed specifically examining this factor, especially in relation to older workers, the literature regarding aging and information processing suggests that paced computer-interactive tasks would be stressful for older persons. There is unambiguous evidence that speed of information processing declines with age [2] . The inherent capabilities of computers and computer networks also affects task complexity. For some tasks such as inventory management, task complexity increases because computers allow extensive and varied information to be rapidly accessed and integrated. This places greater demands on memory and decision-making processes. Information-processing requirements also increase because information retrieval and manipulation requires complicated management of databases. For other tasks such as data entry, information-processing requirements are simplified because the computer imposes tight external control over the type of information presented to the user. In these instances, information-processing requirements are minimized. Evidence, e.g., [9] has been presented which suggests that too much complexity as well as too little complexity may produce stress outcomes. Therefore an important job design issue with respect to computer tasks, especially for older workers given changes in information-processing capacity, is degree of decisional complexity.

2.2 Individual Characteristics

As suggested, a thorough understanding of the potential for stress among older workers must also be based on an analysis of age-related changes in function which are salient to the task situation. One of the most reliable findings regarding the aging process which is critical with respect to computer-interactive tasks is a general slowing in processing as age increases [2]. The size of the age discrepancy in response speed is variable and influenced by factors such as health status, task complexity, response requirements, and practice. However, while practice can reduce the size of the age decrement in response speed, it is unlikely to completely eliminate it. This suggests that computer-interactive tasks such as data entry which are rigidly paced may be inappropriate for older workers.

Other age-related changes in function which have implications for computer-interactive tasks include changes in visual information processing, memory,and decision making. For example, researchers [10] have demonstrated that aging is associated with slower processing of visual information and that older people need more time to read information from displays. Also, older adults have difficulty processing complex and confusing stimuli and allocating attention to task-relevant information. With respect to memory, two age-related changes in memory which are relevant to computer tasks are shortening of the immediate memory span and difficulty in retrieving information from long-term memory (LTM). Considering decision making, although the evidence is less clear, it can be assumed that decision-making and problem-solving abilities decline with age, given the changes in information-processing requirements that support these abilities.

Finally, age-related changes in coping strategies need to be considered. Compensatory actions on the part of older workers typically involve taking more time before making a decision, increased monitoring of action, and exercising increased caution. These types of strategies may not be tolerated for computer-interactive tasks, especially those characterized by tight external control.

Overall the data suggest that changes in information-processing capabilities make computer-interactive tasks potentially stressful for older workers. Thus, there exists a need to develop methodologies which evaluate stress potential.

3. STRESS MEASUREMENT AND AGING

The selection of measures for evaluating stress among older persons needs to take into account: 1) the concern for health in the elderly, 2) the potential for sensitivity of a given measure to stress in the elderly, 3) the need for a measurement system which can reflect momentary as well as prolonged responses to stress, and 4) the requirement that collection of the measures do not involve a significant amount of stress and do not restrict the person from performing the task or become affected by task performance per se.

Also, when interpreting the data, it is important to understand how the aging process affects responses to these measures. For example, the psychophysiological evidence has not been clear with respect to the arousal state associated with advanced age. However, there appears to be consistent data on the reduced ability with advancing age for physiological responses to return to baseline. This suggests that changes in mean levels of response between baseline, task, and recovery periods need to be analyzed in order to better interpret the response of the elderly as a function of hypothesized computer-interactive task stress. Also, to deal with the possibilities of decreased range of response and to better characterize the nature of physiological response by the elderly to potential stress, time-series analysis should be utilized to better focus on patterns of stress.

Other critical issues with respect to stress measurement include the manner in which the measures are taken and the experimental tasks employed. Findings by Eisdorfer [11] indicate that adaptation to the laboratory environment, especially to the measurement recording systems, is especially critical for older adults. Also, in view of a potential for increased arousal in the elderly during learning and the reduced capability for recovery from heightened arousal, sufficient time must be provided between practice and experimental sessions in order to prevent a carryover of stress. Finally, the types of tasks employed is also a factor. Kausler [12] has indicated that elderly people may respond negatively to artificial features of laboratory tasks. Therefore it is important that experimental tasks are realistic and have high face validity.

4. CONCLUSIONS

In sum, it is suggested that some types of computer-interactive tasks, e.g., those characterized by strict pacing or high decisional complexity, may be especially stressful for older adults. This is an important problem as computer and communication technologies are permeating occupational settings, and the number of older people in the population is increasing. To date, research examining stress potential among older persons for these types of tasks is negligible. Therefore, there exists a need to examine the ability of older adults to perform computer-interactive tasks as well as the impact of these types of tasks on the work life of the elderly.

REFERENCES

[1] Taeuber, C. (1984), Older Workers: Force of the Future? in: P. K. Robinson, J. Livingston, J.E. Birren (Eds.), Aging and Technological Advances, New York: Plenum Press, 75-88.

[2] Salthouse, T.A. (1985), Speed of Behavior and its Implications for Recognition, in: J.E. Birren and K. W. Schaie (Eds.), Handbook of the Psychology of Aging, New York: Van Nostrand Reinhold, 400-426.

[3] Cooper, C.L. (1984), Sources of Occupational Stress Among Older Workers, in: P.K. Robinson, J. Livingston, and J. E. Birren (Eds.), Aging and Technological Advances, New York: Plenum Press, 309-220.

[4] Weg, R.B. (1984), Impact of Work and Technology on the Health Status of the Older Worker, in: P. K. Robinson, J. Livingston, J.E. Birren (Eds.), Aging and Technological Advances, New York: Plenum Press, 249-250.

[5] Cox, T., and Mackay, C.J. (1981), A Transactional Approach to Occupational Stress, in: E. N. Corlett and J. Richardson (Eds.), Stress, Work Design, and Productivity, New York: J. Wiley, 91-113.

[6] McGrath, J.E. (1970), A Conceptual Formulation for Research on Stress, in: J. E. McGrath (Ed.), Social and Psychological Factors in Stress, New York: Holt, Rinehart and Winston.
[7] House, J.S., (1974), Occupational Stress and Coronary Heart Disease: A Review and Theoretical Integration, Journal of Health and Social Behavior, 15, 12-27.
[8] Smith, M. J. Cohen, B.G., Stammerjohn, L., and Happ, A. (1981), An Investigation of Health Complaints and Job Stress in Video Display Operations, Human Factors, 23, 387-400.
[9] Caplan, R.D., Cobb, S., French, J.R., Harrison, R.V., and Pinneau, S.R. (1980), Job Demands and Worker Health: Main Effects and Occupation Differences, Ann Arbor: Institute for Social Research, University of Michigan
[10] Cerella, J., Poon, L.W., and Fozard, J.L. (1982), Age and Iconic Read-out, Journal of Gerontology, 37, 197-202.
[11] Eisdorfer, C. (1968), Arousal and Performance: Experiments in Verbal Leaning and a Tentative Theory, in: G. Talland (Ed.), Human Aging and Behavior, New York: Academic Press, 189-216.
[12] Kausler, D.H. (1982), Experimental Psychology and Human Aging, New York: John Wiley and Sons.

Trends in Ergonomics/Human Factors V
F. Aghazadeh (Editor)

THE EFFECTS OF INSTRUCTION ON FINGER STRENGTH MEASUREMENTS: APPLICABILITY OF THE CALDWELL REGIMEN

Valerie J. Berg, Deanna J. Clay, Fadi A. Fathallah, and
Vicki L. Higginbotham

Industrial Ergonomics Laboratory, Human Factors Engineering Center
Department of Industrial Engineering and Operations Research
Virginia Polytechnic Institute and State University
Blacksburg, VA 24061

An experiment was performed to assess the use of the Caldwell Regimen for finger strength measurements. The Caldwell Regimen calls for measurement of static strength using a sustained maximal exertion of four seconds with the average of the first three steady seconds used as the strength measurement, and permits no more than a +/-10% deviation from the average sustained exertion. Maximal pinch strength of thirty subjects was measured using three different techniques: (1) by a sudden maximal contraction, (2) by quickly building to and maintaining a constant maximal exertion, and (3) by slowly (over a period of two seconds) building to and maintaining a constant maximal exertion. The three different techniques of force generation were used with three pinch types: (1) index-thumb pad pinch, (2) three-jaw chuck, and (3) index-thumb lateral pinch. Statistical analysis showed that the different techniques for force generation resulted in different peak pinch strength measurements. Instructions for pinch strength measurements should therefore be precise in indicating the type of force generation desired. As over half of the measurements did not meet the +/-10% criterion for sustained maximal contraction for three seconds, it is recommended that a +/-15% criterion be used with a sustained maximal contraction of two seconds.

1. INTRODUCTION

Most measurements of muscle strength utilize a sudden contraction of maximal amplitude ("peak") to indicate maximal strength. Functional hand force requirements for work or recreational activities normally require a sustained contraction at or below maximal exertion. The peak force measurement is adequate for activities requiring sudden exertion, while a time-integrated score may be superior for tasks requiring a sustained effort.

In the Caldwell Regimen, the subject is instructed to "increase to maximal exertion (without jerk) in about one second and maintain this effort during a four second count" (Caldwell, Chaffin, Dukes-Dubos, Kroemer, Laubach, Snook and Wasserman, 1974). The intention is to attain a continuous maximal strength exertion and to avoid peaks. The Caldwell Regimen requires, (1) a sustained maximal exertion for four seconds, so that three seconds are available for averaging; (2) a transient period of one second before (and after) the exertion; (3) that the mean score of three averaged seconds be used as the strength measurement; and (4) that deviation from the average sustained exertion be no greater than +/-10% during these three seconds.

The research objectives were:

1. Compare maximal forces exerted following three methods of instruction: (a) peak force only (Instruction 1) - a sudden maximal contraction ("peak") force with release of the grip when the subject feels that maximal exertion has been attained, (b) peak force maintained (Instruction 2) - a sudden maximal contraction ("peak") force followed by a sustained MVC for 5.5 seconds, and (c) built-up force maintained (Instruction 3) - gradual increase to maximal exertion over 1 to 2 seconds, followed by a sustained MVC so that the total time for build-up and sustained force is 5.5 seconds.
2. Compare the number of subjects who were able to maintain the force exertion within +/-10% of the average versus those who needed a +/-15% bandwidth.
3. Compare average forces obtained during 2.2 versus 3.3 second intervals of sustained contraction.
4. Determine the effect of instruction on the time required to reach maximal contraction and to initiate sustained force for 2.2 and 3.3 seconds.

2. REVIEW OF LITERATURE

Maximal contraction is influenced by many factors. Maximal exertions in isometric strength testing are reported to be affected by the method of muscle contraction, i.e. force generation (Caldwell and Kroemer, 1973; Kroemer and Howard, 1970; Moudgil and Karpovich, 1969), instructions (Caldwell et al., 1974; Kroemer and Howard, 1970), index of measurement (Kroemer and Howard, 1970), duration of the exertion (Chaffin, 1975; Kroemer and Howard, 1970), subject motivation (Ikai and Steinhaus, 1961; Johnson and Nelson, 1967; Voor, Lloyd and Cole, 1969), time of day (McGarvey, Morrey, Askew, and An, 1984), a loud noise, the subject's own outcry, pharmacologic agents (alcohol, adrenaline and amphetamine) and hypnosis (Ikai and Steinhaus, 1961), knowledge of results (Berger, 1967), and verbal command volume (Johansson, Kent and Shepard, 1983). Other factors cited by Kroemer (1974) to affect motivation (and consequently to affect maximal contraction) include: feedback, ego involvement, noise, incentives, competition, encouragement, spectators, deception by researcher or by subject, and fear of injury (Kroemer, 1974). For this study, each of the applicable areas were considered and controlled according to recommendations revealed in recent literature.

In order to avoid muscle fatigue, strength tests should require force exertions of 10 seconds or less (Kroemer, 1970). Maximal exertion can be maintained for only a brief time before a noticeable decrease in effort is observed (de Vries, 1966; Milner-Brown, Mellenthin and Miller, 1986). The ability to sustain an exertion is dependent on the percentage of maximal force required. An exertion can be sustained longer if the force required is smaller (Caldwell and Smith, 1966; Molbech, 1963; Monod and Scherrer, 1965; Rohmert, 1961).

Rest periods of two minutes for test sessions consisting of fifteen measurements have been reported as adequate (Shanne, 1972, reported in Chaffin, 1975). If only a few measurements are taken, Chaffin (1975) states that thirty seconds should be sufficient. In both cases, the evaluator should verbally assess the subject's need for further rest and statistical analysis should be used to check for fatigue effects (Chaffin, 1975).

The American Society of Hand Therapists' recommendation for the standard position for grip and pinch strength measurements states that the individual "should be seated, with his shoulder adducted and neutrally rotated, elbow flexed at 90 degrees, and the forearm and wrist in neutral position" (Fess and Moran, 1981, cited in Mathiowetz, Kashman, Volland, Weber, Dowe and Rogers, 1985). No significant differences in grip strength have been reported with the wrist in 0, 15, and 30 degrees extension (Pryce, 1980; Kraft and Detels, 1972). Thumb-finger pad (or palmar) pinch with flexed fingers has been reported 92% higher than when the remaining fingers were extended.

The procedures for finger strength testing are not standardized. Instructions, subject position, strength index selection and the regimen for data collection should be standardized to allow comparison. A study by Mathiowetz, Kashman, Volland, Weber, Dowe and Rogers (1985) is the only study on hand and finger strength to report the use of standardized verbal instructions. A review of standardization and protocol for hand strength measurements by Smith and Benge (1985) revealed that in clinical practice and hand rehabilitation, no standardized protocols exist for position, rest periods, verbal instructions, or number of trials. It appears that the same lack of standardization also is prevalent in the industrial setting for finger strength measurements. Specifically, finger strength measurements have not been evaluated in terms of the effects of instruction, index of performance, duration of exertion, or subject position. These evaluations are necessary prerequisites to the formalization of standardized procedures. Standardized procedures should improve the techniques of evaluation, comparison and prediction of the capabilities and limitations of workers.

3. EXPERIMENT

3.1 Subjects

Thirty voluntary untrained subjects (19 males and 11 females) with no test-related physical disabilities participated. The mean age of the subjects was 24 years (range: 20 to 42) with a standard deviation of 4.4 years.

3.2 Testing

Testing sessions lasted 20 to 30 minutes and were conducted in a controlled laboratory environment. The same experimenter was used for all subjects. The experimenter gave instructions with the same voice inflection and intonation for all subjects in order to standardize motivational factors. The experimenter read the same introduction to each subject. All subjects completed a physical fitness questionnaire, and the following anthropometric measurements were collected: height, weight, hand length, finger length, and hand breadth. Hand preference was also recorded (27 subjects were right-hand dominant and 3 were left dominant), and each subject performed strength exertions with the dominant hand.

The experimenter read detailed instructions while demonstrating each pinch (see Figure 1):

1. the thumb-index pad pinch, or palmar pinch (force generated between the pad of the thumb and the pad of the index finger),
2. the thumb-index and middle finger pad pinch, or three-jaw chuck pinch (force generated between the pad of the thumb and the pads of the index and middle middle fingers together),
3. the side, lateral, or key pinch (force generated between the pad of the thumb and the radial side of the middle phalanx of the index finger).

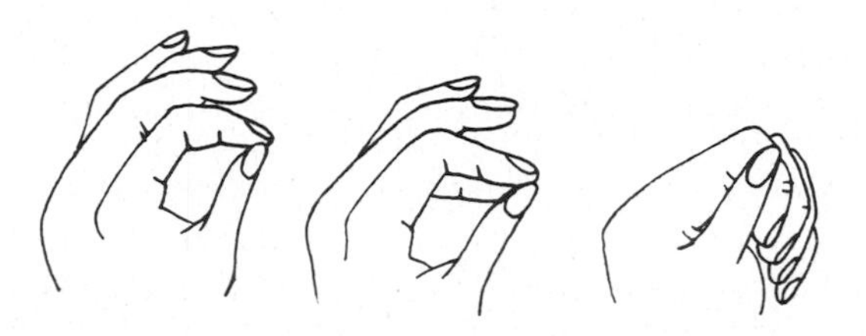

Figure 1. Pinches Tested

Note: The remaining fingers were flexed during all pinches.

Next the subject received an explanation of the three methods of force generation: peak force only, peak force maintained, and built-up force maintained (for details see 1 a, b, c in the research objectives).

The subject was seated in a standardized testing position with the shoulder adducted, elbow flexed at 90 degrees, and forearm and wrist in neutral, as pictured in Figure 2. The subject was allowed to choose the most comfortable wrist angle within a range of 30 degrees dorsiflexion to 30 degrees palmar flexion. The wrist angle chosen was recorded and maintained throughout testing. All wrist angles chosen were between 20 degrees and 30 degrees dorsiflexion (27 of 30 subjects found 30 degrees to be most comfortable).

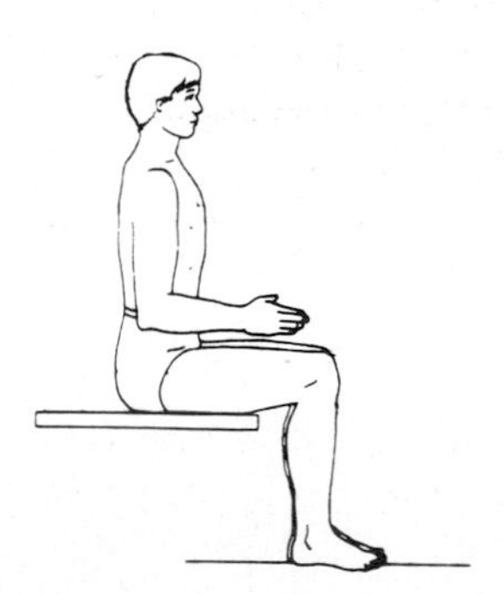

Figure 2. Standardized Testing Position

Rest periods of 30 seconds to 1 minute were provided between trials (exact times depended on computer processing and instruction time). Each subject was informed that additional rest would be provided if requested. No requests were made, indicating that there was no subjective feeling of fatigue.

For each force exertion, the subject placed the fingers comfortably on the gauge, which was held unobtrusively by the experimenter to avoid slippage or dropping of the gauge. The subject's posture and wrist angle were checked before each pinch to insure standardization. The experimenter read the brief standardized instruction before each pinch was performed.

Comments from the subjects were welcomed during the study. There were no complaints regarding the comfort of testing, indicating that subject discomfort did not affect the results. Three subjects repeated one measurement each because the base criterion (+/-15% of the running average over 2.2 seconds) was not attained, but no subjective discomfort or fatigue resulted from these additional measurements.

3.3 Apparatus

Finger pinch forces were collected using a specially designed pinch gauge. The pinch gauge (see Figure 3) consists of two identical aluminum bars held together at their bases with six screws.

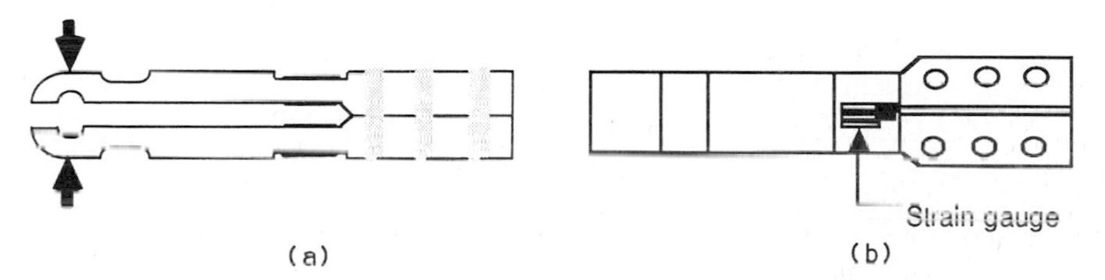

Figure 3. Pinch Gauge Drawing: (a) Side View, (b) Top View

Forces applied at the areas specified in Figure 3a cause a voltage change in four electric strain gauges that are glued on the pinch gauge. Bending of the levers, though less than 1 mm at the finger areas, is proportional to the force applied and changes the impedances in the strain gauges. The voltage signal from the strain gauges (wired in a full Wheatstone Bridge arrangement), is filtered, amplified and sent to the analog-to-digital converter card (A/D) of a computer. A software program was written in BASIC to access the A/D output, and to collect, analyze and store the data.

4. RESULTS

For all 270 accepted trials (30 subjects, 9 trials each), the peak force values and the points in time when the peak occurred were recorded. For trials under instructions 2 and 3, values were also recorded for the mean forces exerted over the 2.2 s interval and the 3.3 s interval. It was also recorded whether or not the mean force values were within the +/-10% or +/-15% bandwidth criteria.

The peak values found in this study are summarized in Table 1. Other studies measuring pinch strength were reported by Hook and Stanley (1986), Jain et al. (1985), and Mathiowetz, Kashman, Volland, Weber, Dowe, and Rogers (1985). Direct comparison of these studies to the present study are difficult due to differences in equipment, pinch type, type of force generation, subject sample, instructions, and subject motivation. The study by Mathiowetz et al. (1985) most closely resembles our experiment. For the three-jaw chuck pinch (pinch type 2) under instruction 2 the present results compared to the Mathiowetz data are shown in Table 2. The values found in the present study are lower than those found in the Mathiowetz study. These differences could be related to discrepancies in any of the above mentioned experimental parameters.

Table 1. Average Strength (in Newtons) Exerted By all Subjects for Each Instruction and Pinch Type

		Males	Females
Instruction	1	86.5	86.3
	2	94.6	65.3
	3	80.2	56.9
Pinch	1	78.2	56.0
	2	84.4	61.1
	3	98.6	67.2

Table 2. Comparison of Strengths (in Newtons) Found in this Study to those Found by Mathiowetz et al. (1985).

Present Study		Mathiowetz et al.	
Males		**Males**	
age 18-30	93.2	age 20-24	118.3
		age 25-29	115.7
Females		**Females**	
age 18-30	65.8	age 20-24	76.5
		age 25-29	78.7

Instruction type had a statistically significant effect on the following dependent measures: peak force value ($F(2,58) = 32.62$, $p < .0001$); time of occurrence of peak force value ($F(2,58) = 82.13$, $p < .0001$); mean force values for the 2.2 s measured interval ($F(1,29) = 4.48$, $.01 < p < .05$) and for the 3.3 s measured interval ($F(1,29) = 4.78$, $.01 < p < .05$). Table 3 gives the results of the Newman-Keuls tests for each measure. Values with different letters (A,B,C) are significantly different at the .05 level.

Table 3. Newman-Keuls Test Results for Instruction

Instruction	Peak Force Value (N)	Time of Occurrence of Peak Force Value(s)	Mean Force Values(N) 2.2 s	3.3 s
1	83.822 A	0.4379 A	-peak only-	
2	77.543 B	0.6795 A	68.609 A	66.624 A
3	71.646 C	2.3012 B	65.254 B	63.213 B

Pinch type also had a statistically significant effect on the dependent measures: peak force value ($F(2,58) = 38.05$, $p < .0001$); time of occurrence of peak force value ($F(2,58) = 3.88$, $.01 < p < .05$); mean force values for the 2.2 s measured interval ($F(2,58) = 37.3$, $p < .0001$) and for the 3.3 s measured interval ($F(2,58) = 39.05$, $p < .0001$). Table 4 gives the results of the Newman-Keuls tests for each measure.

Table 4. Newman-Keuls Test Results for Pinch Type

Pinch	Peak Force Value (N)	Time of Occurrence of Peak Force Value(s)	Mean Force Values (N) 2.2 s	3.3 s
1	70.062 A	1.2573 A	60.267 A	58.322 A
2	75.828 B	1.2116 A	63.474 A	61.998 A
3	87.122 C	0.9498 B	77.053 B	74.435 B

The instruction x pinch interaction had no significant effect on any of the dependent measures.

The difference between the mean force value of the 2.2 s interval (66.89 N) and the 3.3 s interval (64.87 N) was significant, $t = 8.07$, $p < .0005$.

Table 5 shows the percentage of measurements passing the bandwidth criteria (+/-10% and +/-15%) for the 2.2 s and 3.3 s measured intervals.

Table 5. Bandwidth Criteria Comparison

	2.2 s	3.3 s
within +/-10%	34.4%	46.1%
within +/-15%	98.3%	82.2%

5. DISCUSSION

Instructions were found to have a significant effect on the peak force value, on the point in time when the peak occurred, and on the magnitude of the mean force sustained over a given interval. This emphasizes the importance of standardizing instructions which helps not only to maintain reliability within a study, but also allows comparison between studies. In attempting to compare the present study with previously conducted studies, it was uncertain which instructions had been followed in previous studies. Our subjects followed the instructions that were given to them: subjects instructed to peak as quickly as possible peaked earlier and with higher values than when instructed to build up to a peak. Thus it is important to have the instructions clearly define the desired action.

The type of pinch used to generate force was also found to have a significant effect on the peak force, on the point in time when the peak occurred, and on the mean force sustained over an interval. This supports the importance of standardizing the type of pinch used in a study. Pinch types should be selected carefully and should be adhered to throughout a study. Selection of the pinch type should be based on the intended use of the data.

Another factor to consider is the length of the interval during which force will be measured. Most studies test only peak force which is typically greater than a sustained force. A sustained force reflects duration of exertion and is often more applicable to functional requirements. Measurement of a peak or sustained force is dependant on the intended use of the data. For example, data to be used for a real-life application where the sustained force is required should be collected on subjects performing a sustained force.

It a sustained force is measured, its duration is important. It was found that the mean force was higher for the 2.2 s duration than for the 3.3 s duration. The forces measured during the 2.2 s interval were also more stable (able to meet the bandwidth criteria) than those measured during the 3.3 s interval. This suggests changing the requirement of the Caldwell Regimen of finger strength assessment from a sustained interval of 3 s to one of about 2 s.

Another suggested change to the Caldwell Regimen concerns the bandwidth criterion. This study showed that increasing the criterion from +/-10% to +/-15% allowed a greater percentage of the trials to be accepted. In a pilot study, it was found that requiring subjects to sustain a force for 3 s within +/-10% of the mean value resulted in many unacceptable trials which had to be repeated, resulting in subject fatigue.

Recommendations derived from this study for static finger strength measurement are: standardize the instructions; measure a sustained duration of only 2 s; and tolerate measurements within +/-15% of the average sustained force value.

ACKNOWLEDGEMENTS

The authors would like to thank Dr. K. H. E. Kroemer and Ms. M. Susan Hallbeck for their assistance concerning this research.

REFERENCES

References available upon request.

IV

HUMAN–MACHINE INTERACTION

Trends in Ergonomics/Human Factors V
F. Aghazadeh (Editor)

DESIGN OF INTERFACES OF HAND-HELD, TWO-WHEELED DEVICES

H.-J. BULLINGER, W.F. MUNTZINGER, R. ECKERT

Fraunhofer-Institut für Arbeitswirtschaft und Organisation
Stuttgart
F.R.G

Working with hand-held, two-wheeled devices often leads to a high strain for the operator. The strain results mostly from energy-effect related and information-reception related activities of the operator due to work task, enviromental conditions and machine-specific stresses. This contribution is based on research studies and deals with the reduction of the stresses transmitted to the operator by an ergonomic design of the man-machine-interface of hand-held, two-wheeled devices.

1. Introduction

The spectrum of hand-held two-wheeled devices includes numerous machines used for the most varied work tasks, ranging from plowing, cultivating, grading, mowing to snow removal. They have the following features in common :

- drive units of their own (internal combustion engine),
- self-propelled, separate tool drive,
- soil or soil vegetation as their object of work,
- guiding of the machine by the operator walking behind it,
- transmission of guiding and steering forces through handle bars,
- operation of the machine by hand-operated controls.

Particulary in larger-size machines (dead weight above 1500 N), the ergonomic design of the man-machine-interface (MMI) requires special attention, apart from the design ensuring low vibration and low noise, as these machines are mainly used by professionals - and, thus, the man-machine work system constitutes a workstation of its own with up to eight hours per day, dependent of the application involved.

In the course of the technical development of hand-held, two-wheeled machines, the functions of such machines have become more numerous and the associated technology more complex, so that the operator is not only subjected to partly great physiological stresses, but he also has to operate a number of controls, whose operation also results in strains, dependent on the design of these controls. Because such machines are also used along borders of roads with high density of traffic, the safety aspect also plays a decisive role.

This contribution is going to demonstrate approaches to the ergonomic design of MMI´s for hand-held, two-wheeled devices. A mowing machine is used as an example, for which systematic influencing factors and parameters and target criteria for the design of controls will be worked out, taking anthopometric data into account. Recommendations for the design and arrangement of controls will be given and evaluated on the basis of multi-dimensional rating methods. A safety concept for hand-held machines will be demonstrated in addition.

2. Analysis of the man-machine-system

The energy required for performing a work task is drawn by the man-machine work system from man and its work, on the one hand, and from the power supplied by the driving unit, on the other hand. In this process, man´s work is made available to the system in the form of information reception-related activities and energy-effect-related activities. The ergonomic design of hand-held machines aims at minimizing the energy-effect-related activities to be performed by the operator, at the same time fully utilizing the driving power of the machine and ensuring the optimum accomplishment of the work task.

The minimization mentioned above is synonimous with a reduction in the physiological stresses to which the operator is subjected in his work. The total stresses and strains are essentially composed of the following components :

- stresses and strains due to the work task,
- machine-specific stresses and strains,
- environment-specific stresses and strains.

Since almost all energy-effect related components of man´s work enter the system via the MMI, an optimum design of the interface is of correspondingly great importance. This contribution provides more details with a mowing machine as an example, which is regarded as a typical representative of the class of hand-held, two-wheeled machines. These machines are especially used for mowing in heavily cropped terrain, heavily sloped and hilly terrain and road borders. The larger-size types of these machines are now available in variants equipped with hydraulic drives, i.e. hydraulic pump and one-wheel drive through two hydraulic motors, and provided with mechanical or hydraulic power-assisted guides, as distinct from the earlier models with solid drives. The use of power-assisted guides permits steering and maneuvering without major physical exertion, resulting in a reduction of the component of high-strain activities. Figure 1 shows the MMI of a mowing-machine.

3. Anthopometrical data for the design of interfaces

Since the machines treated in this contribution cause heavy mechanical vibrations in vertical and transverse directions during operation, it should be made

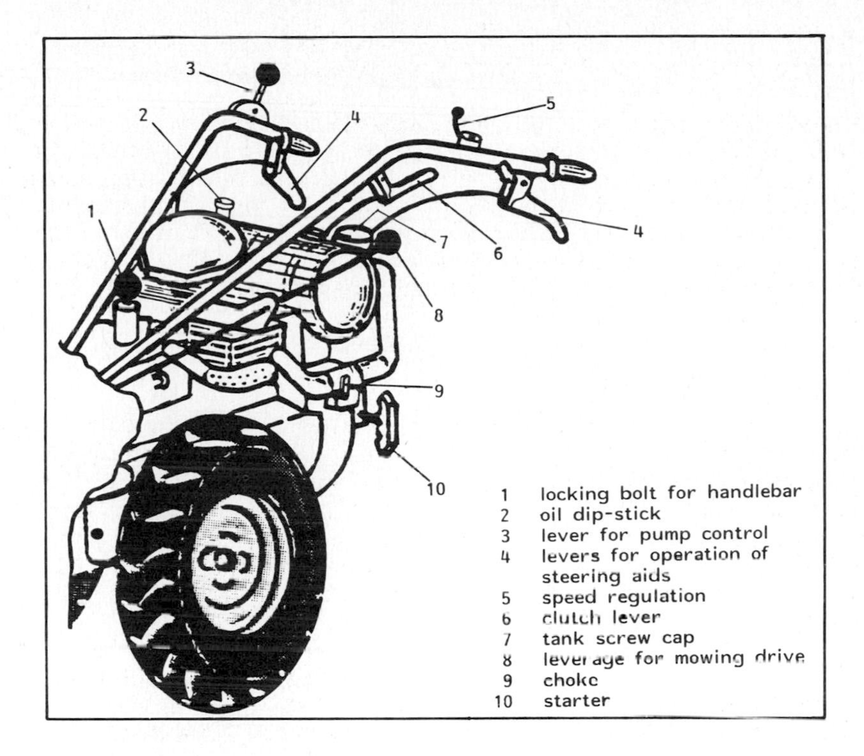

Figure 1
Man-Machine-Interface of a Mowing Machine

sure that a forced, unfavourable posture of the operator will unnecessarily intensify the transmission of vibrations through the handle bar system to the operator. To avoid unfavourable arm postures, the handle height and the angle of the handle axis to the horizontal require particular attention. The machines discussed here are exclusively operated by male operators in the professional field, so that the anthopometrical data of females can be neglected. Based on anthopometric data of the 5th and 95 th male percentile [1] and former studies [2], Figure 2 shows dimensions and angles for the design of handle bars.

Also of great importance is not to exeed the maximum forces to be exerted for operating the controls, see Figure 3. In this connection, particularly the following factors play an important role [3] :

- direction of operation,
- frequency of operation,
- type of clasping the control by hand,
- degree of operator´s practice,
- age and sex of operator,
- path needed to reach the control.

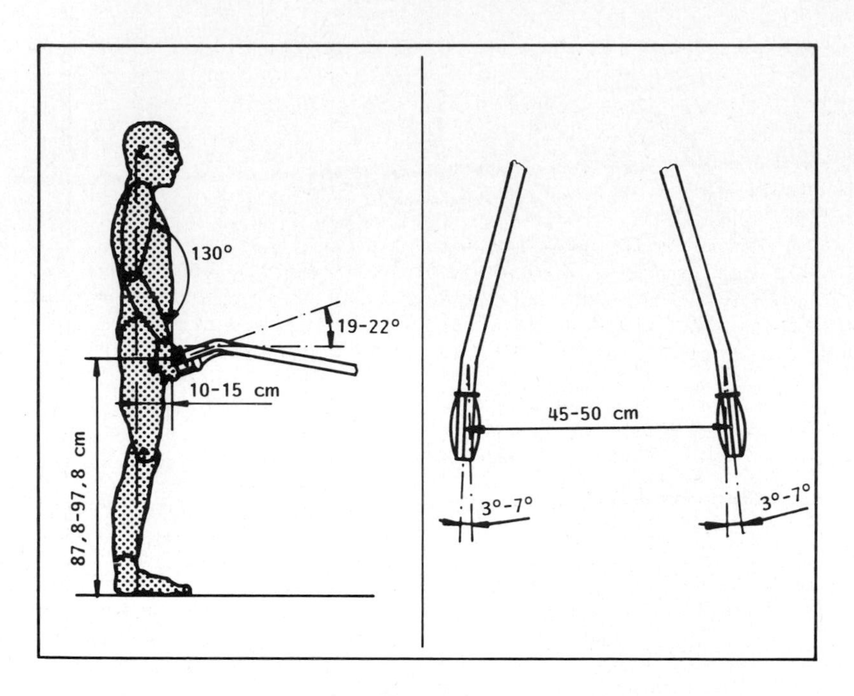

Figure 2
Recommended Dimensions and Angles for the Design of Handle Bars

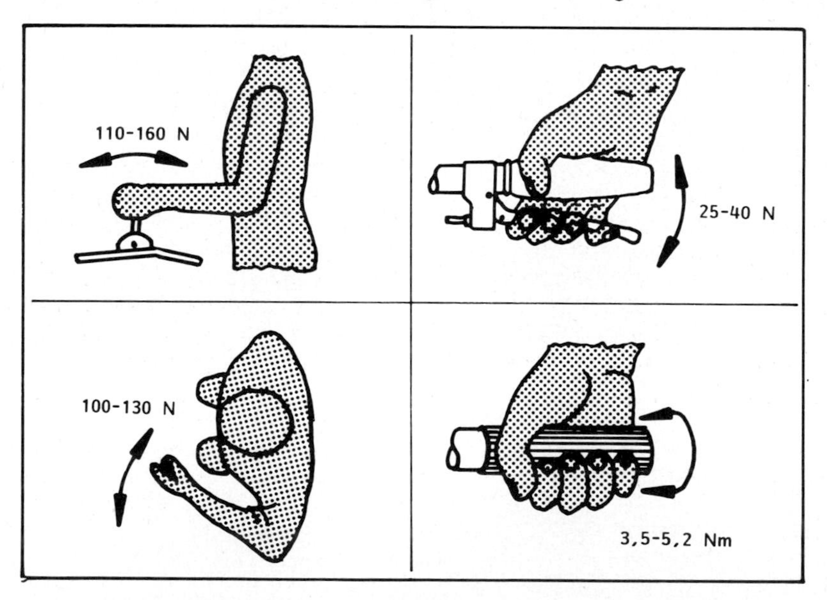

Figure 3
Tolerable Forces and Torque for Operating of Controls

4. Arrangement of the controls

Regarding the arrangement of the controls, it is reasonable to define the hand´s and arm´s reach within which the control to be operated is located. In this connection, the maximum reach of the hand-arm system is decisive, with the position of the operator relative to the machine unchanged. The arrangement of the controls has to be carried on the basis of a score. Therefore, all controls will be rated with a points system, using the frequency of operation and the saftey aspects to the case as criteria. The individual scores thus obtained will be multiplied by a weighting factor and then added. The assignment of the controls to a given reach will then be made on the basis of the amount of the total score as shown in the following example.

The total score calculated in this manner is not only helpful for the arrangement of the controls, but at the same time it is also an indication as to the importance of the controls of a MMI and, therefore, it should be taken into account in the structural design. The larger the total score, the more important is that the control can be easily operated and quickly reached and that it requires low operating forces and operating paths, resp. angles as far as twist-grips are concerned.

Criterion	frequency of operation	safety-related importance	Points
Range	lt. 5 times/per period	no importance	1
	lt. 10 times/per period	low	2
	lt. 20 times/per period	medium	3
	lt. 30 times/per period	high	4
	gt. 30 times/per period	very high	5
Weight.	0.45	0. 55	

Nr. of control (ref. Fig.1)	1	2	3	4	5	6	7	8	9	10
Points Crit. 1	2	1	4	5	3	4	1	3	1	2
Points Crit. 2	3	1	5	5	4	4	1	3	1	1
Σ	0.51	0.20	0.91	1.00	0.71	0.80	0.20	0.60	0.20	0.29

$\Sigma_i = 1/4\,(C_1 {}^* 0.45 + C_2 {}^* 0.55)$ work period = 4 hours

If Σ_i is less than 0.65 arrangement of the control within "indirect zone" is allowable, if Σ_i is greater than 0.65 arrangement of control within "direct zone" is required (refer to Figure 4).

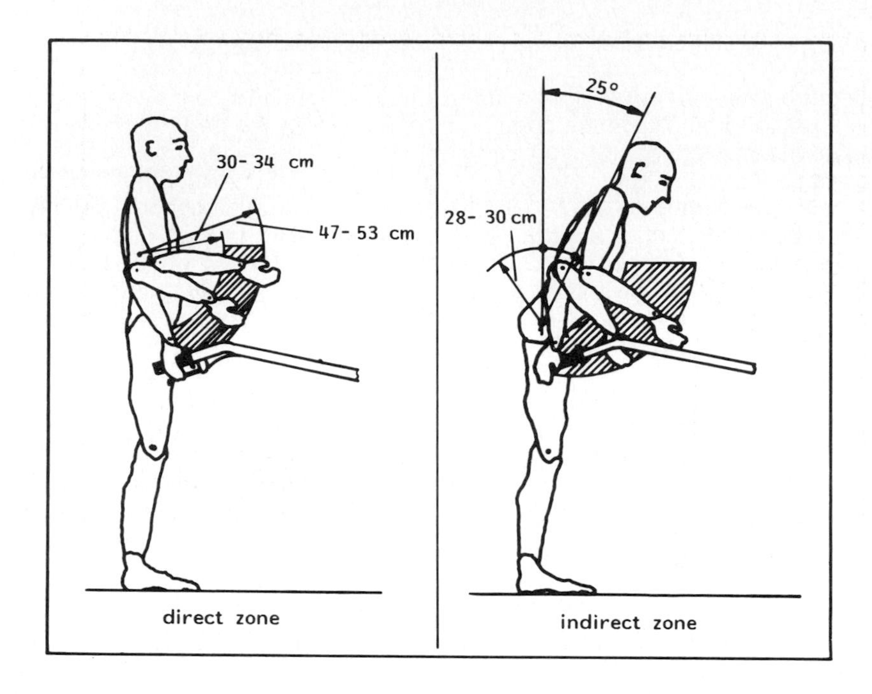

Figure 4
Direct and Indirect Zone for Arrangement of Controls

5. General control design recommendations

The spatial arrangement of the controls is not the only thing that is of importance. Of equal importance is to design the controls in such way that the operator can make a grab movement without visual checking pior to operating a control and that both the grasp movement and the operation of the control conform to the natural movements of the hand-arm system. When designing the contact surfaces of controls, it is important that the movement of the hand relative to the control should be taken into account. Figure 5 is an example of the design of a control for two dimensional movements. The special geometry of the rotationally symmetric control permits both force-oriented or path-oriented movements (contact with the flat of the hand) and sensomotoric movements (contact with finger tips).

The selection of suitable profiles and cover materials for the contact areas of the handle bars is also of importance. On the one hand, this requires convexity along the handle´s longitudinal axis line with the anatomical structure of the hand [4] and, on the other hand the material of the handle must be such that it permits both the transmission of forces and vibration damping.

In handles with levers, the distance between handle and lever must be chosen in such a way that the end-phalanges can clasp the lever, without the hand leaving the handle or the eminence of the hand having to make relative movement to the handle.

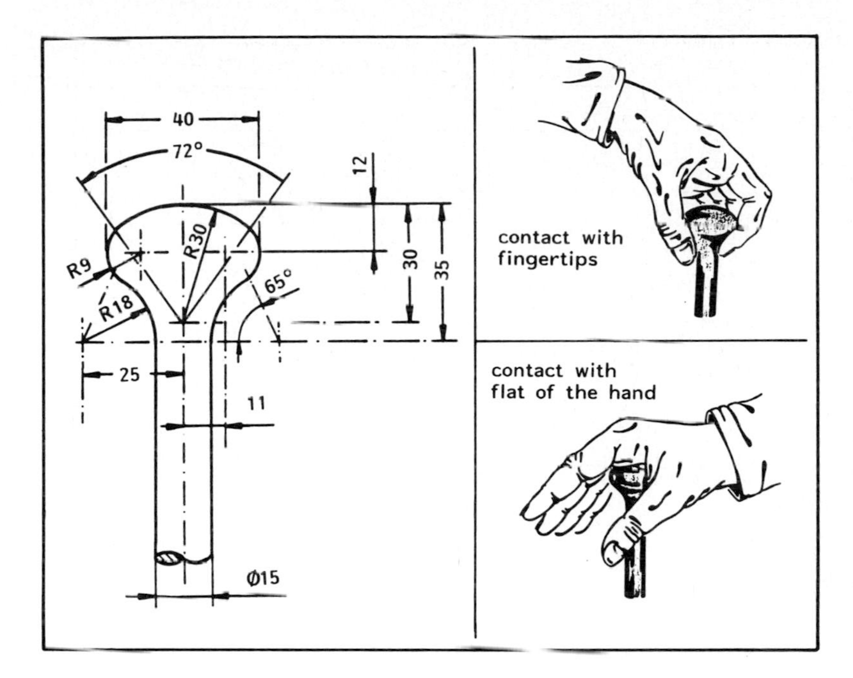

Figure 5
Dimensions of a Two-Dimensional, Multi-Purpose Control

6. Rating of control design alternatives

The rating of design alternatives is a decisive point in the design of interfaces and/or controls. However, before individual alternatives can be rated, it is necessary to define the sequence according to which the individual controls should be selected. In particular, this will be necessary, if specific alternatives of a control rule out individual variants of other controls. The sequence is given by the total scores Σ_i of the individual controls as determined in section 4.

The individual alternatives for the controls are evaluated according to the following quantitative and qualitative criteria :

- operation force F_x, F_y or torqueT (refer to Figure 3),
- operation path x or angle α ,
- length l from normal position of operator´s hand to the point of reaching the control,
- way B of operating the control with both, left and right hand,
- possibilty S of simultaneous operation of other controls,
- compatibility C between direction of operation and effective direction.

A score S_i related to the quantitative or qualitative maximum value will be determined for each criterion according to a rating formula. As described in section 4 the individual scores will be multiplied by the weighting factor $F_{w\,i}$ of the respect-

	Quantitative Criteria			Qualitative Criteria			
Criterion	F_x, F_y, T	x, α	l	B	S	C	Points
Range	0-160N 0-130N 0-5.2Nm	0-±50mm 0-±45°	0- 400mm	impossible	none		1
				difficult	little		2
				possible	medium		3
				easily pos.	high		4
Rating Formula	$1 - \frac{\text{actual Value}}{\text{max. Value}}$			$\frac{\text{given Points}}{\text{max. Points}}$			
Weighting	0.25	0.2	0.15	0.1	0.1	0.1	

$$S_t = \sum_{i=1}^{6} S_i * F_{w_i}$$

ive criterion and then added to obtain the total score S_t. Based on the total score and proceeding by the above mentioned sequence, it will then be possible to select the individual variants for each control and to integrate the best-rated ones into the total concept of the design of the MMI.

7. Safety equipment

The implementation of a safety device is a further important aspect in the design of hand-held machines. The main requirement in this connection is that the drive will immediately stop as soon as the operator fails or looses contact with the machine. This is normally achieved with dead man´s switches or pull-wire switches. These safety systems are inadequate for hand-held machines and they are felt by the operators as a nuissance and, because of this, often bypassed (little acceptance).

The objective, therefore, is to create a safety device which is not immediately obvious to the operator and, still, has a high degree of acceptance. In the concept shown in Figure 6, a circuit coupled to the drive will only interrupt the ignition, when the operator lets go both grips with the machine running. Through the integration of the two pushbutton switches in the grips (mechanical or pneumatic principle), it is achieved that the operator will not see the contact switches as a nuissance. To further improve reliability and safety, emergency stop switches are arranged on both grips in addition.

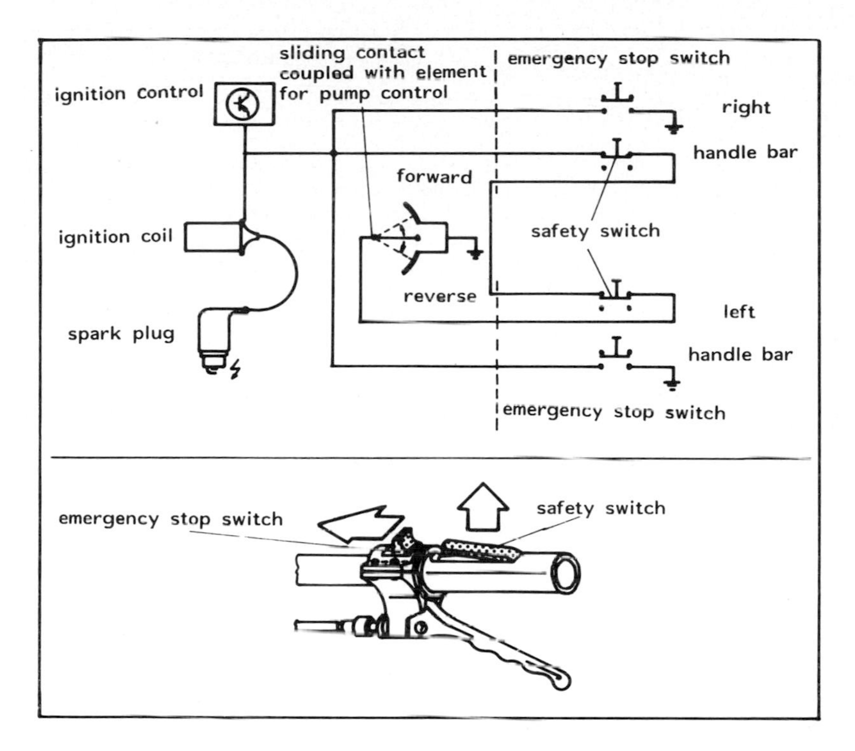

Figure 6
Safety circuit

8. Summary

This contribution shows the importance which attaches to the MMI of a hand-held, two wheeled machine. It illustrates the importance of the ergonomic design of the interface aiming at a reduction in stresses and strains, apart from the development ensuring low-noise and low-vibration machines. It introduces systematic approaches and simple rating methods helping in the development and evaluation of interfaces and their controls for hand-held, two-wheeled machines with due consideration of aspects of anthropometry and saftey.

References

[1] DIN 33402 , Teil 2, Körpermaße des Menschen, Berlin, 1987

[2] Udo Vogt, Ergonomische und sicherheitstechnische Gestaltung von handgeführten Motormähgeräten, Studienarbeit, Institut für Industrielle Fertigung und Fabrikbetrieb, Universität Stuttgart, 1985 (not published)

[3] H.-J. Bullinger, Mensch und Arbeit, Manuskript zur Vorlesung, Lehrstuhl für Arbeitswissenschft, Universität Stuttgart, 1985

[4] H.-J. Bullinger, Einflußfaktoren und Vorgehensweise bei der ergonomischen Arbeitsmittelgestaltung, Habitilationsschrift, Universität Stuttgart, 1978

Trends in Ergonomics/Human Factors V
F. Aghazadeh (Editor)
Elsevier Science Publishers B.V. (North-Holland), 1988

ERGONOMICS OF POWERED HAND TOOLS ON ASSEMBLY LINE WORK

Matti RAUKO, Sakari HERRANEN, Matti VUORI

Technical Research Center of Finland
Occupational Safety Engineering Laboratory
Tampere, Finland

Pneumatic screwdrivers and nutrunners are the most used powered hand tools on assembly line work. The aim of this study was to evaluate the use of pneumatic screwdrivers and nutrunners on assembly line, and produce directions to the proper choice of the tools.

1. INTRODUCTION

Working on assembly line is generally monotonous. The very same work cycles can repeat hundreds of times during work shift. Even more than a half of the total working time may be use of powered hand tools. Constantly repetitive tasks cause muscle strain in upper extremities, which may cause traumas like carpal tunnel syndrome and tennis elbow. In this kind of work, ergonomic and technical specifications of tools are significant.

Only few studies of powered hand tool ergonomics have been published and they have focused attention mostly on noise and vibration measurements and reduction. There has been lack of wide studies of ergonomic and technical specifications of pneumatic screwdrivers and nutrunners for assembly line use.

This study was made to eliminate the lack. The aim of the study was to analyze work stress while using pneumatic screwdrivers and nutrunners, to evaluate the tools from ergonomic and technical point of view and to produce guidelines to the evaluation and the choice of the tools. The study was divided into six parts:

- Work analysis concerning with the use of screwdrivers and nutrunners in assembly line.
- Interview of workers using powered hand tools in their work.
- Laboratory measurements of noise, vibration, grip temperature, trigger force, reaction force and torque repetition capability.
- Creating criteria for the evaluation of pneumatic screwdrivers and nutrunners from ergonomic point of view.
- Evaluation of the tools.
- Producing a guide to the proper choice of powered hand tools for assemblyline use.

2. METHODS

Work analysis consisted of entire body OWAS-SAWO-analysis and hand-wrist posture analysis. So called Ergoprofile method was used to ask how workers suffered the physical stress in their work. Fatigue caused by the use of pneumatic screwdrivers and nutrunners in assembly line were evaluated according to this information.

In the interviews, the workers were asked opinions and experience about their tools; the good and the poor solutions and how they would improve the tools. Interviews and work analysis were made in two cooperation companies in Finland; UPO washing machine factory and SAAB car factory.

48 different tools from seven most common brands of tools in the Finnish markets (Atlas Copco, Bosch, Deprag, Desoutter, Gardner-Denver, Ingersoll-Rand and Uryu) were selected to the laboratory measurements. The tools were operated by the researchers in measurements, so they got also experiences on working with different kinds of screwdrivers and nutrunners.

According to researchers' experience, literature and other information the evaluation criteria, based on 7 characters, was created. The evaluation of the tools was based on the results of laboratory measurements and on using the criteria.

The guide to the choice of powered hand tools for assembly line work were composed according experience and knowledge the researchers got during the study. The frame of the guide grounds on the 7 characters of the evaluation criteria.

3. RESULTS

3.1. Working With the Tools

In average, working on assemblyline in cooperation companies were recovered as physically low loading. Really poor working postures did not occur in the companies according OWAS-SAWO-analysis. Immediate changes were required only in one post, but small changes at the near future were required on almost every post. According to the Ergoprofile method fatigue effected mostly neck and shoulders, lower back and wrists (Fig. 1). One out of five workers felt that the most loading tasks in their work was connected directly to the use of pneumatic screwdrivers and nutrunners. So using of the tools seems clearly to be one of the most remarkable stress factors in assembly line work.

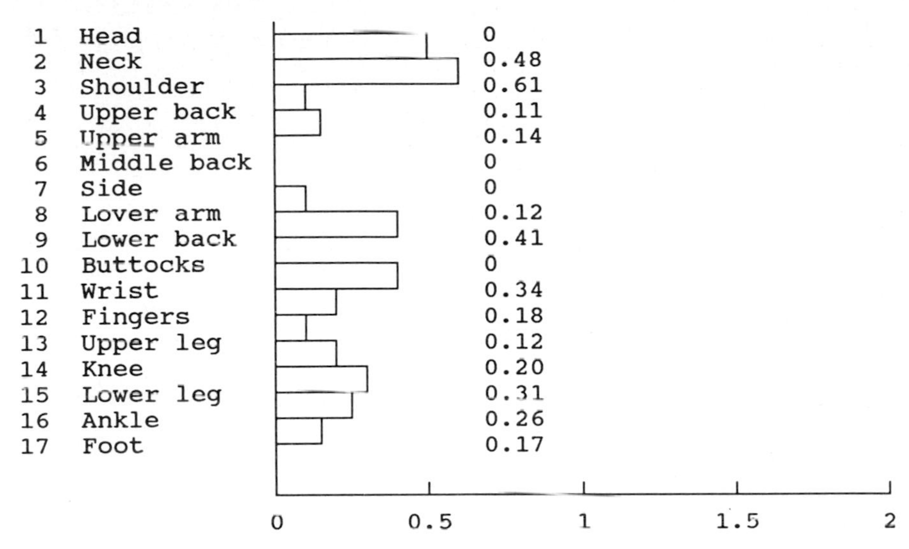

FIGURE 1
Ergoprofile. Mean rates of values given by interviewed workers (n=66).
2 = remarkable pain in this part of the body after work days
1 = minor pain in this part of the body after work days
0 = no pain in this part of the body after work days

Using the tools causes strain especially to the wrists. The worst problem is extreme postures of wrists, occurring a lot while using the tools, even if the work place seems to be otherwise ergonomic well designed. Also reaction force, the counter reaction to the tightening torque, causes strain and shocks to the wrists. The strain can be decreased by improved design of the tools and the posts and by choosing the most appropriate tools for every task, work place and worker.

The major safety hazard while using pneumatic screwdrivers and nutrunners is the rotation of the tool in user's hand caused by reaction force. If that happens, there is the risk to bruise or squeeze one's hand. The most essential thing in prevention of this kind of accidents is fast and reliable function of the clutch and proper shape of the tool and work object (no sharp edges).

3.2. Quality of the Tools

According to the laboratory study, technical and ergonomic properties of the tools were quite good in average (Fig. 2-5). The major shortcomings were high noise levels, poor design of handles, high reaction forces and inconvenient torque setting.

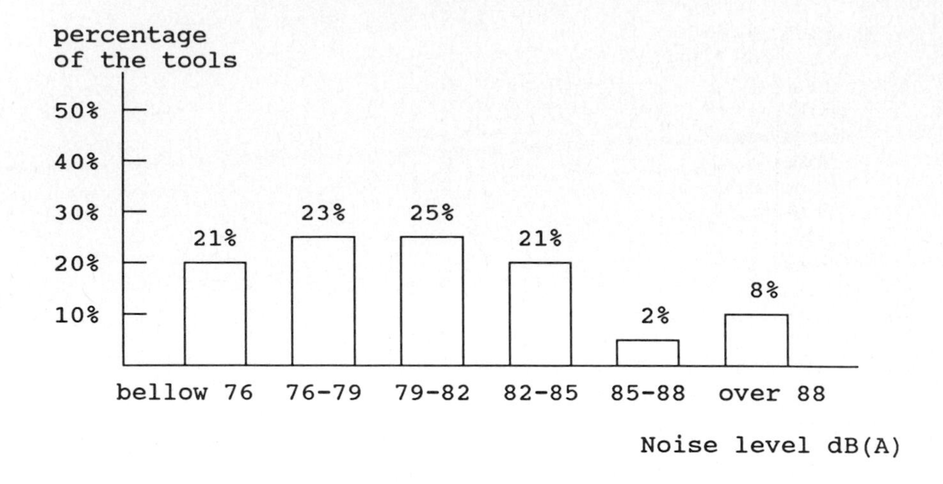

FIGURE 2
Noise level distribution of the tools measured in the laboratory study (n=48)

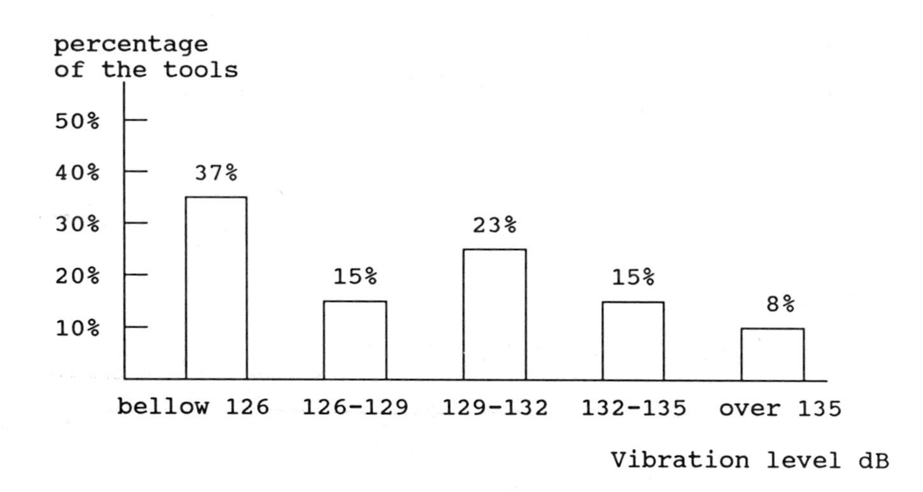

FIGURE 3
Vibration level distribution of the tools measured in the laboratory study (n=48)

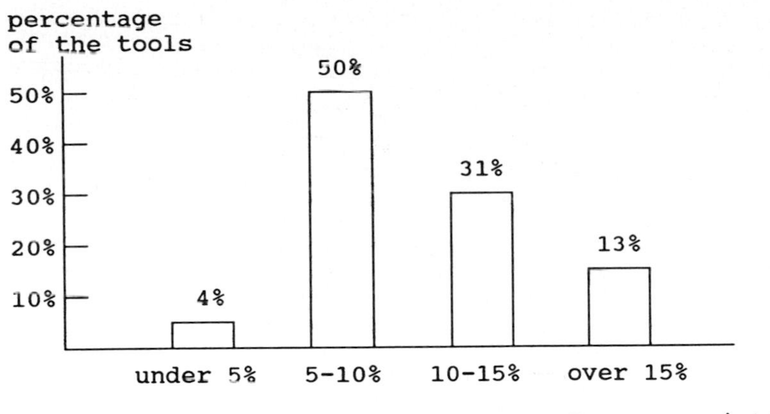

FIGURE 4
Torque variation distribution of the tools measured in the laboratory study (n=48)

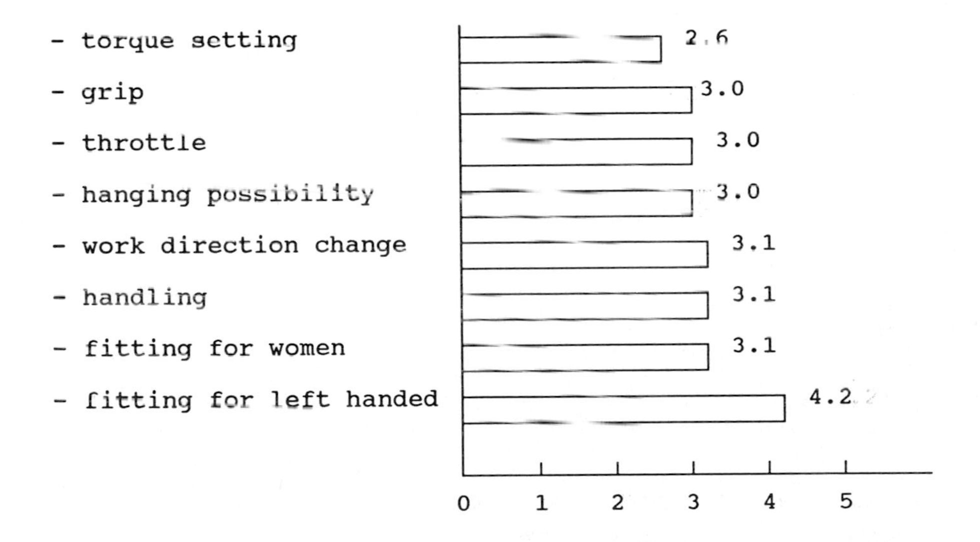

FIGURE 5
Mean rates of grades to properties of the tools evaluated in laboratory study (n=48, grades: 0=very poor, 5=very good)

The interviewed workers were fairly satisfied to their tools (Fig. 6). Least satisfied the workers were to the function of throttle and drive direction change. Also weight and reaction force of strong nutrunners were met as problems. Instead of that the workers did not find vibration or grip temperature of the tools as a problem. In the case of the grip temperature, to this result effected the fact, that most of the workers used some kind of gloves while working.

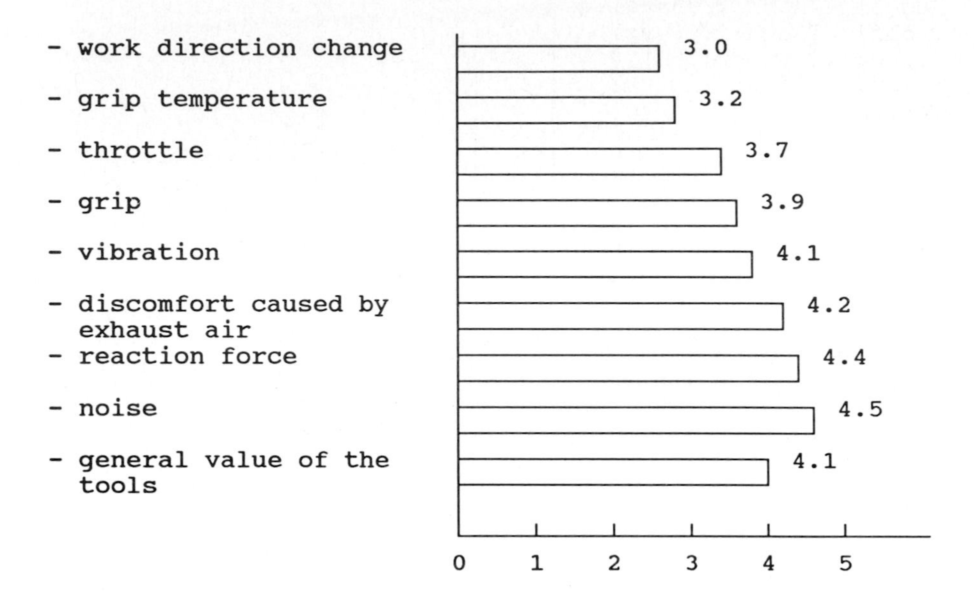

FIGURE 6
Mean rates of grades given by the interviewed workers (n=65) to their tools (grades 0=very poor, 5=very good)

According to the interviews, the workers were more satisfied to their tools than on grounds of results of laboratory study could be supposed. For example to the grip the workers gave average 4.1 (grades 0-5), while in the laboratory study the inspected tools were given mean value 3.0. The reason of the better values given by workers, is partly attention which cooperation companies have focused to the choice of the tools, and partly workers' lack of possibilities to compare their tools to better ones.

3.3. Choosing the Tools

Although the workers were relatively satisfied to their tools, and though according the laboratory study the quality of the tools were quite good, the level of workers' fatigue may be reduced by right choice of the tools. The considerable factors in the choice of pneumatic screwdrivers and nutrunners are:

- Noise and vibration
- Design; dimensions and materials, shape
- Controls; trigger, reverse switch, torque setting
- Handling; weight and center of gravity, fitting for women and left handed, hanging
- Technical quality; speed (RPM), torque variation (production quality), reaction force, reliability, clutch
- Maintenance
- Costs

Further information of choosing procedure of the tools is presented in "Guide for Evaluation and Choice of Pneumatic Hand Tools", composed according the results of the study.

4. CONCLUSIONS

The guide for choice of the tools is major outcome of this study. The guide is intended to the use of industry, when choosing pneumatic screwdrivers and nutrunners for assembly line use, but it may be applied also when choosing other powered hand tools. In the guide is presented:

- chooсing procedure,
- required technical properties and
- required ergonomic properties of the tools.

The guide is so far available only in Finnish, but the English edition will hopefully be published later in 1988.

ACKNOWLEDGEMENTS

This paper is based on research supported by a grant from Finnish Work Environment Fund and completed by Technical Research Center of Finland.

REFERENCES

Herranen, S., Rauko, M., Vuori, M., Paineilmatyökalujen ergonomiset ja tekniset ominaisuudet (Technical Research Center of Finland, Espoo, 1987).

Rauko, M., Paineilmakäyttöisten ruuvin- ja mutterinväänninten ergonomia kokoonpanotyössä (Tampere University of Technology, Tampere, 1987).

Vuori, M., et al, Paineilmatyökalujen arviointi ja valinta opas (Technical Research Center of Finland, Tampere, 1988).

Trends in Ergonomics/Human Factors V
F. Aghazadeh (Editor)

AN INVESTIGATION OF QUANTITATIVE MEASURES OF JOYSTICK CONTROL

Khosrow Behbehani, Sheik N. Imrhan* and George V. Kondraske

Biomedical Engineering
University of Texas at Arlington
Arlington, Texas 76019. U.S.A

Joystick control performance measures obtained from step response and phase plane analysis are investigated to establish simple first order linear relations between them. Results indicate that average movement speed can be predicted from maximum movement speed but other measures should be computed directly. A promising method of quantifying the learning effect emerges which deserves further investigation.

1. INTRODUCTION

Many daily vocational and living activities require tracking using upper extremity. Controlling of lift trucks, operating construction machinery control, and flying sophisticated helicopters are examples of tasks which require upper extremity control using visual feedback. Due to the importance of tracking tasks in various activities, researchers have been trying to analyze them for better understanding, characterization and task performance prediction. Jex and associates [1] were among the first to develop a tracking task to measure the effective operator delay in manual tracking. Recently Eber and Schneider [2] employed a series of tracking tasks to compare the dynamics of the internal model that an operator generates for a tracking task with the actual dynamics of the task. They concluded that operators can achieve accurate tracking without an accurate internal model of the system dynamics. To study pilots' performance capacity, Braune and Wickens [3] used a second order tracking task together with visual and cognitive tests. Most recently, Behbehani et. al. [4] developed tracking test measures using phase plane and step response analysis techniques. The present study explores the association between measures developed in [4] to establish if the magnitude of some measures can be estimated using other measures.

In early 1980's, Kondraske and associates [5] developed a computerized battery of human physical performance tests. Included in the test battery is the Random Step Arm Tracking test. In this test the subject moves the upper extremity in response to a visual step function stimulus which is applied at random intervals. Behbehani and Kondraske [6] implemented methods for extracting quantitative measures of performance from the step tracking test results using step response analysis technique.

---

*Dr. Imrhan is with Indust. Eng. University of Texas at Arlington

In a parallel study, Bhargava et. al. [7] applied the phase plane techniques to the step tracking test results to obtain additional measures of performance. The step response and phase plane measures were further analyzed and modified by Behbehani et. al.[4]. When step response and phase plane analysis are applied to linear deterministic systems the relationship between the results can be established analytically. However, upper extremity response during step tracking test is, in general, nonlinear and stochastic. Hence to establish the relation between the upper extremity step response and phase plane measures statistical techniques must be employed. In particular, it is of interest to explore first order linear regression models relating step response measures to the phase plane measures. Should such simple models with high degree of goodness of fit exist, they can be used to obtain estimates of the step response measures within the same population and provide insight to association between the measures.

2. METHODS

The test equipment has three main components, namely a computer, a video display screen, and a joystick. The computer generates the visual stimulus on the video display and samples the output of the joystick to capture the subjects tracking response. Complete details of instrumentation is given in [5]. The subject sits in front of the video screen and grasps the joystick handle which has a circular disk for fist support. The output of the joystick controls the position of a short vertical line , called follower, on the screen. A long vertical line, called target, appears on screen and jumps laterally at random intervals (2.5 to 4 seconds, uniform distribution). The test instruction to the subject is that once the target jumps, move the joystick as fast as possible to reach the target and align with it after reaching its vicinity. The instruction emphasizes that the speed is of primary interest followed by accuracy. During a test trial the target makes ten jumps and the computer samples and stores the target and follower positions at 60 Hz.

Fifteen normal right-handed subjects were recruited to participate in the test. The age of the six female and nine male subjects ranged form 22 to 36 years with the average value of 27.9 and median of 28 years. The subjects were tested in two sessions separated at least one week but not more than two. In each session, the subjects were tested twice with a few minutes of rest between test trials. The subjects did not receive any payments for their participation in the study.

3. ANALYSIS

The tracking response of the subject is analyzed using step response and phase plane analysis. Detailed explanation of these measures is given in [4]. However, a brief description of their definition is given here for completeness. Figure 1 illustrates the under damped step response of a second order system. The subject's tracking response [e.g. see 4] resembles the response in Figure 1. Thus, measures of speed, accuracy, and coordination can be extracted from the the subject's response. As shown in Figure 1, Movement Time, a measure of upper extremity movement speed, is the total time required for the subject to move the follower from 10% to 90% of the new target position. Percent Overshoot, a combined measure of speed and accuracy, is the maximum follower overshoot expressed as a percentage of the new target position. Settling Time, a measure of coordination and steadiness, is the total time required for the oscillation in the subject response to diminish and for the response to remain within a ±5% of the target position.

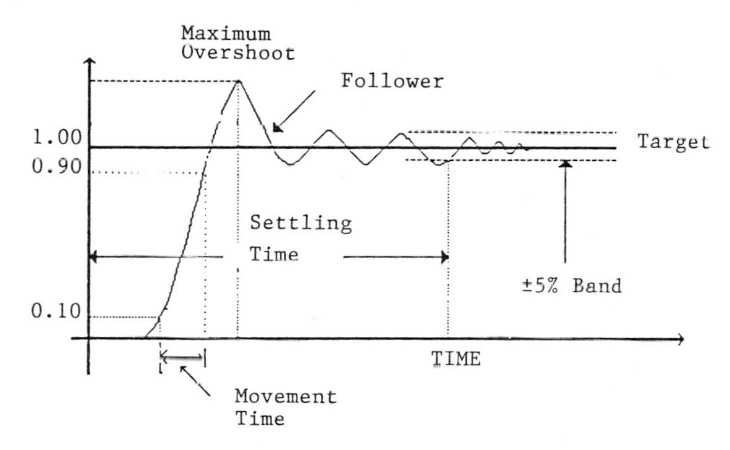

Figure 1. Schematic of follower position in Random Step Tracking Test

Applying the phase plane analysis technique to the subject's tracking response, measures of maximum arm movement speed, Maximum Velocity, and a Coordination Index, can be extracted [4]. Figure 2 is a schematic of the subject's phase plane response. The Maximum Velocity is highest movement speed that the subject generates during the tracking and the Coordination Index is the ratio of the square of the Maximum Velocity over the area circumscribed in the phase plane plot [for derivation see 4]. To account for any impact on the measures due to the direction of arm movement, both phase plane and step response measures are computed for the lateral and medial directions.

To investigate the relationship between the step response and phase plane measures, linear regression and correlation analysis were applied to the results from the two test sessions. Table 1 displays the correlation coefficients for session 1 and 2. The regression analysis was conducted for simple linear first order models only. Higher order or multiple regression models were not considered because the computational effort required for generating such models exceeds that of direct computation of the measures. The R^2 values resulting from regression analysis are displayed in Table 2.

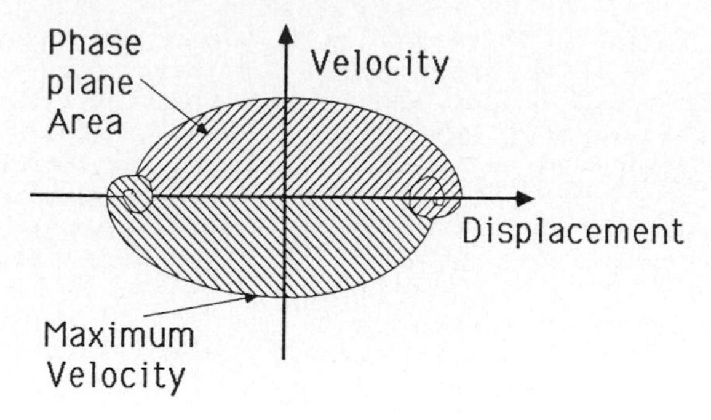

Figure 2. Schematic of a Phase Plane Plot for the Follower Position During Random Step Tracking Test

Table 1. Correlation Coefficients for Step Response and Phase Plane Analysis

	MEDIAL						LATERAL					
	MOVEMENT TIME		PERCENT OVERSHOOT		SETTLING TIME		MOVEMENT TIME		PERCENT OVERSHOOT		SETTLING TIME	
	SESSION 1						SESSION 1					
	TR1	TR2	TR1	TR2	TR1	TR2	TR1	TR2	TR1	TR2	TR1	TR2
MAXIMUM VELOCITY	-0.91	-0.92	0.28	0.26	-0.32	-0.22	-0.93	-0.93	0.23	0.64	-0.21	-0.15
COORDINATION INDEX	-0.5	-0.65	0.29	-0.03	0.15	-0.20	-0.48	-0.64	0.28	0.5	0.03	-0.51
	SESSION 2						SESSION 2					
	TR1	TR2	TR1	TR2	TR1	TR2	TR1	TR2	TR1	TR2	TR1	TR2
MAXIMUM VELOCITY	-0.95	-0.94	0.54	0.5	-0.30	0.32	-0.95	-0.92	0.26	0.80	-0.08	0.4
COORDINATION INDEX	-0.55	-0.60	0.11	-0.07	-0.30	0.09	-0.25	-0.33	-0.17	0.42	-0.21	0.25

TR1 = TRIAL 1
TR2 = TRIAL 2

4. DISCUSSION

Results in Table 1 reveal that only Movement Time is highly correlated to the Maximum Velocity. The level of this correlation remains approximately the same (greater than 0.91) for trials 1 and 2 in both sessions. The high level of correlation between the Movement Time and the Maximum Velocity is expected, as a subject with higher movement speed requires less time to move the follower from 10% to 90% of the target. The R^2 values (Table 2) for the Movement Time and Maximum Velocity are greater than 83% which indicate that a satisfactory estimation of the Movement Time (an aggregate measure) can be obtained from the Maximum Velocity (a single value measure).

Table 2. Percent R^2 Values for Linear Regressional Analysis

	MEDIAL						LATERAL					
	MOVEMENT TIME		PERCENT OVERSHOOT		SETTLING TIME		MOVEMENT TIME		PERCENT OVERSHOOT		SETTLING TIME	
	SESSION 1						SESSION 1					
	TR1	TR2	TR1	TR2	TR1	TR2	TR1	TR2	TR1	TR2	TR1	TR2
MAXIMUM VELOCITY	83.46	84.10	8.13	6.55	10.16	5.06	86.06	85.94	5.41	40.47	4.60	2.26
COORDINATION INDEX	25.46	41.90	8.15	0.11	2.28	4.22	23.50	40.38	7.57	25.10	0.07	26.5
	SESSION 2						SESSION 2					
	TR1	TR2	TR1	TR2	TR1	TR2	TR1	TR2	TR1	TR2	TR1	TR2
MAXIMUM VELOCITY	90.58	88.86	29.36	24.88	8.88	9.98	90.25	85.55	6.73	64.81	0.61	16.04
COORDINATION INDEX	30.07	36.63	1.19	0.46	8.95	0.86	6.52	11.20	2.85	17.62	4.41	6.15

TR1 = TRIAL 1
TR2 = TRIAL 2

Movement Time also exhibits larger correlation with the Coordination Index compared with other measures. The trend in correlation coefficients between these two measures appears to suggest that the level of association between the measures increases from trial 1 to 2 in both sessions. This trend exists in both lateral and medial directions. The increase in correlation coefficients may be due to learning effect, as all conditions were the same from first trial to the second. However, the R^2 values from the regression of these measures are low (below 42%). Thus a simple linear regression can not provide a satisfactory estimate of the Movement Time from the Coordination Index.

Percent Overshoot shows a higher association with both the Maximum Velocity and Coordination Index in lateral compared with the medial direction. In particular, the results suggest that the lateral maximum overshoot in subject's response can be predicted from Maximum Velocity (with correlation coefficient of 0.8 for the trial 2 in the second session) more reliably than the medial overshoot. The corresponding R^2 value for the lateral Percent Overshoot is 64.81% (Table 2), a rough estimate of Percent Overshoot can be obtained. Higher correlation coefficients between lateral Percent Overshoot and Maximum Velocity may, in part, be the result of all subjects being right-handed. Indeed some subjects indicated that they can exert a better control when they move the joystick away form their body.

Although the correlation and R^2 values of lateral Percent Overshoot and Maximum Velocity as well as lateral Percent Overshoot and Coordination Index are low in other trials, they increase from trial 1 to 2 within each session consistently. Such increase may reflect the effect of learning on the subjects response and would be worthy of further investigation. That is, one might be able to quantify the learning rate by monitoring the increase in the correlation between measures. The medial Percent Overshoot does not exhibit a consistent correlation with the phase plane measures.

Settling Time does not show high or consistent correlation with the phase plane measure in either lateral or medial directions. The corresponding R^2 values (Table 2) are also small. Thus it can't be predicted reliably from the phase plane measures.

5. CONCLUSIONS

Among all the step response measures, only the Movement Time has a high correlation with Maximum Velocity and can be estimated from Maximum Velocity using simple linear regression. The results appear to suggest that learning the tracking test increases the correlation between the Movement Time and Coordination Index as well as lateral Percent Overshoot and Maximum Velocity. To establish the statistical significance of this effect, however, a larger number of repeated observations is required. Medial Percent Overshoot as well as lateral and medial Settling Time did not have high correlation with the phase plane measures; thus, they must be computed directly from the test time series data.

ACKNOWLEDGMENTS

This research was supported, in part, by a grant from National Institute on Disability and Rehabilitation, grant number G008300124.

REFERENCES

[1] Jex, HR, McDonnel, JP and Phatek, AV: A "critical" tracking task for manual control research. IEEE Transactions on Human Factors in Electronics, HFE-7, 138-144.

[2] Eberts R, and Schneider W: Internalizing the system dynamics for a second-order system. Human Factors, 27(4):371-393, 1985.

[3] Braune W, and Wickens C: The functional age profile: An objective decision criterion for the assessment of pilot performance capacities and capabilities. Human Factors, 27(6):681-693, 1985.

[4] Behbehani K, Kondraske GV, and Richmond RR: Experimental and Clinical Evaluation of Upper Extremity Visuomotor Control Performance Measures. IEEE Transaction on Biomedical Engineering, In Press.

[5] Kondraske GV, Potvin AR, Tourtellotte WW, and Syndulko K: A computer-based system for automated quantitation of neurologic function. IEEE Transaction on Biomedical Engineering, vol. BME-31(5), pp. 401-414, May 1984.

[6] Behbehani K, and Kondraske GV: New measures of upper extremity response using step response analysis 8th Annual IEEE/EMBS Conference, Ft. Worth; Proceedings, pp. 627-629, 1986.

[7] Bhargava R, Behbehani K, and Kondraske GV: Phase plane analysis of the upper extremity step response. 8th Annual IEEE/EMBS Conference, Ft. Worth, Proceedings, pp. 1596-1598, 1986.

Trends in Ergonomics/Human Factors V
F. Aghazadeh (Editor)

ELBOW FLEXION STRENGTH OF YOUNG FEMALES

Linda J. ROBINSON-EDWIN, Shrawan KUMAR and Hazel CLARKSON

Department of Physical Therapy
University of Alberta
Edmonton, Alberta, Canada T6G 2G4

1. INTRODUCTION

The modern industrialisation has seen increasing number of females in the work force. In the United States, the number of women employees increased at a much faster rate than men workers did (NIOSH 81). Women made up to 39% of the total employment in 1974 compared to 34% in 1964. Most of the increase in women workers was in four major occupations which showed rapid growth, these being services, government, wholesale and retail trade, and manufacturing. Of the total increase only 29% was absorbed in the government and the remainder 71% were either in service industry or trade and manufacturing. Most of the latter category needs physical strength. The changing mix of the working population makes it important that the jobs be designed with both genders in mind. However, most of the work design is based on the data obtained from male workers. Therefore, it was decided to study the isokinetic elbow flexion strength among the female population. This strength could be used for task design by a large number of service industries and manufacturing plants with female operators.

Some of the early work in this area were confined to isometric measurements with objective of determining optimum joint angle for the development of force (Elkins et al 1951). A comparison of the isometric force exerted about the elbow joint in standing versus that in sitting found that the greater mean torque was generated in the sitting position (Provins 1955). The position of the forearm in supination, pronation and midpronation-supination during elbow flexion has also been studied (Elkins et al 1951). The midposition has been found to be most efficient in terms of force production. Despite variation in techniques of measurement by various authors, the typical isometric elbow flexor strength curve has been found to be an ascending-descending curve. The angle of peak tension development was found to vary between 65^0 and 90^0 of flexion, occurring most commonly at 90^0 (Elkins et al 1951). The main factors which determine the shape of the isometric elbow flexor strength curve are the length-tension relationship of muscle modified by the effects of external biomechanical factors.

In an effort to describe the maximum muscular capability during dynamic motion, several authors have derived dynamic elbow flexor strength curves, both in uncontrolled concentric conditions (Singh and Karpovich 1966, Knapik et al 1983) and in velocity controlled concentric conditions (Knapik et al 1983). Though isokinetic strength capabilities may be considered most useful, it has been studied the least. Knapik et al (1983) reported isokinetic elbow flexor curves at the speeds of 36^0, 108^0 and 180^0 per second. They found that the peak torque varied between the angles of 70^0 to 90^0 and the overall shape of the curves was also an ascending-descending one. However, their results were obtained from soldiers. In order to explore this further for its shape and magnitude among women, the current study was conducted.

2. MATERIALS AND METHODS

2.1. Subjects

Twenty-five female volunteers served as subjects in the study. Subjects were screened to ensure that they had no history of systemic pathology or upper limb injury. The subjects anthropometric data are given in Table 1.

Table 1. Anthropometric data of subjects

Variable	Age (years)	Weight (lb)	Height (in)	Fore-arm Length (in)
Mean	23.1	131.5	65.6	16.4
Std. Dev.	2.9	19.5	2.4	.8
Range	19-29	88-167.8	60.8-71.2	14.5-18.2

2.2. Subject positioning

Subjects sat upright with feet flat on the floor. Both upper limbs rested on a cushioned wedge with hands together and touching in the midline, forearms fully supinated and palms up. Both wrists were held in the semi-flexed position as the subjects grasped the effort application bar. The subjects' elbow joint axes, approximately located at the lateral epicondyles, were aligned with the axis of rotation of the input shaft of a Cybex II+ isokinetic dynamometer while the elbows were maintained in full extension. Seat height was adjusted for comfort and ease of movement, by the use of additional seat raises. If necessary, additional foot support was provided so as to maintain a 90^0 angle at hips and knees.

2.3. Experimental procedure

The torque and goniometry channels of the isokinetic dynamometer were calibrated according to the manufacturer's recommendations. The torque channel was calibrated for a full scale deflection of 90 ft.lbs. The subjects' elbows limbs were maintained in full extension while the effort application arm of the machine was aligned parallel with a line joining the elbow joint axis to the point where the hand grasped the grip bar. The input arm was parallel with the forearm within 5^0. This was not considered to be a significant deviation.

General body stabilisation was achieved actively as subjects were required to maintain contact with the seat and the floor or foot support during testing. Stabilisation of the proximal and distal joints was achieved by subjects actively depressing their shoulders and maintaining their wrists in a semi-flexed position. No external stabilisation was applied to the subjects. However, the anterior edge of the testing table was stabilised manually to prevent backward tilting during the test movement.

Isokinetic testing was carried out at two velocities. The 'slow' speed was $60^{o}s^{-1}$ while the 'fast' speed was $90^{o}s^{-1}$. Subjects were randomly assigned the order of testing speed. After being set up, they were each allowed several submaximal trials to familiarise themselves with the equipment. During these trials, the alignment of the subject with the machine was also checked. When properly aligned, the subject was able to complete the required motion smoothly and without experiencing tension or compression in the elbow joint or at the hand. When necessary, subjects were realigned prior to testing.

Three maximal trials of elbow flexion were conducted at each speed. The movement began in full elbow extension and ended in elbow flexion with the subject's hands contacting the chest, chin tipped slightly upwards to avoid contact at this point. Subjects were instructed to pull as hard and as fast as possible without jerking the movement. Verbal encouragement was standardised as a single command given forcibly at the initiation of each repetition, "Now, pull!". During the actual test sequence, two submaximal trials were performed first, each followed by 30 seconds rest. After this, three maximal trials were each followed by 60 seconds rest. Both test speeds were completed in succession. The elapsed angle from initiation of motion was first determined from the height of the goniometer trace of the Cybex (Figure 1). Torque was read from the torque trace at 10° intervals and recorded in ft.lbs. The range of motion achieved by each subject was determined as was the minimum and average range of motion. The starting position of the movement was taken to be the 0° position. The average range of motion achieved by the subjects was 150°. Data reduction was, therefore, performed in 10° increments over the range of 10° to 150° of motion. Torque was then averaged over each of the selected angles and for each speed. The averaged torque data was then converted to percentage of maximum.

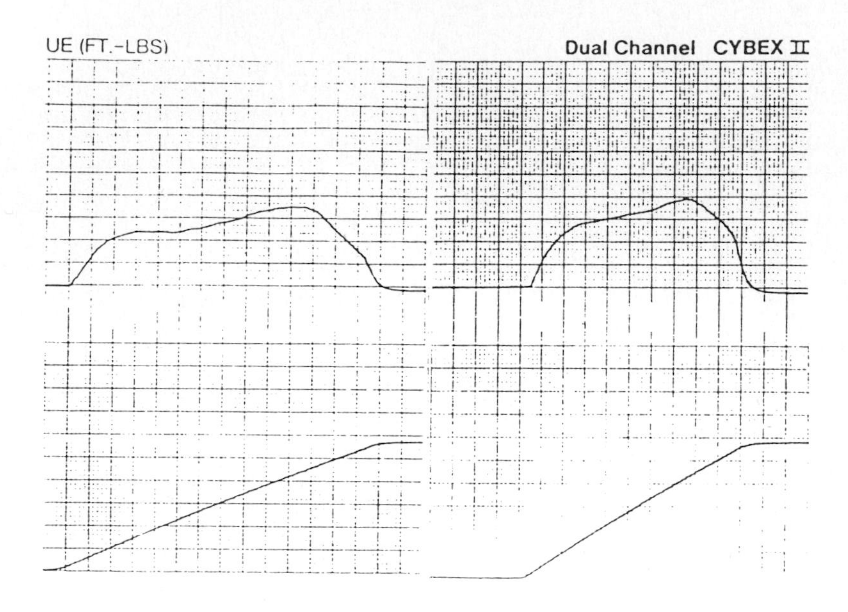

Figure 1. Cybex trace of trial at 60° and $90^{o}s^{-1}$

2.4. Analysis

Statistical analysis of the Cybex data involved the application of a three-way analysis of variance to test for main effects and possible interaction of the factors speed, trial and angle. Factor A, speed, had two levels - $60^{o}s^{-1}$ and $90^{o}s^{-1}$; Factor B, trial, had three levels - trials 1, 2 and 3; Factor C, angle had 15 levels - angles 10° to 150° in 10° increments. The acceptance level of significance was set at .01.

3. RESULTS

The total mean range of motion achieved was $172.76^{o} \pm 9.97^{o}$ while the average individual range of motion varied from 154° to 196°. The first 10° of movement was considered unreliable due to torque overshoot. Data in this range were, therefore, discarded. The elbow flexion capability was analysed and plotted over a range of 10° to 150° of movement (Table 2, Figure 2). The curves at both speeds were ascending-descending. They rose at an almost constant rate to 40° and then at a slower rate, achieving a range of peak torque between 110° and 130° and a distinct peak value at 120°. After this, torque declined slowly at 130° and then rapidly to 150°. Minimum values of torque occurred at 10° and 150° with the value at 150° being approximately three times that at 10° (Figure 1).

The $60^{o}s^{-1}$ curve was separate and distinct from the $90^{o}s^{-1}$ curve and lay above the latter at all points. The analysis of variance showed a highly significant main effect of angle at which torque was measured ($F_{(14,672)} = 122.10$; $p < .000$); there was no signficant main effect of speed. The ABC interaction approached significance at the .01 level, however, no significant interaction effects were found (Table 3).

Table 2. Isokinetic torque exerted by the experimental population at 60^0 and 90^0 per second speed of flexion

Speed	Variable	Torque (ft-lb) at angles in degrees														
		10	20	30	40	50	60	70	80	90	100	110	120	130	140	150
	Mean	9.3	15.1	18.4	19.8	21.5	22.5	24.7	25.5	28.5	30.3	31.6	32	31.6	29.4	25.6
60^0	S.D.	4.9	7.1	10.9	10.1	12.8	10.6	11.4	13	15.6	12	8.9	9.7	10.1	10.4	12.2
	S.E.M.	0.57	.91	1.2	1.1	1.4	1.2	1.3	1.5	1.8	1.4	1.0	1.1	1.17	1.21	1.4
	Mean	6.2	12	15.8	18.3	19.8	21.6	23.6	25.4	27	28.7	29.4	29.4	28.7	25.8	21.4
90^0	S.D.	3.1	5.3	6.2	9.03	8.3	9.4	12	11.2	13.1	10.2	8.6	9.0	9.8	11.1	10.4
	S.E.M.	0.6	1.0	1.2	1.8	1.6	1.8	2	2	2.6	2.05	1.7	1.8	1.9	2.2	2.0

S.D. - standard deviation; S.E.M. - standard error of the mean

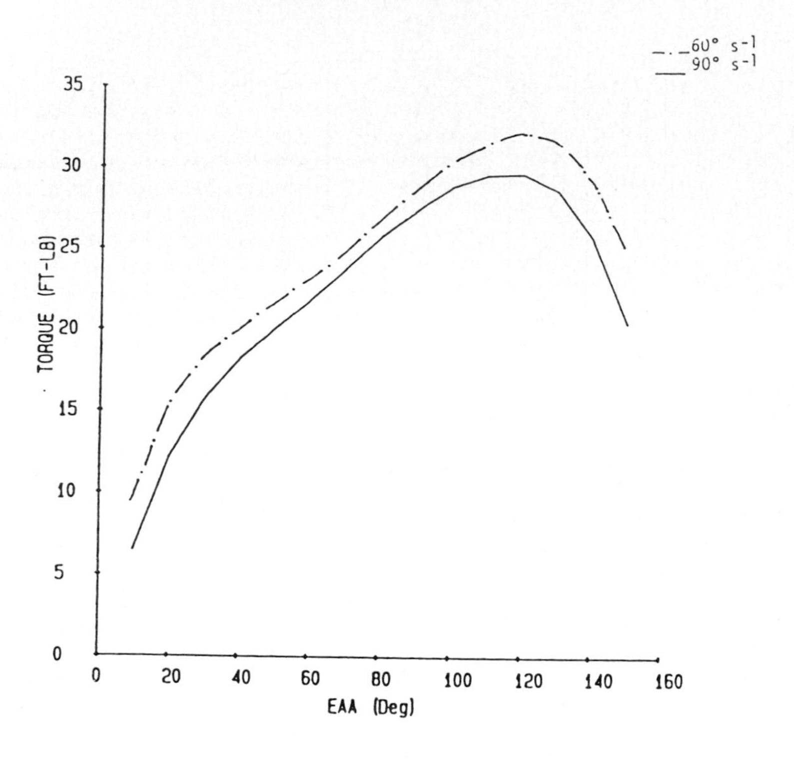

Figure 2. Elbow Flexor Capability Patterns in Ft. Lbs.

Table 3. Summary of 3-way analysis of variance

Source	SS	DF	MS	F	P
BET SUBJ	110229.00	49			
A	3140.0000	1	3140.0000	1.41	0.241
SUBJ W GROUP	107089.00	48	2231.0208		
WITHIN SUBJ	142216.00	2200			
B	125.00000	2	62.500000	1.75	0.180
AB	90.000000	2	45.000000	1.26	0.289
B X SUBJ W G	3432.0000	96	35.750000		
C	95448.000	14	6817.7109	122.10	0.000
AC	581.00000	14	41.500000	0.74	0.731
C X SUBJ W G	37524.000	672	55.839279		
BC	48.000000	28	1.7142849	0.48	0.990
ABC	172.00000	28	6.1428566	1.72	0.011
BC X SUBJ W G	4796.0000	1344	3.5684519		

4. DISCUSSION

The complex interaction of length-tension, angle of pull and force-velocity effects is clearly shown in the shape of the elbow flexion capability plot (Figure 2). This ascending-descending shape is consistent with that described in the literature although there were some important differences. the effect of velocity on force development was not clearly demonstrated by this study. Although a trend was noted for increased velocity of movement to decrease the amount of torque developed, there were no statistically significant differences. The mean shape of the $60^{0}s^{-1}$ and $90^{0}s^{-1}$ curves was similar, particularly when converted to %max torque. However, it is emphasised that the difference between in 60^{0} and 90^{0} per second was not large enough to accentuate this difference. Kumar et al (1988) have shown a significant inverse relationship between the strength and speed. This difference was more pronounced with increase in speed.

In most of the occupations where predominantly the work force is female and the tasks involve elbow flexor strengths, such as packaging, assembly line manufacturing, the loads must be kept low and the range of motion restricted within 30^{0} to 120^{0} of elbow flexion. With prevalent upper extremity disorders like carpal tunnel syndrome and epicondylitis in the work force the safety of workers is of paramount concern. It is emphasised that the maximum torque of only 32 ft-lb was achieved by this group of females. While a proven formula of load reduction for elbow flexors is awaited the magnitude of maximal torques may give the task designers some idea about the capability of such a group. An additional advantage of the elbow flexion capability pattern will be more appropriate design of programs for strengthening the elbow flexor muscles. These characteristics can also be used for designing a proper tool to provide appropriate strength training programs for normals as well as athletes.

REFERENCES

[1] Elkins, E.C., Leden, U.M., Wakin, K.G., Objective recording of the strength of normal males, Arch. Phys. Med. 32 (1951) 639-647.
[2] Knapik, J.J., Wright, J.E., Mawdsley, R.H., Isometric, isotonic and isokinetic torque variations in four muscle groups through a range of joint motion, Phys. Ther. 63 (1983) 938-947.
[3] Kumar, S., Chaffin, D.B., Redfern, M., Isometric and Isokinetic back and arm lifting strengths: devices and measurement, J. Biomech. 21 (1988) 35-44.
[4] N.I.O.S.H., Work practices guide for manual lifting, Department of Health and Human Services, Washington, D.C. 1981.
[5] Provins, K.A., Salter, N., Maximum torque exerted about the elbow joint, J. Appl. Physiol. 7 (1955) 393-398.
[6] Sing, M., Karpovich, P.V., Isotonic and isometric forces of forearm flexors and extensors, J. Appl. Physiol. 21 (1966) 1435-1437.

Trends in Ergonomics/Human Factors V
F. Aghazadeh (Editor)

A COMPARISON OF THE RESULTS OF RECENT STUDIES IN WRIST-TWISTING STRENGTH

Sheik N. Imrhan

Department of Industrial Engineering
The University of Texas at Arlington
Arlington, Texas 76019

Seven relatively recent studies on wrist-twisting strength have been compared to determine their progress in improving one's understanding of this topic. The studies reveal that, for circular screw-type container lids wrist-twisting torque depends mainly on the diameter and surface finish of the lids. The diameter and height influences the type of grip used, and hence torque. Knurls and threads on commercial lids have been ineffective for torque generation, except for very large diameter lids. Enough data does not exist to make strong conclusions concerning relationships between torque and lid height, body posture and different age groups of the population.

INTRODUCTION

The ability to grasp and manipulate objects and to exert forces on them with the hand is a necessary function of all humans, for mastering the environment. With the increase in proportion of the elderly population, many problems hitherto unattended have been brought into sharper focus and are now being approached from an ergonomic standpoint; that is, designing the task, equipment and environment to confirm to the capabilities of the user population. One such problem is the ability to exert wrist-twisting torques (or forces) as in unscrewing or tightening circular lids on food containers and medicine bottle caps, opening and closing faucets, turning door knobs, etc. Recently several investigators have attended to this problem by investigating wrist-twisting strength to determine contributory variables and empirical distributions, and to compare these distributions with torque required to open selected screw-type container lids (Berns, Stuart, Claridge, Wahlstrom and Anell [1]; Berns, [2]; Rohles, Moldrop and Laviana [3]; Loo [4]; Imrhan and Loo, [5 and 6]; and Nagashima and Konz [7]). These researchers, collectively, have recognized the need for data of this type and for an understanding of the underlying biomechanical mechanism of wrist-twisting strength. It is recognized that investigation of the problem has only just begun and that more studies would be needed before ergonomists can state specific guidelines confidently for designing good hand/handle interfaces for tasks requiring great wrist-twisting strength. However the data accumulated so far have shown patterns which are worth reporting.

The purpose of this study was to make a comparative analysis of the results of these recent studies in wrist-twisting strength, indicating (1) where they have identified important variables, (2) the magnitude and direction of the influence of the variables, and (3) where these studies can influence future research.

METHOD

The literature was reviewed to identify all studies which investigated hand strength requirements (wrist-twisting strength) for opening or tightening screw-type lids on containers. An examination was then made of the objectives, methodology and results of these studies to identify similarities and differences which can aid our understanding of wrist-twisting strength. Conclusions were made on how these results can influence future studies to answer questions which these studies have failed to answer. Seven studies (mentioned earlier) were identified which reported objectives and methodologies which made comparison of their results meaningful.

RESULTS AND DISCUSSION

All 7 studies had the following characteristics in common:

1. the measured response was the MVC peak torque which an individual could exert in a counter-clockwise (opening) direction on a circular container lid.
2. only the preferred hand was used.
3. subjects were a "convenience" sample from the elderly population.

Variables which were considered pertained to container lid, task and person:

container lid	- diameter, height (depth), roughness (knurls or threads), and type of material (metal or plastic) for circular screw-type lids
task	- number of repetitions and direction of twist (clockwise or counterclockwise)
person	- other types of hand strength (handgrip and finger pinch), age and anthropometry

Three different types of strength measurement instrument were used to measure wrist-twisting torque - a torque meter manufactured by Snap-On Tools Corp. (U.S.A.), model 6-FU and adapted to have a jar lid attached to its shaft (Rohles et al [3]; and Nagashima and Konz [7]); an Owen-Illinois torque tester manufactured by Owen-Illinois, U.S.A. (Loo [4]; and Imrhan and Loo, [5 and 6]); and a custom designed system in which strain gauges were attached to the container lid (Berns et al [1]; and Berns, [2]). With the exceptions of Imrhan and Loo [5 and 6] none of the other studies attempted to distinguish trends in the torque data as a function of diameter for rough and smooth lids. The trends in the other studies, discussed later, were derived from a reanalysis of these mean data of the other studies.

Biomechanically, wrist-twisting torque (T) for circular screw-type lids can be considered to be the product of the lid diameter (D) and the tangential force (F) exerted at the hand/lid interface; the latter being the product of the handgrip force (G) towards the lid center and the limiting coefficient of friction (μ) at the hand lid interface.

$$T = \mu G D$$

This model assumes that peak torque is recorded when, and if, the hand slips during the effort, or at the point where the subject reaches his maximum effort, if slippage does not occur. It is also assumed that, by the investigator's design or otherwise, the lid, if screwed on to the container, does not unscrew before the subject reaches his maximal effort. The grip force (G) on the lid, for a given individual, is limited by diameter and height (depth) of the lid. This model can be used to explain some of the results of these studies.

Diameter and Lid Height

Not all of the 7 studies explored the effects of all the variables mentioned, and none explored lid height nor type of lid material at different levels. The results, collectively indicate that diameter and surface finish were the two most influential variables on torque. There seems to be a linear trend for smaller lids, in general, giving way to a non-linear trend for larger lids (Imrhan and Loo). The exact limits for the linear trend are not clearly defined because of the different diameters used in the various studies. The available data indicates that the largest range was 28-85 mm (Berns eta al [1]) compared to 27-67 mm for elderly subjects (Rohles et al [3]), 27-48mm for 4 year old children (Rohles et al, [3]), 31-74 mm (Loo [4]; and Imrhan and Loo [5 and 6]). Another difference in the linearity was the different slopes, due probably to differences in experimental conditions, especially body posture and skeletal configuration of the arm while exerting force. From the available data slopes of 0.13 Nm/mm for males and 0.07 for females (Rohles et al, [3]), 0.06 Nm/mm for both males and females combined, (Imrhan and Loo, [6]), and 0.13 Nm/mm for smooth lids, males and females combined (Nagashima and Konz [7]). Imrhan and Loo [5 and 6] found that the nonlinearity at larger diameters was different for rough and smooth lids. For rough lids torque increased monotonically but more slowly towards the higher end of the diameter range. The data of Rohles et al [3] also indicated this trend. In fact, over the entire range tested, the torque-diameter relationship can best be modeled by a geometric curve of the form

$$T = aD^b$$

where a and b are constants.

Because of the lid variable combinations and experimental design used, the study of Imrhan and Loo [6] also revealed that, for smooth lids there was a sharp drop in torque for the 113 mm lid compared to the 74 mm lid in the elderly data - - that is, an increasing - decreasing relationship over the entire range of diameters (31-113 mm). Using a quadratic curve to model this torque - diameter trend, they determined that there was an optimum diameter, at about 83 mm, for torque generation for elderly subjects. This phenomenon can be explained biomechanically. At the very large diameter (113 mm) the handgrip used offers poor mechanical advantage to the finger flexor muscles and the effect of frictional resistance on the rough surface manifests itself. This prevents slippage of the fingers, and increases tangential force and, hence, torque. For the large smooth lid the lesser frictional resistance results in smaller tangential force. For small diameters the finger flexors are in a better mechanical advantage and roughness has less effect (Imrhan and Loo, [6]). The data on children in Loo [4] also indicated a slight increasing - decreasing trend for smooth lids.
The studies of Imrhan and Loo [5 and 6] and Nagashima and Konz [7]

indicated that below 86 mm diameter the effect of knurls and threads on commercially available lids did not enhance torque producing capabilities. The data of Berns, [2] also showed this effect.

Torque Direction and Lid Height

Direction of torque exertion (clockwise versus counter-clockwise) does not seem to affect the magnitude of the torque generated (Rohles et al [3] and Loo [4]). The height (depth) of the lid used would certainly influence the type of handgrip used and, hence, the magnitude of torque generation. In these studies, lid height was kept within a range narrow enough to have little influence in the type of grip used. In other words, it was assumed that lid height was not an influential variable in these studies. The quantitative effect of this variable is, therefore, relatively unknown and needs to be addressed in future studies.

Age and Gender

Even though wrist-twisting strength declined with age in the elderly, the effect was not strong enough to make age a single good predictor of strength. The studies of Rohles et al [3] and Imrhan and Loo [5] showed that the elderly do not constitute a homogeneous group with respect to wrist-twisting strength. Both studies identified 2 groups among males (76 yr and less, and above 76 yr). Among females the former study found 3 groups (62-67, 68-78, and 79-92 yr) while the latter study showed the same 2 groups as for males. In all studies, females exerted weaker torques than males with the female/male ratio being 59, 71 and 77% in 3 of the studies. Such a wide range was due, most likely, to the specific subject samples and experimental conditions in the respective studies.

Anthropometry and Hand Strength

The studies of Rohles et al [3] and Imrhan and Loo [6] were in close agreement with the relationship between wrist-twisting torque and anthropometry or hand-strength. Handgrip and finger pinch were most strongly correlated with wrist-twisting torque. Hand size (length and breadth) were also strongly correlated. However, body weight and height were not always significantly correlated.

CONCLUSIONS

The studies indicated that diameter and surface type are two of the most important variables influencing wrist-twisting strength. While one may expect torque to increase with diameter, the exact mathematical relationship is still not definite because there are possible interactions among diameter, surface type, lid height, and type of handgrip used. More empirical data is needed to establish clear relationships. For the elderly, there seems to be an optimal lid diameter, at about 83 mm, for smooth lids which are difficult to grip when the diameter is very large. For rough lids the relationship between torque and diameter was linear for smaller diameters. The upper limit for this linear trend seems to be about 74-86 mm depending on experimental conditions. Lid height, which influences type of handgrip used, has not yet been systematically investigated, and may prove to be a major factor influencing torque. Apart from torque magnitudes there seems to be differences in trend of the torque-diameter relationships. Again more relevant data are needed to clarify this observation. Body posture and experimental conditions, especially, need to be standardized for testing.

REFERENCES

[1] Berns, T., Stuart, K., Claridge, N., Wahlstrom, G. and Anell, M., The Handleability of Consumer Packaging: summary - part 3. Report no. 54 14 16 46, the Swedish Packaging Research Institute, 1979.
[2] Berns, T., The Handling of Consumer Packaging. Applied Ergonomics, 1981, 12(3), 153-161.
[3] Rohles, F.H., Moldrup, K.L. and Laviana, J.E., Opening Jars:An Anthrompometric Study of the Wrist-twisting Strength of Children and the Elderly, Report no. 83-03, Institute for Environmental Research, Kansas State University, 1983.
[4] Loo, C. H., Predictive Models of Wrist-Twisting Strength. Unpublished Masters Thesis, Louisiana Tech University, 1985.
[5] Imrhan, S.N. and Loo, C., Torque Capabilities of the Elderly in Opening Screw Top Containers. Proceedings of the Human Factors Society, 30th Annual Meeting, 1167-1170, 1986.
[6] Imrhan, S.N. and Loo, C.H., Modelling Wrist-Twisting Strength of the Elderly. Paper submitted to 'Ergonomics' for publication, 1988.
[7] Nagashima, K. and Konz, S., Jar Lids: Effect of Diameter, Gripping Materials and Knurling. Proceedings of the Human Factors Society, 30th Annual Meeting, 672-674, 1986.

Trends in Ergonomics/Human Factors V
F. Aghazadeh (Editor)

THE EFFECTS OF SPATIAL ABILITY ON AUTOMOBILE NAVIGATION

Jonathan F. ANTIN*, Thomas A. DINGUS**, Melissa C. HULSE***, and Walter W. WIERWILLE

Department of Industrial Engineering and Operations Research, Virginia Polytechnic Institute and State Univertsity, Blacksburg, Virginia

A study was performed to evaluate navigation performance associated with the use of two navigation aids: (1) conventional paper map and, (2) moving-map display. The focus of the current paper is the effect of individual differences in spatial ability on navigation. Results showed that although individuals with higher measured spatial ability may be able to navigate more efficiently than those with less ability, there were no differential effects of spatial ability on the type of navigation aid used. These results are examined in the context of the different spatial knowledge representations presented by the two navigation aids.

1. INTRODUCTION

1.1. Automobile Navigation

When driving a familiar route, it can be said that an internal spatial representation or cognitive map exists which allows one to successfully navigate from the given starting point to the desired destination. This construct is implicit in the use of a phrase such as "familiar route." Yet despite sign and landmark cues, and because of problems such as the absence of a direct bird's-eye view of the vicinity when driving, drivers tend to become lost or spend a great deal of time searching for a destination whose location is unknown. Oftentimes drivers have used paper maps or verbal directions from a knowledgeable human to navigate in such circumstances.

In recent years a third choice has been made available - the vehicle navigation aid (VNA, McGranaghan, Mark, and Gould [1]). Although auditory verbal instruction systems have been conceived and interface prototypes favorably evaluated (Streeter, Vitello, and Wonsiewicz [2]), most current and proposed VNAs involve the use of a moving-map display.

---

* Now with the Department of Industrial Engineering, North Carolina State University, Raleigh, North Carolina

** Now with the Department of Psychology, University of Idaho, Moscow, Idaho

*** Now with Applied Cognitive Sciences, Inc., Moscow, Idaho

These systems provide continually updated, graphical map information on the area around the current location of the vehicle and on the specific location and orientation of the vehicle with respect to the displayed map information (McGranaghan, et al. [1]).

1.2. Navigation Aids and Cognitive Mapping

The cognitive map can be viewed at various stages of development along a continuum (Thorndyke and Hayes-Roth [3]). Procedural knowledge is characterized by the ability to navigate within an area based on a cognitive map encompassing landmark locations, and the basic directions and connections (nodes) among key roadways. Survey knowledge is encoded in a more global form and includes such things as the absolute location, size, euclidean distance, and directional relationships among key nodes and links within the area.

One issue regarding these different levels of spatial representation is to what extent will different types of navigation aids bring about different stages of the cognitive map described above. For example, from a paper map can be derived procedural knowledge or survey knowledge, and the level of information provided by a moving-map display will likely depend on the particular design. This could lead to the notion that a particular type of navigation aid can present or aid in the development of a level of spatial knowledge most appropriate for automobile navigation. It was found by Thorndyke and Hayes-Roth [3] that the procedural knowledge gained from navigation experience produced superior performance on an orientation task compared to that produced by the survey knowledge acquired from map study. These results point to the criticality of the procedural level of representation in the process of navigation, and the fact that survey knowledge can be presented and acquired in such a way as to not explicitly include the characteristics of procedural knowledge that are critical for navigation.

1.3. Spatial Ability and Navigation

Individual differences with regard to spatial ability have been shown to be an important determinant of map reading ability. Stasz and Thorndyke [4] found that subjects with high visual-spatial abilities demonstrated better overall map learning than those with lower ability. A subsequent study revealed that a greater proportion of the higher ability subjects were able to use effective attention focusing strategies in studying the map (Stasz [5]). Because of the inherently spatial nature of the act of navigation, it was expected that individual differences with regard to spatial ability would be an important determinant of navigation behavior.

2. OBJECTIVES

It was the objective of the current study to look at the effects of spatial ability on the process of automobile navigation, and to see if spatial ability would differentially affect performance with the different types of navigation aid. Issues associated with the relative effectiveness and efficiency of the conventional paper map and the moving-map display as navigation tools are addressed in Antin, Dingus, Hulse, and Wierwille [6].

3. METHOD

3.1. Subjects

Thirty-two subjects, an equal number of male and female, high and low driving experience volunteers, ranged in age from 18 to 73. High experience subjects were those who drove 10,000 or more miles/year, and low experience subjects were those who drove less than 10,000 but more than 2000 miles/year. Those who drove less than 2000 miles were excluded because it was felt that these individuals were not likely to use navigational aids on a regular basis.

3.2. Apparatus

Apparatus in the experimental vehicle, a 1985 Cadillac Sedan deVille, included an Etak Navigator moving-map display which included several features that were available to the subjects and which might be expected on such a system (McGranaghan et al. [1]). These were the on-screen highlighting of a specific target destination, and allowing the user to view the map information on several different scale levels. In addition to the basic map and vehicle location/orientation information, the system's computer could also present such important items as euclidean distance and direction from the current location to the destination, and a compass.

Other apparatus included conventional paper maps, an AC/DC power conversion system, an IBM Personal Computer with Metrabyte DASH-8 analog/digital converter and PIO-2 parallel port cards, a custom-designed computer interface with push button panels, a Setchel-Carlson CCTV monitor, two videocameras and two videorecorders, and a microphone. A passenger side brake pedal was available to an onboard experimenter in case of emergency. One videocamera was mounted on the hood (on the passenger side) and angled to shoot the driver's face through the windshield. A second camera was mounted on the roof, directly over the subject's head. The hood mounted camera provided information on where the subject's eyes were looking, and the roof mounted camera gave a forward looking view from the point of view of the subject to help determine where the subject was looking outside the vehicle.

Three routes defined by starting points and unfamiliar destination points in the Blacksburg/Christiansburg/Radford area of Virginia were designated as the experimental routes. Each route was selected to match the others in terms of time to complete a route (about twelve minutes), the road types that would likely be traversed on the way to the destination (i.e., four-lane and two-lane highway and residential streets), and overall complexity, to as great an extent as possible.

3.2.1. Spatial ability tests

Three factor-referenced paper and pencil tests (Ekstrom, French, Harmon, and Dermon [7]) were used to measure individual differences with regard to specific aspects of spatial ability judged to be relevant to automobile navigation. Perceptual speed was considered relevant, because navigation success depends on the speed with which relevant cues can be perceived from the environment as well as from any navigation aid. This construct was measured with the Identical Pictures Test. Spatial orientation was considered an important aspect of spatial ability, because bodily orientation is such an important part of navigation. This construct was measured with the Cube Comparisons Test. Thirdly, spatial scanning was measured to determine the subjects' abilities to explore visually a wide or complicated spatial field under time stress. This construct was considered implicitly and directly related to navigation and was measured with the Map Planning Test.

3.3. Procedure

The subject was interviewed to insure that he or she was unfamiliar with the three destinations. The subject then completed a questionnaire concerning demographic and health matters, and was administered the tests of spatial ability. Each subject was thoroughly trained in the use of the moving-map display because of its novelty. The subject was not trained in the use of the paper map, because despite inherent differences in map reading ability which might be expected due to differences in spatial ability and experience, it was assumed that all subjects basically knew how to use a paper map. The final phase of the training procedure allowed the subjects to freely use the moving-map display to actually navigate to a nearby destination whose location was not known; reaching this destination was the training criterion that had to be met for the subject to be permitted to participate in the study.

In an effort to control for the effects of traffic density variation, runs taking place after morning "going-to-work" traffic and before lunchtime were considered low traffic density runs, whereas runs taking place during afternoon "going-home" traffic were considered moderate traffic density runs.

The navigation conditions that were tested included: (1) conventional paper map (2) moving-map display, and (3) memorized route, which served as a baseline of comparison for the two map methods of navigation. The order of presentation of the methods and routes was counterbalanced. In all three conditions the subject was asked to read aloud any street signs or directional signs that were used to navigate; this was used to aid in the classification of glances outside the vehicle in conjunction with the forward looking videotape. Note that the subject was not asked to direct more or less attention to such signs than he or she naturally would, nor was the subject asked to monitor where his or her attention was being allocated, per se.

3.3.1. Paper and moving-map conditions

In the paper map condition, the subject was free to study the paper map for as long as was desired before beginning to drive to the destination; during the drive to the destination the subject was free to use the paper map in any way. This included using the map while driving, and the subject was also instructed that, if necessary or desired, he or she could stop at any safe and legal place to reorient or study the map for as long and as often as he or she wished. The subject drove to the starting point of the route and was instructed to drive to an explicitly specified destination, about 12 minutes away. The subject was requested to drive to the destination as "efficiently as possible." In an effort to allow the subject to navigate naturally, each was allowed to interpret this instruction as they desired. That is, each subject could optimize the route based on time, distance, simplicity, or some other criterion that he or she decided would make the route most efficient. The subject was not informed if a navigation error was committed (i.e., a wrong or missed turn); instead a data run was considered complete when the destination was reached or when 20 minutes had elapsed and the subject clearly could not reach the destination in another five minutes. Identical procedures were followed in the moving-map condition.

3.3.2. Memorized route condition

In the memorized route condition, the subject was shown a route, in detail, on a map which could be studied; the subject was not allowed to look at the map again in the memorized route condition. The subject then drove along the full length of the route, then back again along the same route to the starting point. The next run to the destination without a mistake in following the prescribed route was considered the data run. If errors were committed during a potential data run in this condition, then the data taking ceased, and the subject was informed of the mistake and allowed to continue to drive to the destination. The subject then drove back to the starting point to begin another attempt to complete an error-free run. This process was repeated as many times as was necessary to achieve an error-free run for each subject in this condition.

3.4. Dependent Measures

3.4.1. Time measures

These data included time to study the paper or moving-map display before beginning to drive to the destination (study time), the time taken to drive to the destination (drive time), and the sum of these two (total time). Certain successful data runs were allowed to continue for longer than 20 minutes (up to nearly 24 minutes) if the subject was making steady progress toward the destination, whereas data runs wherein the subject was hopelessly lost were terminated after 20 minutes. So that all runs where the destination was not reached would be scored as taking longer than all runs where the destination was reached, the total time scores of the runs wherein the destination was not reached were assigned the value of 24.0 minutes.

3.4.2. Navigation effectiveness rating

A rating was assigned to each experimental run based on the directness of the routes selected and the presence of navigation errors. A five point rating scale was used; the route was rated a _5_ if a direct route was taken and no errors were committed, whereas the route was rated a _1_ if the destination was not reached. Ratings _2_, _3_, and _4_ were assigned to runs wherein the destination was reached, but indirect routes were taken and/or errors were committed.

3.4.3. Eye movement measures

Each glance while driving (i.e., while the vehicle was in motion) was classified into one of the following categories: (1) roadway-center, (2) roadway-off center, (3) mirrors, (4) signs and landmarks, (5) paper map/moving-map (depending on condition), (6) conventional instruments, and (7) other. A fixation is considered to occur whenever the eyes stop and focus on an image; if there is a change, no matter how slight, there is a new fixation. When a series of fixations are trained on a single object or image in succession, this series of fixations is considered to be a glance. Glances were measured in this study.

4. RESULTS

4.1. Spatial Ability

The scores on each test were standardized so that the scores on one test would be directly comparable to those on the other two, despite initial scale differences. Subjects were grouped with a median split into high and low spatial ability groups based on an equally weighted composite of their standardized scores on the three measures of spatial ability.

Correlation coefficients were calculated between each pair of spatial ability tests. The scores of each test were significantly correlated with each of the other two ($\alpha = 0.05$). The correlations were as follows: between the Identical Pictures and the Map Planning tests, $r = 0.71$, $p = 0.0001$; between the Map Planning and Cube Comparisons tests, $r = 0.45$, $p = 0.01$; and between the Identical Pictures and the Cube Comparisons tests, $r = 0.41$, $p = 0.018$. These results indicate that the three tests measure constructs that are somewhat related (especially the Identical Pictures and Map Planning tests), and that it was logical to group subjects based on a composite of these three test scores.

4.1.1. Time measures

Analyses of variance (ANOVAs) were performed on several of the key time measures including study time, drive time, total time, and average time per glance; as well as the navigation effectiveness ratings. All of these except average time per glance were standardized by route to help control for inherent and unavoidable route differences. The factors of spatial ability, navigation condition, and their interaction were included in each ANOVA; neither spatial ability nor its interaction with navigation condition was significant ($\alpha = 0.05$) for any of the dependent measures tested. However, the interaction of spatial ability with navigation condition was almost significant for the variable, total time, $F(2, 60) = 2.93$, $p = 0.06$.

4.1.2. EYEMAP

The variable, EYEMAP, was the proportion of time spent glancing to the map or map display, depending on navigation condition. An ANOVA was performed on EYEMAP revealing that those in the high spatial ability group spent a lower proportion of time (0.12) glancing to the paper map or moving-map (depending on condition) than did those in the low ability group (0.15), $F(1,30) = 6.75$, $p = 0.0144$. Navigation condition was also significant for this variable, $F(2,60) = 270.34$, $p = 0.0001$. A Newman-Keuls post hoc analysis showed that there were significant differences ($\alpha = 0.05$) among all three navigation conditions on the EYEMAP variable; the means were: 0.0, 0.068, and 0.331 for the memorized route, paper map, and moving-map conditions, respectively.

5. DISCUSSION

5.1. Spatial Ability

Results indicated that subjects with high spatial ability may be able to derive the necessary information from a moving or paper map in less time than those with low spatial ability; this is not surprising. The more interesting hypothesis that spatial ability would differentially affect performance with the two map methods of navigation was not supported.

5.2. Navigation Strategy and Knowledge Representation

With regard to time to reach the destination and the directness and quality of routes selected, the two map methods were not significantly different (Antin, et al. [8]). However, the moving-map display drew a substantially greater proportion of visual attention than did the paper map while driving. The key to understanding these results is in understanding the strategy adopted in the use of each method. The paper map was used to plan, essentially, the entire route from start to finish; that is, a node-to-node sequence was memorized. After this initial study and memorization phase, the paper map was used only as an occasional reference. In contrast, the moving-map display could provide only general route information beyond approximately a one half mile radius of the current location of the vehicle. Despite the additional information and orientation cues provided by the moving-map display, effective use of the system could only be accomplished by continually glancing at the display to acquire important information as it was updated and presented. As a result, the moving-map significantly drew the subject's gaze away from the driving task relative to the norm established in the memorized route condition, as well as in comparison to the paper map.

It seems, then, that the paper map aids in the development of procedural knowledge of an area (for any distance beyond one half mile) to a greater extent than does the moving-map display used in this study, because not all scale levels showed the complete network of streets. This is a feature that is designed to reduce display clutter, and can be expected to be a problem in most moving-map display systems, since a typical moving-map display would likely include an area/detail tradeoff (McGranaghan, et al. [1]). The moving-map display can provide mostly survey information for distances beyond one half mile (e.g., the display in this study indicated current location of the vehicle and its directional relationship to North and the destination, and its euclidean distance to the destination). The paper map provides this information only roughly and only implicitly, but it does provide the complete network of roadways at all times for subjects to study and derive the procedural information so important for navigation.

5.3. Future Research

An avenue of future research should be to find tests of spatial ability which are more sensitive to navigation performance than those used in this study. Even though, as stated above, this study did not show a differential effect of spatial ability on navigation condition, the data did indicate that more research along these lines is warranted. If it could be shown that one type of navigation aid were more beneficial for individuals with more or less spatial ability, then this would be valuable information.

Also, certain design changes could be implemented in the moving-map display which could enhance the ability of this type of system to aid in the development of procedural knowledge at greater distances from the destination. For example, the system could include an on-request detail feature, which would display the complete network of streets on the large area scales. This would likely create undesired clutter, but on a temporary basis it may help to alleviate the problem whereby the moving-map display does not present procedural knowledge.

ACKNOWLEDGEMENTS

This research was sponsored by the Research Laboratories of the General Motors Corporation. The authors express particular thanks to Dr. Kenneth M. Farmer, William Spreitzer, and Richard Rothery of G.M. for their technical and managerial support.

REFERENCES

[1] McGranaghan, M., Mark, D. M., and Gould, M. D. (1987). Automated provision of navigation assistance to drivers. The American Cartographer, 14, 121-138.

[2] Streeter, L. A., Vitello, D., and Wonsiewicz, S. A. (1985). How to tell people where to go: comparing navigational aids. International Journal of Man/Machine Studies, 22, 549-562.

[3] Thorndyke, P. W., and Hayes-Roth, B. (1982). Differences in spatial knowledge acquired from maps and navigation. Cognitive Psychology, 14, 560-589.

[4] Stasz, C. and Thorndyke, P. W. (1980). The influence of visual-spatial ability and study procedures on map learning skill. (Tech Report No. N-1501-ONR). Santa Monica, CA: Rand Corp.

[5] Stasz, C. (1980). Planning during map learning: the global strategies of high and low visual-spatial individuals. (Tech Report No. N-1594-ONR). Santa Monica, CA: Rand Corp.

[6] Antin, J. F., Dingus, T. A., Hulse, M. C., and Wierwille, W. W. (1988). The effectiveness and efficiency of a moving-map display. Manuscript submitted for publication.

[7] Ekstrom, R. B., French, J. W., Harmon, H. H., and Dermon, D. (1976). Manual for kit of factor-referenced cognitive tests. (NR 150 329). Princeton, NJ: Educational Testing Service.

Trends in Ergonomics/Human Factors V
F. Aghazadeh (Editor)

AN ERGONOMIC EVALUATION OF THE FARM TRACTOR

J. M. USHER, F. AGHAZADEH, and R. AZIMULLAH

Industrial Engineering Department
Louisiana State University
Baton Rouge, Louisiana, 70808 U.S.A.

This study examines and compares a group of farm tractors from the viewpoint of ergonomic design. The results indicate that advances have been made in the ergonomic consideration of the tractor design, but that further improvements are still possible, and needed.

1. INTRODUCTION

Farming is a world-wide industry in itself, with many aspects of the trade involving the use of machinery to aid the farmer. The most popular and standard piece of equipment used on farms is the tractor. A tractor is a versatile machine and can function in many roles with the aid of the numerous attachable farming implements. It is because of this versatility and widespread use that the ergonomics of tractor design are so important.

In this study the ergonomic design of the farm tractor is investigated focusing on the ergonomic related features of the tractor. These features encompass the design, placement and type of mirrors, controls, displays, and seating. In addition, the available aids for mounting the tractor and the operating noise levels are evaluated.

2. PROCEDURE

2.1. Tractors

Three tractors were selected from two different manufacturers, Ford and Case International. The first tractor, referred to as Tractor-A, is a Ford model 5610 open-air (i.e. no enclosed cab) tractor. This tractor is a standard 2 wheel drive tractor with 62 HP. Tractor-B, a Case model 1394, was selected as the second tractor because of its similar size (65 HP) and application. The third tractor for study, Tractor-C, is a closed-cab model, the Case 2294. This tractor is a 130 HP, two-wheel drive, model featuring a majority of the luxury features available in tractors. Using these three tractors, the ergonomic differences between tractors from two different manufacturers, and between tractor types (no cab .vs. closed-cab), are studied.

2.2. Methods and Equipment

Since the tractors were not the property of the investigators, the tests involved in the study are those that do not require the operation or alteration of the tractors in any way. All measurements were taken with respect to seat reference point (SRP) with the seat adjusted as far to the rear as possible. Due to the brevity of this report, the tabulated results were not included. Measurements of the force required to operate the foot pedals were obtained using a tension meter. Measurements of the noise level were obtained using a decibel meter.

3. RESULTS AND ANALYSIS

3.1. Field of Vision

The visual field of the tractor driver is an important factor affecting both the safety and quality of their work. The operator requires a clear field of vision both in the front, for steering the tractor, and in the rear, for controlling the attached implements. Improvements in this area come from pro-perly designing the safety frame to provide the maximum possi-ble field of view and employing the use of mirrors to provide a view of the rear of the tractor without body rotation.

The forward field of view for the two open-air tractors, A and B, was clear. No safety frame components were present to cause obstructions. However, neither of these tractors had mirrors and therefore would require the operator to rotate for monitoring operations at the rear of the tractor. The closed-cab tractor had a single mirror located on the interior of the cab. This mirror was rectangular in shape with a convex curvature. These characteristics match those recommended by Sjoflot [1] except for the location and dimensions. Sjoflot specifies a size of 20 x 30 cm and a forward location of 35 - 90 cm in front of the user, while Tractor-C's measured 9 x 20 cm and was located 100 cm in front. This location and reduction in size can lead to reduced performance and, as a result, lack of use. Plenty of room is available in the cab to tolerate an increase in mirror size. The forward view of the closed-cab tractor was reduced due to the framed cab, but not much can be done to change this due to the structural requirements.

3.2. Pedals and Controls

In a survey by Sjoflot's [2], farmers listed the control of implements as one of the most important aspects of tractor design. This concern was accompanied by a grave interest in the layout and operation of both the hand and foot operated controls on the tractor.
Each of the tractors in this study used four standard pedals, the clutch, left brake, right brake and the foot throttle. A

fifth pedal, present on both open-air models, was used to activate the differential lock. On the closed-cab tractor, the differential lock was electronically activated by a rocker switch on the right-hand control panel.

A driver's legs straddle the tractor with each foot resting on a separate floorboard. Pedal location is approximately the same for both of the open-air tractors, with each pedal's pivot point attached below their respective floorboard. Two pedal, a clutch and brake, are on the left-side floorboard and the foot throttle and right brake on the other side. In the closed-cab tractor the driver sits in a compartment much like an automobile where the pedals are located in one cluster under the dash board with the pivot point located above the pedal. Such a placement allows for a more natural movement of the leg in engaging the pedals. According to Phesant and Harris [3], the optimum pedal position to accommodate posture while maximizing the force exerted, is approximately 20 cm below and 74 cm in front of the SRP for both the male and female. The closed-cab model provides this optimum location for all of its pedals except the foot throttle which is set further forward. The pedals for the open-air models are placed below the optimum position, but still in the 38-46 cm recommended range for the forces required. The recommended travel distance for the pedals is 5-15 cm [4]. All the tractors fell within this range (i.e. 6-15 cm), except the travel distance on Tractor-A's clutch (> 18 cm). For each tractor, the force required to activate the pedals fell within the recommended range of 4.5-13.5 kgs of force.

The controls used in the operation of the tractors varied between manufacturers, with a majority controlling standard functions. Tractor-A had the fewest number of controls, with only 11, Tractor-B the next with 19, and Tractor-C, with the closed-cab, the most with 22 controls.

Most controls were of the lever type, with some selectors, knobs and rocker switches. Overall, the types of controls used and the expectation of their action were well coordinated with the function. For each tractor, about 70% of the controls were placed on the right-hand side. This results in an overloading of the users right-hand. The functions performed by the left hand usually involve controls used infrequently, demonstrating that the right hand is doing most of the useful work. Such an overloading of the right hand is most likely an acceptable feature when considering the fact that the tractor driver spend a majority of their time in a twisted posture, with the left-hand on the steering wheel and their upper trunk twisted to the right observing the operation of the implements. This leaves the right-hand the responsibility for operating the controls.

Location of the controls within the operator's reach varied between the three tractors. The distance to each control was calculated using the position values recorded in the study. The reach of both the male and female were then calculated using the fifth percentile values as listed in the anthropometric tables [5]. This would determine if 95% of the users, male or female, would be able to reach the controls.

With the seat set as far back as possible on Tractor-A, only one control was accessible without requiring trunk movement. Movement of the seat forward, half-way, resulted in insignificant change in the reach capabilities. If the seat is moved all the way forward, then operation of the gear shift and throttle lever becomes difficult due to interference with the seat during use. In addition, all control handles except the throttle and engine stop are located below the SRP on Tractor-A. This inability to reach the controls appears to be due only to a lack of planning when compared with the controls placement on Tractor-B.

Tractor-B, from manufacturer 2, was the next best in its placement of controls. Five of the seven frequently used controls can be reached without movement of the trunk and the other two can be reached by tilting the trunk. Some of the non-frequently used controls require movement of the trunk, with the worst being the combining valve and the parking brake. Both of these require that the operator lean over and tilt to the right to activate them. The other non-frequently used controls require either a forward lean or a tilt to the side to reach them, except the PTO engagement lever which requires that the user reach back over the seat to operate it.

The closed-cab tractor provided the best attention to controls placement, with all controls positioned within the reach of 95% of the male and female population. In the closed cab tractor all the controls are located in clusters instead of the sparse arrangement found on the open-air tractors. The frequently used operating controls are all located on the right of the driver in a coordinated grouping. The controls for the cab's interior, are located above the driver on the right hand side, and the driving related controls (i.e. rpm, engine temperature, etc.) on the dashboard.

For each of the tractors, there was interference of the steering wheel with dash mounted controls. The purpose, it seems, is to minimize the size of the dash and therefore, enlarge the area for access. Since most controls on the dash are not frequently used, this tradeoff seems appropriate.

Labeling of the controls was employed using one or more of the three forms; written, numbered or symbolic. The open cab models had the worst labeling, using locations for the labeling that were not visible from the seated position or required a guess as to their corresponding control. Familiarity with tractors is required to decipher the coding, but was standard between the manufacturers studied. The closed-cab model utilized the best labeling scheme incorporating at least two types of labeling with each of a majority of the controls.

3.3. Displays

The displays on a tractor provide information used to monitor the mechanical operation of the tractor. For both of the open-air models, analog displays of the fixed-scale moving pointer type were employed to give information about the engine RPM, fuel level and water temperature. The size of the two tractor's displays were approximately equal and both adequate in design. The RPM display was the larger measuring 8.9 cm square and the others 3.8 cm square. They each utilized black backgrounds with white needles. Tractor-A had yellow numbers and graduation marks while Tractor-B used white. Written and symbolic labels were used on each tractor and color coding was employed to signify ranges. Tractor-A provided 4 warning lamps that would display a back-lit symbol if activated. These provided warning information about low voltage, air filter cleanliness, lights and oil level. Also, contained in the middle of the RPM display was a meter which registers the total hours the tractor has been used, similar to an odometer measuring a car's total mileage.

The closed-cab tractor employs an electronic display utilizing a touch panel keyboard to obtain information about many additional tractor and work functions not normally available, such as the ground speed or the ground-area finished. The information is displayed on a 2.5 x 6.5 cm rectangular lighted LED display. Coolant temperature, fuel level, exhaust temper-ature and voltage level are all displayed using vertical LED linear displays. All displays use written and/or symbolic labeling. Color coding is used on the linear LED displays to define the necessary operating ranges.

The angle of the operator's line-of-sight in viewing the displays was calculated based on the anthropometric measurements for the 5th percentile of the female users and then the 95th percentile of the male users, using both the lowest and highest vertical positions of the seat. The resulting values would then cover the ranges for 95% of the women and 95% of the men. The angle for both open-air tractors fell within the desired range of 10-30 degrees [5]. The closed-cab model yielded angles placing the line of site near the horizontal plane. These values are only for the seat in the lowest position, but should improve with increases in the vertical position of the seat.

TABLE 1: Angles of the Line of Sight in Viewing the Displays.

		Line of Sight (degrees)		
Sex	Position of Seat	Tractor-A	Tractor-B	Tractor-C
Male	Bottom	-18	-29	-3
Male	Top	-22	-30	NA
Female	Bottom	-12	-23	3
Female	Top	-16	-28	NA

The angle of the display from the vertical plane measured 15 degrees for Tractor-B and 21 degrees for Tractor-A. With the seat set in the lowest vertical position, and as far back as possible, only the bottom half of the display on the Tractor-B was visible due to interference with the display case. Any further adjustment of the seat only reduced visibility. The entire display on Tractor-A was visible. This increased visibility was the result of placing the display 14.5 cm further back from the operator. Both these displays used the small tilt angles (from vertical) to eliminate interference from glare. In the closed-cab tractor glare is not as much of a concern due to the presence of tinted glass and overhead protection; therefore, a larger tilt angle is use resulting in easy viewing.

3.4. Seat Design

A tractor's seat design is important to provide for proper posture. Poor working posture is believed to cause increases in onset of operator fatigue and back troubles with continued use. The three tractors studied provided adequate dimensions for the seat bottom and back, with the closed-cab being the largest. Standard adjustment of vertical height and horizontal distance were provided by all manufacturers. For all tractors, adjustment of the seat required the user to lean forward to locate and reach the controls. Both the open-air tractors required that the user walk around to the back of the tractor to adjust the vertical position, but even worse, Tractor-A required that a wrench be used.

In all three tractors, some means of reducing the effects of vibration, and bumpy terrain, on the driver was employed. The open-air tractors used a small pneumatic shock absorber to support the seat and the closed-cab model had some type of electronic isolator. No other adjustments were available except on the closed-cab model which included adjustments for the arm rest, lumbar support, seat bottom tilt angle, and swivel angle. These additional options are the type Sjoflot suggested to improve operator posture and reduce related problems [6]. The ability of the seat to swivel is important in reducing the angle of rotation of the operator while monitoring the functions behind the tractor. The closed-cab tractor seat was covered by a fabric material, while the seats on the open-air models were covered by a vinyl material in order to tolerate exposure to the weather. Seat belts are standard on all three of the tractors, with the closed-cab providing an automatic retraction mechanism for the belt.

3.5. Access Mounting

Due to the function and design of the farm tractor, the mounting of the tractor can be a difficult experience. All the tractors studied provided steps, of the stair ladder type, to aid the operator in mounting the tractor. The recommended distance between steps of this type is 18 - 30 cm [4]. On all the tractors, the height of the first step (44 to 50 cm) exceeded the recommended limit. This is due to the need to maximize the tractor's ground clearance. The height of the other steps, beyond the first, were near the maximum, but still within the recommended range.

Access was only available from the left side on both Tractor-A and the closed-cab tractor, Tractor-C, while Tractor-B allowed access from either side. On Tractor-A, a tool box mounted on the left fender caused interference during access to the seat, and should be relocated. Also on Tractor-A, interference is caused by both gearshift levers located in the center requiring the operator to carefully route the leg or lift it completely over them. Access from the right, on Tractor-B, is met with interference from its gearshift lever, but access from the left is much easier. No handles are provided on either open-air tractors and the natural tendency is to use the steering wheel for leverage. The closed-cab model provides a handrail that runs vertical on the front frame member of the cab. It's location requires a 173 cm reach and therefore, is accessible by 95% of the female population and close to 100% of the male population [5].

3.6. Noise Levels

Noise levels during operation of the farm tractor was listed, by the surveyed farmers, as being of considerable importance [2]. In that survey, 37% of the farmers stated that they experience hearing problems and that less than half were using hearing protection devices.

Noise levels were recorded for both an open-air model, Tractor-B, and a closed-cab model, Tractor-C, operating at idle and at full throttle. The noise levels for the open-air tractor (80 to 90 db) exceed that recommended for prolonged exposure [7]. For the closed-cab model, a 20 db drop from the exterior levels of 88 to 95 db, was attributed to the insulation of the cab. This represents a significant difference and reduces the noise to acceptable levels (i.e. 65 to 75 db). Still, these tractors produce enough noise to make the use of hearing protection a general operating requirement.

4. CONCLUSIONS

Each tractor studied had its advantages and disadvantages from an ergonomic standpoint. When comparing the two open-air tractors, A and B, Tractor-A proved superior in its design for the controls, pedals, and aids for mounting, while Tractor-B provided better design of displays. These results indicate the existence of a difference between manufacturers in their attention to ergonomic design.

The clear leader of this group, in terms of ergonomic design, is the closed-cab tractor, Tractor-C. Just a glance at the interior provides a feeling of comfort and the opinion that its interior-design was well thought out. When compared with the open-air tractor, the closed-cab model proved superior with regards to its ergonomic conformance in the areas of, placement of controls and pedals, the availability of seating adjustments and mirrors, and the design and positioning of displays. In addition, the closed cab tractor provided improvements in access mounting and noise reduction.

The ergonomic improvements that come with the purchase of a closed-cab tractor are not cheap. The approximate cost for adding an air-conditioned cab to an open-air tractor is $6000 on top of the $22,000 price paid for the basic tractor. Other amenities can also be purchased when using a closed-cab, such as an ash tray, cigarette lighter and stereo system. As pointed out by Sjoflot [2] in his survey of Norwegian farmers, a majority of the farmers are willing to pay for a well designed tractor; therefore, the increased cost of the close-cab model should not be a deterrent.

Overall, the tractor manufacturers have made advances in ergonomic design, but still seem to be overlooking certain details which can make a difference in operator performance. Possibly with the continued interest of researchers and the increased exposure in the literature, the manufacturers will take notice of the problems and provide the improvements needed.

REFERENCES

[1] Sjoflot, L., Big Mirrors to Improve Tractor Driver's Posture and Quality of Work, J. Agric. Eng. Res., 1980, 25, pp. 47-55.
[2] Sjoflot, L., The Tractor as a Working-place: A Preliminary Report on a Survey Among Norwegian Farmers and Tractor Drivers, Ergonomics, 1982, pp. 11-18.
[3] Phesant, S. T. and Harris, C. M., Human Strength in the Operation of Tractor Pedals, Ergonomics, 1982, 25(1), pp. 53-63.
[4] Diffrient, N., Tilley, A.R. and Harman, D., Humanscale 7/8/9 Manual, Cambridge, Mass.: The MIT Press, 1981.
[5] Bailey, Robert W., Human Performance Engineering: A Guide for System Designers, Englewood Cliffs, N.J.: Prentice-Hall, Inc., 1982, pp 82-83.
[6] Sjoflot, L., Means of Improving a Tractor Driver's Working Posture, Ergonomics, 1980, 23(8), pp. 751-761.
[7] Eastman Kodak Company, Ergonomic Design for People at Work, Belmont, CA.: Lifetime Learning Publications, 1983, p. 211.

Trends in Ergonomics/Human Factors V
F. Aghazadeh (Editor)

ERGONOMICS OF MECHANICAL TRANSPLANTING

Malcolm E. Wright, Thomas R. Way and Mark D. Miller

Professor, former Research Associate and Research Associate, Agricultural Engineering Department, Louisiana State University, Baton Rouge, Louisiana

A transplanter simulator with five horizontal loading stations was tested in the laboratory to determine the actual and perceived effects of several operating variables on the human operators. Independent variables included eight operator seat positions, three pocket widths for receiving plants, three speeds of operation, and two test durations. Control tests were also done using a prototype of a conventional transplanter. Dependent variables included the percentage of pockets missed, the number of plants laterally mis-positioned in the pockets, and the subjects' perception of difficulty. Best results were obtained when the center of the five loading stations was 60 degrees to the right of the direction the operator faced and the pockets traveled from his left to his right. The best pocket width was 19 cm. A transplanter, duplicating the best laboratory results as closely as practical, was constructed and field tested versus two conventional transplanters at three speeds of operation. Dependent variables included mean plant spacing, the ratio of theoretical to actual plant spacing, the coefficient of variation of plant spacing, and yield. Some significant differences were obtained in these tests but none were dramatic. Operator perceptions favored the experimental planter, indicating a benefit over long planting durations as in commercial planting operations.

1. INTRODUCTION

Transplanting is one of the most labor intensive field operations in the production of vegetables and several other crops. A conventional, one-row, bare-root transplanter has a series of mechanical hands, or "pockets", into which two operators alternately place individual plants. These pockets move vertically downward and deposit the plants in furrows made by those parts of the planter contacting the soil.

Wright and Way (1982) found that the typical rate of uninterrupted transplanting was 31 pockets/min/operator when planting sweet potato cuttings 30 cm apart at 1.2 km/h. Splinter and Suggs (1968) found the major benefit of using multiple loading stations for mechanical transplanting of tobacco occurred when five loading stations were used. Suggs (1979), using the experimental transplanter, found one operator to be as effective as two using a conventional transplanter. Dooley (1983), using an experimental transplanter for forest seedlings, found operators

Approved for publication by the Director of the Louisiana Agricultural Experiment Station as manuscript number 88-07-2072 were able to load plants into 25% more pockets when the pockets traveled horizontally than when moving vertically. Four to six loading stations were necessary to insure continuous pocket fill when operators reached for bundles of seedlings. Dooley predicted a gain in daily production of 15 to 20% for the experimental transplanter compared to conventional transplanting. In work not related to transplanting, McCormick (1957) reported on one-handed visually-controlled positioning movements of humans. The combined accuracy and speed of the movements was best when the target was positioned 60 deg to the right of the direction a subject faced. The overall objective of this study was to determine the effects of operating rate, multiple-loading stations with horizontally-moving pockets, pocket width, and duration of uninterrupted planting on the performance of transplanter operators and on their perceived difficulty. A mechanical transplanting simulator was used and comparative studies were made in the laboratory with a conventional transplanter unit. Also, an experimental transplanter was compared with conventional ones in field studies.

2. EQUIPMENT

A multiple-loading-station transplanter simulator was constructed for labortory use. It had a series of detachable pockets attached to two chains driven by an electric motor and variable speed drive (Fig. 1). Half of each pocket was attached to each chain, leaving a space between halves. Speed was monitored by a tachometer. Three pocket widths - 6.4 cm, 13 cm and 19 cm - were used. The loading-station area was horizontal and plants were loaded into the pockets from above. Sheet metal shields, mounted at the ends of the loading-station area, were used to limit the number of loading stations to five for each pocket width.

A photoelectric system and electric micro-switches were mounted where the pockets left the loading-station area. To detect unfilled pockets, the beam of the photoelectric source and detector shone vertically between the two halves of the pockets. Four micro-switches, connected to TTL devices, were situated such that any object which projected beyond either end of a pocket would close a switch. Artificial plants, 30.5 cm long and similar in size and shape to sweet potato transplants, were used (Wright and Way, 1982). The overall length of each pocket for the simulator was 32.4 cm, allowing 1.9 cm lateral mis-position without a position error. In each test, the operator sat in one of eight positions, designated A through H, as shown in Table 1. The artificial plants were held in a box which was positioned for maximum convenience of the operator.

TABLE 1. THE EIGHT SEAT POSITIONS FOR THE TRANSPLANTER SIMULATOR

Pos.	Diagram	Pos.	Diagram	Pos.	Diagram	Pos.	Diagram
A	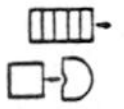	C	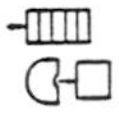	E		G	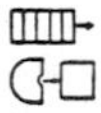
B	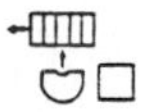	D	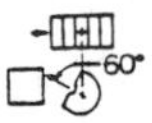 60°	F	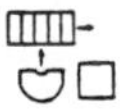	H	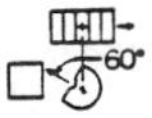 60°

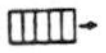

Loading stations and direction of pocket travel.

Seat showing direction operator faces. Plant box.

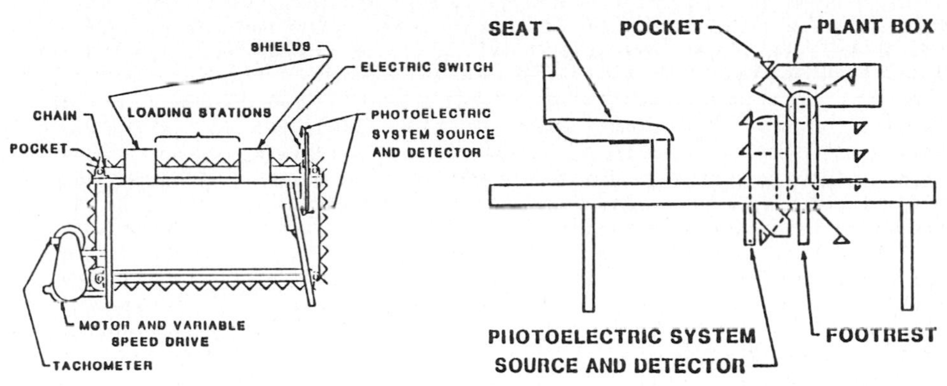

Fig.1. Side View of multiple-loading station transplanter simulation

Fig.2. Side view of conventional transplanting unit.

A conventional transplanting unit was modified for stationary use by removing the soil-contacting parts and attaching support stands (Fig. 2). This unit was driven by the variable speed drive of the simulator. The photoelectric system could be mounted on the conventional unit, when needed, to detect unfilled pockets.

An Apple II Plus microcomputer with a clock was used for data acquisition and analysis. The output of the photoelectric system and the TTL devices was input to the computer. Software recorded current clock time when the photoelectric beam was interrupted or when any of the electric switches closed.

3. PROCEDURE

Sixteen university students, ten male and six female, participated as subjects in this study. Thirteen were right-handed and three left-handed. None had prior experience as transplanter operators.

The study was conducted in three parts, each with several transplanting sessions. During each session, a subject, at the simulator or at the conventional transplanter, attempted to place a plant into each pocket. When the simulator was used, the subject tried to position the plants laterally so neither end of a plant extended beyond the ends of a pocket. The data on unfilled pockets and laterally mis-positioned plants were saved on computer disks.

Latin square experimental designs were used because they are useful in controlling two extraneous sources of variation. The two blocking variables were subjects and the order in which a subject encountered a treatment. A mixed model was assumed for each analysis of variance. The subject was treated as a random factor and the treatment variable for each analysis was treated as a fixed factor. Repeated measures designs were used.

Subjects who participated in the first and third parts of the study did not practice transplanting before beginning the first session. Each participant in the second part obtained transplanting experience in the first part. The study was not intended to compare left-handed with right-handed subjects. Before each session, plants in the box were aligned with one another and with their stems pointing toward the subject. In all tests with the simulator, five loading stations were used.

In Part One of the study, each of eight subjects, all right-handed, performed eight transplanting sessions with the simulator. The seat position of the simulator was different for each session for each subject. The duration of each session was 10 min, pocket width was 6.4 cm, and operating speed was 50 pockets/min.

For Part Two, five transplanter arrangements were used, four with the simulator and one with the conventional transplanter. The four simulator seat positions were A, F, G and H, the ones for which the average numbers of loaded pockets were greatest in Part One. Five subjects participated in Part Two. Four were right-handed and one was left-handed. Three operating rates were used: 40, 50 and 60 pockets/min, in consecutive order for each of the five transplanter arrangements by each of the five subjects. The duration of each session was 20 min and pocket width was 13 cm. After each session, the subject rated his perceived comfort and difficulty.

In Part Three, transplanter simulator seat position H was used with three pocket widths: 6.4 cm, 13 cm and 19 cm. Six subjects participated, four right-handed and two left-handed. The duration of each transplanting session was 20 min. After each session, the subject rated his perceived difficulty.

4. RESULTS OF LABORATORY TESTS

Part One: The treatment independent variable was the seat position of the simulator and the dependent variable was the number of plants loaded in 10 min. The analysis of variance indicated that differences between seat positions did not significantly affect the dependent variable at the 5% level. None of the orthogonal comparisons was significant at the 5% level. The results are presented in Table 2.

TABLE 2. MEAN NUMBER OF PLANTS LOADED INTO SIMULATOR IN 10 MIN

Sim. pos.	No. of plants	Sim. pos.	No. of plants	Sim. pos.	No. of plants	Sim. pos.	No. of plants
H	457	G	446	B	441	D	436
F	452	A	444	C	441	E	431

Part Two: The operating rate had a significant effect on the percentage of pockets loaded per 20 min session for the conventional transplanter ($p = 0.0381$) and for the simulator ($p = 0.0001$). It also had a significant effect on the percentage of mis-positioned plants in the simulator ($p = 0.0001$). The results of Duncan's multiple range test, performed on the means in each of these three sets of tests, are presented in Table 3.

A set of analyses of variance and orthogonal comparisons was performed for each operating rate in Part Two with transplanter arrangement as the independent variable. The dependent variables were the number of plants loaded in 20 min and the percentage of loaded plants that were laterally mis-positioned.

TABLE 3. EFFECTS OF TRANSPLANTING RATE ON OPERATOR PERFORMANCE FOR CONVENTIONAL PLANTER AND SIMULATOR WITH 13 cm POCKETS

Rate in Pockets/min	Mean % of pockets loaded: Conv. Planter	Simulator	Mean % plant mis-positioned in simulator
40	89.9a#	97.5a	4.85a
50	98.0a	93.9a	9.24b
60	83.0b	88.9b	16.8c

#Within each column, means followed by the same letter are not significantly different at the 5% level.

For the 40 pockets/min rate, the transplanter arrangement did not significantly affect the number of plants loaded in 20 min at the 5% level. One orthogonal comparison indicated that the mean number of plants loaded in 20 min was significantly greater for simulator seat positions A, F, G and H than for the conventional transplanter (p=0.0249). Also for the 40 pockets/min rate, the simulator seat position significantly affected the percentage of laterally mis-positioned plants (p=0.0326). The results of Duncan's multiple-range test of these means are shown in Table 4.

The percentage of pockets not loaded was plotted versus the operating rate for each transplanter arrangement in Part Two (Fig. 3). In each case, the percentage increased as the operating rate increased. In all cases, the percentage was greater for the conventional transplanter than for the simulator. The percentage of laterally mis-positioned plants was plotted versus the operating rate for each simulator seat position (Fig. 4). For each, the percentage increased as the rate increased.

Friedman's distribution-free test (Hollander and Wolfe, 1973) was used to analyze the comfort and difficulty ratings for each transplanter operating rate. The transplanter arrangement significantly affected the operators' perceived comfort ratings at the 5% level for the 40 pockets/min rate. The results are presented in Table 5. For the 50 and 60 pockets/min rates, none of the statistical analyses was signficant at the 5% level.

Part Three: The effects of pocket width on the number of plants loaded into the simulator in 20 min and on the percentage of plants laterally mis-positioned during loading was not significant at the 5% level. Friedman's test on the subjects' ratings of difficulty due to pocket width was not significant at the 5% level. The percentage of pockets not loaded was plotted versus the pocket width and showed that this variable decreased as the pocket width increased (Fig. 5).

The data points and linear regression lines of the mean cumulative numbers of plants loaded were plotted versus the time from the beginning of the session. One plot was made for each combination of transplanter arrangement and operating rate in Part Two and for each pocket width in Part Three. The graphs indicated that performance of the operators did not appreciably improve or decline as each session progressed.

TABLE 4. EFFECTS OF SIMULATOR ARRANGEMENT ON MIS-POSITIONED PLANTS AT 40 POCKETS/MIN WITH 13 cm POCKETS

Simulator arr.	Mean % mis-positioned
G	6.44a#
F	4.45b
A	4.33b
H	4.18b

#Means followed by the same letter are not significantly different at the 5% level.

TABLE 5. EFFECTS OF TRANSPLANTER ARRANGEMENT ON COMFORT AT 40 POCKETS/MIN WITH 13 cm POCKETS ON SIMULATOR

Transplanter arr.	Mean comfort rating*
H	4.6a#
F	4.2a
Conv.	4.0ab
A	3.2ab
G	2.2b

*A rating of 1 means very uncomfortable, 5 very comfortable. #Means followed by the same letter are not significantly different at the 5% level.

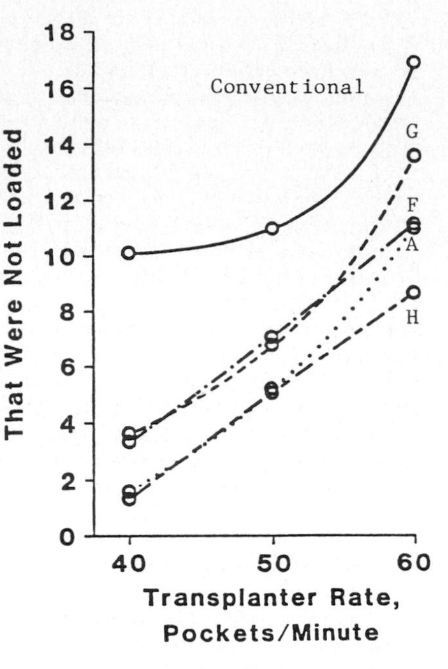

Fig.3. Percentage of pockets not loaded vs. transplanter rate.

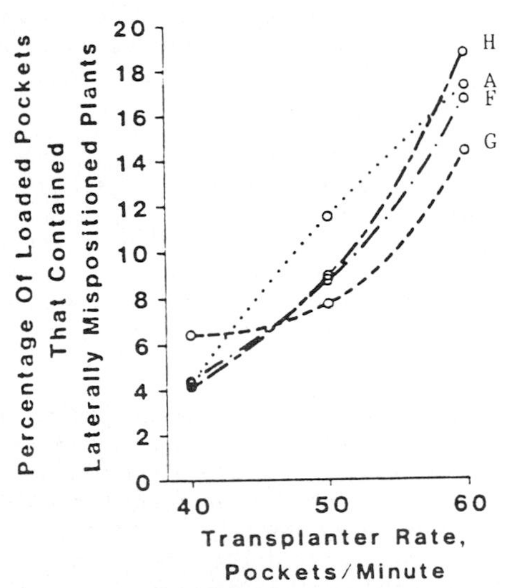

Fig.4. Mis-positioned plants.

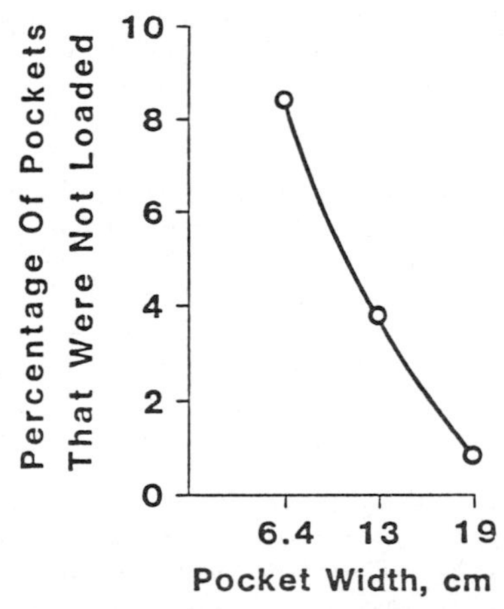

Fig.5. Percentage of pockets not loaded vs. pocket width. Simulator position H at 50 pockets/min.

5. CONCLUSIONS FROM LABORATORY TESTS

(a) Although the differences were not significant at the 5% level, the best transplanter arrangement, by virtue of the majority of the criteria in the study, had five horizontal loading stations and the operator faced so the center of the loading stations was 60 deg to the right of the direction he faced and the pockets moved from his left to his right. The spatial relationships between the seat and the loading stations are detailed in Way (1985). The pocket width was 19 cm. (b) Operating rates above 50 pockets/min effectively reduce an operator's ability to load plants in a transplanter. (c) Lateral mis-positioning of plants in horizontal pockets will increase as operating rate increases above 40 pockets/min. (d) Operators can load significantly more plants into a horizontal multiple-loading-station transplanter than into a conventional planting unit at a typical rate of 40 pockets/min rate. (e) At 40 pockets/min, both the percentage of laterally mis-positioned plants and the comfort of operators using the simulator depended on the position of the operator relative to the transplanter. (f) Laboratory transplanting sessions of 20 min are not long enough to determine effects on operators due to duration of operation. (g) The number of plants loaded in a given time will increase as the pocket width increases.

6. FIELD TRIALS

A conventional transplanter was modified for experimental use and field tested, with sweet potato plants, in 1987, versus two common types of unmodified transplanters. Design constraints, including a maximum width of 102 cm to match minimum row spacings used in conventional field plantings and retrofit of plant receiving parts onto the conventional planter, allowed an approximation only of the best configuration as determined by the laboratory tests (conclusion a above). The experimental transplanter accommodated two operators, seated side by side, with two horizontal loading stations and a plant box directly in front of each. Plants, placed in the pockets, moved toward the center of the machine where they were conveyed downward to the furrow.

In the field test, the three planters, designated Exp, A, and B, were each operated at three speeds, 1.44, 1.60, and 1.76 km/h. The typical speed for commercial transplanting is 1.44 km/h. The test was in a completely randomized block design with four replications for a total of 36 plots. Each plot was one row - 1.2 m x 23 m. Data were taken over a 15 m length within each plot. Dependent variables were: initial plant spacing, ratio of mean to expected spacing, coefficient of variation of spacing, percent survival and potato yield by three grades. Expected plant spacing for the three planters were: Exp = 38 cm, A = 30 cm and B = 30.5 cm.

Several significant differences were machine related. The errors in mean spacing compared to expected spacing were: Exp = +2.9%, A = +10% and B = + 4.2%. The ratios of mean to expected spacings were: Exp = 1.037, A = 1.101 and B = 0.958. Exp and A were different from B but not from each other. There were no significant differences in total yields of potatoes/plot. Mean yields of #1 grade potatoes were: Exp = 17.1 kg, A = 14.4 kg and B = 18.3 kg, where Exp and B were different from A but not from each other.

The mean percent survival decreased from 95.4% to 89.6% over the range of speeds. The coefficient of variation of plant spacing for planters A and B increased with speed but did not increase for the experimental planter.

These results indicate an advantage, though not dramatic, for the experimental planter over the conventional planters in planting uniformity, and in expected plant spacing with no loss of yield. The experimental planter also showed a yield advantage when yields were normalized to average plant spacing. The test plots were not long enough to introduce fatigue effects, even at the highest speed. However, the lack of variation in plant spacing due to speed with the experimental planter seems to be particularly important. A complete redesign of the transplanter to more nearly match the best arrangement of components and operators found in the laboratory tests would potentially yield better results.

REFERENCES

[1] Dooley, J. H. 1983. Transplanter for forest nurseries. TRANSACTIONS of the ASAE 26(6): 1661-1664.

[2] Holland Transplanter Company. Not dated. Catalog No. 50/183. Holland Transplanter Company. Holland, MI.

[3] Hollander, M. and D. A. Wolfe. 1973. Nonparametric statistical methods. John Wiley and Sons., New York, 138-141, 151, 366-371, 373-378.

[4] McCormick, E. J. 1957. Human engineering. McGraw-Hill Book Company, Inc., New York, 314.

[5] Splinter, W. E. and C. W. Suggs. 1968. Simulation studies of human errors in multiple-loading transplanting. TRANSACTIONS of the ASAE 11(6): 844-847.

[6] Suggs, C. W. 1979. Development of a transplanter with multiple loading stations. TRANSACTIONS of the ASAE 22(2): 260-263.

[7] Way, T. R. 1985. Simulated multiple loading station transplanting of sweetpotatoes. M.S. Thesis. Department of Agricultural Engineering, Louisiana State University, Baton Rouge.

[8] Wright, M. E. and T. R. Way. 1982. Unpublished notes. Agricultural Engineering Department. Louisiana Agricultural Experiment Station. Louisiana State University Agricultural Center, Baton Rouge.

V

HUMAN–COMPUTER INTERFACE

Trends in Ergonomics/Human Factors V
F. Aghazadeh (Editor)

EVALUATION OF OCULAR AND MUSCULOSKELETAL SUBJECTIVE DISCOMFORT SCORE RESPONSES FROM VDT OPERATORS

Helmut T. ZWAHLEN and Charles C. ADAMS, Jr.

Industrial and Systems Engineering Department
Ohio University
Athens, Ohio 45701-2979 USA

Subjective discomfort responses measured at eight different times during the workday from two VDT operator studies were evaluated using multiple correlation analysis and other techniques. The results of this evaluation indicate that the responses to some of the subjective discomfort questions are highly correlated and therefore, may be redundant. A revised set of subjective discomfort questions are presented which should make the data collection procedure more efficient in future studies.

1. INTRODUCTION

Several methods have been utilized to measure the effects of the work environment and workload for people operating video display terminals (VDTs). Subjective discomfort responses have emerged as the most viable measures to determine the discomfort and fatigue of VDT operators working in VDT workstations and to develop a quantitative Work-Rest Schedule Model as reported by Zwahlen and Adams [1]. The value of subjective discomfort measures is further enhanced due to the fact that reliable, sensitive and easily obtainable objective measures are not available at the present time and performance measures such as keystrokes per minute and errors appear not to be very sensitive to detect discomfort and early stages of fatigue. Johanssen, et al. [2] suggest that subjective workload measures may be the most reliable method of determining workload because if persons feel loaded, they are loaded, regardless of what any other types of measures might indicate. A similar statement might be made concerning discomfort and/or fatigue. Zwahlen, et al. [3] calculated correlation coefficients for the changes in accommodation for each work session versus the discomfort scores at the end of the same session for the question "hard to see sharply" and found no statistically significant correlation between accommodation changes and the subjective focusing problems. In some instances the two measures were even slightly negatively correlated. It is the objective of this study to evaluate the responses to this questionnaire and to possibly develop a smaller and more efficient revised questionnaire yielding about the same amount of information.

2. METHODS

Study 1 (Zwahlen, et al. [3]) required 6 subjects, all female, to operate a VDT under either a high or low screen viewing situation over two full workdays. Similarly, Study 2 (Zwahlen and Kothari [4]) required 8 subjects, 7 females and 1 male, to perform under the same screen viewing situations while also using either a negative or positive image polarity VDT screen over two full workdays. All fourteen of the VDT operators were experienced typists with normal, uncorrected vision. Each of the VDT operators were paid a base rate with the ability to earn additional incentive pay based upon keystrokes per minutes and errors. The VDT operators were asked to alternately perform either a data entry or file maintenance task on files which consisted of names of chemical compounds, stock numbers with ten characters (eg. Q1M97RS2PM), addresses, phone and bin numbers, prices and an eight digit location code. An Applied Science Laboratories Model 1998 computer controlled eye monitor system collected eye scanning and pupil diameter data unobtrusively and an Apple II computer continuously monitored the typing performance (keystrokes per minute and errors per file) of the subjects. The workstation was well lighted with an Armstrong Tascon lighting fixture and was ergonomically designed with an IBM chair and an adjustable height keyboard. The IBM 3101 VDT used by the VDT operators was fitted with an OEM glare filter.

The VDT operators responded to nine musculoskeletal (Questions 1-9) and nine ocular (Questions 10-18) subjective discomfort questions (initially developed and used by NIOSH), presented sequentially one question at a time always starting with Question 1 on the VDT screen, before and after each of four 90 minute work sessions during a workday. Two of the 90 minute work sessions were conducted in the morning, separated by a 15 minute break, and two of the 90 minute work sessions were presented in the afternoon, also separated by a 15 minute break. The morning and afternoon were separated by a 45 minute lunch break. Figures 1 and 2 present the 18 subjective discomfort questions with a pictorial representation for seven of the musculoskeletal discomfort questions and nine of the ocular discomfort questions. The VDT operators responded to the subjective discomfort questions by pressing the minus (-) key from column 00 to column 72 according to their level of discomfort. The scale was divided into various levels of discomfort using verbal indicators at different points (column 05 "no, not at all", column 20 "only very little", column 36 "somewhat", column 49 "quite a bit", and column 64 "yes, very much so"). The subjective discomfort response data is presented in Zwahlen, et al. [3] and Zwahlen and Kothari [4].

3. ANALYSIS AND RESULTS

Summary measures for the responses to the subjective discomfort questionnaire for both Study 1 and Study 2 are shown in Table 1. F-tests and t-tests performed at the 0.05 level of

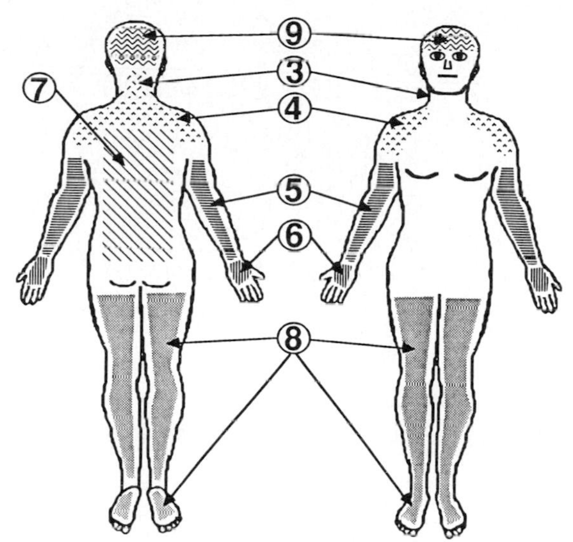

Q1: Do you at this moment feel any form of discomfort in your back, arms or other parts?
Q2: Do you at this moment feel any weariness or general fatigue?
Q3: Do you feel any discomfort, stiffness or soreness in your neck?
Q4: Do you feel any discomfort, stiffness or soreness in your shoulders?
Q5: Do you feel any discomfort, stiffness or soreness in your arms?
Q6: Do you feel any discomfort, stiffness or soreness in your hands?
Q7: Do you feel any discomfort, stiffness or soreness in your back?
Q8: Do you feel any discomfort, stiffness or soreness in your legs or feet?
Q9: Do you have any headache?

Figure 1. The musculoskeletal questions of the subjective discomfort questionnaire presented to the VDT operators one question at a time.

significance to compare the responses from the two VDT studies for the eighteen subjective discomfort questions found only one question (Question 9) statistically different when comparing the studies using the F-test and seven of the eighteen subjective discomfort questions (Questions 5, 9, 10, 11, 14, 15 and 16) differed significantly when comparing the two studies using the t-test.

Tables 2 and 3 present two correlation matrices for the responses to the subjective discomfort questions for Study 1

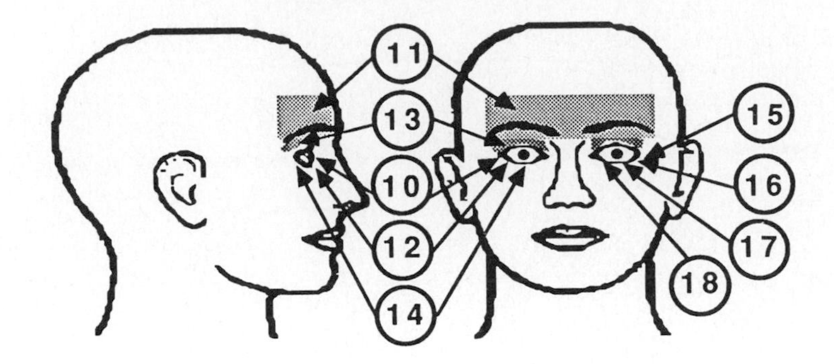

Q10: Do you at this moment feel any form of discomfort in your eyes?
Q11: Do you feel any discomfort or pain above your eyes?
Q12: Do you have any feelings of sand or dryness in your eyes?
Q13: Do you have any feelings of irritation in your eyelids?
Q14: Do your eyes feel sensitive to light?
Q15: Is it hard for you to see sharply?
Q16: Do you blink frequently?
Q17: Are your eyes watering?
Q18: Do you have a dull aching feeling in your eyeballs?

Figure 2. The ocular questions of the subjective discomfort questionnaire presented to the VDT operators one question at a time.

TABLE 1. Summary Measures for the Responses to the Subjective Discomfort Questionnaire for Both Study 1 and Study 2 (Score range 0-72).

STUDY 1 (N=96)						STUDY 2 (N=128)					
Qst. Num.	Mean	Std. Dev.	Qst. Num.	Mean	Std. Dev.	Qst. Num.	Mean	Std. Dev.	Qst. Num.	Mean	Std. Dev.
1	32.7	18.4	10	15.7	14.1	1	32.2	18.4	10	20.3	16.82
2	19.5	15.5	11	10.8	9.5	2	22.1	19.4	11	7.5	10.67
3	33.7	19.2	12	12.5	12.9	3	30.1	22.6	12	10.7	13.65
4	30.1	16.5	13	13.6	13.4	4	28.8	20.5	13	12.1	14.46
5	21.8	14.8	14	15.0	14.6	5	14.4	15.8	14	22.8	18.66
6	11.7	11.6	15	19.5	16.0	6	8.8	11.1	15	25.5	20.22
7	29.9	18.4	16	19.6	17.4	7	34.2	22.6	16	24.7	17.86
8	13.4	12.8	17	13.3	14.2	8	13.4	15.2	17	14.5	15.74
9	8.6	9.8	18	11.8	12.3	9	3.6	6.1	18	13.9	17.55

Qst. Num. -- Question Number Std. Dev. -- Standard Deviation

TABLE 2. Correlation Matrix for the Responses to the Subjective Discomfort Questions From Study 1 (N-96).

Q#	1	2	3	4	5	6	7	8	9	10	11	12	13	14	15	16	17
2	.55																
3	.83	.58															
4	.84	.57	.80														
5	.66	.23	.64	.67													
6	.57	.38	.60	.64	.62												
7	.74	.57	.70	.71	.47	.60											
8	.55	.53	.63	.67	.55	.70	.60										
9	.30	.42	.40	.47	.38	.62	.44	.67									
10	.51	.62	.47	.55	.28	.44	.33	.50	.46								
11	.48	.70	.56	.50	.35	.52	.43	.67	.54	.75							
12	.30	.63	.42	.39	.22	.36	.17	.38	.30	.70	.69						
13	.41	.56	.39	.43	.18	.29	.18	.38	.22	.73	.65	.76					
14	.30	.54	.37	.37	.15	.16	.13	.27	.28	.77	.63	.73	.74				
15	.28	.34	.28	.42	.42	.31	.18	.41	.51	.53	.50	.47	.44	.64			
16	.11	.41	.17	.21	.03	.17	.07	.25	.37	.68	.52	.64	.60	.75	.45		
17	-.0	.38	.10	.07	-.0	.08	-.1	.17	.43	.54	.45	.60	.47	.63	.40	.70	
18	.30	.49	.38	.31	.11	.25	.17	.35	.19	.58	.58	.60	.59	.63	.47	.40	.44

___ Not Statistically Significant at 95% Confidence Level

TABLE 3. Correlation Matrix for the Responses to the Subjective Discomfort Questions From Study 2 (N=128).

Q#	1	2	3	4	5	6	7	8	9	10	11	12	13	14	15	16	17
2	.69																
3	.87	.72															
4	.83	.60	.84														
5	.65	.55	.68	.74													
6	.58	.55	.61	.57	.70												
7	.91	.63	.84	.82	.67	.59											
8	.64	.61	.67	.58	.70	.69	.62										
9	.19	.49	.28	.20	.25	.48	.20	.47									
10	.53	.61	.53	.56	.51	.40	.41	.41	.31								
11	.35	.53	.37	.34	.40	.61	.37	.44	.69	.42							
12	.42	.55	.39	.51	.45	.38	.40	.41	.47	.50	.49						
13	.39	.42	.32	.45	.46	.46	.35	.39	.43	.71	.60	.64					
14	.40	.49	.40	.50	.36	.25	.37	.30	.26	.54	.39	.61	.55				
15	.59	.56	.62	.69	.52	.39	.58	.46	.26	.51	.35	.64	.44	.76			
16	.39	.38	.43	.50	.43	.32	.35	.35	.16	.54	.34	.48	.47	.75	.78		
17	.24	.15	.28	.33	.38	.38	.20	.21	.16	.50	.34	.16	.51	.38	.37	.57	
18	.41	.48	.39	.43	.45	.62	.39	.45	.55	.66	.71	.56	.82	.44	.38	.37	.52

___ Not Statistically Significant at 95% Confidence Level

and for Study 2. Nearly all of the correlation coefficients appear to be statistically significant (141 of 153 cases in Study 1 and 152 of 153 cases in Study 2 are statistically significant at the 95% confidence level), however this could be expected due to the relatively large number of sample observations (N=96 - Study 1; N=128 - Study 2). Thus, it is also important to determine the "practical" significance of the correlation values rather than just the statistical significance. Chatfield and Collins [5] suggest that as a "rough guide", values larger than 0.7 should be considered significant. If values of 0.7 or greater are considered to be of practical significance, then 17 of the 153 correlation values in Study 1 and 15 of the 153 correlation values in Study 2 may be considered to also be practically significant. Comparing the correlation coefficients from the two studies indicates that seven pairs of questions are practically significant in both studies (Questions 1&3-"back, arms, other parts" vs. "neck"; 1&4-"back, arms, other parts" vs. "shoulders"; 1&7-"back, arms, other parts" vs. "back"; 3&4-"neck" vs. "shoulders"; 3&7-"neck" vs. "back"; 4&7-"shoulders" vs. "back"; 10&13-"discomfort in eyes" vs. "irritation in eyelids"; 14&16-"sensitive to light" vs. "blink frequently").

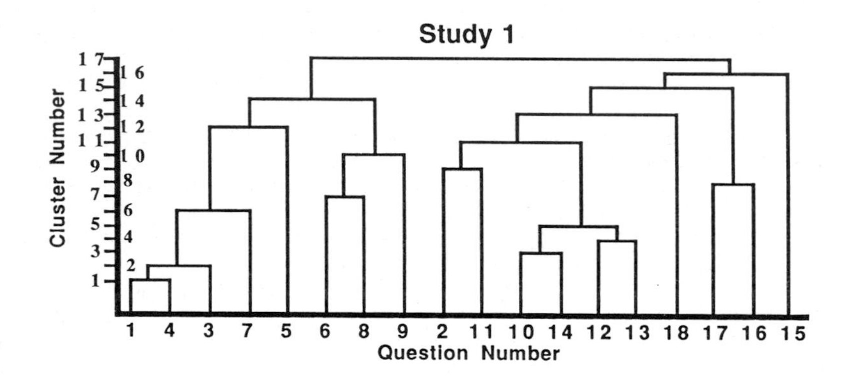

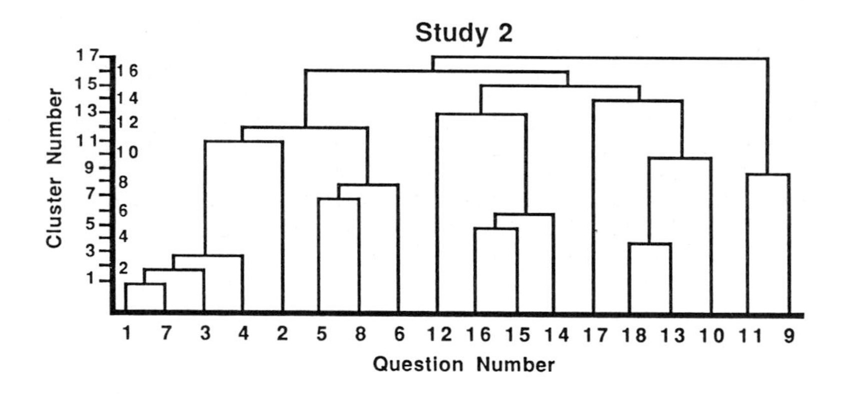

Figure 3. Cluster analysis for both Study 1 and Study 2.

Figure 3 presents dendrograms based on the results of a cluster analysis for both studies. This cluster analysis was performed using the Statpro Statistical Package [6]. Each iteration of the cluster analysis technique which was used gathers together the two items, either questions or previously formed clusters of questions, which have the smallest between-group average distance. The between-group average distance is based upon the correlation coefficient X which is transformed into a distance using the equation X':= (|X|+1)/2. The iterations continue until all of the questions or clusters of questions form one cluster as shown in Figure 2. It is very interesting that in general for both studies, the responses to the musculoskeletal discomfort questions (Questions 1-9) and the ocular discomfort questions (Questions 10-18) do not form clusters until the final iteration of the analysis. This would seem to indicate that the responses to the musculoskeletal discomfort questions generally do not reflect the responses to the ocular discomfort questions and vice versa. In both studies Questions 1-"back, arms, other parts", 3-"neck", 4-"shoulders", and 7-"back", form relatively low level clusters which suggests that they closely correspond to one another. Questions 10-"discomfort in eyes" and 13-"irritation in eyelids" also form relatively low level clusters and closely correspond to each other.

A revised subjective discomfort questionnaire can be developed using the results of the cluster analysis and the correlation matrices. Questions 1-"back, arms, other parts", 3 -"neck", 4-"shoulders", and 7-"back" were all highly correlated and form relatively low level clusters in both Study 1 and Study 2. These questions can be combined to form a single question which should extract about the same information. The question "Do you at moment feel any discomfort, stiffness or soreness in your back, neck or shoulders?" should serve as an appropriate amalgamation of these four discomfort questions. Similarly, Questions 10-"discomfort in eyes" and 13- "irritation in eyelids" form relatively low level clusters and are highly correlated in both of the VDT operator studies. An apt unification of these two ocular discomfort questions is the question "Do you at this moment feel any form of irritation in your eyes or eyelids?". Finally, while Questions 14 -"sensitive to light" and 16-"blink frequently" form a medium level cluster rather than a low level cluster, they are highly correlated in both Study 1 and Study 2 and therefore it may be justifiable to combine these two ocular discomfort questions. A suitable merger of Questions 14 and 16 may be "Are your eyes sensitive to light or do you blink frequently?". The revised subjective discomfort questionnaire containing a total of 13 questions is shown in Table 4.

4. CONCLUSIONS

The evaluation of the responses to the subjective discomfort questionnaire has resulted in a smaller (13 questions) and more efficient discomfort questionnaire which should yield about the same amount of discomfort information. The revised questionnaire could be useful to assess the nature and level of subjective discomfort of VDT operators and could also provide the basis for the use of a quantitative Work-Rest Schedule Model as proposed by Zwahlen and Adams [1].

TABLE 4. Revised Subjective Discomfort Questionnaire

Q1: Do you at this moment feel any discomfort, stiffness or soreness in your back, neck or shoulders?
Q2: Do you at this moment feel any weariness or general fatigue?
Q3: Do you feel any discomfort, stiffness or soreness in your arms?
Q4: Do you feel any discomfort, stiffness or soreness in your hands?
Q5: Do you feel any discomfort, stiffness or soreness in your legs or feet?
Q6: Do you have any headache?
Q7: Do you at this moment feel any form of irritation in your eyes or eyelids?
Q8: Do you feel any discomfort or pain above your eyes?
Q9: Do you have any feelings of sand or dryness in your eyes?
Q10: Are your eyes sensitive to light or do you blink frequently?
Q11: Is it hard for you to see sharply?
Q12: Are your eyes watering?
Q13: Do you have a dull aching feeling in your eyeballs?

REFERENCES

[1] Zwahlen, H.T., and Adams, C.C., Jr., "Development of a Work-Rest Schedule for VDT Work". Paper presented at the Second International Conference on Human-Computer Interaction, Honolulu, Hawaii. In G. Salvendy, S.L. Sauter, and J.J. Hurrell (Eds.), Social, Ergonomic and Stress Aspects of Work With Computers. Amsterdam: Elsevier Science Publishers B.V., 1987, pp. 157-164.

[2] Johanssen, J., Moray, N., Pew, R., Rasmussen, J., Sanders, A., and Wickens, C., "Report of the Experimental Psychology Group". In N. Moray (Ed.), Mental Workload, Its Theory and Measurement. New York: Plenum, 1979, p. 105.

[3] Zwahlen, H.T., Hartmann, A.L., and Rangarajulu, S.L., Video Display Terminal Work with a Hard Copy - Screen and Split Screen Data Presentation. Athens, OH: Ohio University, Dept. of Industrial and Systems Engineering Research Report. Final report distributed by The Report Store, Lawrence, Kansas 66044, July 1984.

[4] Zwahlen, H.T. and Kothari. N.C., The Effects of Dark and Light Character CRT Displays Upon VDT Operator Performance, Eye Scanning Behavior, Pupil Diameter and Subjective Comfort/Discomfort, Athens, OH: Ohio University, Dept. of Industrial and Systems Engineering Research Report. Final report distributed by The Report Store, Lawrence, Kansas 66044, June 1986.

[5] Chatfield, C. and Collins, A.J., Introduction to Multivariate Analysis. New York: Chapman and Hall, 1980, pp. 40-41.

[6] Statpro, New York: Penton Software Incorporated, Multivariate Statistics Section, 1985.

Trends in Ergonomics/Human Factors V
F. Aghazadeh (Editor)

INTER-OCULAR CORRELATION: SPATIAL FACTORS

Charles B. WOODS, Keith D. WHITE, & John H. KRANTZ
Department of Psychology, University of Florida Gainesville, 32611
USA

The effects of different spatial configurations of negative Inter-Ocular Correlation (IOC) on resting binocular vergence positions were measured. Previous work has shown that full field negative IOC stimulations, which may be like the effective binocular stimulation created by viewing a CRT with glare and specular reflections, can affect resting vergence positions. In Experiment 1 negative and zero IOC "grating" stimuli of various bar widths of horizontal or vertical orientation were tested, and were not found to have differential effects on resting vergence position. In Experiment 2, differently sized rectangular areas of negative IOC framed by surroundings of zero IOC (or vice versa) tested the effects of central and peripheral regions of negative IOC retinal stimulation. We found that only when negative IOC comprised a large surrounding frame was resting vergence noticably affected.

1. INTRODUCTION

Vergence eye movements insure that both eyes are appropriately aligned on the object of particular interest. Using both eyes to view an object furthur away or closer to the observer requires the lines of sight of the eyes to diverge or to converge, respectively. One sufficient stimulus for initiating vergence eye movements is retinal disparity [1], the small relative differences between the two retinal images resulting from the geometry of simultaneous visibility of objects at different distances with eyes that are physically separated.

With the aid of vergence eye movements, the visual system engages in binocular matching; a search for the corresponding retinal elements which indicate proper binocular overlay. In the pairing of these retinal elements, certain Inter-Ocular Correlations (IOC's) may exist. In many cases, when an element in one retina is stimulated then a corresponding element in the other retina is similarly stimulated. These are cases when IOC is positive. In some natural settings there are occasions when IOC can be negative; e.g., when a near object occludes one eyes' view of a more distant object. In this instance, stimulation of a retinal element in one eye is not paired with similar stimulation in the other eye. Some binocular models view the visual system as being insensitive to negative IOC, although other models emphasize such sensitivity [2]. A recent study indicated that resting vergence positions (RVPs) adopted while viewing large homogenous fields of negative IOC were substantially different from those adopted while viewing positive IOC or zero IOC (random visual noise which produces random IOC) [3].

In the present study we investigated a set of spatial factors (size, position, spatial frequency, and orientation) which might be important factors in the effects of IOC's on resting vergence positions. In Experiment 1 the spatial periodicity of negative IOC was varied by having subjects view vertical or horizontal "gratings" of varied bar widths, where alternate bars contained zero or negative IOCs. In Experiment 2 the area-dependent effects of negative IOC were measured using differently sized rectangular areas of negative or zero IOC centered in a zero or negative IOC surround.

2. EXPERIMENT 1: SPATIAL PERIODICITY AND ORIENTATION

Methods and Procedure

2.1 Subjects. One experienced and two naive subjects made an equal number of observations of the stimuli. No subject reported any history of strabismus or amblyopia nor wore corrective lenses. Each was found to have a stereoacuity greater than 40 arcsec with a Bausch and Lomb Orthorater.

2.2 Stimulus Generation and Patterns. All zero and negative IOC stimulations were accomplished by using dynamic random element correlograms [4] presented anaglyphically (left eye and right eye stimulations chromatically separated). The circuit used to generate these patterns is described in detail elsewhere [5]. The correlogram stimulus patterns were displayed on an Electrohome RGB monitor whose display screen subtended 12.6 deg horizontal by 8.3 deg vertical (area= 106.2 deg^2) at the 2 meter viewing distance. The stimuli used in Experiment 1 manipulated the spatial periodicity of IOC by varying the spatial frequency of "gratings" containing alternating bars of zero and negative IOC. Eight stimuli were used, 4 vertical gratings and 4 horizontal gratings, both of which varied spatial frequency across 3 octaves. The spatial frequencies used were: 1.75, .88, .44, and .22 cycles/deg for the vertical gratings and 3.25, 1.60, .80, and .40 cycles/deg for the horizontal gratings. Even at the lowest spatial frequencies, our stimuli contained at least 3 full cycles of the grating pattern.

2.3 RVP Measurement. Subjects' RVPs accepted while viewing each of the 8 stimulus patterns were the measures of interest. RVP was assessed psychophysically using a procedure identical to that previously reported [3]. Subjects viewed the correlogram display through beam splitters which allowed dichoptic probes to be optically superimposed upon the center of the display. Left eye and right eye probes were displayed on small black and white monitors placed along separate optical paths but at the same optical distance as the correlogram display. Each probe flashed repetively; on for 120 msec, less than the latency required to initiate vergence eye movements [6], then off for a randomly selected interval of 500-1000 msec (mean 750 msec.). Each probe subtended 9 arcmin horizontally and vertically in a "+" shape, with a stroke width of about 1 arcmin. A computer controlled the presentation of probes for each trial, and an attached joystick allowed the subject to control the probes' movement, allowing both probes to be subjectively aligned. After alignment, the vertical and horizontal coordinates of the probe pair were written to disk. The distance between left eye and right eye probes when subjectively aligned gave a measure of RVP. Since vertical disparity is ordinarilly quite small and not considered a sufficient stimulus for initiating vergence eye movements, and also because our previous study [3] found no evidence that IOC differentially affected vertical RVP's, we will not present the vertical components of RVP from Experiments 1 or 2.

2.4 Data Normalization. Across experimental sessions, the horizontal RVP data were normalized for each subject in the following manner. Each trial in each condition (8 observations per session X 4 sessions = 32 trials) was subtracted from the median RVP value for that condition, giving a difference score for each trial. This normalization helped to eliminate between- subject variability.

All data are presented in Figures 1 and 2. In all 8 graphs normalized convergent and divergent RVP is arbitrarily binned for convicnence of presentation and plotted against number of observations made at that offset, across subjects. A multiplier is given for conversion to arcmin.

2.5 Results

Figure 1 shows RVP data for vertically oriented negative IOC and zero IOC gratings of 4 different spatial frequencies. For all vertical gratings RVPs have been described by their frequency distributions. Each distribution was found to be characteristically bell-shaped or "normal", and found to have peaks at or near zero RVP. RVP distributions measured while viewing vertical gratings were found to be similar for each of the different spatial frequencies, and were considered equivalent.

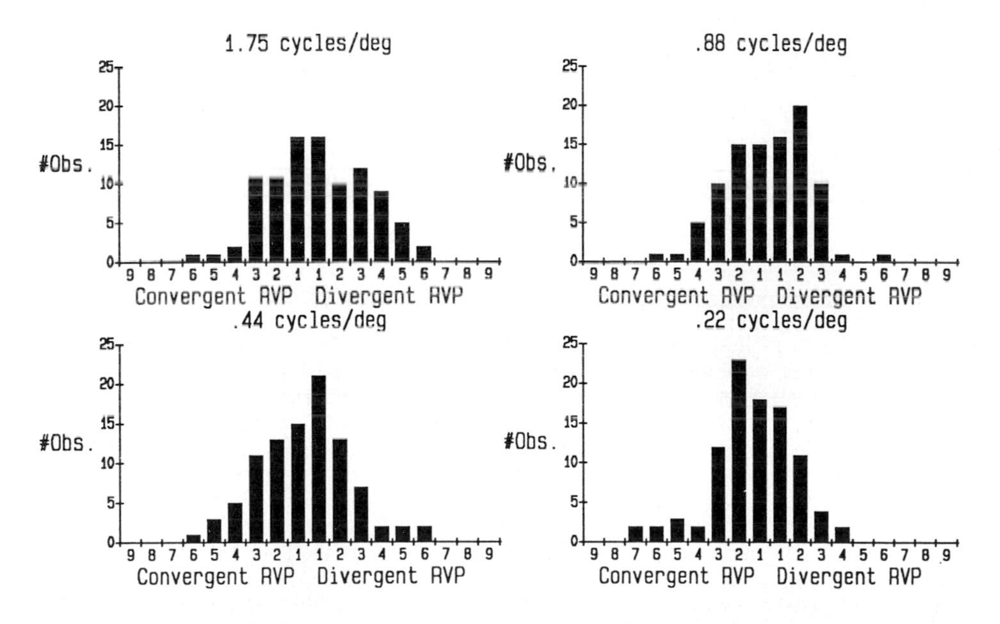

FIGURE 1. RVP frequency distributions for vertical gratings of different spatial frequency. Number of observations are shown in RVP "bins", across subjects. Convergent and divergent RVP can be converted to arcminutes by multiplying bin number times 4.

Figure 2 graphs distributions of normalized RVP acceptable while viewing horizontal gratings. These distributions were found to be quite similar to those of RVP to the vertical grating stimuli in Figure 1. Each can be described as having the same characteristic bell-like shape, and each was found to have its peak near zero RVP.

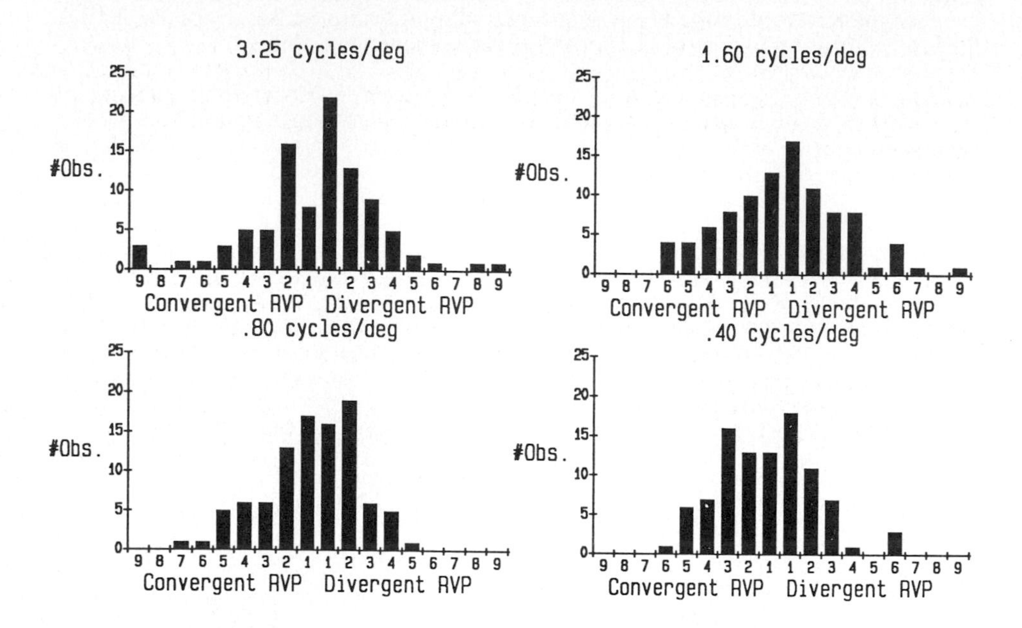

FIGURE 2. RVP frequency distributions for horizontal gratings of different spatial frequencies.

2.6 Discussion

Acceptable RVP 's made while viewing vertical or horizontal grating stimuli of different spatial frequency are well described as bell-shaped distributions. Spatial frequency, at least over the 3 octaves varied in this study, and the 2 orientations (vertical and horizontal) did not differentially affect subjects RVPs.

3. EXPERIMENT 2: CENTRAL AND SURROUNDING STIMULATION

Method and Procedure

3.1 Subjects. Two experienced subjects and 1 naive subject made an equal number of observations in Experiment 2. All subjects met or surpassed the visual criteria of Experiment 1.

3.2 Stimulus Generation and Patterns. The stimulus generation was the same as in Experiment 1. Each pattern used in Experiment 2 consisted of a center/surround arrangment comprising a centrally located rectangular area of one

IOC framed in a surround (the remainder of the display) of the other IOC. The four sizes of central area subtended horizontally by vertically .9 X .6 (area=5.4 deg^2), 5 X 3 (15 deg^2), 7 X 4 (28 deg^2), and 10 X 6 (60 deg^2) of visual subtense. The central area either contained negative IOC in a surround of zero IOC or vice versa, giving 8 total stimulus configurations.

3.3, 3.4 RVP Measurement and Data Normalization were the same as in Experiment 1. In Experiment 2 each data set is based on 8 obervations per session X 6 sessions = 48 trials). The data are also represented here as RVP frequency distributions, and the negative IOC centers in surround zero IOC data and zero IOC centers in surround negative IOC are presented as Figures 3 and 4, respectively.

3.5 Results

Figure 3 shows RVP data for each negative IOC center area size in a surround of zero IOC. In all four cases the distributions peak at small RVPs, with proportionally fewer observations made at greatly convergent or greatly divergent RVPs. Each RVP distribution was approximately bell-shaped or "normal". The differences between the distributions of RVPs measured to the variously sized negative IOC centers in zero IOC surrounds were found to be small enough that the four distributions seem to be equivalent.

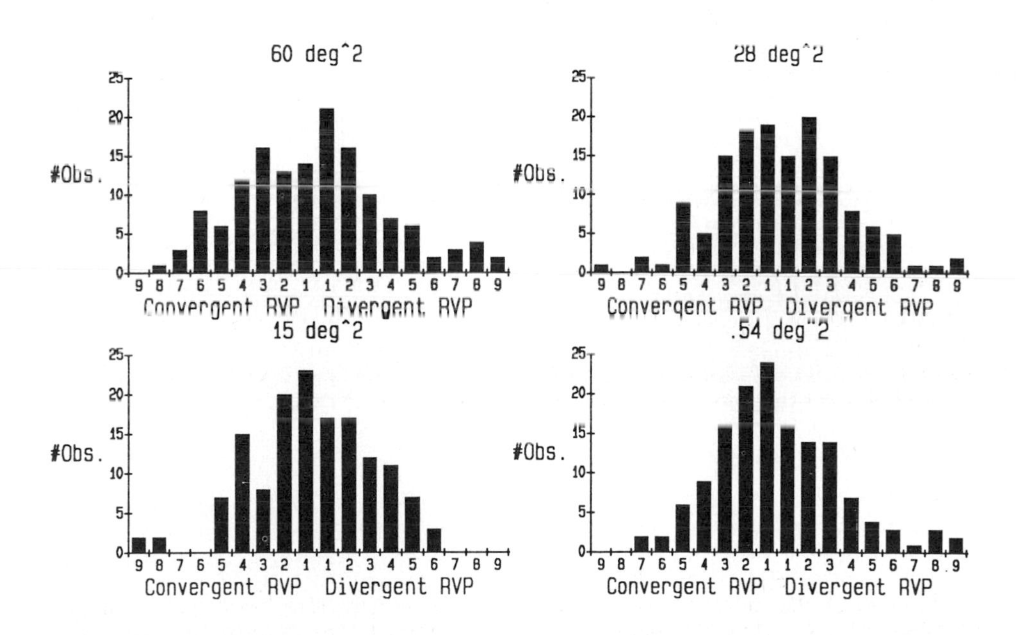

FIGURE 3. RVP frequency distributions for 4 different negative IOC central areas in zero IOC surrounds.

Figure 4 presents the normalized RVP data for each zero IOC center area size in a surround of negative IOC. The bell-shaped distributions which were found to characterize all four data sets in Figure 1 do not accurately characterize every data set in Figure 2. "Normal" adequately describes RVP distributions measured while viewing the 2 larger central areas (60 and 28 deg^2); however, the distributions of RVPs found acceptable while viewing the 2 smaller central areas (15 and .54 deg^2) are too flat to fit this same pattern. As can be seen in Figure 3, the 15 deg^2 data spread considerably more than any data depicted in Figure 3. The .54 deg^2 data in Figure 4 appear bimodal since observations were more frequent at moderate (16-20 armin) convergent and divergent RVPs than at smaller RVPs. It might also be noted that distributions plotted of observations to the center/surround configured stimuli appear to have values more extreme than those to our grating stimuli of experiment 1.

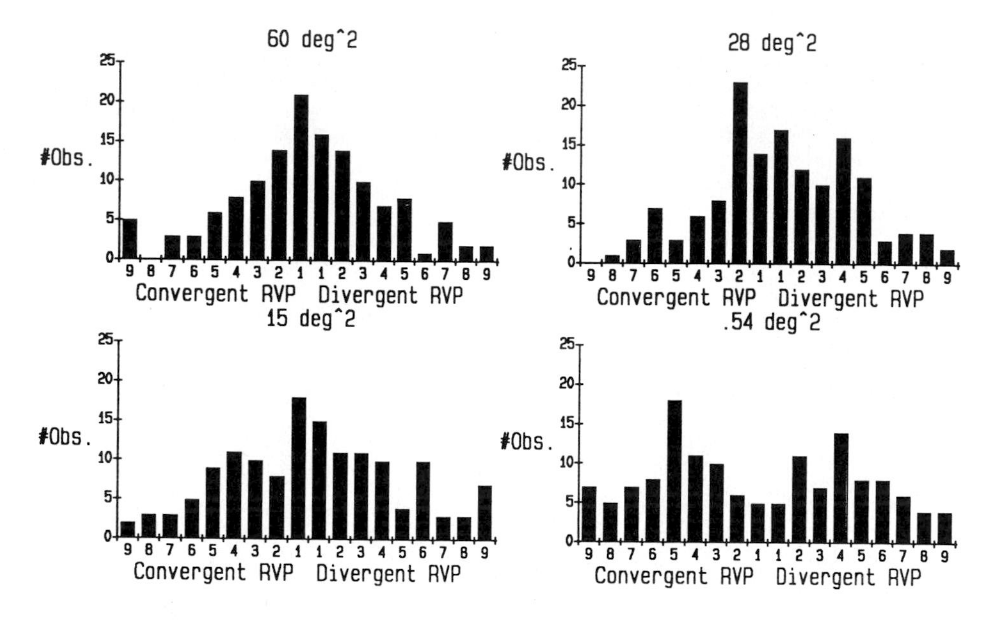

FIGURE 4. RVP frequency distributions for 4 different zero IOC central areas in negative IOC surrounds.

3.5 Discussion

RVPs acceptable while viewing stimuli comprised of negative IOC centers in zero IOC surrounds did not appear to be differentially affected by the central area sizes employed. These distributions are comparable to ones resulting from viewing full-field stimulation (the entire display) of positive or zero IOC [3]. Distributions of RVPs were significantly altered when central areas of zero IOC were 15 deg^2 or less (surrounded by negative IOC). In the most extreme case (.54 deg^2 in Fig.4), the most frequently acceptable RVPs were those approximately 16-20 armin convergent <u>or</u> divergent of those acceptable while viewing any other stimuli.

RVPs seem to be affected by negative IOC only when a substantially large area displays such stimulation. In our earlier study [3], stimulus fields of 106 deg^2 were found to alter RVP. In the present study the inclusion of zero IOC centers covering 60 deg^2 or 28 deg^2 (leaving 46 deg^2 or 78 deg^2, respectively, of negative IOC) abolished the effect on RVP. Only when the centers covered less than 15 deg^2 (leaving 91 deg^2 or 105 deg^2 as negative IOC) was the stimulus effective.

In the present experimental design, stimulus area and eccentricity were unavoidably confounded. We are inclined to suspect that area is the more prominent factor because RVPs were extremely similar between (a) the condition using negative IOC in a 60 deg^2 center (Figure 3), and (b) the condition using negative IOC in a 46 deg^2 surround (Figure 4). For those two stimuli the areas were much more nearly comparable than were the eccentricities, and the data were highly similar.

4. GENERAL DISCUSSION AND CONCLUSIONS

Based on the large body of literature on stereopsis, the best known from of binocular interaction, we might have expected the effects observed in the present study to be strongly dominated by the stimulation of the fovea [4]. This supposition was substantially disconfirmed by our results. In turn this implies for VDT use that one must be concerned not only with the parts of the display to be fixated but also with the substantially peripheral surroundings that are simultaneously visible.

REFERENCES

[1] Westheimer, G. & Mitchell, D. (1969). The Sensory Stimulus for Disjunctive Eye Movements. *Vision Research*, 9, 749-755.

[2] Cogan, A. (1987). Human Binocular Intergration: Towards a Neural Model. *Vision Research*, 27(12), 2125-2139.

[3] White, K.D., Cormack, L.K., Woods, C.B., Krantz, J.H., & Franzen, O. (1987). Binocular Convergence to Various Interocular Correlations. In Asfour (Ed.) *Trends in Ergonomics/ Human Factors IV*. North Holland: New York.

[4] Julesz, B. (1971). *Foundations of Cyclopean Perception*. University of Chicago Press: Chicago.

[5] White, K.D., Cormack, L.K., & Woods, C.B. (1987). 3-D Displays and Eye Movements. In McAllister, D. & Robbins, W. (Eds.) *True 3-D Imaging Techniques and Display Technologies*. SPIE Press. Bellingham, Wa.

[6] Alpern, M. (1972). Effector Mechanisms in Vision. In Kling & Riggs (Eds.) *Experimental Psychology*. Holt, Rinehart, & Winston: New York.

Trends in Ergonomics/Human Factors V
F. Aghazadeh (Editor)

CHANGES IN TEMPORAL INSTABILITY OF LATERAL AND VERTICAL PHORIAS OF THE VDT OPERATORS

T. MAREK[1], C. NOWOROL[1, 2], W. PIECZONKA-OSIKOWSKA[1], J. PRZETACZNIK[1], and W. KARWOWSKI[3]

[1]Institute of Psychology, Jagiellonian University
31-007 Krakow, Golebia 13, Poland

[2]Institute of Mechanization and Energy, Agricultural University
30-149 Krakow, Balicka 104, Poland

[3]Center for Industrial Ergonomics
University of Louisville
Louisville, KY 40292, USA

The main goal of this study was to examine changes in temporal instability of lateral and vertical phorias of the human operators working with visual display terminals (VDTs), and to establish the relationship between the above changes and subjective assesment of visual discomfort. Fifteen female VDT operators participated in the study. Each operator worked for 8 hours on the data-entry task under low and high levels of workload (speed of entry). Temporal instability of the phorias was tested using the Bausch and Lomb Vision Tester. This was done before work and after 3, 5.5 and 8 hours of work. Additionally, the modified Borg scale was used to assess the operators' eye strain. It was observed that the temporal instability of both phorias significantly increased with time.

1. INTRODUCTION

One of the most important components of visual discomfort is disturbance of the muscle balance and the binocular coordination [1]. Sutzliffe [2] and Duke-Elder [3] stressed the need to consider both the accommodation and the convergence components in the eye strain evaluation. Stone [4] pointed out that poor binocular coordination would give rise to a bluring of vision, loss of binocular acuity and stereopsis, and a consequent increase in visual discomfort. Stone and his colleagues [5] have also made reference to the changes in muscle balance under difficult visual tasks.

Work with VDTs has been known to induce conditions of muscle imbalance which may result in either an esophoria or an exophoria. In recent publications Laubli and his colleagues [6], and Voss and her colleagues [7] showed that the phoria (muscle balance) changes while working with VDTs. Some indications of the possible relationships between low contrast and exophoria and between exophoria and subjective visual discomforts have also been found.

Contrary to the above, Dainoff and his colleagues found no changes in heterophorias [8]. Similarly, Gould and Grischkowsky [9], and Woo and his colleagues [10] found no changes in vertical phoria, and in the associated phoria respectively. It follows from the above results that the binocular coordination seems to be a very complex subject. One of the most important factors responsible for the above contradictory results is temporary fluctuation of phoria., i.e. the changes in the muscle balance from one moment to the next.

2. METHODS

2.1 Subjects

Fifteen female VDT operators (office workers 20-28 years of age) participated in the study. Each operator worked for 8 hours entering data into computer under relatively high and low levels of workload. The average pace of work was about 197 and 104 signs per minute, for the high and low levels of workload, respectively. The operators worked in the morning shift with two breaks, first one after 3 hours, and second one 5.5 hours of work. The experimental data was collected four times, i.e.: 1) before work, and 2) after 3, 5.5 and 8 hours of work, respectively.

2.2 Procedures and techniques

The Bausch and Lomb's vision tester was used for examination of temporal stability of lateral and vertical phorias. The tests for near distance (35 cm) was used. Every five seconds the scores from 1 to 15 for the lateral phoria, and from 1 to 9 for the vertical phoria, respectively, were recorded. In total, seven registrations were made. The temporal instability was characterized by the range between maximum and minimum values obtained for all seven registrations. The range of change's in both phorias, lateral and vertical, was registered during a thirty seconds period.

After each registration the modified Borg scale was used for rating of the operator's visual discomfort. The scale had the following values: 0-no eye strain, 0.5-very very weak strain, 1- very weak strain, 2-weak strain, 3- less-than-moderate strain, 4-moderate strain, 5-more-than-moderate strain, 6-less-than-strong strain, 7-strong strain, 8-more-than-strong strain, 9-very strong strain, 10-extreme eye strain.

3. RESULTS

The experimental data was analysed using the analysis of variance (ANOVA). Two factors were taken into account, i.e.: 1) the time of work and 2) the workload level. Figures 1 through 3 show the mean values for temporal instability of lateral and vertical phoria for high and low levels of workload. The Duncan Multiple Range Test on the Means was applied to discover the differences between groups. There was statistically significant increase in temporal instability of lateral and vertical phorias and subjective assessment of eye strain due to time under both work conditions. The values of the F-test for the time of work factor (two way ANOVA) are as follows: 1) temporal instability of lateral phorla, $F= 13.8$ ($p<0 .001$), 2) temporal instability of vertical phoria, $F= 38.4$ ($p< 0.001$), and 3) subjective assessment of eye strain, $F= 135.8$ ($p<0.001$).

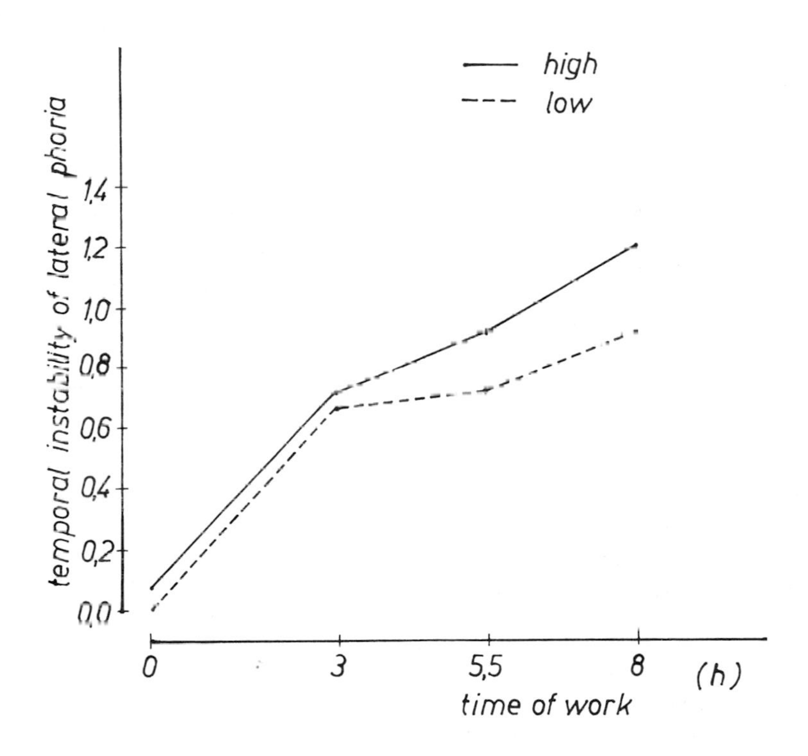

FIGURE 1

Changes of temporal instability of lateral phoria due to time of work. Mean and S.D. values for high level of workload: $m_0=0.07$, $S_0=0.26$; $m_3=0.73$, $s_3=0.80$; $m_{5.5}=0.93$, $s_{5.5}=0.80$; $m_8=1.20$, $s_8=0.56$. Mean and S.D. values for low level of workload: $m_0=0$, $s_0=0$; $m_3=0.67$, $s_3=0.72$; $m_{5.5}=0.73$, $s_{5.5}=0.80$; $m_8=0.93$, $s_8=0.59$.

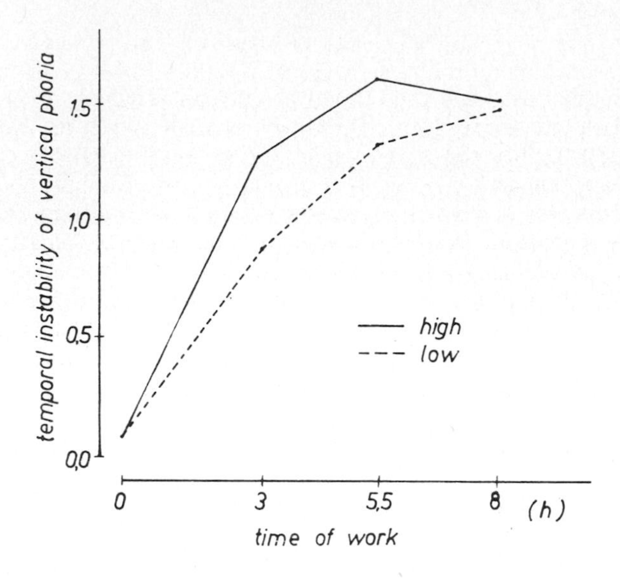

FIGURE 2
Changes of temporal instability of vertical phoria due to time of work. Mean and S.D values for high level of workload: m_0=0.07, s_0=0.26: m_3=1.27, s_3=0.70; $m_{5.5}$=1.60, $s_{5.5}$=0.91; m_8=1.53, s_8=0.52.: Mean and S.D. values for low level of workload: m_0=0.07, s_0=0.26; m_3=0.87, s_3=0.52; $m_{5.5}$=1.32, $s_{5.5}$=0.62; m_8=1.47, s_8=0.52.

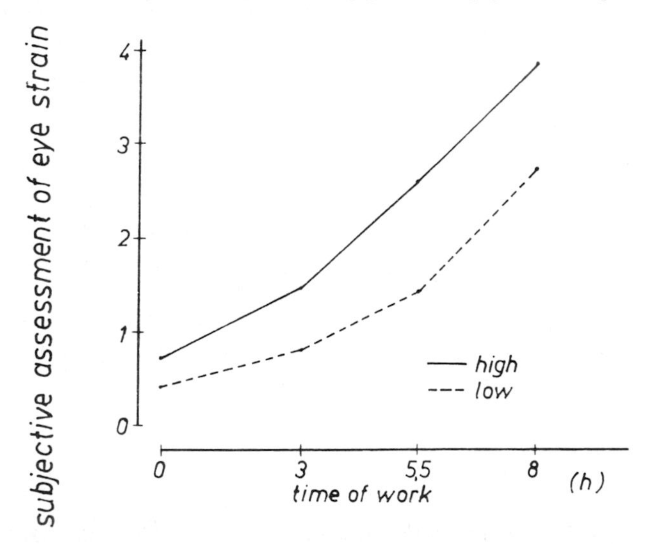

FIGURE 3
Changes of subjective assessment of eye strain due to time of work. Mean and S.D. values for high level of workload: m_0=0.73, s_0=0.90; m_3=1.47, s_3=1.09; $m_{5.5}$=2.60, $s_{5.5}$=1.24; m_8=3.87, s_8=1.06. Mean and S.D. values for low level of workload: m_0=0.40, s_0=0.43; m_3-0.80, s_3=0.68; $m_{5.5}$=1.43, $s_{5.5}$=0.86; m_8=2.73, s_8=1.03.

The results of the Duncan test are shown in Table 1. It can be seen that for both types of phoria, the significant increase of instability appears during the first three hours of work. Contrary to the above, the subjective eye discomfort scores significantly increased in the time interval from three to eight hours of work. In case of the second factor, the workload level, no significant differences between the mean values of temporal instability of lateral phoria were observed. The mean values of vertical phoria differed significantly between the workload levels after the three hours of work. The subjective scores of eye strain differed significantly with respect to the workload at 3, 5.5 and 8 hours of work.

TABLE 1. Differences between mean values m_0, m_3, $m_{5.5}$, m_8, due to time of work (Duncan test) for the temporal instability of both phorias and subjective assessment of eye strain under high and low workloads.

Variables		High level of workload				Low level of workload		
		m_0	m_3	$m_{5.5}$		m_0	m_3	$m_{5.5}$
Temporal	m_0				m_0			
instability	m_3	x			m_3	x		
of lateral	$m_{5.5}$	x	0		$m_{5.5}$	x	0	
phoria	m_8	x	0	0	m_8	x	0	0
		m_0	m_3	$m_{5.5}$		m_0	m_3	$m_{5.5}$
Temporal	m_0				m_0			
instability	m_3	x			m_3	x		
of vertical	$m_{5.5}$	x	0		$m_{5.5}$	x	x	
phoria	m_8	x	0	0	m_8	x	x	0
		m_0	m_3	$m_{5.5}$		m_0	m_3	$m_{5.5}$
Subjective	m_0				m_0			
assessment	m_3	0			m_3	0		
of eye strain	$m_{5.5}$	x	x		$m_{5.5}$	x	0	
	m_8	x	x	x	m_8	x	x	x

x: $p < .001$
0: not significantly different

The correlation coefficients between temporal instability of both phorias and subjective assessment of eye strain were also calculated (see Table 2). It can be seen that the temporal instability of both types of phoria are positively correlated with the subjective assessment of the eye strain. It is also characteristic that under the high workload condition the relationships between temporal instability for both phorias are stronger than under the low workload condition.

TABLE 2. Correlation coefficients between temporal instability of lateral and vertical phorias and subjective assessment of eye strain.

Workload	Temporal instability	Subjective assessment of eye strain
High	Lateral phoria	0.62xx
	Vertical phoria	0.70xx
Low	Lateral phoria	0.49x
	Vertical phoria	0.51x

xx: $p < .01$
x: $p < .05$

The results of this study revealed that the temporal instability increases due to time of work for both lateral and vertical phoria under the high and low workload conditions. In general, both types of temporal instability increase in the first three hours of work. Also, the temporal instability for both phorias are higher under the high than under the low workload conditions, but not all differences are statistically significant. It is characteristic that the dynamics of such an increase for the temporal instability of both phorias differs from that of the subjective assessment of eye strain. While the increase in temporal instability of phorias is most intensive during the first three hours, the subjective scores for the operators' eye discomfort increase in the last five hours of work.

This study also showed that the increase in temporal instability of phorias correlates positively with the subjective assessment of eye strain (see Table 2). In case of the high workload level, the increase in temporal instability of lateral and vertical phorias explains 38 and 49 percent of the operators' visual discomfort, respectively. Under the low level of workload, the instability of lateral and vertical phorias explains 24 percent and 26 percent of the eye discomfort, respectively.

4. CONCLUSIONS

In view of the above results , it can be concluded that the temporal instability of the lateral and vertical phorias is an important factor in the study of visual discomfort of the VDT operators.

REFERENCES

[1] T. Mecaw, Visual fatigue and its measurement, in: J. R. Wilson, E. N. Corlett and I. Manenica (Eds.), New Methods in Applied Ergonomics, Taylor and Francis, London-New York-Philadelphia, 1987, pp. 207-221.
[2] R. L. Sutclife, Visual work, visual comfort and visual efficiency in the wearing industry, J. Psychological Optics, 7 (1950) 16-29.
[3] S. Duke-Elder, System of ophthalmology, in: Ophthalmic Optics and Refraction, Vol. V: Selection IV, Eye strain and visual hygiene, Henry Kempton, London 1970.
[4] P. T. Stone, Issues in vision and Lighting for Users of VDU, in: B. G. Pearce (Ed.) Health Hazards of VDTs, John Wiley and Sons, New York, 1985, pp. 77-88.
[5] P. T. Stone, A. M. Clarce and A. J. Sleter, The effects of task contrast and visual fatigue at a contrast illuminance, Ltg. Res. and Technol.,12, (3) (1980) 144.
[6] T. Laubli, W. Hunting and E. Grandjean, Postural and visual loads at VDT workplace Ergonomics, 24 (1981) 933-944.
[7] M. Voss, K. G. Nymen and U. Bergquist, VDT work and changes in binocular vision-some results, in: Work with Display Units, Stockholm, 1966, pp. 863-866.
[8] M. J. Dainoff, A Happ and P.Crone, Visual fatiguo and occupational stress in VDT poerators, Human Factors. 23 (1981) 423-438.
[9] J. D. Gould and N. Grischkowsky, Doing the same work with hardcopy and with cathode ray tube (CRT) computer terminal, Human Factors, 26 (1984) 323-337.
[10] G. C. Woo, G. Strong, E. Irving and B. Ing, Are there subtle changes in vision after use of VDTs, in: Work with Display Units, Stockholm, 1986, pp. 875-877.

Trends in Ergonomics/Human Factors V
F. Aghazadeh (Editor)

EFFECTS OF MESSAGE TYPE, SCREEN MESSAGE LOCATION, AND RATE OF PRESENTATION ON VDU MESSAGE LEGIBILITY

Stephen J. Morrissey

Trained subjects viewed one of three different message types randomly presented at different locations on a VDU screen. Message types studied were common words and randomly generated nonsense words and alphanumeric strings. Common words were presented at rates of from 4 characters per second to 9 characters per second, and the nonsense words and alphanumeric strings at rates of from 2 characters per second to 7 characters per second.

Common word messages suffered no loss in legibility at any rate of presentation or screen message location. The alphanumeric strings and nonsense words both showed significant drops in legibility when the rate of presentation exceeded 4 characters per second. These two message types were not effected by message location on the VDU screen. It was concluded that if messages are to be presented on a VDU then the rate of presentation should not exceed about 2 characters per second for normal work tasks.

1. INTRODUCTION

One of the most common uses of video display units (VDU's, VDT's) is to present information that changes over time. The rate at which information can be presented to an operator, or the speed at which information can be updated or changed and still read accurately has not been completely determined for VDU's, though there are several recommendations. Bevan (1981) has found that text can be accurately read on a VDU at message presentation rates of from 10-15 characters per second. The proposed guidelines for VDU's (HFS, 1986) suggest that the rate of updating or changing of VDU messages not exceed 2-characters per second. However, this recommendation does not consider screen message location nor different message types.

In a series of studies, Morrissey (1987) and Morrissey and Chu (1987) examined the legibility of different types of messages presented at different locations on a VDU screen with different rates of presentation. They found the rate of presentation was the most important factor determining message legibility. These studies concluded that trained observers could accurately read different types of messages presented on a VDU at a rate of 2-characters per second. However, these studies required subjects to identify any of four different types of messages (common words, nonsense words, number strings, and alphanumeric strings). The more common case in which only one type of message appears was not studied. Thus, their recommendations that information presentation rates not exceed 2-characters per second may be too conservative.

This study examined two questions: First, what is the effect of message type and rate of presentation on message legibility. And, how does the location of the message on the VDU screen effect the legibility of the message?

2. METHODS AND PROCEEDURES

2.1 Independent Variables:

Three independent variables were studied in this research: Message type, message location on the VDU screen, and rate of message presentation in characters per second. The three message types studied were common words of length 4-9 letters, and randomly generated nonsense words and alphanumeric strings of length 2 to 7 characters. All messages were presented for one- second on the VDU screen which gave an effective message presentation rate of 4-9 characters per second for the common words, and 2-7 characters per second for the nonsense word and alphanumeric messages. The differences in message presentation rates were based on the studies of Morrissey (1987) and Morrissey and Chu (1987) which found common words to be least affected by rate of presentation, and nonsense words and alphanumerics the most. Each message type was presented a total of five times at each of five different screen locations, the locations chosen to cover the range of screen message locations found by Morrissey (1987) and Morrissey and Chu (1987) to have the greatest overall legibility. The screen locations studied were a square with corners at the Y coordinates (lines, 1-23) and X coordinates (columns 1-80) 6,20; 6,60; 18,20; and 18,60. The fifth point was set at the center of the screen, 12,40. Characters were white on a black backround, all letters were capitals, and all characters had a visual angle of about 27 minutes of height at the subject viewing location.

2.2 Subjects:

Twelve male and female college juniors and seniors were used as subjects with four subjects randomly assigned to each of the three message types. No restrictions were made on vision, save that if subjects normally wore glasses or contacts, they were to wear them during the experiment. Subjects were allowed to participate only if they felt well, rested, and had not used a VDU in the previous 90-minutes. After subjects arrived, the experiment was completely explained and demonstrated. Subjects were seated at the workstation and a chair and a chin-rest adjusted so the subject was comfortable and the subject's eyes were perpendicular to the center of the VDU screen. The screen intensity and contrast were adjusted as desired by the subject, and subjects were given a training session in which 50 presentations of the particular message type were made. During the training, subjects were corrected when they made a mistake and encouraged to be as accurate as they could. After training, a short rest break was taken then the actual data collection experiment was performed.

3. RESULTS

In evaluating the results, it was found that the common word messages suffered no losses in legibility in any of the test conditions that is, 100 percent accuracy was present in all conditions. Thus, only the data for the nonsense words and alphanumeric strings will be evaluated further. The mean data by message type, width, and screen location are given in Tables 1 and 2, and are plotted in Figure 1. The data for these two message types were evaluated using a three-factor-repeated measures ANOVA. These analyses revealed that only rate of presentation had a significant main effect for the nonsense words ($F(4,72) = 55.0$, $p < 0.0001$) and alphanumeric strings ($F(4,72) = 79.7$, $p < 0.0001$).

The data in Tables 1 and 2 shows very similar responses for both the nonsense words and the alphanumerics, with legibility dropping sharply when the rate of presentation exceeds four characters per second. Tests on the data comparing the accuracy to that of the two and three character per second conditions found that these drops in legibility were significant when the rate of presentation exceeded 5-characters per second ($p < 0.05$).

To further study the influences of the independent variables on message legibility, a stepwise multiple regression analysis was performed. When this was done, it was found that the only variable or combination of variables which had a significant effect on message legibility was message presentation rate. The multiple regression models which best predicted message legibility were found to be:

For Alphanumerics:

Legibility(%) = 145.8 - 15.87Rate(Char/Sec); $R^2 = 0.675$.

For Nonsense Words:

Legibility(%) = 136.9 - 12.8Rate(Char/Sec); $R^2 = 0.613$.

4. DISCUSSION

This study examined how rate of information presentation, the type of message, and its location on a VDU screen affected message legibility, or accuracy with which messages were read. For all message types studied, the rate at which the message was presented was the primary factor that influenced the accuracy with which messages were read. Alphanumeric strings and nonsense words suffered the greatest losses in legibility as the rate of presentation increased. Examination of the data in Tables 1 and 2 shows that the legibility of the alphanumeric strings and nonsense words was very similiar up to a rate of presentation of 5-characters per second. Beyond this rate of presentation, the alphanumeric strings showed greater losses in legibility. This finding is similiar to that reported by Morrissey (1987) and Morrissey and Chu (1987), but the current research provides a better definition of the relationship between rate of presentation and screen location. Also as was reported by Morrissey (1987), for the range of screen message locations studied here, the screen message location did not significantly influence message legibility.

This study found that common words could be read without a loss in accuracy with rates of presentation of up to 9-characters per second. This finding some gives support for Bevan's (1981) finding that textual materials (e.g., common words) can be accurately read at a presentation rate of from 10 to 15- characters per second.

The increased error rate for nonsense words and alphanumeric messages is due to the ambiguity or uncertainty of the message content, and reflects the greater attentional demands needed to determine each character. It is possible that somewhat different results might have been obtained if a more liberal error criterion had been established. In this research, a reply was considered wrong if any part of the message was incorrectly reported. If the criterion of total number of correct character reports to total number of characters presented was used to evaluate performance, the results might have been different. This error criterion was not used in this research as it was desired to study the more probable type of task in which a message was correctly read or not.

Summarizing, this research found that while the 2- character per second maximum rate of change recommended by the proposed VDU standards (HFS, 1986) is somewhat conservative, it is a valid level for a maximum rate of presentation of information on a VDU screen for "normal" use, that is, in cases when the users are not constantly watching the screen and expecting messages at known times.

5. CONCLUSIONS

This study examined how the rate of information presentation and the physical location of a message on a VDU screen affected the accuracy with which different types of messages were read. It was found that only the rate of message presentation in characters per second affected the legibility of the message types studied. Screen message location did not have a significant affect on message legibility. Common word messages were read without any loss of legibility at rates of presentation of up to 9-characters per second. Randomly generated alphanumeric strings and nonsense words suffered large losses in legibility when the rate of presentation exceeded 5-characters per second. These findings give support to recommendations that the rate of presentation of messages on a VDU not exceed 2-characters per second. Although this study found that good legibility is possible at higher rates of presentation, in normal work environments and work tasks, operators will probably not be attending the display as closely as were the subjects in this study; thus the suggestion that the maximum rate of information presentation on a VDU be limited to 2-characters per second is justified.

6. REFERENCES

Bevan, N., 1981: Is there an optimum speed for presenting text on a VDU? International Journal of Man-Machine Studies, 14, 59-77.

Human Factors Society, Inc., 1986: American National Standard for Human Factors Engineering of Visual Display Terminal Work-stations, Revised Review Draft, July, 1986. Santa Monica, Ca.

Morrissey, S.J., 1987: Effects of oblique viewing angle, rate of presentation, screen message location and message type on VDU message legibility. Unpublished technical report.

Morrissey, S.J. and R. Chu, 1987: Legibility of video Display units during off-angle viewing. Behavior and Information Technology, In press.

Table 1: Mean accuracy at screen message location by rate of presentation for the Nonsense Words. Each cell reflect 20 trials.

Y (Row)	X (Column)	Rate of Presentation(Characters/Sec.) 2	3	4	5	6	7
6	20	100	100	95	70	60	25
6	60	100	100	90	80	80	30
12	40	100	100	95	80	70	55
18	20	100	100	100	90	60	35
18	60	100	100	100	85	50	30
Mean, all Locations		100	100	96	81	64	35

Table 2: Mean accuracy at screen message location by rate of presentation for the Alphanumeric Messages. Each cell reflect 20 trials.

Y (Row)	X (Column)	Rate of Presentation(Characters/Sec.) 2	3	4	5	6	7
6	20	100	100	85	85	55	30
6	60	100	100	95	85	60	15
12	40	100	100	95	65	60	15
18	20	100	100	100	70	40	25
18	60	100	100	95	95	50	15
Mean, all screen locations		100	100	94	80	53	20

Trends in Ergonomics/Human Factors V
F. Aghazadeh (Editor)

THE EFFECT OF CRT QUALITY ON VISUAL FATIGUE

Masaru MIYAO*, John S. ALLEN, Selim S. HACISALIHZADE
Stacia A. CRONIN and Lawrence W. STARK

Neurology and Telerobotics Units
University of California
Berkeley, CA 94720, USA

The effects of CRT resolution on visual fatigue and readability were studied. Two kinds of displays with different resolutions and fonts were used. In one experiment the subjects read from each display for one hour. Their reading speed and blink rate were observed during reading. Their eye movement tracking tasks were studied before and after reading. In the second experiment, readability tests with three different character sizes on both displays were conducted and resulting reading eye movements were analyzed. The results show the importance of high resolution screens for readability when undersized characters are used.

1. INTRODUCTION

High resolution video display terminals (VDTs) have recently been introduced to the workplace. The screen resolution is approximately eight times that of "standard" displays. Claims calling for objective substantiation have been made that these high resolution VDTs reduce the amount of eye strain in their users. The purpose of this study was to examine the effect of CRT quality on visual fatigue and to measure the effect of character size and display resolution on readability. A number of methods for measuring eye fatigue quantitatively have been suggested [1, 2 & 3]. But these have met with only moderate success in quantifying such ocular motor changes [4]. This study was conducted with full cognizance of these difficulties. Measurements of performance, eye movement and blink rate were carried out with VDTs of different resolutions. Much thought was given to measuring as many aspects as possible during those measurements.

2. EXPERIMENT I

2.1. MATERIALS AND METHODS

2.1.1 Devices and Task

Two different VDTs were analyzed: (1) a high resolution Laserview™ terminal built by Sigma Design (1664 x 1200 pixels = 2.0 Mpixels) and (2) a standard display terminal (720 x 350 = 0.25 Mpixels). The 19 inch high resolution Laserview terminal and a 12 inch standard terminal were called **H type display** and **S type display** respectively. Both VDTs were reverse screens (dark characters on a light background). The H type CRT had a paper-white phosphor, and the S type CRT had a green phosphor. The angles of the page were approximately 16° and 19° respectively.

---

* Visiting Scholar from School of Medicine, Nagoya University, Japan

Using the Ventura Publisher™ and an IBM PC-AT™ we were able to approximately standardize the character size for the two displays (Table I). Each displayed page was standardized at 73 characters/line and 30 lines/page. Subjects were allowed to assume their preferred viewing distance and seating posture. Lighting and environmental conditions were constant and identical throughout the experiments. The primary visual task consisted of reading aloud from each of the terminals in separate one hour sessions. A contemporary novel titled **Black Jaguar** [5] was used as the reading material. Subjects were asked to read aloud as fast as comfort allowed. The subjects were able to control page changes by themselves.

Table I Character Format of Text (ex: Character "H")

	Dot Matrix	Character Size	Color	CRT Size
H type	12 X 14	(2.3 X 2.7 mm)	white	19 inch
S type	8 X 9	(2.4 X 3.3 mm)	green	12 inch

Table II Experimental Procedure

Smooth Eye Pursuit measured for 10s at each of these frequencies: 0.2 Hz, 0.4 Hz, 0.8 Hz, & 1.0 Hz
Reading Aloud for 1 hour: reading speed recorded for each page
blink rate recorded at three 5 minutes intervals (0 - 5 minutes), (27.5 - 32.5 minutes) & (55 - 60 minutes)
Smooth Eye Pursuit measured for 10s at each of these frequencies: 0.2 Hz, 0.4 Hz, 0.8 Hz, & 1.0 Hz

2.1.2 Subjects

Subjects consisted of ten healthy students: four males and six females with a mean age of 22.7 (range 19-29). Four of the subjects wore glasses. The two parts of the experiment were conducted on separate days with half of the subjects reading from the high resolution CRT first and the other half reading from the standard type first.

2.1.3 Measurement Methods

Ocular motor function was assessed by measuring eye movement tracking before and after reading(Table II). This was accomplished by using the Micromeasurements System 1200™ [6]. By means of an infrared light and a video camera, this system measured the horizontal and vertical positions of the center of the left pupil 60 times a second. Data was stored in a PC-AT. The subject was seated in front of a CRT with his/her head immobilized by a forehead rest and a bite bar. The distance between the target and the eyes was 57cm. The target was a light-spot moving sinusoidally along 20cm of the CRT horizontal axis; the visual angle was 20°.

Blink rate during reading was recorded by the examiner at three five minute intervals: 0-5 min., 27.5-32.5 min. and 55-60 min. The examiner also measured the subject's reading time for each page.

2.2. RESULTS

Reading speed remained relatively constant for both displays during the experiment (Fig.1). The H type reading speeds were insignificantly faster than the S type. In an one hour session, subjects were able to read aloud an average of 29.8±3.9 H type pages (181 words/min.) and 28.9±2.6 S type pages (176 words/min.). H type mean reading speeds were: 122±18s/page for the first three pages, 120±20s/page for the middle three pages, and 119± 22s/page for the last three pages. S type reading speeds for the same intervals were 124±16s/page, 126±18s/page, and 123±18s/page respectively.

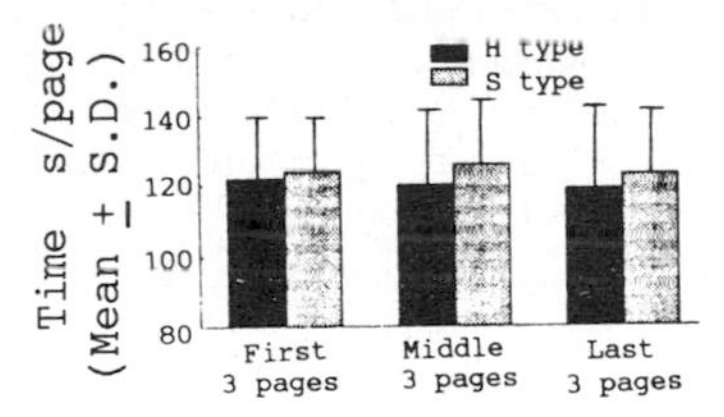

Figure 1 Reading Speed

Reading speed remained relatively constant. The H type reading speeds were insignificantly faster than the standard type.

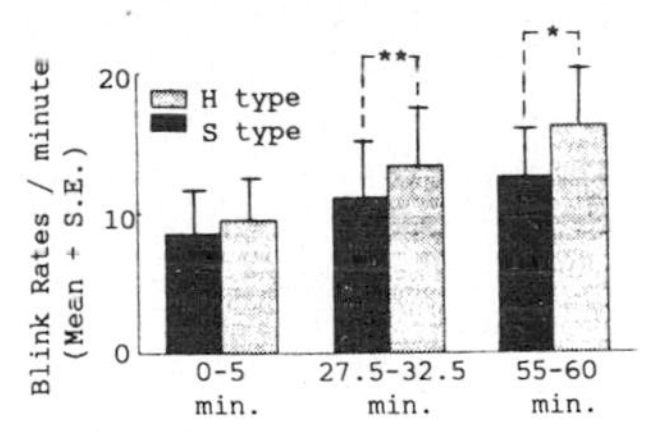

Figure 2 Blink Rates * $p < 0.05$ ** $p < 0.01$

Blink rates were significantly lower for the H type, and insignificantly increased over the one hour reading period.

Blink rates were significantly lower for the H type CRT and increased insignificantly over the one hour reading period for both the H type and S type (Fig.2). H type mean blink rates for the first five min., 27.5-32.5 min., and final five min. were 8.7±3.1 (mean ± S.E.) blinks/min., 11.3±4.0 blinks/min., and 12.3±3.6 blinks/min. respectively. S type mean blink rates for the same five min. intervals were 9.7±3.0 blinks/min., 13.4±4.2 blinks/min., and 16.3±4.8 blinks/min. respectively.

Eye movement tracking of sinusoidally moving targets was evaluated as percent root mean square (RMS) error (Fig.3), and compared statistically (Fig.4). No significant changes in RMS error were found for the H and S display types before and after the one hour reading sessions.

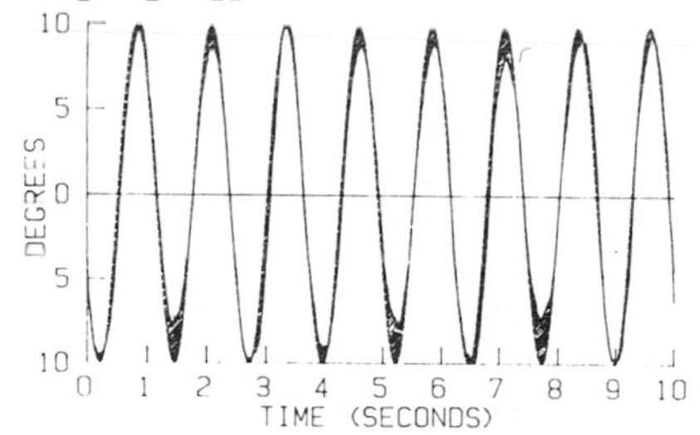

Figure 3 Eye Movement vs. Target

Eye movement tracking of sinusoidally moving targets was evaluated as percent root mean square (RMS) error (shaded area).

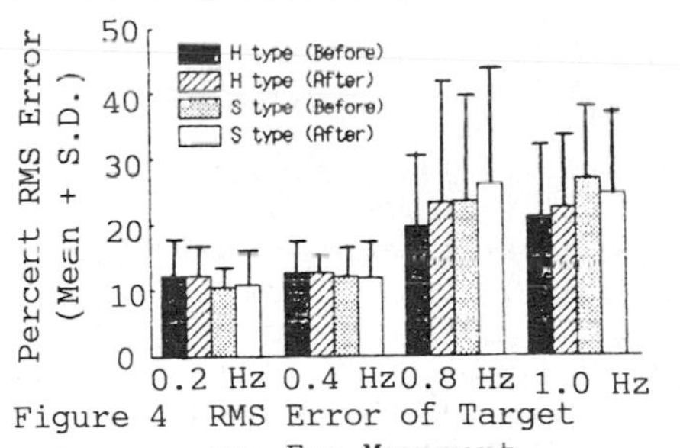

Figure 4 RMS Error of Target vs. Eye Movement

There were no significant differences in percent RMS error tests taken before and after reading

Subjective fatigue was described by all subjects. They complained of considerable eye strain, including dry or watery eyes, after reading from either the H type or S type displays. Two subjects also reported having headaches, one after reading the H type and the other after reading the S type. Half of the ten subjects felt that the H type was less fatiguing. Several mentioned that although they initially preferred the H type CRT because the characters appeared to be "crisper", they felt that after reading for an hour, the S type display was less visually fatiguing.

2.3. DISCUSSION

It is known that it is difficult to measure visual or oculomotor fatigue. Also variability of reading performance, as reflected in large variance in the measurements described above makes quantitative measurement of visual fatigue due to reading an almost impossible task; we were not able to confirm the results in Goussard et al. [4]. In retrospect, one can argue that measurements similar to the ones described above with smaller character sizes and worse resolution may have constituted a more discriminative test due to increased stress. Also it could be argued that measurements under bad reading conditions, such as glare, flicker and lack of contrast would be more effective in the measurement of visual fatigue. However these stressors would be outside of normal task variation and thus less relevant to the general problem. Also one should consider that the subjects complained of much eye strain.

It has been proposed that blink rate is not a measure of fatigue but rather correlates inversely with effort [7,8], especially if performance remains constant. With the H type display, subjects read slightly (but not significantly) faster than with the S type; therefore they might have been able to develop more effort and thus show the significantly lower blink rate observed. The slight (but not significant) increase in blink rates during the experiment may be construed to mean that less effort was necessary as the experiment wore on (even though performance was maintained), secondary to a non-specific training effect or boredom.

Reading aloud was chosen to make sure that the subject read the text. Possibly as a consequence of this, the reading speed was fairly constant, because reading aloud takes about two times longer than reading silently, thus filtering out any possible changes in silent reading speed.

3. EXPERIMENT II

Experiment II was an attempt to compare the readability [9] of medium, small and mini-sized characters displayed on a high resolution CRT to those displayed on a standard type CRT.

3.1. MATERIALS AND METHODS

Using the same devices described in the first experiment, a subject's eye movements were recorded for ten seconds when he/she read from the text, **Black Jaguar**. The text was presented in three different sizes: medium, small and mini on both the H type and the S type CRTs (Table III). Each displayed page of text contained the same number of lines and had approximately the same number of words per line (73 characters/line and 18 lines/page). Horizontal visual angles of each page were 27°, 17° and 12° for medium, small and mini character sizes respectively on H type CRTs. These angles were 24°, 17° and 14° respectively on S type screen.

Subjects consisted of seven healthy females drawn from a student population. The subject was seated in front of a CRT. The distance between the screen and the each subject's eyes was 40 cm.

By observing the return sweep of the subject's pupil on a monitor, the examiner was able to begin recording eye movements after the first line of text was read. The experiment continued for ten seconds. The data was stored in the computer. The order of documents and character sizes presented was randomized for each subject.

3.2. RESULTS

Reading eye movements [9], horizontal eye positions [10] and velocities were studied (Figs.5, 6 & 7). H type reading speeds (characters/s) were 41 ± 13, 45 ± 11 and 44 ± 9 for medium, small and mini characters respectively. S type reading speeds were 42 ± 12, 39 ± 7 and 31 ± 7 for medium, small and mini characters respectively (Fig.8). There were no significant differences in reading speeds for the H type screen. However, in the S type screen, mini letter reading speed was significantly lower than medium size reading speed ($p < 0.05$) and also lower than small size reading speed ($p < 0.01$). There was also a significant difference ($p < 0.01$) in reading speeds with mini-sized characters on S and H type screens.

Intraindividual average peak eye movement velocities while reading (Fig. 7) on H type CRT were $40 \pm 13^{\circ}/s$, $27 \pm 6^{\circ}/s$ and $17 \pm 5^{\circ}/s$ for medium, small and mini characters respectively. On S type CRT they were $37 \pm 11^{\circ}/s$, $25 \pm 9^{\circ}/s$ and $14 \pm 3^{\circ}/s$ for medium, small and mini characters respectively (Fig.9). No significant differences in intraindividual average peak eye movement velocities while reading were detected between H and S type CRTs irrespective of the character size. However, among three sizes there were significant differences ($p < 0.01$) on both H and S type screens.

Average peak return sweep velocities (Fig.7) on H type CRT were $186 \pm 23^{\circ}/s$, $108 \pm 19^{\circ}/s$ and $69 \pm 14^{\circ}/s$ for medium, small and mini characters respectively. On S type CRT they were $158 \pm 13^{\circ}/s$, $108 \pm 37^{\circ}/s$ and $80 \pm 25^{\circ}/s$ for medium, small and mini characters respectively (Fig.10). There was a significant difference in these velocities between H and S type CRTs for medium-sized characters only. No significant differences were detected between H and S type CRTs for small and mini character sizes. However, among three sizes there were significant differences ($p < 0.05$-0.01) on both H and S type CRTs. Note that all velocities (including the ones in the preceeding paragraph) were proportionally decreased by the smoothing algorithms used to process the data; e.g. a 24° return sweep should have a velocity of at least $600^{\circ}/s$ [12].

Number of eye-fixations as defined as the portions of the record where the horizontal velocity of the eye is less than 10% of the average peak velocity of the individual while reading (Fig.7), on H type CRT were 16.4 ± 4.0, 19.3 ± 1.6 and 19.6 ± 2.9 for medium, small and mini characters respectively. On S type CRT they were 15.9 ± 4.5, 20.6 ± 5.8 and 18.6 ± 4.5 for medium, small and mini characters respectively (Fig.11). There were no significant differences among any of these values.

Mean eye-fixation durations (the length of time where the eye is fixated) on H type screen were 114 ± 18ms, 105 ± 15ms and 112 ± 10ms for medium, small and mini characters respectively. On S type screen they were 110 ± 27ms, 115 ± 27ms and 152 ± 35ms for medium, small and mini characters respectively (Fig.12). There were no significant differences among any character sizes on H type screen. However, on S type screen, significant differences ($p < 0.05$-0.01) between mini and the rest two character sizes were observed. Also, there was a significant difference ($p < 0.05$) between two display types with mini size characters. Note that the calculated durations are much less than the actual duration because of the six point smoothing algorithm used; for example, a normal fixation duration for fast reading might be 200-250 ms. However, the significant relative lengthening of the fixation duration of the mini-characters on the S display is real.

Table III Character Size of Reading Text(ex. "H")

	Medium	Small	Mini
H type	18X18dot 3.5X3.5mm	10X10 1.9X1.9	7X8 1.4X1.6
S type	10X10dot 3.0X4.0mm	7X8 2.1X3.2	6X5 1.8X2.0

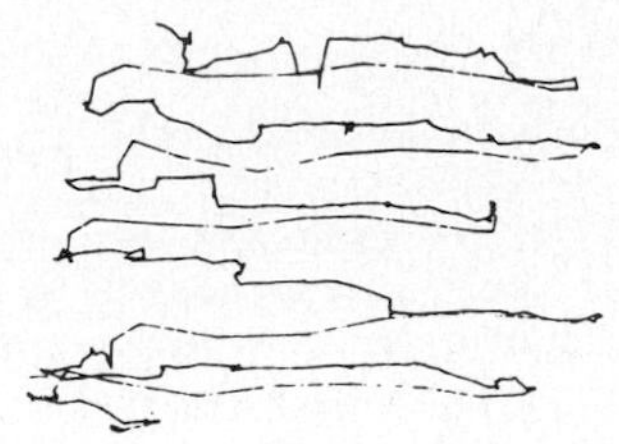

Figure 5 Eye Movement while Reading

Shows eye movement while reading (5.2 lines) for 10s. The vertical scale is shown amplified.

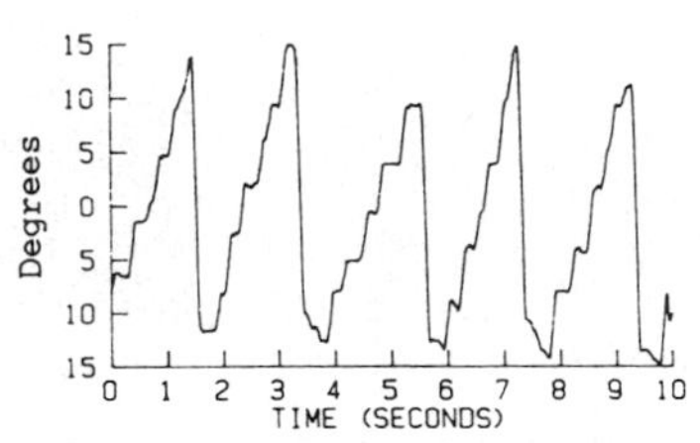

Figure 6 Horizontal Eye Movement vs. Time

Shows relationship between time and horizontal position of eye movements presented in Fig.5.

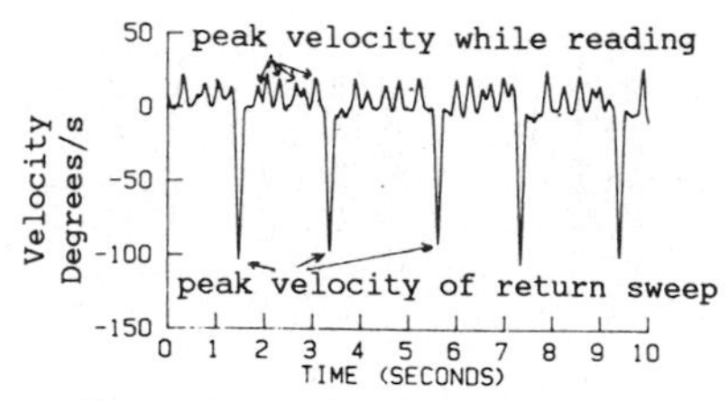

Figure 7 Horizontal Velocity of Eye Movement

Shows relationship between time and velocity of horizontal eye movement (Fig.5). All velocities were decreased by the algorithm used to process the data.

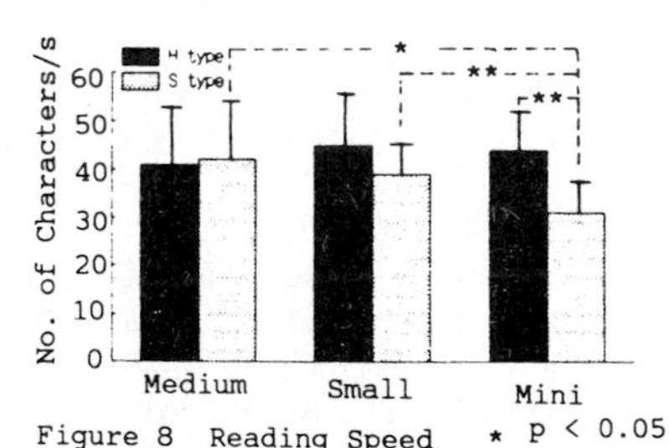

Figure 8 Reading Speed

* p < 0.05

** p < 0.01

Lower reading speed in the S type CRT with mini characters was discerned.

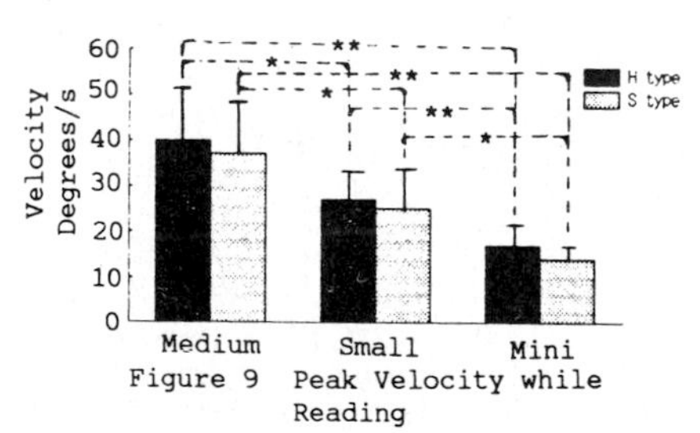

Figure 9 Peak Velocity while Reading

No significant difference between the two displays was noted. Peak velocity increased as visual angle increased.

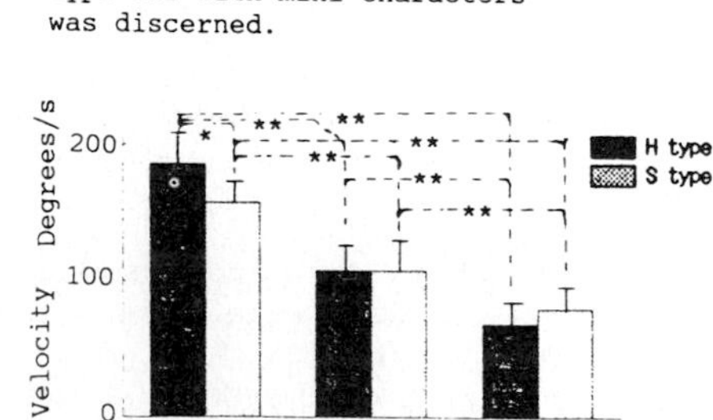

Figure 10 Peak Velocity of Return Sweep

Obvious correlation between visual angle and velocity was shown.

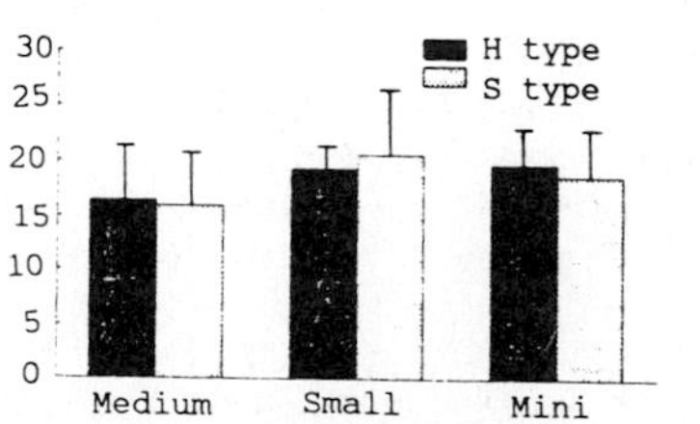

Figure 11 Number of Eye-Fixations

Number of eye fixations while reading remained relatively constant.

Cumulative eye-fixation durations (the total length of time during 10 sec. when the eye is fixated) on H type screen were 1.87±0.70s, 2.02±0.32s and 2.17±0.30s for medium, small and mini characters respectively. On S type screen they were 1.70 ± 0.50s, 2.42 ± 1.07s and 2.76 ± 0.50s for medium, small and mini characters respectively (Fig.13). On S type screen, a significant difference ($p < 0.01$) between mini and medium character sizes was observed. Also, there was a significant difference ($p < 0.01$) between two CRT types with mini size characters.

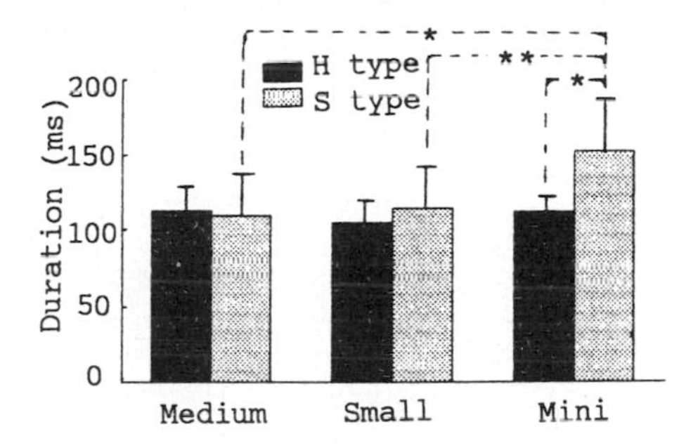

Figure 12 Mean Eye-Fixation Duration

Significantly longer duration for each eye-fixation in S type CRT with mini characters was noted.
Mean + S.D. * $p < 0.05$ ** $p < 0.01$

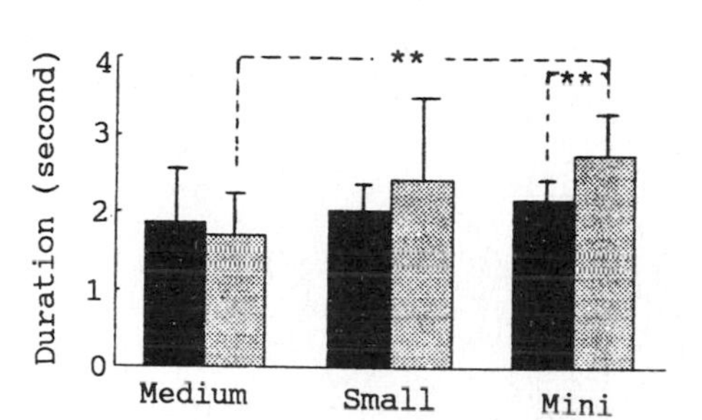

Figure 13 Cumulative Eye-Fixation Duration

Significantly longer fixation time was calculated in 10s reading for S type CRT with mini characters.
Mean + S.D. ** $p < 0.01$

3.3. DISCUSSION

This study suggests the utility of analyzing readability of VDTs using reading eye movements. Reading speed data shows low readability in S type CRT with mini sized characters. ANSI [8] recommends that character height on VDTs be at least 16 and preferably 20 to 22 min. of arc. Measurements described above suggest these values to be too small for the S type CRT. However a look at Table III shows that for the H type CRT a smaller size than suggested may be permissible. Conversely, if it is necessary to use very small characters, a high resolution format should be recommended.

No significant difference between the two types of CRTs was found with respect to the peak velocity while reading and the peak velocity of return sweep. It should be pointed out that the movements pertaining to these values are saccadic. A significant difference was established, however, between the CRTs with mini sized characters as far as the mean and cumulative eye-fixation durations were concerned. This suggests that the slowing down of the reading speed was caused by a perceptual delay.

It is also worth pointing out that the number of fixations during reading does not vary significantly from one CRT to the other or from one character size to the other. The number of lines read during the same time span varied significantly for mini-sized characters on the S-type display. This can be explained as an increase in the number of fixations per line due to difficulty in reading mini characters on low resolution CRT.

Also, the interindividual and intraindividual variabilities in eye-fixation duration in reading different sized characters on both types of CRT are small with the exception of the case of the mini sized characters on S-type CRTs. This might be explained in the following manner. There may be two reasons for fixations: to see better, or to understand better. For this contemporary novel reading matter understandability is not an issue; thus difficulty in early visual processing is the likely culprit. Clearly, the high number of fixations in S-type CRTs with mini characters cannot be due to understanding difficulties which is not a function of readability. From this, it follows that the fixations in S-type displays with mini characters must be due to lower readability.

4. CONCLUSIONS

Measurement of visual fatigue due to reading from different VDTs could not be satisfactorily measured and differentiated. Blink rate changes while reading may provide clues in this respect. Analysis of reading eye movements for displayed texts with different character formats and resolutions demonstrated that the display resolution did not have a significant effect on readability of sufficiently large characters. However, for very small characters, higher resolution improves readability.

ACKNOWLEDGEMENTS

The authors wish to express their gratitude to Shu-Heng Wu, Harold Garland, Munehisa Takeda, Maryse Leroy, Susan Kanno, Steven Stufflebeam, Alan Lee, Corina VanDePol, Paul Owens for their valuable help during the measurements, to Professor Shin'ya Yamada for his encouragement and also to the subjects for their patience. They also thank David Le and Sigma Designs Inc. for long-term loan of the equipment. Selim S. Hacisalihzade was partially supported during this project by the Swiss National Science Foundation grant 5.521.330.615/7.

REFERENCES

[1] Bennett, J. et al. ed. Visual Display Terminals. Prentice-Hall, 1984.
[2] Grandjean, E., ed., Ergonomics and health in modern office. Taylor & Francis, London, 1984.
[3] Scalet, E., VDT health and safety. Ergosyst Assoc. Lawrence, KS,1987.
[4] Goussard, Y., Martin, B. and Stark, L., A new quantitative indicator of visual fatigue. IEEE Trans. B.M.E. 34. 23-29, 1987.
[5] Bonthron, S., Black Jaguar. personal communication, Berkeley,CA,1987.
[6] Shermann, K., Micromeasurements System 1200 Eye Monitor operator's manual. Micromeasurements, Berkeley, CA, 1987.
[7] Kim, W.S., Zangemeister,W. and Stark, S., No fatigue effect on blink rate. Proc.of 20th Annual Conference on Manual Control,337-348,1984.
[8] Wu, S.H. et al., Blink rate as measure of effort in visual task performance, Proc. on Work with Display Units, Stockholm, Sweden, 1986.
[9] ANS for human factors engineering of VDT workstations, Revised Draft. The Human Factors Society, Inc., 1986.
[10] Norton, D. and Stark, L., Eye movements and visual perception. Scientific American, 224, 34-43, 1971.
[11] Jones, A. and Stark, L., Abnormal patterns of normal eye movements in specific dyslexia: Rayner, K. ed., Eye movements in reading. pp.481-491, Academic Press, New York, USA, 1983.
[12] Bahill, A.T. and Stark, L., Trajectories of saccadic eye movements. Scientific American, 240, 84-93, 1979.

Trends in Ergonomics/Human Factors V
F. Aghazadeh (Editor)

VISUAL STRAIN EVALUATION OF VDT OPERATORS USING A LASER OPTOMETER

K.S. LEE*, A.M. WAIKAR* and O. OSTBERG**

* Department of Industrial Engineering, Louisiana State University, Baton Rouge, LA 70803, U.S.A.,
** Ergonomics Lab Plc.
Televerket, 12386 Farsta, Sweden

Effectiveness of eye exercise in reducing eye strain caused by VDT work was investigated. This reduction was compared to the stress alleviation accomplished by using ergonomically designed work station. Three measures, namely increase in accomodation (diopter), subjective rating and task performance were used. Field laser optometer was used to measure changes in eye functions. Five subjects were tested for two days (8 hours/day). Assigned eye exercises were performed for two minutes, every two hours during the breaks, by the subjects while working in fixed design work station.

1. INTRODUCTION

There has been a growing concern about the effect of VDT workstations on eye problems such as burning and itching, eye strain, visual fatigue and blurred vision [1, 2, 3, 4]. The reports of eye strain are widespread, and it affects more than 50% of the operators after an eight-hour work shift [5].

A few studies [6, 7] have questioned whether VDTs are more demanding than intense hard copy reading. The issue remains unsettled but these studies reported that a high percentage of VDT users complained about the eye strain. We also often find ourself rubbing the eyes and/or trying to focus our eyes on an object far away after the long hours of VDT work hoping to reduce eye strain. It is suspected that eye exercises have a potential to reduce eye strain if perfromed regulary. Many people however, have not realized the potential of eye exercises for reducing visual strain. This may be partly due to lack of any systematic study which evaluates the benefit of eye exercises.

While studying microscope operations at a semiconductor plant in the U.S., Emanuel and Glonek [8] recommended eye exercise program to reduce work-related eye strain. Eriksson et al. [9] also suggested intermission gymnastics and relaxation exercises for better health of the worker population. However, none of these studies has examined systematically the effects of eye exercise on eye strain.

The objective of this research was to study the effectiveness of eye exercises at work in reducing the eye strain. If our hypothesis is correct, it may be possible to reduce eye strain to a certain extent by encouraging and suggesting regular, good, inexpensive eye exercise program.

The primary advantages of utilizing exercises to reduce eye strain are low cost and ease of implementation. If eye exercises can sufficiently alleviate eye strain, then it may save temporary myopia and/or eye sight deterioration for a large population engaged in VDT work. Thus, there is a clear need to evaluate the effectiveness of eye exercises.

However, effectiveness in the reduction of eye strain is difficult to be judged in absolute terms unless the reduction is overwhelming. Therefore, the effectiveness of eye exercises should be compared to the effectiveness of a well-known method for a relative comparison. Thus, eye strain of VDT operators while using the ergonomically designed work station was compared to the eye strain of the same operators while using the non-ergonomic (fixed design) work station with eye exercises during their rest breaks.

2. EXPERIMENTAL DESIGN

Each subject worked in the ergonomically designed work station and in the non-ergonomic (fixed design) work station in a randomly assigned order. Ergonomic design included adjustable CRT screen and keyboard, adjustable chair and table, and adjustable footrest. In the ergonomic environment the subjects chose the best table and chair adjustments for maximum physical and visual comfort. The non-ergonomic design included fixed CRT screen and keyboard, fixed table height (72 cm), fixed straight back chair (43 cm high) and no footrest. In the fixed design environment, the subjects performed carefully planned eye exercises.

2.1 Subjects

Five subjects, one male and four females, were used in the study. Their ages ranged from 20 to 32 years. All were college students in good health and had no history of eye or any physical problems. None wore eye glasses and had a 20/20 correctable vision. Each had typing proficiency between 60 to 95 keystrokes per minute. They were paid monetary compensation for their participation in the experiment.

The subjects worked for eight hours in each of the work stations. The subjects were required to rest for at least forty-eight hours to eliminate the effects from the previous experimental participation. Half an hour was allowed for lunch and fifteen minutes each for the two breaks.
In the fixed design work station, subjects performed 2 minutes of assigned eye exercises during each break and lunch time. The exercises were designed from the recommendations of Emanuel and Glonek [8]. A portion of the exercises consisted of focusing on different objects at different distances.

There were also head and eye rotation while focusing on a fixed object.

2.2 Accomodation measurement

Responses based upon human judgements, can vary with time and may also be influenced by psychosocial factors. A few studies [1Ø, 11, 12] reported the use of the Field Laser Optometer as a feasible method to objectively measure altered eye functions, without relying on subjective human judgements.

The Field Laser Optometer (FLO) measures the absolute refractive state of the eye and provides readings in the range of +9 to -3 diopters, with an absolute accuracy of Ø.13 diopters. The FLO contains a 2 milliwatt, He-Ne laser. In the FLO, the beam from the laser is expanded and directed by mirrors onto a white plastic drum. The image of the laser light reflected from the drum is viewed by the observer [13].

To measure the altered eye functions, the subjects were asked to see in the eye tube of the optometer taking a comfortable position. The optometer reading was taken three times in both work stations, before beginning the typing task, after working for four hours and just after the days works (eight hours). Each time, the subject's optometer reading values were recorded in terms of diopters.

2.3 Subjective measurement

The subjects were asked to rate their discomfort using the questionnaire every two hours of VDT work. The questionnaire contained 5 items on eye strain and eye symptoms such as pain above eye, sandy feeling and dryness of eyes, dull feeling and aching of the eye ball, irritation of eye ball and the discomfort of eye in general. The subject responded by pressing the keyboard's "minus-key" so as to produce a line, the length of which represented the subject's felt level of discomfort. The discomfort was rated on the scale from "no, not at all" (score ØØ) to "yes, very much so" (score 5Ø).

2.4 Typing task

The typing task involved two different sessions; file entry, and file maintenance. During the file entry session, the subjects were asked to type a file carefully, but as fast as possible. During the file maintenance session, the subjects were asked to edit a file containing 4Ø-5Ø% wrong characters as fast as possible. Sessions were switched every 2Ø files. The typing speed and errors, including extra and omitted characters, were recorded using a computer program developed at NIOSH. The difficulty in reading and typing for different files was approximately equal.

2.5 Performance measurement

In order to determine the effect of eye strain on performance, the rate of typing speed in key strokes per minute and errors were measured for each work station. Typing rates and errors were averaged for each subject and compared for the two work stations.

Other factors needed to be considered were lighting, temperature, noise, and humidity. To minimize the effect of these extraneous factors, their variation was held to a minimum. Blinds on the window were closed.

3. RESULTS

Figures 1 and 2 show the subjective ratings for 5 different questions about eye strain. It shows that there was significant difference in increase in general eye discomfort between two work stations. It may be noted that the overall eye strain levels continued to climb up despite the the breaks, ergonomic design, and eye exercises.

Figure 3 shows the average level of increase in eye strain and accomodation (diopters) aggregated for all subjects. The subjects reported increased feelings of eye strain and at the same time exhibited increased myopic accomodation when focusing on a 6m distant target. The increase of accomodation was almost the same in both the ergonomic (0.45 diopters) and the non-ergonomic (0.48 diopters) work stations after eight hours of VDT work. However, using the paired t-test, the increase in subjective eye strain rating was found to be significantly smaller (71%) in the non-ergonomic work station (average 8.1) than in the ergonomic work station (ave. 11.4).

These results confirm the hypothesis that eye exercises can reduce eye strain as effectively as the ergonomic design. Due to the unavailabilty of data in the non-ergonomic work station without any eye exercises, it was difficult to evaluate absolute effect of eye exercises on reducing the eye strain. However, it is suspected that the subjects working in the non-ergonomic work station would have experienced higher levels of discomfort without any eye exercise.

There was no significant difference between the two work stations with respect to average typing performance using rate of typing and number of typing errors per file as criteria (figure 4). For the typing entry session, average rate per file was 67 key strokes per minute for ergonomic station versus 65 for the non-ergonomic station. The error rate was 2 errors versus 3 errors per file for the ergonomic and non-ergonomic work stations respectively. For the file maintenance session, the average rate was 48 vs. 47 key strokes per minute respectively but the subjects made, on the average 5 errors in the ergonomic and 4 errors in the non-ergonomic work station.

As shown by figure 5, for both the work stations, the typing rate was lower in the second hour of all the two hour sessions. There was no significant difference in the first hour typing rates for the ergonomic and the non-ergonomic work stations. However, a slight increase was observed in the second hour typing rates of the two hour sessions as the experiment proceeded. For the non-ergonomic work station, this rate increased from 39.5 strokes per minute to 49.5 strokes per minute. The increase was from 46 strokes per minute to 50 strokes per minute for the ergonomic work station. This increase reduced the difference between the overall typing rates in the first and second hours of the experimental sessions.

4. DISCUSSION AND CONCLUSIONS

Reviewing the results obtained for the accomodation and eye strain rating, no significant difference was found between the ergonomic and the non-ergonomic work stations. It was observed that increase was smaller during the morning session in the ergonomic work station than in the non-ergonomic work station. However, this trend was reversed in the afternoon session. This may indicate that the eye strain is somewhat tolerable in the first four hours of VDT work without any eye exercise. However, longer rest breaks and eye exercises may be necessary as the duration of VDT work increases.

The response to the subjective questionnaires for the most part, seems to be similar for both the work stations. However, a slightly higher degree of pain was reported in the non-ergonomic work station. Between the two work stations, the ergonomic station was preferred by all of the subjects. However, when the subjects had completed both the experiments, they indicated that the exercises were beneficial while working in the non-ergonomic work station. It was also suggested that the duration of exercises be increased for optimum benefit.

The performance analysis revealed that the subjects performed slightly better in the ergonomic work station. The rate of typing was approximately 3 percent lower in the non-ergonomic work station. However, statistically there were no significant differences with respect to the rate and errors between the two work stations.

Based on the above findings, it may be concluded that eye exercises assigned to the VDT operators working in the non-ergonomic environment can contribute to reduction of eye strain. The contribution may be as much as that offered by the ergonomic features of the ergonomic work station. However, all subjects indicated that while they had almost no discomfort at the beginning of the day, they ended up with, very much discomfort at the end of eight hours of work. This may imply that any one measure might not be enough to substantially reduce the eye strain of VDT operators. Thus, it is recommended that an appropriate combination of ergonomic features and eye exercises be provided for the VDT operators.

5. REFERENCES

[1] Laubli, Th., Hunting, W. and Grandjean, E., 1981.

Postural and Visual Loads at VDT Workplaces: Lighting Conditions and Visual Impairments. Ergonomics 24: 933-944.

[2] Mourant, R.R., Lakshmanan, R., and Chantadisai, R., 1981. Visual fatigue and cathod ray tube display terminals. Human Factors, 23: 529-540.

[3] Smith, A.B., Tanaka, S. and Halperin, W., Correlates of Ocular and Somatic Symptoms among Video Display Terminal Users. Human Factors 26 (1984) 143-156.

[4] Smith M.J., Health Issues in VDT Work, in Sandelin, J., Bennett, J. and Smith, M.J. (Eds.), Video Display Terminals: Visibility Issues and Health Concerns. (Prentice-Hall, Englewood Cliffs, NJ, (1984) 193-228.

[5] Gunnarsson, E. and Soderberg, I., 1979. Work at a VDT Presenting Textual Information at a Publishing Company: An Inventory of Visual Ergonomics Problems. AMMF 1979:21, National Board of Occupational Safety and Health, Stockholm, Sweden.

[6] Starr, S.J., Thompson, C.R. and Shute, S.J., Effects of Video Display Terminals on Telephone Operators. Human Factors 24 (1982) 699-711.

[7] Starr, S.J., Effects of Video Display Terminals in a Business Office. Human Factors 26 (1984) 347-356.

[8] Emanuel J.T., Glonek R.J., 1974. Microscope Operations: Recommendations for Workplace Layout and Fatigue Reduction. Tech. Report.

[9] Eriksson, M., Enocksson, J., Bjordal, J., Jonsson, B., Jungnell, G., Jalkelov, K., and Lundberg, S., 1983. Microscope Work: An occupational environment survey. Report. The swedish union of salaried employees in hospital and public health services.

[10] Takeda, T., Fukui, Y., Takeo, I., Karasuyama, K., and Kigoshi, T., 1984. A new objective measurement method of visual fatigue in VDT work, Human Factors in Organizational Design and Management, H.W. Hendrick and O. Brown, J.R. (Editors), Elsevier Science Publishers B.V. (North-Holland), pp 193-197.

[11] Ostberg, O. and Takeda, T., Accommodation Performance as a Function of Time of Day and Line of View. Proceedings of the International Scientific Conference on Work With Display Units, Stockholm (1986) 1013-1016.

[12] Zwahlen, H.T., Hartmann, A.L. and Gangarajulu, S.L., Video Display Terminal Work with Hard Copy Screen and a Split Screen Data Presentation. (Ohio University Department of Industrial and Systems Engineering, Athens, OH, 1984). (NIOSH Order No. 83-1775).

[13] Hennessy, R.T. and Richter, O.F., 1979. Field Laser Optometer, Manual, Monterey Technologies, Inc.

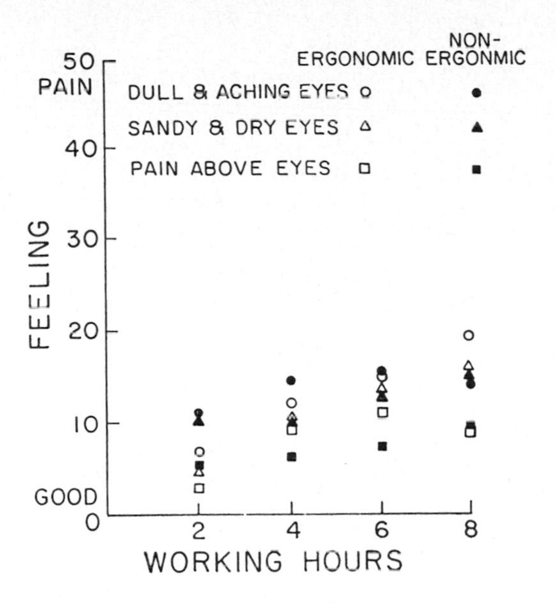

FIGURE 1
Mean increase in ratings of pain, sandy and dry feeling, and dull and aching feeling in eyes for all subjects, during VDT work for two days.

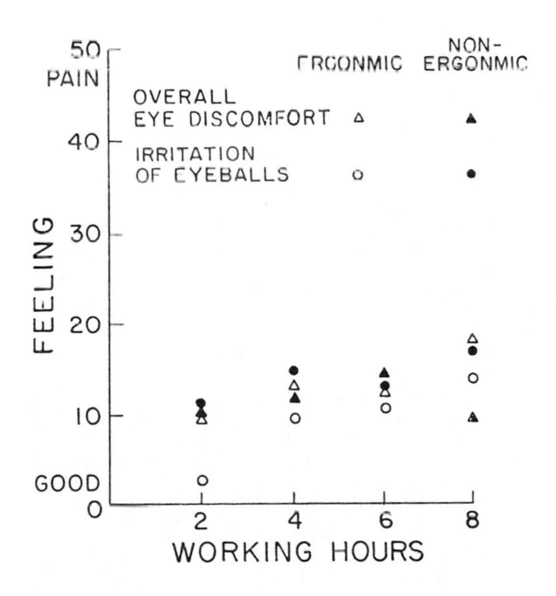

FIGURE 2
Mean increase in ratings of irritation of eye ball and overall eye strain for five subjects, during VDT work for two days.

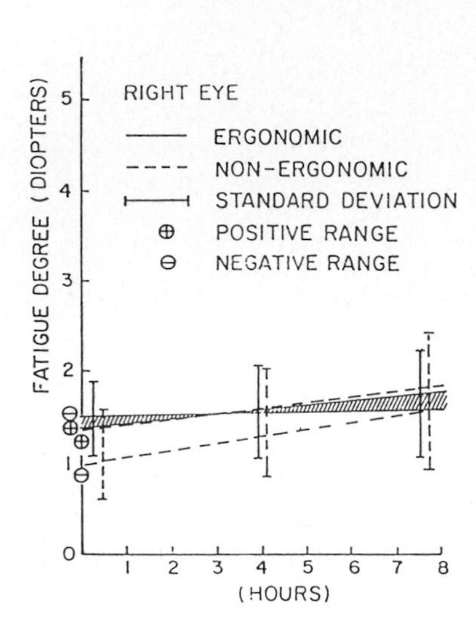

FIGURE 3
Mean increase in accomodation of eye for all subjects during VDT work for two days. For both days, the accomodation baseline was 0.18 diopter (myopic).

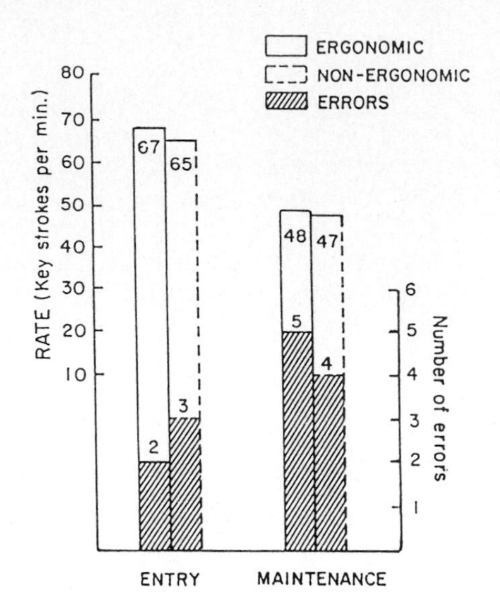

FIGURE 4
Mean key strokes/min. and number of errors per file for a working day for all subjects in ergonomic and non-ergonomic work stations.

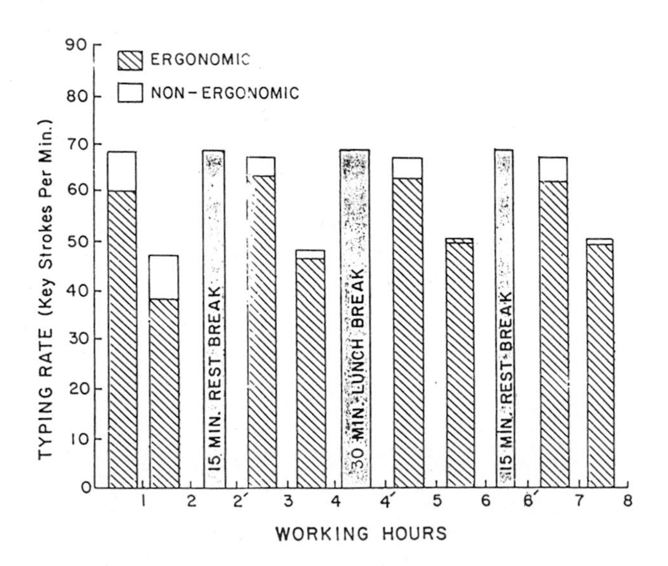

FIGURE 5
Mean key strokes/minute by working hours for five subjects in ergonomic and non-ergonomic work stations.

Trends in Ergonomics/Human Factors V
F. Aghazadeh (Editor)

ERGONOMIC SEATS FOR COMPUTER WORKSTATIONS

Karl H. E. Kroemer

Industrial Ergonomics Laboratory, Human Factors Engineering Center
Industrial Engineering and Operations Research Department
Virginia Tech (Virginia Polytechnic Institute and State University)
551 Whittemore Hall, Blacksburg, VA 24061, USA

In the nineteenth century, physiologists/orthopaedists published theories about "proper" sitting postures and deduced from these recommendations for seat and furniture design. Their concept was that "sitting with an upright trunk means sitting healthily." This idea has remained nearly unchallenged until recently, when various and occasionally contradictory theories about healthy postures while sitting at work have been published. Some of these have led to rather radical proposals for seats at work.

SITTING "UPRIGHT"

In reviewing the literature of about the last 100 years it becomes obvious that postural considerations, e.g. by Cohen, Lorenz and Staffel, were originally based on the concept that the spinal column of the sitting person should be "erect" which actually means curved similar to that of a "healthy normal upright standing" person. This was believed to put the least strain on the spinal column and its supportive structures, including the musculature.

Recommendations were derived for the design and use of seats from this desired erect posture. In particular, school furniture was to be so designed that it facilitated, in fact enforced, an "upright" or "straight" back, with the height of the seat so adjusted that the angles at hips and knees would be close to 90 degrees while the hips were horizontal and the soles of the feet rested on the floor. This "vertical-horizontal" arrangement was also recommended for the arms: the upper arms should hang down and the forearms be horizontal. Pupils were not expected to rest their backs: Cohen's drawing of a school bench shows a wooden protrusion of the front edge of the rear bench poking directly at the lumbar section of the pupil sitting in front. (Akerblom proposed a less painful way for maintaining the lumbar curvature about a century later.)

These ideas of suitable posture and furniture were repeated in the literature, apparently without much questioning or contradiction, until the middle of the twentieth century. In retrospect, this is quite surprising because there was little or no experimental support for the appropriateness of this posture. Also, observing persons sitting freely should have shown then, and does now, that hardly anybody ever maintains this posture if left alone. To follow the recommendation to keep the spine upright or "straight" was also inherently difficult (if not impossible) since the spinal column is, in the lateral view, naturally bent:

there is a physiological forward curve (lordosis) in the lumbar area and in the cervical area, while a backward bent (kyphosis) exists in the thoracic area. Thus, it was and is very difficult to reconcile the fact of three "natural curvatures" in the spinal column with the desire to be upright or straight. Nevertheless, the concept of "sitting upright is sitting well" was generally accepted as true.

EXPERIMENTAL DATA

Modern attempts to quantify strain parameters in the spinal column and trunk were published about the middle of the twentieth century. Most prominent are the experiments measuring muscular tensions (by EMGs), pressures in the intervertebral disks, and intra-abdominal pressure. Other experiments addressed fluid volume and pressure in the lower legs, tissue compression and nerve irritation by the seat pan.

EMG measurements of muscular activities in the trunk show various activities of the six or so major muscle groups pulling down on the upper trunk thus stabilizing the spinal column. Variations in postures, external loads, and backrests or other seat features influence those muscular activities. However, during "normal" sitting these muscle tensions are usually at or below approximately five percent of the maximal voluntary contraction activities, and hardly ever exceed ten percent of MVC. At such low percentages of muscular strength required, variations in muscular exertions hardly allow important conclusions regarding suitable postures. Of course, specific muscular activities, particularly in selected muscles, may have important effects: but such experimental work has hardly started. Intra-abdominal pressures, though possibly bringing some reduction relief of the compression strain of the spinal column, require muscular contractions for pressure maintenance which in turn increases the compression of the spine. Thus, the hypothetical benefits of intra-abdominal pressure are counteracted by the associated muscle tension.

Innovative and courageous measurements of pressure in the intervertebral disks were performed in the 1950s. The published results were used by proponents of the "sitting upright is sitting well" idea to point out pressure benefits associated with this posture. However, careful examination of the published data indicates no clear benefits of sitting erect as opposed to sitting slumped. In fact, leaning on a backrest can reduce disk pressure significantly below the values found while sitting erect without support. Since direct pressure measurements in the disks are difficult, more recently the spinal loading has often been calculated from biomechanical models.

TRUNK-LEG INTERACTIONS

Also in the middle of the twentieth century, the orthopaedic interactions among lower spinal column, pelvic girdle, and thigh and knee angles were investigated. Of particular interest were the angular positions (in the sagittal view) of the pelvic girdle and the adjoined sacrum, upon which rests and with which it is elastically connected the lumbar section of the spinal column. Rotating the pelvic girdle downward in its anterior (forward) section about the ischial tuberosities (resting on the surface of the seat pan) increases via biomechanical interaction the lordosis of the lumbar section of the spinal column. Conversely, rotating the pelvic girdle around the ischial tuberosities so that the anterior section of the pelvis is elevated, flattens out the lordosis towards a kyphotic condition. (These actions on the lumbar spine take place if associated muscles are relaxed; muscle activities can counter the effects.)

A downward tilt of the frontal part of the pelvis can also be facilitated by letting the knees drop below the height of the tuberosities, thus declining the thighs downward. However, muscles that span the hip joint, some in fact also crossing the knee, may play important roles: for example, tightening the posterior muscles running from calves to buttocks generates a force vector at the back which tends to flatten the lumbar lordosis. But if trunk muscles contract to generate counteracting force (according to Newton's Third Law) they may maintain, perhaps even increase the lordosis. "Opening" hip and knee angles may have differing results on the lumbar spine depending on whether muscle tension is existent, or not.

Realization of these anatomic relationships gave rise to recommendations that the seat surface be tilted downward in its front, or tilted upward in its rear section. The first idea was already expressed by Staffel in the nineteenth century, the second was commercially used in the 1950's ("Schneider Wedge") and was recently promoted again (Mandal, balans designs).

The thigh angle may be opened by either leaning back against a large backrest (discussed later) while keeping the thighs essentially horizontal, or by keeping the back more or less upright, but by dropping the thighs. Declining the thighs is facilitated by a steep forward-downward declination of the seat pan which, in turn, makes the buttocks slide down on the seat surface. To counteract the sliding, one can increase the friction between clothing and seat: this pulls uncomfortably on the clothes. Another solution is to contour the seat surface to follow the shape of the seated person's underside. One can thrust the feet on the floor to counteract the sliding force, or alternatively provide support at the knees and lower legs through "shin pads."

BACK SUPPORT

The thinking about whether and how to support the back through a backrest has changed dramatically throughout the last hundred years.

If one sits "upright," a backrest is either unnecessary, or could help to generate the desired form of the spinal column if one presses the back against it, which requires muscular contractions.

Around the middle of this century the idea of a backrest that would conform to and support the lumbar curvature was widely promoted. The associated furniture feature was often called the Akerblom pad. For several decades, such support only in the lumbar region was considered appropriate, as evidenced by the low lumbar board usually attached to the so-called secretarial chair. (The boss usually had a large reclining backrest.)

More recently, two contradictory ideas have been advocated. One is that the muscles of the trunk should be stimulated to maintain (varying) back postures. Such continued but varying muscle contractions are thought to be beneficial. This so-called dynamic sitting does not require a backrest. The opposing idea is to provide a backrest so that one can at least occasionally lean against it to change the posture and relax muscles. If the backrest is tall and so much tilted behind vertical that one reclines on it, the backrest can carry some of the weight of the upper body which consequently needs not be transmitted through the spinal column. Since trunk muscle activities are also diminished, this should reduce the compression loading of the spinal column.

HEAD AND NECK POSTURE

The explosive growth in the use of computer displays at work has brought a large number of complaints about discomfort and pain in the neck and shoulder regions. Many people are forced to hold their eyes in a given position with respect to the computer screen. If the viewing distance and the direction of the line of sight are unsuitable for the individual, he or she must tilt the head and/or bend the neck, possibly even the middle and lower spine, in unbecoming postures. Measurements of the resulting muscle tensions or cervical disk pressure are so far scarce. New experimental data may lead to revised theories about proper head and neck postures, and consequently to ergonomic recommendations for the design of computer furniture.

OBJECTIVE AND SUBJECTIVE EXPERIMENTAL MEASURES

Originally, recommendations for sitting postures were derived from physiologic and orthopaedic theories, without much experimental support. With the development of new measuring techniques and experimental procedures, "objective" dependent variables have been measured, such as disk pressure, muscle tensions (via EMGs), intraabdominal pressures, or skeletal relations either assessed by external observation, by x-rays, or more recently by continuously recording of spine angles.

Given the practical difficulties associated with performing "objective" measures (e.g. disk pressure, muscle tension) and interpreting the results, much emphasis has been placed in recent decades upon "subjective" self-assessments of discomfort, pain, etc. Users and experts have been asked in various ways about their opinions regarding postures, furniture features, adjustments, and other working conditions, particularly in the "computerized office." Learning to elicit and to interpret these subjective statements and to reconcile them with "objective" data for development of furniture design guidelines is an ongoing task.

CURRENT THINKING ABOUT SUITABLE SITTING POSTURES

No "one simple" theory about the proper, healthy, comfortable, efficient, etc., sitting posture at work prevails currently. With the idea abolished that everybody should sit upright, and that furniture should be designed to this end, the general tenet is that many postures are comfortable (healthy, suitable, efficient, etc.) dependent on the individual's body, preferences, and work activities. Consequently, it is thought that furniture should allow many posture variations and hence permit easy adjustments in its main features, such as seat height and angles, backrest position, or knee pads and footrests; and that the computer work station should also allow easy variations in the location of the input devices and of the display. Thus, change, variation and adjustment to fit the individual are the key terms. If any label can be applied to current theories about proper sitting, it is "free posturing."

DESIGNING AND SELECTING ERGONOMIC FURNITURE

No longer guided by the ideological straightjacket of "sitting upright", providing and choosing suitable furniture is again a challenging and rewarding task. Guidance is provided by the following principal criteria:

1. Seat pan

The surface of the seat pan must support the weight of the upper body comfortably and securely. Hard surfaces generate pressure points, which should be avoided by suitable upholstery, cushions, or by providing other surfaces that can elastically/plastically adjust to body contours.

The only inherent limitation to the size of the seat pan is that it should be so short that the front edge does not press into the sensitive tissues near the knee. The height of the seat pan must be widely adjustable, best down to 38 cm (15 in) and up to at least 50 cm (20 in), better 58 cm (23 in) to accommodate persons with short and long lower legs. This adjustment must be easily accomplished while sitting on the chair.

Usually, the seat pan is essentially flat, between 38 and 42 cm (15 and 17 in) deep, and at least 45 cm (18 in) wide. A well rounded front edge is mandatory. Often, the side and rear borders of the seat pan are slightly higher than the central parts of the surface, usually achieved by more compressibility of the inner sections. Figures 1 and 2 illustrate major dimensions of seat pan and backrest.

In the side view, the seat pan is often essentially horizontal, but tilting it slightly backward or forward is usually perceived as comfortable and desirable. Seat pans that are higher in their rear portion and lower at the front facilitate "opening the hip angle". "Semi-sitting" on a distinctly forward-declined seat surface is comfortable for some persons. To counteract the forward/downward thrust along this declined surface, shin pads may be incorporated in the structure. Some inclined seat surfaces are shaped to fit the human underside to counteract slippage.

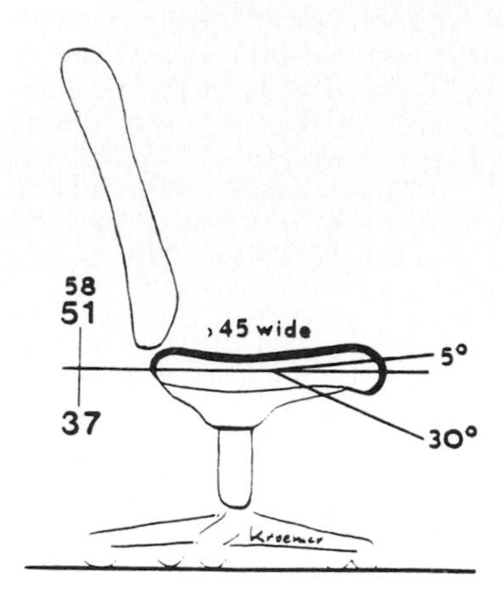

Fig. 1. Essential Dimensions of the Seat.

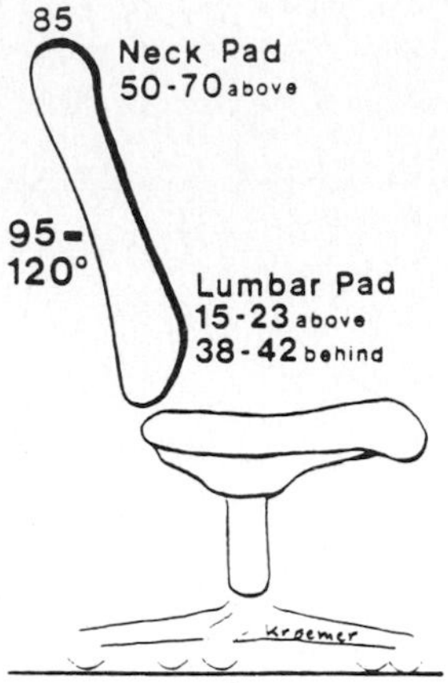

Fig. 2. Essential Dimensions of the Backrest.

2. Backrests

Backrests serve two purposes: to carry some of the weight of upper trunk, arms and head; and to allow muscle relaxation. Both purposes can be fulfilled only when the trunk reclines on the backrest.

The backrest should be as large as can be accommodated at the workplace. This means up to 85 cm (33 in) high, and at least 30 cm (12 in) wide. It should provide support from the head-neck on down to the lumbar region. For this purpose, it is in side view usually shaped to follow the back contours, specifically in the lumbar and the neck regions. An adjustable pad for the lumbar lordosis (e.g. an inflatable cushion) is appreciated by many users. The lumbar pad should be adjustable from 15 to 23 cm (6 to 9 in), the cervical pad 50 to 70 cm (20 to 28 in) above the seat surface.

The angulation of the backrest must be easily adjustable while seated. It should range from slightly behind upright (95°) to 30° behind vertical (120°), with further declination for rest and relaxation desirable. Whether or not the seat back angle should be mechanically linked to the seat pan angle is apparently a matter of personal preference.

3. Armrests

Armrests allow to support the weight of hands, arms, and even portions of the upper trunk and head. Thus, armrests are useful, though often used only for short periods of time. They must be well located and of suitable load-bearing surface. Adjustability in height, width, and possibly direction within rather small limitations may be desirable. However, armrests can also hinder moving the arm, pulling the seat up to a workstation, or getting in and out of the seat. In these cases, having short armrests, or not having armrests at all is appropriate.

4. Footrests

A prevailing need for footrests usually indicates that the height adjustments, particularly of the seat pan, are not sufficient for the seated person. Hence, presence of footrests usually indicates deficient workplace design.

(Some people like to have a footrest, even in form of a rail, at nearly seat pan height so that they can rest their heels on it while extending the legs nearly straightforward. This allows an occasional "straightening of the elevated legs".)

If footrests are unavoidable, they should be so high that the sitting person has the thighs nearly horizontal. Footrests should not consist of a single bar or a small surface (because this severely limits the ability of the sitting person to change the posture of the legs): instead, the footrest should provide a support surface that is about as large as the total leg room available in the normal work position.

5. Design for Motion

People need to stretch, to relax, to move around instead of sitting still. The design of the work furniture, and of the work task, should allow and facilitate "to move the body". Stretching and bending, such as in reaching for a distant object, can be a welcome exercise. Getting up and walking around is still better. Some people even prefer to stand at work, perhaps only for a short time. This requires a workstation where the working height is at or slightly below elbow height while standing. It is often comfortable to put one or the other foot up onto a support surface which should be located at about half knee height.

ERGONOMIC DESIGN OF COMPUTER WORKSTATIONS

The foregoing discussions should have made it clear that several variables combine to determine the "ergonomics of computer workstations". These include psychological and attitudinal as well as organizational conditions. Another important aspect is the physical environment, which includes climate, illumination, and general facility and work space design. A discussion of all these variables goes beyond this text.

Of major importance is the design and use of work equipment, for example data entry devices, such as keyboard, trackball, mouse, touchpad, lightpen, or stick controls. Highly specific, even unusual work tasks and conditions may prevail, and individual preferences may lead to rather unconventional solutions, such as Diffrient's "lounge chair computer workstation".

The human is the most important component of the system since she or he drives the output. Hence, people must be accommodated first: the design of the workplace components should fit all operators, and allow many ideosyncratic variations in working posture. There is no "one healthy upright posture, good for everybody, anytime". Hence, many adjustment features must be considered in an ergonomic workstation. These are schematically shown in Figure 3 for the seat (S), the backrest (B), the footrest (F) if it is needed; for the work surface (W), and the display (D).

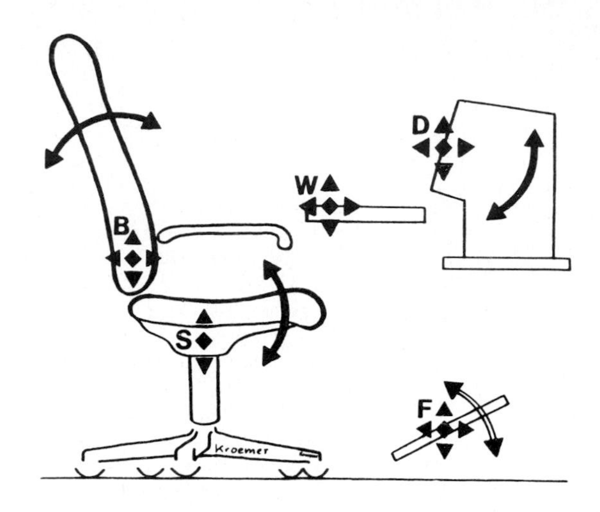

Fig. 3. Adjustment Features of a Computer Workstation.

LITERATURE

Recent overview articles provide rather complete information:

Eklund, J. (1986) Industrial Seating and Spinal Loading. Doctoral Dissertation, University of Nottingham. Linkoeping: University of Technology (ISBN 91-7870-144-9).

Kroemer, K. H. E. (1987) VDT Workstation Design. Chapter in Handbook of Human-Computer Interaction (ed: M. Helander) Amsterdam: Elsevier.

Kroemer, K. H. E. (1987) "Ergonomic" Furniture for Computer Workstations. Workshop Background Paper, Annual Meeting of the Human Factors Society, October 19, 1987. Blacksburg, VA: Virginia Tech, IEOR Department.

Lueder, R. K. (1983) Seat Comfort. A Review of the Construct in the Office Environment. Human Factors 26(3), 339-345.

Tougas, G. and Nordin, M.C. (1987) Seat Features Recommendations for Workstations. Applied Ergonomics 18(3), 207-210.

Trends in Ergonomics/Human Factors V
F. Aghazadeh (Editor)

A KEYBOARD TO INCREASE PRODUCTIVITY AND REDUCE POSTURAL STRESS

Stephen W. HOBDAY

PCD Maltron Ltd
15 Orchard Lane
East Molesey
Surrey, KT8 0BN. England.

A new computer or office machine keyboard has been developed to match hand movements and reduce the postural stress imposed on operators by the Scholes (qwerty) design which engenders fatigue and can lead to pain and disability. The qwerty letter layout has been retained for already trained operators but an alternative efficient new arrangement may be selected. Key tops may be dually or singly designated.

1. INTRODUCTION

As is well known, the physical shape of the Scholes (qwerty) keyboard was established over a hundred years ago within the mechanical limitations of the time. These same limitations also gave rise to the letter layout in which 3 of the 10 most used letters were under the rest position of the fingers. An up, down, or sideways movement had to be made to reach the others before they could be keyed. The delay introduced by this gave sufficient time for the type bar of the previous letter to get out of the way and the arrangement overcame the type bar jamming problem sufficiently for the typewriter to achieve commercial success.

By the mid 1930s most of the mechanical limitations had been overcome and Augustus Dvorak developed a new letter layout with 7 of the 10 most used letters under the fingers. This layout has demonstrated improved performance but has not become widely used. It would seem that the improvement is not enough to justify the retraining effort and it does not address the problems of physical stress associated with the keyboard shape.

The advent of electronic keyboards, mostly in use with visual displays, seems to have highlighted operator stress problems which sometimes occurred with mechanical typewriters. These problems, broadly grouped under the name of Repetitive Strain Injury (RSI), are associated with the over use of particular muscle groups and are found in many occupations. Writer's cramp, Tennis elbow, Cotton picker's arm, are some of the names associated with this type of injury. It also occurs in meat cutting and packing, production and assembly line work in general, to hair dressers and concert pianists. The detail of the injury depends on the actions causing the damage. Medical terms used to describe typical keyboard cases are; Tenosynovitis, Carpal tunnel syndrome, Tendinitis, Peritendinitis, and Epicondylitis. Unfortunately none of these conditions respond quickly to a simple cure and much medical doubt surrounds the best mode of treatment. Prolonged rest seems to be the only effective answer at present. The result is that often the most productive and conscientious operator is forced to stop working altogether, which is a personal career disaster, a severe productivity loss to the employer and a financial loss to the insurance company in meeting compensation claims. Sometimes the onset of RSI from keyboards is rapid (complete disability within a week), but more often it builds up slowly over months or even years.

The symptoms are increasing aches and pains, and / or numbness and tingling in fingers, wrists, arms, shoulders, or neck. There may also be weakness or swellings in hands and arms. These symptoms, when presented, should be regarded seriously and at least some action taken immediately to change the work pattern or load if permanent damage is to be avoided.

Quite apart from clinical levels of RSI damage actually occurring, a surprising number of keyboard operators do put up with mild aches and pains in their daily work. If they complain, they are usually told that they are not sitting properly or have strained themselves in some other activity. While both these comments and others similarly dimissive may be true in some cases, too many occur for them to be ignored. For example, in the offices of the "Engineering Computers" magazine, 14 of the 17 regular keyboard users suffered such pain. So large a percentage can only mean a job related stress. The only common item was the keyboard. Following research in 1974 into the disproportionate amount of ill health in Telex operators in the Australian Postal Service, Professor Ferguson of Sydney University concluded that the stress from keyboard operation was the prime cause [1]. Similar work in Japan [2] and more recently in West Germany [3] has supported this. In 1984 the Australian Public Service reported that 1000 of its 5984 operators were suffering in this way.

The fact that operating a qwerty keyboard can cause pain must mean that an initial stage of tiredness and fatigue has been passed through. From this it follows that to many operators the occupation is at least unpleasant, often painful, and sometimes dangerous. These characteristics of the design can hardly be expected to promote job satisfaction and productivity.

The need for a better keyboard has been realised for many years and alternative designs have been proposed [1,4,5]. Some have been primarily letter layout changes, others have considered the physical problems [6,7,8,9]. In 1976, Professor Ferguson's paper indicating that wrist abduction was the main problem, together with other papers and articles, were brought to my attention by Lillian Malt. From our study of these we decided to try and create a new fully ergonomic design. It was to address the physical stress of wrist abduction, take into account the differing lengths of fingers and be based on movements which can be made easily without strain. It was also to make the most of modern electronics to offer a new letter layout based on a computer analysis of letter use sequences and frequencies. Electronics had made it possible to separate the keys from the printing mechanism opening the way for a three dimensional shape, and permitting key memories to be switched instantly. This allowed the qwerty layout to be selected by already trained operators who would still have the benefit of the new shape.

2. ABDUCTION

The outward turning of the wrists, sometimes to an extreme extent as shown in Fig. 1, requires a sustained muscle tension in the arms to hold the hands in the normal typing position. This tension is partly relieved by slightly raising the shoulders, but this in turn often leads to fatigue and pain across the back.

Fig. 1
Hand abduction at a common keyboard.

By experiment we found that only by splitting the keyboard and separating the two parts by approximately 25cm between centres, could the abduction angles be reduced to zero for average adults, while at the same time keeping forearms nearly parallel in the minimum strain position. Although small angles do occur, this separation also seemed acceptable to children and men with wide shoulders. This design decision was subsequently confirmed by the work of P. Zipp et al [3]. Fig. 2 shows the change of myoelectric activity controlling muscle tension as the abduction angle is varied from 0 to 25 degrees. The operating area shown by the dense shading illustrates the activity needed by the qwerty design. The reduction achieved by the 0 - 5 degree range of the Maltron is clearly shown.

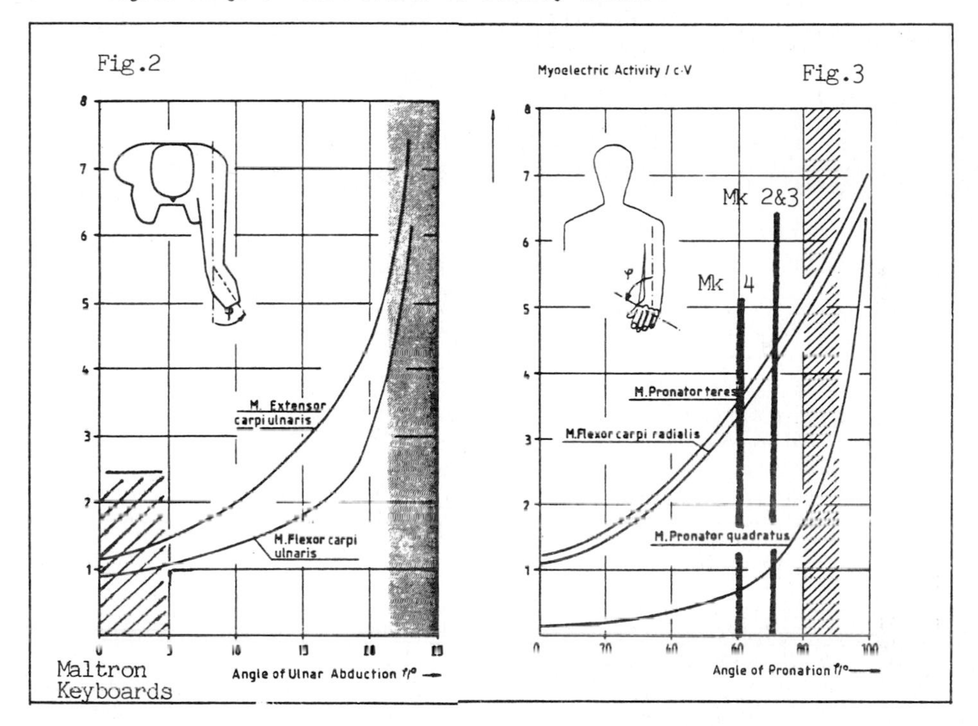

3. HAND SHAPE

The shape of a relaxed hand is easily seen by allowing the arms to hang loosely by the side. In this condition the fingers usually curve through a quadrant. The shorter fingers follow tighter curves so that the line of the ends of the three outer fingers does not make a right angle with the axis of the forearm. The resulting angle is usually about 20 degrees different and is about the same as the inward inclination of the forearms in the minimum strain position mentioned above. The result is that the line of the centre row of keys is nearly straight across the keyboard. If the hand is now raised until the forearm is horizontal, easy finger movements are seen to be arcs of about 90 degrees pivoted on the first knuckle. These movements then define the shape of the keyboard in the front to back direction, but since fingers are not all the same length the key heights and arcs must vary to suit. The division of the keyboard into two key groups with a wide gap in the middle provides an opportunity to consider making more use of the thumbs. Operating with the relaxed hand shape allows thumb movement to cover a significant area without the need to move the fingers from their "home row" key positions.

It was decided to take advantage of this and move the "Return" key to the right thumb alongside the "Space" key so that the long little finger stretch of the qwerty keyboard was no longer needed. This action on the qwerty keyboard, as well as the strain, usually causes loss of finger location and time waste in repositioning the hand. From a logical basis this "Return" key change makes sense since the thumb is immediately available to operate at the end of a word when either a "Space" or "Return" keystroke is needed. Experiment also showed that each thumb could easily access up to 8 or 9 keys. This opened the way for a complete reappraisal of the positions of computer cursor and function keys.

4. PRONATION

Normal keyboard operation requires the hands to be turned to the horizontal palms down position and, as in abduction, this is usually near or at the anatomical limit. Fig. 3 shows the curves of myoelectric activity for the range of movement from palms vertical to horizontal, with the qwerty range shaded. From the point of view of minimum strain, an accordian style keyboard would be the best solution, but such a shape would hardly be acceptable in an office environment. The curves show the importance of reducing pronation as much as possible. In the relaxed hand shape, the shorter outer fingers cause the hands to turn out a little, so the Maltron design uses this effect to reduce pronation to an extent which operators have noticed and welcomed. The 70 degree operating line of the Mark 2 and 3 designs has been drawn on the figure. These designs and the results achieved were the subject of a paper I presented in 1985 [10]. The 60 degree line achieved by the new Mk.4 design is also drawn on the figure. This shows further improvement and a significant reduction of stress from the qwerty levels.

5. KEYBOARD DESIGN

The first design, based on the principles discussed above, was brought into use in 1977 and satisfactory operation confirmed the design philosophy. Trials showed that there was a significant drop in fatigue and an improvement in accuracy. A keyboard of this time is shown in Fig. 4. It was designed to work with an IBM mag card word processor and displayed the Maltron letter layout. The three dimensional shape and the thumb groups are clearly shown.

Fig. 4 Mark 2 Maltron Keyboard

The right hand finger group shows the two rows of keys for the index finger raised to allow for its shorter length. A similar allowance for the little fingers can be seen on the left hand group.

A particular feature is the pair of palm resting pads in front of each finger group. These are not for use during keying since they would restrict the small amount of hand movement needed to key quickly and accurately, but when there is a pause, or "thinking time" is required, the palm may be lowered to the pad without losing finger position registration. This action allows the finger actuating muscles to relax and promotes the flow of blood in them to remove, or at least delay, the onset of muscle fatigue. The pads have proved to be a significant innovation. The centre key of the right hand row of the right thumb group is the "Space" key, with the "Return" key to the immediate left. The white keys are for controlling word processing functions and their positions were determined from an analysis of frequency and action. The centre of the left hand row of the left group is a "Space" key in the qwerty mode which changes to the next most used key "E" when in the Maltron mode with "." adjacent.

Fig. 5 Mark 3 Maltron Keyboard

Since extra keys could not be fitted to the earlier model, the advent of the IBM PC (with number pad and 10 Function keys) called for a new design. The Mk.3 keyboard, developed to be plug compatible with the IBM PC and keyboard compatible clones, is shown in Fig. 5. It can be seen that the essential features have been retained but the moulding has been modified to give a central flat area for the number pad and room for the Function keys along the back. Behind these is a holder for a Function key designation strip which can be easily changed to suit the software in use. To the left of the Function keys is the qwerty - Maltron changeover key to give immediate selection of the preferred letter layout. On this model small red qwerty letters were engraved above the larger Maltron letters. Coloured LEDs at the changeover key showed the set in use. At this time a general philosophy was developed to put all right hand symbols on the right and left hand on the left. This was then extended to cursor keys and other functions. The idea was that actions going forward in the work would be on the right and those going backwards would be on the left. The "Up" and "Left" cursor keys were therefore placed on the left thumb along with "Home" and "Back Space Delete". On the right thumb were the "Down" and "Right" cursor keys together with "End" and "Delete". Placing the "BS Del" key in the left thumb group above the

"E/Sp" key has proved very satisfactory as a conscious error can be corrected immediately. Since the "Control" and "Alt" keys may be needed with a key from either hand, these keys are duplicated and may be selected by either thumb. To make thumb operation of the "Space" and "E/Sp" keys easier, they have been enlarged to double size so that pressing with the side of the thumb, as many operators prefer to do, poses no risk of striking two keys at once.

Pressure from West Germany a few years ago for keyboards with home row heights not more than 30mm above the table, made keyboard manufacturers design short key modules. As these became available, it was decided to redesign the Maltron to use these keys to meet the 30mm requirement as far as possible within our concept and at the same time to see if pronation could be further reduced.

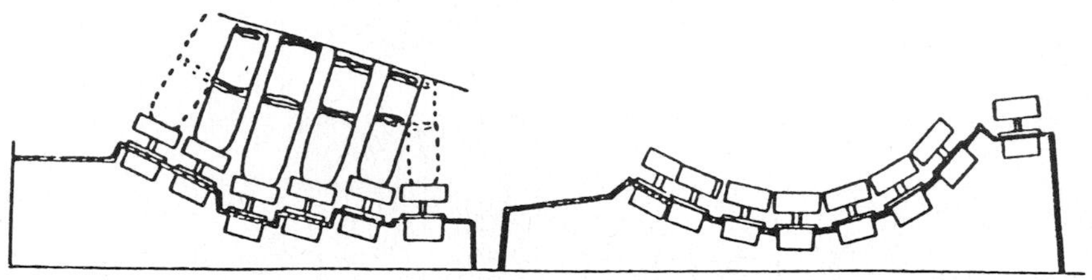

Fig.6 Tilted keys. Fig.7 Arc of movement.

If the relaxed hand and forearm is laid on a table, it can be seen that when the finger ends lie in a vertical plane roughly across the axis of the forearm, the fingers slope outwards at about 20 degrees from the vertical. Because the outer fingers are shorter, the back of the hand shows pronation reduced to the 60 degree region. To take advantage of this and still have the right feeling of comfort at the finger ends, it was decided to tilt the keys outwards to match the slope angle as far as possible. To meet some difficulties in making a moulding, a compromise at 15 degrees was accepted for the 3 outer fingers but with the keys for the index finger raised to suit its shorter length and tilted at 20 degrees. Figs. 6 and 7 show a section along the home row of the right hand and a side view of the little finger outer column of keys. [11] At the same time, to meet the needs of male operators, the key group spacing was increased to 28cm and the key spacing was increased slightly. This design is shown in Fig. 8 where it will be seen that attention has also been given to industrial design aspects of the product to enhance the appearance. Comments from some qwerty operators who found that they needed to look at the letters when adapting to the new shape have been accepted and the old layout is now engraved in equal sized letters coloured yellow, with the Maltron letters in green. When required, either qwerty or Maltron letters alone can be fitted. The electronics will still provide both layouts so that an operator who has learnt to touch type Maltron will still be able to use an apparently qwerty keyboard and vice versa.

In recent years visual problems in connection with VDUs have been studied intensely and any form of glare is known to be a cause of strain. Light reflected from the keyboard to the screen or the operator's eyes, although not in the line of sight, will reduce the screen contrast and contribute to such strain. Also, when the operator looks down at the keyboard, a higher light level from it will cause iris adjustment which has to be changed again as soon as screen viewing is resumed. Since the screen background is usually dark, a dark keyboard will be beneficial in both respects. To meet this, Maltron keyboards, as shown in the figures, are normally made of dark material with a fine grained surface to give minimum reflection.

Fig. 8 Mark 4 Maltron Keyboard

6. QWERTY OPERATION

Since the Maltron shape removes abduction and reduces pronation stress, it can be expected that an improved performance will be obtained from the use of a Maltron keyboard. The flat typewriter style keyboard has been around for so long that a severe cultural shock occurs when an entirely new shape is presented as an office machine keyboard. This is especially so for trained operators who find the different physical layout of key positions to be confusing. Some gentle encouragement may be needed to help to overcome initial reluctance and to practice sufficiently to develop new reflexes. The problem occurs because the sense of key position derived from practice on the flat keyboard is no longer correct. This is a passing phase. It has been demonstrated that in as little as 25 minutes, normal speed can be attained. While this may be exceptional, it can certainly be achieved in a day or so. To help with further improvement, an adaptation training course is supplied with the keyboard. Results show that a 20% improvement in speed is easily accomplished, with a substantial reduction in errors and fatigue as expected.

7. MALTRON OPERATION

The development of a new letter layout is a major task and for the Maltron keyboard was carried out by Lillian Malt in 1976-77. She presented this work at the PIRA Symposium in September 1977. [12] The following extracts from her paper, which may not be readily available, give some idea of the factors considered. "The uneven stretches caused by the diagonal slope of the rows of keys on qwerty result in uneven reach and distance movements, and this together with the letter layout which reinforces language confusions and induces errors, adds to learning difficulties and training time. Of course there are many highly skilled and accurate keyboard operators. They are only a small proportion of the total number of people who learn to use a keyboard and their skill has taken longer to achieve and required greater effort. These difficulties all add to the cost of providing training both in our educational and training establishments, and in industry."

"Siting characters on the keys is a complex matter, and to arrive at an optimum layout many variables require consideration: motion economy principles related to hand and finger movements; finger strength and flexibility; the human neuromuscular structure. All these factors are included, as well as language restraints, such as letter confusions which result in common spelling errors and then appear as common keying errors, and allowance for statistical frequency of letters, single and in combinations of di- and tri-graphs, especially those in the commonest words."

"For accurate keying and for ease of learning, letter layout should take account of cybernetic requirements related to language. Highest source of error in reading and in spelling is located in vowels and vowel graphemes. On the qwerty layout the highest source of error is on the vowels "e" and "i".

Dvorak's layout, mentioned previously, has all the vowels adjacent to one another on the centre (home) row of the left hand. The analysis in his book [4] shows that vowel errors are 49% of the total. So it seems clear that this type of arrangement should be avoided. Again from Lillian's paper:- "If vowels are strategically placed so that they do not appear on adjacent keys, nor on the same finger and same row on the two hands, neural confusion may be avoided. This would provide the best possibility for accurate keying."

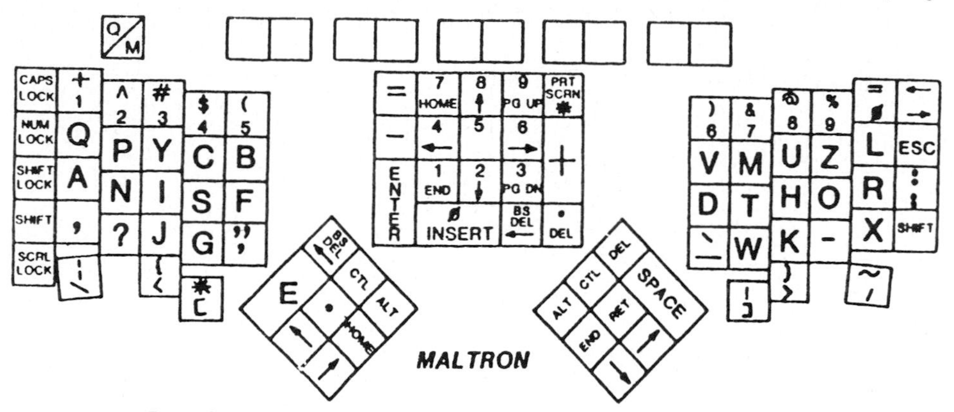

Fig.9 Maltron layout for PC style keyboard.

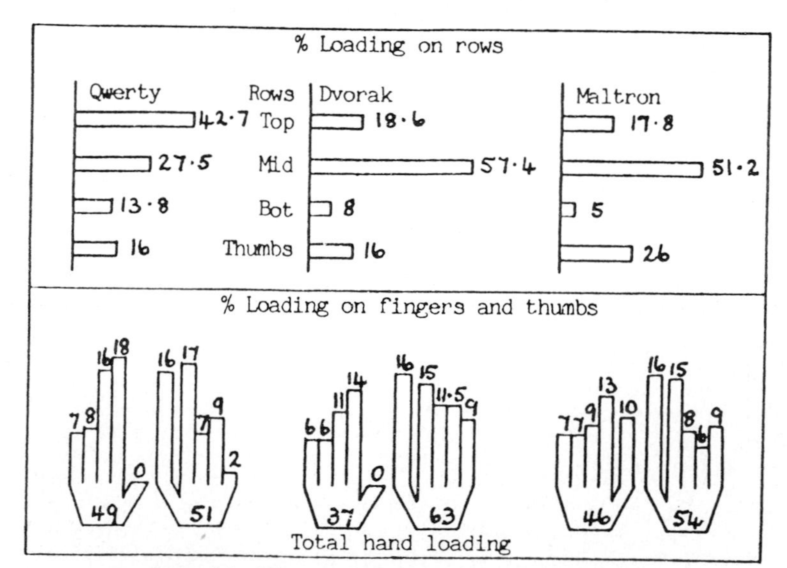

Fig. 10 Finger and Hand Loading Diagram

The Maltron letter layout, developed in accord with the above principles, is shown in Fig. 9. Fig. 10 is a loading diagram showing the amount of work done on the rows and by the fingers and hands for the three different letter layouts. The two operations which slow down keying are; the successive use of the same finger, since a finger cannot be used again as quickly as another one, and a "hurdle" ie. the need to move a finger to the row above or below before striking the key. Based on the figures from an analysis of a million words of English language [13] the table of Fig. 11 shows the number of times each of these occurred for the three letter layouts. The very clear advantage of Maltron can be easily seen. The placing of the letter "E" on the normally unused left thumb is the main reason for the big improvement. The great reduction in Maltron hurdles over qwerty supports the new shape in further reducing the total work load and stress. Trials also indicate that the combined effect reduces learning time to about a quarter of the usual period and errors to one tenth.

Keyboard	SFSU	IR	H	IR	IFSS				% LAW	% CW
					LH	IR	RH	IR		
Qwerty	273450	1	82200	1	12.27	1	13.73	1	43.6	51.9
Dvorak	83700	3.27	3474	23.7	9.33	1.3	7.52	1.8	73.3	86.3
Maltron	24826	11.0	321	256	4.9	2.5	5.50	2.5	77.9	90.5

SFSU — Single finger successive use
IR — Improvement ratio
H — Hurdles
IFSS — Index finger sideways stretch
% LAW — % Letters keyed on home row for all words
% CW — % Letters keyed on home row for the 100 commonest words these = 47% of all language input.

Fig. 11 Table showing improvements due to change in letter layout based on data from Lillian Malt and H. Kucera [12] & [13].

8. CONCLUSION

The Maltron keyboard, when used in either mode, addresses the physical stress problems of keyboard operators and shows a sufficient improvement to justify general adoption. The fact that the design requires only a few hours of adaptation practice by already trained operators to achieve much more comfortable working conditions, and at the same time offers a significant improvement in productivity, should appeal to both operators and employers.

ACKNOWLEDGEMENTS

I have much pleasure in acknowledging Lillian Malt's inspiration in indicating the need for a new keyboard and her considerable work in the creation of a letter layout to take full advantage of the new ergonomic shape we developed. My thanks also to Dr. P. Zipp and Butterworth & Co. (Publishers) Ltd. for permission to use figures from his paper (Figs. 1,2,3).

REFERENCES

[1] Ferguson, D. and Duncan, J., Keyboard Design and Operating Posture, Ergonomics, November 1974, pp 731-744.
[2] Osanai, H., Ill Health of Key-punchers, Journal of Science of Labour, July 1968, pp 367-371.
[3] Zipp, P., Haider, E., Halpern, N., and Rohmert, W., Keyboard design through physiological strain measurements, Applied Ergonomics, Vol. 14.2, June 1983, pp 117-122.
[4] Dvorak, A., Merrick, Dealey and Ford, Typewriting behaviour, American Book Co. 1936.
[5] Chisholm, Alister, The case for revising the layout of typesetting perforator keyboards, British Printer, October 1968, pp 82-87.
[6] International Business Machines Corporation, Improvements relating to Typewriter Keyboards, U.S. Patent No. 1016933, September 1964.
[7] Dodds, Irvine, Keyboard for a typewriter, U.S. Patent No. 3698532, October 1972.
[8] Einbinder, Harvey, Ten finger typewriter keyboards, U.S. Patent No. 4332493, June 1982.
[9] Rose, Michael, Death of the qwerty keyboard? Design World No. 8, 1985, pp 36-43.
[10] Hobday, Stephen W., Keyboards designed to fit hands and reduce postural stress, Proceedings of the Ninth Congress of the International Ergonomics Association, September 1985, pp 457-458.
[11] Hobday, Stephen W., Keyboard, U.K. Patent Application No. 2181096A, April 1987.
[12] Malt, Lillian G., Keyboard design in the electronic era, Printing Industry Research Association, Symposium Paper No. 6, September 1977.
[13] Kucera, H. and Francis, W.N., Computational analysis of present-day American English, Brown University Press, 1967.

VI

WORKPLACE DESIGN

Trends in Ergonomics/Human Factors V
F. Aghazadeh (Editor)
Elsevier Science Publishers B.V. (North-Holland), 1988

ERGONOMIC APPLICATIONS AND CONTROL OF CHEMICAL AEROSOLS

James D. McGlothlin, Paul A. Jensen, Michael G. Gressel, and William A. Heitbrink

Engineering Control Technology Branch, Division of Physical Sciences and Engineering, National Institute for Occupational Safety and Health, Cincinnati, OH 45226.

Ergonomic principles are typically applied in physical environments where physical stress is common. These principles are less commonly applied in the chemical industry to control personal exposure to aerosols. Industrial hygiene principles are commonly applied in the chemical industry to control worker exposures such as improved ventilation, isolation, or process substitution; in manually intensive tasks where the worker is exposed to hazardous aerosols, the principles of ergonomics and industrial hygiene sometimes come in conflict. To address this problem, a method has been developed to incorporate the principles of both ergonomics and industrial hygiene where aerosol exposure sources can be systematically evaluated through real-time sampling and videotaping of work tasks. Application of this method will help to determine the best strategies for reducing aerosol exposure. Examples where this method has been successfully applied are in industries where chemical powders are manually transferred and weighed, and in dental operatories for the control of anesthetic gases during surgery. Two case studies using the industries mentioned above will be presented to demonstrate ergonomic principles for dust control during material transfer and weighing, and for nitrous oxide evaluation and control. The use of this method may determine specific exposure sources and thereby provide engineers, and health and safety practitioners with solutions designed to control exposure including work station design, engineering controls, and improved work practices.

1. INTRODUCTION

Material weighing and transfer of chemical powders is routinely performed in industries all over the world. Traditional dust sampling has shown that ventilation of these operations is partially effective in controlling workers' dust exposure [1]. Unfortunately, traditional

---

[Disclaimer: Mention of company name or product does not constitute endorsement by the National Institute for Occupational Safety or Health.]

methods of dust sampling have not shown which work tasks in the process cause changes in worker dust exposure [2].

Control of nitrous oxide among dental personnel has been a problem since it was identified to be a health hazard [3,4]. To reduce nitrous oxide exposure levels, scavenging systems have been developed to control exposures. In most instances, however, it has been shown that it is very difficult to reduce concentrations below the NIOSH recommended limit of 25 parts per million parts air (ppm) during administration [3]. Part of the problem with control of waste nitrous oxide is determining if such exposures are from ineffective scavenging systems, poor work practices, inadequate general ventilation, or a combination of these variables.

Direct reading instruments which provide continuous data (i.e., real-time sampling devices which monitor airborne aerosol concentrations while workers are performing their tasks) provide a better picture of what activities influence airborne exposure levels. The purpose of this paper is to present information about the utility of integrating real-time instruments [5], with ergonomic task analysis to identify job elements and work processes which cause changes in worker exposures.

2. METHODOLOGY

2.1. Materials Transfer and Weighing.

2.1.1. Process Description.

Environmental measurements and an ergonomic task analysis were conducted in a plastics manufacturing plant which produces a variety of polyvinyl chloride products such as vinyl wallpaper covering. Raw materials for the products in this process are manually transferred and weighed in batches. Following this, the weighed powders are mixed with other ingredients according to recipe and charged into Banbury mixers. After mixing, the plastic products were extruded, milled or calendered into their final form and then shipped to customers. Because some of these raw materials are health hazards, they are transferred and weighed in ventilated work stations (Figure 1). Design engineers from the plant provided a hinged segment of the work platform which could be raised to allow a drum of raw material to be placed inside the booth. An air exhaust plenum formed the back wall of the booth. The primary work tasks included emptying 22.7 kg (50 lb) bags of chemical powder into a fiber drum (approximately 55-gallon capacity) and, using a scoop, the worker transferred the powder from the drum to a small paper bag. The bag was placed on a digital weigh scale and filled according to recipe specifications (in this case 1.06 Kg (2.6 pounds) of powder per bag). Following this the bag was folded closed and placed in a bin behind the worker. This process was repeated until the required number of bags were filled or the fiber drum was empty.

2.1.2. Real-time Sampling for Dust Exposure.

To evaluate respirable dust exposure as a function of depth of scooping from the drum, two types of instruments were used: hand-held aerosol monitors ((HAM), PPM, Inc. Knoxville, TN) and real-time aerosol monitors ((RAM), GCA, Inc. Bedford, MA). The HAM was attached to the worker near the breathing zone, and the two RAMs were used for background dust (i.e., plastic polymer powder): one for dust within the work area, the other placed approximately 20' from the work area. The HAM's and RAMs' analog outputs were connected to an Apple II Plus computer through an AI13 analog to digital converter (Interactive Structures Inc., Bala Cynwyd, PA.). The computer was programmed to store digitized voltages from the converted analog signals. Readings were taken every two seconds and stored on a computer disk. The experiment was terminated when the drum was less than one-third full (about 22 minutes).

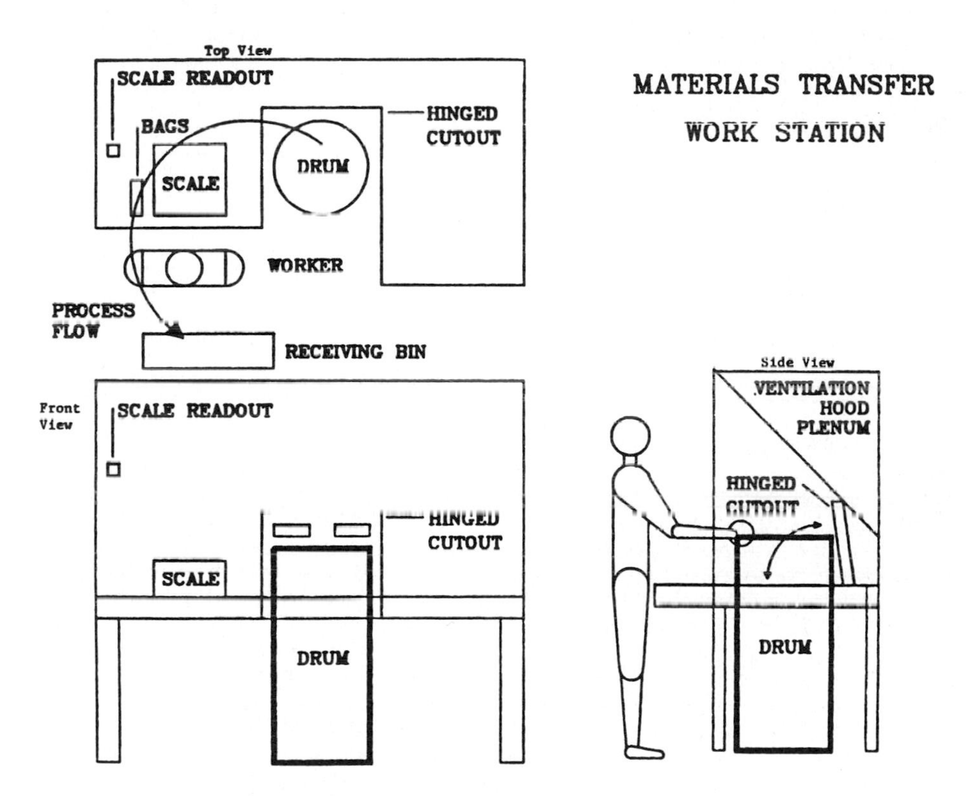

FIGURE 1
Schematic of work station and work process [6].

2.1.3. Ergonomic Evaluation

Material Transfer and Weighing. An ergonomic evaluation of four workers performing a manual weighing and transfer process was conducted using a video camera ensemble and traditional job analysis techniques [7,8]. This analysis included the basic work cycle, job elements in the work cycle, and identification of hazardous elements in the work cycle. Videotapes were played back in the laboratory in slow-motion and stop-action to provide detailed information on work practices, posture, work efficiency, and possible sources of increased personal dust exposure resulting from work activity. The video camera system was synchronized with the internal clock of the computer allowing integration of real-time dust concentration data with work elements for the first three workers. Because of limited resources, the fourth worker could not be measured for dust exposure but could be video taped. Work elements were also examined over time to determine what effect powder depth in the drum had on work practices. The use of the videotaping system in conjunction with real-time instruments provided a detailed look at potential sources of worker dust exposure.

2.2. Dental operatories.

2.2.1. Description of Dental Procedures.

Patients who require dental surgery may be administered nitrous oxide to reduce their anxiety and provide the dentist with better control over the patient. The anesthesia is usually administered through a scavenger nasal mask at a concentration of 40% nitrous oxide and 60% oxygen just before surgery and is reduced in concentration after the major dental work is completed [9]. Most routine dental surgery is performed in one hour or less. Because nitrous oxide concentrations are changing over the course of an operation a direct reading instrument is ideal in determining the changing airborne concentrations.

2.2.2. Real-time Sampling.

During the entire dental operation nitrous oxide was measured and recorded continuously. A Miran IA (Foxboro Analytical Co., South Norwalk, CT) was used to measure the gas concentration levels (ppm). This instrument is designed for field measurement of several gases and vapors, including nitrous oxide. This variable filter, variable pathlength infrared analyzer with a 20.25 meter cell was adjusted to the proper wavelength and pathlength for nitrous oxide (4.47 microns and 8.25 meters, respectively for nitrous oxide concentrations up to 1000 ppm). Nitrous oxide data from the Miran IA was recorded on an analog to digital data logger (Rustrack Ranger, Gulton Industries, Inc., East Greenwich, RI) and uploaded to a portable computer (Compaq 286, Compaq Computer Corporation, Houston, Texas) at the evaluation site.

Ventilation measurements also were taken in the dental operatory which measured 2.4 x 3.2 x 3.0 meters (8 x 10.5 x 10 feet). The air was supplied at 0.0283 cubic meters per second (60 cubic feet per minute) through a 15 cm (6 inch) duct located in the ceiling in the center of the room. Engineering drawings indicated that the supplied air flow should be 0.0519 cubic meters per second (110 cubic feet per minute). Discussions with the building engineer about the ventilation system indicated that the duct work was made of fibrous glass and that the fibrous glass tends to rupture at the "elbows", thus resulting in reduced air supply to the dental suite.

2.2.3. Ergonomic Evaluation.

An ergonomic evaluation of a dentist performing dental surgery was conducted using a video camera ensemble and traditional job analysis techniques as previously described. Analysis included the length of the operation, basic work tasks performed during surgery, and identification of work practices which may cause changes in nitrous oxide concentration. Videotapes were reviewed in the laboratory in slow-motion and stop-action to provide detailed information about the dental procedure. The video camera system was synchronized with the internal clock of the data logger to allow integration of real-time nitrous oxide concentration data with work tasks over the operation period. The use of the video taping system in conjunction with the Miran 1A/Rustrack data logger system provided a detailed look at potential sources of excess nitrous oxide exposure.

2.3. Statistical Methods

For the material transfer and weighing task, sampling data were statistically analyzed to determine which factors affect dust generation. Measurements taken with the HAM were analyzed to resolve whether scooping from the bottom of the drum increased worker dust exposure, and which work activities contributed to this increase. A regression model was developed to fit the dependent variable (i.e., instrument output - a time dependent measure of dust exposure) to the independent variables: bag count (the number of consecutive bags weighed with powder), time spent scooping the powder, time spent weighing the material, and time spent turning to close the bag and place it in the receiving bin [10].

Because of the limited sample size for nitrous oxide exposure, only descriptive statistical analysis was performed.

3. RESULTS

3.1. Material Transfer and Weighing.

Based upon the analysis of these data, a model was developed to predict worker dust exposure. The model elements were relative worker dust exposure, bag count, time spent scooping, time spent weighing, and time spent turning. A plot of the predicted values using the model with these time terms is shown in Figure 2. This figure shows that the worker's dust exposure increased with bag count (a measure of the depth of material in the drum). A plot of these work cycles illustrates an increase in exposure during the scooping activity and a decrease during weighing and turning tasks (Figure 3); suggesting that most of the workers dust exposure is caused by scooping the powder from the drum. Weighing activity is associated with higher dust exposure than turning activity. However, this difference may be an artifact caused by the slow response of the HAMs to high dust exposures during scooping.

The basic job for the materials weighing and transfer task was similar for all four workers. The work cycle for the first three workers took 18 steps to complete, while the fourth worker took 12 steps. Step saving techniques by the fourth worker resulted in less time to perform this task compared to the first three workers. This difference is primarily from the more efficient work layout and motion economy by the fourth worker which eliminated several nonproductive steps. Specific step saving practices included use of both hands (the left hand to grab a bag while the right hand reached and scooped powder material), precision weighing of powder into the bag by careful scoop pouring while inspecting the weight scale, and not folding the bags before putting them into the receiving drum (since the bags were bottom heavy there was no need to fold them to secure the powder).

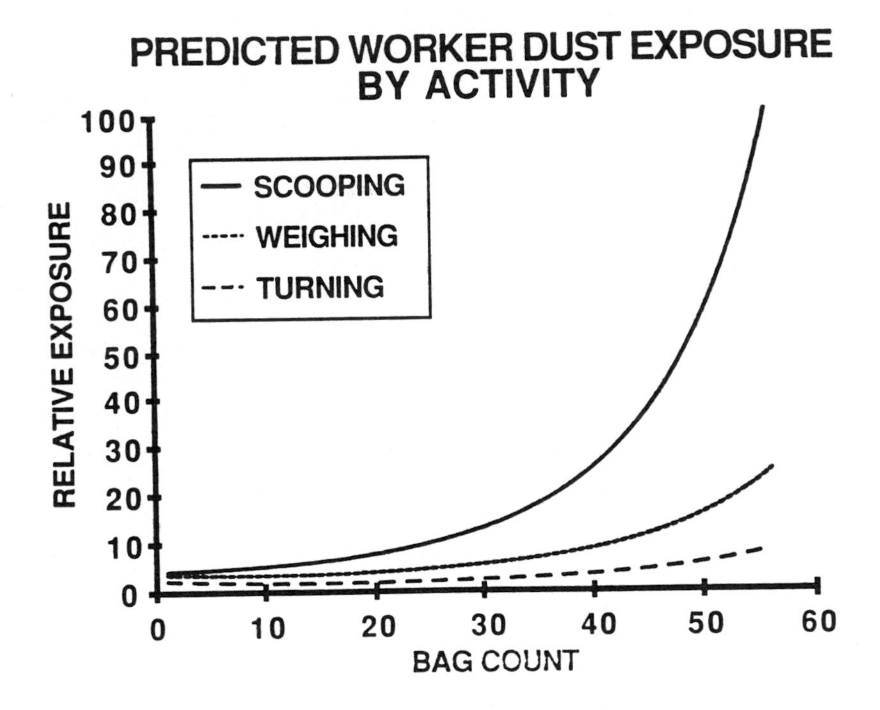

FIGURE 2
Modeled dust exposure of worker as a function of bag count for scooping, weighing, and turning at 8 seconds into an activity [10].

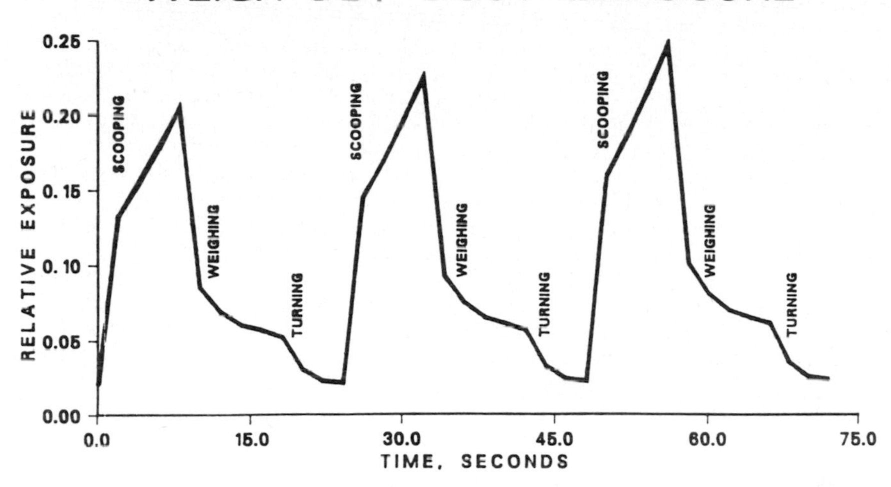

FIGURE 3
Modeled dust exposure of worker for filling bags 51-53 [10].

3.2. Dental Operatories

Nitrous oxide concentrations within the breathing zones of the dentist and dental assistant ranged from 25 ppm at the beginning of the operation to 950 ppm twenty minutes into the operation. Seven liters of gas per minute were supplied to the patient's nasal mask throughout the operation. Nitrous oxide was supplied at 2.5 liters per minute while oxygen was supplied at 4.5 liters per minute. The mixture provided the patient with 40% nitrous oxide and 60% oxygen. During the 45 minute operation, the dentist incrementally reduced the nitrous oxide from 40 to 20 to 10 to 0 percent; the breathing zone concentration of nitrous oxide subsequently decreased to 35 ppm 55 minutes after the operation began.

Work practices of the dentist and the dental assistant were videotaped and analyzed using traditional job analysis techniques. By combining real-time nitrous oxide readings with the videotape analysis, several work elements were found to influence the concentration of nitrous oxide during the course of surgery. These elements included possible deficiencies in the scavenging unit, the amount of nitrous oxide administered by the dentist, the use of a rubber dam (i.e., a barrier placed in the patient's mouth), and the dental assistant's use of the saliva aspirator (reduced nitrous oxide exposure), and air jet (increased nitrous oxide exposure). It also appeared that the patient contributed to the nitrous oxide exposure by talking, coughing, and yawning. Inadequate ventilation also may have influenced the time nitrous oxide stayed in the work environment as well as the possibility of infusion of this gas from other dental offices. Figure 4 shows the changing nitrous oxide concentrations and specific work elements which may have influenced the dentist and dental assistant exposure [11].

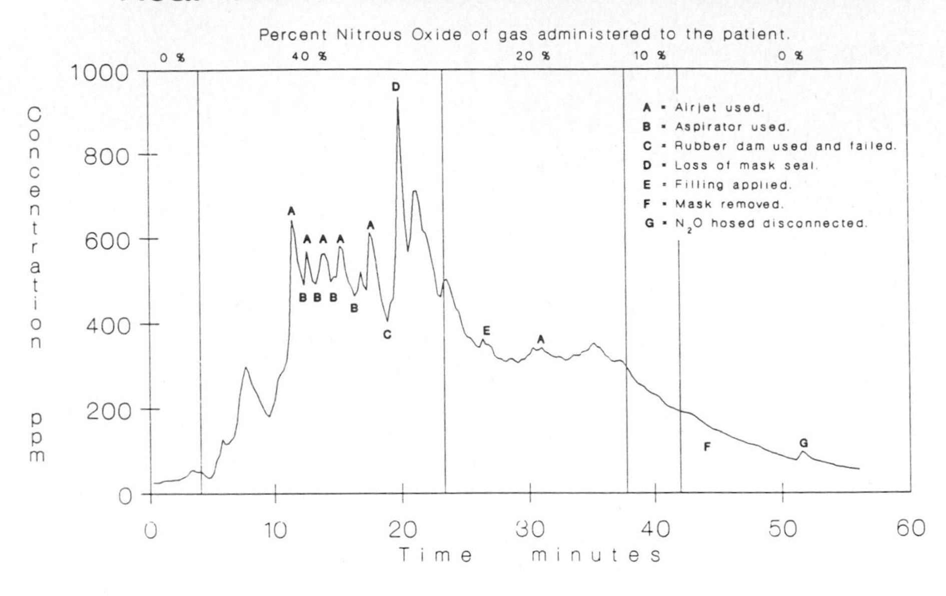

FIGURE 4
Dental operation profile of nitrous oxide over time.

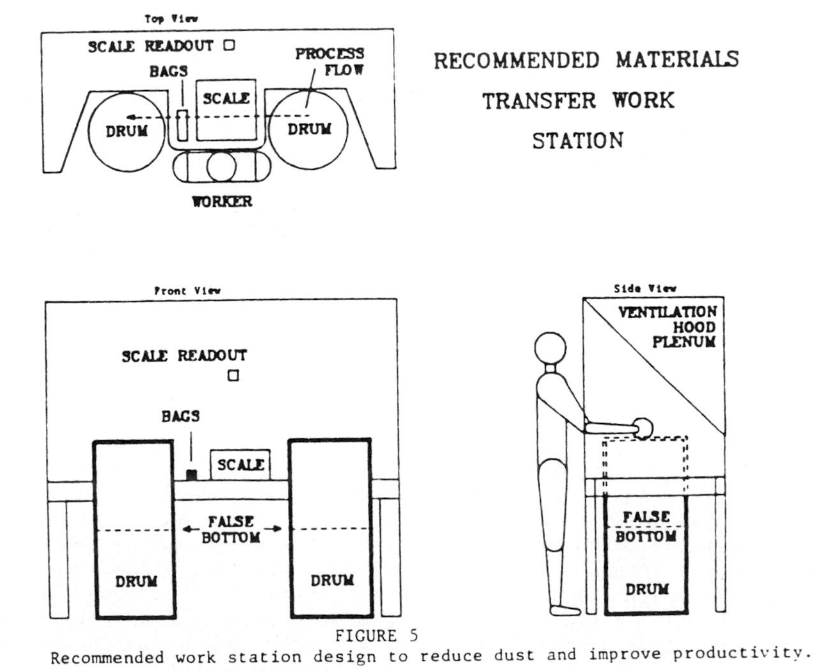

FIGURE 5
Recommended work station design to reduce dust and improve productivity.

4. DISCUSSION

4.1. Materials Transfer and Weighing.

Figures 2 and 3 show that scooping from the drum, and scooping near the bottom of the drum in particular, caused an increase in dust exposure. The real-time data shown in Figure 2 suggest that approximately 36-38 bags of material may be transferred from the 55-gallon drum before dust exposure increases dramatically. Figure 3 shows that scooping material from the drum was the principal source of dust exposure. [The geometric mean for the three personal dust exposures was 57 milligrams per cubic meter of air (mg/m^3) for the empty drums [5]. The permissible exposure limit for nuisance dust is 15 mg/m^3 [12]]. One simple but potentially effective recommendation to control dust during the scooping task is to insert a "false" bottom half way down the drum. Another possibility is to make the drums shorter in depth but larger in diameter. Based on the fourth worker's more efficient work layout, a new work station design was developed (Figure 5). This new work station would take advantage of ventilation controls by locating the drums inside the ventilated hood and incorporates the false drum bottoms to control dust. In addition, the modified layout will provide job symmetry and motion economy to significantly improve productivity.

4.2. Dental Operatories.

This case study showed that the use of a scavenging system does not guarantee a reduction to safe working levels of nitrous oxide. In a Health Hazard Evaluation survey conducted by NIOSH researchers at the same facility in 1979, nitrous oxide levels for personal exposure ranged from 90 - 3500 ppm without scavenging systems [13]. While the scavenging system used during our survey reduced nitrous oxide exposure, it does not appear that scavenging alone will decrease anesthetic gases to the NIOSH recommended exposure limit. Work practices, including the regulation of nitrous oxide by the dentist during the course of the operation, location of the dentist's breathing zone relative to the patient's, the use of the rubber dam, saliva aspirator, and air jet, may influence the amount of waste nitrous oxide dental personnel receive. In addition, the general ventilation including air supply, exhaust, the amount of fresh make-up air, and room size may influence exposure in the operating room and elsewhere.

5. CONCLUSION.

As these studies have shown, real-time instrumentation/data logger systems can be used in a timely manner to identify sources of worker dust and nitrous oxide exposure. For material weighing and transfer, the most important personal dust exposure sources were material scooping and depth of scooping from the drum. Knowledge of the specific task which elevates exposure may be critical with regard to recommendations for work task changes and/or work station modifications. It appears that there are several factors contributing to nitrous oxide exposure including the nitrous oxide supply system, the scavenging unit, work practices, patient interaction, and general ventilation. More research is needed to determine the individual, combined, and interactive effects of these factors. These two case studies illustrate the versatility of systematic evaluations of the worker and work practices, providing an excellent complement to environmental monitoring. Integration of these tools in combination will allow health and engineering professionals to devise more precise strategies to control hazardous chemical exposures in a variety of industries.

REFERENCES

[1] W. McKinnery, W. Heitbrink, Control of Air Contaminants in Tire Manufacturing. NIOSH (DHHS) publication 84-111. Cincinnati, Ohio (1984).

[2] W. McKinnery, W. Heitbrink, The Use of Direct Reading Instruments to Evaluate Engineering Controls and Exposure sources, Paper presented at American Industrial Hygiene Conference. Las Vegas, May (1985).

[3] DHEW (NIOSH) Publication No. 77-140, Occupational Exposure to Waste Anesthetic Gases and Vapors, Criteria for A Recommended Standard. 1977.

[4] DHEW (NIOSH) Publication No. 77-171, Control of Occupational Exposure to Nitrous Oxide in the Dental Operatory. 1977.

[5] M.A. Gressel, W.A. Heitbrink, J.D. McGlothlin, T.J. Fischbach, Advantages of real-time data acquisition for exposure assessment, Applied Industrial Hygiene Journal. Submitted for Publication, 1987.

[6] J.D. McGlothlin, W.A. Heitbrink, M.G. Gressel, T.J. Fischbach. Dust control by ergonomic design. Proceedings of the IXth International Conference on Production Research, Cincinnati, Ohio. pg. 687-694, 1987.

[7] F.B. Gilbreth, Motion Study, Van Nostrand, Princeton, N.J., (1911).

[8] T.J. Armstrong, J. Foulk, B. Joseph, S. Goldstein, An Investigation of Cumulative Trauma Disorders in a Poultry Processing Plant, American Industrial Hygiene Association Journal. 43: 103-116, (1982).

[9] American Dental Association. Dentists' Desk Reference, Materials, Instruments and Equipment, Second Edition. 1983.

[10] M.G. Gressel, W.A. Heitbrink, J.D. McGlothlin, and T.J. Fischbach, Real-time, integrated, and ergonomic analysis of dust exposure during manual materials handling, Applied Industrial Hygiene Journal, Vol 2, No.3, pg. 108-113. May 1987.

[11] J.D. McGlothlin, P.A. Jensen, W. F. Todd, T.J. Fischbach, Study Protocol for "Control of Anesthetic Gases in Dental Operatories." Engineering Control Technology Report, NIOSH, January, 1988.

[12] Title 29, Code of Federal Regulations. Part 1910.1000. Air Contaminants. July 1, 1980.

[13] P.L. Johnson, DHEW (NIOSH) Health Hazard Evaluation Report No. 79-5-564, Children's Hospital Dental Department, Cincinnati, Ohio. 1979.

Trends in Ergonomics/Human Factors V
F. Aghazadeh (Editor)

ERGONOMIC EVALUATION OF CLEANROOM ENVIRONMENTS

Jerry D. Ramsey, James L. Smith and Yeong G. Kwon

Department of Industrial Engineering
Texas Tech University
Lubbock, TX 79409 USA

1. INTRODUCTION

Cleanrooms are widely used in high technology industries. Therefore, special consideration of cleanroom workers is needed due to the work induced stresses from contamination avoidance, clothing requirements, and confinement [1].

2. DESIGN AND METHOD

After completion of a survey to determine potential ergonomic problems in the cleanroom, dominant problems appeared to be garment heat retention, garment vision or breathing interference, and garment design [2]. An experiment was designed to focus on these problems and to use the Texas Tech University environmental chamber to simulate a clean room environment. The temperature of the environmental chamber was set at 72° F with relative humidity maintained at 60%.

Several different sizes of cleanroom garments were borrowed from a microelectronics manufacturing firm. In order to familiarize the subjects with the experimental tasks and procedures, they first performed the simulated tasks in their street clothing. Three sizes of garments were assigned to each subject; under, over, and exact size. The sizes were based on each subject's height and weight which were measured before the experiment. The garment sequence and task sequence were randomly selected for each subject. Subjects did not know the garment sequence or the task sequence. On the fifth trial, a randomly repeated size which was not same size as that of the fourth trial was used.

A series of three tasks required the subjects to use a wide range of body movements which are typically encountered in a clean room. These tasks were:

1. SEATED TASK
The subject played a blackjack computer game for 2 minutes. The task simulated seated work such as writing or data entry and was expected to reveal garment restriction associated with seated tasks.

2. ARM (UPPER ACTIVITY) TASK
The subject placed a cassette box on a shelf located at the subject's shoulder height. The subject was instructed to push the box two feet along the shelf with his or her right hand and then push it back to its original position with the left hand. The subject then removed the cassette box from the shelf using both hands and lowered it to the table. The subject then was instructed to open the cassette box and turn over five cassettes, close the box and place it back on the shelf using both hands. This series of activities was repeated five times. The task simulated reaching and twisting tasks such as the loading and unloading of wafer trays.

3. WHOLE BODY TASK
The subject took the cassette box from the previous work station, walked approximately ten feet to another work station where he or she bent over a table and with both hands released the cassette box, allowing it to go down a slide to approximately two inches above the floor level. The subject then bent down to a squatting position, reached under the table with both hands, and retrieved the cassette box. The subject stood up and walked ten feet back to the original station, placed the cassette box on the table, opened it, turned over five cassettes, closed the box and repeated the entire sequence two more times. This task simulated walking, bending, and squatting tasks similar to those found in clean room processing.

Immediately after completing each task in each size garment, the subject was asked to indicate on a sketch of the human body the degree of restriction for the garment and/or muscle according to the following three point scale: (1) Noticeable (slight) restriction, (2) Minor (moderate) restriction, and(3) Major (considerable) restriction (discomfort).

3. RESULTS

Main independent variables were task, garment size, body region, and sex. The dependent variable frequency of restriction represented a numerical count of the places where restriction was reported. The variable intensity of restriction represented a summation of the degree of restriction values indicated by the three point scale system previously described. Paired t-tests showed no statistical difference between the repeated size data and the original data. This indicated that a learning effect from repeating that task was not significant.

Figure 1 shows the summary chart of task effect. Garment restrictions were significantly different for each task. Task 3 showed the greatest garment restriction and Task 1 the least. The garment restriction data appeared to be a function of the level and range of the active movement (see Table 1.).

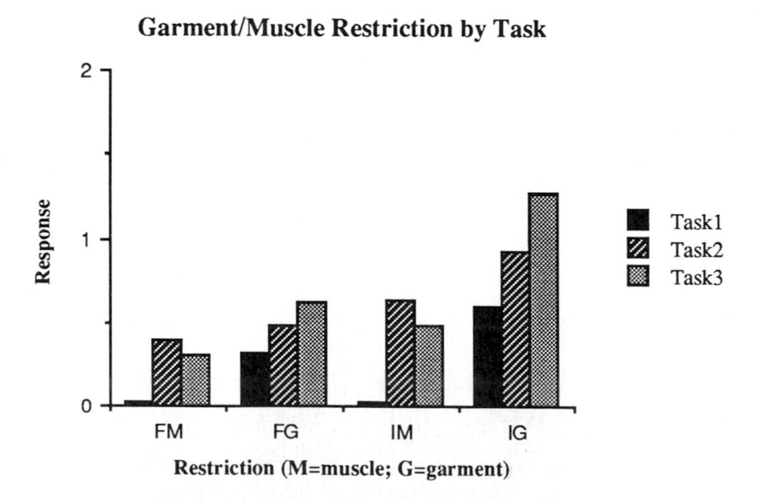

Figure 1. Task effects on garment and muscle restriction.

FREQUENCY OF RESTRICTION						
GARMENT (FM)			MUSCLE (FM)			
MEAN	TASK	GROUP	MEAN	TASK	GROUP	N
.062	3	A	0.40	2	A	320
0.49	2	B	0.31	3	A	320
0.32	1	C	0.03	1	B	320

INTENSITY OF RESTRICTION						
GARMENT (IG)			MUSCLE (IM)			
MEAN	TASK	GROUP	MEAN	TASK	GROUP	N
1.27	3	A	0.64	2	A	320
.093	2	B	0.49	3	A	320
0.60	1	C	0.03	1	B	320

Table 1. Task effects on frequency and intensity of restriction.

Analysis of data with respect to the size variable (n=240) is summarized in Figure 2. For the frequency of restriction for garment, all sizes were statistically different; the undersized garment being most restrictive, followed by the exact-size and the oversize the least restrictive. Since there was no "garment" for the street clothes trial, those means were zero. The intensity of the restriction for garment showed a similar pattern to the frequency of the restriction for garment. However, exact size and over size were not statistically different. This might be due to the non-uniformness of the human body and the wide range of the categorized sizes (see Table 2).

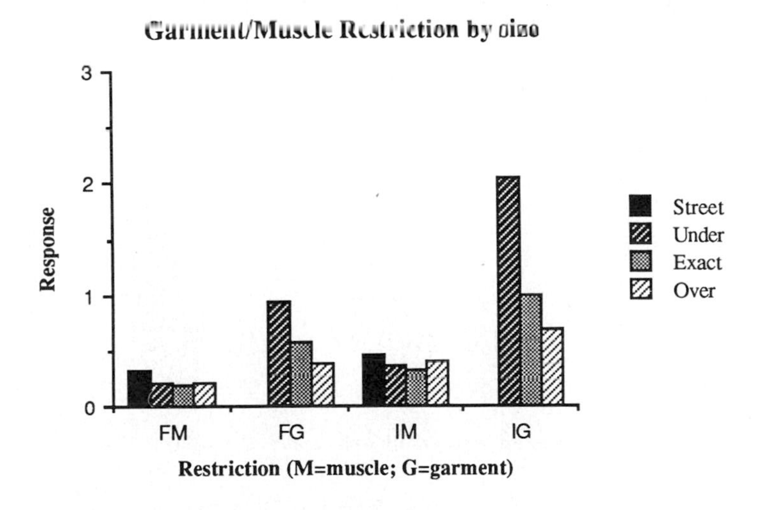

Figure 2. Clothing size effects on garment and muscle restriction.

FREQUENCY OF RESTRICTION						
GARMENT (FG)			MUSCLE (FM)			
MEAN	SIZE	GROUP	MEAN	SIZE	GROUP	N
0.93	Under	A	0.33	Street	A	240
0.58	Exact	B	0.24	Over	AB	240
0.38	Over	C	0.21	Under	AB	240
0.00	Street	D	0.20	Exact	B	240
INTENSITY OF RESTRICTION						
GARMENT (IG)			MUSCLE (IM)			
MEAN	SIZE	GROUP	MEAN	SIZE	GROUP	N
2.04	Under	A	0.45	Street	A	240
0.99	Exact	B	0.41	Over	A	240
0.69	Over	B	0.37	Under	A	240
0.00	Street	C	0.33	Exact	A	240

Table 2. Clothing size effects on frequency and intensity of restriction.

Analysis of data with respect to body region variable (n=240) is summarized in Figure 3. Analysis of the frequency of garment restriction indicated that region 2 (ARMS) had the highest frequency. This was apparently due to the large number of reports of discomfort from the gloves. The region 3 (TRUNK) and region 1 (HEAD) were next highest, but not different from one another. Garment restriction on region 3 (TRUNK) was often related to the crotch and underarms. A similar pattern was reported for the intensity of garment restriction. The mean values were consistently higher (approximately double) and the order was the same except for region 3 (TRUNK). For the intensity of garment restriction the relative restrictions on region 3 (TRUNK) were higher. However, region 3 (TRUNK) was not significantly different from region 2 (ARMS). Region effects are summarized in Table 3.

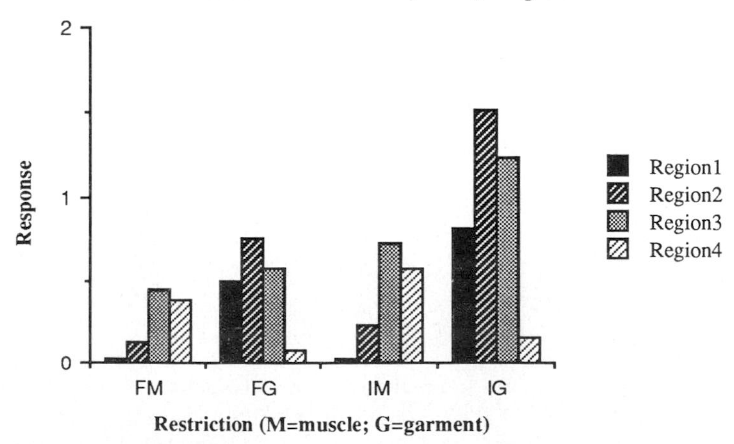

Figure 3. Body region effects on garment and muscle restrictions.

FREQUENCY OF RESTRICTION						
GARMENT (FG)			MUSCLE (FM)			
MEAN	REGION	GROUP	MEAN	REGION	GROUP	N
0.75	2(ARMS)	A	0.45	3(TRUNK)	A	240
0.57	3(TRUNK)	B	0.38	4(LEGS)	A	240
0.50	1(HEAD)	B	0.13	2(ARMS)	B	240
0.08	4(LEGS)	C	0.02	1(HEAD)	B	240
INTENSITY OF RESTRICTION						
GARMENT (IG)			MUSCLE (IM)			
MEAN	REGION	GROUP	MEAN	REGION	GROUP	N
1.52	2(ARMS)	A	0.73	3(TRUNK)	A	240
1.23	3(TRUNK)	A	0.57	4(ARMS)	A	240
0.82	1(HEAD)	B	0.23	2(ARMS)	B	240
0.15	4(LEGS)	C	0.03	1(HEAD)	B	240

Table 3. Body region effects on frequency and intensity of restriction.

In Table 3, four body regions were grouped as follows: region 1 (HEAD) included the head and neck; region 2 (ARMS) included the upper arms, lower arms and hands; region 3 (TRUNK) included the shoulder, underarm, trunk, and crotch; and region 4 (LEGS) included hip, upper leg, lower leg, and feet.

Figure 4 shows the summary chart of the sex effect. Analysis of data with respect to the sex is presented in Table 4. There was a strong and consistently observable relationship for both frequency of restriction and intensity of restriction. Reports of both frequency and intensity for muscle and garment restrictions were much higher for the females than for males. Unisex garments, which are typically designed based on male anthropometric considerations, likely contribute to this situation.

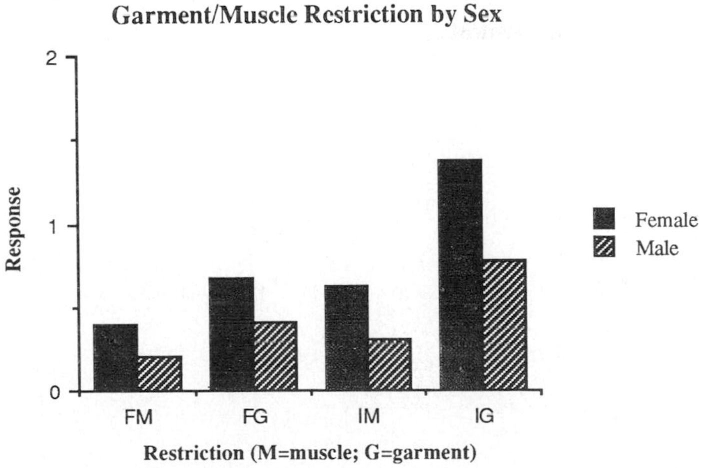

Figure 4. Sex effects on garment and muscle restriction.

FREQUENCY OF THE RESTRICTION GARMENT (FG)			MUSCLE (FM)			
MEAN	SEX	GROUP	MEAN	SEX	GROUP	N
0.67	F	A	0.39	F	A	240
0.41	M	B	0.20	M	B	720
INTENSITY OF THE RESTRICTION GARMENT (IG)			**MUSCLE (IM)**			
MEAN	SEX	GROUP	MEAN	SEX	GROUP	N
1.38	F	A	0.63	F	A	240
0.78	M	B	0.31	M	B	720

Table 4. Sex effects on frequency and intensity of restriction.

4. CONCLUSIONS

As expected, the more active tasks resulted in more muscle discomfort and garment restriction as compared to a seated task. However, the seated cleanroom tasks should not be ignored because sitting for a longer period of time may result in more garment and muscle restriction complaints.

The mean values associated with both frequency and intensity of garment restriction were twice as high for undersize garments as compared to exact size garments. Oversize garments had fewer restrictions than exact size garments, which was expected.

Analyzing of the regions of body where garment interferences and discomfort were observed indicated region problems in the trunk or central body region. High responses of discomfort associated with the head and arms were primarily due to the criticisms of the mask and gloves respectively.

REFERENCES

[1] Czaja, S. J., 1983, "The Role of Ergonomics on Cleanroom Environments", Panel Discussion of Human Factors in Work Environments, Human Factors Society -27th Annual Meeting, 1983.

[2] Ramsey, J. D., Smith, J. L., and Kwon, Y. G., "Ergonomics in the Cleanroom", *Proceedings of the 10th Congress of the International Ergonomics Association*, Sydney, Australia, in print (1988).

Trends in Ergonomics/Human Factors V
F. Aghazadeh (Editor)

ESTIMATION OF AN OPERATOR'S BODY POSTURE IN SAGITTAL AND FRONTAL PLANES

Elżbieta CHLEBICKA

Institute of Production Engineering and Management,
Technical University of Wrocław,
Wrocław, Poland

The results of research on operators' body posture were presented with the help of photogrammetric method in sagittal and frontal planes. A great differentiation was found in a size of the measure of operators' body posture extortion working at different angles within farther range.

1. INTRODUCTION

A body posture assumed during the work performance depends on a kind and technical parameters of a work-place. The body posture may be estimated by subjective and objective methods. A film analysis [1, 2, 3] belongs to the most often applied objective methods. To study the body posture Welon [3] proposed the photogrammetric method which consists in comparison of a working position with a free position on the basis of pictures taken in a sagittal plane. However, it seems that on some work-places a joint film analysis of a body position in a few planes is needed; especially on such work-places where operator's activities include a big work area and besides a forward slope of a body they also cause a side slope of the trunk.
The researches aimed at the estimation of the operator's body posture with the photogrammetric method on the basis of pictures taken in sagittal and frontal planes.

2. MATERIALS AND METHODS

The research program included an analysis of the operators' body posture working in a seated position and performing movements with a right thoracic limb in a horizontal plane in two ranges. The studies concerned male individuals at the age of 20-24, working on a specially constructed measuring station.
A table of 74 cm high on which there were two sensor buttons "start" and "stop" was the main part of this research station. The examined person was to move his hand as quickly as possible from the "start" sensor to the "stop" sensor. The "start" sensor was in a steady position on an edge of the table in the plane of symmetry of the examined person.

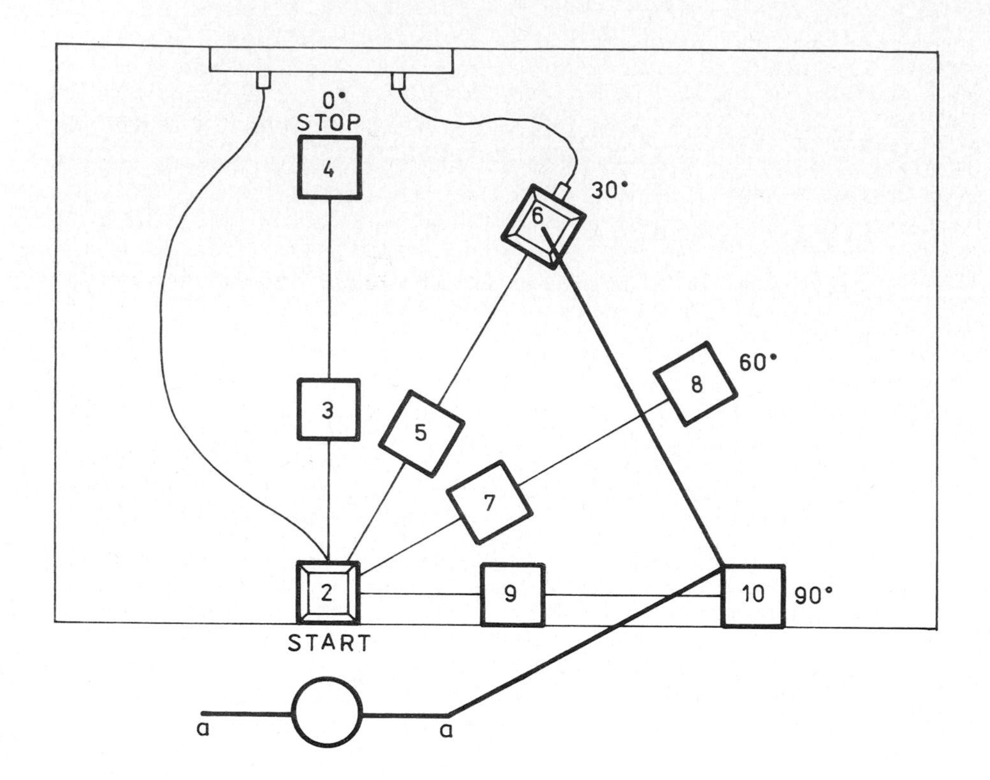

FIGURE 1
Schema of the examined work area.

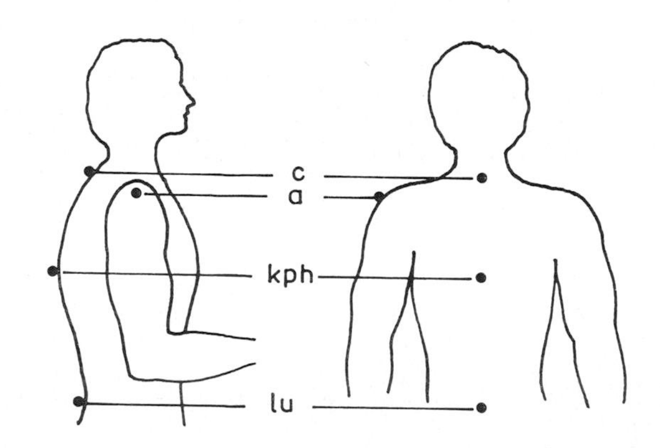

FIGURE 2
Localization of anthropometric points.

The "stop" sensor was placed at angles of 0°, 30°, 60° and 90° in a nearer range of 30 cm and in a farther range of 70 cm (Figure 1). In a moment when the examined person touched the "stop" button, at the same time pictures were taken automatically in the sagittal plane (from one side) as well as in the frontal plane (from a back side). The pictures were taken with two cameras placed at the back of the examined person. On the right side of the measuring station a mirror was placed to take pictures in the sagittal plane. According to the range and angle the body posture of operators was determined in ten working positions (Table 1). At the beginning basic operations were performed which enabled to carry out the main studies. The anthropometric points were marked on a body of the examined person (Figure 2). From the pictures, on the basis of these points, angles of displacement of the whole spine were counted as well as of its thoracic and lumbar parts in the sagittal and frontal planes. In both planes the deviation of the following lines from the perpendicular was measured: cervicale - lumbale one (c-lu), acromion - lumbale one (a-lu), cervicale - kyphoidale one (c-kph) and kyphoidale - lumbale one (kph-lu). On the basis of these parameters a measure of the body posture was calculated separately in the sagittal and frontal planes as a sum of standard differences between the examined parameters in a working position and a free position. Standarization was made by a standard deviation of the free position angles.

TABLE 1
Statistical characteristic of the examined parameters of the operators' body posture in sagittal and frontal planes.

Position		Sagittal Plane				Frontal Plane			
		1	2	3	4	1	2	3	4
1	$\bar{X}$	12,2	17,2	30,2	1,8	0,4	25,2	0,40	0,7
	S	3,7	5,0	5,5	2,9	0,4	2,4	0,8	0,5
2	X	17,7	20,3	39,3	5,8	0,4	24,6	1,0	0,9
	S	2,3	2,4	5,0	1,9	1,4	2,4	3,0	0,8
3	X	18,3	22,0	41,5	14,7	1,2	24,6	1,0	0,9
	S	2,3	2,4	5,0	1,9	0,9	4,0	2,5	0,8
4	X	32,0	40,3	54,7	22,3	-1,4	23,8	-3,2	0,7
	S	6,3	6,9	6,3	4,7	1,3	5,1	2,5	4,0
5	X	16,0	23,2	44,2	5,0	-2,8	22,4	-0,4	-1,8
	S	3,4	6,8	2,6	3,4	1,5	5,4	5,1	1,2
6	X	30,50	42,0	50,2	17,3	0,7	32,0	5,0	5,6
	S	4,5	3,7	7,6	4,1	2,2	3,2	6,0	3,5
7	X	18,3	13,7	42,7	4,3	-2,3	21,4	0,7	-2,2
	S	3,7	5,0	5,8	2,3	1,7	3,5	2,5	3,5
8	X	26,0	33,6	48,3	11,2	4,2	33,6	8,0	6,6
	S	3,7	3,7	4,8	4,0	2,9	2,4	3,7	1,7
9	X	16,3	13,5	41,4	2,3	-3,2	22,0	0,4	-2,2
	S	1,3	1,9	4,8	2,8	2,4	3,7	2,4	1,0
10	X	17,4	15,0	44,8	2,8	-2,8	24,8	0,6	-2,0
	S	1,3	2,3	2,6	3,0	2,9	2,4	4,0	1,6

3. RESULTS

3.1. Analysis of the operator's body posture in the sagittal plane.

Arithmetic means of four analysed angles were presented in Table 1 and arithmetic means of the body posture in Table 2. The calculated angles characterize a slope of the whole operator's body and particular parts of the trunk in the sagittal plane. The largest values of these angles were found in position 4, when the examined person works in a farther range, at an angle of 0°. It is a very inconvenient position for performance of movements in farther range. The work which causes a big frontal slope of an operator harmfully influences the internal organs and it may also cause a rounding of the back. A little bit smaller slope of the operator's body occurs for the angles 30° and 60°. It may be generally assumed that there is a tendency to a decrease of the body frontal slope as an angle of work increases.
It is also confirmed by values of the measure which decrease as an angle of work increases from 0° to 90° (Table 1). The smallest extortion of the body posture occurs during movements performed at an angle of 90° in the horizontal plane. Then a figure of the examined person is the most straight. A slope of the lower part of the spine, i.e. a deviation angle of the kyphoidale - lumbale line from the perpendicular, has the biggest influence on a size of the measure in particular working positions.

3.2. Analysis of the operator's body posture in the frontal plane.

On the basis of pictures taken from a side of the back a side slope of the trunk was estimated (Table 1). According to the range and direction of a thoracic limb movement the calculated angles assume positive or negative values. The greatest deviation of the spine to the right occured during performance of movements at an angle of 60° in a farther range and its greatest deviation to the left occured in a nearer range at an angle of 30°. The calculated values of the measure in the frontal plane confirmed the greatest deviation from the free position during the work in a farther range than in a nearer one except an angle of 90°. The greatest extortion of the body posture occurs during performance of movements with a thoracic limb in the horizontal plane at an angle of 60°. The greatest side slope of examined parts of the spine was found in this position.

3.3. A joint estimation of the operator's body posture in the sagittal and frontal planes.

A general estimation of the operator's body posture was determined on the basis of a sum of the following measures: slope (the sagittal plane) and deviation of a body (the frontal plane) Table 2 . Values of the cumulated measure confirmed a greater extortion of the body posture during the work in a farther range except the movement at an angle of 90°.

TABLE 2
Values of the measure of the operators' body posture extortion.

Position	Body posture measure		
	Sagittal Plane	Frontal Plane	Sum of the measures
2	5,1	1,4	6,5
3	10,7	3,4	14,1
4	21,4	14,2	35,6
5	5,9	15,2	21,0
6	18,8	19,1	37,9
7	5,5	14,7	20,1
8	13,5	34,3	47,8
9	3,9	16,1	20,0
10	4,8	13,8	18,6

During performance of side movements an angle of 90° with a thoracic limb the operator's body posture is subject to the smallest changes. The increase of a movement lenght at this angle does not result in worsening of the body posture. The operator's work in a nearer range at different angles does not cause marked differences in values of the body posture measure. However the influence of the movement direction on a size of the measure is observed during working activities in a farther range. The operator's body posture is loaded to the least degree during side movements an angle of 90° of a thoracic limb. Such working positions should be especially avoided when a frontal slope of the trunk and its side slope occure at the same time. The work in a farther range at an angle of 60° may be the example here (position 8).
The studies carried out for operators working with a thoracic limb at different angles,mainly in a farther range, confirmed the necessity of the body posture analysis in both sagittal and frontal planes. The studies performed in one plane does not give a full information on the body posture extortion.

4. CONCLUSIONS

On the basis of the performed researches the following conclusions may be formulated:

1° Application of the photogrammetric method and the measure of the body posture extortion brought good results in estimation of the body positions assumed by operators during the work.

2° The performed researches confirmed the necessity of application, on some work-places, of the workers body posture analysis, at least in two planes: sagittal and frontal ones.

3° No clear differences were found in a seated body position of operators performing the work of a high-speed character at different angles in a nearer range. While during the work in a farther range a big differentiation occurs in a size of the body posture measure. The least extortion of the body posture occurs for operators working with a thoracic limb in the horizontal plane at an angle of 90°.

REFERENCES

[1] Duncan J. and Ferguson D., Ergonomics 17 (1974) 651.
[2] Eastman M.C. and Kamon K., Human Factors 18 (1976) 15.
[3] Welon Z., Ergonomia 2 (1979) 41.

Trends in Ergonomics/Human Factors V
F. Aghazadeh (Editor)

ERGONOMICALLY DESIGNED CHEMICAL PLANT CONTROL ROOM

R.B. COMBS, Dow Chemical Company, Baton Rouge, Louisiana, 70764-0150 U.S.A.

F. AGHAZADEH, Industrial Engineering Department, Louisiana State University, Baton Rouge, Louisiana, 70803-6409 U.S.A.

This Study provides a generalized development of a computerized chemical plant control room. It combines ergonomic guidelines with computer control technology to provide an efficient man/machine interface.

1. INTRODUCTION

The main purpose of this study was to design a parallel interface computerized process control room which will reduce the problems associated with serial information displays. Parallel displays depict all parts of the process continuously on large screens or panels. Serial displays present process information sequentially on video display monitors. The problems with serial displays include difficult operator training, lack of overall process display continuity and reduction in response time to cascading alarms.

Inspection of existing plant control room installations utilizing serial display devices by this author have found that additional strip chart recorders are often installed during the first year of operation. This indicates a high level of display saturation occurs during major process upsets which require more information than is readily available on the serial display.

Another problem associated with serial interfaces is response time [1]. Sequential monitoring of information depicts a small portion of the overall system. Repetitive screen changes are required to determine the cause and magnitude of the process upset. After the cause is determined, the operator must correctly key in the appropriate codes to correct the problem. If the operators are new or have not had problems in this area recently, precious time may be lost as they review the screen to determine which codes are necessary. If a strip chart recorder is not dedicated to the system, additional time is lost while the appropriate configuration parameters are keyed in to trend the proper time frame and then make a decision.

Serial displays take longer to learn and require constant use to maintain familiarity [2]. If a portion of the process does not require frequent observation, it may be weeks before a display is viewed for trouble-shooting or training purposes. New operators indicate serial displays require a great deal of time for proficiency because one must be aware of all process interactions to make necessary corrections before the alarm state is reached. Operators take great pride in reducing the number of alarms which occur on their shift and believe a perfect performance would result in elimination of alarms.

Masking of alarms can occur because serial displays typically show the alarms in the sequence in which they occured. If two unrelated alarms occur, one would go unanswered until the primary alarm is resolved. False alarms cause concern because an operator must know which additional displays must be selected to verify that an alarm is false. If the alarm is ignored, serial information displays will conceal the problem until it drives other process areas into the alarm state.

If a piece of critical process equipment shuts down and causes disturbances in the downstream process, the alarms will build up in queue and remain unanswered until the primary problem is resolved. Additionally, serial displays will mask disturbances unless they reach an alarm state and are acknowledged. These disturbances can be detrimental to product quality and personnel safety.

The objective of this study was to design a parallel control room interface to reduce the response time required during major process upsets, decrease the amount of training required for operation, and increase overall process continuity.

2. MODULE DESIGN

An ergonomic parallel display module has been designed to decrease operator response time during a process upset. Studies by Zwaga [2] have determined the hierarchy of activities in a process control room. These activities are integrated into a control module utilizing anthropometric variables of shoulder height, elbow height, fingertip height, hip breadth, thumb tip reach, extended thumb tip reach, shoulder breadth, forearm to forearm breadth, and eye height for 5th percentile female and 95th percentile male population.

The module housing is sized to allow an operator unrestricted access in the horizontal and vertical planes. Maximum width is dictated by 5th percentile female back width and shoulder - elbow length and is 993 mm. Minimum width is 95th percentile male shoulder breadth and this is 533 mm. The module width is selected to fall within these boundaries at 914 mm (fig. 1).

The total height of the module is 2743 mm (fig. 2). The upper 914 mm contains a color video monitor (fig. 3) which

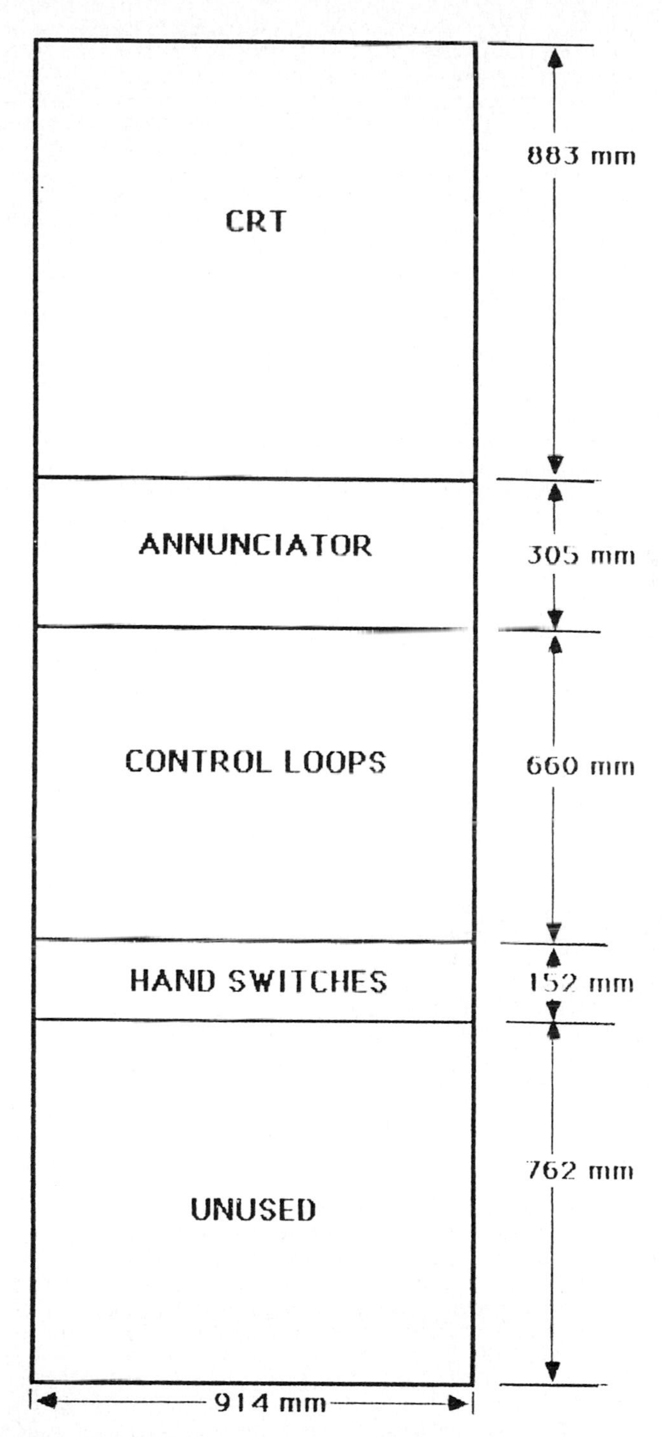

FIG. 1 CONTROL MODULE (FRONT VIEW)

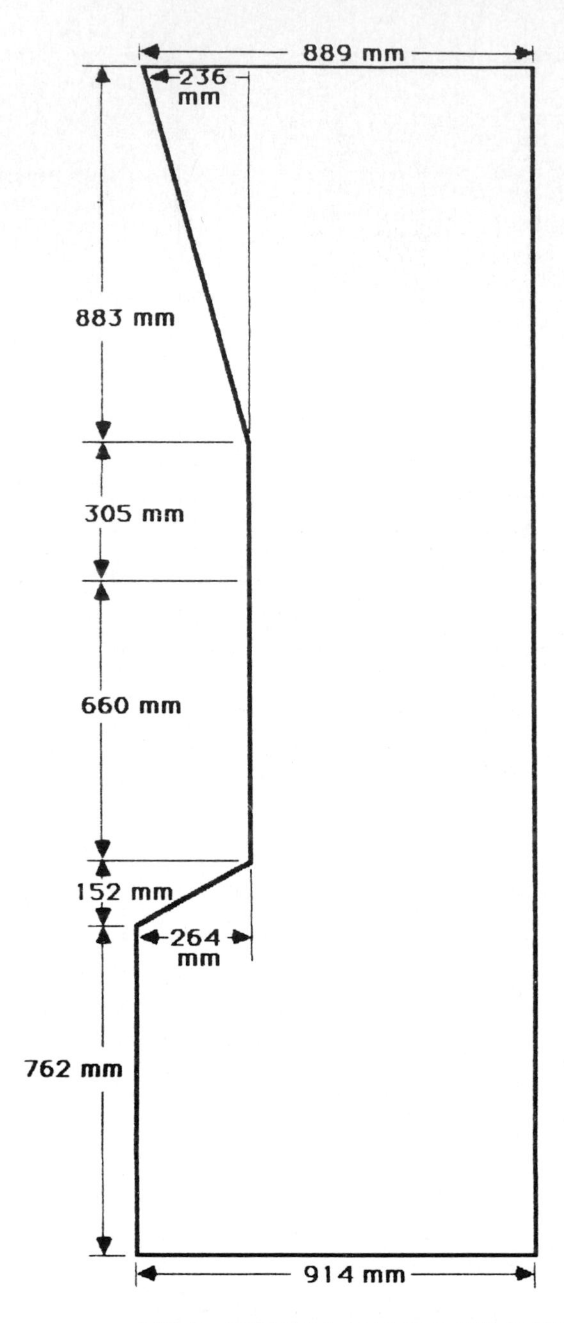

FIG 2. CONTROL MODULE (SIDE VIEW)

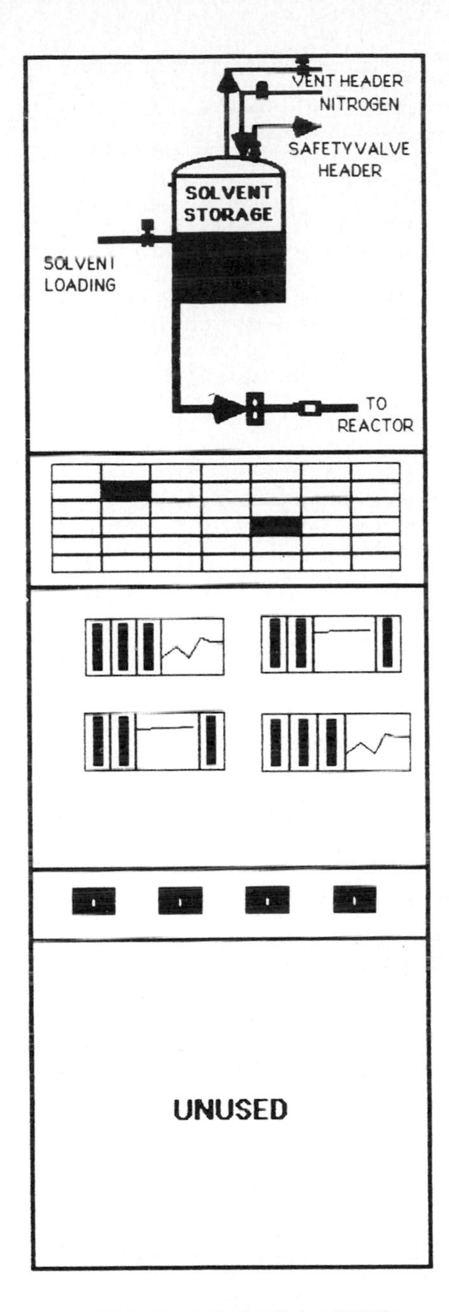

FIG 3. MODULE LAYOUT

pivots along the horizontal axis for ease of viewing. There is a light pen attached to allow selection of individual screens or to show alarm setpoints on individual process parameters. The default screen contains a graphic diagram of the equipment controlled by the module. Each display shows interconnecting lines between adjacent modules. All process flows lead to the appropriate destination on each module display to give the impression of one continuous process interconnected across all display screens. The back of the module is designed to be easily opened to give access to the internals for simplified maintenance. Alarmed areas of the display are programmed to change to a red or yellow color when entering the alarm condition to give visual warning of the location of process problems.

The next 305 mm of display space are within the prime viewing limits for the selected population of all American citizens between ages 18 and 65. This area contains a multiple window annunciator for alerting operators of alarm conditions. Each tile is clearly labelled with a descriptive message which is free from confusing abbreviations. Tile dimensions are at least 25 mm high by 50 mm wide. The top row is reserved for process critical alarms and descending rows will contain lower priority alarms. All labels are grouped to correspond to the position of the equipment the tile references on the video display monitor. The light source for the annunciator consists of two separate light bulbs. If one bulb fails, the tile will remain lit. A test switch is provided on the annunciator to allow all bulbs on the annunciator to light up simultaneously to determine if any are faulty and need replacement. An acknowledge button is also placed in several convenient locations for ease of operator access.

Tone generators ranging from 500 to 3000 hertz are mounted in the top of each module. These tone generators have warble style frequency modulators capable of phasing from one to three seconds. Multiple module control rooms would have the lowest tone and lowest phase modulation at the beginning of the process and the highest tone and phase modulation at the end of the process. This would give each module a distinctive sound signature to assist operators in determining the initial location of the alarm.

When an alarm condition occurs, the annunciator tile will flash quickly and the tone generator will sound until the operator strikes the acknowledge button after determining the cause of the alarm. This will slow the flashing of the light and silence the tone. When the instrument goes out of the alarm conditions, the light will turn off and the tile will be dark.

The next section of the module is the most accessible for hand manipulation of controls. Therefore, it contains the individual process controllers and strip chart recorders. These are grouped in chunks of five or less and are arranged to simulate the same configuration as is depicted in the graphic process display. Each chunk has individual loop controllers which monitor the status of the selected process variables. Groups of these controllers combined with multi-pen strip-chart recorders will allow complete control of each piece of process equipment.

The area located from 762 mm to 914 mm above the floor in the module contains the hand switches required for manual switching of redundant systems. Switches are also used for process critical electric valves and emergency reaction snuffing systems. These switches have self contained authentication lights to allow operators to recognize when the function is complete after switch activation. The switches are grouped according to their location in the graphic display. The bottom 762 mm of the module is not used because it is difficult to see or operate this area from a normal upright position. This area may be used for storage of test equipment or tools.

Any process which requires more than one module may arrange them in a row or group them in a semicircle (fig. 4). The net result of using these modules is focusing of operator attention immediately in the direction of the process problem. A new operator with no process experience would still be attracted to the proper module and would know what general area of the control board needs attention. Operators can scan the length of all modules immediately to see if other problems are beginning to build up because of a particular process disturbance. Continual scanning of all process graphic displays enhances operator familiarity with all parts of the plant. In the event of a total computer failure, the individual controllers will keep the plant running until the problem can be solved or the plant is shut down safely. The system is also easily configured so the control scheme can expand as the plant expands.

3. CONCLUDING REMARKS

Accepting computerized systems with their inherent advantages does not imply acceptance of serial information interfaces. This work has designed a parallel interface process control module which is compatible with the fifth and ninety-fifth percentiles of the working population. The module is designed to allow manual, individual loop, and computer control of the chemical process. The individual modules can be clustered for more complex processes and are easily configured for plant expansions.

REFERENCES

[1] Kragt, H. "A Comparative Simulation Study of Annunciator Systems." I. Chem. E. Symposium Series No. 90, Pergammon Press: Institution of Chemical Engineers, 1984.

[2] Zwaga, H.J.G. "Evaluation of Integrated Control and Supervision Systems. I. Chem. E. Symposium Series No. 90, (2), pg 133 - 146 Pergammon Press: Institution of Chemical Engineers, 1984.

FIG 4. MODULE CONFIGURATION

Trends in Ergonomics/Human Factors V
F. Aghazadeh (Editor)

AN EFFECTIVE METHODOLOGY TO THE DESIGN OF WORK ENVIRONMENT FOR A FAST-FOOD RESTAURANT

Sergio E. Martinez, Development Professor,
Industrial Engineering Department
Florida International University

Khokiat Kengskool, Assistant Professor,
Industrial Engineering Department
Florida International University

And

Juan C. Valdes, Senior Student,
Industrial Engineering Department
Florida International University

This research project introduces a methodical approach to safety in the design of a fast-food restaurant. The motivation of this research project on fast-food restaurants was the frequent source of controversy and the current critical reviews received by both communities and government agencies. Some of the adverse reaction to the locations of fast-food restaurants are: traffic and parking congestion, loitering, litter from disposable containers, noise, odor, bright lights, and decrease in property value.

INTRODUCTION

The main objective of a fast-food restaurant is to serve customers a quick and inexpensive meal. Fast-food restaurants thrive on high volume and fast turnover. Approximately three out of five families eat out at least once a week and the trend is upward in the United States. Fast-food franchises presently account for thirty per cent of the total sales at all eating places [1].

Typical fast-food restaurants have a number of different characteristics that can be labeled as possible sources of criticism. Problem areas encountered are: inadequate locations, zoning regulations, parking areas, drive-thru areas and vehicular and pedestrian congestion [1]. These problem areas can be properly addressed utilizing an effective safety methodology in the design approach to the work environment. Fast-food restaurants can also have either a positive or a negative impact on their surroundings.

CASE EXAMPLE:

This is an example of a typical fast-food restaurant located in a commercial zone area. This is an ideal location for everyone concerned because a fast-food restaurant is a good source of attracting people into an area. However, if this restaurant does not have an adequate size parking lot to accommodate its customers and/or the drive-thru window is not properly located or staffed to handle volume at peak hours it could create traffic and parking congestion. This could imposibilitate adjacent land owners from have an easy access to their places of business.

The approach towards an effective methodology in this design are: choosing a proper and safe location, zoning regulations, provide an adequate size parking lot area, and safety measures in the different work area locations. The methodological steps in the design of a work environment for a fast-food restaurant, if adhered, will address the current critical reviews from both the communities and government agencies.

A PROPER AND SAFE LOCATION

The first approach in the utilization of an effective methodology is choosing an appropriate and safe location. A proper location for a fast-food restaurant is an accessible area with at least one major artery leading to it as it is illustrated in the site plan of this design, Figure 1. Commercial or business type districts and certain type of industrial and manufacturing areas are most desirable. However, residential districts may be chosen if the restaurant can be made compatible with existing uses.

Continuing with the effective safety methodology, once a location has been determined, environmental tests as required by local regulations must be performed before any ground clearance takes place. The purpose of these environmental tests are to determine whether there is any type of archaeological value in the land to be excavated. Another test to be performed for safety reasons is to determine the presence of any type of toxic gases in the area that could be of any physical hazard. The elevation of the ground is another safety factor in determining the location of a fast-food restaurant due to the possibility of a flood in the area.

Although zonification has no safety implications, it is a principal factor in the decision making policies when choosing a location. Planning and Zoning officials are concerned with the nuisance effects of fast-food operations on the the public. Many communities have adopted specific zoning regulations for treating fast-food restaurants. Since zoning treatment in general is not consistent from jurisdiction to jurisdiction, specific zoning provisions for treating fast-food restaurants cannot be drawn. In each case the local zoning regulations must be carefully checked. Many zoning ordinances do not even distinguish between fast food restaurants and other restaurants or commercial uses.

PARKING LOT AREA

The second effective approach to the design of a fast food restaurant is the parking lot area. Fast-food restaurants generate a lot of traffic. This is why the parking lot area in the present design is set back far enough, as shown in the site plan, so that cars do not back into pedestrians walking in the sidewalk or disrupt the flow of traffic on the street. The entrances and exits in this design are properly constructed and located, not to inhibit the traffic flow and increase the incidence of traffic accidents.

For the safety and convenience of customers wide and perpendicular parking which yields greater space utilization has been implemented in the design [2]. This provides the customers with betters than 2/1 customer/parking space ratio. There are two reserved handicapped parking spaces as required by the municipality where this restaurant is located, as shown in Figure 1. An easy access ramp was implemented in front of the main entrance of the restaurant for the convenience of handicapped customers. Another safety measure in the design of this parking lot was a high coefficient asphalt covering to help prevent possible slipping when the ground is wet. The minimum illumination level has been set at one foot-candle, this will give customers the safety of having ample lighting at night. An added safety measure is a video camera for surveillance of the parking lot.

The implementation of a drive-thru window in a fast-food restaurant is a popular item with those customers who do not have the time to get off their cars. Drive-thru windows exemplifies what fast-food restaurants thrive on, to produce a high volume and fast turnover. An appropriate safety measure for drive-thru windows is locating them at the rear of the restaurant. Keeping the drive-thru traffic as distant as possible from pedestrians and as far away from the street as possible gives the customers and the adjacent property owners some assurance that measures have been taken to prevent any type of accidents, as shown in Figure 1.

DINING ROOM DESIGN

One of the safety measures that can be taken in the dining room area is the utilization of a cohesive type of floor covering. Carpeting or cohesive tile are the two basic alternatives available. Although carpet is less slippery than tile it is also more costly, both for first cost and for maintenance.

Safety approved plate glass is recomended if glass has been implemented in the design. For the safety and protection of the customers there should be no hidden obstacles in the design of the dining room area. Additional safety requirements in the dining room design are to eliminate any sharp edges on the furniture and table supports should be designed for customer's comfort and not to block the customer's pathway. When the customers are in queue, a sturdy channeling pipe is safer than a chain for the customers in case they want to lean on it. The serving area requires proper illumination of about 50 to 75 foot-candles so that customers will get a good view of their order. Bathrooms for both man and women must have facilities for the handicapped as required by local regulations.

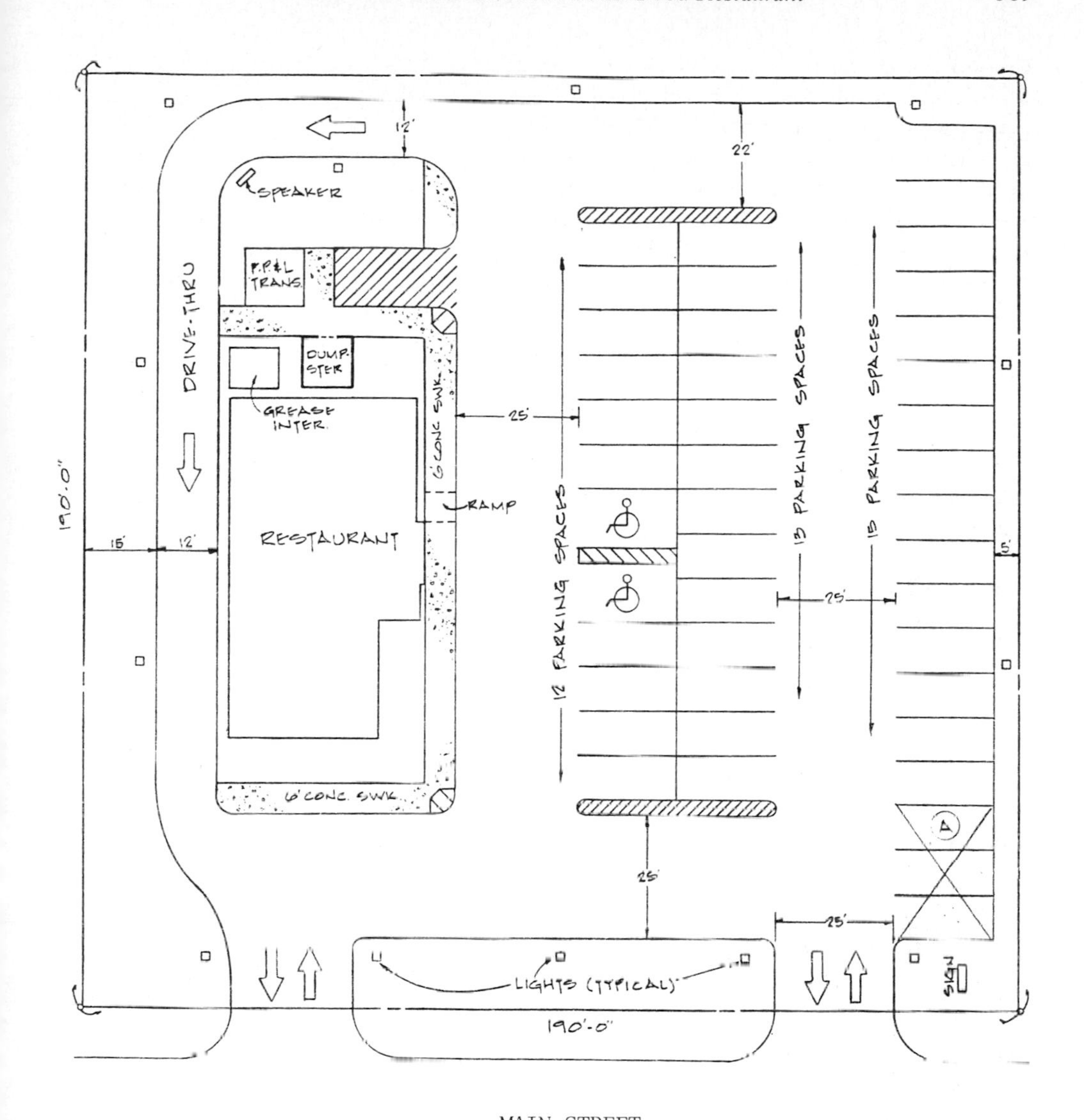

MAIN STREET

Figure 1
Site Plan Case Example
- Total Area 36,100 square feet
- Parking Lot capacity 40 cars

Note: Area (A) should not be implemented as parking spaces because of possible traffic and parking congestions.

KITCHEN AREA

The kitchen area in a fast-food restaurant basically consist of the cooking area, crew room, manager's office and the storage areas.

The cooking area must be very safety oriented in fire protection because it is a likely place for a fire to start. The cooking area as well as the whole restaurant must have an overhead sprinkler system for fire protection. Depending on the size of the broiler the fire protection code will dictate the size of exhaust required.

The crew room should be implemented in the design for the employees recreation and safety storage. The manager also requires an office where employee's files and the safe are kept.

Two types of storage areas are required in a fast food restaurant; dry food storage area for goods that must be kept fresh and a freezer for meat. The size of the food storage area depends on the amount of sales of the restaurant. An additional rear exit is required by local codes. This additional exit can be conveniently located adjacent to the storage area which can be utilized for the loading and unloading of goods into the restaurant.

TIME STUDY OBSERVATION

In order to avoid parking and traffic congestions, an adequate number of cashiers must be available to handle customers during peak hours. A time study observation was performed on an average worker of a similar size fast-food restaurant as in this design during peak hours. The observed readings were recorded using a stopwatch utilizing the continous movement method so no time would be omitted. The length of time per element is simply the difference between the starting time and ending time [3].

The findings of this time study observation was that an average cashier should take 85 to 90 orders per hour. With four registers in operation a similar restaurant should handle an average of 320 to 360 an hour.

CONCLUSIONS

The usefulness of this design approach to the work environment for a fast-food restaurant is validated through the current critical reviews from the communities and government agencies involved in the design of the workplace and work environment for a fast-food restaurant. The utilization of similar methods have proven to be successful by some of the major fast-food franchises.

REFERENCES

[1] Mc Allister, Anttoinette, Zoning for Fast-Food and Drive-In Restaurant. Chicago: American Society of Planning Officials, 1976

[2] Tompkins, James A. and White, John A., Facility Planning. John Wiley & Son, Inc., 1984.

[3] Konz, Stephan, Work Design: Industrial Ergonomics Second Edition John Wiley & Sons, Inc., 1983.

Trends in Ergonomics/Human Factors V
F. Aghazadeh (Editor)

HUMAN ISSUES IN AUTOMATED (HYBRID) FACTORIES

Anil MITAL
Lalji J. GEORGE

Ergonomics Research Laboratory
Mechanical and Industrial Engineering Department
University of Cincinnati
Cincinnati, OH 45221-0072, U.S.A.

Many manufacturing industries in the United States have undergone tremendous changes in the past few years. The most profound of the new changes is the change in the level of automation. As the levels of automation have increased, so have the manner in which routine jobs are handled. This paper reviews the changes in the role of human operators, caused by increased automation, and discusses some of the relevant issues such as training, supervision, work measurement, and safety.

1. INTRODUCTION

Reduced production costs, improved product quality and enhanced worldwide competitiveness are the goals engineers wish to accomplish by automating processes and assemblies. As the drive for increased levels of automation is accelerating, the automatic factory (completely automated factory) is becoming a reality. The expectation is that all manufacturing industries will, one day, become automatic and intelligent. However, the transition from the traditional manual (labor intensive) factory to an automatic factory, even though widespread, is far from over. One can easily come across examples of the following three categories of factories depending upon the level of automation:

1. Manual factory (labor intensive and almost no automation),
2. Automated factory (has elements of both the manual and automatic factory; some functions are automated while others remain manual), and
3. Automatic factory (all functions and elements are automated and no human intervention is necessary for satisfactory operation).

The majority of progressive manufacturing industries fall in the second category, also known as the hybrid factory. Depending upon the level of automation, the preponderance is either in favor of manual functions or automated functions.

While the role of humans performing labor intensive functions is no different today than it has been in the past (for instance, perform all manufacturing processes, such as painting, welding, and assembly, and handle materials, by either conveyors or carts), the role of humans involved in automated functions (for instance, supervision, computer programming, training, response to contingencies, etc.) most certainly is.

The purpose of this paper is to review the changing nature of the factory and discuss some of the related human issues. In particular, attention is drawn to the following four issues:

1. Supervision,
2. Training,
3. Work Measurement, and
4. Safety

2. THE PRESENT DAY FACTORY

As mentioned earlier, the drive to automate is very pervasive. There is hardly any factory in the United States which has not been touched by the automation tidal wave. While the majority of present-day factories cannot be classified as manual factories, they have yet to become factories of the future (automatic factories). The overwhelming majority are hybrid factories (factories which utilize a combination of manual and automatic processes for the manufacture or assembly of a product). The levels of automation in hybrid factories vary a great deal. The degree of automation could vary from stand-alone islands of automation to partially automated processes. In a typical hybrid factory, the functions that generally are automated are materials handling, storage and warehousing, painting, welding, minor assemblies or sub-assemblies, palletizing, and packaging. Some aspects of manufacturing process itself, such as turning, may also be automated.

3. SUPERVISION

While the need for supervision in manual factories is extensive, it is minimal in automatic factories. Since a hybrid factory is a combination of manual and automatic factories, the supervision needs are also a mix of the two kinds. The functions for manual components are the same as the functions for manual factories (inspection, maintenance, etc.) and functions for automated components are the same as for automatic factories (overseeing the process and equipment, fault diagnosis, response to breakdowns, etc.). The skill level required by operators in such an environment is the broadest, requiring both the information processing skills as well as the basic manual skills. In many respects, therefore, the supervisory needs in hybrid factories are not only the broadest, but most difficult as well. The levels of skill and training requirements are also considerably greater than either for manual or automatic factories.

4. TRAINING

Training of personnel in automated factories is a key issue. It is important, therefore, to determine the skill levels that are required both by the operators and maintenance technicians. Another key consideration is the delineation of responsibilities between operators and technicians. This "fine line" is very evident in the start-up stages, where operators are "babysitting" the automated process.

Due to a lack of knowledge of the process, the operators may be totally dependent upon the maintenance technicians to fix even routine problems, such as re-orienting a part in a feeder, ejecting a completed product assembly from a work station, etc. With adequate training and instruction of the operators, this problem can be minimized. In operator selection, training and asignment, the following should be considered:

1. Operators should be assigned responsibilities based on competence levels. For example, operators with high mechanical and electronic aptitudes can be assigned the task of overseeing robot operations, whereas personnel with lower aptitudes can be assigned routine tasks, such as routine process monitoring or inspection tasks.

2. It should be recognized by management that during the start-up phase, an operator over-dependence on maintenance and engineering personnel will exist. This can be minimized by monitoring operators initially "on the job" and giving them enough leeway in fixing routine problems until a "comfort level" at which he can perform independently is reached. A compromise has to be reached as to how much this leeway should be determined since an operator error would result in extensive damage to a manipulator or peripheral equipment.

3. A significant problem in operator training may be caused if the class room training is not supplemented with hands-on experience. This is critical since problem recognition and the speed with which the operator fixes random problems is dependent on continual hands-on experience.

5. WORK MEASUREMENT

Work measurement has traditionally been concerned with the measurement of human labor. Human labor and capital constitute the two legs of production. The product of the human labor could be easily defined and measured in the past. Today, people have more sophisticated functions to perform and the product of their labor is difficult to measure.

The goals of traditional work measurement are however still applicable today. They are: 1) the better use of labor resources through balanced work loading, improved manpower planning and labor cost control and 2) better use of capital assets through soundly based product planning, better material forecast and standards for plant and machinery.

To achieve these goals, work measurement practioners try to establish a fair and acceptable time (standard) for a specified job which may be achieved by trained qualified workers applying themselves with reasonable effort.

The first stage in a work measurement application is the analysis of the work into its constituent elements, the size of which varies with the type of measurement technique employed. The second stage, measurement, is accomplished by one of the three main techniques: direct observation timing with performance rating, work sampling or predetermined time systems. Element time can then be reassembled and included in a data bank.

The accuracy of these systems have to be periodically audited to reflect changes in the process, the technology and the competitive environment.

Work measurement in the automated factory is a fairly complex issue. From this standpoint the following are critical:

1. Performance evaluation and measurement of operation specialists and maintenance specialists in jobs that are not clearly defined.

2. Establishment of incentive schemes for these personnel.

3. Establishment of a standard data base for operations performed by these personnel.

5.1. Performance evaluation and measurement

Ideally, direct labor in an automated factory accounts for the efficient operation of a manufacturing process (% uptime of the automated equipment). The question now arises as to how this uptime should be used in evaluating personnel. Through a database developed from historical data, factors affecting uptime can be determined, e.g. the time of the shift, vendor equipment specification, start-ups, etc. Other factors that can be used to evaluate personnel are the average quality of output, their performance in diagnosing and responding to emergency situations, etc. Through work sampling, the types of breakdown can also be categorized to provide further information on both the personnel and the process.

5.2. Incentive schemes

Since the predictable workload of operators and maintenance technicians in an automated environment is a relatively small percentage of the total time, incentive schemes can be based on groups of personnel working in combination to achieve process uptime, production throughputs, and quality products. The bottom line that should be considered from a direct labour standpoint are:

1. Is the production process producing at the level of production that is called for, based on minimum downtime of equipment and processes?

2. Can the production process, if required, meet the capacity requirements it was designed for, based on timely operator interventions, efficient process control?

From an indirect technical standpoint, the considerations would be:

1. Has enough flexibility been designed into the process to ensure quick changeovers for the variety of product mix?

2. Can the existing process be innovatively improved from a standpoint of quality of product, increased throughput and less downtime?

5.3. Establishment of a standard data base for work standards

Since the automated factory is a relatively recent occurrence, very little data exists for typical work procedures that are performed by personnel in these environments. An initial data base can be established by vendor data (calibration procedures for robots, preventative maintenance on AS/RS system). However, by and large standard data for automated processes can only be established through experience.

6. SAFETY

Safety of personnel in automated factories is also a major concern. As a result of accidents and injuries due to automated equipment, automation safety has received repeated attention in the past few years. In an automated factory environment, with a large number of automated components in operation simultaneously, safety becomes critical. This is especially true in the "bring up" of automated systems, such as robot assembly lines, when manufacturing personnel might inadvertently step into the work envelope of a robot during application, teaching or debugging. It is critical that personnel involved in maintenance, installation and operation understand the unpredictability and capabilities of an automated device, such as a moving robot arm, in case of a malfunction such as an encoder or resolver breakdown. Thus, a safety awareness needs to be created and emphasized repeatedly. The key to automation safety is to understand the most effective way for protection and still have easy access to the automated work area for intervention and maintenance purposes.

Some of the safety measures that can be implemented are:

1. Hard hats for all personnel entering the manufacturing area for protection from material flow on overhead conveyors.

2. Safety mats, safety shields and light curtains strategically placed to protect personnel from robot arms, mechanical fixtures and feeders, shuttles and moving material handling devices, such as conveyors. A key switch can be provided to override these devices for maintenance purposes.

3. Emergency power-off buttons and emergency power-off chords installed at strategic locations to deenergize moving devices (It is advisable to have an operator monitor the maintenance technician during the period he works within the work envelope. The operator should have easy access to the emergency power-off button during this period).

4. Strobe lights can be used to indicate the work station status:

Color of Strobe	Conditions
Amber	Parts low condition of feeders, magazines.
Blue	Parts out condition of feeders, magazines.

Red	Malfunctions, such as part jams, problems in product infeed, outfeed, improper clamping, etc. Operator intervention required.
Green	Maintenance in progress.
White	Station operational.

It should also be realized that safety can not be compromised to accelerate installation and production schedules, as an equipment related injury could have serious implications and cause set backs on production schedules and employee and management morale.

7. SUMMARY

This paper discusses some of the human issues relevant in automated factories. Better understanding of these issues will lead, through research, to efficient solutions.

Trends in Ergonomics/Human Factors V
F. Aghazadeh (Editor)

SYSTEM RELIABILITY WITH HUMAN COMPONENTS

C. M. Klein and J. A. Ventura

Department of Industrial Engineering
University of Missouri-Columbia
Columbia, MO 65211

ABSTRACT

Complex system reliability has received a great deal of attention over the years. With more and more technological advancements, systems are becoming more involved and consist of more subsystems than ever before. This is especially true in the area of military aircraft. The determination of aircraft reliability, and hence availability and maintainability is critical in determining mission capability and war-time readiness. However, it seems that with the technological advancements, the concern of human performance is not taken into account when determining the reliability of an aircraft system. It is the purpose of this paper to reintroduce the human aspect into aircraft system reliability, discuss its implications, and present a method for its determination.

1. INTRODUCTION

In dependability, there are a number of measures to indicate the performance of equipment. Three of the most important are: pointwise availability - the probability that the system will be operable at specified instant of time; reliability - the probability that the system will not fail during a given interval of time; and interval availability - the expected fraction of a given interval of time the system will be operable.

The probability that a system successfully performs in a given interval of time is called "system reliability", or sometimes called "probability of survival". Generally, the system reliability must be built in at the design level of a project. This reliability of the system is dependent on the individual component reliabilities. However, the actual reliability of the system once it is operable must be determined and is affected by the period of time and age of the system, material, etc. Hence, system reliability can be viewed in 2 different stages: during the design phase, and after the system is operating.

For a large system, attempts should be made to divide the overall system into suitable subsystems which can be handled conveniently one at a time. Every attempt should be made to keep the subsystems independent. Once the subsystems are determined the state space of each subsystem is reduced either by merging states or by truncating very low probability states. The subsystems are then combined into a complete system and the required reliability measures can be evaluated.

As Leemis [11] notes, the determination of system reliability measures usually involves four elements. These are: an environment, a system state vector, a state transition mechanism, and a system structure. The environment generally includes elements external to the system, but that affect the performance of the system. However, even though the environment affects the system, the system has no influence on the environment. The system state vector identifies the distribution of the initial state and the potential states the system can assume. It is assumed that a state is not binary, operating or not operating, but can have gradual degradation of performance. The transition mechanism is simply the rate or rates at which components of the system change states. Lastly, the system structure is used to describe the relationships of the states with one another, i.e. series, parallel, etc.

As can be seen from above, the keys to viewing any system are the components in the system and their relationship. Therefore, introducing the concept of human components is not difficult. The difficulty comes in assessing the reliability of the human component and determining where and when it has an effect on the system.

It should be noted, however, that in most situations involving aircraft and aircraft design, the aspect of human reliability does not enter into the evaluation of system reliability or system availability. It is the purpose of this paper to develop a model to determine the system reliability of an aircraft. This reliability is based on the individual component reliabilities which include human components such as pilot reliability and human related maintenance and maintainability. Methods for determining the reliability of the human components as well as the system reliability will be given.

2. AIRCRAFT RELIABILITY

The key difference between aircraft reliability and general systems reliability is the measure of performance. Generally, when looking at systems reliability, the measure of performance is the reliability of the system itself or the availability of the system. Aircraft reliability, however is concerned with the probability of starting and completing a mission without interruption. It may be possible to have good aircraft reliability even though the aircraft system reliability is low. Concerns that must be addressed in determining aircraft reliability are airplane readiness, maintenance times, down time, unscheduled maintenance, diversions, priority impedence, and environment. In general, aircraft reliability will be viewed as

$$R_A = 1 - \frac{\text{mission interruptions}}{\text{total missions attempted or scheduled}}.$$

The probability of mission interruption is given by

$$I_A = 1 - R_A.$$

Hence, to determine aircraft reliability the entities of component reliability, system reliability, maintainability, efficiency of maintenance, priority impedence, and environment must be carefully considered.

It is easily seen that a large portion of aircraft reliability is dependent on some sort of maintenance action. However, classically the aspect of human reliability has not been considered in these actions. This concept will be explored in the next section. Another key part to aircraft reliability is reliability of the system itself.

Systems reliability can be modeled with great generality (see Barlow & Proschan [1]) through the use of and combinations of simpler systems such as series and parallel configurations. A common type of complex system is the k out of n system. For this system to be operating at least k of its n components must be operating. The reliability function for this system is quite difficult to find except when the probability that component i is functioning properly, p_i, is the same for all n components. That is, $p_1=p_2=\ldots=p_n=p$. In this case the system reliability R_s is given by

$$R_s = \sum_{i=k}^{n} \binom{n}{i} p^i (1-p)^{n-i}$$

A more general method of finding systems reliability is to view the structure of a system as a network and then apply some standard network programming techniques to the problem. In the network the arcs represent the components, and flow from a source node to a sink node indicates that the system will operate successfully. Through the use of the concepts of minimal cuts it is possible to find the exact system reliability, but once again this is a tedious task.

Due to the difficulty in finding the reliability of a complex system from component data in closed form it is necessary to approach the problem from a different perspective. One such perspective is the use of simulation (see Ingerman [9], Jennings and Harlan [10], and Dietz [6]). Based on the concepts presented above it is possible to construct a simulation procedure for aircraft reliability that considers the effects of human reliability. This will be discussed in section 4.

3. HUMAN COMPONENTS

As noted in the previous section, the consideration of the human component in the determination of system reliability is virtually ignored. In most textbooks and articles dealing with system reliability the aspect of the human component is never considered. This is in part due to the fact that many systems studied do not have a human component or are designed to make the human component as noncritical as possible.

This is not the case in terms of aircraft reliability. Even though most studies on aircraft reliability do not consider the human component, it is impossible to deny its existence and affect an aircraft reliability. The most obvious human component is the pilot. However, in most modern aircraft today, the pilot can virtually be viewed as a redundant system once the plane is airborne. The most critical time for pilots is in takeoffs and landings. Errors during these times are generally critical.

The human component that has the largest effect on aircraft reliability is that which is contained in the maintenance aspects. Errors in maintenance may not be critical in terms of system reliability, but in terms of aircraft reliability they can cause a system to be unreliable and reduce the total number of missions that can be flown. They can however, also be very critical as was illustrated by the DC-10 accident at O'Hare Airport in Chicago, in 1979.

Another concern in maintenance considerations is the number of actions needed to repair a failed component. For example, assume a hydraulic pump malfunctions in a landing gear. Although the pump is bad, in order to repair it, several service actions may be necessary and the time to complete service can be long. Since there are many actions that must be taken, the chance for errors increases which results in potentially longer service times.

In order to model human error, the work of Rook [13] is generally used. His model can be used to compute the probability of no function failures over all independent types of tasks. Based on Rook [13], let

$$F_{ki} = q_{ki} Q_{ki}$$

be the occurence probability of function failure resulting from the kth error mode of the ith operational task. Q_{ki} is the conditional probability that if mode k error of the ith operation occurs, function failure will occur also, and q_{ki} is the probability that the ith task arises in an error of the kth mode.

The overall probability of no function failure over n independent tasks can be shown to be

$$R_T = \prod_{i=1}^{n} (R_{nfi})^{s_i}$$

where $(R_{nfi})^{s_i}$ is the probability of no failure over all s_i ith type independent tasks with s_i being the number of times the ith type of task is performed. This model seems to work well for repetitive manual systems.

For time-continuous reliability modeling of human components it is possible to follow classic reliability modeling. Following Dhillon and Singh [5], define $h_e(t)$ to be the time-dependent human error rate, where

$$h_e(t) = - \frac{1}{R_e(t)} \frac{dR_e(t)}{dt} .$$

Here $R_e(t)$ is the human reliability at time t. By rearranging and integrating the above equation, the following form of $R_e(t)$ is found.

$$R_e(t) = \exp\left[-\int_0^t he(t)dt\right].$$

This reliability expression holds as long as $h_e(t)$ is described by some statistical distribution. Regulinski and Askren [12] showed that the Weibull distribution gives a very good fit to human error data.

As can be seen from the above discussion and from recent literature (see, for example, Dhillon and Rayapati [4]), the modeling of human reliability is a known technique and one that can be implemented. It is essential in determining aircraft reliability that the human component be assessed also. This human component enters into the model in terms of personnel on the aircraft and, especially, the personnel involved in maintenance actions.

As was seen in section two, aircraft reliability can be modeled in terms of its components. Therefore, the consideration of human reliability in the system can be seen as the addition of components. Although adding human components is easy, the determination of the system and aircraft reliability, as noted in section 2, is difficult. Therefore, a simulation approach to determine performance measures will be taken.

4. RELIABILITY SIMULATION MODEL

The simulation model designed for this system is based on the work of Leemis [11] and his general lifetime model. In the simulation model five major elements are considered. As were described in section 1 these are: environment, system state vector, state transition mechanism, system structure, and the addition of performance measures.

The environment element is very important. Environmental concerns have an effect on mission readiness, ability to complete a mission, and on human performance. The environment element is linked to the state transition mechanism by the proportional hazards model of Cox and Oakes [2] when the distributions are the same, and by the mixture model of Everitt and Hand [7].

The state element describes the current condition of the system and is given by an n vector where n is the number of components. The ith element of the vector represents the state of the ith component and can assume a finite number of values. These values allow the modeling of multistate systems which reflect gradual degradation not just operating and failed.

The state transition mechanism allows items to change states. This can either be by degradation, or improvement. This in turn allows the modeling of both scheduled and unscheduled maintenance. This is done by defining an E matrix for each component. The E matrix is a time dependent generator that determines the state of each component based on the hazard functions for the components.

The system structure element describes how the components of the system relate to each other. Therefore, the system structure allows the modeling of very simple series or parallel systems, but also complex systems.

The performance measures are problem dependent but allow the determination of system reliability, aircraft reliability, mean time to repair, mean time between maintenance actions, mean down time, and average number of missions.

The hazard functions of each component are used to generate variates for simulation by using the cumulative hazard function H(t),

$$H(t) = \int_0^t h(\tau)\, d\tau.$$

The variates are generated by using a unit exponential variate as an argument in the inverse cumulative hazard function (see Griffith [8] and Devroye [3]).

Each of these elements are used to do a Monte Carlo simulation to estimate the measures of performance. Through the flexibility of the simulation model and the inclusion of the environment element it is fairly easy to model and simulate an aircraft, its maintenance procedures, and its human components. The flexibility of the model allows the human components to be added and viewed as any other component of the system. This also allows the modeler to estimate the effects of human performance and to test hypotheses or changes of design or reliability.

5. CONCLUSIONS

The aspect of human reliability is not new. However, human performance and its effects on systems have been classically relegated to the area of human factors. When an aircraft designer is asked about the human component of his system, he will generally respond by discussing how the cockpit was designed so that every instrument was this or that. Rarely, if ever, are human performance considered in the design stages of aircraft. However, in this paper it has been shown in terms of aircraft reliability, the ability of an aircraft to start and complete a mission, that human performance can affect the reliability greatly. This comes more from maintenance considerations than on board personnel, but both should be considered.

Since human performance can be viewed as another component of the entire system, a general structure can be used to model aircraft reliability. However, to determine the measure of performance is quite difficult and estimates must be used. A simulation model based on four major elements was discussed. This method makes finding the measures of performance relatively easy.

ACKNOWLEDGEMENTS

The authors would like to acknowledge and thank Dr. L. Leemis for his valuable input and help on the simulation model.

REFERENCES

1. R. Barlow and F. Proschan, Statistical Theory of Reliability and Life Testing: Probability Models, Holt, Rinehart and Winston, Inc., 1975.

2. D. Cox and D. Oakes, Analysis of Survival Data, Chapman and Hall, New York, 1984.

3. L. Devroye, "The Analysis of Some Algorithms for Generating Random Variates With a Given Hazard Rate", draft, McGill University, 1984.

4. B. S. Dhillon and S. N. Rayapati, Reliability and Availability Analysis of On Surface Transit Systems, Microelectronics and Reliability 24, 1029-1033, 1984.

5. B. S. Dhillon and C. Singh, Engineering Reliability: New Techniques and Applications, John Wiley & Sons, New York, 1981.

6. D. C. Dietz, "Translating Aircraft Reliability and Maintainability into Measures of Operational Effectiveness", Proceedings of the Aircraft Systems, Design and Technology Meeting, The American Institute of Aeronautics and Astronautics, Dayton, Ohio, 1-5, 1986.

7. B. Everitt and D. Hand, Finite Mixture Distributions, Chapman and Hall, New York, 1981.

8. W. Griffith, "Multistate Reliability Models", Journal of Applied Probability, 17, 735-744, 1980.

9. D. Ingerman, "Simulation Models: Analysis and Research Tools", Proceedings 1978 Annual Reliability and Maintainability Symposium, IEEE, 184-189, 1978.

10. H. A. Jennings and R. W. Harlan, "Aircraft Operations and Logistics Simulation", 1980 Proceedings Annual Reliability and Maintainability Symposium, IEEE, 320-326, 1980.

11. L. Leemis, "Stochastic Lifetimes: A General Model", PhD Thesis, Purdue University, 1984.

12. T. L. Regulinski and W. B. Askren, Mathematical Modeling of Human Performance Reliability, in Proceedings of Annual Symposium on Reliability, IEEE, 5-11, 1969.

13. L. W. Rook, Reduction of Human Error in Industrial Production, Report No. SCTM 93-62(14), Sandia Laboratories, Albuquerque.

Trends in Ergonomics/Human Factors V
F. Aghazadeh (Editor)

SELF-REPORTS IN ERGONOMICS: AGREEEMENT BETWEEN WORKERS AND ERGONOMIST

Kathryn WOODCOCK WEBB, P.Eng.

Centenary Hospital,
Scarborough, Ontario, Canada
M1E 4B9

An ergonomic survey of workplace and task conditions was conducted among ten workers in a task and compared to the review by an ergonomist. Inter-worker agreement and agreement of the workers with the ergonomist varied across subject area. While some items and sections were well reported, the shortcomings would discourage use of self-reported data without corroborative objective data. Further study is encouraged.

INTRODUCTION

Self-reports by incumbents is a common data collection technique for the application of ergonomics in the field. Perceptions of a few or many aspects of task and workplace conditions may be surveyed by ergonomist, manager, or worker using instruments found in references, original instruments, or interviews. The object of the survey may be to determine if a problem exists, to gather evidence to support proposed changes, to assess the effectiveness of a change, or to compare between different situations. However, very little evidence exists to support or discourage the direct use of self-reported data. A few studies have explored reliability of self-reports of individual factors or their correlation with objective measures, however a number of others have used the subjective measure as the criterion when evaluating the appropriateness of an objective index. There is no evidence that prescribes that self-reports are sufficiently reliable or valid to be used for full ergonomic surveys. The present study was initiated to respond to this deficiency.

An original Survey of Workplace and Task Conditions was developed for this study, based on the person-task-environment model. Ten sections explored a range of factors (sensory, attention, knowledge, posture, reach, force, manipulation, energy, environment, social) in 224 questions. About half of the items were response scale format (5cm scale read as five equal intervals) and half boxes to check as applicable. Following external review by a consulting ergonomist, pilot testing, and revision, data was solicited in a large urban hospital's kitchen. Ten respondents reported on one particular task - plating meals on the belt-line - which is the data on which the present analysis is based. As is often the case with field experiments, the size of the sample was restricted by the size of the workforce. The task was observed and measured by the consulting ergonomist according to normal applied ergonomics data gathering techniques. The ergonomist had no knowledge of workers' responses. The workers' responses were compared among workers and with the ergonomist.

RESULTS

Overall

Clustering around the mean or degree of agreement among the workers was used as an indicator of reliability for the response scale items. Overall, 68 (70%) of these items' responses had a 95% confidence interval of +/-1.0 scale unit or narrower around the sample mean ("clustered"). Of these, 12 (12% of the total) had a confidence interval of only 0.5 scale unit in either direction from the mean ("very clustered"). Among boxes requesting a check if applicable, at least 75% of the respondents gave the same response, either checking or not, in the case of 75% of the check-box items. Different patterns were observed depending on whether the condition was judged present by the ergonomist. Among not-present conditions, 80% (57 of 71) of items had over 75% of respondents in agreement; among present conditions, only 60% (26 of 39) of items have such agreement.

'Validity' of the workers' responses was interpreted as agreement between workers and ergonomist. In the case of response scale items, this was a lack of significant difference between the worker sample and the ergonomist. Unfortunately, the choice of significance level must be arbitrary. Items significant at $p<.01$ are quite certainly different from the response of the expert and thus not "valid" in the context of this investigation. Those which show no difference at $p<.05$ are probably not different from the ergonomist. Agreement was noted between the answers of the workers and the ergonomist in up to 80% of the response scale items (using the liberal criterion of difference only significant at $p<.01$). Using a more skeptical perspective, 54% of the items did not differ significantly at $p<.05$. Considering only the clustered items (as if only clustered items were sufficiently 'reliable' to consider), 53% were not significantly different from the ergonomist at $p<.01$, as were 35% at $p<.05$. On the check-box items, the overall proportion of the sample agreeing with the ergonomist was 69%. This varied from 35% for conditions present to 86% for conditions not present.

Sensory Factors

Items in the section about sensory factors elicited good inter-worker agreement in both formats (85% of check box items had at least 75% of workers in agreement). The response scale items elicited some exaggeration, but check box items showed understatement of conditions, mainly failure to identify conditions that were present. The difficulty of hearing signals and discriminating them correctly was overstated by workers. However, workers understated the noise level in the Environment section and did not agree with the ergonomist's evaluation of poor hearing, all too usual in a workforce of average age in the 40's, with mean 8 years experience in a noisy environment. The difficulty balancing was slightly overstated ($p < .05$) by workers, although the task involved no obvious challenges to balancing, and workers did not allege, for example, riding on moving equipment. While the ergonomist identified six present impediments to vision that were not cited by the majority of the workers, workers cited a problem not identified as present by the ergonomist.

Attention Factors

Inter-worker agreement on scale format items about attention factors was significantly better than the rest of the questionnaire and good among check boxes: 73% of items. Workers underestimated the frequency of response required, the speed of incoming cues (both $p < .01$), and the concentration required ($p < .05$). They also underestimated ($p < .02$) the degree to which they needed to assimilate incoming cues into a decision about action. The most significant difference of the entire questionnaire (at $p < .001$) was the subjective judgment of shift length, which was rated worse, too long, by the workers. Among check box items, the workers had a good degree of agreement with the ergonomist among not-present conditions, but over 75% of workers failed to cite four of the conditions present. Workers were less critical of certain aspects of the overlap between successive cues for action: they did not report cues incoming during the previous action or the need to ignore demands for action temporarily, or the absence of an information buffer. They also did not report distraction from other peoples' activity. Except for shift length, all worker responses understate demands or problems identified by the ergonomist.

Knowledge Factors

Although all but one of the workers' responses to the scale questions were clustered, they were not in good agreement with the ergonomist, tending to overstatement. The pattern appeared to be that of making the task appear very mentally demanding and the performance of others deficient. In the minority, two items understated the task demands: knowledge of results was rated better by workers than the ergonomist, and although the memory needed for the task was strongly exaggerated, workers understated how often it was called upon. In contrast to overrating of, say, adverse environments, this overrating of knowledge demands may be thought of as less a criticism than a way to express pride in the mastery of a task. The pattern of overstatement here appears to have more to do with self-esteem.

Postural Factors

The workers' responses to response scale items were reasonably clustered (8 of 10 within +/- 1 scale unit), but overstated several problems and demands compared to the ergonomist. Workers exaggerated the discomfort during the task but did not indicate body parts bothering all the time, and overestimated both the potential for stress-relieving postural change during the task and the opportunity to leave the workspace. Workers overstated difficulty of ingress/egress and transit through the work area, possibly in reference to other job tasks. This latter suspicion is supported by the workers' high rating of seat stability, despite the ergonomist's observation that seating was not used during the task. The workers had little agreement on the check box items (the worst in the questionnaire, with 55% of items lacking agreement), and poor agreement with the ergonomist, at best an average agreement with the ergonomist on factors present. This pattern signifies exaggeration of postural demands. Notably, over 75% of workers (contrary to the ergonomist) cited excessive use of postures holding arms out at the sides, holding head bent, and standing on one foot. In view of these erroneous reports, it is unusual that workers failed to recognize stooping and upper body leaning postures. It is also unusual that corresponding body part discomforts were not identified in this or the Manipulation section, as would have been expected (Corlett and Bishop, 1978). The postural section was the worst reported of all with regard to check box items.

Reach Factors

There was remarkably little agreement among workers to the items in this section, all response scale format. Two of the five non-clustered items had plausible interpersonal variability explanations, however in one case the range allowed by the ergonomist overlapped the 95% confidence interval by only 0.2 scale units, that is, effectively not at all. In this case, the workers' range was understated compared to the ergonomist's report. The majority of responses showed reasonable agreement with the ergonomist, with significant understatement of frequency of reaching ($p<.01$, and for dynamic reaching, $p<.02$) despite exaggeration ($p<.05$) of the disorder of object grouping - a major factor in the reach requirement of a task.

Force Factors

The majority of scale responses (9 of 14) were not clustered, but among check boxes 72% of items had over 75% of workers in agreement. The non-clustered items in this section did not appear to be particularly subject to interpersonal differences; only three of nine non-clustered items related to a range allowed by the ergonomist. Six (of 14) scale items disagreed with the ergonomist, and eight (of 21) check box items disagreed. While four items weakly overstate the strength demands and musculoskeletal risk, both frequency of force application and object distance from the body responses understate demands, the latter significant at $p<.01$. The ability to apply force to an object close to the body considerably reduces both workload and risk of injury.

Manipulation Factors

The Manipulation section responses to scale items were not well clustered (only 4 of 13) but 74% of check box items elicited agreement among workers. Once again, the role of interpersonal variability was tested and failed to explain the disagreement except in one case. Peculiarly, worksurface height, perhaps the most obvious opportunity for interpersonal variability to affect agreement, elicited exactly unanimous response from workers, in fact agreeing exactly with the ergonomist's mid-range assessment. Agreement with the ergonomist was found in all but two (of 13) scale items. However, the disagreement with the ergonomist was only seen at the weakest level of significance. Significantly different scale items overstated the experience needed to attain the requisite skill, and reported use of a more demanding style of object grip. Check box responses were mainly understatements; ten (of 19) check box items had over 75% of workers disagreeing with the ergonomist. The majority of these disagreements are workers' non-reporting of the physical discomforts predicted by the ergonomist. Despite the strong exaggeration of discomfort in the postural section, the individual body part check boxes did not elicit the expected reports of discomfort, as was found in the postural section, where workers did not indicate that any body parts bothered all the time. The exception was the report of headache (contrary to the ergonomist). Workers incorrectly identified error potential in control shape and 'other', and failed to identify the error potential arising from object position.

Energy Factors

Responses to Energy items were reasonably well-clustered and well-matched with the ergonomist for both item formats. The one non-clustered item does correspond to interpersonal variability anticipated by the ergonomist. Both scale items significantly different from the ergonomist are overstatements. The strongly significant ($p<.01$) item is an underestimate of the frequency of performing different tasks, which overstates both the demands of monotony and an inability to use alternative muscle groups at different work through the same shift, thereby relieving fatigued muscles. It is implausible that the workers actually do not recognize changing from one task to another, thus this seems to reinforce the earlier comment about the confusion between job and task focus. Among the check boxes, the workers did not cite the inability to keep up, resulting in error as did the ergonomist.

The energy section (physical workload, fitness and rest) seems to agree with other studies finding subjects' estimates of physical workload to be accurate (eg. Borg, 1978). This was a task rated as moderately demanding, and the same agreement may not be found for very heavy or very light workload tasks. Workers mainly exposed to less-demanding work may overstate physical demands compared to the more accurate reports of physical labourers (Ilmarinen et al, 1979).

Environmental Factors

Although interworker agreement was good, the workers understated three of the four response-scale environmental items. A lack of fresh air was significantly exaggerated, but the noise level was understated (both $p<.01$), as were heat and humidity (at $p<.05$). Interworker agreement to environmental check box items was slightly worse than average, however this section was the only one in which workers agreed with the ergonomist on a clear majority of items, whether present (71%) or not (83%). As with the scale items, the disagreement with the ergonomist related mostly to understatement of conditions: interestingly, the failure to recognize the unmistakeably present factors of hot surfaces and distracting odours. The latter understatement seems peculiar in view of the complaint about fresh air. Contrary to the ergonomist, the workers identified slippery floors as a factor, perhaps thinking of other tasks in their job which could take them to the dishroom, or perhaps generalizing from conditions that were often present but not evident on the day of observation by the ergonomist.

Although in the surveyed task the workers downplayed slight environmental problems, this task environment is not extreme and a pattern of understatement of problems may not be generalizable to more extreme conditions. The workers' overstatement of a lack of fresh air seems to be contradicted by the failure to cite the bothersome odours which were present. However, temperature, noise, hot surfaces and odours are created directly in the task environment, hence may seem less out of the workers' control. Workers, lacking the ergonomist's knowledge of the levels of environmental conditions that cause performance and health detriment, may perceive an unpleasant condition as something more serious, particularly if they feel powerless to effect change, as is often the case with ventilation. Workers may downplay problems arising from factors under their control or associated with their own choice, as groups of construction workers underestimated risks particular to their own trade (Zimolong, 1985).

Social Factors

The workers agreed with one another and the ergonomist to a reasonable degree, tending to understate the negatives of the social environment perceived by the ergonomist, particularly the value of their task to the organization and co-workers. Workers were more satisfied with their own skill than predicted by the ergonomist. Workers attributed to their peers a lower quality expectation than did the ergonomist. This appears to extend the pattern identified in the knowledge section, of elevating one's own performance and undermining ones peers'.

Excessive demands or oppressive conditions could degrade one's job satisfaction, as suggested by Wilson and Grey (1986); inversely, an unpleasant organizational climate can cause one to become less tolerant and more critical of conditions. The generally positive reports about social factors and tendency to overstate adverse demands and conditions contradicts these theories. Despite the lengths pursued to ensure confidentiality, it is possible that social factors were not reported honestly and are actually more negative. If this were the case, social dissatisfaction could be to blame for the negative reports of some conditions, particularly where workers have overstated problems but have not conspicuously exaggerated.

DISCUSSION

Interpersonal Differences?

Interpersonal differences are a legitimate reason why workers would not agree among themselves. However, the majority of the not-clustered items were not those identified as potential ranges by the ergonomist. Furthermore, the majority of potential ranges identified did not materialize in the present sample. Of 29 items on which workers' responses were not clustered, eight (27%) had interpersonal difference potential specified by the ergonomist. Of 12 items on which the workers closely clustered, there were two Items In which the ergonomist anticipated interpersonal variability. In addition, another 11 items had anticipated interpersonal variability, but on which the workers' responses were clustered. Thus, less than one-third of the non-clustered items were linked to expected interpersonal variability factors, and no interpersonal variability was found in over half (13 of 21) of the items where it was judged possible.

Exaggeration?

One reservation about workers' self-reports is the potential for exaggeration. The proportion of response scale items on which the workers' response overstated a problem or task demand was actually close to 50% of the items of significant disagreement at both $p<.05$ and $p<.01$. Among box questions on which at least 25% of the workers disagreed with the ergonomist's judgment, workers overstated problems in approximately 13 to 14% of the items, to the same approximate degree regardless of whether the condition was present; understatement was almost solely the failure to identify present conditions. It appears to be as legitimate to worry that workers will overlook problems, as was seen extensively in the check-box format.

Not Promising?

On the basis of the present data, it would be risky to recommend use of worker self-reports about general task and workplace conditions. Neither inter-worker agreement (reliability) or agreement with the ergonomist (validity) are adequate in all topic areas. However, it may be satisfactory to pursue self-reports about certain aspects of the person-task-environment interaction, for instance the energy demands. Attention factors could be surveyed and expected to be slightly understated, likewise knowledge factors could be discounted for the tendency to self-flattery. However, it would appear to be ill-advised to solicit self-reports of postural factors without corroborating objective evidence. This finding demands further investigation, as there is no practical source of information about bodily comfort (EMG's not being a simple and convenient field tool), and success has been claimed for postural discomfort surveys used in a number of studies by Corlett and several colleagues, and others.

Inter-worker disagreement and disagreement with the ergonomist were both more prevalent about present conditions than about not-present conditions or all conditions. This suggests that workers have difficulty consistently recognizing conditions. This would be less surprising in free response than with conditions suggested by the questionnaire. It appears that the workers are less able or less ready than the ergonomist to identify the conditions queried by the instrument. This could have significant implications applied to tasks with more conditions present, that is, more problematical tasks. It would be interesting to experiment with worker training in basics of ergonomics as it may affect this finding.

The premise of this criterion validation exercise has been that the ergonomist's assessment is the correct one. However experienced, the ergonomist cannot report the absolute truth, merely the closest approximation based on the observable aspects of the task and workplace. If an ergonomist's data gathering exercise would have been accepted as a sole rationale for action and a self-report is to substitute for this, then the ergonomist's report is a legitimate criterion. However, even as the criterion, it merely stands for and does not become the truth. While the literature has indicated, for example, that awkward postures are predictive of discomforts (Corlett and Bishop, 1978) and the ergonomist logically deduces the nature of the discomforts induced by the task, it is difficult to accuse that the workers' reports of their own sensations are wrong. Perhaps the ergonomist's report of a condition being present was based on an unrepresentative day of observation, and the workers' reports are based on a generalization of eight years' experience.

SUMMARY

This initial exploration merits further investigation in larger work groups and different types of tasks, particularly testing the relationships in more adverse environments and more problematic tasks. It appears that over-reliance on self-reported data should not be encouraged unless a pattern of better reliability and validity is found on more extensive review. While certain aspects of an ergonomic review are within the abilities of workers to report well, others are badly reported. A lack of certainty about which factors are which impairs use of both overall ergonomic surveys and surveys of individual parts of the person-task-environment system. Corroborating objective data should be used.

REFERENCES

BORG, G.1978. Subjective aspects of physical and mental load. *Ergonomics* **21**(3):215-220

CORLETT, E.N. AND BISHOP, R.P. 1978. A technique for assessing postural discomfort. *Ergonomics* **19**:175-18

ILMARIMEN, J., KNAUTH, P., KLIMMER, F., AND RUTENFRANZ, J. 1979. The applicability of the Edholm scale for activity studies inindustry. *Ergonomics* **22**(4):369-376

WILSON, J.R. AND GREY, S.M. 1986. Perceived characteristics of the work environment. in Brown, O. and Hendrick, H.W. (eds.) *Human Factors in Organizational Design and Management* Elsevier-North Holland: Amsterdam

ZIMOLONG, B. 1985. Hazard perception and risk estimation in accident causation. in Eberts, R.E. and Eberts, C.G. (eds.) *Trends in Ergonomics/Human Factors II* Elsevier-North Holland: Amsterdam

Trends in Ergonomics/Human Factors V
F. Aghazadeh (Editor)

AN ETHNIC ANTHROPOMETRIC SURVEY AS AN EDUCATIONAL TOOL

Don E. MALZAHN, Jeffrey E. FERNANDEZ,
and Chol-Hong KIM

Department of Industrial Engineering
The Wichita State University
Wichita, KS 67208 (USA)

International students can contribute to the multi-ethnic anthropometric data base. The collection of anthropometric data from 101 Korean female garment workers and the ergonomic design of several workstations were conducted as part of a Masters Thesis in Human Factors. Since many international students have close ties with either a sponsoring organization or organization that they have worked for, they may have access to a subject population. Sound planning and documentation can reduce the problems inherent in a cross-cultural study.

1. INTRODUCTION

This study, involving the collection of anthropometric data of Korean female garment workers and the ergonomic design of several work stations, was conducted as part of a Masters Thesis in Human Factors. Masters students in human factors from other countries may find sound thesis topics in the country of their origin. This requires them to demonstrate their competence in the environment in which they will eventually practice the profession. Simultaneously, these studies will provide access to an international anthropometric data base. The incorporation of this information in the design of products manufactured in the United States will increase these products international market appeal. The rest of the world has ready access to the data collected on the United States and European populations.

If an object or a space is intended for human use, its shape and dimensions should be derived from the characteristics of the human body and senses [1].

Anthropometric data can depend upon various factors such as race, age, sex, and individual characteristics. There are wide differences in body size and shape among races, subraces, and national or ethnic groups. Racial difference may vary from each other either in overall size (as measured by stature or weight) or in body proportion [2]. For example, Western people are usually taller than Orientals but may be different in body proportion.

There has not been much engineering anthropometric collected data for the Korean population [3]. The rapid industrialization of the Korean economy will require the effective application of appropriate anthropometric information.

2. PREPARATION

2.1. History of Garment Industry in Korea

The textile industry was once the driving force behind Korea's industrialization, growing at an annual rate of nearly 20 % (real value added) between 1970 and 1978. It expanded a mere 4.5 % during 1979 - 1982 [4]. The export of textiles in 1983 was 6 billion dollars, making Korea the second largest net exporter of textiles in the world. The industry also retained its pre-eminence among the various manufacturing subsectors in its share of trade (27 % of total merchandise export in 1982), employment (26 % of the industrial work-force in 1982) and value added (16 % in 1982).

The textile industry's employment is decreasing (from 427,000 employment in 1978 to 383,000 employment in 1982). The labor costs of the Korea textile industry compare favorably with those of other countries (Korea 6.4 % of total cost, Japan 14.6 % and U.S. 21.5 % in 1983) [4]. A steady increase in labor cost (25 % per year in the seventies) has jeopardized Korea's relative labor cost advantage.

2.2. Arrangement with Company

Many international students have significant levels of industrial experience. The relationship between students and their previous employers can provide the basis for accumulating an ethnic anthropometric data base.

It is difficult to find industrially employed subjects in the United States. Access to this data requires that either the subjects be paid for data collection on their own time or time off from the job must be contributed by the firm. Neither of these alternatives are usually available to the Masters level graduate student.

The former employer of the graduate student performing the study described in this paper, was very receptive to a proposal for an anthropometric survey. Data collection was permitted during normal work time.

Communication with Company

Six months before the survey was conducted, contact with the firm was established through telephone and letters. Initial contact with the firm was followed by a brief (less than one page) general proposal. Shorter written documents reduce the frequency of communication errors. As much control as possible was retained by the supervising faculty so that the educational experience could be managed.

Extent of Communication

It was important to communicate with every level of management in the company. The chief executive officer and staff of the company were provided with the kind of information that they needed in order to make a commitment. Those who were in direct supervision of the employees needed more specific kinds of information; dates, times, and specific subject selection process. Communication with the subject was also important and incorporated in the data collection procedure.

2.3. Checking the Equipment

All equipment needed for the survey was determined and the equipment which was not available in Korea was placed on loan from the department. Customs checks were considered in selecting the equipment to be transported to Korea. The equipment taken by the student were a GPM anthropometer and a Jamar adjustable dynamometer. Equipment borrowed in Korea consisted of a video recording system.

2.4. Measurement Rehearsal

Since the survey was conducted in Korea, there was no way to supervise the survey. Therefore, the measurement rehearsal was performed in the presence of the directing faculty members so that they were confident that the data collection procedure was appropriate. The standardization of data collection procedures is essential if the resulting information is to become part of a standard anthropometric data base. The dress rehearsal provided an opportunity for supervising faculty to develope confidence in the student's ability to handle the problems that may be encountered in data collection.

3. PROCEDURE

3.1. Subjects

Subjects were selected from a list of all employees provided by the company. A total of 101 female subjects between the ages of 17 and 26 and in good health were identified. The list of subjects was returned to the company.

Each subject was told the purpose of the study and her role in providing information relevant to the improvement of her workplace. A brief description of each of the measurements was provided with illustrations and Korean language text [5]. The large illustration of the appropriate positioning for each measurement allowed the subjects to easily assume the appropriate posture.

While the measurement was being conducted, each subject was treated very carefully. The nature of each measurement was communicated verbally followed by a request for the subject's consent.

3.2. Measurement

<u>List of the Measurement</u>

1. Height
2. Standing Eye Height
3. Standing Shoulder Height
4. Elbow Rest Height
5. Knuckle Height
6. Sitting Height
7. Sitting Eye Height
8. Sitting Shoulder Height
9. Knee Height
10. Popliteal Height
11. Elbow Rest Height
12. Thigh Clearance
13. Buttock-knee Length
14. Buttock-popliteal Length
15. Shoulder-elbow Length
16. Forearm-hand Length
17. Hand Length
18. Hand Breadth at Metacarpale
19. Hand Breadth at Thumb
20. Hand Thickness at Metacarpale III
21. Foot Length
22. Foot Breadth
23. Weight
24. Grip Strength

3.3. Documentation of the Measurement

The data collection procedure was documented with photographs and video tape. Several still photographs were taken of the experimental setting and a video recording was made of one complete data collection session. This provided the faculty with assurance that the desired procedures were followed.

3.4. Design Application

Although some automation is in place in the industry, it is till quite labor intense. Approximately 7,000 female workers work 10 net hours per day each, six days a week, or a total of 21,840,000 labor-hour/year. The sewing operations are performed in the seated position and finishing operations such as ironing and packing are performed in the standing position. Some tasks are performed over 600 times per day. Good biomechanical design of the workstation is essential.

The student and company personnel selected six tasks for possible design based on the anthropometric data. Each of the tasks was photographed and and relevant dimensions were noted. After the student returned to campus, three tasks were selected with the assistance of the faculty advisor. Since there was no opportunity to get additional measurements, it was essential that the tasks be well documented. By keeping the tasks simple, the data collection was more straightforward.

The development of several smaller task designs required the student to use a wider range of the data collected and increases the probability that the designs will be implemented. A single large task design may have exposed the process to additional risk because of the complexity of the problem, missed data, and difficulty in communication.

3.5. Schedule of the Student

Followings represent the schedule of student for his thesis accomplishment.

1) Contact with company : Jan. 1987 to Mar. 1987

2) Checking Equipment and Measurement Rehearsal : Feb. 1987 to Apr. 1987

3) Data Collection : Jun. 1987 to Jul. 1987

* Five subjects per day and one hour per each subject

4) Data Analysis : Aug. 1987 to Oct. 1987

5) Design Application : Oct. 1987 to Nov. 1987

6) Result Documenting : Nov. 1987 to Dec. 1987

4. SUMMARY

1. International students can contribute to the multi-ethnic anthropometric data base.

2. Anthropometric data from 101 Korean female industry workers was collected in a well documented manner and can included in the engineering anthropometric data base.

3. The human engineering problems in countries undergoing industrial development provide ample opportunity for students to demonstrate their capability. The sound documentation of their solutions provided by a thesis can be beneficial when the student seeks employment.

4. Sound planning can overcome the potential produced by a student performing a majority of his thesis work in another country.

REFERENCES

[1] Pheasant, S., Bodyspace: Anthropometry, Ergonomics and Design Taylor & Francis Co., 1986.

[2] Chapanis, A., Ethnic Variables in Human Factors Engineering, The Johns Hopkins University Press, 1975.

[3] Park, K. S., Human Engineering, Korean Edition, Young-Ji Publication Co., 1986.

[4] IBRD, Korea: Development in a global context. Publication of The International Bank for Reconstruction and Development/ The World Bank, 1984.

[5] Kim, Chol-Hong, "Anthropometric Measures: The Korean Female and The Workplace" (1987), Unpublished Masters Thesis, The Wichta State University, Wichita, Kansas.

Trends in Ergonomics/Human Factors V
F. Aghazadeh (Editor)

THE EFFECTS OF FLOOR TYPES ON STANDING TOLERANCE IN INDUSTRY

M.S. REDFERN, D.B. CHAFFIN

Center for Ergonomics, University of Michigan, Ann Arbor, MI 48109-2117

It is recognized by studies in many industries that standing tolerance is affected by the type of floor used. This study investigated the effects of different flooring conditions on workers required to stand for long periods of time using a psycho-physical approach. Nine different flooring conditions were used including concrete, seven types of mats, and a visco-elastic shoe insert. Questionnaires administered at the end of the workday were used to monitor the perceived fatigue and pain associated with the various floors. The results of the study showed that there was a significant difference between the ratings of the floor for perceived hardness, overall body fatigue, and leg fatigue. Specific discomfort areas of the body were also rated by the workers with the results showing a significant difference between the flooring conditions on the different areas of the body. In evaluating the results, the study showed that workers who are required to stand for prolonged periods of time experience significant levels of fatigue and discomfort in different areas of the body. It appears that the effectiveness of a floor in relieving this fatigue is a function of it's hardness and it's depth before bottoming out as well as flooring dynamic properties.

1. Introduction

In many industries, workers are required to work while standing, walking, and/or carrying loads. Lower extremity discomfort and fatigue from forced long term standing and walking is a problem often identified, but seldom documented. In 1983, it was reported by the American Podiatric Association that 83 percent of the industrial work force in the United States had foot or lower leg problems such as discomfort, pain or orthopedic deformities. Bousseman, *et al.* [1] showed in the laboratory that long term standing is a direct cause of pain and discomfort. Morgora [2] showed that the incidence of low back pain was highest in those workers who stood regularly every working day for periods of more than four hours.

It is difficult to define the problem of standing an walking tolerance because the cause and effect relationship is so elusive. There are no models or methods to directly measure the effects of long term standing and walking. It is interesting and insightful to note the attempts of the workers to alleviate the problem. It is common to see a worker put cardboard, a mat, or a piece of rubber on the floor to try to reduce the perceived stress and tiredness in his/her legs. Flooring and shoe manufacturers are also beginning to become sensitive to this problem. Flooring companies now sell various mats and floor coverings that they claim help to alleviate tiredness and fatigue in the legs due to prolonged standing and walking.

2. Objective

The purpose of this research project was to investigate the effects of different flooring conditions on workers that are required to stand while performing their job. Questionnaires were used at the end of the workday to monitor the perceived fatigue and pain experienced by the workers when on the various floors.

3. Methods

A total of nine flooring conditions were used including concrete, seven types of mats, and a viscoelastic shoe insert. Table 1 describes these conditions. Hardnesses of the floor conditions were measured by a stress-deformation analysis using an Instron machine.

Table 1 : Flooring conditions used in the flooring study.

1) thin - 1/16" rubber mat
2) medium - 1/4" rubber mat
3) thick - 3/8" rubber mat
4) Tri - hard mat with Tri-laminate padding underneath
5) Notri - hard mat without any padding
6) Concrete - bare concrete floor
7) A - viscoelastic mat
8) Insert - viscoelastic shoe inserts
9) Uneven - soft, uneven surface mat

The first three floors (1, 2 and 3) are mats made of the same material but of varying thicknesses, with the thicker ones creating more "give" under the feet of the workers. The material used, a rubber material, is fairly stiff. The next three floors (4, 5 and 6) were the controls. The hard mats with and without the tri-laminate padding, a soft foam rubber, (4 and 5) are a standard in the surveyed plant. Most of the workers who stand throughout the day put these mats down instead of standing on concrete. The final three floors (7, 8 and 9) are other commercial methods of attempting to reduce leg fatigue. The viscoelastic floor (7) is an experimental mat supplied by a company that produces primarily vibration reduction materials. The insert (8) is a viscoelastic shoe insert made of a similar material as the mat (7) but made to fit inside the shoe. The same principle is involved regarding shock absorption. It is also claimed by the manufacturer to reduce low back pain by reducing shock. The last floor, the uneven surface (9), is a new commercial mat designed to reduce fatigue. The theory behind it (as put forth by Brantingham [3]) is that the uneven surface allows slight movement of the ankles thereby improving blood circulation in the muscles of the lower legs. The mat is also soft relative to the others.

Questionnaires were used to monitor the perceived fatigue and pain associated with the various floors. Fourteen workers, all of whom stand throughout their shift, were asked to complete the questionnaire to evaluate the particular floor he/she was using. The floors were rotated between subjects every two weeks until every subject had evaluated every floor condition. The hardness and elastic properties of the floors were measured using an Instron compression tester to get an objective measure of these parameters to compare to results of the questionnaire.

4. Results

The first three questions on the questionnaire asked the workers to rate the floors according to: 1) perceived hardness of the floor, 2) overall tiredness of the body at the end of the day, and 3) tiredness in the legs at the end of the day. Table 2 is a summary of the results of this section.

Table 2: Mean and Standard Deviations of ratings for the perceived hardness of the floor surfaces.

	Perceived Hardness		Overall Tiredness		Leg Tiredness	
	Mean	S.D.	Mean	S.D.	Mean	S.D.
1) 1/16" (thin)	3.5	1.4	3.5	0.7	3.7	0.8
2) 1/4" (med)	3.4	1.2	3.1	1.0	3.1	1.0
3) 3/8" (thick)	2.4	1.2	2.1	1.2	2.1	1.2
4) Hard mat w/ Tri-lam.	2.2	1.2	2.4	1.3	2.5	1.2
5) Hard mat w/o Tri-lam.	4.7	0.5	3.8	0.8	3.9	0.9
6) Concrete	4.8	0.4	4.2	1.0	4.5	0.8
7) Viscoelastic mat	4.0	0.7	3.8	0.7	3.7	0.7
8) Viscoelastic shoe insert	2.5	1.4	2.4	1.1	2.0	1.1
9) Uneven mat	1.9	1.0	3.3	0.9	3.2	1.0

The results of the study showed that there was a significant difference between the ratings of the floors for perceived hardness, overall body fatigue, and leg fatigue. When the data were stratified by each floor type, the correlations between the three indices ranged from .60 for the concrete floor to .97 for the thick rubber mat. In general, the ratings of perceived hardness correlated very well with the measured hardness of the floors with the one exception of the uneven mat (9) which was perceived as the softest mat. The general correlations between perceived hardness and either leg tiredness or body tiredness were both about .50. A major difference seen between the perceived hardness ratings and these general tiredness ratings was that the uneven mat (9) was rated as the softest floor, however the tiredness rating was 3.3 which was fifth in rank and just below the thin rubber mat (1). The uneven mat also was fifth in rank for the leg tiredness ratings.

Specific discomfort areas of the body were also rated by the workers. These areas were divided as follows: Feet, Ankles, Shank, Knee, Thigh, Hips, Lower Back, and Upper Back. The results showed a significant difference between the flooring conditions on the different areas of the body. The concrete floor (6) and the hard mat (4) consistently had the highest discomfort ratings, while the lowest ratings were shared by the thick rubber mat (3), the tri-laminate mat (5) and the shoe insert (8). F - tests between the mean discomfort ratings for each floor showed that they were significantly different ($p<.05$) for the foot, ankle, shank, knee, upper back and lower back. The others, the leg and hip, were not significantly different ($p<.05$) between floor types.

When looking at the ratings for the different parts of the body, the feet had the highest discomfort rating followed by the ankle. In fact, the general discomfort ratings involved in standing tend to decrease as the distance from the floor is increased. For example, the feet ratings are higher than the ankle, the shank are higher than the ankle, and so on.

5. Discussion

In evaluating the results, the study showed that workers who are required to stand for prolonged periods of time experience significant levels of fatigue and discomfort in different areas of the body. Also, the floor surfaces on which they stand during their shift affect the workers' perceptions of this tiredness and discomfort. It appears that the effectiveness of a floor in relieving this fatigue is a function of it's hardness and it's depth before bottoming out. A floor can be too soft, however. This is evidenced by the fact that the uneven mat (9) was relatively soft and had low perceived hardness levels, but had relatively high tiredness ratings. The dynamic properties of the floor may also be important. These properties reflect how the floors absorb and transmit forces to the body when the foot hits the ground. The non-linear dynamic properties such as in the viscoelastic material also have some effect.

Another observation from this study is the high correlation between leg tiredness and general tiredness. This indicates that the flooring is not only affecting the legs, but the entire body in some way. The discomfort ratings showed that the lower legs were most affected by the type of floor used. It seems that the body segments closer to the floor had the higher ratings. In other words, the discomfort diminishes as the distance away from the floor.

6. Conclusions

This research has shown that workers who are required to stand for prolonged periods of time experience significant levels of fatigue and discomfort in different areas of the body. Also, the floor surfaces on which they stand during their shift affect the workers' perceptions of this tiredness and discomfort. It appears that the effectiveness of a floor in relieving this fatigue is a function of it's hardness and it's depth before bottoming out. Further research needs to be done to determine the effect of flooring on the human body. Investigations into both the physiologic and biomechanical influences should be undertaken. Better methods to identify and quantify the effects of long term standing are also necessary to continue this work.

Acknowledgements

The authors wish to thank Ford Motor Company, Body and Assembly Operations Division for their support of this research.

[1] Bousseman, M.; Corlett, E.N. and Pheasant, S.T. "The relation between discomfort and postural loading at the joints.", Ergonomics **25**(4): 315-322, 1982.

[2] Morgora, A "Investigations of the relation between low back pain and occupation.", Industrial Medicine, **41**(12): 5-9, 1972.

[3] Brantingham, C.R.; Beckman, B.E.; Moss, C.N.; and Gordon, R.B. "Enhanced venous pump activity as a result of standing on a variable terrain floor surface.", J of Occupational Medicine **12**(5): 164-169, 1970.

Trends in Ergonomics/Human Factors V
F. Aghazadeh (Editor)

"WALKING CONDITION AND FLOOR TYPE EFFECTS ON KINEMATIC VARIABLES OF GAIT "

M.A.EL-NAWAWI*, S.B.MOSHREF* AND A.A.ZEINELABIDIEN**

* Professor,Dept. of Mech.Eng.,Al-Azhar University,Cairo,Egypt
** Res. Assoc.,Dept. of Mech.Eng.,Al-Azhar University,Cairo,Egypt

The characteristic differences in gait pattern of normal individuals appear to be minor, especially on even ground. Such differences increase significantly in case of lower extremity amputees as well as patients suffering from various pathologic ailments of lower extremities. In this investigation, one patellar tendon bearing below-knee modular prosthesis with a solid-ankle cushion-heel (SACH) foot is selected as test prosthesis. The experiments are performed on different types of flooring (laboratory floor, sand and asphalt) at three free levels of walking condition (slow, normal and fast). Variables such as floor type and walking condition are statistically analysed to determine their effects on several temporal and distance factors of gait (actual walking speed, cadence, step length differences, swing phase time differences, stance phase time differences and double support time differences). The significant effects are determined using the F-test statistic at the 0.01 and 0.05 levels of significance. Interesting results and conclusions regarding the type of prosthesis, walking condition and floor type are introduced.

1. INTRODUCTION

Gait studies are usually conducted to develop methods and means to help the disabled and to allow patients with various pathologies to be helped in the most effective manner. Special attention is directed in such studies to obtain design information relating to prosthetic aids. Investigations into temporal and distance parameters looking into aspects such as step and stride length, duration of foot contact, swing or variations in velocities or unsymmetries of walking [1-5] gain special interest, while others investigate variations in the force patterns together with some aspects of body kinematics [6-10].

Gait deviations such as differences in step lengths, stance and swing times of both legs as well as times consumed in transporting the body weight completely between legs result in the inconsistency observed in the pattern of walking.

Pathologic ailments of lower extremity, amputation level, orthosis and prosthesis alignment and design, floor type and walking condition represent the main factors affecting gait deviations.

The research under consideration is conducted in order to evaluate the floor type and walking condition effects upon the unsymmetry of walking of an amputee wearing a conventional PTB-B/K prosthesis with SACH foot. The kinematic approach is adopted in this study, where the actual walking speed, cadence, step length difference, swing phase time difference and stance phase time difference between sound and prosthetic legs are monitored together with the double support time difference.

2. EXPERIMENTAL WORK

Experiments were carried out using one male subject age 32, 179cm height with a unilateral below-knee amputation of traumatic origin.

Three experimental walking tracks: asphalt (as), laboratory floor (la) and fine grain sand (sa), 1 m. width and 20 m. long each are selected as the walking surfaces for the purpose of this investigation. In addition, three levels of walking condition, namely slow (S), normal (N) and fast (F) are taken as the free levels assumed by the subject during each run.

A complete stride occured between successive heel strikes of the same foot. The line between the ipsilateral heel strikes provided a measure of stride length and also represented the walking direction or line of progression. The forward distance travelled between consecutive footsteps determined step length measured along each line of progression.

To correlate the stride times with stride lengths, special foot inserts for each shoe is recommended. Each foot insert is provided with four switches connected in parallel at the outer contour of the heel area and two other switches at the toe region. Such construction makes the simple flexible strip contacts (switches) more effective during walking, especially, on the fine grain sand track. The two switch groups in each foot insert are connected to two channels of an oscillographic recorder (8-channel Nihon-Kohden Polygraph). The recorder pen deflection is activated upon setting on the switch in any of the prescribed foot areas. This aciton lasts as long as the switch points are in contact. The contact and release events for each of the heel and toe of both feet are marked on the polygraph recording paper.

The experimental runs are performed on asphalt, laboratory floor and fine grain sand tracks at three free levels of walking condition: slow, normal and fast. The subject is made familiar with the walking track and condition through a number of training trials conducted before the experimental runs.

3. RESULS AND DISCUSSIONS

The independent variables under investigation being the three floor types and the three walking condition levels are statistically analysed to study their effects on the prescribed kinematic variables. The significant effects are determined using the F-test statistic at the 0.01 and 0.05 levels of significance. The following discussion of results highlights the main findings obtained by the analysis of variance.

3.1. Effect of Floor Type

The analysis of variance results showed that the floor type has a highly significant effect at the 0.01 level upon the actual walking speed (V), cadence (C), stance phase time difference between the sound and the prosthetic legs (Stance pd). Such significance is also found at 0.01 level for the floor type upon the double support time difference (Support td), where the double support time difference considered in this study is the difference between the time consumed in transporting the body weight completely from sound to prosthetic leg and the time consumed in transporting the body weight completely from prosthetic to sound leg. The floor type is also shown to affect the swing phase time difference (Swing pd) at the 0.05 level of significance. No significant effect is shown for the floor type upon the step length difference (Step ld). Fig (1). Shows a plot of means and standard deviations of actual walking speed versus the floor type. The laboratoy floor and asphalt tracks yield nearly the same values of actual walking speed while the fine grain sand track yields the lowest values. Similar findings are given for the floor type effects on Cadence. The fine grain sand track gives the amputee a feeling that the prosthetic limb is not adequately stable during walking. This feeling affects the step length values of both sound and prosthetic legs.

Fig. (2) illustrates a plot of means and standard deviations of both swing and stance phase time differences versus the floor type. The highest values of swing and stance phase time differences are recorded when the amputee walks over asphalt track. The fine grain sand track yields the lowest values for swing and stance phase time differences while the laboratory floor track results fall in between.

The standard deviation values indicated in the figure reflect the narrow margin obtained for the swing and stance phase time differences for asphalt and laboratory floor results. Under all combinations of floor type and walking condition, the amputee tends to have a longer prosthetic step length than sound step length, longer prosthetic swing phase than sound swing phase and a shorter prosthetic stance phase than sound stance phase. These observations are maintained throughout all experimental runs, whether the amputee started moving with the sound or the prosthetic leg.

Such findings are in agreement with those obtained by Murray et al. [11] under all three walking conditions. Hoy et al. [12] findings for sound and prosthetic step length of amputee children are similar to the prescribed results. However, when a solid ankle cushion heel (SACH) foot prosthesis is used in their study,only the results obtained under the fast walking condition level showed similarity with the present work findings.

The increase of the prosthetic step over the sound one may be explained by the reluctance or inability of the amputee to spend a longer support time on the prosthetic single limb. Such behavior forces the subject to shorten the prosthetic single limb support time and allows a longer time interval for sound single limb support. The longer prosthetic swing phase time compared to the sound swing phase time is due to the fact that the amputee experiences some difficulty in accelerating the swing of the prosthesis because of the relatively short lever arm of the residual limb. The reduced values of axial compression forces transmitted to the stump in case of sand walking track allows the amputee to prolong the prosthetic

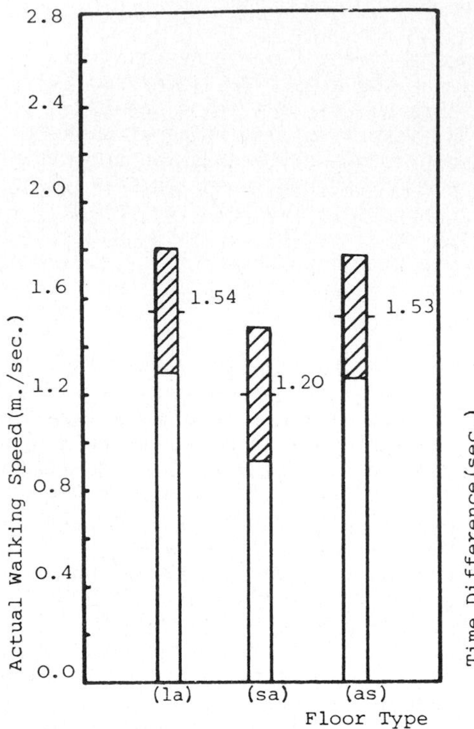

Fig. (1) Plot of means & standard deviations for actual walking speed versus the floor type.

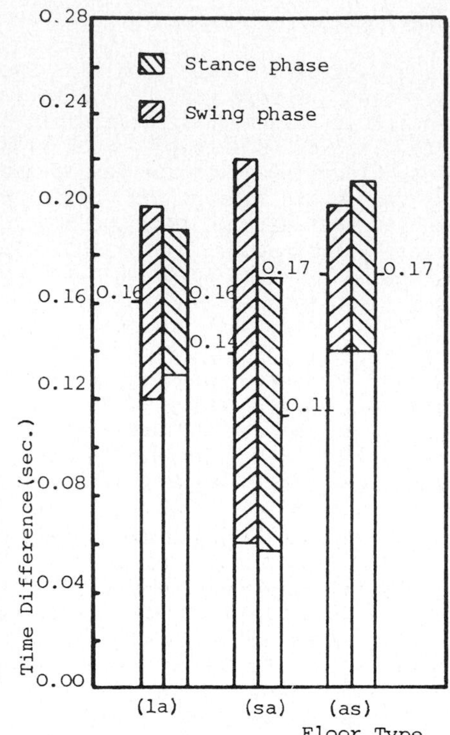

Fig. (2) Plot of means & standard deviations for swing & stance phase difference between sound & prosthetic legs versus floor type

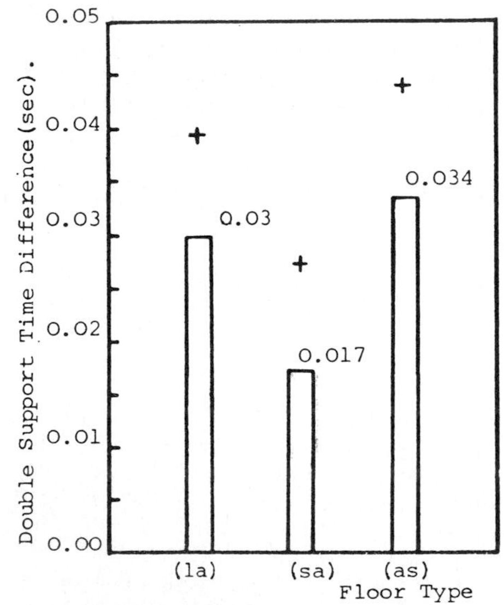

Fig. (3) Plot of means & standard deviations for double support time differences vs the floor type (points indicate standard deviations from mean).

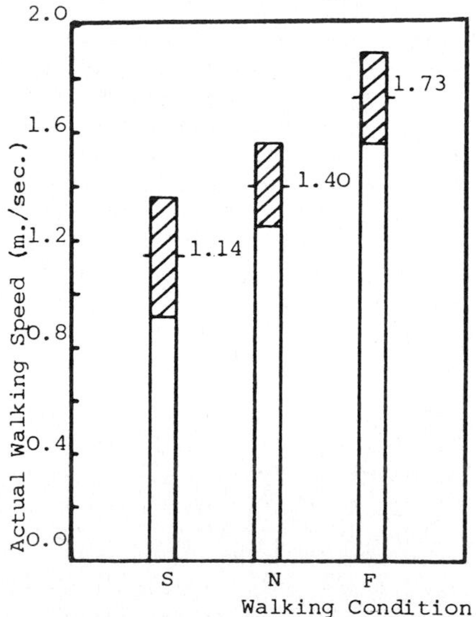

Fig. (4) Plot of means & standard deviations for actual walking speed vs the walking condition.

stance, hence reducing the time differences in stance and swing [13 and 14].

Fig. (3) illustrates the means and standard deviations of double support time difference versus the floor type. The least double support time difference value is observed when the subject walks over the fine grain sand track followed by the laboratoy floor track. The asphalt track yields the highest value for double support time difference. The support time difference is the result of subtracting the load transport time (sound to prosthetic) from the load transport time from prosthetic to sound. This order is maintained throughout all laboratory floor and asphalt data findings, while reversed in some fine grain sand track data.

During walking and immediately after prosthetic heel rise, an increasing load is applied to the toe rubber portion of the prothesis. This load deflects the toe rubber upwards. The net result is a storage of an amount of energy in the toe rubber portion. During the last stage of the prosthetic stance phase and just befor toe-off, the amputee begins to carry part of the load on the sound leg to reduce the weight upon the prosthesis. The toe rubber portion re-expands and gives back some of the energy previously stored in it. The toe rubber portion does not provide more than partial support for the body during the last part of the toe-off. In contrast to the strong calf muscles of the sound leg, the toe rubber portion cannot forcibly deliver substantial energy for push off [15]. Such difference may provide an interpretation to the extended time interval in transporting the body weight completely from prosthetic to sound leg, as obtained in asphalt and laboratory floor track data findings. The reversed order given in some fine grain sand readings as well as the small differences between the pushing off times of prosthetic and sound legs may be interpreted by the looseness of the fine grain sand structure.

3.2. Effect of Walking Condition

The analysis of variance results indicate that the walking condition has a highly significant effect at the 0.01 level upon the actual walking speed, cadence, swing phase time difference and stance phase time difference between the sound and the prosthetic legs. Such significance is also found for the walking condition effect upon the double support time difference. No significance is indicated for the walking condition effect upon the step length difference.

The fast speed level of the walking condition yields the highest values of actual walking speed and cadence (e.g. Fig (4)).The order of increase, as shown in the figure, indicates a proportional relationship between the actual walking speed and the walking condition.

The increase of the prosthetic step over the sound step under all levels of walking condition is maintained throughout all experimental runs. The step length difference also shows a proportional increase in relation to the three gradually increased levels of walking condition, inspite of the absence of statistical significance of the effect.

The consistency of increase of prosthetic swing over the sound swing and the decrease of prosthetic stance in comparison with the sound stance under all walking condition levels are observed. Furthermore, both swing and stance phase, either sound or prosthetic, maintained a gradual

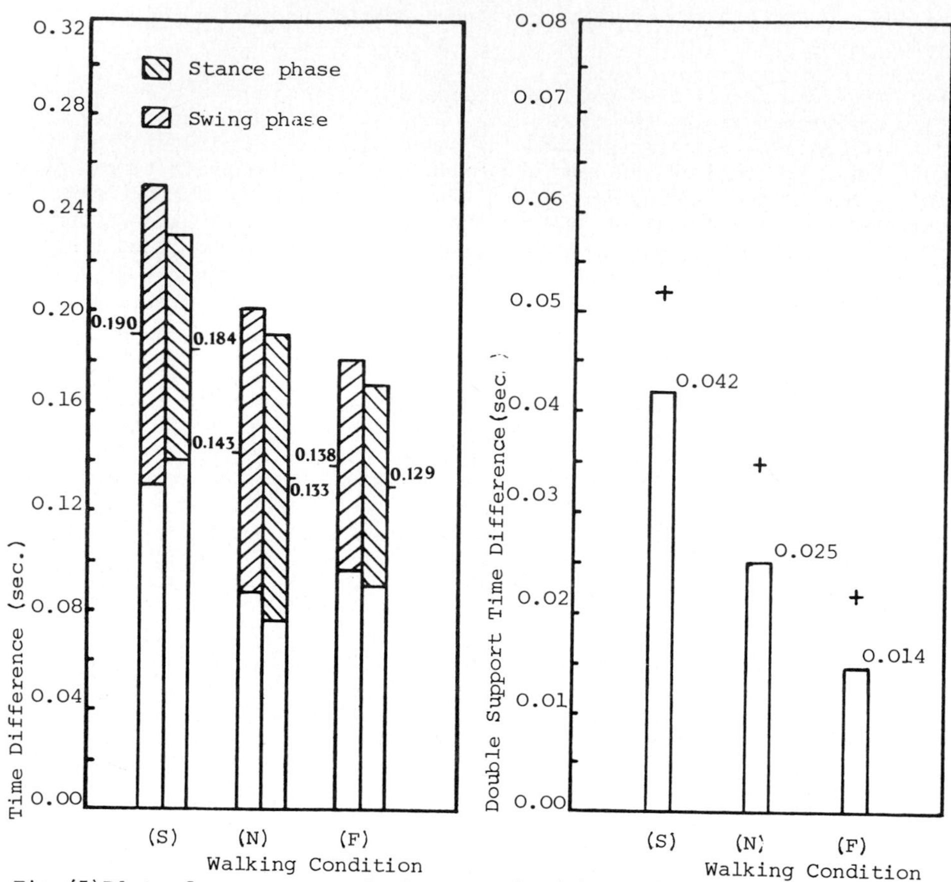

Fig. (5) Plot of means & standard deviations for swing & stance phase difference between sound & prosthetic legs versus walking condition.

Fig. (6) Plot of means & standard deviations for the double support time difference versus walking condition.

decline with the increase of walking condition level. However, the swing and stance phase time difference results indicate inverse relationship to the walking condition (Fig. 5). The highest swing and stance phase time difference values are obtained at the slow level of walking condition (S), while the lowest values are observed at the fast level (F).

The step length, the swing and stance phase difference results obtained in this investigation differ from those reported by Murray et al. [11] and Hoy et al. [12]. The results obtained in this investigation, as well as others [16] and [17], suggest that the different circumstances under which each study is conducted, the variation in the prosthesis type and the means of fitting to the wide population of amputees affect the consistency in the trend of the step length difference, swing and stance phase time differences behaviour under the different levels of walking condition.

Figure 6 shows that the double support time difference maintains a gradual decline with the increase of walking condition level.

3.3. Floor Type by Walking Condition Interaction Effects

The ANOVA results show a highly significant effect at the 0.01 level for the floor type by walking condition interaction upon cadence, swing phase time difference and stance phase time difference. No significance is reported for the floor type by walking condition interaction upon the actual walking speed, step length difference and the double support time difference.

The highest actual walking speed and cadence values are observed when using the laboratory floor track, under normal and fast levels of walking condition, while their highest values are observed with asphalt track under the slow level. On the other hand, the lowest values of actual walking speed and cadence under the three levels of walking condition are consistently yielded when walking on the fine grain sand track.

Similarly, the highest step length difference values under the three levels of walking condition are given by the fine grain sand track. The asphalt track yields the lowest values of step length difference at the three levels of walking condition, while the laboratory floor track results fall in between.

The highest swing phase time difference and stance phase time difference are observed on the asphalt track under the fast level of walking condition while their highest values are given with the laboratory floor track under the normal level. Their lowest values are obtained using the fine grain sand track under the normal and fast levels of walking condition.

The double support time difference values at the three levels of walking condition are observed at the highest with the asphalt track, while at the lowest with the fine grain sand track.

4. CONCLUSION

Based on the previously discussed results, the following conclusions could be drawn:
1) The floor type has a highly significant effect at the 0.01 level upon the actual walking speed, cadence, stance phase time difference between sound and prosthetic legs. Such significance is also found at 0.01 level for the floor type upon the double support time difference. No significance is indicated for the floor type upon the step length difference.

The variations in the walking surface structure and the ankle joint capabilities of sound and prosthetic legs cause the main abnormalities during walking. The fine grain sand track hence yields the highest step length difference between the sound and prosthetic legs.

2) The walking condition significantly affects the actual walking speed, cadence, swing phase time difference and stance phase time difference between the sound and prosthetic legs. Such significance is also found at the same level (0.01) for the walking condition upon the double support

time difference. No significance is shown for the walking condition upon the swing phase time difference.

The results obtained in this investigation, suggest that the different circumstances under which each study is conducted, the variation in the prosthesis type and means of fitting to the wide population of the amputees affect the consistency in the trend of the step length difference, swing phase time difference and the stance phase time difference behaviour under the different levels of walking condition.

3) A highly significant effect at the 0.01 level is observed for the floor type by walking condition interaction upon cadence, swing phase time difference and stance phase time difference. No significance is reported for the floor type by walking condition interaction upon the actual walking speed, step length difference and the double support time difference. The increase of the prosthetic step over the sound one may be explained by the reluctance or inability of the amputee to spend a longer support time on the prosthetic single limb. Such behaviour forces the subject to shorten the prosthetic single limb support time and allows a longer time interval for sound single limb support.

4) The longer prosthetic swing phase time compared to the sound swing phase time is due to the fact that the subject experiences some difficulty in accelerating the swing of the prosthesis because of the relatively short lever arm of the residual limb.

5) The type of floor most frequently used by the subject in his daily activities appears to influence the different parameters under investigation. Such statement is supported by the findings given in the asphalt track results in particular and in the laboratory floor results in general.

REFERENCES

[1] Grieve D.W., Gear R.J.: Relationships between length of stride, step frequency, time of swing and speed of walking for children and adults. Ergonomics 9, pp 379-399, September 1966.

[2] Godfrey C.M., Jousse A.T., Brett R, Butler J.F.: Comparison of some gait characteristics with six knee joints. Orthot. Prosthet. 29, pp 33-38, September 1975.

[3] Grieve D.W.: Gait patterns and speed of walking. Biomed. Eng. 3, pp 119-122, 1968.

[4] Foley C.D., Quanbury A.O., Steinke T.: Kinematics of normal child locomotion - a statistical study based on TV data. J. Biomech. 12, pp 1-6, 1979.

[5] Grieve D.W.: The assessment of gait. Physiotherapy 55, pp 452, 1969.

[6] Seireg A. and Arvikar R.J.: The prediction of muscular load sharing and joint forces in the lower extremities during walking. J. Biomech. 8, pp 89, 1975.

[7] Bresler B. and Frankel J.P.: The forces and moments in the leg during level walking. ASME Trans. 72, pp 27, 1950.

[8] Elftman H.: The measurement of the external force in walking. Science 88, pp 152-153, 1938.

[9] Elftman H.: Forces and energy changes in the leg during walking. Am. J. Physiol. 125, pp 339-356, 1939.

[10] Winter D.A.: Overall principle of lower limb support during stance

phase of gait. J. Biomech 13 (11), pp 923-927, 1980.
[11] Murray M.P., Mollinger L.A., Sepic S.B. and Gardner G.M.: Gait patterns in above - knee amputee patients - hydraulic swing control versus constant friction knee component. Arch. of Phys. Med. Rehabil. 64 (8), pp 339-345, 1983.
[12] Hoy M.G., Whiting W.C. and Zernicke R.F.: Stride kinematics and knee joint kinematics of child amputee gait. Arch. of Phys. Med. Rehabil. 63 (2), pp 74-82, 1982.
[13] Steffens H.P., Engelke K.W. and Boenick U.: Experimental analysis of the loads acting on below-knee modular prostheses on outdoor walking surfaces. Biomedizinische Technik, Band 22 (1-2), pp 8-12, 1977.
[14] Boenick U., Steffens H.P., Zeuke R. and Engelke E.: Experimental analysis of the forces and moments acting on above-knee and below-knee modular prostheses on outdoor walking grounds. A report of a conference covered by the international society for prosthetics and orthotics (ISPO), pp 153-160, 1977.
[15] Murphy E.F.: Lower - extremity components. American Academy of Orthopaedic Surgeons, Orthopaedic appliances atlas, Ann. Arbor, Mich., Vol. 2, Chapt. 5, 1952.
[16] El-Nawawi, M.A., Moshref, S.B. and Zeinelabidien, A.A., "Kinematic Variables Study of Type of Ankle in Below-Knee Prosthesis.", Proceedings of the Third International Conference on Production Engineering Design and Control (PEDAC) Alexandria Univ., Alexandria, Egypt, pp 245-254, 1986.
[17] El-Nawawi, M.A., Moshref, S.B. and Zeinelabidien, A.A., "Kine[illegible]tic Variables Study of Type of Ankle in Below-Knee Prosthesis.", Proceedings of the Annual International Industrial Ergonomics and Safety Conference, pp 1075-1086, Miami, Florida, U.S.A., 1987.

Trends in Ergonomics/Human Factors V
F. Aghazadeh (Editor)

EVALUATION OF OFFICE CHAIRS A VALIDATION STUDY

Ram R. BISHU, David J. COCHRAN and Michael W. RILEY

Department of Industrial and Management Systems Engineering
University of Nebraska-Lincoln
Lincoln, Nebraska, USA.

Five typical office chairs were evaluated in a field experiment involving the office secretaries of the College of Engineering and Technology at the University of Nebraska-Lincoln. The evaluation procedure used a general comfort rating scale, a body part discomfort rating scale, and a chair feature evaluation checklist. The differences among the chairs were consistently observed in all the dependent measures, thereby reiterating the effectiveness of the evaluation procedure. It appears that the body part discomfort ratings of the back regions are critical in chair comfort.

INTRODUCTION

Current recommendations on chair design have been derived from chair evaluation studies performed over the last two decades. A wide range of methods have been used for evaluating chairs. For example Oxford [1] compared the chair against the anthropometric data and principles of chair design. Jones [2] used fitting trials to adjust the chair to the operator. Experimental evaluations have ranged from those in laboratory setting [3] to those in field setting [4]. More recently Drury and Coury [5] recommended a three stage evaluation procedure for the absolute evaluation of a single chair, namely 1) evaluate the chair against published dimensional recommendation; 2) use fitting trials to allow the subjects to fit the chair to their body; and 3) evaluate comfort directly at the work place using a known methodology. A part of this methodology has been used in the evaluation of a forward sloping chair [6] and in the field evaluation of office chairs [7].

As Branton [8] has noted, seating is a means to an end and not an end itself and therefore the task to be performed is an important consideration in chair evaluation and recommendation. Further, the fact that comfort is a subjectively defined personal sensation, the final criterion in chair selection should be the perceived seating comfort for the user population. This criterion stresses the importance of having a reliable and valid scale for measuring chair comfort. Published studies on chair and seating comfort indicate the use of a variety of comfort scales ranging from the two point adjuctive pairs (eg, comfortable / uncomfortable) by McLeod et al, [4] to the eleven point scale used by Shackel et al [9]. Similarly the use of a number of body part discomfort scales such as the three point scale used by Grandjean [3] and the five point scale developed by Bishop and Corlett [10] has also been reported in the literature. Evaluators have generally supplemented the chair comfort measurements with some subjective measures of the chair features [9].

OBJECTIVES

The objectives of this investigation were to evaluate and compare five office chairs using the method suggested by Drury and Coury [5], and to develop a model for predicting chair comfort using the general comfort rating, body part discomfort rating scales and the chair feature check list.

METHOD

Four office chairs, from a set of currently popular brands, were obtained from a local furniture maker. A typical university classroom chair was chosen as the fifth chair. The details of the chairs are given below:

Chair No. 1: United Chair Model S/22
Chair No. 2: Krueger Model DONAP-1
Chair No. 3: Hon Model 5820 Code V
Chair No. 4: Kinetic Furniture Model 105/902
Chair No. 5: Class room chair

Five secretaries of the College of Engineering and Technology at the University of Nebraska-Lincoln participated in this experiment. The order of assignment of chairs to the secretaries was randomized.

For the data collection, the subjects were requested to sit on each of the experimental chairs for an entire work day and perform their daily work. A two stage evaluation procedure for measuring the chair comfort was followed. The general comfort rating (GCR) scale developed by Shackel et al [9] and body part discomfort scale [10] were used for measuring comfort. These scales were administered every hour and their responses recorded. At the end of the work day the chair feature check list [9] was administered to assess some of chair's features.

The responses on the body part discomfort scales were combined to give a set of dependent variables, as shown below:

1. Body part discomfort frequency (BPDF): The percentage of body part discomfort responses which were non zero at any measurement time.

2. Body part discomfort severity (BPDS): The mean scale value of all non zero responses at any measurement time. The non zero responses to the various body parts were combined to yield the following body part discomfort severity measures: BPDS neck, BPDS arm, comprising the shoulder, upper and the lower arms; BPDS back, comprising the upper, middle and the lower back; BPDS buttocks, and BPDS leg, comprising the thighs and the legs.

3. General Comfort Rating (GCR). The scale value of the overall comfort rating at any measurement time.

RESULTS

A series of ANOVAs were performed on this data. For the general comfort rating the Time and the Chair effects were significant and so was the Subject x Chair interaction. Figure 1 shows the plot of GCR for the different chairs. A clear difference in the overall comfort patterns is seen among the five tested chairs. According to a range test, the ranking of chairs in increasing order of discomfort was chairs 4, 2, 3, 1, and 5 respectively.

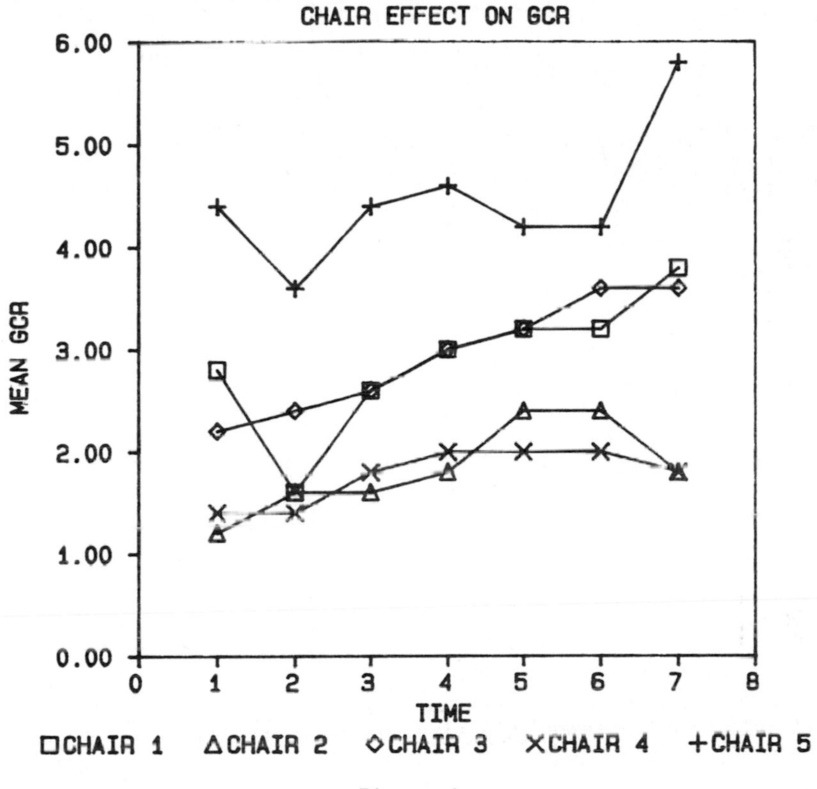

Figure 1

For the body part discomfort frequency, the Time, the Chair, and the Subject x Chair effects were significant. Figure 2 shows the plot of this. Time effect is expected given that people tend to become tired at the end of the day. The differences among the chairs are demonstrated here as well. The ranking of the chair in order of increasing discomfort was 2, 1, 4, 3, and 5.

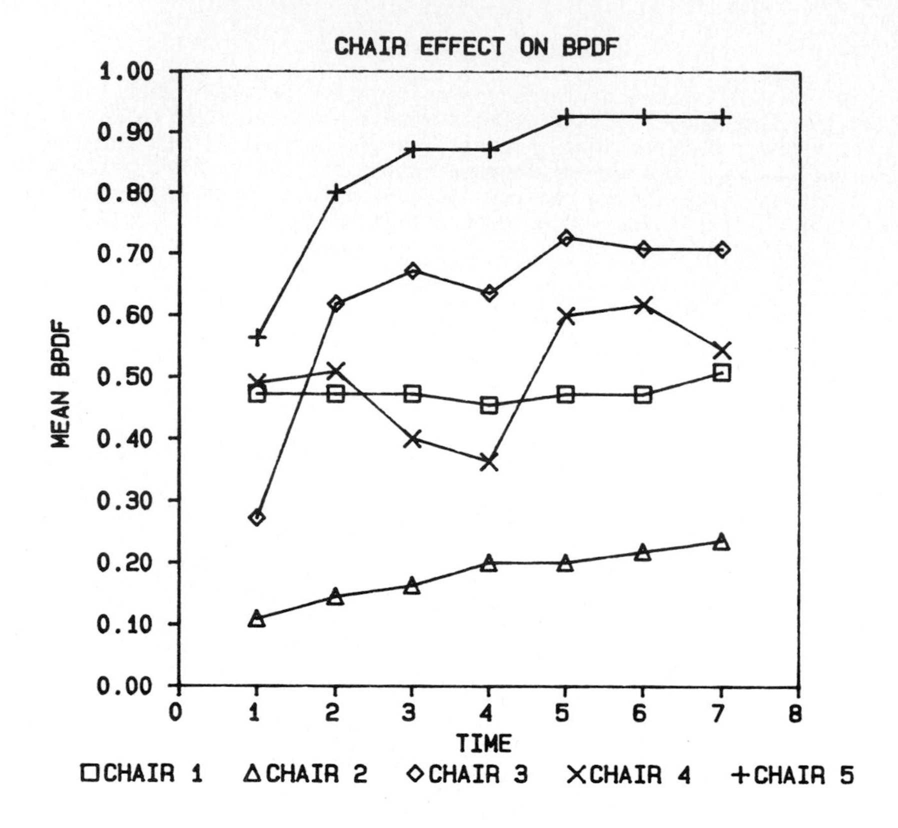

Figure 2

Among the five BPDS variables described above, BPDS neck and BPDS arm were not analyzed further. It was reasoned that the discomfort on these regions would be task and arm rest dependent, two factors not controlled in this investigation. The Time and the Chair effects were significant for BPDS back as well as the Time x Chair and Subject x Chair interactions. Figure 3 shows the plot of these effects. As in the previous cases the differences among the chairs are observed here as well. The ranking of the chair in increasing order of discomfort was 2, 4, 3, 1, and 5. Figure 4 shows the effects of chair and time on BPDS leg. The Time, Chair, and Subject x Chair effects were significant. The rank order of chairs in increasing order of discomfort, was 2, 4, 3, 1, and 5.

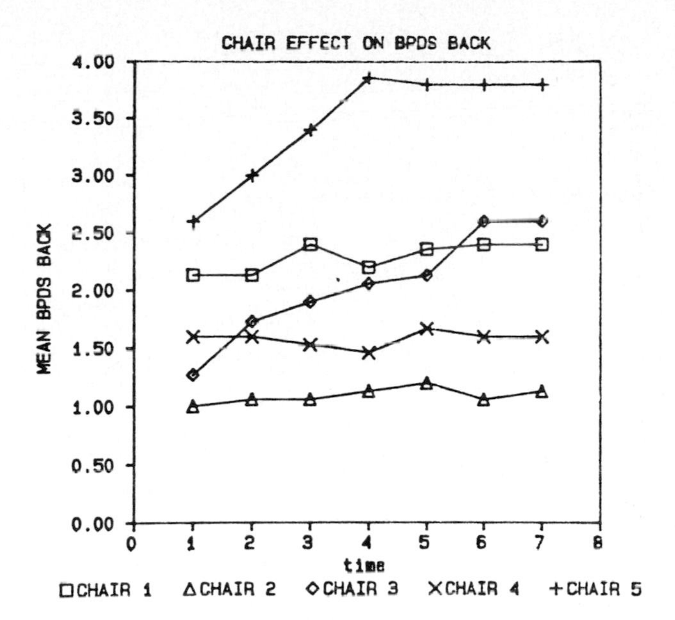

Figure 3

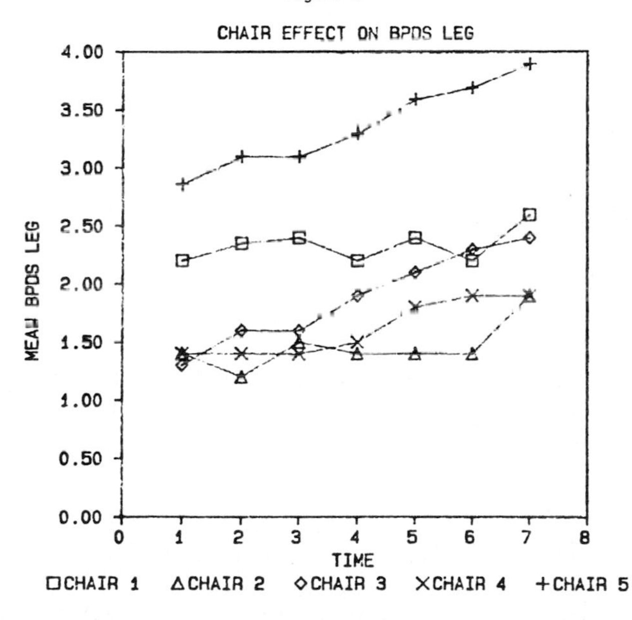

Figure 4

The results of the chair feature checklist is shown in Figure 5, with a mean value plus or minus one standard deviation for each of the experimental chairs. Large variability seem to exist in subjects' perception of the chair features. However it is observed that for the comfortable chairs (Numbers 2, 1 and 4) most of the responses are around the mean value of the scale with less variation, while the responses on chairs 1 and 5 seem to be more dispersed, and away from the optimum value.

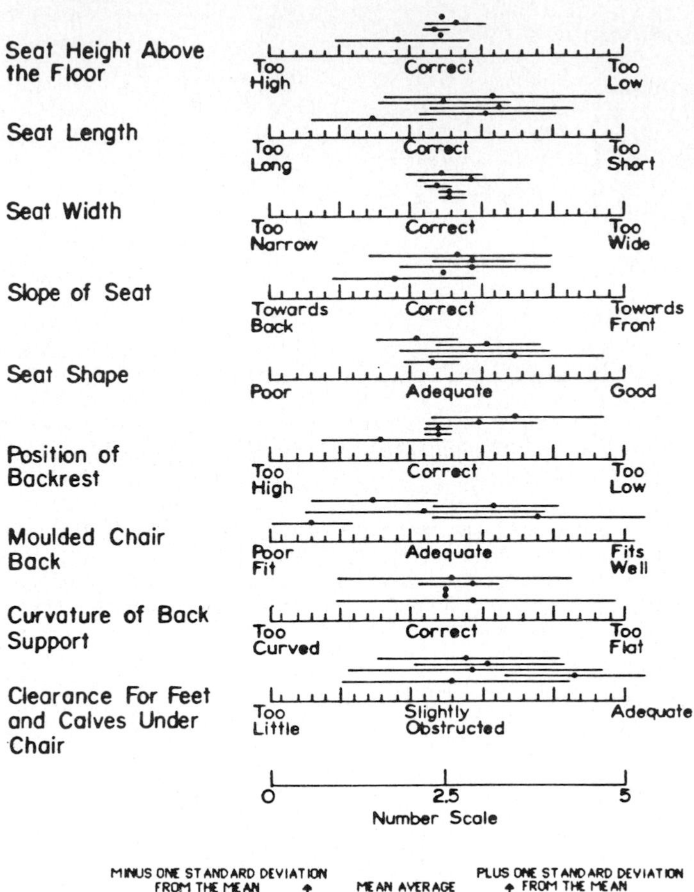
CHAIR FEATURE CHECKLIST
FEATURES:
Seat Height Above the Floor
Too High
Correct
Too Low
Seat Length
Too Long
Correct
Too Short
Seat Width
Too Narrow
Correct
Too Wide
Slope of Seat
Towards Back
Correct
Towards Front
Seat Shape
Poor
Adequate
Good
Position of Backrest
Too High
Correct
Too Low
Moulded Chair Back
Poor Fit
Adequate
Fits Well
Curvature of Back Support
Too Curved
Correct
Too Flat
Clearance For Feet and Calves Under Chair
Too Little
Slightly Obstructed
Adequate
0
2.5
5
Number Scale

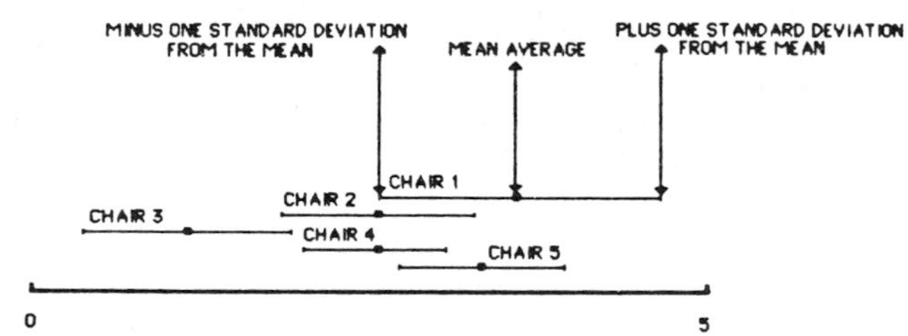
MINUS ONE STANDARD DEVIATION FROM THE MEAN
MEAN AVERAGE
PLUS ONE STANDARD DEVIATION FROM THE MEAN
CHAIR 1
CHAIR 2
CHAIR 3
CHAIR 4
CHAIR 5
0
5
THIS THE RATING SCALE WITH A RANGE OF 0 TO 5
THE SCALE IS PROVIDED WITH THE FIGURE

Figure 5

DISCUSSION AND CONCLUSION

This study validates the chair evaluation methodology described by Drury and Coury [5]. Drury and Francher [6] had used this method in evaluating a forward sloping chair and more recently Burri et al [7] had used the same in evaluating ten ergonomic chairs.

This investigation has shown that the evaluation procedure using the overall comfort rating, the body part discomfort rating and the chair feature check listing gives reliable and consistent results. Considering the fact that Figures 1 to 4 show identical differences among the chairs tested here reiterates the usefulness of this simple, and yet effective methodology for chair evaluation. As seen by the Tukey's range test on the various mean discomfort ratings the overall ranking of the chairs in increasing order of discomfort was chairs numbers 2, 4, 1, 3, and 5 respectively.

It is reasonable to expect that the overall comfort experienced in a chair is a complex function of the time of sitting, the task to be performed, and the comfort or the discomfort felt in the different body parts while sitting. The overall comfort was measured by the general comfort rating scale, while the Bishop Corlett [10] scale was used to measure the discomfort levels at the various body parts. It appears that among the five severity factors tested here, BPDS back predicts the overall comfort level best (r^2 = .85). Further if it is assumed that the perceptions of the chair features are related to comfort or discomfort felt in the various body parts, the BPDS back, in turn, is best predicted by length, width, shape of the chair and the shape of the back support (r^2 = .86). Other chair features do not seem to affect the extent of discomfort on back. It should be noted that the task was not a controlled factor here, but given that all the subjects were trained office secretaries of engineering departments, it is reasonable to assume that they would have had similar tasks.

Two other noteworthy findings are found from the Subject X Chair interaction for all the dependent variables and the Time X Chair interaction for BPDS back. Consistent subject differences and the Subject x Chair interactions indicate that individual preferences may be an important issue in deciding the best chair for a user task environment, among a set of reasonably good chairs. The Time x Chair interaction for the BPDS back is an interesting result, suggesting that longitudinal studies, and not short evaluation sessions, may be needed to give a reliable comparison among chairs.

In summary it can be said that:

1. The "GCR - BPD - CFC" evaluation procedure compares the chairs reliably and consistently.

2. Subject preferences may be an important factor in deciding on what is the best chair.

3. Among the nine features in the chair feature check list, the seat length, seat width, seat shape, and molded chair back were the important features.

REFERENCES

[1] Oxford, H. W., 1969, Ergonomics, 12.2, 140-161, Anthropometric data for educational chairs.

[2] Jones, J. C., 1969, Ergonomics, 12.2, 171-181. Methods and results of seating research.

[3] Grandjean, E., Hunting, W., Wotzka, G., and Scharer, R., 1973, Human Factors, 15, 3, 247-255. An ergonomic investigation of multipurpose chairs.

[4] McLeod, P., Mandel, D. R., and Malvern, F., 1980, The effects of seating on human tasks and perceptions, H. R. Poydar (ed.)., Proceedings of the Symposium Human Factors and Industrial Design in Consumer Products, Medford, Mass: Department of Engineering Design, Tufts University.

[5] Drury, C. G., and Coury, B. G., 1982, Applied Ergonomics, 13.3, 195-202. A methodology for chair evaluation.

[6] Drury, C. G., and Francher, M., 1985, Applied Ergonomics, Evaluation of a foreward sloping chair, 16.1, 41 - 47.

[7] Burri, G. J., Czaja, C. J., Drury, C. G., and Helander, M. G., 1987, A field evaluation of office chairs, In the Proceedings of the 31st Annual Meeting of the Human Factors Society, Santa Monica, CA, pp 1121 - 1122.

[8] Branton, P., 1969, Ergonomics, 12.2, 316-327, Behaviour, Body Mechanics and Discomfort.

[9] Shackel, B., Chidsey, K. D., and Shipley, P., 1969, Ergonomics, 12.2, 269-306, The assessment of chair comfort.

[10] Corlett, E. N., and Bishop, R. P., 1976, Ergonomics, 19, 175-182. A technique for assessing postural discomfort.

Trends in Ergonomics/Human Factors V
F. Aghazadeh (Editor)

DOUBLE CURVATURE BACKREST: RESULTS OF A PILOT STUDY

Anil MITAL

Ergonomics Research Laboratory
Mechanical and Industrial Engineering Department
University of Cincinnati
Cincinnati, OH 45221-0072, U.S.A.

A pilot study was conducted to test the hypothesis that if the area of contact between the lower back and the backrest can be maximized, the resulting stresses and the body fatigue will be minimized. Using four volunteers, 3 males and 1 female, four different backrests which maximized the area of contact, one for each person, were built in-house and tested against a standard (control) backrest. EMG and subjective fatigue data were collected on all four subjects. No significant differences between the standard backrest and the ones built for testing the hypothesis were found. The hypothesis, thus, could neither be rejected nor accepted. However, the study convinced the author to undertake a future investigation in which professionally built backrests, which maximized the area of contact, are used to test the hypothesis.

1. INTRODUCTION

Supporting the lower back area during seated work is considered essential for reducing musculoskeletal stresses and body fatigue. The backrest supports the lower back both while leaning forward, during work, and leaning backward, during relaxation. In the former case, if designed properly, the backrest supports, or is intended to support the lumbar region while in the later case it is intended to relax the back muscles (right trapezins, right latissimus dorsi, and left and right sacrospinalis - Lundervold (1958)).

While a slightly curved posture relaxes the muscles, an upright posture reduces disc pressures by keeping the spinal column in its natural shape. A periodic change in disc pressure, however, is essential in meeting the nutritional needs of the intervertebral discs (Kramer (1973)). Since individuals frequently lean forward and backward, while seated and working, the disc nourishment problem is not so acute. The general objective of a good backrest therefore is to reduce disc pressure by assisting the spinal column maintain its natural shape while at the same time pushing the trunk slightly forward, so that the weight of the trunk is in balance and back muscles are relaxed. It seems that both these goals, reduction in the disc pressure and reduction in the strain on the back muscles, are impossible to achieve simultaneously since a backrest supporting a forward bend cannot assist in holding the spinal column in its natural shape. Thus it appears that the problems of the sitting posture cannot be alleviated without frequently changing the posture itself.

The purpose of this pilot study was to test a new kind of backrest which assisted not only in keeping the spinal column upright but in allowing the trunk top to bend forward for reducing disc pressure. It was hypothesized that if the area of contact between the lower back and the backrest was maximized, the resulting stresses (disc pressure) and the body fatigue (muscle strain) would be minimum. The backrest tested had double curvature along the spine and along the back (figure 1).

2. BACKREST CONSTRUCTION

In order to produce double curvature backrests, casts of individuals' backs were made. The casts were produced by having the subject lie prone on the ground. A thin cotton sheet was used to cover the lower back region of the subject. The cotton sheet was then covered with a layer of clear polyethylene wrap to provide a surface on which the material, used to cast the backrest, could set. The polyethylene layer was then covered with paste wax to allow easy removal of the cast. Once the paste wax application was complete, cast material (ordinary automative body putty) was applied. The putty was applied to about 19 mm thickness and was spread from about 5 cm above the coccyx to the lower third rib. After approximately 10 minutes, when the putty had set sufficiently, the cast was removed and allowed to set completely. Once completely set, any irregularities (holes, etc.) were covered with modeling clay. A plaster cast (38 mm thick) was then made from the putty cast. Once set, the irregularities of the plaster cast were sanded off and it was coated with lacquer to provide a smooth surface.

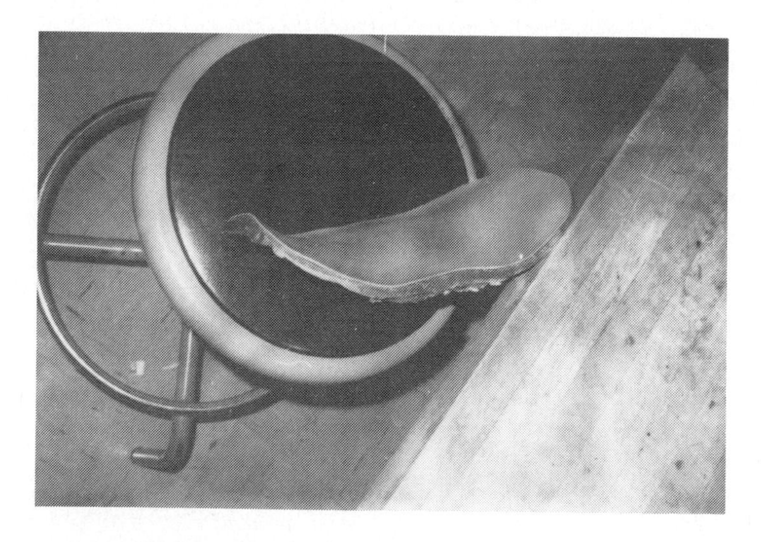

Figure 1. Double Curvature Backrest.

Once the plaster cast was ready, a dam was built around it with modeling clay and the surface itself was coated with paste wax for easy separation. Fiberglass resin and cloth were then applied in layers on the plaster until a thickness of approximately 9.5 mm was obtained. Twenty-four hours were allowed for the fiberglass to set. Once set, the fiberglass casts were separated, cleaned and smoothened. Finally, the fiberglass casts were covered with padded vinyl about 2 mm thick.

3. EXPERIMENTAL PROCEDURE

In order to determine the muscular strain resulting from the double curvature backrests, they were mounted, one at a time, on a standard office chair. The adjustable backrest of this office chair was the control backrest. The angle of the backrest in both cases, double curvature and standard, was adjustable and such that the backrests were constantly in contact with the lower back. In case of the double curvature backrest, this ensured that the natural shape of the spine in the lower thoracic region and lumbar region was maintained. For each subject, the double curvature backrest, that was unique to his/her back, was used; the standard backrest was the same for everyone.

The EMG activity of the subjects in the back muscles (right and left erector spinae) at the L3 level were recorded with the help of bipolar surface electrodes. The electrodes were placed approximately 3 cm from the midline of the back (Colombini et al. (1986)).

Perceived comfort was evaluated using a subjective questionnaire based on Shackel et al. (1969), Drury and Coury (1982) and Congleton et al. (1985). Each person completed the questionnaire, while seated, for each of the two backrests. The response to a question was either poor, adequate, or excellent. Using a random order, subjects sat in the chair with either the double curvature backrest or the standard backrest. The duration of experiment was one hour per condition (2 conditions per subject).

4. RESULTS AND DISCUSSION

The EMG activity of the back muscles remained unchanged with time for both backrests. The effect of time, as determined by the analyses of covariance, was insignificant ($p \geq 0.05$). The mean EMG values for the two backrests were also not different from each other (at the 5% level of significance). The muscle strain in both cases, double curvature backrest and standard backrest, thus, was the same.

Since the double curvature backrest ensured that the lower spine region, lumbar and lower thoracic regions, maintained its natural shape, the disc pressure was also presumed to be lower compared to the standard backrest in which case the upper edge of the pelvis rotated backward (backward rotation of the pelvis leads to greater disc pressures (Nachemson (1974)).

The double curvature backrest, thus, minimized both the disc pressure and the muscle strain.

Table 1 Subjective Questionnaire Responses for the Standard Backrest

	POOR	ADEQUATE	EXCELLENT
Upperback	0	3	1
Midback	0	1	3
Lowerback	0	3	1
Upperarms	0	2	2
Breathing	0	1	3
General	0	2	2
Width	0	4	0
Height	0	3	1
Horizontal Radius	0	2	2
Vertical Radius	0	2	2
Flexibility of Spine	1	3	0
Flexibility of Arms	0	4	0
Ability to Relax	2	2	0
Comfort	2	2	0
Ability to Provide Support			
Without: Stiffness	2	0	2
Numbness	1	0	3
Soreness	1	0	3
Pain	1	0	3
WEIGHTED TOTALS	-10	0	28
OVERALL WEIGHTED TOTAL	18		

NOTE: Weight = -1, for Poor; 0, for Adequate; 1, for Excellent.

Tables 1 and 2 summarize the answers to the subjective questionnaire for the standard and double curvature backrests, respectively. The overall comfort rating was 18 for the standard backrest and -5 for the double curvature backrest. It was felt that the low comfort rating for the double curvature backrest was caused primarily by a lack of padding, compared to the standard backrest, as well as the poor quality of fabrication. In spite of these shortcommings, the double curvature backrest was able to provide support to the back with no perceived pain. The experience with the double curvature backrest suggests that if a professionally built backrest is employed, the subjective comfort rating for it will be higher than the rating for the standard backrest.

5. CONCLUSION

A double curvature backrest was tested in this pilot study. The EMG data indicated there was no difference in the muscle strain between the double curvature backrest and the standard backrest which permits greater forward bending. The close contact between the double curvature backrest with the back ensured that the spine retains its natural shape. The disc pressure, thus, was minimized. Subjective comfort ratings suggest tha a longer study, with professionally built double curvature backrests, should be conducted.

Table 2 Subjective Questionnaire Responses for the Double Curvature Backrest

	POOR	ADEQUATE	EXCELLENT
Upperback	1	2	1
Midback	2	2	0
Lowerback	1	2	1
Upperarms	0	2	2
Breathing	0	1	3
General	1	3	0
Width	1	3	0
Height	1	2	1
Horizontal Radius	1	3	0
Vertical Radius	2	2	0
Flexibility of Spine	2	2	0
Flexibility of Arms	0	4	0
Ability to Relax	4	0	0
Comfort	1	3	0
Ability to Provide Support			
Without: Stiffness	4	0	0
Numbness	1	0	3
Soreness	1	0	3
Pain	0	0	4
WEIGHTED TOTALS	-23	0	18
OVERALL WEIGHTED TOTAL	-5		

NOTE: Weight = -1, for Poor; 0, for Adequate; 1, for Excellent

REFERENCES

[1] Colombini, D., E. Occhipinti, C. Frigo, A. Pedotti, and A. Grieco, Biomechanical, Electromyographical and Radiological Study of Seated Postures, The Ergonomics of Working Postures (Eds.: N. Corlett, J. Wilson, and J. Manenica) (1986) 331-344.

[2] Congleton, J.J., M.M. Ayoub and J.L. Smith, The Design and Evaluation of the Neutral Posture Chair for Surgeons, Human Factors 27 (1985) 589-600.

[3] Drury, C.G. and B.G. Coury, A Methodology for Chair Evaluation, Applied Ergonomics 13 (1982) 195-202.

[4] Kramer, J., Biomechanische Veranderungen im Lumbalen Bewegungssegment, Hippokrates, Stuttgart (1973).

[5] Lundervold, A., Electromyographic Investigations During Typewriting, Ergonomics 1 (1958) 276-233.

[6] Nachemson, A., Lumbal Intradiscal Pressure - Results from In-Vitro and In-Vivo Experiments With Some Clinical Implications, 7, Wiss. Konf., Deutscher Naturforscher and Aerzte, Springer, New York (1974).

[7] Shackel, B., K.D. Chidsey, and P. Shipley, The assessment of Chair Comfort, Ergonomics 12 (1969) 269-306.

VII

ENVIRONMENTAL STRESSES

Trends in Ergonomics/Human Factors V
F. Aghazadeh (Editor)
Elsevier Science Publishers B.V. (North-Holland), 1988

COMPARISONS OF AIR VS. LIQUID MICROENVIRONMENTAL COOLING FOR PERSONS PERFORMING WORK WHILE WEARING PROTECTIVE CLOTHING

Phillip A. Bishop

The University of Alabama
Tuscaloosa, AL, USA

Sarah A. NUNNELEY, John R. GARZA, and Stefan H. CONSTABLE

USAF School of Aerospace Medicine, VNC
Brooks AFB TX, USA

Work productivity of personnel performing physical labor is compromised by the thermal burden imposed by the use of certain types of protective clothing. One strategy for increasing work capacity under these conditions involves supplying personal cooling during scheduled work breaks. Previous findings [Constable, S., et al. Aviat. Space and Environ. Med. 58(5):495, 1987] suggest that work capacity may be doubled in hot environments with this approach. This study compared the physiological responses of personnel wearing the U.S. military chemical protective suit whose torsos were cooled using a closed-circuit liquid cooling system, with their responses to an open circuit cool air system. Eight subjects (5M, 3F) worked repetitive cycles consisting of 30 minutes of treadmill walking at ≈40% of VO_2 max (≈ 500 watts) followed by 30 minutes of rest in environmental conditions of $38^o/26^o/43^o$ C (dry bulb, wet bulb, globe). Liquid cooling was supplied to a vest at 0.8 - 1.0 liter per minute with an inlet temperature between 10° and 15° C. Air cooling was provided to a chest and middle back distribution system at approximately 500 liters per minute at vest inlet temperatures from 15° to 20° C. A small amount of this air was diverted to the face mask. Work-rest cycles were repeated for four to six hours. Selected means and standard errors are shown below. None of the differences shown between liquid and air cooling were statistically significant.

	Liquid	Air
work time (min)	143 (9)	143 (9)
skin temp (^{o}C)	28.4 (1.1)	29.7 (0.8)
peak rectal temp. work (^{o}C)	38.2 (0.2)	38.0 (0.2)
minimum rectal temp rest (^{o}C)	37.7 (0.1)	27.4 (0.1)
sweat prod. (ml/min)	10.4 (1.9)	8.9 (1.2)
sweat loss (ml/min)	6.0 (1.2)	5.9 (0.4)

*P. Bishop was sponsored by AFOSR/AFSC, United States Air Force under contract F49620-85-C-0013.

The reduction in rectal temperature during rest (-0.6 C) was identical for both systems. Each system offers certain advantages and both systems substantially increased work capacity in comparison to rest without cooling. However, whereas a relatively stable cyclic state was achieved with cooling, early cumulative fatigue could not be fully mitigated.

1. INTRODUCTION

In some industrial and military situations, personnel must perform physical labor in toxic or hazardous environments. These situations require that protective clothing be worn. Because of the inherent design of many protective garments, material clo and im/clo values are often high. If ambient temperatures are warm and metabolic heat production is high, the protective suit can significantly compromise the body's normal cooling mechanisms, and the resultant heat storage can seriously restrict total work capacity.

One of the most effective strategies for ameliorating this problem is the employment of a personal cooling system. Approaches to personal cooling have often employed either closed-circuit liquid, or open-circuit air systems. Both liquid and air systems offer certain particular advantages. In an earlier review, Nunneley [1] pointed out a general practicability for the liquid approach. More recently, other researchers [2, 3] have made some initial comparisons between liquid and air personal cooling for the relief of heat stress. Wissler [4] has addressed some of the theoretical considerations of these systems with the use of a complex physiological model. From previous research it might be concluded that the employment of chilled air systems may be preferred for certain applications.

Improvements in technologies involved with the man-machine interface of these systems have improved their physiological cooling capacities. This is of critical importance and each system requires individual, laboratory evaluations and comparisons that precisely simulate the projected application. Recently, Constable et al. [5] have shown a U.S. Air Force developed, liquid cooling system to be effective when used in an intermittent fashion. Intermittent cooling (while subjects rested) was investigated in the present study because it is much simpler and more practical than attempting to continuously cool individuals while they are active and mobile. The "ambulatory" approach, which requires the user to carry the entire cooling system while working, is not currently practical on a large scale basis. The purpose of our study was to directly compare the physiological efficacy of intermittent liquid and air microenvironmental cooling for the reduction of heat stress in subjects performing heavy work under thermally stressful conditions. This type of comparison should be valuable in providing information that would facilitate the application of the intermittent cooling concept.

2. METHODS

2.1 Subjects

Subjects for this study were five male and three female volunteers.

Informed consent was obtained from all subjects. The mean subject physical characteristics are presented in Table 1. Surface area was calculated from a Dubois equation nomogram 6 .

Table 1. Physical Characteristics of Subjects and Energy Costs. Means ± (s.e.);(N = 8)

Age (yrs)	Height cm	Weight (kg)	Surface Area (m^2)
36 (3.0)	174 (3.0)	71.7 (3.8)	1.85 (.07)
Max $\dot{V}O^2$ (l/min)	**Work $\dot{V}O_2$ (l/min)**	**% Max $\dot{V}O_2$ (%)**	**Max HR (bt/min)**
3.62 (.33)	1.42 (.09)	39.1	184 (9.6)

2.2 Work Task

The work task used for all tests consisted of walking 1.34 m/s (3 mph) at either three or six percent grade. The selected work load elicited approximately 40% of maximal aerobic capacity when the subject was wearing the protective ensemble. Subjects performed the exercise in an environmental chamber where the conditions were $38^0/24^0/44^0$ C for dry, wet, and black globe temperatures, respectively. The work/rest cycles consisted of 30 minutes of walking followed by 30 minutes of sitting. Personal cooling was applied only during the periods of rest. Rest periods were conducted inside the chamber with minimal exposure to radiant heat. These work/rest intervals were based upon pilot work at the School of Aerospace Medicine which suggested that this ratio was effective in minimizing cumulative heat storage, while maximizing work time, under these specific conditions.

2.3 Clothing

The protective clothing (PC) worn was the U.S. Armed Forces Chemical Defense Ensemble, which consists of a protective mask, hood, charcoal impregnated jacket and trousers overgarments, and rubber gloves with cotton liners. Although the ensemble includes rubber overboots, these were omitted, primarily for safety concerns associated with the treadmill walking. A military fatigue shirt and trousers and cotton socks were worn underneath the PC. The insulative value for this ensemble was clo 2.55 and im/clo 0.28. The air cooling system also supplied 80 liters per minute of cool air to the protective mask. Also, a light cotton undershirt was normally worn under the air cooling vest.

2.4 Cooling Systems

The air cooling system has been described elsewhere [2]. Air was supplied to the vest from a U.S. Army designed cooling unit which produced 15°C air at approximately 500 liters per minute. Outlet air was directed through a flow meter for measurement purposes. The vest received 85% of the air flow with the remainder directed to the face mask. Subjects who felt uncomfortably cool were permitted to vent some of the chilled air directly to the environment, thus reducing the potential cooling power delivered to the body. Subjects were seated during all cooling sessions.

The liquid cooling system consisted of a vest with 48 meters of tygon tubing in a slip liner covering approximately 0.5 m of body surface area (ILC Dover, Frederica, DE). The vest was snug fitting and extended from the shoulders to the umbilicus, also covering approximately 24 cm of the upper arm. This cooling garment is essentially the top half of the one worn by U.S. NASA astronauts during extra-vehicular activities. Water in a mixture with five percent propylene glycol was supplied at a rate of 0.8-1.0 liters per minute to the vest. The coolant inlet temperature ranged from 10° to 15° C and was delivered by a specially designed cooling system developed by the U.S. Air Force.

Pre and post, nude, and fully clothed body weights along with fluid intake (ad libitum during rest) were measured. Rectal, chest, and thigh temperatures were recorded every 30 seconds by a custom, computerized data collection system.

2.5 Statistical Analysis

Mean differences in the physiological responses to liquid and air cooling were tested for significance using one-way repeated measures ANOVA. An alpha of $p < .05$ was used in significance testing.

3. RESULTS AND DISCUSSION

The primary focus of this research was to compare two forms of intermittent personal cooling. The feasibility of using this approach was initially characterized at the USAF School of Aerospace Medicine using liquid microenvironmental cooling under stressful conditions [5]. The relatively high specific heat of water as a cooling medium implies that this system will have a higher potential capacity for physiological cooling than air. Alternatively, other researchers have empirically noted the large potential capacity of an air cooling system to remove body heat under thermally stressful conditions [7]. Data shown in Figures 1 and 2 suggest that in practice both systems performed well, at least under these conditions. In fact, there was a trend for the air cooling to attenuate distortions in body core temperatures and heart rates slightly more relative to liquid cooling from cycle to cycle. Although, the physiological significance of this trend may be small, it can be seen that both systems greatly reduced the amount of heat stress incurred compared to control conditions (i.e., no cooling). Work capacity was at least doubled, when auxiliary cooling was added. It is important to note that the number of data points was decreased after two hours in the control condition and beyond four hours when cooling was employed.

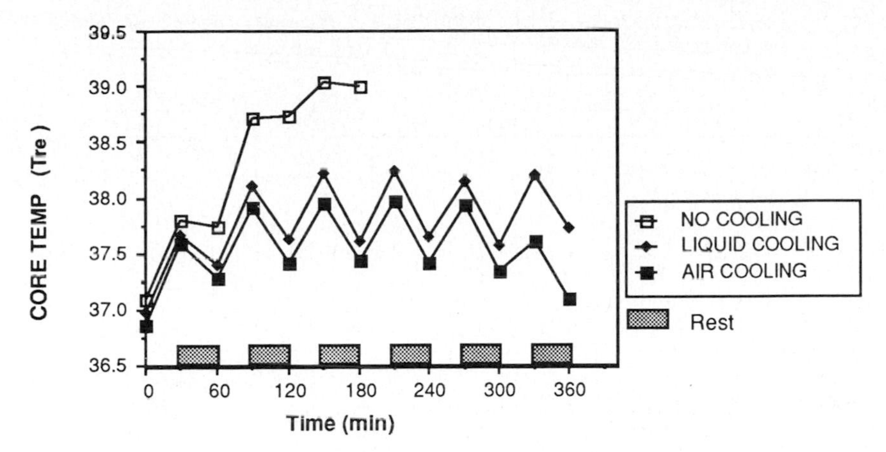

Figure 1. Mean Rectal temperature (Tre) response to intermittent work

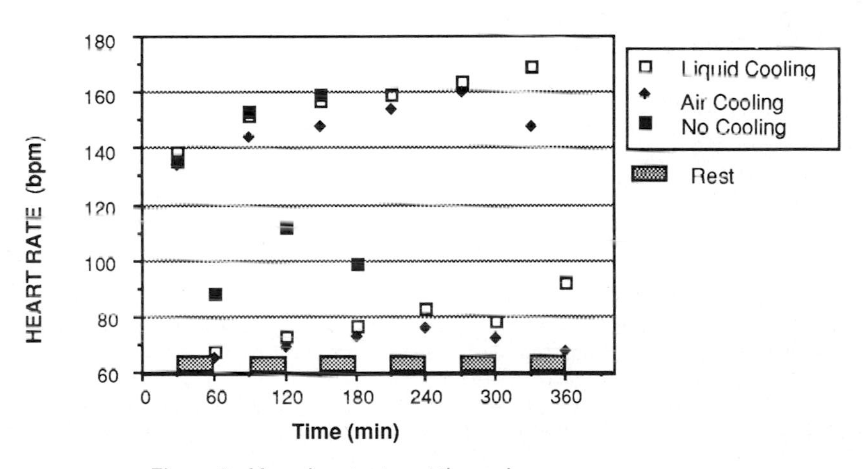

Figure 2. Mean heart rates at the end of each work and rest cycle

A statistical comparison was made of certain physiological parameters after the last work and rest cycles (see Table 2). The only significant difference between the cooling perturbations was observed for thigh skin temperature. It appears that the air cooling was slightly more effective using this measure. Overall, air cooling seemed to attentuate the physiological stress as well as the liquid cooling approach.

Table 2. Final Physiological and Rating of Perceived Exertion (RPE) Observations. Means ± (s.e.) of subjects.

	Final "Work"		Final "Resting"	
	Liquid	Air	Liquid	Air
Rectal (T_{re} °C)	38.2 (0.1)	38.0 (0.2)	37.7 (0.1)	37.4 (0.1)
Chest (T_{sk} °C)	37.4 (0.2)	36.9 (0.3)	28.3 (0.9)	29.4 (0.7)
Thigh (T_{sk} °C)	37.2 (0.3)	36.8 (0.2)	36.9 (0.3)*	35.6 (0.3)
Heart Rate (bt/min)	164 (5.8)	156 (7.5)	86 (4.5)	74 (6.7)
RPE Scale (6-20)	13.6 (1.0)	14.9 (0.8)	--	--

*Indicates significant difference between liquid and air cooling, $p < .05$.

It would have been expected that the evaporative cooling component would have been significantly higher with air cooling. In some manner, the air system more than compensated for its inherently low convective cooling capacity. Surprisingly, there appeared to be little difference in the effects of the type of cooling on sweating rates or sweat evaporation (see Figure 3). Although most of the subjects did, in fact, sweat less when air cooling was used and evaporated a higher percentage of their perspiration (66% evaporated for air cooling vs. 57% for liquid cooling) this was not the case for every subject. The lack of a demonstrated statistical difference between cooling approaches for these two variables is difficult to explain. Personal observations made by the investigators indicated that the subject's skin and clothing was drier at the end of the experiment when air cooling was used. Furthermore, most of the subjects indicated that they felt drier with the air cooling. There is some error in these measurements, as some individuals sweated so profusely that there was obvious "drippage" onto the floor as the garments became soaked through. Further investigation regarding sweat production and loss is warranted.

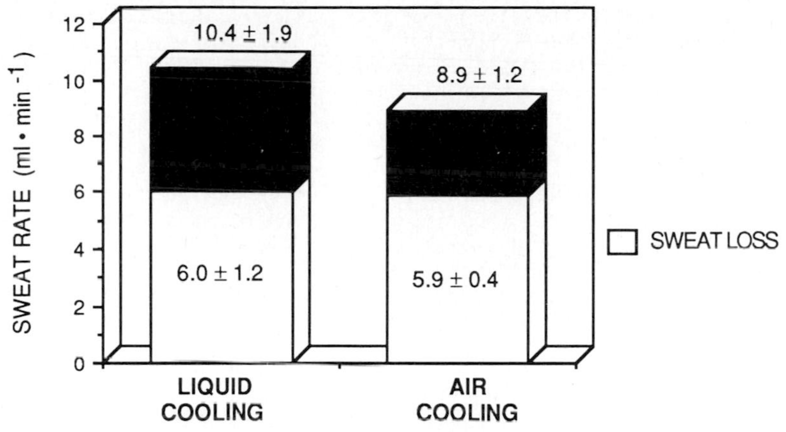

FIGURE 3. Sweat production and loss for liquid and air cooling ($\overline{X}$ ± s.e.)

Shapiro et al. [2] compared air-cooled and water-cooled systems in contrived "tank crews" performing various work tasks under hot-wet and hot-dry conditions. Under hot-wet conditions, the air-cooling produced significantly lower heart rates and higher sweat evaporation rates than the water-cooling. In hot-dry conditions, only the sweat evaporation rate was significantly higher for air-cooling. Epstein et al. [3] also compared air and water cooling of resting subjects in a hot-dry environment. They found that both systems produced similar sweat rates. The observed strain index (product of heart rate, change in rectal temperature and sweat rate) was 25% lower (more favorable) with the air-cooled system. Statistical differences were not reported.

These two previous studies and the present study employed different exercise regimens and cooling systems. Though there were large differences in experimental design, the differences among physiological responses to air and liquid systems were relatively small. The most obvious discrepancy between our findings and previous findings was the small difference in sweat evaporation between the two systems.

4. CONCLUSIONS

Under the conditions studied here, air and liquid cooling systems were equally effective in reducing body temperatures during intervals of rest. Air cooling was preferred by most subjects and appeared to produce drier clothing. Although the data showed little difference in sweat evaporation rates between the two systems, we conclude that both liquid- and air-cooled systems have good potential for application to problems involving work in heat while wearing protective clothing.

REFERENCES

[1] Nunneley, S. A. Water cooled garments: A review. Space Life Sciences (1970) 2:335-360.

[2] Shapiro, Y., K. B. Pandolf, M. N. Sawka, M. M. Toner, F. R. Winsmann, and R. F. Goldman. Auxiliary cooling: comparison of air-cooled vs water-cooled vests in hot-dry and hot-wet environments. Aviation, Space, and Environmental Medicine (1982) 53(8):785-789.
[3] Epstein, E., Y. Shapiro and S. Brill. Comparisons between different auxiliary cooling devices in a severe hot/dry climate. Ergonomics (1986) 29(1):41-48.
[4] Wissler, E. H. Simulation of fluid cooled or heated garments that allows man to function in hostile environments. Chemical Engineering Science (1986) 41:1689-1698.
[5] Constable, S. H., P. A. Bishop, S. A. Nunneley, and T. Chen. Personal cooling during resting periods with the chemical defense ensemble. Aviation, Space, and Environmental Medicine (1987) 58(5): 495.
[6] Dubois, D., and E. F. Dubois. Clinical calorimetry X. A formula to estimate the surface area if height and weight are known. Archives Internal Medicine (1961) 17:683.
[7] Pimental, N. A., H. M. Cosimini, M. N. Sawka, and C. B. Wenger. Effectiveness of an air-cooled vest suing selected air temperature and humidity combinations. Aviation, Space, and Environmental Medicine (1987) 58:19-24.

Trends in Ergonomics/Human Factors V
F. Aghazadeh (Editor)
Elsevier Science Publishers B.V. (North-Holland), 1988

BIOPHYSICAL EVALUATION OF HANDWEAR FOR COLD WEATHER USE BY PETROLEUM (POL) HANDLERS

W.R. SANTEE[1], T.L. ENDRUSICK[1] and L.P. WELLS[2]

[1] U.S. Army Research Institute of Environmental Medicine
Natick, MA USA 01760

[2] U.S. Army Natick Research, Development and Engineering Center
Natick, MA USA 01760

This study addresses the question whether a prototype petroleum, oil and lubricant (POL) handler's glove provided adequate protection for cold weather use. A pumping simulator fitted with an actual fuel pump handle which operated a timer was constructed. Subjects wore the new Extended Cold Weather Clothing System (ECWCS) uniform and the prototype gloves. Subjects entered the chamber and stood quietly for 10 min, then began a pump/rest cycle consisting of 6 min of "pumping" followed by a 1 min rest. The simulator timer operated only when the pump lever was fully compressed against a resistance of 5.7 kg. Measurements consisted of finger temperatures for the pumping and non-pumping hand, and total pumping time for a thirty minute period. The initial test environment was -34.4°C(-30°F) and 2.2 $m \cdot s^{-1}$ (5 mph) "wind". Initial results indicated that endurance time in the -34.4°C environment barely exceeded the 10 min baseline. The temperature was raised to -28.9°C (-20°F) for all subsequent tests. Final results show a mean "pumping" time of 17 min (n=6). The prototype glove was determined to meet the basic 6 min pumping requirement at -28.9°C.

1. INTRODUCTION

Even in the absence of effects on core body temperature (hypothermia), the physical damage and reduction in morale associated with cold injury is of considerable significance [13,15,16]. Loss of personnel due to cold injury is a significant manpower consideration [7,9,13,15]. Industrial petroleum, oil and lubricant (POL) handlers working under adverse climatic conditions may encounter the same problems of exposure and cold injury, but the military situation may be compounded by the rapid mass deployment in a military emergency of personnel lacking any prior experience in cold weather operations. In addition, under combat conditions, fueling facilities may be temporary, and shelter for workers limited. The military scenario therefore represents a worst-case situation relative to industrial requirements [9,15].

Discomfort and cold injury are experienced initially in the extremities of individuals engaged in outdoor cold weather activities. Therefore, adequate hand

and footwear assume a disproportionate importance in the selection of cold weather clothing. This report concerns the testing of handwear for a specific task, cold weather POL (petroleum, oil and lubricant) handling. In the process of presenting the specific results from the study, the basic principles and limitations of handwear testing and selection will be presented.

As a first step in comparing different clothing items to determine gross relative potential insulation, the Biophysics branch, USARIEM uses a heated copper hand model. Our regional controlled model measures the insulation of different zones of the handwear, thereby isolating areas of inadequate insulation or construction without the difficulty or exposure to potential discomfort or injury of human subjects. Physical models can eliminate deficient handwear as a first step in selecting adequate handwear.

Model testing is however, entirely static and the surface temperature is uniform and maintained at a single constant set point. For example, a tight fit on the model may compress the insulation while a looser fit traps more dead air, increasing the effective insulation. Actual surface temperatures vary by hand region, and change as muscular activity and blood flow change the energy input into various hand regions. When the subject is actually performing a task, such as pumping fuel or tightening a pipe connection, the stresses and general variability increases relative to simply maintaining one rigid position.

In a dynamic state, handwear performance is dependent on the performance of the total uniform system worn in the given environment. When the body is subjected to general cold stress, a primary response is to maintain central body temperature by restricting heat transfer to the peripheral extremities by the mechanism of vasoconstriction [5]. Hence, if the overall uniform system provides insufficient insulation to conserve the core body temperature, the hands and feet will receive less heat via blood flow and drop in temperature.

The uniform worn for the test was a new Extended Cold Weather Clothing System (ECWCS). The uniform system [9] is layered; starting with polypropylene underwear, an inner clothing layer of trousers and shirt, a fiber-pile jacket, polyester batting liners and outer shell trousers and parka made of a "breatheable" polytetrafluoroethylene (PTFE) laminate. The outer shell is waterproof, providing protection from moisture and fuel spills. Vapor barrier boots, winter weight socks and a balaclava or pile cap complete the uniform system.

A single, static model determination of dry insulation should not be the entire basis for determining if military handwear provides adequate cold protection. The preceding anthropomorphic variables can only be determined by testing with human subjects. After selection and development of prototypes based on model testing results, the next testing phase is human testing in an environmental chamber. Subjects are instrumented and perform specified tasks in a controlled set of environmental conditions. Chamber testing provides a quantitative measurement of human performance under a specific set of conditions.

During the chamber tests, many functional aspects of the handwear are determined because of the dynamic nature of actual wear as opposed to static testing. However, the limited, controlled activities in the chamber do not completely predict the adequacy of the handwear for actual use in the field. In actual field use, the handwear is subject to more varied stress, more environments, wetting of insulation by sweat or water, soiling and general wear. To ensure that the handwear is exposed to extreme conditions and use, it is necessary to issue a number of gloves for several months to personnel at several locations. If the human chamber tests are by-passed and a prototype is tested only by limited field issue, it is possible that the inadequacy of the prototype will be exposed at the cost of multiple cold-injuries. The other principal shortcoming of outdoor field tests is that the physical environment is neither predictable, nor controllable; whereas in a chamber the environment can be specified and replicated on demand.

Whenever possible, a testing program for protective clothing, whether handwear or complete environmental suits, should follow the following sequence; static testing with a biophysical model, dynamic human tests in a controlled environment, and limited field issue.

A consideration in handwear design is the difference between gloves and mittens. In a traditional glove, all five hand digits are separated. In a traditional mitten, only the thumb is isolated. The prototype POL four finger glove is a compromise that attempts to combine some of the heat conserving advantages of mittens with the increased dexterity of gloves. The prototype glove is a Navy design. The "glove" is actually a three layer ensemble; a neoprene moisture and lubricant resistant outer shell, an inner liner of foam insulation and finally a standard five-finger wool glove liner. An alternative liner with micro-fiber insulation was also developed for the test, but rejected when fitting tests determined that the pattern delivered fit the fingers too snugly.

2. METHODOLOGY

This study consisted of two phases; preliminary studies and the full scale chamber study. The preliminary phase was improvised, utilizing available "cold chambers", a field portable data logger, and staff as volunteer subjects. The objectives of the preliminary studies were primarily to determine the best chamber temperature for the formal test and to develop procedures for running the simulated pumping.

The results of the pilot studies and a field test report [2] lead to the selection of an initial test chamber temperature of -34.4°C (-30°F). A combination of a cool dressing room 15°C (60°F) and the 10 min baseline period was intended to prevent sweat from accumulating in the clothing and to dissipate any excess heat stored prior to entering the chamber.

Seven male subjects were recruited after receiving a detailed orientation on test procedures and potential hazards and signing an informed consent volunteer statement. The tests were organized into a sequence of events; harnessing and dressing, chamber entry, 10 min chamber baseline, 30 min pump rest cycles, chamber exit and recovery and undressing. Skin surface temperatures were collected at 11 points; including the nail beds of the middle and little fingers on both the left and right hand. Subjects were encouraged to wear only the components of the ECWCS uniform necessary to be comfortable until just prior to entering the chambers to avoid sweating or excess heat storage.

Only one performance criterion was specified; completion of six minutes of simulated fuel pumping [17]. To determine if sustained pumping was possible, the protocol specified that subjects would "pump" for a cumulative 6 min, rest for 1 min, then repeat the cycle until 30 min of activity had elapsed or the subjects were removed. A subject was removed from the chamber when his finger reached a temperature of 5°C, a monitor removed the subject or the subject voluntarily removed himself.

A pumping simulator was designed and assembled for the chamber testing. To eliminate differences in subject height, the pump handle mount was designed to be adjustable for both angle and height. Each pump was mounted on a separate stand with start and stop signal lights and an audible alarm. Fuel pump handles are typically not standardized, but the most common in the military are 3.8 cm (1.5 inch) size. One 2.5 cm (1 inch) fuel handle was too small to fit a hand with the POL glove inside. A special kit that adapted a 1 inch handle to a D-handle operating lever was used. The switch on each pump was connected by cable to a central timing console that had a digital display of total elapsed time and cumulative pumping time. The effort required to compress the pump handle to operate the switch was 5.5-5.9 kg.

Subjects entered the test chamber, and assumed a position facing the pump stands with their back to the chamber air flow. The instrument leads were connected to the data acquisition system and the baseline period began. The initial test condition was -34.4°C (-30°F) and 2.2 $m \bullet s^{-1}$ (5 mph). Four pump "configurations" were tested; the improvised D-handle remote kit, the normal uninsulated handle, a "field-improvised" insulated handle, and a simulation of a pump fitted with a hold-open device. Pump handles were insulated by taping two thin layers (approx. 3.5 mm each, uncompressed) of closed cell foam to the lever and handle of the nozzle. The method of insulation could be considered a simple, field expedient improvisation. The final condition was simulated by simply gripping the handle without operating the lever, equivalent to operating the pump with a hold-open device.

3. RESULTS

Formal testing was conducted over a 5 day period. During the first two days of testing, when test chamber air temperatures were at -34.4°C (-30°F), the majority of test subjects reached the 5°C (41°F) limit in less than 10 min. Subjects had complained to monitors regarding the pre-test conditions, but did not demonstrate shivering or other overt signs of cold stress. Dressing room temperatures were raised from 15°C (60°F) to 21.1°C (70°F) for the third test day. The chamber air temperature was increased to -28.9°C (-20°F) for the remainder of the tests.

The total chamber time (10 minute baseline plus pumping) and actual pump operation time for all test days are summarized in Table I. Repeated measures ANOVA analysis for test days 3,4,5 for n=5 determined no significant differences between the three different pump configurations.

Table I. Mean chamber endurance and pumping operation times (-28.9° C)

Day	pump	endurance	s.d.	pumping time	s.d.	n
3	1.5" pump	33:08	5:55	17:05	5:38	6
		(33:06)	(6:37)	(16:54)	(6:16)	(5)
4	insulated	27:51	11:34	14:23	9:26	5
5	"hold-open"	26:45	9:54	14:48	8:40	5

The modified 1 inch pump handle, which utilized a bolt-on modification which allowed the use of a large D-handle was considered a significant improvement by the majority of the subjects. Test measurements are insufficient to quantitatively support the subjects' opinions. Only 3 subjects were able to simulate pumping with the grip modification kit, but this was interpreted as a function of the thermal environment rather than the stress of pumping.

Two subjects complained about the increase in the grip diameter on the insulated handle and attributed the reduction in their pumping performance to the increased grip size. They also expressed an opinion that the insulation was an improvement in terms of heat loss through the hand.

The test results (Figure 1) clearly demonstrate the oscillation in right hand skin temperatures due to the alternating pump and rest cycles approximating a 6-7 min recovery for most subjects.

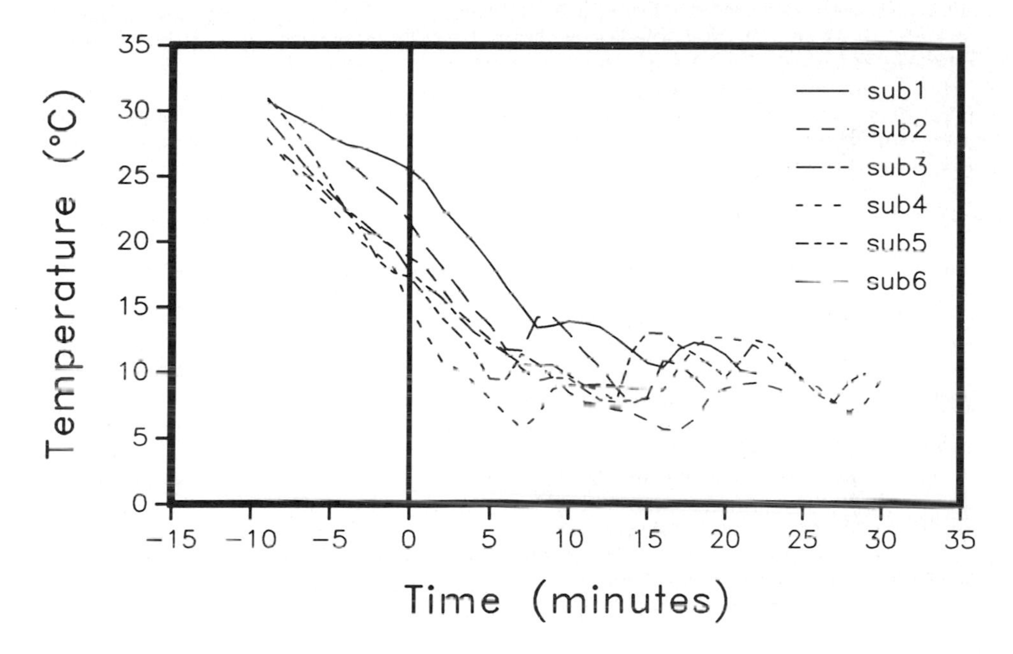

Figure 1. Middle finger temperature (°C) versus time (min)

In both test environments the neoprene shell and foam liner became less flexible when exposed to the cold, even while being worn by the subjects. As a result the gloves were effectively bulkier and more difficult to manipulate during pumping in the cold than at room temperature.

The subjects were all oriented with their backs to the wind, so facial exposure was reduced relative to field conditions. By properly utilizing the snorkel parka, the subjects had little trouble with cold in the face or nose regions. Some frost accumulated inside the front of the hood for some subjects. No subject complained of cold feet.

4. DISCUSSION

The pattern of oscillating finger temperatures is described as a "pseudo-hunting" reaction because of the similarity to the fluctuations in surface temperature in cold stressed appendages caused by alternating periods of vasoconstriction and vasodialation [6,14]. The pattern demonstrated by the test subjects is caused by a similar variation in blood flow. Tightly squeezing the pump lever reduces blood flow to the fingertips and during the rest period blood flow increases to the hand. These alternating flow rates mimic the hunting reaction also referred to as cold-induced vasodilation (CIVD). The situation is complicated however by the muscle heat generated by the pumping activity which may offset or moderate the effects of reduced blood flow.

The test conditions were more rigorous than normally experienced in the field. Subjects were required to stand for 10 minutes prior to starting the simulated pumping action. Subjects were required to fully compress the pump lever to operate the timer and to use only one hand to pump. During current actual field operations, pump operators emerge from a heated shelter and begin pumping immediately. A glove which passes the minimum requirement under the stated test should be expected to perform adequately under actual field conditions in an equivalent environment.

Of interest was the extremely rapid finger cooling observed for the majority of subjects on the first two test days (-34.4°C). In our preliminary report, the difference was associated with the difference in chamber temperatures. The results agreed with a field report [2] conducted at -39°C(-38°F) and preliminary work in our lab that indicated that pumping performance would decline at -28°C relative to previous work at -20°C. Heat loss is directly related to the product of the combined heat transfer coefficients and the temperature gradient. A combination of a large thermal gradient, psychological reaction to the intense cold ("cold anxiety") and accompanied adrenergic response, followed by the abrupt onset of vasodilation could have caused an excessively rapid temperature drop. During the test period, we assumed that was the case and moderated pre-conditions to reduce subject "cold anxiety." However, when the data were plotted, it becomes apparent that the initial finger temperatures were too low for the majority of subjects. The most reasonable explanation is that finger temperatures were already lower for some subjects prior to entry into the chamber by excessive sympathetic drive.

It was anticipated that the initial 15°C (60°F) dressing room condition would be temporarily uncomfortable for some subjects. It was assumed that any discomfort would rapidly disappear when the subjects dressed. What was not anticipated was that some subjects apparently became sufficiently chilled to initiate vasoconstriction in the extremities and further that the response persisted even after the subjects were dressed in arctic levels of insulation. It is also unusual that none of the observers noted mechanical thermogenesis (shivering); an overt response to excess heat loss. An early pre-chamber onset of vasodilation due to dressing room conditions appears to be the primary factor that caused the lower initial finger temperature. For two subjects, those effects apparently also occurred on the last test day when the dressing room was set at 18.3°C (65°F).

One consideration is that the onset of vasoconstriction is more rapid and persistent in certain populations [8,10,11,12]. Our limited data and other studies (Endrusick, in preparation) suggests that some subjects displayed those traits. Smoking habits may also affect the rate of hand rewarming [3].

As indicated above, the test and pre-test conditions were more rigorous than currently experienced by POL handlers. At -28.9°C (-20°F), the average pumping time was 17:05 min. At -34.4°C (-30°F), 2 subjects were able to exceed the minimum pumping time requirement. The wind chill temperature with a 5 $m \cdot s^{-1}$ wind speed at -28.9°C is -31.5°C (-24.7°F); although the effects of wind chill were reduced by the clothing and the orientation of the subjects. On the basis of the preceding, a lower effective temperature limit of -31.7°C (-25°F) was recommended for the POL glove.

The chamber testing revealed several points which were not determined during testing with the static physical model. The primary factor was the stiffening of the shell and foam liner. Also of importance was the difficulties encountered in donning and removing the gloves when worn with cold weather clothing. The cuffs of the prototype handwear are difficult to fit over the outer layers of arctic clothing. Dressing and rapid removal or replacement would be enhanced by larger cuffs. Other types of liner insulation may overcome the problems related to the cold-induced stiffening of the foam liner.

The design of the handle is a very important consideration in the selection of cold weather POL equipment. The 1 inch (2.5 cm) pump handle was too small to allow the insertation of a hand wearing a medium prototype POL glove. The 1.5 inch pump handles varied in terms of whether the fuel flowed through the handle. This would affect both possible convective heat loss to fuel flowing through the metal handle and the dimensions of the handle.

5. CONCLUSIONS

The results indicate that all test subjects (n=6) were able to exceed the basic 6 min pumping requirement at -28.9°C (-20°F). Several subjects were able to exceed the requirement at -34.4°C (-30°F) despite the cool pre-conditions. A lower effective temperature limit of -31.7°C (-25°F) was recommended for the POL glove. However, the pre-test conditions clearly influenced the performance of subjects and must be taken into consideration when establishing operating limits.

ACKNOWLEDGEMENTS

The authors would like to acknowledge the following people for their contributions to this study: Dr. T. Lon Owen, Mr. Joseph M. McGrath, CPT M. Whittle, Mr. George Scarmoutzos, Jr. and Ms. Laura B. Myers.

REFERENCES

[1] Barger, M., Outfitting the foot soldier: US Army working hard on ideas but moving slowly to field them. Armed Forces Journal May (1987) 60-61.

[2] Brinson, W.H. and J.L. Bailey, Report (USA Cold Regions Test Center, FT. Greely,AK, 1987).

[3] Cleophas,T.J.M., J.F.M. Fennis and A. van't Laar, Finger temperature after a finger-cooling test: Influence of air temperature and smoking. Journal of Applied Physiology 52 (1982) 1167-1171.

[4] Endrusick,T.L., Tolerance time and finger temperature responses while wearing three types of handwear in moderate and extreme cold-dry environments (1988, in preparation).

[5] Keatinge,W.R., The effect of general chilling on the vasodilator response to cold. Journal of Physiology 139 (1957) 497-507.

[6] Lewis, T., Observations upon the reaction of the vessels of the human skin to cold. Heart 15 (1930) 177-208.

[7] Hamlet,M.P., An overview of medically related problems in the cold environment. Military Medicine 152 (1987) 393-396.

[8] Iampietro,P.F., R.F. Goldman, E.R. Buskirk and D.E. Bass, Response of Negro and white males to cold. Journal of Applied Physiology 14 (1959) 798-800.

[9] McCraig, R.H. and C.Y.Gooderson, Ergonomic and physiological aspects of military operations in a cold wet climate. Ergonomics 29 (1986) 849-857.

[10] Miler,D. and D.R. Bjornson, An investigation of cold injured soldiers in Alaska. Military Medicine 127 (1962) 247-252.

[11] Newman, R.W., Cold acclimatization in Negro Americans. Journal of Applied Physiology 27 (1969) 316-319.

[12] Rennie, D.W. and T. Adams, Comparative thermoregulatory responses of Negros and white persons to acute cold stress. Journal of Applied Physiology 11 (1957) 201-204.

[13] Steinman, A.M., Adverse effects of heat and cold on military operations: history and current solutions. Military Medicine 152 (1987) 389-392.

[14] Vanggaard, L., Physiological reactions to wet-cold. Aviation, Space, and Environmental Medicine 46 (1975) 33-36.

[15] Vaughn, P.B., Local cold injury-menance to military operations: A review. Military Medicine 145 (1980) 305-311.

[16] Whayne,T.F. and M.E. DeBakey, Cold Injury, Ground Type (Washington,DC: U.S. Government Printing Office, 1958).

[17] Winsmann, F., Evaluation of POL handwear and insulating pump handles. Report (Military Ergonomics Division, USARIEM, Natick, MA, 1978).

Trends in Ergonomics/Human Factors V
F. Aghazadeh (Editor)

THE ROLE OF TEXTILE MATERIAL IN CLOTHING ON THERMOREGULATORY RESPONSES TO INTERMITTENT EXERCISE

Ruth NIELSEN[1*] and Thomas L.ENDRUSICK[2]

[1] National Institute of Occupational Health
S-17184 Solna, Sweden
[2] U.S. Army Research Institute of Environmental Medicine
Natick, MA 01760-5007, USA

The physiological effect of different textile materials used in the underwear of an ensemble were studied in the development of over heating or chilling in humans during intermittent exercise in a cold environment. Underwear prototypes manufactured from five different fiber type materials were tested as a part of a standardized clothing system on eight male subjects. Testing occured in a climatic chamber at $T_a=5^oC$, $T_{dp}=-3.2^oC$ and $V_a=0.32$ $m \cdot s^{-1}$. The test consisted of a twice repeated procedure of 40 min cycle exercise (54% of $\dot{V}O_2max$) followed by 20 min of rest. Differences were found in both the amount of non-evaporated and evaporated sweat with the five different underwear configurations No significant difference could be detected in responses of esophageal temperature, skin temperatures, skin wettedness, and onset time of sweating. It was concluded that the textile material used in underwear in a normal work garment has a small influence on the insensible heat disipation during intermittent exercise in a cool environment.

1. INTRODUCTION

Working conditions in many areas have been improved parallel to the technological developments of the society. However, exposure to the cool or cold work environment is still a daily reality for many industrial and outdoor workers. After-exercise chill is a common reason for thermal discomfort in these environments [1, 2]. The phenomenon is especially relevant for dressed people performing intermittent exercise, as after-exercise chilling is related to the interplay between the physiological reactions of the human body and the clothing system.

Commonly used clothing ensembles for cold environments comprise two or more clothing layers: underwear, possibly inner layers, and an outer clothing layer. The main part of the skin surface is not in contact with the ambient environment, but with the micro environment under the clothing and the underwear itself. Thus, underwear has a special function in relation to the sensation of the fabric-to-skin interface and may also be of importance for the resulting micro environment over the skin. Advertisements for thermal underwear, claim benefits of different textile

* The experimental part of this work was done while R.Nielsen held a National Research Council-USARIEM Research Associateship.

material in both insulating power, vapor transmission and wicking ability "..will not absorb moisture but allows it to wick away from the body into the outer garments where it evaporates, keeping the body dry, warm and comfortable" [3]. From this it seems possible that proper use of underwear could minimize or eliminate the development of after-exercise chill.

The thermal resistance of an undergarment is quite small as compared to the overall resistance of a total garment. In the literature (4), the addition of thermal underwear to a clothing system is supposed to add little extra warmth or protection for the wearer, and in terms of differences in intrinsic thermal resistance these differences are insignificant as long as the fit and design remain the same [5].

The latent heat dissipation from the skin to the environment takes place mainly as diffusion resulting from a difference in vapor pressure between the skin and the environment [6]. The diffusion is restricted by the resistance of the clothing layers. Water/sweat can also be transported in the textile fibers themselves, on the surface of the fibers or by capillary action in the yarn [6]. The transport of water by capillary action is negligible in natural fibers; however, textiles made of polypropylene and treated polyester fibers, that do not take up water in the textile fibers, have a considerable ability to transfer water by capillary action [7]. In natural fibers the water transport in the textile fibers themselves is most important. There are important differences in the water/sweat absorbing ability of textile fibers [8]. Textile fibers will absorb and desorb water when the humidity around them changes, and reach an equilibrium with their environments. Wool is able to absorb and retain significantly more water/sweat in the fibers than any other textile and thus, it takes longer time before the air in the textile cavities is exchanged with water during sweat bursts. Also the ab-/desorption processes mean that a heat flow takes place between fibers and air. By measurements on a thermal manikin, release of 20 $W \cdot m^{-2}$ by a light woolen garment when changing from a 5% to a 70% environment has been shown [5]. This has the same effect as an increase of 0.9 clo in the thermal insulation or walking into a room with a $6^{0}C$ higher air temperature. The release of heat was considerably less when the same ensemble was made of cotton/polyester (11 $W \cdot m^{-2}$), and even smaller when it was made of polyester/polyamide (7 $W \cdot m^{-2}$).

From a theoretical point of view the difference in water absorbing and water transporting ability of textiles such as wool, cotton, polyester and polypropylene, can be expected to produce both differences in wetting-time of underwear and other clothing layers, and differences in energy exchange caused by condensation and evaporation processes, and thus contribute to the degree of sweating during exercise, the after-exercise chill and general cold discomfort. The purpose of the present study was to investigate if underwear manufactured from different textile materials caused different thermoregulatory responses on persons performing intermittent exercise in an environment resulting in both periods of both sweating and chilling.

2. MATERIALS AND METHODS

2.1. Garment Description

Underwear manufactured in a 1-by-1 rib knit from 5 different fiber type materials (cotton (F1), wool (F2), polypropylene (F3), and polyester with two different kinds of surface treatment (Capilene=F4 & Thermax=F5)) were evaluated. Measurements of fabric thickness were performed on samples of cloth [9]:F1=0.94 mm, F2=0.97 mm, F3=0.84 mm, F4=0.79 mm, F5=0.69 mm.

All five underwear prototypes were tested as a part of a typical, standardized clothing system on human subjects. The clothing system was comprised of a two-piece long-sleeved/long-legged underwear ensemble, a Battle Dress Uniform (BDU) shirt and trousers (50% cotton/50% nylon), woolen socks, gym shoes, and woolen gloves. Before any testing was done, all clothing systems were laundered and air-dried 5 times without the use of any detergent.

2.2. Subjects

Eight healthy males volunteered for the present series of experiments. They had an average ($\pm$s.d.) age of 25 $\pm$3.7 years, weight of 71$\pm$7.2 kg, height of 175 $\pm$3.3 cm, DuBois surface area (A_{Du}) of 1.86$\pm$0.104 m^2, $\dot{V}O_2max$ of 3.46 $\pm$0.637 l $O_2 \cdot min^{-1}$, and percentage body fat of 14 $\pm$4.6 %.

2.3. Experimental Protocol

Conditions were designed so as to mimic real-life situations in which sweating and after-exercise chill would develop, and where this type of clothing would normally be worn. Testing occured in a climatic chamber at an air temperature (T_a=T_g) of 5.2 $\pm$0.31^{0}C, a dew point temperature (T_{dp}) of -3.2 $\pm$0.48^{0}C (~54 % rh), and an air velocity of 0.32 $m \cdot s^{-1}$.

The clothing was stored in the antechamber at an air temperature of 29^{0}C and 20 % relative humidity (0.8 kPa) at least two hours before the experimental procedure began. The dressing of the subject also took place in this antechamber. Each subject reported to the laboratory at the same time of the day for all experiments. After arrival he was weighed in the nude and then instrumented with chest electrodes for heart rate (HR), thermocouples for esophageal and skin temperatures (calf, thigh, chest, lower back, upper back, upper arm, forearm, hand, and forehead) Each piece of clothing was weighed on a balance (Sauter, model K12), put on the subject and when he was completely instrumented, a dressed weight was recorded (Sauter, model KR120). Upon entering the test environment the subject was instrumented with dew point sensors [10] on the skin underneath the garment (back, chest, thigh) before he mounted a cycle ergometer placed on a Potter balance (model 23B). Approximately 10 min after entering the test chamber the subject began the 2-hour test. The test comprised a twice repeated bout of 40 min cycle exercise (60 r.p.m.; 2.0 $\pm$0.40 kp) followed by 20 min of rest. Each subject always exercised at the same exercise intensity, that had been choosen so it would approximate 55% of his $\dot{V}O_2max$. Esophageal, skin and air temperatures, as well as dew point temperatures at the skin and in the ambient air were monitored on a HP200 computer every minute during the test and stored for analysis. Changes in body weight were sampled from the Potter balance every 20 seconds on a HP85 computer and HR was recorded every 10 min. $\dot{V}O_2$ and $\dot{V}CO_2$ were measured by open circuit spirometry using an automated system (Sensormedics Horizon MMC) during the last 5 min of the first exercise and rest period, respectively. Two minutes after cessation of the test the subject left the test chamber and undressed in the antechamber. Nude body and individual clothing component weights were recorded immediately after the subject had undressed.

3. CALCULATIONS AND STATISTICAL METHODS

Metabolic energy production (M) was calculated from the measurements of oxygen consumption ($\dot{V}O_2$) as [11]

$$M = (0.23RQ+0.77)\cdot\dot{V}O_2\cdot k\cdot 60\cdot A_{Du}^{-1} \qquad (W\cdot m^{-2})$$

in which RQ is the respiratory exchange ratio, $\dot{V}O_2$ is the oxygen consumption in $l\ O_2\cdot min^{-1}$, and k is the energy equivalent of oxygen ($5.873\ W\cdot h\cdot l\ O_2\cdot min^{-1}$).

Mean skin temperature ($\bar{T}_{sk}$) was calculated as an area-weighted average of measurements (modified from [11]):

$$\bar{T}sk = 0.05T_{hand}+0.07(T_{forearm}+T_{upper\ arm}+T_{head})+0.20T_{calf}+0.19T_{thigh}$$

$$+0.175(T_{chest}+(T_{upper\ back}+T_{lower\ back})/2) \qquad (^{o}C)$$

Evaporative heat loss from the skin (E_{sk}) over the total experimental period was determined from the continuous monitoring of weight loss on the Potter balance corrected for weight of respiratory water loss (E_{res}) [12] and metabolic weight loss [13]. No dripping took place, because all excessive sweating was absorbed in the clothing. Total non-evaporated sweat loss (Sw_{ne}) was measured as the difference between clothing weight before and after the experiment. Total sweat loss (Sw_{tot}) was calculated as the sum of E_{sk} and Sw_{ne}.

Vapor pressures at the skin surface and in the ambient air were determined from the local dew point temperature recordings using the Antoine equation.
Local skin wettedness, w, on back, chest and thigh was calculated as:

$$w = 100\cdot(P_{sk}-P_a)/(P_{ssk}-P_a) \qquad (\%)$$

where P_{sk} is the vapor pressure at the skin surface obtained from the dew point sensor, P_{ssk} is the saturated vapor pressure at the local skin temperature and P_a is ambient water vapor pressure.
An average skin wettedness for thigh and torso area was estimated using each local skin surface area's fraction of the total body surface area:

$$w = 100\cdot(0.175\cdot w_{chest} + 0.175\cdot w_{back} + 0.19\cdot w_{leg})/0.54 \qquad (\%)$$

Repeated-measures analysis of variance (ANOVA) was used to determine whether the factor 'textile material' had any significant effect on physiological reactions or sweat accumulation in the clothing. In the event that ANOVA reveiled significant main effect, Tukey's critical difference was calculated and used to locate significant difference between means. Data are presented as mean $\pm$s.d. All differences reported are significant at the $P<0.05$ level. A paired t-test was used to test if there was any difference in physiological reaction in the first and second test period.

4. RESULTS

Work intensity was at average 64 $\pm$11.6 $W\cdot m^{-2}$ during the 40 minute cycling periods. Metabolic energy production (M) did not vary between clothing systems, and were at average (n=40) 344 $\pm$51.2 $W\cdot m^{-2}$ during exercise and 65 $\pm$11.6 $W\cdot m^{-2}$ during rest.

Core temperature as represented by T_{es} was not influenced by the fiber material of the underwear worn (Figure 1). T_{es} averaged 36.6 $\pm$0.19^{0}C for all tests in the first minute of exercise. After 20 minutes of exercise a steady-state T_{es} value of 37.4 $\pm$0.18^{0}C was reached. During the rest period T_{es} decreased quickly to reach a value of 36.8 $\pm$0.20^{0}C in the last minute. The course of T_{es} over the second exercise/rest period was similar to the course over the first period and similar temperature values were measured at the end of the two periods.

Mean skin temperature ($\bar{T}_{sk}$) did not vary between types of underwear worn (Figure 1). The average values (n=40) at the beginning of the exercise and at the end of each period were 31.0 $\pm$0.61, 30.9 $\pm$1.01, 29.1 $\pm$0.62, 30.3 $\pm$1.01 and 28.7 $\pm$0.63 ^{0}C. All values except for F5 during exercise and F2 at rest were significantly lower at the end of the second period than the corresponding values in the first period.

Average skin wettedness (w) did not show any difference between underwear conditions (Figure 1). Average values (n=40) at the beginning of the exercise and at the end of each period were 7 $\pm$5.2, 60 $\pm$11.3, 41 $\pm$13.1, 62 $\pm$10.6 and 45 $\pm$14.4 %. Wettedness were slightly higher ($p<0.05$) in the second exercise and rest period compared at the first period.

Onset of sweating was considered to take place when the dew point sensors at the skin recorded an increase in vapor pressure, and it began 9 $\pm$2.74 min after the start of the exercise. No difference could be detected

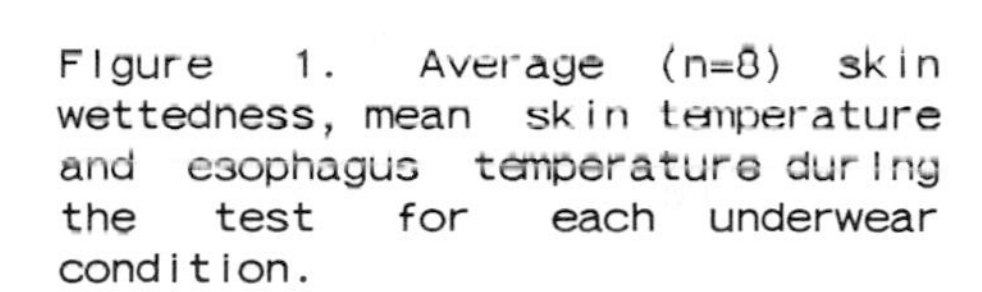

Figure 1. Average (n=8) skin wettedness, mean skin temperature and esophagus temperature during the test for each underwear condition.

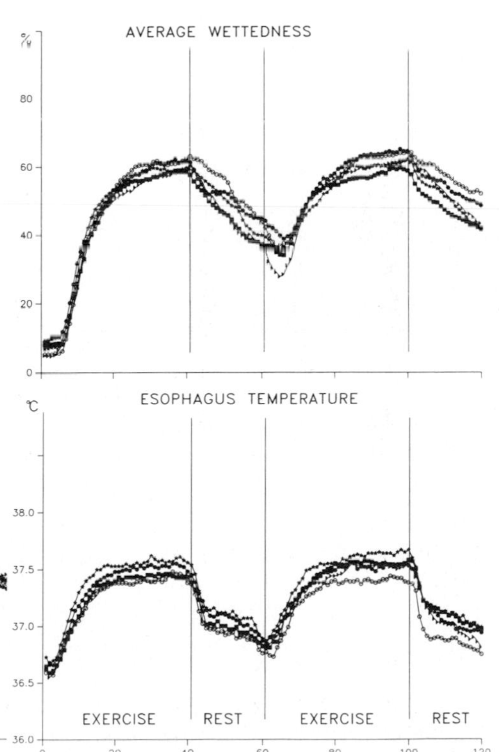

between the five different underwear conditions; nor was there any significant difference in the time to onset of sweating between the first and second exercise period.

Total sweat loss (Sw_{tot}) shown in Figure 2 was significantly lower with F4 underwear compared to F1, F2 and F3 underwear and with F5 than with F2 underwear. Evaporation of sweat (E_{sk}) totalled over the test period was significantly higher with F3 compared to F4 and F5, and higher with F2 than with F5. The underwear material had a significant effect on the amount of non-evaporated sweat (Sw_{ne}) absorbed in the clothing ensemble worn (Figure 2). More sweat was absorbed when either F1 or F2 underwear were worn compared to F3, F4 or F5 underwear (F1 >F3, F4, F5; F2 >F3, F4, F5). Considering the underwear alone, both F1 and F2 underwear contained more sweat than all three types of man-made fiber underwear (F1 >F3, F4, F5; F2 >F3, F4, F5). There was no difference between the amount of sweat absorbed in F3, F4 and F5 underwear; nor was there any difference between the amount of sweat absorbed in F1 and F2 underwear, respectively. There was no difference, and thus no underwear effect, in the amount of sweat found in the BDU-jacket, and the BDU-trousers showed only a difference between F2 and F4 conditions (F2>F4). Socks, shoes and gloves contained after all experiments very little moisture (3, 6 and 6 g, respectively).

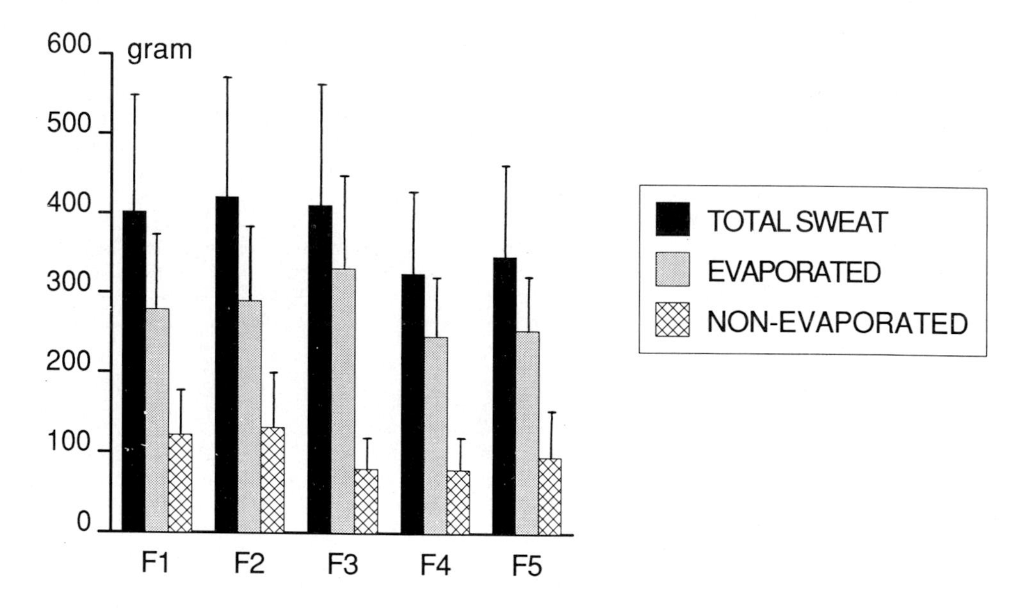

Figure 2. Total amount of sweat, evaporated sweat and non-evaporated sweat with the five clothing systems (mean $\pm$sd).

5. DISCUSSION:

Different fiber type material in the underwear of a clothing system has an effect on thermoregulatory responses during intermittent exercise in a cold environment. In the present study differences in sweat loss, evaporation of sweat and amount of non-evaporated sweat absorbed in the underwear were found. No differences were detected in chilling at rest, but with a longer rest period differences may develop.

It was decided to use intermittent exercise rather than one continuous exercise period followed by a period of rest. It was expected, and confirmed, that a dampening or wetting of the clothing system occured over the course of the first period and this changed the course of the thermophysiological responses in the next period. Initially $\bar{T}_{sk}$ decreased slightly; but after onset of sweating it increased due to insufficient evaporation of sweat as shown by the increase in skin wettedness. In the first rest period $\bar{T}_{sk}$ decreased significantly and was at the onset of the second exercise period $2^{o}C$ lower than at the start of the test. $\bar{T}_{sk}$ increased during the second exercise period and reached a steady state $0.5^{o}C$ lower than the steady state level in the first exercise period. As other variables were constant this can probably be related to the dampening of the clothing system, that causes a lowering of the insulation value of the clothing system [14, 15]. This is supported by a similar lowering of $\bar{T}_{sk}$ in the second rest period and a slightly increased skin wettedness in the second period. During the first rest period skin wettedness showed that a drying out of the clothing took place, but after 20 minutes of rest the microenvironment was still humid (w=41%). Except in more extreme environments, the course of T_{es} is determined by the work level [16]. The clothing only influences the thermophysiological responses at the body surface.

No differences were found in skin wettedness and $\bar{T}_{sk}$ due to different fiber type material of the underwear indicating that the humidity conditions in the microenvironment between skin and underwear were similar. Therefore, the observed differences in wet heat exchange must be caused by differences in the thermal processes taking place within the clothing system. The amount of sweat absorbed in the wool and the cotton underwear was as predicted from the textile characteristics being larger than the amount absorbed in the man-made textiles. These were all made of hydrophilic fibers, where the surface was made hydrophobic by a chemical treatment. The differences in total sweat production are difficult to explain. Possible mechanisms are hypothesized below using sweat production with F4 as a reference.

Absorption of sweat in wool (F2) over the skin surface results in an incomplete evaporation and cooling of the skin, plus liberation of sorption heat at the skin. Both tend to increase $\bar{T}_{sk}$. The absorbed sweat is transported in the wool fibers further into the clothing where it evaporates and decreases the temperature between the clothing layers. This creates a steeper temperature gradient over the underwear and a greater dry heat flow. The tendency to increase $\bar{T}_{sk}$ may increase sweat production, and could be the cause of the greater sweat production observed.

The increased sweat production with polypropylene underwear (F3) is found as increased evaporation. This may be explained by sweat being wicked through the underwear, and evaporated further into the clothing. The incomplete evaporation at the skin tend to increase $\bar{T}_{sk}$ resulting in a higher sweat production. However, a similar response could be expected for F4 and F5 provided the information supplied by the manufacturer is correct. The difference between polypropylene and the treated polyester fibers could be related to the slight difference in cloth thickness, but as the difference between F3 and F4 are small, this cause is not likely.

6. CONCLUSIONS

In conclusion, different textile material in the underwear of a clothing system has an effect on thermoregulatory responses during intermittent exercise in a cold environment. However, thermal underwear does not keep its wearer completely dry as often claimed. In periods where sweating occur skin wettedness increases to similar levels independent of material of the underwear, and at rest mean skin temperature and skin wettedness decreases to the same values. However, the textile material of the underwear making the contact to the skin absorbs and contains less sweat when man-made hydrophilic fibers are used.

ACKNOWLEDGEMENT
The authors express their appreciation to the volunteers whose participation made this study possible, and to L.Stroschein, T.Guerra, R.Oster, G.Newcomb, J.Bogart, L.Levine, G.Sexton and C.Levell for their help.
The views, opinions and/or findings in this report are those of the authors and should not be construed as official Department of the Army position, policy, or decision, unless so designated by other official documentation.

REFERENCES:

[1] Nielsen,R., Clothing and thermal environments. Appl.Ergonomics 17 (1986) 47-57.
[2] Pugh,L.G.C., Clothing insulation and accidential hypothermia in youth. Nature 209 (1966) 1281-1286.
[3] Cabela's mid-winter catalog 1986-87.
[4] Fonseca,G.F., Heat transfer properties of ten underwear-outerwear ensembles. Textile Res.J. 40 (1970) 553-558.
[5] Olesen,B.W., and Nielsen,R., Thermal insulation of clothing measured on a movable thermal manikin and on human subjects. Technical report 7206/00/914 (Technical University of Denmark. 1983).
[6] Mecheels,J., Kleidung aus chemiefasern für heisse klimabedingungen. Textil-industrie 72 (1970) 859-66.
[7] Pontrelli,G.J., Partial analysis of comfort's Gestalt. In Hollies, N.R.S., and Goldman,R.F. (eds), Clothing Comfort (Ann Arbor Science Publishers, Inc., 1977).
[8] Newburgh,L.H.,Physiology of heat regulation (Saunders, Philadelphia, 1949).
[9] ASTM: D1774-64. Standard test method for measuring thickness of textile materials.
[10] Graichen,H., Rascati,R., and Gonzalez,R.R., Automatic dew point temperature sensor. J.Appl.Physiol. 52 (1982) 1658-1660.
[11] Gagge,A.P., and Y.Nishi. Heat exchange between human skin surface and thermal environment (In: "Handbook of Physiology" edited by Lee, D.H.K., Bethesda, M. 1977, chapter 9).
[12] Fanger,P.O., Thermal Comfort (McGraw-Hill, Inc., New York, 1970).
[13] Kerslake,D.McK., The stress of hot environments (Cambridge University Press, England, 1972).
[14] Hall,J.F., and Polte,J.W., Effect of water content and compression on clothing insulation. J.Appl.Physiol. 8 (1956) 539-545.
[15] Holmer,I., Heat exchange and thermal insulation compared in woolen and nylon garments during wear trials. Text.Res.Inst. (1985) 511-18.
[16] Nielsen,M., Die regulation der Körpertemperatur bei Muskelarbeit. Skand.Arch.Physiol. 79 (1938) 193-230.

Trends in Ergonomics/Human Factors V
F. Aghazadeh (Editor)

A METHODOLOGY FOR ASSIGNING VARIABLE RELAXATION ALLOWANCES: MANUAL WORK AND ENVIRONMENTAL CONDITIONS

Andris FREIVALDS and Joseph H. GOLDBERG

Department of Industrial & Management Systems Engineering
The Pennsylvania State University
207 Hammond Building
University Park, PA 16802

This paper develops a physiological basis for assigning variable relaxation allowances for manual work and environmental conditions. The resultant allowances are compared to the currently accepted International Labour Office (ILO) recommended values. The ILO values compare favorably to the physiological standards only on average. There are large deviations for particular conditions which cannot be accounted for with the ILO formulations. It is therefore, recommended that the proposed allowances be used in industry.

1. INTRODUCTION

The normal time for an operation is the time that a qualified operator would need to perform the job if the operator worked a standard performance level. An allowance is added to the normal time to account for interruptions and delays so as to provide a fair and readily maintainable standard. These allowances are classified as personal needs, fixed and variable relaxation and unavoidable delays. The personal needs allowances are typically assigned 5% of the normal time, the fixed relaxation allowance 4%. For the variable relaxation allowances (VRA) a variety of schemes have been developed with the International Labour Office [1] being most typically accepted. These were developed through consensus agreements between management and workers across many industries and have not been directly substantiated. On the other hand, since the 1960's much work has been done in developing specific ergonomic standards for the health and safety of the U.S. worker. The purpose of this paper is therefore, to examine how well these standards compare and to propose a more comprehensive scheme of VRA's to account for current ergonomic standards.

2. DEVELOPMENT OF A PHYSIOLOGICAL BASIS FOR MANUAL VRA's

2.1. Muscle Strength and Fatigue Considerations

In industrial activity, one is most interested in measuring fatigue during work and determining recovery after completion of the work. The most immediate result of muscle fatigue is the reduction in muscle strength. Rohmert [2] observed the following:

1. Reduction in maximum strength occurs only if the static holding force exceeds 15% of maximum strength.

2. The longer the static muscular force lasts, the greater the reduction in available strength.

3. There are no individual or muscular differences if the forces are normalized to the individual's maximum strength for that muscle.

4. Recovery is a function of the degree of fatigue, i.e. the same percent decrease in maximal strength will require the same amount for recovery.

Based on Rohmert's [2] observations, Freivalds and Kaleps [3] quantified the relationship between maximum holding time and maximum holding force as follows:

$$T = \frac{1.197}{\left(\frac{f}{100} - .15\right)^{.618}} - 1.2083 \qquad (1)$$

where T = maximum holding time (min)
f = holding force (lbs)

The holding force (f) is normalized by 100 lbs, the average of the three basic standardized strengths utilized in lifting: arm lifting strengths, leg lifting strengths, and torso lifting strengths collected on 1522 industrial male and female workers [4]. Thus based on Eq. 1, the required VRAs for infrequent lifts (e.g. one lift per minute) of short duration (e.g. .1 minute) are presented in Table 1. For more frequent lifts metabolic considerations would become predominant.

Table 1. Variable Relaxation Allowances (%) for Manual Work Based on Various Physiological Considerations

				Metabolic	
Load (lbs)	ILO	Strength	Lift	Push/Pull	1 Std. Dev.
5	0	0	1.3	0	4.7
10	1	0	2.0	0	5.9
15	2	0	2.9	0	7.7
20	3	.1	4.1	.4	13.1
25	4	.4	5.5	1.0	16.5
30	5	.7	7.0	2.1	20.1
35	7	1.2	8.9	3.2	23.9
40	9	1.8	10.9	4.8	27.9
45	11	2.7	13.1	6.5	32.0
50	13	3.8	15.3	8.5	36.2
60	17	7.2	20.5	12.6	44.7
70	22	13.1	25.8	17.8	53.5

2.2. Metabolic Considerations

Metabolic considerations become very important for high rates of activities. They are also important in determining rest allowances for posture considerations. Metabolic models have been developed that predict expected energy expenditure for various activities [5], [6]. These can be used to predict and compare energy expended for three basic postures:

$$\dot{E}_{SIT} = .0105 \text{ BW} \tag{2}$$

$$\dot{E}_{STAND} = .0109 \text{ BW} \tag{3}$$

$$\dot{E}_{AWKWARD} = .0129 \text{ BW} \tag{4}$$

where $\dot{E}$ = energy expenditure (kcal/min)
BW = body weight (lbs)

In addition to the basic posture, the model assumes that other work, load lifting or hand assembly, would be added on top of the basic posture.

For lifting, pulling, pushing and other manual material handling tasks, there are specific equations to predict energy expenditure for a task given specific frequencies, loads and distances [6]. These are fairly complicated equations, also including various other task and individual factors. Eliminating some factors and averaging across others one can arrive at the following three simplified equations:

$$\dot{E}_{LOW\ LIFT} = [142\ (32-H_1) + (2.08\ L_U+.8L_D)\ (H_2-H_1)]/10000. \tag{5}$$

$$\dot{E}_{HIGH\ LIFT} = [22.8(H_2-32) + (3.22\ L_U+1.03L_D)\ (H_2-H_1)]/10000. \tag{6}$$

$$\dot{E}_{PUSH/PULL} = D\ (.2+.027L)/100. \tag{7}$$

where:

D = distance load pushed or pulled
H_2 = upper lifting height (in)
H_1 = lower lifting height (in)
L = effective force pushed or pulled (lbs)
L_U = load lifted up (lbs)
L_D = load lifted down (lbs)
$\dot{E}_{LOW\ LIFT}$ = energy expenditure for a low lift ($H_1<H_2\leq 32$ in) (kcal)
$\dot{E}_{HIGH\ LIFT}$ = energy expenditure for a high lift ($H_2>H_1\geq 32$ in)(kcal)
$\dot{E}_{PUSH/PULL}$ = energy expenditure for a push or pull (kcal)

For intermediate heights, (i.e. $H_2>32$ in, $H_1<32$in) a combination of low and high lifts must be used:

$$\dot{E}_{LIFT} = \dot{E}_{LOW\ LIFT}\ (H_1,H_2=32) + \dot{E}_{HIGH\ LIFT}\ (H_1=32,H_2) \tag{8}$$

In addition the energy expenditure for the basic posture as well as the number of times a lift or push/pull is performed must be accounted for.

Thus the final energy expenditure equation is:

$$\dot{E}_W = (\dot{E}_{\substack{SIT\\STAND\\AWKWARD}} \times T + N \times \dot{E}_{\substack{LIFT\\PUSH\ PULL,\ ETC.}}) / T \tag{9}$$

where $\dot{E}_W$ = energy expenditure for work
T = time observed (min)
N = number of lifts or push/pulls per time T

The total amount of rest required for any given work activity can be calculated from an equation developed by Murrell [7]:

$$R = \frac{(\dot{E}_W - S)}{(\dot{E}_W - 1.5)} \tag{10}$$

where R = rest time (% of total time)
S = a standard working rate (kcal/min)

The standard working rate is accepted to be about 1/3 of maximum physical work capacity and is approximately 4.67 kcal/min for a combined male and female population [8]. Also converting the rest time from percent of total time to a percent allowance, one obtains:

$$VRA = 31.5\ (\dot{E}_W - 4.67) \tag{11}$$

2.3. Simulation and Discussion

It was decided to simulate a variety of lifting, pushing and pulling tasks using different loads, heights and frequencies. Loads were identical to those in the ILO table (Table 1). A combination of H_2 and H_1 were used ranging from 0 to 60 in. in 10 in. increments; the only constraint being that H_1 was less than H_2. Frequencies used were 1, 2, 4 or 8 per minute. These are all values to be typically expected in industry, resulting in a total of 1186 combinations. The average VRA based on muscle fatigue across different heights and frequencies for lifting and pushing/pulling are tabulated in Table 1.

A comparison of the VRAs as calculated from the two various approaches indicates similar trends. The average VRAs based on lifting and the average VRAs based on push/pull straddle the ILO recommended values. In fact, average values of the two would almost be superimposed on the ILO values. The increase with load is relatively linear with each additional increment in load requiring a similarly large increment in energy expenditure. The VRA based on muscle fatigue deviates most, showing very small values for low force values consistent with the low fatiguability of slow twitch motor units and then rising sharply for large force values, consistent with the high fatiguability of the fast twitch motor units. Thus, perhaps the best overall approach is to be conservative, ie. to use the larger of the two allowances for any given situation.

A more important factor is the large variability in VRAs as shown by the

+1 standard deviation. This variability in VRAs is due to various heights and frequencies used in the simulations which are not accounted for in the ILO recommendations. These variations can be up to three times the mean VRA value. Thus the task factors of frequency of lifting and height of the lift are critical and have just as much effect on the VRAs as load. For the most accurate application of VRAs one should calculate VRAs based on all three task factors rather than using an average value as recommended by ILO.

3. DEVELOPMENT OF A PHYSIOLOGICAL BASIS FOR ENVIRONMENTAL CONDITION VRAs

3.1. Environmental Considerations

Excessive heat is a stress to which most workers are exposed at one time or another as created by the demands of the particular industry. During these conditions there occurs a heat exchange between the body and the environment which can be modeled by:

$$S = M \pm C \pm R - E \tag{12}$$

where M = heat gain of metabolism
C = heat gained or lost by the body due to convection
R = heat gained or lost by the body due to radiation
E = heat lost by the body through evaporation of sweat
S = heat storage (or loss) of the body

For thermal neutrality, S must be zero. If the summation of the various heat exchanges across the body result in a heat gain then resulting heat will be stored in the tissues of the body with a concomitant increase in core temperature and a potential heat stress problem.

Many attempts have been made to combine the physiological manifestations of these heat exchanges along with environmental measurements into one index. Perhaps the best index is the Heat Stress Index (HSI), originally developed for industry by Haines and Hatch [9] and Belding and Hatch [10] and later modified McKarns and Brief [11]. The index is based on a comparison of the amount of sweat that has to be evaporated (evaporation required to maintain thermal equilibrium = E_{REQ}) with the maximum amount of sweat which can be evaporated in the specific climatic conditions (maximum evaporative capacity = E_{MAX}). For a clothed worker evaporation required (E_{REQ}) is determined from the heat balance equation:

$$E_{REQ} = M \pm R \pm C \tag{13}$$

where

$$C = .756\ V^{.6}\ (T_{DB}-95)\ \text{Btu/hr} \tag{14}$$

$$R = 17.5\ (T_W-95) \tag{15}$$

and

$$T_W = [(T_G + 40)^4 + (1.03X0^8 V^{.5}(T_G-T_{DB})]^{.25}-460 \tag{16}$$

with M,C,R in Btu/hr, T in $^{\circ}$F and V in ft/min. Maximum evaporative capacity (E_{MAX}) is given by McKarns and Brief [11] as:

$$E_{MAX} = 2.8\ V^{.6}\ (42-P_A)\ \text{Btu/hr} \tag{17}$$

E_{MAX} is limited to 2400 Btu/hr. which represents the evaporation of one quart of sweat per hour, which, is considered approximately the maximum a person can produce for eight hours a day.

The ratio

$$HSI = 100{*}E_{REQ}/E_{MAX} \tag{18}$$

indicates the level of heat stress, with a value of 100 being considered the maximum that can be tolerated for eight hours a day.

Allowable exposure time (AET), in hours, can be calculated from the rate at which the body stores the excess heat gain:

$$AET = W\ H_c \Delta T/(E_{REQ}-E_{MAX}) \tag{19}$$

where W = weight of worker (kg)
H_c = heat capacity of human body, .827 Btu/lb/$^{\circ}$F
ΔT = increase in core temperature ($^{\circ}$F)

Typically suggested maximum increases in core temperature have been on the order of 1.8°F [12]. Similarly the minimum recovery time (MRT) can be calculated from the rate at which the body loses the stored heat during recovery at resting conditions (M = 357 Btu/hr).

$$MRT = WH_c\ \Delta T/(E_{MAX} - E_{REQ}) \tag{20}$$

A relaxation allowance (RA) would then simply be:

$$RA = \frac{MRT}{AET} \times 100 \tag{21}$$

A comparison between the RA developed here and the ILO recommended values can be made, although there are difficulties in applying the ILO general specifications. The Kata thermometer, used by ILO, is essentially an alcohol-filled thermometer developed by Hill [13]. The bulb is heated in warm water until the column rises into an upper reservoir and is wiped dry. The dry instrument is suspended in an air stream and the fall of the column is measured with a stopwatch. The cooling time is a function of air velocity and air temperature and can be expressed in terms of cooling power. Thus the dry Kata can be considered to be a measure of both radiation and convection. A wet Kata uses a wet cotton wick fitted around the bulb to express the cooling power under conditions of humidity and can be considered as a measure of maximum evaporation. One must assume that the ILO values of Kata cooling power include both the dry and wet Kata measurements, in which case the ILO cooling power (CP) can be expressed as:

$$CP = \pm R \pm C - E_{MAX} \tag{22}$$

As a result Eq. 19 and Eq. 20 becomes respectively:

$$AET = WH_c \Delta T/(M-CP) \quad (23)$$
$$MRT = WH_c \Delta T/(CP-M) \quad (24)$$

Two more assumptions must be made. The maximum allowable energy expenditure for an eight hour day is 1269 Btu/hr for males and 952 Btu/hr for females [8]. Should the energy expenditure exceed these values then relaxation allowances for manual work must be used. For the present, an average metabolic energy expenditure of 1112 Btu/hr for a combined male and female population will be used. Second, the surface area of the skin for an average human is 20 ft^2 [10] , of which 60% becomes an effective area for heat exchange when fully clothed [14].

The ILO cooling power values are given in milicalories/cm^2/sec and can be converted to Btu/hr by multiplying by the conversion factor of 13.272 to obtain Btu/ft^2/hr and by the effective surface area of the worker of 12 ft^2/hr. The resulting values are tabulated in Table 2 and are used in Eqs. 23 & 24 in place of cooling power (CP) to calculate the required allowable exposure time and minimum recovery time.

Table 2. Values of Current ILO and Proposed VRA as a Function of Cooling Power

ILO Cooling Power		Allowance	
(in millicalories/cm^2/sec.)	Btu/hr.	ILO	Proposed
16	2548	0	0
14	2230	0	0
12	1911	0	0
10	1593	3	0
8	1274	10	0
6	956	21	26
5	796	31	72
4	637	45	170
3	478	64	524
2	319	100	∞

3.2 Discussion

There are large discrepancies between the ILO VRAs and proposed VRAs, with the proposed VRAs showing an exponential increase toward larger values than the ILO VRAs (Table 2). This large discrepancy can be explained on the basis of several factors. First of all, the ILO standards do not provide a clear statement of the environmental conditions involved. The Kata cooling power is not identified as wet, dry or combined and does not include radiation. Secondly the individual environmental factors (such dry bulb temperature, wet bulb temperature and wind velocity) are not identified. Thirdly, the ILO allowances do not account for other individual characteristics, such as the level of physical work, that may change the allowances. Metabolic energy expenditure can vary over a very large range and consequently alter the VRA over a similarly large range. Thus, the proposed, straightforward method is recommended for use in industry.

4. CONCLUSIONS

The ILO variable relaxation allowances, were based on work done by the Personnel administration, Ltd. in London based on consensus agreements between management and workers across many industries. They have not been clearly presented and directly substantiated in the literature. Based on comparisons with proposed VRAs as developed from more current ergonomic standards and guidelines, it appears that there may be serious deficiencies in the ILO allowances were comparable only on average. For specific cases they could differ by as much as a factor of three. For environmental conditions, the allowances differed sharply and could not be even reconciliated on an average basis. The only justification for the ILO allowances is that these have been used for many, many years with perhaps general satisfaction and that their formulations are very easy to use. However, it is recommended that a more current and accurate approach be used as specified in the above paper.

5. REFERENCES

[1] International Labour Office: Introduction to Work Study, (ILO Geneva, 1957).

[2] Rohmert, W., Ermittlung von Erholungspausen fur statische Arbeit des Menschen, Int. Z. Angew. Physiol., 18, (1960), 123-140.

[3] Freivalds, A. and Kaleps, I., Modeling of dynamic active neuromusculature responses, submitted to Journal of Biomechanics, (1987).

[4] Chaffin, D. B., Freivalds, A. and Evans, S.R., On the validity of an isometric biomechanical model of worker strengths, IIE Transactions, 19 (1987) 280-288.

[5] Aberg, V., Elgstrand, K., Magnus, P. and Lindholm, A., Analysis of components and prediction of energy expenditure in manual tasks, The International Journal of Production Research, 6, 1968, 189-196.

[6] Garg, A., Chaffin, D. B., and Herrin, G. D., Prediction of metabolic rates for manual materials handling jobs, American Industrial Hygiene Association Journal, 39, (1978) 661-674.

[7] Murrell, K.F.H., Human Performance in Industry, (Reinhold, New York, 1965).

[8] Bink, B., The physical working capacity in relation to working time and age, Ergonomics, 5, (1962) 25-28.

[9] Haines, G. F. and Hatch, T. F. Industrial heat exposure-evaluation and control. Heating and Ventilation, 49, (1952) 93-104.

[10] Belding, H. S. and Hatch, T. F. Index for evaluating heat stress in terms of resulting physiological strains. Heating, Piping and Air Conditioning, 27, (1955) 129-136.

[11] McKarns, J. S. and Brief, R. S. Nomographs give refined estimate of heat stress index. Heating, Piping & Air Conditioning, 38, (1966) 113-116.

[12] NIOSH, Criteria for a Recommended Standard to Occupational Exposure to Hot Environments, (U.S. Dept. HEW, HSM 72-10269, 1972).

[13] Hill, L. The Science of Ventilation and Open Air Treatment. Part I, (Med. Res. Coun. Spec. Rept., No., 32, London, 1919).

[14] AIHA. (American Industrial Hygiene Association). Heating and Cooling for Man in Industry, (AIHA, Akron, OH, 1975).

Trends in Ergonomics/Human Factors V
F. Aghazadeh (Editor)

A SUBJECTIVE EVALUATION OF A MICROCLIMATE COOLING SYSTEM

Fariborz TAYYARI

Department of Industrial Engineering
Bradley University
Peoria, Illinois 61625

A microclimate cooling system was tested in this study that consisted of a vest with 54 small ice bags inserted in the pockets sewn onto the inside of it. The test was performed in an environmental chamber using four young male subjects performing a walking task under a thermal condition of approximately 40° C and 75% relative humidity. The results were evident of the significance of the cooling system in reducing the subject's heat stress.

1. INTRODUCTION

In many occupations, often times, physical activities are performed in environments with adverse thermal conditions. Heat stress may not only deteriorate the worker's performance, but also can lead to a heat disorder. Heat stress, associated with such environments, is the total heat load imposed by the internal and external sources. The internal sources of heat stress consist of the body-generated metabolic heat (which is directly affected by the physical workload), body temperature, as well as the state of heat acclimatization. The external sources include the thermal conditions of the work environment, that is, air temperature, radiant heat, humidity and air movement.

Although a feeling of heat discomfort may be experienced, no health damage risk has been reported to be associated with lower levels of heat stress. But, if the heat stress exceeds the worker's heat tolerance limit, adverse health effects can occur. The most critical heat disorders are heat stroke, heat exhaustion, heat cramps and prickly heat (heat rash). Explanation of these disorders and their preventive measures and/or treatments can be found in most articles and books written on the heat stress subject [e.g., 1, 2, 3].

When performing physical activities is potentially dangerous to safety and health, some safeguard provisions would be necessary. To reduce the heat stress of individuals the application of the following techniques can be very useful:

1.1. Administrative Techniques

* Screening and selecting personnel suitable for the work conditions.

* Provision of a heat acclimatization period before assigning personnel to their work [4].

* Provision of a suitable work-rest schedule [5, 6, 7].

* Providing and encouraging drinking water to prevent dehydration [1, 3].

* Personnel rotation--from one shift to another and/or from one work area to another.

1.2. Engineering Techniques

* Modifying work methods and redesigning the equipment and work space to reduce the amount of work energy requirements.

* Providing air conditioning systems to control air temperature, humidity and air circulation.

* Insulating and shielding against radiant heat.

1.3. Personal Clothing

* Using appropriate clothing and protective equipment, including personal cooling systems.

However, there are situations in which neither the administrative techniques are sufficient nor application of engineering techniques are economical and/or feasible. Therefore, many investigators concentrated on the development and applications of microclimate cooling systems as a feasible and economical alternative to substitute the environmental (macro-climate) cooling.

There are a large number of reports regarding the design, evaluation and/or improvement of various types of microclimate cooling systems ranging from partial- to whole-body cooling. So far as literature concerns, three major categories of cooling systems have been developed. They are **air-ventilated** [e.g., 8, 9], **water-circulated** [e.g., 10, 11] and **ice-cooled** [e.g., 12, 13, 14, 15]. In addition, some investigators (e.g., 16, 17, 18, 19] have attempted localized cooling systems (i.e., cooling head, neck, and/or wrists).

2. OBJECTIVES

This study was conducted to evaluate the effectiveness of a prototype microclimate cooling system [15] in reducing heat stress as measured by subjective responses of individuals to working in a hot environment. The tested cooling system is an ice-cooled jacket that has been tailored in such a way to open and close in front (to be worn as a sleeveless vest) and on both sides (to be worn as a poncho). The specifications of this system are given in Table 1.

Table 1. Specifications for the Microclimate Cooling System

Specification	Description
Type of System	Adjustable Vest
Inner Shell Material	840 Denier Nylon
Outer Shell Material	Vinyl Nylon
Insulation Filling Material	Velour Fabric (Blanket)
Thickness of Insulation Filling	5 mm (3/16 inch)
Number of Pieces	1
Number of Pockets	54
Attachment of Ice Bags	Replaceable
Size of each Pocket	8 x 10 cm (3.2 x 4 in)
Size of each Ice Bag	7.5 x 10 cm (3 x 4 in)
Ice Contact Area on the Torso	0.4 m2 (approximately)
Coolant Material	Water
Weight of each Ice Bag	70 gm (2.5 oz)
Total Weight of Coolant	3.78 Kg (8.32 lbs)
Heat Absorption Capacity	506 W (450 Kcal)
Weight of System without Coolant	0.76 Kg (1.68 lbs)
Total Weight of the System	4.54 Kg (10 lbs)

3. METHODOLOGY

To evaluate the microclimate cooling system, four male college students of good physical condition were selected to serve as the subjects of the study. The criteria for selecting the subjects was passing a physical examination as well as a treadmill test. A treadmill was used to simulate the desired physical workloads. In each experimental session, the speed of treadmill for the subject was set at a level equivalent to the desired metabolic rate [20]. An environmental chamber was used to establish and control thermal conditions of 39.6° C dry-bulb temperature and 75% relative humidity.

All subjects were heat acclimatized during a seven-day period, prior to the experimentation, under an environmental thermal condition of about 39.7° C dry-bulb and 35.4° C wet-bulb. The subjects walked on the treadmill at a metabolic rate of 250, 300 and 350 W on the first, second and third days, respectively, and 400 W on the last four days of acclimatization. The walking task during the heat acclimatization and experimental sessions was scheduled to be performed intermittently for 15 minutes followed by a resting period of 5 minutes for a total of 95 minutes per session. However, the following criteria were established for the early termination of any experimental session, in which:

* the subject's heart rate exceeded 190 minus his age,
* the subject's rectal temperature exceeded 39.2° C, or
* the subject became unable to continue the walking task or developed any sign of heat disorders.

At the end of each 15 minute walk-bout, the subject was asked to rate his subjective "feeling of discomfort" and "thermal sensation." Then, these responses were quantified according to the scales developed by Kamon [14]. The scales with a slightly change are shown in Table 2.

Table 2. Scales of Subjective Responses

Feeling of Discomfort	Thermal Sensation
0: Very Comfortable	0: Cool
1: Comfortable	1: Neutral
2: Slightly Uncomfortable	2: Slightly Warm
3: Uncomfortable	3: Warm
4: Very Uncomfortable	4: Hot
5: Extremely Uncomfortable	5: Very Hot

4. RESULTS AND DISCUSSION

The microclimate cooling system, which was investigated in this study, revealed itself to be significantly effective means of reducing heat stress of individuals working in hot environments. None of the subjects were able to complete the 95-minute experimental sessions without using the cooling system. The average tolerated exposure time in this case was 73 (± 14) minutes. But when the subjects were provided with the cooling system, they easily completed the sessions and were still able to continue their walking task. This phenomenon by itself was a strong evidence of the effectiveness of the cooling system. However, this claim was further supported by the subjective responses.

After 55 minutes of exposure, the average subjective rating of their *feeling of discomfort* was 4.00 (very uncomfortable) when the subjects did not use the cooling system and 2.75 (between slightly uncomfortable and uncomfortable) when they used the system. Their average rating of the *thermal sensation* was 4.5 (between hot and very hot) without the cooling system and 2.5 (between slightly warm and war) when the cooling system was utilized.

Based on the results of this study, it can be concluded that both the employee and employer in a hot industrial setting would benefit from the application of microclimate cooling systems. They minimize the employee's risk of developing heat disorders by removing his metabolic heat by conduction. This increases the worker's heat tolerance limit which results in an improved work productivity.

REFERENCES

[1] Astrand, P.O. and Rodahl, K. (1977). ***Textbook of Work Physiology: Physiological Bases of Exercise.*** (2nd ed.), McGraw-Hill Book Company: New York.

[2] Lahey, J.W. (1984). What to do when the heat's on. *National Safety News*, 130(3): 62-64.

[3] McArdle, W.D.; Katch, F.I. and Katch, V.L. (1986). ***Exercise Physiology--Energy, Nutrition, and Human Performance.*** (2nd Edition), Lea & Febiger: Philadelphia.

[4] Kuhlemeier, K.V.; Miller, J.M.; and Dukes-Dobos, F.N. (1976). *Assessment of Deep Body Temperature of Workers in Hot Jobs.* National Institute for Occupational Safety and Health (DHEW), Publication No. 77-110.

[5] Murrel, K.F.H. (1965). ***Human Performance in Industry.*** Reinhold Publishing Corporation, New York.

[6] Ergonomics Guides (1971). Ergonomics Guide to Assessment of Metabolic and Cardiac Costs of Physical Work. ***American Industrial Hygiene Association Journal***, 32(8): 560-564.

[7] ACGIH (1987). ***Threshold Limit Values and Biological Exposure Indices for 1987-1988.*** American Conference of Governmental Industrial Hygienists: Cincinnati, Ohio.

[8] Crockford, G. W. and Lee, D. E. (1967). Heat-Protective Ventilated Jackets: A Comparison of Humid and Dry Ventilated Air. ***British Journal of Industrial Medicine***, 24(1): 52-59.

[9] Aurora. D. D. K. (1970). ***Physiological Evaluation of an Air Cooled Shirt Utilizing Dynamic Insulation.*** M.S. Thesis (unpublished), Kansas State University, Manhattan, Kansas.

[10] Billingham, J. (1962). Heat Exchange between Man and His Environment on the Surface of the Moon. ***Journal of British Interplanetary Society***, 17: 297-300.

[11] Burton, D. R. and Collier, L. (1964). ***The Development of Water Conditioned Suits.*** Technical Note ME-400, Royal Aircraft Establishment, Farnborough.

[12] Van Rensburg, A. J.; Mitchell, D.; Van Der Walt, W. H.; and Strydom, N. B. (1972). Physiological Reactions of Men Using Microclimate Cooling in Hot Humid Environments. ***British Journal of Institute of Industrial Medicine***, 29: 387-393.

[13] De Rosa, M. I. and Stein, R. L. (1976). ***An Ice-Cooling Garment for Mine Rescue Teams.*** Report of Investigations 8139, U. S. Dept. of the Interior: Bureau of Mines, Pittsburgh, PA.

[14] Kamon, E. (1983). ***Personal Cooling in Nuclear Power Stations.*** Final Report RP-1705 to EPRI, Palo Alto, CA.

[15] Tayyari, F. (1986). ***Design and Evaluation of a Micro-Climate Cooling System Using a Vest with Ice Bags.*** Ph.D. Dissertation (unpublished), Texas Tech University, Lubbock, Texas.
[16] Ramsey, J. D. and Coleman, A. E. (1972). Localized Cooling to Reduce Heat Stress. ***Proceedings of the Human Factors Society: 16th Annual Meeting***, 205-206.
[17] Nunneley, S. A.; Troutman, S. J.; and Webb, P. (1971). Head Cooling in Work and Heat Stress. ***Aerospace Medicine***, 42(1): 64-68.
[18] Williams, B. A. and Chamber, A. B. (1971). ***Effect of Neck Warming and Cooling on Thermal Comfort.*** NASA SP-302, pp. 289-294.
[19] Shvartz, E.; Ben-Mordechai, Y.; and Magazanik, A. (1976). Neck and Back Cooling in a Hot Environment. ***Israel Journal of Medical Sciences***, 12(8): 796-799, 1976.
[20] Tayyari, F. (1987). Estimating Energy Expenditure in Level Walking. ***Trends in Ergonomics/Human Factors IV***, (Shihab S. Asfour, editor), Part A, pp. 285-290, Elsevier Science Publishers B.V. (North-Holland).

Trends in Ergonomics/Human Factors V
F. Aghazadeh (Editor)

AN EFFICIENT METHOD OF VERIFYING HEAT STRESS PROBLEMS IN INDUSTRY

David C. Alexander and Leo A. Smith

Department of Industrial Engineering
Auburn University
Auburn, Alabama 36849 USA

1. INTRODUCTION

In process industries each summer, there are many urgent requests to resolve heat stress problems. The requests begin in June, and multiply with rising temperatures. The ergonomist or hygienist has only a few months to gather information, perform analyses, and recommend solutions.

Typically, tho solutions are not implemented in time to alleviate problems during the current season. This timo lag results in additional heat exposure and frustration for those working in the heat. At the same time, the ergonomist is working at a frenzied pace to gather and analyze data, and to develop solutions.

2. PROBLEM STATEMENT AND CONCEPT

The traditional approach to the analysis of heat stress problems is to gather environmental data (temperatures, humidity, air flows) along with detailed metabolic and work/recovery data. This approach allows extensive analysis and detailed conclusions.

However, this analysis may not be required or warrented. The purpose of most heat stress studies is to (1) determine if a problem exists and (2) propose workable solutions. Time spent to prove that problems exist takes away from the time to work on solutions. Typically, limited information is required to determine whether there is a problem or not. Then, if there is a problem, work can begin on a solution.

By collecting and using the minimum amount of information for each study, the ergonomist will be able to perform more studies within the summer season. This has the obvious advantages of (1) preventing heat illnesses, (2) completing more studies, and (3) saving valuable time for the ergonomist.

The point is that information has a cost to obtain and/or to analyze. In the field of heat stress, there are many forms of information - environmental, metabolic, and work/recovery data. Without good protocols, it is easy to feel a need to see "just a little more data". The unnecessary data collection costs in terms of time to perform a study, and in the performance of fewer studies.

Examples of inexpensive information include most of the environmental data, i.e., wet, dry and radient temperatures, air flow, and humidity. Information with moderate cost is heart rate, skin temperature, and estimates of work/recovery cycles, work hours, and clothing effects. High cost information includes measured metabolic data and work/recovery cycles.

By examining this information, one can see that initial diagnoses using the inexpensive information are the preferred way to begin a study. Then, if and when additional information is needed, it can be obtained. This can be called JIT (Just-In-Time) information.

Since selective use of information is important, this paper uses the concept of triage to recommend a fast and efficient method of verifying that a heat stress problem exists. Triage is a method of dividing battlefield casualties into three groups: those who are beyond help, those who do not need immediate help, and those for whom immediate help will make a difference. The limited medical resources on the battlefield are concentrated on those in the third group.

In a similar manner, triage can be used to categorize heat stress problems into three areas:

1. those where a solution is obviously necessary,
2. those where a solution is not necessary, and
3. those where more information is needed to determine whether a solution is necessary or not.

In the first case, the ergonomist can begin working on a solution sooner, rather than using a detailed analysis to prove that a problem indeed exists. In the authors' experience with industry, it is not uncommon to find that the effort required to prove that a problem exists is as expensive as the eventual solution to that same problem.

In the second case, the limited resources of the ergonomist can be used elsewhere, with little affect on the safety and health of those in the organization.

In the third case, just enough information is collected to determine whether a solution is required or not. The collection of this information can be thought of as a series of steps that simply narrows the area of uncertainity between known problems and the area where no problems exist. When it becomes known that a problem indeed exists, then solutions are developed. If it becomes known that no problem exists, the problem is dropped. The goal is to learn whether to pursue problem solutions or just to drop the problem altogether.

Figure 1 provides a picture of what this might look like. The area of uncertainity is narrowed with each additional piece of information. The rate and degree of the narrowing of the uncertainty band will vary with the information already available, and with the new information collected. It is not usually uniform.

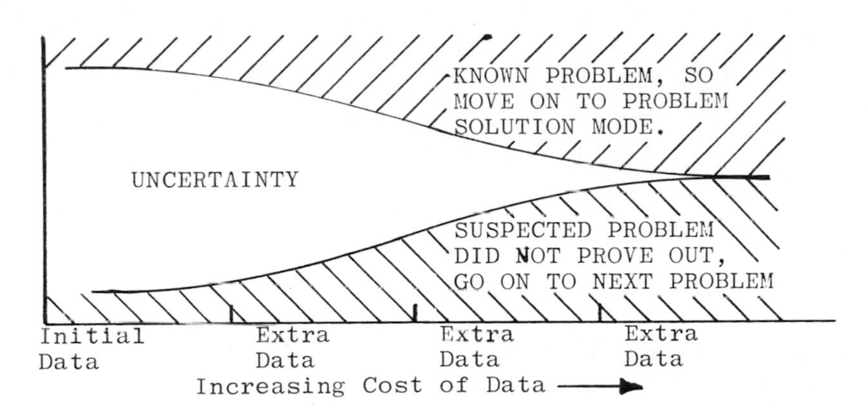

FIGURE 1. Use of Additional Information to Verify Problems

3. METHODOLOGY

3.1 STEP 1 - APPROXIMATED WBGT

The methodology advocated herein suggests using the minimum amount of data initially and that this data be obtained in the least costly manner. The initial data necessary to verify heat stress problems includes the information needed to calculate the wet-bulb globe-temperature and some estimate of the metabolic requirements for the task. These data, along with knowledge of the task's work/recovery regimen may be compared to the "heat stress alert limits" recommended by NIOSH [1] to evaluate the severity of the problem.

Figure 2 presents the NIOSH recommended limits for unacclimatized workers. A similar set of curves has been recommended for use in situations involving individuals known to be fully heat acclimatized to the task environment. If the intersection of the task WBGT and workload data falls above the appropriate work/recovery regimen curve then a problem likely exists.

The WBGT formula for indoor and outdoor work applications are:

$$\text{Indoors:}\quad WBGT = 0.7(Tnwb) + 0.3(Tg)$$
$$\text{Outdoors:}\quad WBGT = 0.7(Tnwb) + 0.2(Tg) + 0.1(Ta)$$

where Tnwb is the unaspirated wet bulb temperature, Tg is the black globe temperature, and Ta is the air temperature.

Numerous options are available to obtain these temperatures ranging from simple thermometers appropriately enclosed in wetted wicks and six inch copper spheres painted black to electronic WBGT meters with data display and/or printout devices. For convenience, purchase of one of the electronic

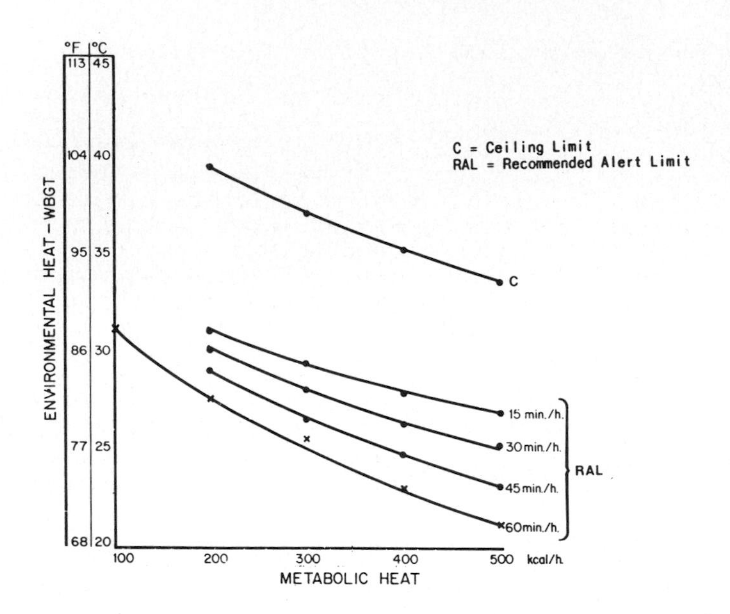

FIGURE 2. Recommended Heat Stress Alert Limits for Heat Unacclimatized Workers

WBGT meters is a wise investment if the plant anticipates having to investigate several heat stress related complaints.

The metabolic data can be much more difficult and time consuming to obtain. In this initial step it should be estimated using data available in various texts (for example see Alexander [2] p. 198, or the procedure suggested by NIOSH [1] p. 60. Likewise, complete knowledge of the work/recovery regimen requires time consuming task analysis and work measurement procedures. In this step in the investigation estimates only are necessary and may be obtained from talking with workers and their supervisors or by cursory observation.

Determining the intersection of the temperature, estimated metabolic levels, and work/recovery regimen intersection on Figure 2 provides a quick and simple estimate of the degree of the problem. If there is a question about the metabolic load, simply go to a higher level when reading the figure. Similarly, if one is concerned that the work/recovery regimen data may be wrong, assume a "worst case" situation. Use of the NIOSH recommended levels for unacclimatized workers (Figure 2) is also recommended at this point just to be on the safe side.

If this evaluation indicates that no real problem exists, the study may be terminated. Also if a serious problem is indicated, the ergonomist may choose to immediately go on to a problem solution phase. However, if step 1 indicates a possible problem or a borderline situation, then one should go on to the next step in which the metabolic levels and the work/recovery regimen are more accurately determined.

3.2 STEP 2 - DETAILED WBGT

In this step the ergonomist should perform a more detailed evaluation of the metabolic demands of the task. Since the metabolic level is an important contributor to heat stress, studying it closer will provide additional insight. Variation in estimates of metabolic levels of as little as 15 percent can make a significant difference in the results obtained from the use of Figure 2. More detailed metabolic estimates may be obtained by more careful evaluation of published data or by estimation from heart rate data obtained from one or two workers using simple wrist palpation or a "home exercise" heart rate monitor. Various tables and graphs of heart rate data vs. energy expenditure data are presented in standard ergonomics texts. Again one must recognize that this data, while superior to that used in step 1, is still only an estimate of the metabolic load since the heart rate data, while simple to obtain, is quite sensitive to the thermal environment experienced by the worker.

A close look at the worker population is appropriate at this time. Exactly what is the composition of the work force - male/female, age, degree of physical fitness? How are workers selected? What is the work experience and level of safety training typically brought to this job by the workers? Can they change jobs if they request a transfer? Are other comparable jobs available, i.e., would someone feel forced to stay on the job in spite of perceived health hazards? A careful evaluation of the percentage of workers who could be expected to be acclimatized to the job should be made. Perhaps the use of "unacclimatized" worker curves in Figure 2 is too conservative except for newly assigned workers.

Another area to investigate at this time is the available health data. Have there been heat related medical incidents and, if so, does the seriousness of these incidents correlate with the estimates of the severity of the job heat stress obtained in step 1? If the correlation is low, then additional fact finding is definitely needed.

Another action that should be taken at this time is to estimate the air velocity in the work area and to evaluate the possible contribution of the types of clothing and personal protection equipment worn by the workers on the heat load they experience. Information relative to these and other factors are discussed by NIOSH [1] and Smith and Ramsey [3].

3.3 STEP 3 - APPLICATION OF THE HEAT STRESS INDEX

If after completing step 2 the ergonomist is still uncertain as to the extent of the potential severity of the heat stress situation it becomes appropriate to use the calculation power and detail of the Heat Stress Index (Belding and Hatch [4]). The use of the HSI requires knowledge of metabolic levels equal to that obtained in step 2 but more specific knowledge of the environmental conditions. Thus some additional time must be devoted to data collection.

The HSI offers the advantage that one can easily see the degree of the problem - the more the calculated value of the HSI exceeds 100, the more serious is the problem. Also using the HSI one can calculate allowable exposure time and minimum recovery time estimates using the appropriate data for both the work and rest area environments and activities.

To conveniently perform HSI calculations the ergonomist needs to write a program to manipulate the equations. This requires a time investment, but provides the capability to quickly perform "simulations" to test the sensitivity of the level of the heat stress problem to changes in the environmental and metabolic levels. This evaluation can offer considerable insight into what actions might best be taken to control the heat stress situation.

3.4 STEP 4 - PHYSIOLOGICAL MEASUREMENT

Often manipulation of the HSI will provide enough information to determine the severity of the problem and suggest the actions necessary to effect a solution. If doubts still remain as to the significance of the problem, and thus the capital investment justified to resolve it, a detailed work physiology and task activity analysis should be performed as a final assessment step. The time invested up to this point may be as high as 40 manhours; nevertheless, the additional time required for carrying out metabolic and complete task analysis measures can run as high as 4 to 10 weeks of the ergonomist's time. Thus step 4 should be viewed as an "only if absolutely necessary" activity.

During step 4 detailed measures of the heart rate and energy expenditure of the workers during task performance and recovery are obtained. These measures allow assessment of the degree and intensity of the work and the resulting stress. Recognizing the potential variability in the observed responses, one should monitor several people working on all the suspect tasks. Additionally, the ergonomist should determine baseline fitness levels for each of the subjects. At the same time, a detailed task analysis should be completed using appropriate job analysis and work measurement procedures to obtain an accurate description of all task activities and the time required to perform each.

The detailed data should be discussed with plant medical, industrial hygiene, and industrial engineering personnel to obtain a consensus view of the extent of the heat hazard and suggestions for its resolution. This level of analysis is appropriate when the question is very specific, when the solution implemention costs are very high, or when there are many people exposed to the heat stress.

4. AN EXAMPLE

A 48 year old female performs a job in a textile mill that requires her to work in three different areas in the plant.

Data pertinent to her task is given below.

	Area 1	Area 2	Area 3
Working heart rate	120	105	87
Globe Temperature	88	82	75
Air Temperature	81	82	75
Natural wet-bulb	78	70	66
Air movement (fpm)	50	100	50
Personal Protective Equip't	Rubber apron and gloves	Rubber gloves	

Her work time in each area is
Area 1: 7:30 - 10:00 AM; 10:45 - Noon; 1:30 - 2:30 PM
Area 2: 10:00 - 10:45 AM; Noon - 1:30 PM
Area 3: 7:00 - 7:30 AM; 2:30 - 3:00 PM

4.1 STEP 1 - APPROXIMATED WBGT

WBGT = .7 NWB + .3 GT
Area 1 - .7(78) + .3(88) = 81 F; Area 2 = 73.6 F;
Area 3 = 68.7 F

At this point, Area 1 is a concern, especially if the work load is heavy. A description of the work tasks to estimate the metabolic load (in Kcal/hr) is needed, and normally would be available. From the heart rate data, estimate a metabolic load of 350 Kcal/hr. The work in Area 1 lasts for 2.5 hours each morning.

Based on the Figure 2, there is concern with continuous work at 350 kcal/hr at a WBGT of 81 F. Note that if the work were only 30 min/hr, or if the work load were less than 225 Kcal/hr, there would be less cause for concern.

Since the task is still in the uncertainty region, go on to Step 2.

4.2 STEP 2 - DETAILED WBGT

A closer look at the work/recovery regimen shows that the task is not performed all day, and may be self-paced.

The work population includes females and older workers (above 45). No information is known about the availability of other jobs. No information about health problems is known. The air velocity in that area is low. Part of the body is covered by impervious clothing, thus restricting evaporative cooling somewhat.

Using this information with adjustments indicated in Smith and Ramsey [3], the WBGT can be adjusted by - 4 F. The concern is more serious now. Therefore, going to Step 3 is appropriate.

4.3 STEP 3 - APPLICATION OF HEAT STRESS INDEX

After estimating the additional data needed to use the HSI [4], its application reveals serious problems. The HSI for area 1 is far above the index of 100 where problems can begin.

The "Allowable Exposure Time" (or AET) is determined to be approximately 1/2 hour which is much less than the 2 1/2 hour exposure actually encountered. This indicates a serious health concern.

It is time to move into a problem solution mode. There is no need to use detailed metabolic measures and work task analysis to continue to evaluate the potential seriousness of this situation.

6. BENEFITS AND CONCLUSIONS

The major benefit of this approach is a more effective use of limited ergonomist's time, as shown in Figures 3 and 4. The other benefits include effective use of data, a more timely response to requests, and a methodology that can be understood and used by technicians.

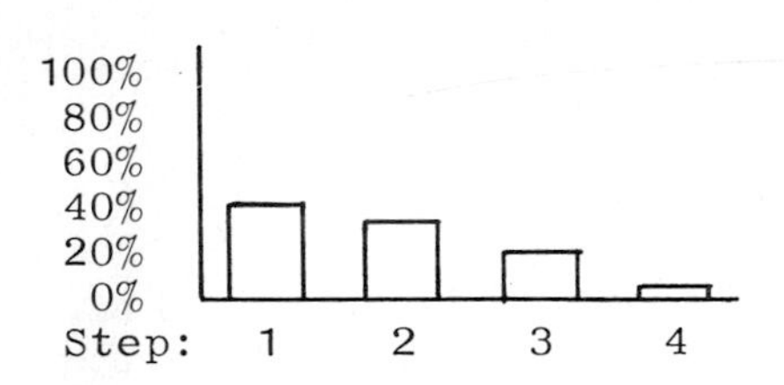

FIGURE 3. % of Problems Verified at each Step

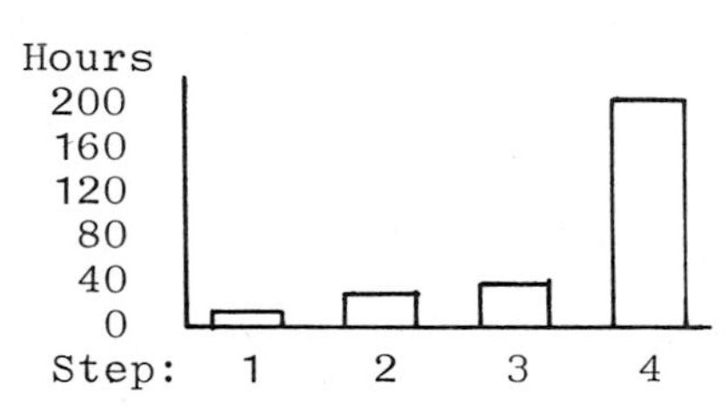

FIGURE 4. Cumulative Time to Complete Steps

Through the use of this methodology, the ergonomist will be able to deal with more heat stress problems in a given season. Coincidentally, the problems will be resolved faster in the operations areas, thus easing the burdens of heat stress for operating personnel.

REFERENCES

[1] NIOSH, Occupational Exposure to Hot Environments: Revised Criteria 1986, U.S. Department of Health and Human Services, DHHS (NIOSH) publication No. 86-113.
[2] Alexander, David C., The Practice and Management of Industrial Ergonomics, Prentice-Hall, Inc., Englewood Cliffs, NJ, 1986.
[3] Smith, James L. and Ramsey, Jerry D., "Designing Physically Demanding Tasks to Minimize Levels of Worker Stress," Industrial Engineering, May, 1982, p. 50.
[4] Belding, H. S. and Hatch, T. F., "Index for Evaluating Heat Stress in Terms of Resulting Physiological Strain", Heating, Piping and Air Conditioning, Vol. 27, August, 1955, pp. 129-135.

Trends in Ergonomics/Human Factors V
F. Aghazadeh (Editor)

THERMAL CHARACTERISTICS OF SURGICAL GOWN MATERIALS

Robert E. SCHLEGEL, Sharon D. CHEATWOOD and Brad SCHMIDT

School of Industrial Engineering
University of Oklahoma
Norman, Oklahoma 73019

A study was conducted to evaluate the heat transfer characteristics of three different surgical gown materials: 50% cotton/50% polyester, 40% polyethylene/60% woodpulp, and a microfilter layer between two layers of polypropylene. Thermal characteristics of the surgical environment were simulated in an environmental chamber while four male subjects performed a manipulative task under two portable Operating Room lights. Results of the study indicated that there was no significant difference between the 50/50 woven cloth and the 40/60 nonwoven material, but that both of these gowns produced a greater skin temperature increase than the microfilter layer material.

1. INTRODUCTION

Surgical gowns provide a bacterial and moisture barrier between operating room personnel and the patient. The traditional surgical gown is a nondisposable woven cotton blend. In many operating rooms there has been a transition from the nondisposable gown to a disposable gown made of a nonwoven fabric. Many OR personnel prefer the cloth gowns which are considered more comfortable, cooler, and more durable. However, the nonwoven gowns are often preferred by hospital administration because they reduce the hospital overhead involved in cleaning, sterilization and repair.

Regardless of the type of material, OR gowns should provide sufficient heat transfer that the body surface temperature remains within comfortable limits. Some surgeons have been hesitant in adopting the disposable gowns because of the discomfort due to warmth of the gown while operating. This factor has appeared constantly during informal interviews with surgeons. Product managers and sales representatives for disposable gowns are aware of this problem.

A study was conducted to evaluate the heat transfer characteristics of three different surgical gown materials: 50% cotton/50% polyester, 40% polyethylene/60% woodpulp, and a microfilter layer between two layers of polypropylene. Thermal characteristics of the surgical environment were simulated in an environmental chamber while subjects performed a manipulative task under two portable Operating Room lights.

2. BACKGROUND

The human thermoregulatory system consists of three control mechanisms: heat transfer by blood flow, secretion of sweat, and shivering. Through the heat transport function of the blood, the blood vessels, especially the capillaries, act as distributors of heat, picking up heat from warm tissues and distributing it to cooler tissues. By using the blood for transfer, heat can be moved from the body's interior to areas of the skin that are cooled by the environment. When clothing impedes the heat transfer from the body, the body surface becomes warmer. This warmth is transferred to the interior of the body by the blood stream and is indicated by higher skin and core temperatures.

The temperature of the skin provides a quantitative measure of the heat that is transferred by the blood to and from the body's interior. Mean skin temperature is an important measure of human response to the thermal environment. The skin temperature may be measured at various body sites with thermistors. Ramanathan's [1] weighted mean skin surface temperature formula is recommended by Mitchell and Wyndham [2] for its simplicity and small number of site readings. Four body sites are used: the chest, upper arm, medial thigh and lower leg. The formula is as follows:

$$T_{surface} = .3\ [t_{chest} + t_{arm}] + .2\ [t_{thigh} + t_{leg}]$$

The indoor climate comfort zone is 68 to 72°F (20 to 22°C). Grandjean [3] has shown that if test subjects are asked to report when they feel really comfortable, the range is narrow, perhaps 4 or 5°F (2 or 3°C). The range of thermal comfort depends on many variables including the properties of the clothing being worn and the level of physical exertion.

The properties of clothing which determine the rate of heat transfer are the resistance to dry heat loss through convection and radiation and the resistance to evaporative heat loss through the transfer of evaporated sweat [4]. If the property of the fabric does not allow proper body heat exchange to take place, thermal discomfort is experienced. This sensation of discomfort can range from mere annoyance to one of pain according to the extent to which the heat balance is disturbed. Some of the important side effects of thermal imbalance are weariness, sleepiness, decline in performance, and increased liability for errors [3].

Clark and Cena [5] noted the following three factors which determine the radiant heat loads experienced by clothed men in sunshine:

(1) posture relative to the direction of the heat source,
(2) reflection and transmission properties of the fabric, and
(3) insulation between the site of radiation absorption and the skin surface.

These three factors must also be taken into consideration when working under overhead lights.

In a normal situation, humans can restore a correct balance of heat exchange by modifying the environment. In the operating room, however, this ability is lessened. Once in a sterile gown and gloves at the operating table under overhead lights, a surgeon cannot remove clothing if he becomes too warm. This warmth can be experienced even though the room temperature away from the overhead lights is 68 to 73°F (20 to 23°C) and the relative humidity is 40 to 60%.

3. SURGICAL GOWN MATERIALS

Operating room clothing should be made of nonlinting material, must constitute an effective bacterial barrier and must be comfortable. Operating room fabrics should be made of nonflammable materials. This is of special concern in operations where cauteries and lasers are in use. Also, the fabrics should not allow for static buildup.

In order to be an effective bacterial barrier, the fabric must be resistant to air-borne and liquid-borne contamination. Resistance to air-borne contamination is achieved by taking into consideration two important factors. First, bacteria are approximately 2 microns in size. Therefore, the interstitial spaces that result from the manufacturing process should be less than 2 microns. Second, bacteria cannot travel a path consisting of three consecutive right angles. Thus, the fabric should be processed so that a *tortuous path* exists from one fabric surface through to the other fabric surface. Resistance to liquid-borne contamination is achieved by using fabrics that are moisture repellant where the surface tension of the fabric is greater than the surface tension of liquids.

Three types of gowns were tested in this study, a woven cloth gown and two different types of nonwoven gowns. All gowns were extra large in size. This size is purchased by many hospitals because it fits a large number of body types. The gowns are the wrap-around type that fasten at the neck and provide a "sterile back" for the person wearing the gown. Knitted material is used for cuffs that hug the wrist. Detailed descriptions of the gowns follow.

(1) The woven gown was made of a 50% cotton/50% polyester blend that is treated to prevent static buildup. The front of the gown from shoulder to hip has an extra layer of fabric made of 100% cotton. This extra layer is also found on the forearm area of the gown. The gown is manufactured using a standard weave at 90° angles and a thread count of 180 threads per inch (Cloth 180). The fabric is classified as a Class II fabric in terms of flammability. This is intermediate flammability with a time of flame spread from 4 to 7 seconds [6]. The bacterial efficiency of the fabric (Cloth 180) is less than that of Cloth 280 which has a Moist Bacterial Barrier filtration of 93.7%. After a new gown is washed and sterilized three times, this is reduced to about 26% [7]. This percentage is the same as for Cloth 140 which is not recommended as a bacterial barrier fabric.

(2) One disposable nonwoven gown was made of 40% polyethylene and 60% woodpulp. This fabric is purchased by several medical companies. The fabric manufacturing process is the spun-laced process in which a web of polyethylene is mechanically bonded to the woodpulp by water jets. Small needle-like holes are punctured straight through the fabric with the water jets. Air gaps between the strands of polyethylene and the woodpulp also exist. This accounts for the fabric's Moist Bacterial Barrier filtration of 64.9%. This fabric is a Class III fabric in terms of flammability. This is highly flammable with a flame spread of less than four seconds. A rapid and intense burning is characteristic of this class. Such textiles are considered dangerously flammable and are recognized by the trade as being unsuitable for clothing because of rapid and intense burning [6]. This fabric is also not recommended for use with Class IV lasers [8]. In addition, precautions should be taken if this fabric is used where a cautery might fire and ignite the fabric.

(3) The other disposable gown was made of a layer of microfilter material between two layers of spunbonded polypropylene. The layers are point heat bonded. The fabric, because of the manufacturing process, has 80 to 100 consecutive right angles of microfilter or *tortuous path* from one fabric surface to the other. The fabric has a Moist Particle Filtration Efficiency of 96.9%. This fabric is a Class I fabric with a flame spread greater than seven seconds [6]. The fabric melts before it ignites.

4. METHODOLOGY

Four male subjects from the same geographical area, between 19 and 26 years of age, participated in the study. Each subject wore each of the three gowns while performing a simulated surgical task for sixty minutes. The order in which the subjects wore the surgical gowns was counterbalanced. Body surface temperature was measured every two minutes to indicate how much heat transfer was restricted by the surgical gown. If more heat was retained by the gown, the body surface temperature would be higher than if the heat was transferred away from the body.

An environmental chamber was used to maintain the levels of the control variables as follows:

(1) room temperature - 70°F (21.1°C)
(2) relative humidity - 50%
(3) air velocity - 25 cycles per hour
(4) constant heat radiation from surgical lights.

Surgical lights were used to illuminate the task. The lights were placed at the head and foot of the "operating table". Two subjects stood on either side of the "patient". Five thermistors were plugged into each of two tele-thermometers. Four thermistors were attached to each subject using micropor tape to minimize the heat transfer barrier effect. Two thermistors, one per telethermometer, were used to measure the ambient room temperature. The clothing (scrub suit, booties, and surgical cap) worn by each subject was the same type of clothing worn by OR personnel (Table 1).

Table 1. Clothing Worn By Test Subjects.

Briefs

Socks (ankle length)

Shoes (tennis shoes)

Scrub Suit Top (short sleeved and collarless)
50% cotton/50% polyester
180 thread count

Scrub Suit Pants (loose, fitted at waist)
same fabric as top

Surgical Cap (nonwoven, loose fitting)

Surgical Mask

Shoe Covers (disposable, fabric unknown)

Surgical Gowns (loose fitting)
(1) 50% cotton/50% polyester
(2) 40% polyethylene/60% woodpulp
(3) polypropylene/microfilter

Laytex Gloves

5. RESULTS

Baseline skin temperatures were collected during the first ten minutes of the task. The average of these readings was subtracted from each of the last five readings at the end of the 60-minute trial. The average increases for the three gowns are presented in Table 2. Analysis of variance verified a significant difference between the gowns ($p < 0.0001$).

A Duncan multiple range test indicated that there was no significant difference between the 50/50 woven cloth and the 40/60 nonwoven material, but that both of these gowns produced a greater skin temperature increase than the microfilter layer material. This gown allowed heat to be transferred from the subject to the atmosphere better than the other two gowns.

Table 2. Average Increases in Mean Skin Temperature.

Gown	Temperature Increase
50/50 Woven Cloth	1.88°F (1.045°C)
40/60 Nonwoven	2.01°F (1.115°C)
Microfilter Layer	0.48°F (0.266°C)

To evaluate the effects of minor fluctuations in the air temperature, a Pearson correlation was computed between the room temperature and the skin temperature. A correlation coefficient of 0.21 leads to the conclusion that the room temperature did not have a significant effect on the skin temperature of the subjects.

6. CONCLUSIONS

The average increases in body surface temperature for the different types of surgical gown materials were not the same. The gown with the microfilter layer allowed surface body heat to be transferred better than the 50/50 woven cloth and 40/60 nonwoven gowns. Furthermore, the microfilter fabric has the highest bacterial barrier efficiency and the best flame retardance.

It is recommended that further studies be performed to examine other activities in the operating room environment. These activities may include holding clamps, handing instruments to the operating surgeon, and helping the operating surgeon get prepared for the operation. In this study, the surgical lights were a control variable. A changing radiant heat load should also be investigated.

Standards for the use of fabrics in the Operating Room should be established requiring materials of the highest quality available on the market. There should be no compromise in the areas of bacterial efficiency, particle generation, flammability and comfort when these standards are developed. It is hoped that the information provided by this study will be of some value in the development of these standards with respect to thermal comfort.

REFERENCES

[1] Ramanathan, N.L. (1964). A New Weighting System for Mean Surface Temperature of the Human Body. *Journal of Applied Physiology, 19,* 531-533.

[2] Mitchell, D. and Wyndham, C.H. (1969). Comparison of Weighting Formulas for Calculating Mean Skin Temperature. *Journal of Applied Physiology, 26,* 616-621.

[3] Grandjean, E. (1982). *Fitting the Task to the Man, An Ergonomic Approach.* London: Taylor and Francis, Ltd.

[4] Holmer, I. and Elnäs, S. (1981). Physiological Evaluation of the Resistance to Evaporative Heat Transfer by Clothing. *Ergonomics, 24,* 63-74.

[5] Clark, J.A. and Cena, K. (1978). Net Radiation and Heat Transfer Through Clothing: The Effects of Insulation and Color. *Ergonomics, 21,* 691-696.

[6] Consumer Product Safety Commission (1982). *The Govmark Book of Reprints of Flammability Standards and Flammability Test Methods of Textiles, Plastics and Other Materials Used in Home and Contract Furnishings.* 1st Edition, Bellmore, New York: The Govmark Organization, Inc.

[7] Domin, M.A. (1985). Reducing Risk in Surgery: The Role of Barrier Fabrics. Oklahoma City, Oklahoma: Seminar at the Park Suite Hotel.

[8] American National Standards Institute (1984). *Laser Safety in the Health Care Environment,* ANSI Z-136.3 Draft.

Trends in Ergonomics/Human Factors V
F. Aghazadeh (Editor)

THERMAL COMFORT SIMULATION TESTS FOR IMPROVAL OF HOT ENVIRONMENT IN A GLASSWORK

Dr. László BANHIDI, László FABó(1),
Dr. József SZERDAHELYI(2)

(1)ÉTI, Hungarian Institute for Building Science
(2)OMI, Hungarian Institute for Working Health

The workers in a Hungarian glasswork, producing individual blown products, are often exposed to 60 oC globe temperature. A design institute was commissioned to improve the comfort conditions by the application of ventilation system. In-situ thermal and physiological measurements followed by laboratory simulation of the hot environment and radiant heat sources were involved in the research. 16 glasswork employees were invited in the laboratory for tests. Different ventilation and air cooling alternates were simulated. The tests resulted subjective thermal comfort values and skin temperatures that were consequently used for evaluation of ventilation alternates, and for selection of optimal parameters. The paper presents the methods of the in-situ measurements and laboratory simulation tests and the results.

1. INTRODUCTION

In a Hungarian glasswork the workers often experience extremely hot environment, mostly in summertime, partly due to heat radiation, partly to excessive indoor air temperature. In winter thermal comfort problems are caused by asymmetric radiation and cold air draught.

This environment adversely influences the comfort and the performance of the workers, for many of them leave the factory, causing manpower management problems, being difficult to find properly trained replacement.

Following different attempts to solve the problem the management contracted a design institute to design a ventilation system for the improvement of the comfort conditions. The ÉTI and OMI institutes were involved to do the necessary research and investigations as subcontractors.

The investigations consisted of two stages:
- measurements at the site in summer and winter,
- laboratory tests under simulated summer conditions.

2. MEASUREMENTS AT THE SITE

The thermal comfort investigations carried out at the site included:
- recording the subjective thermal comfort sensation of the workers on individual test questionnaire sheets,
- measurements of the thermal comfort parameters of the indoor environment.

2.1. Subjective thermal comfort tests

To reveal the subjective judgement of the thermal comfort conditions and reflects of the workers test questionnare sheets were distributed among the workers. The test sheet had the form below:

Personal thermal comfort test sheet

....
code number and date

Job description:
Age: Sex:
For how long have you been employed in the present job:
Typical physical activity of your work: standing, sitting, walk
Are you sensitive to changes in weather:
cold front : high, medium, low
hot front : high, medium, low
Your sensitivity to draught : high, medium, low
How can you tolerate hot : well, medium, not
How can you tolerate cold : well, medium, not
Do you perspire when hot : heavily, medium, slightly
Do you smoke? : yes, no (....cigarettes/day)
Do you drink alcoholic beverages : regularly, occassionally, not
Do you take any medicin : regularly, occassionally, not
if you do, what:......................................
Any health problems: high/low blood pressure, heart, lung, bile, kidney, thyroid, gland, neurosis, ulcer, allergy, rheumatic complaints, skin desease, other:
....

Are you satisfied with labour conditions: yes, no
What disturbs you most : in winter temperature,
heat radiation,
ventilation system,
draught,
in summer temperature,
heat radiation,
ventilation system,
draught
Do you use the fans: yes, no
where do you put them: ....
why there: ....
Which ventilation did you find best, and why:................................
Which air flow direction do you prefer: to your face, to your back, from above, from below, or:...
Do you change the ventilation rate during the working hours:
in winter: do you change it yes, no

how:..
in summer: do you change it yes, no
how:..
Which period of the year do you find most unpleasant:.........................
why:..
Do you find the air dry wet,
Describe your working clothing
in winter:..
in summer:..
Mark your most frequent position on the sketch below /oven no., or cooling belt no. should be noted/

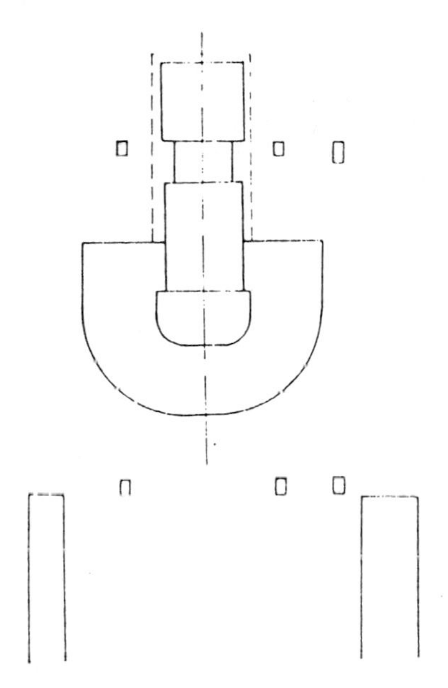

Comments on the questionnaire:
a./ The subjects were not required to give their names, (code numbers were used for identification), following our routine in similar tests, that proved to contribute to relaxing test stress and obtaining more reliable replies.
b./ First group of the questions concerned the general state of health, habits and other factors of the subjects influencing the thermal comfort sensation.
*c./ The next fundamental question concerned the workers' satisfaction with the working conditions, investigating the order of importance of disturbing effects.
d./ The following group of questions targeted the use of the current ventilation system.
e./ In the next section the workers' choice was asked of the best ventilation system among those tried during the recent years.
f./ The next question concerned the direction of the ventilation air flow best preferred by the workers.
g./ Finally the ventilation demand and clothing were surveyed in different seasons.

The subjects were asked to mark their most frequent position on the sketch at the bottom of the sheet. /Very few of them have a permanent working place in the workshop/.

As it can be seen, the questionnaire ensured the possibility to obtain all necessary data for preparing the experimental program and drawing some conclusion.

2.1.1. Evaluation of the replies

213 of the distributed 300 questionnaire were returned that gives a fairly good 71% response ratio. Not all of the returned sheets were completely filled, some of the questions remained without answer or, in some cases, the reply was not evaluable. The first main facts of the test were the following:

180 men and 33 women returned the filled questionnaire.
46 men (34%) and 16 women (61%) were satisfied with the labour conditions, while dissatisfied were 112 men (66%) and 10 women (39%).

The above percentage values, however, are somewhat deceiving for two reasons:
- 18% of the replies were not evaluable,
- the number of the female workers was relatively low, and they work at the worst place, on the platform in front of the oven.

Since the response ratio to the question concerning discomfort effects were 90% for men and 85% for women, it was reasonable to assume that those who gave not evaluable reply or no reply at all might belong to the dissatisfied.

As regards the four discomfort sources listed the replies showed the following distribution:
In winter 132 men and 25 women mentioned only one discomfort source,29 men and 2 women mentioned 2, three factors were referred to by one woman and all the four by only one man. In winter 132 men and 25 women mentioned only one discomfort source, 29 men and 2 women mentioned 2, three factors were referred to by one woman and all the four by only one man.

Distribution of the replies as regards the different listed discomfort sources was the following:

Winter complaints are due to: temperature : 24 men, 6 women
heat radiation : 19 men, 1 woman
ventilation : 18 men, 2 women
draught :135 men, 23 women

Winter complaints are due to: temperature : 24 men, 6 women
heat radiation : 19 men, 1 woman
ventilation : 18 men, 2 women
draught :135 men, 23 women

To the question concerning changing the adjustment of the ventilation the replies were:
In winter 107 men and 10 women regularly used the adjustment possibility of the current mobile ventilation system, while 43 men and 12 women did not.
In summer 99 men and 16 women regularly used the adjustment possibility of the current mobile ventilation system, while 42 men and 5 women did not.
Most unpleasant season of the year from indoor thermal comfort aspect in the glasswork was found to be:
- Spring: by 14 men and 1 woman
- Summer: by 129 men and 21 women
- Fall: by 11 men and no woman
- Winter: by 22 men and 3 women.

Summerizing the results it was found that:
a./ More than 60% of the workers is dissatisfied with the working conditions.
b./ Winter complaints are due in 70% to draught,
in 13% to working conditions,
in 8.5% to heat radiation,
in 8.5% to ventilation system.
c./ Shares of the above discomfort factors in summer complaints are 25%, 39%, 27% and 9%, respectively.
d./ The following percentage of the workers used the adjustment possibility of the present ventilation system:
in winter: 68%,
in summer: 71%.
e./ From indoor thermal comfort aspect summer was said to be the most unpleasant season by 75% of the workers and only 12% said it was winter (spring and fall were claimed by the rest).

From the above findings the following conclusions might be drawn:
- from thermal comfort point summer was the most unpleasant period of the year in the glasswork,
- summer discomfort complaints are generally due to hot indoor climate,
- in winter most discomfort is caused by draught,
- the possibility of adjusting the ventilation system to local personal comfort is used by 70% of the workers.

2.2. Measurements of thermal comfort data at the site

Neglecting the detailed description of measured parameters, methods and instrumentation the following summary is given of this experimental stage.

Thermal and air flow conditions (air- and globe temperatures, heat radiation, air velocity, relative humidity) were mapped at different frequented working places in the workshop.

Of the results we should like to point out two extreme values measured at the site:
- air temperature: 55 - 60 oC,
- outdoor air temperature available for ventilation on the hottest days: 32 oC.

3. LABORATORY TESTS UNDER SIMULATED SUMMER CONDITIONS

3.1. Test arrangement

Summer conditions were to be simulated in the Microclimate Laboratory of the Hungarian Institute for Building Science. The air conditioning system of the test room of the laboratory is designed to produce max. 40-45 oC air temperature, that is well below the required 55-60 oC globe temperature, therefore supplementary individual radiating panels had to be used.

Since the radiant heat load had two sources at the original site, one was the orifice, the other the external envelop of the ovens, the 5*3 m test room was arranged accordingly. The oven orifice was simulated by a Solectronic gasfired infrared radiant screen of 950 oC surface temperature. The surface temperature of the ovens was simulated by 5 pcs 1kW electric line radiators mounted on one end wall of the test room. The flue gases from the gas appliance were exhausted from the test room. By permanent operation of the air conditioning system and by setting the supplementary radiant heat sources at the required distances from the test zone the summer indoor conditions measured at the site and extreme conditions as well could be simulated.

Workers from the glasswork were invited to take part in the tests as subjects.

Glass blowing work could not be simulated, therefore the subjects were requested to perform other manual work, assembling and disassembling propelling pencils, standing at one place with their back to the radiant heat sources, and from time to time go closer to the radiant heat sources as if drawing liquid glass from the oven.

Different ventilation methods were simulated by combinations of perforated suspended ceiling, simulating flow conditions of the general ventilation at the site, and dampers that directed part of the supplied air into the required direction locally, in concentrated jets. The working place desk was positioned to ensure that the jet hit the subject frontally from above.

Three workers were tested simultaneously in the test room where the following microclimate parameters were maintained:
- 50-60 oC Vernon globe temperature,
- 40-42 oC sheltered air temperature,
- 34-36 oC air temperature,

and different air velocities as measured at head level in the ventilation flow directed downward:
0; 0.3; 0.5; 1.0; 1.5; and 2.0 m/s.

Tests were also carried out to study the thermal comfort benefit of reduced heat radiation from the envelop of the ovens. In this case the Vernon globe temperature was reduced to 40-44 oC, and the sheltered air temperature to 36-38 oC, while keeping the air temperature unchanged. This was achieved by sheltering part of the radiant heat sources.

The test subjects, 3 persons every day, arrived at the Microclimate Laboratory at about 9 in the morning and, having completed the tests, left about 2 or 3 pm. The tests took 4.5-5.0 hours.

The tests had three stages:
- preliminary examinations in the adaptation room,
- simulation tests in the test room,
- post-test examination in the adaptation room.

3.2. Test method

The following data were measured and recorded:
Preliminary tests: recording the psychical and physiological condition of the subjects by filling the daily pre-test sheet, measurement of skin temperature, pulse, blood pressure, response time, etc.

Simulation tests: thermal and flow parameters of the indoor climate, physiological data /skin and core temperature, moisture loss by perspiration, etc./, subjective thermal comfort sensation data /sensation scales were used for temperature and draught/.
Post-test examination: the same data were recorded as before the test completed with the subjects' opinion on the tested ventilation versions.

4. EVALUATION OF THE LABORATORY TESTS

From the tests more than 10,000 data were available for evaluation and determination of certain limiting values of the ventilation acceptable from thermal comfort aspect. Omitting the detailed presentation of data processing, the principles, findings and conclusions are given below.

The tests were based on two assumptions as to the temperature of the supply air, that were used alternately:
- some cooling is available (adiabatic cooling by humidifying), that can reduce the dry bulb temperature of the supply air by about 5 oC,
- outdoor air is supplied without any treatment, and blown to the workers.
In this respect the tests showed that:
a./ A +27 oC supply air temperature, that assumes a 5 degree cooling of the outdoor air, improves the subjective thermal comfort sensation from "very hot" to "acceptably hot" if the air velocity reaches 1.0 m/s.
b./ In the same case the same improvement can be achieved at 50-70% of the subjects, at 0.5 m/s velocity, by sheltering the radiant sources.
c./ In case of +32 oC supply air temperature at least 1.5 m/s air velocity is necessary for a general "acceptably hot" vote, whereas 1.0 m/s reduces the number of this vote to 50%.
d./ Sheltering the radiant sources and providing 1.0 m/s air velocity yields "acceptably hot" vote of 50-70% of the subjects.

On the course of the basically thermal comfort tests physiological examinations were also involved to determine certain functional changes of the human organism, that are of distinguished importance in case of physical work under thermal exposure. The frequency of pulse and the variation of blood pressure were measured in order to determine loads on heart and circulation. For the determination of the heat, water and salt balance of the body, skin and core temperatures and the perspiration were also measured, respectively.

The simulation test was divided into 30 minute periods. Climatic conditions were constant during a period. This time lapse proved sufficent to achieve steady state conditions in pulse frequency and core temperature, and allowing for establishing variation trends of the other physiological parameters.

Initial mean value of pulse frequency was found to be 80.5/min, (with 56 and 110/min extremes), that practically showed no change by the end of the tests: the mean was 80.1/min, with 56 and 100/min extremes. Most of the measured pulse frequencies fell far from the 110/min limit of physiological acceptability. Similar findings characterize the variation of the blood pressure.

The average skin temperature (tsk), measured at the end of each 30 min period showed characteristic changes. At 27 oC supply air temperature the increase of the air velocity from 0.5 to 1.0 m/s reduced "tsk". Sheltering the radiant sources alone and together with an increase in the air velocity resulted further reduction. At a constant air temperature and sheltering the reduction of the air velocity to 0.5 m/s causes a sudden increase of the skin temperature. Ceasing the sheltering at 32 oC supply air temperature further increases "tsk", that was reduced again by a repeated use of sheltering.

Under the climatic and labour conditions of the tests the core temperature showed minimal fluctuation within the acceptable 1.0-1.2 oC range of variation.

The average quantity of the perspiration loss was 683 gr, but in a number of cases exceeded or almost reached 1000 gr. The results showed the degree of themal exposure, proved the acclimatization ability of the tested persons, and the existance of likely significant heat balancing reserves.

It can be generally stated on the basis of the physiological measurements that the findings and conclusions agree with those from the thermal comfort tests. Against the adverse effects of the high air temperature and infrared radiation primarily sheltering and air velocities higher than 1.0 m/s can protect. On the base of the laboratory simulation these should be applied in the glasswork at the site.

5. PRACTICAL UTILIZATION OF THE TEST RESULTS

Findings and conclusions of the research were utilized in the design and installation of a new ventilation system that now operates. The parcentage of workers satisfied with thermal comfort conditions increased to 80-85%, a fairly good value in this respect.
The whole research program can be regarded as an example of readily utilizing thermal comfort research achievements and experimental technique to solving practical problems.

REFERENCES

/1./ Thermal Comfort Study of the Glasswork of Ajka ÉTI Study, Budapest, 1983. rep.no.:5233 (in Hungarian)

/2./ Bánhidi, L.: Survey of Laboratories Investigating Aspects of Thermal Comfort CIB Working Paper Publ., no.:89, 1985., pp 1-34.

/3./ Bánhidi, L.: Possibilities of Measuring Microclimate Parameters and Human Thermal Comfort Proc. of 5th Conf. on Thermogrammetry and Thermal Engineering, Budapest, 1987. pp. 127-134.

VIII

EFFECTS OF NOISE AND VIBRATION ON PERFORMANCE

Trends in Ergonomics/Human Factors V
F. Aghazadeh (Editor)

ASPECTS OF CONSENSUS STANDARDS DEVELOPMENT IN BIODYNAMICS

John C Guignard, MB, ChB*

GB Associates (New Orleans)
824 Kent Avenue
Metairie, LA 70001-4332, USA

A continuing effort to develop and refine consensus standards for the measurement, evaluation and control of human exposure to mechanical vibration and shock (impact) is being made internationally by the International Organization for Standardization (ISO) Subcommittee ISO/TC 108/SC 4 (Human exposure to mechanical vibration and shock); and at the national level in many countries. In the United States, the counterpart effort is the responsibility of ANSI (American National Standards Institute) Working Group S3-39 (S2), which operates under the aegis of the Acoustical Society of America. Existing ISO and American national standards, some now at the stage of draft or under revision, include guidelines for the evaluation of human exposure to whole-body vibration in the range 0.1 to 80 Hz, and to hand-transmitted vibration in the band 8 to 1000 Hz; and cognate standards which define biodynamical terminology, coordinate systems and descriptive terms for force and motion inputs to the human body, human driving-point mechanical impedance and transmissibility models, human impact exposure, and the taxonomy of motion- and vibration-sensitive human activity and task performance. This communication reviews current progress and some problems of defining criteria and standards for human vibration and shock exposure.

1. INTRODUCTION

Standardization of system components had its origins in ancient times; and modern principles of engineering standardization had become well established by the middle of the last century, for such varied ends as the facilitation of international trade and the rapid and efficient repair, maintenance, and interchange of military materiel. The present century has seen the development of industrial, national and international standards for the measurement, evaluation and prevention of hazards to man and the environment from adverse physical and chemical agents.

Beginning in 1964, a continuing effort to develop and refine consensus standards for the measurement and control of human exposure to mechanical vibration and shock (transient acceleration or impact) has been made internationally by the ISO (International Organization for Standardization), which was established in 1948 and now numbers more than 70 member nations. An important aim of this effort has been to unify standardization activity within the global biodynamics community and to establish

---

* Convenor, ISO/TC 108/SC 4, Working Group 1 (Biodynamic terminology).

an intelligible consensus in place of a chaotic plethora of conflicting standards and guidelines previously issued, often with a very meager scientific or technical justification, by a multitude of individual industries and agencies.

1.1. Current Activities of ISO/TC 108/SC 4 and its US Counterpart

The drafting and preparation (for ISO ratification) of standards in the field of biodynamics are the prerogative of Subcommittee 4 (Human exposure to mechanical vibration and shock: identified within ISO as ISO/ TC 108/SC 4) of ISO Technical Committee 108 (Mechanical vibration and shock). At national level in the United States, a counterpart effort is the responsibility of ANSI (American National Standards Institute) Working Group S3.39(S2), which meets and conducts its business under the aegis of the Acoustical Society of America (ASA) Standards Secretariat*. Corresponding national efforts, and participation in the consensus standardization process at international level, are conducted by the actively interested member bodies (the national standardization institutes) of the respective countries which support the work of ISO. The standards developed or adopted for promulgation by ANSI in the United States are commonly the basis for formulating industry and military standards, including relevant sections of Military Standard MIL-STD-1472 [1].

It is important to note that ISO and ANSI standards are consensus standards; and, accordingly, are not mandatory. Moreover, they are issued with a finite lifetime upon the expiration of which they are subject to review and, if necessary, redrafting, before being voted upon in committee with a view to their reissue. Standards such as those reviewed below for the evaluation of human exposure to vibration provide guidelines without attempting to set binding limits (although in some member countries ISO or national standards may be incorporated into legislation).

Existing ISO and American national standards, some in the process of revision, include guidelines for measuring and limiting human exposure to whole-body vibration in the range 0.1 to 80 Hz; and to hand-transmitted vibration at frequencies encompassing the range 8 to 1,000 Hz. Cognate ISO standards, some presently under revision or at the stage of draft, define biodynamic terminology; coordinate systems and descriptive terms for force and motion inputs to the human body; human driving-point impedance and vibration transmissibility models; human impact exposure; and the taxonomy of motion- and vibration-sensitive human activity and task performance. Perhaps of particular interest to the ergonomics and human factors engineering communities, a draft standard on reference postures for describing mechanical vibration and shock inputs to the human body is currently (1988) being voted upon within ISO/TC 108/SC 4 to determine its future course of development as an international or adjunct standard. A list (not exhaustive) of related ISO and some American National Standards is provided in the bibliography at the end of this paper [Appendix A].

---

* Inquiries about American national standardization efforts (including US participation in ISO/ TC 43 and TC 108 activities) in acoustics and mechanical vibration, or concerning the availability of cognate ANSI and ISO standards, should be addressed to the Standards Manager, Acoustical Society of America Standards Secretariat, 335 East 45th St, New York, NY 10017-3483.

2. BIODYNAMICS STANDARDIZATION ACTIVITY IN THE US

With the exception of military standards development, American national standardization activity in biodynamics currently falls within the purview of ANSI working group S3.39(S2) on human exposure to mechanical vibration and shock. That group is the accredited United States counterpart (Technical Advisory Group [TAG]) to ISO/TC 108/SC 4 on the same topic. The working group has been responsible for the preparation of American National Standards issued by ANSI [see bibliography] on the evaluation of human exposure to whole-body vibration and to hand-transmitted vibration; and on related topics already mentioned. Reports of cognate standardization activity in the United States and at international level are reported periodically in the **Journal of the Acoustical Society of America.**

An important function of S3.39(S2) and similar working groups (often depending substantially on the initiative of the chair and individual expert members) is the maintenance of liaison with other national (including industrial) and international standards-writing groups, whose fields inevitably overlap, in order to minimize the likelihood of confusing and wasteful duplication or conflict of consensus standardization efforts. In the area of biodynamics and biomechanical anthropometry, for example, the work of ISO/TC 108/SC 4 and its US counterpart overlaps with the activities of ISO TC 22 (on road vehicles); TC 43 (acoustics); TC 159 (ergonomics); and certain other ISO technical committees.

In the United States, military standardization is the prerogative of the US Department of Defense, which establishes its own committees for the purpose. The current status and activity in military standards development in such areas as human engineering design criteria (MIL-STD-1472 series, which includes biodynamics standards), anthropometry of military personnel, and human factors engineering at large, is described in reference [2].

3. INTERNATIONAL STANDARDIZATION ACTIVITY IN BIODYNAMICS

Through ANSI and its corresponding counterpart groups, the United States is a participating member of International Organization for Standardization (ISO) Technical Committee ISO/TC 108 on mechanical vibration and shock; and of TC 108's several subordinate groups, including Subcommittee ISO/TC 108/SC 4 on human exposure to mechanical vibration and shock. The technical work of ISO/TC 108/SC 4, which meets approximately annually, is carried out by six working or ad hoc groups, covering such areas as biodynamic terminology (WG 1); human exposure to whole-body vibration; (WG 2); human exposure to hand-transmitted vibration (WG 3); human exposure to impact (WG 4); biodynamic modeling (WG 5); and human exposure to repetitive shocks (bumps) superimposed on vehicle ride vibration (ad hoc). Collectively, these groups have produced several ISO standards or drafts reviewed below [or listed in the bibliography]; and are revising these as well as developing newly-proposed standards as ISO-approved or proposed work items within the subcommittee's scope. Historically, it has been a common practice for ANSI in the United States (and corresponding national standardization bodies in other member countries of ISO) to adopt or adapt these standards as national standards. Current activity of ISO/TC 108/SC 4 includes the revision or de novo drafting of the following and other biodynamic standards.

4. NOTES ON CERTAIN ISO BIODYNAMICAL STANDARDS BEING REVISED OR AT THE STAGE OF DRAFT

4.1. International Standard ISO 2631 (revision)

International Standard ISO 2631, "Guide to the evaluation of human exposure to whole-body vibration", was reissued in 1978 [3] as an attempt to achieve an international consensus regarding the measurement and evaluation of human exposure to whole-body vibration in the frequency range 1 to 80 Hz. Its precursor, ISO 2631-1974 had superseded a confusing multiplicity of guidelines, many based on meager, biased or unreliable data, that had been written in many different countries for particular industries or applications [4].

ISO 2631-1978 prescribed standard methods of measuring and evaluating human exposure to whole-body vibration disturbing or potentially harmful to man at work or in transportation; and it provided guidelines for restricting human exposure to such vibration in the range 1 to 80 Hz. That band was selected having regard to several practical considerations, including current engineering applications; the function and limitations of instrumentation for evaluating dynamic environments impinging on seated or standing man; and the availability of substantial data concerning the human frequency-response to whole-body vibration which are suitable for incorporation into a consensus standard [4, 5]. Several countries, including the United States*, have adopted national standards incorporating essentially the same guidelines.

The central concept and quantitative formulation embodied in ISO 2631-1978 was the "Fatigue/Decreased Proficiency [FDP] Boundary". This was an adjustable guideline (essentially a human frequency-response function based on laboratory data reflecting the principal human resonance characteristics and correlated discomfort or performance decrements) for limiting human exposure to whole-body vibrational accelerations in the spinal (**z**) and transverse (**x, y**) axes respectively.

An important tenet was that the standard should be most conservative (ie, the tolerable acceleration boundary for any given combination of frequency, exposure time and direction should be lowest) in the frequency bands of the principal body resonance phenomena (falling mainly in the three octaves of the spectrum from 1 to 8 Hz): for those biomechanical phenomena are strongly correlated with physiological stress, disruption of human activity and task-performance, and discomfort; and also, highly probably [5, 6], with chronic disorders of occupational origin associated with severe whole-body vibration exposure on a daily basis.

The criterion for applying the "FDP" boundary is prevention of decrements in task-performance or disruption of human activity by whole-body vibration. The standard provided a guide to acceptable levels of root mean square (rms) acceleration at the point of input to man (typically his seat or the floor or deck on which he stands) as a function of vibration frequency and notional cumulative exposure time (computed on a daily basis using a standardized formula). The text of the standard,

---

* American National Standard ANSI S3.18-1979 is based on the same data and essentially differs only editorially from the ISO standard. The basic human vibration frequency response function of ISO 2631-1978 is similarly enshrined in the pertinent section of MIL-STD-1472 [1].

which must be carefully consulted in order to interpret or apply the numerical guidelines legitimately, contains instructions for weighting the putatively acceptable rms acceleration values when other criteria (eg, occupational safety and health; passenger ride) apply; when the spectrum or level of vibration varies substantially for different periods of time during the exposure being evaluated; or when the vibration is interrupted during the exposure or is otherwise markedly nonstationary), as frequently happens during daily occupational exposures.

It has been demonstrated experimentally that this standard is protective (perhaps overly so) of human performance of various tasks during z-axis sinusoidal vibration in the range 2 to 32 Hz for exposures up to 8 hours [6]; and field data indicate that, in working and military vehicles, for example, the ride is judged by independent criteria to be unsatisfactory or marginally acceptable for operational purposes when the applicable guidelines of ISO 2631-1978 are substantially exceeded (at least, for short-term exposures). The standard can provide a convenient basis for the comparison and evaluation of vehicle ride data [8].

Nevertheless, this standard as presently promulgated has been the focus of cogent and justifiable criticism on several grounds [6, 9]. For example, the acceleration-weighting procedure recommended for different criteria of application (eg, safe exposure; ride comfort) is based on the arguable assumption that there is a simple hierarchical relationship between vibration-induced discomfort, performance decrement and occupational health hazard. Moreover, the weighting procedure for calculating notional daily exposure time has been based on very meager data or on unproven theory concerning the time-dependence of human response to whole-body vibration (particularly when physiologically mediated effects on long-term performance and health are involved [6]).

Moreover, the 1978 standard ignored demographic and other cicumstantial factors that may strongly influence human response to vibration on a group or individual basis; and made no statement regarding population percentiles supposedly at risk or to be protected from whole-body vibration nuisance or hazard.

Accordingly, current work of ISO/TC 108/SC 4 is directed towards revision and elaboration of the standard; and also to extending its scope to the frequency decade below 1 Hz. The band 0.1 to 1 Hz is associated specifically with motion sickness [10] and gross disruption of human activity and task performance in ships [11, 12], aerospace and land-borne vehicles. Part of the intent of a standard such as ISO 2631-1978 and its revisions [11, 13] is to encourage the collective international acquisition of further and better experimental and field data which can be used to formulate more detailed and reliable predictive biodynamic models and, ultimately, more soundly based and useful standards.

ISO 2631 continues under active revision with the intention of reissuing it in the form of several sections having specific applications. Some revised sections have recently been ratified by member countries of ISO as current International Standards [11, 13] while others are presently at the stage of draft. Moreover, supplementary standards have been issued or are being developed for particular situations or applications, such as fixed buildings and offshore structures, ships, rail transportation and so on; or to formalize predictive biodynamical models of human response to whole-body vibration within a specific frequency range [see list in Appendix A]. A counterpart draft United Kingdom standard [14] has endeavored to address some of the widely acknowledged conceptual as well as factual deficiencies in ISO 2631-1978; and to anticipate improvements to be made in the international standard when the draft is eventually ratified and reissued in full. An interesting and potentially important innovation [14] has been the concept of "vibration dose value": this can be used to prescribe risk from noxious effects of vibration, although it is recognized that there is still a lack of epidemiological and supporting data to establish causal and consequent does-response relationships between whole-body vibration and occupational disease or injury [5, 6, 14].

4.2. Draft International Standard ISO 8727

This draft standard [15] on biodynamic coordinate systems is presently (1988) being circulated internationally for comment and vote. It represents the fruition of an effort originating in the United States to establish a reliable and rationally based hierarchical system of anatomical, instrumentational, and reference orthogonal coordinate systems for use in biodynamics research and applications, where there is a need to define precisely the vibrational or impact force and motion inputs to the human body or its analog models; and to measure the distribution of inertial forces within the body. The keystone of this document is the establishment of a standard system of right-handed orthogonal coordinate systems originating in, and orientated with reference to radiographically determinable anatomical landmarks in the bony skeleton. The anatomical coordinate systems (established for the head and neck; the pelvis, considered as the principal point of entry for whole-body vibration; and the hand, considered as the primary receiver of hand-transmitted vibration [16]) can be referenced to external basicentric and geocentric systems; and in turn serve as the frame of reference for inertial data recorded using external man-mounted bioinstrumentation.

4.3. Draft Proposal ISO 5805

This draft proposal on the terminology of mechanical vibration and shock affecting man is a revision (now being circulated internationally for comment and vote) of ISO 5805-1981 [17]. It is intended to supplement the international standard terminology of mechanical vibration and shock, ISO 2041, which has itself recently undergone revision for reissue by ISO/TC 108.

4.4. Draft Proposal (ISO Work Item WI 01-93)

This draft proposal for a standard taxonomy of vibration- and motion-sensitive human activity and performance [19] is a new proposal (now being circulated internationally for comment), intended to supplement ISO 2631 (revision) and to be of general value in providing a framework for describing and predicting the effects of vibration and low-frequency motion on human activity and work, productivity or mission effectiveness.

4.5. Draft Proposal (ISO Work Item WI 05-94)

This draft proposal on a standard combined model for human whole-body mechanical driving-point impedance and transmissibility includes the revision of ISO 5982 and ISO 7962 [see Appendix A]. It is currently being circulated internationally by ISO/TC 108/SC 4 for comment. The biodynamic model is intended to provide a standard basis for specifying and comparing inertial responses of man and human analogs to actual test or computer-simulated vibrational force and motion inputs; and to serve as a standard human analog model for predicting the human biodynamical response to measured or predicted mechanical inputs (eg, from vehicle ride).

4.6. Proposed ISO Work Item on Standard Reference Postures

A new proposal [20] on standard reference postures for describing force and motion inputs to man is under consideration by ISO for adoption either as a new work item or as an annex to an existing ISO standard or draft proposal. It is intended to provide a standard basis for preventing or resolving ambiguity in defining human posture and orientation with respect to vibrational and impact force and motion inputs. Some existing standards and published reports in biodynamics, for example, refer to vibration affecting subjects who are seated, reclining or recumbent, without making clear the range or orientation of the posture reported. Yet these factors are known to have a strongly modifying effect the transmission and distribution of vibrational forces within the living body [4, 5]. The draft proposal defines the limits of sitting, reclining, recumbent, and other functional postures in terms of ranges of the angles standing between the human torso or major body segments and the principal axes of external biodynamic coordinate systems.

5. REFERENCES

[1] United States Military Standard MIL-STD-1472 (series: currently C), Human Engineering Design Criteria for Military Systems, Equipment, and Facilities, Washington: US Department of Defense, 1974; 1985.

[2] US Department of Defense: Human Factors Standardization Document Program Plan (Revision 5), 20 August 1987.

[3] International Organization for Standardization [ISO], Guide to the Evaluation of Human Exposure to Whole-body Vibration, Geneva: ISO, International Standard ISO 2631-1978.

[4] Guignard, J C, Vibration, in: Guignard, J C & King, P F, Aeromedical Aspects of Vibration and Noise, AGARDograph AG-151, Neuilly-sur-Seine, France: NATO/AGARD, 1972, Part 1, pp 2-113.

[5] Guignard, J C, Vibration, in Patty's Industrial Hygiene and Toxicology, Vol 3B, Theory and Rationale of Industrial Hygiene Practice: Biological Responses, Cralley, L J & Craller, L V (eds), New York: Wiley, 1985, Chapter 15.

[6] Wasserman, D E, Human Aspects of Occupational Vibration, Advances in Human Factors/Ergonomics - 8. Amsterdam &c: Elsevier, 1987.

[7] Guignard, J C, Landrum, G J & Reardon, R E, Experimental Evaluation of International Standard (ISO 2631:1974) for Whole-body Vibration Exposures, University of Dayton Research Institute, Dayton, Ohio, Final Report to the National Institute for Occupational Safety & Health, UDRI-TR-76-79, December 1976.

[8] Stikeleather, L F, Hall, G O & Radke, A O, A Study of Vehicle Vibration Spectra as Related to Vehicle Dynamics, Society of Automotive Engineers, New York, SAE Paper 720001, January 1972.

[9] Oborne, D J, Whole-body Vibration and International Standard ISO 2631: A Critique, Human Factors, 25, 55-69, 1983.

[10] Guignard, J C & McCauley, M E, The Accelerative Stimulus for Motion

Sickness in Man, in: Crampton, G H (ed), Motion and Space Sickness, Boca Raton, Florida: CRC Press (in preparation), Chapter 14.
[11] ISO, Evaluation of Human Exposure to Whole-body Vibration - Part 1: General Requirements, Geneva: ISO, International Standard ISO 2631/1-1985.
[12] Bittner, A C, Jr & Guignard, J C, Human Factors Engineering Principles for Minimizing Adverse Ship Motion Effects: Theory and Practice, Naval Engineers Journal, 97 (4), 205-213; Discussion, Ibid, 97 (5), 107-111, 1985.
[13] ISO, Evaluation of Human Exposure to Whole-body Vibration - Part 3: Evaluation of Exposure to Whole-body z-axis Vibration in the Frequency Range 0.1 to 0.63 Hz, Geneva: ISO, International Standard ISO 2631/3-1985.
[14] British Standards Institution [BSI], Draft British Standard, Guide to the Evaluation of Human Exposure to Whole-body Mechanical Vibration, BSI Document 86/71027, 1986.
[15] ISO, Standard Biodynamic Coordinate Systems, Geneva: ISO, Draft Proposal ISO/DP 8727*.
[16] ISO, Guidelines for the Measurement and Assessment of Human Exposure to Hand-transmitted Vibration, Geneva: ISO, International Standard ISO 5349-1986.
[17] ISO, Mechanical Vibration and Shock - Human Exposure - Vocabulary, Geneva: ISO, Draft Proposal ISO/DP 5805* (revising ISO 5805-1981).
[18] ISO, Vibration and Shock - Vocabulary, Geneva: ISO, International Standard ISO 2041-1975 (under revision).
[19] ISO, Mechanical Vibration and Shock - Human Exposure - Standard Taxonomy of Motion- and Vibration-sensitive Human Activity and Performance, ISO, Draft Proposal ISO/DP XXXX* (ISO Work Item WI 01-93).
[20] ISO, Mechanical Vibration and Shock - Human Exposure - Standard Reference Postures, ISO Document pending approval as a work item, ISO/TC 108/SC 4/WG 1 N77A (1987, amended 1988).

6. APPENDIX A: SUPPLEMENTARY LIST OF COGNATE STANDARDS AND DRAFTS

6.1. International (ISO) Standards and Drafts*

ISO 1503, Geometric orientations and directions of movement.
ISO/DIS 2631/2, Evaluation of human exposure to vibration and shock in buildings (1 to 80 Hz).
ISO/DIS 2631/4, Evaluation of crew exposure to vibration on board sea-going ships (1 to 80 Hz).
ISO 4865, Analog analysis and presentation of vibration and shock data.
ISO/DIS 5347, Methods for the calibration of vibration and shock pickups.
ISO/DIS 5348, Mechanical mounting of accelerometers (seismic pickups).
ISO 5353, Earth-moving machinery - Seat index point.
ISO 5982, Mechanical driving-point impedance of the human body.
ISO 6165, Earth-moving machinery - Basic types - Vocabulary.
ISO 6897, Guide for the evaluation of the response of occupants of fixed structures, especially buildings and off-shore structures, to low-frequency horizontal motion (0,063 to 1 Hz).
ISO 7096, Earth-moving machinery - Operator seat - Transmitted vibration.
ISO/DIS 7962, Vibration and shock - Mechanical transmissibility of the

---

* Dates of some documents are omitted because they are at the stage of draft or draft proposal and are accordingly ephemeral.

human body.
ISO/DIS 8041, Human response vibration measuring instrumentation.
ISO 8661, Pt 1, Measurement of vibrations in hand-held power-driven tools - Part 1: General.
ISO 8661, Pt 2, Hand-held power-driven tools - Measurement of vibrations - Part 2 (Chipping hammers).

6.2. ISO Publications on Standards Writing and Issuance

[Note: These publications are issued by the ISO Secretariat in Geneva, Switzerland. In the United States they are obtainable through ANSI.]

ISO Catalogue [of techical committees and standards], 1987.
ISO Directives for the techical work of ISO, 1985.
ISO Rules for the drafting and presentation of international standards, 1986.

6.3. American National Standards

ANSI S3.18-1979, Guide for the Evaluation of Human Exposure to Whole-body Vibration [follows closely the guidelines in ISO 2631-1978. Also identified as ASA 38-1979].
ANSI S3.29-1983, Guide to the Evaluation of Human Exposure to Vibration in Buildings [cf ISO/DIS 2631/2].
ANSI S3.32-1982, Mechanical Vibration and Shock Affecting Man - Vocabulary [cf ISO/DP 5805].

7. APPENDIX B: KEY TO ABBREVIATIONS

ASA	Acoustical Society of America
ANSI	American National Standards Institute
ISO	International Organization for Standardization
ISO/DIS	ISO Draft International Standard
ISO/DP	ISO Draft Proposal
ISO/TC../SC..	ISO Subcommittee
ISO/TC..	ISO Technical Committee
WG..	ISO Working Group
XXXX	ISO/WG document not yet assigned a DP number.

8. APPENDIX C: BIBLIOGRAPHY (SUGGESTIONS FOR FURTHER READING ON HUMAN EXPOSURE TO VIBRATION)

Gierke, H E von, Nixon, C W & Guignard, J C, Noise and Vibration, In: Foundations of Space Medicine and Biology, Calvin, M & Gazenko, O G (eds), Joint USA/USSR Publication, Washington, DC, National Aeronautics & Space Administration; and Moscow: Academy of Sciences of the USSR, Volume II, Book I, 1975, Chapter 9.
Guignard, J C, Cralley, L J, Cralley, L V & Bittner, A C, Jr (eds), Human exposure to vibration, New York: John Wiley & Sons, Inc (in preparation). [Chapter 20 will address standards.]
International Labor Office (ILO). ILO Code of Practice, Protection of Workers Against Noise and Vibration in the Working Environment, Geneva: ILO, 1977.

Trends in Ergonomics/Human Factors V
F. Aghazadeh (Editor)

COMBINED EFFECTS OF WHOLE BODY VIBRATION AND NOISE ON PERFORMANCE IN A MULTIPLE REACTION-TIME-TASK

Gert NOTBOHM and Eckhard GROS

Institute of Occupational Medicine
University of Duesseldorf
Duesseldorf, FRG

Performance in a five-choice-reaction-task was studied under six conditions of environmental stress: white noise of 75 dBA resp. 100 dBA alone and in combination with sinusoidal whole body vibration of 4 resp. 8 Hz. Reaction time was shortest with combined stress of 8 Hz vibration and 100 dBA noise, whereas accuracy of answers was best with noise alone. Theoretical implications of these findings and the importance of subjective evaluation of the experimental situation are discussed.

1. INTRODUCTION

In several industrial branches, whole body vibration and noise belong to the most common stressors at the workplace. Nevertheless, research on the combined effects of these factors on human performance has been very limited up to now. In an extensive literature review, only twelve publications covering this field were found.

Four types of tasks were used in these studies: simple reaction-time tasks /1,2,3,4/, vigilance tasks /5,6/, tracking tasks /1,2,3,4,7,8/ and arithmetic tasks /1,2,9,10,11,12/. For none of these different kinds of performance, the review yielded unanimous results about the interaction of whole body vibration and noise. Reduced performance by addition of a second stressor was as well reported as improved performance or no change at all. Obviously, the effect of combined whole body vibration and noise depends on a variety of experimental factors, such as specific features of the task, criterion of performance (correct answers or speed), intensity levels of the stressors, vibration frequency and waveform, duration of experimental exposure etc.

Those studies, however, which compared different intensity levels both of single and combined stressors give support to an intriguing suggestion: the effect of combined stressors seems to depend heavily on the intensity levels applied, thus augmenting the effect of the single stressor in one case and antagonizing it in another case. For example, in a complex counting task it could be demonstrated that performance under noise of 65 dBA became worse when vibration was added, whereas performance under noise of 100 dBA became better when combined with whole body vibration /11/. The same was true for tracking performance under 60 dBA vs. 100 dBA noise /4/, but when 110 dBA were applied instead of 100 dBA, the combined effect of vibration and noise changed from subtractive to additive /3/. With different intensity levels, other authors, too, reported less impairment of performance by combined stressors of high intensity than by single stressors or by combined stressors of lower intensity levels /2,5,12/.

Thus, there is much evidence that the interaction of vibration and noise is quite complex rather than just additive and that the intensity levels chosen play a dominant role in it, although the results in detail are still controversial, and the underlying mechanisms are far from being understood. Further investigation of this interaction is important because of its practical consequences for performance in working situations as well as for its theoretical implications: such an interaction can only be explained by a common - presumably cognitive - factor influenced by vibration as well as by noise.

In this study, we intended
- to explore systematically the single and combined effects of whole body vibration and noise on human performance by using two levels of white noise (75 dBA and 100 dBA) alone and in combination with vibration,
- to extend the range of vibration frequencies studied up to now in this field by comparing the effects of 4 Hz and 8 Hz vibration in combination with noise,
- to investigate the interaction of these stressors on performance in a five-choice-reaction-task which was not used before in this field and which we consider more typical for usual working situations than arithmetic tasks and less dependent on physical interference than tracking tasks,
- to assess subjective evaluation of the experimental situation as a possible source of variation in performance.

Referring to the results of former studies /4,11/, we expected for both vibration frequencies that performance would be best with 75 dBA noise alone and worst with 75 dBA noise and vibration. With the combination of vibration and 100 dBA noise, performance should be better than with 100 dBA noise alone.

2. METHODS

2.1. Subjects

20 male university students were volunteered for participation in the experiment. They ranged in age from 19 to 33 years. As determined by standard audiometric methods, all Ss had normal hearing within the frequency range of 1000 to 6000 Hz. They had undergone a physical examination for general health and spinal troubles and received incentive pay for their participation.

2.2. Apparatus

Sinusoidal vibration stimulation was presented by a hydraulic Toni MFL vibration exciter DAL 25/100. With both frequencies, 4 Hz and 8 Hz, acceleration was 3 m/s^2. Ss sat in a comfortable chair with back rest, which was mounted on top of the shake table, and were restrained by a lap belt. During the trials, acceleration was constantly monitored at the seat of the chair.

The noise exposure was produced by a Brüel & Kjaer type 1405 noise generator and passed bilaterally to Universum Stereo 200 headphones worn by the subject. White noise levels of 75 dBA and 100 dBA were applied. As the noise emission of the vibration exciter did not exceed 70 dBA, the noise level of 75 dBA was sufficient to mask all ambient noise.

2.3. Task

On the shake table, the subject's console was mounted bearing a determination device (DTG 2000 from Zak). It consisted of five coloured buttons on a horizontal panel and a glass plate on a vertical panel. Behind the glass plate, lights in the same five colours flashed in random sequence controlled by a Zak Bioport micro-computer. There were five lights for each colour irregularly distributed over the plate. The flashing rate depended on the subject's reactions, i.e. each pushing of a button elicited the next light stimulus.

Each single reaction time, total amount of reactions per trial and scores of correct and wrong answers ("hits" and "errors") were automatically computed by micro-computer. For statistical procedures, mainly mean reaction time per trial and score of errors per trial were used.

2.4. Procedure

On the day of physical examination, Ss were asked to complete several questionnaires to assess habitual dimensions of personality, and the tendency of interference was tested by the Stroop colour-word-test.

The experimental session for each subject started with a training phase consisting of three trials of the reaction task lasting six minutes each. Ss were instructed to react as quickly and as accurately as possible during all trials. After training the subjects took a short rest and then repeated the task in eight trials of six minutes each. The first and last of these trials served as control and were performed with no vibration and without any machinery noise in the vicinity of the testing area. In the remaining six trials, each subject was exposed to different stressful conditions in random order. The six experimental conditions were: white noise of 75 resp. 100 dBA alone and in combination with sinusoidal whole body vibration of 4 resp. 8 Hz. In the intervals between two trials lasting six minutes, subjective assessment of the perceived vibration effects and of the subjective evaluation of the situation and the performance were registered by questionnaire.

The entire experimental session took about three hours for each subject. Ss were randomly assigned to experimental sessions in the morning or in the afternoon.

3. RESULTS

3.1. Reaction time and errors

Fig. 1 shows the mean **reaction time** for all experimental conditions. Control 1 and 2 are plotted separately because of the distinct differences in performance. Analysis of variance for reaction time data yielded no significant effect of the experimental conditions ($F(7,133) = 1.86$, $p < 0.08$). One-tailed t-tests according to our hypotheses, however, revealed an outstanding performance in the 8 Hz/100 dBA-condition: in this condition, reaction time was significantly shorter compared with 8 Hz/75 dBA ($p < 0.004$), with 100 dBA noise alone ($p < 0.046$), and with the first control

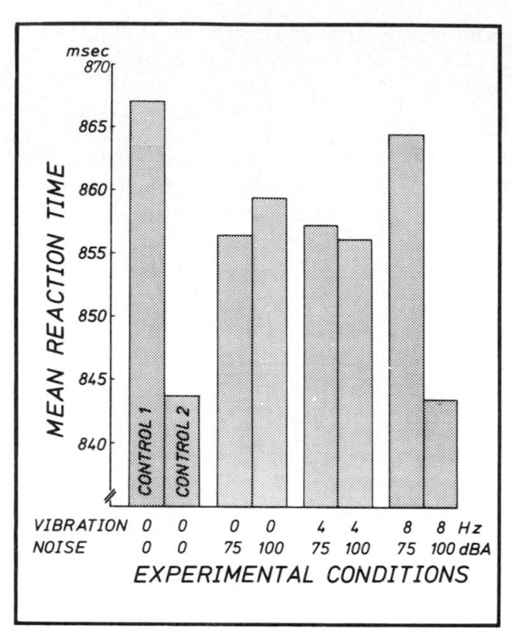

FIGURE 1
Mean reaction time for all experimental conditions (in msec)

($p < 0.001$). With regard to 4 Hz vibration, no significant differences between experimental conditions could be found.

For the amount of **errors** in reaction, ANOVA for repeated measures clearly showed a significant effect of the experimental conditions ($F(7,133) = 4.14$, $p < 0.004$). In fig. 2, the means of errors in the controls and the stress conditions can be seen. Contrast calculations clarified that in the first control, errors were significantly less compared with all combinations of whole body vibration and noise. In the second control, errors were significantly less compared with the 4 Hz/75 dBA- and with the 8 Hz/100 dBA-condition. Further, errors in the 75 dBA-noise- as well as in the 100 dBA-noise-condition were significantly less compared with all conditions of combined stressors.

For further investigating the prominent role of the 8 Hz/100 dBA-combination, we calculated one-tailed t-tests for the differences in **hits** between this condition and all other experimental conditions. Two significant effects resulted: in the 8 Hz/100 dBA-condition, the score of hits was higher than in the 8 Hz/75 dBA-condition ($p < 0.023$) and in the 4 Hz/75 dBA-condition ($p < 0.003$).

For clarifying any learning or practising effects, we finally calculated analyses of variance not for the experimental conditions, but for the experimental **trials.** There was no significant effect of trials, neither for reaction time ($F(7,133) = 1.30$, $p < 0.2549$) nor for errors ($F(7,133) = 1.90$, $p < 0.0738$). Contrast calculations showed that only the results of the controls differed considerably from some experimental trials. Reaction time in the second control was significantly shorter than in the first, whereas errors were significantly less in the first control.

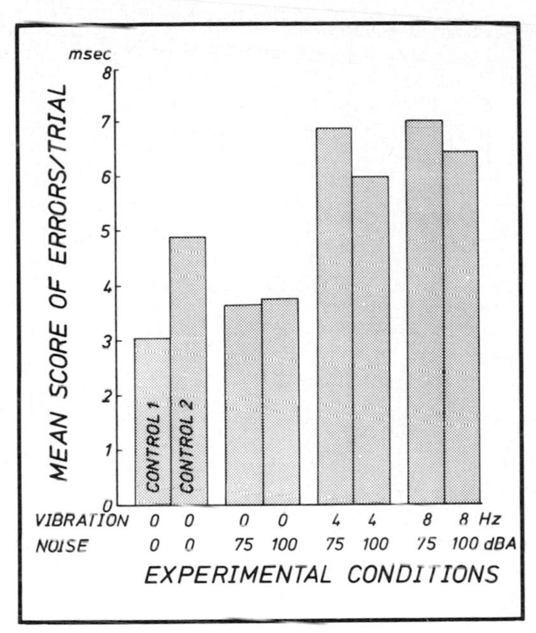

FIGURE 2
Mean score of errors for all experimental conditions

3.2. Subjective evaluation

From the subjective data, which were collected between trials by questionnaire, here we will only mention a few which are closely related to the performance data.

After trials involving exposure to vibration, Ss were asked to rate how tolerable or intolerable they experienced the vibration, and they were to mark their judgement on an analogous scale of 10 cm length, which was labeled "as bearable as possible" at the left and "as unbearable as possible" at the right. Thus, the distance of the subject's mark from the left side gave a measure of increasing intolerance of the vibration.

For this variable, ANOVA for repeated measures showed a highly significant effect of experimental conditions ($F(3,57) = 9.09$, $p < 0.0001$). Contrast calculations proved that in the 4 Hz/75 dBA- as well as in the 4 Hz/100 dBA-condition, vibration was experienced significantly less endurable than in both the 8 Hz/75 dBA- and 8 Hz/100 dBA-condition.

Further, with the following item Ss were asked to attribute their performance to external or internal factors /13/ after each trial :

"To which factors do you attribute your performance in the last trial ?
Please take your performance as 100 % and decide to which extent **each** of the following factors has contributed to your performance :
- the difficulty of the momentary situation ___ %
- strength of will ___ %
- ability or exercise ___ %
- casual circumstances (luck or bad luck) ___ %"

For the mean percentage in these attributional variables, ANOVAs for repeated measures yielded the following effects of experimental conditions :
- for "situation": $F(7,133) = 0.80$, $p < 0.5896$
- for "will" : $F(7,133) = 7.34$, $p < 0.0001$
- for "exercise" : $F(7,133) = 7.66$, $p < 0.0001$
- for "luck" : $F(7,133) = 2.95$, $p < 0.0067$

Contrast calculations revealed exactly the same pattern for the variables "exercise" and "will": in **all** combined stress-conditions, the percentage of performance attributed to these internal factors was significantly less than in the controls or noise-alone conditions. For the variable "luck" the results were less systematical: only in the 4 Hz/75 dBA-condition, performance was significantly less attributed to "luck" than in most of the other conditions.

3.3. Tendency of interference

From all the tests of personality dimensions which we applied to the Ss, here we only present findings with the Stroop test, as it had been established in former work /14/ that the tendency of interference affects performance under noise as a major moderating variable. With the German form of the Stroop test /15/, several basic functions of cognitive performance can be measured, but for our purpose, the variable INT was most interesting which can be described as an indicator for the individual tendency to be easily disturbed by task-irrelevant stimuli (at low scores) or to keep concentration selectively on relevant information (at high scores).

From table 1 can be seen that high scores in this variable correlate negatively with the amount of errors, especially in conditions of combined stress. The correlations with the amount of hits are significantly positive for nearly all experimental conditions. Ss with high selectivity towards stimuli obviously tend to perform better in this reaction task, and they still do so under stressful conditions.

TABLE 1. Correlations between selectivity towards divergent stimuli and performance measures in all experimental conditions

	experimental conditions							
	control 1	control 2	0 Hz/ 75 dBA	0 Hz/ 100 dBA	4 Hz/ 75 dBA	4 Hz/ 100 dBA	8 Hz/ 75 dBA	8 Hz/ 100 dBA
reaction time	0.54	0.55*	0.59**	0.43	0.60**	0.34	0.49*	0.41
errors	-0.33	-0.25	-0.23	-0.63**	-0.41	-0.46*	-0.58**	-0.59**
hits	0.55*	0.55*	0.58**	0.47*	0.61**	0.39	0.53*	0.47*

* p < 0.05 ** p < 0.01

4. DISCUSSION

For **reaction time**, we could not find significant differences neither between the two combinations of 4 Hz vibration and noise nor between these combinations and noise alone. So far, our hypothesis of an interaction between vibration and noise derived from literature could not be confirmed. For vibration of 8 Hz, however, we established a subtractive effect when combined with the higher noise level of 100 dBA. This finding agrees with studies on other types of performance reporting subtractive effects of 6 Hz vibration combined with noise of 100 dBA /4,11/. It may be hypothesized that in vibration of 4 Hz, the mechanical impact of whole body vibration is too dominant to allow the noise to interfere with vibration in this particular task. Therefore, further investigation of these stressors should apply an extended range of vibration frequencies in order to study the fine mechanisms of stressor interaction more thoroughly.

Accuracy of answers which was used as a criterion of performance in most of the studies cited above clearly showed more detrimental effects of all combinations of stressors compared with noise alone. This result does not confirm former findings of subtractive effects of combined stress which were obtained with other kinds of performance. Obviously, the interactive effects of vibration and noise also depend on the specific demands of the task. Thus, for generalizing experimental results to working situations, type of task and criterion of performance should be carefully suited to the characteristics of the specific workplace.

Learning effects during experimental sessions cannot be totally excluded. Reaction time in the second control was significantly shorter than in the first. However, for accuracy of answers the relation was reverse. As analysis of variance showed no significant sequence effect and as the stress conditions were applied in random order, differences in performance between these conditions can plausibly be attributed to this independent variable and not to exercise.

Summarizing our findings on performance data, there is still no distinct answer to the question how whole body vibration and noise combine in their effect on performance. On the contrary, their interaction appears to be more and more complex, as type of task and vibration frequency prove to be important variables, too. But at least, there is growing evidence that this interaction can take very different directions depending on a variety of circumstances. This gives support to the hypothesis that the effect of vibration on performance is not merely mechanical. Both stressors, vibration as well as noise, seem to influence cognitive or motivational factors, and this aspect we wanted to assess with our subjective data.

Subjective factors which may contribute to indivdual performance under stress conditions had not been intensively studied in this field up to now. With the Stroop test we found high correlations between **tendency of interference** and individual performance in most experimental conditions. This finding suggests that a considerable part of variation in performance under combined stress is due to habitual dimensions of personality. Besides tendency of interference, related features may prove to be important. Further exploration of this aspect may explain some of the inconsistency in experimental findings stated in the beginning.

The same conclusion can be drawn from our results concerning **momentary subjective evaluation** of the experimental situation and its influence on performance. Intolerance of vibration was very distinctive among the vibration conditions, but these differences did not reflect the differences in performance measures. For the assessment of causal attribution, however, the results corresponded exactly with the amount of errors in experimental conditions: a higher score of errors in the combined stress conditions was subjectively mirrored in a minor estimation of self-forced performance or vice versa.

These computations only compared means of subjective items in different experimental conditions. In a next step, we will explore if these subjective data allow to identify styles of evaluation which are related to performance under single and combined stress. Further study of such relations between cognitive processes and performance may promise more insight in the function of human factors in working situations.

REFERENCES

/1/ Grether, W.F., Harris, C.S., Mohr, G.C., Nixon, C.W., Ohlbaum, M., Sommer, H.C., Thaler, V.H. and Veghte, J.H., Effects of combined heat, noise and vibration stress on human performance and physiological functions. Aerospace Med., 43 (1971) 1092-1097.
/2/ Grether, W.F., Harris, C.S., Ohlbaum, M., Sampson, P.A. and Guignard, J.C., Further study of combined heat, noise and vibration stress. Aerospace Med., 43 (1972) 641-645.
/3/ Harris, C.S. and Sommer, H.C., Interactive effects of intense noise and low-level vibration on tracking performance and response time. Aerospace Med., 44 (1973) 1013-1016.
/4/ Sommer, H.C. and Harris, C.S., Combined effects of noise and vibration on human tracking performance and response time. Aerospace Med., 44 (1973) 276-280.
/5/ Hughes, J., An investigation into some aspects of human sensitivity to the combined effects of noise and vibration. Human Response to Vibration Conference, University of Sheffield (1972)
/6/ Rao, B.K.N. and Ashley, C., Effect of whole body low frequency random vertical vibration on a vigilance task. J. Sound Vibr., 33 (1974) 119-125.
/7/ Harris, C.S. and Shoenberger, R.W., Combined effects of noise and vibration on psychomotor performance. AMRL-TR-70-14, Aerospace Medical Research Laboratory, Wright-Patterson AFB, Oh. (1970)
/8/ Innocent, P.R. and Sandover, J., A pilot study of the effects of noise and vibration acting together; subjective assessment and task performance. Human Response to Vibration Conference, University of Sheffield (1972)
/9/ Ioseliani, K.K., Effect of vibration and noise on ability to do mental work under conditions of time shortage. Environmental space sciences, 1 (1967) 144-146.
/10/ Harris, C.S. and Sommer, H.C., Combined effects of noise and vibration on mental performance. AMRL-TR-70-21, Aerospace Medical Research Laboratory, Wright-Patterson AFB, Oh. (1971)
/11/ Harris, C.S. and Shoenberger, R.W., Combined effects of broadband noise and complex waveform vibration on cognitive performance. Aviat. Space Environ. Med., 51 (1980) 1-5.
/12/ Sandover, J. and Champion, D.F., Some effects of a combined noise and vibration environment on a mental arithmetic task. J. Sound Vibr., 95 (1984) 203-212.
/13/ Weiner, B. and Kukla, A., An attributional analysis of achievement motivation. Journal of Personality and Social Psychology, 15 (1970) 1-20.
/14/ Gros, E., Lärm, Schlaf und Leistung (Pahl-Rugenstein, Köln, 1985)
/15/ Bäumler, G., Farbe-Wort-Interferenztest (FWIT) nach J.R. Stroop (Hogrefe, Göttingen-Toronto-Zürich, 1985)

Trends in Ergonomics/Human Factors V
F. Aghazadeh (Editor)

EFFECTS OF MECHANICAL VIBRATION ON PERIPHERAL BODY TEMPERATURE AND THE HEAT EMISSION PATTERN OF THE HAND

Elsayed Abdel-Moty, M.S.
Tarek M. Khalil, Ph.D., P.E.

Department of Industrial Engineering
and Department of Neurological Surgery
University of Miami
P.O.Box 248294
Coral Gables, FL. 33124

Sinusoidal mechanical vibration was applied, in a controlled laboratory environment, to the right hands of seven subjects. Three levels of vibration frequencies were selected. These were: 5 Hz, 20 Hz, and 100 Hz. These frequencies were applied in random order to each subject. Each frequency was applied for a period of 30 minutes at a level of 1 g in the horizontal direction. The change in the peripheral fingers temperature was digitally monitored over this period of time. Also, the effects of vibration on the Heat Emission Pattern of the hand was studied through the use of Liquid Crystal Thermography. Results and their implications in the design of man-machine systems are presented. The advantages and limitations of using Thermography to evaluate occupational and musculoskeletal stresses are addressed.

1. INTRODUCTION

The physiological responses due to mechanical vibration have been studied by many investigators. Hyperventilation, increased heart rate, change in blood chemistry, reduced blood flow, change in tracking performance, feelings of fatigue and discomfort under certain vibrating conditions are but some of the symptoms associated with this type of industrial stress [2, 3, 4, 5, 6]. Of special interest to ergonomists and human factors experts are the hand-held vibrating tools. Numerous reports have linked hand-held vibrating tools with many diseases including reduced blood flow, decreased skin temperature, loss of grip strength, neuritis, decalcification, and cysts of the radial and ulnar bones [16]. Although most of these pathologies are not disabling, advanced cases of exposure to vibration have been known to lead to gangrene of the finger tips [13]. Despite the guidelines by safety specialists regulating the exposure time and levels of vibration, symptoms still develop. Since symptoms develop in varying periods of time [7], early detection could be an effective prevention strategy for the avoidance of adverse effects. The main objective of this study was to investigate the effects of low and high frequencies of sinusoidal mechanical vibration on peripheral body temperature and the heat emission pattern of the hand. The utility, value, and limitations of using Thermography in the evaluation, documentation and early detection of vibration-induced symptoms were studied.

Thermography is the technique of mapping the distribution of the body's surface heat. Two non-invasive thermographic systems are being employed in clinical practice. These are the infrared telemetry system and the liquid crystal system. The skin temperature recorded using these systems depends mainly on the changes in the local blood flow [1, 9]. Liquid crystal thermography utilizes cholesteric liquid crystals which change color with small temperature variations. The flexitherm system, which was used in this study (Flexi-Therm, Inc., N.Y.), uses a flexible sheet imbedded with liquid crystals and is attached to the elastic side of a transparent cassette. The cassette is then inflated with air and when pressed against the body surface, it molds to the skin. The system is equiped with an instant camera that provides instant colored photographs. Each thermographic image displays a 7-color temperature scale. Each of the seven colors represents a temperature, with the black color representing the coolest and the dark blue representing the warmest. Each of the other colors (brown, orange, yellow, green, blue) represents an incremental temperature change of about 1 oC respectively. Thermographic images are evaluated by comparing both sides of the body for each individual. In clinical applications, a thermogram is considered normal only if a series of photographs show high degree of bilateral symmetry [15].

Thermography was used in the past for the detection of breast cancer and circulatory disorders. Recent applications, even though lack sufficient consistency and justification, include diagnosis of skeletal disorders, documentation of soft tissue injuries, detection of peripheral arterial dysfunction, and evaluation of repetitive-stress trauma [9, 10, 14]. There are many less familiar applications in non-medical fields such as in chemistry, agriculture [8], and industry [11, 12]. This study investigates the use of thermography in the evaluation of an industrial factor, namely mechanical vibration.

2. METHODS

Seven college students participated in this study. The sample consisted of 4 males and 3 females. The average age of the sample was 22.3 yr (SD=4.3 yr) and the average body weight was 140.4 lb (SD=24.3 lb). Each subject was advised not to smoke or drink coffee for at least 12 hours before experiment time. Before the start of the experimental procedures, each subject was asked to wash the hands with soap and water. Hands were then dried, cooled, and skin temperature was allowed to stabilize for 10 minutes. The subject was then asked to place the palms of both hands on the Flexitherm. As soon as the Heat Emission Pattern stabilized, an instant photograph of the temperature pattern of both hands was taken. The subject was then seated on an adjustable chair. A digital temperature sensor was attached to the thumb of the right hand. The subject was instructed to gently grasp to a vibrating machine handle (Figure 1). Vibration was then applied at the predetermined frequency for a continuous period of 30 minutes. This bi-directional angular vibration was generated at a level of 1 g (9.8 m/sec*sec). In order to insure repeatability and consistency in performance, each subject maintained a vertical line on the screen of a CRT in a defined position (tracking task). This procedure was employed to insure continuous use of the hand-held control. Thumb temperature was continuously monitored and was recorded at the end of each minute. Immediately following the task performance period, a second thermogram was taken. The same procedure was repeated on the following two days for each of the remaining vibration frequencies.

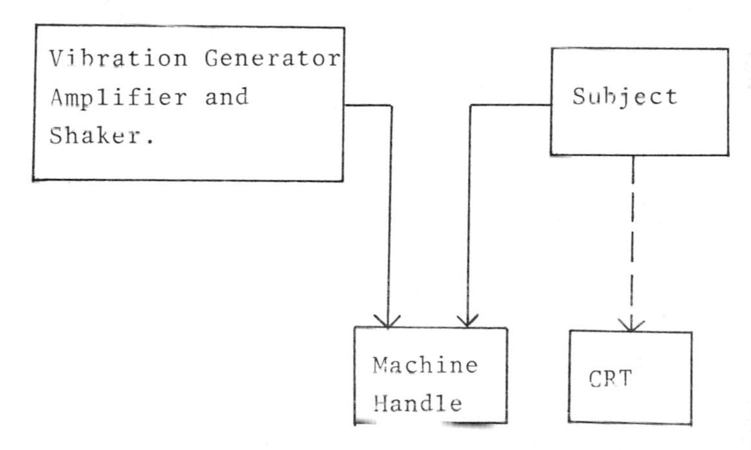

Figure 1. Simplified Diagram of the Experimental Set-Up.

3. RESULTS

Since the body's surface temperature varies between individuals, and over time within the same subject, only changes in thumb temperature were analyzed. The mean values of the change in temperature (temperature at minute "t" - initial temperature) were plotted for each minute of the task performance period (Figure 2). Each profile represents the response to a vibration frequency. This figure shows that the physiological responses to the various vibration frequencies were different, with 100 Hz showing an abrupt and more rapid increase in temperature than 5 Hz or 20 Hz. Statistical analysis have shown that the increases in the temperature were significant for all vibration frequencies. The average increase in thumb temperature (over the entire 30 minutes period) was 4 degrees (SD = 1.3) when vibration was applied at 5 Hz. The correspon-

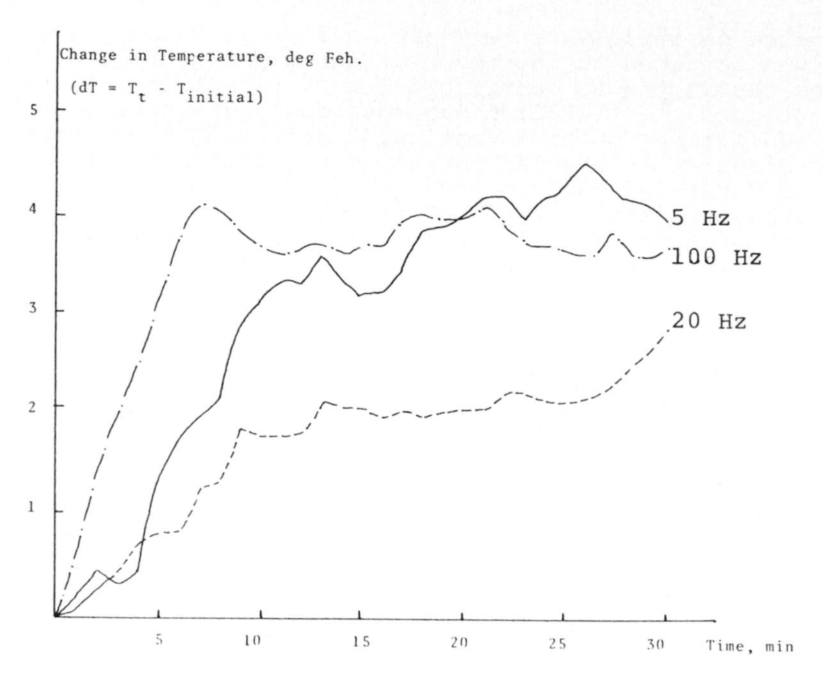

Figure 2. Accumulative Change in Thumb Temperature for Each of the Vibration Frequencies.

ding value for the 20 Hz and the 100 Hz were 2.9 (SD = 0.98) and 3.5 (SD = 0.62) degrees respectively. It can be seen also that long-term application (at least 30 minutes) of the body's resonance frequency (5 Hz) and higher frequencies (20 Hz and 100 Hz) show similar increases in local temperature. Another important finding was that the female subjects experienced statistically larger changes in thumb temperature than the male subjects.

Since the heat patterns vary between individuals, the thermograms were interpreted by comparing the temperature of a single location of the hand before and after the task performance period. In order to analyze each thermographic image, the heat pattern of the hand was divided into five regions (Figure 3). The change in color of each region was translated into quantitative terms by using the color code/scale of the thermographic cassette. Findings are presented in Table 1. In this table a (+) represents an increase, a (-) represents a decrease, and a (0) represents no change in the color image and consequently in temperature. Similar to the findings from analyzing the change in thumb temperature recorded digitally, the female subjects (subjects #1, #3 and #4) showed larger changes in the heat pattern of the hand due to vibration. The increase in the thumb temperature, as recorded digitally, was not consistently manifested in the thermographic images of the thumb (region #2). This could be due to the different hand grasping techniques by the subjects which in turn could result in difference

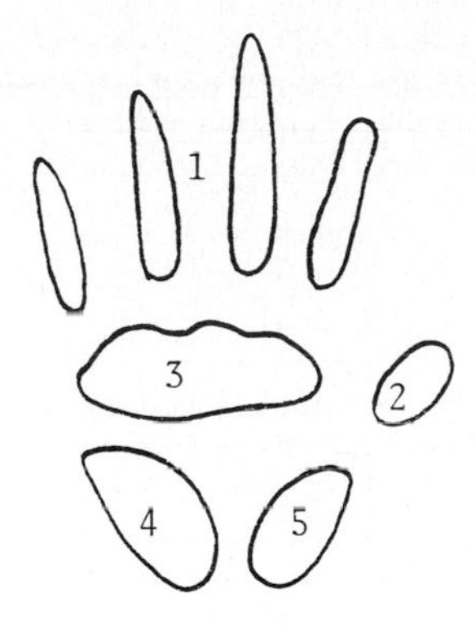

Figure 3. Definition of the 5 Regions Identified in the Thermographic Image of the Right Hand. Numbers are in Reference to Table 1.

TABLE 1

Changes in Regional Temperature as Shown in the Thermographic Images of the Right Hands

		Subject #						
FREQUENCY	REGION	1	2	3	4	5	6	7
5 Hz	1	-3.9	-1.1	0	+3.9	+0.5	+0.5	-1.3
	2	-3.9	-1.1	+2.2	+1.4	+0.5	+0.5	-1.3
	3	-2.5	-1.1	-3.9	0	-0.5	+0.5	-0.5
	4	+1.4	0	+3.9	+2.5	-0.5	+0.5	-0.5
	5	-2.5	-1.1	-5.0	-3.9	-0.5	+0.5	-0.5
20 Hz	1	-4.1	+1.6	+3.6	-1.3	+1.3	-1.6	0
	2	-1.2	+1.6	+1.6	-1.0	+0.5	-1.6	+1.1
	3	-2.5	0	+1.1	-1.3	-1.0	-1.6	0
	4	-5.5	0	+1.6	+1.6	+1.8	0	0
	5	-3.9	0	0	+1.3	+1.8	0	+1.1
100 Hz	1	+3.9	0	+3.4	+5.0	-1.1	+1.6	+1.6
	2	+2.6	0	-3.9	+3.9	-1.1	+1.6	+1.6
	3	-3.6	0	+1.4	+2.5	-1.1	-1.6	+1.6
	4	+0.5	0	-1.3	+5.5	0	0	0
	5	-1.1	0	-2.5	+3.9	-1.1	-1.1	+1.6

in pressure on any one location of the hand. There was a variable change in the heat of certain regions of the hand after the 30 minutes period. This, again, reflects the varying amount of grasp subjects had to apply to the vibrating machine handle.

4. CONCLUSIONS

Results of this study show that mechanical vibration of the hand at the body's resonance frequency of 5 Hz produces the highest average temprature change over the task period. Results also suggest that vibration at 100 Hz produces a quick rise in hand temeprature in the first 5 minutes of performance. This frequency could be useful in some medical application where rapid heating of body tissue is required.

The evaluation of the change in heat patterns using thermographic images was inappropriate for this study due to several factors. First, it was difficult to describe the images in a quantitative manner, and hence to accurately monitor the changes in temperature. Secondly, this technique requires elaborate preparation procedures. It was found to be sensitive to environmental changes in temperature and humidity. Thirdly, the effect of factors other than vibration (such as grasping techniques which determine the amount of hand pressure on the vibrating handle) could overshadow the actual effects of vibration on body heat.

Before it can be recommended for further use, we recommend that the application and interpretation of liquid crystal thermography in assessing this and other types of occupational stresses should be done with extreme caution.

REFERENCES

[1] Fischer, A.A., Tissue Compliance Recording. A Method for Objective Documentation of Soft Tissue Pathology, Arch Phys Med Rehab 62 (1981) 542.
[2] Greco, E.C., Khalil, T.M., Moty, E.A., Vibration Induced Synchronization in Myoelectric Activity. 35th ACEMB Meeting (Philadelphia, PA: 1982) pp. 197.
[3] Hornick, R., Vibration. In Bioastronautics Data Book (2nd ed.). Cited in E.J. McCormick and M.S. Sanders: Human Factors in Engineering and Design, 5th ed. (New York, McGraw-Hill Book Co.: 1982).
[4] Khalil, T.M., Ayoub, M.M., Work-Rest Schedule Under Normal and Prolonged Vibration Environment. AIHA Journal, March, 1976.
[5] Koradecka, D., Peripheral Blood Circulation Under the Influence of Occupational Exposure to Hand Transmitted Vibration. In: D. Wasserman and W. Taylor (eds.), Proceedings of the International Occupational Hand-Arm Vibration Conference (Cincinnati, Ohio: 1977)

[6] Leatherwood, J., Dempsey, T., Clevenson, S., A Design Tool for Estimating Passenger Ride Discomfort Within Complex Ride Environments. Human Factors 22 (1980) 291-312.
[7] Leonida, D., Ecological Elements of Vibration (Chiper's) Syndrom. In: D. Wasserman and W. Taylor (eds.), Proceedings of the International Occupational Hand-Arm Vibration Conference (Cincinnati, Ohio: 1977)
[8] Merritt, T.W., A Review of the Use of Thermography in the Evaluation of Occupational Stress. Ergonomics IE News XVII (1983) 1-3.
[9] Studebaker, E., Champlint, K., Liquid Crystal Thermography and Venous Physiology Applied to IV Catheterization. National Intervenous Therapy Association 2 (1984) 11-15.
[10] Thurston, N.M., Kent, B., Jewell, M.J., Blood, H., Thermographic Evaluation of the Painful Shoulder in the Hemiplegic Patient. Physical Therapy 66 (1986) 1376-1381.
[11] Tichauer, E.R., Thermography in the Diagnosis of Work Stress due to Vibration Implements. Proc Int'l Occup Hand-Arm Vibration Conf (NIOSH Publication 77-170: pp. 160-168)
[12] Tichauer, E.R., The Objective Corboration of Back Pain Through Thermography. J Occup Med 19 (1977) 727-731.
[13] Walton, K., The Pathology of Raynaud's Phenomenon of Occupational Origin. In: W. Taylor (ed.), The Vibration Syndrom (New York: Academic Press, 1974)
[14] Wexler, C.E., Lumbar, Thoracic and Cervical Thermography. J Neural Orthop Surg, Nov (1979) 37-41.
[15] Wexler, C.E., An Overview of Liquid Crystal and Electronic Lumbar, Thoracic and Cervical Thermography. Tarazana, CA, (Thermographic Service Inc., 1981)
[16] Williams, N. Biological Effects of Segmental Vibration. J Occup Med 17 (1975) 37-39.

Trends in Ergonomics/Human Factors V
F. Aghazadeh (Editor)
Elsevier Science Publishers B.V. (North-Holland), 1988

PILOT SUBJECT EVALUATION OF WHOLE-BODY VIBRATION FROM AN UNDERGROUND MINE HAULAGE VEHICLE

Thomas G. BOBICK, Sean GALLAGHER, and Richard L. UNGER

Bureau of Mines, U.S. Department of the Interior, Pittsburgh Research Center, P.O. Box 18070, Pittsburgh, PA 15236

The U.S. Bureau of Mines has developed an in-house test facility to evaluate the effects of whole-body vibration levels experienced by underground mobile equipment operators. Vibration data were collected from a coal haulage vehicle via a uni-axial accelerometer attached to the machine frame. The data were analyzed and processed so a computer-controlled vibration platform could duplicate the vibration signals. Six men (36.0 yr of age ± 6.9 SD) participated in a pilot study to evaluate the effects of shock and whole-body vibration on heart rate (HR), blood pressure (BP), back extensor strength, stature, manual dexterity, and subjective discomfort. The subjects were exposed to vibration for 30 min periods while seated in a typical operator's seat (with the backrest angle at 90° or 130°) that was plain steel or treated with 2 in of foam padding. To control for diurnal variations in stature, HR, and BP, each subject repeated the same protocol on a separate day, but without the vibration. Results indicated that vibration significantly increased the HR, systolic BP, mean BP,and the overall subjective discomfort rating ($p < .05$). Also, the number of times subjects reported discomfort increased significantly during vibration ($p < .05$) and when seated in the steel seat ($p < .005$). The seat back angle had no significant effect on any of the dependent measures.

1 INTRODUCTION

The presence of shock and vibration while operating mobile equipment is a growing concern in the underground mining industry. Minimal research has been conducted on the exposure of underground equipment operators to whole-body vibration, or to the design of appropriate seating. Even recent models of mobile mining equipment have seats that are only a bent steel plate. Other seats may have padding, but the materials can wear out quickly in the harsh underground environment. Also, the seat is usually attached directly to the machine frame since vibration isolation systems are difficult to install because of space limitations. Thus, the operator is subjected to almost constant vibration and shock loading during equipment operation.

During some informal tests in a low-seam coal mine, Bureau researchers found that while riding for 30 min in a typical personnel vehicle, they were exposed to nearly 35 pct of the 8-hr exposure limit set by the International Standards Organization [1] for whole-body vibration (WBV). Also, a past Bureau research program [2] conducted a limited evaluation of mobile underground coal mine equipment operators to WBV. These data indicated that between 33 pct and 39 pct of the operators are exposed to vibration levels that exceeded the ISO fatigue-decreased proficiency level (intended to preserve human working efficiency) and that 7 pct to 14 pct exceeded the exposure limit (intended to protect workers from physical injury or disease caused by daily exposure at work).

Various research studies and reviews have indicated that WBV can affect the musculoskeletal system [3-7], the cardiovascular system [4-6, 8-11], and gastrointestinal system [4, 6, 12]. In addition to these physical and physiological effects, other studies have investigated the effects of WBV on performance tests ([8] mentions six studies and [12] reviews another eight) and on subjective evaluations of WBV exposure [9, 13-15].

The purpose of this project is to determine the effects of WBV on mobile underground mining equipment operators. This paper presents the initial results of pilot subject testing. Data from this research program will be used to recommend changes in seat design for underground mining equipment.

2. METHOD

2.1 Subjects

Six healthy men (36.0 yr of age ± 6.9 SD) volunteered to participate in a pilot study that examined the effects of vibration, seat back angle, and the presence or absence of foam padding on various physical, physiological, and performance measures. The test subjects were all employees of the Bureau's Pittsburgh Research Center. They received a thorough physical examination and graded exercise tolerance test [16] prior to participating to insure good health. Subjects were advised of the nature of the investigation and signed an informed consent form before participation.

2.2 Experimental Design

The independent variables in this investigation were (1) presence or absence of random, broad-band vibration, (2) seat back angle of 90° or 130°, and (3) presence or absence of foam padding material on the seat pan and back. Dependent measures included back strength, manual dexterity, stature, heart rate (HR), systolic, diastolic, and mean blood pressures (BP), and subjective discomfort. Test conditions were randomized and performed in a counterbalanced fashion to control for bias due to order of testing.

2.3 Apparatus

Figure 1 presents a schematic of the equipment used in this experiment. Subjects sat in a test seat, which was equipped with an adjustable backrest, to which padding could be easily installed. One of the configurations tested was a duplicate of a typical operator's seat. The adjustable seat was mounted on a hydraulically powered, computer-controlled shake table. Heart rate was obtained using a Beckman* Dynograph Recorder, Model R-511A. Blood pressures were acquired with a Narco Scientific Adult/Pediatric Non-Invasive Blood Pressure Monitor. Back extensor strength was measured with a strain gage-type load cell manufactured by the Prototype Design and Fabrication Company, used in conjunction with the Model ST-1 Force Monitor. Manual dexterity was measured with the Stromberg Dexterity Test. Stature was measured by carefully positioning the subjects against a surface that was inclined at 10° to which an anthropometer was attached. While the subjects were seated in the various experimental conditions, they wore stereo headphones through which "pink noise" was played to mask ex-

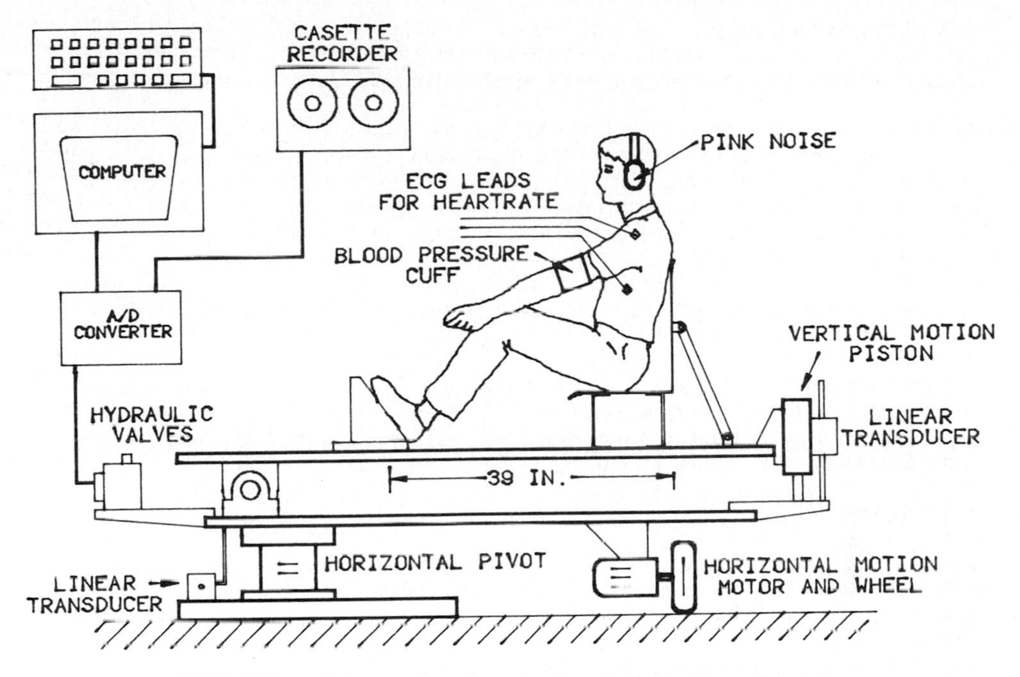

FIGURE 1. Schematic of experimental test equipment.

traneous auditory signals. The foam material was manufactured by Dynamic Systems, Inc. and was the brand designated as Pudgee. Lastly, the subjects completed a subjective discomfort form three times during the test period. Figure 2 provides a schematic of the body divided into different areas on which the discomfort estimate was based. Subjects rated their discomfort on a 7-point scale, from just noticeable (1) to moderate (4) to severe (7).

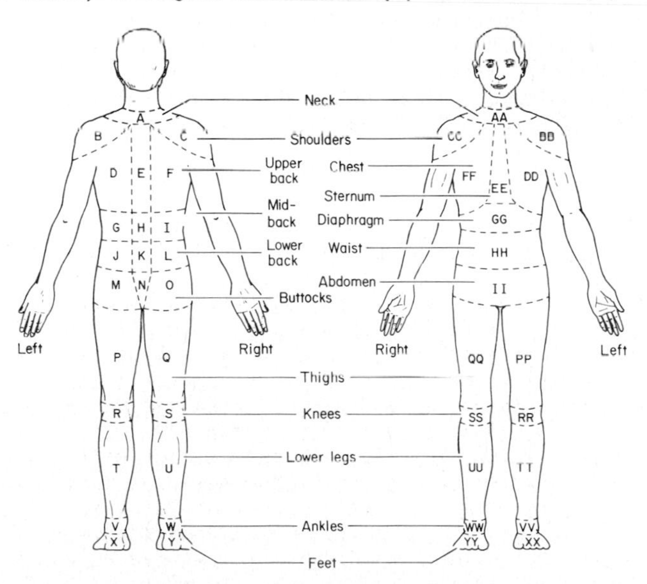

FIGURE 2. Body areas for subjective discomfort evaluation.

2.4 Experimental Task

Prior to testing, the subjects were instructed as to the experimental protocol and how to fill out the subjective discomfort form. Additionally, they practiced the Dexterity Test and became familiar with providing a maximum voluntary isometric back exertion via a test-retest procedure that followed the Caldwell Regimen [17]. After these practices, the before-testing data were collected; these included the average of three stature measurements, one back extensor strength test, and a timed dexterity test.

The subject was then positioned in the seat. Before beginning the 30-min test period, each subject was instructed to sit quietly so resting heart rate and blood pressure levels could be obtained.

During testing, heart rate was collected during the last ten sec of every min. The final 25 HR values were averaged and taken as the mean value for that test configuration. Blood pressures were collected every 5 min during programmed pauses in the vibration cycle. Systolic, diastolic, and mean values were averaged at the end of the test. Every 10 min (at min 9, 19, and 29) the subjective discomfort form was completed by the subjects.

Immediately after the 30-min test period, stature, back strength, and dexterity measurements were collected for comparison with the before-test measurements. A 30-min break was provided for the subject to attend to personal needs and to recover by relaxing in a reclined posture before beginning the next experimental condition.

2.5 Data Treatment

The results of data collected for all nine dependent measures were analyzed using a 2 x 2 x 2 (vibration or not x seat back angle x seat material) analysis of variance with repeated measures (ANOVR) statistical package [18]. The back strength, dexterity, and stature measurements were normalized for each test condition. As a follow-up to these analyses, a 2 x 2 x 4 (vibration or not x before and after testing x the four tests, as administered) ANOVR was run on the back strength and dexterity. For these analyses, the data were normalized for the highest back strength and quickest dexterity time measured in both test days. Critical alpha levels were 0.05 in all cases.

3. RESULTS

3.1 Performance and Physical Data

The upper portion of Table 1 summarizes the results for the two performance measures (back strength and dexterity) and the one physical measurement (stature). Initially, these measurements were normalized to the data collected prior to each different test configuration. None of the independent variables significantly affected these three dependent measures.

Subsequently, the data for back strength and dexterity were normalized across both testing days. None of the independent variables (vibration-no vibration, before-after, and the four test conditions) had a significant effect on the back strength; there was, however, a significant interaction ($F_{1,5} = 7.818$, $p < .05$) between the vibration and the before-after conditions on the normalized back strength. The nature of this interaction

TABLE 1 -- SUMMARY OF EFFECTS FOR ALL TEST CONDITIONS

Dependent Measures	Independent Variables		
	Vib'n-No vibration	90° - 130°	Ab - Steel
Back Strength	n.s. (0.507)	n.s. (0.680)	n.s. (0.714)
Dexterity	n.s. (0.302)	n.s. (0.320)	n.s. (0.052)
Stature	n.s. (0.571)	n.s. (0.358)	n.s. (0.737)
Heart Rate	$F_{1,5}$=8.481, p=.033 vib'n incr's HR	n.s. (0.977)	n.s. (0.790)
Systolic BP	$F_{1,5}$=12.664,p=.016 vib incr's Sys BP	n.s. (0.286)	n.s. (0.849)
Diastolic BP	n.s. (0.061)	n.s. (0.150)	n.s. (0.320)
Mean BP	$F_{1,5}$=11.849,p=.018 vib incr's Mean BP	n.s. (0.159)	n.s. (0.141)
No. of Subjective Discomfort Eval'ns	n.s. (0.078)	n.s. (0.562)	$F_{1,5}$=6.950,p=.046 st incr's number
Overall Subjective Discomfort Rating	$F_{1,5}$=8.908, p=.031 vib incr's rating	n.s. (0.710)	$F_{1,5}$=26.075,p=.004 st incr's rating

was that while back strength decreased following the test conditions, it was decreased by a greater amount when the subjects were vibrated.

Concerning the normalized dexterity data, one of the three independent variables had a significant effect. The order of administration of the dexterity tests had a significant effect on the time it took the subjects to complete them ($F_{3,15}$ = 8.953, $p < .05$). Using Duncan's multiple-range test to determine significant differences between the mean times for the four test conditions indicated that the time to complete the first test of the day was significantly longer than subsequent tests ($p < .05$). The times to complete the dexterity test on the second, third, and fourth trials of the day were not significantly different from one another ($\underline{p} > .05$). Additionally, the fastest completion time (for all subjects) always occurred on the second test day, regardless of whether it was a vibrating or non-vibrating condition.

3.2 Physiological Data

As shown in Table 1, heart rate, systolic BP, and mean BP were all significantly increased during the vibrating test day ($p < .05$). Although heart rate was not significantly affected by either the angle of the seat back or the presence or absence of foam material, there was a significant interaction between these two independent variables on the mean heart rate of the subjects ($F_{1,5}$ = 7.723, p = .039).

3.3 Subjective Discomfort Data

The lower portion of Table 1 presents the subjective discomfort data. The number of times subjects reported discomfort was significantly affected by whether they were seated in the untreated (steel) seat or treated with foam padding ($F_{1,5} = 6.950$, $p = .046$). Additionally, the overall rating of discomfort was significantly affected by both the vibrating test condition ($F_{1,5} = 8.908$, $p = .031$) and, again, whether the subjects were sitting in the steel seat or the treated one ($F_{1,5} = 26.075$, $p = .004$). Neither the number of times the subjects reported discomfort nor the overall discomfort rating were significantly affected by the angle of the seat back.

4. DISCUSSION

4.1 Performance and Physical Data

None of the independent variables had a significant effect on the back strength of the test subjects. However, when the back strength data were normalized for the effects across the entire day (as opposed to just each test period), there was a significant interactive effect between the vibrating condition and the measurements collected before and after the testing ($F_{1,5} = 7.818$, $p = .038$). Even though back strength is presently not being affected by the vibration, it would be interesting to investigate the possible effects of longer periods of vibration (perhaps 60 min) on back strength. Additionally, most of the controlled laboratory studies investigate the effect of sinusoidal vibration. In this particular study, however, the vibration spectra is broad band; this may be a confounding factor in the back strength measurements.

Similarly, the dexterity test was not significantly affected by any of the independent measures. The analysis of the normalized times across both testing days indicated that the mean time to complete the dexterity test decreased slightly for each subsequent evaluation. A learning process seemed to be occurring, perhaps offsetting any real changes in dexterity and reaction time caused by the vibrating condition. Interestingly, the data showed a trend that subjects performed better on the dexterity test after sitting on the steel seat. This may have been due to an increased arousing effect, which was reported on by Kjellberg and Wikstrom [12].

Similar to the back strength and dexterity data, none of the independent variables had a statistically significant effect on the stature measurements. Changes in stature due to vibration have been reported [19, 20]; however, the magnitude and even the direction of change have been shown to be frequency dependent [19]. Also, in most other studies examining changes in stature, subjects have experienced sinusoidal WBV. It may be that random, broad-band vibration exposure has less effect on the stature of an individual. Changes to the protocol may increase the likelihood of establishing an effect from vibration on stature measurements.

4.2 Physiological Data

Three of the four physiological dependent measures were significantly ($p < .05$) affected by the vibrating condition. Hasan [8] conducted a thorough review of the literature related to the biomedical effects of whole-body vibration. He describes conflicting reports on the effects of vibration on the cardiovascular system. In general, most investigators have found no alteration in the heart rate or mean blood pressure at low frequencies of vibration. After an initial burst of activity, the HR and BP reverted back to normal or even slightly subnormal levels with continued exposure. However, a slight increase in HR and BP has been observed in subjects at a vibration frequency of 5 Hz [8]. The fact that the shake table generated a broad-band random signal may be the explanation for the significant increases in the heart rate, and in systolic and mean blood pressures. The diastolic blood pressure was also elevated, but did not achieve statistical significance.

None of the four physiological measures were affected by the seat back angle, or the presence/absence of the foam material. This was surprising since the hypothesis was that HR or BP would be affected by these two variables. There was, however, a significant interaction ($p < .05$) between these two independent variables and their effect on the HR of the subjects.

4.3 Subjective Discomfort

The last two categories of the dependent variables were associated with the subjective discomfort of the subjects. As expected, the number of times the subjects reported discomfort was increased significantly when seated in the steel seat. The number of times discomfort was reported was also increased for the vibrating condition, although not significantly. Both the vibrating condition and the steel seat significantly increased the overall rating of discomfort felt by the subjects during the testing. It is quite surprising to see that the seat back angle did not have any effect on either of the discomfort categories. This type of information will be valuable for the purposes of designing a favorable work station. Other angles will be investigated in future research studies.

5. SUMMARY OF PILOT TEST RESULTS

The findings of this pilot subject testing are summarized below:

1. Heart rate and two of the three blood pressures were significantly increased by the vibrating condition.
2. The steel seat had a significant effect on both of the subjective discomfort categories. The vibrating condition affected only the overall discomfort rating.
3. The posture of the subject (seat back angle) had no significant effect on any of the nine dependent variables.
4. Back strength, dexterity, and the stature measurements were not significantly affected by any of the independent variables.

6. RECOMMENDATIONS FOR FUTURE TESTING

The ultimate goal of this research project is to determine the effects of WBV on mobile underground mining equipment operators, and to develop improved seating designs for this equipment. The results of the present study have raised several points to be considered in future testing:

1. The low-seam underground environment often requires the equipment to be operated while drivers are lying on their side or stomach. While there were no significant effects on any of the dependent measures from the two seat back angles investigated in the present study, future research should investigate more supine postures.
2. The 30-min vibration period was chosen based on estimates of the duty cycle of underground coal haulage equipment. A longer vibration period

or shorter rest period may be more appropriate to provide an accurate description of the whole-body vibration effects.
3. Any new seat configurations considered for future equipment designs will have to consider the space limitations of the operator compartments. Electromyographic data will be collected (during the existing pauses to measure BP) to determine whether localized fatigue is occurring in the erector spinae, trapezius, or other muscles.
4. Additional performance tests should be included in the test protocol to increase the external validity. A tracking task and/or a reaction time test will be used to better assess the effects of WBV on operating proficiency.
5. It is important to know whether mobile underground mining machine operators are experiencing compression of the intervertebral disks from WBV. The current measurement procedure, however, does not permit the level of precision that is needed. An improved stature measurement device is needed for future research studies.

FOOTNOTE

*Reference to specific brands does not imply endorsement by Bureau of Mines

REFERENCES

[1] International Standards Organization (ISO), Guide for the evaluation of human exposure to whole-body vibration, (1974), ISO 2631-1974.
[2] Remington, P.J., D.A. Anderson, and M.N. Alakel, BuMines Contract No. J0308045, March 1984, (BBN Report No. 5616), 98 pp.
[3] Carlsoo, S., Applied Ergonomics, (1982) 251-258.
[4] Seidel, H., R. Heide, Int. Arch. Occup. Environ. Health, (1986) 1-26.
[5] Wilder, D.G., B.B. Woodworth, J.W. Frymoyer, and M.H. Pope, Spine, (1982) 243-254.
[6] Wilder, D.G., J.W. Frymoyer, and M.H. Pope, Automedica, (1985) 5-35.
[7] Chaffin, D.B. and G.B.J. Andersson, Occupational Biomechanics. (John Wiley and Sons, New York, 1984).
[8] Hasan, J., Work-Environment-Health, (1970), 19-45.
[9] Soule, R.D., Vibration. Chap 26. in: The Industrial Environment-Its Evaluation and Control. (U.S. Dept of HEW, NIOSH, U.S. G.P.O., 1973).
[10] Guignard, J.C., Evaluation of Exposure to Vibrations. Chapter 13. in: Patty's Industrial Hygiene and Toxicology. Volume III: Theory and Rationale of Industrial Hygiene Practice, (John Wiley and Sons, New York, 1979).
[11] Helmkamp, J.C., E.O. Talbott, and G.M. March, Am. Ind. Hyg. Assoc. J., (1984) 162-167.
[12] Kjellberg, A. and B.O. Wikstrom, Ergonomics, (1985) 535-544.
[13] Meister, A., D. Brauer, N.N. Kurerov, A.M. Metz, R. Mucke, R. Rothe, H. Seidel, I.A. Starozuk, and G.A. Suvorov, Ergonomics, (1984) 959-980.
[14] Oborne, D.J. and P.A. Boarer, Ergonomics, (1982) 673-681.
[15] Weaver, L.A., Professional Safety, (1979) v. 24, no. 4, pp. 29-37.
[16] American College of Sports Medicine, Guidelines for Graded Exercise Testing and Exercise Prescription. (Lea and Febiger, Philadelphia, 1980).
[17] Kroemer, K.H.E., H.J. Kroemer, and K.E. Kroemer-Elbert, Engineering Physiology. (Elsevier Science Publishers, Amsterdam, 1986).
[18] Games, P.A., G.S. Gray, W.L. Herron, and G.F. Pitz, Behavioral Research Methods and Instrumentation, (1980) 467.
[19] Corlett, E.N., J.A.E. Eklund, T. Reilly, and J.D.G. Troup, Applied Ergonomics, (1987) 65-71.
[20] National Aeronautics and Space Administration, Anthropometric Source Book, Volume I: Anthropometry for Designers, NASA Ref. Pub. 1024, 1978.

Trends in Ergonomics/Human Factors V
F. Aghazadeh (Editor)

SHIPBOARD EVALUATION OF MOTION SICKNESS INCIDENCE

Alvah C Bittner, Jr, PhD and John C Guignard, MB, ChB

Analytics, Inc
2500 Maryland Road
Willow Grove, PA 19090

GB Associates (New Orleans)
824 Kent Avenue
Metairie, LA 70001-4332

Seakeeping trials of a United States Coast Guard (USCG) cutter in March 1984 included an evaluation of seasickness in two work-stations. Analyses including correlation of questionnaire and ship motion data led to these conclusions: (1) seasickness was characterized by at least two functionally independent factors, identified as F1 (symptomatic General Motion Illness) and F2 (Retching-Vomiting); (2) both factors showed similar carry-over and location effects (although F1 was strongly associated with the vertical component of the ship's motion, while F2 was associated with large transverse relative to vertical components of motion); and (3) validity of future sea trials requires (a) multiple-score scaling of motion sickness, (b) control of subject crew movement about the ship during observation, and (c) avoiding steaming patterns that induce extraneous carry-over effects.

1. INTRODUCTION

The maintenance of readiness and the accomplishment of naval operations in rough seas is of growing concern to modern strategists (Kehoe et al, 1983; McCreight & Stahl, 1985). This concern has led to seakeeping trials of new vessels which address performance, safety, and well-being in moderate to severe sea conditions (eg, Baitis, Applebee & McNamara, 1984; Olson, 1977; Wiker et al, 1979). This paper reports an investigation of the incidence and implications of seasickness experienced during seakeeping trials of a medium endurance United States Coast Guard (USCG) cutter.

Seakeeping trials of this ship were conducted by the David Taylor Naval Ship Research and Development Center (DTNSRDC) for the USCG. In earlier reports, DTNSRDC has reported the overall plan of those trials, as well as operational aspects of crew and ship performance measured at sea (Applebee & Baitis, 1984; Baitis, Applebee & Meyers, 1984). Because of exceptional incidences of seasickness noted previously by the USCG in the mission-critical communications support center (CSC) and the communications center (CC), we were requested to evaluate the problems in those work-stations. We were also called upon to to consider the implications of our findings for the conduct of future seakeeping trials.

We have earlier confirmed the exceptional incidence of crew seasickness when working in the CC and CSC; and recommended the application of five human factors engineering principles for minimizing the incidence of adverse effects (Bittner & Guignard, 1985; 1987). The present report is concerned specifically with analyses of unpublished seasickness and associated data collected from CC and CSC personnel during the DTNSRDC/USCG seakeeping trials.

1.1. Seakeeping Trials

Seakeeping trials took place on 6 and 7 March 1984 during which data were collected under experimental steaming conditions. On each experimental day, it had been intended that the ship would steam an octagonal course so as to obtain data at a variety of headings (ideally, in a constant sea state). However, deviations from the planned order of "legs" (segments) of the octagonal course occurred on the second day for operational reasons. It had also been intended that every 40-minute leg of the octagon was to contain two 20-minute segments during which the ship maintained course and speed while its anti-roll fin stabilization system was switched on and off sequentially. Ideally, this steaming pattern (with and without fin stabilization) was to be done at the maximum permissible speed within physical and human operational constraints (eg, avoidance of slamming, excessive deck wetness, etc). This artifice had been intended to permit an appraisal of the action of fin stabilization but the developmental status of the ship's stabilization system precluded the attempt (Baitis et al, 1984).

Variations in sea conditions during the trials were substantial. Significant wave heights varied from 3 to 15 feet (Sea State 5), which, in conjunction with the course changes, led to wide variation in associated conditions aboard the ship (Applebee & Baitis, 1984). The latter variations provided for the data base conditions to allow correlation with observed crew response.

1.2. Purpose

This report evaluates variations in motion sickness experienced by the personnel on duty in the CC and CSC during seakeeping trials. One intermediate goal was to evaluate the relationship of motion sickness incidence and development with measured ship motion variables. Another goal was to evaluate those relationships with regard to the conduct of future seakeeping trials.

2. METHOD

The subjects were 16 USCG officers and enlisted personnel who, during the seakeeping trials, were assigned to duties in the CC or CSC.

2.1. Questionnaires

Two questionnaires were administered to the subjects during the seakeeping trials: (1) Motion Sickness Symptoms (MSS) and (2) Prewatch Status (PS) described in an earlier report (Bittner & Guignard, 1987). The MSS evaluated on a 0-to-3 scale the range of signs and symptoms of motion sickness listed in Table 2.

2.2. Ship Motion Measurements

Electronic instrumentation was deployed by DTNSRDC observers at fixed locations to measure and record the ship's motion in response to sea, wind and heading variations. The physical data acquired included both "transverse" (athwartship) and "normal" (ship's vertical) root mean square (rms) acceleration values, which have been reported previously (Applebee & Baitis, 1984).

2.3. Procedure

The MSS and PS questionnaires were administered daily to all observed personnel before they went on watch. This pre-watch administration was done after breakfast and had been preceded by at least a 4-hour rest period with the opportunity for sleep. The MSS also was administered at the end of every 20-minute octagon segment. Typically, subjects were followed over their staggered 4-hour watch during which about half an octagon (6 to 9 20-minute segments) would be steamed. The data were later correlated with the corresponding transverse and normal rms acceleration data recorded at a relatively central point near to the crew's mess (Applebee & Baitis, 1984).

3. RESULTS

Data analysis was conducted in two phases. First, a factor analysis was conducted of the 354 MSS questionnaires collected over all observation days and subjects using an approach similar to that adopted in prior time-course investigations (eg, Sampson _et al_, 1983). This served to reduce and concentrate the data prior to more detailed evaluation. In the second phase, relationships between factor and other variables were explored by means of regression analyses. The results of the two phases of analysis are presented sequentially below.

3.1. Phase I: Factor Analysis of Symptomatology

Table 1 presents the intercorrelations, means ($\bar{X}$), and standard deviations (SD) for symptoms across the 354 questionnaires with ratings on a 0-to-3 scale (0 = symptom not present to 3 = severe). Interestingly, the mean ratings of all symptoms typically averaged less than unity (1 = mild), although ratings of 3 (severe) were observed in some cases for all symptoms. Part of the explanation for the low averages may have been that many sufferers, with increasingly severe symptoms, removed themselves from the areas of observation to less stressful locations (eg, sleeping bunks). This selective reduction resembles an effect seen in epidemiological studies of adverse human responses to other stressful conditions such as whole-body vibration (Spear _et al_, 1976).

Principal factor analysis (PFA) with iterated communalities revealed two factors which together essentially explained the variation of the correlation matrix (Dixon, 1981). Of a total of 55.2% of the variance which could be explained maximally by up to six factors, 38.9% was explained by the first factor (F1) and 47.8% by the first two factors (F1 and F2) combined. Table 2 shows the rotated factor loadings and communalities for the observed symptoms listed in Table 1. Examining Table 2, it may be seen that F1 exhibits its highest loading (0.83) on variable 1 (identified with general feelings of illness); and, except for "Sweating", exhibits positive loadings across the spectrum of elicited symptoms. This pattern suggests empirical characterization of F1 as "General Motion Illness". On further examination, we found that F2 exhibits its highest loadings (0.78 and 0.56, respectively) on variables 7 (identified with retching) and 8 (frank vomiting). F2 is accordingly

TABLE 1. Symptom* intercorrelations, means and standard deviations (N = 354)

	1	2	3	4	5	6	7	8	9	10	11	$\bar{X}$	SD
1	1.00											.85	.78
2	.44	1.00										.67	.76
3	.78	.47	1.00									.65	.82
4	.44	.37	.33	1.00								.66	.76
5	.03	-.09	-.03	.14	1.00							.22	.49
6	.39	.36	.44	.36	.09	1.00						.80	.78
7	.10	.12	.17	.07	.25	.13	1.00					.01	.14
8	.19	.18	.21	.14	.16	.07	.47	1.00				.02	.17
9	.71	.55	.74	.23	-.06	.36	.12	.12	1.00			.90	.78
10	.63	.48	.63	.45	.01	.68	.17	.17	.52	1.00		.95	.85
11	.61	.51	.56	.44	.10	.48	.21	.16	.56	.68	1.00	.59	.70

* Symptoms listed in Table 2.

TABLE 2. Rotated factor loadings (pattern) for symptoms

Symptom Number & Name	Factor 1	Factor 2	Communality
1. General Ill Feelings	.828*	.041	.688
2. Dizziness	.618*	.063	.386
3. Nausea	.817*	.071	.673
4. Headache	.485*	.116	.249
5. Sweating	-.012	.331*	.109
6. Sleepiness	.585*	.099	.352
7. Dry Heaves (Retching)	.107	.777*	.615
8. Vomiting	.156	.557*	.335
9. Stomach Awareness	.766*	-.012	.587
10. Fatigue	.806*	.121	.664
11. Difficulty Thinking	.758*	.162	.600

* Loadings greater than 0.33.

characterized as "Retching-Vomiting" (ie, the clinically observable, frank manifestations of seasickness). It is noteworthy that these manifest responses, although positively associated, occurred relatively independently of the subjects' symptomatic judgements of general feelings of illness. This is consistent with spontaneous comments by some crew members, as well as earlier clinical observations of the disassociation of the magnitude of illness from the manifestations of retching and vomiting (T G Dobie, personal communication, 1984).

The results of the factor analysis are collectively consistent with a motion sickness syndrome in which retching and vomiting are relatively disassociated from the magnitude of general illness.

3.2. Phase II. Correlation of Seasickness and Motion Variables

The correlational analyses in this section are concerned with changes in the motion sickness factors (F1 and F2) which were associated with motion variables over the two days of seakeeping trials. Independent motion variables included: Common logarithmic (LOG) transformations of both the ship's normal and transverse rms acceleration values; as well as five pseudodichotomous (1 = present; 0 = not present) variables inindicating subject location during a particular test leg. It should be noted that the rms transformations were selected on the a priori expectation that subjective change would relate most closely to the logarithms of those variables (cf Bittner & Chatfield, 1980). The five pseudodichotomous variables were designated shipboard locations in the CC (PD1), CSC (PD2), Mess (PD3), Bunk (PD4), or undifferentiated (PD5). Figure 1 illustrates schematically the locations of the CC, CSC and Mess areas in relation to the ship's effective center of rotation (ECR): not shown are the Bunk areas dispersed about the ship (Chatterton & Braithwaite, 1978).

The respective lag score (F1B or F2B), as measured at the end of the preceding period, was included as an independent variable to provide for assessment of change in F1 or F2. This controlled for individual and other differences observed during the analyses of the prewatch status variables (eg, sleep). This also provided for an integrated autocorrelative model, with exponential growth and decay (Box & Jenkins, 1976). Moreover, this procedure was in accord with recommended statistical methods for the assessment of change (Cronbach & Furby, 1970). Variables were selected as a minimal comprehensive set of those available, based on prior investigations (Bittner & Guignard, 1985; Lawther & Griffin, 1985; Wiker et al, 1979).

Independent variables were evaluated in a forced sequence during regression analyses of factors F1 and F2. Based on a structured model (Kenny, 1979) of seasickness, the rms ship motion variables were first removed, followed by the location variables; and subsequently by the respective lagged motion sickness score. This sequence was used in anticipation of substantial colinearity between the independent variables based upon earlier analytic and empirical work (Applebee & Baitis, 1984; Wiker et al, 1979). This structured analysis was used to provide control for the colinearity of variables which otherwise might obscure the attribution of effects.

F1's Relationship to Motion Variables. BMDP2R (Dixon, 1981) stepwise regression analysis, following the procedure detailed above, was performed on the formal trial data for F1 summarized in Table 3. The

TABLE 3. Correlations, means and standard deviations for ship and sickness variables (N = 226)

		1	2	3	4	5	6	7	8	9	10	11	$\bar{X}$	SD
1	LOG(RN)	1.0											-1.07	.20
2	LOG(RT)	.12	1.0										-1.10	.30
3	PD1	.01	.00	1.0									.27	.42
4	PD2	-.04	-.11	-.48	1.0								.44	.50
5	PD3	-.02	.13	-.16	-.27	1.0							.08	.28
6	PD4	-.08	.01	-.11	-.18	-.06	1.0						.04	.20
7	PD5	.08	.04	-.28	-.46	-.16	-.11	1.0					.21	.41
8	F1B	.11	.07	.04	-.27	.14	.15	.12	1.0				.12	.92
9	F2B	.21	.16	-.07	-.11	.35	-.05	-.00	.07	1.0			.01	.82
10	F1	.15	.08	.09	-.22	.09	.05	.10	.85	.08	1.0		.12	.93
11	F2	-.03	.14	.03	-.08	.19	-.04	-.04	.02	.54	.05	1.0	.01	.89

TABLE 4. Variable coefficients for F1 and F2 factor loadings

Variable	F1-Equation	F2-Equation
Coefficient	0.342	-0.359
LOG (rms-normal)	0.225	-0.592
LOG (rms-transverse)	0.091	0.307
PD1 (CC)	0.116	0.167
PD2 (CSC)	0.036	0.031
PD3 (Mess)	-0.090	-0.019
PD4 (Bunk)	-0.351	-0.058
PD5 (Other)	0.000	0.000
F(1) Before	0.868	-
F(2) Before	-	0.618

resultant multiple correlation was both substantial in amount (RSQ = 0.734) and very highly significant [$F(7,218) = 85.99$; $p <$ E-11]*. The stepwise increment in RSQ due to: LOG(rms-Normal) was 0.024 [$F(1,218) = 19.35$; $p <$ E-04]; LOG(rms-transverse) was 0.003 [$F(1, 218) = 2.71$; $p = .10$]; Location (PD1-PD5) was 0.046 [$F(4,218) = 9.43$; $p <$ E-06]; and F1B was 0.661 [$F(1,218) = 541.92$; $p <$ E-11]. Table 4 summarizes the model coefficients for F1.

The model developed in this analysis describes the functional effects of location and sea state on factor F1. The coefficients for the rms acceleration variables as well as the regression analysis both point to the dominant role of rms vertical motion in the development of F1 (cf Table 2). These motion effects can be evaluated cumulatively in combination with other variables. Predicted increases and decreases in General Motion Illness are consistent with those manifested in previous empirical research (Bock & Oman, 1982; Guignard & McCauley, 1982; McCauley _et al_, 1976).

F2's Relationship to Motion Variables. The BMDP2R stepwise regression analysis similarly was applied to the data for F2 summarized in Table III. The resultant multiple correlation was of moderate magnitude (RSQ = 0.317) but very highly significant [$F(7,218) = 14.45$; $p <$ E-10]. The stepwise increment in RSQ due to: LOG (rms-Transverse) was 0.019 [$F(1, 218) = 5.97$; $p = .015$]; LOG (rms-Normal) was 0.002 [$F(1, 218) = 0.54$; $p > .46$]; Location (PD1 - PD5) was 0.033 [$F(4,218) = 2.63$; $p = .35$]; and F2B was 0.264 [$F(1,218) = 84.12$; $p <$ E-10]. Table 4 summarizes the model coefficients for factor F2.

The model developed in this analysis describes the effects of motion and location on factor F2. In contrast with the results for F1, LOG(rms-Transverse) was found to be the primary motion predictor for this variable. LOG(rms-Normal), however, was included in the model because its coefficient was significant in Table 4 [$F(1,218) = 6.82$; $p < .01$]. Also in contrast with the analysis of F1, location effects were not directly found to be significant in the regression analysis. However, the significant monotonic [RHO = 1.00; $p < .017$] relationship between the F1 and F2 coefficients argued for their inclusion. As for F1, the F2 location coefficients again point toward the CC and CSC as particularly provocative while the Mess and Bunk areas were relatively unprovocative. The relative effects of rms-Transverse and rms-Normal motions distinguish the models for F1 and F2.

4. DISCUSSION

This research evaluated motion sickness experienced during seakeeping trials of a medium endurance USCG cutter. This evaluation was concerned first with the relationships between seasickness and ship motion; and second with the implications of the results for the planning and conduct of future seakeeping trials.

4.1. Motion Sickness Evaluation

Two motion sickness factors were followed over the course of the seakeeping trials: General Motion Illness (F1) and Retching-Vomiting (F2).

---

* E-n denotes the -nth power of 10.

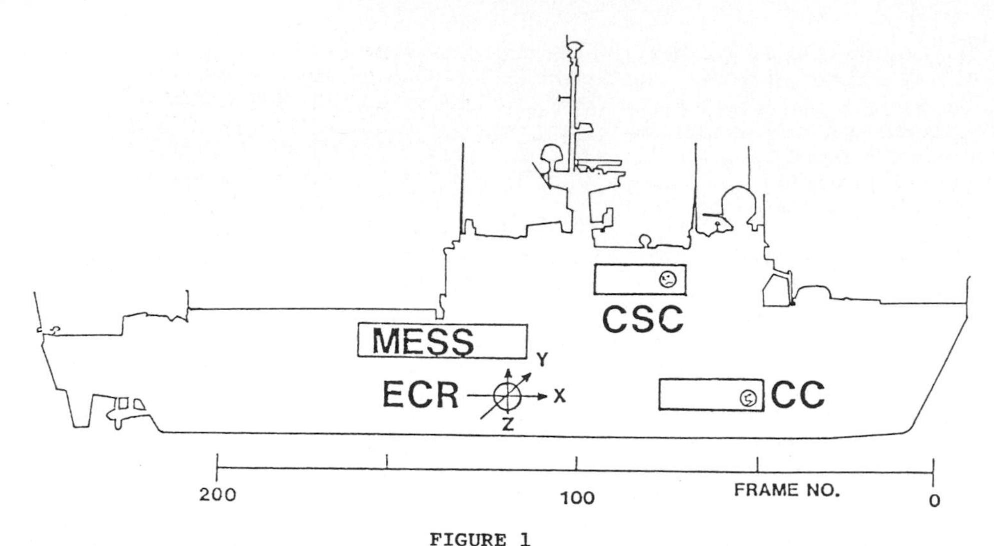

FIGURE 1
Relative locations of the ship's communications center (CC), communications support center (CSC), mess, and effective center of rotation (ECR).

Factors F1 and F2 exhibited similar functional responses to the within-day variables (cf Table 4). First, comparing lagged coefficients, it can be seen that both F1 and F2 have positive weights (0.868 and 0.618, respectively) for their corresponding values (F1B and F2B). These weights provide for time-course increases in motion sickness which are consistent with those seen in laboratory experiments using single-frequency and complex motions (McCauley et al, 1976; Guignard & McCauley, 1982). In addition, these weights predict an exponential recovery, when moving from a more to a less provocative location, which is in keeping with earlier research (Bock & Oman, 1982). Second, comparing the coefficients for Location (PD1 - PD5), an ordered relationship is apparent. This order is consistent with previous observations and an earlier developed first principle for minimizing motion sickness: namely, locate critical stations near the ship's effective center of rotation) (Bittner & Guignard, 1985). The lagged and location model coefficients taken together are consistent with earlier analyses.

The coefficients for the transformed motion variables in the F1 and F2 models differ substantially, as pointed out earlier. For F1, the coefficient for LOG(rms-Normal) and LOG(rms-Transverse) accelerations were both positive, with the former dominating the latter. Indicating that vertical accelerations are paramount in determining F1, this is consistent with previously published empirical research (eg, Lawther & Griffin, 1985) and with our own earlier analysis (Bittner & Guignard, 1985, p 207). For F2, the transformed coefficients for Normal and Transverse motions were of opposite sign (-0.593 and +0.307, respectively). This result indicates that F2 was more strongly associated with the mechanical conditions that probably increase complex angular head movements. Consequently, this result supports principles II and III of Bittner and Guignard (1985) for minimizing motion sickness (minimize head movements; align operator with a principal axis of the ship's hull). The differences in the F1 and F2 models consequently point toward a two-factor picture of seasickness while supporting earlier posited principles for its prevention.

4.2. Implications for Seakeeping Trials

This research has implications for planning future seakeeping trials, particularly with regard to scaling and experimental (trial) design.

Scaling. Motion sickness symptoms were factor analysed as a direct test of the long-held assumption that symptom severity is unidimensional. This assumption is implicit in the single-score rating systems which have been in use for some two decades under diverse conditions giving rise to a motion sickness syndrome (eg, Graybiel, et al, 1968; Lackner & Graybiel, 1983; Lawther & Griffin, 1985; Wiker et al, 1979). The results of our factor analysis have revealed that the unidimensional assumption is overly simplistic: there are at least two motion sickness dimensions (F1 and F2). Of practical significance, these two factors have been shown to be differentially associated with motion conditions. Specifically, rms-Normal motions predominate for F1 and large rms-Transverse (relative to rms-Normal) for F2. Our results argue for multiple-score scaling of motion sickness symptomatology in future seakeeping trials.

Experimental Trial Design. Considerable care was required in performing the correlational analyses because of the problems associated with the design of the sea trials to obtain the present data. The octagonal steaming pattern and movement of the crew about the vessel induced undesirable correlations between the independent variables. The sequential steaming pattern, in particular, led to colinearity of current and lagged ship motion variables because consecutive octagon legs differed by only 45 degrees of heading. It is of interest that this colinearity was mathematically predictable a priori using the Standard Ship Motion Program (SMP) (cf Applebee & Baitis, 1984). For assessment of the gross mechanical response of the vessel, such a correlation does not present a serious problem. However, the colinearity of the current and lagged motion variables induces an artifactual correlation between the current human response and lagged variables. Consequently, the separation of the direct from carry-over ship motion effects on the human response presented a statistical problem which was the result of the steaming pattern. The complexities of analyzing human response results obtained during octagon steaming trials have been alluded to previously (Wiker et al, 1979). We compensated for the artifactual correlation problem by the structured analysis described earlier.

The movement of subject personnel about the ship also complicated the analysis of the results. Over the study, slightly more than one third of crew responses were taken in various locations outside the CC and CSC (cf Table 3). Augmenting this problem were a number of sequences of crew response interrupted because of unscheduled rest or recovery in their bunks. These interruptions led to partial confounding of the various independent variables (eg, location; and individual physiological state). These variables could only be separated by structured analysis.

The human response results could be improved in future seakeeping trials by selecting alternative steaming patterns as well as by controlling subject movement. Analytically, the SMP or a similar model could be used to predict vessel responses over alternate patterns. Motions predicted from this model could in turn be used to estimate the resulting subject responses by means of a human response approximation (eg, the F1 model from this report or the Relative Motion Sickness Index from

Bittner & Guignard, 1985). The resulting ship motion and corresponding human responses could then be estimated for candidate steaming patterns; the degree of colinearity previewed; and a candidate pattern selected. However, for any selected pattern, the uncontrolled movement of the observed crew within the ship could still weaken the results unless anticipated in the experimental design. Previous seakeeping studies have sometimes required their observed crew to remain at specific workstations for the duration of an entire trial (eg, Wiker et al, 1979). Such an approach, unfortunately, does not provide for evaluation of seasickness growth and recovery under typical operational conditions. As an alternative, "balanced-for-residual-effects" designs (Cox, 1958) would compensate for these limitations by systematic variations in crew location. We recommend careful attention to steaming pattern and crew movement in future seakeeping investigations.

5. ACKNOWLEDGMENTS

The data reported here were collected while the authors were staff members of the Naval Biodynamics Laboratory, New Orleans. This work was performed for DTNSRDC under an agreement with the USCG. Opinions are the authors' and are not necessarily those of the supporting agencies.

6. REFERENCES

Applebee, T A and Baitis, A E. Seakeeping Investigation of the U.S. Coast Guard 270-Ft Medium Endurance Class Cutters: Sea Trials Aboard the USCGC Bear (WMEC 901), Report No DTNSRDC/SPD-1120-01, Bethesda, MD: David W Taylor Naval Ship Research and Development Center, August 1984.

Baitis, A E, Applebee, T R & McNamara, T M. Human factors considerations applied to operations of the FFG-8 and LAMPS MK III, Naval Engineers Journal, 96 (4), 191-199, 1984.

Baitis, A E, Applebee, T R & Meyers, W G. U.S. Coast Guard 270-ft medium endurance class cutter fin stabilizer performance, Report No DTNSRDC/SPD-1120-02, Bethesda, MD: David W Taylor Naval Ship Research and Development Center, October, 1984.

Bittner, A C, Jr & Chatfield, D C. A signal detection theory function and paradigm for relating sensitivity (d') to standard and comparison magnitudes, Proceedings of the 1980 International Conference on Cybernetics and Society, Boston, Massachusetts, 8-10 October 1980, 978-984.

Bittner, A C, Jr & Guignard, J C. Human factors engineering principles for minimizing adverse ship motion effects: Theory and practice, Naval Engineers Journal, 97(4), 205-213; discussion: ibid, 97(5), 107-111, 1985.

Bittner, A C, Jr & Guignard, J C. Magnitude estimation of motion sickness in an operational environment. Presented to the Acoustical Society of America at its 111th meeting, Cleveland, Ohio, 16 May 1986. J Acoust Soc Am Suppl 1, 79, S86-S87 [Abst]. 1986.

Bittner, A C & Guignard, J C. Shipboard evaluation of motion sickness incidence and human factors engineering problems. Presented to the Fourth International Meeting on Low Frequency Noise & Vibration, Umeå, Sweden, June 9-11, 1987. Proceedings, 4-1-1 [Abst]. 1987.

Bock, O L & Oman, C M. Dynamics of subjective discomfort in motion sickness as measured with a magnitude estimation method, Aviation,

Space & Environmental Medicine, 53, 773-777. 1982.
Box, G E P & Jenkins, G M. Time series analysis: Forecasting and control (Revised edn). San Francisco: Holden-Day. 1976.
Campbell, D T & Stanley, J C. Experimental and quasi-experimental design for research, Chicago: Rand-McNally. 1966.
Chatterton, H A & Braithwaite, T. The design of the United States Coast Guard 270-foot medium endurance cutter. New York: Society of Naval Architects and Marine Engineers. Paper no 11, presented at the Spring Meeting, STAR Symposium, New London, CT, 26-29 April 1978.
Cox, D L. Planning of experiments, New York: John Wiley & Sons, 1958.
Cronbach, L J & Furby, L. How to measure change - Or should we? Psychological Bulletin, 74, 68-70, 1970.
Dixon, W J (Ed), BMDP statistical software, Los Angeles: University of California Press, pp 359-387. 1981.
Graybiel, A, Wood, C, Miller, E F & Cramer, D B. Diagnostic criteria for grading the severity of acute motion sickness. Aerospace Medicine, 39, 453-457. 1968.
Guignard, J C. Vibration, in: L V Cralley & L J Cralley, Eds, Patty's industrial hygiene and toxicology, 2nd edn, Volume 3B, Biological Responses, New York: John Wiley & Sons, pp 653-724. 1985. Chapter 15.
Guignard, J C & McCauley, M E. Motion sickness incidence induced by complex periodic waveforms, Aviation, Space & Environmental Medicine, 53, 554-563. 1982.
Harman, H H, Modern factor analysis, 3rd edn. Chicago: University of Chicago. 1976.
Kehoe, J W, Brower, K S & Comstock, E N. Seakeeping, Proc US Nav Inst, 109, 63-67. 1983.
Kenny, D A, Correlation and causality, New York: John Wiley & Sons. 1979.
Lackner, J R & Graybiel, A. Etiological factors in space motion sickness, Aviation, Space & Environmental Medicine, 54, 675-681. 1983.
Lawther, A & Griffin, M J. The motion of a ship at sea and the consequent motion so sickness amongst passengers. Ergonomics, 29, 535-552. 1986.
McCauley, M E, Royal, J W, Wylie, C D, O'Hanlon, J F & Mackie, R R. Motion sickness incidence: Exploratory studies of habituation, pitch and roll, and the refinement of a mathematical model. Goleta, Calif: Human Factors Research, Inc, Technical Report No 1733-2. Arlington, Va, Office of Naval Research. April 1976.
McCreight, K K & Stahl, R G. Recent advances in the seakeeping assessment of ships, Naval Engineers Journal, 97 (4), 224-233, 1985.
Olson, S R. A seakeeping evaluation of four naval monohulls and a 3400 ton SWATH, Center for Naval Analysis, Washington, DC. Memorandum 77-0640. 1977.
Reason, J T & Brand, J J. Motion sickness, London, New York & San Francisco: Academic Press. 1975.
Sampson, J B, Cymerman, A, Burse, R L, Maher, J T & Rock, P B. Procedures for the measurement of acute mountain sickness, Aviation, Space & Environmental Medicine, 54, 1063-1073. 1983.
Spear, R C, Keller C A & Milby, T H. Morbidity studies of workers exposed to whole body vibration, Archives of Environmental Health, 31(3), 141-145. 1976.
Wiker, S F, Kennedy, R S, McCauley, M E & Pepper, R L. Susceptibility to seasickness: Influence of hull design and steaming direction, Aviation, Space & Environmental Medicine, 50, 1046-1051. 1979.
Williams, R E. Comments on Bittner & Guignard (1985), Naval Engineers Journal, 97(5), p 109. 1985.

Trends in Ergonomics/Human Factors V
F. Aghazadeh (Editor)
Elsevier Science Publishers B.V. (North-Holland), 1988

RETROFIT NOISE CONTROL MODIFICATIONS FOR AN UNDERGROUND MINE HAULAGE VEHICLE

Thomas G. BOBICK

Bureau of Mines, U.S. Department of the Interior, Pittsburgh Research Center, P.O. Box 18070, Pittsburgh, PA 15236

Diesel-powered haulage vehicles used in underground mining pose a significant noise overexposure problem for the operators. Full-shift dosimeter surveys indicated that load-haul-dump (LHD) vehicle operators had noise exposures that ranged from 160 to 470 percent of the allowable. Retrofit treatments were installed on a typical LHD machine during two 3-day work periods. Three acoustical instrumentation systems indicated that the noise level at the operator's location was reduced by 4.5 to 5.2 dBA. These reductions will increase the permitted time of vehicle operation by 87 to 106 percent. The total cost for the abatement materials amounted to $325. This project indicated that low-cost modifications can be effective in reducing the noise generated by diesel-powered LHD vehicles.

1. INTRODUCTION

1.1 Development of the Diesel-Powered Haulage Vehicle

Mobile equipment is used in all types of mining to haul ore from the mining face, where the material is removed from the formation, to a dumping point, where it is loaded into or onto a conveyance that takes the mined product to the surface for further processing. During the late 1940s, the capacity of drive train components (transmission, torque converter, axles) had been sufficiently increased to permit their incorporation into a diesel-powered vehicle that would be rugged enough to survive the demanding conditions underground. The development of a highly mobile and maneuverable ahulage vehicle with a large bucket capacity was considered by many in the mining industry to be a major technological advance (Gambill, 1978; Wenberg, 1978).

During the 1960s, the use of diesel-powered equipment steadily increased because its versatility and mobility contributed to dramatic increases in production at many mining operations. Increases in the average annual production ranged from a modest 24 percent (Wenberg, 1979) to an impressive 187.5 percent (Olsen, 1980). There are always trade-offs, however. Along with the increase in size and productivity, a corresponding increase in noise generation also occurred.

The purpose of this project was to develop retrofit modifications to reduce the noise exposure of operators of LHD equipment.

2. NOISE PROBLEM OF DIESEL-POWERED HAULAGE VEHICLES

One of the first research studies (Bolt Beranek and Newman, 1975) to define the extent of the noise problem with diesel equipment in underground mining was sponsored by the Bureau of Mines and conducted by Bolt Beranek and Newman Inc (BBN). This study indicated that approximately 18,500 underground miners were exposed to noise from an estimated 4000 diesel-powered machines.

A project that followed directly from the 1974 study was a hands-on noise control project (Bolt Beranek and Newman, 1977) that demonstrated the effectiveness of retrofit noise control modifications on a load-haul-dump (LHD) machine. The modified LHD was a Wagner Model ST-5A (Scoop-Tram), and was equipped with a 5-cu yd capacity bucket. This vehicle was powered by a 180-hp, 8-cylinder Deutz air-cooled diesel engine (model F8L-714).

2.1 Noise Reduction of Diesel-Powered Mining Equipment

Large diesel-powered mining/construction machines have a number of noise generating sources and a variety of paths which the sound travels along to reach the operator's ears. The various sources should be rank-ordered in terms of overall effect to the operator; related to this is the importance of attenuating the dominant sources first. The relative importance of the sources depends on the physical condition of that particular machine. The design of a vehicle (and its operating condition) will determine which noise sources are the most hazardous to the operator's hearing. When treating any machine, some diagnostic testing must be conducted on the machine to determine which components will contribute most to the operator's noise exposure.

3. NOISE CONTROL PROJECT

The experimental work was conducted by the author when employed by the Noise Branch, Pittsburgh Technical Support Center, Mine Safety and Health Administration (MSHA). The modification work was initiated by a written request for assistance from a salt mining company in central New York via the Albany subdistrict office, MSHA. After preliminary evaluation surveys were conducted on five cited LHD vehicles, a meeting was held to discuss future plans. Discussions centered on how applicable the retrofit treatments developed by BBN for a Wagner ST-5A LHD would be in quieting the cited vehicles.

Although the BBN project had successfully reduced the operator's noise level, the noise control modifications were developed for a specific LHD. Since the treatments had been fabricated by BBN, essentially they were a unique item. Neither the LHD manufacturer nor its exclusive parts/service company had experience in developing noise abatement hardware for their diesel products. The initial estimate by the LHD manufacturer for the costs of designing, fabricating, and installing the noise-control treatments on an ST-5A LHD was approximately $20,000. The mining company felt that this was excessive for modifications that had not yet been proven on a routine basis.

In an attempt to show that noise control modifications could be developed for LHDs fairly routinely, the Noise Branch (specifically the author) agreed to develop, install, and evaluate acoustical modifications for one of the company's five units. The machine chosen for the modification project had been cited with a full-shift dosimeter value of 297 percent.

3.1 Methods and Materials

The retrofit program was conducted in the company's main underground shop during two separate 3-day work periods. Each work period consisted of

(a) a partial-shift baseline survey, (b) two days of developing and installing noise control modifications, and (c) a partial-shift follow-up survey.

The modifications developed during the initial work period consisted of acoustically treating a metal barrier fabricated for the operator's side of the engine compartment. The mining company made it in two pieces, mainly for easy installation. Unfortunately, the middle of the barrier presented a line-of-sight leak for the airborne noise from the engine. This opening was blocked off with acoustical fiberglass that had a backing of lead material.

The inside of the two-piece metal barrier was treated with vibration damping material and acoustical fiberglass. Both materials were also installed on the underside of the existing engine hood; the opposite side of the engine compartment was untreated because of concerns about potential overheating.

The underside of the transmission compartment hood (located directly in front of the operator) was treated with acoustical fiberglass only. Insufficient clearance prevented installing the damping material. Instead, ordinary conveyor belting was attached to the top of the hood to reduce its vibration through the addition of weight.

The steering column passes through a slot in a metal panel located directly in front of the operator. This slotted opening is also a line-of-sight path for the transmission noise. A piece of heavy, flexible material was attached to the perimeter of the steering column guide and extended toward the floor to partially block this opening. At the completion of this modification work, the first follow-up survey was conducted.

The second work period was conducted 10 weeks later. The lead backing on the material covering the middle of the two-piece barrier had degraded noticeably. During the second modification period, the original two-piece barrier was changed to completely eliminate the noise leak in the middle. Some minor leaks still existed along the barrier edges, but they were sufficiently eliminated with proper maintenance.

Because only moderate temperature increases occurred during the first test period, an enclosure was fabricated for the opposite side of the engine compartment. The enclosure was constructed from a framework of angle iron that had ordinary conveyor belting bolted to it. Acoustical fiberglass was bolted to the inside of the enclosure for noise absorption. The entire system was then attached to the LHD with only four bolts.

Finally, a panel located directly in front of the operator (immediately to the left of the steering column slot) was removed, cleaned, and treated with vibration damping material and acoustical fiberglass. Both materials were used to maximize the noise-barrier and noise-absorption characteristics of the metal panel that shields the operator from the transmission noise.

4. DATA DISCUSSION

At the time of the modification work, three separate instrumentation systems were used to collect the acoustical data during the two before-and-after evaluation surveys. These included using a hand-held sound level meter (SLM) to collect short-duration measurements (grab samples) when the stationary vehicle was revved to full throttle (maximum-rev tests). The noise from the LHD unit was also tape recorded at the same locations that the SLM data were collected during the maximum-rev tests. Finally, partial-shift dosimeter measurements were collected while the vehicle conducted normal production work. Table 1 summarizes the acoustical data collected with the three separate systems for both work periods. Figure 1 is a schematic of the LHD vehicle showing the locations where grab samples and tape recordings were collected (operator's position and at three locations along the operator's side of the engine compartment: (1) near the torque converter compartment, (2) middle of engine compartment, and (3) near the cooling fan).

Referring to table 1, the results obtained at the operator's location with the three instrumentation systems were very consistent. Comparing the untreated condition with the end of the second modification period, the sound level meter (Type 2 measurement system) measured a 4 to 5 dBA reduction (101-101.5 to 96.5-97 dBA); the tape recordings (Type 1 system) measured a 5.2 dBA reduction (from 100.5 to 95.3 dBA); and, the equivalent noise levels calculated from the partial-shift dosimeter dosages indicated a 4.6 dBA reduction (from 100.0 to 95.4 dBA). A 5-decibel decrease in the overall noise level will result in a doubling of the machine's permitted operating time.

Figure 2 provides the one-third octave band analyses of the before- and after-modification conditions, measured at the operator's position. This figure shows that the modifications developed in this project were effective from 100 Hz to 10,000 Hz. Except for the 160-, 315-, and 500-Hz bands, the reduction in each one-third octave band was greater than 4 dB, and as much as 8, 10, or 11 dB in the higher frequency bands (greater than 2,000 Hz).

Approximately 10 weeks after the second modification period, a return visit was made to demonstrate the effectiveness of the noise control treatments for representatives from three mining companies and five different MSHA offices. During the demonstration/evaluation visit, both engine compartment barriers were removed to show the ease of installation and removal, and to determine the increase in the noise level at the operator's position when removed. SLM data were collected along both sides and at the operator's position when the barriers were removed; all other noise control treatments remained on the LHD. The SLM data are presented in Table 2.

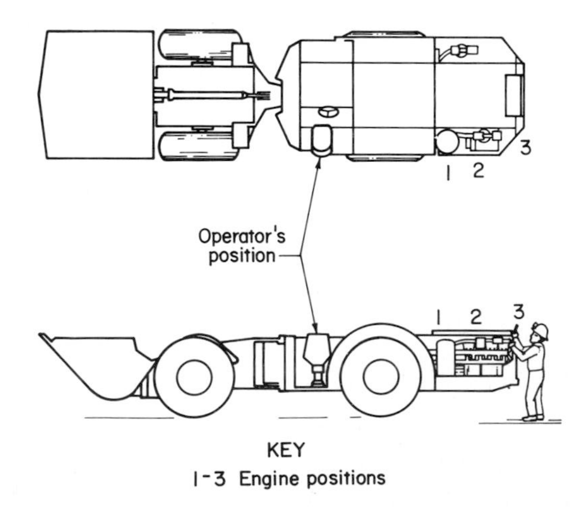

Figure 1. Schematic of load-haul-dump vehicle showing operator's location and positions along diesel engine where acoustical data were collected.

Comparing the SLM measurements from Table 2 for the operator's position (when both barriers were in place) with the SLM measurements from Table 1 for the operator's location (at the completion of the second work period) indicates an average increase in the noise level of about 0.5 dBA. Since the tolerance of the SLM is ± 2 dBA, these data indicate that the operator's noise level remained virtually the same over the ten-week period.

The reduction obtained with the barrier installed on the operator's side of the engine compartment ranged from 6.8 dBA (106.0 to 99.2 dBA) for the tape recorded data, and 6.3 dBA (106.8 to 100.5 dBA) for the SLM data; these values (measured 3 ft from side of barrier) are presented in Table 1. Figure 3 presents the one-third octave band spectra for the untreated and treated (at the end of the second work period) two-piece metal barrier.

TABLE 1. SUMMARY OF THE NOISE CONTROL MODIFICATIONS TO THE TEST LHD AND RESULTING ACOUSTICAL DATA COLLECTED WITH THREE INSTRUMENTATION SYSTEMS

Modifications	Stationary, Max Rev, SLM[1], dBA	Stationary, Max Rev, Tape[2], dBA	Dosimeter Dosages	Equivalent Noise Level
First Work Period (Completely Untreated)	Opr Position 101-101.5 3 ft from engine Avg is 106.8	Opr Positon 100.5 3 ft from engine Avg is 106.0	94% & 105% Avg is 99.5% (120 min exposure)	100.0 dBA
First Work Period (Treatments listed in footnote 3)	Opr Position 98 3 ft from engine Avg is 102.5	Opr Position 99.9 3 ft from engine Avg is 102.9	85% & 90% Avg is 87.5% (145 min exposure)	97.7 dBA
Second Work Period (Treatments listed in footnote 4)	Opr Position 99 3 ft from engine Avg is 102.5	Opr Position 98.8 3 ft from engine Avg is 101.7	79% & 83% Avg is 81% (128 min exposure)	98.0 dBA
Second Work Period (Treatments listed in footnote 5)	Opr Position 96.5-97 3 ft from engine Avg is 100.5	Opr Position 95.3 3 ft from engine Avg is 99.2	72% & 73% Avg is 72.5% (164 min exposure)	95.4 dBA

[1]Sound level meter data [2]Tape recorded data

[3]Engine hood and two-piece barrier for operator's side of diesel engine treated w/vibration damping mat'l and acoustical fiberglass. Underside of transmission hood trt'd w/fiberglass; conveyor belting attached to top for mass loading. Steering column slot in front of opr blocked w/barrier mat'l.

[4]Modifications listed in footnote 3 still in place; some deterioration had occurred to lead-backed acoustical mat'l that blocked opening in two-piece barrier (made by mining company) for opr's side of diesel engine.

[5]Changed two-piece barrier to one-piece design (closed off opening in the middle). Installed an enclosure (made of scrap conveyor belt) on opposite side of engine. Treated inside of enclosure with acoustical fiberglass. Panel located to the left of steering column slot (in front of operator) was treated with vibration damping material and acoustical fiberglass.

TABLE 2. SOUND LEVEL METER DATA (dBA) COLLECTED ON FOLLOW-UP SURVEY OF THE COMPLETELY TREATED LHD, MACHINE STATIONARY, MAXIMUM REV CONDITION

	BOTH BARRIER AND ENCLOSURE IN PLACE			
Opr position	97-97.5			
Position 1	99-100	Position 4	103	(103.4, RTA[1])
Position 2	101	Position 5	104	(104.1, RTA)
Position 3	100-101	Position 6	101-102	(101.9, RTA)
[Avg of the 3]	[100.3]	[Avg of the 3]	[102.8]	(103.1, RTA)
	ONLY CONVEYOR BELT ENCLOSURE REMOVED			
Opr position	97-97.5			
Position 1	99.5-100	Position 4	106	(106.5, RTA)
Position 2	101	Position 5	110	(109.0, RTA)
Position 3	101-102	Position 6	106-106.5	(105.8, RTA)
[Avg of the 3]	[100.8]	[Avg of the 3]	[107.4]	(107.1, RTA)
	BARRIER & ENCLOSURE REMOVED--ALL OTHER TREATMENTS IN PLACE			
Opr position	97.5- 98			
Position 1	103 -104	Position 4	106	(No RTA data)
Position 2	104.5-105	Position 5	109.5-110	(No RTA data)
Position 3	105	Position 6	106	(No RTA data)
[Avg of the 3]	[104.4]	[Avg of the 3]	[107.2]	(No RTA data)

[1]Real time analysis of tape recorded data; provided for comparison only.

The noise reduction achieved with the conveyor belt enclosure ranged from 4.0 dBA (107.1 to 103.1 dBA) for the tape recorded data to 4.6 dBA (107.4 to 102.8 dBA) for the SLM data; these data are presented in the right-most columns of the upper part of Table 2. The acoustical reduction due to the treated conveyor belt enclosure is shown in figure 4.

Most importantly, Table 2 shows that even with the engine barrier and enclosure removed (all other treatments in place), the noise level at the operator's location (97.5-98 dBA) increased only slightly (approximately 0.5 to 1 dBA for the operator's data in Table 2 and 1 to 1.5 dBA for the operator's data in Table 1). These slight increases indicate that the most important noise treatments were those installed around the transmission. Thus, the transmission is the primary noise source for the LHD operator.

4.1 Retrofit Labor and Material Costs

Any discussion of retrofit noise control modifications would be incomplete without some mention of the costs of materials used, and the total number of hours needed to develop, install, and evaluate them. Table 3 provides a breakdown of costs (1979 values) for the materials used to modify this LHD. Cost of materials was approximately $325.

The total time spent on the noise control project amounted to 32 work days. Approximately 85 percent was used by MSHA personnel developing, installing, and evaluating the modifications. The remaining 15 percent was used by mining company personnel accompanying the MSHA researchers and assisting in developing some of the treatments during both work periods.

An important point, which has been shown to be accurate from previous noise-control projects, is that as the modification work is repeated on subsequent equipment, the development and installation time will decrease.

5. CONCLUSIONS

The retrofit noise control modification project developed for this specific LHD vehicle has shown the following:

1. Retrofit noise control modifications for diesel-powered LHD vehicles can be developed that are effective, and are of minimal cost.
2. The transmission is the primary noise source for the operator of this specific LHD.
3. The noise control treatments resulted in a 4.5- to 5-dBA decrease in the noise measured at the operator's position. (Each 5-dBA decrease in overall noise level doubles the vehicle's permitted operating time.)
4. The two-piece metal barrier provided a 6- to 7-decibel reduction for the operator's side of the diesel engine.
5. The conveyor belt enclosure provided a 4.5-decibel reduction on the noisier side of the diesel engine (opposite the operator).
6. The partial-shift dosimeter data indicated an equivalent noise level of 95.4 dBA for the operator of the quieted LHD. Assuming a full-shift noise dosage of 132 pct is permitted by the MSHA inspectors, the quieted LHD could operate for a total of 299.7 min (5 hr).
7. Finally, assuming a typical work shift of 6.5 hr and a permitted dosage of 132 pct, the noise level at the operator's location would have to be reduced an additional 2 dBA (to 93.5 dBA) to achieve compliance.

TABLE 3. BREAKDOWN OF MATERIAL COSTS

MATERIAL	TOTAL USED	UNIT COST	TOTAL COST
Absorptive fiberglass, 1 in thick (Insul-Quilt, Insul-Coustic, Inc)	31 sq ft	$ 2.00/sq ft	$ 62.00
Damping material (E.A.R. C-1002, E.A.R. Corp)	24 sq ft	$ 7.00/sq ft	$168.00
Conveyor belting	25 sq ft	$0.50/sq ft est.	$ 12.50
Angle iron, 1.5 in	14 ft	$ 1.60/ft	$ 22.40
Sheet metal, 14 ga.	11 sq ft	$24.00 per 4 x 8 ft sheet	$ 8.25
Metal studs, 1/4 by 1 in (used to install acoustical mat'l)	103	$ 0.10 ea	$ 10.30
Rubber cover buttons	103	$ 0.20 ea	$ 20.60
Miscellaneous (Nuts. bolts, duct tape, etc)		approx	$ 20.00
		TOTAL:	$324.05

REFERENCES

[1] Bolt Beranek and Newman Inc, U.S. Bureau of Mines Contract No. H0346046, 1975, NTIS No. PB 243-896/AS, 225 pp.

[2] Bolt Beranek and Newman Inc, U.S. Bureau of Mines Contract No. H0262013, 1977, NTIS No. PB 288-854/AS, 76 pp.

[3] Gambill, T.L., Proceedings, Fall Meeting, Society of Mining Engineers (AIME), Diesels in Underground Coal Mines, 1978, pp. 3-5.

[4] Olsen, R.W., Mining Congress Journal, v. 66, no. 7, 1980, pp. 19-24.

[5] Wenberg, R.V., Proceedings, Fall Meeting, Society of Mining Engineers (AIME), Diesels in Underground Coal Mines, 1978, pp. 25-29.

[6] Wenberg, R.V., Mining Congress Journal, v. 61, no. 1, 1979, pp. 21-25.

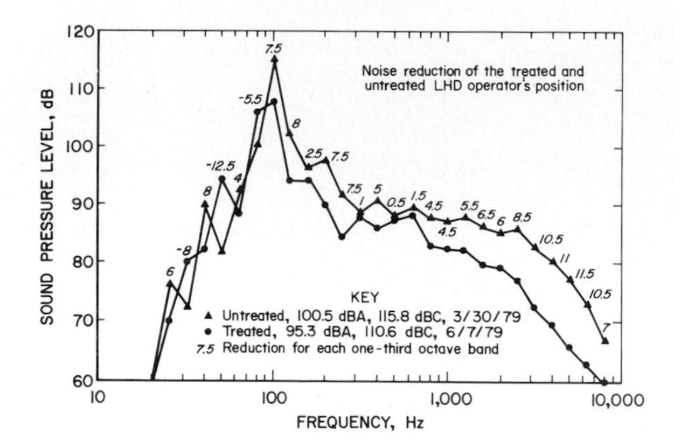

Figure 2. One-third octave band spectra of the untreated and treated LHD vehicle, measured at the operator's position.

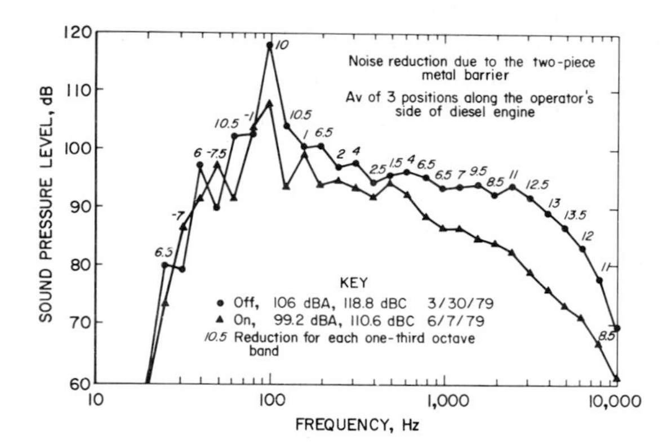

Figure 3. One-third octave band spectra measured 3 ft from untreated and treated two-piece metal barrier installed on opr's side of diesel engine.

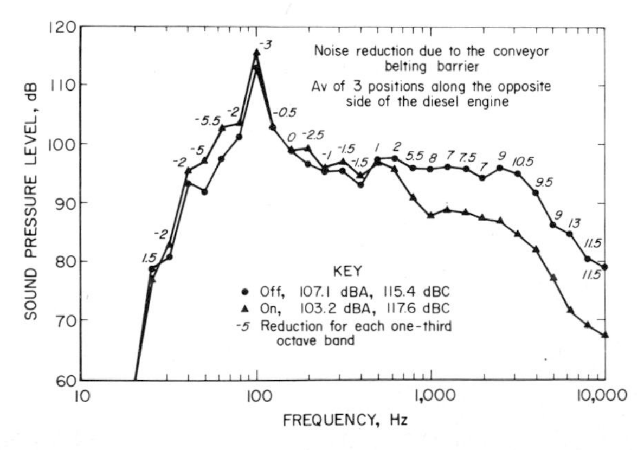

Figure 4. One-third octave band spectra (3 ft from conveyor belt enclosure). After installation; then removed (but all other tr'tments in place).

Trends in Ergonomics/Human Factors V
F. Aghazadeh (Editor)

VIBRATION AND NOISE PROBLEMS ASSOCIATED WITH THE GARDENING EQUIPMENT

K. LEE, F. AGHAZADEH and A. WAIKAR

Department of Industrial Engineering
Louisiana State University
Baton Rouge, Louisiana, U.S.A. 70803-6409

This paper presents the results of a pilot study to investigate vibration and noise problems associated with the selected gardening equipment. Four different criteria, namely grip strength, coordination skill level, caligraphy skill, and subjective evaluation were used to evaluate vibration effect on the human body. Noise produced by these selected equipment was measured in a repair room and on the field to compare the levels with OSHA's permitted level. Effect of vibration was investigated by conducting another experiment involving lawn mowing in the field. Eight subjects performed the mowing task for an hour to evaluate vibration effect. In certain cases the permissible noise level was surpassed. Subjects showed slight grip strength reduction and were affected by the tremors.

1. INTRODUCTION

Many individuals spend several hours in mowing or trimming lawn at home. It is not unusual for persons, who engage in commercial mowing, to spend more than six hours per day, seasonally. Even though there has been substantial improvement in reducing or controlling vibration and noise of the gardening equipment, the level of vibration and noise remain a problem.

There are several ways in which vibration can affect task performance [1]. The most significant adverse effects are a) temporary reduction of visual acuity, b) mechanical interference with control manipulation, and c) general discomfort which causes fatigue and reduces motivation. Other ways in which vibration may influence performance are by its effects upon the central nervous system and its effects upon physiological processes such as the heart rate or the respiration rate.

This investigation was conducted to see if the vibration of the lawn mowers causes any discomfort and whether it affects the performance of the operators. The noise level of the selected gardening equipment was also tested for the possible adverse effects on the users.

2. VIBRATION EFFECT EXPERIMENT

An experiment involving lawn mowing was conducted on the field to measure the effect of vibration on the human body.

2.1. Subjects

Eight male volunteers participated in this study. Their ages ranged from 19 to 24 years. They were college students in good health and had no history of musculoskeletal problems. None of them were involved in commmercial lawn mowing.

2.2. Measurements

Four different measurements were conducted. They were measurements of coordination performance, the subjects' overall physical pain and discomfort, calligraphic skill and grip strength.

Coordination Level

The coordination skill level was tested by measuring the time taken by the operators to perform the tasks as well as the accuracy in performing the tasks. The tasks consist of threading a needle and tracing straight lines (vertical and horizontal lines) and a curved line (sinusoidal curve).

Caligraphy Skill

The subjects were asked to write down their names the way they usually do it. They were also asked to draw circles and vertical lines within two parallel horizontal lines.

Grip Strength

The subjects' right and left hand grip-strength were measured using a grip strength tester (Model 78Ø1Ø: Lafayette Instrument Co.). This test was performed at fifteen minute intervals.

Subjective Ratings

The subjects were asked to rate their sensations of pain and discomfort in their arms and hands at the end of fifteen minute intervals. A simple feeling scale which indicates the degree of pain and discomfort and the degree of shakiness feeling was used as follows:

Degree of Pain	Feeling
1- 5	Not at all
6-15	Very little
16-24	Somewhat
25-33	Quite a bit
34-50	Very much

Degree of Shakiness	Feeling
1- 5	Not at all
6-15	Very little
16-24	Somewhat
25-33	Quite a bit
34-50	Very much

2.3. Procedure

The equipment used in the experiment was a self-propelled lawnmower in good mechanical condition. The subjects were asked to cut grass for a total period of one hour. Before any cutting of grass took place, the subjects were asked to perform the coordination, calligraphy and grip strength tests. They were also tested every 15 minutes during their lawn mowing task. The tests took 2 to 3 minutes each time.

Even though the size of the grass cut by the subjects was not monitored, the grass size was, on the average, short. The height of the grass cut by the subjects performing this experiment was on the average from 2 to 5 inches above the cutting plane. This experiment was performed in a field which had no slope, or natural obstructions such as trees, furrows or grooves.

3. NOISE TEST

Four different types of equipment (sixteen machines) were measured for noise. Several different measurements of noise of gardening equipment were performed to see if the noise from the equipment falls within the limit recommended by OSHA. The measurements were taken at two hardware stores. The type of equipments used were: mowers, tillers, edge cutters and weeders. The measurements were taken at a height of 5.7ft above the machine and measured inside and outside the work shop. Several factors listed below were taken into consideration:

1) The speed of the engine, low and high speeds.

2) The age of the machine, the manufacturing date.

3) The type of the engine, and the muffler.

4) The measurement site.

4. RESULTS

The vibration and noise were tested in two different experiment. The results are presented in the separate sections.

4.1. Vibration Effects

Figure 1 shows the average levels of discomfort of the values aggregated for all the subjects. This figure shows that the average levels of discomfort increased as time progressed during the experiment. However, the discomfort and hand-arm shakiness feeling level were not significantly high.

Figure 2 shows the average levels of grip strength for the left and right hands aggregated for all the subjects. The strength levels for both hands decreased as time progressed in the experiment. This may be due to an increase in the muscular fatigue of the operators. Another finding from this test is the fact that the left hand strength decreased in a larger proportion than the right hand strength. This may be due to the fact that all the subjects were right-handed. They all had higher grip strength and larger endurance in their right hand than in their left.

It can be seen from Figure 3 that the subject's coordination performance did not decrease. This can be attributed to the short working period. A similar trend was found from the needle-threading test and the caligraphy test. There was not significant evidence of a decrease either in the operator's psychomotor coordination performance or in the ability to write due to vibration and/or muscular fatigue caused by an hour of cutting grass with a powered lawn mower.

However, it took more time for the subjects to follow the path of the sine wave as time progressed in the experiment. Figure 4 shows the average time that it took for the subjects to perform the task following a pre-determined path after 15, 30, 45, and 60 minutes of the mowing experiment. This may be due to the fact that as time progressed in the experiment, the subjects became more and more fatigued; so, an increase in the subjects concentration and attention was needed in order for them to perform the given task.

4.2. Noise Levels

Table 1 shows the noise Level of the tested gardening equipment. The average noise level outside was 87 dB for 11 tested machines which is below the OSHA's [2] 8 hour permissible limit (90 dB) but two machines surpassed the 4 hour limit (95 dB). Inside noise level was 90 dB which is the same as OSHA's 8 hours permissible limit but 38% of them surpassed the limit for 4 hours. Table 2 shows that the noise level exceeded OSHA limit when the noise was measured at 1.57m above the machines. Also it was found that in an open area the noise level decreases by an average of 8 dB.

It was found that the age of the machines, the muffler size, and the engine speed affected the noise level. For example, machine A that was made in 1986 had a noise level of 85 dB outside the work shop, and the same type machine, which was made in 1981 had a noise level of 90 dB.

The machines that had larger mufflers, had lower noise levels. The machines performing at low speed resulted in a lower noise level, than at high speed. At high speed, the noise level increased by an average of 10 dB.

Table 1. Noise Level of Gardening Equipment

Name		Manufacturing Date	Noise Level	
			Inside	Outside
Mower	1	1981	92 dB	88 dB
	2	1981	95	90
	3	1981	92	87
	4	1986	94 *	87 *
			85 **	80 **
	5	1986	90	85
Edge-Cutter	1	1982	95	92
	2	1985	94	90
	3	1986	96	92
Tiller	1	1986	90 *	84 *
			80 **	76 **
	2	1986	87 *	85 *
			77 **	75 **
Weeder	1	1986	98	96
	2	1986	98	97
	3	1987	94	86

* at high speed
** at low speed

5. DISCUSSIONS

This study presents a few possible ways to measure the effect of vibration caused by gardening equipment on the human body. However, due to the brief and simplified nature of this pilot study any significant results, other than the grip strength and a slight decrease in feeling of discomfort could not be obtained. Even it was not clear whether the decline in grip strength was the result of vibration or the result of physical work.

The results do, however, imply that long hours of mowing work could result in some adverse effects on the subjects. It still remains unanswered whether extensive use of gardening equipment for several years may cause any adverse effects. It is recommended that a further study on subjects who use gardening equipement very frequently and over a long period of time should be undertaken.

The gardening equipment noise may not cause excessive adverse effects to the ordinary home owner, since it is not used very frequently, and the duration of each use is short. But this equipment can cause hearing problems to the commercial mowers or repair persons because of the close range between them and the machine and also because of the long duration of operation. So it is recommended that they wear protective hearing devices.

Table 2. Noise Level of the gardening equipment at 1.57m from the equipment.

Name		Manufacturing Date	Noise Level
Mower	1	1981	97 dB
	5	1986	97
Edge-Cutter	1	1985	98
	3	1986	101
Weeder	1	1986	106
	3	1986	100

ACKNOWLEDGEMENTS

The authors would like to thank Mr. M. Juarez and Mr. A. Qundes for their help in collecting data.

REFERENCES

[1] Tempest, W. 1976, Infrasound and low frequency vibration, Academic Press, New York.

[2] McCormick, E. and M. Sanders, 1982, Human Factors in Engineering and Design, McGraw Hill, 5th Ed.

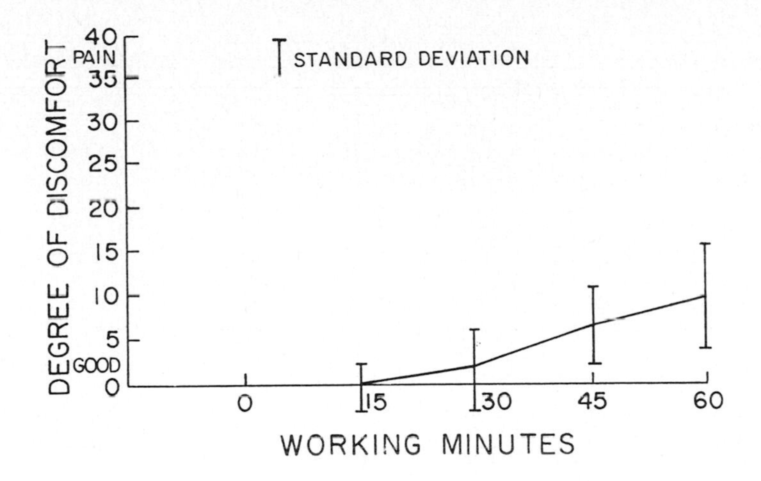

FIGURE 1
The average levels of discomfort reported by all subjects after lawn-mowing.

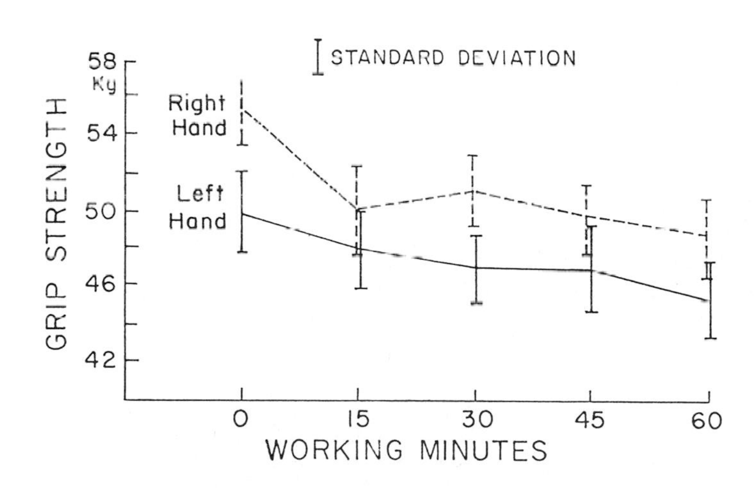

FIGURE 2
The average levels of grip strength for left and right hands for all the subjects after lawn-mowing.

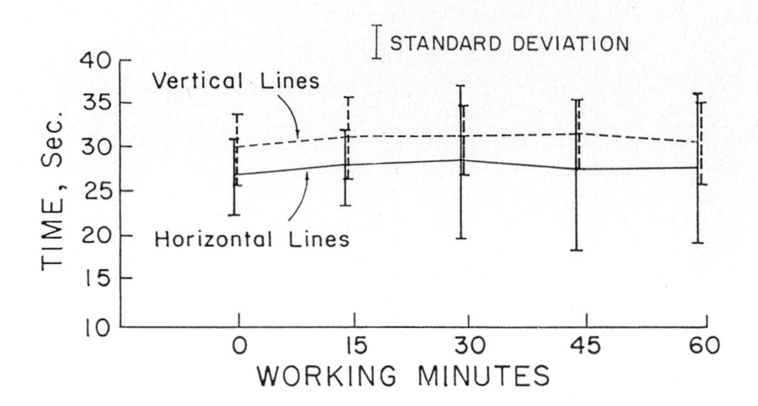

FIGURE 3
Average time taken to trace straight patterns for all subjects.

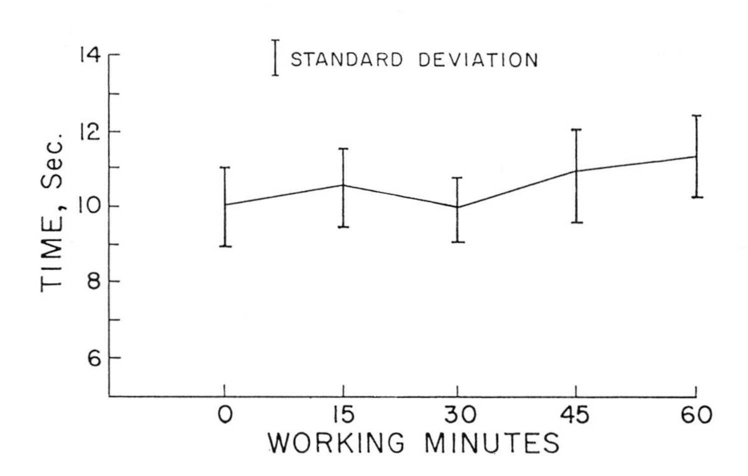

FIGURE 4
Average time taken to trace a curved line pattern for all subjects.

IX

SAFETY MANAGEMENT AND INJURY CONTROL

Trends in Ergonomics/Human Factors V
F. Aghazadeh (Editor)

JOB LOAD AND HAZARD ANALYSIS: A CO-OPERATIVE APPROACH TO IDENTIFY AND TO PREVENT ERGONOMIC AND OTHER HAZARDS AT WORK

Markku MATTILA

Occupational Safety Engineering
Tampere University of Technology
Tampere, Finland*

The aim of this study was to improve occupational health care by means of systematic workplace investigations titled Job Load and Hazard Analysis. The method is a simple job hazard analysis. Each of five items is rated on a three-point scale. Information is gathered by observations, interviews, and a worker questionnaire. Occupational health and safety personnel and worker representatives are assessing the problems cooperatively, whereafter the measures are developed. The method was tested in building industry. It produced an overall analysis of occupational hazards, and it increased the number and quality of proposed preventive measures. The new method was evaluated to be clearly superior to previous practices. The study showed that occupational health care and safety professionals can effectively prevent health impairment even in difficult settings like construction work.

1. INTRODUCTION

The Finnish Occupational Health Care Act obliges employers to provide occupational health care for employees to prevent health impairment caused by work. Workplace investigations, a new concept first defined in the act, entail an analysis of the health hazards inherent in work and evaluation of their impact on health. The objectives of the present study were: 1) to develop a simple, systematic method for the analysis of work load and hazards to be used in occupational health care and in occupational safety, 2) to test the method as part of the occupational health care procedures of a construction firm.

2. MATERIALS AND METHODS

2.1. Job Load and Hazard Analysis

The purpose of the method is to provide a summary of the loads and hazards of jobs and to provide the basis for other measures. Tha identification of hazards is first made through preliminary job analysis, in which five factord are considered: i) chemical hazards, ii) physical hazards, iii) physical work load, iv) mental stress, and v) accident risk. Each item is assessed on a three-point scale according to occurance and relevance to workers' health [1].

---

* P.O. BOX 527, 33101 Tampere, Finland

The information was gathered and the analysis was done: i) by occupational health care and safety personnel at the site, through worker observations and interviews, and ii) by workers themselves on a questionnaire form (in which the items were illustrated with several typical examples).

The findings are discussed job by job by the co-operation team, which consists of those participating in safety inspection (occupational health care personnel, the safety manager, worker representatives, and supervisors) and the line manager. The worker's own assessments were analyzed before the meeting, and the results were presented to the group. The group made a joint assessment of the hazards and stress of every job. After the meetings of the co-operation team and on the basis of these assessments, the occupational health care personnel or safety personnel reached their conclusions idependent and made the proposals for the future. The stages of the procedure are presented in Figure 1.

To determine interrater reliability the scores of two experienced raters, the nurse and the safety engineer, were compared. The two simultaneously rated 32 jobs. The reliability coefficient varied from 0.87 (mental stress) to 0.95 (work load).

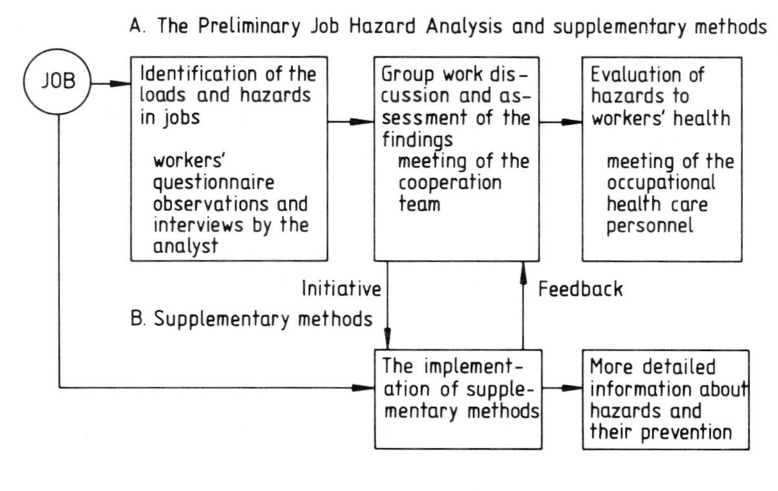

FIGURE 1

The structure of the method [1].

2.2. Subjects and Setting

The method was tested in this study, as part of the actual occupational health care in the building construction industry, for a 2,5 year period. Three building sites were chosen: construction sites for a theatre, an industrial plant, and an apartment house.

The jobs were analyzed regularly at two or four months intervals. The information collected thus applied to different seasons and different construction stages.

3. RESULTS

3.1. Loads and Hazards of Jobs

The analysis yielded an overall assessment of job loads and hazards (fig 2), which was summarized from the results for all the analyses made. It illustrated the essential loads and hazards of different jobs and revealed the most critical jobs. It helped direct occupational health professionals' attention to jobs most in need of their efforts, and it served as the basis for planning the content of the occupational health care program.

Object of analysis (job occupation, worksite)	Type of load or hazard														
	Chemical hazards			Physical hazard			Physical work load			Mental stress			Accident risk		
	0	1	2	0	1	2	0	1	2	0	1	2	0	1	2
1. Cleaner (social facilities)		▲		▲				▲		▲				▲	
2. Cleaner (lumber)		▲			▲			▲		▲				▲	
3. Lumber boy	▲				▲			▲			▲				▲
4. Concrete layer		▲				▲			▲		▲				▲
5. Cement worker			▲		▲				▲		▲			▲	
6. Construction worker (other)		▲			▲			▲			▲				▲
7. Picker			▲			▲			▲		▲				▲
8. Surveyor		▲			▲			▲				▲		▲	
9. Timberman (supporting structures)	▲					▲			▲		▲				▲
10. Timberman (other structures)	▲				▲			▲			▲				▲
11. Repair worker		▲				▲		▲			▲			▲	
12. Warehouse worker	▲			▲			▲				▲		▲		
13. Crane driver	▲				▲			▲			▲			▲	

Rating scale:
0 = no load and/or hazard ▲
1 = some load and/or hazards; may effect health ▲
2 = much load and/or many hazards; definite effect on health ▲

FIGURE 2

An overall assessment of some jobs at the theatre construction site.

3.2. The Occupational Health Care Program

Another result of the study was the new occupational health care program compiled by the occupational health professionals. The program specified what type of occupational health care was needed for every job studied and also contained proposals for future preventive measures. Some examples are presented in table 1.

Table 1. Part of the occupational health program compiled as a result of the study.

Jobs	Hazards to be informed by occupational health professionals	Medical Examinations			Preventive Measures
		Initial examinations	Due to specific hazards	Additional examinations	
1. Concrete layer	Noise, vibration, risk of skin disease, poor work postures, ergonomic problems in the whittling of concrete molds, accident risks	When the worker enters the occupation	-	Examination due to noise, vibration, and dust exposure: examination of the skin and musculoskeletal system	Safety boots, ear protectors, work experience, to avoid working alone
2. Timberman for supporting structures	Noise, vibration, poor work postures, accident risks	When the worker enters the occupation	Periodic medical examination due to noise exposure	Examination due to vibration exposure, examination of the musculoskeletal system	Ear protectors
3. Bricklayer	Exposure to plaster, poor work postures, accident risks	When the worker enters the occupation	-	Examinations due to exposure to chemicals contaided in plaster, examination of the musculoskeletal system	Protective gloves, improved ergonomics, placement of noisy work stations

3.3. Proposals for prenventive measures and their implementation

The preventive character of the workplace investigations in occupational health care was improved. Proposals aimed at preventing hazards were made for all jobs involving a considerable physical work load or other hazard with a definite effect on workers health. Altogether 68 proposals were made after four inspections at the theatre construction site (table 2). Most of them concerned the workers and their personal protective equipment. Of all the proposals 37 % were implemented on the site. Many proposals were realized during or just after the safety inspection. Some of the improvements required management decisions.

The workplace investigations had an impact on management at the construction site. The minutes of supervisors' weekly metings included 132 occupational safety issues, ie, 3.2 issues per meeting, during the period of 8 months at the theatre construction site. The safety supervisor took an active part in the meetings and usually decided which issues were discussed at each meeting. In this way line management was informed of the safety problems identified during the investigation process and became involved in making decisions.

Table 2. Proposals for preventive measures made after the four inspections at the theatre construction site and their implementation.

Object	Proposals (N = 68) %	Implemented measures for proposals (N = 25) %
1. Technology	19	38
2. Method, organization production, control	28	21
3. Workers	53	39
Total	100	100

3.4. Benefits and Costs

After the study the medical examination program became more detailed and comprehensive. The participants themselves evaluated the new method as being clearly superior to the former practice. Acceptance in the line organization was the only aspect evaluated to be less with the new method than with the old one. Workers especially praised the new method over the old one.

The implementation costs of this systematic method were evaluated on the basis of the time necessary for its completion. The reseacher's development of the method is not included here. The new method required more work than the old one but for a medium-sized or large company the use of the new method proved not to be an essential economic question: occupational health and safety personnel in any case spend this much time to analyze work conditions. The real issue is how time can be used the most efficiently.

4. CONCLUSIONS

The systematic method for workplace investigations proved to function well in practice. It clearly improved the quality of occupational health care. It provided a comprehensive overview of the hazards in different jobs, it gave an illustrative basis for the consideration of preventive and medical measures, it enhanced the content of occupational health care program, and it increased the number and quality of proposals to line management for preventive measures.

The method can be characterized by three concepts: job analysis, worker involvement, and group problem-solving. In occupational health care and in occupational safety, job analysis makes it easy to connect information about hazards on the job with information about man's health [2]. It offers many possibilities to study interactions between man and the characteristics of work.

Worker involvement through the questionnaire and interviews during inspections and through the inclusion of worker representatives in the co-operation group was fruitful. It emphasized workers' central role as the object of occupational health care and guaranteed that all the problems recognized by workers themselves not only came to light but also were taken into account in the analysis. This approach also formed a new cooperative forum in the organization and thus helped improve the safety climate.

After this study was finished, the method has been used at new construction sites with different production techniques, as well as at other industries. A current application concerns a construction repair job. The aim is to develop a new ADP-program for the method, suitable for personal computers, so as to enable use of all the data processing possibilities at the workplace level.

The information system gathered by Job Load and Hazard Analysis offers new possibilities to identify safety problems in advance, before the problematic circumstances actually exist. The utilization of this information system for prevention also requires further study.

ACKNOWLEDGEMENTS

The Finnish Work Environment Fund financed the study. The Occupational Health Care Centre of Tampere and Pertti Kivi, MD, gave their consent and assistance. The author wishes to thank all the organizations and persons whose help made the study possible.

REFERENCES

[1] Mattila, M.K., Job load and hazard analysis of workplace conditions for the occupational health care, British Journal of Industrial Medicine (1985) 656-666.

[2] McCormick, E.J., Job Analysis: Methods and applications (AMACOM, New York, 1979).

Trends in Ergonomics/Human Factors V
F. Aghazadeh (Editor)

EFFECTIVE GROUP ROUTINES FOR IMPROVING ACCIDENT PREVENTION ACTIVITIES AND ACCIDENT STATISTICS[&]

Ned Carter[§] and Ewa Menckel

National Institute of Occupational Health
Social Psychology Unit
Solna, Sweden

A combination of two group routines, an accident investigation group and a review group, was tested for three years at one company. The investigation group assisted foremen in their accident investigations. The review group discussed, monthly, accidents which had occurred since the previous meeting and was responsible for checking on the implementation of prevention measures. Improvements in safety activities and accident statistics were noted following the first year and at one- and two-year follow-ups. The accident investigation group routine was then tested at two companies and the review group routine at an additional two companies. Improvements in prevention activities were noted. All four companies reported decreases in number of accidents as compared to the preceding year and reductions in accident severity were noted at companies using the review group routine. The routines, in combination and individually, were begun with minimal instruction and assistance and each was positively received by foremen and personnel involved. The findings suggest that the group routines could be of benefit to other companies.

1. INTRODUCTION

Menckel and Carter (1) presented data from a case study in which a combination of two group routines, an accident investigation group and a review procedure, was associated with improvements in safety activities and accident statistics at a large commercial bakery. An investigation group consisting of the union's head safety representative, a company representative in safety matters and the company nurse was created to assist foremen in their accident investigations, ideally, within a few days of the accidents. In addition, a delegation of the company safety committee reviewed and discussed during their monthly meetings all accidents which had occurred since the previous meeting and was responsible for checking on the implementation of prevention measures. Following the start of these new group routines, accidents were investigated more quickly and there were increases in reported near accidents as well as proposed countermeasures. Decreases in reported accidents and accident severity were also noted. The improvements were maintained at one- and two-year follow-ups.

& This study was financed by a grant from the Swedish Work Environment Fund.
§ Present address: Department of Occupational Medicine, University Hospital, Uppsala, Sweden.

The present paper describes an evaluation of the accident investigation group routine at two companies and of the review group routine at an additional two companies.

2. METHODS

A proposal was made to the safety committees at each of four companies described below, that an accident investigation group and a review group be tested for one year. The safety committees at two breweries, 136 and 450 employees, respectively, decided to test the accident investigation group. The safety committees at a tire manufacturer (1480 employees) and a warehouse (900 employees) decided to test the review group. All four companies were similar to the bakery in that they maintained an on-site health office, had departmental union safety representatives, had access to a safety engineer, conducted quarterly factory-wide safety investigations and had a safety committee which met quarterly. Problems with delays in accident reporting and low employee interest in safety activities were also similar to those at the bakery.

After the implementation of the procedures, some problems in the operation of the accident investigation groups were noted. The bases of the problems were delays in notifying the group that an accident had occurred and conflicting demands on group member´s time which interfered with their attendance at group investigations. The investigation group routine required about 45 minutes/investigation. The review procedure was implemented as planned, although accident reports were reviewed quarterly rather than monthly as in the original study. The review procedure usually required no more than 20 minutes of meeting time.

3. RESULTS

3.1. Safety Activities

Measures of safety activities included time from accident occurrence to formal accident investigation, number of proposals for countermeasures and participation of union safety representatives in investigations. Data from the year during which the group routines were tested were compared with data from the preceding two years. Accidents at the breweries were investigated more quickly (Figure 1). A small increase in number of proposals for countermeasures was noted at the warehouse (Figure 2) and at all four companies there were increases in union safety representatives´ participation in accident investigations.

3.2. Accident Statistics

Measures of accident statistics included number of reported near-accidents, number of accidents and mean number of days absent per accident. All companies reported fewer accidents than for the preceding year and for the breweries and the warehouse, the results represent all-time lows. Accident statistics for the warehouse, which tested the review group, are shown in Figure 3. Accident severity had been stable for the two preceding years but decreased by 50% during the year the review procedure was tested. A closer analysis revealed that there had been no absences longer than one month as compared to five cases per year for the previous two years. Mean

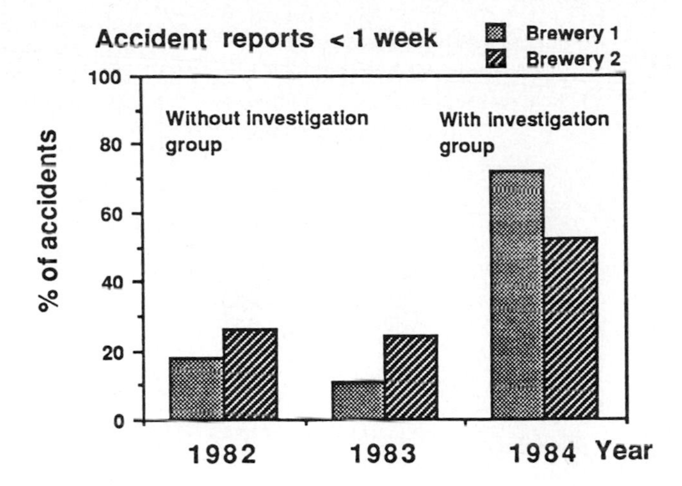

Figure 1. Accident reports (%) completed within 1 week of accident occurrence at two breweries without (1982-83) and with (1984) an investigation group.

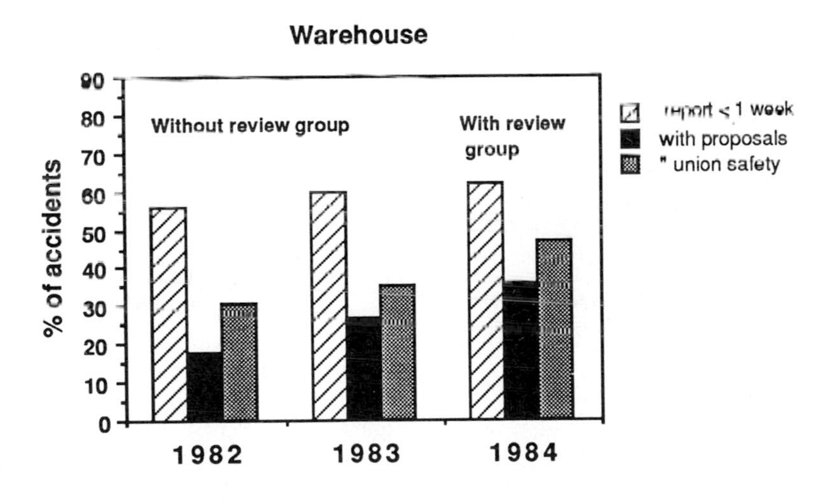

Figure 2. Accident reports (%) from a warehouse, completed within 1 week of occurrence, containing proposals for countermeasures and in which a union safety representative participated, without (1982-83) and with (1984) a review group.

number of says absent per accident at the tire manufacturer was reduced from 21 to 16. At this site the reduction occurred for the categories of accidents causing absences "less than one week" and "longer than one month".

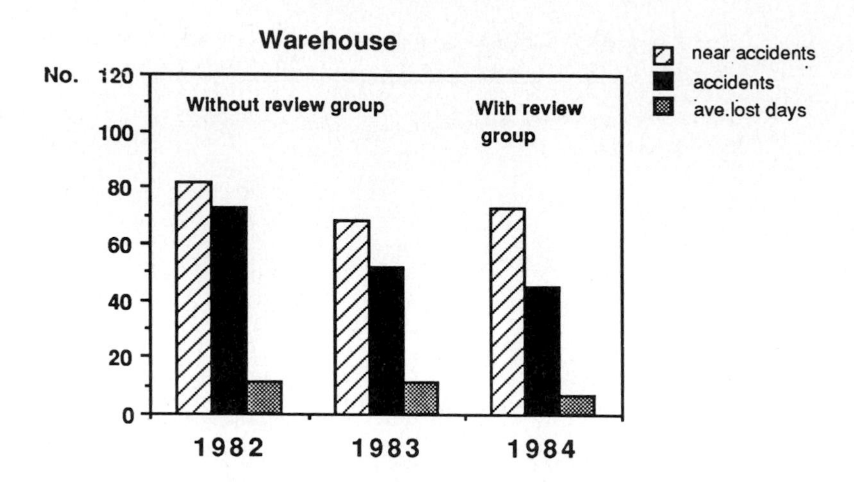

Figure 3. Accident statistics at a warehouse without (1982-83) and with (1984) an accident report review routine.

3.3. Social Validation

Interviews with participants revealed that the review routine received positive comments in that it was wasy to initiate and apply. The review procedure reportedly permitted an overview of accident risks and deficiencies within the safety organization. The safety committees at the tire manufacturer and the warehouse decided to continue using the review routine. Despite difficulties implementing the accident investigation group routine, participants reported that taking part in investigations had been beneficial to them in their safety related activities. The participants planned to continue to use the investigation group procedure on specific occasions. All participants noted that the study had motivated discussions concerning which accident data should be accumulated and how that data could be used in planning accident prevention activities.

4. CONCLUSIONS

Both group routines were positively received at the companies and were associated with positive changes on measures of prevention activities and accident statistics. The fact that the company safety committees determined which routine would be tested and whether it would be continued increases the validity of the routine and the results in that such a decision is a prerequisite for the application of the majority of accident prevention routines.

The exact mechanisms for the positive findings are uncertain. The role of the groups seems however to be to increase the likelihood that accidents are investigated rapidly and systematically, that countermeasures are discussed and that there is follow-up on the implementation of countermeasures.

Since the routines were tested separately rather than in combination, a detailed comparison with the results reported by Menckel and Carter (1) is precluded. The present results indicate however that each of the two group routines can benefit employees and company safety organizations in their efforts to improve accident prevention.

REFERENCE

(1) Menckel, E. and Carter, N. (1985), The development and evaluation of accident prevention routines: A case study. Journal of Safety Research, 73-82.

Trends in Ergonomics/Human Factors V
F. Aghazadeh (Editor)

EVALUATION OF SAFETY ATTRIBUTES IN ACCESS SYSTEMS

Häkkinen,K., Pesonen, J., and Väyrynen, S.

Department of Safety
Institute of Occupational Health
Laajaniityntie 1, SF-01620 Vantaa, Finland

Safety is a multidimensional property of a system which usually cannot be expressed by any single yardstick. Besides stable and objective variables, also dynamic and subjective criteria must be integrated in the evaluation. Thus information about passive physical reality must be complemented with data on active product-user situations and on the opinions of users.

A traditional method in the evaluation of access systems is to check that the structural details meet the requirements set by standards. New techniques were developed in order to better cope with our conception of the true nature of safety. The essential features of the analysis are (i) a system model, (ii) accident scenarios,(iii) analysis of physical properties, (iv) study of observed manifestations in simulated use and (v) opinions of users.The procedures were applied to portable ladders and access systems of tractor cabs.

INTRODUCTION

Access systems have the objective of providing human access from one level to another. Practically always there is some risk involved when such a system is in operation. About 20 % of occupational accidents are related to tracks for human movement. Injuries with mobile machinery occur to a large extent during access and egress tasks. Some 20-50 % of the accident cases are usually classified in this category, the proportion depending on the type of machine applied.(1,2,3)

Accidents due to slipping and falling and safety aspects of ladders and stairs are classical topics of industrial safety practice. However, recommendations and provisions are usually based on rule-of-thumb knowledge rather than on any rational or scientific approach. This notion has been expressed also for engineering and architectural design of stairways (4). Due to the apparent simplicity of many access injuries, detailed investigations are often omitted.

In any safety evaluation of an engineering detail, mere mechanical information on dimensions and components is not enough . There must be some knowledge available related to an active product-user interaction. Moreover, safety can never be measured purely objectively. Safety is a property connoted also with the values and opinions of people . Thus safety assessment should include experiences on use and the subjective comments of users.(5,6)

A traditional method in the evaluation of access systems is to check that the structural details meet the specific requirements set by standards (2,3). New techniques were developed in order to better cope with the true nature of safety. The procedures have been applied in safety studies of portable ladders and access systems of tractors.

Portable ladders are access systems used extensively at work. Usually one ladder serves many distinct access purposes, and thus the safety features of the system are to a large extent determined by situations and procedures of application. Obviously there is also a temptation to label the user as the culprit in an accident, without a closer examination of the other components of the access system. However, when looking over design details of ladders, one can easily find characteristics either promoting or preventing accidents (7,8).

About 20 % of tractor accidents occur during access or egress. Here the access system is a subsystem of another system,the tractor. Several requirements for the design have been set by standards and specific studies (2,3,9).

SYSTEM MODELS

Safety attributes and their mutual relationships are often poorly understood by specialists in engineering and production , but also by safety practitioners. Understanding of the complex system can be improved when the system models are constructed by the experts who have detailed technical knowledge of the system and also of the safety attributes involved .

System models focus on the production objectives of the system, on the input required and output expected. Thus they can direct thoughts to alternative ways of attaining the same objective, e.g. substituting of ladders by fixed stairways (Figure 1). One can also visualize relationships between technical, human and environmental characteristics and the accident output. The complex entirety can be seen from the details sometimes overemphasized due to single serious events or due to specific specializations, e.g. materials engineering, dominating in working groups. In spite of the incompliance of every single model, system models when originating from knowledge and experience can improve our safety picture of the access system in concern.

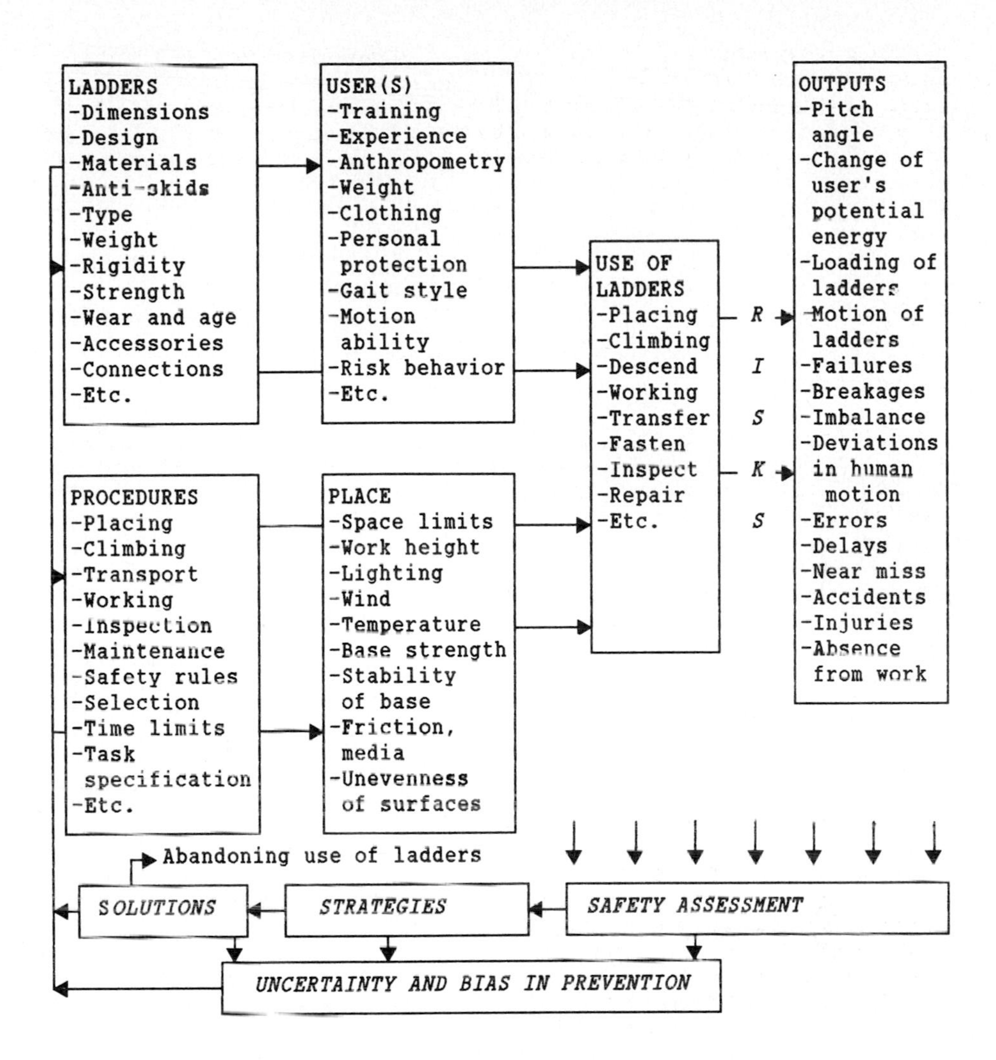

FIGURE 1. A system model for the evaluation of the safety attributes in the access systems based on the use of portable ladders (10).

ACCIDENT SCENARIOS

There is no doubt that safety problems lie there where accidents tend to happen. When accident data are available, they can be arranged properly to outline typical accident scenarios. All accidents, even near-accidents, represent a valuable source of data, in spite of the unreliability of most reporting systems.

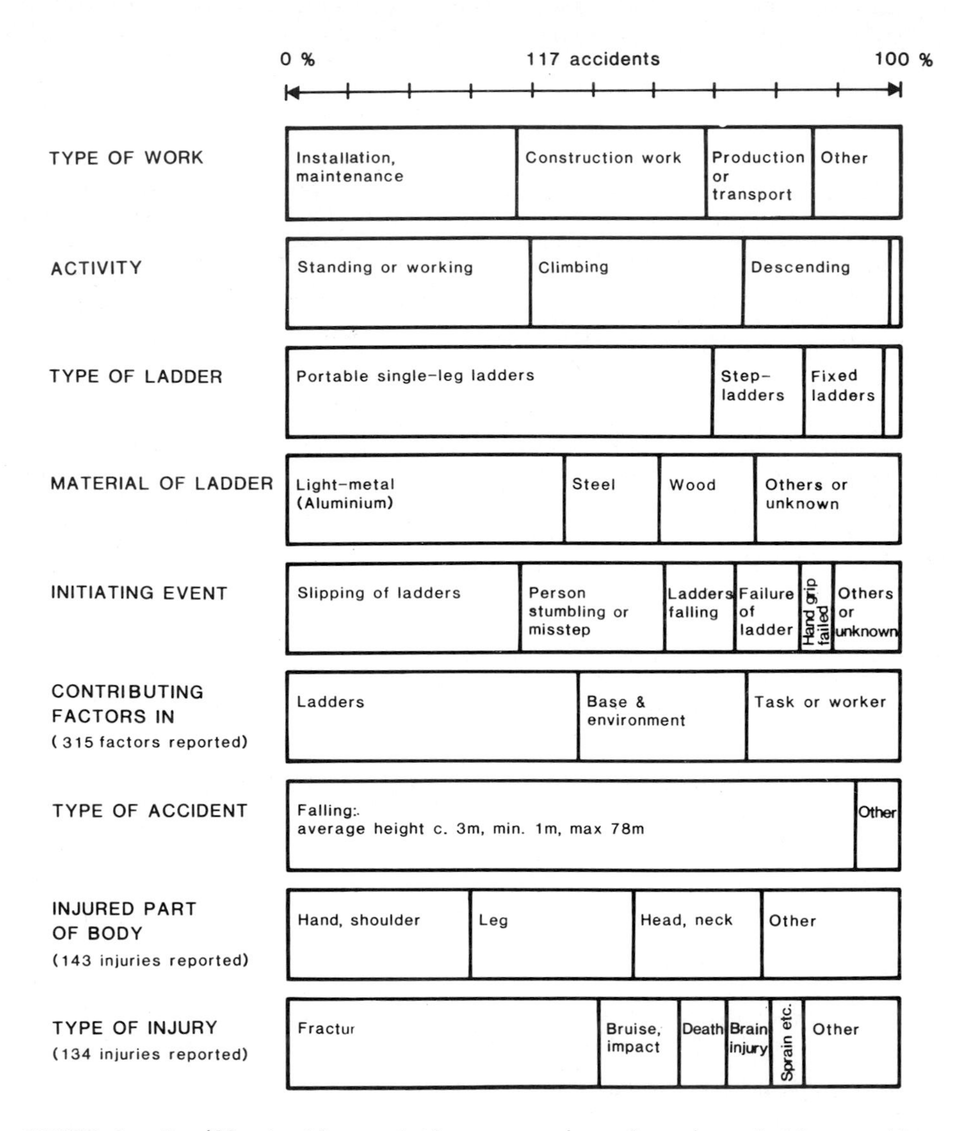

FIGURE 2. An illustration of the scenarios of serious ladder accidents occurred in Finland during the years 1977-1985 (10).

Utilizing the reports of serious accidents from the past ten years, an attempt was made to find out what a typical ladder accident is like: where it happens, how, why and to whom. Figure 2 contains a summary of the characteristics of the 117 serious accidents included in the investigation reports of our labor inspectorate.

A ladder accident most frequently takes place in installation and maintenance work or in construction work. The person is usually standing, working or climbing up on a portable light-metal pitching ladder when the incident starts. The ladder either slides away under him or he loses his balance, falling down from an average height of three meters. The resulting injuries include fractures or bruises on hands, legs or head.(10)

In serious accidents, the contributing factors were most often found in the ladders (46%). The most common factor induced by a ladder was insufficient bottom support (15%). Factors resulting from the base and surroundings rose to 27%. Of factors connected with the task and the worker (25%), the most common one was lack of suitable tools (9%).

PHYSICAL PROPERTIES

Physical characteristics have a paradoxical role as safety attributes. Safety standards often give exact guidance values but the relevance of many of them for safety can be questioned. A safety assessor can focus all his attention to variables easily measured with a tape measure without analyzing the practicality of the complex entirety of the non-measurable hazards present in patent or latent form. Nonetheless, some dimensions can be important safety measures per se, and they are useful precisely because they are clearly quantitative.

The four main components of the access system of a tractor are (i) steps (height, arrangement, size), (ii) handholds (type, arrangement, size), (iii) doors (size, shape, hinging) and (iv) internal dimensions of cab (9).

In our study altogether 26 measurable physical variables were identified. The properties have varying importance for safety. When the optimal and allowance limits for these measures and the weights of importance were defined, evaluation programs based on the calculation of evaluation matrices could be run to yield a ranking list of commercial solutions.(3)

Some mandatory requirements for physical attributes exist in standards. Regardless of our opinion on how correct or practicable they are, we must take them as given. The compliance with standards can be checked automatically by establishing rules in the evaluation program. The set of rules governing tractor access systems according to the Finnish regulations is demonstrated in Figure 3 in the format in which they were included in the computerized evaluation.

Lightyear

SUBJECT:TRACTORS
VERSION:PROPERTIES

RULE NAME	RULE
STEP HEIGHT	Step height MUST BE AT MOST 550 (ELIMINATION RULE)
STEP WIDTH	Step width SHOULD BE AT LEAST 250 (WEIGHT = 75)
STEP DISTANCE	Step distance SHOULD BE AT MOST 300 (WEIGHT = 75)
STEP DEPTH	Step depth SHOULD BE AT LEAST 150 (WEIGHT = 75)
DOOR HEIGHT	Door height SHOULD BE AT LEAST 1350 (WEIGHT = 50)
DOOR WIDTH	Door width MUST BE AT LEAST 250 (ELIMINATION RULE)

FIGURE 3. The set of rules governing the evaluation of tractor access systems according to Finnish regulations (dimensions are in mm).

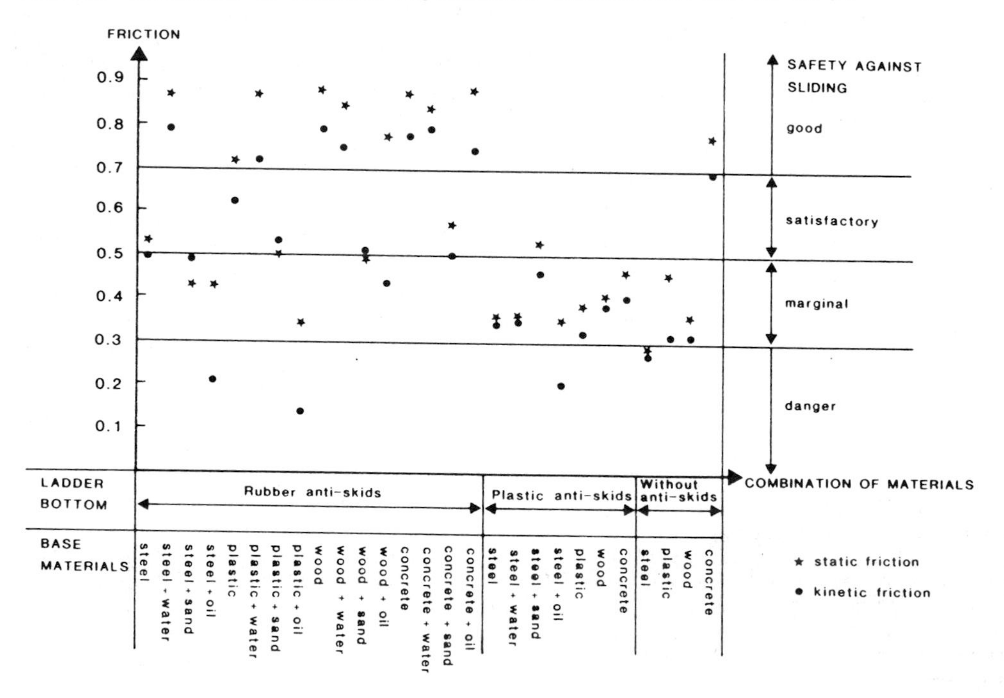

FIGURE 4. Static and kinetic friction coefficients obtained in the slide tests of an aluminum ladder in combination with various floor and medium materials, and an estimate of the slide risk (10).

Accident reports revealed that the incidents originating from the slipping of a ladder were common. Prevention should therefore be focused on the control of slipping. Unfortunately, current standards concentrate on strength requirements for ladder structures and anti-skid effectiveness is not covered, although it is a safety criterion of major importance(10).

The safety margin against slipping was determined as the difference between the required value of friction and the actual friction in slipping conditions. The required friction is dictated by real horizontal and vertical forces during climbing and descending. The slide tests to measure actual friction were performed by using several material combinations of ladder foot, medium and floor (Figure 4). In general, rubber pads at the bottom end prevent sliding rather well, whereas slipping took place easily when plastic pads or bare metallic ends were used. On very slippery surfaces, e.g. oily floors, slipping occurs abruptly with all anti-skids without prior warning.

MANIFESTATIONS IN USE

An access system is not only a static mechanical structure. There is always a dynamic interaction between human beings and materials when access systems are in operation. Thus safety analysis must contain some information on this interaction. Information on usage can be obtained by collecting evidence of real usage situations at worksites, or by arranging simulated usage tests.

In the study of simulated access into tractor cabs, variables such as the time share of a three-point contact, patterns of body postures during access, and the time of access are of importance (3). In ladder access tests, a meaningful number of certain disturbances, e.g. omitting a rung, changes in gait patterns, and inadequate hand grips, could be observed (10).

Some variables, though important, proved to be controversial safety criteria. E.g. it is considered positive from the safety standpoint that a three-point contact can be maintained as long as possible during access into a cab (2). However, the situation cannot be assessed as good from an ergonomic viewpoint, when access design compels an operator to maintain the three-point contact during the entire access (compare mountain climbing).

Ladder tests revealed some important deviations among users from the recommended practices. In a sample of 147 settings the average pitching angle when setting the ladder was 66 degrees, while a pitch angle of about 75 degrees is often recommendded in safety standards to achieve optimum balance. The tendency to use lower than recommended pitch angles was also found in user tests reported by Irvine and Vejvoda (11).

Further, users prefer to keep their hands on siderails rather than on rungs, whereas rungs are preferred by safety manuals. It may be appropriate to look at the requirements for ladder design, procedures for ladder use and training of users once again in the light of these observed manifestations in ladder use. E.g., the siderails of ladders are often sharp-edged and inconvenient as hand grips. Would it be a safer approach to design siderails for gripping as well, because people nonetheless seem to want to hold on the siderails ?

Simulation is useful because variables can be controlled, tests can be repeated and accurate measurements can be conducted. But performance in a simulation setting can be quite different compared to extreme environments and difficult tasks in real life. In spite of the efforts to produce disturbances and deviations in simulation studies, simulated use always tends to concentrate on normal and smoothly running situations. Therefore simulation studies should be complemented by authentic usage experience including accident data.

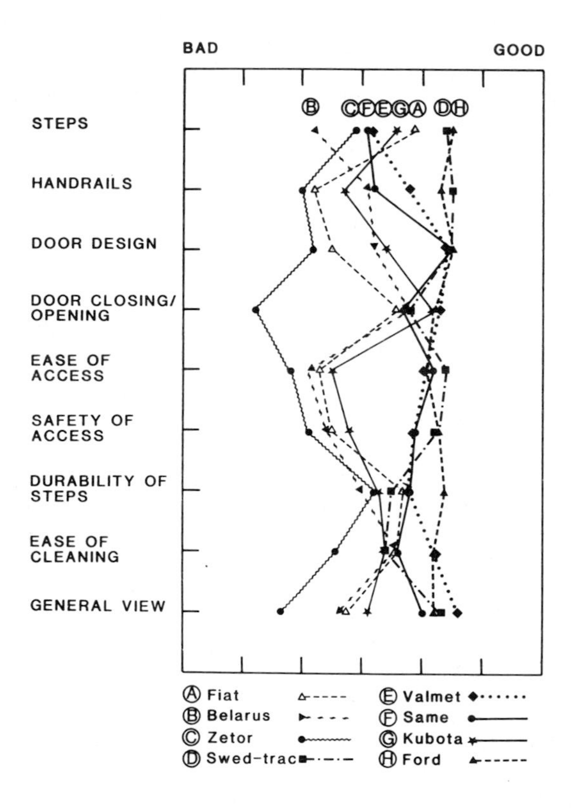

FIGURE 5. The profiles of opinions given by 10 participants of the simulated access tests for the access systems of the eight tractors (3).

OPINIONS OF USERS

Safety is a property having both objective and subjective contents (6). The opinions of users reflect the subjective side - the values and viewpoints of people as to what is safe and what is not. The access systems considered safe to enter a tractor cab in the fifties are now regarded as unsafe; opinions also vary from place to place.

Most of the 20 maintenance men participating in the ladder study regarded slide prevention and robustness as the most important characteristics for safety. Based on the accident figures these properties have a crucial meaning for safety. User opinions on various ladder marks were much in accordance with the structural safety evaluation conducted for the same set of ladders (10). Experienced users seem to have a realistic view of the relevant safety properties.

Figure 5 presents the profiles of opinions concerning access systems given by the ten participants for the eight test tractors studied. Again, it is evident that some solutions are commonly considered good and others not. Some properties , such as durability and ease of cleaning proved to be difficult to assess during a short-term experiment, as wide deviations in opinions indicate. Apparently field experiences are needed for more stable assessments.

DISCUSSION

Safety cannot be measured by any single yardstick. Therefore multiple measures must be developed for safety evaluations. These measures should cope with the true nature of safety; thus an objective analysis of physical characteristics has to be complemented by a study of subjective factors. Besides stable design factors, also dynamic phenomena manifested in the operation of the system should be integrated into the analysis.

In this presentation a set of techniques was demonstrated for two distinct types of access systems. The multidimensional nature of safety was tackled by site simulation and participation of operators. Obviously this approach produces a better measure of safety than a standard review of design details and dimensions. An analysis of accident data is a necessary component for a safety evaluation,as it reflects the degree to which the hazards and risks are manifested in accident outcomes. System modelling contributes to the understanding of the whole safety problem, and helps to relate production, hazards, human factors and prevention with each other.

A computerized expert support system including the elements shown above can be constructed by modern software packages or by using commonly used programming languages. In the projects reported in this presentation, a Lightyear(C) decision modelling software and a BASIC programme written by one of the authors were used. The tools developed can be used in access systems design, in purchasing decisions and in devising safety procedures and standards.

REFERENCES

(1) Cohen,H.H.,Templer,J.,Archea,J., An Analysis of Occupational Stair Accident Patterns. Journal of Safety Research 16(1985) ,171-181.

(2) Couch,D.B., Fraser,T.M., Access systems of heavy construction vehicles: Parameters, problems and pointers. Applied Ergonomics 12(1981):2, 103-110.

(3) Häkkinen,K., Väyrynen,S.,Pesonen,J., Access to and from a tractor cab. Ergonomics 1988. In print.

(4) Fitch,J.M.,Templer,J.,Corcoran,P., The Dimensions of Stairs. Scientific American 231(1974), 82-90,92

(5) Brown,R.,Green,C., Precepts of Safety Assessment. J.Operational Research Society 31(1980) ,563-571.

(6) Driedger,G., Die sicherheitstechnische Bewertung von Mensch-Maschine -Systemen.VDI-Verlag GmbH, Düsseldorf, 1984

(7) Goldsmith,A., The Manufacturer's Role in Preventing Ladder Accidents. Hazard Prevention 1985:Sept./Oct., 26-29.

(8) McIntyre,D.R., The Effects of Rung Spacing on The Mechanics of Ladder Ascent. Ph.D.Dissertation, University of Oregon, 1979.

(9) Bottoms,D.J., Barber, T.S., Chisholm, C.J.,Improving Access to the Tractor Cab. J.Agric.Engng.Res. 24(1979),267-284.

(10)Häkkinen, K., Pesonen, J., Rajamäki, E.: Experiments on the safety of portable ladders. Journal of Occupational Accidents 1988. In print.

(11)Irvine, C., Vejvoda, M.: An investigation of the angle of inclination for setting non-self supporting ladders. Professional Safety 1977:7, 34-39.

Trends in Ergonomics/Human Factors V
F. Aghazadeh (Editor)

PITFALLS IN THE USE OF ACCIDENT DATA TO FOCUS ERGONOMIC ATTENTION

Kathryn WOODCOCK WEBB, P.Eng.

Centenary Hospital
Scarborough, Ontario, Canada
M1E 4B9

Use of accident databases for safety measurement and research is jeopardized by confusion between inherent risk and risk control. Separation of these factors is essential. Behaviour modification and quality assurance methods can be used to generate a management safety control index for each work unit. Discounting accident experience by the management safety control factor points to inherent risk and hence research needs for the remainder. A descriptive and not cause oriented accident database would provide researchers with data for study, separately or in conjunction with management safety control standards and performance records. Focussing management measurement on safety control standards should clarify understanding of performance as well as improve control.

INTRODUCTION

Accident data is not actually needed for a basic accident prevention program. This illusion is based on a misconception that only accidents reveal hazards in the work. Ergonomic techniques, in particular task analysis, focus on person-task-environment mismatches which create the potential for error and injury. Existing knowledge may be applied in the workplace based on analysis of the work, subject to the risk management formulas or policy in force in the organization. On that basis, hazard reduction measures may be prescribed. Such management tools as supervision, Quality Assurance activities or Performance Audits should monitor the implementation and continued function or application of prescribed hazard reduction measures.

Most accidents result from well-known risks (Grondstrom, 1980). It is interesting to speculate that these are primarily of the probabilistic sort common to the ergonomist's person-task-environment system; deterministic hazards are more often controlled. Investigations usually have little difficulty establishing the spatial and temporal cause and effect relationships leading to the accident. When a person-task-environment mismatch is involved in an accident, the practitioner usually can recognize the ergonomist's insight, but to wait for accidents to investigate so that practitioners can appreciate the potential consequences of the hazardous mismatches already predicted by researchers is a costly learning experience. Though investigation could be used to establish culpability (for failure to apply or enforce application of readily available knowledge), it is fashionable to insist that affixing blame is not an object of accident investigation. Thus, cause-oriented investigation of individual accidents has little worthwhile to offer. While "the next [ferry] accident is less likely to be caused by bow doors being left open ... waiting for an accident to occur is an expensive way of doing ergonomics studies" (Buck, 1988).

Managers and those who evaluate managers (including other managers and company directors, the labour inspectorate, insurers, and workers' representatives) assume that aggregated accident data can be used to monitor how diligently and effectively management is preventing accidents and where management needs to be more diligent. Unfortunately, basing action on aggregated data delays attention to problems; in many workplaces, some problems may never accumulate data to levels that would capture the attention of responsible parties. Almost any hazards that could be detected by database analysis would probably have been obvious without elaborate cross-tabulations, and factors not foreseen at the time of system design or data entry could not be discerned from the data: the human brain is more efficient for this free form pattern recognition and problem solving (Kletz, 1976).

It should be noted that a common motive for accumulation and 'analysis' of accident data is not actually to discover patterns but to 'prove' them: a safety manager/-nurse/workers' representative may wish to provoke official recognition of a problem, and usually budgetary support of some corrective measure. Recognition that a certain factor is a problem in a certain workplace should not require the manipulation of figures. To draw that conclusion from a database requires that the factor be probed in collection of the raw data. This 'witchhunt' approach results in consciously or unconsciously biased investigation and suggestive interrogation guaranteed to overstate the problem or, if the factor is incriminating to the principal in the accident, to discourage honesty or deter reporting. If a problem is sufficiently obvious that detailed data can be solicited about it for inclusion in the database, perhaps the responsible approach is to address the problem directly.

While management oriented people use accident data to indicate the effectiveness of management's efforts to control risk, research attention is often turned to certain firms or fields in proportion to overall accident incidence. Scientists in both psychological and physical sciences fields assume that accidents are an indication of a need for more research into risk reduction. This arises from a belief that all research findings are immediately applied in the field. However, what looks like an epidemic and untapped research topic is often a failure to apply available knowledge, as often due to unawareness of its existence as to an unwillingness to invest in it. To remedy this requires more efforts spent by researchers (or by intermediaries, the specialist consultants) on transfer of new knowledge. Re-researching previously-studied fields in search of new explanations to old phenomena is a waste of research efforts. (Not wasteful is study by social scientists and management scientists interested in the process of and impediments to transfer of research knowledge.)

Obviously, both risk control and inherent risk factors are interwoven and inseparable in the numerical measures of accidents. In the field there will be a level below which the most diligent manager could not eliminate risk. For example, an operation which can only be performed by using a toxic chemical can minimize but not eliminate the risks of the chemical. Accidental injuries and occupational illnesses which occur despite the application of all available knowledge are undoubtedly due to some residual hazards in the job, task, or workplace. The scientist must discover a less toxic process or increasingly effective controls, for example ways to prevent release of the chemical, new protective gear, or inspection techniques. Management's performance cannot be measured by the total number of accidents: surely a poor management in a low risk industry does not deserve praise while the best management in a high risk industry receives criticism.

When seeking the top priorities for accident prevention research, ergonomists and other researchers need to deduct the proportion of accidents that occur due to management's failure to implement existing knowledge to control hazards. Unfortunately, there is no means to separately quantify the risk inherent in an operation. What is required is an index measuring management control alone, which would provide a synthetic means to separate management control from the accident frequency Index. When the residual risk is identified, it is appropriate to initiate research to prescribe preventive and protective measures. This separation approach is the key to the proposed model of accident data analysis. It may appear that a simple approach at the firm level to measurement of management control is to compare accident rates to firms in the same industry, but there are always confounding differences in business volumes, workforce demographics, clientele, products, technology, geography, and other factors. Also, this option is unavailable for comparisons across industries.

Although the following model requires further specific development, safety practitioners and researchers should nevertheless consider the contamination issue in the use of accident data, and should begin to work on ways to isolate the factor of interest.

A MODEL OF ACCIDENT DATA ANALYSIS

Figure 1 illustrates the relationship between the total accident experience and management safety control and inherent risk components.

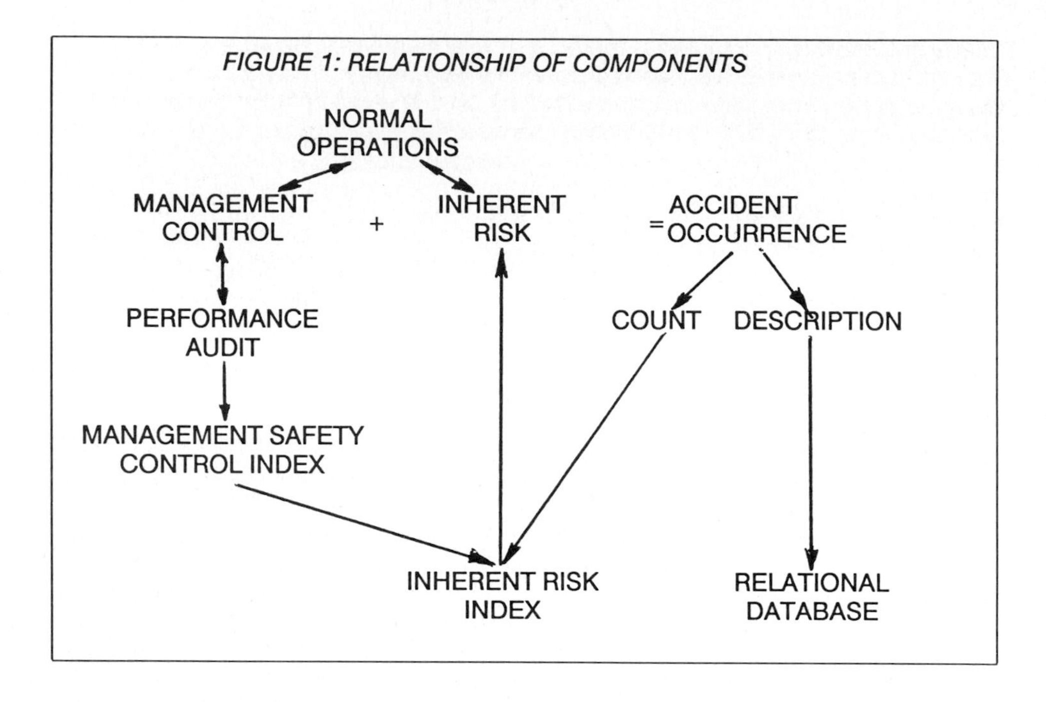

Management Safety Control Index

If each work unit or department had a list of standards and criteria which described 'safe operation in this unit', a numerical safety performance score could be assigned. Few would argue with the principle that such a score measures the effectiveness of management control: legislation confirms that management is responsible for the safe and unsafe behaviour of individual workers. The behavioural modification approach to safety (promoted by Komaki et al 1978; Sulzer-Azaroff 1978; Smith et al, 1978; extended and replicated by others) monitors normal performance, not accidents. Similar to quality assurance techniques, these approaches typically establish certain behavioural standards and criteria, incorporate means for observing and taking observations of performance, and incorporate performance feedback and initial and on-going instruction. The means used to monitor behaviour would be ideally suited to development into a management safety control index.

Difficulties with such a measurement method are mainly the complexity of developing standards and criteria in each unit, and in the manipulation of standards and criteria to overstate management safety control. Neither difficulty should be insurmountable. Quality assurance standards and criteria are a mainstay of most industrial operations. The Canadian health care industry recently introduced institution-wide QA programs to consolidate existing QA activities and encourage their development of those in areas where they had been lacking. While not a trivial undertaking, it has been possible for most institutions to achieve respectable results, and it is expected that these programs will improve and evolve over time. If improvement in safety performance and safety measurement is expected, it is this type of commitment that management safety indices will require. Less time spent by managers, workers' safety representatives, and safety practitioners investigating and identifying familiar causes for accidents will free up time for committees and individual effort to develop the necessary safety standards and criteria.

Development of the standards and criteria internally should not jeopardize their validity. Joint labour-management committees, called for in most health and safety legislation, could monitor the standards and criteria, with the assistance of such external audit agencies as labour inspectorate, insurers, and consultants.

Inherent Risk Component

The measure of the risk inherent in an operation or industry could be thought of as the component left in the accident index after management safety control has been deducted. If management is performing at a 100% level, then the accident rate is entirely due to inherent risk requiring new scientific knowledge. If management is performing at the 60% level, then 40% of accident experience is attributable to management shortcomings and 60% must be due to inherent risk. It is important to avoid allocating individual accidents to one pile or the other, as this introduces the element of bias; the management safety control index must be thought of as the full measure of management safety control. If it is suspected that it is outdated or inadequate for this task, it should be updated or modified. Ideally, as the management safety control performance tended to zero, some minimum inherent risk would apply, however it is likely that this would be an infrequent situation and would most usually exist in low-risk industry sectors.

Descriptive Data

The actual descriptive data from the database has not yet been mentioned. The preceding indices require only a knowledge of normal operations (for the management safety control index) and simple normalized counts of accidents (for computing inherent risk). Other than the date and location of the accident and the name and occupation of the injured person, the content of individual data records are widely varied. Most are based on one or more of several models of accident causation. As discussed, such probing into the accident's cause does nothing more than seek to determine culpability which, aside from supporting or defending a legal action, has little value. Legal counsel would prefer to have a simple record of facts without an opinionated elaboration of perceived 'cause' of the reasons for the event. One of several accident descriptive models (eg. Shannon and Manning, 1980; Laughery, 1984) would be far better suited than the causation models.

The database of accident descriptions would serve as a resource for researchers pursuing the priorities identified through the calculated inherent risk. Alone, it is a source of epidemiological data, though of statistical value only when consolidated to generate a database of sufficient size. Combined with a review of individual managers' management safety control standards and criteria and performance scores, the relational database would permit non-application of available knowledge to be confirmed or ruled out. The researcher could then proceed to further study the data, observe the operations of interest, conduct pure or applied scientific experiments or through other means develop a solution to the problem.

SUMMARY

The proposed safety measurement system depends on a sequence of steps:

1. Responsible job, task, and workplace design respecting current scientific findings - by practitioners

2. Development of standards and criteria that describe safe work unit operation, and the measurement of management safety control through establishment and maintenance of monitoring practices - by practitioners

3. Accumulation of accident occurrence data and accident description records - by practitioners

4. Determining an inherent risk index by 'discounting' accident occurrence data to a proportion equivalent to the management safety control index - transition from practitioners to researchers

5. Auditing management safety control standards and criteria periodically, using internal and external auditors - by practitioners

6. Review of inherent risk index for priorities, and undertaking of risk reduction research using the accident description data or other methods - by researchers

The proposed system involves more effort in clearly establishing what safety will look like, and separating the proportion of accidents due to sub-optimal control. Researchers will largely continue their work in the same way but in areas more appropriately prioritized according to the needs of diligent managers. Practitioners will substantially change their approach to safety measurement. They will monitor their performance by observation of normal operations and almost eliminate the attention to individual accidents that is currently preeminent in accident prevention.

It is difficult for a typical safety practitioner to visualize the abandonment of the well-established custom of accident investigation, yet it is a significant investment in time that could often be spent more profitably in developing and monitoring standards and criteria. In fact, some accident investigation will continue. The management safety control audit process will likely incorporate analysis of a proportion of randomly selected accidents regardless of severity as a quality assurance activity to ensure that policies and prescribed preventive measures have been followed. In addition, detailed study of individual severe (eg. long-term or permanent disability or fatality) accidents may be desired as a preparation for whatever external inquiries may be made. Workers' representatives may wish to review a larger proportion of accidents for signs of violation of policy that indicate poor supervision, and will legitimately agitate for management follow-through. These should all be secondary methods, *in addition to* the on-going supervision and monitoring of prescribed preventive measures, and never as the sole means to discover such violations.

Accident data will always be weak due to evasion of reporting, different legal and customary thresholds of reportability, unrepresentativeness of accidents of different severities, and other factors. Conscious planning in data collection and research design, however, will maximize its utility. While some of the shortcomings of accident data must be accommodated, others must be rectified. The most important among these is the clarification of what the data enumerates - risk management or inherent risk - and the establishment of separate measures to monitor both.

The clarification process will benefit not just clarity but also the performance of the control and the research. As managers and workers enunciate the characteristics of safe work, put in place training and proper conditions, and monitor and feed back performance, the focus of their accident prevention efforts moves, from regulations, slogans, statistics, and hope, to meeting the specific requirements of their own situation. There is no doubt that occupational safety will improve as a result. As researchers spend less time repeating their own research and the research already done by others, more time is available to research new areas, or to focus on the transfer of knowledge.

REFERENCES

BUCK, L. 1988 Human error at sea. Forum, *Communique* Newsletter of the Human Factors Association of Canada, January

GRONDSTROM, R. 1980 Serious occupational accidents - an investigation of causes. *Journal of Occupational Accidents* **2:**283-289

KLETZ, T.A. 1976 Accident data -- the need for a new look at the sort of data that are collected and analysed. *Journal of Occupational Accidents* **1**:95-105

KOMAKI, J., BARWICK, K.D., AND SCOTT, L.R. 1978. A behavioural approach to occupational safety: pinpointing and reinforcing safe performance in a food manufacturing plant. *Journal of Applied Psychology* **63**(4):434-445

LAUGHERY, K.R. 1984b *Safety Measurement Systems* - advanced workshop. International Conference on Occupational Ergonomics: Toronto

PIMBLE, J. AND O'TOOLE, S. 1979. Analysis of accident reports. *Ergonomics* **25**(11):967-979

SHANNON, H.S. AND MANNING, D.P 1980 The use of a model to record and store data on industrial accidents resulting in injury. *Journal of Occupational Accidents* **3**:57-65

SMITH, M.J., ANGER, W.K, AND USLAN, S.S. 1978. Behavioural modification applied to occupational safety. *Journal of Safety Research* **10**(2):87-88

SULZER-AZAROFF, B. 1978. Behavioural ecology and accident prevention. *Journal of Occupational Behaviour Management* **2**:11-44

Trends in Ergonomics/Human Factors V
F. Aghazadeh (Editor)
Elsevier Science Publishers B.V. (North-Holland), 1988

SHIFTWORK AND SAFETY: A REVIEW OF THE LITERATURE AND RECENT RESEARCH RESULTS

Jon A. WAGNER

Mining Engineer, Twin Cities Research Center
Bureau of Mines, U.S. Department of the Interior
Minneapolis, MN 55417

Of recent concern to many companies is the effect of shift scheduling practices on accident rates and accident patterns. For this reason, the Bureau of Mines reviewed extensive literature on the subject of shiftwork and safety in an attempt to characterize the cause-and-effect nature of this phenomenon for various work tasks and job environments. In addition, the Bureau examined 10 years of accident data for a group of iron ore mining companies that operated on the same shift rotation schedule. This paper summarizes some of the more important findings. Of special interest are the roles of particular workdays of the night shift and age of workers on the timing and frequency of accidents. In general, older workers tended to have a greater share of their accidents toward the end of the night shift series and during the last few hours of a night shift, compared to younger workers.

1. INTRODUCTION

1.1. Rationale

The objective of this paper is to summarize some of the factors affecting the safety of shiftworkers and analyze these factors in an actual case study. The data presented will implicate some factors as being significantly important for the worker population studied.

1.2. Temporal Factors and Shiftwork Safety

As summarized below, there are six major time-related factors that can affect the safety of shiftworkers.

1.2.1. Fatigue

Fatigue may be defined as the loss of physical, mental, or emotional energy due to extended time on a task without adequate rest. It is important to consider the type of task being performed when analyzing safety performance for fatigue effects. For instance, a foreman may experience "stress overload" and mental fatigue during a day of extremely active supervision. Conversely, a haulage truck driver may experience "stress underload" and boredom during a day of normal haulage operations. Both workers can show performance decrements toward the end of a work shift, but due to very different reasons.

1.2.2. Night Work

Potential safety problems with night work are logistical and biological. First, work during hours of darkness often includes reliance on artificial illumination which can be inadequate. At the very least, the appearance of the outside environment undergoes a significant change. Second, night work opposes the human being's natural sleep-wakefulness cycle [1]. Unless a person is given adequate time to partially adjust to night shifts, work will be performed during an "alertness trough" during which the person would normally be sleeping.

1.2.3. Exposure

Accident trends during work hours are not meaningful unless the exposure or rate and type of activities are known. For example, if accident data show that an accident peak occurs during the final hour of the shift, it must be ascertained if the workers are indeed doing the same tasks, at the same rate, and under the same conditions as the earlier hours, before a fatigue effect can be deduced. Scientists researching the accident rate at a British coal mine noted such a peak, but observed that the miners would make a mad dash out of the mine at the end of the shift, thereby sustaining injuries [2]. In this case, the nature of the hazard exposure changed between intermediate hours of the shift and the last hour of the shift.

1.2.4. Time of Day

Previous research has noted the existence of circadian, or "about 1 day" rhythms in human performance, including error commission [3-4]. Most performance rhythms correlate with core body temperature which, for a day-active person, rises throughout the daylight hours and drops sharply from approximately 9 p.m. to 6 a.m. [5]. The low-temperature trough in the early morning hours (2 a.m. to 6 a.m.) corresponds with deep sleep and extremely impaired alertness. Apart from the early morning hours, the human body experiences an alertness dip about 12 hours earlier, known as the postprandial or postlunch dip. The severity of this alertness decrement may be mediated by the timing and composition of the lunch-time meal [6-7].

It must be noted that time-of-day variations in accident rates can be due to reasons other than circadian rhythms of performance. For instance, exposure levels often change in parallel with light levels. To illustrate, a mine haulage truck driver may be sufficiently aroused during the day, which should prevent any accidents arising from lack of alertness or inattention. However, there might indeed be accidents occurring, due to increased traffic in the mine pit during day shift hours (including foremen, mechanics, electricians, labor crews, pump crews, etc.). Conversely, fewer accidents may occur during hours of darkness because of lower traffic density.

1.2.5. Shiftworker Dischrony

Shiftworker dischrony, or the internal desynchronization of the body's internal circadian rhythms, can be described as occupational jet lag. This pervasive fatigue manifests itself as a general malaise, lack of energy, or overall depression [8], due to the changing of shifts.

Laboratory research has shown that some physiological functions may require up to 16 days to become "reentrained" to a new schedule which is 8 hours out of phase with the old schedule [9-10]. However, not all physiologic functions become reentrained at the same rate; hence, shiftworker dischrony results. In human terms, this period of adjustment is characterized by mental and physical fatigue, sleep disruption, a likelihood of impaired performance, and possibly increased susceptibility to disease [11]. The ideal epidemiological approach to determining the effects of shiftworker dischrony is to compare the accident records of rotating shiftworkers versus permanent day workers, with all other conditions being identical. Unfortunately, the small number of studies on this topic are not conclusive, with few implicating a dischrony effect [12-13].

1.2.6. Reentrainment After Shift Change

One factor that has been overlooked by most accident researchers is the effect of shift change on accident causation. When a worker works a night shift after a series of day or afternoon shifts, three adverse performance consequences are possible: (1) work may be attempted without an adequate prior period of sleep, (2) work may be performed during the alertness trough, a time when the body is normally geared toward sleep, and (3) internal desynchronization of circadian rhythms, or dischrony, may occur.

It may be hypothesized that as successive night shifts are worked, sleep becomes more regular during the day, the alertness trough realigns itself with the new sleeping hours, and circadian rhythms gradually move in phase with each other and with the new schedule. Unfortunately, this process has not been shown to occur perfectly in the lives of shiftworkers [14]; however, it seems certain that these processes play a vital role in how a person feels and performs, especially during the first week of night work. An examination of the distribution of accidents ocurring on the night shift, with respect to how many night shifts have been worked successively, would be a key in determining the effect of shift change (and shiftworker dischrony) on accident causation.

1.3. Other Factors and Shiftwork Safety

There are a number of other factors that can influence the effects of temporal factors on safety performance, including nature of work, supervision, age, experience, personality, coordination, mental state, alcohol intake, drug usage, and nutrient intake. Obviously, how a shiftworker interacts with hazards in the workplace may depend a great deal on these personal and organizational factors.

One such factor that has attracted particular attention is age. Theoretically, the age of the worker may affect when, how, and why an accident occurs. It is known that, with age, significant psychological and physiological changes occur. Accuracy of risk perception improves [15], reaction time slows down, agility decreases [16], and accidents may have more severe consequences [17-18]. Lifestyle priorities generally change from maximizing social time to protecting health [19]. In addition, previous research has noted that middle-aged shiftworkers experience physiological changes that lead to increased sleep disturbances, early awakenings, and decreased flexibility of circadian rhythms [20-22], often around the age of 45 [23]. An analysis of age effects on accidents should therefore consider if older workers suffer a greater number of fatigue-related accidents compared to younger workers. This fatigue factor would show itself as an increase in accidents during the last few hours of a shift, as well as during the last few days of a shift cycle.

2. METHOD

The remainder of this report deals with results of research on the possible role of the temporal and age factors identified above in accident causation. A case study of accidents in northern Minnesota and Michigan iron mines was chosen because of the following special features. First, the accident and injury data necessary to perform the analysis are available in the Denver Safety and Health Technology Center accident and injury database, which is compiled by the Mine Safety and Health Administration (MSHA). By law, all lost-time injury accidents must be reported to MSHA in a format that provides detailed accident information. In addition, the Bureau of Mines has developed the Accident Data Analysis (ADA) program, which allows the user to perform specific types of sorts and risk analyses [24]. Given the availability of the data and the database management program, a thorough analysis of actual mine accidents was deemed feasible.

Second, during the period from 1975-84, the iron mines of the U.S. Lake Superior region were all similar in terms of equipment, mining technique, mineral processing, manpower requirements, haulage methods, and management methods. The mines and associated plants considered for this study are all taconite operations, which mine and process low-grade iron ore. These operations can be reasonably grouped together to form one large pool of accident data that could be analyzed in aggregate.

Third, and most important, all of the taconite mines and plants operated with the same basic 20-day rotation schedule for the 10-year period, despite minor fluctuations in demand and production. This schedule, which completes its cycle after 20 working days, allows the researcher to match accident records with the exact shifts within which the accidents occurred. By knowing the date and time of an accident, one can look at a calendar and determine the day of the week and the corresponding work shift (either day, afternoon, or night duty). Most often, the work shift was already specified in the accident reports. For rotating shiftworkers, the day of the week likewise corresponds with a particular workday in a series of same shifts. For instance, an accident occurring on a Saturday at 5 a.m. is known to have occurred during the second workday in the series of seven night shifts (see figure 1).

This paper focuses only on night shift accidents because workers on the night shift were known to be exclusively on the same rotational schedule depicted in figure 1. An analysis of accidents suffered by night shift workers would therefore not include workers on other types of shifts or rotational patterns. On the other hand, afternoon shift accident victims were an unknown mixture of the standard shift rotators and day-to-afternoon rotators. Likewise, day shift accident victims were a mixture of standard rotators, day-to-afternoon rotators, and straight day workers. Thus, night shift accident analysis gives the purest indication of schedule-related accident causation factors.

Another important point about night shift accidents is that the night

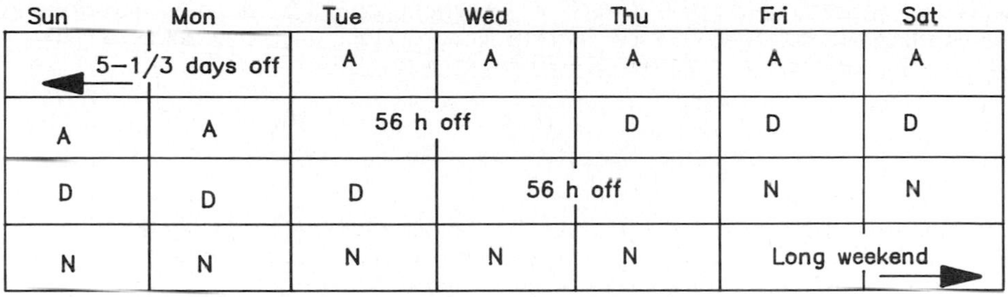

Sun	Mon	Tue	Wed	Thu	Fri	Sat
← 5-1/3 days off		A	A	A	A	A
A	A	56 h off		D	D	D
D	D	D	56 h off		N	N
N	N	N	N	N	Long weekend →	

KEY A=Afternoon shift, 3–11 p.m. D=Day shift, 7 am–3 p.m.
N=Night shift, 11 p.m.–7 a.m.

Figure 1. – Standard 20-day rotation schedule used in taconite operations, 1975–1984. Note: Night shift begins at 11 p.m. on the previous day.

shift was purely production oriented; only those workers necessary for production and operation maintenance were part of crews that rotated through night shifts. Therefore, the workforce remained at constant levels throughout the series of night shifts, and produced output at a constant rate, as established by production planners. The only variations in workforce during night shifts would have been due to absences that usually required replacements. As a result, fluctuations in work demands and workforce levels most likely were small and probably had a minimal effect on the relative accident incidence rates between night shifts.

The accident data from 10 taconite operations were analyzed. A total of 6,546 accidents were reported during the 10-year period. Of these, 814 occurred on the night shift. During this time, the taconite industry employed approximately 14,000 workers. It is not known how many workers rotated through the night shift. The chi-square statistic was used to test the null hypothesis that the number of accidents did not vary significantly ($p<.05$) in relation to time of day or day of week.

3. RESULTS AND DISCUSSION

For the sake of brevity, only those results dealing with night shift accident patterns will be discussed in this paper. (Those interested in obtaining a full set of results may contact the author or order the appropriate Bureau of Mines publication [25]).

3.1. Time of Day

Figure 2 shows the overall time-of-day plot for night shift accidents during the 10-year period. The first hour of the shift is relatively safe, with most workers involved with instruction, preparation, and travel to work sites. The major spike occurs during the midnight hour; most jobs are being started and workers are beginning their work routines. After this, the frequency level steadily declines, except for the 1 a.m. to 3 a.m. disparity. A coffee break during the 1 a.m. hour may partially account for the relatively low accident level during that hour. Note that 61 pct. of the accidents occurred before midshift.

Figures 3, 4, and 5 present results for three age groups of the miner population. In the <=29-year-old group (figure 3), the accident level peaks during the second hour and declines through the remainder of the shift. Effects of cumulative fatigue and/or the trough in the circadian rhythm of alertness (4 a.m. to 6 a.m.) are not evident in these accident data. Note that 63 pct. of the accidents occurred before midshift.

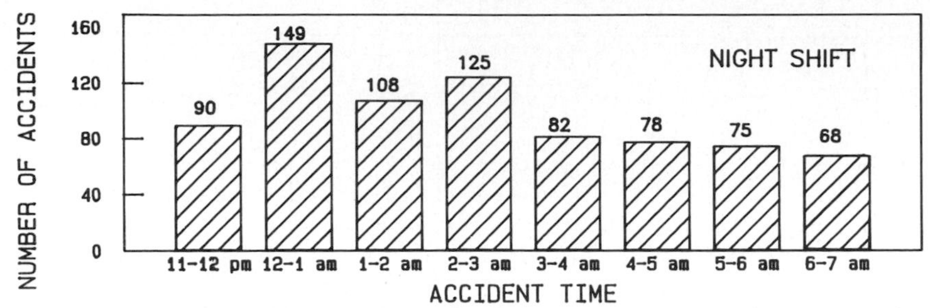

Figure 2. – Accidents on night shift, by time of day. Some accidents are not included because they occurred pre– or postshift or at unknown times. The data are significant at the 0.05 level. Sixty–one pct. of the accidents occurred before midshift.

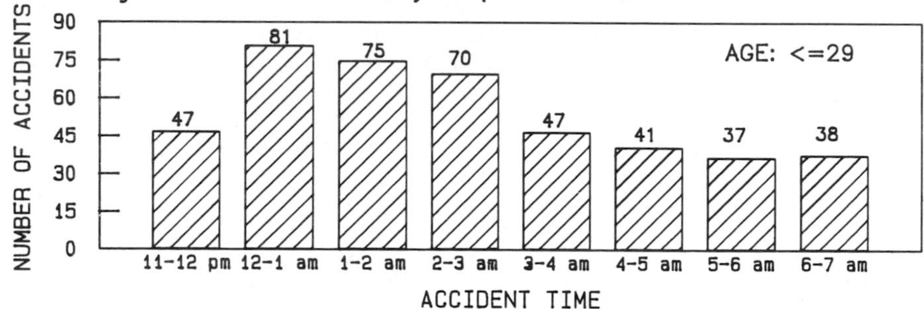

Figure 3. – Night shift accidents for the <=29 year–old–group, by time of day. Some accidents are not included because they occurred pre– or postshift or at unknown times. Sixty–three pct. of the accidents occurred before midshift; data are significant at 0.05 level.

In figure 4, the 30-39 age group shows two accident peaks, one during the midnight hour and one during the 2 a.m. hour. The unusually low accident level at 1 a.m. cannot be fully explained by a 10-minute coffee break. Note that 64 pct. of the accidents occurred before midshift.

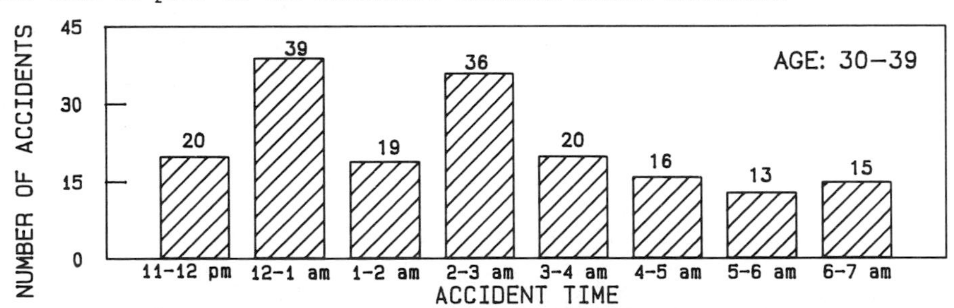

Figure 4. – Night shift accidents for the 30–39 age group, by time of day. Some accidents are not included because they occurred pre– or postshift or at unknown times. Sixty–four pct. of the accidents occurred before midshift; data are significant at the 0.05 level.

Of special interest is the graph for the >=40-year-old group (fig. 5). Here a bimodal distribution can be seen, with peaks at both the midnight hour and at 5 a.m. However, the data are not statistically significant. It is possible that these older workers are either more careful (through experience) at the beginning of a shift and/or are more prone to suffer from the effects of cumulative fatigue or lack of alertness toward the end of a shift. This postulated decreased alertness may be due to circadian rhythm dischrony, cumulative sleep deficit, work during the alertness trough, or some combination thereof. Note that only 51 pct. of the accidents occurred before midshift.

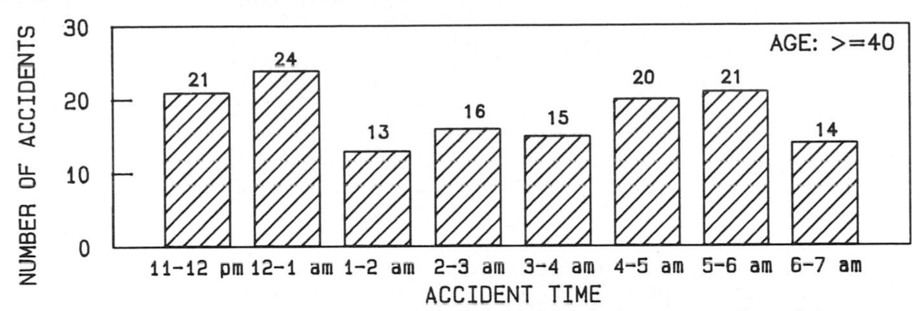

Figure 5. – Night shift accidents for the >=40 year-old group by time of day. Some accidents are not included because they occurred pre- or postshift or at unknown times. Only 51 pct. of the accidents occurred before midshift; however, data are not significant at 0.05 level.

3.2. Day of Shift

Figure 6 shows the overall accident frequency for successive workdays on night shift. The first night shift (Friday) shows a relatively high accident level, followed by lower accident levels possibly due to an adaptation to night shift. However, Monday's shift (beginning Sunday evening) shows an accident spike, followed by a gradual tapering. It is possible that seven night shifts in succession do not provide for the successful reentrainment of the body's circadian rhythms if the social weekend, during mid-series, is allowed to interfere. Social, family, religious, and/or activities during daylight hours may limit sleep-taking behavior and disrupt circadian rhythms.

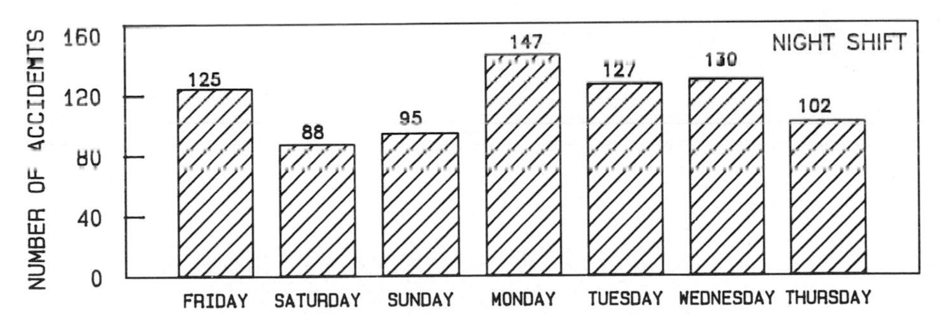

Figure 6. – Accidents on night shift, by day of week. Data are significant at 0.05 level.

Figures 7, 8, and 9 show the night shift accident frequencies by day of shift for the population, broken into three age groups. The youngest group, <=29 years, shows an accident pattern similar to the overall pattern (fig. 7). The middle age group, 30-39 years, has an extremely low accident level on Saturday, plus the familiar accident spike on Monday (fig.8). The rest of the week shows a tapering off, though not as steep as the under-30 age group. In fact, a slight increase can be seen for Wednesday (fig. 8).

Figure 9 shows the accident frequency pattern for the >=40 age group. There is a relatively flat distribution until the spike on Wednesday. Notably absent is a Monday spike and a high initial accident level. The data, however, do not yield statistical significance.

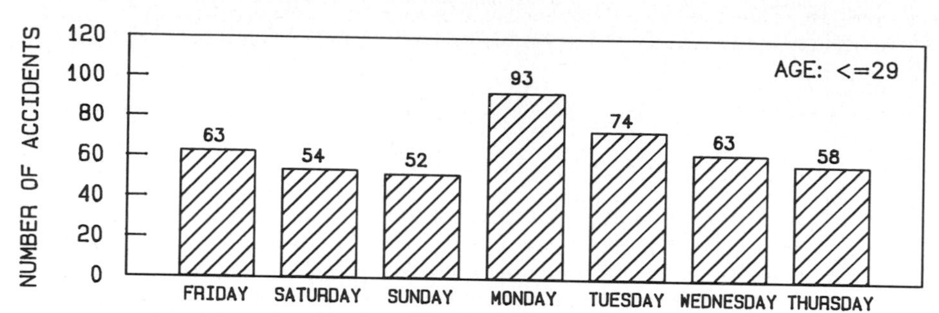

Figure 7. – Night shift accidents for <=29–year–old group, by day of week. Data are significant at 0.05 level.

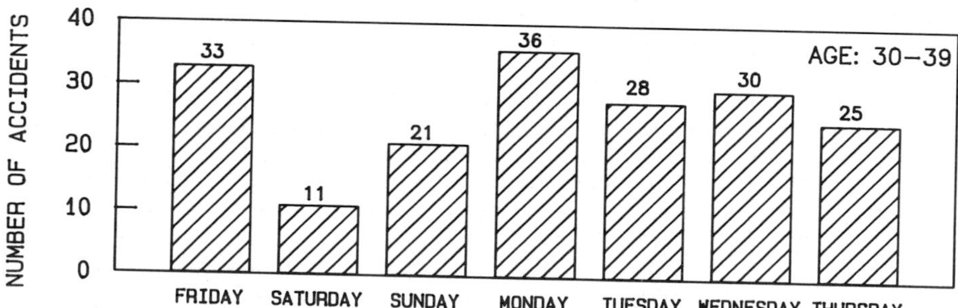

Figure 8. – Night shift accidents for 30–39–year–old group, by day of week. Data are significant at 0.05 level.

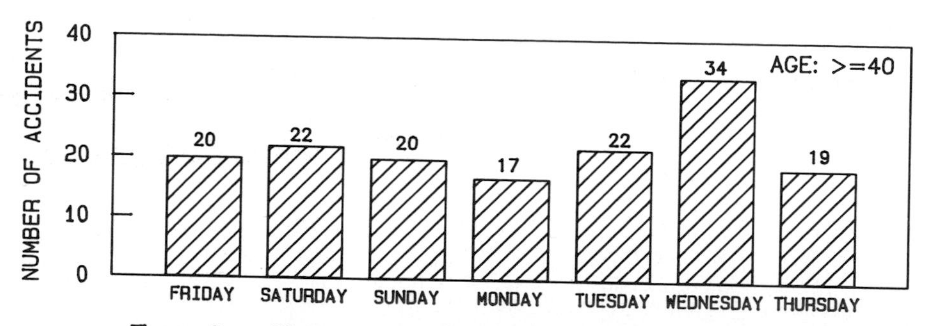

Figure 9. – Night shift accidents for >=40–year–old group, by day of week. Data are not significant at 0.05 level.

Why does the accident pattern change with increasing age? It may be that as a worker matures, his or her coping behaviors increase and lifestyle priorities change, such that more emphasis is given toward getting adequate sleep at a consistent time of day, especially on the social weekend. This will tend to decrease the Monday spike. However, as a worker enters his or her 40's, the natural ability to obtain day sleep and to shift body rhythms begins to decline. As a result, sleep deficit may accumulate toward the end of the work week, as suggested by the Wednesday spike. However, the return to normal frequency levels for the >=40-year-old workers on Thursday, the last day, is unexplained. Anecdotal information suggests that there may be a "second wind" phenomenon in anticipation of time off at the end of the shift sequence.

4. CONCLUSIONS

The data presented in this report show that the night shift accident frequencies for this mining workforce are dependent upon the combined effects of time of day, day of shift, and age. In general, younger workers seem prone to having more of their accidents early in the shift series, immediately after the social weekend, and in the first few hours of a night shift, compared to older workers. Conversely older workers, especially those 40 or older, seem prone to having more of their accidents toward the end of the night shift series and during the last few hours of a night shift, compared to younger workers. It should not, however, be concluded that older workers are comparatively more safe or less safe than younger workers. The data presented in this report could not be normalized to account for population sizes in the various age groups.

The collective results on shiftwork and accident levels reported in this study point to the need for more extensive human factors research on the possible interaction between shiftwork, temporal factors, and age in relation to work performance decrements and accident risk. Given that many industries are now dealing with continuous operations plus an aging workforce, this need appears clearly defined.

REFERENCES

[1] Minors, D.S. and Waterhouse, J.M., Circadian Rhythms and the Human (Wright, Bristol, 1981).
[2] Colquhoun, W.P., private communication, University of Sussex, Brighton (1986).
[3] Ehret, C.F., Applied Chronobiology: Helping Shift Workers Beat Circadian Vertigo, Logos 1, Argonne Nat. Lab. (U.S. GPO, Washington, DC., 1983).
[4] Colquhoun, W.P., Accidents, Injuries, and Shift Work, in: Rentos, P.G. and Shepard, R.D., (eds.), Shift Work and Health, Dept. HEW (NIOSH), pub. No. 76-203, (GPO, Washington, D.C., 1976) p. 162.
[5] Colquhoun, W.P. (ed.), Biological Rhythms and Human Performance (Academic Press, London, 1971) p. 41.
[6] Folkard, S. and Monk, T.H., (eds.), Circadian Performance Rhythms, in: Hours of Work (Wiley, Chichester, 1985) pp. 41-42.
[7] Craig, A., Baer, K., and Diekmann, A., The Effects of Lunch on Sensory Perceptual Functioning in Man, Intl. Archives of Occup. Environ. Health, No. 49 (Springer Verlag, 1981) pp. 105-114.
[8] Ehret, C.F., Future Perspectives for the Application of Chronobiological Knowledge in Occupational Work Scheduling, Invited Testimony for the Investigations and Oversight Subcommittee of the Committee on Science and Technology (U.S. House of Rep., Washington, D.C., Mar. 24, 1983) p. 4.

[9] Klein, K.E. and Wegmann, H.M., The Effects of Transmeridian and Transequatorial Air Travel on Psychological Well-Being and Performance, in: Chronobiology: Principles and Applications to Shifts in Schedules, NATO Advanced Study Inst., Series D: Behavioral and Social Sciences, No. 3 (Sijthoff and Noordhoff, The Netherlands, 1980) pp. 339-352.
[10] Klein, K.E. and Wegmann, H.M., Circadian Rhythms of Human Performance and Resistance: Operational Aspects, in: Nicholson, A.N., (ed.), Sleep, Wakefulness, and Circadian Rhythm, AGARD Lecture Series No. 105 (NATO, 1979) pp. 2:1-2:9.
[11] Smolensky, M.H., Paustenbach, D.J., and Scheving, L.E., Biological Rhythms, Shiftwork, and Occupational Health, in: Crailey, L. and Crailey, L., (eds.), Patty's Industrial Hygiene and Toxicology, 2nd ed., v. 3B: Biological Responses (Wiley, 1985) pp. 267-269.
[12] Smith, M.J., Colligan, M.J., Frockt, I.J., and Tasto, D.L., Occupational Injury Rates Among Nurses as a Function of Shift Schedule, J. Safety Res., v 11, No. 4 (1979) pp. 181-187.
[13] Social Statistics of the Statistical Office of the European Communities, Erhebung uber sitz und Art der durch Arbeitsunfalle verursachten Verletzungen (Eisen-und Stahlindustrie), Report No. 5/6 (1971) pp. 144-172.
[14] Akerstedt, T., Adjustment of Physiological Circadian Rhythms and the Sleep-wake Cycle to Shiftwork, in: Folkard, S. and Monk, T.H., (eds.), Hours of Work (Wiley, Chichester, 1985) pp. 186-188.
[15] Finn, P. and Bragg, B.W.E., Perception of the Risk of an Accident by Young and Older Drivers, Accident Analysis and Prevention, v. 18, No. 4 (Pergamon Press, Aug. 1986) pp. 289-298.
[16] Welford, A.T., Ageing and Human Skill (Oxford Univ. Press, 1958) pp. 117-128.
[17] Buck, P.C. and Coleman, V.P., Slipping, Tripping, and Falling Accidents at Work: A National Picture, Ergonomics, v. 28, No. 7 (July 1985) pp. 950-955.
[18] Vernon, H.M. and Bedford, T., A Study of Absenteeism in a Group of Ten Collieries, Ind. Fatigue Res. Board, Rep. No. 51 (HMSO, London, 1928) pp. 39-42.
[19] Klein, M., Personal communication about unpublished shiftworker survey data, Sleep Physiology Laboratories, Lincoln General Hospital, Lincoln, NE (1987).
[20] Torsvall, L., Akerstedt, T., and Gillberg, M., Age, Sleep, and Irregular Workhours, Scand. J. Work and Environ. Health, v. 7 (1981) pp. 196-203.
[21] Pavard, B., Vladis, A., Foret, J., and Wisner, A., Age and Long Term Shiftwork With Mental Load: Their Effects on Sleep, J. Human Ergology, v. 11, supplement (1982) pp.303-309.
[22] Monk, T.H. and Folkard, S., Individual Differences in Shiftwork Adjustment, in: Folkard, S. and Monk, T.H., (eds.), Hours of Work (Wiley, Chichester, 1985) pp. 228-230.
[23] Akerstedt, T., Interindividual Differences in Adjustment to Shift Work, in: Colquhoun, W.P. and Rutenfranz, J., (eds.), Studies of Shiftwork (Taylor and Francis, London, 1980) pp. 121-130.
[24] Bowers, E.T., Using ADA (Accident Data Analysis) in Mine Safety Research, Bureau of Mines OFR 72-86 (1985) 96 pp.
[25] Wagner, J.A., Temporal Factors in Mining Accidents: A Case Study of the U.S. Lake Superior Iron Ore Industry, Bureau of Mines Report of Investigation (in press, 1988).

Trends in Ergonomics/Human Factors V
F. Aghazadeh (Editor)
Elsevier Science Publishers B.V. (North-Holland), 1988

PREVENTION STRATEGIES ADOPTED BY SELECT COUNTRIES FOR WORK-RELATED MUSCULOSKELETAL DISORDERS FROM REPETITIVE TRAUMA

Vern Putz-Anderson
National Institute for Occupational Safety and Health
Division of Biomedical and Behavioral Science
4676 Columbia Parkway, Cincinnati, Ohio U.S.A.

Abstract

Within the last 10 years reports of tendinitis and related soft tissue injuries to the upper limbs have increased to an extent that several countries have developed and implemented national programs aimed at preventing such problems. Yet information about these programs has not been widely circulated. One reason for the limited public attention may be a failure to recognize the ubiquitous nature of these disorders. This is most likely reinforced by the distinctive expressions used by various countries to refer to these conditions. In the United States, for example, the term cumulative trauma disorder is most often used. In Sweden the label occupational cervicobrachial disorder is used. In Great Britain and Australia the most common label is repetitive strain injury. Just as each of these countries has adopted special terms for describing such work-related disorders of the musculoskeletal system, each country has also formulated a plan for responding to these disorders, which ranges from standard-setting to information dissemination. Select features of these programs are presented. In addition, this paper identifies two recent trends in the workplace that may account for the growing incidence of work-related musculoskeletal disorders.

Terminology and Background

Several distinctive labels have appeared in the occupational health literature of different countries over the last 10 years to identify work-related disorders stemming from repetitive trauma to the upper limbs. The common term in Great Britain and Australia is repetitive strain injury (RSI). In the United States the term cumulative trauma disorder (CTD) is often used, and in Sweden these conditions are referred to as ergonomically-related injuries, or by the anatomical descriptive label: occupational cervicobrachial disorder (OCD). Collectively, these conditions are identified here as simply: work-related musculoskeletal disorders.

Although these terms are seldom found in textbooks, the underlying medical conditions are commonly included under one or more of the topics describing nonarticular rheumatism, psychogenic rheumatism, degenerative joint disease, osteoarthritis, or overuse injury. To further complicate matters, any one of at least 30 separate diseases from the International Classification of Disease Codes (ICD) may be cited as the diagnosis of a given work-related musculoskeletal disorder (Table 1). Moreover, terms such as cumulative trauma disorder, or repetitive strain injury are not included in the ICD list.[1]

Work-related musculoskeletal disorders are normally distinguished from rheumatoid arthritis by the absence of a systemic infection agent. In general, work-related musculoskeletal disorder are local or regional afflictions primarily affecting the soft-connective tissues of one or more structures of the upper limbs. Since the disorders may also be caused by non-occupational factors, a job analysis is usually necessary to identify the work component, or trauma. Usually the source of the trauma, however, is not that evident, but rather arises out of the interaction between a worker and a job. Hence, a job is considered to be high-risk for musculoskeletal disorders if the task poses demands that either exceed the capacities of the worker or requires work motions that are biomechanically unsound. Moreover, it is the cumulative and protracted action of these situations that yields a chronic trauma disorder that distinguishes it from the simpler acute form of accidental injury.

The history of work-related musculoskeletal disorders can be traced to the late 1800's.[2] It was the early to mid 1970's, however, before reports first began to regularly appear in the international scientific literature. Before 1970, the occupational and safety practices of most industrialized countries were focused on preventing traumatic injuries and illnesses from exposure to physical and chemical agents. Contemporary interest in work-related musculoskeletal disorders is due to a combination of factors that includes changes in the nature of work and the composition of the work force. Two of the most significant trends contributing to the increase in reports of musculoskeletal disorders are the mechanization of work and increased employment in the service and information sectors.

ICD Code	Description of Disorder
354	Mononeuritis of Upper Limb
354.0	Carpal Tunnel Syndrome
354.1	Other lesion, medial nerve
354.2	Lesion, ulnar nerve
354.3	Lesion, radial nerve
443.0	Raynaud's Syndrome
715	Osteoarthritis and allied disorders
719	Other and unspecified disorders of joints
723.3	Cervicobrachial syndrome
726.1	Rotator cuff syndrome
726.32	Lateral epicondylitis
727	Other disorders of synovia, tendons and bursa
727.0	Synovitis and tenosynovitis
727.03	Trigger finger
727.04	Radial styloid tenosynovitis (de Quervain's disease)
727.2	Specific bursitis often of occupational origin
727.4	Ganglion or cyst of synovium, tendon or bursa
728	Disorders of muscle, ligament and fascia
729	Other disorders of soft tissue
840	Sprains/strains of shoulder and upper arm
841	Sprains/strains of elbow and forearm
842	Sprains/strains of wrist and hand
955	Injury to peripheral nerves of shoulder and upper limb
959	Injury, other and unspecified
959.2	Shoulder and upper arm
959.3	Elbow, forearm and wrist
959.4	Hand

Table 1. ICD Codes for Upper Limb Cumulative Trauma Disorders

Changes in Work Patterns and the Work Force

Since work has become more mechanized and segmented, the locus of work has shifted from the level of the back, where lifting was a main stress, to the level of the eye, where the hands and arms are the most vulnerable structures. In the modern workplace, the worker is typically seated doing routine repetitive motions. Mass production is no longer unique to the factory, but has been successfully implemented in the modern white collar environments. This is characteristic of work found in many service and data processing industries. Recent employment statistics indicate that more people in the U.S. work in fast food restaurants than in the traditional steel and auto industries.[3]

Perhaps the biggest change in jobs is the rate at which work is performed. Assembly lines and automation have increased the pace of production, while reducing the physical effort and complexity of the work activity. As a consequence, a worker may repeat a stereotyped motion many thousands of times per day. Each repetition may cause little or no tissue damage. It is the cumulative affect of the repeated trauma over months or years that is responsible for many of the symptoms of pain, inflammation and restricted motion that are characteristic of work-related musculoskeletal disorders.[4]

Just as the work process is changing, a similar change is occurring in the composition of the work force. With the increase in life expectancy and decrease in fertility rate, the work population in many countries is growing older. In the U.S., the median age in 1984 was 31.2, up from 28 years in 1970.[5] Older workers have fewer accidents than less experienced co-workers, but their musculoskeletal systems are less resilient to daily exposures of repetitive trauma. Similarly, as the work force expands and diversifies, jobs and tools that were designed for a specific work population become sources of biomechanical stress for workers who have different physical attributes and strengths. Industries that have not considered some form of intervention program to alter the design of work and tools to fit individual workers, may also be contributing to the rising incidence of work-related musculoskeletal disorders.[6]

There have also been changes in peoples' tolerance for occupational injury and disease. In the past, a worker had two options when faced with unsafe or physically-demanding work: stay and endure, or seek other employment. As workers become better educated, they no longer accept the prospects of injury, pain or potential disability as the risk of employment. Workers are more prone to seek redress for injury and illness from disability and compensation systems.[7] Hence, changing expectations about health have challenged the widely held belief that aches and pains are the normal consequences of the degenerative process of aging and the accepted costs of manual labor. Furthermore, for certain highly repetitive-assembly work and food processing jobs, the younger workers are as much at risk as the older workers. In such jobs, the onset of work-related musculoskeletal disorders is measured in months, not years. The older worker may be indirectly protected as a result of self-selection and seniority that provide relief from jobs that induce chronic-musculoskeletal discomfort.

Recent data indicate that the incidence of tendinitis, strained ligaments, and nerve entrapments have increased for all age groups.[8] Moreover, this is occurring at modern worksites where the jobs are considered safe. The resulting health consequences and the loss of productivity stemming from these changes have not escaped the attention of leading occupational and safety practitioners from various countries. There is a growing consensus in many countries that such musculoskeletal disorders are a costly and largely preventable occupational health problem.[9] In fact, several countries have introduced prevention programs even though they have been unable to determine the national incidence of these disorders.

Lack of National Incidence Data

In designing a prevention program, it is important to have an understanding of the scope of the problem. National health priorities need to be established; resources need to be allocated. Usually, one of the first goals of a health program is to introduce or upgrade a national surveillance system to track major categories of occupational injuries and illnesses. Even in countries where adequate surveillance systems are in place, the data may be difficult to interpret because of confounded classification systems and overlapping diagnostic labels.

In the U.S., for example, following the passage of the Occupational and Safety Act of 1970, a comprehensive reporting system was established that contained a category for "Disorders Associated with Repeated Trauma." This category is used not only to record musculoskeletal disorders, but also "noise-induced hearing loss" and "vibration effects."[a] To separate hearing loss from tendinitis problems, it is necessary to refer to the Supplementary Data System (SDS) (OSHA Form No.101) to obtain estimates of national figures for musculoskeletal problems. The SDS system obtains much of its data from the workers' compensation systems of a large sample of participating states. This database was used by NIOSH investigators to generate incidence rates of repetitive trauma disorders for a number occupations and industries. Although the data are from 1979 statistics, the average industry-wide incident rate for that year was nearly 10 claims per 100,00 workers.[10] Rates varied from a high of 38 claims per 100,000 workers to a low of 2. For select worksites, NIOSH investigators have found incidence rates for repetitive motion disorders to range between a high of 58% of the employees in one plant to a low of 3% of the employees.[11]

More recent incidence data are now available from select states. California, Wisconsin, Ohio, North Carolina, and New Hampshire have reported significant increases in work-related musculoskeletal disorders, such as carpal tunnel syndrome. For example, in Ohio between 1980 and 1984 there was a 3-fold increase in the number of worker's compensation claims

---

[a] This category was formed in the 1978 revision of the reporting requirement of Public Law 91-596 (OSHA Act).

(WC) for cumulative trauma disorders.[12] Some of the increase may be due to an extension of WC eligibility to include injuries and illnesses with no lost time. In New Hampshire, WC claims for musculoskeletal disorders increased from 12.9% of all claims in 1976 to 20.5% in 1986.[13] In North Carolina, the reported cases of all musculoskeletal disease more than doubled between 1982 (7.7%) and 1984 (17.3%). Current estimates by the U.S. Department of Labor for inflammation of the joints, which includes tenosynovitis, account for 20% of illnesses reported each year.[14] In the U.S., these conditions have increased to the extent that they have become one of the major negative side effects of modern technology.

In Sweden, ergonomically-related injuries accounted for about 20% of all occupational injuries reported in 1982.[15] Their reporting system is sufficiently detailed that the nature of the hazardous condition, the part of the body injured, and the industries most at risk can be identified. For example, in 1982 most of the ergonomic injuries were attributed to unusual strenuous movements or working postures affecting the shoulder-arm region of workers employed in the processing of food. Sweden has one of the most comprehensive occupational health surveillance systems for assessing national incidence levels of these disorders.[16] One factor that may have encouraged more inclusive reporting is their use of the general rubric "ergonomically-related disorders" as a catch-all phrase for categorizing work-related musculoskeletal disorders. This category is sufficiently broad so as to reduce the inevitable diagnostic confusion surrounding International Classification of Disease Codes or specialized trauma definitions.

Current national statistics are unavailable for Australia. However, data obtained from the New South Wales Workers' Compensation Commission (Australia) showed that for the period of 1978 to 1982, the incidence of repetitive strain injury for females increased from 526 cases to almost 1500 cases. In the early 1970's, reports from Australia documented the first cases of a form of musculoskeletal disorders stemming from repetitive, low trauma exposure. What was then described as a telegraphist's cramps is now known as repetitive strain injury. In 1984, the Medical Journal of Australia referred to these conditions as "the new industrial epidemic."[17]

In the early 1960's, the Industrial Survey Unit of the Arthritis and Rheumatism Council in Great Britain reported that rheumatic diseases were second only to respiratory diseases in accounting for lost work in Britain.[18] Evidence at that time supported the view that occupational factors were related to the etiology of some chronic rheumatic diseases, such as tenosynovitis, bursitis, and osteoarthrosis.[19] In 1985, Britain's occupational reporting system was revised.[b] Tenosynovitis and similar conditions were excluded from this revision because of the acknowledged difficulty in diagnosing and defining these conditions. As a result, there are no current national statistics for Great Britain on those individuals seeking benefits for repetitive strain injuries.

---

b. "Reporting of Injuries, Diseases and Dangerous Occurrences Regulation." (Riddor) 1985.

International Perspective: Some CTD Prevention Strategies

From the literature, four countries were identified as having distinct national strategies for work-related musculoskeletal disorders. They are Sweden, Australia, Great Britain, and the United States. Other countries may also have prevention or health promotion activities concerned with these disorders, but the information was either unavailable outside of the country or incomplete. In contrast, almost every country has some form of publicized standards or guidelines for the prevention of back injuries from lifting or materials handling. The literature, however, documenting control procedures for lifting is extensive and beyond the scope of this paper.

Although it is too early to evaluate the success of any of the musculoskeletal strategies, it is instructive to compare the different approaches followed by each country in recognition of the rising incidence rate of work-related musculoskeletal disorders. In general, each country has taken an approach that reflects the scope of the perceived problem and the unique social and economic status of the country. The recent experience of Great Britain has been somewhat different, and points to particular problems in establishing legal-based guidelines without standard definitions and diagnostic criteria. Below is a brief summary of each country's efforts.

Sweden in 1982 issued a regulation with the title "Ordinance concerning work postures and working environment."[20] This ordinance, prepared by the National Board of Occupational Safety and Health in Sweden, represents the first attempt by any country to provide legislation for controlling work-related musculoskeletal disorders. The ordinance consists of seven sections, each specifying one or more provisions concerning work postures, working environment and physical load. Rather than account for every work contingency, the provisions were developed to indicate general principles for the implementation of the Work Environment Act in Sweden. The following excerpt from Section 1 of the ordinance illustrates its content and spirit.

> Work must be designed so as to avoid unnecessarily fatiguing or otherwise strenuous work postures and working movements. Efforts must be made to enable the person doing the work to vary his work posture and working movements. If there is little opportunity of variation, the person doing the work must be given suitably disposed breaks.

Although the ordinance does not contain an ergonomic equivalent of a "threshold limit value," or similar quantitative values for controlling biomechanical stress at the workplace, it does provide a set of useful recommendations and explanations for good ergonomic work practices. The ergonomic ordinance represents one element of Sweden's comprehensive occupational safety and health program that was initiated in 1978 as the Work Environment Act. This act provides a framework for the operation of the National Board of Occupational Safety and Health, which provides the central administrative authority in Sweden for matters of occupational safety and health. The ergonomic ordinance has been used as a basis for information, discussion, and citations. An implementation program has been introduced that reflects the philosophy of the Swedish work environment legislation that "active health and safety should be conducted at company level as the employer's responsibility in cooperation with the employees."[15]

Australia, in a similar fashion, developed a "Model Code of Practice." The code was a response to a "near epidemic of musculoskeletal problems" among several high risk occupations in their country that included packers, sorters, machinists and keyboard operators.[21] In early 1980, a tripartite special commission was formed to draft a report on ways to prevent repetitive strain injury (RSI), the Australian equivalent of work-related musculoskeletal disorders. Worksafe Australia was released in 1986, published for the National Occupational Health and Safety Commission by the Australian Government Publishing Service.

The objective of the Australian Model Code of Practice was to provide a framework for minimizing the risk of, and managing RSI. The proposed prevention strategy consists of recommendations covering eight main elements or factors, including the following: work organization, job and task design, task variation/work pauses, work adjustment periods, workplace and environment design, technology selection and equipment design and education and training.

The philosophy and content of the Australian Code of Practice parallels the Swedish ordinance. For example, Item 1 of the Australian Code stresses the need for work "to be organized so that employees are able to regulate some of the pressures related to their work." Likewise, the Swedish ordinance suggests that "efforts must be made to arrange work in such a way that the employee can influence his own working structure," (Work Environment Act, Ch. 2, Section 1, Subsection 2).

Great Britain has a long legislative history covering the health and safety of workers. Early legislation focused on high-risk work groups. This legislation included: the Woollen and Worsted Textiles Regulation of 1926, the Jute Regulation of 1948, and the Pottery Regulation of 1950. Each of these regulations addressed the general issue of safe manual materials handling. The regulation warned managers not to employ workers in ways that would cause injury to them. In some cases actual work or load limits were specified. The forerunners of the existing Health and Safety Act of 1974 (HSAWA) was the Factories Act of 1961 and the Offices, Shops and Railway Premises Act of 1963. HSAWA did not replace prior regulations, but did change the manner of enforcement and responsibility. More important, HSAWA reflected a shift in the philosophy of regulation from detailed prescriptions to one of broad legal obligations, supported by codes of practice, standards and guidance material. This was intended to allow for a broader range of methods that can be customized to meet the needs of different industries.[22]

It is significant that repetitive strain injuries to the upper extremity are not included in the current reporting system of Great Britain. Nevertheless, musculoskeletal conditions are recognized as a significant occupational health problem by the Medical Division of Britain's Health and Safety Executive. This group is actively engaged in a program of research designed to clarify the epidemiology of these disorders and associated occupational factors. Despite the lack of surveillance data and the absence of specific legal language defining work limits or standards, a "guidance note" on tenosynovitis has been developed to provide information on identifying and controlling repetitive strain injury. Advisory leaflets are also being developed for employees. As more information about the condition is obtained, it is anticipated that prevention programs will be developed and disseminated. This would be in the form of a "code of practice, backed up by framework guidance, the latter encompassing general principles."[22] Each industry could then develop the prevention program that would most closely meet their needs.

In the **United States,** the National Institute for Occupational Safety and Health (NIOSH) completed in 1985 a Proposed National Strategy for Preventing Musculoskeletal Injuries.[23,24] This document was introduced, discussed, and revised at a national symposium held in cooperation with the Association of Schools of Public Health.[c] Unlike the Australian and Swedish documents, the U.S. musculoskeletal document focussed on the development of a plan or strategy for preventing work-related musculoskeletal disorders, rather than providing detailed guidelines, recommendations for intervention, or enforceable standards. This key distinction also reflects the division of responsibilities among the U.S. occupational safety and health organizations as provided by Public Law 91-596.

Public Law 91-596, enacted in 1970, established three organizations concerned with occupational safety and health in the federal government. They include: the Occupational Safety and Health Administration (OSHA) in the Department of Labor, responsible for setting standards, inspections and compliance; the National Institute for Occupational Safety and Health (NIOSH) in the Department of Health and Human and Services, responsible for research, information dissemination, training and developing recommendations or guidelines for OSHA's consideration; and the Occupational Safety and Health Review Commission, a three-member judicial panel whose principal function is to hear appeals from employers who wish to contest citations issued by OSHA.

Within this framework, NIOSH developed a musculoskeletal strategy that was directed primarily at research and informational needs. The strategies included: (1) better surveillance and diagnostic information for identifying the disorders, (2) applied research to develop predictive models, (3) expansion of the role of the public and private sectors to define and

---

[c]. The musculoskeletal strategy was one of five occupational health strategies prepared and presented in 1985 at a national symposium held in Atlanta, GA.

implement ergonomic interventions, and (4) the development and dissemination of user-oriented guides to labor, management and the OSHA field staff. The guides are designed to provide methods for detecting and controlling work-related musculoskeletal disorders.

As noted in the U.S., there are no special ordinances or OSHA standards for cumulative truama disorders.[25] Hence the prevention of these disorders, as reflected in the NIOSH strategy document, will occur voluntarily when both management and workers recognize that it is to their mutual benefit, from both a health and economic standpoint, to cooperate in reducing work-related injury and lost time. As indicated, one way to facilitate that process is to provide the necessary information about the cause and prevention of of these disorders. An example of this is NIOSH's recently completed manual on Cumulative Trauma Disorders.[26] A main conclusion of this user-oriented guide was that musculoskeletal disorders can be controlled at the workplace by using ergonomic principles to redesign jobs and tools. Successful interventions, however, usually require the combined participation of plant managers and workers.

Conclusion

By the mid-1980's, almost every industrialized country had instituted some type of program to protect the safety and health of their workers. These programs, however, typically do not address work-related musculoskeletal disorders affecting the upper limbs. Countries that have taken steps to prevent these disorders, either by developing a specific national strategy or by disseminating manuals or guidance documents, have increased the awareness of these problems within their country. This is the first step toward prevention. From an international perspective, the different labels and overlapping terms used by different countries to refer to these disorders may have impeded critical scientific interchange and research. As an example, work-related musculoskeletal disorders are not even included as a group in the amended list of occupational diseases provided by the International Labor Organization. Moreover, of the 4,000 standards published by the International Organization of Standards, only those on seating and vibration are directly related to the prevention of these disorders.

One solution is to hold an international meeting with a specific agenda that will address the following main issues:

1. The need to clarify the redundant and poorly defined terminology.
2. The need to identify or define the etiology of these disorders.
3. The need to develop common diagnostic standards and outcome variables for surveillance systems.

This meeting should include researchers, practitioners, and government health officials from interested countries. If interest is sufficient, the committee could serve as an international advisory body to promote clinical and scientific consensus related to (1) definitions of work-related musculoskeletal disorders, (2) diagnostic criteria for musculoskeletal disorders, and (3) the development of uniform surveillance methods and outcome criteria.

REFERENCES

1. The International Classification of Diseases, 9th Revision, DHHS Publication No. (PHS) 80-1260, U.S. Government Printing Office. Vol 1, 1980.

2. Ramazinni, B., 1771, In Wright, W. (trans.): The Diseases of Workers. Chicago, University of Chicago Press, 1940.

3. "Golden arches not blast furnaces symbolize the current economy." Newsweek, February 16, 1981.

4. Radin, E.L., Mechanical aspects of osteoarthrosis. Bulletin on the Rheum. Diseases, 26:7, 862-865, 1976.

5. Hoffman, M.S. (Editor) The World Almanac and Book of Facts. Pharos Books, New York, N.Y., p. 218, 1987.

6. Silverstein, B.A., Fine, L.J., and Armstrong, T.J., Hand wrist cumulative trauma disorders in industry. Br. J. of Ind. Med. 43, 779-784, 1986.

7. Hadler, N. M., Medical ramifications of the federal regulation of the social security disability insurance program. Annals of Internal Medicine. 96 (5), 665-669, 1982.

8. Armstrong, T.J., Fine, L.J., and Silverstein, B.A., Occupational Risk Factors: Cumulative Trauma Disorders of the Hand and Wrist. Final Report on NIOSH Contract #200-82-2507, Cincinnati, Ohio, 1985.

9. Armstrong, T.J., Ergonomics and cumulative trauma disorders. Hand Clinics. 2(3):553-565, 1986.

10. Jensen, R.C., Klein, B.P., Sanderson, L.M. Motion related disorders traced to industries, occupational groups. Monthly Labor Review. Sept. 13-16, 1983.

11. Habes, D.J., and Putz-Anderson, V., The NIOSH program for evaluating biomechanical hazards in the workplace. J. of Safety Research. 16(2):49-60, 1985.

12. Tanaka, S., Seligman, P. Halperin, H., Thun, M. Use of worker's compensation claims data for surveillance of cumulative trauma disorders. In press: J. of Occupational Medicine. 1988.

13. Schwartz, E., Use of workers' compensations claims for surveillance of work-related illness - New Hampshire. MMWR, 36,(43):713-720, 1987.

14. Bureau of Labor Statistics: Occupational Injuries and Illnesses in the United States by Industry - 1985. U.S. Government Printing Office, Washington, D.C. 1987.

15. Nilsson, B., Bjurvald, M., Stjernberg, K., Implementation programme of the Swedish ordinance concerning work postures and working movements. Ergonomics. 30:(2)431-436, 1987.

16. National Board of Occupational Safety and Health. Ordinance Concerning Work Postures and Working Movements, (Stockholm: National Board of Occupational Safety and Health) AFS, (6) 1982

17. Ferguson, D., The "new" industrial epidemic. The Medical Journal of Australia. 318-319, March, 1984.

18. Anderson, J.A.D., Occupation as a modifying factor in the diagnosis and treatment of rheumatic diseases. Current Medical Research and Opinion. 2(9):521-528, 1974.

19. Anderson, J.A.D., Rheumatism in industry: a review. British J. Industrial Medicine. 28:103-121. 1971.

20. Danielson, G., Edstrom, R., and Lindh, G., Ordinance concerning Work Postures and Working Movements, National Board of Occupational Safety and Health. Stockholm, Sweden. 1984.

21. National Occupational Health and Safety Commission. "Repetitive Strain Injury (RSI): A Report and Model Code of Practice." Worksafe Australia. Australian Government Publishing Service, Canberra, 1986.

22. Edwards, F.C., Prevention of musculoskeletal injuries in the workplace National Approaches to safety standards: Great Britain. Ergonomics. 30(2):411-417, 1987.

23. Proposed National Strategies for the Prevention of Leading Work-Related Diseases and Injuries, Part 1. Washington, D. C.: the Association of Schools of Public Health under a cooperative agreement with the National Institute for Occupational Safety and Health, 1986. (NTIS (ID # 87 114 740).

24. Bucsela, J. [Editor] Summary of Proposed National Strategies for the Prevention of Leading Work-Related Diseases and Injuries. In press: American Journal of Industrial Medicine, 1988.

25. U.S. Department of Labor, OSHA Safety and Health Standards (29 CFR 1910), OSHA 2206, General Industry Standards, 1976.

26. Putz-Anderson, V. [Editor] Cumulative Trauma Disorders: A Manual for Musculoskeletal Disorders of the Upper Limbs, Taylor and Francis, 1988.

Trends in Ergonomics/Human Factors V
F. Aghazadeh (Editor)
© Elsevier Science Publishers B.V. (North-Holland), 1988

ROBOTS AS AN ERGONOMIC CONTROL TO MUSCULOSKELETAL INJURIES IN THE WORKPLACE

A.M. GENAIDY*, G. RONCINI*, R. DAWOOD*, AND A. MITAL**

*Department of Industrial Engineering
Western Michigan University
Kalamazoo, Michigan 49008, U.S.A.

**
Ergonomics Research Laboratory
University of Cincinnati
Dept. of Mechanical and Industrial Eng.
Cincinnati, Ohio 45221-0072, U.S.A.

Physical tasks account for a large number of musculoskeletal injuries such as back injuries and cumulative trauma disorders. Engineering controls should be employed to eliminate and/or minimize these injuries. One such control is the use of robots. The main objective of this study is to examine the capabilities and performances of both humans and robots.

1. INTRODUCTION

Statistics reported by researchers, federal agencies, and insurance companies show that physical tasks performed by humans account for a large number of workplace injuries [1-7]. Back injuries and cumulative trauma disorders (CTDs) are examples of workplace injuries.

Approximately, about 35% of all workers' compensation claims are related to back injuries [8] and an estimated $14 billion is paid annually in direct financial compensation [9]. The direct costs may be as much as four times this amount [6]. Based on estimates derived from the Bureau of Labor Statitics [10] about 1 million workers suffered back injuries in 1980.

CTDs are being recognized as a leading cause of economic burden on compensation systems, loss of productivity, and human suffering [11]. More than half of the workers in the United States have jobs with potential for the CTDs [7]. Major categories include construction, services, manufacturing, and clerical [12]. Habes and Anderson [13] reported that with the proliferation of assembly-line techniques, the increasing tempo of production and the widespread use of vibrating and air-powered tools, CTDs have become a fact of industrial life.

Reduction in the number of musculoskeletal injuries associated with physical tasks has, therefore, become a major of many governmental agencies, insurance companies, as well as researchers.

2. OBJECTIVES

NIOSH [1] reported that lifting tasks may fall within three categories based on the concept of action limit (AL) and maximum permissible limit (MPL). Lifting tasks above MPL are viewed as unacceptable to most workers and require engineering controls. Lifting tasks below AL represent nominal risk to most industrial workers. Lifting tasks between AL and MPL are unacceptable without administrative or engineering controls. NIOSH cited mechanical aids as one of the engineering controls. According to this guide, mechanical aids may encompass: conveyors, cranes and hoists, industrial trucks, hooks, bars, rollers, jacks, platforms, and trestles. One sort of mechanical aid that is not reported in the NIOSH guide is industrial robots. Industrial robots have the potential to replace humans in injurious environment specially the ones leading to back injuries and CTDs.

To ensure the success of industrial robots in injurious work environments, they should be designed to simulate the motions performed by humans in various manual operations. The main objective of this research is to examine human and robot capabilities and performances. Such knowledge would help designers to build future robots that simulate the functional abilities of human operators.

3. BACKGROUND

3.1. Definition

The Robot Institute of America defined a robot as "a reprogrammable, multifunctional manipulator designed to move material, parts, tools, or specialized devices through variable programmed motions for the performance of a variety of tasks" [14].

3.2. Applications

Robots have been implemented successfully in various industries [14-15]. Materials handling is one of the most common applications of industrial robots where robots are used to advantage in handling heavy or fragile parts as well as parts that are very hot or very cold. Further, programmable assembly systems utilizing robots can effectively automate low to medium volume and batch assembly operations since assembly operations are a major use of labor in manufacturing. Currently, few programmable assembly systems are in operation in production.

3.3. Human Vs Robots

Paul and Nof [16] compared the capabilities of a robot and a human operator as defined by methods-time measurement "MTM" [17]. The results of the comparison are summarized in table 1. The reach, move, grasp, and position elements were only considered in the analysis. Paul and Nof pointed out that the MTM elements RC, MA, G1C, G2, AND G4 are impossible to perform by a robot. On the other hand, the authors defined the following elements as suitable for the robots while highly inefficient and may be impossible to perform by a human operator: (1) spatial reach defined as move to an absolute position and orientation in space such as storage grid, (2) spatial move defined as move an object to an exact postion and orientation in space, and (3) measure grasp defined as combined grasp and measurement of an object. It should be noted that robot work elements can be translated into program statements. Accordingly, the robot can perform a given operation.

Table 1. Comaprison of MTM and Robot Motions [16]

Element	Case	MTM Description	Robot Equivalent
Reach	A	Reach to object in fixed location or to object in other hand or on which hand rests	Move end-effector directly to given position and orientation
	B	Reach to single object in location which may vary slightly from cycle to cycle	Move to a position close to object; refine position by touch or force feedback
	C	Reach to object jumbled with other objects in a group so that search and select occur	Highly inefficient for robot
	D	Reach to a very small object or where accurate grasp is required	Move to a position close to object;re-establish position of object by vision, refine position by touch or force feedback

Table 1. Cont'd

Element	Case	MTM Description	Robot Equivalent
Reach	E	Reach to indefinite location to get hand in position for body balance or next motion or out of way.	Indefinite movement to get out of way is equivalent to class A for robot, otherwise serves no purpose
Move	A	Move object to other hand or against stop	Cannot move an object against a stop without deccelera- tion (use class C instead)
	B	Move object to approximate or indefinite location	Move object to some given position with given acceleration and decceleration
	C	Move object to exact location	Move object close to given position and orientation; move to final position by force or vision feedback
Grasp	G1A	Small pickup-small, medium or large grasp by itself, easily grasped	Close hand
	G1B	Very small object or object lying close against a flat surface	Locate surface by force, close hand
	G1C1	Interference with grasp on bottom and one side of nearly cylindrical object. Diameter larger than 1/2"	Free object from interference, relocate it and close hand
	G1C2,3	Interference with grasp on bottom and one side of nearly cylindrical object. Diameter less than 1/2"	Impossible for robot
	G2	Regrasp	Impossible for robot

Table 1. Cont'd

Element	Case	MTM Description	Robot Equivalent
Grasp	G3	Transfer grasp	Position object in other hand, release (possibly when two hands are available
	G4	Object jumbled with other objects so search and select occur	Highly inefficient for robot
	G5	Contact, sliding, or hook grasp	Close hand while in motion

Element	MTM Definition Class of Fit	Symmetry	Robot Equivalent
Position	1-loose, no pressure required	S SS NS	Perform initial insertion of object into opening; then insert completely
	2-close, light pressure required	S SS NS S	Insert edge of object into opening; align object; then insert completely
	3-exact, heavy pressure required	SS NS	Insert edge of object into corner of opening; insert edge; align object; then insert completely

Based on the aforementioned discussion of MTM elements versus robot capabilities, Paul and Nof developed a method called robot time and motion (RTM). RTM consists of eight elements categorized into reach, stop on error, stop on force or stop on touch, move, grasp, release, vision, and process time delay (table 2). The standard time of the whole operation is the sum of the time values assigned to the various elements making the operation. RTM differs from MTM in that the elements are based on the physical parameters of the robot, maximum torque, resolution, sensors, etc. leading to an exact method of predicting task times. MTM, on the other hand, take human variability into account in providing estimates of elements times.

Wygant and Donaghey [18] reported that RTM does not provide a convenient procedure to convert estimated times for manual tasks into robot performance time estimates. Wygant [19]

Table 2. Description of RTM Elements

Element	Symbol	Description
REACH	Rn	Describes the motion of an empty hand to a position. Since a robot may not always reach directly, it may have to move through a series of (n-1) intermediate points before it reaches the final position.
STOP ON ERROR	SE	Describes the manipulator being brought to rest within a given position error balance.
STOP ON FORCE and STOP ON TOUCH	SF ST	Describes the manipulator being brought to rest by force or touch sensing.
MOVE	Mn	Describes the motion of a loaded hand to a position. Mn is identical to Rn except that its time is increased depending on the load.
GRASP	GR	Describes closing the fingers. This element is used to grasp an object in a given position. It also describes closing the hand.
RELEASE	RE	Describes opening the fingers.
VISION	VI	Describes the robot obtaining visual input. It is usually used to identify and locate objects and their features.
PROCESS TIME DELAY	TI	Specifies unavoidable delays delays during which the robot must wait.

developed a method termed ROBOT MOST which is based on the concept of the MOST predetermined motion time system [20]. ROBOT MOST can be used to predict robot times for various operations. Wygant and Donaghey [19] developed a computer program that can be used in selecting a robot which is capable of performing a given task as specified by parameters such as maximum horizontal reach, maximum load, repeatability, and whether or not continuous path motion is required. The program hosts a data base that includes operating specificatons of 334 individual robot models.

Nof et al [21] pointed out that human operators and robots differ in numerous characteristics within the context of job design. These characteristics are: (1) robots are unaffected by social and psychological effects; (2) for humans, training and retraining are more difficult; (3) humans possess a set of accumulated skills and experience; (4) robots do not have any significant individual differences; (5) robot abilities can be structured for a particular task to a greater degree than can a human' abilities.

Paul and Nof [16] compared the performances of both human operator and a robot for a pump assembly task. The robot performance required more than eight times the estimated human performance performance time. Wygant and Donaghay [17] reported that the total time for drilling a 1/2" hole in casting was 0.22 and 0.50 min for humans and robots, respectively. Based on unpublished results by the authors, it was found that the average times for assembling two squared blocks were about 10.01 sec and 2.95 sec for robots and humans, respectively.

4. DISCUSSION

The literature search conducted in this study reveals the following points: (1) Robots require no learning on the job while humans learn on the job; robots can be taught, through programming, to do the job more consistently than humans; and (2) Robots are slower than humans, however, they can work more than one shift a day ; if humans do work more than one shift, then they will be overstressed; with the advancement of technology, a robot may overcome this limitation. These findings may assist in making a decision toward selecting a robot as opposed to a human operator. Having the advantage of no learning, long hours of operation, no absenteeism, and more importantly the elimination of work injuries strongly support the use of robots in the workplace. An ergonomic control of this sort should be the first choice of the industrial manager as well as practitioners of safety and health. This is very true in industries where there is a heavy use of hand tools such as knives at high repetitions, often associated with excessive manual force and unnatural postures of the various parts of the upper extremity. Meat and poultry industries are typical examples of such an environment. It is recommended that the documented work methods of a manual operation be used in conjunction with a selected or designed robot for its programming to do specific jobs, thus eliminating the CTDs.

In sum, it is concluded from this study that research is warranted to (1) design robots having human skills and capabilities, (2) develop robot time prediction models for various tasks, and (3) to develop a knowledge base of different types of robots to fit the robot to the task.

5. REFERENCES

[1] National Institute for Occupational Safety and Health,

1981. Work Practices Guide for Manual, Cincinnati, Ohio.
[2] Chaffin, D. B., Herrin, G. D., Keyserling, W. M., Foulke, J. A., 1977. Pre-Employement Strength Testing. DHEW (NIOSH), Publication No. 77-163, Cincinnati, Ohio,
[3] Snook, S. H., 1978. Ergonomics, 21: 963.
[4] Ayoub, M. M., Bethea, N. J., Deivanayagam, S., Asfour, S. S., Bakken, G. M., Liles, D. H., Mital, A., and Sherif, M., 1978. Determination and Modeling of Lifting Capacity. DHEW (NIOSH), Grant 5 RO1-OH-00545-02.
[5] Ayoub, M. M., Asfour, S. S., Bakken, G. M., Bethea, N. J., Liles, D. H., and Selan, J., 1983. Effects of Task Variables on Lifting Capacity. DHHS (NIOSH), Grant No. 5RO-OG-00798-04.
[6] Asfour, S. S., Khalil, T. M., Moty, E. A., Steele, R., and Rosomoff, H. L., 1983. Back Pain: A Challenge to Productivity Proceedings of the VIIth International Conference on Production Research, Windsor, Canada.
[7] Putz-Andersen, V. (Editor), August, 1987. Cumulative Trauma Disorders Manual for the Upper Extremity. NIOSH, Cincinnati, Ohio.
[8] National Safety Council, 1983. Accident Facts. Chicago.
[9] Taber, M., 1982. Reconstructing the Scene, Back Injury. Occupational Safety and Health, 51: 16.
[10] Bureau of Labor Statistics, 1982. Back Injuries Associated With Lifting. U. S. Department of Labor, Bulletin 2144, Washington, D. C.
[11] Kesley, J. L., White, A.A., Pastides, H., and Bisbee, G. E., Jr., 1979. J. of Bone and Joint Surg., 61A (7): 959.
[12] Bureau of Industrial Economics, 1984. U. S. Industrial Outlook: Prospects for Over 3000 Industries. U. S. Department of Commerce, Washington, D. C.
[13] Habes, D. and Putz-Andersen, V., 1985. Journal of Safety Research, 16 (2): 49.
[14] Engelberger, J. F., 1980. Robotics in Practice. American Management Association, Saranac Lake, New York.
[15] Tanner, W. (Editor), 1983. Industrial Robots: Volume II/Applications. Robotics International of SME, Society of Manufacturing Engineers, Dearborn, Michigan.
[16] Paul, R. P. and Nof, S. Y., 1979. International Journal of Production Research, 17 (3): 277.
[17] Maynard, H. B., Stegmerten, G. J., and Schwab, J. L., 1948. Methods-Time Measurement. McGraw-Hill, New York.
[18] Wygant, R. M. and Donaghey, C. E., 1987. Proceedings of the Third International Robotic Systems Education and Training Conference, Society of Manufacturing Engineers, Detroit, Michigan.
[19] Wygant, R. M., 1986. Robots vs Humans in Performing Industrial Tasks: A Comparison of Capabilities and Limitations. Unpublished Ph. D. Dissertation, University of Houston, Houston, Texas.
[20] Zandin, K. B., 1980. MOST Work Measurement Systems. Marcel Dekker, New York.
[21] Nof, S.Y., Knight, J. L., Jr., and Salvendy, G., 1980. AIIE Transactions, 12 (3): 216.

Trends in Ergonomics/Human Factors V
F. Aghazadeh (Editor)

Unsafe Behavioral Responses of Agricultural Equipment Operators to Interruption of Machine Function

R. Lewis Hull

Human Factors and Product Safety Engineering Research
Advance Technology, Inc.
Wichita, Kansas

The typical operator response to task interruption with agricultural equipment is indecision and learning unsafe behavior if experience does not indicate adverse results or if equipment has not been designed to assure safe user performance. To understand the behavioral responses common to agricultural equipment operators, this research examines three areas: 1) the occasion upon which a response occurs; 2) the response itself; and 3) the consequences. With agricultural equipment, task interruption most often involves the functional aspects of the equipment. This can provoke an operator into taking an action, which may not readily be perceived to be hazardous, to expedite job completion. All too often the consequences of the action results in injury. Therein, there exists the probability that given behavior can be predicted to occur at a given time based upon an understanding of classical operator responses to defined situations.

1. INTRODUCTION

Many generalizations relating to accident causation with agricultural equipment may imply that a machine operator is a hapless victim of his own actions. However, the machine contribution to accidents and resultant injuries is a significant factor in the extremely high agricultural injury and death rate as discussed by McKnight (1984), Sevart and Hull (1985), Hull (1987), and Purschwitz and Field (1987). Impetus for this discussion is to further the study of Duncan (1981) concerning operator behavior in response to interruption of machine function, and the assessments of Aherin and Murphy (1987) contained in, "Impact of Operator Training on Reducing Losses". The objective is to ascertain some elements of human factors analysis that determine the extent to which the operator of a machine can control hazardous situations, if typical behavioral responses are considered in the design of agricultural equipment.

There are numerous variables of behavior on which the probabilities of operator responses are dependant. The stimulus of task interruption is, of course, the important independent variable for this discussion. Other important variables are found in such areas as motivation and emotion. An operator of a machine "learns" to respond by his own previous behaviors as he "learns" to respond to the machine and the environment of use. Interruptions because of equipment malfunction will occur and the human factors design specialist must direct the operator's behavior with proper equipment design strategies. The practical objective to the study of variables of operator behavior is to develop enough knowledge to predict the variations, within limits, if the nature of certain factors upon which variability depends are changed. The problem then is to classify groups of variables of behavior so that the one variable under study can be controlled within limits that allow evaluation. Smith (1979) offers some assistance with functional behavior classification;

> "all the sensory or perceiving functions can be considered under the category of perceptual skills (vision, audition, tactile discrimination, kinesthetics); all "doing" functions under the category of motor skills; and "judging" functions under diverse categories like behavior and stress responses, and fatigue."

Therefore, the difficulty is to analyze basic behavior and classify functional response in terms which parallel hazardous actions of agricultural equipment operators which result from task interruption.

2. AGRICULTURAL EQUIPMENT FUNCTIONAL INTERRUPTION

Agricultural equipment operators often are confronted with work task interruptions because of a machine system malfunction. Many times these interruptions occur because the machine system fails to process crop being harvested when the material becomes clogged in the mechanism thus requiring the operator to remove the blockage. An operator has no choice but to respond to this type of functional problem with some action before the work task can be completed. The operator, through "learning" or response acquisition, will discover that the procedure can be expedited using the machine power to assist in clearing the blockage. Since the functional interruption of the machine was remedied, since the operator was "careful", and since no accident transpired there is motivational reinforcement to continue a practice which agricultural specialists know to be extremely hazardous.

It is necessary then for agricultural equipment designers to recognize that functional interruptions are characteristics of the machine system with which the operator must contend. The equipment design must provide an operator with acceptable options for safely correcting a system malfunction. The question of operator behavior related to interruption of agricultural equipment function is non-exclusive of the question of operator feedback from the reinforcing motivation of task completion as generated by the machine system. An interactive machine system that assures the achievement of the system goals must depend upon successful system performance which, in turn, relies upon positive, productive interactions between the operator, the equipment, and the environment in which the activity takes place. Operator activity in response to interruptions will depend on both the equipment design and prior operator behavior.

3. OPERATOR BEHAVIORAL RESPONSES

The behavior of an agricultural equipment operator while interacting with a machine system can most often be explained; when a response occurs and is then reinforced, the probability of that particular response occurring again, under similar stimuli, is increased. However, for most agricultural equipment operators, this type of reinforced behavior can have both positive and negative consequences. When an operator responds to the interruption of machine function, the reinforcing reward for a particular action will most often be work task completion, which will be considered by the operator as positive, but significance must be placed on possible adverse consequences. A relation between behavior and task completion can be implied, when an agricultural equipment operator with appreciable knowledge about a machine system, possibly inferred from experience, does not respond adequately. Importance must be given to behavioral performance based upon incentives. Within this context of task interruption, the concepts of motivation, frustration and conflict are important considerations when an inadequate response occurs. The process of operator decisions which lead to inadequate responses, that results in unsafe conditions, must be countered by human factors stratagem to assure safe operational responses are reinforced.

When an operator is making a decision concerning different actions, the ultimate decision will more likely be made based upon an action that can maximize the goal objectives, that is, as the operator understands those choices. Busemeyer (1985) distinguishes among three classes of decisional situations as being decisions made under conditions of certainty, risk, and uncertainty. "Certainty" decisions have a single known outcome based upon each action. "Risk" decisions have a set of multiple outcomes based upon each action and the possibility of each outcome is known. While, "Uncertainty" decisions have a set of multiple outcomes based upon each action and the possibility of each outcome is not known. The decisional situation for an operator of agricultural equipment during functional interruption is somewhere between the risk and the uncertainty class of decision. Hasher and Zacks (1984) combine with this to explain that; when people make decisions under conditions of partial uncertainty, they usually rely upon alternatives to which they have been frequently exposed to in the past, or upon past experiences in similar situation.

4. CONSEQUENCES OF OPERATOR RESPONSES

The human factors approach to hazard analysis of functional interruption with agricultural machine systems can be based on two ruling concepts: 1) the intrinsic hazards of machine operation originate emergencies which can lead to accidents with different levels of injury severity; 2) the accidents themselves are dynamic events which create their own operational hazards that must be detected and controlled to determine the course of, and limit the injury severity of, an accident. In contrast to some approaches to accident analysis and research, based largely on the search for physical defects and worker negligence, the hazard and risk control approach views accidents and emergencies as dynamic events involving progressive injury potential which requires incorporation of human factors concepts which seek to limit the severity of injury resulting from an accident. Agricultural equipment accidents are basically operational in character and are in part manageable, if not preventable. Near and minor accidents may reflect a high degree of operator skill in controlling an emergency situation where an operator has "learned" to cope with hazardous equipment and avoid injury. The machine and situational feedback effects have marked influences on operator performance even in accidents with severe injury potential.

5. HUMAN ENGINEERED MACHINE SYSTEMS

Human factors ideology must accept the evolutionary role of technology and social process in understanding the basic psychological capacities and behavioral responses of man. Human engineering applies human factors theory to the design of machines and other technological systems. Human engineering when combined with system control has the objectives to: 1) differentiate system components and system controls; 2) to specialize component and controller activities; and 3) to establish a hierarchical dominance and subordination of controls in performing different functions with respect to maintaining integration of the system, reacting to environmental conditions, and generating specialized activities. Human factors theory has no alternative than to assume that safety will satisfy system requirements for the operator. Human factors methods emphasize the preventive approach, seeking to design the conditions and the processes of operational control of machines so as to detect, monitor, and reduce the risk associated with hazards before accidents occur. Within these concepts, the objective is to analyze conditions which influence agricultural equipment operators to respond to functional interruptions of equipment with unsafe behavior.

The more complex a machine system: 1) the more difficult the operation; 2) the greater the need for operational "learning"; 3) the less precise and safe the final level of performance; 4) and the more specialized the operational performance must be. With the fact that control performance is specialized, in relation to the particular task situation and to the particular machine, operator ability may not be determined primarily by system "learning". That is, the potential of transferring or generalizing skill performance from one machine to another may not always be determined by operational "learning" on a particular machine or task situation. Rather, it is determined by similarity among critical feedback factors of prior "learning". The critical feedback factors are crucial to the human factors design of the control operations as the machine is applied to particular situations.

When control functions are assigned to man in a control system it is certain that, unlike other system elements of control, a human controller can introduce many undesirable response features. The most obvious of these is response delay consisting of perceptual time, discrimination time, and reaction time. Equally as apparent is the unpredictability of motivation, fatigue, and intellectual distractions, unique in the human element and requiring considerable attention for safe system performance. Safety performance for the operator is determined in the same way as other aspects of the system. Safety performance is inherently dynamic.

Hazards in machine operation are also dynamic and operational. Hazards occur as an intrinsic aspect of machine performance which reflect interactions between the operational design of the machine, the control situation, and the specialized performance of an operator. Performance must be directed by developing techniques of detecting, monitoring, and regulating machine hazards to eliminate or reduce injury. There are no generalities of safe performance with machines in controlling hazards except to provide the operator with a full understanding of the design and operation of a machine which is already designed for minimal risk of injury. Therefore, through the human factors design of the machine, the operator must be provided with circumstance to "learn" the necessary skill to meet the performance requirements of using the machine efficiently, as well as safely.

6. SYSTEM DESIGN

Agricultural machine systems have four major components: 1) a functional or operational component which cuts, processes, separates, or performs some other operation; 2) a control component which relates to the operator's motor activities; 3) an actuator or power component which converts operator control into output with some magnification or transformation of time, range, or force of operator movement; and 4) an integrative component which links the operator's control action with the operational action so as to produce the performance of the machine. The design of the functional components of a machine are significant for safe operation of a machine. To insure safety, operational components must be designed to fit the objectives and environment related to the specific function of the device and to provide effective sensory feedback to the operator. Christensen and Howard (1985) reference Ramsey's "Accident Sequence Model" which illustrates some of operator feedback considerations when designing system and control concepts.

1) Perception and Recognition - The hazard or hazardous condition may or may not be perceived and even if perceived may not be recognized as a hazard.

2) Decision Making - After recognizing a hazard, the individual has to decide whether or not to avoid the hazard. Is he willing to take the risk, even if he knows what it is? Obviously, attitude and motivational considerations come very much into play at this stage.

3) Capability - Even after deciding to avoid the hazard (i.e., not take the risk), the individual must be capable of developing and executing the required avoidance responses.

A machine system interactively alters the feedback that an operator gets from his own motions. This process determines specialization of performance, physiological regulation, and understanding of the machine system.

Human factor inquiry concerning behavior consist essentially of investigating the time, range, and force feedback properties and compliances between motor response and sensory processes. In normal behavior, the individual receives two or more types of feedback. Perceptive feedback is that produced by a movement itself and sensed kinesthethetically, tactually, visually, or in other ways. Operational feedback is the perceived effect of the body movement on the environment. When a machine is used, the individual receives feedback from the machine that is different from that produced by body movement or from the environmental effect. A human factors principle for effective performance in both reinforced and non-reinforced behavior is that perceptive and operational feedback effects should be as compliant as possible with body movement in time, range, and force properties. Visual, auditory, tacitly, and kinesthetic perception are dependent on dynamic motor control of sensory input and receptor sensitivity, especially in different aspects of attention and sustained motivated attention. What this means is that all aspects of perception and attention in work are active behaviors controlled and guided by motor-sensory feedback processes. Problems of vigilance, monotony, perceptual loading, noise interference, and the job hazards and stress connected with them can be approached best from a human factors point of view by understanding the dynamic motor-sensory mechanisms of perception.

7. OPERATIONAL WARNING EXAMPLE

The operational safety warning or "Caution Limit" is the final "limit" where the designer can exhibit direct control over an operational hazard at the time of functional interruption. It is the communication link between the knowledge which the designer has concerning the hazard of a machine system and the knowledge which the operator needs to control the situation and avoid this hazard. Baker and Aherin (1987) explain:

> "The primary purpose of product safety communication is to motivate the product user to follow recommended safe work practices in the utilization of the product so as to minimize the potential of injury."

Rogers (1975) describes three crucial components to accomplish this communication link between the designer and the operator of a machine system:

1. The magnitude of noxiousness of the depicted event and personal relevancy of the event.

2. The probability of the occurrence of the event.

3. The efficacy of the protective response.

Safety warnings or "Caution Limits" must incorporate human factor design principles as should other components of a machine system that influence behavioral motivation.

A warning is essentially a notice that a hazard exists as an operational condition requiring system interaction for avoidance. For the designer, after a warning has been designed for system requirements, some method such as behavioral response analysis, must assess the effectiveness of that warning. That is, even though the warning may be perceived by an operator, there is no assurance that the perception will be followed by appropriate cognitive action. At the cognitive level, cues, provided by the warning, must interact with such factors as past experience, attitudes, and perceived urgency. Robinson (1977) expands these ideas in the study titled, " Human Performance in Accident Causation: Towards Theories on Warning Systems and Hazard Appreciation", with the reasoning:

> "Progress in reducing accidents has been materially impeded by two problem: 1) the "motivational" attitudes towards the human, as opposed to the performance limitation view; and 2) the lack of human performance theory, models, and data applicable to accident causation."

Then continuing;

> "We see two distinct types of failure causes: 1) the failure of the warning to capture the operator's attention; and 2) the failure of the operator to execute the proper response."

Even though a warning is incorporated into a system design, even though a "skilled" operator recognizes a hazard, and even though the operator is capable of responding, the behavioral motivation for cognitive action can be hidden within the prior "learning" experience.

Safety messages can be most effective when presented at a time and under conditions where the operator understands the importance of altering operational patterns. Without this understanding, a safety message and repetition of a safety message will likely have a diminutive effect. Warnings not presented timely and under conditions of understanding will sometime be ignored to accomplish task completion and develop a pattern of an unsafe behavior in response to a particular stimulus such as task interruption. Possibly the best approach for assuring a safe response is to suspend the operational processing of information by encroaching upon the mechanism of attention. Thus, an effective warning for safe operator performance when confronted with functional interruption of agricultural equipment must: 1) be recognized by the operator; 2) invoke a desirable response; 3) and become an operational behavior. An important consideration for directing the conduct of an operator is that once a functional interruption occurs a search through memory will occasion an appropriate action based upon a learned and practiced response.

Warning signs should provide the operator with means, to identify, interpret, and the procedure to follow, to avoid hazards of the system. The characteristics of a warning should be determined by such factors as: 1) the accuracy of identification required; 2) the time available for recognition or other response; 3) the distance and location at which the warning must communicate; and 4) the criticality of the warning for safe performance. Cunitz (1981) relates, "The ultimate test of the effectiveness of a warning involves determining whether or not a warning actually changes the behavior of the people who are to be warned." A warning must give the operator the information (hazard level, hazard identification, procedure, and consequences) needed for the safe operation of the equipment. However, warnings must do much more in the areas of emotion, motivation, incentive, and stimulus consistent with a behavioral response which is intuitive toward hazard avoidance. Through the application of basic human factors principles and methods the necessary information of a warning can be conveyed so as to ensure that interpretation and response is timely, safe, and motivational toward task completion.

8. SUMMARY

In understanding unsafe behavioral response when functional interruptions occur, some assumptions and interpretations regarding the underlying potential cause of accidents and factors that contribute to their occurrence and severity must be formulated. Emergency and accident situations must be anticipated by the designer of agricultural equipment. However, hazards must not be left by the designer, and then make it the operators responsibility to anticipate accident situations. Accidents are operational and at some level involve system failure or inadequacy of control to meet the developing accident situation. Operational weaknesses of the system, in most cases, cause accidents. Accidents are not a single event, but rather dynamically developing events with progressive deterioration and loss of operational control of a machine or the situation. Accidents develop as specialized events in relation to the human factors considerations of agricultural equipment design within a given situation where an operator must respond correctly to avoid injury. It is the human factors understanding of unsafe behavioral responses of agricultural equipment operators to the interruption of machine function that will reduce the number of deaths and serious injuries.

REFERENCES

1. Aherin, R. A. and Murphy, D. J., "Impact of Operator Training on Reducing Losses", ASAE Paper Number 87-5528, American Society of Agricultural Engineers, December, 1987.

2. Baker, L. D. and Aherin, R. A., "Creating Effective Product Safety Messages", ASAE Paper Number 87-5515, American Society of Agricultural Engineers, December, 1987.

3. Busemeyer, J. R., "Decision Making Under Uncertainty: A Comparison of Simple Scalability, Fixed Sampling, and Sequential Sampling Models", Journal of Experimental Psychology, March, 1985.

4. Christensen, J. M., and Howard, J. M., "Human Factors Concepts and Methods for Agricultural Engineers", American Society of Agricultural Engineers Seminar, December, 1985.

5. Cunitz, R. J., "Psychologically Effective Warnings", Hazard Prevention, Volume 17, Number 3, May/June 1981.

6. Duncan, J. R.., "Operator Behavior: Responses to Interruptions of a Machine Function", Technical Memorandum No. 536, June, 1981.

7. Hasher, L. and Zacks, R. T., "Automatic Processing of Fundamental Information: The Case of Frequency of Occurrence", American Psychologist, December, 1984.

8. Hull, R. L., "Human Factors of Risk Reduction in User/Equipment Misinteraction", ASAE Paper Number 87-5002, American Society of Agricultural Engineers, June, 1987.

9. McKnight, R. H., U.S. Agricultural Equipment Fatalities, 1975-1981: Implications for Injury Control and Health Education, Doctor of Science Thesis, John Hopkins University, Baltimore, Maryland, 1984.

10. Purschwitz, M. A. and Field, W. E., "Scope and Magnitude of Injuries in the Agricultural Workplace", ASAE Paper Number 87-5514, American Society of Agricultural Engineers, December, 1987.

11. Robinson, G. H., "Human Performance in Accident Causation: Towards Theories on Warning System and Hazard Appreciation", Proceedings of the Third International Systems Safety Conference, October, 1977.

12. Rogers, R. W., "A Protection Motivation Theory of Fear Appeals and Attitude Change", The Journal of Psychology, 91, 1975.

13. Sevart, J. B. and Hull, R. L., "A Statistical Evaluation of Agricultural Related Deaths", ASAE Paper Number MCR 85-103, American Society of Agricultural Engineers, April, 1985.

14. Smith, K. U., Human-Factor and Systems Principles for Occupational Safety and Health, National Institute for Occupational Safety and Health, U. S. Department of Health, Education, and Welfare, September 1979.

Trends in Ergonomics/Human Factors V
F. Aghazadeh (Editor)
Elsevier Science Publishers B.V. (North-Holland), 1988

FACTORS REPORTED IN AMPUTATION AND OTHER INJURY CASES AT MECHANICAL POWER PRESSES BETWEEN 1976 AND 1984

John ETHERTON

Division of Safety Research
National Institute for Occupational Safety and Health
Morgantown, West Virginia

This paper presents findings of injury factor analyses using a new NIOSH data base which contains information on cases of amputation and other injury which occurred at mechanical power presses. This data base provides more specific information regarding this source of injury than is currently available from workers' compensation claim cases. This data base incorporates injury reports collected in accordance with OSHA regulation 1910.217(g)(1). One of the analyses in this paper determined frequency of safeguard type reported for 2184 cases of power press injury. Of the eight different categories of safeguard type, two-hand devices ranked first with 29% of the cases and fixed barriers ranked second with 25% of the cases. A trend diagram of the proportion of cases reporting a two hand device as the safeguard showed it to be increasingly important, growing from slightly over 20% of the cases in 1976 to over 40% in 1984.

1. BACKGROUND

When workers' fingers are accidentally amputated their manual skills and productive capacity are irretrievably diminished. Preventing such losses is preferable to compensating for them. The financial compensation received by workers who suffer an amputation varies with the severity of amputation and with the nature of any liability action which is pursued. For instance, the worker compensation insurance payments received in 109 serious amputation cases in 1977 in Minnesota totaled $406,146 [1], a figure not including medical care and rehabilitation costs. As a result of liability action, this total worker compensation amount for one state has been exceeded in certain individual amputation cases.

In 1979, the Bureau of Labor Statistics (BLS) determined that about 21,000 workers in the United States sustained an amputation and about 95% of these injuries involved the amputation of a finger or fingers [2]. About ten percent of these involved press type machines. The National Institute for Occupational Safety and Health (NIOSH) has reported that the 151,000 employees who work at mechanical power presses continue to risk having a hand or part of a hand amputated on the job [3].

One injury prevention strategy for presses has been to set standards for the level of performance of press system parts that are critical to safety (controls, safeguards, maintenance). If followed voluntarily or due to enforcement, the combination of requirements in a safety standard should prevent amputations. Voluntary safety standards for mechanical presses were first available in the U.S. in the early part of this century. The general recognition that workers at mechanical power presses were at high risk of crushing and amputation injuries led the Occupational Safety and Health Administration (OSHA) to adopt the voluntary ANSI B11.1 standard [4] as regulation in 1971. In a 1974 standards change, OSHA deleted the no-hands-in-die rule which had been in the voluntary standard and established a data collection process intended to identify details of injuries incurred at presses as an aid in evaluating the standard. In the 1974 rulemaking, sweep-type devices were scheduled to be phased out over a two-year period because they were considered ineffective and also could be a source of injury. In 1978, a notice of proposed rulemaking on OSHA's machine guarding standards, Subpart O, resulted in comments on changes toward performance language. In 1979, reference to sweep-type devices was removed from the standard as part of a standards cleanup project. In the 1980's attention has focused on proposed rule changes to allow presence sensing device initiation (PSDI) for mechanical power presses. In 1987, a NIOSH Current Intelligence Bulletin suggested ways for employers to reduce risks to mechanical power press operators by supplementing the requirements on palm buttons and foot controls in the existing OSHA power press standard [3]. In addition, the ANSI B11.1 standard is routinely revised or reaffirmed following an ANSI five-year update policy. To evaluate how effective the level of performance of a particular design or use requirement has been in actually preventing amputations at presses, data on design and use factors in injury cases is needed.

One of the press safety issues faced by standards setting organizations is the question generally referred to as "hands-in-die" operation. There are two alternative safety approaches to loading press dies: 1) design a method for loading and removing work from the die so that no normal exposure of the hands to the point of operation is needed (hands-out-of-die, or HOOD), or 2) permit exposure to the point of operation only if appropriate safeguards are installed to prevent the closure of the dies from occurring while the hand is present (hands-in-die, or HID).

For technical feasibility reasons, the second alternative was chosen by OSHA in 1974 occupational safety standards rulemaking. Detailing the reasoning for this approach, OSHA explained that it lacked sufficient injury surveillance information for comparing which of these approaches could achieve greater progress in preventing amputations at mechanical power presses [5]. While leaving exposure to hazard (i.e., method of feeding) up to the employer, special safeguarding requirements were added to the standards for hands-in-die operation. Also, requirement 1910.217(g)(1), "Reports of injuries to employees operating mechanical power presses," was added: 1) to acquire more reliable data and 2) to evaluate the effectiveness of the new standard. The information specified for reporting to fulfill this requirement is shown in Table 1.

In 1977, OSHA did followup investigations on 50 injury cases as a quality control measure. Personnel who were responding to the reporting requirement did so to the extent that the technical information required was available to them.

Table 1. The data required in reports

Employer's name, address and location of the workplace (establishment).
Employee's name, injury sustained, and the task being performed (operation, set-up, maintenance, or other).
Type of clutch used on the press (full revolution, part revolution, or direct drive).
Type of safeguard(s) being used (two hand control, two hand trip, pullouts, sweeps, or other). If the safeguard is not described in this section, give a complete description.*
Cause of the accident (repeat of press, safeguard failure, removing stuck part or scrap, no safeguard provided, no safeguard in use, or other).**
Type of feeding (manual with hands in dies or with hands out of dies, semiautomatic, automatic, or other).
Means used to actuate press stroke (foot trip, foot control, hand trip, hand control, or other).
Number of operators required for the operation and the number of operators provided with controls and safeguards.

* For more detail on safeguard types see [6,7].
** The "cause" stated in these reports is the employer's own explanation of cause. This explanation is useful for formulating hypotheses for studies to evaluate causal associations. Others may have different cause explanations for a case. This data system does not contain alternative cause explanations of injury cases.

2. THE DATA BASE

A computerized data base of information on reports to OSHA about power press injuries has been developed at NIOSH. The purpose of the data base is to provide details on mechanical power press accidents which are not routinely reported in compensation cases. The information in this data base will be useful in developing hypotheses for studies on the effectiveness of different types of safeguarding on mechanical power presses under various operating conditions.

An injury case in this data base is defined as one which has occurred because a finger, hand or other body part was caught between two dies (in the point of operation) when the press dies closed as they were designed to be closed. A continuum of severity (nip, crush, amputate, fatality) is represented. The severity of an amputation depends on the amount of the body exposed (in the point of operation) at the instant of die closure.

The data base currently contains coded information on over 2500 injury reports at mechanical power presses, all submitted to OSHA between 1976 and 1984. The file is maintained in an IBM 4361 mainframe at NIOSH's Division of Safety Research, Morgantown, West Virginia. Previously published findings from this set of injury reports were univariate tabulations of factors in the reports [8,9]. Interactions between variables (such as employer-reported cause of injury when a two-hand device was the intended safeguard) are reported here for the first time.

3. ANALYSIS

3.1. Effectiveness of Two-Hand Devices as a Safeguard

Part of OSHA's first request for information from the data base was to determine what is known about injury cases when the safety distance for two-hand devices were based on the 1.6 m/s hand speed constant now in the standard. NIOSH research on hand speed and the after reach hazard when two hand devices are the safeguard on power presses has led OSHA to consider revising Standard 1910.217(c)(3)(vii)(c). In this standard, a value of 1.6 m/s specifies the speed a hand will achieve when an employer is determining how far to locate two-hand devices from the press die. In 652 injury cases, a two-hand device was reportedly used. Cases were ranked by employer-reported cause (Figure 1). "Repeat of press" ranked first (25%) and "failure of press" ranked second (17%) out of 7 cause categories. These are failure modes which a two-hand device cannot protect against. The distance of two-hand devices from the point of operation was not specified in the injury reports.

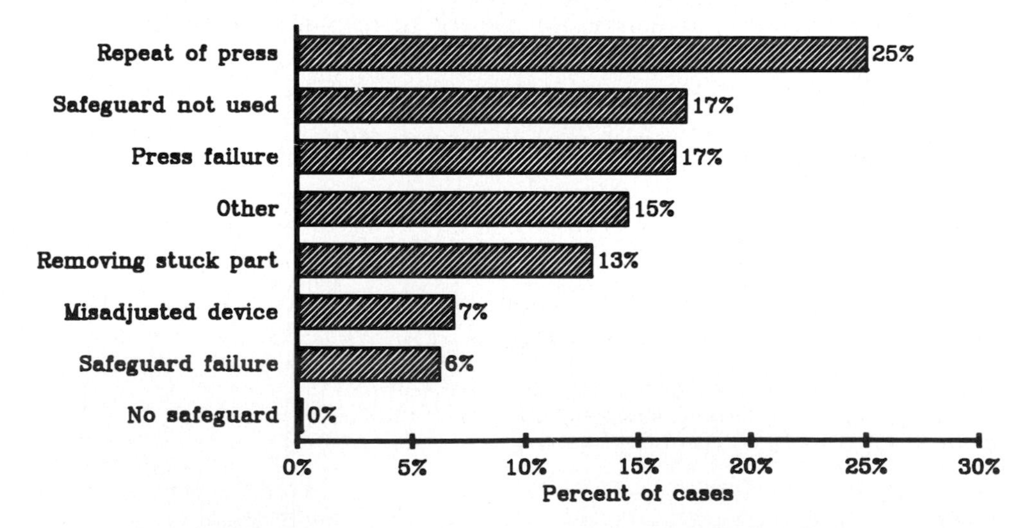

Figure 1. Percentages of reported cause of injury within 652 normal operation injury cases with two hand device reported as the safeguard.

Trend analysis of the proportion of cases reporting a two-hand device as the safeguard shows that the two-hand device has apparently been increasingly important as a safeguard, growing from slightly over 20% of the injury cases in 1976 to over 40% in 1984 (Figure 2).

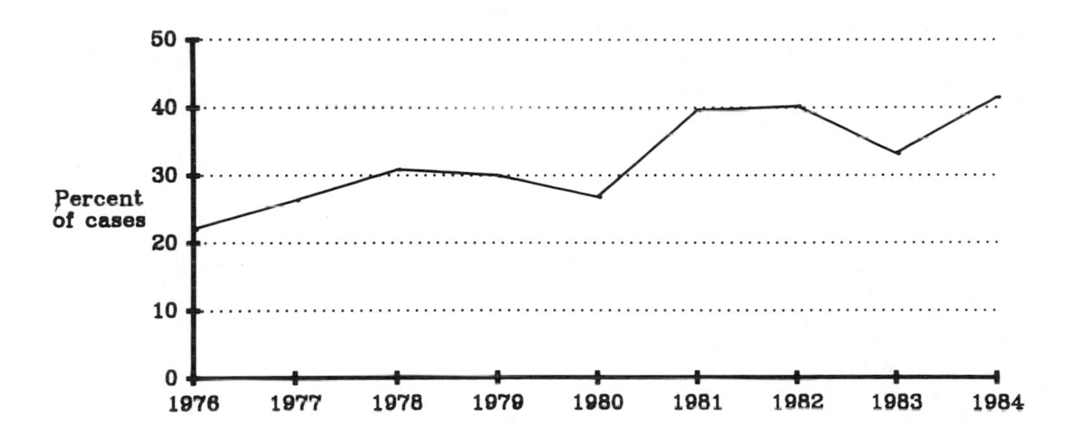

Figure 2. Trend between 1976 and 1984 in the percentage of normal operation injury cases with two hand device reported as the safeguard.

3.2. Production Status When Injury Occurred

Injury during normal operation was much more common than during setup (Figure 3). However, the relative severity between these categories has not yet been examined[1].

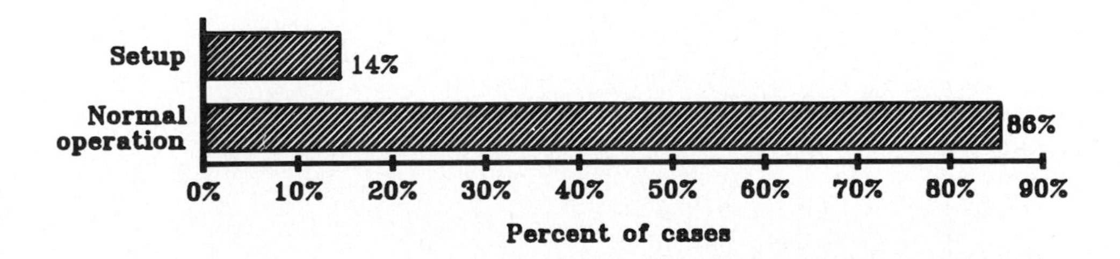

Figure 3. Percentages of reported task within 2574 injury cases.

3.3. Feeding Method

About as many injuries were reported where the feeding method was hands-in-die as were reported for no-hands-in-die operations (Figure 4).

The design options for achieving no-hands-in-die operations (handtools, manual loading away from the dies, automatic feeding) are also shown in Figure 4.

This comparison suggests the need to focus amputation prevention efforts on the use of effective safeguarding, even for no-hands-in-die operations. It also suggests that employers carefully consider equipment and human reliability problems if they choose to operate hands-in-die. Over the years 1976 to 1984 the injury frequency trend has been slightly increasing when hands-in-die is the feeding method (Figure 5). In both HOOD and HID cases, OSHA's safeguarding rules should have been followed.

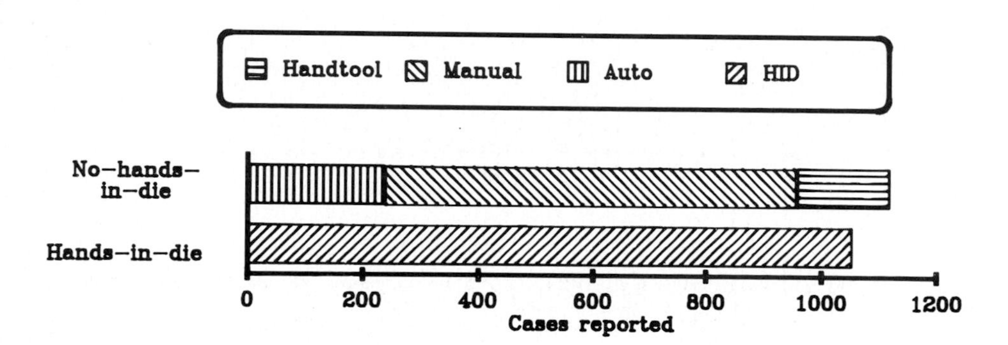

Figure 4. Frequency of hands-in-die and no-hands-in-die as a task design factor within 2172 cases of injury during normal operation.

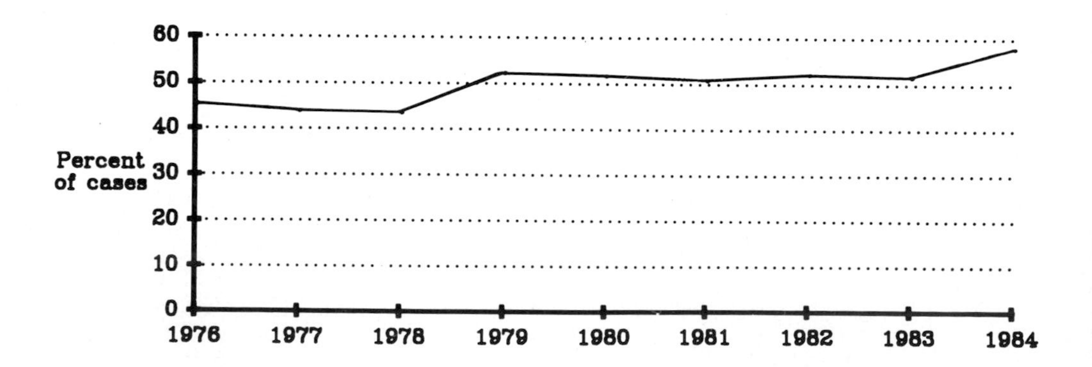

Figure 5. Trend between 1976 and 1984 in the percentage of normal operation injury cases with the method of feeding designed as hands-in-die.

3.4. Type of Safeguard

Frequency of safeguard type reported for 2184 cases of normal operation power press injury were ranked (Figure 6). Of the eight different categories of safeguard type, two hand devices ranked first with 29% of the cases and fixed barriers ranked second with 25% of the cases. It has been suggested that safeguards should be compared only with respect to the particular kind of application, e.g. blanking, short run, automatic, small pieces, in order to understand which safeguards are appropriate to which applications [7].

3.5. Means Used to Actuate Press Stroke (Controlling the Moment When Hazardous Motion is Initiated)

When a foot control was being used to initiate downward motion of the ram, safeguards in three categories accounted for 79% of the injury cases (Figure 7). These were: fixed guard (33%); pullout/restraint (29%); and no safeguard (17%). Misadjusted or misused fixed guards and pullout/restraint devices can increase amputation risk when foot controls are used on presses [3,8].

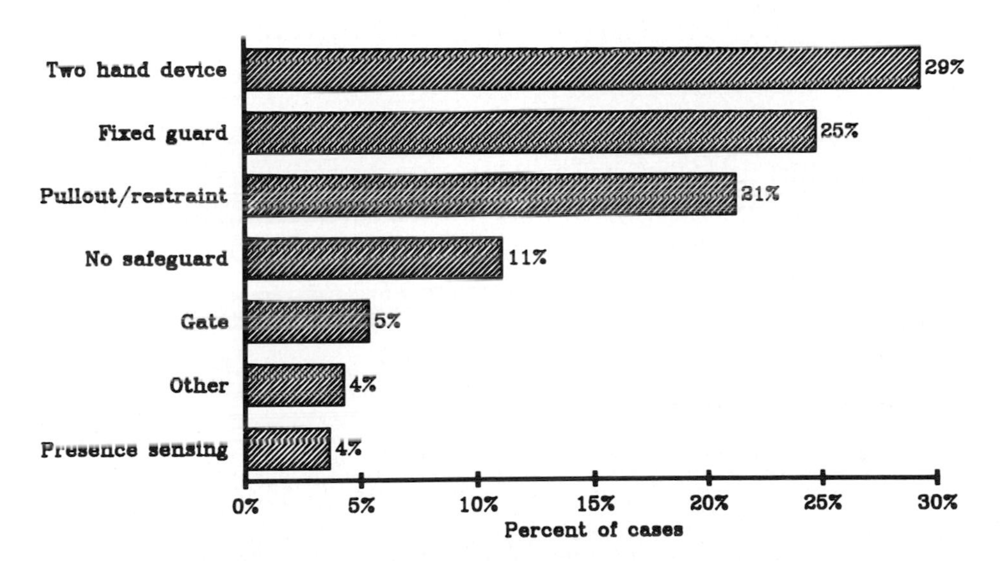

Figure 6. Percentages of reported safeguard used within 2184 normal operation injury cases. (40 reports indicated redundant safeguards)

4. CONCLUSIONS

4.1. Two-Hand Devices

The data base indicates growing importance of the two-hand device as a factor in injury cases. Further study of the causal relationships in such cases is needed. In particular, collecting information on distance

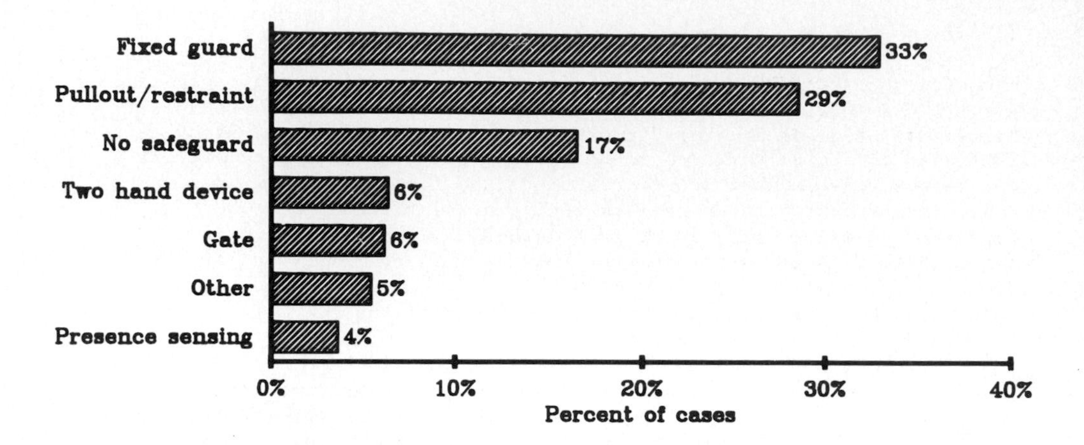

Figure 7. Percentages of reported safeguard used within 1495 normal operation injury cases with foot control reported as the actuating means.

between palm buttons and press dies in injury cases is requisite to evaluating the effectiveness of protection with two hand devices.

4.2. No-Hands-In-Die

Directly comparing the efficacy of HID task designs with the HOOD alternative still cannot be done because these methods have not been separately enforced on large populations. The OSHA rule in effect while these data were being collected allowed any employer to use the hands-in-die alternative whenever convenient. Manufacturers who have voluntarily enforced a HOOD policy since 1974 might provide safety research data toward comparing the effectiveness of HOOD with the current standard permitting HID.

4.3. Safeguard Effectiveness

Comparing the effectiveness of different safeguarding systems is not possible using only the data in this injury information system. A source of exposure data for different safeguards is needed. A method for comparing the effectiveness between safeguarding types is proposed here. This method is to measure the ratio between two proportions: the proportion of injury cases in which X is the reported safeguard compared to the proportion of all press operations which use safeguard X.

To do this two tasks must be done. First, the reliability of the OSHA data needs to be checked to ensure that the first proportion is accurate. Second is collecting a representative sample of the proportion of press operations where different safeguards are used. This could be done by cooperating safety organizations.

4.4. Continuing the Data Collection

This current surveillance system provides the only readily available means of monitoring injury trends associated with particular safeguarding systems and feeding methods for mechanical power presses. It should, therefore, be continued. The effect of underreporting on the results presented may or may not be a problem and should be assessed.

The value of this reporting system for monitoring the effectiveness of OSHA's 1910.217 standard, as pressworking technology evolves, is evidenced by the following excerpts from cases in the system:

1. PSDI - While setting up a press on August 13, 1984, a male employee sustained an amputation of the distal two phalanges, middle finger, right hand. The reported means used to actuate the press stroke was a proximity sensor which operated the press on extraction from the field. A more detailed report (by phone or on site) of this incident from a safety research perspective would have been useful in the current rulemaking on PSDI.

2. Robotics for Press Feeding - While reaching into a die to dislodge a stuck part on August 7, 1980, a female employee's left index finger was amputated in the middle phalange. The reported means for feeding this press was a robot. The press cycled when the robot was pushed back to a position where it normally engaged an interlock to actuate the press. General misconceptions of robotic technology make it unclear whether this was a robot according to the definition adopted by the Robotics Industry Association in 1984. This is, however, an example of how the reporting system can permit monitoring of the effectiveness of the standard as more workers program, operate, and maintain robot fed press systems.

The analyses reported in this paper establish the need to take the next step toward getting the data required to compute injury incidence rates for presses using each type of safeguarding method. Determining incidence rates requires collecting data on proportions of presses where each type of safeguard is applied. Such an injury incidence rate would give workers, management, and standards-setting organizations one measure of safeguarding effectiveness, and such research would be important for promoting the maximum use of machine guarding technology.

ACKNOWLEDGMENTS

The author acknowledges the assistance of Carrol Burtner and Ray Souryal, OSHA/Office of Safety Standards Development in obtaining copies of the injury reports, and, Mike Moll, Pat Cutlip, and Nick Kornick, NIOSH/Division of Safety Research, Data Analysis Section, in the development of this report. The author also acknowledges the many other farsighted safety professionals who have contributed since the early 1970's to the evolution of this injury surveillance system.

FOOTNOTES

[1] The importance of performing a severity comparison is suggested by a recent report on fatalities [10] which showed 12.7% of UAW

skilled trades workers killed on the job were diemakers, a trade closely associated with setting up mechanical power presses.

REFERENCES

[1] Olson, D. and Gerberich, S., Traumatic Amputations in the Workplace, Journal of Occupational Medicine, 28(7) (1986) 480-485.

[2] McCaffrey, D.P., Work-Related Amputations by Type and Prevalence, Monthly Labor Review 104 (Sept) (1981) 35-41.

[3] Injuries and Amputations Resulting from Work With Mechanical Power Presses, Current Intelligence Bulletin No. 49, (DHHS (NIOSH) Publication No. 87-107, Cincinnati, 1987).

[4] American National Standard for Machine Tools-Mechanical Power Presses-Safety Requirements for Construction, Care, and Use. (American National Standard Institute, Inc., New York, 1970).

[5] Federal Register, Vol. 39, No. 233, (U.S. Government Printing Office, Washington, DC, Dec. 3, 1974) pp. 41841-41846.

[6] Mechanical Power Presses, Part 1910.217, Code of Federal Regulations, (U.S. Government Printing Office, Washington, DC, 1986).

[7] Etherton, J., The Use of Safety Devices and Safety Controls at Industrial Machine Workstations, in: Human Factors Handbook, Salvendy, G. (ed.) (John Wiley and Son, Inc., New York, 1987).

[8] Trump, T. and Etherton, J., Foreseeable Errors in the Use of Foot Controls at Industrial Machines, Applied Ergonomics, 16(2) (1985) 103-111, .

[9] Pizatella, T. and Moll, M., Simulation of the After-Reach Hazard on Power Presses Using Dual Palm Button Actuation, Human Factors 29(1) (1987) 13-22.

[10] Mirer, F., Occupational Fatalities Among UAW Members (International Union, United Automobile, Aerospace, and Agricultural Implement Workers of America, Detroit, Michigan, 1987).

Trends in Ergonomics/Human Factors V
F. Aghazadeh (Editor)
Elsevier Science Publishers B.V. (North-Holland), 1988

COMPARATIVE HAZARDOUSNESS OF METALWORKING MACHINES

Roger C. JENSEN and John R. ETHERTON

Division of Safety Research
National Institute for Occupational Safety and Health
Morgantown, West Virginia

In order to reduce the frequency of machine-related injuries, research resources made available for machine safety research need to be committed to those types of industrial machinery where it can benefit the most workers. Consequently, a priority plan for machine safety research is being developed. The plan is to compare various types of machines using multiple indices that reflect the need for research. This paper describes one of these indices, a hazard index based on the ratio of number of injuries to number of machines. Results indicated that shears and slitters are relatively hazardous, and apparently warrant in-depth investigation to determine what can be done to reduce injuries from such machines. However, the place of these machines on an machine research priority list will also consider other factors.

1. INTRODUCTION

Many of the most severe occupational injuries occur when workers come in contact with industrial machine hazards. About 54 percent of occupational amputations result from contact with a machine and 60 percent of the work-related amputations occur in manufacturing industries [1]. Data such as these indicate that the approach to machine safety used in the United States has only been partially successful in controlling the hazards of the many machines used in manufacturing industries. There is an apparent need to increase the use of present day engineering methods for safeguarding machines, as well as a need for research to extend the state-of-the-art.

The U.S. Public Health Service agency responsible for conducting and supporting industrial safety research is the National Institute for Occupational Safety and Health (NIOSH). Since the establishment of NIOSH in 1971 it has recognized the importance of allocating research resources to projects that have potential for preventing large numbers of severe injuries.

An example of this concern is found in the machine safety research area. In the early 1970s NIOSH contracted for an assessment of the need for research and improved standards for safeguarding personnel at metalworking and woodworking machines. The contractor was Bendix Launch Support Division. The contract report [2] ranked 29 different metalworking machines based on a composite index which included for each machine category:

1. An index of average injury frequency rate.
2. An index of average injury severity.
3. Average compensation paid per case.
4. Total number of the machines in use.
5. Total number of workers exposed.
6. An index reflecting the proportion of the machines which are unguarded.
7. An index reflecting the number of hazards inherent to the machine.

The composite index, based on multiplying unweighted values of the seven components, found mechanical power presses ranked as the number one priority, with a composite index 3.7 times that of the second highest priority machine category, lathes. Mechanical power presses had the highest index values of: 1) injury frequency, 2) injury severity, and 3) workers exposed. It was second in average compensation paid per case. These indices explain why the NIOSH Division of Safety Research has considered work at power presses the type of metalworking machine operation most in need of safety related improvements. Many of the results from NIOSH research projects on power press safety were included in the recent publication entitled "Injuries and Amputations Resulting from Work with Mechanical Power Presses" [3].

Since 1975 a trend toward increasing use of numerically controlled machines and computer integrated manufacturing systems may have changed patterns of exposure to machine hazards. Consequently, NIOSH has undertaken a project to develop a more current priority plan to guide its machine safety research program. The plan will include a priority list of various types of machines using multiple quantitative indices that reflect the need for safety research as well as anticipated improvements in injury prevention that could result from research. One of the indices being considered is an index of hazardousness for groups of similar machines based on the ratio of number of injuries to number of machines. This paper describes sources of data for this index and presents values of the index for several categories of machines used in the metal processing industries.

2. METHODS

Data for number of injuries involving defined machine groups were obtained from workers' compensation data from the year 1983. These data were part of the Supplementary Data System (SDS), an occupational injury and illness record system established by the U.S. Bureau of Labor Statistics. The SDS is described more completely in another article [4]. For this analysis the search strategy described below was to select cases that met specified criteria for industry and machine.

Data for the number of machines were obtained from a survey conducted by American Machinist Magazine in 1983 [5]. The survey used a stratified sampling approach to obtain representative data for different size employers. Because the American Machinist survey was limited to metal processing industries, the search for injury cases was also restricted to these same industries. The comparisons were also restricted to five general machine groupings for which data were available in both sources. Materials handling equipment, such as conveyors, were not listed in the American Machinist survey. Industrial robots were not listed in either source.

Table 1 shows the machine categories used in this analysis. It lists the word descriptors used in the SDS records on the left, and the descriptors used in the American Machinist survey in the middle column. The right column shows the number of machines in each category. Note that the SDS descriptors are indicative of similarity in the general operations performed at the point-of-operation of machines. However, machines within the same SDS classification may be used for quite different metal processing operations. For instance, within the "shears, slitters, and slicers" category, shears cut flat plate whereas slitters cut wide coils of thin metal into narrower strips.

TABLE 1
Number of Machines in U.S. Metalworking Industry in 1983

SDS Descriptor[1]	American Machinist Descriptors[2]	Number
Shears, slitters, & slicers	NC shearing machines	755
	Plate & sheet shears: mechanical	22,921
	Plate & sheet shears: hydraulic	5,784
	Bar, angle, & rotary shears	6,073
	Other power operated non-NC punch & shear	3,802
	Coil processing systems	10,852
		50,187
Presses (not including printing presses)	NC punching machine	5,468
	Non-NC punching machines	27,783
	Mechanical power presses	198,973
	Hydraulic presses (not forging)	53,292
	Pneumatic presses	17,656
		303,173
Saws	Cutoff & sawing machines	169,107
Drilling, boring, & turning machines	NC turning machines	33,352
	Non-NC turning machines	332,327
	NC boring machines	5,064
	Non-NC boring machines	40,462
	NC drilling machines	7,993
	Non-NC drilling machines	281,453
		700,650
Casting, forging, welding machines	Die casting machines	8,095
	Forging machines	19,613
	Electric arc-welding equipment	313,704
	Electric resistance-welding equipment	62,940
	Energy beam processing[3]	936
	Gas welding machines (not hand)	5,367
	Welding robots	484
		411,139

NC: numerically controlled
1: Supplementary Data System [4]
2: from American Machinist [5]
3: electron beam, laser welding

Table 2 lists the industry categories included in this analysis. The list includes the two-digit Standard Industrial Classification (SIC) code, and the word description of the industry. SIC codes are based on the product manufactured [6]. The American Machinist Survey [5] included all of these industries. The search of SDS cases was restricted to these same industries.

TABLE 2
Industrial Categories Included in Analysis

SIC	Industry
25	Furniture and Fixtures
33	Primary Metal Industries
34	Fabricated Metal Products, except machinery and transportation equipment
35	Machinery, except electrical
36	Electrical and Electronic Machinery, Equipment, and Supplies
37	Transportation Equipment
38	Measuring, analyzing, and controlling instruments; photographic, medical and optical goods; watches and clocks
39	Miscellaneous Manufacturing Industries

SIC: Standard Industrial Classification code number, 2-digit.

The SDS search identified cases that met three requirements. First, the cases were part of the 1983 reference year SDS data set. Second, the industry of employment was within one of the 2-digit categories listed in Table 2. Third, the source of injury was one of the machine categories listed in the left column of Table 1.

3. RESULTS

Results from the search of 1983 workers' compensation records of the 30 states that participated in SDS revealed a total of 10,065 cases involving the five machine groups within the specified industries. These injury cases included all types of injury in which these machine groups were listed as the source of injury. Two machine types included in this analysis but not in the 1975 Bendix priority ranking are slitters and die casting machines. Table 3 shows for each machine group: the number of injury claims, the machine population, and the hazardousness index (1000 x the ratio).

TABLE 3
Machine Groups Ranked According to Comparative Hazardousness Index

Machine Groups	Injuries	No. Machines	Index
Shears and slitters	1,014	50,187	20.2
Saws	2,252	169,107	13.3
Presses (not printing presses)	3,461	303,172	11.4
Drilling, boring, and turning	2,213	700,650	3.2
Welding, casting and forging	1,125	411,139	2.7
	10,065	1,634,255	

4. DISCUSSION

These results show that the shears and slitters group is relatively hazardous with respect to number of injuries for machine population. Examples of these two machine types are shown in Figures 1 and 2. These two machine categories are actually quite different from the perspective of safety engineering because of differences in the nature of worker involvement in loading, set-up, and maintenance operations. This points to a need for more specific classification categories in the SDS coding system. Such data would make it possible to examine injury data on slitter machines and slicer machines in order to compare their relative hazardousness as well as to identify differences in common accident scenarios involving the different types of machines.

The results of this analysis indicate a need for more in-depth investigation to determine the circumstances associated with injuries involving industrial machines generally, and for slitters and slicers in particular. Such information would help extend the state-of-the-art for minimizing the risks involved in using these machines. At the present time, however, there are some recognized hazards and worker-machine interactions according to various authorities. Known hazard areas at shears include points between the blades and under the hold-down clamps. Protecting helpers from falling material at the rear of the machine is also an important consideration. A potentially fatal practice in operating slitters is hand insertion of paper or filler material into the recoiler. This is prohibited in the ANSI B11.14 safety requirements for coil-slitting machines/systems [7], but better guidelines for implementing the standard may be needed.

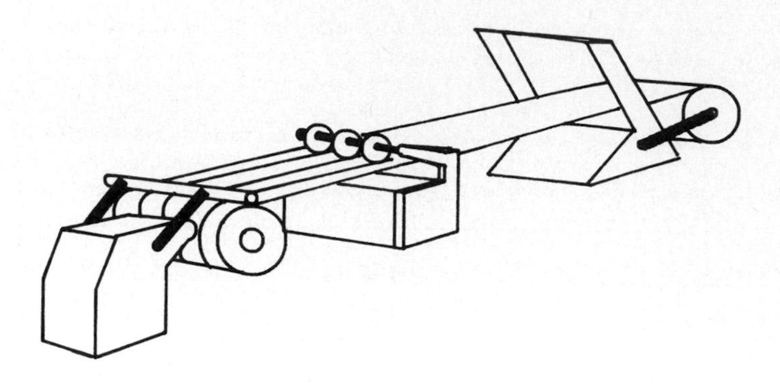

FIGURE 1
Example of a coil slitter machine.

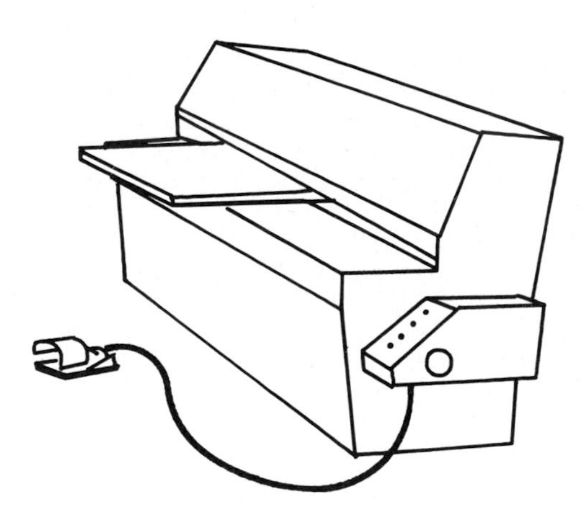

FIGURE 2
Example of a sheet metal shear machine.

The comparative hazardousness index presented here has several limitations. First, it only applies to metal working machines used in metal processing industries. Second, it does not consider injury severity. Third, it does not directly indicate need for research. Consequently, the place of these machines on a research priority list will not be determined by this index alone. Other indices will be developed to account for other aspects of concern such as number of workers potentially impacted, technical difficulty of making the machines safer, severity of injuries, and potential injuries prevented if the research is successful and can be applied effectively. Specific needs for research to improve machine safety standards and human factors guidelines should become apparent in this priority setting process.

REFERENCES

[1] McCaffrey, D.P., Work-related amputations by type and prevalence, Monthly Labor Review 104 (1981) 35-41.

[2] Bendix Corporation, Machine Guarding - Assessment of Need. DHEW Publication Number (NIOSH) 75-173. National Institute for Occupational Safety and Health, Cincinnati, Ohio (1975).

[3] NIOSH. Injuries and Amputations Resulting from Work with Mechanical Power Presses. DHHS Publication No. 87-107, National Institute for Occupational Safety and Health, Cincinnati, Ohio, (1987).

[4] Jensen, R.C., How to use workers' compensation data to identify high risk groups. Chapter 12, pp. 364-403, in: Handbook for Occupational Safety and Health, L. Slote, editor, John Wiley & Sons, New York (1987).

[5] American Machinist Magazine. The 13th American Machinist Inventory of Metalworking Equipment 1983. American Machinist (November 1983) 113-144.

[6] Office of Management and Budget. Standard Industrial Classification Manual, 1972. U.S. Government Printing Office, Washington, D.C.

[7] ANSI B11.14 Committee. Safety Requirements for Construction, Care, and Use of Coil-Slitting Machines/Systems, ANSI B11.14 Standard. American National Standards Institute, New York (1983).

Trends in Ergonomics/Human Factors V
F. Aghazadeh (Editor)
© Elsevier Science Publishers B.V. (North-Holland), 1988

Safety Perceptions and Use Patterns For ATVs and Other Motorized Recreational Vehicles

Edward W. Karnes*, S. David Leonard**,
Theresa Schneider*, Walter Pedigo*, David Krupa*, and Ed Madigan*

*Metropolitan State College, Denver, Colorado
**University of Georgia, Athens, Georgia

Three surveys involving 3622 persons were performed in order to determine the characteristics of use of ATVs and other recreational vehicles. Topics were the frequency of vehicle use, knowledge of safety practices, adherence to safety practices, perception of risk of injury, accident occurrence, and perception of need for training in use of the vehicles. The data indicate that both risk perception and knowledge of safety practices are low. Despite the high injury rate for ATVs, they are perceived as less risky than many other recreational vehicles. It was concluded that more adequate information about the basis for risks should be presented.

1. INTRODUCTION

Safety issues concerning all-terrain vehicles (ATVs) have become important for both the general public and human factors professionals due to the high frequency of deaths and catastrophic injuries associated with their use. Unfortunately, heretofore, information about the use of these vehicles and the general public's perception of the risks involved has been rather sparse. The studies reported here were designed to provide such information.

ATVs were introduced by Honda in 1970 as a motorcycle-type vehicle with three wheels. The distinctive characteristics of ATVs are cycle-type handlebars, a long seat, and large, low-pressure, balloon-type tires. During their early years sales of ATVs were modest. Advertising often focused on utility uses and consisted of radio ads and print ads presented mostly in enthusiast magazines. In the early 1980s several other manufacturers introduced their versions of three-wheeled ATVs, and Suzuki introduced the first four-wheeled model.

Coincident with the entry of other manufacturers into the market, advertising efforts increased and the type of ads changed. Network television (TV) ads became prominent, portraying the vehicles as "fun" recreational vehicles appropriate for all family members, including children. The emphasis on women and children riders distinguished ATV ads from ads provided for similar recreational vehicles such as trailbikes and motorcycles. ATVs were shown as truly all-terrain vehicles by scenes showing riding on snow, grass, sand, dirt, mud, rocky areas, and through water. The ads often portrayed "adventuresome" and carefree uses of the vehicles by riders of all ages. Some of the ads even showed dangerous riding practices such as riding without helmets, performing "wheelies," and jumping, despite the fact that some manufacturers warned against such practices. Presumably the ads were successful in that sales of ATVs increased dramatically, soon exceeding those of both trailbikes and snowmobiles.

With the increase in the popularity of ATVs came an even more dramatic and disproportionate increase in accidents resulting in deaths and catastrophic injuries. The public concern about this resulted in an investigation, given top priority, by the U.S. Consumer Product Safety Commission (CPSC) in 1984 to evaluate dangers associated with the use of ATVs in the United States. A review of the concern, governmental actions, and injuries and deaths associated with ATVs through 1986 was presented earlier (Karnes, Leonard, Schneider & Rachwal, 1987). Recent data provided by CPSC show no decline in the rate of approximately 20 deaths and 7,000 estimated hospital-treated injuries per month associated with ATVs. For the time period 1982 through June 8, 1987, there have been a total of 789 deaths and 271,700 hospital-treated injuries (Schachter, 1987). Many of the dead and injured have been youngsters. For example, CPSC data have shown that 44% of both deaths and hospital-treated injuries involved children less than 16 years of age and 20% of the victims were less than 12 years of age (Newman, 1986). The magnitude of the problem is revealed in a reanalysis of CPSC death and injury data by Downs (1987), who noted that over the estimated eight-year lifetime of a three-wheeled ATV, two of every three to five ATVs driven by youngsters (less than 16) will send someone to a hospital emergency room, and deaths involving child riders of three-wheeled ATVs are at least twice as likely as for older riders.

The concern over dangers associated with riding ATVs has mostly focused on the three-wheeled ATVs, but the question of differences in safety between three- and four-wheeled versions remains open. Newman (1986) reported that the risk of injury per 100 vehicles in use for three and four wheelers was 5.1 and 3.3, but that the risk of death was the same. Manufacturers, particularly Honda, have maintained that the accident, death, and injury data are misleading and that the safety of three-wheeled ATVs is no different than four-wheeled ATVs. In fact, reanalyses of CPSC injury data performed for Honda (Heiden Associates Inc., 1986) indicated that four-wheeled ATVs were involved in slightly more tip-overs and overturns than their overall average involvement in all injuries and in a substantially above average proportion of injuries caused by flipping-forward accidents. McCarthy (1987) opined that a possible explanation for the higher accident rates with the three-wheeled ATVs is that it may be an artifact of age (i.e., the fleet of three wheelers is older than the fleet of four wheelers, and accident rates for older vehicles of any type are usually higher than newer vehicles because of rider behaviors and vehicle malfunction rates).

Differences in death and injury rates between ATVs and other motorized recreational vehicles have been viewed by some as evidence of the extreme danger posed by ATVs. For example, injury data provided by CPSC for 1985 had shown that the injury rate for ATVs was roughly twice that of mini/trail bikes and three to five times that of snowmobiles. However, Verhalen (1986) presented a reanalysis of the injury data that showed essentially no differences when injury rates were adjusted by estimated frequency of use of the various vehicles. The estimates were that ATVs were used ten times as frequently as snowmobiles and two to three times as frequently as trailbikes. These estimates were based on the testimony of seven witnesses (four ATV dealers and three users) before the CPSC public hearings in 1985 and 1986. Subsequently, the General Accounting Office (GAO) (1986) issued a report concluding that Verhalen's analysis was based on unsubstantiated estimates and could not be considered statistically valid. Nevertheless, the issue of relative usage of ATVs in relation to other recreational vehicles has been raised as an important factor in evaluating the injury and death rates associated with the ATVs.

Manufacturers of ATVs have consistently responded to the issue of ATV safety by denying that their products are unreasonably dangerous. They have contended that the causes of accidents resulting in injuries and deaths to riders of their vehicles are due to rider carelessness and misuses of the vehicles. Cited as major misuses by the ATV manufacturers are riding on public roads, riding with passengers, riding under the influences of alcohol and drugs, riding without helmets, and children riding without parental supervision. The CPSC 1986 Task Force report on regulatory options for ATVs provided some support for the ATVs manufacturers' position by reporting that many of the accidents were the result of foreseeable misuses by ATV operators.

In direct opposition to the ATV manufacturers' contention that their vehicles are safe if not misused, many safety and engineering professionals have contended that it is not rider misuses but rather defects in the design of ATVs and the conduct of the manufacturers that makes them unreasonably dangerous. The term "rider active" used by ATV manufacturers to describe the operational characteristics of their vehicles has been reviewed by critics as euphemism to hide the harsh reality that the vehicles are both "tricky" and dangerous. Frequently cited as design defects that cause dynamic instabilities and complex operational characteristics include short wheel bases, high centers of gravity, solid rear axles, and lack of suspensions (on some models). The conduct of manufacturers has been questioned in terms of: (1) failures to provide adequate and durable warnings and appropriate safety information; (2) deceptive promotion that creates false and unrealistic expectations of safety and ease of operation; (3) failures to provide formal hands-on training; and (4) failures to exercise adequate control over dealer representations to consumers.

Criticisms of the position taken by ATV manufacturers that rider misuses are the major causes of accidents and deaths are abundant. For example, David Schpeltzer (1986), in response to the 1986 report of the CPSC All-Terrain Vehicle Task Force, noted that ATVs are manufactured and sold primarily for recreational purposes, and that very few consumers that use the machines expect to encounter anything other than a good time. He further commented,

> The briefing package also contains references to the immaturity and lack of judgment of children operating ATVs. Here again, such behavior on many occasions was and is reasonably foreseeable. It should be kept in mind when referring to risk taking propensities of the operator that many consumers, especially children and parents of children, do not have an appreciation of the risks they will be taking or the risks they have taken, given the absences of, or inadequacies of, the warnings and instructions, the extensive marketing and promotion of the vehicles, and the lack of an effective hands-on training program . . . The report seems to shift the responsibility for the injuries and deaths from the makers and promoters of this very dangerous product to the user (Schpeltzer, 1986).

Much litigation has occurred as a result of ATV-related injuries and deaths. A wide variety of issues has been examined by human factors and safety consultants. Some prominent topics involved are consumer expectations and risk perceptions; adequacy of warnings and safety information on the vehicle, in manuals, obtained from dealers, and in advertising and promotional materials; effects of advertising on consumer expectations; age requirements and limitations; unique handling characteristics of ATVs, and transfer of training from operating two-wheeled recreational vehicles; formal training requirements; and adequacies of injury data bases. Unfortunately, a paucity of empirical data is available to support or refute many of the conflicting opinions presented by experts.

The objective of the studies presented here is to provide survey data about consumers' knowledge of safety requirements, risk perceptions, use of protective equipment, and frequencies of use for various recreational vehicles. The results of three separate studies are reported. Because of the disparities in sizes for age, sex, and experience groups, unweighted means methods were used in all statistical analyses. All significant differences (reliable) noted in the results are significant at the .01 level unless otherwise noted.

2. Experiment 1

2.1 Method

A total of 1493 subjects included 734 students at Metropolitan State College in Denver, Colorado, 382 high school students from a suburban Denver high school, and 377 students from two state universities and one liberal arts college in Georgia. They were surveyed in intact classes and told that their participation was voluntary. They responded to a questionnaire that included items relating to riding experience and frequency of use (months and hours per year) for five recreational vehicles: three-wheel ATVs, four-wheel ATVs, trailbikes, snowmobiles, and motorcycles.

2.2 Results and Discussion

The percentage of persons with riding experience for three-wheel or four-wheel ATVs, trailbikes, snowmobiles, and motorcycles were 33%, 28%, 16%, and 23%. Calculation of the mean hours of use was made difficult by the fact that a small percentage in each group (about 5%) produced estimates that exceeded any reasonable likelihood of use (indeed, some estimates exceeded the actual hours per year). It was assumed that these individuals were trying to indicate high usage; therefore, to reduce distortions without discarding these presumably high users, the hours-per-year data were categorized into nine sets as follows: 1-25, 26-50, 51-75, 76-100, 101-200, 201-300, 301-500, 501-1000, and over 1000. Midpoints of the first eight categories and 1000 were used in calculating grouped-data means for each vehicle type.

Single-factor unweighted means ANOVs were performed to compare the subpopulations with respect to usage of each vehicle. The motorcycle was the only vehicle for which a reliable difference among subpopulation means was found. Colorado college students estimated greater motorcycle use than did either Georgia college students or Colorado high school students. For vehicle-usage analyses, these subgroups were combined.

Means were calculated for three categories of ATV use, those who used only three-wheelers (50.61), those who used only four-wheelers (238.13), and those who used both (144.24). The weighted mean for hours of use by all ATV categories was 150.67. Means for trailbikes, snowmobiles, and motorcycles were 70.06, 39.80, and 114.88 respectively. Comparisons among the vehicle groups (with the three ATV categories combined) indicated that among riders of these vehicles, ATVs were ridden reliably more than the other vehicles, and motorcycles were ridden reliably more than trailbikes and snowmobiles which did not differ significantly. The ratios of means for hours of use from this survey suggest that ATVs are ridden 2.15 times as frequently as trailbikes and 3.79 times as frequently as snowmobiles. These ratios are different than those used by Verhalen (1986), and indicate that more extensive data concerning frequency of use needs to be obtained for reliable adjustments to death and injury rates.

3. Experiment 2

The area of concern in this experiment was the extent to which individuals wear protective devices, specifically helmets, while riding recreational vehicles, and what conditions might affect the use of helmets. In particular, we were concerned with the question of whether helmets were worn more frequently with some vehicles than others. A further question was whether or not ATV riders constitute a subpopulation that differs from non-riders of ATVs in their attitudes towards safety.

3.1 Method

A total of 987 subjects were drawn from schools in the metropolitan Denver, Colorado area and included 627 adult college students, 250 high school students, aged 15 to 17, and 110 junior high school students, aged 10 to 12. They were surveyed in intact groups and participation was voluntary. The questionnaire included items for riding experiences (single or double), accident experiences (alone, driving with a passenger, or as a passenger), injury experience (severity and type of treatment), and frequency of helmet use. Frequency of helmet use was rated on a four point scale ranging from never to always. Added to the vehicles listed previously were bicycles.

3.2 Results and Discussion

Single-factor unweighted means ANOVs showed no significant differences in frequency of helmet use among the various age groups for any of the vehicles. Thus, for other analyses the data from these groups were combined. Percentages of persons with riding experience for bicycles, three-wheel ATVs, four-wheel ATVs, trailbikes, snowmobiles, and motorcycles were 99, 37, 26, 60, 34, and 69.

Because not all subjects rode each type of vehicle, to compare relative usage of helmets across vehicles it was necessary to form a subset of subjects for each pair of vehicles and perform within subjects' t tests. From these tests it was determined that in terms of reported usage: 1) helmets are worn reliably less frequently on bicycles than on the five motorized vehicles; 2) reported helmet usage is not reliably different among three-wheel ATVs, four-wheel ATVs, and snowmobiles; 3) helmet usage is reliably less frequent for ATVs and snowmobiles than for two-wheeled motorized vehicles; and 4) helmets are worn reliably more frequently on motorcycles than on trailbikes.

It is known from CPSC and ATV manufacturers marketing data that most persons do not use helmets when riding ATVs. This behavior may be explained two ways. Either people do not regard ATVs as inherently dangerous, or ATV riders are primarily individuals who are risk-takers by nature or training or both. If ATV riders fit in this category of heedless risk-takers, then the percentage of non-riders of ATVs who use helmets on other vehicles should be greater than the percentage of ATV riders who use them. There were no reliable differences between ATV riders and non-riders in their use of helmets on any of the other vehicles. The relationship of accident experience to vehicle use will be discussed in conjunction with Experiment 3.

4. Experiment 3

Because the data in Experiment 2 involved some instances of multiple t tests, it was decided to evaluate both rider behavior and estimates of risk and difficulty of operating various vehicles by a between subjects technique. Thus, subjects in this experiment were asked about only one vehicle.

4.1 Method

A total of 1142 adult college students from Metropolitan State College provided information about one of each of six different motorized recreational vehicles: three-wheeled ATV, four-wheeled ATV, trailbike, snowmobile, Odyssey (a Honda dune-buggy vehicle), and jet ski (a motorized water vehicle). Group sizes for the aforementioned vehicles were 294, 170, 181, 176, 151, and 170.

The questionnaire for each vehicle was identical with the following exception: the three-wheel ATV, four-wheel ATV, and trailbike questionnaires included items for single and passenger-riding experience, accident experience, and injury experiences. All six questionnaires required subjects to indicate riding experiences. Three questions required rating scale responses relating to perceptions of risk, chances of tipping over or overturning (i.e., vehicle stability), and difficulty of learning to operate the vehicle. Subjects indicated their responses by using a seven-point rating scale where one always referred to extremely low and seven extremely high. A fourth item required subjects to indicate their perception of the necessity for formal training by using a five-point rating scale from 1 = absolutely not necessary to 5 = absolutely necessary. Two final open-ended questions required subjects to write narrative responses concerning what an operator should and should not do to safely operate the vehicle.

4.2 Results and Discussion

Percentage of persons with riding experience on three-wheel ATVs, four-wheel ATVs, trailbikes, snowmobiles, jet skis, and Odyssey were 31, 28, 62, 36, 25, and 0. Each of the four rating scale questions was evaluated by two-factor between subject ANOVs with six levels of vehicle types as one-factor and two levels of sex and two levels of experience (rider and nonrider) as the other factors. All main effects were significant for each question. Except for the second question (stability perceptions), none of the interactions was significant. For question 2 the interaction of vehicle type and sex was reliable.

Three-wheel ATVs were perceived as involving reliably higher risk than four-wheel ATVs, snowmobiles and Odyssey ($p < .04$). Trailbikes were perceived as reliably more risky and jet skis as less risky than the other five vehicles. For all vehicles, males perceived less risk than females, and persons with riding experience perceived less risk than persons without riding experience.

Chances of tipping over or overturning were perceived as not reliably different for three-wheel ATVs, four-wheel AVTs, and snowmobiles. Those three vehicles were perceived of as involving reliably less chance of overturning than Odyssey, jet skis or trailbikes which showed no reliable differences in stability perceptions. The Sex X Vehicle Type interaction was accounted for by the fact that females perceived ATVs as being more stable and trailbikes as being less stable than males, but no reliable sex differences were found for the other vehicles.

Four-wheeled ATVs were reliably perceived as being easiest to learn to operate with three-wheeled ATVs and snowmobiles next easiest but not reliably different from one another. Odyssey and jet skis did not differ reliably from one another, but both were perceived as reliably easier to learn to operate than trailbikes, which were considered the most difficult of all. Males and riders rated the vehicles as easier to learn to operate than females and nonriders.

The necessity of hands-on training was perceived as reliably greater for trailbikes than the other five vehicles. Males and riders perceived less need for formal training on the vehicles than females and nonriders.

Responses to the open-ended questions regarding knowledge of safety practices were divided into responses of riders and nonriders and sorted into 13 categories. Two major findings emerged from these data. One finding was that ability to express knowledge of specific safety practices or requirements was severely limited. Poteniating the concern with this finding was the fact that frequenceis of response for the categories did not differ reliably between riders and nonriders.

Because of certain specific dangers expressly related to the design of ATVs, riders should be aware that carrying passengers and riding on paved surfaces are extremely dangerous. Further, when riding any unstable or open vehicle, use of a helmet is a very important safety measure. Currently all ATVs provide warnings that say "wear helmets," "read owner's manuals," "don't carry passengers," and "don't drive on roads." The percentages of experienced riders of ATVs who stated these as necessary safety practices were 30%, 11%, 9%, and 5%. Among nonriders of ATVs the percentages were 28%, 11%, 7%, and 7%. An especially important warning that is presented in some owner's manuals is to always keep your feet on the foot pegs. Only 1% of experienced riders stated that practice; 0.5% of nonriders stated that practice. If one is to contend that adequate warnings have been provided on the vehicles or, at least, in owners manuals, and therefore, the reason that drivers disregard warnings is irresponsibility on their part, these data provide no support. Rather, they suggest that the warnings have not been adequately presented. Whether this is because they are too inconspicuous, or the advertising and word-of-mouth information presented by salesmen and others is too contradictory, cannot be established from this study. Clearly, however, the message hasn't been received by the individuals who need it most.

Questions on accident occurrence included in Experiments 2 and 3 enabled us to determine the relative frequency and circumstances of accidents for the various vehicles. The percentages of persons having accidents during single riding and passenger riding were 46% and 54% for ATVs, 58% and 42% for trailbikes, and 68% and 32% for bicycles. The difference was not reliable for ATVs but was for both trailbikes and bicycles. In addition, the accident frequencies while riding double were reliably greater for ATVs than for trailbikes or bicycles. These findings are particularly relevant because carrying passengers on ATVs is so common (Karnes, et al., 1987). Data obtained in the present studies indicate that passenger riding is common on other recreational vehicles as well. The percentage of riders who reported carrying or riding as passengers on three-wheel ATVs, four-wheel ATVs, trailbikes, snowmobiles, and motorcycles were 80, 85, 87, 90, and 96 respectively. Passenger riding was reliably greater on motorcycles than the other off-road recreational vehicles. A sex difference was also found in riding double. More females indicated that they had ridden double than males on all vehicles except bicycles, and in all cases the difference in favor of females was reliable.

5. Conclusions

The results of these surveys show that ATVs are in common use among young adults, teenagers, and even preteenagers, and that they are perceived as being safer than two-wheeled motorized vehicles. The fact that ATVs were rated lower in all categories of risk is the most reasonable explanation for the finding that helmets are less frequently worn while riding on ATVs than while riding other vehicles. It should be noted that the data in these surveys were collected after some widely disseminated TV publicity and attendant newspaper reports concerning the dangers associated with ATVs. Lacking similar data for prior years, we don't know what influence this publicity had, but it surely didn't increase the image of safety. Certainly our results are consistent with frequently stated opinions of consumers' perceptions of ATV safety and stability. For example,

> A big selling point for ATVs. Whether you're buying for fun-riding or work-riding, rest assured that it will be safe-riding. With the additional wheel (or wheels), ATVs offer a measure of stability that can never be found in a conventional motorcycle. Such stability allows the most inexperienced rider to climb aboard an ATV and have the ride of his or her life. This feature can be used to lure over the older folks, or to convince a protective mother to let her young son give that little ATC 70 a whirl around the parking lot. (Staff, Motorcycle Dealernews, 1983).

We can only speculate why public perception of risk for ATVs is not greater, but their innocuous appearance and promotion as a "fun" vehicle and no indication in advertising or, until recently, the news media, of dangers involved has not educated the public to the fact that ATVs require special and complex psychomotor and judgmental skills, and that these skills may involve negative transfer from other vehicles, such as bicycles, that we have all ridden. In particular, the immaturity of children's psychomotor and judgmental skills may account for the large number of deaths and injuries among riders under 16 years old.

The finding that ATV riders have no greater appreciation than non-riders of the safety requirements is a matter of great concern. If injuries and deaths are to be reduced, ATV riders must appreciate the degree and severity of the risks involved in riding ATVs. Safety information and warnings must be provided in a form that will alert and educate users to facts that the vehicles are not as easy and as safe to ride as they may appear to be. Safety information in owners' manuals must provide explanations of the reasons for specific safety practices and the consequences of failure to comply, especially for safety practices where the reader's prior knowledge is likely to interfere with acceptance of the warning (e.g., riding on paved surfaces and riding with passengers).

One of the most serious problems from a safety standpoint is passenger riding. Karnes and Leonard (1986) have shown that children and adults do not have an inherent appreciation of dangers associated with passenger riding on ATVs.This lack of appreciation is undoubtedly related to the present finding that carrying passengers is a common practice on all recreational vehicles. That was not a surprising finding since, with the singular exception of ATVs, at least some models of the motorized-recreational vehicles included in Experiment 2 are designed to accommodate passengers. ATVs are unique as recreational vehicles because the long seat invites passenger use and the dangers associated with passenger riding are so great. Carrying passengers on ATVs is extremely dangerous for several reasons: the passenger may interfere with weight shifting required to turn the vehicle or prevent tipovers or flipovers; the passenger adds more weight to the rear of a vehicle that is already rear heavy; and the vehicle provides a unique danger of having the operator's or passenger's foot being caught by the rear tires. The present studies and the results reported by Heiden Associates (1986) have shown that the presence of passengers on ATVs increases the risk of accident and injury. Since users are unaware of these dangers when carrying or riding as passengers, warnings need to be conspicuously and forcefully presented. Also, explanation of the dangers needs to be provided in owners' manuals to promote user acceptance of the warnings since, for most people, prior knowledge is not likely to facilitate understanding of the reasons for a warning that provides only a simple prohibition against passenger use. Most certainly existing warnings that are accompanied by explanations in owners' manuals such as "the vehicle load limit and seating configuration do not safely permit the carrying of a passenger," will tend to mislead rather than inform the reader of the dangers associated with riding double.

REFERENCES

[1] Downs, T. (1987, September). **Risk of death or injury involving children riding three-wheeled ATVs.** Unpublished manuscript, Houston, Texas.

[2] General Accounting Office (GAO) (1986, November). **Concerns about staff memorandum relating to all-terrain vehicles.** U.S. General Accounting Office, Human Resources Division, Washington, D.C. 20548.

[3] Heiden Associates, Inc. (1986, February 14). **Draft evaluation of CPSC preliminary report on ATV-associated injuries.** Prepared for Somers, Hall & Verrastro, P.C. (Confidential: Attorney-Client Work Product), Washington, D.C.

[4] Karnes, E.W. & Leonard, S.D. (1986). Consumer product warnings: Reception and understanding of warning information by final users. In W. Karwowski (Ed.), **Trends in Ergonomics/Human Factors III** (pp 995-1003). Elsevier Science Publishers B.V. (North-Holland).

[5] Karnes, E.W., Leonard, S.D., Schneider, T., & Rachwal, G. (1987). Factors involved in perceptions of risk in riding all-terrain vehicles. In S.S. Asfour (Ed.), **Trends in Ergonomics/Human Factors IV** (pp 473-482). Elsevier Science Publishers B.V. (North-Holland).

[6] McCarthy, R.L. (1987, June 4). **Deposition of R.L. McCarthy.** Frank Cusimano, Jr. et al. (plaintiffs) vs. American Honda Motor Company, Inc., et al., No. 499782, Superior Court of State of California, County of San Diego.

[7] Newman, R. (1986, September). **Analysis of all-terrain related injuries and deaths.** U.S. Consumer Product Safety Commission, Directorate for Epidemiology, Division of Hazard Analysis (EPHA), Washington, D.C. 20207.

[8] Schacter, L. (1987, August 11). **Update of ATV deaths and injuries.** Memorandum to Commissioners. U.S. Consumer Product Safety Commission, Washington D.C. 20207.

[9] Schpeltzer, D. (1986, September 10). **Recommendations on ATV briefing package.** Memorandum to Leonard DeFlore. U.S. Consumer Product Safety Commission, Directorate for Compliance and Administrative Litigation, Washington, D.C. 20207.

[10] Staff (1983, August). All-terrain vehicles: still the hottest show in town. **Motorcycle Dealernews**, p. 12.

[11] Verhalen, R.D. (1986, June 13). **ATV project: Comparative injury tables.** Memorandum to Nick Marchica, U.S. Consumer Product Safety Commission, Washington D.C. 20207.

Trends in Ergonomics/Human Factors V
F. Aghazadeh (Editor)

IMPROVED SAFETY DIRECTIONS: SAFER WORK WITH GRINDING MACHINES

Anne Seppälä

Department of Occupational Safety
Institute of Occupational Health
Laajaniityntie 1, 01620 Vantaa, Finland

Accident statistics have revealed that severe accidents happpen with hand-held grinding machines. The safety content of manufacturers' manuals for grinding was studied. The contents were compared with the Finnish standards for safe grinding. It was found that the manuals contained many pitfalls and even errors in the text and figures. A new guide for safer grinding was drafted. Special attention was paid to the contents and way of presenting the information. The guide was tested with vocational school students. It was found that the students who read the guide got significantly better scores both in theoretical and practical tests, as compared with those who had not read the guide or had received only the normal vocational instruction for using the grinders. The results demonstrate the value of porper educational material and guidance in the formation and adoption of safe working habits.

1. ACCIDENTS WITH MANUAL GRINDING

It has been noted that serious accidents may occur in the use of hand-held grinding machines. The Finnish National Board of Labour Protection keeps a special register on serious accidents (absence from work for at least one month). 63 serious accidents due to working with grinding machines were documented between 1977-1984. The total number of registered serious accidents at that time was 3328.

Of the 63 grinding accidents, 27 cases occurred in connection with a hand-held grinding machines. 19 of these were analysed as to their main causes (Table 1).

Table 1: Main causes of serious accidents with hand-held grinding machines (n=19)

Cause	n
Grinding method	8
Spark spray	7
Wheel guard	6
Choice of the wheel	4
Flanges and nuts	2
Test use	2

These factors suggest that the accidents caused by hand-held grinding machines are essentially related to lack of knowledge and incorrect performance of the worker.

2. CURRENT MANUALS

Manuals and sales promotion leaflets were obtained from the dealers of grinding machines. As to the safety contents of the sales promotion brochures, they were often statements about the technical safety factors of the machines, such as safety throttle, anti-vibration motor, prevention of the spreading of dust, built-in silencer. It seemed obvious that technical safety was a good commercial quality of the machines.

A content analysis of the actual instruction manuals was carried out to get a view of the safety items of current grinding manuals. 13 manuals were obtained from the dealers of the hand-held grinding machines. The contents were compared with the effective Finnish safety standards for hand-held grinding machines.

18 demands for the safe use of hand-held grinding machines were defined on the basis of the standards. On the average, 10 demands were mentioned in the manuals. The best manual had 15 references to such demands, the poorest one only 6.

The most often mentioned demands were (frequency):

- choice of grinding tool 13
- use of wheel guard 12
- assembling the wheel 11
- use of eye protectors 11

The least mentioned demands were:

- correct working methods
- grinding angle 5
- cutting 3
- hand grip and press 7
- noise hazards 4
- risk of fire 4

The least mentioned demands are closely related to the actual accidents that have occurred with grinding machines. Some manuals even contained errors in the text or figures. These errors may be relevant to the operation safety of the machines, e.g. mixing flanges and blotters or specifications of straight and depressed-center wheels.

3. TESTING A NEW SAFETY GUIDE FOR MANUAL GRINDING

3.1. Contents of the guide

A new safety guidebook was drafted on basis of the standards, accident analysis and content analysis of the current manual.

The guide contained the following 11 specifications:

- choose a correct grinding tool
- inspect the tool
- choose correct flanges and nuts
- fix the wheel
- inspect the hold of the wheel guard
- test the equipment
- protect the schoundings against the spark spray
- use personal safety devices
- see to it that the equipment is in good condition
- store the machine and wheels properly.

Each of the specifications contained 1-7 central safety statements, and was illustrated by a related picture. The information was expressed as direct instructions of what to do, what is the safe way of handling, etc. The statements were usually in the form of positive directives.

3.2. Subjects and methods

The safety guide was tested with students from a vocational school. A group of first-year students from a metal working department formed the experimental group, which read the guide.

Two control groups were not given the safety guide. These were: a group of first-year students in the automobile assembly lines and a metal working line; and a group of second-year students from the metal working lines. The first control group and the experimental group had a similar knowledge base. The first-year groups had not received any formal vocational education in the use of grinding machines.

Both of the groups had, however, used grinding machines, mostly at home or in junior high school.

The second control group had received instruction at the vocational school in the use of grinding machines and had practised their use.

A questionnaire concerning the main safety specifications in the use of grinding machines was given to both of the first-year groups, as well as to the second-year students, at the beginning of the first semester. After about three months, the guide was given to the experimental group. A second survey with the same questionnaire was carried out within a week from the hand out. The second-year students participated only in the first survey.

3.3. The results

The results of the questionnaire are shown in Table 2.

Table 2: Means and standard deviations of the experimental and control groups in the questionnaire survey on safety knowledge regarding the use of hand-held grinding machines.

Groups	I survey			II survey		
	N	Mean	Standard deviation	N	Mean	Standard deviation
Experimental group	41	4.88	2.50	36	9.58	2.06
1st-year controls	38	5.42	2.36	28	7.89	1.93
2nd-year controls	28	6.46	1.91	-	-	-

In the first survey, both of the control groups got slightly better results than the experimental group. The difference of the means was significant between the second-year controls and the experimental group ($t=2.97$, p .01), but not between the first-year controls and the first-year experimental group ($t=0.99$, n.s.). The results indicate that the second-year students had received more teaching and had more experience in the use of grinding machines than the students who were at the beginning of their actual vocational training.

In the second survey, the experimental group was better than both of the control groups. The mean of the experimental group was significantly higher than that of the first-year controls ($t=3.37$, p .001), and of the second-year controls ($t=6.26$, p .001). Reading of the safety guide thus significantly improved the safety knowledge of the students.

3.4. A practical test

A practical test was also arranged for the first-year groups. The students were given a grinding machine and a selection of various wheels and flanges. They were asked to simulate a grinding process by starting from the choice of the wheels and flanges, and proceeding to the working methods and storing of the equipment. The performance was observed and rated according to the safety specifications stated in the new safety guide.

The experimental group performed correctly about 52 % of the tasks compared with 43 % by the control group. The experimental group paid special attention to the spark spray, correct grinding methods, use of safety goggles, and storing the equipment. As regarde accident potential, these activities are essential. The control group was especially poor in noticing the dangers of spark sprays and controlling the fitness of the wheel to the wheel guard.

4. CONCLUSIONS

Reading the safety guide significantly improved the knowledge and performance of the students. The gain was positive even when compared with the knowledge of the students who had received the vocational training in the use of grinding machines. This suggests that traditional educational methods do not emphasize safety aspects as much as would be desirable.

Manuals instructing in the use of various tools and machines should also contain relevant, essential safety specifications for safer use of the equipment. Safety would thus be an integral part of the teaching of students in the use of tools and machines.

Trends in Ergonomics/Human Factors V
F. Aghazadeh (Editor)

OCCUPATIONAL HAZARDS INVOLVED WITH CHIMNEY SWEEPING

Paavo TIITTA

Tampere University of Technology
Occupational Safety Engineering
P.O.Box 527
SF-33101 Tampere
Finland

The aim of this study was to determine the accident risks of chimney sweeps, to produce information on their tasks, mortality and incapacity for work for the use of the health services and to produce suggestions to improve the occupational safety of chimney sweeps. The theoretical basis of the study were stress-strain model, accident investigation models based on hazard danger model and risk behavior thesis. The material was based on data from statistics, a questionnaire, interviews and visits to the work sites. The results showed extremely high accident indicators and classifications that differed from the usual, and quite apparent selection in occupation because of one's health.

1. INTRODUCTION

In the chimney sweeps' occupation there is a high risk of accidents and different illnesses. Insurance companies in Finland have rated this occupation as having one of the highest premiums paid because of the number of potential accidents. Studies have reported conditions such as irritation eczemas, chronic coughing and infections of respiratory organs. Both Sweden and Denmark have shown an excess of the incidence of death from cancer is this occupation [2], [3], [4]. The deaths from cancer among chimney sweeps were noticed as early as 1775 in England [6].

Illnesses are connected to the soot dust and especially to the compounds of vanadium, sulphur and polycyclic aromatic hydrocarbons in the soot. However, the amount of metals and other carcinogens differ largely depending on the fuel used (fuel oil, gas oil, wood, coal, waste products). Due to the burning of oil, the geographical origin and the refining process of the fuel as well as the burning reaction affect the composition of the soot. The acid burning gases have led to rapid corrosion of safety railings and ladders.

In total, there are about 1200 chimney sweeps in Finland. Of them 400 are municipal, while the rest are, for the most part, self employed. A small number of chimney sweeps work either part time or for private contractors. Because they are few in number, little attention has been paid to the occupational safety of chimney sweeps. However, the outdoor work, changing working environment, the types of wages, working in different places and many hazardous substances in work involve special features to decrease the working safety. Also new tasks like the maintenance of air shafts have not become known to the health services.

The study was started on the initiative of the chimney sweeps' labor unions and was carried out by Tampere University of Technology and the Institute of Occupational Health. The main objectives of the study were:

(i) to define the occupational accidents,
(ii) to give a total picture of different tasks and to produce basic data for the mortality, early retirement and incapacity for work,
(iii) to produce suggestions to improve the occupational safety of chimney sweeps.

2. MATERIAL AND METHODS

To define the occupational accidents, the study material was taken from both statistics of the insurance companies and the occupational accident register held by the National Board of Labor Protection. The accident rate and ratio and the number of lost working days were taken from the statistics. For the data from the occupational accident register 19 variables were classified to clarify the course of events leading to the injuries. The accident data were taken from the years 1980 - 1984, when 236 accidents had occurred. In handling the accident notification forms sent by the insurance companies and the State Accident Office, the National Board of Labor Protection sorts out accidents which have led to at least three days' incapacity and occupational diseases in which the disability caused by the disease is at least 10 % regardless of the length of absence.

To analyze the accident data, a model was formed, based on the Finnish accident study model [1] and the hazard-danger model (Figure 1). According to the model used an accident can occur when the worker and the cause of the accident are in the same place at the same time. The cause of an accident is a force, energy, material or construction that has an influence on the normal way of working. Some determining factors of accidents can be for example the subjective experience of risks, or risk behavior when estimating the danger.

The data for the mortality study and incapacity for work were taken from the statistics kept by the Statistical Centre of Finland. Also a questionnaire of occupational accidents, subjective illnesses and working conditions and the use of personal protectors was made as well as visits to the work sites. The total number of questionnaires received was 329, and the total of 60 work sites were investigated in different seasons in order to gain insight into constructional, hygienic and ergonomic factors of the work.

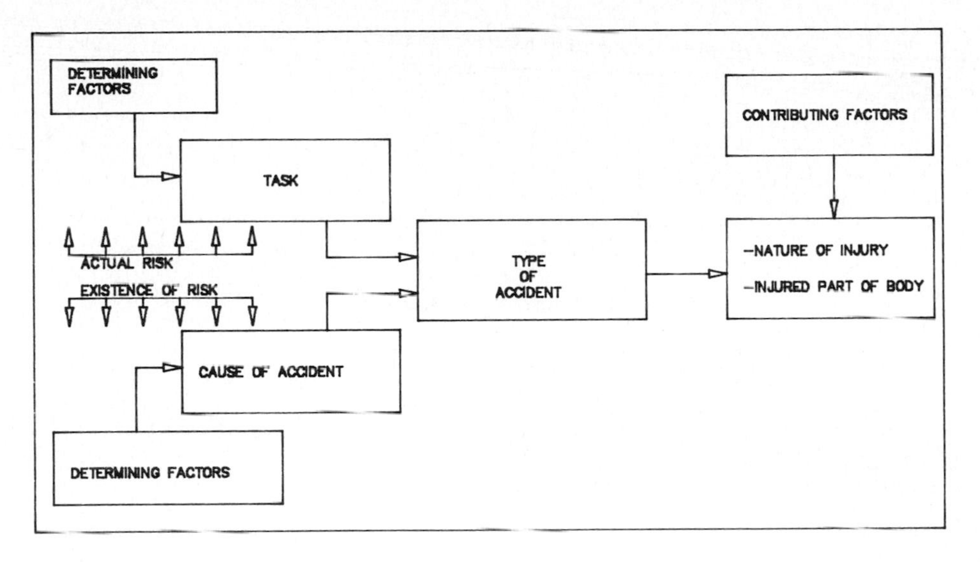

FIGURE 1
The model used in the accident study.

3. RESULTS

The accident study results showed extremely high accident indicators and classifications that different from the usual. The accident frequency (accidents causing at least three lost days per one million workhours) was almost three times greater and the accident ratio (accidents causing at least three lost days per one thousand workers) over three times greater than in all industrial accidents in 1984. Over 60 per cent of the accidents were due to slipping, tripping and falling over, falling from high working or passage platforms or falling from a height in connection with the collapse or fall of the platform. In 1984 these types of accidents made up 26 % of all industrial accidents [5]. In building, the most hazardous industry in Finland, the percentage was 33. Both in all industrial branches and in building the type of accident most common was striking against objects, while in the chimney sweeps' occupation this came after the causes mentioned above with a percentage of 16. The main classifications of chimney sweeps' accidents are seen in Figure 2.

The statistical study of the mortality and incapacity for work among chimney sweeps with the information obtained from the questionnaire and the interviews showed quite an apparent selection of occupation because of one's health. In numbers, the main reasons for both mortality and incapacity for work were circulatory illnesses and musculoskeletal diseases, but statistically the differences were small. During the visits to the work sites the ergonomics were noted to be poor. The temperature of the working environment varied a great deal and the working postures were similar. Not enough attention has been paid to the maintenance of chimneys, stoves, boilers and air shafts in architectural planning.

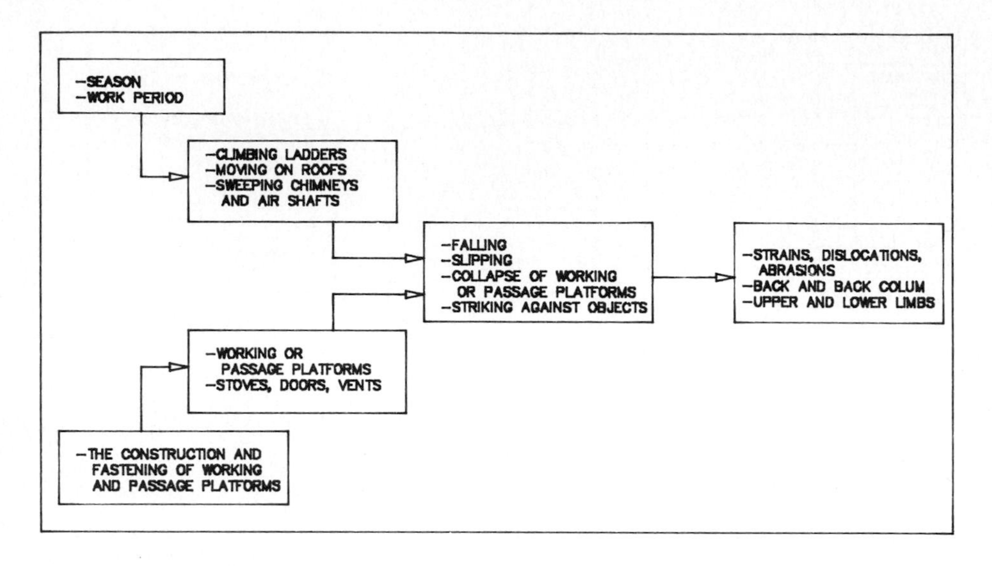

FIGURE 2
The main classifications of chimney sweeps' accidents.

4. DISCUSSION AND CONCLUSIONS

In this study the risks were seen as determining factors of both accidents and occupational illnesses. As a result the accident indicators (accident rate and ratio) differed clearly from the statistics. The model used in the accident study gave results that were seen both in the questionnaire and visits to work sites while observing the constructional safety.

Though one must take into consideration that the study material was not extensive, it was possible to point out ways of improving work safety. The most important improvements in this were in the areas of safety inspections, training in estimating the risks of the occupation, safety concepts at the design stage of both chimneys and air shafts and the construction regulations, especially on safety railings and ladders. More research is needed to determine the effectiveness of statistical data in occupational accident prevention, the improving of the informative value of the statistics and the integration of this technique with the behavioral approach to occupational safety.

As results, mortality, occupational illnesses and diseases and early retirement were seen both in the statistics and in the questionnaire and the interviews with the workers. An evident selection because of one's health is to be seen in the occupation, and an epidemiological study of chimney sweeps should be undertaken. Also the concentration of dust in the air shafts should be studied. When comparing with the results of other foreign studies one must consider the differences in chimney sweeps' work and working conditions between countries as well as differences in attitudes in protective matters.

5. ACKNOWLEDGEMENTS

This paper is based on the author's thesis, submitted of Tampere University of Technology in partial fulfillment of the requirements for the M.Sc. degree. The author wishes to thank Professor Markku Mattila, Tampere University of Technology, and L.Sc. Helena Mäkinen, Institute of Occupational Health, for support, interest and guidance during the work. The study was supported by a grant from the Ministry of the Interior.

6. REFERENCES

[1] Centre for Industrial Safety, Investigation of an accident (in Finnish). Helsinki 1987. 31 pages.

[2] Gustavsson, P., Gustavsson, A. & Hogsted, C., Mortality among Swedish chimney sweeps (in Swedish). Arbete och Hälsa 1986:15. 26 pages.

[3] Hansen, E.S., Mortality from cancer and ischemic heart disease in Danish chimney sweeps: a five year follow up. American Journal of Epidemiology 117(1983)2, pp. 160 - 164.

[4] Hansen, E.S., Olsen, J.H. & Tilt, B., Cancer and noncancer mortality of chimney sweeps in Copenhagen. International Journal of Epidemiology 11(1982), pp. 356 - 361.

[5] National Board of Labor Protection, Industrial Accidents 1984. Official Statistics of Finland, Tampere 1985. 158 pages.

[6] Wade, L., Occupation and cancer. Archives of Environmental Health 9(1964)3, pp. 365 - 374.

Trends in Ergonomics/Human Factors V
F. Aghazadeh (Editor)

SCALE VALUES FOR WARNING SYMBOLS AND WORDS

S. David Leonard
University of Georgia

Edward W. Karnes and Theresa Schneider
Metropolitan State College

Previous research on the differences in signal words by Leonard, Matthews, and Karnes, [1] found perception of risk was not affected by which signal word was used. Because attentional factors were minimized and because some organizations do have prescribed meanings for signal words in terms of the amount of hazard involved, it is possible that there might be differences in attention paid to signal words. Both ranking and rating of the hazardousness of situations using six signal words and six symbols demonstrated that the population stereotypes of risk do differ.

1. INTRODUCTION

A significant factor in promoting safe behavior is the effectiveness of warnings. In order for warnings to be useful it is necessary that they be seen, understood, and followed. We will not discuss the perceptual aspects, despite the fact that a short trip down one of our highways indicates that ergonomic considerations of visual perception are commonly ignored. Our concern is primarily with the last of these factors, that of compliance with warnings. However, it seems likely that the factor of understanding will strongly influence that of compliance, and attending to warnings may be influenced by factors other than basic perceptual phenomena.

There is some controversy in the literature regarding the effectiveness of warnings. For example, Dorris and Purswell [2] found very little compliance with certain warnings, and McCarthy, Finnegan, Krumm-Scott, and McCarthy [3] concluded from an extensive review of the literature that there was no evidence of effectiveness of warning signs. In addition, McCarthy, Robinson, Finnegan, and Taylor [4] suggested that a plethora of warnings could be counterproductive by competing with one another for attention. The implication is that warnings may be omitted. It is, of course, possible that warnings may be less effective than they should, as pointed out by McCarthy, Horst, Beyer, Robinson, and McCarthy [5]. They compared a warning with an instruction label and found that when the word "warning" was used but fewer details were placed on the label more errors were made than when the label "general information" was used. Indeed, in our own work examining warnings used on all-terrain vehicles (ATVs) we have found instances in which warnings can be misleading. Consider the statement, "Do not ride on road. Vehicle does not conform to federal motor vehicle safety standards." This might easily be interpreted to mean, "It is not legal to ride this on the road (i.e., don't get caught)." This warning would probably not deter the rider from riding on other paved surfaces which pose the same problems of unstable handling. We must, therefore, keep in mind the necessity of clarity of expression in any sort of sign, as was so ably shown by Chapanis [6].

Contrary to the view that warnings have no effects, there is evidence that warnings can be effective. Both Wright, Creighton, and Threlfall [7] and Godfrey, Allender, Laughery, and Smith [8] present evidence that concern for warnings is greater when the perceived hazard is greater. Further, Godfrey, Rothstein, and Laughery [9] presented evidence for the effectiveness of warnings in naturalistic settings, although compliance decreased when the cost (in time or effort expended) was high. In line with these results Leonard, Matthews, and Karnes [1] found both rated perception of risk and ratings of the likelihood of compliance were affected by the presence of consequences of failure to comply. A simulation study described by Karnes and Leonard [10] resulted in some subjects refusing to continue after having read the appropriate warning.

Previous research (Leonard, Matthews, and Karnes, [1]) has suggested that certain segments, at least, of the populace show minimal differentiation between the words caution, warning and danger when attached to warning signs. A related result using different signal words was found by Wogalter, Godfrey, Fontenelle, Desaulniers, Rothstein, and Laughery, [11]. A possible explanation for this result is that individuals know, or think that they know, the consequences of various hazards. Thus, when a hazard is mentioned, they attach no differential meaning to the signal words. Some evidence for this interpretation was found by Leonard, Matthews, and Karnes [1] in a significant interaction between perception of risk for specific items and whether or not consequences were presented. Ratings increased significantly for items which might seem innocuous, such as a car battery, but for which serious consequences, such as a possible explosion, were presented. Items rated as more serious without consequences presented didn't change appreciably when consequences were given.

In sum, it seems that people are likely to attend to warnings to the extent that they think they are dealing with something that really is dangerous. The problem is what to do about truly dangerous objects that are not generally perceived as dangerous.

One might assume that it is appropriate to include consequences on all warnings, and that this would take care of the apparent lack of information contained in the signal words. Although it is clearly good ergonomic practice to include consequences in warnings, we felt that obtaining more basic information about signal words would be worthwhile from two standpoints. First, in the Leonard, Matthews, and Karnes [1] experiment attentional factors were at a minimum, because all subjects were directed to examine the warnings. But some organizations do have prescribed meanings for signal words in terms of the amount of hazard involved, such as those promulgated by ANSI, FMC Corporation, and various military and governmental groups. Because the general public is likely not aware of, much less familiar with, these standards, it is useful to see how well they conform to population stereotypes. Second, it is conceivable that there might be differences in attention paid to a warning, if the signal word generates a higher perception of risk.

Two experiments were conducted to scale the apparent risk associated with a set of words and symbols frequently used on warning signs.

2. EXPERIMENT 1

Because various standards call for using the words caution, warning, and danger for different levels of hazard, it was decided to determine the rank order of these items in terms of their population stereotypes. In addition, several other warning type words and several symbols commonly used in warnings were examined.

2.1. Method

2.1.1. Subjects

Subjects were 83 male and 112 female undergraduate students at a large state college in Colorado with ages ranging from 16 to 49. Another 7 subjects were discarded for failure to use the ranking procedure.

2.1.2. Materials and Procedure

Six words and six symbols were arranged in five different random orders. The symbols used were the radiation symbol, the skull and crossbones, the "do not" symbol of a slash across a circle, a triangle with an exclamation point inside, an illustration of a hand in gears, and a hand receiving an electric shock. The last two symbols were taken from the FMC Handbook [12]. The words were "attention," "be careful," "caution," "danger," "deadly," and "warning."

Subjects were asked to assume that they saw a warning sign with one of the words or symbols on top of it and to rank the signs from 1 to 12 in order of hazardousness.

2.2. Results

The correlation between male and female rankings was .97; therefore, only the overall rankings are reported in Table 1. In general, it appeared that words or symbols that were more specific in terms of consequences were ranked higher. One interesting finding was the relatively low rank (fifth overall) of the symbol for radiation. It seems likely that this resulted from a lack of knowledge of the meaning of the symbol. Anecdotal support for this notion was the fact that several individuals asked what that symbol was during and after the experimental sessions.

These results provide support for the notion that information about the consequences of an act has the strongest effect on perception of risk.

3. EXPERIMENT 2

Because the differences found could have been forced by the ranking procedure, it was deemed useful to confirm the results by using a rating procedure. In addition, it seemed worthwhile to test the hypothesis that the radiation symbol was not known by some of the respondents.

3.1. Method

A total of 90 male and 130 female undergraduate students were drawn from two institutions in Georgia. Their ages ranged from 17 to 51.

Table 1

Overall Means of Rank Order Scale and Rating Scales

Presented as Words				Presented as Symbols			
Item	Mean Rank	Symbol	Word	Item	Mean Rank	Symbol	Word
Deadly	2.32	4.93	4.95	Skull & Crossbones	2.42	4.87	4.72
Danger	4.16	3.53	3.45	Electric Shock	4.63	3.97	3.69
Warning	6.44	2.70	2.53	Radiation	6.11	3.47	4.30[a]
Caution	7.31	2.34	2.21	Hand-in-Gear	6.40	3.15	2.78
Be Careful	9.59	1.70	1.69	Slashed Circle	8.93	2.44	2.33
Attention	9.86	1.72	1.77	Exclamation Point	9.78	1.93	1.81

a - Presented as a word in the Word condition.

Two sets of ratings were obtained. One stimulus set, the Symbol set, used the same items as were used in the ranking procedure. The second set, the Word set, was the same except that the word radiation was used, instead of the symbol. Both sets were presented in the same order. Only one order was used, inasmuch as no order effect was found in the ranking task which was presumed to be more sensitive to order effects. Subjects were instructed to use a rating scale with 5 indicating a hazard that could be life-threatening and 1 representing a minimally hazardous event.

3.2. Results

As in the ranking procedure the male and female results corresponded quite closely; therefore, only the combined data are presented. It may also be seen in Table 1 that the ratings corresponded quite closely with the rankings. A correlation of -.97 was obtained between the mean rank order and the mean ratings for the Symbol group. Such a correlation is reliable at the .01 level. This supports the notion that the ranking data were not merely an artifact of procedure.

There was also a close correspondence between mean ratings in the symbol and word conditions, $r = .97$. Thus, the difference between the mean ratings for the word radiation and the symbol is all the more obvious. A chi-squared statistic computed using the frequency of each rating category for the word radiation and the radiation symbol indicated the difference was significant, $\chi^2_{(4)} = 27.18$, $p < .001$.

One other analysis was performed on the symbol group data. It compared the responses of subjects to the symbol signals with their responses to the word signals. A total score was obtained for each subject for the ratings of all six words and for the ratings of all six symbols. The mean of the total ratings for symbols was 2.96 greater than the mean for words. This difference was significant, $t_{(132)} = 8.12$, $p < .001$.

4. GENERAL DISCUSSION

It seems apparent that in an abstract sense there are differences in the weights assigned by the public at large to different signal words and symbols. Further, in a loose sense, the population stereotype fits the order of seriousness proposed by sign-makers (e.g., FMC, [12]), that is, danger, warning, and caution are ranked in that order. This suggests that, despite the findings of Leonard, Matthews, and Karnes [1] and Wogalter, et al. [11] we should not necessarily use only one warning signal word. One might be tempted to use the signal word deemed most serious in all cases, just to be sure of getting the user's attention, given the findings of Wright, Creighton, and Threlfall [7] and Godfrey, et al. [8] that concern for warnings is greater when the perceived hazard is greater. We would reject such an approach, because it would likely work for only a very limited time followed by extinction. In addition, if something like the skull and crossbones were used in a wide variety of situations, the extinction would probably generalize to poisons and reduce its effectiveness in warning about them. On the other hand, if limited to extremely hazardous conditions where the dangers might not be appreciated by the casual user, such use might be condoned. For example, on an ATV the U-Haul corporation used a skull and crossbones on either side of the word "warning" with the statements, "ATCs flip every which way-see users guide," and, "ATCs corner strangely-see users guide." Because potential riders would very likely consider these warnings as involving serious consequences, the level of risk perception associated with the symbol would not undergo much extinction. But use at a cross-walk, for example, would erode the general usefulness of the symbol.

On a global basis it might also seem that the use of symbols would be better than words, because the symbols were rated as involving more hazard. This is not necessarily the case for two reasons. First, the symbols and words were not based on the same concepts. Indeed, in the direct comparison the word radiation produced greater ratings of hazard-ousness than the symbol. Further, several subjects indicated that they had not comprehended some of the other symbols, such as the pinch point of the gears. It is likely that the symbols are rated higher because most of them included either a directly observable or or a well-known consequence. That is, most college students (and most of the population at large) have learned that the skull and crossbones means poison, and they have also learned that many poisons are fatal. It is interesting to note that the word deadly was rated even higher in hazardousness than the skull and crossbones. Clearly, an experiment to determine whether sym-bols have some inherent advantage over words would be of interest. And symbols whose meaning can be universally recognized need to be devised.

A basic problem is that many hazards are highly improbable on any one occasion, but they have a high enough aggregate probability to justify precautions being taken. An example is the use of a hammer, as described by Dorris and Purswell [2]. It is likely that the ultimate answer to warnings about these hazards will be some form of training taken at a very early stage of life. Otherwise, extinction of the fear will have taken place, and it will be difficult to shape the appropriate behavior. Education seems to have effects as seen in the declining number of smok-ers and the increasing number of seat belt wearers in our society. One area in which ergonomics professionals can help is in the development of training programs and in devising symbols that display consequences.

ACKNOWLEDGEMENTS

The authors thank Drs. John E. Hesson, G. William Hill, IV, and Harvey B. Milkman for assistance in obtaining subjects, as well as Dr. Stuart B. Katz and Mr. Hajime Otani for constructive criticism of the manuscript.

REFERENCES

[1] Leonard, S.D., Matthews, D., and Karnes, E.W. (1986). How does the population interpret warning signals? **Proceedings of the Human Factors Society--30th Annual Meeting.** Human Factors Society: Santa Monica, CA, 116-120.

[2] Dorris, A.F. and Purswell, J.L. (1977). Warnings and human behavior: Implications for the design of product warnings. **Journal of Products Liability**, Vol. 1, 255-263.

[3] McCarthy, R.L., Finnegan, J.P., Krumm-Scott, S., and McCarthy, G.E. (1984). Product information presentation, user behavior, and safety. **Proceedings of the Human Factors Society--28th Annual Meeting.** Human Factors Society: Santa Monica, CA, 81-85.

[4] McCarthy, G.E., Robinson, J.N., Finnegan, J.P., and Taylor, R.K. (1982). Warnings on consumer products: Objective criteria for their use. **Proceedings of the Human Factors Society--26th Annual Meeting.** Human Factors Society: Santa Monica, CA, 98-102.

[5] McCarthy, G.E., Horst, D.P., Beyer, R.R., Robinson, J.N., and McCarthy, R.L. (1987). Measured impact of a mandated warning on user behavior. **Proceedings of the Human Factors Society--31st Annual Meeting.** Human Factors Society: Santa Monica, CA, 479-483.

[6] Chapanis, A. (1965). Words, words, words. **Human Factors**, 7, 1-17.

[7] Wright, P., Creighton, P., and Threlfall, S.M. (1982). Some factors determining when instructions will be read. **Ergonomics**, 25, 225-237.

[8] Godfrey, S.S., Allender, L., Laughery, K.R., and Smith, V.L. (1983). Warning messages: Will the consumer bother to look. **Proceedings of the Human Factors Society--27th Annual Meeting.** Human Factors Society: Santa Monica, CA, 950-954.

[9] Godfrey, S.S., Rothstein, P.R., and Laughery, K.R. (1985). Warnings: Do they make a difference? **Proceedings of the Human Factors Society--29th Annual Meeting.** Human Factors Society: Santa Monica, CA, 669-673.

[10] Karnes, E.W., and Leonard, S. D. (1986). Consumer product warnings: Reception and Understanding of warning information by final users. In W. Karwowski (Ed.) **Trends in Ergonomics/Human Factors III.** Amsterdam: Elsevier. 995-1003.

[11] Wogalter, M.S., Godfrey, S.S., Fontenelle, G.A., Desaulniers, D.R., Rothstein, P.R., and Laughery, K.R. (1987). Effectiveness of warnings. **Human Factors**, 599-612.

[12] FMC Corp. (1985). **Product Safety Sign and Label System.** FMC: Santa Clara, CA.

Trends in Ergonomics/Human Factors V
F. Aghazadeh (Editor)
Elsevier Science Publishers B.V. (North-Holland), 1988

SAFETY ISSUES RELATING TO AGRICULTURAL MACHINES MODIFIED FOR DISABLED OPERATORS

Terry L. WILKINSON and William E. FIELD*

Breaking New Ground Resource Center
Agricultural Engineering Department
Purdue University
West Lafayette, Indiana USA

In 1986, the authors conducted a study for the National Institute of Handicapped Research** to evaluate modifications made to self-propelled agricultural machines intended for use by operators with physical disabilities. It became apparent to the authors that little consistency existed between the modifications evaluated with the exception that all were functional and did the task desired. Numerous safety and ergonomic concerns were observed which had the potential for causing injury, unnecessary risk, and operator discomfort.

1. INTRODUCTION

Agricultural production has the highest disabling injury rate of any occupation in the United States. It is estimated by the National Safety Council that in 1986, 170,000 farm and ranch related injuries were reported of which at least 3400 resulted in permanent disability [1]. In addition to work-related disabling injuries, farmers and ranchers also experience permanent disabilities from motor vehicle accidents, recreational activities, accidents in the home, and diseases such as arthritis and stroke. The Breaking New Ground Resource Center at Purdue University, established to assist farmers and ranchers with physical disabilities, estimates that approximately 560,000 American farmers and agricultural workers are hindered in the completion of essential work-related tasks due to a variety of physical handicaps.

One of the most serious barriers that this population must cope with is the operation of agricultural machines essential in most phases of crop production. Consequently, there have been many attempts to modify the more frequently used machines, such as tractors and combines. It will be the purpose of this paper to discuss the potential hazards associated with these, often unique, modifications and present general suggestions for improving their safety and function.

2. METHODS

The Breaking New Ground Resource Center was established in 1979 and has assembled an extensive collection of information on work-site modifications to make agricultural operations more

* Terry Wilkinson is a doctoral research assistant in agricultural safety and health, and William Field is Professor of Agricultural Engineering at Purdue University.

** Research sponsored by the Department of Education/National Institute of Handicapped Research. Grant Number G008535172.

accessible for farmers and ranchers with physical disabilities. Many of the ideas relate to the access and operation of agricultural machines and include such items as manlifts and control modifications. Some of these devices have successfully solved the problem of operating equipment but often presented problems and hazards with respect to their design and operation. Recognizing these problems and the potential for serious injury, the authors conducted a study in 1986, sponsored by the National Institute of Handicapped Research, to evaluate manlifts and control modifications that would have potential for use on various types of self-propelled agricultural machines. In choosing the machines that would be evaluated from those previously identified, the following criteria were used: quality of construction, workmanship, adaptability, safety, and function. The authors identified 29 modified agricultural machines that met the criteria and provided the basis for completion of the study.

A standardized evaluation form was developed to record data obtained from each machine. Data was collected on techniques being employed for accessing agricultural machines, control modifications (primarily modifications to the brake and the clutch pedals), and accessories (grip bars, seating, transfer devices). The form was designed to obtain the following information during the evaluations: materials used in construction, quality of construction, type of power system used on lifts, types and location of controls, accessibility and ease of use, ease of service and maintenance, adaptability to other machines, and potential hazards the modifications presented to the operator.

The modifications evaluated represented a diversity of techniques being employed to transfer a mobility-impaired operator from the ground to the operator's station. These modifications included: powered chairlifts (Figure 1); platform lifts with the operator standing (Figure 2), sitting or in a wheelchair (Figure 3); sling lift; and a tractor designed to carry the operator in his or her wheelchair.

The results of the overall evaluations were assembled and published by the authors to provide a resource for designers and builders of these modifications, farmers, and rehabilitation professionals [2].

Figure 1. Powered Chairlift Mounted on a John Deere 4440 Tractor

Figure 2. Platform Lift Mounted on a Ford TW-20 Tractor

Figure 3. Wheelchair Lift Mounted on an International Harvester 186 Tractor

3. SAFETY AND ERGONOMIC ISSUES

Hazards are present on machinery if they are: unrecognized, recognized but ignored, or recognized but a means to remedy the hazard is not feasible or practical. Many of the hazards and ergonomic problems observed during the evaluations of modified agricultural machines appeared to be present because of the following key factors:

1. There was a lack of plans, standards, guidelines, or documented experience to follow when designing these modifications on agricultural machines for use by operators with physical handicaps;
2. The majority of the modifications were made by farmers or local craftsmen without the benefit of available rehabilitation engineering information or the assistance of a professional engineer;
3. The desire of the farmer to put function over safety, or comfort.

Off-highway vehicles are modified infrequently compared to highway vehicles designed to assist those with physical disabilities. Tormoehlen and Field [3] found that there were no directly applicable standards to follow when modifying agricultural machines. Wilkinson [4] examined various guidelines and principles used by several of the modification designers and found they lacked consistency. Adaptive devices and wheelchair lifts in the automotive industry are generally designed to follow the Veterans Administrations established standards. It was determined that some of the standards on lifts and control modifications could be used as a reference when making modifications to agricultural machines [5,6]. It was also determined that adaptive devices and lifts designed for automobiles could not be readily adapted to agricultural machines.

4. CONCERN OVER EXPOSURE TO PRODUCT LIABILITY LITIGATION

There have been some agricultural machine modifications designed by engineers or manufacturers which have resulted in one-time only ventures because of the concern over the potential exposure to product liability litigation. There have been other engineers and machinery manufacturers who have refused to make needed modifications because of the product liability issue. Some of the major concerns with respect to product liability include:

1. The possibility of the disabled operator suffering an injury or a secondary injury while mounting or operating modified agricultural machines;
2. There have been no studies conducted to determine the safety record of disabled farmers who use modified agricultural machinery;
3. There have been no studies or testing conducted to determine the level of risk that the modifications present to operators with physical handicaps;
4. The abilities and capabilities of each disabled operator are uniquely different and require specific design considerations for each individual.

Product liability is not the only barrier that keeps manufacturers from making modifications for disabled agricultural operators. A second major obstacle is the existence of a wide diversity of designs, sizes, makes, and models of agricultural machines which makes it extremely difficult and expensive for a manufacturer to make modifications commercially available. A third obstacle is the limited market for modified agricultural machinery as compared to the demand for adaptive devices and lifts on automobiles and vans.

5. RESULTS OF HAZARD IDENTIFICATION AND ERGONOMIC EVALUATION

The major hazards identified on the 29 modified agricultural machines are described in the following sub-sections.

5.1. Power Systems

Hazardous power systems were identified on 45 percent of the machines evaluated. The most common of these were machines equipped with lifts powered by electric winches that utilized a flexible steel cable to lift the operator or lift assembly on which the operator rode. This practice is not recommended by the manufacturer of nearly all portable winches which were incorporated into these designs. The other hazardous power system was found on lifts that operate off the machine hydraulics which required that the tractor or combine be running for the lift to operate. Using a manlift while the machine engine is running exposes the operator to the hazard of potentially starting the machine while it is in gear, accidentally shifting the transmission into an operating position throwing the operator off, or trapping the operator if the engine should fail preventing use of the lift system.

5.2. Pinch Points

Pinch points were present on over 90 percent of the machines evaluated. They were most common on platform lifts that were operated in a standing position. The operator could get his hands, fingers, arms, feet or toes caught and possibly suffer serious injury. This problem is especially acute with respect to operators with spinal cord injuries who cannot sense touch on their lower extremities and might not be able to detect an injury occurring.

5.3. Lack of Protective Railings

The lack of protective railings was primarily noted on platform lifts which are operated in a standing position (see Figure 2). Nine of 10 standing platform lifts evaluated lacked protective railings. The operators of these lifts have limited mobility and a slight movement or a slip could cause them to fall. This is even a greater concern considering the distance (between 4-9 feet) of vertical lift required to access these machines.

5.4. Lack of Protective Devices on Chairlifts

Thirteen chairlifts were evaluated during the study (see Figure 1). A sudden unexpected movement could cause an unrestrained operator to lose his balance and fall out of the lift seat. Armrests, footrests, and seat belts are devices that could be used to secure the operator to the seat to prevent a fall and possible injury. Four of these chairlifts lacked any type of protective device. The most common device used were armrests placed on one side of the seat.

5.5. Restricted Access To Machine

The placement of the lifts on 65 percent of the machines evaluated restricted access to the operator's station. This situation forced an able-bodied farmer to either use the lift to access the agricultural machine (noted in 31 percent of the cases), or blocked one side of the machine from access (noted in 34 percent of the cases). This situation violates an existing American Society of Agricultural Engineers (ASAE) standard regarding the requirement for two methods of access and egress.

5.6. Poor Switch Placement

If a lift is left in a raised or partially raised position, the disabled operator could not access the machine by himself. The placement of a switch on the machine that the operator could use to lower the lift when he is on the ground would remedy this problem. It is equally important to have access to a control switch near the operator's station. Poor switch placement was found on 72 percent of the machines evaluated.

5.7. Clutch and Brake Modifications

Clutch and brake modifications were mounted on 65 percent of the machines (brake modifications shown in Figure 4). Clutch levers that would be difficult to remove were found on 55 percent of the machines while 34 percent of the brake levers were found to be difficult to remove. This feature is important because on most agricultural operations more than one person will operate a machine. If modified controls are in the way, they may prevent an able-bodied operator from having full control of the machine. A hazard found with the modified clutch controls involved 52 percent of the cases being designed without a means to lock it in the disengaged position. Clutch modifications need to be designed in this manner to allow the operator to freely operate other controls with his/her hand.

Figure 4. Modified Brake Controls Mounted on a Ford 4610 Tractor

6. CONCLUSIONS

1. Designers and builders of agricultural machine modifications for disabled farmers have used a variety of concepts most of which were not documented or tested to ensure reliability and safety. There will continue to be safety and ergonomic-related problems until a documented set of guidelines, recommended practices and/or standards are developed for agricultural machines that will be modified for operation by those with physical handicaps.
2. Automotive wheelchair lifts and adaptive equipment are not readily adaptable to self-propelled agricultural machines. The standards and guidelines used for these devices have potential applications to modifications made on agricultural machines to improve reliability and safety.
3. The power system on a manlift will most likely be an electric winch despite the warnings by most manufacturers of these devices that these units are not suitable for this application.

4. Lifts will most likely lack adequate protective railings to prevent falls and improve stability during transfer. The presence of pinch points on lifts were common and exposed operators to the potential for serious injury. The mounting of a lift on a machine will also likely restrict access to the operator's station by able-bodied operators.
5. The strategic placement of switches on modified agricultural machines to insure easy access by a disabled operator will most likely not be considered.
6. Clutch and brake modifications will most likely be designed to be left in place when an able-bodied operator uses the machine which might hinder usual operation of the machine.

7. RECOMMENDATIONS

1. Considering the number of farmers and agricultural workers that are continually added to the ranks of those with physical handicaps, evaluation of various types of agricultural machines should be continued to identify new design concepts that provide the greatest accessibility for those with various types of physical limitations.
2. Research into the safety and durability of these modifications need to be conducted to determine the level of potential risk they present to operators with physical handicaps. An accident reporting system should be established to monitor injuries and identify problem areas.
3. A procedure for testing lifts and control modifications that will be mounted on self-propelled agricultural machines should be developed.
4. Consideration should be given to compiling guidelines and applicable standards related to agricultural machine modifications. Eventually, a set of recommended practices, guidelines and/or standards for the modifications of agricultural machines for operators with physical handicaps should be developed. The possible standard organizations involved would be the American Society of Agricultural Engineers (ASAE) and the Society of Automotive Engineers (SAE).

8. SUMMARY

Farmers/ranchers and agricultural workers with physical handicaps face many barriers with one of the most serious being the operation of agricultural machines. Machines equipped with man-lifts and control modifications helps remove this major vocational barrier. Regardless of how essential the operation of agricultural machines is to the successful farm, any modifications made to these machines should be designed to insure the health and safety of the disabled operator and others.

To date, most of the modifications to agricultural machines have been homemade by farmers who do not have the assistance of professional engineers or rehabilitation engineering professionals. In addition, these individuals have few plans, standards, guidelines or documented experiences to use when designing necessary modifications. In some cases, the desire of the farmer or rancher to put function over safety resulted in some modifications which were poorly constructed with respect to materials used and workmanship, hazardous to operate, and designed without consideration of basic ergonomic principles. The benefits to this unique population, of all the recent advances in rehabilitation technology, have yet to be realized.

REFERENCES

[1] Accident Facts 1987 Edition. National Safety Council, Chicago, Illinois, 1987.

[2] Wilkinson, T.L., and Field, W.E. Modified Agricultural Equipment: Agricultural Equipment Manlifts for Farmers and Ranchers with Physical Handicaps. (Breaking New Ground, Purdue University, West Lafayette, Indiana, 1987)

[3] Tormoehlen, R.L. and Field, W.E. Safety Implications of Physical Handicaps For Operators of Agricultural Equipment. West Lafayette, Indiana: Purdue University, Department of Agricultural Engineering, June 1983. (ASAE Paper No. 83-5045)

[4] Wilkinson, T.L. Evaluation of Self-Propelled Agricultural Machines Modified for Operators with Serious Physical Handicaps. Masters dissertation, Purdue University, December 1987.

[5] Veterans Administration. Standard Design and Test Criteria For Safety and Quality of Automatic Lift Systems For Passenger Motor Vehicles. Report No. VAPC-A-7708-3, June 28, 1977.

[6] Veterans Administration. Standard Design and Test Criteria For Safety and Quality of Special Automotive Driving Aids (Adaptive Equipment) For Standard Passenger Automobiles. M-2, Part IX, G-9, March 1978.

Trends in Ergonomics/Human Factors V
F. Aghazadeh (Editor)

AGE RELATED DIFFERENCES IN INJURIES IN THE HOME

Sheik N. Imrhan

Department of Industrial Engineering
The University of Texas at Arlington
Arlington, Texas 76019

Tabulated data published by the National Electronic Injury Surveillance System have been analyzed to determine trends and relationships across age groups for products associated with a relatively high estimated frequency of injuries but not necessarily being the cause of the injuries. The data indicated an upward trend in injury rate over the years for all age groups, with only a few exceptions. For the 6 most hazardous products, the 0-4 yr and 65+ yr age groups had the greatest injury rate. The interpretation of the data is rendered difficult because causative factors have not been identified and because injury rate is affected by the frequency of use of these products, another unknown factor.

INTRODUCTION

Accidents in the home are more frequent than reported data indicate. Very often they are due to incompatible relationships between people and systems; for example an elderly person falling from steep stairs designed for capabilities adequately possessed by adults but which decline beyond a certain age. Though many accidents and injuries are not usually reported to a hospital or other system which accumulates data for subsequent summarization or analysis, the available data may still be very useful. Even when the exact causes of injuries are not known, data which denote an association between injury type (or product) and frequency or severity of injury may still help in efforts to understand accident occurrence across products and age groups.

Of particular interest is the elderly. With their declining cognitive, physiological and physical capacities (Lawton [1], Welford [2], and Fozard [3]), environments and systems with which they formerly interacted successfully may now pose difficulties. Some of these declining capacities, utilized often for accomplishing their daily living activities, are related to visual and auditory perception, psychomotor skills, memory, learning, muscular and cardiorespiratory strength and endurance (Lawton [1], Welford [2] and Fozard [3]). Age related trends in the frequency of occurrence of injuries in the home may reveal areas of difficulties in cases where the elderly and other age groups actually use certain products. Of course, such data are not comprehensive; they would not indicate problem areas where the elderly avoided using certain products for which they perceived the task as being too difficult or not absolutely necessary. Reported data, on a large scale, do not usually contain this level of detail.

One of the most important sources of product related injuries has been the National Electronic Injury Surveillance System (NEISS). The NEISS Data Highlights [4] represents tabulated data from 5,939 hospitals in the U.S. and its territories that reported having emergency departments. The NEISS tabulated data indicate that a particular product was associated with an injury. It does not state whether the product actually caused the injury. The actual number of reported cases of injuries has been used to derive an estimated number for the total population of the U.S. and its territories. Products associated with an annual estimate of less than 10,000 emergency department injuries for the whole U.S. population were not included in the NEISS data tables.

The purpose of this study was to determine age related differences in injuries in the home from an analysis of recently published NEISS tables of injury data. The data from the elderly age group (65+ yr) was used as a standard for comparison, and the analysis focused on those products which were associated with a relatively high incidence of injuries in the elderly. An arbitrary value of 30 injuries per 100,000 persons over 65 yr in 1982 was used. Twelve such products were identified, but data were not available for 4 of them in 1978.

METHOD

Tabulated data for the 12-month periods ending in September for the years 1978 (or 1979) to 1982 were used. Data prior to 1978 were not directly comparable because of some important differences in the reporting procedure. The figures used were estimates, not actual number of injuries, per 100,000 persons within each of the following age groups: 0-4, 5-14, 15-24, 25-64, and 65+ yr. The estimates were made by inflating the actual number of accidents from 5,939 hospitals to represent the whole population. These hospitals were treated as a sample representative of institutions with emergency treatment department in the U.S. and its territories.

RESULTS AND DISCUSSION

The NEISS data were analyzed and interpreted in two main ways: (1) the trend in estimated number of injuries across age groups for products associated with higher frequency of injuries, and (2) chronological trend from 1978 (or 1979) to 1982, in injury frequency for high incidence products. For most products, the trend in the estimated number of injuries across age groups was similar from year to year. An increase in estimated injury frequency over the years occurred for all age groups, in general, though not in the same proportion. A typical example is for 'beds' which had a relatively large increase from 1978 to 1982 (Table 1).

Table 2 shows the products for which the estimated number of accidents per 100,000 persons 65 yr and over exceeded 30. Because of the similarity of trend across years, mentioned earlier, the latest year, 1982, which had the highest figures and which was typical, was selected for illustration (Table 2).

For the 12 most hazardous home products (or sets of products) given in Table 2, 7 showed an approximate U-shaped distribution of estimated

Table 1. Increase in injuries associated with beds over 5 years for different age groups.

	Estimated number of injuries per 100,000 persons per age group				
	Age Group				
YEAR	0-4	5-14	15-14	25-64	65+
1978	355	86	29	22	59
1979	423	101	42	25	74
1980	399	89	52	27	106
1981	558	104	48	33	116
1982	603	105	51	31	127

Table 2. Home products with estimated number of injuries exceeding 30 per 100,000 elderly for 1982.

Product	Estimated number of injuries per 100,000 persons per age group				
	Age group (yr)				
	0-4	5-14	15-24	25 64	65+
Stairs, steps, ramps and landings	685	276	367	334	446
Chairs, sofas and sofa-beds	568	94	67	69	132
Beds (excl. water-beds)	603	105	51	34	127
Bathtubs and non-glass shower structures	131	34	34	34	69
Tables	806	105	60	48	68
Carpets and rugs	52	17	16	18	62
Ladders and stools	27	19	24	65	56
Desks cabinets, shelves and footlockers	221	94	56	46	50
Power home workshop saws	5	13	35	47	44
Porches, balconies, open side floors, etc.	109	42	29	27	35
Lumber and paneling	91	97	100	87	35
Unpowered cutlery and knives	28	58	113	85	31

number of injuries over age (defined by age groups). The 0-4 yr and 65+ yr groups had the largest estimates, with the other age groups having much smaller estimates. These relationships must be viewed with caution, however, when making comparisons across age groups because of the greater likelihood for injuries on 0-4 yr and 65 yr to be reported and treated at hospitals, compared to the other age groups. The estimates for these 2 age groups may therefore be exaggerated; this would be especially true for minor injuries. Thus for beds in 1982, the estimates, based on reported number of cases, for the elderly is 3.7 times (127/34) that for the 25-64 yr group but the true number of injuries was probably much less than the factor of 3.7. Nevertheless, another point of view would be to regard reported injuries as severe enough to make a product hazardous. Thus the estimates in the NEISS tables, though not having the same proportion to the true unknown number of injuries in all age groups, can be regarded as an index of the hazard associated with a particular home product. Thus beds may perhaps be regarded as about 3.7 times as hazardous to 65+ yr old compared to 25-64 yr olds.

The 6 most hazardous home products for the elderly (1-6 in Table 2) seemed to be those whose usage required skillful performance in walking, climbing, sitting from a standing posture, standing from a sitting posture and maintaining equilibrium. These tasks require mainly the ability to perceive visual stimuli, coordination of complex movements to maintain the body in static and dynamic equilibrium, speed of reaction, muscular strength and attention - the capacities for which there has been a steady decline beyond 65 years of age.

From 1978 (or 1979) to 1982 the percentage increase in the estimated number of injuries (Table 3) for most hazardous products (Table 2) for almost all age groups was much greater than the corresponding increase in population for each age group. Overall, the 5-14 yr, 15-24 yr and 25-64 yr groups had the smallest percentage increases - - 49, 54 and 52%, respectively; while the 0-4 yr and 65+ yr groups had the largest increases (90% and 81% respectively). By comparison, the percentage changes in population within age groups, for the 0-4, 5-14, 15-24, 25-64 and 65+ yr groups, from 1978 to 1982 were 9.8, -4.7, 8.1, 8.3 and 9.6%, respectively (U.S. Bureau of Census [5]). Only two products showed a decrease in the estimated number of injuries - ladders and stools, and power home workshop saws, and both were for the 15-24 yr group. Of all the products, beds seem to be the most hazardous, having not only a very large estimated number of injuries (Table 2) but also a very large percentage increase over the years.

Table 3. Percentage increase in estimated number of injuries in different age groups from 1978-1982.

Product (1978-1982)	Percentage increase (%) Age group (yr)				
	0-4	5-14	15-24	25-64	65+
Stairs, steps, ramps and landings	53	37	28	35	55
Beds (excluding water-beds)	136	21	78	45	115
Bathtubs and non-glass shower structures	71	97	114	72	88
Tables	44	27	63	62	133
Carpets and rugs	378	68	58	43	97
Ladders and stools	15	6	-2	15	8
Porches, balconies, open side floors, etc.	80	69	65	48	58
Unpowered cutlery and knives	94	97	123	132	130
Product (1979-1982)					
Desks, cabinets shelves and footlockers	26	30	26	47	97
Lumber and paneling	117	78	99	104	116
Chairs, sofas and sofa beds	30	28	28	20	40
Power home workshop saws	31	22	-13	5	38

CONCLUSIONS

The results of this study indicate that injury rate due to home products are increasing steadily over the years and, with very few exceptions, for all age groups. It is possible that the rate of increase over years, as indicated by the data, may be different from the true, unknown, injury rate because of imperfections of the data gathering system. However, the relatively large number of hospitals sampled - 5,939 - gives some strength to the level of validity of the data. The fact that the data were derived from reported injuries to hospitals implies that the actual number of injuries (reported plus unreported) would most likely be several times the published estimates for different age groups, and that this factor may be different for different age groups. The products associated with the greatest estimated number of injuries for the elderly were also those which had relatively smaller estimates for the 5-14, 15-24 and 24-64 yr groups. The data in the NEISS tables do not provide information concerning the cause of injuries, and do not reflect the relative frequency of contact by persons of different age groups, with particular products. Hence a product with a low reported number of injuries for the elderly, say, may be due largely to the fact that that age group refrained from using that product, for some particular reason(s). The data, however, shows the association between reported injury frequency and product by age group. Severity of injuries has not been documented by age group and, hence, cannot be analyzed in that manner.

REFERENCES

[1] Lawton, M. P., The Impact of the Environment on Aging and Behavior; in Birren, J.E. and Schaie K. W. (Eds.) Handbook of the Psychology of Aging, New York, Van Nostrand Reinhold, 1977 pp. 276-301.

[2] Welford, A. T., Signal, Noise, Performance and Age, Human Factors, 1981 23 pp. 97-109.

[3] Fozard, J. L., Person-Environment Relationships in Adulthood: Implications for Human Factors Engineering, Human Factors, 1981 23 pp. 7-27.

[4] National Electronic Injury Surveillance System Data Highlights, Directorate for Epidemiology, Consumer Product Safety Commission, Washington 1978, 1979, 1980, 1981, and 1982.

[5] U.S. Bureau of Census, Current Population Reports, 1985.

Trends in Ergonomics/Human Factors V
F. Aghazadeh (Editor)

AGING AND ENVIRONMENTAL SAFETY

Paul K. H. KIM

School of Social Work
Louisiana State University
Baton Rouge, Louisiana, U.S.A.

Introduction: Graying America

America today joins the developed countries on the continents of Europe and Asia which are characterized as "graying nations". The median age of the U.S. population is expected to increase by 23% in the 20 year span between 1980 and 2000; namely, from 30.4 to 37.3 years of age. In 1985, older Americans who were 65 years old or older numbered 28.5 million, representing about 12% of the U.S. population. This indicates an increase of 11% or 2.8 million more older persons during the five years since 1980. Further, this figure will increase to 35 million people, representing about 13% of the total population, at the turn of the century (AARP, 1986). Today, it is also estimated that about 5600 Americans attain the age of 65 every day (Harper, 1987). Moreover, American older women outnumber their counterparts today with a sex ratio of 147 to 100, and this figure will increase to a higher ratio of 251 to 100 for older persons 85 and older (AARP, 1986), which is the fastest growing segment of the U.S. population. Additionally, the minority elderly, including Blacks, Hispanics, Native Americans (Indians), and Asian-Pacific Islanders will outgrow the 1980 figure by 50%, which is a faster rate than that of their white counterparts, whose increase will be about 35% by the year 2000 (U.S. DHHS, 1986).

Unfortunate concomitances of the advancement of age can be characterized by poor physical and mental health, poverty, age discrimination, etc. Health conditions of older persons are depicted as 30.6% in fair or poor health; 44.2% having activity limitations; and 38.2 and 14.8 days a year of restriction and bed disability, respectively (National Center for Health Statistics, 1980).

Environment, living quarters and the neighborhood community in particular, play a significant part in the process of growing old in America, simply because the environmental conditions affect behaviors of older persons. In fact, environmental impact on human aging is so serious, that a sub-discipline of gerontology, called "Environmental Gerontology", has been established, and research and training activities are burgeoning today (Lawton, 1980).

The Objective of the Paper:

This paper is designed to discuss an ecological theory of environmental gerontology, based on which to examine environmental problems facing older Americans and to bring a public awareness, particularly in environmental engineers, to the issue. By having a more aggressive and systematic research and development undertaken by ergonomics engineers, the quality of the American elderly will be enhanced, and perhaps the

United States will reach one step closer to a Haven for the Elderly.

The World Dictionary defines environment as "all the conditions, circumstances, and influences surrounding, and affecting the development of an organism or groups of organisms". According to Lewin's Field Theory (1951), environment is a significant functional part of human behavior, i.e., Human Behavior = f (P x E), meaning that human behavior is a function of the interface between person and environment. The environment (fields of organisms) that surrounds human beings is so powerful that it is considered to be the most influential factor on either normal or abnormal human growth and development. Namely, most schools of social work subscribe a functional school of thought which emphasizes environmental impact on socially objectionable behaviors, including social problems. Economic poverty, social injustice, even mental illness, etc., are related to "problems of living" caused by environmental givens, rather than individual fault or inferiority. Behaviorists support this notion with a contention that stimulus produces response, the "S-R" theory.

On the other hand, the humanistic school of psychology focuses on individual freedom of choice, in that the ultimate decision-maker as to the "external push" (environmental givens or stimulus) is the right and capability of the individual who faces the situation. This implies then that environment is to be perceived, interpreted, and internalized by the person, and that that person defines the value of the environment and behaves as he believes, irrespective of external conditions or circumstances. Briefly, this means that personal freedom in determination is much stronger than environmental reasoning, though "freedom and reasoning" are perhaps interrelated and complementary to each other.

Lawton (1980) classifies environment as follows:

1. Personal environment: Consisting of significant individuals in the life of a person, including parents, spouse, children, friends, and others closely associated with.

2. Group environment: Force that influences an individual to act as part of a group, such as group pressure, social norms, and perhaps cultural mandates.

3. Suprapersonal environment: The aggregate of individuals in physical proximity which is usually expressed as the average or the mode. Examples of such environmental characteristics are "the average age of neighbors, their average income, the most frequent ethnic group in a housing project."

4. Physical environment--the natural or man-made environment: Empirically measurable as numbers, grams, centimeters, seconds, etc.

Despite the fact that all types of environment listed above have an impact on older persons in one way or another, the focus of this paper is on physical environment. Unlike social scientists who study the dynamics of human behavior vis-a-vis human relations and socioeconomic and cultural conditions, ergonomists perhaps take the perspectives of physical environment, particularly in terms of physical safety factors of homes (living quarters), community (neighborhood), and work place. In view of the majority of older persons retired from the work force, this paper is designed to concentrate on physical aspects of (1) the home environment where one spends most of his/her time as he/she grows older, and (2) situations of immediate neighborhoods including areas where needed services are available to older persons. Of an estimated 29 million older Americans today, about 27.5 million (95%) are affected by those two environments, by the simple virtue of living in their own homes, apartments, and groups homes.

<u>Ecological Theory of Aging</u>

Lawton and Nahemow (1973) expand Lewin's field theory, and "person" in particular. They specify the set of a person's "competence" as the domains of biological health, sensorimotor functioning, cognitive skill, and ego strength. These personal characteristics are more or less internal, and it has been empirically documented and proven that human capabilities gradually reduce as one advances in age. On the other hand, Lawton and Nahemow advocate Murray's concept of "environmental press", or pressures or influences, more or less external to the person which force individuals to react. Pressures include physical and social conditions such as conditions of neighborhoods, social and political systems, etc.

Combining "competences" and "environmental press" together, Lawton and Nahemow posit what is called "ecological theory of aging". This theory explains human aging in general, and physical and mental conditions of the elderly in particular, as a product of interface between the person's physical, psychological (cognitive), and mental capabilities, and his/her perceptions of environmental conditions. Depending on a person's past life style and perhaps his/her competencies, given environmental situations are viewed differently in terms of degree of worth and interest of the person who faces them in his/her life. The theory further suggests that the person who is physically and mentally capable of dealing with his/her environment, can cope with a varying degree of environmental pressures: Namely, the person with a higher competence can handle a stronger press in the environment, and vice versa. Subsequently, based on this theory, one can establish an "adaptation level" which could be interpreted as an indicator of a person's life and problem coping skills in a specific time and place.

The ecological theory of aging hypothecates that competent elderly can cope with a strong environmental push to a certain degree yet if environmental press is too weak, he/she develops maladaptive behaviors. The opposite situation also creates a negative affect on older persons, in that those with lower competencies cannot cope with a stronger environmental press, and when/if environmental push endures, they develop a maladaptive behavior as well. Furthermore, it suggests that "the less competent the individual, the greater the impact of environmental factors on that individual," known as "environmental docility hypothesis" (Lawton, 1980, 14).

As such, one significant implication of this theory is that, the optimum adaptation level for older persons can be established by maintenance and enhancement of "competencies" in conjunction with the development and modification of "environmental press". Although these two important domains of older persons are affected by research on and services of social and health care programs, as well as social sciences and physical engineering, health professions and psychology are more or less focused on the issues related to "competencies", while sociology, political science, urban planning, and disciplines of physical engineering, concentrate on environmental issues. The profession of social work is unique in that it brings both domains together and provides evaluative measures on the products of respective disciplines. Improving competence alone cannot assure older persons' optimum quality of life, and vise versa. According to the theory, both should be interfaced and complementary to each other. Along the same line of reasoning, the theory may suggest that environment should fit the needs of the individual with rich resources and safety, and subsequently the individual should be able to maintain and even develop new competencies.

Environmental Problems facing Older Persons

It should be remembered that as people get older, they unwittingly develop physical and mental impairments. About 85% of American elderly have one or more chronic illnesses, and 50% of them must limit their activities. The old-olds (75 years old and over) experience mental disorders (such as Alzheimer's and other types of dementia) in their later life more than any other age group.

For the treatment and deterrence of physical and mental illnesses, American older persons consume a disproportionate number of prescription drugs, about 25 percent of the total of such medications dispensed (Williamson, et al., 1980). An unfortunate complication for one who has several chronic conditions is the need for multiple drugs, which may cause adverse drug reactions unless carefully controlled. Thus, among those who are hospitalized, about 5% are being treated for adverse drug reaction and the elderly are over-represented among such patients. Mental problems--depression or confusion, which are prevalent among older persons, are common results of adverse drug reaction (Watson, 1982). To illustrate the point further: 35% to 40% of all tranquilizers are prescribed for older persons. Unless taken with extreme care, older persons are prone to become addicted to Librium and, if withheld, many develop an emotional dependency and physical withdrawal symptoms; Mellaril, a strong tranquilizer, reduces physical activity to the point at which mental atrophy occurs (Butler and Lewis, 1982).

It follows, then, that the advancement of human age is bound to bring physical and mental weaknesses, and consequently requires social as well as environmental resources with which older persons can maintain their lives to the fullest potential. For this reason, perhaps, any design or plan for the development and/or modification of the physical environment to better meet the needs of the elderly seems to be more realistic and less risk oriented. The rest of this paper, therefore, focuses on physical environment problems facing older persons today in the United States--homes and communities in particular.

Home Environment:

As one gets older, he/she spends more time at home. Seventy percent (70%) of all housing units headed by the elderly are owner-occupied homes, and 80% of them are free and clear of mortgage. The realization of an American dream to have a home is their pride, so that despite the fact that housing often has deficiencies and is sub-standard, they desire to keep it because of its "symbolic value" (Lawton, 1980). Moreover, upon retirement with a fixed income, the elderly cannot afford to buy another home of the same size and specificity. Unfortunately, it is reported that 21.3% of all accident victims are older persons, and about two-thirds of them occur around the house. In view of the fact that elderly women live longer than their male counterparts, more women than men experience accidents in old age. Accident in older women cause 4 days of disability per year, as compared to one day for older men (Hendricks & Hendricks, 1981). Thus, accidents are the 4th leading cause of death for older persons, with women having a 59% higher risk than men.

According to the U.S. Consumer Product Safety Commission (1985), elements within the household physical environment that may cause accidents for older persons include:

1. **Electrical Cords:** Standard 18 gauge extension cords can carry 1250 watts and require care if one uses several appliances attached to a single cord, as it can become overheated and catch fire. Extension cords stretching from wall plugs across hallways may cause a person to trip, possibly resulting in broken bones. Moreover, continued use of damaged (as well as frayed or cracked) cords shock and cause fire.

2. **Rugs, Runners and Mats:** It is reported that in 1982, over 2500 older persons were treated in hospital emergency rooms for injuries that were blamed on rugs and runners around the house. For too often falls result in chronic pain or even fatal injuries for the elderly.

3. **Heaters:** Many homes occupied by older persons were built before the 40's and 50's, and are not energy efficient, requiring portable heaters, gas and/or electric, to keep them warm. Such heating appliances installed too close to furniture and flammable materials such as curtains and rugs may cause extensive house fires. In addition, older persons are vulnerable to being suffocated with carbon monoxide poisoning created by old and/or nonexistent ventilation systems and/or unhealthful indoor air polluted by appliances that use gas or kerosene.

4. **Clothing fires:** Older persons account for a significant proportion (70%) of deaths resulting from clothing fire. The height of the kitchen range and types of portable heaters (either gas flame or electric) naturally constitute dangers for older person. Poor vision heightens the potential dangers of such appliances still further.

5. **Low lighting and glare:** Light fixtures on the ceiling require older persons to use a ladder or stool to change bulbs, making for a difficult and dangerous task. Unfortunately, in 1982, more than 1500 elderly were treated in hospital emergency rooms for fall injuries resulting from standing on such items. Moreover, in the event the person replaces the bad bulb with an incorrect size, heat may be trapped in and result in fire.

6. **Chimney:** A clogged chimney can cause a poorly-burning fire as well as poisonous fumes and smoke coming back into the house, or perhaps even spars ashes to a serious fire.

7. **Hot water temperature:** High temperatures (more than 120 F degrees) cause burns sustained with hot water from the faucet. With age, it becomes more difficult for the older person to distinguish between hot and cold.

8. **Dark areas in the house:** In the dark, older persons easily become confused and disoriented. They may easily run into walls and/or trip over other household objects and fall, often times resulting in broken bones. Hip fractures sustained by falls are very common in elderly women.

9. **Medication:** Medication bottles are not only child-proof but also adult-proof. Older persons who have arthritis of the fingers experience difficulty in opening medication bottles. Although instructions for opening bottles are generally found on the top of the cap,

small and/or worn-out letters are difficult to be discerned by older persons whose eye sight is poor. The elderly who have to take three or four medications every day for their chronic diseases experience additional difficulty, in that the time and dosage of medications are often incorrectly kept. Thus, drug overdose or under-dose is not unusual among the elderly, and subsequently their disease is not properly treated, so that many elderly must be readmitted to the hospital.

10. Bedding fire: It is reported that 42% of deaths related to mattress and bedding fires were attributed to older persons. Such a fire is caused by smoking in bed, over heated electric blanket, and heating pads.

11. Power tools: In 1982 more than 5200 older persons were treated in hospital emergency rooms for injuries sustained by power tools used at home. Wrong size fuses, improperly connected ground wires, and subsequent electric shocks, missing guards from power tools, etc. are contributing factors to injuries related to power tools.

12. Flammable and volatile liquids: Vapors of flammable and volatile liquids, such as gasoline and kerosene, not only are toxic if inhaled, but also travel through the air and if contact is made with heat or flame, can result in fire.

13. Stairs without handrails and lighting systems: Handrails support older persons as they climb up and down stairs and can prevent them from a tumbling fall on the stairs. Loosened handrails and uncomfortable grips may contribute to accidents in older persons.

14. Narrow surfaces of, and loose carpet on, the stairs: Even a small difference of step surfaces and heights of the stairs can cause a fall. Protruding nails, steps edges which are blurred or hard to see, deep pile and/or patterns of dark colored carpeting, etc. make older persons vulnerable to falls.

Community Environment:

Just as anyone else, older persons wish, and have to go out of their homes for shopping, visitations, church, banking, etc. Lawton (1980) identifies some community activities of the elderly in an urban area as follows:

Activities	Median %	Modal distance
Visitation of children	98	less than 10 blocks
Grocery shopping	87	less than 3 blocks
Physician's Office	86	more than 20 blocks
Shopping (other than grocery)	70	
Church	67	less than 6 blocks
Bank	64	less than 6 blocks
Visitation with friends	61	less than 6 blocks
Visitation with relatives	57	
Beauty/Barber shop	40	
Restaurant	31	more than 20 blocks
Park	30	less than 3 blocks
Clubs/Meetings	29	more than 20 blocks
Entertainment	19	more than 20 blocks
Library	18	

Thus, older persons are most frequently out in the community for visitations, shopping, physician's visits, church activities, and banking. Except doctor's offices, the elderly commonly use community facilities closer to their living quarters, within about 10 blocks. As much as they can, the elderly try to situate themselves at a closer distance to their children, friends, and relatives.

Lawton also presents from the U.S. Bureau of the Census' Report on Annual Housing Survey (1973-1976), older persons' perceptions of their environmental situations summarized as follows:

Perceptions	Exists
Street traffic	37.2%
Street noise	24.6
Industrial activities	20.1
Airplane noise	19.5
Inadequate street lighting	18.5
Neighborhood crime	15.5
Streets need repair	14.3
Trash, litter	14.2
Street impassable (snow, mud)	7.9
Odors, smoke	7.9
Run-down houses	7.4
Abandoned structures	5.6
	Inadequate
Public transportation	32.8
Shopping	15.8
Medical resources	12.1
Police protection	8.4
Fire protection	4.5
Schools	2.4

According to the report above, the top five environmental presences that concerned urban older persons are street-related (noise, traffic, and needed repair), neighborhood crime, and trash/litter. Therefore, these conditions plus industrial activities and odors/smoke in their community are most frequently blamed as reasons why the elderly wish to move.

An Ergonomic Approach to Environmental Problems facing Older Persons

Ergonomics as "humanside" engineering is a long overdue science. As much as human beings enjoy the products of physical engineering which have contributed immensely to the convenience and comfort of living conditions and to the improved quality of life, they suffer as well, with inadvertent by-products. The time has come to consider the human aspects of engineering; time not for the sake of engineering, but for the betterment of human life.

One aspect of the ergonomic approach to alleviating human sufferings is prevention, in which engineers not only re-evaluate their inventions in the past, but test the safety of new products, particularly relative to old age. As systems theoreticians contend, the preventive approach to problem-solving is much cheaper and much less tragic than any after-maths. To further illustrate the point: Engineers today should learn from human experience--predicaments facing older persons in this case, and invent preventive measures which can be applied to existing products used in homes and communities. In addition to public education, it is imperative to invent preventive apparatus which warn older persons to possible dangers. Because even warnings may not be enough, safety apparatus may be necessary to cut off functions of electricity, power tools, etc. before any danger occurs. What this implies is that, it is wise to have simple gadget built into the electrical system which cuts off electricity automatically when the elderly person, or anyone for that matter, uses bulbs of higher than the capacity of the wire; heat detector systems that trigger water sprinkler systems when unexpected fire is detected in a kitchen and on a bed; fuel lines that cut off automatically when heat is trapped; sensory lighting systems to light areas when a person is present; etc. In the community, the traffic system has a built-in structure that controls traffic flow. Unfortunately, streets today are more speed-oriented than human oriented, and older persons, who walk slower than younger persons, are extremely scared of fast moving traffic.

Another ergonomic approach to human safety is multidisciplinary. This approach calls for a close working relationship between physical engineering and human service professions. The latter disciplines would provide the former with the impact of physical engineering on human behaviors and social issues, and develop and implement policy and program strategies by which ergonomic products reach to older persons regardless their social economic status. This approach is designed to encourage cooperation between disciplines and to improve the community environment in which older persons, and all residents in that matter, can enjoy safety. We have a law that mandates sidewalk cracks be repaired and home-accident insurance purchased, but we have no law to guarantee safe streets; we have a law not to disturb our neighbors with noise, but there is also a law to permit factories to create noise and air pollutions.

It follows, then, that physical engineering and the profession of social work must work together to minimize human miseries. There must come a time in which one hand appreciates what the other hand does, and vice versa. Through this mutual effort, people, young or old, shall have "physical Walden Two" at least.

BIBLIOGRAPHY

AARP (American Association of Retired Persons). (1986). **A Profile of Older Americans.** Washington, DC: Author.

Butler, R.N. & M.I. Lewis. (1982). **Aging and Mental Health** (3rd ed.). St. Louis, MO: C.V. Mosby.

Harper, M.S. (1987). **Federal supports in mental health.** A lecture delivered at the Center for Life Cycle and Population Studies of the Louisiana State University, Baton Rouge, Louisiana.

Hendricks, J., & C.D. Hendricks. (1981). **Aging in Mass Society: Myths and Realities.** (2nd ed.). Cambridge, MA: Winthrop Publishers.

Lawton, M.P. (1980). **Environment and Aging.** Monterey, CA: Brooks/Cole Publishing Co.

Lawton, M.P., & L. Nahemow. (1973). Ecology and the aging process. In C. Eisdorfer & M.P. Lawton, eds., **Psychology of Adult Development and Aging,** Washington, DC: American Psychological Association.

Lewin, K. (1951). **Field Theory in Social Science.** New York, NY: Harper & Row.

Murray, H.A. (1938). **Explorations in Personality.** New York, NY: Oxford University Press.

National Center for Health Statistics. (1980). **Health in the United States: Chartbook.** Washington, DC: U.S. Government Printing Office.

U.S. DHHS. (1986). **Aging America: Trends and Projections.** Washington, DC: U.S. Government Printing Office.

U.S. Consumer Product Safety Commission. (1985). **Safety for Older Consumers.** Washington, DC: Author.

Watson, W. (1982). **Aging and Social Behavior: An Introduction to Social Gerontology.** Belmont, CA: Wadsworth.

Trends in Ergonomics/Human Factors V
F. Aghazadeh (Editor)

MANAGEMENT OF CARPAL TUNNEL SYNDROME

B. Thomas Harter, Jr., M.D.*, Kathleen C. Harter, M.D. and
Frank W. Archer, M.B.A.

Information about the treatment of carpal tunnel syndrome secondary to work-related activity in a patient population seen over a three year period is discussed.

1. INTRODUCTION

Carpal tunnel syndrome, one of several upper extremity conditions included under the broad heading of cumulative trauma disorders, is the most common peripheral compression neuropathy, and is one of the leading causes of impairment for the production worker.[1-4] It is the opinion of these authors that conservative medical management of cumulative trauma disorders, including carpal tunnel syndrome, involves both the treatment of the worker and the modification of the work environment. Surgery should be considered only when the condition is severe, or when conservative treatment has proved ineffective. A surgical operation should be considered prophylactic and not therapeutic; the symptoms may be treated successfully by surgery, but the underlying etiology is not corrected. Frequently, the worker has permanent occupational restrictions after surgery.

2. CLINICAL EXAMINATION

Classic symptoms include numbness and tingling of the affected hand, including all of the fingers with the exception of the ulnar side of the ring finger and small finger. Parasthesias are often nocturnal and awaken the patient from sleep. Upper extremity pain is very common and can radiate from the hand to the shoulder or neck. Weakness and thenar atrophy are signs of more advanced stages of the disease. Clinical signs indicative of carpal tunnel syndrome include a positive Phalen's test, a positive Tinel's sign, and decrease in light touch, vibration, and two-point discrimination. Because the differential diagnosis is quite variable, a thorough history and physical examination with particular attention to the nervous system is mandatory to rule out other causes.[5]

3. NERVE CONDUCTION STUDIES

Most of the symptoms of carpal tunnel syndrome, and the majority of physical findings are subjective in nature. However, because carpal tunnel syndrome is considered a compensable disease and often work related [6,7], objective findings are extremely important to allow the physician to make correct definitive decisions concerning treatment. Nerve conduction studies provide this information. Following electrical stimulation

* Hand Surgery, 4006 Dupont Circle, Louisville, Kentucky 40206-6048

of a nerve, a response from sensory and motor nerve fibers can be recorded. A motor latency of the median nerve greater than 4.3 milliseconds or a sensory latency of the median nerve greater than 3.6 milliseconds is abnormal.[8] Historically, elevated distal latencies were often considered to be an indication for surgery. More recently there has been a trend in the treatment of carpal tunnel syndrome away from surgical exploration [7,9,10], even for patients with markedly abnormal nerve conduction studies. This study demonstrates the effectiveness of non-surgical treatment for carpal tunnel syndrome in patients who have objective findings on nerve conduction studies with elevated distal sensory and motor latencies.

4. CONSERVATIVE TREATMENT

During the past three years, a number of patients have been examined and diagnosed as having work-related carpal tunnel syndrome. The diagnosis was based on abnormal nerve conduction studies, as well as clinical signs and symptoms. The patients often presented with chief complaints of pain, numbness and tingling that they attributed to the type of work that they performed. Some have undergone surgery, but the majority have had conservative treatment. Criteria for selection in this study involved evaluation of motor and sensory latencies in the affected hand, with the median motor of 4.3 milliseconds or greater, or the median sensory of 3.6 milliseconds or greater. Patients must have received conservative treatment long enough that at least two nerve conduction studies were performed. Generally, repeat neurologic testing is required every four to six months depending on the severity of the initial problem. Both the latency between stimulus and response, and the size of the amplitude of the response are measured. Latency is generally reported directly in milliseconds, but can also be recorded as a function of distance and is termed velocity. Injury sufficient to cause changes in the nerve increases the latency and slows the velocity. Generally, the sensory latencies were performed in antidromic fashion and motor latencies were performed orthodromically.

Conservative treatment on all patients included patient education with a video type and brochure explaining the anatomy, causes, and symptoms of carpal tunnel syndrome, as well as how modifications in activities might lead to improvement in symptoms. The limitations of surgery were also explained. Patients were instructed to wear night splints with the option of wearing them while working if this allowed for improvement in symptoms. All patients were given Vitamin B-6, 100 mg. daily. Where appropriate, patients were instructed and assisted in making job changes or changes in personal activities.

A total of 131 patients qualified for inclusion in the study. Of this total, 98 patients (75%) did not undergo surgery, and responded satisfactorily to conservative treatment. The remaining 33 patients (25%) had surgery. Of those with surgery, 19 (58%) had surgery immediately because of atrophy, extremely abnormal nerve conduction studies, or failure to respond to conservative treatment rendered by their referring physician. Fourteen (42%) patients tried conservative treatment for more than three months. Satisfactory improvements were not realized, and consequently surgery was performed. Thus, conservative treatment was not successful in 10.7% of the total cases.

Table I provides a comparison of median motor and median sensory latencies of the initial examination in comparison with the last examination.

TABLE I - Patients Receiving Conservative Treatment with Two or More Electromyograph Examination

EMG READING	NO. OF PATIENTS	INITIAL AVERAGE READING Median Motor	Motor Sensory	LAST AVERAGE READING Median Motor	Motor Sensory
Under 4.5	20	4.33	3.81	4.32	3.71
4.5 to 4.99	21	4.66	3.87	4.35	3.73
5.0 to 5.49	18	5.15	4.15	4.48	3.80
5.5 to 5.99	16	5.70	4.55	4.72	3.87
6.0 to 6.49	8	6.09	4.69	5.76	4.36
6.5 to 6.99	4	6.68	5.75	5.18	4.69
7.0 to 7.49	4	7.24	6.08	6.26	5.44
7.5 to 7.99	4	7.75	N/R	6.20	N/R
8.0 to 8.49	0	-	-	-	-
8.5 to 8.99	1	8.60	5.84	6.80	4.56
9.0 to 9.49	0	-	-	-	-
9.5 to 9.99	0	-	-	-	-
10.0 to 10.49	1	10.40	N/R	9.40	N/R
10.5 to 10.99	1	10.80	N/R	5.00	N/R
	98				

N/R= No Response

5. CONCLUSION

Patients having carpal tunnel syndrome with objective findings of nerve conduction study abnormalities can frequently be treated successfully by conservative methods. Patient motivation and cooperation are essential for a satisfactory recovery.

REFERENCES

[1] Armstrong, T., and Chaffin, D., Journal of Occupational Medicine, 21 (1982), 481.
[2] Hymovich, L., and Lindholm., Journal of Occupational Medicine, 8 (1966), 573.
[3] Jensen, R., Klein, B., and Sanderson, L., Monthly Labor Review, September (1983), 13.
[4] Chaplin, E., and Kasdan, M., Journal of Plastic and Reconstructive Surgery, 75 (1985), 722.
[5] Chaplin, E., Kasdan, M., and Corwin, H., Hand Clinics, 2 (1986), 513.
[6] Cannon, L., Bernacki, E., and Walter, S., Journal of Occupational Medicine, 23 (1981), 255.
[7] Kasdan, M., and Janes, C., Journal of Plastic and Reconstructive Surgery, 79 (1987), 456.
[8] Jebsen, R.H., Archives of Physical Medicine and Rehabilitation, 48 (1967), 185.
[9] Ellis, J., Azume, J., and Folkers, K., Research Communication in Chemical Pathology and Pharmacology, 17 (1977), 175.

[10] Wolaniuk, A., Vadhanavikit, S., and Folkers, K., Research Communication in Chemical Pathology and Pharmacology, 41 (1983), 501.

X

CONTROL OF BACK INJURIES

Trends in Ergonomics/Human Factors V
F. Aghazadeh (Editor)

COMPREHENSIVE BACK INJURY PREVENTION PROGRAM: AN ERGONOMICS APPROACH FOR CONTROLLING BACK INJURIES IN HEALTH CARE FACILITIES

John W. AIRD, Pirjo NYRAN, Gwenn ROBERTS

Health Care Occupational Health and Safety Association,
150 Ferrand Drive, Don Mills, Ontario. Canada M3C 1H6

An ergonomics approach to comprehensive back injury prevention has been developed to address the perceived shortcomings of current approaches that have exhibited only limited effectiveness. This paper provides an overview of the multi-faceted components of the approach, and reports preliminary results of its effectiveness in controlling back injuries. Retrospective case studies of three health care facilities that have adopted this approach were employed for this pilot investigation.

1. INTRODUCTION

The problem of musculoskeletal overexertion injuries, particularly back injuries, in the health care industry is enormous and widespread. Of all 1987 lost time injury claims* in the health care industry in Ontario, 40.6% were back injuries. The majority of these (90.9%) were strains or sprains resulting from overexertion. Nursing staff suffered 68.6% of these back injuries which appear to be related primarily to the demands of the job rather than personal characteristics of the employees [1,2,3]. Physical demands associated with particular patient handling tasks have been shown to be very high [4,5,6].

Solutions for how to address the problem of back injuries are less clear. The effectiveness of training has been questioned [7,8]. Preplacement selection programs are only aimed at newly hired employees, and are limited by the size of the available worker pool and human rights requirements. Redesigning the job to better match the capabilities of the typical workforce is likely the most effective prevention strategy. However, the alternatives for changing the design of equipment or the workplace are often impractical, prohibitively expensive, or unavailable. Changes to staffing levels are largely out of the control of a publicly funded facility.

An ergonomics approach to comprehensive back injury prevention has been developed, based on the rationale that the predominant underlying factor behind overexertion musculoskeletal injuries is a mismatch between the capabilities of the employees and the demands of the job. There are four essential groups of interacting factors, as illustrated in Figure 1, that must be addressed in order for the program to be effective. These factors are the anthropometric characteristics, capabilities and fitness of the employees, the specific job demands, the training of staff to carry out job demands safely, and management support required to ensure that these three factors are addressed.

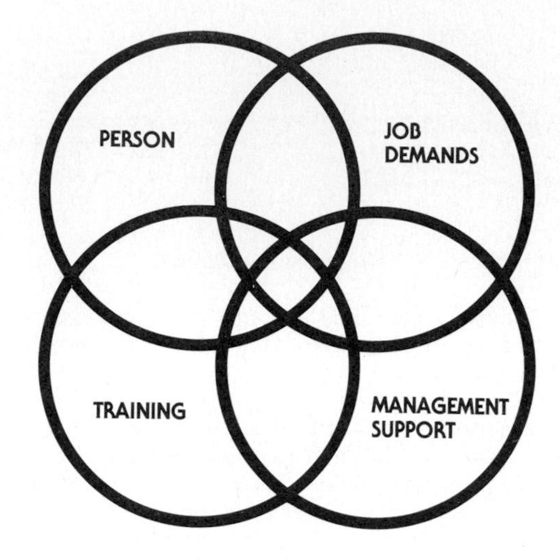

FIGURE 1

Interactive Factors Essential for an Effective Program

Programs aimed at assessing and influencing the capabilities and fitness of the person to meet the demands of the job include pre-placement assessments, back care education, fitness promotion, pausegymnastics and rehabilitation/return to work programs. The effectiveness of such programs in reducing back injuries is dependent on voluntary participation and application of the principles by participants; however, these are factors over which management has little control. If the job demands greatly exceed the capabilities of the workers, such programs will not eliminate the mismatch. Therefore it is essential that the comprehensive injury prevention program include task analysis mechanisms aimed at reducing the demands. Interventions can include changes in procedures, organization of the work, equipment, workplace design, and in staffing levels. Employees also must be trained in the procedures and skills required to utilize the existing resources in the most safe and effective way to carry out their duties. Finally, there must be upper management commitment and support. A study of seven geriatric hospitals evaluating the effectiveness of two program strategies suggested that the management control systems had a greater effect on reducing back injuries than the education program [10]. Policies and procedures must be developed and clearly communicated to all levels of staff, who must be held accountable for fulfilling these responsibilities. If there is lack of compliance, then it is important to identify why the staff are unable to follow policies and procedures, and corrective action must be taken. It may be that the policies or procedures are impractical or the workload pressures may encourage individuals to take shortcuts.

This approach to comprehensive back injury prevention recognizes differences between health care facilities. The programs implemented will vary depending upon the available resources, and the limitations of equipment and workplace design. Facilities must implement the strategies for prevention which will have the greatest impact on improving the match between the employees and the jobs in their institutions.

A pilot project was initiated to assess the effectiveness of this ergonomics approach to back injury prevention.

2. METHODOLOGY

A retrospective case study approach was adopted for the pilot investigation. A hospital, a home for the aged and a hospital linen service that have adopted an approach to managing back injuries corresponding to our ergonomics approach, were selected. Profiles of the program components functioning in the three facilities, and the dates they were initiated, were developed through telephone interviews.

The effectiveness of the programs in reducing back injuries was analyzed using lost time injury claim information contained on the System 57 data base of the Ontario Workers' Compensation Board. Strain and sprain injuries were analyzed by occupational group and part of body. The majority of such injuries result from overexertion situations (95.0% of all strain or sprain back injuries result from overexertion situations provincially), although it is recognized that a small percentage of the injuries may result from other situations such as falls. The results of an analysis not included in this paper confirmed that the majority of strain and sprain injuries suffered in the three facilities were due to overexertion situations.

Injury frequency rates were calculated where paid hours of work were made available from the facilities, to provide an estimate of relative risk of injury and to normalize the data for changes in the size of the worker population. The frequency rates were calculated using the following standard formula.

$$\frac{\text{Number of injuries} \times 200{,}000}{\text{Paid hours of work}} = \text{approximation of the percentage of employees suffering an injury over the year}$$

3. CASE STUDIES

3.1 Nursing Areas in a Hospital

In response to the high number of back injuries, one of Ontario's largest hospitals initiated activities to upgrade its existing back program in 1983. All head nurses participated in a core group leader training program in 1984 to provide them with the expertise to develop appropriate policies and procedures for their departments and to provide job-specific training. Additional lifting devices were also purchased for all units.

The major components of the back injury prevention program were initiated facility-wide in 1985. The pre-placement assessment program included a medical examination and an assessment of lifting skills. The orientation program included an assessment of body mechanics and lifting skills on the job site. An assessment of lifting skills was also carried out with all injured staff returning to work, and education on general back care and lifting techniques was provided monthly. Responsibilities and accountabilities for the programs were established at the supervisory levels. No formal mechanism was established for analyzing the job demands or for problem solving.

In response to the high number of back injuries occurring in orthopaedic wards, which represent four of the 44 nursing wards, a more intensive training and problem-solving program was initiated. This involved primarily on-the-job training and problem solving sessions with small groups co-ordinated by the core group leader. Injury statistics for these wards were provided by the facility for 1985-87 only.

A number of components were added to the program in 1987. A general fitness program was made available to all staff, conducted by a qualified fitness instructor during the lunch hour. A special exercise program was made available to staff who had experienced back problems. A modified work program was initiated to help individuals who had been off work for a long period of time, return to work. Injured employees are now assessed for functional abilities, first by the occupational health physician and then by the trained core person on the job site, before returning. This latter assessment includes a comprehensive assessment of lifting skills. The facility has also drawn on outside resources to assist with task analysis to address job design problems, and is examining the possibility of developing internal resources to carry out these functions.

The lost time injury statistics for the entire facility are presented in Table 1. The number of lost time back injuries for all nursing units, and for the four orthopaedic wards only, are presented in Table 2. Paid hours of work were not available for the nursing areas, prohibiting the calculation of frequency rates.

TABLE 1

Number and frequency of strain or sprain lost time injuries for all occupational groups in a hospital.

	TOTAL		BACKS	
YEAR	NUMBER	FREQUENCY	NUMBER	FREQUENCY
1984	93	3.21	71	2.45
1985	109	3.52	77	2.49
1986	107	3.37	79	2.49
1987	104	3.18	65	1.99
% change 1984-87	+ 11.8	- 0.1	- 8.4	- 18.8

TABLE 2

Number of back injuries for all nursing areas and orthopaedic nursing areas only in a large hospital.

YEAR	ALL NURSING	ORTHOPAEDIC NURSING
1984	48	—
1985	56	21
1986	42	15
1987	39	10
% change		
1984-87	- 18.7	--
1985-87	- 30.4	- 52.4

The results outlined in Table 1 indicate a facility-wide decrease in both the number (-8.4%) and frequency (18.8%) of back injuries from 1984 to 1987. Over this same time period, the total facility frequency rate for all types of injuries remained relatively constant (-0.1%). This suggests an improvement associated with the facility-wide program initiated in 1985. The higher percentage change in the frequency of back injuries (-18.8%) compared to the number of back injuries (-8.4%) illustrates the value of normalizing the data for changes in staff population. There was an even greater reduction in the number of back injuries in nursing (-18.7%) over the same time period as shown in Table 2. Additional components initiated in nursing included core group leader training for all head nurses and the purchase of mechanical lifting devices. The more intensive program conducted in the orthopaedic wards showed a reduction in back injuries (-52.4%) even greater than that seen across nursing in general (-30.4). This supports the additional value of incorporating program components which address job-specific problem solving. The results also reflect the more concerted and visible support provided by head nurses in these areas. The effectiveness of the programs initiated in 1987 would not be expected to be reflected in this statistical analysis.

3.2 Nursing Areas in a Home for the Aged

The administrator in a medium-sized home for the aged attributed the increasing accident frequency and cost to the high incidence of back injuries. In 1985, extensive back care education was provided to all levels of staff. Additional mechanical lifting devices were purchased and assessment of residents was initiated, which included an assessment of every lifting situation in advance by in-house trainers. Such an assessment considers the characteristics and capabilities of the patient and the care-giver, and environmental factors including the availability and design of equipment. This information was recorded on the patient care plan and communicated through a logo system so that all staff were clearly aware of the procedures. All staff were held accountable on an on-going basis for carrying out correct policies and procedures, and educational sessions were provided on a regular basis.

More detailed individual claims information was available which permitted a specific analysis of overexertion strain and sprain injuries resulting from patient handling situations. The results of the statistical analysis are outlined in Table 3.

TABLE 3

Number and frequency of lost time injuries for nursing areas in a home for the aged

	TOTAL		BACKS	
YEAR	NUMBER	FREQUENCY	NUMBER	FREQUENCY
1984	3	6.9	3	6.9
1985	6	13.3	4	8.8
1986	4	8.7	0	0
1987	4	8.2	4	8.2
% change 1984-87	+ 33.3	+18.8	33.3	+ 18.8

The results show no back injuries occurred in 1986 following the initiation of the program in 1985. However, there were four back injury claims in 1987. Analysis of the four back injuries by the director of nursing revealed that two back injuries were caused when staff did not follow procedures, even though the logo system identified use of the mechanical lift. The third employee sustaining a back injury was identified as an abuser of the WCB system and had a long history of injuries unrelated to work. No specific contributing factor could be identified for the fourth back injury. The director of nursing in all cases followed up on these injuries and appropriate disciplinary measures were taken where necessary.

These results illustrate the danger of using only a single statistical measure, such as the number of lost time injuries, to assess the effectiveness of a program. A more indepth analysis suggests that when the program is followed, the risks of suffering a back injury have been reduced. It also illustrates the limitation of using lost time injury statistics to evaluate programs in smaller populations where a few injuries can greatly bias the results.

3.3 A Hospital Linen Service

An ergonomic assessment of the workplace design and layout, procedures and physical demands of the tasks was carried out in 1984 in response to senior management's concerns about the high number of back injuries in a hospital linen service. A number of strategies were initiated in 1985 as a result of the assessment to address the identified problems. Several jobs were re-designed. Working levels were adjusted to better match the characteristics of the employees. Jobs that were traditionally rotated on a weekly basis were rotated every two hours. Intensive back education for all employees was initiated. Job specific training and re-orientation after injury were also provided. As it was identified that many of the job demands could not have been designed out, a pausegymnastics program to break the monotonous work routine was established. Twelve employees were trained as pausegymnastics leaders. The program was run twice daily and was used as a warm-up exercise at the beginning of the shift. Occupational health services were contracted in 1986 to provide prompt treatment and early rehabilitation. The results of the statistical analysis are outlined in Table 4.

TABLE 4

Number and frequency of strain or sprain lost time injuries in a hospital linen service.

	TOTAL		BACKS	
YEAR	NUMBER	FREQUENCY	NUMBER	FREQUENCY
1984	26	10.1	10	3.9
1985	26	9.5	16	5.9
1986	20	6.6	11	3.6
1987	16	5.4	5	1.7
% change 1984-87	- 38.5	- 46.5	- 50.0	- 56.4

Since the implementation of the programs in 1985, there has been a 56.4% reduction in the frequency of back injuries and a 46.5% drop in the frequency of strains and sprains in general. These results support the effectiveness of the multi-faceted ergonomics approach. The current pilot investigation was not designed to demonstrate the impact of the rehabilitation and return to work program established in 1986. Such a program would be expected to reduce days lost and recurrence of injuries rather than the total number of injuries.

4. DISCUSSION

The results of the three case studies provide preliminary support for the effectiveness of the multi-faceted ergonomics approach to back injury prevention. The greater reduction in back injuries in the orthopaedic wards compared to nursing in general also suggests that on-the-job instruction and problem solving is an effective strategy for addressing some of the job-specific factors. This observation confirms the conclusions of previous studies [1,2,3] which have pointed to the need to address job-related factors. The large reduction in injuries in the hospital linen service indicates that the principles of the program can be effectively applied to different health care facilities.

Interpretation of these preliminary results is limited. Harber et al. [2] noted that by looking at lost time injuries data only, the prevalence of occupational back pain is greatly underestimated. There is a need for more in-depth analysis to determine the effectiveness of this ergonomics approach. The days lost due to injuries and the recurrence of injuries also need to be examined. The effectiveness of programs targeted at particular subgroups of employees, such as preplacement, orientation and return to work programs, need to be separated from the analysis of the total employee population. It is not possible to separate the relative impact of each component of the programs, though the hypothesis is that a multi-faceted approach will address the shortcomings of any one approach, and therefore be more effective. Finally, lack of objective, qualitative information to evaluate the programs was another limitation of the methodology. Despite these limitations, there would appear to be sufficient support for the effectiveness of this approach to the prevention of back injuries to warrant further investigation.

FOOTNOTES AND REFERENCES

* Based upon lost time injury claims for 1987 up to and including September 30, 1987, drawn from the System 57 data base of the Ontario Workers' Compensation Board.

[1] Videman, T., Nurminen, T., Tola, S., Kuorinka, I., Vanharanta, H. and Troup, J.D.G., Spine, 9(4), 1984, 400-404.

[2] Harber, P., Billet, E., Gutawski, M., SooHoo, K., Lew, M. and Roman, A., Journal of Occupational Medicine, 27(7), 1985, 518-524.

[3] Venning, P.J., Walter, S.D. and Stitt, W., Journal of Occupational Medicine, 29(10), 1987, 820-825.

[4] Dehlin, O. and Lindberg, B., Scandinavian Journal of Rehabilitation Medicine, 7, 1975, 65-72.

[5] Gagnon, M. and Lortie, M., A biomechanical approach to low-back problems in nursing aides, in: Asfour, S.S., (ed.), Trends in Ergonomics/Human Factors IV (North-Holland, Amsterdam, 1987) pp.795-802.

[6] Khalil, T.M., Asfour, S.S., Marchette, B. and Omachonu, V., Lower back injuries in nursing: A biomechanical analysis and intervention strategy, in: Asfour, S.S., (ed.), Trends in Ergonomics/Human Factors IV (North-Holland, Amsterdam, 1987) pp.811-821.

[7] Dehlin, O., Berg, S., Andersson, G.B.J. and Grimby, G., Scandinavian Journal of Rehabilitation Medicine, 13, 1981, 1-9.

[8] Stubbs, D.A., Buckle, P.W., Hudson, M.P. and Rivers, P.M., Ergonomics, 26(8), 1983, 767-779.

[9] Snook, S.H., Campanelli, R.A. and Hart, J.W., Journal of Occupational Medicine, 20, 1978, 478-481.

[10] Wood, D.J., Spine, 12(2), 1987, 77-82.

Trends in Ergonomics/Human Factors V
F. Aghazadeh (Editor)

SUDDEN-MOVEMENT/UNEXPECTED LOADING AS A FACTOR IN BACK INJURIES

Terrence J. STOBBE and Ralph W. PLUMMER

Industrial Engineering Department
West Virginia University
Morgantown, West Virginia 26506

INTRODUCTION

Back injuries have been, and will continue to be, a significant problem in the industrial workplace. While there are many reasons for this, the fact remains that our understanding of the phenomena is limited. The cause most commonly ascribed to these injuries is manual material handling, and within that broad class of activities, the most frequent cause is generally thought to be lifting. Numerous studies of the manual-material-handling/lifting problem are cited in the "Work Practices Guide for Manual Lifting" (NIOSH, 1981), and need not be repeated here. They all focus on lifting as the root of the problem, and indeed, it is one of the roots. More recently, studies in the mining industry by Stobbe (1983), and Unger (1983), have supported this conclusion.

The injury data cited in much of this prior research has been based on Worker's Compensation Form type accident/injury reports. At their best, these reports provide sketchy details about the accident situation. To really understand the details of what happened in a back injury situation, we need much more information.

Another more recent set of studies by Stobbe (1986), and Smith (1986) suggested that sudden energy release may be an important factor in these injuries. Another study done in the mining industry directly confirmed the importance of sudden-movement and unexpected-loading as a causative factor in back injuries (Bobick, 1987). While these data do not negate the importance of prior work, they do suggest that the study of industrial back injuries must be expanded beyond the narrow confines of lifting injuries. They also suggest that basing our conclusions about injury patterns on Employer's First Report of Injury type data will sometimes be misleading.

This paper reports the results of the further investigations into the sudden-movement/unexpected-loading related injuries.

METHODS

Developing a more in-depth understanding of the industrial back injury phenomena requires that the researcher obtain information that is not readily available from the standard accident report or employer's first report of injury. In most cases, the additional information is only available from the injured person. The two logical choices for obtaining the information are asking the person to complete some type of questionnaire and interviewing the person. Blind completion of a questionnaire has the advantage of allowing access to a large number of subjects quickly, but suffers from the problem of misunderstanding of the question by the subject and of the response by the researcher. An interview on the other hand takes much more researcher time, which means less data can be collected; it does however, reduce the chances of either form of misunderstanding because the researcher can ask additional clarifying questions specific to topics that appear to be confusing. In view of the above, we elected to conduct this research using interviews to gather the additional data.

The research was conducted in three industries: underground coal mining, a hospital complex, and a tire manufacturing. A dozen coal mines were involved, while only one hospital and one tire plant were involved. The injured persons interviewed were selected based on the criteria that they were suffering from a low-back injury, were available to be interviewed, and were willing to be interviewed. Not all persons who were injured during that data collection period were interviewed.

The interviews took place either in an office provided by the host facility, or at the underground mines; many of the miners were interviewed at their work site. The interviews were conducted in private, and none of the information is shared with the sponsoring companies.

The interview itself took from forty minutes to an hour to complete, and consisted of fifty-four questions in four categories: 1) questions designed to determine the who-what-where-when-why-how of the accident and its associated events; 2) questions designed to assess the biomechanical and environmental factors associated with lifting injuries; 3) questions which investigated the examination, treatment, and care received by the injured persons; and 4) questions designed to determine the injured person's historical exposure to physically stressful work and leisure activites. Since participation was voluntary, the interviewees had the option of not answering any of the questions. Initially, there was some fear that the injured would be reluctant to discuss the injuries, but the only questions which provoked non-responses were those relating to physically stressful activities not a part of their current job.

The coal mines which are participating in the research are typical of the range of work environments commonly encountered in underground mining. The mines are spread over four states, and vary in height from 54 to over 96 inches. Both diesel and electric mines are represented as are longwall and continuous miner operations, and union and non-union mines. Injured persons were interviewed independent of their job title -- thus the group includes newly hired laborers, experienced miners and mechanics, and supervisory and administrative personnel.

The hospital complex includes a modern 550 bed hospital, with an affiliated long-term patient care facility. In this facility, only health care personnel with the job of nurse, nurse's aide, or attendant were interviewed. The tire plant was a large plant in the Southern United States.

At the tire plant, personnel were interviewed without regard to job title, and a variety of occupations are represented in the data.

RESULTS

The results described in this paper relate only to sudden movement and lifting, and their relative contribution to back injuries in the locations studied. The data collected in the project are much broader in scope and will be reported elsewhere.

The first table summarizes the activity the person was engaged in at the time the injury occurred. It is specific to the injured persons actions when injured, not to the workplace activity. There are a number of empty cells in the table because some activity categories were not relevant to all of the industries. Some categories deserve clarification: walking is movement between points without transporting anything; carrying refers to walking or stepping while also moving some type of material, but not including lifting, pushing, or pulling; enter/exit refers to entering or leaving a mine vehicle, or climbing/stepping on or off a vehicle; riding refers to sitting in a mine vehicle which is in operation, as a driver, operator, or passenger; bending refers to situations in which the person had adopted a stoop or squat posture to access equipment for maintenance or other purposes, and the prolonged difficult posture resulted in back pain; and balancing patient describes a situation in which the health care professional is walking along with an ambulatory patient for the purpose of assisting the patient in case they get weak or dizzy.

TABLE 1

Injured Person's Activity At the Time of Injury
(expressed as percent)

ACTIVITY	INDUSTRY UNDERGROUND COAL MINING	HEALTH CARE	TIRE AND RUBBER PLANT
LIFTING	30	39	18
PUSH/PULL	19	13	41
WALKING	16	9	9
CARRYING	9	4	14
REACHING	3	4	14
BENDING	3	4	4
BALANCING PATIENT	--	22	--
ENTER/EXIT	6	--	--
RIDING	11	--	--
OTHER	3	5	0

The first thing to note is that the data suggests an interesting distribution of injury causes. Lifting, commonly thought to be the major factor in back injuries accounts for less than one-third of the cases. Grouping lifting, pushing, pulling, and carrying together into the broader category of material handling accounts for almost sixty percent of the cases, This still leaves alarge percentage of the cases due to other causes.

There is considerable causal activity inconsistency across industries which is probably due to inherent differences in their work environments. The inconsistency decreases when the lifting and push/pull categories are combined. This is significant because these two categories account for most of the situations in which a person is consciously applying the large forces commonly associated with back injuries.

Table 2 is an analysis of the forces applied during the lifting injuries. If the data is pooled across industries, almost forty percent of the lifting injuries occurred while lifting light loads (<30 lbs), less than thirty percent occurred lifting moderate loads (31-50 lbs), and about one-third occurred while lifting heavy loads (>50 lbs). There is again some inconsistency between industries, with health care being especially biased toward the heavy category. These are patient lifts, and even with two or three persons involved, the load exceeds 50 lbs.

TABLE 2

WEIGHT OF OBJECT BEING LIFTED OR CARRIED AT TIME OF INJURY
(data expressed as percent)

	INDUSTRY		
OBJECT WEIGHT	UNDERGROUND COAL MINING	HEALTH CARE	TIRE AND RUBBER PLANT
0-10 LBS	21	0	13
11-30 LBS	18	10	50
31-50 LBS	36	20	25
> 50 LBS	25	70	12

The third table analyzes sudden movement and unexpected loading as factors in back injuries. These factors are a direct contrast to the lift/push/pull situation in which the person expects to be loaded. In the case of sudden or unexpected loading, the neuromuscular system has no time in which to prepare the appropriate set and magnitude of muscular contractions needed to prevent injury. Typical examples include slipping and tripping, getting bumped unexpectedly, applying force to an object and having it suddenly break loose, or conversely be pushing or pulling an object and having it stop unexpectedly, having a patient fall and grab the assisting nurse, and having a load which is being lifted or carried shift suddenlt. Table 3 suggests these kinds of events are a highly significant cause of back injuries.

When viewed across industries, there are again some inconsistencies. Health care has the lowest frequency of these injuries, probably because a high percentage of their injuries occur while dead-lifting patients. Their sudden movement injuries occured when walking patients suddenly fell grabbing them for assistance, and while walking and carrying or falling out of a chair.

TABLE 3
SUDDEN MOVEMENT AS A FACTOR IN BACK INJURIES
(expressed as a percent)

INDUSTRY	SUDDEN MOVEMENT DID OCCUR	SUDDEN MOVEMENT DID NOT OR MAY NOT HAVE OCCURRED
UNDERGROUND* COAL MINING	54	40
HEALTH CARE	39	61
TIRE & RUBBER MANUFACTURING	68	32

* 6% of these injuries resulted from falling roof

The last table is a more detailed analysis of the sudden movement/unexpected loading injuries. Again, as in Table 1 there are a number of blank cell entries because not all activity categories relate to all industries. Riding refers to injuries which occured while an occupant of a mine vehicle which either was involved in a collision, or which ran over an uneven spot in the roadway or track causing the person to be "bounced around". The distinction between slip and trip is based on surface slipperiness as opposed to unevenness. Entry/exit refers to a person entering or leaving a mine vehicle, with the injury usually resulting from stepping down blindly onto an uneven surface and losing their balance. It should be noted that very few of the slips, trips and entry/exit injuries resulted in a person actually falling to the ground.

TABLE 4
CAUSE OF SUDDEN MOVEMENT OR
UNEXPECTED LOADING RELATED INJURY
(expressed as percent)

CAUSE DESCRIPTION	INDUSTRY: UNDERGROUND COAL MINING	HEALTH CARE	TIRE AND RUBBER PLANT
Slipped While Walking	11	0	14
Tripped While Walking	10	21	21
Load Shifted Unexpectedly	13	49	0
Struck by or Jerked by Something	4	10	21
Lift/Push/Pull With Sudden Force Release	16	0	8
Jerk on Object That Does not Move	11	10	36
Riding on Vehicle	20	--	--
Entry/Exit	13	--	--
Other	2	10	0

While we might reasonably assume that most of these sudden movement injuries occur due to slips and trips, in fact they account for just under one-fourth of these injuries. The balance occur for a variety of reasons. In health care, the load (usually a walking patient) shifts unexpectedly. In the other two industries, a combination of exerting force on something which suddenly moves, exerting force on a moving object which stops suddenly, and attempting to move something that is struck by jerking on it are the other main causes. In the mining industry, vehicle related events are a significant cause of sudden movements as well.

DISCUSSION

This paper has reported an overview of back injury data collected by the interview process in three industries. It has focused on the relative relationship between lifting and sudden-movement/unexpected-loading. The data reported show that the two categories are at least equally important as causes of industrial back injuries. This result is consistent with other data reviewed by the authors which shows that a large percentage of contact injuries are also the result of the unexpected -- or perhaps more correctly the result of a lack of forethought. Controlling injuries of this type requires a combination of workplace design that minimize the liklihood and effects of sudden-movement/unexpected-loading events, tool design and selection that also minimizes them, and training which emphasizes the severity of effect that can result from failing to think through a job.

The data obtained from this approach to studying back injuries also shows that this more detailed approach to the study of back injury events provides a better understanding of the problem, and leads to different and broader conclusions about the causes and cures of the industrial back injury dilema than are available from the more traditional approach which studies employer's first reports of injury or their equivalent.

REFERENCES

Bobick, T, Stobbe, T, and Plummer, R. "An Analysis of selected Back Injuries Occurring in Underground Coal MIning", US Bureau of Mines, IC 9145, July, 1987.

NIOSH, Work Practices Guide For Manual Lifting, Technical report 81-122, National Institute for Occupational Safety and Health, 1981.

Smith, G, "Injuries as a Preventable Disease: The Control of Occupational Injuries From the Medical and Occupational Health Perspective", W.H.O. Conference on Musculoskeletal Injuries in the Workplace, May, 1986.

Stobbe, T., Plummer, R., " Analysis of Coal Mining Back Injury Statistics", US Bureau of Mines IC 8948, August, 1983.

Stobbe, T., Bobick, T., and Plummer, R., "Musculoskeletal Injuries in Underground Mining", Ann of ACGIH, V 14, 1986.

Unger, R., and Connelly, D., "Materials Handling Methods and Problems in Underground Coal Mines", US Bureau of Mines IC 8948, August, 1983.

Trends in Ergonomics/Human Factors V
F. Aghazadeh (Editor)

PATIENT HANDLING DEVICES: AN ERGONOMIC APPROACH TO LIFTING PATIENTS

Bernice D. OWEN

University of Wisconsin-Madison, School of Nursing
600 Highland Avenue
Madison, Wisconsin 53792

An ergonomic approach recommended for decreasing back pain in nurses is the use of mechanical devices for lifting patients. There is a paucity of data available about the use and acceptance of patient handling devices. Questionnaires were sent to a random sample of 175 long-term care facilities; 120 (69%) were returned and usable for data analysis. Gait belts, lift sheets, hydraulic lifts, and bathtub lifts were most frequently used. Common problems were lack of skill in using devices, time involved, lack of available devices in patient's rooms, and patient fear. Patient characteristics important for use of devices included physical and mental abilities such as comatose status and combative activities. Nurses believed use of patient handling devices could help to decrease back pain. There were no significant differences in responses according to size (bed capacity) of long-term care facility; lift sheets were used significantly more often in facilities where majority of patients were dependent.

1. INTRODUCTION

Patient handling tasks appear to play an important part in the back pain problem of nurses. Data consistently reveal that acts of lifting/transferring patients are the most frequent precipitating factors perceived to be contributory to back pain (Harber et al., 1985; Jensen, 1985; Owen, 1985). A link has been established between exposure to heavy load handling (patient handling) and risk of back injury; this link has been supported by biomechanical models (Chaffin & Andersson, 1984). It appears that the layout and design of work surfaces for patient care are such that manual lifting and transferring of patients cannot be done without high levels of postural stress. For example, some patient beds cannot be adjusted to accommodate the anthropometric characteristics of patient handlers.

Many researchers have concluded an ergonomic approach is best for reducing back stress (Bell, 1984; Harber et al., 1985; Owen, 1987; Stubbs et al., 1986; Troup, 1981). They believe nurses should not be expected to adjust to the patient environment and one avenue to prevention could be the use of mechanical devices to assist with patient handling tasks. There is however, a paucity of data about the use or effectiveness of patient handling devices (PHD). Bell (1984) studied hoists (United Kingdom) and found them rarely used and when used it was primarily for transferring patients to the bathtub. The principal reasons given for non-use were interference with the goals of care for independence of the patient and the time needed for use. For non-use he concluded, "the factors are very complex involving the overall needs of individual patients, the actual ward environment, staffing levels, staff attitudes and levels of knowledge" (p. 207).

From discussion with clinical nurses it appears they have a negative attitude toward the use of PHDs. Common responses for non-use are: time, patient fears, devices are not accessible, and lack of knowledge or skill. Torma-Krajewski (1986) studied lifting tasks and concluded staff attitudes were negative toward PHD's and the devices actually interfered with lifting.

The purposes of this study were to: 1) identify what PHDs are currently used and frequency of use, 2) determine which PHDs are used with what patient handling tasks, 3) delineate the problems, if any, that patient handlers have with use of PHDs, 4) find out patient characteristics that appear to be important when considering use of PHDs, and 5) determine opinions of nursing personnel concerning PHDs. Data were collected from long-term care facilities (nursing homes) because Jensen (1987) indicated the highest percent of disabling back injuries for nursing personnel occurred in these settings.

2. METHODS

2.1. Sample

There are 485 licensed long-term care facilities in Wisconsin. A random sample of 175 was secured from a listing published by the Wisconsin Department of Health and Social Services. Questionnaires were sent and 70% (123) were returned. Three were not usable due to inadequate data.

All respondents were nurses; the majority (77%) held the title of Director of Nursing. Ten percent were head nurses or staff nurses; the remaining held titles such as staff development coordinator.

Bed capacity of the facilities ranged from 20 to 620. Fifty-seven (48%) had less than 150 beds. Two had more than 600 beds; these were county-run facilities. In addition to bed capacity, respondents cited the percent of patients that required total, partial, or no assistance with activities of daily living and nursing care. There were 58 homes that had one-third dependent and two-thirds partial or self-care patients; 57 homes had two-thirds dependent and one-third partial or self care patients (status unknown for five homes).

2.2. Instrument and Procedure

The questionnaire consisted of six sections. First was a letter explaining the purpose of the study and process of selection of sample. Section 2 elicited data such as bed capacity. Third listed the PHDs (See Appendix A for description of devices): hydraulic lift (e.g., Hoyer), bathtub lift, lift sheet, gait/transfer belt, trapeze, turn table, patient roller, sliding board, bed scale, rope ladder, inflatable lift, other; the respondent used a Likert scale (0-4) to note frequency of use of devices. Next the respondent matched the devices with tasks such as lifting patient up in bed. In Section 5 open-ended questions related to problems with use of patient handling devices and characteristics of patients indicating need for a device. Likert scales were used in Section 6 for data collection on opinions concerning PHDs.

3. RESULTS

3.1. Frequency of Use

Most frequently used devices were gait/transfer belt, bathtub lift, lift sheet, and hydraulic lift. Sixty-four percent (n = 76) used gait/transfer belts very often; 59% (n = 71) placed bathtub lift in the very often category; lift sheet was used very often by 52% (n = 62); and hydraulic lift by 34% (n = 41). Rope ladder was rarely used and then in only one facility; turn table was rarely used but was present in five facilities. The devices used primarily for lifting or transferring patients in a horizontal position were not used frequently in any of the long-term care facilities (Table 1).

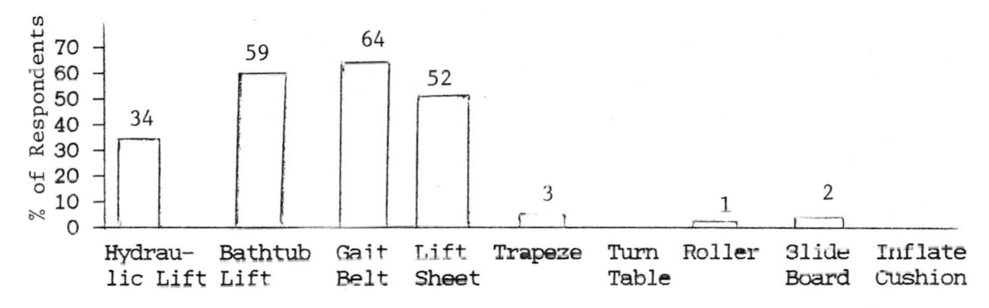

TABLE 1. Percent Who Use Various Patient Handling Devices Very Often

The facilities were grouped by bed capacity to determine differences in use of PHDs by size of institution. The frequency of use categories were collapsed into two groups: never, seldom and sometimes equal one category, the other was often and very often. There were no significant differences between those with a bed capacity of 150 or less (n = 57) and those with over 150 patients (n = 62) in frequency of use of hydraulic lift (Chi-Square = .4, df = 1, p = .53), bathtub lift (Chi-Square = .01, df = 1, p = .91), gait belt (Chi-Square = .00, df = 1, p = 1), or lift sheet (Chi-Square = 1.2, df = 1, p = .26). Frequency of use was also studied in relation to number of dependent vs partial/self-care patients. There were no significant differences between self-care status and hydraulic lift (Chi-Square = 1.7, df = 1, p = .19), gait belt (Chi-Square = .03, df = 1, p = .86), or bathtub lift (Chi-Square = .74, df = 1, p = .38). Significant differences were found with lift sheet (Chi-Square = 7.7, df = 1, $p < .005$); those with two-thirds dependent patients used lift sheet more frequently.

3.2. Devices Used with Patient Handling Tasks, Problems Encountered, and Patient Characteristics Important for Use of PHDs.

Most PHD's were for transferring patients from bed to chair, chair to bed, to and from bathtub, and lifting up in bed. Common problems were: time involved, staff experience, and lack of availability (Table 2).

Patient characteristics listed most frequently were: inability to support own weight, obesity, and weakness (Table 3). Others were unconscious state, confused or combative behavior, those in pain, and spastic conditions. Many, however, added comments to emphasize need for professional judgment in use of devices with certain patient characteristics. For example, caution was expressed with use of hoists and lift sheets for obese patients because hoists have weight limits and sheets may tear, or patients in pain may have pain intensified by tight slings or belts.

Problem	Hydraulic Lift	Bathtub Lift	Gait Belt	Lift Sheet	Trapeze	Sliding Board
Too time consuming	34% (of responses)	0	25%	5%	0	0
Unsafe, unstable device	18%	0	0	0	0	0
Patient fearful or says "no"	12%	5%	4%	0	0	4%
Need for maintenance	12%	4%	0	0	0	0
Staff have lack of experience	10%	4%	17%	5%	0	8%
Requires 2 people	10%	4%	0	25%	0	4%
Takes too much space	6%	0	0	0	0	0
Uncomfortable for patient	5%	0	10%	0	0	5%
Can't use if want to increase independence of patient	15%	0	0	0	0	0
Lack of availability (in room)	4%	0	10%	0	10%	10%
Patient must have strength and cognitive ability	0	0	0	0	22%	0

TABLE 2. Problems Encountered with Use of Various Patient Handling Devices

Hydraulic lift. All facilities which use this device (93%) use it for transferring patients from chair to bed and bed to chair. Of these, 37% also use it for transferring patients on and off the toilet and 32% into and out of bathtub. Seventy-six percent mentioned at least one problem. Time involved in positioning the patient, arranging the canvas slings, and in transferring the patient was considered a problem by 34%. Eighteen percent stated the device was unsafe or unstable, especially for obese, combative, or comatose patients. Other problems were: the patients were fearful or refused use of this device, for safety it required two people, staff lacked experience and/or knowledge, equipment took up too much space for storage and for use. Patient characteristics were: lifts should be used with patients who can not support their own weight (74%), were obese (43%) or weak (34%). Few stated combative/confused behavior (4%), in pain/arthritis (4%) or spastic conditions (2%). Seven cautioned about using hoists if patients had frail skin.

Bathtub Lift. Eighty-six percent of facilities have this device; 5% mentioned problems of patient fear or refusal to use the lift and the device requires two people for use. Frequent characteristics were inability to support own weight (75%), weakness (50%), and those in pain/arthritis (32%).

Gait Belt. These were used by 86% of the facilities. All stated it was used for transferring from bed to chair and vice versa. Of these 52% also mentioned on and off toilet transfers and 19% reported chair to tub transfers. Most common problem cited (25%) was the time needed for correct placement of the belt. Knowledge/skill of staff may be a problem (17%) because correct location and tightness of fit are essential for patient safety and comfort. The availability of a belt at the time of transfer is important (8%). Thirteen indicated how they adamately were in favor of gait belts: "gait belts are a must," "we have decreased back injuries by 75%," and "we have a policy our nursing assistants must wear a gait belt so it is always available to put on the patient." Five spoke adamately against gait belts: "we've had broken ribs from these," and "significant skin problems occur when using belts." Patient characteristics for use were combative/confused behavior (48%) and spastic conditions (49%). Fifteen cautioned against gait belt use with patients who were very obese, had rib or abdominal problems, frail skin or pendulous breasts.

Patient Characteristics	Hydraulic Lift	Bathtub Lift	Gait Belt	Lift Sheet	Trapeze	Sliding Board
	%	%	%	%	%	%
Can't support own weight	74	75	35	75	21	32
*Obese	43	14	10	62	30	9
Weak	34	50	29	52	0	5
Unconscious	8	0	5	57	0	9
*Combative/Confused	4	8	48	48	0	3
*Arthritis/Pain	4	32	6	43	0	0
Spastic	2	8	49	60	0	0

*Keen professional judgment needed for use of any patient handling device

TABLE 3. Patient Characteristics Important for Use of Patient Handling Devices

Lift sheet. Staff used lift sheets for lifting patients up in bed in 85% of the facilities. Of these, 68% also used them for turning patients in bed and 15% for transfers from bed to cart and vice versa. The need for two people was the most common problem (25%); the amount of time was cited by 5% and also staff need knowledge/skill with this device (5%). Some feared tearing of the sheets during a lift with an obese patient or when the material was worn thin. This was the most frequently listed device for use with all categories of patient characteristics: for those patients who could not support their own weight (75%), were obese (62%) and those with spastic conditions (60%).

Bed trapeze. Sixty-five percent reported use and primarily for lifting patient up in bed; 42% also stated assisting patient onto bedpan. Twenty-two percent reported the patient could help with repositioning in bed. A problem cited by 25% was many patients do not have strength and/or cognitive ability to use this device. In addition, 10% stated the trapeze was generally not available on enough beds in their facilities.

Characteristics listed were cannot support own weight but must have upper body strength (21%) and obesity (30%).

Patient roller. This was used by 5% for transferring patients from bed to cart and vice versa. All stated problems of discomfort to patient and availability of the device within close proximity.

Sliding board. Transfer from bed to chair tasks were completed using this board (27%). Another 8% use the board for transferring from bed to cart and vice versa. The most frequent problem cited (10%) was lack of availability in close proximity to patient's room. Staff lacked knowledge/experience with this device (8%), and a few felt patients were fearful of the board or it was uncomfortable for them. This was used with patients who could not support their own weight (32%).

The remaining four devices were used by few or no facilities: Bed scale. Three percent used a bed scale to suspend a patient while making the bed. The two problems given were: lack of staff experience and use of scale was time consuming. Rope. Not used by any facilities. Inflatable cushion. Four percent stated use for turning patients while in bed. No problems were listed. Turn table. Three percent used this device for pivoting when transferring from bed to chair but all stated a problem if the patient was not cooperative or have cognitive ability. Other. None were listed. Few characteristics were listed for roller, bed scale, or inflatable cushion.

3.4. Opinions

Forty-nine (41%) indicated "yes definitely" to the statement "I feel the fatigue from repetitive lifting of patients could be reduced with increased use of patient handling devices"; there were 45 (37.5%) who stated "yes"; 17 (14%) circled "maybe"; 9 (7.5%) said "probably not" and none indicated "no." There were no significant differences between these responses by size of institution (Chi-Square = .92, df = 3, p = .82), or by self-care status (Chi-Square = 3.7, df = 3, p = .29).
The second opinion solicited was in response to "I feel back problems in nursing could be reduced by increased use of patient handling devices." For "yes definitely" there were 54 (45%) responses; for "yes" there were 45 (35%); 15 (12.5%) indicated "maybe"; and 9 (7.5%) stated "probably not" and no one stated "no." There were no differences between these responses by size of institution (Chi-Square = 2.3, df = 3, p = .51), or by self-care status (Chi-Square = 2.5, df = 3, p = .47).

4. DISCUSSION

The most frequently used PHDs were gait belts, draw sheets, bathtub lifts and hydraulic lifts. The size of the nursing home was not important for determining frequency of use of PHDs. Those nursing homes that had twothirds dependent patients, however, used the lift sheet significantly more often than those homes where the majority of patients were of partial/self-care status.

Transferring patients from bed to chair and vice versa were the most frequent tasks listed; the PHDs cited most often for use with these tasks were the hydraulic lift and gait belt. In general, the devices used primarily for lifting/transferring patients out of bed in a horizontal position were not frequently used in these nursing homes, e.g., roller and sliding board. This probably reflects the emphasis placed on transferring patients into a chair so there is little need for carts/stretchers in nursing homes.

Inability to support own weight was the most frequently cited patient characteristic indicating need for PHDs such as hydraulic lift, bathtub lift, lift sheet, and sliding board. Spasticity and combative/confused status were the most frequent characteristics mentioned for gait belt.

More problems were cited with the hydraulic lift than with any other PHD; the amount of time needed for use of this device was most frequently listed. Bell (1984) found time was an important factor; with a bed to chair transfer, the manual technique was executed in 20 seconds, the hoist took 295 seconds (which did not include the time it took to locate the hoist and transfer it to the patient's room). The unsafeness and instability of the lift were also frequently mentioned. Comments concerning safety included: two staff were needed so hoist did not tip, combative behaviors facilitated a tip over, the weight limits stated in the instruction manual should be reduced, and frequent maintenance was required to ensure safety.

Time was also a problem with gait belts. Additional problems included lack of staff knowledge/skill, comfort of patient, and lack of immediate availability. No mention was made of physical comfort/safety of the nurse. In the laboratory, Gagnon, Sicard and Sirois (1986) studied safety by estimating the forces at L5/S1 and assessing mechanical work and energy transfers in a task of raising a "patient" from a chair using three different methods (with the hands, with the forearms behind the patient's back at the shoulder level and with a gait belt). "The method requiring a belt to lift the patient was found to be considerably more strenuous for the spine and also to require a larger amount of work..." (p. 407). Because so many unsolicited comments were made in favor of gait belts, further clinical study should be done in evaluating their safety. If they are found to be biomechanically acceptable and comfortable for the nurse, a vigorous training program to increase the knowledge/skill of the nurse may decrease patient discomfort and skin problems. Also, increased skill may decrease the time problem. Another factor is cost; cost is not prohibitive as compared to hydraulic lifts and bathtub lifts.

The respondents overwhelmingly (80%) held the opinion that back problems in nurses could be reduced by increased use of PHDs.

These findings provide insight into availability and frequency of using PHDs, into problems experienced by nurses in nursing homes, and into patient characteristics that should be considered when determining need for PHDs. It is imperative, as stated by a number of nurses, to use good professional judgment when using PHDs with the lifting of patients.

REFERENCES

[1] Bell, F. (1984). Patient lifting devices in hospitals. London: Groom Helm.

[2] Chaffin, D., & Andersson, G. (1984). Occupational biomechanics. New York: John Wiley & Sons.
[3] Gagnon, M., Sicard, C., & Sirois, J. (1986). Evaluation of forces on the lumbo-sacral joint and assessment of work and energy transfers in nursing aides lifting patients. Ergonomics, 29, 407-421.
[4] Harber, P., Billet, E., Gutowski, M., SooHoo, K., Lew, M., & Roman, A. (1985). Occupational low-back pain in hospital nurses. Journal of Occupational Medicine, 27(7), 518-524.
[5] Jensen, R. (1987). Disabling back injuries among nursing personnel: Research needs and justifications. Research in Nursing & Health, 10, 29-38.
[6] Jensen, R. (1985). Events that trigger disabling back pain among nurses. Proceedings of the Human Factors Society 29th Annual Meeting, 799-801. Santa Monica, CA: Human Factors Society.
[7] Owen, B. (1987). The need for application of ergonomic principles in nursing. In S. Asfour (Ed.), Trends in Ergonomics/Human Factors IV. North Holland: Elsevier Science Publishers.
[8] Owen, B. (1985). The lifting process and back injury in hospital nursing personnel. Western Journal of Nursing Research, 7, 445-459.
[9] Stubbs, D., Baty, D., Buckle, P., Fernandes, H., Hudson, M., Rivers, P., Worringham, C., & Barlow, C. (1986). Back pain in nurses: Summary and recommendations. Guildford, England: University of Surrey.
[10] Torma-Krajewski, J. (1986). Occupational Hazards to Health Care Workers. Northwest Center for Occupational Health and Safety, September.
[11] Troup, D., Lloyd, P., Osborne, C., and Tarling, C. (1981). The handling of patients: A guide for nurse managers. Kempshott, Great Britain: J.B. Shears & Sons Ltd.

APPENDIX A - Description of Patient Handling Devices

(1) Hydraulic Lift - A mobile hoist used to transfer patient by use of slings attached to patient and hoist.
(2) Bathtub Lift - Hydraulic/air pressure lift (in shape of chair seat) attached to bathtub. Patient is placed in tub from lift seat.
(3) Lift Sheet - A sheet (may be draw sheet) is placed under the patient's shoulders and buttocks. Usually used in a two person maneuver to reposition/lift patient within the bed.
(4) Gait/Transfer Belt - A belt placed snuggly around the abdomen of the patient to enable the nurse to grasp the handles to transfer the patient.
(5) Trapeze - An attachment to the bed that enables patient to grasp a bar overhead to reposition self in bed.
(6) Turn Table - The patient stands on the pivot board, the handler turns the patient approximately 90° and lowers him to the chair.
(7) Roller - A rigid, board-like apparatus with rollers so patient can be rolled in a horizontal position.
(8) Board - Board-like structure with slippery surface to facilitate transfer from bed to cart.
(9) Bed Scale - Many are of a board-like surface that slips under the patient. Patient is lifted off the bed to be weighed.
(10) Rope Ladder - Soft material formed in rope shape, attached to end of bed so patient can pull self to sitting position.
(11) Inflatable Cushion - Apparatus placed under lower back/buttocks and inflated to raise the hips of patient for bedpan use.

Trends in Ergonomics/Human Factors V
F. Aghazadeh (Editor)
Elsevier Science Publishers B.V. (North-Holland), 1988

HOSPITAL BED DESIGN AND OPERATION - EFFECT ON INCIDENCE OF LOW BACK INJURIES AMONG NURSING PERSONNEL

David NESTOR

Division of Safety Research
National Institute for Occupational Safety and Health
Morgantown, West Virginia

The hospital services industry employs approximately six million people and reports an average injury rate twice that of other service industries. Forty percent of the job-related back injuries to hospital employees are accounted for by the nursing service. Of the nurses leaving the profession due to illness forty percent do so because of back pain. An estimated eighty percent of compensatory back pain injuries in nursing personnel are attributable to patient handling. The majority of patient handling tasks are performed while the patient is either in or close to the hospital bed. No studies have reported the effect of hospital bed design/operation on the incidence of low back pain among nursing personnel. This pilot study was initiated to determine the effect of existing hospital bed design/operation characteristics on the incidence rates of low back pain among nursing personnel in the hospital setting.

Data were collected at three hospitals employing a total of 1,133 full-time equivalent nursing personnel. Hospital bed design/operation characteristics were surveyed using a data collection tool specifically developed for this study. The characteristics included: time to raise bed from lowest to highest position, and location of bed height control device. Retrospective data on back injuries reported by nurses on nine wards in three hospitals were obtained for a three year period. A measure of the ratio of specific ward personnel back injury incidence rates to overall hospital nursing service personnel back injury incidence rates was developed to account for differences in the three hospitals. The results of this pilot study showed that there were differences in ratios of low back injury when controlling for bed design and operation. Recommendations for hospital bed design and operation are discussed.

1. INTRODUCTION

The hospital services industry employs approximately 6 million people. During the period from 1982 to 1985 the industry reported an average injury rate of 3.2 lost time cases with 53.5 lost work days per 100 full-time employees.[1,2] In comparison with other industries, the injury rates of the health services industry were twice those of other service industries and were comparable to blue collar workers.[3,4] In a 1980 study, injury rates for nurses and nurses aides ranked just behind truck drivers, warehousemen and loggers, and ahead of construction laborers, carpenters, welders, sawyers and various other so-called high-hazard occupations.[4] The Bureau of Labor Statistics (BLS) data show that in 1985 the health services industry had an injury rate of 6.8 (up from 6.0 in 1984) per 100 full-time employees. These same data show that the manufacturing industry experienced an injury rate of 4.4 (down from 4.5 in 1984) per 100 full-time employees.[5]

One injury category which contributes heavily to the lost time rate for the health services industry is musculoskeletal injuries, specifically, low back injuries. The prevalence of low back pain (LBP) and its occupational significance among persons in the nursing profession have been documented.[3,6-23] A survey of job-related back injuries in a hospital in Delaware revealed that 43 per cent of the employees in the nursing service accounted for 67 per cent of the injuries due to lifting.[13] Another study showed that 40 per cent of all nurses who leave the profession through illness, do so because of back pain.[24]

An estimated 73 to 81 percent of compensatory back pain injuries in nursing personnel are attributable to patient handling.[23,25] Patient handling tasks associated with low back injuries are lifting patients in bed, moving beds, assisting patients into and out of beds and lifting patients from beds to a litter, wheel chairs, etc.[10] Many of these and similar patient handling tasks exceed the Action Limits (AL) and Maximum Permissible Limits (MPL) for lifting as determined according to the National Institute for Occupational Safety and Health (NIOSH) Work Practices Guide for Manual Lifting.[26,27]

Manual handling of patients is a skilled activity which requires time, practice and application in order to be accomplished in a safe and comfortable manner.[28] If these requirements of time, practice and appropriate application of patient handling tasks are not achieved, then injuries may occur. Faverge's analysis of injuries demonstrated that man-machine and man-system relationships played an important role in safety.[29] Besides physical strain, other factors may predispose workers to injuries including: inadequate knowledge of the work process, mental anxiety due to lack of experience, unusually stressful situations, inappropriate use of equipment and poor design of equipment and equipment controls.[29]

The hazards associated with the design and operation of certain types of hospital beds have been documented.[30-32] Among the hazards are patient falls caused by uneven bedframe positioning and operation, patient fatalities caused by crush injuries associated with pedestal-style beds with a "walk-away" down control, patient fatalities associated with bed siderail design and malfunction, patient and staff injury due to poor hospital bed wheel, frame and control design and operation, and patient and staff injury and death associated with hospital bed electrical fires.[30-33] Evaluations of hospital beds and patient handling devices used by nursing personnel during patient handling tasks have been developed according to patient comfort, electrical safety, mechanical safety, equipment repair and purchase cost considerations with little, if any, consideration of nursing personnel capabilities or limitations.[30,31,34-45] Only two unpublished studies evaluated both the design and the operational characteristics of hospital beds as they relate to nursing personnel.[46,47]

The interaction of patient, nurse and hospital bed is time dependent and labor intensive for nursing personnel. The patient spends upwards of 90 percent of his or her hospital stay in or near the hospital bed.[46] Nursing personnel performing direct patient care spend approximately 50 percent of their working time dealing directly with the patient while he or she is in or close to the hospital bed.[48] The problem encountered when manual patient handling tasks are being performed is that it is difficult to make a compact manageable load out of a jointed, maybe paralyzed, overweight or unpredictable human being (patient).[49] In addition to dealing with an unmanageable load, many of the patient handling tasks are asymmetric-type lifts that place additional compressive and shear force loads on the nurse's trunk and musculoskeletal system. The unexpected loading of the nurse's musculoskeletal system (e.g. when a patient slips) during a manual patient handling task often causes considerable risk of injury to the lifter. In a recent study examining trunk loading and expectation of loading, average muscle forces for the unexpected loading condition exceeded those in the expected conditions by nearly two-and-a-half times. The peak muscle forces developed during the unexpected loading condition were, on average, 70 percent greater than the peak muscle forces developed in the expected loading condition.[50] Many of the manual patient handling tasks performed by nursing personnel involve similar loading due to the patient's unexpected resistance to movement, poor positioning of the patient in bed or by the patient becoming unable to assist during a transfer which renders the patient totally dependent upon nursing personnel for support and safety.

Various forms of intervention have been recommended for reducing risk of back injury. Employee training is the most common recommendation. A few authors have discussed and proposed pre-employment evaluations, behavior modification and ergonomics programs.[51-58] The use of patient handling devices has also been proposed as a mechanism for reducing exposure of nursing personnel to stressful manual patient handling.[59] Studies have indicated that mobile hoists tend to be under utilized by the nursing staff because of inadequate training on the use of the devices, complex operation and inability to use the devices in confined spaces within the hospital rooms, bathrooms and hallways.[59,60] The use of other patient handling devices such as slings and roller boards has not been studied

No studies have reported the effect of hospital bed design and operation on the incidence of low back pain among nursing personnel. The purpose of this pilot study was to determine the effect, if any, of hospital bed height cycle time and nursing control location upon the incidence rates of low back injury among nursing personnel in the hospital setting.

2. METHOD

This pilot study was conducted at two Veteran's Administration (VA) medical center hospitals and one state medical center. VA medical center "A" has 66 acute care beds and 126 extended care beds. This facility has an average of 184 full-time equivalent employees in their nursing service. VA medical center "B" has 216 acute care beds and no extended care beds. This facility has an average of 146 full-time equivalent employees in their nursing service. The state operated medical center "C" has 452 acute care bed and no extended care beds.

Their nursing service averages 803 full-time equivalent employees. Each nursing service of all three hospitals includes registered nurses, licensed practical nurses, nursing aides and nursing orderlies.

Hospital bed design and operation characteristic data utilized in this study were obtained through the use of a hospital bed evaluation form designed specifically for this project. Information abstracted from the bed evaluation form was used to identify the independent variables, bed height cycle time and location of nursing controls. A representative model of each bed used by the study facilities was examined and evaluated using this survey instrument.

The hospital bed height high/low cycle times and the locations of the nursing controls on the beds were selected for evaluation in this study. The cycle times and the locations of nursing controls were selected because the height of the bed must be altered prior to initiating a manual patient handling task in order to achieve maximum efficiency and safety. The speed at which the bed height adjustment can be made and the accessibility of the nursing controls may determine when or if the nursing personnel utilize these design features. For example, if the bed cycle time was long and the available time to perform the manual patient handling task was limited due to staffing, patient load, work pressure etc., the nurse may decide to forego bed height adjustments and take the risk of performing the patient handling task at an other than optimal, safe height. This same situation may also occur if the nursing controls were not conveniently located and do not allow for rapid access and operation. On the other hand, if the cycle time was short and the nursing controls were easily accessible, the nurse may find it convenient to perform the height adjustment thereby making the work station safer for the manual patient handling task.

3. DATA COLLECTION

Injury information used for calculating low back injury (LBI) incidence rates for this study was abstracted from the Occupational Safety and Health Administration (OSHA) Form 101 for each facility for the calendar years 1984 through 1986. Additional detailed accident/incident information concerning the VA facilities was taken from VA Form 2162-Report of Accident. Similar detailed accounts of accidents/ incidents for the state medical center hospital were abstracted from the hospital's insurance carrier accident reporting forms.

Annual LBI incidence rates for 1984-1986 for each hospital in the study were determined by calculating the rate of low back injuries per 100 full-time equivalent (FTE) employees of nursing service.[27,59] The number of full-time equivalent employees of nursing service for each hospital was calculated for the calendar years of 1984, 1985 and 1986. A low back injury incidence was defined as any musculoskeletal injury to the low back, with or without lost time, during the study year. No distinction was made between "new" and "repeat" incidents of low back injury. Incidence rates for the medical/surgical (including neurologic/neurosurgical, orthopedic and general surgical) and psychiatric wards were also calculated. The medical/surgical wards were chosen for evaluation due to their high patient care acuity. These wards were generally occupied by the more dependent patients that required a higher level of patient care, nursing management and more skilled manual patient handling tasks. On the other hand, the psychiatric wards of the study facilities had a lower degree of patient care acuity. These patients are usually ambulatory and required a minimal amount of manual patient handling. The data from the psychiatric wards were used as control/reference information for nursing personnel having patient contact but not having a high frequency of exposure to manual patient handling tasks.

The LBI incidence rates for nursing personnel were calculated as follows:

$$\text{Nursing Service Incidence Rates} = \frac{\text{Number of LBI in Nursing Service}}{\text{100 Full-Time Nursing Employees}}$$

$$\text{Nursing Ward Incidence Rates} = \frac{\text{Number of LBI on Nursing Ward}}{\text{100 Full-Time Ward Nursing Employees}}$$

Specific ward incidence rates were compared to hospital rates. From this comparison, a ward:hospital ratio was calculated for each of the two types of wards.[61] This ratio was calculated as follows:

$$\text{Ward/Hospital Ratio} = \frac{\text{Incidence Rate Ward}}{\text{Incidence Rate Hospital}}$$

This ratio was used as the dependent variable for analyses. The two independent variables were bed height high/low cycle times and the locations of nursing control.

The following operational definitions were used for this study. The bed height high/low cycle time was defined as the time, in seconds, required to cycle the bed from the minimum height to the maximum height. The cycle time for each bed was measured with a stop watch and recorded in seconds.

The location of the nursing controls on the bed was defined as the physical location of the nursing control panel used when adjusting the bed height. The three most common locations of nursing controls are as follows:

- Foot Bed Frame
- Side Bed Frame
- Side Rail

4. RESULTS

Analysis of the data compiled from the survey instrument revealed that in facility "A", 50% of the medical/surgical and psychiatric beds were manual hand-crank operated and 50% were electric motor driven. All controls were located at the foot of the bed. These two bed types required an average of 38 seconds to cycle from the lowest to highest bed height positions. Bed height adjustments ranged from 27" low position to 40" high position (including 6" mattress). The ward/hospital low back injury incidence ratios for this facility were the highest of the three study hospitals.

Facility "B" medical/surgical beds were 100% electric motor driven with primary nursing controls located on the side bed frame. Secondary bed height and nursing controls were located at the foot of the bed. Bed height adjustments ranged from 27" low position to 40" high position (including 6" mattress). The cycle time for this group of beds was 22 seconds. The psychiatric beds used for comparison were fixed height (15 inches) without adjustments or controls.

Facility "C" medical/surgical and psychiatric beds were 100% electric motor driven with primary nursing and patient controls located in the bed siderail. Secondary bed height and nursing control lock-out mechanisms were located at the foot of the bed. Cycle time for this group of beds was 20 seconds. Bed height adjustments ranged from 23" low position to 43" high position (including 6" mattress).

Low back injury incidence rates per 100 full-time equivalent (FTE) employees were calculated for hospital nursing service, medical/surgical wards and psychiatric wards for each facility for the years 1984, 1985, 1986 (Table 1).

Table 1 - LBI Incidence Rates/100 FTE's

	Facility A			Facility B			Facility C		
Year	H	M/S	P	H	M/S	P	H	M/S	P
1984	7.5	18	0.7	4	7.7	0	4.1	7.4	0.4
1985	4	11.7	0.5	4.8	10.7	0	5.1	9.7	0.3
1986	8.5	14.5	0.8	4.7	6.6	0.5	5.0	8.0	0.3

H = Hospital Nursing Service
M/S = Medical/Surgical Wards
P = Psychiatric Ward

Ward/hospital low back injury incidence ratios for the three facilities during the years 1984, 1985 and 1986 were calculated and are shown in Table 2. A simple average of these ratios was compared to the bed cycle time (Table 3) and nursing control location (Table 4).

Table 2 - Ward/Hospital Low Back Injury Incidence Ratios

	Facility A		Facility B		Facility C	
Year	Med/Surg	Psyc	Med/Surg	Psyc	Med/Surg	Psyc
1984	2.4	.09	1.9	0	1.8	.10
1985	2.9	.12	2.2	0	1.9	.06
1986	1.7	.09	1.4	.11	1.6	.06

Table 3 - Bed Hi-Low Height Cycle Time vs. Average Ward/Hospital LBI Incidence Ratio

Facility	Cycle Time	LBI Incidence Ratio
A	38 seconds	2.3
B	22 seconds	1.8
C	20 seconds	1.8

Table 4 - Nursing Control Location vs. Average Ward/Hospital LBI Incidence Ratio

Facility	Control Locations	LBI Incidence Ratio
A	Foot	2.3
B	Side Bedframe*/Foot	1.8
C	Side Rail*/Foot	1.8

* Primary control location

5. DISCUSSION

This pilot study was conducted as a pre-cursor to a larger study of the effect, if any, of hospital bed height cycle times and location of nursing controls on the incidence rates of low back injury among nursing personnel in a hospital setting. The results of this study suggest that an increased incidence of low back injury among nursing personnel might be partially explained by long bed height cycle times and nursing controls located at the foot bed frame. The one facility which uses beds that demonstrate a bed height cycle time of 38 seconds and had nursing controls located at the foot of the bed had the highest ward/hospital low back injury ratios in the study. This may suggest that the long bed cycle time and awkward location of nursing controls may be a factor to consider when evaluating the incidence of low back pain among nursing personnel using this type of bed. Facilities "B" and "C" had bed height cycle times of 20-22 seconds and nursing controls located on the bed side frame or siderail. These two facilities had similar ward/hospital low back injury incidence ratios lower than facility "A". These data suggest that a hospital bed with a height cycle time of approximately 20 seconds and nursing controls located on the bed side frame or siderail may be most useful for encouraging nursing personnel to use hospital bed design and control characteristics properly, prior to and during manual patient handling tasks in the hospital setting.

Medical/surgical ward/hospital low back injury ratios were higher than overall hospital rates and psychiatric ward/hospital ratios for all three facilities. This finding suggests that the high degree of patient care acuity of the medical/surgical patients, regardless of the bed type, may expose the patient handler to more demanding manual patient handling tasks resulting in a higher incidence of low back injury among these nursing personnel.

6. CONCLUSIONS

Guidelines for hospital bed design and control location may be developed utilizing the findings of this descriptive pilot study, the American Hospital Association published recommendations [62] and the unpublished works of Solomon and Gandette [46] and Carpman and Grant [47].

Suggested guidelines include:

- Siderail, bed sideframe located nursing controls with secondary height controls and lock-out mechanisms located at the foot of the bed.
- Electrically operated beds with a bed height high-low cycle time of 20 seconds or less.
- Minimum bed height of 20" (including 6" mattress).
- Maximum bed height of 43" or higher (including 6" mattress).
- Minimum bed length 73".
- Maximum bed length of 87".
- 5" diameter, double ball bearing, swivel type casters with crossed locking system and mid-frame mounted foot "lock-unlock" mechanisms.
- Lighted, large graphic patient control symbols with international meanings that patients can operate with either hand.
- Visible, contrasting siderail release mechanisms.
- Momentary "down" height controls.
- Removable head/foot boards.
- 6" interspring mattress which conforms to Flammability Standard DOC FF-4-72 and California Technical Bulletin #117. (Refer to Solomon & Gandette for additional criteria)[46]
- IV and traction equipment receptacles.
- Two or more electric motors for independent operation of bed function should be completely enclosed, overload protected with reset button and insulated from metal parts of the bed.

The findings of this descriptive pilot study and results of published and unpublished works cited in this document should serve as a starting point for future research in developing safe, ergonomically designed hospital beds.

ACKNOWLEDGEMENTS

The author sincerely appreciated the very constructive comments on the draft of this article from Roger Jensen, Catherine Bell, Tim Pizatella, Roger Nelson, Ph.D., Ann Myers, Sc.D. and Carol Stuart-Buttle. Special thanks to Pollyanna Fiorini for her attention to detail in preparing this manuscript.

REFERENCES

[1] The Bureau of National Affairs, Inc. Occupational Safety and Health Reporter, 14(25):494, November 22, 1984.

[2] The Bureau of National Affairs, Inc. Occupational Safety and Health Reporter, 16(35):644, November 19, 1986.

[3] Magora, A., Investigation of the Relations Between Low Back Pain and Occupation. Industrial Medicine, 39(11):465-471 and 39(12):504-510, 1970.

[4] Omenn, G. and Morris, S. Occupational Hazards to Health Care Workers: Report of a Conference. American Journal of Industrial Medicine, 6:129-137, 1984.

[5] The Bureau of National Affairs, Inc. Occupational Safety & Health Reporter, 16(25):628, November 19, 1986.

[6] Magora, A., Investigation of the Relations Between Low Back Pain and Occupation. Part IV Medical History & Symptoms. Scand J Rehab Med, 6:81-88, 1974.

[7] Troup, J., et al. Back Pain in Industry: A Prospective Survey. Spine, 6:61-69, 1981.

[8] Back Pain, Office of Health Economics, White Crescent Press Ltd., England, No.78, July, 1985.

[9] Cust, G., Pearson, J. and Mair, A., The Prevalence of Low Back Pain in Nurses. International Nursing Review, 19:169-179, 1972.

[10] Harber, P., Billet, E. and Gutowski, M., et.al. Occupational Low Back Pain in Hospital Nurses. Journal of Occupational Medicine, 27:518-524, 1985.

[11] Videman, T. Nurminen, T. Tola, S., Kuorinka, I. and Vanharanta, H. Low-Back Pain in Nurses and Some Loading Factors of Work. Spine, 9(4):400-404, 1984.

[12] Stubbs, D., Buckle, W. Hudson, M., Rivers, P., Worringham, C. Back Pain in the Nursing Profession. I Epidemiology and Pilot Methodology. Ergonomics, 26(8):755-765, 1983.

[13] Hoover, S. Job Related Back Injuries in a Hospital. AJN, pp 2078-2079, December, 1973.

[14] Prezant, B., Demers, P. and Strand, K. Back Problems, Training Experience and Use of Lifting Aids Among Hospital Nurses. Trends in Ergonomics/Human Factors IV, Part B, pp 839-846, 1987.

[15] Pines, A., Skulkeo, K., Pollack, E., Peritz, E. and Steif, J. Rates of Sickness Absenteeism Among Employees of a Modern Hospital. The Role of Demographic and Occupational Factors. British Journal of Industrial Medicine, 42:326-335, 1985.

[16] Douglas, B. Health Problems of Hospital Employees. Journal of Occupational Medicine, 13(12):555-560, December, 1971.

[17] Klein, B., Jensen, R. and Sanderson, L. Assessment of Worker's Compensation Claims for Back Strains/Sprains. Journal of Occupational Medicine 26:443-448, 1984.

[18] Dehlin, O., Hedenrud, B. and Horal, J. Back Symptoms in Nursing Aides in a Geriatric Hospital. Scand J Rehab Med, 8:47-53, 1976.

[19] Ferguson, D. Strain Injuries in Hospital Employees. The Medical Journal of Australia, 1:376, February, 1970.

[20] Jensen, R. Disabling Back Injuries Among Nursing Personnel: Research Needs and Justification. Research in Nursing & Health, 10:29-38, 1987.

[21] Owen, B. The Lifting Process and Back Injury in Hospital Nursing Personnel. Western Journal of Nursing Research, 7:445-459, 1985.

[22] Stubbs, D., Rivers, P., Hudson, M. and Worringham, C. Back Pain Research. Nursing Times, pp 857-858, May 14, 1981.

[23] Pines, A., Cleghorn de Rohrmoser, D. and Pollack, E. Occupational Accidents in a Hospital Setting: An Epidemiological Analysis. Journal of Occupational Accidents, 7:195-215, 1985.

[24] Bangs, M. Back Injury Information Gap. Nursing Times, page 1726, November 7, 1974.

[25] Jensen, R. Events that Trigger Disabling Back Pain Among Nurses. Proceedings of the Human Factors Society-29th Annual Meeting, 1985, pages 799-801.

[26] National Institute for Occupational Safety and Health, Work Practices Guide for Manual Lifting. DHH(NIOSH) Pub.# 81-122. Superintendent of Documents, U.S. Government Printing Office, Washington, D.C.

[27] Torma-Krajewski, J. Analysis of Injury Data and Job Tasks at a Medical Center. Trends in Ergonomics/Human FActors IV, Part B, pp 863-874, 1987.

[28] Haynes, S. Counting the Cost of Sickness. Nursing Times, pp 50-51, August 15, 1984.

[29] Faverge, J. Analyse de la Securite du Travail en termes de facteurs de risque. Revue Epidemiologique et Sante Publique, 25, pp 229-241, 1977.

[30] Inspection of Electric Beds. Health Devices, The Emergency Care Research Institute, Vol. 1, June 1971.

[31] Electric Beds. Health Devices, The Emergency Care Research Institute, Vol. 2, January, 1973.

[32] Electric Beds. Health Devices, The Emergency Care Research Institute, Vol. 15, November, 1986.

[33] Health Devices Alerts - Hazard Update 10-347. Health Devices, The Emergency Care Research Institute, March 13, 1987.

[34] Reizenstein, J. and Grant, M. From Hospital Research to Hospital Design. Patient and Visitor Participation Project, Office of Hospital Planning, Research and Development, University of Michigan, Ann Arbor, 1982.

[35] Bergstrom, D., Ellwood, P. and Grendahl, B. Rehabilitation Experts Confer with Manufacturers on Design of Hospital Beds. Hospitals, JAHA, 39:101-106, December 1, 1965.

[36] Robertson, J. And So To Beds...Patient Support Systems. Nursing Mirror Supplement, pp ii-xvii, May 24, 1979.

[37] Walsh, T. and Hanks, T. A Study of the Dimensions and Problems Associated with Equipment Malfunction and Accidents in Hospitals. California Hospital Association, 1972.

[38] Spencer, S. The Development of a New Hospital Bed. The Australian Nurses Journal, 5(1):36-38, 1975.

[39] Norton, D. By Accident of Design, E & S Livingstone, London, 1970.

[40] Winn, F. Evaluation System Benefits Bed Purchase Decision. Hospitals, JAHA, 43:98-99, May 16, 1969.

[41] Scott, J.C. Design of Hospital Bedsteads. Nursing Times, pp 1351-1352, October 6, 1967.

[42] Roberts I. Design of Hospital Beds. Nursing Times, pp 632-634, May 13, 1966.

[43] Shaw, A. and Fisher, J. Falls from Hospital Beds as a Result of Poor Mattress Design: A Case History. Journal of Medical Engineering & Technology 3(6):301, November, 1979.

[44] Garfin, S. and Pye, S. Bed Design and its Effect on Chronic Low Back Pain-A limited Controlled Trial. Pain, 10:87-91, 1981.

[45] Guidelines for Construction and Equipment of Hospital and Medical Facilities (1984 ed.), HHS Office of Health Facilities, Division of Facilities Resources, Rockville. Maryland.

[46] Solomon, L. and Gaudette, R. Adult General Hospital Bed and Patient Room Furniture Evaluation (unpublished report). Office of Hospital Planning, Research and Development, University of Michigan, Ann Arbor, 1984.

[47] Carpman, J. and Grant, M. Hospital Patient Room Furnishings Mock-ups (unpublished research report #25). Patient & Visitor Participation Project, Office of Hospital Planning, Research and Development, University of Michigan, Ann Arbor, 1984.

[48] Methods of Studying Nurse Staffing in a Patient Unit a Manual to Aid Hospitals in Making Use of Personnel. NTIS Doc.#HRP-0900595, U.S. Dept. of Health, Education and Welfare, Chapter IV, May, 1978.

[49] Fordham, M. Medical Ergonomics, Ergonomics, 30(3):551-562, 1987.

[50] Marras, W., Rangarajulu, S. and Lavender, S. Trunk Loading and Expectation. Ergonomics, 30(3):551-562, 1987.

[51] Morris, A. Program Compliance Key to Preventing Low Back Injuries. Occupational Health and Safety, 3, 44-47, 1984.

[52] Snook, S. Campanelli, R. and Hart, W. A Study of Three Preventive Approaches to Low Back Injury. Journal of Occupational Medicine, 20(7):478-481, July, 1978.

[53] Burgel, B. and Gliniecki, C. Disability Behavior. AAOHN Journal, 43(1):26-30, 1986.

[54] Raistrick, A. Nurses with Back Pain - Can the Problem be Prevented? Nursing Times, pp 853-856, May 14. 1981.

[55] Daly-Gawenda, D., Kempinski, P. and Hudson, E. Pre-Employment Screening-Its Use and Usefulness. AAOHN Journal, 34(6):269-271, 1986.

[56] Gates, S. and Starkey, D. Back Injury Prevention-A Holistic Approach. AAOHN Journal, 34(2):59-62, 1986.

[57] Wood, D. Design and Evaluation of a Back Injury Prevention Program Within a Geriatric Hospital. Spine, 12(2):77-82, 1987.

[58] Scholey, M. Patient Handling Skills. Nursing Times, pp 25-27, June 27, 1984.

[59] Hollis, M. Safe Lifting for Patient Care. (2nd Ed.) Blackwell Scientific Publications, Boston, Mass, 1985.

[60] Bell F. Patient-Lifting Devices in Hospitals. Croom Helm, London, England, 1984.

[61] Kleinbaum, D., Kupper, L. and Morganstein, H. Epidemiologic Research, Chapter 8, Lifetime Learning Publications, California, 1982.

[62] Carpman, J., Grant, M. and Simmons, D. Design That Cares. Chapters 6 & 10, AHA Publication 043180, AHA Publishing, Inc., Chicago, Ill., 1986.

Trends in Ergonomics/Human Factors V
F. Aghazadeh (Editor)
Elsevier Science Publishers B.V. (North-Holland), 1988

WHICH SUBPOPULATIONS OF THE MINING INDUSTRY ARE AT A HIGHER- OR LOWER-THAN-AVERAGE RISK FOR BACK INJURY PROBLEMS?

Shail J. BUTANI

Twin Cities Research Center
U.S. Bureau of Mines
5629 Minnehaha Avenue South
Minneapolis, MN. 55417

The U.S. Bureau of Mines in 1986 conducted a probability sample survey to measure the characteristics of the U.S. mining industry workforce. This paper discusses the importance of collecting demographics data for the purposes of analyzing existing injury and illness data. In particular, it shows how demographics survey data are utilized in identifying subpopulations of the 1986 mining workforce (by sex, age, experience at present job, and job title or occupation) that exhibited a disproportionately higher- or lower-than-average number of work-related back injuries, as well as lost workdays due to back injuries. These injuries in 1986 accounted for 19 pct of all mining incidents and 29 pct of all lost workdays.

The results of the analysis show that of all the subpopulations studied, continuous miner and related machinery operators in the coal industry had the highest risk for back injuries and related lost workdays, about four times the average. For both coal and metal and nonmetal sectors (metal, stone, sand and gravel, and nonmetal), the workers in the age group 30 to 39 were at a higher than average risk, whereas those age 50 and over were the least prone (below average) to back injury problems.

1. INTRODUCTION

One of the primary objectives of the U.S. Bureau of Mines is to conduct research in the area of health and safety of the Nation's miners, aimed at reducing the incidence rate of work-related injuries and illnesses in the domestic mining industry. In order to reduce the overall incidence rate, the Bureau needs to know which groups or subpopulations of the workforce exhibit higher-than-average incidence rates.

To identify the high risk groups, information about the injured victims and about the workforce are required. Present regulations permit the Mine Safety and Health Administration (MSHA) to collect information on all mine injuries requiring medical attention. Hence, a database containing, among other things, information on the occupation, age, sex, and job experience of the injured workers is available. Unfortunately prior to 1986, a similar database did not exist for the total miner population. The total population characteristics are crucial pieces of information which are required to determine the relative injury measures (e.g., injury rates and indices), in order to compare and contrast various subpopulations or groups. Hence, the demographics survey was conducted by the Bureau in 1986 to provide information about populations at risk and thus aid the research in pinpointing the hazardous segments of the population. Table 1 illustrates this analysis for total number of incidents and back injuries; it also compares the risk for workers in the coal sector, relative to that for workers in the metal and nonmetal sector.

TABLE 1

Basic Statistics On Workers, Total Incidents, And Back Injuries For 1986 Coal And Metal and Nonmetal[1] Mining

Sector	No. of Workers[3]	Total Incidents[2]		Total Lost Workdays[2]		Back Injuries[2]		Lost Workdays Due to Back Injuries[2]	
		No.	Rate[4]	No.	Rate[5]	No.	Rate[6]	No.	Rate[7]
Coal	146,395	12,962	8.9	309,215	211	2,640	1.8	93,141	64
Metal & Nonmetal	167,389	9,067	5.4	131,889	79	1,624	1.0	33,880	20

[1]Metal and Nonmetal includes metal, stone, sand and gravel, and nonmetal industries.

[2]Source: 1986 MSHA Injuries/Illnesses/Fatalities database.

[3]Source: 1986 Demographics Survey; excluding job title category office workers.

[4]Rate = (No. of Incidents/No. of Workers) X 100.0

[5]Rate = (No. of Lost Workdays/No. of Workers) X 100

[6]Rate = (No. of Back Injuries/No. of Workers) X 100.0

[7]Rate = (No. of Lost Workdays Due to Back Injuries/No. of Workers) X 100

According to an article published in the January 1988 issue of the journal titled "Occupational Health and Safety News Digest" [1]: "(1) At least 20 pct of all work injuries are to the back; (2) about 400,000 disabling work injuries to the back occur each year; and (3) about 40 pct of all recorded absences from the workplace are due to backaches." The data in table 1 show that 19 pct of all mining incidents in 1986 were to the back, accounting for 29 pct of all lost workdays. Hence, back injuries are also a major problem facing the mining industry. The purpose of this paper is to use demographics survey data to analyze work-related back injuries separately for the workers in the coal and metal and nonmetal sectors in relation to the attributes of sex, age, experience at present job, and job title or occupation.

2. DATA COLLECTION METHODOLOGY

The demographics survey was based on probability design theory and methodology; it utilized a two-stage stratified probability sample design [2]. The questionnaire was designed to collect information on the following characteristics: occupation, principal equipment operated, work location at mine (underground, surface at underground, surface, plant or mill, office), experience at present job, experience at present company, total mining experience, job related training, age, sex, race, and education.

Based on the survey's design criteria, a sample of 3,708 establishments was selected from a universe of 18,350 mining establishments in the United States. The initial mailing of questionnaires took place in March 1986, and the respondents were contacted again by followup mailings and/or telephone calls. The data collection phase was completed in September 1986 with a high degree of success; the survey achieved an overall usable response rate of over 75 pct. As part of the data collection phase, all the returns were reviewed and edited for completeness and accuracy of the data; the inconsistencies were reconciled. The returned questionnaires contained information for about 40,000 workers and represent the data source for the population estimates.

The estimates of reliability (i.e. variances, standard errors, and coefficient of variations) for the population estimates were computed using a random variance group technique [3].

The data for back injuries statistics were obtained from the 1986 MSHA database which is based on a census.

3. INJURY ANALYSIS METHODOLOGY

As mentioned in the introduction, the objective was to analyze the number of back injuries, as well as the number of lost workdays due to this injury, separately for coal and metal and nonmetal mining sectors, in order to identify subpopulations at high or low risk. For each sector, the back injuries were analyzed separately by sex, age, experience at present job, and job title or occupation.

The first part of the analysis tested the null hypothesis that the population distribution of an attribute is the same as the distribution of workers incurring back injuries. Example: For each age group in the coal sector, the proportion of nonoffice workers (identified by job title or occupation) is equal to the proportion of workers incurring back injuries. The reason for excluding office workers from the analysis was to account for some of the obvious difference in job risk. The same type of hypothesis was also tested for the number of lost workdays due to back injuries. All hypotheses were tested at the significance level of 0.05. The null hypothesis was tested by using the Pearson chi-squared test for the goodness-of-fit [4] corrected for the survey design effect. Since the Pearson chi-squared test is based on simple random sample, the correction for the design effect is essential because the demographics survey used a complex survey design (i.e., two-stage stratified). The methodology used in this paper was developed by Rao and Scott [5].

The second part of the analysis involved identifying those subpopulations having higher- or lower-than-average risk for back injuries. This required testing the null hypothesis that for a subpopulation (e.g., age group 40 to 49 in the coal sector) the percentage of total workers in that subpopulation (23.8-table 2) is equal to the percentage of workers incurring back injuries (23.6-table 2). The null hypothesis assumes that the difference between 23.8 and 23.6 is not statistically significant at a preassigned significance level, for this analysis it is 0.05. Since the injury data were based on a census, they were not subject to any sampling error. Hence, the task of testing this hypothesis was accomplished by constructing a 95 pct confidence interval for the proportion of workers for each subpopulation category. If the proportion of injuries fell outside the interval, then the difference was significant. The same test was also applied for lost workdays due to back injuries. That is, if the proportion of lost workdays fell outside the 95 pct confidence interval for percentage of workers, then the difference was significant. The significant differences are shown by an asterisk (*) in tables 2 and 3. In the case of coal workers, age group 40 to 49, the proportion of back injuries was not significant while the proportion of lost workdays was significant (table 2). That is, workers in this subpopulation were subjected to average risk for number of back injuries but the severity of these injuries as measured by lost workdays was higher than average.

TABLE 2
Distribution of Back Injuries and Related Lost Workdays (LWDS), and Workers
By Sex, Age, Job Experience, and Occupation For 1986 Coal Industry

Attribute	No. of Injuries	Pct[1] Injuries	Pct[1] LWDS	Pct[1] Workers[2]	95 Pct[1] Confidence Limit Lower	95 Pct[1] Confidence Limit Upper	I_{INJ}[3]	I_{LWDS}[4]
Sex[5]								
Male	2,592	98.2*	97.9	97.7	97.4	98.0	1.01	1.00
Female	48	1.8*	2.1	2.3	2.0	2.6	.78	.91
Age[6]:								
15 to 23	52	2.1	1.0*	2.6	2.0	3.2	.81	.38
24 to 26	133	5.3	3.7*	5.5	4.8	6.2	.96	.67
27 to 29	287	11.4*	9.6	10.2	9.5	10.9	1.12	.94
30 to 34	633	25.1*	26.7*	21.9	20.8	23.0	1.15	1.22
35 to 39	544	21.6*	21.3*	19.8	18.8	20.8	1.09	1.08
40 to 49	593	23.6	27.0*	23.8	23.0	24.6	.99	1.13
50+	275	10.9*	10.7*	16.2	15.6	16.8	.67	.66
Job Experience[7]:								
0> to 1	501	20.6	19.5	19.6	18.3	20.9	1.05	.99
1> to 2	337	13.9*	14.9	15.8	14.7	16.9	.88	.94
2> to 3	221	9.1*	8.6*	10.3	9.5	11.1	.88	.83
3> to 5	381	15.7	17.6	16.5	15.3	17.7	.95	1.07
5> to 10	641	26.4	23.6*	25.9	24.9	26.9	1.02	.91
10> to 20	322	13.2*	13.7*	10.7	9.9	11.5	1.23	1.28
20>	29	1.2	2.0*	1.3	1.1	1.5	.92	1.54
Occupation[8]:								
Beltpersons	155	6.0*	7.1*	3.6	3.2	4.0	1.67	1.97
Continuous Miner, Related Machine Operators	418	16.3*	18.3*	4.4	4.1	4.7	3.70	4.16
Drillers, Roof Bolters	276	10.7*	11.4*	7.6	6.8	8.4	1.41	1.50
Heavy & Mobile Equipment Operators	248	9.7*	9.8*	18.9	17.9	19.9	.51	.52
Laborers, Miners, Utility Persons	329	12.8*	12.3*	15.9	15.6	16.2	.81	.77
Managers	131	5.1*	5.5*	11.7	11.1	12.3	.44	.47
Shop Persons	562	21.9*	20.2	20.1	19.3	20.9	1.09	1.00
Shuttle Car, Scoop Operators	289	11.2*	11.1*	7.5	6.9	8.1	1.49	1.48
Others	161	6.3*	4.3*	10.4	9.7	11.1	.61	.42

*Significant at the .05 level.
[1]Due to rounding, sum of percentages may not equal to 100.0
[2]Excluding job title category office workers.
[3]I_{INJ} = Pct Injuries/Pct Workers.
[4]I_{LWDS} = Pct LWDS/Pct Workers.
[5]Excludes data for those workers whose sex was unspecified.
[6]Excludes data for those workers whose age was unspecified.
[7]Excludes data for those workers whose job experience was unspecified.
[8]Excludes data for those workers whose job title or occupation was unspecified.

TABLE 3

Distribution of Back Injuries and Related Lost Workdays (LWDS), and Workers
By Sex, Age, Job Experience, and Occupation For 1986 Metal and Nonmetal Industry

Attribute	No. of Injuries	Pct[1] Injuries	Pct[1] LWDS	Pct[1] Workers[2]	95 Pct[1] Confidence Limit Lower	95 Pct[1] Confidence Limit Upper	I_{INJ} [3]	I_{LWDS} [4]
Sex[5]:								
Male	1,587	97.7*	98.4*	97.3	97.1	97.5	1.00	1.01
Female	37	2.3*	1.6*	2.7	2.5	2.9	.85	.59
Age[6]:								
15 to 23	128	8.1*	6.4	5.8	5.2	6.4	1.40	1.10
24 to 26	136	8.6*	5.9*	7.2	6.8	7.6	1.19	.82
27 to 29	175	11.1*	8.9	9.0	8.4	9.6	1.23	.99
30 to 34	307	19.5*	18.2*	15.7	15.3	16.1	1.24	1.16
35 to 39	281	17.8*	17.9*	15.3	14.9	15.7	1.16	1.17
40 to 49	320	20.3*	25.1*	23.5	23.0	24.0	.86	1.07
50+	231	14.6*	17.7*	23.6	22.9	24.3	.62	.75
Job Experience[7]:								
0> to 1	411	28.5*	28.0*	20.0	18.9	21.1	1.43	1.40
1> to 2	196	13.6*	12.6	12.5	11.7	13.3	1.09	1.01
2> to 3	123	8.5*	8.4*	9.2	8.7	9.7	.92	.91
3> to 5	170	11.8	10.1*	12.4	11.8	13.0	.95	.81
5> to 10	295	20.4*	20.1*	23.4	22.3	24.5	.87	.86
10> to 20	207	14.3*	17.0	16.9	15.8	18.0	.85	1.01
20>	41	2.8*	3.8*	5.7	5.3	6.1	.49	.67
Occupation[8]:								
Heavy & Mobile Equipment Operators	285	18.1*	18.7*	27.7	26.7	28.7	.65	.68
Laborers, Miners, Utility Persons	345	21.9*	23.4*	12.0	11.1	12.9	1.83	1.95
Managers	55	3.5*	3.3*	11.8	11.4	12.2	.30	.28
Plant Operators	307	19.5*	15.5*	16.8	16.2	17.4	1.16	.92
Shop Persons	442	28.0*	30.2*	19.3	18.7	19.9	1.45	1.56
Others	142	9.0*	8.8*	12.5	12.2	12.8	.72	.70

*Significant at the .05 level.
[1]Due to rounding, sum of percentages may not equal to 100.0
[2]Excluding job title category office workers.
[3]I_{INJ} = Pct Injuries/Pct Workers.
[4]I_{LWDS} = Pct LWDS/Pct Workers.
[5]Excludes data for those workers whose sex was unspecified.
[6]Excludes data for those workers whose age was unspecified.
[7]Excludes data for those workers whose job experience was unspecified.
[8]Excludes data for those workers whose job title or occupation was unspecified.

Additionally, in order to compare and contrast various groups or subpopulations, a relative measure called index was computed for back injuries and for lost workdays. The index for back injuries (I_{INJ}) is simply the ratio of percentage of back injuries to the percentage of workers; similarly the index for lost workdays (I_{LWDS}) is percentage of lost workdays divided by the percentage of workers. An index of 1.00 indicates an average risk; the larger the index is, the higher the risk is for that group. For example, an index of 1.15 for back injuries (I_{INJ}) means this group of workers was at a 15 pct higher-than-average risk.

4. RESULTS

4.1 Overall Distributions

For each of the four attributes, two chi-squared tests were performed, one for back injuries and one for lost workdays. Thus, eight different chi-squared tests were performed for each sector, or a total of 16 for the coal and metal and nonmetal sectors. Fifteen of the sixteen chi-squared tests rejected the null hypothesis that the overall distribution of workers is the same as the distribution of back injuries or lost workdays due to back injuries. The only chi-squared test that did not reject the null hypothesis was for the lost workdays distribution by sex in the coal sector.

4.2 Subpopulations

Sex--The results (table 2 and 3) indicate that females are incurring proportionately fewer back injuries than males. Although all the differences are significant with the exception of lost workdays for the coal sector, these differences for all practical purposes are too small to warrant any action.

Age--The relative measures by age differ somewhat between the coal and metal and nonmetal sector. In both sectors, however, the workers in the age group 30 to 39 were anywhere from 8 to 24 pct higher-than-average risk while the workers in the age group 50 and over were at a considerably lower than average risk. These workers (50+) incurred proportionately about one-third (1.00 minus 0.67) less back injuries and about one-fourth to one-third fewer days lost from work due to back injuries.

Experience at present job--For this attribute, the relative measures are vastly different for the two sectors. In the coal industry (table 2): (1) the group of workers who suffered a disproportionately high number (about 25 pct) of back injuries as well as lost workdays due to back injuries were those with more than 10 years but less than or equal to 20 years of experience, and (2) workers with more than 20 years of job experience had significantly higher (54 pct) lost workdays from back injuries relative to their numbers in the population but the risk for back injuries was about average (I_{INJ} = 0.92). The data for this subpopulation, however, should be interpreted with caution as there were only 29 back injuries. In the metal and nonmetal sector (table 3): (1) the workers with 1 year or less experience were at a considerably (40 pct) higher-than-average risk, and (2) the workers with more than 20 years job experience had the safest record.

Job title or occupation--The indices for this category varied vastly for coal (table 2). The continuous miner and related machinery operators had an unusually high risk, suffering about four times as many back injuries and related lost workdays as their numbers in the population. Also at high risk were beltpersons; drillers and roof bolters; and shuttle car and scoop operators. In the metal and nonmetal industry (table 3), the laborers, miners, and utility persons had the highest risk, about twice the average. Also, the shop workers in this sector had about 50 pct higher than average risk for back injuries as well as related lost workdays. Although the plant operators had proportionately 16 pct more back injuries, the severity of these injuries as reflected by lost workdays was a little less than average. In both sectors, the heavy and mobile equipment operators, and managers were well below the average risk for back injury problems.

5. CONCLUSIONS

The above analysis indicates the importance of collecting and utilizing the demographics data for purposes of identifying subpopulations of the mining industry that are exhibiting higher- or lower-than average risk for back injuries. This information will be used to improve and expand mine health and safety research, and to customize training and safety programs for specifically identified demographic sectors of the mining population. It is hoped this work will encourage additional research along these lines.

REFERENCES

[1] Mansfield, Gall. Work Accidents Cost Billions in Time, Dollars. Occupational Health & Safety News Digest, V. 4, Jan.'88, pp 1-3.

[2] Cochran, W. G. Sampling Techniques. John Wiley and Son, Inc., NY, 1966.

[3] Wolter, Kirk. Introduction to Variance Estimation. Springer-Verlag, NY, 1985.

[4] Siegel, Sidney. Nonparametric Statistics For the Behavorial Sciences. McGraw Hill Book Co., Inc., NY, 1956.

[5] Rao, J. N. K., A. J. Scott. The Analysis of Categorical Data From Complex Surveys. Chi-Squared Tests for Goodness of Fit and Independence in Two-Way Tables. J. Am. Stat. Assoc., V. 76, June 1981,pp 221-230.

XI

OCCUPATIONAL BIOMECHANICS AND STRENGTH MEASUREMENT

Trends in Ergonomics/Human Factors V
F. Aghazadeh (Editor)

ERGONOMICS, HUMAN RIGHTS AND PLACEMENT TESTS FOR PHYSICALLY DEMANDING WORK

R.D.G.Webb and D.W.Tack

Humansystems Incorporated
Guelph, Ontario, Canada

ABSTRACT: This paper outlines emerging Human Rights issues in Canada related to ergonomics in general and placement tests for physically demanding work in particular.

1. INTRODUCTION

Ergonomics considers people in a systems design context. Just as electronic and mechanical components need to be carefully integrated into any occupational system, so do human beings. The purpose of this integration is to benefit both individual well being and organizational effectiveness but although these two goals are inseparable, they do not always coincide. The role of ergonomics within the systems design context is to ensure that task demands match the abilities (actual or potential) of the person performing the work. The foundation for this is a detailed analysis of existing or anticipated demands using a range of observational and simulation techniques and the application of relevant life and behavioral science data to the issues that arise.

Although the priorities may vary, three strategies can be combined to ensure the best fit between task demands and user characteristics. These are the design of equipment and workplace layout, improvement of skill and knowledge through learning (training), and placement of people in accordance with their abilities and preferences.

Although individuals will be the end users, this approach is usually based on group data. Target groups have to be identified and an acceptable range defined for relevant characteristics. Commonly, a range accomodating 90% of the target group is used, working to the 5th percentile and/or the 95th percentile depending on the interaction between the task and the user characteristics in question. If the consequences of exclusion are severe, or if there is little additional cost involved, a greater proportion of the target population will be included by using the 1st or 99th percentile. People with characteristics falling outside this range may then be identified and, depending on the costs and the consequence, either excluded or the work modified to take account of special cases.

It is important to note that while the balance between design, placement and training may vary from case to case, it is difficult to imagine a situation in which, implicitly or explicitly, some aspect of each strategy will not be present. Conventionally, ergonomics gives first priority to physical design changes but because economic and technological constraints usually limit the range of people with which any design can cope, placement and/or training will be needed. While the role of training is to bring the skills of individuals within the range of the design criteria, the role of placement is to ensure selection of the best person for the job, the best job for the person or, for unavoidable risks, to screen out people who will be vulnerable. It is also important to note that the balance can be changed according to priorities. Emerging Human Rights decisions may be changing the perception of what constitutes an appropriate balance between these strategies.

2. HUMAN RIGHTS

A major issue in Human Rights legislation is to ensure that individuals are not unfairly treated as a result of employment practices based on questionable assumptions about group characteristics. In practice, interpretation of the legislation depends on the precedents set in specific cases and the structure of the institutions set in place to pursue the legislation.

In Canada, federal and provincial jurisdictions have each have separate Human Rights codes and Human Rights Commissions with the responsibility to cover other areas as well as employment (for example education, goods, services and accommodation). Aspects of employment covered include recruitment, hiring, promotion, and dismissal as well as the terms and conditions of employment (such as hours, pay, shiftwork, and performance evaluation). Protected categories of people may also vary. For example the Ontario Human Rights Code covers race, ancestry, place of origin, colour, ethnic origin, citizenship, creed, sex, handicap, age, marital status, family status, and record of offenses.

The need to balance the concerns of employer and employee is generally recognized as is the principle of equal employment opportunity based on a merit principle. For instance, within the terms of any collective agreement, employers may:

- Define specific employment needs according to business priorities.
- Require job related qualifications and experience.
- Hire, promote and deploy the most suitable person for specific positions.
- Establish measurable standards for evaluating job performance.

On the other hand, it is reasonable that employees/applicants should:

- Be considered and compete equally for jobs for which they are qualified.
- Have required skills, experience and education clearly outlined.
- Be informed of duties and performance expectations.
- Be advised of shortcomings and permitted an opportunity to effect improvement in job performance.

Application of the Canadian Human Rights Act (1976) has been affected by the U.S. Civil Rights Act (1964) and its interpretations, in particular the landmark Griggs v Duke Power in 1971. This case established that once adverse impact on a protected category has been shown, the burden of proof shifts to the employer to demonstrate that the requirement is related to job performance. The current Canadian position may be summarized by the ruling of the Supreme Court of Canada in another case; Ontario Human Rights Commission et al. v The Borough of Etobicoke (1982) concerning the requirements for a bona fide occupational qualification.

> *...must be imposed honestly, in good faith and in the sincerely held belief that such limitation is imposed in the interests of adequate performance of the work involved with all reasonable dispatch, safety and economy, and not for ulterior or extraneous reasons aimed at objectives which could defeat the purpose of the Code. In addition it must be related in an objective sense to the performance of the employment concerned, in that it is reasonably necessary to assure the efficient and economical performance of the job without endangering the employee, his fellow employees and the general public. (C.H.R.R., D/783, 6894).*

Another important case is that of Action Travail des Femmes v. Canadian National (1984) concerning systematic discrimination against the hiring of women for railway work. This case, among other things, concluded that the use of a mechanical aptitude test validated elsewhere had to be validated again for the jobs in question and that validation should be based on a detailed job analysis. Furthermore, tests used for women should be validated for women. It also concluded that abilities that were being demanded that were much higher than those actually required and that much of the knowledge could be acquired on the job. In passing, the tribunal commented on other work sample selection tests being used (such as carrying an 83 lb piece of equipment) as being unrepresentative of the manner in which the work was actually carried out.

3. ESTABLISHING DISCRIMINATION

Different categories of discrimination relating to employment exist: adverse impact, disparate treatment, and failure to accommodate to the special needs of employees. All three categories are of potential interest to ergonomists.

Adverse impact occurs when an apparently neutral test, such as height or strength, results in the disproportionate rejection of a protected group. An example would be the tendency of strength tests to reject more women than men. It is important to note that it is not necessarily unacceptable for a test to have adverse impact, provided the employer can show that the test is a bona fide job requirement.

Disparate treatment occurs when different criteria are systematically applied to different protected groups even though the groups are actually similar in relevant characteristics and responsibilities. An important feature in many such cases has been to show intent to discriminate. Intent need not be malicious but merely based on misplaced stereotypes. An example might be the routine exclusion from driving jobs of people with a hearing impairment.

Failure to make reasonable accommodation occurs when an employer refuses to make allowances for the special needs of a member of a protected group (such as religious observance, or a physical disability). This requirement is not yet fully established in Canada. (In 1986 and 1987, the Canadian Human Rights Commission recommended that the Canadian Human Rights Act be amended to make it a discriminatory practice to refuse to make reasonable accommodation for the special needs or obligations related to a prohibited ground of discrimination.) The case of the Ontario Human Rights Commission and O'Malley v Simpsons-Sears (1985) found that intent is not an essential element of such discrimination. This case decided that where there is an adverse effect on the complainant (in this instance on religious observance), the requirement does not have to be struck down, but the effect on the complainant must be considered. Reasonable steps should be taken to accommodate the complainant without undue expense or interference with business operations.

Although there are exceptions, a general pattern is becoming established both in the U.S. (Hogan and Quigley 1986) and Canada (Fitzsimmons-LeCavalier 1984, Cronshaw 1986). First, prima facie evidence of adverse impact is required. Under the 4/5ths rule (Equal Employment Opportunities Commission 1978), a selection rate for any group which is less than 80% of the rate for the group with the highest rate of acceptance has been accepted as evidence of adverse impact or disparate treatment. More precise tests of statistical significance (at the $p<.05$ level) and, where statistical significance exists, of practical significance have been used in some cases (Fitzsimmons-LeCavalier 1984). Also the characteristics of the groups being compared are being more closely scrutinized for distorting factors such as the "chilling effect" on potential applications once a discriminatory practice is known in the community.

Once evidence of adverse impact has been established, the burden of proof shifts to the employer to show that the test is job related. This evidence should be based on a before-the-fact analysis of representative job demands with objective scores of on-the-job performance. Furthermore, the employer needs to show both a reasonable business necessity through the use of utility analyses (eg Cronshaw and Alexander 1985) and that no equivalent alternative exists that would reduce the adverse impact.

A feature of the literature is the notion of "reasonableness". Interpretations of what is reasonable vary from case to case but there is a noticeable trend towards the expectation of more detailed technical knowledge about characteristics of target populations and job demands, the utility of various options in the balance of design, placement and training, and the complexities of validating selection tests (Cronshaw 1986, Hogan and Quigley 1986). The trend is away from using assertions from expert witnesses (for instance experienced supervisors) towards a requirement for objective data.

4. ERGONOMICS AND HUMAN RIGHTS

This outline illustrates some of the conceptual and philosophical similarities between Human Rights legislation and ergonomic principles and practice. Both focus on people. Both have the problem of extrapolating between the needs of individuals and estimated characteristics of target groups. Both have the problem of balancing the needs of a system (usually as perceived by the employer/manager) and the needs of individuals on which the system depends. These similarities represent both an obligation and an opportunity. The obligation arises from a professional commitment to provide relevant expertise in cases where it is needed. The opportunity is to have a wider influence on the application of ergonomics since the impact of a precedent setting legal finding will oblige conformity across an entire community. There is little doubt that Human Rights professionals see ergonomics as having the potential to make several contributions.

One of the biggest of these contributions may be to make it quite clear that there are choices in the balance between the strategies of design, placement and training that every job represents and that judicious changes in that balance can significantly alter accessibility to jobs for protected groups or individuals with particular handicaps. More specifically, ergonomics can help with:

- o Information about the degree of accommodation required to modify working conditions for individuals (eg placement of people with particular handicaps in suitable work).
- o Data on relevant biomechanical and/or anthropometric characteristics of target populations and the overlap between them eg sex differences in strength and endurance.
- o Analysis techniques to determine sensory, cognitive and physical demands of essential duties.
- o Design modifications able to expand the potential labour pool.
- o Measurement techniques to determine levels of ability among individuals
- o Equivalence of job demands for cases related to employment equity in rates of pay.
- o Verification or otherwise of population stereotypes in instances of disparate treatment.
- o Overlaps between abilities for different groups such strength in men and women.

A quick review of tribunal and court actions in 1986 related to the federal Canadian Human Rights Commission (which covers federal government staff, Crown Corporations such as Canada Post, interprovincial transportation, broadcasting, and telephone companies and banks) illustrates the potential. Highlights for that year included claims turning on stereotypical assumptions about the physical performance abilities of asthmatics, the ability a person with hearing impairment to drive a truck, the ability of diabetics to work as train drivers, the use of mechanical aptitude tests validated on men in the selection of women, the relationship between age and suitability as a bus driver. In the U.S., there have been many cases in which height and or weight have been challenged as suitable job qualifications (Hogan and Quigley 1986) particularly for jobs as police officers and firefighters. In many cases, the requirement has been shown to be insufficiently substantiated by job demands. For example, in the U.S. case of Mieth v Dothard (1976) (described by Hogan and Quigley (1986)) which concerned selection tests for police officers, the only skill to which height was definitely related was that patrol cars could not be adequately driven by people shorter than 5ft or taller than 7 ft.

5. PLACEMENT TESTS FOR PHYSICALLY DEMANDING WORK

Placement testing for physically demanding work is only one example of the application of ergonomics in this area. Factors related to performance ability or screening for risk of injury have been widely discussed in the literature either directly (Arnold et al. 1982, Ayoub 1983, Campion 1983, Griffin and Troup 1984, Hogan and Quigley 1986, Keyserling et al. 1980, NIOSH 1981, Pytel and Kamon 1981) or indirectly (Ayoub et al 1980, Chaffin and Andersson 1984, Garg et al. 1980, Kumar 1986, McGill and Norman 1985). It is possible to identify many indicators of the ability to perform physically demanding work other than body height and weight. For example: callisthenics (Sharkey 1981), work samples (Arnold et al 1982), static strength tests (Chaffin et al 1978) and dynamic strength tests (Pytel and Kamon 1981). The content validity requirement to include both strength and endurance has also received greater recognition (eg NIOSH 1981, Campion 1983)

Because of the inevitability of adverse impact for any test designed to screen people for physically demanding work whether for performance ability or reduced risk of injury, there is every indication that placement testing for physically demanding work is likely to receive increasing scrutiny under Human Rights legislation (eg Hogan and Quigley 1986) yet most attention has come from industrial/organizational psychology (Cronshaw 1986)..

The issues which are likely to receive particular attention relate to the validity of test procedures in general (construct, content and criterion validity). More particularly, the question of sex differences will result in a call for greater concentration on validation and normative data for women and in general, the suitability of normative data bases will be questioned. Not only will evidence be required that the data base is representative of the target labour pool, but cut-off points for women identifed by extrapolation from strength or endurance data based on all-male samples are likely to be challenged. Similarly, concurrent criterion validation studies based on men (as many existing work populations will be) will be closely scrutinized. In the absence of data showing the equivalence of male and female performance, equivalence is unlikely to be assumed. Simple formula such as female strength values being 60% of that of males are unlikely to be easily accepted since wide differences in overlap have been shown depending on the part of the body being tested and the posture adopted (Pheasant 1983, Laubach 1976). The data in Arnold et al 1982 shows that, for work samples, overlaps in performance between the sexes can be as much as 85% (using a sledgehammer); even lifting 50 lb bags overlapped 41%.

If top down selection is used, then aggressive challenges may meet assumptions that higher scores relate to practically significant differences in performance ability with utility to the employer. For tests developed on similar jobs, validation for the target organization will probably be expected with differences in work practices, production pace, and even factors such as work posture to be considered since even small changes in object size with no change in object weight can make an appreciable difference in job demands (Webb and Handyside 1982).

The role of work experience and training is also likely to become an issue. It is sometimes suggested that women approach strength and endurance testing differently from men because they are less experienced in manual handling. The implication of this is that, given men and women of equivalent potential, the women will have somewhat lower test scores and thereby be denied the opportunity to acquire the necessary experience.

Finally, there are several relevant statistical issues: for example the strength of correlation coefficients and the issue of practical vs statistical significance. Generally speaking it is rare to find correlation coefficients of over 0.6 in the literature on psychological tests (Nunnally 1978) even though a correlation coefficient of 0.7 is desirable. Nevertheless, for groups, the use of tests with quite low correlations will result in an improvement in group performance thereby allowing the argument of business necessity despite inaccuracy in particular instances. However, where physical measures and work samples are involved then correlations higher than 0.7 are possible (eg Arnold et al. 1982) and may even be expected.

6. CONCLUSIONS

Ergonomics and Human Rights appear to be following parallel but, by and large, independent routes with only occasional transfers of knowledge and philosophy. Despite, or perhaps because of, the pitfalls, there may be considerable mutual benefit from an increased interaction: for ergonomics to make a technical contribution in specific cases, and for Human Rights to act as philosophical counter balance during the development and application of ergonomics in many areas.

7. REFERENCES

ACTION TRAVAIL DES FEMMES v. CANADIAN NATIONAL (TD 10/84) 1984. *Decision rendered by the Tribunal under the Canadian Human Rights Act, August 22nd, 1984.*

ARNOLD J.D., RAUSCHENBERGER J.M., SOUBEL W.G., GUION R.M. 1982. *Validation and Utility of a Strength Test for Selecting Steelworkers.* Journal of Applied Psychology, 67.5,588-604.

AYOUB, M.M, MITAL, A., ASFOUR, S.S., BETHEA, N.J., 1980. *Review, evaluation and comparison of models for predicting lifting capacity.* Human Factors, 22, 257-269.

AYOUB M.A., 1983. *Design of a pre-employment screening program.* In: "Ergonomics of Workstation Design" (Ed T.O. Kvalseth). Butterworth: London.

CAMPION M.A. 1983. *Personnel selection for physically demanding jobs: review and recommendations.* Personnel Psychology, 35, 527-550.

CANADIAN HUMAN RIGHTS COMMISSION, 1987. *Annual Report: 1986.* Ottawa, Ontario.

CRONSHAW S.F., ALEXANDER R.A., 1985. *One answer to the demand for accountability: Selection utility as an investment decision.* Organizational Behavior and Human Decision Processes, 35, 102-118.

CRONSHAW, S.F. 1986. *The status of employment testing in Canada: a review and evaluation of theory and professional practice.* Canadian Psychology, 27, 2, 183-195.

EQUAL EMPLOYMENT OPPORTUNITY COMMISSION, 1978. *Uniform Guidelines on Employee Selection Procedures.* Federal Register, 43, 38290-38315.

FITZSIMMONS-LECAVALIER P., 1984. *The use of statistics as evidence in individual and systemic human rights cases.* Research and Policy Branch, Canadian Human Rights Commission, Ottawa, Ontario.

GARG A., MITAL A., ASFOUR S.S.A. 1980. *A Comparison of Isometric Strength and Dynamic Lifting Capacity.* Ergonomics, 23, 13-27.

GRIFFIN A.B., TROUP J.D.G., 1984. *Tests of lifting and handling capacity: Their repeatability and relationship to back symptoms.* Ergonomics, 27, 3, 305-320.

GRIGGS et al. v. DUKE POWER CO. 1971. United States Court of Appeals. October Term 1970. Pages 424-436. 401 U.S. 424.

HOGAN J., QUIGLEY A.M. 1986. *Physical Standards for Employment and the Courts.* American Psychologist, 41, (11), 1193-1217.

HOGAN J.C. 1980. The State of the Art of Strength Testing. In "Women, work and health: challenges to corporate policy." (Eds: D.C. Walsh and R.H. Egdahl). Springer-Verlag: New York.

KEYERSERLING W.M., HERRIN G.D., CHAFFIN D.B., 1980. *Isometric strength testing as a means of controlling medical incidents in strenuous jobs.* Journal of Occupational Medicine, 22, 332-336.

KUMAR, S. 1986. *Isometric, isokinetic strength and maximum acceptable lift.* Human Factors Assoc. of Canada, 19th Annual Meeting, 115-118.

LAUBACH L.L. 1976. *Comparative muscle strength of men and women: a review of the literature.* Aviation, Space and Environmental Medicine, 47, 534-542.

MCGILL, S.M., NORMAN, R.W. 1985. *Dynamically and statically determined low back moments during lifting*. Journal of Biomechanics, 18, 12, 877-885.

MIETH V DOTHARD 1976. *418 F.Supp. 1169 (D.Ala. 1976)*

NATIONAL INSTITUTE FOR OCCUPATIONAL HEALTH AND SAFETY (NIOSH) 1981. *Work practices guide for manual lifting.* Technical Report 81-122. U.S. Department of Health and Human Services: Cincinnati, Ohio.

NUNNALY J.C. 1978. *Psychometric Theory.* McGraw-Hill, New York.

ONTARIO HUMAN RIGHTS COMMISSION et al. v. SIMPSONS-SEARS LIMITED 1985. File No.:17328. (Supreme Court of Canada, December 1985).

ONTARIO HUMAN RIGHTS COMMISSION et al. v. THE BOROUGH OF ETOBICOKE, 1982. Volume 3, Decision 164, C.H.R.R. D/781. (Supreme Court of Canada, May June 1982.)

PHEASANT, B.T. 1983. *Sex differences in strength - some observations on their variability.* Applied Ergonomics, 14.3, 205-211.

PYTEL, J.L., KAMON, E. 1981. *Dynamic test as a predictor for maximal and acceptable lifting.* Ergonomics, 24, 663-672

SHARKEY, B.J. 1981. *Fitness for Wildland fire fighters.* The Physician and Sports Medicine, 9, 4, 93-102.

WEBB R.D.G., HANDYSIDE J. 1982. *Brickhandling: A case study.* Applied Ergonomics, 13, 191-194.

Trends in Ergonomics/Human Factors V
F. Aghazadeh (Editor)

A MULTIVARIATE ANALYSIS OF DIRECTIONAL MOVEMENT TIME

Don E. MALZAHN, Jeffrey E. FERNANDEZ, Robert J. MARLEY, and Jalaluddin DAHALAN

Department of Industrial Engineering,
The Wichita State University
Wichita, KS 67208-1595

An experiment using 100 able-bodied adults was performed to determine the influence of isometric muscular forces upon hand-arm movement times. Canonical analysis showed that muscular force had a negligible effect upon movement time within a typical workspace. Other variables such gender and reaction time were shown to influence movement time.

I. INTRODUCTION

Reaction and movement times are variables that play an important part in describing motor performance. Many factors have been shown to be associated with hand-arm movement time. For example, studies have documented the effect of age upon movement time [1,2]. Gender has also been shown to influence movement time [3,4,5]. Hand preference can influence movement time for complex tasks [6,7].

The question of movement direction was addressed by Kattan and Nadler [8]. They found significant within-subject differences in movement time with regards to motion direction. The direction of differences was the same for both superior and inferior hands [4]. Isometric muscular force has been proposed as a key factor in movement time differences among individuals [1,9].

The primary objective of this study was to determine whether muscular force, or strength, accounted for within-subject differences in movement time with greater strength associated with shorter movement times. As part of an ongoing experiment on functional human performance, the present study assessed four measures of isometric muscular force that corresponded to four different directions of hand-arm movements. These measures were analyzed in terms of their effect upon the movement times in each direction.

2. METHODS

2.1. Subjects

Subjects for the study consisted of 100 able-bodied adults.

This group was evenly divided between males and females (50 each). For both males and females, there were 26 individuals in the 20 to 29 year old group, 14 in the 30 to 39 year old group, and 10 in the 40 to 49 year old group.

2.2. Apparatus

The apparatus used in this study was the Available Motions Inventory (A.M.I.) test package [10]. The A.M.I. has been shown to be an effective and reliable instrument for identifying the functional abilities of the upper extremities and thereby useful in individual specific task design and modification [11,12].

2.2.1. Devices

The devices used in this study were the Reaction-Reach subtest of the A.M.I. (see Figure 1) and the Applied Force subtest (see Figure 2). The Reaction-Reach subtest was fixed to a horizontal panel positioned directly in front of the seated subject at one inch above the seated elbow height. The panel is symmetrical with regards to testing either the left or right hand. Four different measures of movement time for both hands were measured: LR--Lateral Reach (center-front to side); LM--Lateral Move (side to center-front); TR--Transverse Reach (center-front to top); and TM--Transverse Move (top to center-front). All movements were 12 inches in distance and all switches were 3.5 x 3.5 inches in size.

The Applied Force subtest was also fixed to a horizontal panel positioned directly in front of the subject. Force was measured by a lever arm connected to a strain gauge. Four measures of directional force are measured in this manner: AFL--Applied Force Lateral (center to side pull); AFM--Applied Force Medial (side to center pull); AFPull--Applied Force Pull (straight pull towards body); AFPush--Applied Force Push (straight push away from body). Each of these measures was repeated for the both the superior and inferior hand.

2.3. Procedures

Hand superiority was empirically derived by normalizing the performance scores for both hands on all A.M.I. tasks. The hand with the highest overall mean z-score was determined to superior and, likewise, the hand with the lower mean z-score was classified as inferior [7].

2.3.1. Directional Movement Time

This procedure called for the subject to initially depress a start switch corresponding to the particular movement direction. Following a verbal "ready" cue, the subject reacted to an auditory stimulus (80 db, 4K Hz tone) and moved as quickly as possible to depress the appropriate target switch. For each direction, two trials were performed and the timings for each trial were averaged. Two times were observed: 1) reaction time to the stimulus and 2) movement time to the target. This

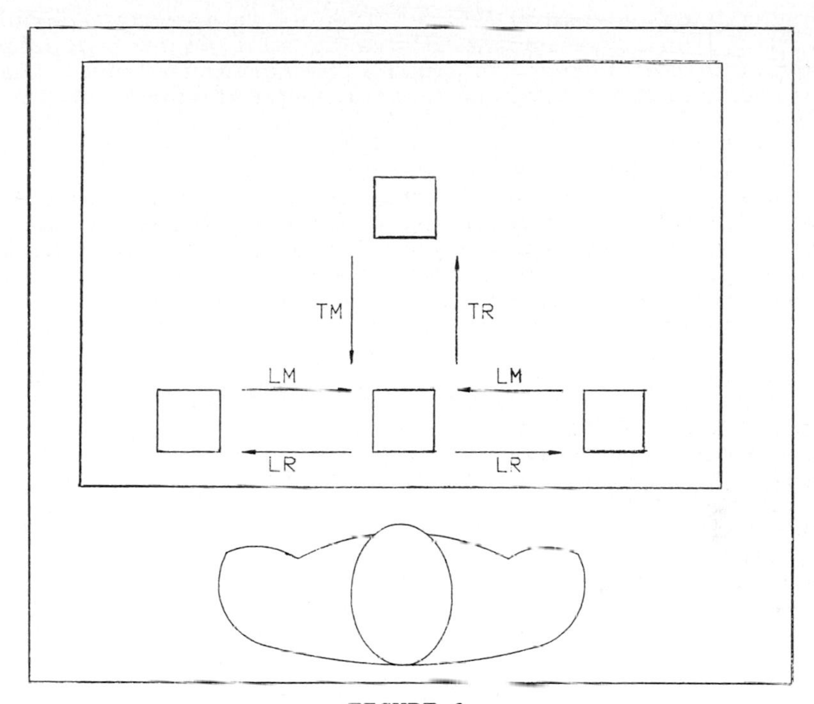

FIGURE 1
The A.M.I. Reaction-Reach subtest module.

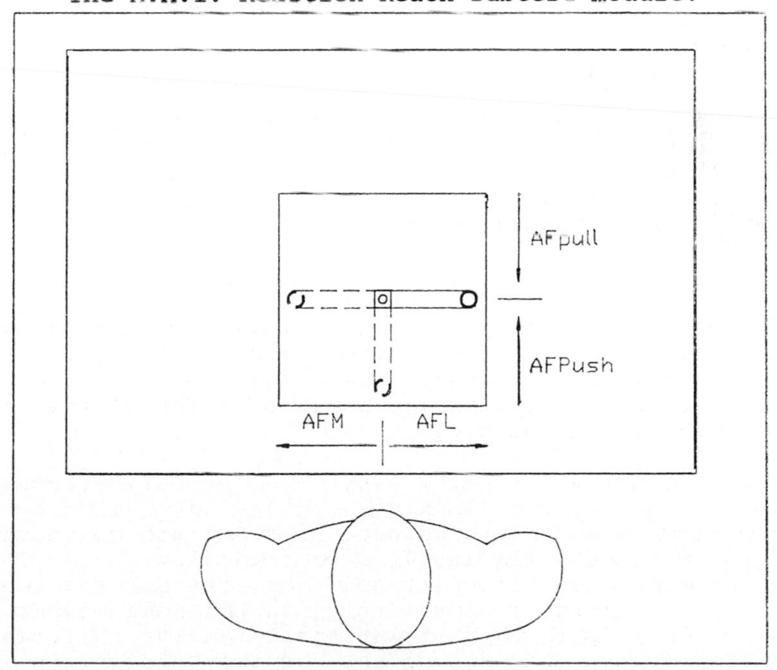

FIGURE 2
The A.M.I. Applied Force subtest module.

procedure was repeated for both hands in the four directions. It should be noted here that due to experimental constraints, reaction times were not paired with movement times. Rather, the four reaction times were combined for a mean reaction time (MRT).

2.3.2. Directional Applied Force

Each subject was asked to apply force to the lever arm in the direction which corresponds to a reaction/movement direction stated above. Two trials were performed in each of these directions and the mean score was recorded. This procedure was repeated for both the superior and inferior hands.

3. RESULTS AND DISCUSSION

The data were analyzed using the SAS statistical package (SAS Version 5.16, 1986). Figure 3 shows the mean movement times paired with the corresponding strength measures. The primary data analysis was done using the Multivariate Tests option of PROC REG. This procedure performed canonical analysis upon the four strength measures and the four movement times. Previous research had shown that movement time performance profiles reveal no interaction with hand preference [4], therefore only the results of the superior hand were used in this analysis.

Results from this procedure indicate that directional, isometric muscular force is not related to hand-arm movement time. The maximum canonical correlation coefficient extracted was $R^2 = 0.1076$. This coefficient was considered not meaningful. Thus interpretation of the canonical weights and structure coefficients were not attempted. Significance of these canonical variates was given in Wilk's Lambda = 0.8256 with $F(16,272) = 1.102$, $p = 0.3524$.

A detailed examination of the regression parameters, in Table 1, further shows that none of the parameters for the strength variables show a significant relationship with the movement times. Only gender of the subject and mean reaction time are significantly related to movement time which agrees with previous studies [3,4,5]. Also from Table 1, it is clear that age group had little effect. Note that this sample attempted to approximate the industrial population and no subject over 60 years of age was included.

Precise explanation for the within-subject differences in movement time is still lacking. This study failed to find a significant relationship between strength and movement time unlike Era, et al. [1] and Viitasalo and Komi [9]. The studies just mentioned, however, examined strength and movement times at the neuromuscular level. On the other hand, the current study, like that of Kattan and Nadler [8], concentrated on functional measures of the body members moving in space. At this level, the influence of strength seems to have given way to other variables.

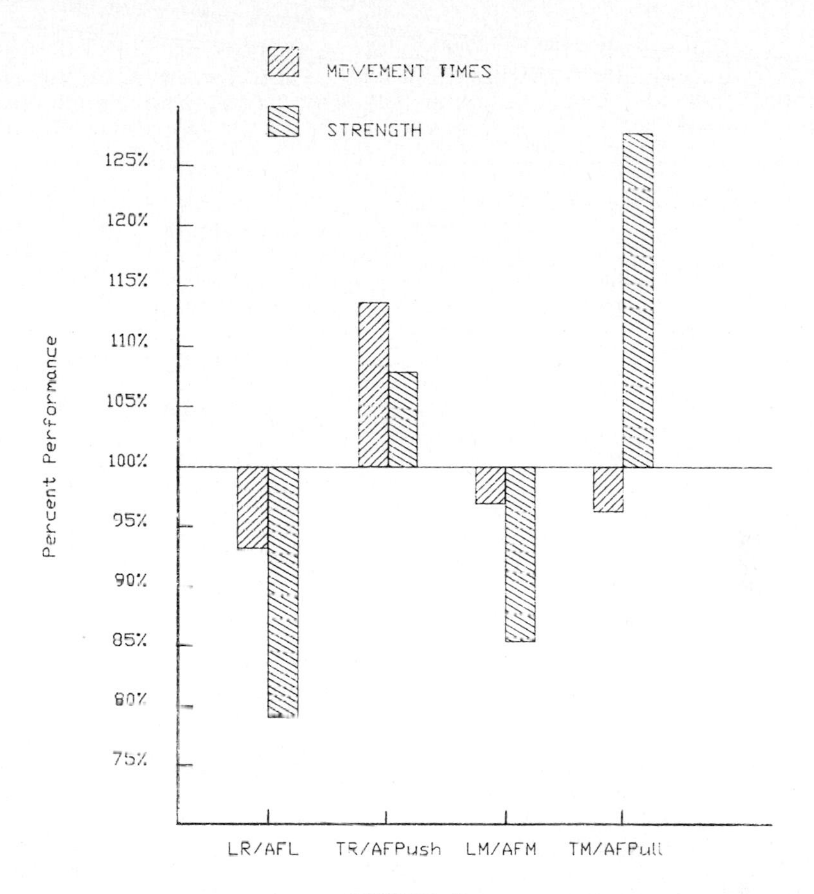

FIGURE 3
Mean movement times and associated strength measures expressed as percentage of mean performance.

4. CONCLUSION

Based upon the results of this experiment, it appears that isometric muscular strength is unrelated to movement time within the constraints of a typical workstation. It is clear from reviewing previous research that there are within-subject differences in movement time with regards to motion direction. However, this study has shown that isometric strength in a particular movement direction does not contribute to these differences. This is despite the between-subject relationship that seems to be evident in the means reported in Figure 3.

TABLE 1
Significance of Non-Intercept Regression Parameters Between Movement Time and the Strength, Gender, Age, and Reaction Time Variables.

Dependent Movement Times	Independent Variables	T for H_0 Parameter = 0	Prob.
Lateral Reach	AFL	1.181	0.241
	AFM	0.179	0.858
	AFPush	1.880	0.063
	AFPull	-1.548	0.125
	Gender	-2.828	0.005 *
	Age	-0.420	0.676
	Mean Reaction	5.043	<0.001 *
Transverse Reach	AFL	1.178	0.242
	AFM	0.295	0.769
	AFPush	1.864	0.066
	AFPull	-1.605	0.112
	Gender	-2.540	0.013 *
	Age	-0.434	0.665
	Mean Reaction	5.180	<0.001 *
Lateral Move	AFL	0.460	0.647
	AFM	0.531	0.597
	AFPush	1.367	0.175
	AFPull	-0.740	0.462
	Gender	-2.154	0.034 *
	Age	-0.148	0.882
	Mean Reaction	5.781	<0.001 *
Transverse Move	AFL	0.475	0.636
	AFM	-0.162	0.871
	AFPush	1.243	0.217
	AFPull	-0.160	0.873
	Gender	-2.362	0.020 *
	Age	-0.157	0.876
	Mean Reaction	4.223	<0.001 *

* indicates significance at 0.05 level

Future research in this area may concentrate on identifying the physiological, psychological or biomechanical foundations of these differences. In terms of industrial task design where movement time may play a critical role, isometric muscular force is a variable that need not necessarily be considered.

REFERENCES

[1] Era, P., Jokela, J., & Heikkinen, E. (1986). Reaction and movement times in men of different ages: A

population study. Perceptual and Motor Skills, 63, 111-130.

[2] Welford, A.T. (1980). Motor Skill and Aging. In, Nadeau, C.H., Halliwell, W.R., Newell, K.M., & Roberts, G.C., (Eds.), Psychology of Motor Behavior and Sport: 1979. Champaign, IL: Human Kinetics.

[3] Hodgkins, J. (1963). Reaction time and speed of movement in males and females of various ages. Research Quarterly, 34(3), 335-343.

[4] Marley, R.J., Malzahn, D.E., & Fernandez, J.E. (1987). Potential factors in movement time: Implications for functional evaluation of individuals with disabilities. In, Asfour, S.S. (Ed.), Trends in Ergonomics/Human Factors IV. North-Holland: Elsevier.

[5] Wrisberg, C.A., Hardy, C.J., & Beitel, P.A. (1982). Stimulus velocity and movement distance as determiners of movement velocity coincident timing accuracy. Human Factors, 24(5), 599-608.

[6] Annett, J., Annett, M., Hudson, P.T.W., & Turner, A. (1979). The control of movements in the preferred and non-preferred hands. Quarterly Journal of Experimental Psychology, 31, 641-652.

[7] Rahimi, M., Malzahn, D.E., & Musa, H. (1985). Hand preference and difference patterns using the Available Motions Inventory (AMI), Human Factors Society 29th Annual Meeting, Baltimore, pp. 167-171.

[8] Kattan, A., & Nadler, G. (1969). Equations of hand motion path for work space design. Human Factors, 11(2), 123-130.

[9] Viitasalo, J.T., & Komi, P. (1981). Interrelationships between electromyographic, mechanical, muscle structure and reflex time measurements in man. Acta Physiologica Scandinavica, 111, 97-103.

[10] Malzahn, D., and Kapur, R. (1980). An ability evaluation system for persons with physical disabilities. Santa Monica, CA: Proceedings of the Human Factors Society 24th Annual Meeting. 114-118.

[11] Malzahn, D. (1984). Functional evaluation for task modification using the Available Motions Inventory. In Halpern, A.S., and Fuhrer, M.J. (Eds), Functional Assessment in Rehabilitation. Baltimore: Brookes.

[12] Rahimi, M. and Malzahn, D. (1984). Task design and modification based upon physical ability measurement. Human Factors, 26(6), 715-726.

Trends in Ergonomics/Human Factors V
F. Aghazadeh (Editor)

PREDICTION OF MAXIMUM DYNAMIC STRENGTH FROM MULTIPLE REPETITIONS WITH A SUBMAXIMAL LOAD

T. L. DOOLITTLE
Department of Environmental Health SC-34
University of Washington
Seattle, Washington 98195

Karl KAIYALA
Training Division
Seattle Fire Department
Seattle, Washington 98104

The concept that muscular strength (a single maximum effort, *aka* 1-RM) and muscular endurance (maximum repetitions with a submaximal load) are related on a continuum, has been discussed in the literature for many years. Until recently, however, there has been no mathematical equation to express this relationship. Building upon the work of Berger, the authors derived a theoretical equation that relates the number of repetitions possible to the fraction of 1-RM being lifted. The purpose of this investigation was to test a variant of that equation for its efficacy in predicting 1-RM from the number of repetitions performed with a submaximal load. Subjects performed as many repetitions possible with loads representing 40 to 90 per cent of their 1-RMs. Regression analyses comparing actual and estimated 1-RM values resulted in R-squares ranging from 0.76 to 0.96.

1. INTRODUCTION

Maximum dynamic strength has been defined operationally as the maximum weight that an individual can lift through a full range of motion once, with the demand being sufficient that the lift cannot be immediately repeated. This phenomenon is frequently termed "one repetition maximum" or "1-RM". The determination of 1-RM is routinely done in strength training and athletic situations, but not without concerns that minimize its effectiveness in most industrial or clinical settings. Determining 1-RM requires a trial and error approach which can be time consuming as well as fatiguing. The potential contaminant of fatigue dictates that the measure be confirmed after the individual has had time to fully recover (usually on a subsequent occasion) further exacerbating the time requirement. Risk to injury is present during the determination of 1-RM. It is relatively low, however, if the individuals being tested and those doing the testing are experienced. Since this is the case in strength training and athletic situations, the risk of injury is tolerable. Such is not the case in most industrial testing situations, and the risk of injury it too high to be acceptable. This has led to the use of isometric or isokinetic tests from which inferences are made regarding the individual's dynamic strength capacity.

The concept that muscular strength (a single maximum effort) and muscular endurance (maximum repetitions with a submaximal load) are related has been implied by the literature since before the turn of the century. Logan and Foreman [6] formalized the concept in 1961 by describing the phenomena as the "Strength-Endurance Continuum". The mathematical relationship, however, remained unknown. While Logan and Foreman described the continuum for dynamic strength, others [4,8] attempted to quantify the relationships with isometric measures. In part no doubt, because of the greater ease of measurement in that mode.

Rohmert [8] observed that maximum isometric force could be maintained for only a few seconds while 75% of maximum could be held for about 21-sec, 50% for 60-sec, and 15% for extended periods of time. Caldwell [4] in essence replicated these findings, and in addition found that the relationships were applicable to females as well as males.

Berger [2], while not acknowledging it as a "strength-endurance continuum", derived an equation for predicting 1-RM values for chin and dip exercises from the number of repetitions performed. The equation was of the general form:

$$\text{1-RM} = \beta \times \text{no. of chins} + \text{body weight} \qquad (1)$$

with four different values for beta, adjusted for body weight. The beta values were 4.66, 4.85, 5.88, and 6.82 for body weights of 130, 150, 170, and 190 pounds respectively. Berger indicated that for other body weights a beta coefficient should be interpolated. Berger and Medlin [3] demonstrated the effectiveness of the equation by extrapolating the beta coefficients for individuals weighing from 80 to 216 pounds and employing it with junior high school boys. For both the college age men and the younger cohort the predictive accuracy when compared with actual measures of 1-RM for both chins and dips was impressive. Correlation coefficients ranged from 0.86 to 0.95.

2. METHODS

Since Berger [2] found a single equation to be effective with two dissimilar exercises, it appeared reasonable that the quantitative relationship should apply to the fractional utilization of 1-RM and the repetitions possible for other lifting movements.

2.1. Equation Derivation

A theoretical equation that relates the number of repetitions possible in a lifting regimen to the fraction of 1-RM being lifted was derived. This was accomplished by acknowledging that in both the chin and dip exercise the mass being lifted is the body weight. Dividing both sides of the equation by the mass yielded equation (2).

$$1\text{-RM} * M^{-1} = \beta * M^{-1} * R + 1 \qquad (2)$$

Utilizing Berger's beta and mass values a mean coefficient (0.034) was determined. The inverse of 1-RM divided by mass is, obviously, the fraction of the individual's maximum that is being lifted. Thus, equation (3), as previously reported [5], expresses the percentage of maximum (that any given mass will be) as a function of the number of repetitions.

$$\text{Percent of Max} = (0.034*R + 1)^{-1} * 100 \qquad (3)$$

The purpose of this investigation was to test a variant of equation (3) for its efficacy in predicting 1-RM from the number of repetitions performed with a submaximal load. The equation was:

$$1\text{-RM}_{est} = L*(0.034*R + 0.966) \qquad (4)$$

where: $1\ RM_{est}$ = estimated one maximum repetition
L = load lifted
R = number of repetitions

2.2. Validation

Forty-two adult male volunteers, ranging in age from 23 to 50 (mean = 34.5), served as subjects in the first cohort. They were volunteers from 90 firefighters who had been selected at random for a study pertaining to entry level firefighter strength and performance standards. All were healthy and considered fit for the arduous job of fire fighting. Inspection of their average strength scores will verify that they possessed average or above levels of strength for men of their age.

The inclined press exercise was performed. This is essentially a shoulder to reach (S-R) lift in ergonomics terminology. To standardize the lifting action it was performed while seated with the back bearing on a padded surface inclined at 70 degrees. A barbell was pressed from the level of the sternal notch to full arms reach overhead in a vertical plane passing through the shoulder joints. A pronated grip was employed with the hands approximately shoulder width apart.

One repetition maximums (1-RM) were carefully determined for each subject. The subjects then were randomly assigned to perform repetitions at 90, 80, 70 or 60 per cent of their 1-RM. They were instructed to perform as many repetitions as possible without pauses between executions. Each subject was actively coached and encouraged until momentary muscular failure (inability to complete a full repetition correctly) occurred.

Utilizing equation (4), estimated 1-RM values were calculated from the number repetitions performed and the load lifted. A comparison of mean values are contained in Table 1. The R-squared from regression analysis between actual and estimated 1-RM values was 0.85, with a standard error of prediction of ±3.11 kg.

TABLE 1

INCLINED PRESS STRENGTH SCORES
Mean (kg) ± Std. Dev.

COHORT	N	ACTUAL	ESTIMATED	r
MALE	42	56.6 ±7.6	56.4 ±8.0	.92
FEMALE	56	33.5 ±6.4	31.5 ±6.3	.92
COMBINED	98	43.4 ±13.4	42.1 ±14.2	.98

A second cohort of 56 female volunteers repeated the procedures except that they performed their maximum repetitions with loads approximating 85, 70, 55 or 40 per cent of their 1-RM values. The women were comprised of a group of prospective firefighters, each of which was seeking an evaluation as to her potential for becoming able to meet the entry level strength requirements of the fire department. They ranged in age from 18 to 36 (mean = 25.1). All were deemed to be healthy, and their scores indicated they were above average in upper body strength for females of their age. Correlations between actual and estimated 1-RMs resulted in an identical R-squared of 0.85, and standard error of prediction of ±2.42 kg.

Fifty-five of the females in the second cohort also were tested for 1-RM and maximum repetitions with the biceps curl exercise. The curl represents a knuckle to shoulder (K-S) lift, in ergonomics terminology, that is accomplished solely with elbow flexion. In order to insure that the agonist muscle group was isolated, the exercise was performed with buttock and shoulder blades in contact with a vertical surface. Utilizing the same equation, estimated 1-RM values were computed and compared with the actual 1-RMs (Table 2). The resulting R-squared was 0.76, with a standard error of prediction of ±2.62 kg.

TABLE 2

BICEPS CURL STRENGTH SCORES
Mean (kg) ± Std. Dev.

COHORT	N	ACTUAL	ESTIMATED	r
FEMALE	55	28.4 ±4.6	27.5 ±5.2	.87

3. DISCUSSION

These findings agree very favorably with those reported for isometric contraction. As noted above, Rohmert [8] and Caldwell [4] found that maximum isometric force could be maintained for about 2-seconds. A duration that closely approximates that required to execute one repetition of the lifting movements utilized in this investigation. They further reported that 50% of maximum could be held for about 1-minute. At 2-seconds per repetition for dynamic lifting, one would complete 30 repetitions in one minute. This rate is very close to that employed in continuous dynamic lifting as employed in this study. If 30 is used for R in equation (4) the estimated 1-RM will be 1.986 times the load. Likewise, 11 repetitions will demonstrate that the load is 75% of 1-RM, and require approximately 22-seconds which is virtually identical to the 21-seconds found by Rohmert.

A plausible physiological rationale for the similarity between the isometric and dynamic strength-endurance relationships, is the relative tension in the muscle and associated reduction in blood perfusion. The notion that there exists a relaxation phase during continuous rhythmic dynamic lifting, which allows blood perfusion, during lowering is probably unfounded. The eccentric tension required, in lowering a mass under control, may well be sufficient to preclude blood perfusion, and yield the similarity in results. Lending credibility to this rationale is the preliminary experience of the authors that indicates that while only 10 or 11 repetitions can be executed in succession without any pause at 75% of 1-RM, at a rate of 6 repetitions per minute 30 repetitions can be performed with the same mass. The approximately 8-seconds of rest between executions appears to allow sufficient blood perfusion to delay the onset of fatigue.

Astrand and Rodahl [1] in the latest edition (which was published while this work was in progress) of their classic text, assert that a load corresponding to 90-95 per cent of 1-RM should be able to be lifted 3 times, while those of 85, 75, and 65 per cent should be able to be lifted 6, 10 and 15 times, respectively. While they do not cite experimental evidence to support their assertion, all of the relationships are in close agreement with those expressed by equation (4). Pollock et al []indicate that the average individual should be able to complete 12 to 15 repetitions at 70% of 1-RM, and that the competitive athlete ought to be capable of 20 to 25 repetitions. Again they cite no experimental evidence. This study is in complete agreement with their assertion for the average individual, however, it does not support the notion that the trained person has the magnitude of advantage that they suggest. The endurance decrement, when the load is expressed as a percentage of maximum, appears to be independent of absolute strength. This is supported by the findings of this study, as well as those of Caldwell [4] in his investigation with isometric efforts of males and females.

3.1. Limitations

Berger's [2] original equation was based on a maximum of 26 repetitions, which probably was a function of the minimal percentage of 1-RM that could be tested. The minimum percentage attempted in this study was 40% of 1-RM. The maximum repetitions performed by any individual was 60, with 8 combinations yielding 26 or more. It would appear safe to state that equation (4) will yield valid estimates for loads with up to 40 to 50 repetitions, provided the individual is adequately motivated. Empirical evidence gained by utilizing this approach with industrial workers (in clinical settings) indicates that some individuals will quit after 25 to 30 repetitions even when it appears that they could have continued. In these instances the workers jobs were not on the line so the motivation was minimal. Thus, if a load is too light rendering to many repetitions possible motivation may be a significant factor in obtaining valid results.

Two other considerations also limit the extent to which the equation may apply. The first is the relative mass of the body segment being lifted to the total mass (i.e. body segment plus load). Since it has been stated by numerous authors that one can work a full shift exerting 15 to 22 percent of one's max, theoretical calculations were made estimating repetitions down to 10% of 1-RM. For a one arm biceps curl with an assumed 1-RM of 22.7 kg plus a forearm mass of 2 kg, and taking lever length into consideration, the number of repetitions possible with a 2.3 kg load is reduced from 266 to 185. The factor that potentially would cause this is that the 2.3 kg load which appears to be 10% of 1-RM is nearer 14% of 1-RM when limb torque is figured in the computation. In contrast, a load of 20 kg (apparent 89%) is reduced from 5 repetitions to 4.78 which is trivial. With a 11.35 kg load (apparent 50%) the repetitions are reduced from 30 to 27.7, which is still within the error of estimate. A full investigation of the effect of segment mass awaits further research, at present it is the author's contention that for low loads it may have a significant impact. In the same manner, the present equation should not be used with squat lifting. The similarity of results with the biceps curl exercise, and the previous findings of Berger [2], however, do indicate that the equation may be extended to other forms of two arm lifting.

The second consideration that limits the use of the equation with low percentages of 1-RM and high numbers of repetitions, relates to the metabolism supporting the activity and the muscle fiber types being recruited. While a full discussion of this concept is beyond the scope of this paper, it is sufficient to note that short burst strength activities primarily are accomplished with slow twitch oxidative (so called SO or Red) fibers. In contrast, long lasting aerobic activities are accomplished primarily with fast twitch glycolytic (so called FG or White) fibers. While knowledge about how and when the transition occurs and the degree of overlap in fiber function in the human is rudimentary, it appears the shift in emphases takes place at between one and four minutes. That is activities of less than one minute of all out effort are thought to be supported primarily by SO fibers, while those all out efforts that can be sustained for greater than four minutes are thought to be accomplished primarily with FG fibers. Therefore, the further one moves out on the strength-endurance continuum beyond 30 repetitions (approximately one minute of all out effort) the less accuracy that can be expected from the equation. As noted above, the results appear to be within tolerable limits to about 40 to 50 repetitions.

3.2. Gender

The results with the females replicated almost exactly those of the males, as was the case with Caldwell's [4] subjects. Granted the absolute strength scores were higher for the males, the endurance scores at the same relative percentages could be explained by the same equation. Unlike Caldwell's subjects where "the weakest male was 22 lb stronger than the strongest female", there was overlap between the cohorts in this investigation. Since the statistics where virtually identical for the two cohorts on the inclined press, they were combined into a single analysis. As may be noted in

Table 1, the standard deviation was higher for the combined group reflecting the greater range. The R-squared was 0.96, and the standard error of prediction was ±2.82 kg. These findings indicate that equation (4) is equally applicable to both genders and may be employed with mixed groups.

4. CONCLUSION

Predicated upon the findings of this study, the following conclusion appear warranted: Equation (4) adequately describes the mathematical relationship of the "Strength-Endurance Continuum" espoused by Logan and Foreman [6] and may be employed for estimating 1-RM for dynamic lifting with submaximal loads using both arms. Further, it may be employed effectively with either or both sexes.

REFERENCES

[1] Astrand, P. & Rodahl, K. Textbook of Work Physiology (3rd Ed). New York: McGraw-Hill, 1986.
[2] Berger, R., Res. Quart., (1967) 330.
[3] Berger, R. & Medlin, R., Res. Quart., (1969) 460.
[4] Caldwell, L., J. Engn. Psych, (1963) 155.
[5] Doolittle, T. & Kaiyala, K. A Generic Performance Test for Screening Firefighters, in: Asfour, S., (ed), Trends in Ergonomics/Human Factors IV (North-Holland, Amsterdam, 1987) pp. 603-610.
[6] Logan, G. & Foreman, K., The Physical Educator, (1961) 103.
[7] Pollock, M., Wilmore, J. & Fox, S. Health and Fitness Through Physical Activity. New York: John Wiley & Sons, 1978.
[8] Rohmert, W., Intern. Z. Agnew. Physiol., (1960) 123.

Trends in Ergonomics/Human Factors V
F. Aghazadeh (Editor)

DISTORTION OF TRUE BODY WEIGHT AND STATURE FROM SELF ESTIMATION

Sheik N. Imrhan

Department of Industrial Engineering
The University of Texas at Arlington
Arlington, Texas 76019

Victorine Imrhan

Hood General Hospital
Grandbury, Texas

Rose Pin

Department of Industrial Engineering
The University of Texas at Arlington
Arlington, Texas 76019

A comparison was made between measured and self-estimated values of body weight and stature in a sample of 261 college students. The results indicated that they can estimate more accurately than the general population (from previous studies). The estimates were not dependent on gender nor recency of previous measurement, but weight was slightly dependent on frequency of previous measurements.

INTRODUCTION

To be able to predict a person's body weight and stature by simply asking him/her to state (estimate) them would be of immense value to an investigator. The cost in time and effort would become redundant especially in cases where these are the only two measurements of interest. A few studies have been performed to estimate the accuracy of self reported body weight and stature. They dealt mostly with the assessment of nutritional and health status. Overall the results seem to indicate that individuals, in general, can estimate their own body weight and stature accurately and that errors of estimation often show significant correlations with identifiable person characteristics, such as gender, obesity, etc. Errors have been analyzed for related factors in an attempt to find predictable patterns which can aid in improving the accuracy of predicting true body weight and stature from self-estimations. Stewart [1] has shown that subjects seem to have digit preference, thus introducing errors of estimation. Some observers have found that, in the general population, heavier persons tend to underestimate their body weight, with lighter men tending to overestimate their weight (Stewart, [1]; Pirie, Jacobs, Jeffery and Hannan [2] Schlichting, Hoilund-Carlsen and Quaade, [3]; and Wing, Epstein, Ossip and LaPorte, [4]). Some of the results of these studies seem to be sample dependent. For example, whereas Pirie et al [2] found that heavier women (over 170 lb) in a large sample (n = 3407) tended to underestimate their weight more than men (over 190 lb). Tell et al [5] found no significant difference in errors between heavier men and women Dwyer et al [6]. The latter's sample was 146 adults in a weight-loss program. Dwyer have found that girls, especially heavier ones, tend to underestimate their body weight more than boys. However, Huenemann et al [7] found that 16 year old boys estimate more accurately than 16 year girls. Another factor of concern is the type of scale people are accustomed to using, but Pirie et al [2] ascribes errors due to differences in scales as random instead of systematic. Whereas, amount of clothes people usually wear when weighing themselves may be a possible source of systematic error. One study, Tell et al [5], failed to confirm this. They found little difference between those who usually weigh in street clothes and shoes (average error = 6.1 lb or 2.9%) and those who usually weigh without clothes (average error = 6.4 lb or 3.2%). Recency of weighing was, however, found to be influential. Tell et al [5] found that those who had a lapse of less than 15 days before they last weighed were more accurate in their self estimation than those who had a longer lapse (average error of 4.2 lb versus 7.4 lb. Stature is usually overestimated by men, especially shorter men, while women tend to underestimate it slightly (Pirie et al, [2]).

The purpose of this study was to determine the accuracy of self-estimation of body weight and stature in a specific population which has heretofore not been investigated - college students. It is postulated here that this population, by virtue of its level of education and motivation to succeed in life, is probably more aware of its stature and weight and should be able to estimate them more accurately than the general population. The study also attempted to determine factors associated with errors in estimation and hence derive regression equations which can predict true stature and weight from their estimated values accurately.

METHODS

Two hundred and sixty-one students between the ages of 18 and 45 years (mean = 23.0 yr) participated in the study voluntarily. They were told that the purpose of the study was to determine the relationship between estimations of their own stature and weight and measured values. They completed a questionnaire requesting their age, sex, self-estimate of stature and body weight, recency of self-measurement of stature and body weight, and frequency of these measurements, over the previous 12 months. They were told to use any measurement unit they preferred. Immediately afterwards their stature and body weight were measured accurately. Stature was measured, without any shoes with a Pfister anthropometer (to the nearest mm) according to the method in NASA reference publication: 1024. Body weight was measured without shoes but with light clothing; jackets and other heavy clothing were removed. A portable LaFayette weighing scale (model 01150) was used; it measured to the nearest 0.5 lb.

RESULTS AND DISCUSSION

Two main statistics were used to measure and assess the accuracy of self-estimation of stature and body weight:

1. the mean percentage error (MPE) - the average error over all subjects, where a subject's error is computed from

$$\frac{\text{measured value - observed value}}{\text{measured value}}$$

2. the mean absolute percentage error (MAPE) - the mean absolute error.
3. the mean absolute error (ME).

Whereas MPE indicates overall direction (algebraic sign) of the errors of estimation, MAPE and AE indicate overall magnitude, ignoring direction.

Overall, students estimated both body weight and stature very accurately (MPE for weight = 0.35% (0.6 lb), and for stature - 0.63%(-1.1 cm lb); MAPE = 1.95%(2.92 lb) and 1.11%(1.92 lb), respectively. These figures indicate a slight underestimation for weight and a slight over estimation for stature. These figures are more accurate than those in the previously mentioned studies - MPE of 5%, 1.6 and 3.1%, and 2.9%, in the studies of Charney et al [8], Palta et al [9] and Tell et al [5], respectively. Males (n = 187) and females (n = 74) underestimated their weight almost equally (males = 0.35% and females = 0.34%), whereas females overestimated their stature slightly more (-0.7%) then males (-0.6%), though the difference was not statistically significant. These observations differ from those of previous studies which showed differences in estimation between males and females. This is probably due to the fact that the sample of this study contained few persons with extremes of weight and stature and, hence, few who would estimate under the pressure to confirm to the investigators' expectations (Tell et al [5]). Moreover college students probably know their weight and stature more accurately than the general population.

In the sample of 261 students, 111 estimated their weight within 2.0 lb, 205 within 5.0 lb, and 56 above 5.0 lb. For stature, 85 students estimated within 20 cm, 213 within 4 cm, 253 within 5.0 and 8 above 5.0 cm.

Analysis of the data according to frequency of measurements within the previous year, indicated that those who weighed at least 2-3 times per month (n = 85) were, overall, more accurate (MAPE = 1.6%) in their body weight estimations than those who didn't weigh within the last 12 months (n = 13; MAPE = 3.2%) and those who weighed between 1-6 times within the last 12 months (n = 153; MAPE = 2.1%). For stature their was no relationship, a plausible observation in light of the fact that stature changes very little in adulthood. Analysis according to recency of measurements indicated that those who weighed within the last 6 months were more accurate than those who didn't weigh within the last 12 months. There were no significant relationships between accuracy of estimation and recency in measuring height.

The data in this study also failed to indicate sharp differences in the accuracy of estimation among students of different weight and stature categories.

For predicting true weight (TW) or stature (TS) from self estimated weight (EW) or Stature (ES), respectively, only 2 equations seem to be necessary.

1. $TW = 1.25 + 0.995\ EW \quad (R^2 = 0.98)$
2. $TS = 7.37 + 0.952\ EH \quad (R^2 = 0.95)$

CONCLUSIONS

The results of this study indicate that college students can estimate their weight and height more accurately than the general population. They can estimate stature a little more accurately than weight; the latter seems to be slightly dependent on frequency of measurement within, at least, the previous 12 months. This indicates that if investigators are satisfied with the magnitude and direction of errors in this study, they can eliminate the time and cost of performing the measurements on this population, by merely asking them to estimate their own weight and stature. Accuracy of measurements does not appear to be gender dependent. Both stature and weight are independent of recency of measurement, but weight seems to be slightly dependent on frequency of measurement.

REFERENCES

[1] Stewart, A.L., The Reliability and Validity of Self-reported Weight and Height, J. Chronic Dis., 1982 35 pp. 295-301.
[2] Pirie, P., Jacobs, D., Jeffery, R., and Hannan, P., Distortion in Self-reported Height and Weight Data, J. Am. Diet. Assoc., 1981 78 pp. 601-606.
[3] Schlichting, P., Hoilund-Carlsen, P., and Quaade, F., Comparison of Self-reported Height and Weight with Controlled Height and Weight in Women and Men, Int. J. Obes., 1981 5 pp. 67-76.
[4] Wing, R.R., Epstein, L.H., Ossip, D.J., and LaPorte, R.E., Reliability and Validity of Self-reported Observers' Estimates of Relative Weight, Addict. Behav., 1979 4 pp. 133-140.
[5] Tell, G.S., Jeffery, R.W., Krammer, F.M. and Snell, M.K., Can Self-reported Body Weight be Used to Evaluate Long-term Follow-up of a Weight-loss Program, J. Am, Diet, Assoc., 1987 87 pp. 1198-1201.
[6] Dwyer, J.T., Feldman, J.J., Seltzer, C.C., and Mayer, J., Body Image in Adolescents: Attitudes Toward Height and Perception of Appearance, J. Nutr, Educ., 1969 1 pp. 14-22.
[7] Huenemann, R.L., Shapiro, L.R., Hampton, M.C., Mitchell, B.W. and Behnke, A.R., A Longitudinal Study of Gross Body Composition and Body Conformation and Their Association with Food and Activity in a Teenage Population: Views of Teen-age Subjects, Am. J. Clin. Nutr., 1966 18 pp. 325-338.

Trends in Ergonomics/Human Factors V
F. Aghazadeh (Editor)
Elsevier Science Publishers B.V. (North-Holland), 1988

REPEATABILITY OF STATIC AND ISOKINETIC MAXIMUM VOLUNTARY BACK STRENGTH EXERTIONS

Sean GALLAGHER

Bureau of Mines, U.S. Department of the Interior, Pittsburgh Research Center, P.O. Box 18070, Pittsburgh, PA 15236

Twelve subjects (mean age 36 ± 8 SD) performed static and dynamic back strength exertions in standing and kneeling postures using an isokinetic dynamometer. A test-retest criterion was employed that required two maximum voluntary contractions (MVCs) to be within 10 percent of one another for each of twelve experimental conditions. The higher of these two values was accepted as the true maximum exertion. Data were analyzed to determine (1) the number of trials required to satisfy the test-retest criterion, and (2) at which trial the MVC occurred for each experimental condition. An analysis of variance with repeated measurements (ANOVR) demonstrated that dynamic back strength measurements required significantly more trials to satisfy the test-retest criterion ($F_{1,11} = 5.837$, $p < .05$) than did the static exertions. The posture assumed (standing or kneeling) had no effect on the number of exertions required ($F_{1,11} = 0.258$, $p = .622$). Analysis of the static exertions indicated that back angle (i.e., trunk flexion of 22.5°, 45.0°, or 67.5°) had no effect on the number of repetitions required ($F_{2,22} = 1.082$, $p = .356$). Speed of contraction (i.e., 30°/s, 60°/s, or 90°/s) was a significant factor in number of trials required in the isokinetic tests ($F_{2,22} = 6.073$, $p < .01$); the 90°/s condition required significantly more trials than the two slower dynamic speeds. The average trial upon which the true maximum was observed was lower for static (2.39 trials) tests than dynamic (2.82 trials) tests; however, this difference was not significant ($F_{1,11} = 4.199$, $p = .065$). The results of this study indicate that fast isokinetic contractions are not as easily repeatable as slower isokinetic or static contractions. This may be due to the affects of rapid acceleration and perhaps an increased potential for a variable jerking motion during the strength measurement. The results of this study indicate that the use of a test-retest criterion may provide better estimates of both isometric and isokinetic strength.

1. INTRODUCTION

Strength has been shown to be a complicated and highly variable function [1]. Even with well-standardized measurement methods, the standard deviation on repeated tests on the same subject is on the order of ±10 percent or even higher. Some of the difficulties associated with assessment of muscular strength include control of the body parts to be measured and those to be excluded, controlling the posture of the subject, and control of the motivation level of the subject [1,2]. Muscular strength has also been shown to vary considerably throughout the course of the day [1]. In an attempt to reduce the variability in reported strength measurement procedures, Stobbe and Plummer [3] proposed that a test-retest criterion be used whereby two maximum voluntary contractions (MVCs) must be within 10 percent of one another, with the higher value taken as the true MVC. This paper describes results of maximal static and dynamic back exertions using the aforementioned criterion.

2. METHOD

2.1 Subjects

Subjects in this investigation were twelve healthy underground miners (mean age of 36 years ± 8 SD). The participants were paid volunteers, all of whom passed a thorough physical examination and graded exercise tolerance test. The subjects were given an explanation of the requirements of the study, along with the potential risks and benefits of participation. All subjects read and signed an informed consent form prior to the start of testing.

2.2 Apparatus

Back strength was measured using a modified isokinetic dynamometer [4]. This device was modified to allow assessments of both standing and kneeling back strength measurements through use of a platform that could be raised or lowered. In both postures, the subject's pelvis was stabilized through the use of restraining straps. Data on torque production and trunk angle was collected on-line using an A/D converter hooked to a microcomputer.

2.3 Procedure

Prior to the start of testing, subjects performed 5 minutes of light exercise on a bicycle ergometer, and executed a series of five trunk stretching exercise as a warm-up. Subjects then performed a series of practice trials at the various speeds and trunk flexion angles used in the study. During the actual experiment, all subjects performed at least two back extension exertions in twelve different conditions. Two postures were examined: kneeling and stooped. In each of these postures, three static and three dynamic exertions were performed. The static tests were performed at trunk flexion angles of 22.5°, 45.0°, and 67.5°. The dynamic tests were performed at isokinetic speeds of 30°/s, 60°/s, and 90°/s. A test-retest criterion was used where the peak torque of two MVCs had to be within 10% of one another; the higher of the two values was accepted as the true MVC. The subject rested at least two minutes between exertions.

2.4 Data Analysis

The dependent variables of interest in this investigation were: (1) the number of trials needed to obtain a true MVC according to the test-retest criterion; and (2) the trial on which the true MVC occurred. Separate 2 x 2 (static or dynamic contraction x standing vs. kneeling posture) analysis of variance with repeated measures (ANOVR) were performed to examine whether velocity of contraction or posture had an effect on the number of trials needed to obtain a true MVC or on the trial upon which the maximum contraction occurred. For static contractions, a univariate ANOVR was done to examine whether trunk angle had an effect on the number of trials required or on the trial on which the maximum contraction occurred; similarly, for dynamic contractions an ANOVR was performed to see if the speed of contraction influenced either the number of trials needed to obtain the true MVC, or the trial upon which the MVC occurred. Data on the amount of torque produced by these subjects is beyond the scope of this paper and will be presented elsewhere.

3. RESULTS

3.1 Number of Trials Required for MVC

The 2 x 2 analysis of variance demonstrated that dynamic exertions required significantly more trials to satisfy the MVC criterion than did static exertions ($F_{1,11} = 5.837$, $p < .05$). On the average, 2.39 trials were needed in the static conditions, while 2.82 trials were necessary in dynamic situations. The posture assumed during the exertions did not affect the number of trials required to achieve a true MVC ($F_{1,11} = 0.258$, $p = .62$).

The trunk angle assumed during exertions in static conditions did not affect the number of repetitions ($F_{2,22} - 1.082$, $p = .36$). However, in dynamic tests, speed of contraction did influence the number of exertions needed to satisfy the MVC criterion ($F_{2,22} = 6.073$, $p < .01$). A Duncan Range Test (alpha = .05) indicated that while the 30°/s (2.46 trials) and the 60°/s (2.67 trials) isokinetic conditions were not significantly different from one another, the 90°/s contraction required significantly more trials to establish a true MVC (3.50 trials).
Table 1 summarizes the data pertaining to the number of trials necessary to satisfy the MVC criterion. Examination of this table shows that for the majority of tests, only 2 exertions were required to satisfy the MVC criterion. Static exertions never needed more than four exertions; however, as many as seven repetitions were needed in the dynamic tests.

3.2 Trial on Which MVC Occurred

Table 2 summarizes the data pertaining to the trial upon which the MVC actually occurred during these tests. There was a trend that suggested that the true MVC occurs after more trials during dynamic tests; however, this trend was not significant ($F_{1,11} = 4.199$, $p = .065$). Posture did not influence the trial upon which the true MVC occurred ($F_{1,11} = 0.765$, $p = .400$). The angle of the trunk during static tests did not affect the MVC trial ($F_{2,22} = 1.085$, $p = .355$). In dynamic tests, there was a trend that suggested that the true MVC occurred after more trials; however, once again, this trend did not achieve significance ($F_{2,22} = 2.952$, $p = .073$).

Inspection of Table 2 shows that the true MVC occurred most often on the second exertion, whether the exertion was static or dynamic. The second most frequent trial where the true MVC occurred was on the first exertion. Together, the first and second exertions accounted for over 75% of all the maximum contractions. Rarely did the true maximum occur after the third trial.

TABLE 1.- Number of back strength exertion trials required to satisfy the test-retest criterion for the test sample (n=12)

Condition	Number of Trials Required 2	3	4	5	6	7
Standing Static 22.5°	7	4	1			
Standing Static 45.0°	8	1	3			
Standing Static 67.5°	10	2				
Standing Dynamic 30°/s	7	4	0	1		
Standing Dynamic 60°/s	7	4	0	0	1	
Standing Dynamic 90°/s	5	2	2	2	1	
Kneeling Static 22.5°	9	1	2			
Kneeling Static 45.0°	9	1	2			
Kneeling Static 67.5°	9	3				
Kneeling Dynamic 30°/s	9	2	0	1		
Kneeling Dynamic 60°/s	8	4				
Kneeling Dynamic 90°/s	3	3	4	1	0	1
Total	91 (63.2%)	31 (21.5%)	14 (9.7%)	5 (3.5%)	2 (1.4%)	1 (0.7%)
Totals for:						
Static Tests	52 (72.2%)	12 (16.7%)	8 (11.1%)			
Dynamic Tests	39 (54.2%)	19 (26.4%)	6 (8.3%)	5 (6.9%)	2 (2.8%)	1 (1.4%)

4. DISCUSSION

4.1 Number of Trials Required for MVC

The evidence of this study suggests that static back exertions are generally more repeatable than are dynamic contractions. However, this difference appears to be primarily attributable to the increased number of trials required at the fastest isokinetic speed (i.e., 90°/s), with an average of 3.5 trials necessary to satisfy the criterion. The 30°/s condition (average of 2.46 trials) and the 60°/s condition (average of 2.67 trials) did not differ substantially from the static tests (average of 2.39 trials). The average number of trials required during isometric strength tests in this study (i.e., 2.39 trials) is very close to the 2.43 trials/subject-test combination reported by Stobbe and Plummer [3] in their tests of various isometric strengths. Posture did not have a significant effect of the number of trials required, nor did the angle of the trunk during static exertions.

The greater variability in dynamic strength repeatability is reflective of the increased complexity of dynamic strength. Acceleration and

TABLE 2.- Trial upon which the MVC occurred for each of the experimental conditions (n=12)

	Trial on Which MVC Occurred				
Condition	1	2	3	4	5
Standing Static 22.5°	5	5	2		
Standing Static 45.0°	4	7	1		
Standing Static 67.5°	4	6	2		
Standing Dynamic 30°/s	6	5	1		
Standing Dynamic 60°/s	3	5	4		
Standing Dynamic 90°/s	3	2	4	2	1
Kneeling Static 22.5°	2	7	2	1	
Kneeling Static 45.0°	3	7	1	1	
Kneeling Static 67.5°	5	7			
Kneeling Dynamic 30°/s	3	6	2	1	
Kneeling Dynamic 60°/s	8	4			
Kneeling Dynamic 90°/s	3	3	3	3	
Total	44 (30.6%)	66 (45.8%)	25 (17.4%)	8 (5.6%)	1 (0.1%)
Totals for:					
Static Tests	23 (31.9%)	39 (54.2%)	8 (11.1%)	2 (2.8%)	
Dynamic Tests	21 (29.2%)	27 (37.2%)	17 (23.6%)	6 (8.3%)	1 (1.4%)

deceleration of the trunk can cause considerable variability in force output and is apparently more difficult for subjects to duplicate accurately [5]. However, it is noteworthy that, on the average, the slower isokinetic speeds in this study did not differ substantially from the static strength tests in terms of the number of trials required to achieve the repeatability criterion. It has been shown recently that most lifting activities occur at relatively slow trunk speeds [6]. Therefore, it may be advantageous when examining back strength with regard to materials-handling capabilities to concentrate on slower dynamic contractions, due to the decreased variance and increased external validity of the measurement.

4.2 Trial Upon Which MVC Occurred

The true MVC occurred on either the first or second trial over 75% of the time in this study. In static tests, the first two trials accounted for more than 85% of all MVCs; however, in dynamic exertions, the first two trials accounted for only about two-thirds of the MVCs. However, it is interesting to note that the second exertion was consistently the one where the MVC occurred most often, no matter whether the test was static or dynamic. This finding suggests that subjects require some learning before being able to produce a maximum contraction in the various experimental conditions, even though the subjects had performed several practice trials at different speeds on the dynamometer before the experiment began.

As can be seen from Table 2, only about 30% of the MVCs occurred on the first exertion (whether the contraction was static or dynamic). This implies that a single strength exertion may not be truly indicative of the maximum force capabilities of a test subject. This possibility should be considered by researchers when examining the strength characteristics of individuals.

4.3 Evaluation of the Test-Retest Criterion

There is a great deal of inherent variability in the measurement of human strength capabilities. Any method of testing that can be used to increase the reliability of strength data, therefore, deserves to be scrutinized by the thoughtful researcher. The test-retest criterion recommended by Stobbe and Plummer [3] would appear to be one useful method of helping to control for the conspicuous variability experienced when assessing muscular strength. The data presented in this paper supports the statement by these authors that the use of single and dual test criteria will underestimate some portion of the subject population's strength. The test-retest criterion appears to be a method which, while requiring an additional time investment, reduces the chance that underestimates of strength capabilities will occur. Other methods that merit further study in obtaining consistent and reliable strength data are monitoring heart rate during and after strength tests to assess subject motivation [1], and the use of verbal encouragement during exertions [7].

5. CONCLUSIONS

Based on the results of this study, the following conclusions were drawn:

Static back strength tests require fewer repeated trials to satisfy a test-retest criterion where peak strength observed must be ± 10% of that seen in another exertion. Slow dynamic contractions (30°/s and 60°/s) required a similar number of repetitions as the static tests; however, fast dynamic contractions (90°/s) required more repeated trials to satisfy the criterion.

The true maximum voluntary contraction was usually observed on either the first or second exertion. However, the maximum was observed to occur more frequently on the second exertion than on any other trial.

Using a test-retest criterion appears to be a useful method of assuring that underestimates of subject or population strength do not occur.

REFERENCES

[1] Astrand, P. O., and K. Rodahl., Textbook of Work Physiology (New York, NY, 1977).
[2] Kroemer, K. H. E., Kroemer, H. J., and Kroemer-Elbert, K. E., Engineering Physiology (Amsterdam, Elsevier Science Publishers, 1986).

[3] Stobbe, T. J., and R. W. Plummer. A Test-Retest Criterion for Isometric Strength Testing. Paper in Proceedings of 28th Annual Meeting of the Human Factors Society (San Antonio, TX, October 22-26, 1984), pp. 455-459.
[4] Marras, W. S., A. I. King, and R. L. Joynt., Spine (1984) 176.
[5] Chaffin, D. B., and G. B. J. Andersson., Occupational Biomechanics (New York, John Wiley and Sons, 1984).
[6] Marras, W. S., and P. E. Wongsam. Archives of Physical Medicine and Rehabilitation (1986) 213.
[7] Johansson, C. A., B. E. Kent, and K. F. Shepard. Physical Therapy, (1983) 1260.

Trends in Ergonomics/Human Factors V
F. Aghazadeh (Editor)

Strength Testing May Be An Effective Placemont Tool For The Railroad Industry

Paul B. McMahan

Manager, Safety Research Division
Association of American Railroads
50 F Street, N.W. #7516
Washington, D.C. 20001

Job-related physical tests may be an effective tool for placing workers in physically demanding jobs in the railroad industry. The Association of American Railroads (AAR) has completed an initial evaluation of the potential benefits of job placement based on job-related physical tests. AAR researchers conducted a job analysis of track laborers. This analysis found that tasks involving the manual handling of materials were frequently associated with over-exertion injuries. The researchers then conducted a biomechanical analysis of these tasks. Based on this biomechanical analysis, they designed isometric strength tests which "simulate" the requirements of the job. These are tests of the whole-body strength capability of workers. They focus on measuring the strength of the back, shoulder and arms. Comparison of the measured strength capability of track laborers with their past safety performance showed that weaker individuals had more injuries. Thus job-related physical tests may be an effective pre-employment job screening and placement tool.

INTRODUCTION

Many railroad jobs continue to require better-than-average strength and fitness. Manual tasks involving handling heavy materials such as railroad ties, timbers and track parts; operating switches, couplers, and hand brakes; and using heavy hand and portable power tools are performed regularly by several crafts. Over-exertion injuries of all types account for over 50 percent of all lost time injuries. The direct costs of these injuries averages over $250 million annually. And sprains and strains of the lower back account for more than half of the over-exertion injuries—25 percent of all lost time injuries. Trackmen, yardmen, road trainmen, and carmen continue to be involved in the majority of the cases. These crafts accounting for more than 60 percent of the lost time injuries but only 35 percent of the man hours worked.

The need for more effective job placement techniques is evident. This research was conducted to evaluate the potential for application of strength-based screening tests in pre-employment screening and job placement in the railroad industry. Trackmen were selected for study because they are required to perform a large number of manual tasks with potential for over-exertion. And the industry is likely to employee a number of these individuals in the future.

JOB ANALYSIS

Track laborers, especially section or extra gangs, are required to perform a wide-ranging and large number of track maintenance tasks. These include the installation and maintenance of all track structures and other numerous general maintenance activities. They use a large variety of hand and portable power tools—ranging in weight from 4-200 pounds. Materials handled by hand range in weight from 10-500 pounds. Team lifting is very very important.

On-site observations and structured interviews with subject matter experts revealed several "manual" processes common to track laborers. (See Table 1.) These processes may be performed **without** the benefit of materials handling aids and power tools—depending upon the availability of such equipment.

Table 1. General Processes For Track Laborers

1.	Replacing Switch Points
2.	Replacing Switch Stands And Timbers
3.	Replacing Frogs And Switch Timbers
4.	Replacing Rail Sections
5.	Highway Grade Crossing Rehabilitation
6.	Distribution Of Ballast From Work Trains
7.	Distribution Of Materials For Rail Laying System Gangs
8.	General Maintenance Activities
9.	Spot Rail Realignment And Surfacing
10.	Spot Tie And Timber Renewal
11.	Loading And Unloading All Track Materials

We found that the potentially most stressful tasks within these processes involved loading, unloading and carrying of various materials by hand. This includes switch points(340 lb), switch stands(100 lb), switch timbers(324 lb), kegs of spikes(200 lb), rail grinders(75 lb), rail drills(200 lb), rail saws(60), bags of rail anchors(150 lb), crossing timbers(480 lb),.and other materials(20-100 lbs). Workers use team lifting and simple hand tools, such as tie tongs, to handle some of the heavier of these materials.

BIOMECHANICAL ANALYSIS

The objective of strength-based job screening is to reduce the frequency and severity of overexertion injuries. Accomplishment of this objective requires a thorough understanding of two interactive components of the job, the workplace and the worker. Biomechanical analysis provides a common scientific basis for understanding both the physical work requirements and the physical abilities of the worker. It helps us rank order the strength requirements of the various tasks involved in a job. It identifies the muscle groups which limit performance on each task. And it predicts the percentage of adult working populations that could be expected to perform each task.

We used the two-dimensional static biomechanical model developed by the University of Michigan's Center for Ergonomics [1] to evaluate the loading, unloading and carrying tasks performed by trackmen. We made video recordings of these tasks to document the methods used, and digitized posture data from the video recordings. We then used the biomechanical model to identify those tasks falling between the NIOSH Action Limit (AL) and Maximum Permissible Limit (MPL) [2] for male strength. These tasks provided the basis for design of the strength tests. Figure 1 shows a scatter plot of the hand locations for these tasks. The obvious clustering in Figure 1 indicates three zones which represent the range of task situations which are most stressful.

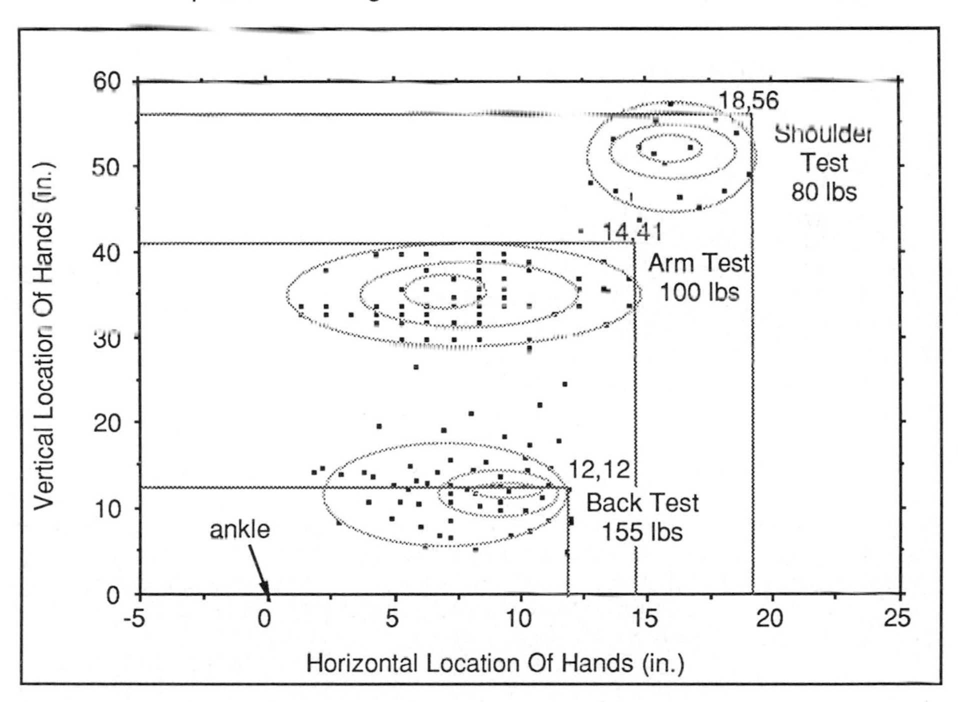

Figure 1
Illustration of Hand Locations for Tasks Analyzed and Corresponding Parameters for Isometric Strength Tests

The biomechanical analysis showed that the back, shoulder and arm muscles limit performance in the respective zones indicated in Figure 1. The Figure also shows the parameters we selected for three isometric strength tests to "simulate" the tasks analyzed.

Each of the tests is two-handed and symmetric. The handles of our test apparatus were set at 12 horizontal inches from the ankle of the subject and 12 vertical inches from the floor for the back test. The corresponding settings for the arm and shoulder tests were (14,41) inches and (18,56) inches respectively. The test apparatus measures the static force applied, in pounds.

We followed the recommendations published in the "Ergonomic Guide for the Assessment of Human Static Strength" [3] during the administration of the strength tests. Subjects were permitted to use their preferred posture. We gave repeated trials, up to five with sufficient rest, for each of the three strength tests until we obtained three trials with measurements within 5% of each other. We then computed a "score" for each of the three strength tests by averaging the three trials.

MEASURED STRENGTH AND INJURY HISTORY

We completed the strength assessment of eighty incumbent trackmen and compared their retrospective safety performance (i.e. reported injury history) with their measured strength capability.

Weaker individuals had significantly more injuries in the past. Figure 2 shows that persons with a prior back injury had an average measured back strength of 162.9 pounds. Whereas, those with no prior injury at all had a average measured back strength 221.2 pounds. Similarly, persons with a prior shoulder injury had an average measured shoulder strength of 119.0 pounds. And those with no prior injury had a average measured shoulder strength of 153.8 pounds. These differences between previously injured and uninjured individuals are statistically significant (α=.05) for both back and shoulder strength. However, the differences between these individuals were not statistically significant for measured arm strength. Persons with a prior arm injury had an average measured arm strength of 109.7 pounds. Whereas, those with no prior injury at all had a measured arm strength 128.4 pounds.

Unfortunately we are uncertain of the cause of the poor performance of persons with prior injuries. That is, we are uncertain whether they performed poorly on our tests because they are indeed inherently weaker or because they were injured previously. Because none of our subjects had experienced lost-time injuries within one year of our tests, we might assume that they were fully recovered from any previous injuries. Thus, it would seem that weaker individuals indeed do have more injuries. Prospective analysis of the safety performance of these individuals is underway to confirm this finding.

Back Strength Vs Past Injury Type

Type Injury	N	Mean	Std.
Back	12	162.9	49.7
Other	27	204.6	48.1
None	41	221.2	56.7

Comparison	Mean Diff	Sig.
Back vs None	58.3	yes
Back vs Other	41.7	yes
Other vs None	16.7	no

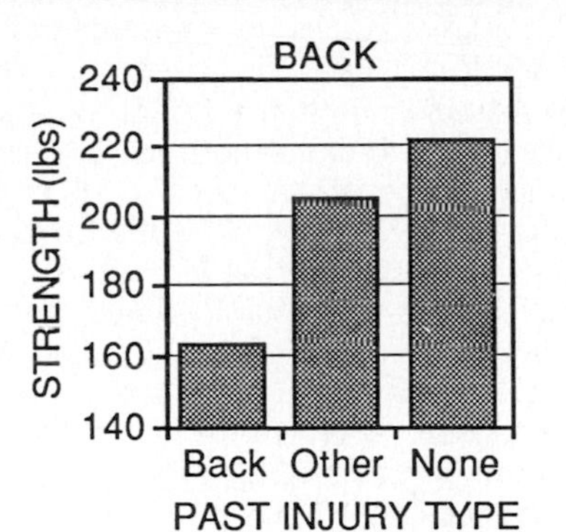

Shoulder Strength Vs Past Inj.Type

Type Injury	N	Mean	Std.
Shldr	8	119.0	36.2
Other	31	144.4	45.2
None	41	153.8	43.8

Comparison	Mean Diff	Sig.
Shlder vs None	34.8	yes
Shlder vs Other	25.4	no
Other vs None	9.4	no

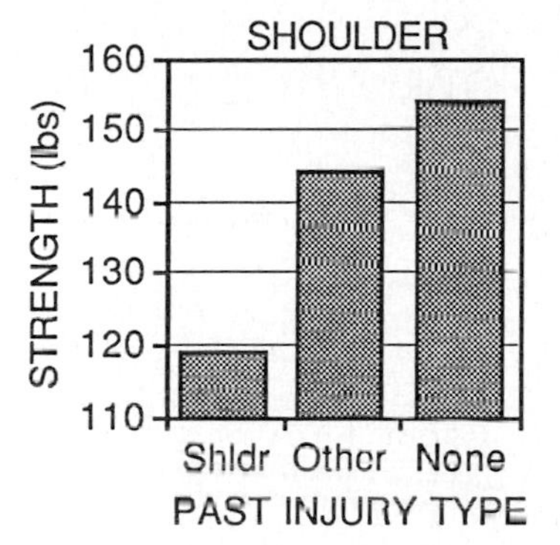

Arm Strength Vs Past Injury Type

Type Injury	N	Mean	Std.
Arm	5	109.7	15.2
Other	34	125.7	36.2
None	41	128.4	29.3

Comparison	Mean Diff	Sig.
Arm vs None	18.7	no
Arm vs Other	16.0	no
Other vs None	2.7	no

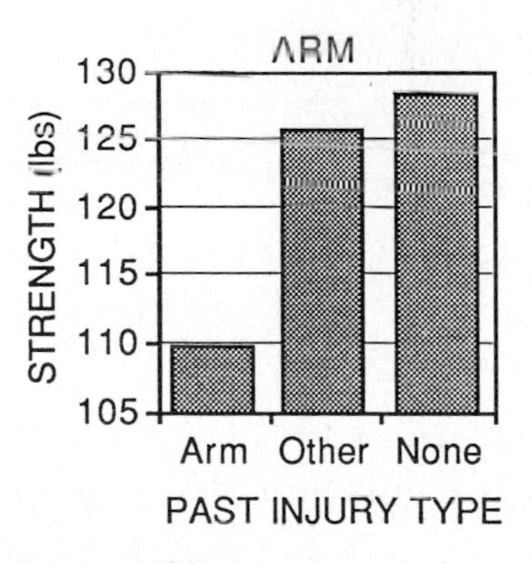

Figure 2
Measured Back, Shoulder and Arm Strength Versus Past Injury Type

LOAD-STRENGTH RATIO

One way to compare a worker's capability with the requirements of a task is to examine the load-strength ratio. This is simply the ratio of the load required by task to the measured force exertion capacity (strength) of the worker. Figure 3 shows the hypothetical relationship between the load-strength ratio and injury rates. The ratio is less than 1 when the worker's capacity exceeds the requirements of the task. The ratio is greater than 1 when the requirements of the task exceed the capacity of the worker. We expect that injury rates will rise as the ratio gets larger. It is generally recommended that a person not be required to exert more than 70 percent of his maximum capability for continuous work. This is the same as saying that the load-strength ratio should not exceed 0.7.

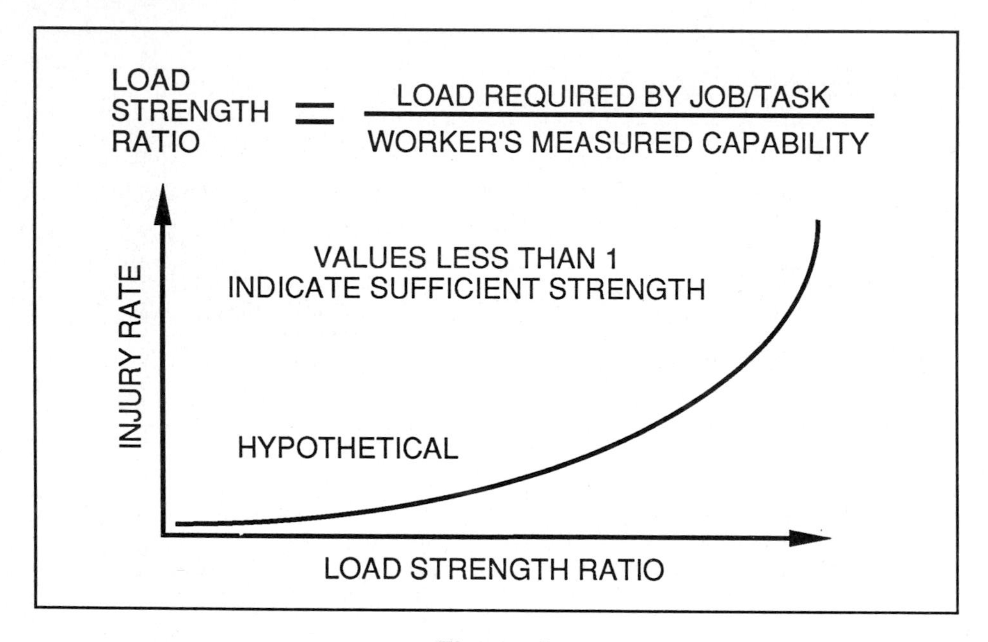

Figure 3
Hypothetical Relationship Between Load Strength Ratio And Injury Rate

INJURY HISTORY VERSUS LOAD-STRENGTH RATIO

The maximum loads required by the tasks simulated by our strength tests were 155 pounds for the back test, 80 pounds for the shoulder test, and 100 pounds for the arm test. Using these values we computed the load-strength ratios for each subject for each of the strength tests. Figure 4 shows the distribution of the employees we tested by load-strength ratio group and the related injury history for back, shoulder and arms.

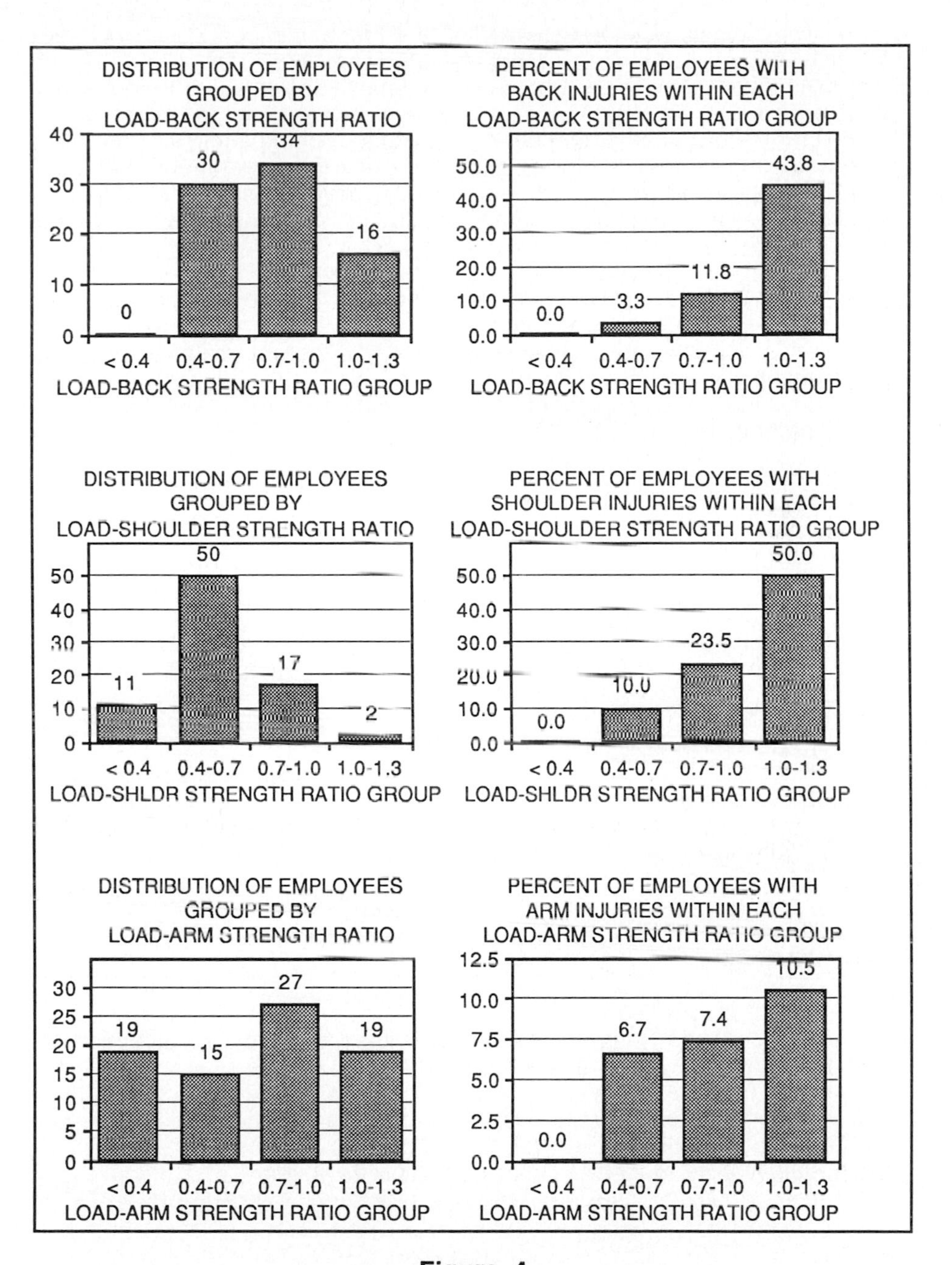

Figure 4
Load-Strength Ratio For Back, Shoulder and Arm Versus Past Injury Type

Although the sample sizes are small, Figure 4 tends to support the hypothesis that injury rates will rise as the load-strength ratio rises. The back strength capability of twenty percent (16) of the employees tested was less than the task requirement of 155 pounds. That is, their load-back strength ratios exceed 1.0. And Figure 4 shows that 43 percent of this group had experienced prior back injuries. This is four times the injury rate of persons whose measured back strength was greater than the task requirement. Similar results are evident for shoulder and arm injuries. All of these results are limited by the small sample sizes.

Figure 4 also tends to support the recommendation that the load-strength ratio should not exceed 0.7 for continuous work. Over 91 percent of all the prior back injuries experienced by this group of employees occurred to persons whose load-back strength ratio exceed 0.7. Similarly, 50 percent of the shoulder injuries and 80 percent of the arm injuries occurred to persons whose load-strength ratios exceed 0.7.

SUMMARY

This research gives strong **indication** of the potential benefits of job-related strength tests for pre-employment screening and job placement. Extrapolation of the results suggests that about 90 percent of back injuries could be prevented by selecting individuals whose capacity exceeds job requirements. That is, a load-strength ratio limit of 0.7 would be recommended for continuous work. But we cannot justify extrapolation of the findings of this research due to the limited number of employees tested. In addition, prospective analysis of the safety performance of the tested employees is needed to confirm the finding that weaker individuals have significantly more injuries and that strength testing can be an effective means for identifying these individuals.

REFERENCES

[1] Chaffin, D.B. and Garg, A., A Biomechanical Computerized Simulation of Human Strength. AIIE Trans. March 1-15, 1975.

[2] National Institute of Occupational Safety and Health, Work Practices Guideline for Manual Lifting, Department of Health and Human Services (NIOSH) Publication No. 81-122, March 1981.

[3] Chaffin, D.B., Ergonomic Guide for the Assessment of Human Static Strength, American Industrial Hygiene Association, J. 36:505-510, 1975.

Trends in Ergonomics/Human Factors V
F. Aghazadeh (Editor)

CHANGES IN POSTURAL STABILITY, PERFORMANCE, PERCEIVED EXERTION AND DISCOMFORT WITH MANIPULATIVE ACTIVITY IN A SUSTAINED STOOPED POSTURE

R. Wickstrom, A. Bhattacharya and R. Shukla*

Biomechanics-Ergonomics Research Laboratory and
*Biostatistics-Epidemiology Laboratory
Department of Environmental Health
University of Cincinnati Medical Center
Cincinnati, OH 45267-0056

Postural stability tests, Borg's Rating of perceived exertions and modified Corlett-Bishop discomfort ratings were obtained from eleven males aged 28-43 years following a five-minute tests of manipulative performance at standing work stations requiring erect versus stooped trunk postures. Statistically significant ($p < .05$) increases in perceived exertion and discomfort and a decrease in work performance were demonstrated following the activity requiring a stooped work posture in comparison to the same activity performed with the trunk erect. A decrease in postural stability was significant ($p < .05$) only for the group of subjects who experienced noticeable discomfort in the low back area during the stooped posture activity. The high discomfort group also reported significantly higher levels of perceived exertion. These results suggest that quantification of postural sway and psychophysical measures of exertion and discomfort may provide better characterization of standing work activities.

1. INTRODUCTION

Many industrial tasks require the workers to adopt abnormal stooped postures -- leading to backache, fatigue, reduced performance and postural instability. Because relatively few jobs in industry today require high energy expenditure or work in extreme physical environments, the constraints imposed by posture on work activity has received increased attention. Corlett et al. [1] note that the barrier beyond which a posture will be maintained is an unacceptable level of pain. Therefore, difficult postures which cause unacceptable levels of pain may impair worker performance.

Kvalseth [2] notes that workplace designs that require standing operators are most appropriate where the following activities occur frequently:

1. Handling of heavy loads,
2. Work requiring a high degree of mobility in the workplace,
3. Extended reaches and moves of substantial magnitude,
4. Work requiring manual downward forces of substantial magnitude.

Kvalseth [2] stressed that tasks requiring a high degree of body stability and equilibrium or fixed body postures for an extended time period are best designed for sitting operators.

Standing work postures which are repeated on a daily basis over a period of years have been shown to cause short and long-term effects on the cardiovascular and musculoskeletal system of the worker. von Wely [3] demonstrated a high degree of correlation between working postures and the complaints received at the industrial health clinic. He noted that the probable site of pain from work in a stooped posture was in the lumbar region and erector spinae muscles. Cardiovascular problems are also common in occupations that require extensive standing postures. When the worker stands in a static position for an extended period of time, blood and tissue fluids tend to accumulate in the legs resulting in venous pooling, swollen legs, painful feet, and the eventual formation of varicose veins. Although sitting may decrease the physiological load on the cardiovascular system, prolonged sitting is not without its own health disadvantages.

Working in stressful static work postures can potentially effect postural stability by several mechanisms:

1. Following standing exercise and the assumption of static standing postures, blood tends to pool in the legs and may deprive both the heart and the brain of oxygen needed to function properly.
2. Fatigue may produce metabolic end products like lactic acid which may produce local and central effects on the sensory motor system which contribute to the control of postural equilibrium.
3. Abnormal standing work postures may directly modify joint and muscle proprioceptors which provide afferent input for the control of postural stability.

Therefore, to maximize worker performance and well being, it is necessary to first characterize the physical and psychological effects of standing work activities and postures.

Several investigators have attempted to relate physiological response to activity to the perception of workload [4,5]. Borg's Self-Rating Scale [5] of perceived exertion is scaled from 6-20. Multiplying the activity rating by a factor of 10 has been shown by Borg [5] to correlate to absolute heart rate levels for dynamic whole body efforts. Corlett and Bishop [4] developed a self-rating chart of postural discomfort which recorded the distribution of discomfort in the body and its change during the work period. The use of this discomfort chart in a study involving spot welders provided direct evidence of the benefits of ergonomic changes.

Although several experimenters have attempted to quantify posture and manual lifting at work, no one has evaluated the effect of work posture and activities on postural stability. For over a century, postural sway has been keenly studied in the fields like neurology because of its importance in the global screening of sensory motor systems which control posture [8]. The force platform has been traditionally used for quantifying the movement pattern of the body's center of pressures (CP). A graphical presentation of the movement pattern (x versus y) of the CP is generally called a stabilograph [9]. Yagoda [10] has performed a theoretical analysis of human postural stability based on biomechanical considerations in order to substantiate that human postural stability is a criteria which should be included in worker protection policies. Because postural stability may be affected by standing work activities, this study was designed to compare the stooped versus erect work posture with respect to postural stability, perceived exertion, postural discomfort and performance.

2. MATERIALS AND METHODS

The multi-axis biomechanics force platform used in this study was validated previously by Bhattacharya et al. [9]. This system was used in conjunction with a custom-developed pattern recognition algorithm. The high frequency response (400 Hz) strain-gauge type six component force platform (Model OR6-3, Advanced Mechanical Technology, Inc., Newton, MA) quantitates changes in body sway by measuring the reaction forces at the feet (Fx, Fy, Fz, Mx, My, Mz) and then calculating the location and movement pattern of the body's center of pressure, based on the equations provided in Bhattacharya et al. [9]. The parameter chosen to characterize the body sway was the mean radial deflection of the body center of pressure from the mean value of center of pressure coordinates during the 30 sec test procedure.

The Minnesota Rate of Manipulation Turning Test was chosen as the manual dexterity activity based on its extensive use in pre-employment screening for assembly line tasks requiring rapid manipulation and sorting of parts. The test requires the subject to pick up a series of cylindrical blocks from a rectangular board with one hand, shift the block to his other hand, and place the block back into its hole upside down. The standard test height was modified in this study to an ergonomically recommended work height [11] of 5-10 cm below elbow height for Test 1 (trunk erect), and 38 cm for Test 2 (requiring a stooped posture). A five minute time period was chosen because static postural efforts have been shown to cause fatigue for slight static efforts or abnormal postures maintained for 5 minutes or more [11].

Borg's Rating of Perceived Exertion [5] consists of a scale of 6 to 20, with corresponding descriptive terms ranging from very, very light (7) to a very, very hard (19).

The standard Bishop-Corlett Chart [4] consists of a body diagram in which different areas may be ranked from most to least painful. This chart was modified for the purpose of this study by including an analogue scale for each body part which ranged from just noticable to intolerable discomfort. Responses were weighted numerically from zero (no discomfort) to 10 (intolerable discomfort). Only the low back discomfort rating was considered in the statistical analysis because the low back was the only area of significant discomfort reported during the stooped activity.

Experimental Protocol

Eleven male volunteers (ages 28-43 years) were assessed for a 30-second baseline postural stability test during quiet standing on the platform in bare feet, with eyes closed, heels together, and feet at a 30° angle. Following baseline postural sway measurements, the subjects were given the standardized instructions for the Minnesota Rate of Manipulation Turning Test. After a brief practice period at an ergonomically recommended work height of 5-10 cm below the elbows, the subjects were instructed to work as quickly as possible at this work height for 5 minutes. Immediately following this five-minute activity, the 30 sec sway test with eyes closed was repeated. Upon completion of the sway test, the subjects were asked to rate their perceived exertion for the five minute activity using Borg's Whole Body Scale of Perceived Exertion. The subjects were also asked to rate their postural discomfort using a modified Corlett-Bishop discomfort scale for each body part. The number of blocks turned during the five minute time period was used as an index of performance. The work height was then adjusted to a 38 cm work height and the same five minute protocol and data collection was carried out for the lower work height.

3. RESULTS AND DISCUSSION

Statistical analysis consisted of paired t-tests (or a nonparametric analogue of paired t-test) and linear regression of the measured parameters (see Table 1). When the data for all eleven subjects was analyzed collectively for differences in the mean radius of sway (RMean), rate of perceived exertion (RPE), low-back discomfort (LBD), and performance, the following trends were identified:

1. Increased ($p < .001$) RMean of sway following the stooped posture activity compared to the control of normal sway.
2. Increased ($p < .05$) RMean of sway following the erect posture activity compared to the control of normal sway.
3. Slightly greater RMean of sway following the stooped posture activity versus the erect posture activity which was not statistically significant.
4. Increased RPE ($p < .005$) during the stooped posture activity versus the erect standing activity.
5. Increased LBD ($p < .05$) during the stooped posture activity versus the erect standing activity. The low back was the only body area of significant discomfort noted by the eleven subjects, and low back pain only occurred with the stooped posture activity.
6. Decreased ($p < .005$) manipulative performance by 8.3% for the stooped posture activity versus the erect posture activity.
7. Significant ($p < .005$) positive correlation between LBD and RPE ($r = .83$).
8. Nearly significant ($p = .062$) positive correlation between LBD and RMean following the stooped posture activity ($r = .58$).

Based on the positive correlation between LBD and the dependent variables RPE and RMean of sway following the stooped posture activity, the eleven subjects were arbitrarily divided into high ($LBD>2$) and low ($LBD\leq 2$) discomfort groups based on low back discomfort scores during the stooped activity.

Within the high discomfort group, the results were similar to the pooled results for all eleven subjects except the RMean of sway following the stooped posture activity was significantly increased ($p < .05$) over the RMean of sway following the erect posture activity. The RMean of sway of the erect posture activity was not significantly different from the control. Within the low discomfort group, the only significant difference for the stooped posture versus erect posture activity was an increased RPE for the stooped activity. There were no significant differences in RMean of sway between the stooped posture versus erect posture activity.

TABLE 1
SUMMARY PARAMETERS FOR POSTURAL SWAY, EXERTION, LOW-BACK DISCOMFORT AND PERFORMANCE

	ALL SUBJECTS (n=11)		HIGH DISCOMFORT GROUP (n=6)		LOW DISCOMFORT GROUP (n=5)	
VARIABLE	Mean(SD)	P	Mean(SD)	P	Mean(SD)	P
MEAN RADIUS SWAY (cm)						
Stoop (S)	.719(.220)	--	.858(.203)	--	.552(.074)	--
Erect (E)	.637(.186)	--	.615(.124)	--	.644(.256)	--
Control (C)	*.478(.106)	--	.530(.112)	--	.416(.059)	--
(S) vs (E)	.082(.270)	.339	.243(.163)	.015	-.112(.250)	.375
(S) vs (C)	.241(.169)	<.001	.328(.175)	.006	.136(.090)	.028
(E) vs (C)	.159(.186)	.018	.085(.139)	*.208	.248(.210)	.058
RATE OF PERCEIVED EXERTION						
Stoop (S)	12.55(1.97)	--	13.67(1.75)		11.2(1.30)	--
Erect (E)	8.91(1.58)	--	9.83(1.17)		7.8(1.30)	--
(S) vs (E)	*3.64(1.91)	*.003	3.83(2.23)	*.034	3.40(1.67)	.01
LOW BACK DISCOMFORT						
Stoop (S)	3.09(3.27)	--	5.33(2.73)	--	*0.40(0.89)	--
Erect (E)	*0.00(0.00)	--	*0.00(0.00)	--	*0.00(0.00)	--
(S) vs (E)	3.09(3.27)	.011	5.33(2.73)	.005	*0.40(0.89)	.374
PERFORMANCE (# BLOCKS)						
Stoop (S)	476.3(59.30)	--	459.5(65.43)	--	496.4(50.10)	--
Erect (E)	436.6(83.28)	--	411.3(95.13)	--	467.0(62.64)	--
(S) vs (E)	39.7(40.80)	<.005	48.2(34.6)	<.005	29.4(49.00)	<.15

*Nonparametric analysis (Wilcoxon)

In comparing the parameters for the high versus low discomfort groups, the RMean of sway and RPE were significantly higher ($p < .05$) for the high discomfort group. Despite a trend towards greater reduction in performance for the high discomfort group, the differences were not statistically significant.

In conclusion, poor work station design which promotes stooped working postures may have a significant affect on body sway, perceived exertion, low back discomfort, and performance. The rate of perceived exertion and body sway effects appear to be positively correlated to the level of discomfort. Since heart rate was not monitored during the experiment, it is unclear as to whether the perception of perceived exertion is related to the perception of discomfort or a whole body cardiovascular response. The body sway correlation with discomfort may be the result of localized muscle fatigue of postural muscles which control balance.

Since postural balance is an essential requirement of standing work activity, it is important to monitor the effect of work station design on body balance, discomfort, and perceived exertion. The force platform analysis of body sway has many potential applications within the field of ergonomics; however, further research is needed to determine the true relationship between postural sway, local and whole body fatigue, and discomfort.

ACKNOWLEDGEMENT

The authors appreciate the technical assistance and advice provided by Vernon Putz-Anderson, Ph.D., National Institute for Occupational Safety and Health, Cincinnati, OH. We would also like to acknowledge the assistance of Ms. Barbara Howard and Ms. Diane Gorman in the manuscript preparation.

REFERENCES

[1.] Corlett, E.N. and Manenica, I., The effects and measurement of working postures, Applied Ergonomics 11.1 (1980) 7-16.
[2.] Kvalseth, T.O., Work station design, in: Alexander, D.C. and Pulaf, B.M., (eds.), Industrial Ergonomics -- A Practitioner's Guide (Industrial Engineering & Management Press, Norcross, GA, 1985).
[3.] van Wely, P., Design and disease, Applied Ergonomics 1 (1970) 262-269.
[4.] Corlett, E.N. and Bishop, R.P., A technique for assessing postural discomfort, Ergonomics 19-2 (1976) 175-182.
[5.] Borg, G., Psychological aspects of physical activities, in: Larson, L.A. (ed.), Fitness, Health and Work Capacity (Macmillan, New York 1974) 141.
[6.] American Industrial Hygiene Association, Work Practices Guide for Manual Lifting (NIOSH Publication 81-122, Cincinnati, 1983).

[7.] Rohmert, W. and Mainzer, J., Influence Parameters and Assessment Methods for Evaluating Body Postures, in: Ergonomic interventions to prevent musculoskeletal injuries in industry, American Conference of Government Industrial Hygienists (Lewis Publishers, Chelsea, MI, 1987).
[8.] Romberg, M.H., Manual of nervous diseases in man, London Syndenham Society (1853) 395-401.
[9.] Bhattacharya, A., Morgan, R., Shukla, R., Ramakrishanan, H.K. and Wang, L., Non-invasive estimation of afferent inputs for postural stability under low levels of alcohol, Annals of Biomedical Engineering 15 (1987) 533-550.
[10.] Yagoda, M.P., A basic theoretical analysis of human postural stability, Mechanics Res. Commun. 1 (1974) 347-352.
[11.] Grandjean, E., Fitting the task to the man (Taylor & Francis, London and Philadelphia, 1986).

Trends in Ergonomics/Human Factors V
F. Aghazadeh (Editor)

A COMPARISON BETWEEN ISOKINETIC TRUNK STRENGTH AND STANDARD STATIC STRENGTH TESTS

J.W. YATES

Exercise Physiology Laboratory
University of Louisville
Louisville, KY 40292, U.S.A.

and Waldemar KARWOWSKI

Center for Industrial Ergonomics
University of Louisville
Louisville, KY 40292, U.S.A.

ABSTRACT

Isokinetic strength of the trunk extensors and flexors was measured on eight males, in addition to four static strength tests: arm, stooped back, leg, and composite. Concentric trunk extension (CTE) and trunk flexion (CTF) strength was measured at speeds of 15, 30, 60, 90, and 120 degrees/sec. Eccentric trunk extension (ETE) and trunk flexion (ETF) strength was measured at speeds of 15, 30, and 60 degrees/sec. CTE strength ranged from a high of 351 ± 63 Nm at 15 degrees/sec to a low of 254 ± 95 Nm at 120 degrees/sec. CTF strength changed very little with speed and averaged 154 + 33 Nm. Eccentric torque exceeded concentric torque from 12 to 24% over the range of velocities tested. Correlations between TE and TF strength over the different velocities generally exceeded 0.84. Correlations between concentric and eccentric contractions were somewhat lower, but always exceeded 0.64. Neither concentric nor eccentric isokinetic contractions correlated well with static strength. These data suggest that concentric and eccentric strength of th back extensors and trunk flexors are highly related over a range of contractions velocities, but are not highly correlated with static strength.

1. INTRODUCTION

In the late 1960's and early 1970's several studies were published which suggested that the rate of low back injuries was greater in "heavy" than in "light" industries [1,2,3]. Chaffin and coworkers [4,5] extended this by showing that the incidence of low back injuries was three times as great in individuals who demonstrated less static strength than required by the job. Since this correlation was found, static strength tests have been used for preemployment screening. The reasoning behind this testing was to eliminate individuals with insufficient strength to perform the job and hence reduce low back injuries.

Recently the validity of using static tests to predict success during dynamic work has been questioned [6,7,8]. This has led to dynamic strength testing, primarily of the trunk extensor and flexor muscles. However, the relationship between dynamic back strength and incidence of low back injury has not yet been established. The purpose of this study was to investigate the relationship between dynamic trunk strength and static strength tests which have been shown to be related to the incidence of low back pain [4,5]. In addition, the ratio between trunk extensor and trunk flexor strength, and the relationship between concentric and eccentric trunk strength was investigated.

2. METHODS AND PROCEDURES

2.1. Subjects

Eight males participated as subjects in this experiment after signing an Informed Consent. Physical characteristics for the subjects are shown in Table 1.

Table 1. Age, physical characteristics, and static strength [N] of the subjects (n=8).

Variable	Mean	S.D.	Range
Age (years)	26.8	4.1	22-32
Height (cm)	176.3	5.8	167.6-182.9
Weight (kg)	72.8	14.1	49.9-90.7
Arm Strength	395.3	131.1	123-549
Stooped Back	1095.4	144.2	933-1193
Leg Strength	1631.0	191.5	1293-1907
Composite Strength	1420.7	269.8	1037-1938

2.2. Equipment

All static strength measurements were made using a SM-500 Interface load cell connected to an Apple-Isaac 91A data acquisition system. A 12-bit, A-D converter sampled the load cell at a rate of 200 samples/second for a period of five seconds. The raw A-D values along with the calibration values were saved on floppy disks. Peak force was immediately printed on a printer.

Dynamic back strength was measured on a Kin-Com isokinetic dynamometer. The Kin-Com is similar in function to a Cybex except it allows both concentric and eccentric isokinetic muscle contractions at velocities ranging from zero to 210 degrees/second. The system is controlled by an

IBM-PC with a 20 Meg hard-disk drive. The data analysis system allows the raw data to be output as well as other manipulations such as averaging force over a range of motion.

2.3. Static Strength Measurements

Static measurements of arm strength, stooped back strength, leg strength, and composite strength were made according to the procedures described by Ayoub et. al. [9]. For each of these efforts the highest of two trials were recorded. When the two trials differed by more than 10%, a third or fourth trial was performed as necessary. The greatest force exerted during a 5 second contraction was chosen as the MVC.

2.4. Dynamic Trunk Strength Measurement

Both concentric and eccentric dynamic trunk extension and trunk flexion strength were measure on a Kin-Com isokinetic dynamometer. Concentric strength was measured at 15, 30, 60, 90, and 120 degrees/second, while eccentric strength was measured at 15, 30 and 60 degrees/second. The reported torque values represent the average torque produced by the subject over the range from 70 to 100 degrees. All trunk strength measurements were made with the subject in a sitting position.

The axis of rotation of the Kin-Com lever arm was adjusted to approximate the L5-S1 vertebra of the spine. The connector pad was adjusted so that it rested on the upper portion of the sternum during trunk flexion and approximately midway on the scapula during trunk extension. The length of the lever arm was stored on the computer in order to convert the force produced into a torque value. Each contraction was performed from approximately 50 degrees trunk flexion (90 degrees = upright sitting) to approximately 115 degrees trunk extension. Eccentric contractions were alternated with concentric contractions of the same speed. The number of contractions performed at each speed was determined by the repeatability of the subject. When the subject performed two closely matched contractions as determined from the graphical display, he proceeded to the next speed. Testing always proceeded from the slowest speed of movement to the fastest speed.

2.5. Data Analysis

Pearson product correlation coefficients were calculated to determine if any relationships existed between maximum dynamic trunk strength and static strength.

3. RESULTS AND DISCUSSION

3.1 Results

The results of the isometric strength tests are shown in Table 1. Dynamic trunk extension and flexion strength is depicted in Figure 1. Static strength tests were generally highly correlated among themselves, butstatic strength was not highly correlated with dynamic strength. For example, the correlation between the static stooped back lift and the static leg strength was r=0.86, while r=0.87 was the correlation between static stooped back and static composite lift. However, the correlations between static stooped back strength and dynamic back extension strength were generally low and ranged from r = -0.25 to r = -0.62. On the other hand, correlations between dynamic trunk extension strength over the tested speeds (i.e. torque produced at 15 degrees/sec was highly correlated with the torque produced at 30 degrees/sec) ranged from r = 0.68 to r =0.92. These same relationships were consistent across all static and dynamic strength tests.

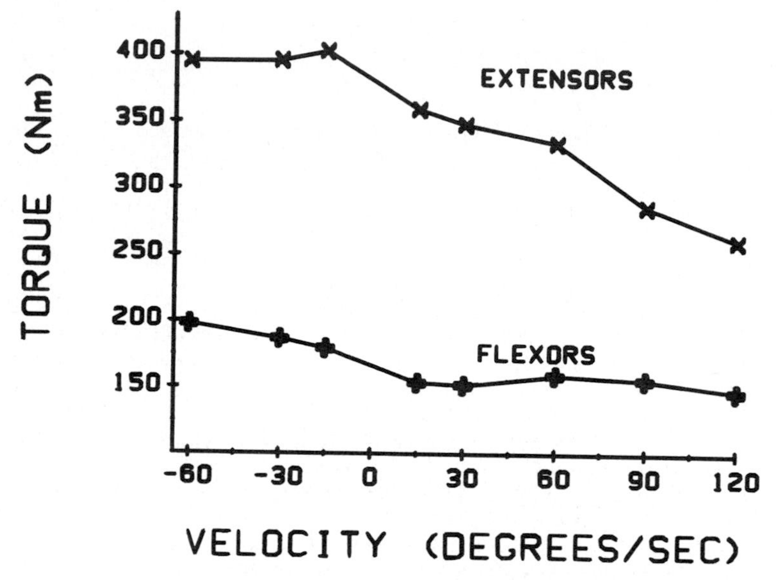

Figure 1. ISOKINETIC STRENGTH OF TRUNK EXTENSORS (X) AND TRUNK FLEXORS (+) FOR VARIOUS MOVEMENT SPEEDS. NEGATIVE VELOCITIES REPRESENT ECCENTRIC CONTRACTIONS WHILE POSITIVE VELOCITIES REPRESENT CONCENTRIC CONTRACTIONS.

Table 2 shows the ratio of trunk extensor to trunk flexor strength for both concentric and eccentric contractions. For both types of contractions the ratio decreased with increasing speed of movement. This was due to the fact that trunk flexor torque remained relatively constant as the speed of movement increased, while trunk extensor torque decreased (see Figure 1). Eccentric contractions were not performed at speeds faster than 60 degrees/second due to the increased possibility of injury.

The ratio of eccentric strength to concentric strength, shown in Table 3, seems to be slightly higher for the trunk flexors than for the trunk extensors. However, the number of subjects in this study is probably too low to make any generalizations.

Table 2. Ratio of trunk extensor to trunk flexor strength for various speeds of movement.

	Speed of Movement (degrees/second)				
	15	30	60	90	120
Concentric	2.34	2.29	2.09	1.83	1.77
Eccentric	2.24	2.12	2.00	----	----

Table 3. Ratio of eccentric contractions to concentric contractions at various speeds of movement.

	Speed of Movement (degrees/second)		
	15	30	60
Trunk Flexors	1.17	1.23	1.24
Trunk Extensors	1.12	1.14	1.18

3.2 Discussion

During the last ten years, instruments capable of measuring dynamic trunk strength have become commercially available, and a great deal of information on the dynamic strength of the trunk flexor and extensor muscles has been collected. However, the relationship between dynamic trunk strength and low back injuries has not yet been established. The correlations between dynamic trunk strength and static strength tests found in this study were negative and low. The work of Chaffin and his colleagues [4,5] suggests that static strength tests are predictive of the incidence of low back injuries. While the findings of this study do not make it possible to predict the relationship between dynamic trunk strength and incidence of low back injuries, they do suggest that caution is advised.

The above findings suggest the possibility that even though the low back is the most often injuried component of the lifting chain, it is not the weak link. It is possible that weakness in another area of the lifting chain results in an overuse of the back, which then results in injury. For example, the strength of the shoulder muscles have been shown to be highly predictive of static tray lifting strength [10]. An inability to generate sufficient force from the shoulders could result in more stress on the back and legs and increase the chance of injury. Given the above findings we believe that one should be cautious in the use of dynamic trunk strength in the prediction and prevention of low back injuries.

This study has several limitations. The subjects were college graduate students, not industrial workers, and the number of subjects was low. In addition, the methods used to test dynamic back strength may also have been a factor. The Kin-Com back testing unit tests the subject in a sitting position, hence the pelvis is stabilized so that only the trunk extensor and flexor muscles were used during the activity. This is logical if one desires to truely know the strength of the flexors or extensors alone.

However, posture has been shown to influence back extension torque [11]. Langrana and Lee [8] found that standing extension strength was 18% greater than sitting extension strength while standing flexion strength exceeded sitting flexion strength by 97%. Since the static strength tests are performed while standing, it is possible that the correlations were influenced by testing dynamic strength while sitting. Cybex trunk testing units measure strength with the subject standing. It remains to be seen whether the same relationships hold for this unit.

A survey of the literature suggests that the subjects of this study were stronger than in other studies [11,12,13,14]. For example, literature values for trunk extension at 30 degrees/second range from a low of 176 Nm [11] to a high of 253 Nm [8]. Subjects in this study generated an average torque of 359 Nm at the same speed. Similar differences exist at higher velocities of movement as well. The differences cannot solely be attributed to testing in the sitting versus standing position. Langrana and Lee [8] reported that back extension in the standing position exceeded that produced in the sitting position. However, the relationship between back extension torque and velocity was similar between this study and others.

The ratio of eccentric torque to concentric torque ranged from 1.17 to 1.24 for the trunk flexors and 1.12 to 1.18 for the trunk extensors for this study. These values are very similar to the 1.24 and 1.20 reported by Reid and Costigan [15] for the trunk flexors and extensors, respectively. Eccentric contractions have recently been shown to have advantages over concentric contractions when diagnosing anterior pain in the knee [16]. A concentric isokinetic evaluation of the knee revealed no differences between the involved and noninvolved side. However, the ratio of eccentric torque to concentric torque was less than one for the involved side while the non-involved side was always greater than one. It needs to be determined if a similar relationship exists in patients with low back pain.

The ratio of the trunk extensors to flexors was considerably larger for this study than those reported previously [13,15]. Reid and Costigan reported a concentric value of 1.32 (at a speed of 25 degrees/second) and an eccentric value of 1.29. The values found for these subjects were 2.29 and 2.12 for concentric and eccentric contractions, respectively. A portion of this difference can be explained by the testing position. When sitting, the trunk extensors are stretched as compared to a shortened position for the trunk flexors. Therefore, it is likely that the sitting position shortened the trunk flexor muscles so that they generated less torque compared to the extensor muscles.

4. CONCLUSIONS

Within the limitations of this study the data presented here suggest that dynamic trunk extension and flexion strength is poorly correlated with static strength tests commonly used for preemployment screening. When tested in the sitting position, the ratio of trunk extensor to trunk flexor strength ranges from 2.34 to 1.77; this ratio decreases with increasing speed of movement. In normal subjects, eccentric strength exceeds concentric strength for both the trunk extensors and flexors (range from 12% to 24%).

ACKNOWLEDGEMENTS

This project was supported by grant RO3 OHO2229 from the National Institute for Occupational Safety and Health of the Centers for Disease Control and by a Graduate Research Council Grant (UHSC: 120-85) from the University of Louisville.

REFERENCES

[1] Kosiak, M., Aurelius, J.R., and Harfield, W.F., Backache in Industry, J. Occup. Med., 1967, 8, pp. 51-58.

[2] Magora, A., Investigation of the Relation between Low Back Pain and Occupation. Part I, Ind. Med. Surg., 1970, 39, pp. 21-37.

[3] Rowe, M.L., Low Back Pain in Industry: A Position Paper, J. Occup. Med., 1969, 11, pp. 161-169.

[4] Chaffin, D.B., and Park, K.S., A Longitudinal Study of Low Back Pain as Associated with Occupational Lifting Factors, Am. Ind. Hyg. Assoc. J., 1973, 34, pp. 513-525.

[5] Chaffin, D.B., Herrin, G.D., Keyserling, W.M., and Foulke, J.A., Pre-employment Strength Testing in Selecting Workers for Materials Handling Jobs, NIOSH Technical Report, DHEW(NIOSH) Publication No. 77-163, 1977.

[6] Kamon, E., Kiser, D., and Landa-Pytel, J., Dynamic and Static Lifting Capacity and Muscular Strength of Steelmill Workers, Am. Ind. Hyg. Assoc. J., 1982, 43, pp. 853-857.

[7] Kroemer, K.H.E., An Isoinertial Technique to Assess Individual Lifting Capability, Human Factors, 1983, 25, pp. 493-506.

[8] Langrana, N.A., and Lee, C.K., Isokinetic Evaluation of Trunk Muscles, Spine, 1984, 9, pp. 171-175.

[9] Ayoub, M.M., Bethea, N.J., Deivanayagam, S., Asfour, S.S., Bakken, G.M., Liles, D., Mital, A. and Sherif, M. Determination and modeling of lifting capacity. Final Report, HEW (NIOSH) Grant No.5 R01 OH-00545-02, September, 1978.

[10] Yates, J.W., Kamon, E., Rodgers, S.H., and Champney, P.C., Static Lifting Strength and Maximal Isometric Voluntary Contractions of Back, Arm and Shoulder Muscles, Ergonomics, 1980, 23, pp. 37-47.

[11] Gallagher, S., Marras, W.S., and Bobik, T.G., The Function of Trunk Muscles during Back Strength Exertions in Standing and Kneeling Postures, in: Asfour, S.S., (ed.), Trends in Ergonomics /Human Factors IV, (North-Holland, Amsterdam, 1987) pp. 27-34.

[12] Hause, M., Fujiwara, M., and Kikuchi, S., A New Method of Quantitative Measurement of Abdominal and Back Muscle Strength, Spine, 1980, 5, pp. 143-148.

[13] Davies, G.J. and Gould, J.A., Trunk Testing Using a Prototype Cybex II Isokinetic Dynamometer Stabilization System, J. Ortho. Sports Phys. Ther., 1982, 3, pp.164-170.

[14] Thompson, N.N., Gould, J.A., Davies, G.J., Ross, D.E., and Price, S., Descriptive Measures of Isokinetic Trunk Testing, J. Othro. Sports Phys. Ther., 1985, 7, pp. 43-49.

[15] Reid, J.G. and Costigan, P.A., Trunk Muscle Balance and Muscular Force, Spine, 1987, 12, pp. 783-786.

[16] Bennett, J.G. and Stauber, W.T., Evaluation and Treatment of Anterior Knee Pain Using Eccentric Exercise, Med. Sci. Sports Exer., 1986, 18, pp. 526-530.

Trends in Ergonomics/Human Factors V
F. Aghazadeh (Editor)

THE EFFECT OF STARTING POSITION AND SPEED OF MOVEMENT ON MAXIMUM STRENGTH IN DYNAMIC STRENGTH TESTING

Ehsan Asoudegi

Department of Industrial Engineering
Northern Illinois University
DeKalb, IL 60115-2854

Terrence Stobbe and Majid Jaraiedi

Department of Industrial Engineering
West Virginia University
Morgantown, WV 26506-6101

Dynamic strength testing is another dimension in the field of pre-employment screening. Most manual work requires the use of dynamic strength. That is to say, most industrial work is not static, but requires dynamic qualities for such activities as lifting, pushing, and carrying. Because velocity and acceleration are involved in these activities, these factors must be measured and their effects on human strength determined. The accuracy of dynamic strength testing depends on dynamic factors such as speed of movement and starting position. This paper examines the effect of starting position and speed of movement on the maximum dynamic strength.

1. INTRODUCTION

1.1 Background

Work injuries are a major source of incapacitation, suffering and cost to workers as well as expense to employers. The National Safety Council (1985) in <u>Accident Facts</u> estimated that in 1984 the workforce of the United States experienced 1.9 million work-related injuries. For government, industry, insurance companies, labor, and researchers, the reduction of work injuries remains a significant concern.

Previously, injuries have been prevented through such practices as careful worker selection, good training procedures, and designing the job to fit the worker (Snook and Ciriello, 1972). The obvious means to reduce work injuries is to design a job without excessive lifting, lowering, pushing, pulling, carrying, holding, etc. Ergonomic job design is the primary way of preventing work injuries. Good job design is especially important because it offers a more permanent engineering solution by reducing workers' exposure to hazards, which consequently reduces medical and legal worker-selection problems and makes it easier to select appropriate replacement for absent workers. Designing the job to fit the worker can reduce injuries by as much as one-third according to Snook, 1978. Thus, although job design is not a complete solution, it is significantly more effective than merely selecting the worker for the job or training the worker to fit the job. However, good job design requires a knowledge of worker capabilities and limitaions, and, according to Snook, Campanelli and Hart, 1979, careful selection and training of workeres should also be used.

Automation and the mechanization of jobs have decreased human strength demands somewhat, but manual power is still the primary energy source in many occupations. For example, jobs requiring muscular strength, such as materials handling or maintenance work, often are not practical to automate totally. Workers whose strengths are not well matched to the muscular requirements of their jobs are at greater risk of sustaining injuries than their co-workers who are better matched (Chaffin, 1975; Chaffin et al., 1977; Snook et al., 1979). When manual handling tasks require more strength or endurance than a worker can exert without excessive stress, injuries occur. Chafin et al. (1978) concluded that a systematic strength assessment and job placement program can significantly reduce injuries. Matching the requirements of the job and the muscular attributes of worker would benefit not only the worker, but also industry.

1.2 PRE-EMPLOYMENT STRENGTH TESTING

The assessment of human strength has been of considerable interest to ergonomists and researchers. Many questions have been raised regarding what type strength tests are effective and how many tests should be performed.

Chafin et al. (1978) investigated the potential effectiveness of pre-employment strength testing in reducing the incidence and severity of musculoskeletal and back problems in materials handling jobs. In the study, strength tests were given to 551 employees before assignment to new jobs, where they were monitored for 18 months. Chaffin found that a worker's liklihood of sustaining a back injury or musculoskeletal illness increases when job lifting requirements approach or exceed the isometric strength capability demonstrated by the individual in a simulation of the job. The study suggests that industry should implement pre-employment programs based on the strength performance criterion.

Traditionally, pre-employment strenth-testing programs designed to reduce work injuries have consisted of isometric (static) testing. Statis strength is defined as "the maximal force muscles can exert isometrically in a single voluntary effort" (Roebuck, Kroemer, and Thompson, 1987). A strength-testing standard recommending the use of static tests for measurement of human strength was developed in 1972 (Caldwell, et al., 1974) and adopted several years later as an "Ergonomics Guide for the assessment of Human Static Strength" by the American Industrial Hygiene Association (Chaffin, 1975). However, since people work dynamically, isometric tests may not provide accurate information on industrial work because a static test can not truly simulate a dynamic task. For example, the lifting of a 25 pound box from the floor to waist level is not static because moving the box requires acceleration. For this manual task, static testing is not a precise form of evaluation because acceleration changes the true load, and the body movement changes the worker's posture and the contribution of the different muscle groups.

The alternative for providing a more accurate measure of strength is dynamic testing. By simulating the dynamic tasks required in industrial work, dynamic strength testing may provide a better measure of the strength needed to perform the tasks. Dynamic testing can provide sensitive detection of muscle weakness specific to some part of the range of motion or some functional contraction speed. By creating a better simulation, such as that provided by dynamic strength testing, allows more accurate matching between the force requirements of the job and the worker's strength. This match would result in fewer injuries and in the end, fewer expenses for industry in jobs requiring human strength.

The accuracy of dynamic strength testing depends on dynamic factors such as speed of movement and starting position, and control of subject posture (which is fixed during isometric strength testing). This paper examines the effect of starting position and speed of movement on the maximum dynamic strength through an extensive isokinetic study of dynamic knee movement. The knee was chosen because its uniaxial planer simplifies measurements. Also, measurements of knee strength are more repeatable than measurements of back strength, and the knee is less susceptible to injury than the back.

2. METHOD

2.1 Subjects

Ten healthy adult male subjects consisting of students and faculty at West Virginia University volunteered to participate in this study. Their ages ranged from 20 to 36 years, with a mean of 26.6 years. Their heights ranged from 66 to 74 inches, with a mean of 69.40 inches. Their weights ranged from 120 to 185 pounds, with a mean of 158.60 pounds. Subjects with similar physical backgrounds (without athletic training) were chosen to eliminate possible muscle-training effects on the measured strength.

Eliminating large variations in the subjects' anthropometric characteristics increases the homogeneity of the data. Therefore, when possible, subjects with similar anthropometric characteristics were selected. This allowed a consistent distance from the center of knee rotation to the point at which the KIN-COM arm contacted the subject's knee (moment arm). Keeping the moment arm reasonably constant minimized the effect of mechanical advantage on the results of this study.

2.2 Testing Device (KIN-COM)

The Kinetic Communicator Exercise System is a hydraulically-driven, microcomputer-controlled dynamometer for the test, measurement, and rehabilitation of human joint function. The KIN-COM user performs a movement, or a series of movements, against a resistance that the machine provides via a rotating lever-arm system. The machine-controlled movement modes include isokinetic, semi-isotonic (lever-arm speed is continually adjusted to maintain constant resistance), and passive joint movement. The unit can induce concentric, eccentric, or isometric contractions of the muscles involved. The KIN-COM is controlled through feedback loops that monitor the position and speed of the lever arm and the force being exerted by the user. A strain gage bridge is used to measure force. A bar-encoded shift measures position and speed. This machine compares favorably with other dynamometers such as the CYBEX II. For most clinical and muscle-strength research applications, the KIN-COM appears to provide acceptable and valid measurements, and it is at least as accurate as other available dynamometers (Asoudegi, 1987).

2.3 Testing Procedure

After completion of a health questionnaire, each subject was examined for left knee muscular disorder by a physical therapist. Subjects with past knee injuries or cardiovascular problems were excluded from the study. A consent form was signed by these subjects who were qualified by the physical therapist to participate in this knee study.

Before starting the experiment, each of the subjects, was acquainted with the equipment and experimental procedure. To start the exercise, the subjects were positioned supine with the pelvis strapped to the table of the exercise unit. Each subject was required to perform a minimum of nine tests representing all possible combinations (Figure 1) of speed (0 degree/second, 60 degrees/second, 120 degrees/second)

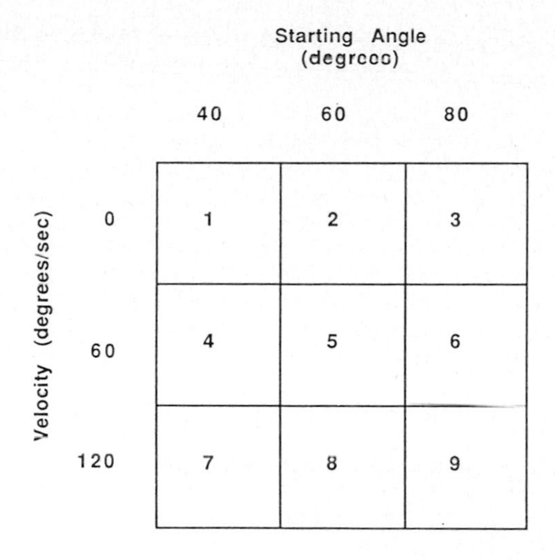

Figure 1 Strength testing combination.

and starting position (40 degrees, 60 degrees, 80 degrees of knee flexion; Figure 2). The sequence of these combinations was randomized for each subject. Each test consisted of four consecutive maximal concentric and eccentric knee movements resisted by the KIN-COM dynamometer.

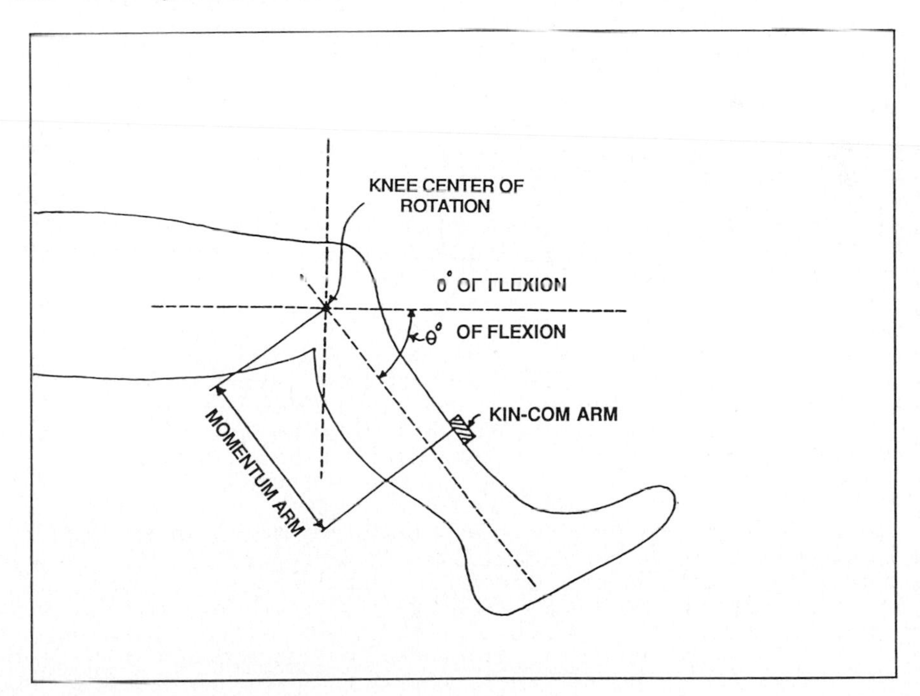

Figure 2 Side View of Subject's Knee During the Test

To eliminate the effect of fatigue, each subject was given a three minute rest period between each test. During each test, subjects were instructed verbally to exert as much force as they could. No feedback was given regarding their performance.

2.4 Experimental Design

The objective of this study was to investigate the effect of starting angle and speed of motion (independent variables) on maximum observed torque. To accomplish this, an orthogonal polynomial model was used with the speed of movement and the starting angle as the two independent variables and the maximum torque as the dependent variable.

3. DATA ANALYSIS AND RESULTS

To test for the effect of the starting position and speed of movement on the maximum torque, the following polynomial model was proposed:

$$y_i = \beta_0 + \beta_1 x_{1i} + \beta_2 x_{2i} + \beta_3 x_{1i} x_{2i} + \beta_4 x_{1i}^2 + \beta_5 x_{2i}^2 + \varepsilon_i \quad (1)$$

Where:

$\beta_\emptyset, \beta_1, \beta_2, \beta_3, \beta_4, \beta_5$ are parameters
Dependent variable Y represents the maximum observed torque (newton x meter)
Independent variable X_1 represents the speed of movement (degree/second)
Independent variable X_2 represents the starting angle (degree)
ε_i is the random error term $N(\emptyset, \sigma^2)$

The two independent variables X_1 and X_2 were deliberately selected at equispaced levels so that the proposed model could be transformed to an orthogonal polynomial model.

The maximum torque for each subject during strength testing is recorded for each combination of starting angle (4Ø degrees, 6Ø degrees, 8Ø degrees) and speed of movement (Ø degree/second, 6Ø degrees/second, 12Ø degrees/second).

Since X_1 and X_2 are at equispaced levels, the polynomial model (1) can be transformed to an orthogonal polynomial model such as:

$$y_i = \alpha_0 + \alpha_1 z_1(x_{1i}) + \alpha_2 z_1(x_{2i}) + \alpha_3 z_1(x_{1i}) z_1(x_{2i}) + \alpha_4 z_2(x_{1i}) + \alpha_5 z_2(x_{2i}) + \varepsilon_i$$

Where $Z_1(X_1)$ and $Z_1(X_2)$ are polynomials of degree one in X_1 and X_2. Also $Z_2(X_1)$ and $Z_2(X_2)$ are polynomials of degree two in X_1 and X_2. The results of statistical analysis, shown in table 1, were used to evaluate a series of null hypotheses.

Table 1

Results of orthogonal polynomial analysis for the effect of starting position and speedof movement on the maximum strength.

Source of Variation	SS	df	F Ratio (Phase I)	F Ratio (Phase II)	Regression Coefficients	Estimated Regression Coefficients
(Intercept)	5201775	1	2015.30	2037.18	a_0	240.41
(Speed of movement)	10073	1	3.92	3.95	a_1	-11.60
(Starting Angle)	120064	1	46.51	47.02	a_2	44.73
(Speed of Movement) (Starting Angle)	1882	1	.70	-	a_3	-8.40
$(\text{Speed of Movement})^2$	2960	1	1.15	-	a_4	3.78
$(\text{Starting Angle})^2$	576	1	.22	-	a_5	-11.70
Error	214235	83	-	-		
Total	5551505	89	-	-		

These hypotheses were evaluated in two phases. In the first phase, only the following null hypotheses were tested:

$$H_0: \; j - \emptyset \qquad \text{for } j = 3,4,5$$

Against alternative hypotheses:

$$H_A: \; j \neq \emptyset \qquad \text{for } j = 3,4,5$$

From the results in Table 1, we conclude that the parameters α_3, α_4, and α_5 are statistically insignificant.

The additive property of an orthogonal polynomial makes it possible to pool the sum of squares for these parameters with the error sum of squares without recomputing the coefficient of significant terms. Also, the degree of freedom for each parameter can be added to the degree of freedom for the error term.

In the second phase, the following null hypotheses were tested:

H_O: $_j$ = Ø for j = Ø,1,2

Against alternative hypotheses:

H_A: $_j$ ≠ Ø for j = Ø,1,2

Results from the second phase indicate that parameters α_0,α_1,and α_2 are statistically significant (P<.Ø5).

From the tested hypotheses, it can be concluded that speed of movement and starting angle have a significant effect on the maximum observed torque.

4. SUMMARY

Determining the effect of the starting angle and speed of movement on maximum measured strength was a major objective of this study. The starting position determines which part of the range of motion needs to be measured to determine a subject's maximum strength.

An orthogonal polynomial model with maximumm strength as its dependent variable and starting angle and speed of movement as its independent variables was used to test for the effect of starting position and speed of movement on maximum strength. This study showed that the speed of movement and starting angle significantly affect the maximum measured strength.

The resulting polynomial model indicates that the maximum measured strength increases as the starting angle increases. Also, the results show that the maximum measured strength increases as the speed of movement decreases, with the highest value at a speed of Ø degree/second (isometric). If isometric strength testing is used to assess human strength during dynamic activities, it will tend to over-estimate the subject's strength. This over-estimation may be a serious problem for a person assigned to a dynamic job based on isometric strength tests. The worker is at an increased risk of injury while performing the job because his true strength capability (dynamic strength) is lower than the measured isometric strength.

5. REFERENCES

1. National Safety Council, Accident Facts, Chicago, Illinoia, 1985.

2. Snook, S.H. and Ciriello, V.M., "Low Back Pain in Industry." Amer. Soc. of Safety Eng. J., 17(4):17-23, 1972.

3. Snook, S.H., "The Design of Manual Handling Tasks." International Ergonomics Society Lecture - 1978, Bedfordshire, England, 1978.

4. Snook, S.H., Campanelli, R.A., and Hart, J.W., "A Study of Three Preventive Approaches to Low Back Injury." J. Occ. Med., 1979.

5. Chaffin, D.B., "Ergonomics Guide for the Assessment of Human Static Strength." Amer. Indus. Hyg. Assoc. J., 36(7):July, 1975.

6. Chaffin, D.B., Herrin, G.D., Keyserling, M.K., and Foulke, J.A., "Pre-Employment Strength Testing." NIOSH Technical Report, CDC-99-74-62, May, 1977.

7. Chaffin, D.B., Herrin, G.D., and Keyseling, W.M. "Pre-Employment Strength Testing." Journal of Occupational Medicine, 20:403-408, 1978.

8. Roebuck, J.A., Kroemer, K.H.E., and Thompson, W.G., Engineering Anthropometry Methods, John Wiley and Sons, New York, New York, 1975.

9. Caldwell, L.S., Chaffin, D.B., Dukes-Dobos, F.N., Kroemer, K.H.E., Lauback, L.L., Snook, S.H., and Wasserman, D.E., "A Proposed Standard Test Procedure for Static Muscle Strength Testing." J. of Am. Ind. Hyg. Assoc., 35:201, 1974.

10. Asoudegi, Ehsan, Comprehensive Study of Isokinetic Strength Testing, Ph.D. Thesis, West Virginia University, Morgantown, 1987.

Trends in Ergonomics/Human Factors V
F. Aghazadeh (Editor)

INFLUENCE OF WORKLOAD AND ANTHROPOMETRIC MEASURES ON MUSCULOSKELETAL COMPLAINTS

Ryszard PALUCH and Anna ŁAZOR

Institut of Organization and Management
Technical University of Wrocław
Wrocław, Poland

The subjective feeling of pain of the musculoskeletal system has been studied in a sample of operators from a departmenf of non-ferrous metals smelter. For determination of influence of work, age, body height and weight on appearance of pains the analysis of variance has been used. The main factor contributing to pain was found to be work load, and in a lesser degree age, body height and weight. Anthropometric characteristics "per se" contribute to pain in a minor degree but their influence on appearance of pain is considerable in case of a unsuitably designed work space.

1. INTRODUCTION

The causes of complaints and disorders of the musculoskeletal system are complex. Apart of genetic conditioning the causes initiating complaints are sought in external factors such as heavy physical work, constrained body postures, repeatability of movements, factors of physical environment (e.g. vibration, temperature) et.c.

Independently of profesional factors in the motoric system complaints are found caused by the wear of the organism and ageing of tissues. The frequency of their appearance increases with age and the degenerative processes begin about the 30 year of life. That phenomenon is distincly visible in lumbar spine and lower back complaints (Hedberg et al. 1981, Brinckmann 1985, Hettinger 1985, Koeller et. al 1986). Apart of work factors and age in the process of "wear and tear" of intervertebral discs and other elements of the motoric system there participates also the so-called internal or hidden load decided by anthropometric characteristics of the individual. The most important of them are body height and weight. Their magnitudes are not indifferent to the developed forces and moments acting upon the particular structures of the musculoskeletal system (Tichauer 1978, Klausen 1985, Dul and Hildebrandt 1987).

However, there are numerous reports on the lack of relationship between complaints of the musculoskeletal system and anthropometry (Pedersen et al. 1975, Ferguson 1976, Kuorinka and Viikari-Jantura 1982, etc.).

The aim the present study is to answer the question whether, apart of workload, such traits as age, body height and weight, contribute also to the increase of musculoskeletal complaints.

2. MATERIAL AND METHOD

The study was carried out on a sample of 87 male non-ferrous metals plant operators (age: $\bar{x}$ = 34,2, s = 11,1; body height: $\bar{x}$ = 173,8, s = 6,9; body weight: $\bar{x}$ = 76,0, s = 10,4).
The activities of the metallurgists under study included assistance in loading of furnaces, collecting samples, supervising the melting process, casting, collecting and packing of casts.

Local complaints of the musculoskeletal system have been estimated on the basis of subjective feeling of pain, in a five-degree scale. The operators were asked to rate their pain and mark it on a schematic diagram ot the human body, with 18 body regions specified. The examination were carried out twice: before and after work (Paluch and Piesiewicz 1986).

For determination of reasons of pain occurence in the particular body regions the analysis of variance has been used. The independent variables were: age, body height and weight, and work (the increase of pain felt between the first and second examination), and the dependent variable - pain. The materials were analysed in age samples equal in number (19-28, 29-32, 33-59 years), body height (160-170, 171-177, 178-195 cm), and body weight (57-74, 75-81, 82-105 kg); the traits were considered in three schemes respectively:

1. age, body height, work.
2. age, body weight, work,
3. body height, body weight, work.

In the analysis effects of the main factors and effects of interaction on intensity of feeling of pain were used. The significance of hypotheses ($p<0,05$) have been tested by means of Fisher's-Snedecor's F test.

3. RESULTS

For all the body regions significant relations have been found between analysed factors and the pain felt (Fig. 1). The main factor contributing to pain was work load, in a lesser degree age, body height and weight.

Work load, except of the buttock's region, is the main factor, and in the neck region, shoulders and calves the only factor explaining the causes of the of the discomfort felt by the operators. Age for forearms and feet, and body height for forearms and the buttock region as main factors, and age-body weight for thighs, and body height-body weight for the right forearm as effects of combined acting, are the pain causes "per se".
In the remaining significant relations anthropometric traits decide on pain together with work, and the trait which particularly often manifests itself in the interaction with work is body weight.

The results can be, in fact, reduced to the conclusion that the shorter, lighter and younger operators feel more discomfort than the taller, heavier and older ones.

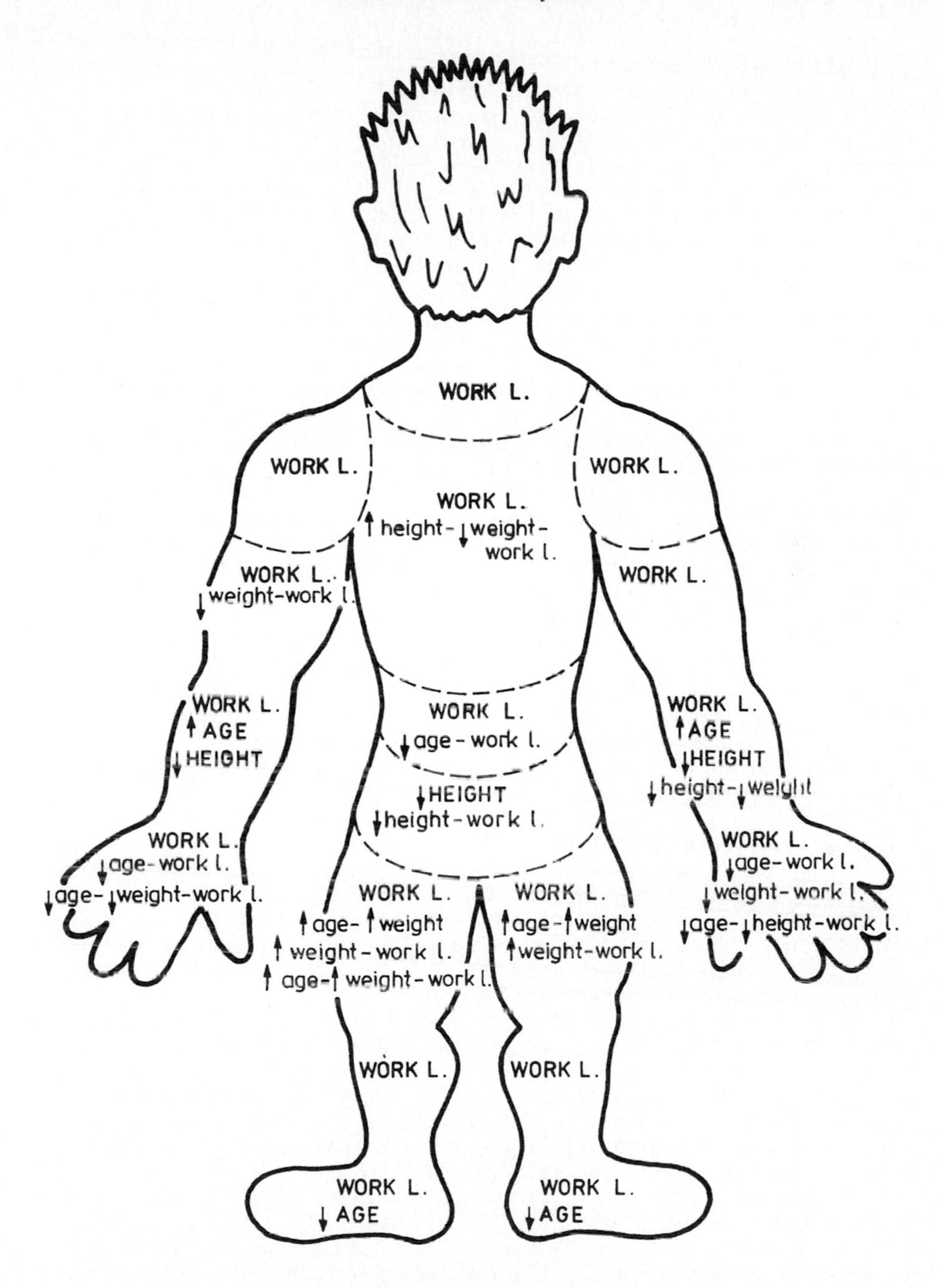

FIGURE 1
Main and interactional relations between the traits studied and pain in the musculoskeletal system.
↑Relation positive: pain increases with increasing trait value. ↓Relation negative: pain increases with decreasing trait value.

4. DISCUSSION

It is supposed that the extensity and intensity of pain related to anthropometric properties of the operators is influenced by work conditions, particularly activities including a considerable component of static work, as e.g. distant reaches, too high or to low working spaces and dynamic activities, as e.g. frequent bending. Particularly meladjustment of spatial conditions to body measures of the operator wil produce additional loads caused by a constrained posture.

Body height contributes to pain in upper limbs and trunk, not appearing as a factor of discomfort in lower limbs, whereas, body weight co-decides on pain in almost all the body regions. The increase of pain with lower body height in forearms, in the right hand and buttocks should be connected with distant reaches. Again, the increase of pain in the upper limb, with a lower body weight of the operator, results probably from the relation of lower force with lower muscle mass.

There is a significant increase with age of pain in forearms and thighs, and again a decrease in the lumbar region, in hands and feet. The intuitive reasoning that pains in the motoric system increase with age seems to be sufficient and, in general, is unopposed. However, the reverse phenomenon that pains decrease with age does not confirm with our experience and arises opposition. Silverstein et al. (1985) explain the negative relation with age und years of the job on hand wrist CTDs by selective processes. This hypothesis is somewhat doubtful because there is, above all, the problem why the selection concerns only the distal sections of extremities? It affects hands and not forearms, in which pain increases with age. Similarly in feet, where pain decreases with age, but shows an increasing tendency in calves, and in thighs a significant increase of pain is recorded in the older age group. This phenomenon may by explained by the worse working technique and lower skill in manual activities in younger operators (Karnes et al. 1986), and a greater "care" for the feet in older operators. It can be noticed that older workers - as far as possible - more often use chairs than younger ones. Also by some degenerative processes at distal parts of extremities.

The age distribution evidences that the group of operators under study is influenced by selective factors. More than a half (52 %) of the operators are placed in the age interval 18-30 years, and the remaining ones between 31-59 years. However, in practice, selection does not concern body height and weight. There are only slight shifts in the distributions toward a shorter stature and lower body weight. It is supposed, that selection acts rather toward elimination of individuals not resistant to termal stresses and considerable physical loads.

The analysis of variance has shown the complexity of factors deciding of pains in the motoric system. This complexity depends not only on simultaneous, significant acting of several factors but also on various interactions of these factors. E.g. in forearms, hands and thighs the pain is a resultant of numerous factors of various intensity and very complicated interaction configurations. This complexity is additionally intensified by the "anisotropy" of acting of anthropometric characteristics. A greater body weight may contribute to the increase of pain felt in a certain region and to the decrease in another region. Together with a lower body weight, and in interaction with other factors, there is an increase of pain in the upper limb (left arm, right forearm and both the hands). Again, together with a higher body weight there is an increase of pain in the left and right thigh. The case is more complicated when the anisotropy of factors (acting in interaction) influences pain and concerns correlated traits. E.g. together with higher stature but with a lower body weight there is an increase of pain in the pectoral section of the spine. In the population studied the linear correlation coefficient betwwen body height and body weight was r = 0,381.

It seems that in the light of mentioned above complex relationships "work-anthropometric characteristics-discomfort" such studies should be considered the results of which rendered no relation between work and discomfort of the motoric system (Pedersen et al. 1975, Kuorinka and Viikari--Juntura 1982), between age and discomfort (Chaffin and Park 1973, Pedersen et al. 1975, Kuorinka and Koskinen 1979) and between body height and weight and discomfort (Ferguson 1976).

Besides the divergences between "dose" and "effect" may by only apparent as effects of physical loads are modified by cumulative and diffuse effects (Grandjean 1980, Dainoff 1984), by mental stress caused by dissatisfaction of work (Van Wely 1970, Luopajärvi 1985) as by individual susceptibility of the operator.

FINAL REMARKS

Anthropometric characteristics "per se" make a minor causing for musculoskeletal complaints of the operators, and manifest itself as discomfort and pain only in disadvantageous working conditions.

The analysis of variance "separating" to some extent the pains of the musculoskeletal system, resulting from the process of work itself, from pains caused by maladaptation of the work place to individual characteristics of the operator, exposed the significance of proper designs of working space of the workstand.

The results presented here are not of universal value as they concern specific working conditions, as well as, definite anthropometric characteristics.

The analysis of variance revealed numerous, statistically significant relations between load and pain, not indicated by analysis of linear correlations (Paluch and Piesiewicz 1986). For example, despite of lack feeling discomfort between the right and left extremities, the analysis of variance has proved the existence of functional cross-asymmetry (right upper extremity and left lower extremity) in reactions to load.

REFERENCES

1. Brinckmann, P., Pathology of the vertabral column, Ergonomics, 28 (1985) 77-80.
2. Chaffin, D.B. and Park, K.S., A longitudinal study of low-back pain as associated with occcupational weight lifting factors, American Industrial Hygiene Association Journal, 34 (1973) 513-525.

3. Dainoff, M.J., Ergonomics of office automation - a conceptual overview, in: Proceedings of the 1984 International Conference on Occupational Ergonomics, Toronto, II (1984) 72-80.

4. Dul, J. and Hildebrandt, V.H., Ergonomic guidelines for the prevention of low back pain of the workplace, Ergonomics, 30 (1987) 419-429.

5. Ferguson, D., Posture, aching and body build in telephonists, Journal of Human Ergology, 5 (1976) 183-1986.

6. Grandjean, E., Fitting the task to the man, Taylor and Francis Ltd., London, (1980).

7. Hedberg, G., Björkstén, M., Ouchterlony-Jonsson, E. and Jonsson, B., Rheumatic complaints among Swedish engine drivers in relation to the dimensions of the driber's cab in Rc engine, Applied Ergonomics, 12 (1981) 93-97.

8. Hettinger, T., Occupational hazards associated with diseases of the skeletal system, Ergonomics, 28 (1985) 69-75.

9. Karnes, E.W., Freeman, A. and Whalen, J., Engineering work standards for warenhouse operations: effects performance ratings, age, gender, and neglected variables, in: Karwowski W., (ed.), Trends in Ergonomics/Human Factors III (North-Holland, Amsterdam, 1986) pp. 535-543.

10. Klausen, K., Physique and manual working capacity, Ergonomics, 28(1985) 99-105.

11. Koeller, W., Muehlhaus, S., Meier, W. and Hartmann, F., Biomechanical properties of human intervertebral disc subjected to axial dynamic compression - influence of age and degeneration, Journal of Biomechanics, 19(1986) 807-816.

12. Kuorinka, I. and Koskinen, P., Occupational rheumatic diseases and upper limb strain in manual jobs in a light mechanical industry, Scandinavian Journal of Work, Environment and Health, 5, suppl. 3, (1979) 39-47.

13. Kuorinka, I. and Viikari-Juntura, E., Prevalence of neck and upper limb disorders (NLD) and work load in different occupational groups. Problems in classification and diagnosis, Journal of Human Ergology, 11 (1982) 65-72.

14. Luopajärvi, T., Interaction of work load and functional capacity, in: Proceedings of the 9th Congress of the IEA, Bournemouth, (1985) 955-957.

15. Paluch, R., and Piesiewicz, A., Relationships between subjective symptoms and objective work load, in: Karwowski, W., (ed.), Trends in Ergonomics/Human Factors III (North-Holland, Amsterdam, 1986) pp. 1185-1193.

16. Pedersen, O.F., Pedersen, R. and Staffelt, E.S., Back pain and isometric back muscle strength of workers in a Danish factory, Scandinavian Journal of Rehabilitation Medicine, 7 (1975) 125-128.

17. Silverstein, B., Fine, L., Armstrong, T., Joseph, B., Buchholz, B. and Robertson, M., Cumulative trauma disorders of the hand and wrist in industry, in: Proceedings of the 9th Congress of the IEA, Bournemouth, (1985) 955-957.

18. Tichauer, E.R., The biomechanical basis of ergonomics, (John Wiley and Sosn, Inc., 1978).

19. Van Wely, P., Design and disease, Applied Ergonomics, 1(1970)262-269.

Trends in Ergonomics/Human Factors V
F. Aghazadeh (Editor)

EFFECTS OF HANDLE LENGTH AND BOLT ORIENTATION ON TORQUE STRENGTH APPLIED DURING SIMULATED MAINTENANCE TASKS

S. DEIVANAYAGAM and Tom WEAVER

Department of Industrial Engineering
Tennessee Technological University
Cookeville, TN 38505

This paper will describe an experimental investigation undertaken to study the torque strength capabilities of individuals in simulated maintenance task conditions. The primary objective was to assess the variability in torque strength due to length variations of the wrench handle.

1. INTRODUCTION

Historically, the supportability and maintainability considerations have had a low priority during the development stage of weapon systems. Most often, these considerations have been postponed to the later stages of design and development of the system. Therefore, any maintainability related deficiencies in the design generally become obvious only late in the life cycle of the system. At that point in the system's life, considerable financial and other resources are spent to rectify the deficiencies. Alternately, no change is made in the design in order to save this cost. The result of not changing the design usually results in a higher total cost of the system expended during maintenance and support functions. For example, estimates show that approximately 35 percent of the total lifetime cost of weapon systems in the inventory of the United States Air Force is spent on maintenance and support functions alone.

The United States Air Force is currently making an effort to reduce the cost of maintenance and support without any reduction in the readiness and effectiveness of the weapon system. The weapon systems are required to have the maintainability considerations embedded within the design features of the system. If a quantitative approach is introduced to maintainability, the variables such as time, effort, and expense required to repair the system must be accounted for in this approach.

The Air Force Human Resources Laboratory and the Air Force Aerospace Medical Laboratory have initiated a program to develop a computer aided design tool. This tool is conceived to be in the form of a computerized biomechanical man model of a maintenance technician. The CREW CHIEF model, among other things, requires an ergonomic database describing the maintenance technicians' body size and strength capabilities in order to properly simulate these characteristics for modeling purposes. Since the use of a handtool such as a ratchet wrench is an important requirement for successful performance in a wide variety of maintenance tasks, a series of experimental investigations were conducted to gather data on maximum torque strength that a person can apply using a ratchet wrench.

2. EXPERIMENTAL STUDY

One particular phase of the study proposed the use of three ratchet wrenches of varying lengths. The relationship between torque exertions and wrench handle lengths was examined as an approximation to the classical definition. It is the belief of the experimentors that the linear relationship between torque values and increases in perpendicular distance from the axis of rotation does not hold true in real life situations due to task related as well as biomechanical variables. The primary objective of the experiment is to identify and analyze the observed differences between the actual and ideal relationships between torque values and the change in moment arms due to wrench handle lengths. An extensive experimental investigation was designed and conducted to investigate several aspects of the torque strength capabilities. This paper describes only that part of the study and the results relating to the objective mentioned above.

The following experimental variables and their respective levels were considered in the study.

A. Bolt Head Orientation - three levels; Vertical, Facing, and Transverse. Vertical bolt orientation was defined as the condition in which the bolt head was placed vertically up. Facing bolt orientation corresponded to the bolt head being placed horizontal with the axis of rotation of the bolt through the midsaggital plane of the body. Transverse bolt orientation was defined as the bolthead placed in a horizontal position with the axis of rotation of the bolt perpendicular to the midsaggital plane of the body.

B. Wrench Position - four levels; 0, 90, 180, 270 degrees. In all bolt head orientations the clockwise increasing of angles was maintained. For vertical bolt head orientation the 0 degree wrench position was defined as that position in which the handle extended farthest away from the subjects. For the facing and transverse bolt head orientations the 0 degree wrench position was defined as the position in which the handle extended vertically upwards.

C. Wrench Length - three levels; 5.5 inches, 7.5 inches, and 12.0 inches. This distance was measured from the center of the socket to the middle of the gripping point of the wrench handle. Three different sizes of commercially available "Snap-On" brand ratchet wrenches were used.
The comparisons of torque values were conducted at three wrench positions per bolt orientation thus producing nine comparisons. A total of 27 exertions per subject were required for the purpose of comparing the wrench length to torque values in these nine locations. Only clockwise torque was examined in the study.

During the torque session, the bolt head was positioned at a vertical height corresponding to 60 percent of the subject's vertical reach. Vertical reach was defined as the maximum height which a subject could reach with the right hand clenched while standing flat footed. Horizontal distance from the bolt head was maintained at 50 percent of the subjects grip length. Grip length was measured by placing two poles in the subject's hands and having the subject hold the two 3/4 inch diagonal, 4 feet long poles in a vertical orientation. The subject was instructed to move the right pole to the center line of the body while keeping the right arm extended. The left pole was brought in to the center of the chest and the horizontal distance between the poles was measured. During the test, the left foot was required to be at 50 percent of the grip length; however, the right foot could be placed up to two feet behind this line depending upon individual preference. The subject's body was centered about the bolt with the bolt head located along the midsaggital plane.

3. APPARATUS

A torque dynamometer with a strain gage load cell and a bridge amplifier was used to gather torque data. The load cell was attached to the bolt such that any torque exerted on the bolthead was sensed by the load cell. The output of the load cell was directly proportional to the force exerted on the bolthead and was amplified using a Honeywell Accudata bridge amplifier. The resultant analog voltage was fed to a microcomputer equipped with a PC Mate Techmar Analog/Digital Converter. The analog signal was sampled at the rate of 10 per second and was digitized. The results were printed immediately in tabular form along with a set of statistics for the experimentor to decide on the acceptability of the test. The accepted data was recorded on a floppy disk for further analysis by the experimentor. Figure 1 shows the schematics of the data acquisition system.

4. SUBJECTS

Six male and six female volunteers were selected for participation in the study. The subjects ranged in age from 20 to 29. All subjects were considered to be in good to excellent physical health and no one with a previous history of back injury, hernia, or other questionable health conditions was selected. A set of 22 anthropometric and strength measurements on each subject was taken during the first session of the experiment. All measurements were performed in accordance with the procedures mentioned in references Ayoub et. al.[1], Webb Associates [2], and McDaniel, et. al.[3]. The summary statistics of the anthropometric and strength data are given in Table I.

5. PROCEDURE

Torque exertions were divided into three sessions with one session per bolt orientation. Subjects were required to schedule sessions at least three days apart in order to exclude any effects of fatigue. All torque exertions were randomized for each session. Subjects took mandatory rest breaks of two minutes minumum between exertions.

The subject was required to exert his or her maximum torque on the bolt head with the wrench for a period of four seconds at the end of which time a buzzer sounded signalling the end of data collection. Torque values were sampled and digitized at 100 millisecond intervals. The first second was considered to be the time required to develop to the maximum strength. The subject was instructed to maintain this maximum torque for the subsequent three seconds. The average maximum torque was calculated taking into account only the torque values during the last three seconds.

The acceptability of the data resulting from a given exertion was evaluated by an automatic range calculation during the exertion. The average torque from one to four seconds plus 10% of the average was calculated as the upper limit of the range while the average torque minus 10% of the average was calculated as the lower limit of the range. A maximum of three sampled data points was allowed outside of this range. If it exceeded three data points, the exertion was not accepted and was repeated.

Another requirement concerned the ratio of the peak torque value during the first second and the mean torque from one to four seconds. The range of tolerance for this ratio was from 0.8 to 1.2. If the ratio was greater than 1.2, the subject had an instantaneous, non-smooth peak during the first second. If the ratio was less than 0.8, the subject had not achieved the maximum torque within the alloted time. Exertions were repeated up to two times. After the third exertion, the exertion closest to the given criteria was accepted.

6. RESULTS

Table II shows the torque strength values for the three bolt head orientations, three wrench lengths, and males and females averaged for the six subjects. The results indicate that while the torque strength values increased with increasing handle lengths, the increase is not linearly related. For example, Figures 2, 3, and 4 show the torque strength versus wrench handle length relationships at 180 degrees wrench positions for the three bolt orientations. The data for males and females are combined in these figures. The solid lines represent the actual data and the broken lines represent the ideal relationship if the torque strength at 7.5 inches length is taken as the reference. It can be seen that when the wrench handle length increased from 7.5 inches to 12 inches the resultant increase in torque strength falls short of ideal in all cases. This must be due to the fact that the longer hand requires the upper extremity to be placed in a less advantageous biomechanical configuration. The only deviation from this general trend is observed at facing bolt orientation and 180 degree wrench position. In this situation the decrease in torque strength that must have occured ideally due to the 2 inch reduction in handle length (comparing small to medium) did not happen. It is suspected that the shorter handle length brought the grip position just below the midchest height which is considered a preferred height for the application of maximum force in the right to left direction.

7. CONCLUSIONS

In conclusion, it can be said that the variation in wrench handle lengths, while affecting the torque strength values, does not seem to bear any direct relationship. As the number of subjects tested in this pilot study was small, we are not in a position to make definite conclusions. A more detailed study has been recently concluded with the objective of further investigating this relationship. The data is currently being analyzed.

REFERENCES

[1] Ayoub, M.M., et. al., Establishing Physical Criteria For Assigning Personnel To Air Force Jobs (Texas Tech University, Lubbock, Texas, September 1982)

[2] Webb Associates, Anthropometric Source Book, Volume I: Anthropometry for Designers, NASA Reference Publication 1024 (Yellow Springs, Ohio, July 1978)

[3] McDaniel, Joe W. et. al., Weight Lift Capabilities of Air Force Basic Trainees, AFAMRL-TR-83-0001 (Wright-Patterson Air Force Base, Ohio, May 1983)

[4] Deivanayagam, S., Development of Prediction Models For Human Torque Strength, Final Report Submitted to Southeastern Center for Electrical Engineering Education, St. Cloud, Florida, (The University of Texas, Arlington, Texas, 1985)

TABLE I. ANTHROPOMETRIC AND STRENGTH MEASUREMENTS

MALES			FEMALES	
AVERAGE	STANDARD DEVIATION		AVERAGE	STANDARD DEVIATION
106.67	9.43	INC WT LIFT 6 FT (LB)	53.33	11.06
177.90	5.00	STATURE (CM) 165.33	165.33	11.38
93.63	2.57	SITTING HEIGHT (CM)	87.40	4.51
146.67	36.99	LIFT HOLDS 40 LB (SEC)	51.00	18.67
92.70	3.60	TROCHANTRIC HEIGHT (CM)	88.40	6.46
48.65	2.41	TIBIALE HEIGHT (CM)	45.40	3.42
140.00	18.26	INC WT LIFT ELBOW HT (LB)	81.67	14.62
36.03	1.15	ACROMION-RADIALE LENGTH (CM)	31.90	1.95
26.33	1.22	RADIALE-STYLION LENGTH (CM)	23.52	1.98
166.67	19.72	INC WT LIFT KNUCKLE HT (LB)	121.67	16.75
35.22	1.36	BICROMIAL BREADTH (CM)	31.32	2.50
34.10	1.06	BI-TROCHANTRIC BREADTH (CM)	33.33	1.51
42.67	15.02	LIFT HOLDS 70 LB (SEC)	4.50	5.06
6.72	0.55	LATERAL MALLEOLUS HEIGHT (CM)	6.88	0.76
76.77	7.41	WEIGHT (KG)	61.37	6.46
83.33	4.41	ELBOW HEIGHT LIFT (LB)	40.17	6.59
60.33	2.82	GRIP LENGTH (CM)	49.78	2.84
215.32	5.95	VERTICAL REACH (CM)	198.15	12.69
113.33	18.17	ONE HAND PULL (LB)	62.83	14.92
270.50	70.65	38 CM VERTICAL LIFT (LB)	122.00	49.91
49.00	7.06	GRIP STRENGTH RT (KG)	28.54	4.79
48.88	7.67	GRIP STRENGTH LT (KG)	27.13	4.49

TABLE II. AVERAGE TORQUE STRENGTH FOR MALES AND FEMALES

VERTICAL						
Wrench Length	Small		Medium		Large	
sex	m	f	m	f	m	f
0°	-	-	-	-	-	-
90°	12.8	11.7	18.7	15.0	27.6	22.7
180°	13.4	10.1	20.0	14.6	26.8	21.4
270°	9.1	6.5	11.5	10.7	15.8	13.8
FACING						
Wrench Length	Small		Medium		Large	
sex	m	f	m	f	m	f
0°	11.5	9.0	16.3	12.1	22.5	17.7
90°	15.4	12.2	24.4	15.5	36.4	22.0
180°	14.0	9.7	15.8	12.0	23.3	17.3
270°	-	-	-	-	-	
TRANSVERSE						
Wrench Length	Small		Medium		Large	
sex	m	f	m	f	m	f
0°	10.0	8.8	14.0	12.7	21.6	19.3
90°	-	-	-	-	-	-
180°	13.5	12.7	25.8	17.0	35.6	25.5
270°	20.4	13.5	34.1	20.9	45.6	26.9

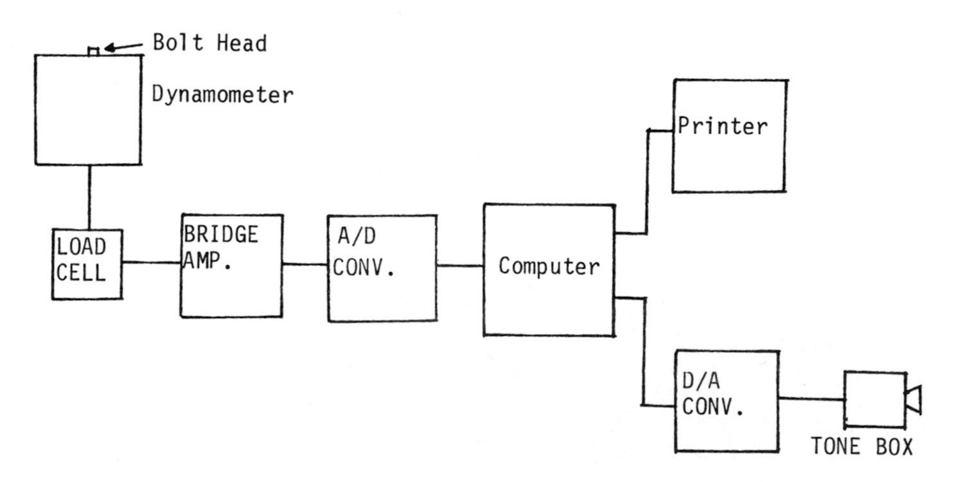

FIGURE 1
SCHEMATICS OF THE DATA ACQUISITION SYSTEM

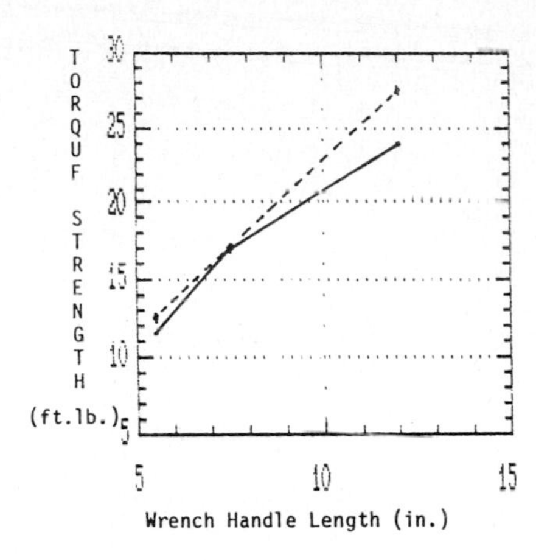

FIGURE 2

Wrench Length vs Torque Strength
Vertical Bolt Orientation 180°

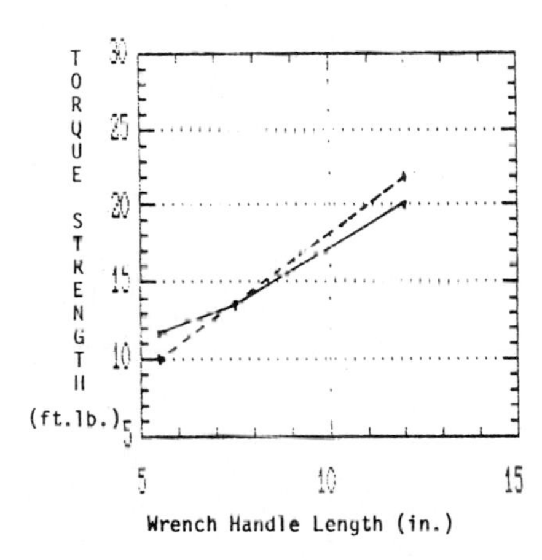

FIGURE 3

Wrench Length vs Torque Strength
Facing Bolt Orientation 180°

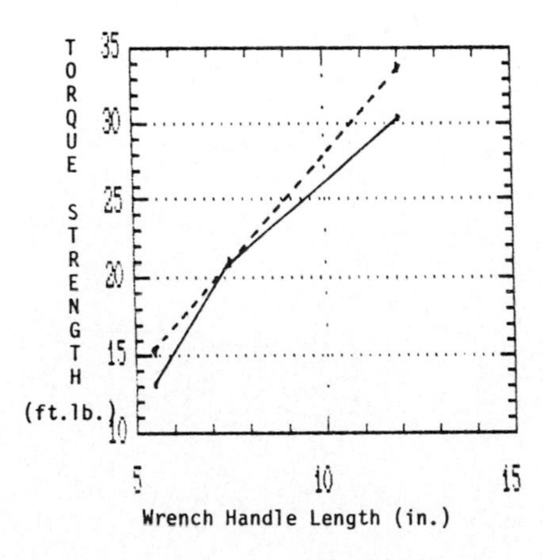

FIGURE 4

Wrench Length vs Torque Strength
Transverse Bolt Orientation 180°

XII

MANUAL MATERIALS HANDLING

Trends in Ergonomics/Human Factors V
F. Aghazadeh (Editor)

THE PSYCHOPHYSICAL APPROACH: THE VALID MEASURE OF LIFTING CAPACITY

Jeffrey E. FERNANDEZ
Industrial Engineering Department
The Wichita State University
Wichita, Kansas 67208

M. M. AYOUB
Industrial Engineering Department
Texas Tech University
Lubbock, Texas 79409

A laboratory experiment was conducted to study the differences in lifting capacity determined by the physiological and the psychophysical approaches, and to study the effect of psychophysical lifting capacity when measured before and after the eight hour lifting sessions. The psychophysical lifting capacity, physical work capacity, and physiological lifting capacity were determined of twelve male subjects. The subjects handled the load for several eight hour sessions, after which they again estimated their psychophysical lifting capacity. Results indicate that the physiological lifting capacity overestimates the psychophysical lifting capacity at the low frequencies, but in fact underestimates the psychophysical lifting capacity at the high frequencies. Psychophysical lifting capacity before and after the eight hour sessions showed slight but insignificant increases.

1. INTRODUCTION

The psychophysical, physiological and biomechanical approaches are used to measure individual's lifting capacity. Research over the last two decades reveals numerous factors influencing lifting capacity, and limitations, advantages and disadvantages of these approaches. Presently there seems to be an ongoing controversy of the validity of the psychophysical approach.

The psychophysical approach estimates an individual's lifting capacity by quantifying their subjective tolerance to the stresses of manual material handling [1]. The physiological approach is concerned with the physiological stresses on the body, and the biomechanical approach determines the forces imposed upon the musculoskeletal system.

Comparing the psychophysical and the physiological approaches Garg and Ayoub [2] concluded that the lifting capacity estimated by the physiological approach overestimated the lifting capacity estimated by the psychophysical approach at the lower frequencies but underestimated at the higher frequencies. However, Mital [3] concluded that the lifting capacity estimated by the physiological approach overestimated the lifting capacity estimated by the psychophysical approach over the frequency range of 1 to 12 lifts per minute as shown in figure 1.

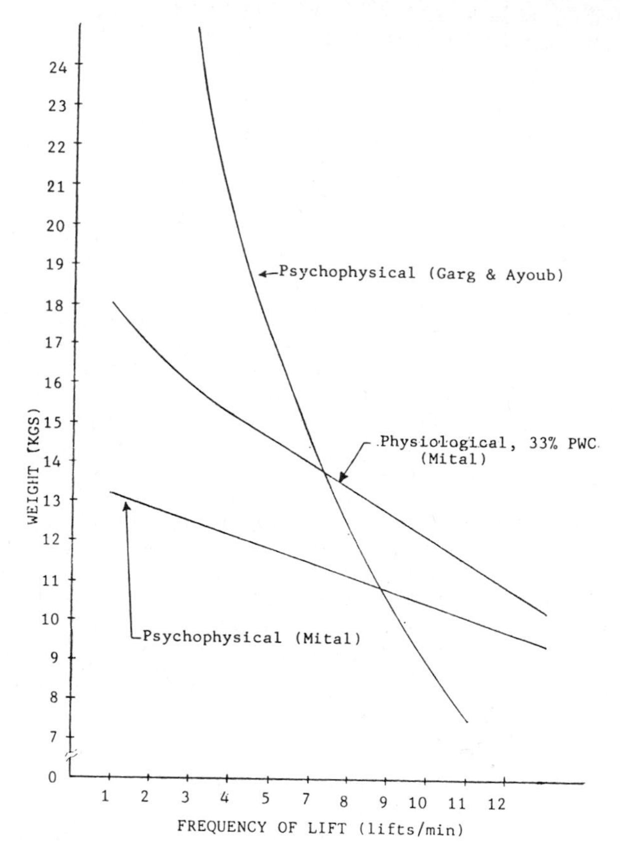

Figure 1. Comparison of the psychophysical and physiological lifting capacity as concluded by Garg and Ayoub [2] and Mital [3].

Various studies have been conducted utilizing the psychophysical approach: studies that looked into the effect of varying some of the factors affecting lifting capacity, and verification studies that were conducted to confirm if the lifting capacity estimated by the psychophysical approach did, in fact, change when lifting for extended periods.

This study was undertaken to document any difference in lifting capacity determined by the physiological and the psychophysical approaches, and the effect of lifting capacity estimated by the psychophysical approach when measured before and after the eight hour lifting sessions.

2. METHODS AND PROCEDURES

2.1 Subjects

Twelve male subjects from the student body at Texas Tech University participated in this study. A summary of the subjects data is presented in Table 1. The subjects were financially reimbursed for participating in the experiment.

TABLE 1

Summary of the subjects data

	Mean	s.d.
Age (years)	24.75	3.98
Weight (pounds)	173.47	24.79
Height (inches)	70.64	1.79

2.2 Experimental Design

The experiment was divided into 5 parts. Experiment 1 was lifting for 25 minutes to determine the psychophysical lifting capacity, experiment 2 was determining the lifting physical work capacity (PWC), and experiment 3 was lifting using the physiological approach. Experiment 4 was lifting for eight hours, with and without weight adjustments. Lastly, experiment 5 was the repetition of experiment 1, but after experiment 4 was completed.

The experimental design was a randomized complete block design with subjects as blocks. The order of the tests were randomized for each subject within each experiment.

2.3 Familiarization Period

Each subject had to undergo a familiarization period. The familiarization period consisted of a 4 day, one hour per day program. During the first four days the subject was required to lift from floor to knuckle height using the psychophysical methodology. The duration of the lifting task was 40 minutes. The subject adjusted his load until he felt that the load represented his maximum weight of lift. The frequency of lift for the four day familiarization sessions was randomly selected from either 4 or 6 lifts per minute [4].

2.4 Psychophysical Lifting Phase (experiment 1)

The subjects were asked to lift from floor to knuckle height (30"), a 18 x 11.5 x 12" box with handles at 2 and 8 lifts per minute.

The starting weight was randomly set at relatively heavy or relatively light. Subjects were allowed to adjust the weight of load to the maximum that they could lift without strain, or discomfort, and without becoming tired, weakened, overheated or out of breath [5]. The adjustment period lasted for 25 minutes. The final weight at the end of the period was considered the "maximum acceptable weight of lift (MAWOL)".

2.5 Physical Work Capacity (PWC) Determination (experiment 2)

A submaximal test as described by Kamon and Ayoub [6] was used to reduce risks to the subjects. The subjects were asked to refrain from eating, smoking or consuming carbonated drinks during the lifting sessions. A mouthpiece was inserted in the subject's mouth and a nose-clip on his nose so that respiration would occur only through his mouth. The resting heart rate was measured with the subject resting in a sitting position. This was known as the Resting Heart Rate (RHR); the Maximum Heart Rate (MHR) was calculated by the formula 220-Age [7]. The Heart Rate Range (HRR) was then determined from MHR-RHR [8].

The PWC was calculated by extrapolating the regression line of oxygen consumption and heart rate at three steady states to predict the maximum oxygen consumption at maximum heart rate [6].

Oxygen consumption was assumed to be proportional to percentage HRR [9]. The protocol called for the second and third workload to be 50% and 65% of VO_2 maximum respectively, therefore the loads were adjusted so that 50% and 65% of HRR were achieved.

The lifting range used was floor to knuckle height; the lifting frequency was kept constant at 2, 6 and 8 lifts per minute. The oxygen consumption (liters per minute) and heart rate (beats per minute) were recorded throughout the test. Steady state had to be achieved before the next workload was applied; this was usually obtained in 4 to 5 minutes.

2.6 Physiological Lifting Phase (experiment 3)

In this section of the study the oxygen consumption was measured at various weight of lifts for the frequencies: 2 and 8 lifts per minute. The physiological approach states that an individual can work at 33% of his/her PWC for an eight hour period; therefore, knowing an individual's PWC, the load (weight of lift) was varied and oxygen consumption was measured at steady states. A pocket calculator was utilized to determine whether the load should be increased or decreased; after necessary interpolation, the weight at 1/3 PWC was determined.

2.7 Eight hour Lifting Phase (experiment 4)

The starting weight for the eight hour sessions was the MAWOL at 2 and 8 lifts per minute, determined in the 25 minute bouts (experiment 1). There were two variations in the eight hour sessions: the weight could be adjusted and the weight was onstant [4].

2.8 Repeat Psychophysical Lifting Phase (experiment 5)

This 25-minute bout experiment was conducted after the eight hour lifting sessions were completed, at 2 and 8 lifts per minute. The instructions and protocol was the same as the initial psychophysical lifting phase (experiment 1).

3. RESULTS AND DISCUSSION

A summary of the MAWOL and the lifting capacity using the 1/3 PWC criteria for the physiological approach is presented in Table 2. For the calculation of the physiological and psychophysical lifting capacity at 2 lifts per minute, the PWC at 2 lifts per minute was utilized; in the same manner, the PWC at 8 lifts per minute was utilized for the lifting capacity at 8 lifts per minute.

TABLE 2

Summary of the Psychophysical and Physiological Approaches in Determining Lifting Capacity

	Freq.	Lift. Cap.(lbs)		PWC	
	(1/min)	Mean	s.d.	Mean	s.d.
Psychophysical	2	53.75	13.14	31%	9.97%
Psychophysical	8	30.42	6.13	38%	6.53%
Physiological	2	67.38	33.02	33%	16.17%
Physiological	8	21.65	14.20	33%	21.37%

When the percentage PWC was calculated for the psychophysical lifting capacity using the PWC at 6 lifts per minute, instead of PWC at 2 and 8 lifts per minute, the percentage PWC was 25.8 and 40.9 at 2 and 8 lifts per minute respectively. Figure 2 shows the lifting capacity based on the psychophysical and physiological approaches. The lifting capacity based on the physiological approach overestimated the lifting capacity based on the psychophysical approach at 2 lifts per minute by 25.36%. However, at 8 lifts per minute, the lifting capacity based on the physiological approach underestimated the lifting capacity based on the psychophysical approach by 28.83%. This is contrary to Mital's [3] results which stated that the lifting capacity based on the physiological approach overestimated the lifting capacity based on the psychophysical approach over the frequency range of 1 to 12 lifts per minute. The results of the present study support the conclusions of Garg and Ayoub [2].

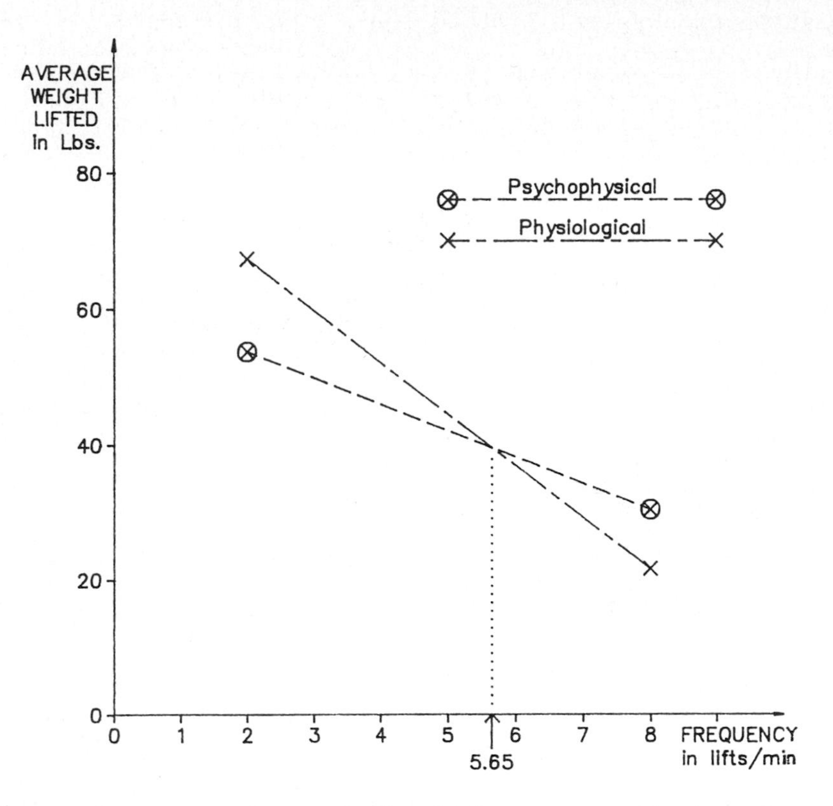

Figure 2. Comparison of the lifting capacity based on the psychophysical and physiological approaches

The straight line fit in Figure 1 is because only two frequencies were considered in this study. Actually there is no linear fit over the whole frequency range; at the low frequency from 0.1 to 2 lifts per minute the curve is negatively exponential, but from 2 to 12 lifts per minute it is approximately linear [10]. From Figure 1, at 5.65 lifts per minute the lifting capacity based on the two approach coincide, which is similar to the conclusions of Garg and Ayoub [2].

At the low frequency of 2 lifts per minute the physiological system is not stressed enough; therefore, the physiological approach is a gross overestimation of the lifting capacity. At low frequencies the biomechanical approach is more appropriate, while at high frequencies the physiological approach is more appropriate. Therefore, utilizing the psychophysical approach to estimate lifting capacity is appropriate over the entire frequency range when compared to utilizing the physiological or the biomechanical approaches.

Figure 3 shows the lifting capacity based on the psychophysical approach adjusted for the 8 hours, with the lifting capacity based on the physiological approach. The MAWOL utilized in this figure was the weight at the end of hour 8 at 2 and 8 lifts per minute. It is evident from Figure 3 that at the low frequency the physiological approach still overestimates the psychophysical approach; also, at the high frequency the physiological approach still underestimates the psychophysical approach. The frequency where both approaches estimate the same weight is 7.07 lifts per minute.

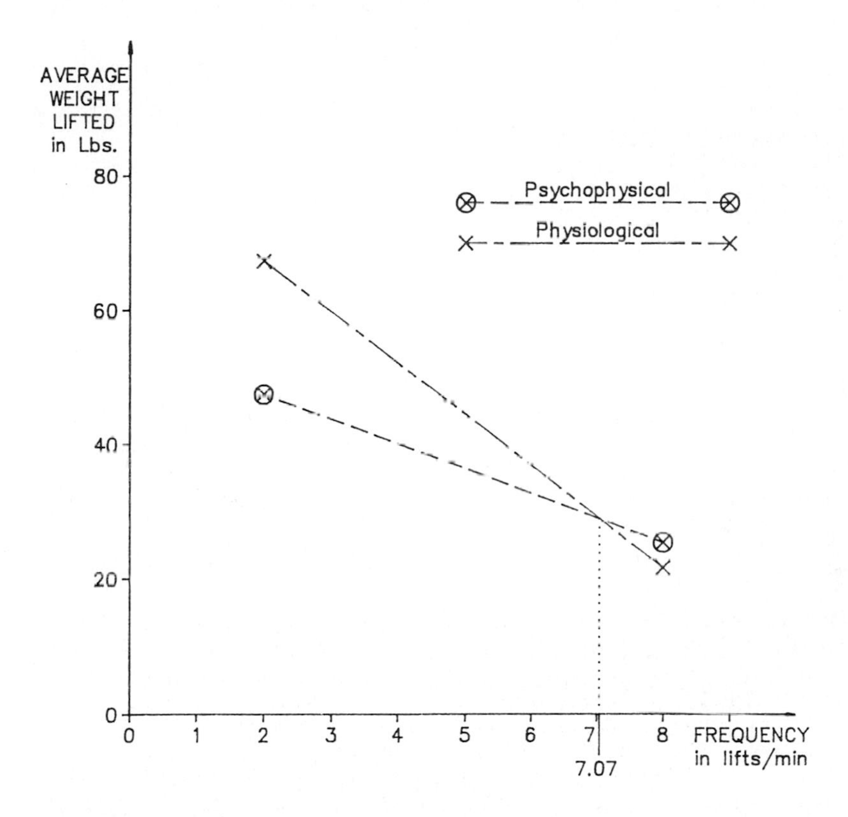

Figure 3. Comparison of the lifting capacity based on the adjusted psychophysical and physiological approaches

Table 3 presents a summary of MAWOL, before and after the 8 hour sessions. It is observed from the table that there was an increase of 0.54% for MAWOL at 2 lifts per minute and an increase of 4.17% for MAWOL at 8 lifts per minute. These increases were not significant at $\underline{p} < 0.05$. This indicates that the subjects were relatively accurate and comfortable with their initial estimates and did not find any need to make any significant adjustments to their repeated lifting capacity estimates (post MAWOL).

TABLE 3

Summary of MAWOL, before and after the 8 hour sessions (N=12)

	Mean	s.d.	Min. value	Max. value
S1	53.375	13.139	34.50	74.50
S2	30.417	6.133	21.50	39.25
PS1	53.667	8.636	37.50	65.25
PS2	31.687	4.609	22.25	36.75

where S1 = MAWOL, 2 lifts/min.
S2 = MAWOL, 8 lifts/min.
PS1 = Post MAWOL, 2 lifts/min.
PS2 = Post MAWOL, 8 lifts/min.

4. CONCLUSIONS

The following are the conclusions drawn from this study:

1. The physiological lifting capacity overestimates the psychophysical lifting capacity at the low frequencies, and in fact does underestimate the psychophysical lifting capacity at the high frequencies. The psychophysical approach seems to be a valid measure of lifting capacity across the lower and moderate lifting frequency range, however at the higher frequencies the physiological method appears to be a more reasonable approach.

2. Lifting capacity estimated by the psychophysical approach was relatively consistent as the subjects arrived at the same estimate in repeated trials. The psychophysical capacity estimated in an hour or more could be a better estimate of lifting capacity than the 25 minute bout [4].

REFERENCES

[1] Ayoub, M.M. (1987). The Problems of Manual Material Handling. In Asfour, S.S., (Ed.), Trends in Ergonomics/Human Factors IV, North-Holland, New York, NY.

[2] Garg, A., and Ayoub, M. M. (1980). "What Criteria Exist for Determining How Much Load can be Lifted Safely", Human Factors, 22(4), pp 475-486.

[3] Mital, A. (1985). "A Comparison between Psychophysical and Physiological Approaches Across Low and High Frequencies Ranges",Journal of Human Ergology, 14(2), pp 59-64.

[4] Fernandez, J. E. (1986). Psychophysical Lifting Capacity Over Extended Periods. Unpublished Ph.D. Dissertation, Texas Tech University, Lubbock, TX.

[5] Snook, S. H. (1978). "The Design of Manual Handling Tasks", Ergonomics, 21(2), pp 963-985.

[6] Kamon, E., and Ayoub, M. M. (1976). "Ergonomic Guide to Assessment of Physical Work Capacity", American Industrial Hygiene Association, Akron, OH.

[7] Astrand, P. O., and Rodahl, K. (1977). Textbook of Work Physiology, Second Edition, McGraw-Hill Book Co., New York, NY.

[8] Intaranont, K. (1983). Evaluation of Anaerobic Threshold for Lifting Tasks, Unpublished Ph.D. Dissertation, Texas Tech University, Lubbock, TX.

[9] deVries, H. A. (1980). Physiology of Exercise for Physical Education and Athletics, Third Edition, Wm. C. Brown Company Publishers, Dubuque, IA.

[10] Ayoub, M. M., Selan, J. L., Karwowski, W., and Rao, H. P. R. (1983). "Lifting Capacity Determination", Proceeding: Bureau of Mines Technology Transfer Symposia, Pittsburg, PA, and Reno, NV.

Trends in Ergonomics/Human Factors V
F. Aghazadeh (Editor)

SYSTEM COMPARISON OF TWO ADVANCED METHODS FOR MEASURING ANGULAR DISPLACEMENT OF TORSO

Issachar Gilad*, Don B. Chaffin+, Mark Redfern+ and Seong N. Byun+

*Faculty of Industrial Engineering & Management, Technion - Israel Institute of Technology, Haifa 32000, Israel.

+Center for Ergonomics, The University of Michigan, Ann Arbor, MI 48109, U.S.A.

A comparative study was conducted to evaluate two different methods used to measure trunk kinematics during the lifting task. The following two methods were compared in a laboratory study: 1) Opto-electronic detection method using a Selspot camer system. This system requires complex laboratory setup and needs a high level of technical expertise. 2) A new miniature electronic inclinometer method, which is based on a high Tech. detection electrolytic potentiometer, enabling direct measurement and is low in equipment and software requirements. Quantitative assessments of the trunk motion of male and female subjects while performing dynamic lifting are presented and discussed. The comparison revealed that with care in calibration the two systems displayed similar torso angle measurements for a large variety of test conditions. Cross correlation between the angle estimates show consistent agreement in angle trajectory over time at T5, less consistency was obtained at C4 and L5. Factors affecting the performance of the two systems are discussed.

1. INTRODUCTION

The quantitative assessment of the configuration of the vertebral column is an important diagnostic procedure, and can contribute to the understanding of postural reaction to physical work stress, particularly during manual lifting. There is general agreement that about 8 out of 10 workers will experience back pain at some time during their working careers (Snook, [7]). The National Safety Council stated in its 1984 Accident Facts that the number of reported low back injuries in industry has been increasing. It has also been established that the physical activity most frequently associated with the onset of low back pain sypmtons is lifting (NIOSH, [6]). Because manual lifting is so prevalent in industry, e.g. 30% of workers are estimated to be required to lift loads and because it is often the cause of serious musculoskeletal disability, a special detailed evaluation of lifting tasks is often warranted (Chaffin and Andersson, [9]). A careful review of the literature regarding the hazards of manual materials handling revealed that many facets of the problem, particularly lifting, still remain inadequately researched (Herrin et al, [5], and Drury, [3]). Recent biomechanical evaluations of lifting have indicated that small changes in spinal column configurations while lifting can cause major changes in spinal column forces (Tichauer et al, [8], and Anderson et al, [1]). These findings create a need for the development of improved methods of measuring the spinal column configurations during lifting tasks.

The objective of this study was to evaluate the new inclinometer device. Based on the collective user experiences of the preceding systems, certain criteria for selection of an improved system resulted. The new measurement device was evaluated by comparison to a computerized spot detection system. The two methods were compared in a set of controlled lifting tasks (Gilad et al. [4]).

2. **METHODS**

To compare the two systems, measurements of the angles of three gross divisions of the vertebral column were used to describe postural changes during controlled lifts. The divisions chosen were: the cervical region at the C4 level, the thoracic region at the T5 level, and the lumbar region at the L5 level.

The Selspot System

The system employed to measure, and to compare the kinematics of torso curvature was the Selspot computerized spot detection system, with one detection camera. This system has several characteristics not found in other systems for spinal analysis. It includes software to control the entire measurement process from calibration through result presentation. This strongly integrated system produces accurate estimates of the geometric values of spinal kinematics during task performance, with each spot located with about $\pm$ 0.5 cm accuracy. To obtain estimates of the 2-D spatial coordinates of a specific landmark on the human body, a small infrared light emitting diode (IR LED) is attached to a known body segment location. The image of the LED is focused by a lens system onto a semiconductor plate. Signals are obtained from this semiconductor plate which are related to the two dimensional coordinates of the LED.

The Miniature Inclinometer System

The new Miniature Inclinometer system employed to measure the kinematics of torso curvature is based upon small angle detectors manufactured by Spectron, see Figure 1. The detectors are single axis electrolytic resistance potentiometers. They require an AC excitation signal, and provide a proportional voltage output as the unit is tilted relative to the vertical gravity vector. They were specifically designed by the manufacturer to control the gradient output while being subjected to a variety of motions and vibrations.

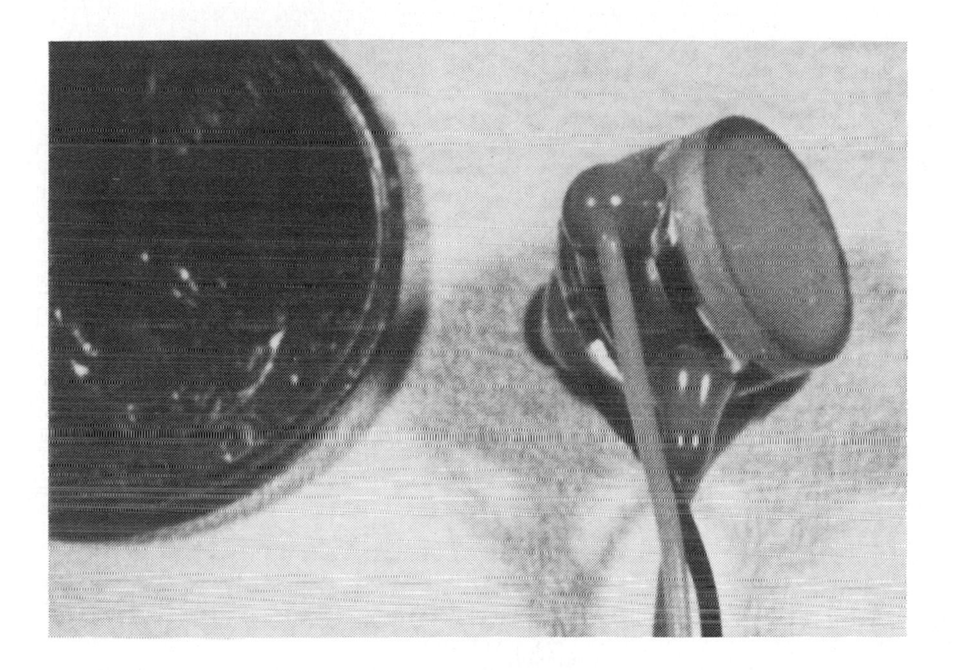

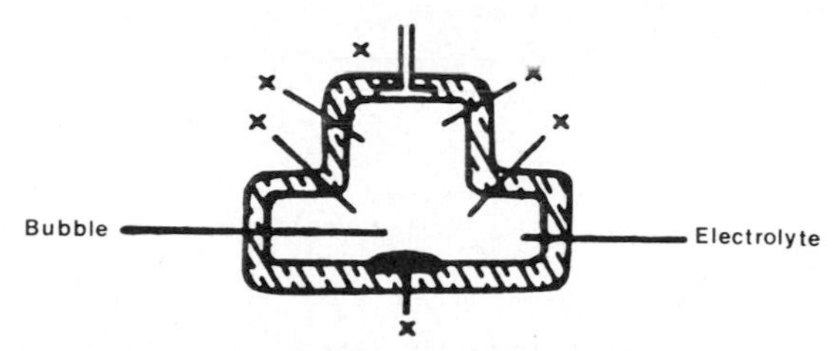

Cross sectional view showing position of bubble in chamber, X indicates bubble contact with 6 of 8 walls

Figure 1: The Spectron Miniature Inclinometer Detector and a simplified cross section of chamber.

The detector is a one piece glass enclosure approximately one centimeter in diameter. The internal platinum contacts and external terminals are sealed into the glass to prevent electrolyte leakage. The internal geometry is designed so that when the chamber is partially filled with electrolyte, a bubble results which maintains direct surface contact with six of the eight walls, this containment of the bubble gives stability to the performance of the detector as it is rotated.

The procedure of lifting tests was recorded by a video camera and screened on a video monitor. Information about the assigned parameters for each test is simultaniously super imposed with the actual lifting motions on the monitor. This on-line video control set-up enables to relate each set of measured geometrical data (x,y and angular values) to the performance of task during its motion picture. The video control set-up helped especially during evaluation of odd values when observed on the data output files and during the process of analyzing the different phases of the lifting act. To define the exact location of the reference points on the video monitor, reflective photographic posts were used, they are seen on the monitor as small bright light sources, easy to follow. Figure 2 illustrates the components of two measuring systems together with the video control set-up and the computer.

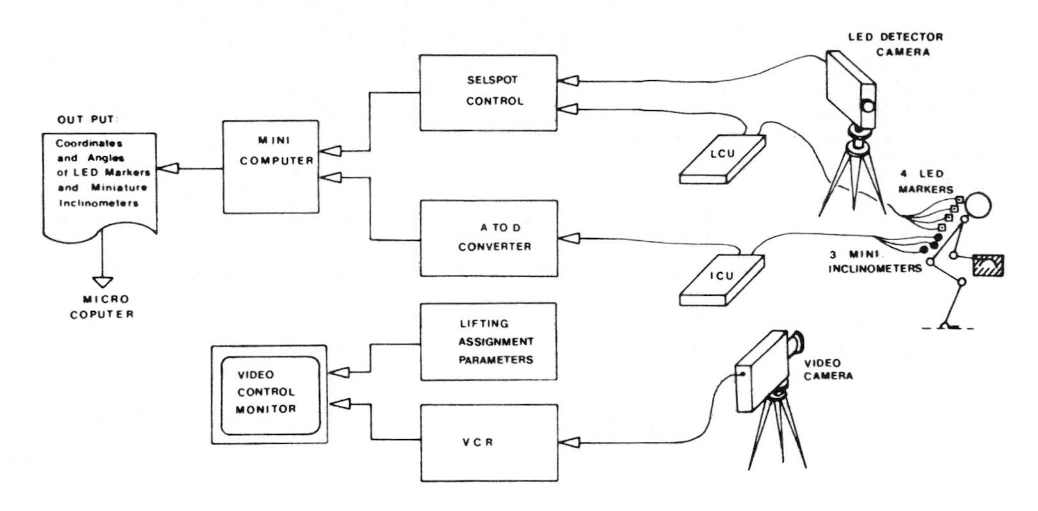

Figure 2: Components diagram of the two detection systems and the video control set-up.

The comparison of the systems' abilities to detect torso kinematics accurately was accomplished by having two subjects perform lifting tasks. The tasks were of varying magnitudes, from different horizontal locations, to assigned heights (Floor-F, Knuckle-K, Shoulder-S, Reach Height-RH), in three lifting postures (Straight Back-SB, Flexed Back-FB and Free Lifting). A factorial design was delineate for this comparative study, based on the variables: (1) Subject gender, (2) lifting posture, (3) lifting heights. The dependent variables were: (1) Horizontal distance (2) lifting weights, as they were related to subjects' authropometry and masuclar capacity for the assigned variables.

Male and female paid volunteers, one of each, were elected to participate in the methods comparison study. Their present and past health history was checked and found to be healthy with no medical limitations of the musculoskeletal system. Subjects of both gender were selected because of anatomical differences between the sexes, which lead to biomechanical differences in the lifting act. Tichauer [8] asserted that in the female pelvis, the hip joints are located further forward than in the male. It favors a forward tilt of the female pelvis and causes a difference in "resting" lumbosacral angle between the two body types. Because of the biomechanics of lifting, the change in curvature of the spine when loaded, is postulated to be greater in women than in men.

3. RESULTS

Spinal configuration data were obtained simultaneously from both the Selspot system and the new inclinometer system for each subject during each lifting task. Time series correlation coefficients for angle changes of torso movements as calculated during dynamic lifting of the female subject, for the two methods are listed in Table 1. Graphs of the angular data, such as seen in Figure 3, were drawn to compare the various angle trajectories in relation to torso motion over time. The motion as seen in the graphs are divided into five phases. The first phase is the normal standing posture before beginning the task. The second phase is while the subject reaches to begin the lift. The third phase includes the lift from the original location to the destination location. The fourth phase occurs as the subject returns the weight to the original location. The fifth phase is the return of the subject back to standing posture.

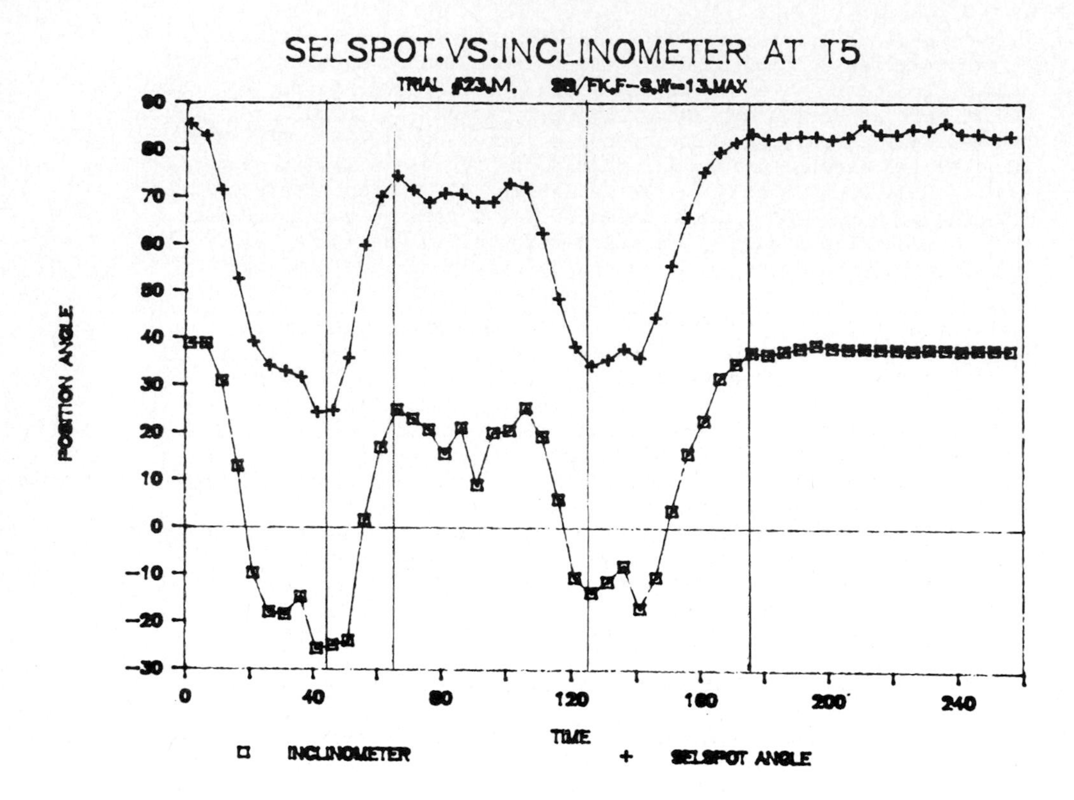

Figure 3: Graphs of data obtained simultaneously from both, Selspot and miniature inclinometer systems from the Thoracic Spine at T5. The graph lines demonstrate the angular displacement at the thoracic region of a male subject while performing Straight Back lift, Floor to Shoulder vertical height, 13 kg weight, from Maximal reach distance.

TABLE 1

Time series Correlation Coefficient for angle changes of torso movements during dynamic lifting, on measured by Selspot and miniature inclinometer systems, for female subject.

Trial No.	Lumbar Spine at L5	Thoracic Spine at T5	Cervical Spine at C4
1-6 (SB.F-K)	0.95-0.89	0.98-0.97	0.75-0.35
7-12 (FB,F-K)	0.94-0.92	0.97-0.95	0.97-0.98
13-18 (FREE,F-K)	0.96-0.79	0.98-0.96	0.90-0.76
19-24 (SB,F-S)	0.91-0.79	0.98-0.97	0.80-0.44
25-30 (FB,F-S)	0.95-0.91	0.99-0.97	0.97-0.93
31-36 (FREE,F-S)	0.96-0.87	0.98-0.97	0.87-0.77
37-42 (SB,F-RH)	0.97-0.84	0.98-0.92	0.90-0.76
43-48 (FB,F-RH)	0.96-0.95	0.98-0.96	0.97-0.95
49-54 (FREE,F-RH)	0.94-0.88	0.98-0.96	0.95-0.72
55-57 (SB,K-S)	0.68-0.25(1)	0.95-0.06	0.79-0.56
58-60 (FB,K-S)	0.88-0.77	0.97-0.96	0.82-0.62
61-66 (FREE,K-S)	0.91-0.81	0.97-0.93	0.77-0.60
67-69 (SB,K-RH)	0.71-0.22(2)	0.92-0.55	0.88-0.59
70-72 (FB,K-RH)	0.90-0.78	0.97-0.95	0.95-0.90
73-78 (FREE,K-RH)	0.89-0.84	0.96-0.93	0.95-0.81
79-81 (SB,S-RH)	0.72-0.28	0.52-0.37(3)	0.80- -0.25(3)
82-87 (FREE,S-RH)	0.66-0.20(5)	0.67-0.37(4)	0.73- -0.17(4),(5)

Errors obtained in LED estimates, of the Selspot system, for trials:

(1) trial #55 (2) trial #67 (3)trial #79 (4) trial #82 (5) trials #82, #85 and #87.

4. DISCUSSION

There are a few possible reasons for the differences between the two systems measures at given points of locations on the torso, these are listed below:

1. Difference in system references: The Selspot system, which has the capability to estimate body configurations in 3-dimensional space by using two cameras simultaneously, was configured with only one camera in this experiment. It was therefore limited to analysis of movement in one plane (i.e. and sagittal plane). Since this system is limited to measurement in a single plane, estimates of a specific motion may be affected by unrealized, instantaneous out-of-plane motions. The inclinometer system, however, is not constrained in the same manner, this system measures angles relative to gravity. In other words, the Selspot plane of measurement is referenced to the camera position while the inclinometer plane of measurement is referenced gravity and may not be sensitive to lateral or torsional motions as much as the Selspot.

2. Adjacent body segment motion: System measurement function of a motion can be affected by movement of adjacement body segment. For example, head movement in frontal and/or tansverse plane can cause differences of measurement between Selspot and Inclinometer systems at C4 and L5, rather than T5, where possible skin motions will more likely occur.

3. Lifting dynamics: Under the condition that a lifting task should be done within the same time, lifting postures which have different torso movement amplitude in sagittal plane can generate different magnitude of acceleration of deceleration at a reference point. That may result in difference of measurement performance. Evaluation of the electromechanical component, reveals that the Inclinometer system is supposed to be more erroneous than the Selspot sytem when acceleration or deceleration effect is high.

4. Difference in initial conditioning: Though care was taken to calibrate both systems at the start of the experiment to reduce optical distortion, LED signal errors, and minimize the alignment error, minor calibration differences may still cause disagreement in system performances.

A quantitative comparison of the two systems is provided compactly in Table 2. The criteria chosen, and remarks, are from the researcher's experience in using these systems.

TABLE 2

Quantivative Comparison of Selspot and Inclinometer Techniques

Criteria	Selspot	Inclinometer
Angle measurement method	Collects coordinates and uses mathematical transformation	Collects analog angle signal and then digitizes
Cost	$35,000 - $50,000 including minicomputer	$8,000-$12,000 including microcomputer
Software requirement	Relatively complex	Relatively simple
Calibration	Measured angles are not largely affected by calibration	Simple initial calibration is important to collect accurate data
Reading range	360 degrees with absolute reference frame	120 degrees with relative reference frame
Motion effect in measurement	No effect by a motion in frontal/transverse plane	Effects with a motion in frontal/transverse plane
Interference with motion	Marker can be obscured by another body segment	None
Environmental factor	Excess light in background	None
Complexity of operation	Skill required to operate computer and electronic system	Same
Ease of attachment	Fairly easy to attach to subject	Same
Data repeatability	Depends upon consistent alignment of markers on subject	Depends upon maintaining same angle as reference point from trial to trial
Speed effect	Low	High
Easy to find joint of motion	Easily shown by stick diagram	No diagram
Memory space to store data	Depends upon sampling rate, but high	Same
Effort required to analyze	Depends upon data and facilities, but high	Same

ACKNOWLEDGEMENTS

This work was partially supported by research gifts from both GenCorp and Owens Corning Fiberglass Corporation. The electronic inclinometers were provided by Hoggan Health Industries, and the Selspot System from NIOSH DSR contract 210-81-3104.

REFERENCES

[1] Anderson, C.K. and Chaffin, D.B., A Biomechanical Evaluation of five Lifting Technique, presented at the International Conference of Occupational Ergonomics, (1984.)

[2] Chaffin, D.B. and Andersson, G., **Occupational Biomechanics**, John Wiley & Sons Pub., New York, 111-146, (1984).

[3] Drury, C.G., (Ed.) Safety in Manual Materials Handling, **DHEW, (NIOSH)** Publication No. 78-185, (1978).

[4] Gilad, I., Chaffin, D.B., Redfern, M., Byun, S.N., A System Comparison of Two Methods for Measuring Angular Displacement of Torso During Dynamic Lifting, Tech. Report, 1986-A. Center for Ergonomics, University of Michigan (1986).

[5] Herrin, G.D., Chaffin, D.B. and Mach, R.S., Criteria for Research on the Hazards of Manual Materials Handling, NIOSH Contract Report, CDC 99-74-118, (1974).

[6] NIOSH (National Inst. for Occupational Safety and Health), Work Practices guide for manual lifting, NIOSH Tech. Report Pub. No. 81-122, U.S. Dept. of Health and Human Services, (1981).

[7] Snook, S.H., The Design of Manual Handling Tasks, **Ergonomics** 21, 963-985 (1978).

[8] Tichauer, E.R., Miller, M. and Nathan, I.M., Lordosimetry: a New Technique for the Measurement of Postural Response to Material Handling, **American Industrial Hygiene Assoc. J.**, 1, 1-12, (1973).

Trends in Ergonomics/Human Factors V
F. Aghazadeh (Editor)

TRANSFER FUNCTIONS OF TRUNK MUSCLE EXTENSIONS DURING MOTION

William S. Marras, Ph.D.

Industrial and Systems Engineering Department
The Ohio State University
Columbus, OH 43210

This study has observed and quantified the reaction of the trunk extensor muscles when the trunk is moving at typical lifting velocity rates. In this study 45 male subjects between the ages of 17 and 61 were tested for their ability to exert torque about their low back under static and dynamic sagittally symmetric lifting conditions. Trunk muscle electromyography (EMG) was used to monitor the activity of the trunk muscles. Generally, back muscle EMG activity and trunk torque decreased (at different rates) as the trunk velocity increased. However, during an actual lift trunk torque requirements remain constant regardless of lift velocity. In order to deal with this situation, this study has developed transfer function profiles of EMG per unit torque which describes how trunk loading changes with velocity. Correlations between EMG and torque were also found to change as a function of velocity. These results are used to discuss how static lifting models might be adjusted to account for added trunk load due to trunk velocity.

1. INTRODUCTION

Motion studies which assess the ability of workers to perform manual materials handling (MMH) tasks have traditionally focused on two ways of approaching the evaluation. One approach involves the analysis of a workers ability to handle a load that is moving at a given rate outside the body (Aghazadeh and Ayoub; 1985). The other approach isolates the movement of a body segment and evaluates the response of body structures to motion and torque applied around that segment (Marras et al. 1984). Segment analyses which focus on the lumbrosacral junction (L5/S1) have appeared with increasing frequency in the literature recently (Marras et al. 1984; 1985; Andersson et al. 1986). These studies have focused upon the reaction of forces internal to the trunk in response to external loading. These internal forces are of particular interest since they represent a restorative moment to the forces imposed external to the body. Since the distance (from L5/S1) at which these internal forces act is very small compared to the distance at which external forces are applied, these internal forces are often very large and become the primary loading sources of the spine during work.

Previous studies (Marras et al., 1984; 1985) have investigated the activity of the internal force generating structures of the trunk in response to torque produced about L5/S1 during motion (constant angular velocity). These studies investigated the activities of 10 trunk muscles and intra-abdominal pressure in response to trunk velocities which varied from static exertions to the maximum velocity each subject could produce with the trunk while still being able to generate a torque about L5/S1. These studies showed that the activities of all these trunk structures changed dramatically as trunk velocity increased, particularly at the higher velocities.

A recent study by Kim and Marras (1987) investigated the L5/S1 angular velocities which subjects exhibited when asked to lift at slow, medium and fast speeds. The mean velocities exhibited by these subjects ranged from about 15 deg/sec for slow lifts to 28.1 deg/sec for fast lifts. These velocities are much slower than the range investigated by Marras et al. (1984, 1985).

The objective of this study was to investigate the response of the internal force generators of the trunk when the trunk was moving at velocities which were representative of MMH lifting velocities. Since the back extensor muscles would represent the main agonist internal forces involved in this motion, only the latissimus dorsi and erector spinae muscles were observed in this experiment.

Previous studies have revealed that as trunk velocity increases the trunk muscle activities and available torque production decrease. However, during MMH the torque requirements remain a function of the distance of the load from L5/S1. Thus, changes in trunk angular velocity can influence the internal loading of the trunk. This analysis is unique in that the trunk muscle responses were considered as a function of the electromyographic activity (EMG) per unit torque over the velocity range. In other words, a transfer function (output compared to input) was defined to represent the relative nature of EMG with respect to trunk torque production.

2. METHOD

2.1. Subjects

Subjects in this experiment consisted of 45 healthy males who had a negative history of low back disorder. Subject age ranged from 17 to 61 years (mean = 31 S.D. = 8.6). Mean subject height and weight were 179.3 cm and 79.8 kg respectively. The subject populations represents a mix of students, faculty and laborers.

2.2 Design

The experimental design consists of a 3 (angle) x 4 (velocity) design with repeated measures where velocity and angle were completely crossed. Trunk angle was defined at three levels (0 deg, 22.5 deg, and 45 deg), with the 0 deg angle corresponding to a normal standing posture. The 22.5 and 45 deg angles relate to forward flexion angles of the trunk relative to the 0 deg upright posture. The velocity variable levels were defined in terms of trunk angular velocity about L5/S1. Four fixed velocity levels were defined (0 deg/s, 15 deg/s, 30 deg/s, and 90 deg/s). The 0 deg/s condition represents an isometric or static exertion. The other velocity conditions consisted of isokinetic angular velocities.

Five dependent measures were defined in this experiment and consisted of: (1) the maximum voluntary torque that a subject could exert about the L5/S1 junction during the return from a flexed trunk posture (this moment is similar to that experienced by the back link during a backlift), (2) the integrated EMG of the right latissimus dorsi muscle (LATR), (3) the integrated EMG of the left latissimus dorsi muscle (LATL), (4) the integrated EMG of the right erector spinae muscle (ERSR), and (5) the integrated EMG of the left erector spinae muscle (ERSL).

Once the subject was strapped into the experimental reference frame, he was required to produce a maximal voluntary torque against the axis arm of the dynamometer. The exertion started with the trunk at a 60-deg forward angle and culminated past the 0-deg angle for isokinetic exertions. The first 15 deg of motion permitted the subject to attain an isokinetic velocity state. Static exertions were recorded with the dynamometer axis arm frozen in set positions.

2.3 Apparatus

A reference frame was constructed which position a Cybex II isokinetic dynamometer in line with the L5/S1 junction of the back in the sagittal plane. This frame was constructed so that trunk motion will be isolated to that of L5/S1 in this plane of the body. Muscle activity was recorded via surface electrodes glued over the muscles of interest. The EMG activities of these muscles were amplified with miniature preamplifiers placed at the muscle site. These EMG signals were monitored, filtered, further amplified, and "integrated" prior to analog-to-digital conversion and recording on a microcomputer. The dynamometer and potentiometer (trunk angle) signals were recorded in a similar fashions. Figure 1 shows the configuration of this experimental apparatus.

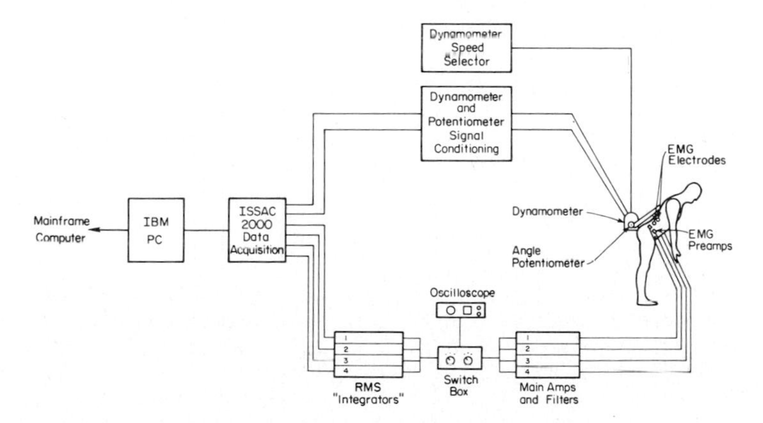

Figure 1: Experimental Apparatus

2.4 Data Processing

All dependent measures were evaluated as a function of window values which represent the mean activity of the variable as the trunk passed through the experimental angles at each experimental velocity. Each EMG signal was normalized with respect to each subject's maximum EMG activity (pretest) and resting EMG activity (2nd pretest) observed at each experimental angle.

3. RESULTS

Univariate analysis of variance was used to test the statistical significance of each dependent variable. Each dependent variable exhibited a statistically significant response as a function of velocity, angle and velocity-angle interaction (these F values are not reported here due to space contraints but are available from the author). Figures 2, 3 and 4 show the activities of the latissimus dorsi muscles, erector spinae muscles and torque as a function of velocity. The latissimus dorsi muscles and torque decrease as a function of increasing velocity. It is also significant to point out that trunk torque capability decreases by 0.55% of maximum torque for every deg/sec increase in trunk velocity. However, figure 3 shows that the erector spinae muscles respond differently. In this case the greatest amount of muscle force produced throughout an exertion was exhibited under the 15 deg/sec condition.

When the window values of the dependent variable activities were considered as transfer functions dramatically different behaviors were observed. Figures 5 and 6 show the EMG-torque transfer functions as a function of velocity and angle for the latissimus dorsi and erector spinae muscles respectively. The surfaces of these figures represent equal torque productions. Therefore, in each figure as trunk velocity and trunk angle increase greater EMG activity is required to produce a given amount of torque with the back.

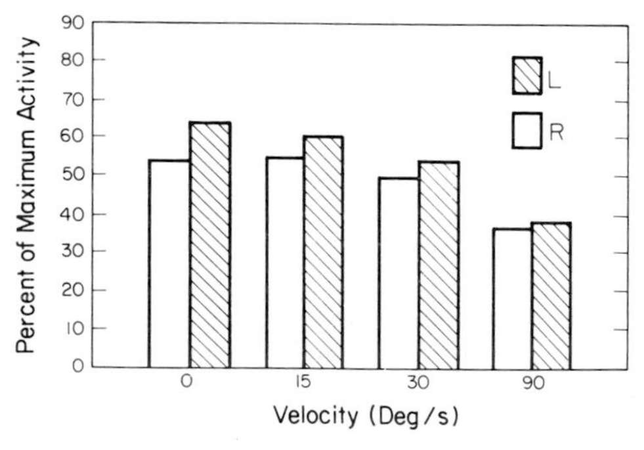

Figure 2: Mean EMG Activity of LATR and LATL.

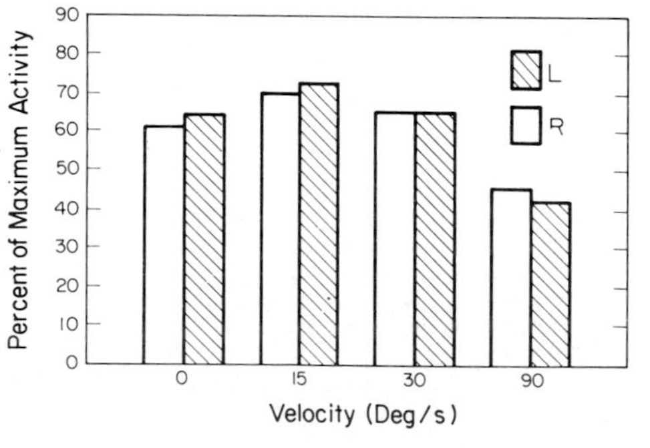

Figure 3: Mean EMG Activity of ERSR AND ERSL.

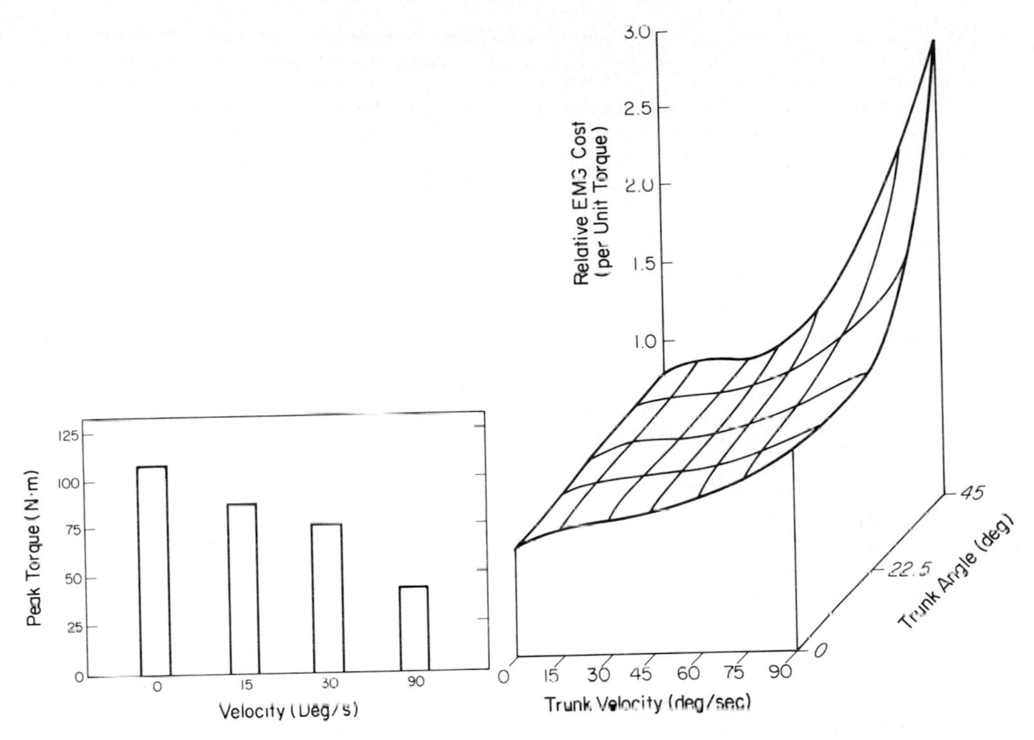

Figure 4: Torque production as a function of velocity

Figure 5: Mean EMG activity per unit torque for latissimus dorsi muscles.

Finally, the relationship between EMG and torque was investigated as a function of velocity. This relationship is shown in Figure 7. The figure indicates that the correlation coefficient between EMG and torque increases as velocity increases for erector spinae muscles but remains relatively constant over velocity for the latissimus dorsi muscles.

4. DISCUSSION

These results suggest that several significant changes occur in the behavior of the internal and external force generation of the trunk when angular velocity about L5/S1 is present compared with static conditions. This information could be used as a basis to adjust static biomechanical models so that they are more reactive to realistic (motion) MMH conditions. Several considerations should be made based upon these findings.

First, a reduction in torque genration capacity about L5/S1 is present when motion is introduced into the experimental task. This reduction is also dependent upon trunk angle. The reduction of torque is most apparent at the 45 deg trunk angle. This is particularly significant for MMH since this often represents the starting point of a dynamic lift. Under true dynamic conditions (non-isokinetic), this initial motion phase would involve acceleration of the trunk to overcome inertia. These results show that motion affects torque capacity significantly at this point in a lift. This may help explain why many injuries occur at the beginning of a lift.

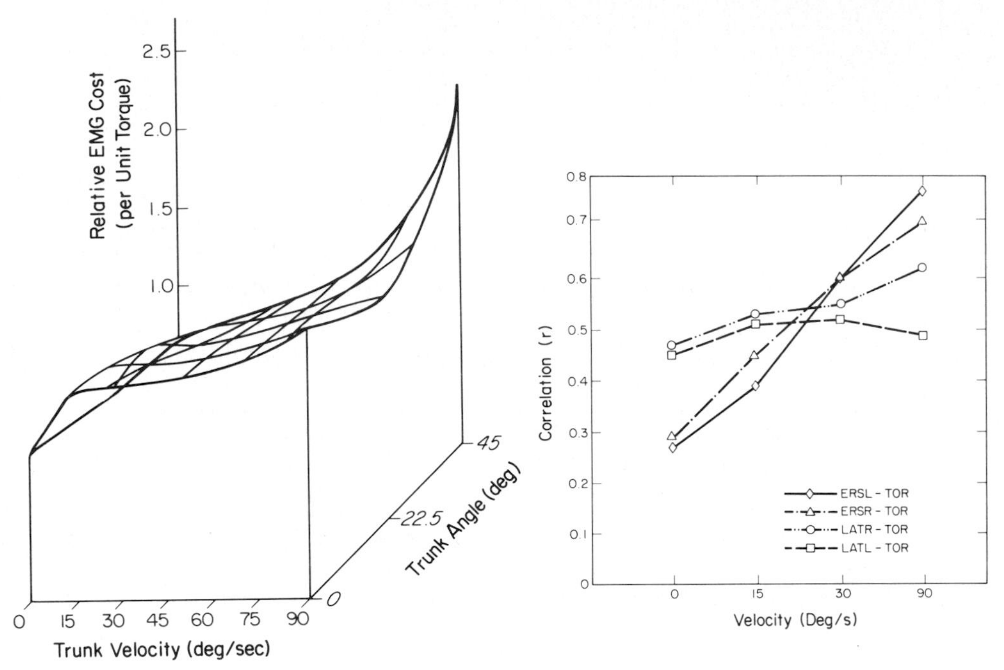

Figure 6: Mean EMG activity per unit torque for erector spinae muscles.

Figure 7: Correlation between EMG and torque over velocity conditions.

These results have also quantified the torque loss experienced by the trunk as trunk velocity increases. Generally, over all angles, a 0.55% reduction of maximum torque is experienced for every deg/sec increase in trunk angular velocity. A recent study by Marras and Wongsam (1986) has reported that the mean velocity of the trunk about L5/S1 was about 36 deg/sec for normal healthy subjects lifting light loads under leg lift and back lift conditions. Therefore, using this velocity reductions factor their dynamic capability at this speed would be at about 80% of that predicted by static evaluations.

Second, the cost of motion to the back muscles has been described via transfer functions. Figures 5 and 6 indicate that the internal forces needed to produce a given amount of external torque vary greatly as a function of trunk velocity and angle. This information can be used to adjust biomechanical models which account for the internal loading of the trunk during MMH. For example, the static transverse plane model described by Schultz and Andersson (1981) may be adjusted by weighing the latissimus dorsi and erector spinae muscle contributions in their trunk force and momement equations by the transfer function value (adjusted for trunk angle and velocity) described in this study. Marras et al. (1986) has also shown in a simple model of internal loading how the compressive load on L5/S1 can more than double by including just 90 deg/sec of trunk velocity compared to a static model.

Third, an interesting change in correlation between trunk force and erector spinae EMG activity occurs when trunk motion is increased. This suggests that, with further research, the EMG signal may be meaningful in interpreting true dynamic MMH activities provided that L5/S1 motion characteristics are documented. Further correlations among trunk loading factors are discussed in Marras et al. (1987).

Finally, the contribution of the erector spinae muscles, which represent a fairly large internal force of the trunk, have been delineated in this study. Previous static analyses have defended static interpretations of the biomechanical system by claiming that a static analysis represents a "worst case" situation. Figure 3 has shown that the greatest amount of force generated by these muscles occurs at slow velocities not under static conditions. It is also interesting to note that the static responses of this muscle are similar to responses at velocities of 30 deg/sec. However, the torque generation capability of the trunk is decreased at 30 deg/sec and previous research (Kim and Marras 1987) have shown that subjects lifting considerable loads judge 30 deg/sec as a fast velocity. Hence, this study shows that some interesting changes occur in the behaviors of the internal structures at slow velocities and more research should be performed to investigate the effects of these changes on the biomechanical system. In particular, issues such as asymmetric loading, muscle coactivations and non-isokinetic motion must be explored.

REFERENCES

[1] Aghazadeh F. and Ayoub M.M. A comparison of dynamic- and static-strength models for prediction of lifting capacity. Ergonomics 28(10), 1409-1417 (1985).

[2] Anderson, C.K., Chaffin, D.B., and Herrin G.D. A study of lumbrosacral orientation under static loads. Spine 11(5), 456-462 (1968).

[3] Kim J.Y. and W.S. Marras. Quantitative trunk muscle electromyography during lifting at different speeds. International Journal of Industrial Ergonomics 1(3), 219-229 (1987).

[4] Marras, W.S., King, A.I., and Joynt, R.L. Measurements of loads on the lumbar spine under isometric and isokinetic conditions. Spine 9(2), 176-188 (1984).

[5] Marras, W.S., Joynt R.L. and A.I. King. The force-velocity relation and intra-abdominal pressure during lifting activities. Ergonomics 23(3), 603-613 (1985).

[6] Marras W.S. and Wongsam P.E.. Flexibility and velocity of the normal and impaired lumbar spine. Archives of Physical Medicine and Rehabilitation 67, 213-217 (1986).

[7] Marras, W.S., Wongsam, P.E., and Rangarajulu, S.L. Trunk motion during lifting: the relative cost. International Journal of Industrial Ergonomics 1(2), 103-113 (1986).

[8] Marras, W.S., Rangarajulu, S.L. and Wongsam, P.E. Trunk force development during static and dynamic lifts. Human Factors 29(1), 19-29 (1987).

[9] Schultz, A.B. and Andersson G.B. Analysis of loads on the lumbar spine. Spine 6(1), 76-82 (1981).

Trends in Ergonomics/Human Factors V
F. Aghazadeh (Editor)

SUBJECTIVE JUDGEMENT OF LOAD HEAVINESS AND PSYCHOPHYSICAL APPROACH TO MANUAL LIFTING

Waldemar KARWOWSKI and Anthon BURKHARDT

Center for Industrial Ergonomics
University of Louisville
Louisville, KY 40292, USA

The objective of this study was to examine human perception of load heaviness, and to determine the meaning of seven linguistic values, i.e.: 'very light', 'light', 'less-than-medium', 'medium', 'more-than-medium', 'heavy', and 'very heavy' as descriptors of the load lifted. These linguistic categories chosen for seven standard boxes of 10, 15, 25, 35, 45, 55, and 60 lb were also compared to the psychophysically-selected loads under the same lifting conditions. Since a significant proportion of subjects chose the 'very heavy' or 'heavy' weight categories as the maximum acceptable loads for an 8-hour work shift, it was suggested that the instructions used in the *rating of acceptable load method* be changed.

1. INTRODUCTION

Four different approaches have been used in the past to develop ergonomic guidelines for manual lifting of loads. These approaches are epidemiological, biomechanical, physiological and psychophysical [1,2]. The psychophysical approach [3,4] led to development of the widely used technique of acceptable rating to determine load-handling capacities of industrial workers. Reliability of this technique was recently investigated by Mital [5] and Karwowski and Yates [6], and some of the limitations with regard to its applicability in industrial settings were discussed by Griffin et al. [7] and Snook [8].

The psychophysical approach to setting limits for manual lifting calls for subjective adjustments of the work-load (usually the weight of load lifted) to the maximum level that a worker can sustain over an 8-hrs day without straining him/herself or without becoming unusually tired, weakened, overheated or out of breath [3].

As pointed out by Gamberale [9], the above rating of acceptable load (RAL) is based on the assumption that workers are able to indicate with some accuracy the highest workloads tolerable to them under given working conditions, and that such workloads are below the loads leading to manual handling injuries. So far, the only proof that such assumption is valid, are the findings that when workers are asked to subjectively rate the degree of physical effort or strain in their jobs, low back pain appeared significantly more frequently in those who believed their work to be harder [10].

2. OBJECTIVES

A study by the U.S. Department of Labor (1982) showed that 35% of workers injured while performing manual lifting tasks thought that the loads handled at the time of accident were "too heavy" for them. Although the meaning of "heavy" or "very heavy" and other categories of load heaviness cannot be clearly determined, mainly due to large differences in human strength capacities, the RAL method which is based on ill-defined subjective perception of the muscular effort, attempts to determine human lifting capacity in representative industrial situations. As pointed out by Karwowski and Ayoub [12], human perception of the maximum acceptable weight of load lifted is by its very nature imprecise and inexact. Moreover, it is not clear whether the RAL method is employing the scaling of sensory or perceptual stimuli [9,13].

The main objective of this study was to evaluate human perception of load heaviness, and specifically to determine the meaning of seven linquistic variables as subjective descriptors of the load lifted, namely: 'very light' (VL), 'light' (L), 'less-than-medium' (LM), 'medium' (M), 'more-than-medium' (MM), 'heavy' (H), and 'very heavy' (VH) categories. In addition, the maximum acceptable weights of load determined using the RAL method were compared with the weights corresponding to the chosen categories of load heaviness.

3. METHODS AND PROCEDURES

3.1 Subjects

Nineteen males, college students, participated as subjects in the laboratory experiment. All subjects had some previous experience in manual handling, and were given an additional training in manual lifting over a period of two weeks (4 hours a week) before the start of data collection. The subjects did not know about the true purpose of the study, and until the second part of the experimentation , they were not aware of the rate of acceptable rating (RAL) method. Age, body weight, and anthropometric dimensions of the subjects are shown in Table 1.

The isometric strength measurements (arm, shoulder and back) were made according to the procedures described in [6], while the isokinetic strength tests were done following the procedures outlined by Pytel and Kamon [15]. The subjects were paid for their work on an hourly basis. Each subject read and signed an informed consent form prior to beginning of the project.

TABLE 1. Age and physical characteristics of the subjects [N = 19].

Variable	Mean	S.D.	Range
Age [years]	26.0	2.1	20 - 33
Body weight [lbs]	181.0	27.0	146 - 237
Height [cm]	179.0	8.3	167 - 198
Static arm strength [lbs]	90.4	13.6	65 - 111
Static shoulder strength [lbs]	137.5	25.9	73 - 181
Static back strength [lbs]	221.3	45.0	117 - 300
Dynamic arm strength [lbs]	69.3	16.7	37 - 105
Dynamic shoulder strength [lbs]	43.9	7.5	34 - 59
Dynamic back strength [lbs]	151.1	31.2	71 - 200

3.2 Procedures

Seven standard, unmarked containers (36x36x34 cm) with the weights of 10, 15, 25, 35, 45, 55 and 60 lbs were used. Each box was coded, and neither the code nor the actual weight in the box was not known to the subjects. The subjects were asked to lift the container from 24 inches above the floor to 48 inches height with the frequency of two lifts per minute. Both the frequency and height of lifting were chosen in order to minimize the potential effects of the relevant biomechanical and physiological stresses.

A total of thirty minutes was allowed for the subject to make a determination as to which of the above seven load heaviness categories best described the weight of the container. All subjects were encouraged to take as much time as they thought they needed to make the best judgement. Each subject was asked to lift each of the seven containers in a random order and express his opinion about the perceived load heaviness. In the second part of the experiment, the rate of acceptable load (RAL) method was used. The subjects were given standard instructions devised by Snook [3], and were asked to select the maximum weight of load lifted that they felt should be acceptable to them for an 8-hour shift in a representative work situation. The task variables were the same as in the first part of the experiment, and thirty minutes were allowed for the adjustment period.

4. RESULTS AND DISCUSSION

4.1 Results

The results of the first part of the study are illustrated in Figure 1. It can be seen that twelve subjects (about 63%) considered a 60.0 lb box as 'heavy', while another (30%) felt that such container was 'very heavy'. Sixty three perecent of the subjects thought that a 55 lb load was 'heavy'. The average maximum acceptable weight (MAW) of load determined using the RAL method was 55.9 lbs (S.D. = 9.9 lbs), with the range between 41.2 lbs and 77.3 lbs. Nine subjects selected MAW values above the 55.0 lb load, while ten of them selected MAW values between 41.2 and 51.6 lbs. The number of subjects who chose MAW values in the

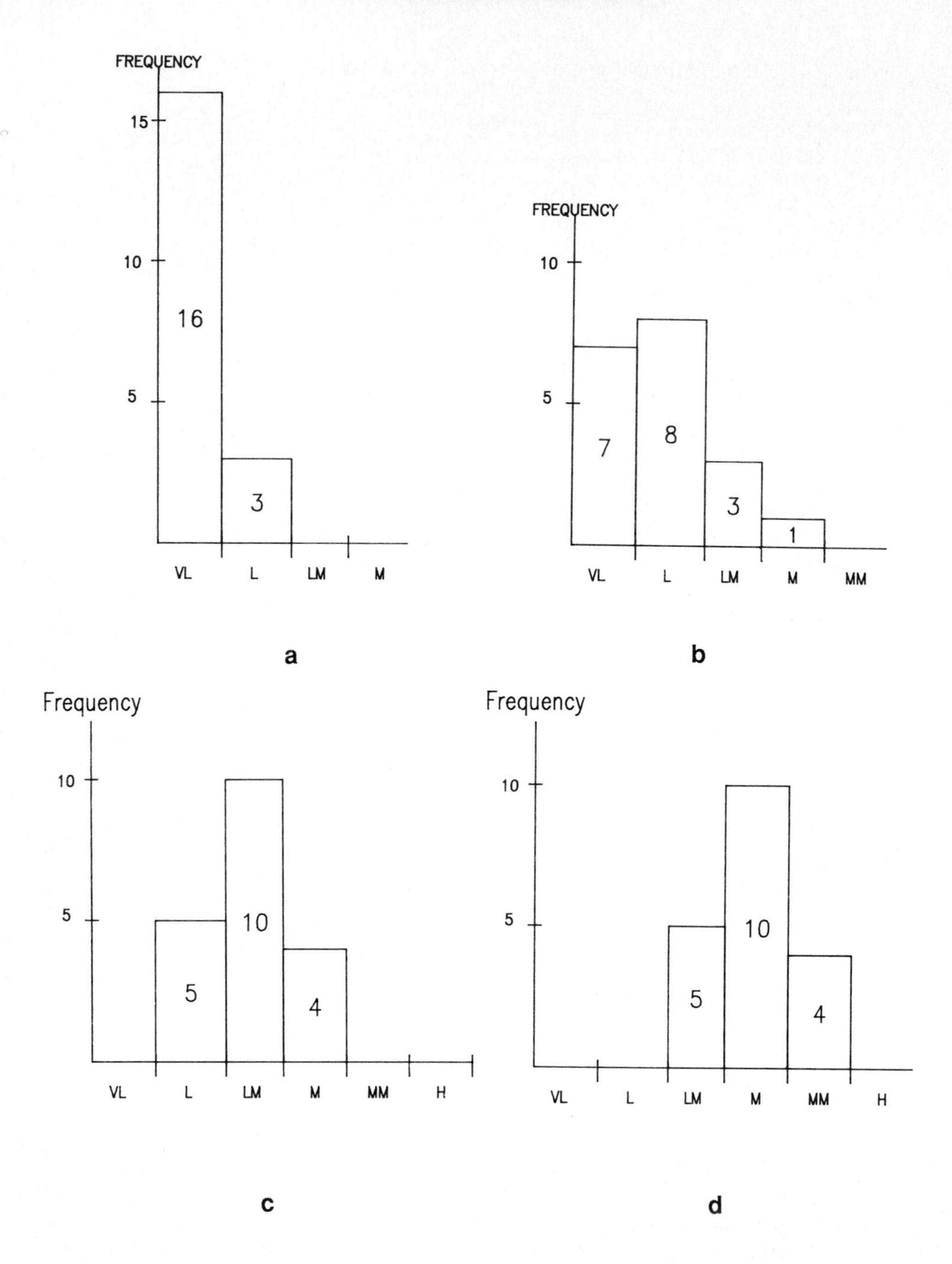

FIGURE 1. Frequency chart for the load heaviness categories associated by the subjects with: a) 10 lbs box, b) 15 lbs box, c) 25 lbs box, and d) 35 lbs box.

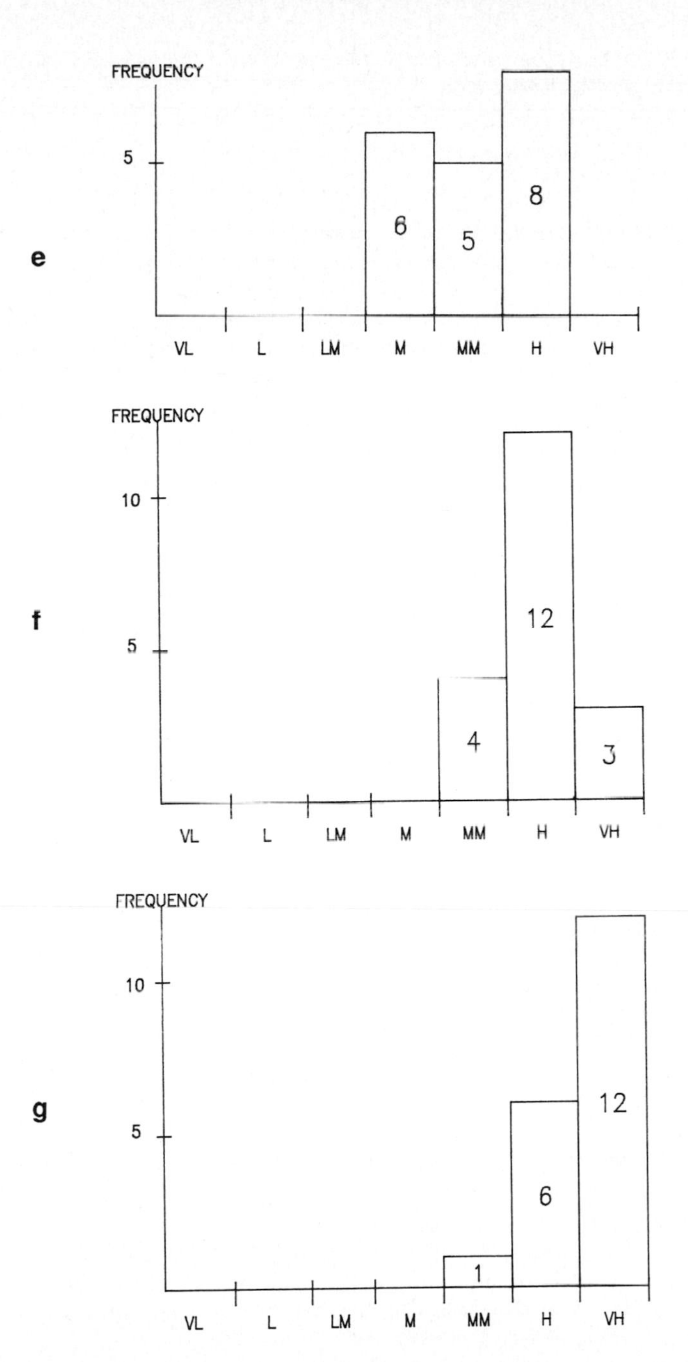

FIGURE 1 (continued). Frequency chart for the load heaviness categories associated by the subjects with: e) 45 lbs box, f) 55 lbs box, and g) 60 lbs box.

'55-60 lb range', '60-65 lb range' and a 'greater than 65 lb range' was three in each of the above categories, respectively.

4.2 Discussion

A comparison of the selected maximum acceptable weights (MAWs) with the subjective categories of load heaviness revealed the following observations. Out of the twelve subjects who considered a 60.0 lb box as 'very heavy, four subjects selected the MAW values above the 60.0 lbs, while eight chose MAW values below that level. In addition, out of the six subjects who felt that a 60 lb container was 'heavy', two and four subjects selected the MAW values above and below that level, respectively. Overall, about 63% of the subject population selected the loads which were in their opinion either 'heavy' or 'very heavy' as the maximum acceptable for an 8-hour shift.

5. CONCLUSIONS

This study shows that application of the RAL method may lead to overestimation of the true lifting capacity for the infrequent lifting tasks. The significant proportion of the experienced student handlers selected loads judged by them as 'very heavy' or 'heavy', as the maximum acceptable weights to be lifted every thirty seconds during an 8-hour workshift. Therefore, in order to increase reliability of the psychophysical lifting norms, it is suggested that the rating of acceptable load method be modified by changing the instructions given to the subjects.

Also, in view of the above results, extreme care should be exercised when applying the current psychophysically-based lifting norms in industry, and the pre-employment strength testing should always be used to verify the individual's capacity for manual lifting tasks.

REFERENCES

[1] Garg, A., and Ayoub, M. M., Human Factors, 22 (1980) 475.
[2] National Institute for Occupational Safety and Health, Work Practices Guide for Manual Lifting, US Department of Health and Human Services, (1981).
[3] Snook, S. H., Ergonomics, 21 (1978) 963.
[4] Ayoub, M. M et al., Human Factors, 22 (1980) 257.
[5] Mital, A., Human Factors, 25 (1983) 485.
[6] Karwowski, W., and Yates, J. W., Ergonomics, 29 (1986) 237.
[7] Griffin, A. B., Troup, J. D. G. and Lloyd, D. C., Ergonomics, 27 (1984) 305.
[8] Snook, S. H., Ergonomics, 28 (1985) 323.
[9] Gamberale, F., Ergonomics, 28 (1985) 299.
[10] Magora, A., Industrial Medicine, 39 (1970) 465, 504.
[11] OSHA, U.S. Department of Labor, Washington D.C., (1982).
[12] Karwowski, W. and Ayoub, M. M., Ergonomics, 27 (1984) 641.
[13] Foreman, T. K., et al., Ergonomics, 27 (1984) 1283.
[14] Pytel, L., and Kamon, E., Ergonomics, 24 (1981) 663.

Trends in Ergonomics/Human Factors V
F. Aghazadeh (Editor)

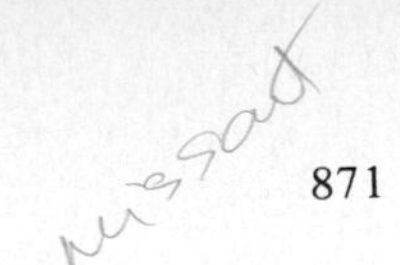

PSYCHOPHYSICAL AND PHYSIOLOGICAL RESPONSES TO ASYMMETRIC LIFTING

Arun GARG and J. BANAAG

Department of Industrial and Systems Engineering
University of Wisconsin--Milwaukee
Milwaukee, Wisconsin U.S.A. 53201

Maximum acceptable weights and heart rates were measured on eight male college students for lifting in the sagittal plane and at three different angles of asymmetry (30, 60, and 90°). The subjects lifted a box from the floor to an 81 cm high table and from an 81 cm high table to a 152 cm high table at a rate of 3, 6 and 9 lifts min^{-1} for a period of 1 hour. The maximum acceptable weights decreased and the heart rate increased with an increase in the angle of asymmetry ($p < 0.01$). Correction factors of 9, 15 and 21% for maximum acceptable weights at 30, 60 and 90° of asymmetric lifting are recommended.

1. INTRODUCTION

Manual materials handling is believed to be the primary cause for work related musculoskeletal injuries, especially back injuries. One of the strategies for reducing the extent of these injuries is ergonomic job design. Several studies have provided guidelines for determining allowable loads for two-handed, symmetrical lifting in the sagittal plane [1, 2, 3]. Twisting of the spine while lifting is common in workplaces. However, little is known about acceptable loads under asymmetric lifting conditions because of the lack of studies on such tasks.

The objective of the present study was to determine maximum acceptable weights and heart rates for two-handed, repetitive asymmetric lifting tasks of one hour duration. Two-handed symmetric lifting tasks were also studied to provide a basis for comparison.

2. METHOD

2.1. Subjects

Eight paid volunteer male college students acted as subjects. They were all judged to be in good physical health and claimed never to have had any musculoskeletal or cardiovascular problems. Their mean age, body weight, stature, and static arm, torso and leg strengths are summarized in table 1.

2.2. Asymmetric Lifting Tasks

Asymmetric lifting was studied in 30, 60 and 90° lateral planes to the right, while the subjects kept their feet in the sagittal plane. In

Table 1. Summary of anthropometric and strength measurements from eight male subjects

Variable		Mean	Standard Deviation	Range
Age	(years)	24.3	1.9	22 - 28
Body weight	(kg)	78.3	8.3	60.0 - 86.6
Height	(cm)	179.3	5.3	171.0 - 187.5
Arm strength	(N)	320.6	91.2	156.9 - 460.8
Back strength	(N)	370.6	102.9	186.3 - 500.0
Leg strength	(N)	1003.9	262.7	598.0 - 1500.0

addition, symmetric lifting was studied in the sagittal plane (0°) for comparison purposes. A 51 x 38 x 25 cm (length x width x height) box with a false bottom and cushioned handles was placed perpendicular to one of these planes and the subject was required to lift the box using both hands from the floor to a height of 0.81 m (bench height) without moving his feet, then take a step, if needed, and finally place the box on a 0.81 m table in front of him. Two research assistants lowered the box to the floor. The same procedure was used for 0.81 m to 1.52 m (about shoulder height) lifts. The subjects lifted the box using whatever body posture they found to be most comfortable. The horizontal distance between the ankles and the center of the box was not controlled to simulate real-life lifting conditions.

2.3. Maximum Acceptable Weights

A psychophysical method [2] was employed to determine the maximum weight of the load acceptable to the subject for a work duration of 1 hour. The instructions given to the subjects were the same as those used by Snook and Irvine [4]. The subjects were asked to adjust the weight of the box by adding or subtracting lead shots to the maximum amount that they could lift comfortably at a rate of 3, 6 or 9 lifts min^{-1} for a duration of 1 hour.

2.4. Heart Rate

An exersentry cardiotachometer with three built-in electrodes was used to measure the heart rate. A portable cassette tape recorder was attached to a belt and was connected to the cardiotachometer to record heart beat (R wave) at rest (standing) and from the 13th to 15th min, 28th to 30th min, 43rd to 45th min, and from the 58th to the 60th min from the beginning of work.

3. RESULTS

3.1. Maximum Acceptable Weights

Maximum acceptable weights and heart rates are summarized in figures 1 and 2. Maximum acceptable weights decreased with an increase in the angle of asymmetry of lifting for all three frequencies and two height levels (figure 1). The subjects selected heavier weights for the 0.81 to 1.52 m height than for the floor level to 0.81 m height at all four angles of asymmetry and three frequencies (figure 1). Also, maximum acceptable weight decreased with an increase in lifting frequency for all four angles of asymmetry and two lifting heights (figure 1). A mixed model analysis of variance of maximum acceptable weight with repeated measures (where angles, frequency and height level were considered as fixed factors and subjects as random factors) showed that there were significant differences ($p < 0.01$) between the four angles, the three frequencies and the two height levels.

3.2. Heart Rate

Heart rate increased with an increase in angle of asymmetric lifting (figure 2). On the average, heart rate increased by about 6 beats min^{-1} while the maximum acceptable weight decreased by 5.9 kg at 90° as compared to 0°. The heart rates for the 0.81 m to 1.52 m height were consistently lower than those for the floor level to 0.81 m height (figure 2), though heavier weights were selected at the higher height.

There was an increase in heart rate with time for a given lifting task. For example, for lifting from the floor to bench height at 9 lifts min^{-1} and at 90° of asymmetry, the heart rate increased from about 124 beats min^{-1} at 15 minutes from the beginning of work to about 135 beats min^{-1} at the end of 60 minutes (figure 3). However, all heart rates at 60 minutes of work were less than 130 beat min^{-1} [5], except those for 9 lifts min^{-1} (figure 3).

4. DISCUSSION

A more useful analysis is to normalize the maximum acceptable weights in lateral planes with those in the sagittal plane. An analysis of variance of the percentages of maximum acceptable weight showed that while there were significant differences between angles ($p < 0.01$), frequency and height level had no significant effect ($p \geq 0.05$). The observed decreases were 9, 14, and 21% in maximum acceptable weight at 30, 60 and 90° of asymmetric lifting. These decreases are very similar to those reported by Garg and Badger [6] for occasional lifting. Thus, the existing data in the literature on the maximum acceptable weights for the symmetric sagittal-plane lifting can be multiplied by the above percentages to obtain equivalent values for asymmetric lifting.

5. CONCLUSION

This study has shown that the maximum acceptable weights for repetitive asymmetric lifting for one hour were significantly lower and the heart rates were significantly higher than those for symmetrical sagittal-

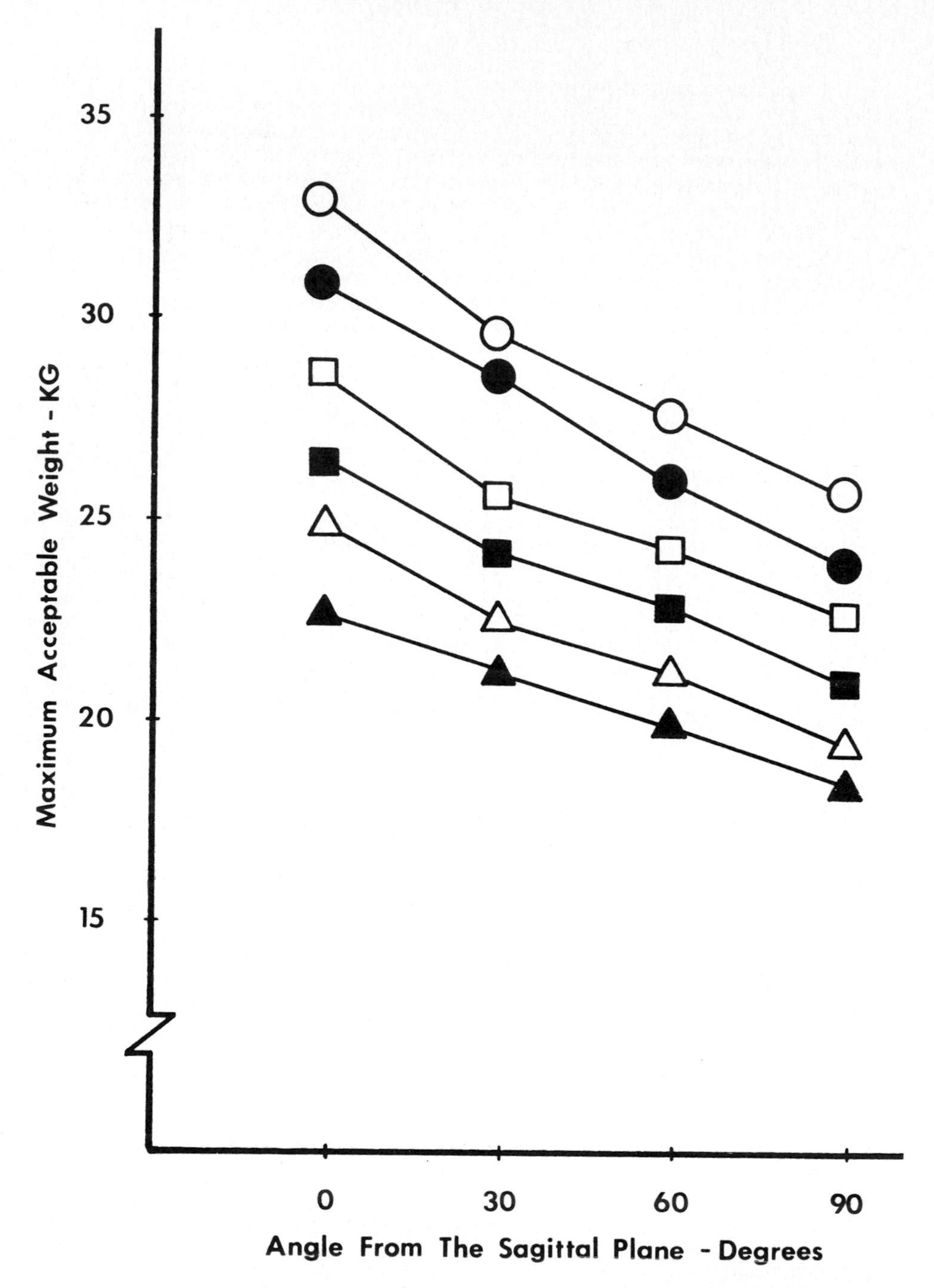

Figure 1. Effect of asymmetric lifting on maximum acceptable weight for lifting from floor to 0.81 m height (●, ■, ▲) and from 0.81 m to 1.52 m height (O, □, Δ) at 3 (●, O), 6 (■, □) and 9 lifts min^{-1} (▲, Δ).

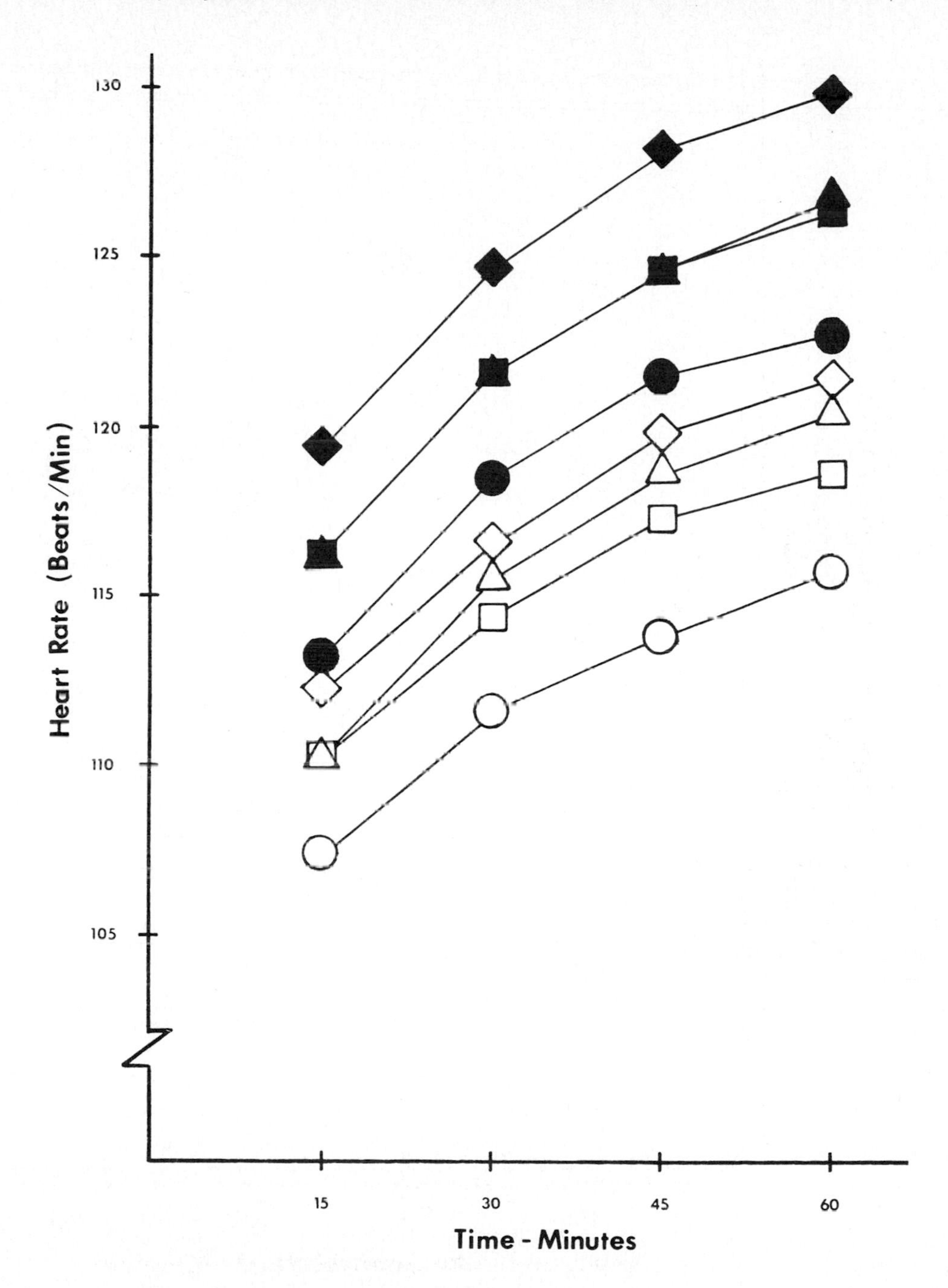

Figure 2. Heart rates for symmetric and asymmetric lifting as a function of time. Heart rates for lifting from floor to bench height (●, ■, ▲, ■,) and from bench height to 1.52 m height (O, □, Δ, ◇), at 0° (●, O), 30° (■, □), 60° (▲, Δ) and 90° (◆, ◇).

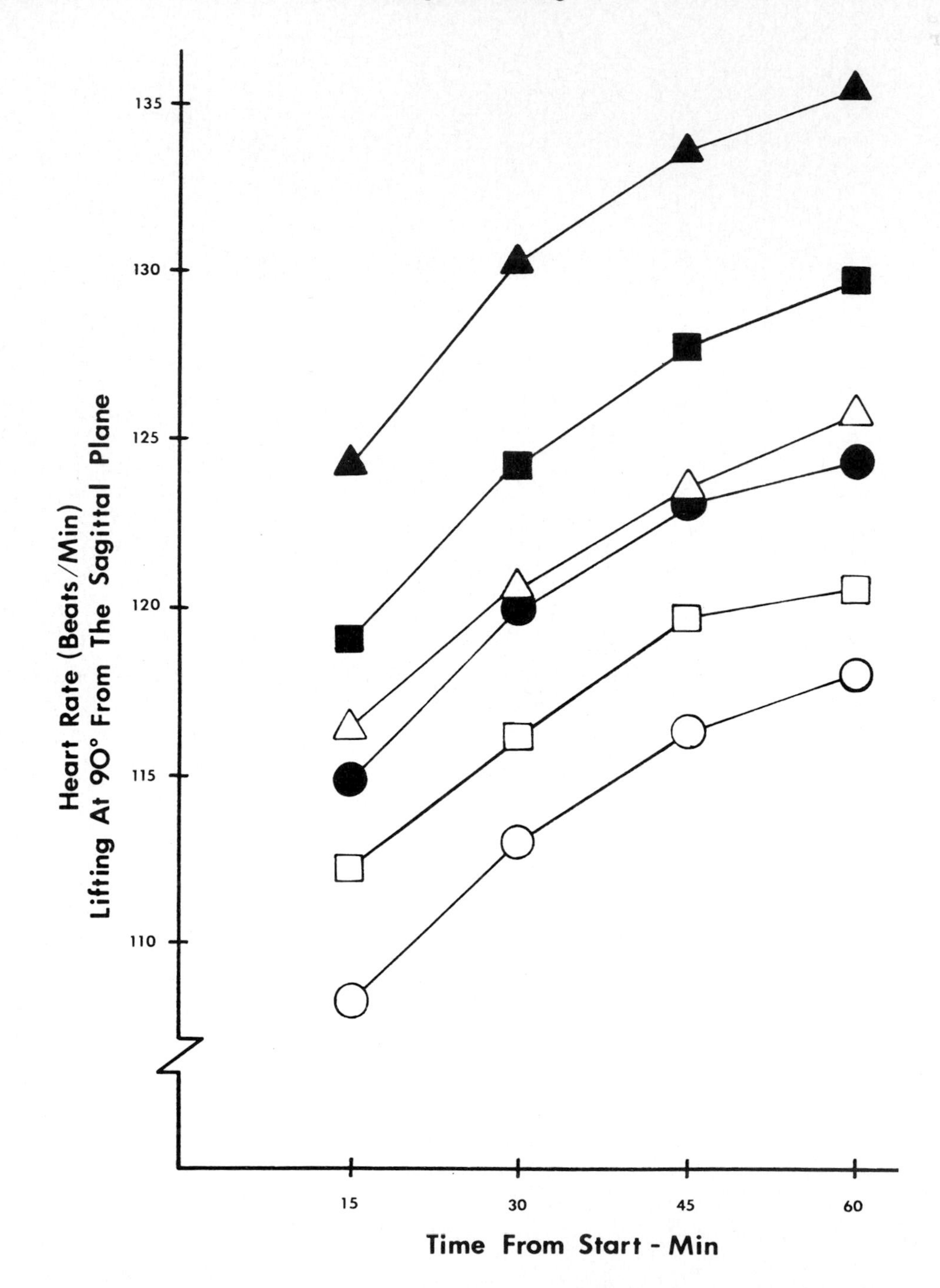

Figure 3. Effects of time and frequency on heart rate for lifting at 90° of asymmetry from floor to bench height (●, ■, ▲) and bench height to 1.52 m height (O, □, Δ) at 3 (●, O), 6 (■, □) and 9 lifts min^{-1} (▲, Δ).

plane lifting for all three frequencies and two height levels studied. Correction factors of 9, 14 and 21% for maximum acceptable weight at 30, 60 and 90° of asymmetric lifting are recommended to modify the existing data on the symmetric sagittal-plane lifting.

REFERENCES

[1] AYOUB, M. M., MITAL, A., BAKKEN, G. M., ASFOUR, S. S., and BETHEA, N.J., 1980, Development of strength and capacity norms for manual materials handling activities: the state of art. Human Factors, 22, 271-283.

[2] SNOOK, S. H., 1978, The design of manual handling tasks. Ergonomics, 21, 963-985.

[3] U. S. Department of Health and Human Services, 1981, Work practices guide for manual lifting. DHHS (NIOSH), Publication No. 81-122, Cincinnati, Ohio.

[4] SNOOK, S. H., and IRVINE, C. H., 1967, Maximum acceptable weight of lift. American Industrial Hygiene Association Journal, 27, 322-329.

[5] A.I.H.A. Technical Committee, 1971, Ergonomics guide to assessment of metabolic and cardiac costs of physical work. American Industrial Hygiene Association Journal, 32, 560-564.

[6] GARG, A., and BADGER, D.B., 1986, Maximum acceptable weights and maximum voluntary isometric strengths for asymmetric lifting, Ergonomics, 29, 879-892.

Trends in Ergonomics/Human Factors V
F. Aghazadeh (Editor)

DYNAMIC BIOMECHANICAL MODEL FOR ASYMMETRICAL LIFTING

H.C. Chen and M.M. Ayoub

Department of Industrial Engineering
Texas Tech University,
Lubbock, Texas 79409

A three dimensional biomechanical model and its associated representation tools had been developed to investigate the unbalanced loading on human body when performing non-symmetrical liftings. "ExpertVision" system by Motion Analysis was used to collect displacements of predefined targets on human body to describe the body movements and these data were used to predict the forces and moments acting on major body joints. Using linear programming techniques, the compression and shear forces were estimated on L5/S1 disc. The model was validated by comparing the model predicted force components and measured ground reactive ones collected by a "Kistler" force platform. Load distribution ratios between hip joints was also estimated by the use of force platform. Comparisons between symmetrical and non-symmetrical liftings involving 90 degrees right turn, as well as the effects of three different weight levels, were made in the current study.

1. INTRODUCTION

When manual material handling (MMH) tasks such as container palletizing is performed, more body movement, i.e. lateral bending and/or twisting, will be involved than when performing sagittal lifting. These tasks can be classified as asymmetric load handling. Under symmetric load handling condition, the stresses on the musculoskeletal system are equalized bilaterally. Due to the nature of asymmetrical lifting condition, unbalanced loading on the back muscles will take place, producing concentrated stresses on certain back muscle groups (NIOSH, 1981), and can be the cause of several orthopedic deformities (Hoogmartens & Stuyck, 1978). Evidence also shows the higher degree of severity of stresses during asymmetrical lifting than during symmetrical lifting. The relationship between the measured myoelectric activities of back muscles and the measured disc pressure for both nonsymmetrical and symmetrical lifting was investigated by Örtengren, et al. (1981). The results showed that at certain level of muscle activity the asymmetric lifting produces higher disc pressure than does the symmetric lifting. Kumar (1980) found significantly different relationships between the measured intra-abdominal pressure and electromyography (EMG) when performing lifting tasks in different planes. In order to compare the degree of stressfulness of asymmetrical lifting with symmetrical lifting, a dynamic biomechanical model was developed to study the particular human lifting motions across three dimensional space and to predict loads imposed on the musculoskeletal system. Of particular interest are the lower spinal loads which require more sophisticated and complicated models to predict the internal muscle exertion in the trunk.

2. The Lifting Tasks

Five male paid volunteers participated in the study. Each was asked to perform two types of lifting tasks, symmetric and asymmetric, with a total of 9 trials. For each type of task, the subjects were asked to perform lifting at three different weight levels with both feet on a force platform. Three other trials were performed for non-symmetric tasks with only the right foot on force platform.

3. The Video and Force Data Collection

The "ExpertVision" motion analysis system was used to record the displacement profiles of predefined body landmarks on the subjects. When the subject assuming the starting lifting position, a fixed reference axes for the system was defined with its origin set to the right of and behind the subject. The subject faced the positive x direction with y axis to his left and parallel to his frontal plane. According to right hand rule, the z axis was defined as positive toward the head. The "Kistler" force platform was used to collect ground reactive forces and moments, and the data were recorded through "DASH-16" AD converter installed in a MS-DOS microcomputer. The position of the landmarks and reactive forces and moments were collected simultaneously in real time and were synchronized by the use of a simple circuit design.

4. Data Reduction and Filtering

An indirect method was used to collect the desired body landmarks due to the complexity and difficulty of 3D data collection. A stick with two targets with predefined distance, which were actually tracked, was used to extend a body target if necessary. The desired body landmark was then estimated by the relationship of landmark and the two reference points on the stick. Both the final video data and force data were filtered by a digital low-pass filter routine as described in Winter (1979) before input to the biomechanical model.

5. Model Design and Development

5.1 The Biomechanical Model

A dynamic biomechanical lifting model containing 9 links of body segment equivalents was developed to predict the stresses on the following two categories of the human body:

(1) All major joints (ignoring the wrist joints)
(2) L5/S1 disc surface.

The model estimated the force and moment components about a predefined system axis for each joint as described in the model. The dynamic effects of body segment motions and only the trunk rotational effect were taken into account by the model. On the estimation of lower lumbar loads, a ten muscles trunk model developed by Schultz, et al., (1981 and 1982) was modified and used in the current study. All the force and moment components associated with L5/S1 disc were adjusted and calculated about a sub-coordinates system along its surface and with its origin situated at its center through the lifting process. The 3D biomechanical model was also designed to accommodate both symmetric and asymmetric, with 90 degree twist to the right, lifting conditions.

5.2 The Anthropometric Data

The model used a nine segment human body representation. The nine segments are lower arms, upper arms, trunk, upper legs and lower legs. The segment weight, segment length and center of mass locations were estimated as ratios to the body weight and body height (Chaffin and Andersson, 1984; Winter, 1979), respectively. Other necessary properties of body segments such as moment of inertia, radius of gyration were obtained from previous investigators' findings.

5.3 The Kinematics and Kinetics

With the body landmarks coordinates estimated using the Motion Analysis System, the kinematics of motion, were calculated. Using these kinematics, the kinetics such as forces and moments were calculated using free body diagrams. Assuming the body is in dynamic equilibrium, the instantaneous reactive forces are computed by adding the dynamic effects to the gravity (static) effect of segment weights at each joint. Thus, for any link j:

Reactive force = Static forces
+ Linear inertial force at CM of segment
+ Reactive force from j-1 joint

The moments at the joints were computed by adding the linear acceleration effects to the static effects, plus the inertial effects of the segment rotational moment. Then

Mj = Mj-1 + Static moment
+ moment due to linear acceleration effects
+ moment yield by reactive forces from adjacent joint j-1
+ inertial moment due to rotation of link

where:
Mj = Reactive moment vector acting about joint j
Mj-1 = Reactive moment vector from j-1 joint.

The force and moment components were calculated starting with the upper extremities. The estimated components then transmitted through subsequent segments to the ankle joints which were defined as ground level.

5.4 The Load Distribution Ratio at Hip Joints

When the loads were transmitted to the trunk, the reactive forces and moments were first estimated about the midpoint between the hip joints. The loads which the right hip joint sustained was expected to be greater than the one by the left hip joint. These load distribution ratios were estimated by the ratios of collected reactive vertical forces of right foot to the sum of model predicted ground forces at two feet. The model then was fine tuned using these ratios to find the loads at hip joints.

5.5 The Trunk Model

The second section of this model estimated the external loads acting on the center of L5/S1 disc and , in turn, predicted the internal muscle forces, compressive and shear forces on disc surface. A modified three dimensional static model introduced by Schultz and Andersson (1981) which included ten major trunk muscle equivalents spanning the lumbar region was implemented. Referring to Figure 1 for the schematic diagram of the trunk model, an imaginary plane was assumed cutting through the body at L5 level and parallel with the surface of the L5/S1 disc. A coordinate sub-system was defined with its origin at L5/S1 center, and coordinate directions were selected as: Y axis positive to the left, X axis positive anteriorly, and Z axis positive superiorly. The x and y axes were contained in the transverse cutting plane while the Z axis was perpendicular to cutting plane through the center of the L5/S1 disc. All the external forces and moments were assumed to act at the L5/S1 center and about this coordinate system. It was assumed that this cutting plane is stiff enough to maintain the orientations of muscle groups relative to center of L5/S1 disc constant. Following the requirements to keep the upper body in stable condition, the Newtonian equations for force and moment equilibriums must be met. Thus, for force components, referring to Figure 1 for the force notations and orientations:

$$Fx = (Ir+Il) * \sin \beta - (Xr+Xl) * \sin \delta + Sa$$

$$Fy = (Ll-Lr) * \sin \gamma - Sr$$

$$Fz = C + P - (Er+El) - (Ll+Lr) * \cos \gamma - (Rl+Rr) - (Ir+Il) * \cos \beta - (Xr+Xl) * \cos \delta$$

and for moment components

$$Mx = (Rl-Rr) * Yr + (El-Er) * Ye + (Ll-Lr) * \cos \gamma * Ye + (Il-Ir) * \cos \beta * Yo + (Xl-Xr) * \cos \delta * Yo$$

$$My = (Er+El) * Xe - (Rr+Rl) * Xr + (Lr+ll) * \cos \gamma * Xl - (Ir+Il) * \cos \beta * Xo - (Xr + Xl) * \cos \delta * Xo + P * Xp$$

$$Mz = (ll-Lr) * \sin \gamma * Xl + (Ir-Il) * \sin \beta * Yo - (Xl-Xr) * \sin \delta * Yo$$

The solutions for the above six equations could be obtained by using a linear programming technique with "minimum compression force on the spine" as the objective function. Schultz, et al., (1981 and 1982) found that such techniques produced good agreement between computed muscle tension and measured magnitude of EMG data. Due to the limits of muscle strengths, it was assumed that the muscle contraction intensities not exceed a reasonable level such as 100 N/cm^2 (Schultz, et al, 1981).

5.6 Determination of L5/S1 Disc Orientation

During the process of asymmetric lifting, the trunk will experience flexion and rotational motions. The tilted angle of L5/S1 disc to the system x axis was adjusted as a function of the angles between trunk and thigh (El-Bassoussi, 1974). Two imaginary lines, (1) connecting midpoint of shoulder joints and midpoint of hip joints, and (2) connecting midpoint of hip joint and midpoint of knee joints, were used to describe the flexion of the trunk and upper legs. The rotational angle of cutting plane was determined by the projection angle of the hip on transverse plane formed by rotating y axis located on L5/S1 disc, and the fixed y reference axis for whole system.

5.7 Estimation of Medial Rotational Effect

The model considered only the medial rotational effect of the trunk link. The trunk rotation angle was estimated at center of mass level and was computed taking a ratio of relative rotation angle of shoulder to the one of hip girdle. The ratio was taken as the location of C.M. of trunk from the hip to the trunk length.

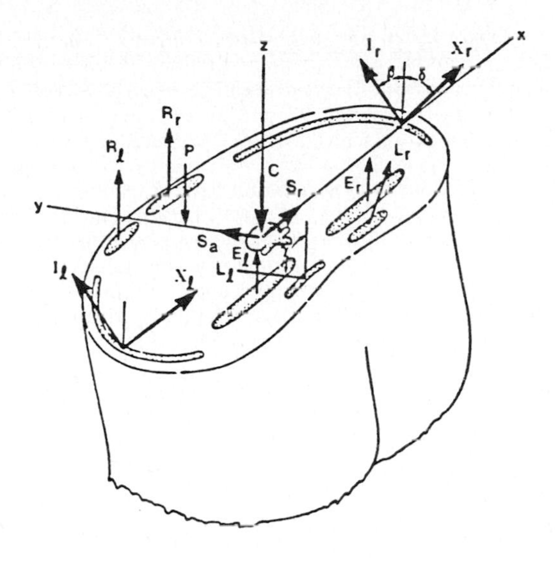

Rr = Right Rectus Abdominus
Rl – Left " "
Lr = Right Latissimus Dorsi
Ll = Left " "
Ir = Right Internal Oblique
Il = Left " "
Xr = Right External Oblique
Xl – Left " "

Er = Right Erector
El = Left "
P= Intra-abdominal Pressure

C = Compression
Sr = Right lateral shear
Sa = Anterior Shear

Figure 1. The ten muscles trunk model (Schultz, 1981)

5.8 The Intra-abdominal Pressure

The trunk model also took the intra-abdominal pressure into account using the estimation equation developed by Andersson, et al. (1984). The values of IAP's, as a function of resultant moment at L5 level, were calculated for every body posture during the course of the lift. The acting area of this pressure is estimated as 0.465 of the product of trunk width and depth at L5/S1 level (Schultz, et al. 1982). The varying lever arm of IAP was expressed as a function of angles of the trunk and thigh from the vertical (El-Bassoussi, 1974).

6. Results and Discussions

6.1 Gross Model Validation

The sum of model predicted reactive forces at ankle joints was used to compare against the measured ground reactive forces. These comparisons were made for both symmetric and asymmetric type of liftings at each force component level. The results are shown in Figure 2. The vertical forces followed a consistent pattern and the difference at the end of the lift was possibly due to the subject's "drop box". The other two force components were quite smaller with some possible noise presented.

6.2 Effects of Weights on Spinal Compression and Shear

The subjects used the psychophysical weight selection criterion to determine their individual one time maximal weight of lift. This weight was used to classify the individual's percentile capacity from the lift tables presented by Snook (1978). Using these tables, the capacity at six lifts per minute was selected as the heaviest weight for the subject. Two other lower weights were selected. Figure 3 showed the spinal loads profiles produced by three different weights. Generally speaking, the higher weight did yield higher values of compression and shear forces than the ones of lower weight for both lifting conditions.

6.3 Comparison of Asymmetric with Symmetric Liftings

The differences of spinal loads, i.e. compression and shear forces, had been studied for both asymmetric and symmetric lifting tasks. In spite of different weight levels used, the compression yielded by symmetric tasks were higher than the one by asymmetric tasks for a certain weight. However, the shear forces, anterioposterial and lateral, were higher for asymmetric tasks than for symmetric. After examining the lifting motions through stick diagram, it was found that the lifting postures were very similar for both tasks in the starting phase of lifting and the profiles of spinal loads appeared to be similar. At the last stage of lifting, with more body twisting for asymmetric lifting, the differences of spinal loads were more significant. The time histories of lateral shear in symmetric liftings were found oscilating around a certain value as compared to zero value assumed in sagittal models in previous studies. It was believed that the lifting performed not in sagittal plane in reality.

6.4 Model Structure and Output Presentation

The three dimensional dynamic biomechanical model, the linear programming routine and the associated text and graphic presentation were written in Microsoft Quickbasic and integrated as a package. Dynamic three dimensional stick diagrams, lifting motion history in a single plane, kinematic and kinetic data can be viewed by using the presentation section.

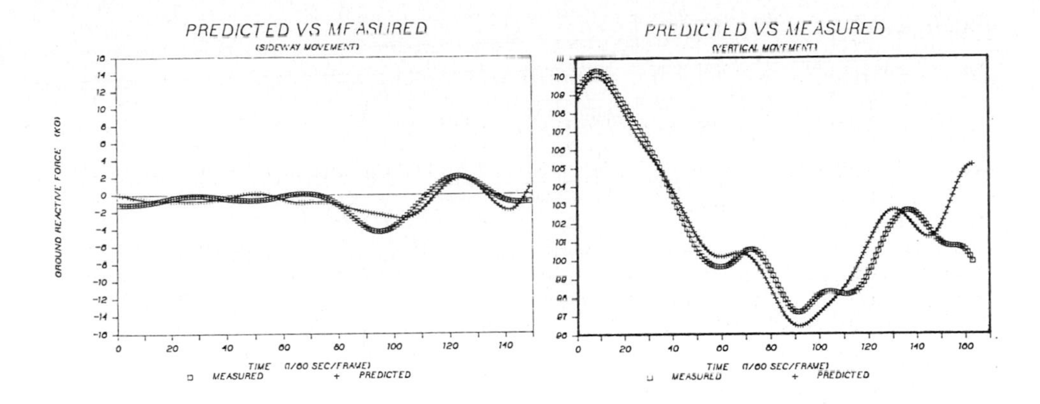

Figure 2. Comparison between predicted values and measured in vertical and sideways movements.

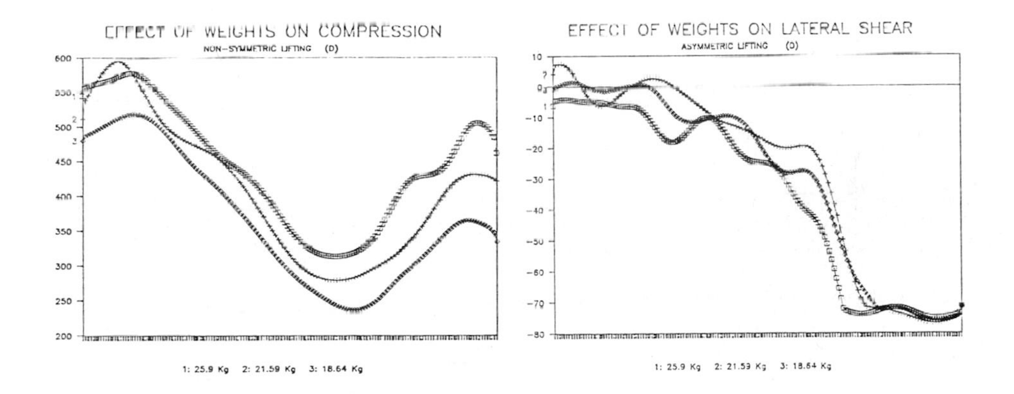

Figure 3. The compression and lateral shear yielded by three weights under asymmetric lifting condition.

7. References

[1] Anderson. C.K., Chaffin, D.B., Herrin, G.D., and Matthews, L.S., A Biomechanical Model of the Lumbosacral Joint During Lifting Activities, J. Biomechanics, v. 18, no. 8, (1985) 571-584.

[2] Chaffin, D.B., and Andersson, G.B.J., Occupational Biomechanics, (John Wiley & Sons, 1984).

[3] Chen, H.C., Biomechanical Identification of Musculoskeletal Stresses During Asymmetric Lifting: A Dynamic 3-D Approach, unpublished Ph D. thesis, (Industrial Engineering), Texas Technology University, Lubbock Texas, 1988.

[4] El-Bassoussi, M.M., A Biomechanical Dynamic Model for Lifting in the Sagittal Plane, unpublished Ph.D. thesis (Industrial Engineering), Texas Technology University, Lubbock, Texas, 1974.

[5] Hoogmartens, M.J., and Stuyck, J.A., Vibration Electromyography As a Differential Measurement of the Postural Tone in the Left and Right Spinal Muscles, Biomechanics VI, (University Park Press, 1978).

[6] Konz, S., Dey, S., and Bennett, C., Forces and Torques in Lifting, Human Factors, 15(3), (1973) 237-245.

[7] Kumar, S., The Physiological Cost of Three Different Methods of Lifting in Sagittal and Lateral Planes, Ergonomics, 23, (1980) 987-993.

[8] Muth, M.B., An Optimization Model for the Evaluation of Manual Lifting Tasks, thesis, North Carolina State University, Raleigh, (1976).

[9] National Institute for Occupational Safety and Health, A Work Practices Guide for Manual Lifting, Tech. Report No. 81-122, U.S. Dept., of Health and Human Services (NIOSH), Cincinnati, OH, (1981).

[10] Ortengren, R., Andersson, G.B.J., and Nachemson, A.L., Studies of Relationships Between Lumbar Disc Pressure, Myoelectric Back Muscle Activities, and Intra-abdominal (Intragastric) Pressure, Spine, vol. 6, no. 1, Jan/Feb. (1981).

[11] Schultz, A.B., Andersson, G.B., Haderspeck, K., Ortengren,R., Nordin, M., and Bjork, R., Analysis and Measurement of Lumbar Trunk Loads in Tasks Involving Bends and Twists, J. Bio., vol. 15, no. 9, (1982) 669-675.

[12] Schultz, A.B., Andersson, G., Ortengren, R., Haderspeck, K., Nachemson, A., Loads on the Lumbar Spine, Validation of A Biomechanical Analysis by Measurements of Intradiscal pressures and Myoelectric Signals, J. Bone and Joint Surg., 64-a, (1982) 713-720.

[13] Schultz, A.B., and Andersson, B.J.G., Analysis of Loads on the Lumbar Spine, Spine, 6(1), (1981) 76-82.

[14] Snook, Stover H., The Design of Manual Handling Tasks, Ergonomics, VOL. 21, no. 12, (1978) 963-985.

[15] Winter, D.A., Biomechanics of Human Movement, (John Wiley & Sons, Inc., 1979).

Trends in Ergonomics/Human Factors V
F. Aghazadeh (Editor)

THE VALIDITY OF PREDETERMINED MOTION TIME SYSTEMS IN SETTING WORK STANDARDS FOR MANUAL MATERIALS HANDLING

A.M. GENAIDY [*], A. MITAL [**], M. OBEIDAT [*] and S. Puppala [*]

[*] Department of Industrial Engineering
Western Michigan University
Kalamazoo, Michigan 49008, U.S.A.

[**] Ergonomics Research Laboratory
Dept. of Mechanical and Industrial Eng.
University of Cincinnati
Cincinnati, Ohio 45221-0072, U.S.A.

In recent years, predetermined motion time systems (PMTS) have been growing in popularity as an industrial engineering tool to set time standards. This may be attributed to the release of the United States Air Force MIL STD 1567A, the development and spread of microcomputers, and the subjectivity of time studies. The main objective of this study is to evaluate the use of PMTS in setting work standards for manual materials handling operations.

1. INTRODUCTION

Predetermined motion and time systems (PMTS) have been growing in popularity in recent years. This could be attributed primarily to two factors. First, the release of the military standard MIL STD 1567A by the United States Air Force has enhanced to a great extent the spread of PMTSs. The military standard states that type I standards representing 80% coverage of all touch labor are to be established using a recognized and accurate technique to reflect an accuracy level of 10% at a 90% confidence at the operational level. Second, computer-based PMTSs offer improved accuracy as well as greater speed in determining and applying standards. According to Shell [1], such systems are particularly appropriate where short cycle, repetitive operations are involved, or where high productivity losses or excess labor complaints occur.

The main objective of this research is to evaluate PMTS in setting work standards for manual mateials handling (MMH).

2. REVIEW

2.1. Definition

A PMTS is defined as an "organized body of information, procedures, techniques, and motion times employed in the study and evaluation of manual work elements. The system is expressed in terms of the motions used, their general and specific nature, the conditions under which they occur, and their previously determined performance times" [2].

2.2. Assumptions and Limitations

An allowed time in PMTS is determined by (1) listing the basic motions required for a task, (2) associating the time values with the listed basic motions, (3) adding the time values to obtain the base time, and (4) adjusting whatever allowances are needed for the operation [3].

In essence, PMTSs are based on two fundamental assumptions. These are: (1) an accurate, universal time value can be assigned to each basic motion, and (2) the addition of individual basic motion time values will be equal to the time for the whole task "the sum of the parts is equal to the whole" [3].

Nadler [3] questioned the validity of the additivity assumption on which PMTSs are based. He cited the following mathematical example to illustrate his viewpoint. Suppose that (1) three basic motions have equal average time (t) and standard deviation (sd), and (2) there is no interaction among the t's. The coefficient of variation (cv) for a basic motion is equal to (sd/t). The additivity assumption states that T = t1 + t2 + t3 = 3t where T is the overall time to perform the three motions in sequence. The standard deviation (SD) for T is

$$SD = (sd_1^2 + sd_2^2 + sd_3^2)^{1/2} = 3^{1/2}\ sd.$$

The coefficient of variation (CV) for T is

$$CV = (SD/T) = 3^{1/2}\ sd/3\ t = sd/3^{1/2}\ t,$$

or in general form,

$$CV = sd\ /\ t\ n^{1/2}\ .$$

As the number of basic motions increases, the coefficient of variation of T will decrease. Nadler concluded from this example that the accuracy of standard time based on a PMTS will get better as the job gets longer. For shorter cycle jobs, the usefulness of PMTSs for setting time standards depends on the desired accuracy.

Nanda [4] noted that (1) most PMTSs are proposed as a solution to the estimation of the mean performance time for a given work cycle comprised of a specific series of basic motions, and (2) no solution is given to the estimation of the variance of the performance time for a given work cycle comprised of a specific series of basic motions under manual and particularly under machine paced conditions. Nanda pointed out that extensive tables of the expected time values of of basic motions are available from a number of different sources. However, these tables make no mention of the variance of the time values of basic motions or their correlation with other basic motion times. Thus, Nanda suggested to test the assumption that the covariances are independent of the specific nature of the adjacent elements.

Sanfleber [5] reported that the accuracy of PMTSs must focus, among other things, on the following points: (1) possible inaccuracies due to the pecularities of a particular system, e.g., not exactly definable motion elements, unclear rules for the work analysis, insufficient consideration influential factors or elemental times with non-uniform performance levels, and (2) errors may result from incorrect application of a given system, e.g., when the work analyst does not recognize all motion elements in the course of the analysis or overlooks or misinterprets certain work complications.

2.3 Time Study Assessment

Ghiselli and Brown [6] (1967) pointed out that the atomistic conception of human behavior is open to serious criticism. According to these authors, the atomistic notion of human behavior means that each individual movement of a total complex task can be considered as a separate and independent unit; if one movement is eliminated then the total time to complete the task is reduced in proportion to that which was required in the execution of that movement as part of the complete cycle. Ghiselli and Brown conducted experiments aimed at disproving the atomistic concept. Subjects were presented with six keys arranged in the following fashion:

A D F
B C E

and were instructed to tap them in a certain sequence, the time needed to make the movement between each pair of keys being recorded. Some of the movements were eliminated in a second sequence, and times of movements again were recorded. Each set of records was taken after the subjects had mastered the subject and had reached their maximum speed of movement. The results showed that the original sequence required 2.42 sec. The elimination of two movements that required 0.71 sec was expected to reduce the total time to 1.71 sec (2.42-0.71=1.71). The actual time required to perform the second sequence was 1.81 sec. All but one of the movements retained in the second sequence required more time than they had required in the original sequence. In a second experiment, opposite results were obtained. The elimination of some of the movements reduced the time much more than was expected. The time required to perform one of the remaining movements in the new sequence was reduced significantly.

Abruzzi [7] conducted a series of experiments to test the

independence assumption of PMTSs. The results indicated that while in some instances independence among basic motions could be achieved, in other instances independence could not obtained. Abruzzi commented on his findings by stating that (1) the relationships among elements largely depend on the operator, (2) the number and magnitude have an important effect on the independence of elements. Abruzzi suggested that as a rule of thumb, independence is much more likely when the shortest element lasts at least five hundredths and the median component lasts at least ten hundredths.

Nadler and Wilkes [8] and Nadler and Denholm [9] conducted experiments aimed at testing the additivity assumption of PMTSs. Based on their results, the authors concluded that the assumption was not valid within the experimental conditions reported.

Hall [10] reported the results of an experiment which provided evidence that the basic motions are related. Hall stated that the removal of one basic motion because of its undesirability may well be reflected in other elements. Hall concluded that if production is to remain constant, variations in time of the individual basic motions must be and are compensated for by either an increase or decrease in the time of the other basic motions which compose the cycle. This compensation may occur among the basic motions of one hand or between basic motions of both hands.

Schmidtke and Stier [11] carried out seven series of experiments aimed at testing several assumptions of PMTSs. Schmidtke and Stier questioned that none of the existing PMTSs considered the effect of the direction of motion on the time values assigned to basic motions. Their results showed that (1) the time for a motion at an angle of about 145o with the front plane of the body and over a distance of 40 cm is about 20% greater than at an angle of about 55o, and (2) vertical motions are performed about 10% more quickly than horizontal motions. The results of another series of experiments showed that a difference of about 60% in motion time for comparable arm motions without weight and with a weight of 1.5 kg, respectively. Schmidtke and Stier concluded that there are few grades within the weight classes of existing PMTSs, the lowest of which ranges from 0 to 1.5 kg in the MTM system and from 0 to 0.9 kg in the Work Factor system. In a third series of experiments, the authors tested whether or not the times required to perform a sequence of motion elements are directly dependent on each other. This was achieved by setting two targets of 3 mm and 30 mm in diameter next to each other and opposite to a single target of 30 mm in diameter at a distance of 36 cm. The subject experimented upon were asked to touch in turn either of the two targets lying side by side with a contact pin, and to start these motions from the single target. The results indicated that the time for the backward motion from a target of 30 mm or 3 mm in diameter is dependent on the size of the target to an extent that after having touched a target of 3 mm the backward motion takes an average of 22% longer than the way back from the target of 30 mm in diameter. Schmidtke and Stier further pointed out that existing PMTSs overlooked the effect of repetition frequency on motion time. They asked their subjects to do an assembly at a frequency of about 2 sec and then later to do it at a frequency of about 1 min. During the series of assembly work interrupted by different activities and carried through at a frequency of 1 minute, the increase in time amounted to between 35 and 62%. The last objection of Schmidtke and Stier to PMTSs is that the time values of basic motions are not constant. They reported that when omitting a basic motion from an operation, the time per operation is not reduced by the time value of this omitted basic motion, but the remaining time values of basic motions increase by an average of about 10%. They concluded that the time values of basic motions are not mathematical quantities which may be added or subtracted without hesitation.

A series of comments appeared in the Journal of Industrial Engineering following the publication of the study reported by Schmidtke and Stier [11]. Sellie [12], Taggart [13], Honeycutt [14], and Bailey [15] suggested that Schmidtke and Stier [11] did not apply properly MTM, Work Factor, and BTM systems in arriving at the standard times for some of the tasks performed in their experiments. Moreover, some of these authors pointed out that Schmidtke and Stier have insufficient knowledge of the various PMTSs. Davidson [16] indicated that while the research by Schmidtke and Stier provides a basis for rejecting the scientific validity of certain PMTSs, an inquiry into the operational validity of PTSs was not made.

Schmidtke and Stier [17] responded to the comments raised about their earlier publication. The main emphasis was to disprove some of the comments made by Sellie [12], Taggart [13], Honeycutt [14], and Bailey [15].

Konz and Rode [18] investigated the control effects of weights of 0.12, 1.25, 2.38, and 3.25 lbs on hand-arm movements in the horizontal plane. Their results showed that (1) time increased 6% per lb, (2) the subject weight has no effect, and (3) cross body movements should be given 7% additional time. Konz and Rode compared their experimental results with those of MTM, Work Factor, and BMT PMTSs. They arrived at the following conclusions: (1) MTM gives inadequate importance to weight; the slope of the time-weight relationship is too flat, and no time values were assigned to any weight less than 2.5 lb, a conclusion reached earlier by Schmidtke and Stier [11], (2) BTM also gives inadequate emphasis to weight, providing an increase of 0.018 sec/lb instead of the 0.026 lb/sec found in Konz and Rode's data, and (3) Work Factor seems to have an appropriate slope for males but its interval which is 5 lb is too big.

Buffa [19], Buffa [20], and Buffa and Lyman [21] pointed out that the expected value of the sum of three random variables x, y, and z connected by a joint density function f(x,y,z) is equal to the sum of their separate expected values, i.e., E(x+y+z) = E(x) + E(y) + E(z). This is true regardless of interaction between the variables. Buffa and Lyman [21] conducted an experiment to determine whether or not additivity of basic motions holds true for the overall cycle time predictions despite interactions among the basic motions. Time measurements were made in a light assembly task requiring 16 basic motions in the complete cycle and 10 basic motions in the incomplete cycle for 16 male subjects. The results indicated that the total incomplete cycle times predicted from data obtained in the complete cycle did not differ significantly from times actually measured even though there was evidence of interactions among the basic motions and the variables. Buffa and Lyman concluded that the additivity assumption seem to be valid where several basic motions are involved. Other research work by Buffa [18-19] reached similar conclusions.

Nanda [22] conducted three experiments aimed at testing the uniqueness, additivity, and independence of basic motions of PMTSs. These assumptions were defined in Nanda's research as follows: (1) uniqueness means that the work elemental time developed in the context of one task can be applied in the context of work cycles of different tasks, (2) independence means that the time for a work element is independent of work elements preceding and following it, and (3) additivity implies that the mean time to perform a given work cycle comprised of a specific series of basic work elements, is equal to the sum of the mean times of the individual work elements. The test of uniqueness was based on the hypothesis that u1 = u2 = u3 = ... = uk where u's represent the mean time for the same element in each of the K different experimental sequences. In some instances, Cochran's test of homogeneity of variances was also carried out. The tests of independence and additivity were based on the hypothesis that the sum of the variances of individual elements in a work cycle is equal to the cycle variance. A correlation test between work elements was also performed by constructing a covariance matrix within a period of one cycle. For each covariance, a product moment correlation was then computed and the null hypothesis was tested. From the covariance matrix, Wilk's multivariate test for independence was also performed for each experiment as discussed by Abruzzi (1952). The results of the statistical analyses revealed the following:

(1) The analysis of variance showed no significant difference between mean times of work elements developed in different sequences of work cycles.
(2) The correlation tests showed no consistent pattern; variations were primarily due to subjects. There appear to be occurrences of correlation between transport and

release, specifically in cases where release of parts is involved rather than a contact release.

(3) Additivity was valid for a high percentage of subjects even though some elemental performance times did show some correlation.

According to Nanda, the rejection of Wilk's multivariate test of independence in every experimental case, as a basis for rejecting the validity of elemental mean time systems, seems to be an improper conclusion for the operational basis. Nanda concluded that the assumptions of unique, independent, and additive mean elemental times are valid.

Sanfleber [5] conducted experiments to investigate whether the additivity of basic motions can be considered sufficiently accurate in practice. The findings showed statistically significant effects resulting from the overall pattern of motion. However, the effects of such motion influences differed in character from one person to another, not only in their absolute amount but frequently also in their trend. This means that for one group the introduction of a certain change in the motion cycle resulted in extended and for another in decreased times for the individual elemental times considered. Consequently, the effects of motion influences sometimes eliminated each other when the test results of several persons were engaged. Sanfleber concluded that generally the effects of motion influences which occurred, even though they were found to be statistically significant, were of little practical importance. However, motion influences could be of greater importance in practice when the motion pattern was such that a distinct rhythm could be developed.

3. DISCUSSION

Based on the literature search conducted in this study, it is not clear whether the assumptions of PMTS are met. There were two opposing views regarding this issue. One group of investigators is in support of the validity of PMTSs assumptions (e.g., uniqueness, independence, and additivity), while the other group raised serious doubts with respect to the validity of these assumptions. The problem is more severe when there is weights involved in operations performed in different planes of motion. The effect of weight handled and motion planes on time standards is not well addressed in the various PMTSs. Furthermore, the majority of PMTSs do not take into consideration the effect of various task variables (height of lift, container size, etc.) and worker variables (age, sex, body weight, anthropometric variables, etc.) on time standards. It is, therefore, concluded that research is warranted to further examine PMTSs under different job conditions and operators of various characteristics.

4. REFERENCES

[1] Shell, R.L. (Editor), Work measurement: principles and practice (Institute of Industrial Engineers, Industrial Engineering and Management Press, Technology Park/Atlanta, Norcross, Georgia, 1986) pp. 191.
[2] Industrial Engineering Terminology (Institute of Industrial Engineers, Industrial Engineering and Management Press, Norcross, Georgia, 1983).
[3] Nadler, G., Work design (Richard C. Irwin, Homewood, Illinois, 1963).
[4] Nanda, R., J. of I. E., XVIII (1), 120-122, 1967.
[5] Sanfleber, H., I. J. of Prod. Res., 6 (1), 25-45, 1967.
[6] Ghiselli, E.E. and Brown, C.W., Personel and industrial psychology (McGraw-Hill, New York, 1948).
[7] Abruzzi, A., Work, workers, and work measurement (Columbia University press, Morningside Heights, New York, 1956).
[8] Nadler, G. and Wilkes, J.W., Advanced Management, XVIII, 20-22, 1953.
[9] Nadler, G. and Denholm, D.H., 1955. J. of I. E., VI, 3-4, 1955.
[10] Hall, N.B., Jr., J. of Appl. Psyc., 40 (2), 91-95, 1956.
[11] Schmidtke, H. and Stier, F., J. of I. E., XII (3), 182-204, 1963.
[12] Sellie, C., J. of I. E., XII (5), 330-333, 1961.
[13] Taggart, J.B., J. of I. E., XII (6), 422-427, 1961.
[14] Honeycutt, J.M., Jr., J. of I. E., XIII (3), 172-179, 1962.
[15] Bailey, G.B., J. of I. E., XII (5), 328-330, 1961.
[16] Davidson, H.O., J. of I. E., XIII (3), 162-165, 1962.
[17] Schmidtke, H. and Stier, F., J. of I. E., XIV (3), 119-124, 1963.
[18] Konz, S. and Rode, V., AIIE Transactions, 4 (3), 228-233, 1972.
[19] Buffa, E.S., J. of I. E., VII (2), 217-223, 1956.
[20] Buffa, E.S., J. of I. E., VIII (6), 327-333, 1957.
[21] Buffa, E.S. and Lyman, J., J. of I. E., 42 (6), 379-383, 1958.
[22] Nanda, R., J. of I. E., XIX (5), 235-242, 1968.

Trends in Ergonomics/Human Factors V
F. Aghazadeh (Editor)

THE COMPARISON OF THE ARM CRANK AND CYCLE EXERCISE TESTS FOR THE PREDICTION OF THE AEROBIC AND ANAEROBIC WORK PERFORMANCE IN MANUAL MATERIAL HANDLING TASKS

Veikko LOUHEVAARA and Pentti TERÄSLINNA

Department of Physiology
Institute of Occupational Health
Helsinki, Finland

Department of Health, Physical Education and Recreation
University of Kentucky
Lexington, Ky, USA

Physiological responses including gas exchange, blood lactate, heart rate, blood pressure, and perceived exertion were studied in submaximal and maximal tasks of manual parcel handling, and compared to the corresponding responses obtained during arm crank and cycle exercise. The purpose of the comparisons was to establish how well the arm crank and cycle exercise tests would predict the actual aerobic and anaerobic work performance during parcel handling. The subjects were 21 healthy male sorters of postal parcels. The physiological responses to manual handling of postal parcels substantially differed from those to arm cranking but they were almost equal to those obtained during cycling. The cycle exercise test can be considered to be appropriate and more valid than the arm crank exercise test for the prediction of the aerobic and anaerobic work performance in manual material handling tasks that require in addition to lifting, continuous moving and carrying of loads.

1. INTRODUCTION

Studies on aerobic and anaerobic responses to manual material-handling have mainly been focused on lifting tasks, and the responses have been compared with various techniques and frequencies of lifting and load weights (1, 2, 3, 4, 5).

In spite of the high research activity on the physiology of manual material-handling (6, 7) there is a lack of information on tasks requiring in addition to lifting, continuous moving and carrying of loads. Furthermore, knowledge is scarce about aerobic and anaerobic responses to material-handling in relation to upper body exercise with an arm crank ergometer (8) as well as to leg exercise with a cycle ergometer.

Consequently, the purpose of the present study was (i) to compare the physiological responses to the parcel handling and sorting to those obtained during arm crank and cycle exercise, and (ii) to establish how well the arm crank and cycle exercise tests would predict the actual aerobic and anaerobic work performance during parcel sorting.

2. MATERIAL AND METHODS

2.1. Subjects

Twenty-one healthy male sorters of postal parcels volunteered for the study (Table 1). Their mean (±SD) work experience at this job was 10.3±6.2 years.

Table 1. Subjects' physical characteristics. Maximal oxygen uptake ($\dot{V}O_2max$) and maximal external power output were assessed during cycle and arm crank exercise, and during parcel sorting. The values given are the means±SD for 21 subjects.

Age (years)	33.3 ± 5.9
Height (cm)	178.4 ± 7.2
Weight (kg)	78.3 ± 12.7
Body fat[a] (%)	18.4 ± 5.3
$\dot{V}O_2max$ ($l\ min^{-1}$)	
Cycling	3.24 ± 0.44
Arm cranking	2.54 ± 0.32
Parcel sorting	2.18 ± 0.78
Max power output (W)	
Cycling	274 ± 9
Arm cranking	155 ± 8
Parcel sorting	<10

[a]According to Durnin and Rahaman (9)

2.2. Test procedures

The work test simulating manual sorting of postal parcels and incremental maximal arm crank and cycle exercise tests were carried out in the laboratory at temperate conditions.

The cycle exercise test started with sitting at rest on the cycle orgomotor for 4 minutes. The first external work load was 50 W, and the load was increased by 50 W every fourth minute. The pedalling rate was kept at 50 revolutions per minute. The subjects were asked to continue pedalling until exhaustion. At each submaximal work load the exercise was interrupted after 3 minutes for 30 seconds while blood pressure was measured, a sample of venous blood was taken, and the rate of perceived exertion (RPE) was asked (10). Thereafter the exercise was continued at the preceeding work load to the end of the period of 4 minutes. Blood pressure was measured and venous blood sample was taken also before exercise at rest and immediately after the maximum.

The procedure during the arm crank exercise test was similar to the cycle exercise test with the exception of body posture and external work loads. The first load was 25 W, and it was increased by 25 W every forth minute. The subjects stood with no torso restrains during arm cranking including the pauses for 30 seconds.

The parcel sorting test was carried out in a simulated workplace constructed in the laboratory with equipment supplied by the Central Post Office. One hundred parcels with standardized weights and sizes were made for the sorting test. The mean weight of the parcels was 5.1 kg, the range being 0.5 - 17.5 kg. Each parcel had a postal code which was either 01260 or 01620 (equal numbers of each). The parcels were sorted according to the postal code from a parcel container onto two trollies at a distance of 2.5 m.

The procedure and measurements of the sorting test were similar with the exercise tests. At the beginning of the sorting test the container was full of parcels in a random order. After an initial rest period of 4 minutes sitting, the subjects sorted parcels with free style at the following work rates: slow, habitual, accelerated (piecework rate), and maximal. The work rate was increased every fourth minute. After 3 min work at slow, habitual and accelerated work rate the sorting was interrupled for 30 seconds for blood sampling and blood pressure and RPE measurements. Habitual, accelerated and maximal work rates were selected individually by the subjects. The slow work rate was fixed at 3 parcels min^{-1}.

2.3. Equipment and measurements

In the exercise and work tests, gas exchange (respiratory frequency, f; tidal volume, V_T; minute ventilation, $\dot{V}_E$; oxygen uptake, $\dot{V}O_2$; carbon dioxide output, $\dot{V}CO_2$; and respiratory exchange ratio, R) was continuously measured by a Morgan Exercise Test System. Lactic acid concentration in venous blood was analyzed emzymatically (11).

Heart rate (HR) was recorded every minute with a Sport Tester PE 3000 Monitor. The systolic blood pressure (SBP) was measured with the conventional auscultatory technique.

3. RESULTS

During slow sorting of 3 parcels min^{-1} the mean (±SD) $\dot{V}O_2$ was 1.03±0.25 l min^{-1}, and at maximal work rate (17±8 parcels min^{-1}) 2.18±78 l min^{-1}. At the comparable mean $\dot{V}O_2$ levels of about 1 and 2 l min^{-1} the external work loads during arm cranking were 25 and 100 W, respectively. During cycling the corresponding work loads were 50 and 150 W, respectively (Fig. 1).

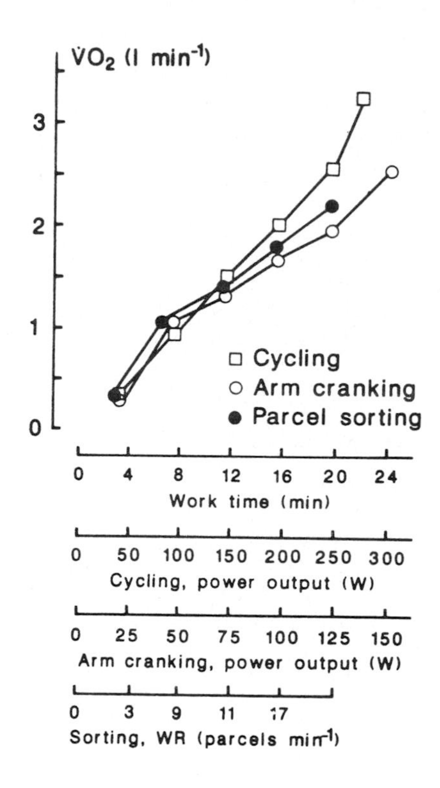

FIGURE 1
The oxygen uptake ($\dot{V}O_2$) in relation to external power output during cycle and arm crank exercise and to work rate (WR) during parced sorting. The values are means for 21 subjects.

According to the analysis of variance and the student's t-test at the mean $\dot{V}O_2$ level of about 1 l min^{-1} all tested responses (V_T, f, $\dot{V}_E$, R, La, HR, SBP, and RPE) with the exception of f and SBP, were significantly lower during parcel sorting than during arm cranking. Howerer the differences in responses to parcel sorting and cycling were minor. Only in SBP the difference appeared to be statistically significant at the 5 % level (Table 2).

Table 2. The comparison of the physiological responses to cycling, parcel sorting and arm cranking at the mean $\dot{V}O_2$ level of about 1 l min^{-1}.

	Cycling	p(t)	Parcel sorting	p(t)	Arm cranking	p(F)
$\dot{V}O_2$ (l min^{-1})	0.96 ± 0.11	NS	1.03 ± 0.25	NS	1.09 ± 0.16	NS
f (breaths min^{-1})	17.2 ± 4.2	NS	17.6 ± 3.1	NS	19.2 ± 4.3	NS
V_T (l)	1.27 ± 0.29	NS	1.19 ± 0.28	<.001	1.47 ± 0.32	<.01
$\dot{V}_E$ (l min^{-1})	21.2 ± 4.8	NS	20.4 ± 3.4	<.001	27.4 ± 4.4	<.001
R	0.76 ± 0.09	NS	0.77 ± 0.11	<.001	0.87 ± 0.09	<.001
La (mmol l^{-1})	1.4 ± 0.5	NS	1.7 ± 0.7	<.001	2.7 ± 0.8	<.001
HR (beats min^{-1})	92 ± 14	NS	91 ± 14	<.001	106 ± 18	<.01
SBP (mmHg)	151 ± 13	<.05	138 ± 14	NS	142 ± 14	<.05
RPE	0.4 ± 0.5	NS	0.3 ± 0.3	<.05	0.7 ± 0.7	<.05

Table 3. The comparison of the physiological responses to cycling, parcel sorting and arm cranking at the mean $\dot{V}O_2$ level of about 2 l min^{-1}.

	Cycling	p(t)	Parcel sorting	p(t)	Arm cranking	p(F)
$\dot{V}O_2$ (l min^{-1})	2.05 ± 0.12	NS	2.18 ± 0.78	NS	1.97 ± 0.24	NS
f (breaths min^{-1})	22.1 ± 3.6	NS	25.2 ± 6.0	NS	29.0 ± 4.5	NS
V_T (l)	2.24 ± 0.48	<.01	1.81 ± 0.43	<.01	2.11 ± 0.30	<.01
$\dot{V}_E$ (l min^{-1})	48.0 ± 4.8	NS	46.5 ± 21.1	<.05	60.6 ± 8.5	<.01
R	0.96 ± 0.07	<.05	0.91 ± 0.07	<.001	1.04 ± 0.07	<.001
La (mmol l^{-1})	2.0 ± 0.6	<.05	2.7 ± 1.6	<.001	5.5 ± 1.3	<.001
HR (beats min^{-1})	135 ± 18	NS	139 ± 30	<.01	154 ± 21	<.05
SBP (mmHg)	181 ± 20	<.001	158 ± 24	NS	158 ± 19	<.001
RPE	2.8 ± 2.1	NS	2.4 ± 1.3	<.05	3.9 ± 1.9	<.05

At the mean $\dot{V}O_2$ level of about 2 l min^{-1} each response during parcel sorting except f and SBP still remained significantly lower when compared to arm cranking. Furthermore, during parcel sorting SBP, V_T and R also remained significantly smaller than during cycling. On the other hand, at this $\dot{V}O_2$ level La during parcel sorting increased significantly higher than during cycling (Table 3).

4. DISCUSSION

At the mean $\dot{V}O_2$ levels of about 1 and 2 l min^{-1} physiological responses to the arm crank exercise considerably differed from those to parcel sorting and cycle exercise. The responses to parcel sorting and cycling were almost equal particularly at the $\dot{V}O_2$ level of about 1 l min^{-1}, showing that this type of sorting included a large component of dynamic muscle work. However, during parcel sorting V_T and SBP values constantly remained lower than during cycling. This trend in V_T was probably due to the static work of the upper body, limiting free motion of the thorax. In the continuous material-handling tasks the reduction of V_T may cause hypoventilation and disturb gas exchange. In the present analysis the development of insufficient gas exchange during parcel-sorting may be seen in lower R values at the mean $\dot{V}O_2$ level of about 2 l min^{-1} when compared to the cycle exercise. The unexpected SBP responses to parcel sorting and exercise tests remain unexplained.

The present results indicate that the cycle exercise test can be considered to be appropriate and more valid than the arm crank test in predicting the aerobic and anaerobic work performance of a parcel sorter. In all probability this was due to the dominant effect of the large lower body skeletal muscle mass involved in the manual handling and sorting of postal parcels.

5. CONCLUSIONS

The following conclusions can be reached:

1. The physiological responses to manual handling and sorting of postal parcels substantially differed from those to arm cranking but they were almost equal to those obtained during cycling.

2. The cycle exercise test can be considered to be appropriate and more valid than the arm crank exercise test for the prediction of the aerobic and anaerobic work performance in manual material handling tasks that require in addition to lifting, continuous moving and carrying of loads.

ACKNOWLEDGEMENTS

This study was supported by the Central Post Office and the Ministry of Finance. The authors wish to acknowledge Päivi Piirilä, M.D., Olli Korhonen, M.D., Susanna Salmio, Maija Jokinen, Sirpa Leino and Eeva Tolin for their skillful assistance during the laboratory experiments.

REFERENCES

(1) Snook, S.H., 1978, The design of manual handling tasks. Ergonomics, 1,963-985.
(2) Petrofsky, J.S., and Lind, A.R., 1978, Metabolic, cardiovascular and respiratory factors in the development of fatigue in lifting tasks. Journal of Applied physiology: Respiratory, Environmental and Exercise Physiology, 45, 64-68.
(3) Petrofsky, J.S., and Lind, A.R., 1978 a, Comparison of metabolic and ventilatory responses of men to various lifting tasks and bicycle ergometry. Journal of Applied physiology: Respiratory, Environmental and Exercise Physiology, 45, 60-63.
(4) Garg, A., and Saxena, U., 1979, Effects of lifting frequency and techniques on physical fatigue with special reference to psycho-physical methodology and metabolic rate. American Industrial Hygiene Association Journal, 40, 894-903.
(5) Peacock, B., 1980, The physical workload involved in parcel handling. Ergonomics, 23, 417-424.
(6) National Institute of Occupational Safety and Health, 1981, Work practices guide for manual lifting, DHHS (NIOSH) Publication no. 81-122, Cincinnati, OH.
(7) Troup, J.D.G., and Edwards, F.C., 1985, Manual handling and lifting. An information and literature review with special reference to the back. (London: Her Majesty's Stationary Office), p. 22-28.
(8) Sawka, N.M., 1986, The physiology of upper body exercise. Exercise and Sport Sciences Reviews, 14, 175-211.
(9) Durnin, J.V.G.A., and Rahaman, N.M., 1967, The assessment of the amount of fat in human body from measurements of skinfold thickness. British Journal of Nutrition, 21, 681-689.
(10) Borg, G., Ljunggren, G., and Ceci, R., 1985, The increase of perceived exertion, aches and pain in the legs, heart rate and blood lactate during exercise on a bicycle ergometer. European Journal of Applied Physiology and Occupational Physiology, 54, 343-349.
(11) Karlsson, J., Jacobs, I., Sjödin, B., Tesch, P., Kaiser, P., Sahl, O., and Karlberg, B., 1983, Semi-automatic blood lactate assay: experiences from an exercise laboratory. International Journal of Sports Medicine, 4, 45-48.

Trends in Ergonomics/Human Factors V
F. Aghazadeh (Editor)

DOES THE ADVANTAGE OF LORDOTIC POSTURE OVER STRAIGHT BACK POSTURE TRANSLATE INTO EXTRA LIFTING CAPACITY?

Anil MITAL

Ergonomics Research Laboratory
Mechanical and Industrial Engineering Department
University of Cincinnati
Cincinnati, OH 45221-0072, U.S.A.

This study was undertaken to determine if the biomechanical advantage of lordotic posture over straight back posture translates into additional lifting capacity. Sixteen males voluntarily participated in the study. The maximal psychophysical lifting capacity of each subject was determined for three different box sizes for the floor to 32" height level. Two different lifting postures were employed by the subjects: lordotic and straight back. The results indicated no difference in the lifting capacity for the two postures. The changes in lifting capacity with box size were as expected. It was concluded that the biomechanical advantage of the lordotic posture does not translate into additional lifting capacity.

1. INTRODUCTION

The configuration human body assumes during manual lifting has always been of interest to researchers interested in preventing overexertion injuries and low back pain. Many different lifting postures have been studied over the years. In particulr, stoop (straight legs), squat (straight back), and free-style (semi-squat) postures have received attention in previous scientific investigations (Brown(1971), Garg and Herrin (1979), Kumar and Magee (1982), Leskinen et al. (1983), Kumar (1984)).

The present state of the art suggests that for manual lifting tasks, the free-style posture might be the best. Between stoop and squat postures, squat posture is biomechanically less stressful provided the load is compact enough to be lifted from between the legs (Garg and Herrin (1979), Leskinen et al. (1983)). Bulky containers, on the other hand, can produce greater spinal moments if a squat posture is employed. The metabolic costs associated with the squat posture have also been reported to be higher (Garg and Herrin (1979), Kumar and Magee (1982), Kumar (1984)). Among the three common lifting postures, squat, stoop, and free-style, the free-style method of lifting is metabolically least expensive (Brown (1971)).

Generally it is accepted that there is no single correct method to lifting loads. The NIOSH work practices guide for manual lifting (1981) acknowledges that instructions to "correct" lifting postures should be avoided until their complexities are well understood. However, both the National Safety council (1971) and the International Labour Office (1967) recommend Kinetic lifting posture. The Kinetic lifting posture is basically a straight back, not vertical back, and bent knees posture which takes into account the positioning of the feet and body momentum of the individual while lifting. The benefits of Kinetic lifting when the lift is entirely vertical are however questionnable.

In recent study, Hart (1985) compared several different postures and concluded that lordotic posture (back is curved inside by the same extent as it is when a person stands up straight) minimizes spinal moments as compared to straight back, or kyphotic, and bent knees posture. The obvious question this conclusion leads to is: Does this advantage (reduced spinal moment) result in enhanced lifting capacity? The purpose of this study was to answer this question.

2. METHODS

2.1. Subjects

Sixteen males voluntarily participated in this study. Their age ranged between 20 and 23 years. The height and body weight of the subjects ranged between 1.69 m and 1.9 m and 60.9 kg and 99 kg, respectively. All subjects were in good physical condition. Prior to participating in the experiment, all subjects were instructed in the lifting procedure. Work clothes were worn during the experiment.

2.2. Experimental Design

A randomized complete block factorial design was used. Subjects were the blocks. The independent variables included the box size (12 inches, 18 inches, and 24 inches long in the sagittal plane; all boxes were 8 inches high and 12 inches wide) and posture (straight back - bent knees posture and curved back (lordotic) - bent knees posture). The maximum acceptable weight of lift for a single lift was the response variable. Lifting was performed in the floor to 32 inch height region. The climatic factors were controlled (temperature 21° - 22°c; relative humidity 45% - 55%).

2.3. Experimental Procedure

The psychophysical methodology was used by the subjects to determine their 'maximum acceptable weight of lift' for single lifts for each of the six treatment combinations (3 box sizes x 2 postures). In order to determine the maximum acceptable weight of lift, subjects were started, randomly, with either a very light weight or very heavy weight and were permitted to make adjustments to it (addition or subtraction of weight). As many adjustments as needed were permitted. However, at least 15 minutes elapsed between successive trials. The final adjusted weight was measured. Data for only one combination was collected on any given day. Several subjects however participated on the same day. The order of treatments was randomized for each subject.

3. RESULTS

The data were analyzed by carrying out an analysis of variance (ANOVA) on the maximum acceptable weight of lift. Table 1 shows the results.

Table 1 Analysis of Variance with Weight as the Dependent Variable

Source	DF	SS	F-Value	p(%)
Model	20	39155.79	12.87	0.01
Error	75	11413.05		
Total	95	50568.84		
Subjects	15	35920.88	15.74	0.01
Posture	1	3.97	0.03	87.22
Box Size	2	3038.16	9.98	0.01
Posture*Box Size	2	192.78	0.63	53.36

As shown in Table 1, the affect of posture was insignificant. The effects of subjects and box size, as expected, were highly significant. The interactive effect between posture and box size was also not significant.

Table 2 Overall Means, Standard Deviations and Ranges of the Maximum Acceptable Weight of Lift (Kg)

Variable	Mean	Standard Deviation	Range
Posture			
Straight Back	51.00	11.00	25.00-72.00
Curved (Lordotic) Back	51.18	10.05	29.51-69.46
Box Size			
12 inches	54.00	10.01	32.69-71.96
18 inches	51.47	9.90	27.47-68.55
24 inches	47.79	10.84	24.97-69.46

Table 3 Overall Means and Standard Deviations of the Maximum Acceptable Weight of Lift (Kg) for Various Posture-Box Size Combinations

Posture	Box-Size	Mean	Standard Deviation
Straight Back	12"	53.96	11.48
	18"	50.49	10.52
	24"	48.38	10.98
Curved Back	12"	54.07	8.68
	18"	52.46	9.48
	24"	47.27	11.04

Tables 2 and 3 show the overall means for posture and box size and for the six treatment combinations. As evident from these tables, posture effect is non-existent.

4. DISCUSSION

The purpose of this work was to determine if the biomechanical advantage of the lordotic posture results in greater lifting capacity compared to the straight back posture. The intent was not to verify Hart's (1985) results. And for that reason, no EMG data were collected. During the experiment, subjects were however asked to verbally indicate which posture they considered less strenuous.

The lack of difference in the liftng capacities for the two postures tends to suggest that stress on the spine is not related to the trunk flexion moment. The advantage of the lordotic posture, which leads to lower trunk flexion moments (Hart (1985)), therefore, were not translated into increased lifting capacity.

The subjects, unanimously, indicated that the lordotic posture was more stressful. In their opinion, they had to make an extra effort to curve the back inside. It was felt that if they had to lift repeatedly (frequent lifting), maintaining inside curvature would be impossible.

The results and subjective opinions obtained in this study leave one with the impresion that advantages of the lordotic posture, at least at this stage of awareness, may not lead to any practical gains in terms of increased lifting capacity.

5. CONCLUSIONS

This study investigated the effects of lordotic and straight back postures on psychophysical lifting capacity. Since no posture effect was observed, it was concluded that the biomechanical advantage (lower trunk flexion moment) of the lordotic posture does not translate into extra lifting capacity.

REFERENCES

[1] Brown, J.R., Lifting As An Industrial Hazard, Labour Safety Council of Ontario, Ontario Department of Labour, Canada.
[2] Garg, A. and G.D. Herrin, Stoop or Squat: A Biomechanical and Metabolic Evaluation, AIIE Transactions 11 (1979) 293-302.
[3] Hart, D.L. Effect of Lumbar Posture on Lifting, Ph.D. Dissertation, West Virginia University, Morgantown, West Virginia, U.S.A. (1985).
[4] Himbury, S., Kinetic Methods of Manual Handling in Industry, Occuptional Safety and Health Series, No. 10, I.L.O., Geneva, Switzerland (1967).
[5] Kumar, S., The Physiological Cost of Three Different Methods of Lifting in Sagittal and Lateral Planes, Ergonomics 27 (1984) 425-433.

[6] Kumar, S. and D.J. Magee, Energy Cost of Lifting in Sagittal and Lateral Planes by Different Techniques, Proceedings of the VIIIth Congress of the International Ergonomics Association, Inter-Group, Tokyo, Japan (1982) 644-645.
[7] Leskinen, T.P.J., H.R. St. Elhammer, I.A.A. Kuorinka and J.D.G. Troup, A Dynamic Analysis of Spinal Compression with Different Lifting Techniques, Ergonomics 26 (1983) 595-604.
[8] National Institute for Occupational Safety and Health, Work Practices Guide for Manual Lifting, Cincinnati, Ohio, U.S.A. (1981).
[9] National Safety Council, Human Kinetics and Lifting, National Safety News (1971) 44-47.

XIII

WORK PHYSIOLOGY

Trends in Ergonomics/Human Factors V
F. Aghazadeh (Editor)

ON DEVELOPMENT OF AN INDEX OF PHYSICAL FITNESS

Suebsak NANTHAVANIJ, Ph.D. and Howard GAGE, Ph.D.

Department of Industrial and Management Engineering
New Jersey Institute of Technology
Newark, NJ 07102

A laboratory experiment was conducted to investigate the possibility of using parameters derived from a person's heart rate profile for an index of physical fitness. Ten male subjects performed a physical activity which involved moving loads between two work stations. Heart rate was continuously monitored and several parameters were derived from its profile such as: the difference between working and normal heart rates, the ratio of working to normal heart rate, and the recovery decay rate. Statistical comparisons of individual parameters between two different test days did not show any significant differences (at the 0.05 level). Some parameters indicated the potential for being used in an index of physical fitness since they reflected the difference among subjects and were partly correlated with their physical conditions.

1. INTRODUCTION

A physiological definition for "Physical Fitness" is still not clear. In sports, atheletic competition represents the classical test of physical fitness or performance capacity. To be more specific, comparing between two competitors with similar atheletic skill, a person who is more physically fit will have a better chance of winning. Performance capacity can be measured objectively (in terms of time or distance) or subjectively (as in gymnastics, figure skating, or diving). It is the combined result of the coordinated exertion, and integration of a variety of functions (Astrand and Rodahl [1]).

In industry, physical fitness may be defined as an ability of a person to handle and maintain vigorous industrial activity without fatigue (or more appropriately, less fatigue) for a prolonged period. Many factors contribute to a physically fit body performing physical work. They include an ability of the body to deliver nutrients to the working muscle fiber, oxygen uptake, cardiac output, etc. (Astrand and Rodahl [1]).

Physical work load may be assessed either by measurement of oxygen uptake during operation, or by indirect estimation of oxygen uptake on the basis of heart rate recorded during work performance. Although an analysis of oxygen uptake provides a direct assessment of physical work load, the use of heart rate has gradually gained in interest among researchers since it can be accomplished relatively easily. Current technology enables the researcher to study heart rate via a wireless portable heart rate monitor which functions as both electrical transmitting and receiving units. This type of instrument also allows subjects to move freely without interference from the cables.

Grandjean [2] stated that for light work, heart rate increases quickly to a so-called working level, which remains relatively constant for the duration of work, and returns to normal a few minutes after cessation. With more strenuous work however, the heart rate continues to increase until either work has ceased, or the worker is forced to stop due to exhaustion. These phenomena imply that, when working at the same level, the heart rate profile of a more physically fit person will follow the former pattern while that of a less physically fit person will follow the latter. This implication suggests that differences in heart rate porfiles may be used to indicate relative levels of physical fitness.

Several recent research findings were reported regarding the use of changes in heart rate in physical work load assessment. A variety of factors which could influence these changes were investigated including: work pace, (Salvendy and Knight [3]; Das [4]), work procedures, (Ilmarinen and Louhevaara [5]), environment, (Gertner et al [6]; Smith et al [7]), apparel (Riley et al [8]), and load characteristics, (Genaidy et al [9]). Results of these findings will be discussed under Conclusions. Some heart rate characteristics used as criteria for physical work load assesment included: maximum heart rate, heart rate response, and differences between working and normal heart rates. However, it was noticed that each of these characteristics was studied on a qualitative basis; hence the conclusions drawn were limited due to variations in physical condition of the subjects.

This paper describes an attempt to quantify the level of physical fitness using key parameters derived from a person's heart rate profile. Initially, as many parameters as possible were investigated. These included: normal heart rate, working heart rate, difference between working and normal heart rates, ratio of working to normal heart rate, and recovery decay rate. The objective of this study was twofold. Firstly, the repeatability of those aforementioned was tested. Secondly, the differences among subjects for each of the individual parameters were correlated with personal history factors such as: exercise habits, smoking habits, and current physical status.

2. EXPERIMENTAL DESIGN

Ten male subjects participated in this research study. Their ages ranged between 21and 29 years. Certain selected physical characteristics such as age, body weight, and body height were recorded as summarized in Table 1. Additionally, smoking and exercise habits were also noted.

TABLE 1
Summary of Biographical and Certain Physical Characteristics Data

	Age (yr)	Weight (hg)	Height (m)
Mean	23.10	75.93	1.79
Std. Dev.	2.64	12.77	0.08

Subjects were required to wear a special plastic belt and wristwatch on their bodies. The plastic belt functioned as a heart rate transmitter while the wristwatch was a receiver. The belt was worn around the subject's chest just below his arm pits. It housed two conductive rubber electrodes which collected heart rate signals and transmitted them to the receiver. The wristwatch received the signals continuously and saved them in its memory at five second intervals.

The experiment used in this study simulated a repetitive physical task usually found in an industrial environment. It involved carrying loads from one place to another at a normal work pace. Two work stations were set up in the Human Factors Laboratory. The distance between them was 7.8 meters. The task was to move sixteen boxes from one work station to another. The weights of boxes varied between 9.07 and 15.88 kg, with mean and standard deviation values of 14.43 and 2.62 kg., respectively. Experimental procedures consisted of the following steps:

The experiment started with the subject resting in a seated position for two minutes and then walking around for another two minutes to warm up. He began his task by moving one box at a time from the first work station to the second at a work pace of 2.33 km/hr. After all sixteen boxes were relocated, he then moved them back following the same procedure. This task lasted approximately eight minutes. After it was finished, the subject walked around for two minutes before returning to his seat. He would then sit and rest for another eleven minutes until his heart rate returned to its normal level. The entire experiment lasted twenty five minutes and was repeated once more on another test day. In summary, the experiment involved the following five steps.

1. Rest (2 minutes)
2. Pre-work (2 minutes)
3. Work (8 minutes)
4. Post-work (2 minutes)
5. Rest (11 minutes)

During the experiment, heart rate was continuoulsy monitored. At the end of each step, time markers were entered into the wristwatch memory storage by the experimentor. Then the heart rate data stored in the memory storage of the receiver was downloaded to the computer and saved on computer diskette for further analysis.

3. RESULTS

Figure 1 shows a plot of the heart rate profile for one subject. The various time marks represent changes from one experimental step to another. Specifically, time marks numbered 1,2,3, and 4 indicate the end of initial rest, the end of pre-work, the end of work, and the end of post-work, respectively. As can be seen, this heart rate profile agrees with those described by Grandjean [2]. A person's heart rate increases rapidly as soon as he starts his pre-work and work activities. When the heart rate reaches its elevated level, it remains relatively unchanged until the cessation of work. Then, it immediately decreases to its normal level even though physical activity (post-work) is continued at a lighter level for a certain period of time.

The plots in Figure 2 show the heart rate profiles of two subjects, one who smokes regularly and the other who is not a smoker and exercises regularly. It is clear that although both subjects exhibit similar heart rate profiles, the smoker has both higher resting and working heart rates than the non smoker. Nevertheless, the heart rate profiles of individual subjects show remarkable similarity when compared on two different test days (Figure 3).

The heart rate profiles shown in Figures 1-3 also indicate that rate of recovery of heart rate can be expressed in terms of the following exponential model.

$$Y = ab^X$$

where: Y = Heart rate data (beats/minute)
X = Recovery time (seconds)
a,b = Parameters

Exponential regression was performed to obtain a heart rate recovery model from all heart rate profiles. It was found that the coefficient of correlation ranged between 0.70 and 0.97, with a mean of 0.88.

Key parameters derived from the heart rate profile can be summarized as follows.

1. R1 (Initial resting heart rate): The average of the heart rate data taken during the first two minutes.

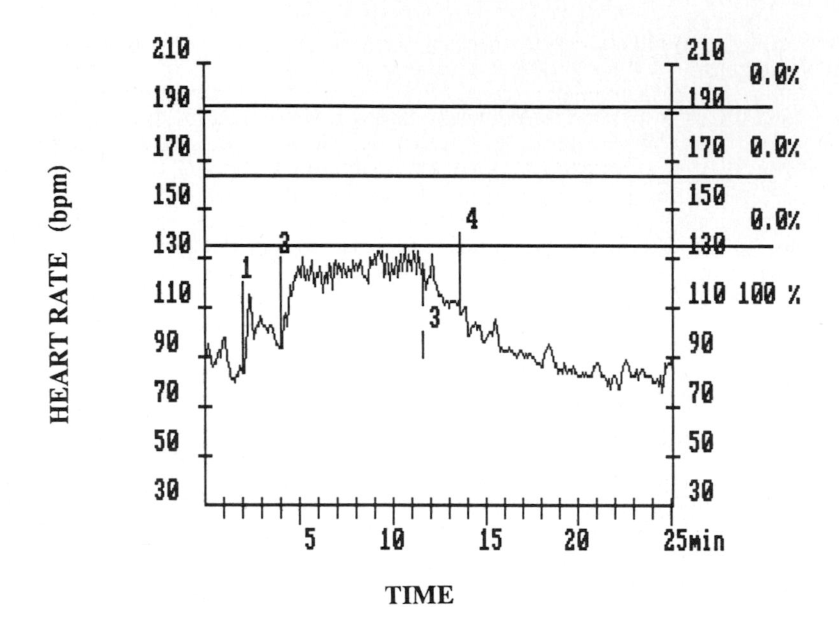

FIGURE 1
A Plot of Heart Rate (bpm) Against Time (Second)

2. W (Working heart rate): The average of the heart rate data taken during the last two minutes of work.
3. R2 (Final resting heart rate): The average of the heart rate data taken during the last two minutes of the experiment.
4. W-R1: The difference between the working and the initial resting heart rate.
5. W/R1: The ratio of the working to the initial resting heart rate
6. RDR (Recovery decay rate): The ratio of the natural logarithm of slope to the natural logarithm of intercept, as determined from the plot of heart rate data taken during four minutes after the cessation of work against recovery time.

The respective means and standard deviations for these parameters are given in Table 2 for both test days 1 and 2. Statistical analysis was performed using a one-way ANOVA to compare between the two test days. The results did not indicate any significant differences (at a level of 0.05).

4. CONCLUSIONS

The results of this study clearly indicate the possibilities for deriving an index of physical fitness from a person's heart rate profile. Though the small subject sample does limit the use of these results, several interesting facts are demonstrated. All parameters investigated show remarkable repeatability. Due to the ease of use of the portable monitoring unit, an indirect assessment of physical performance via heart rate profiles provides an alternative for researchers to investigate the relative physiological cost of work as it is being performed.

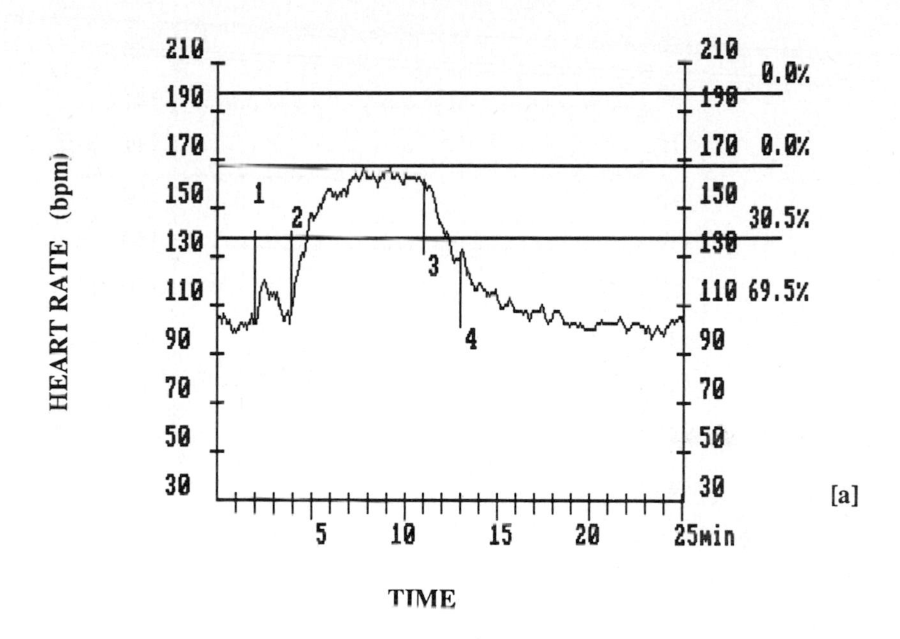

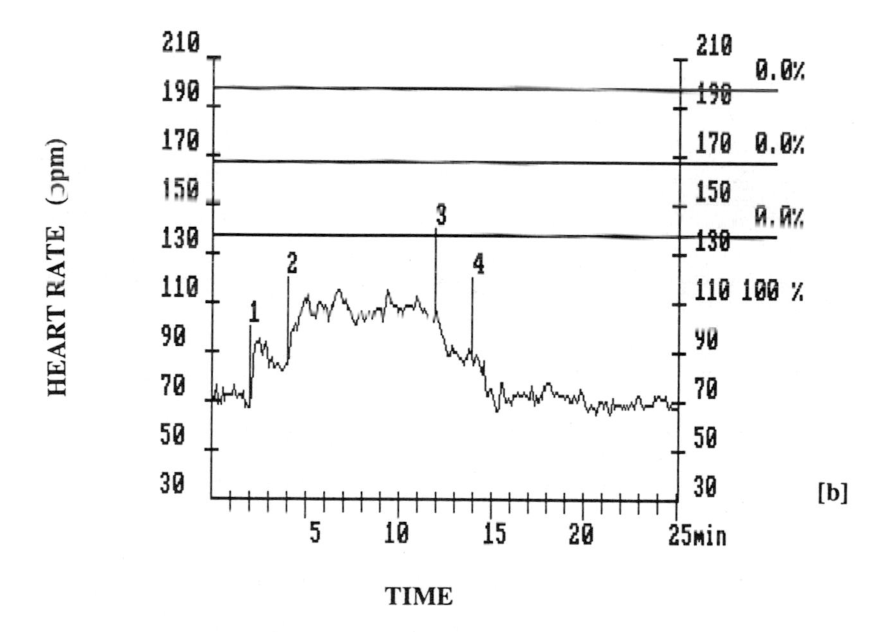

FIGURE 2
Heart Rate Profiles of a Smoker (a) and a Non-Smoker (b) under the same Physical Work Load

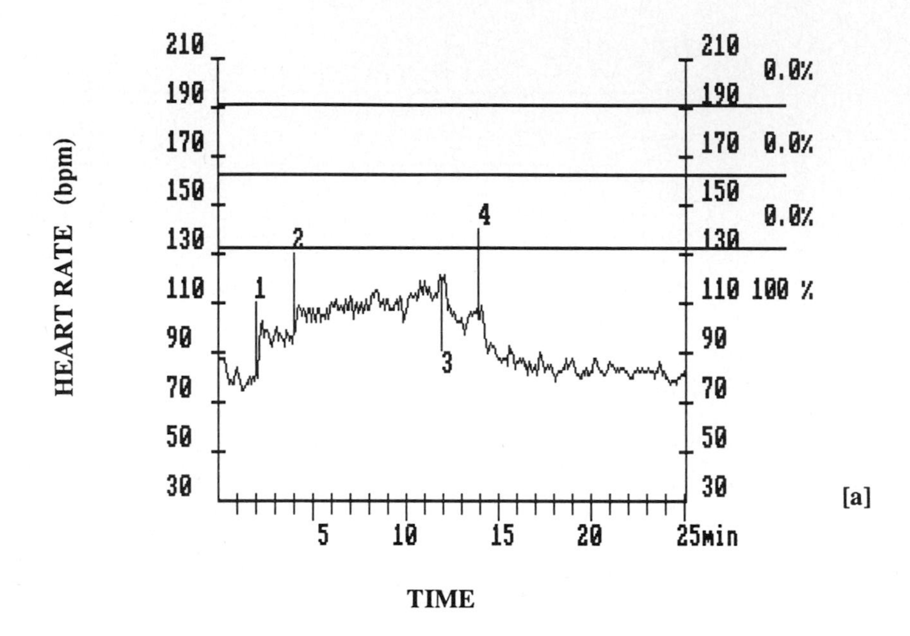

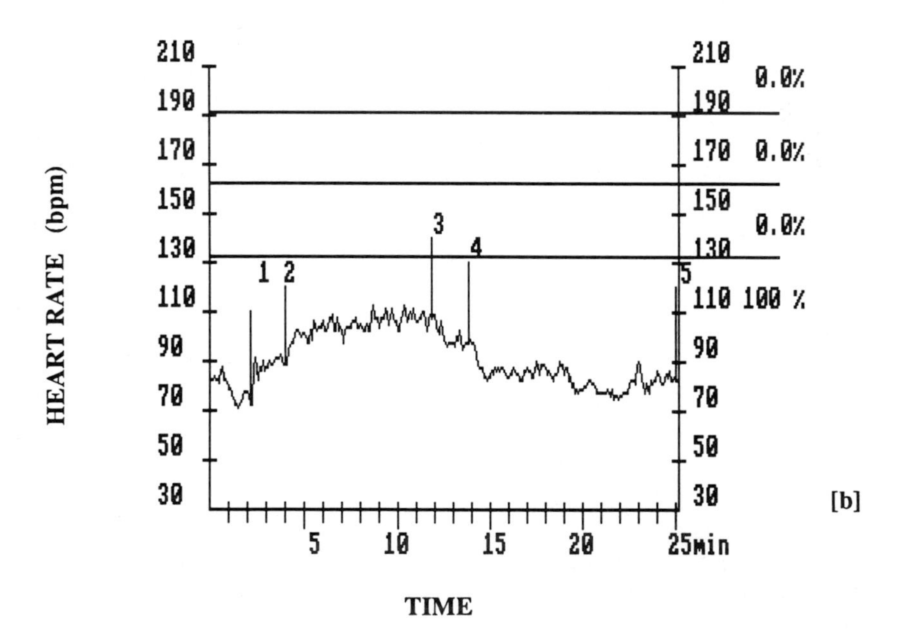

FIGURE 3
Heart Rate Profile of One Subject from Test Day 1 (a) and Test Day 2 (b)

TABLE 2
Summary of Heart Rate Parameters for Test Day 1 and Test Day 2

	TEST DAY 1				TEST DAY 2			
	Min	Max	Mean	Std. Dev.	Min	Max	Mean	Std. Dev.
R1 (bpm)	61	103	**87**	13.01	71	98	**84**	9.30
W (bpm)	104	163	**125**	16.97	104	142	**120**	13.38
R2 (bpm)	57	106	**86**	14.03	70	102	**86**	11.60
W-R1 (bpm)	24	61	**39**	9.41	26	60	**36**	9.39
W/R1	1.24	1.71	**1.46**	0.13	1.32	1.73	**1.43**	0.12
RDR ($x10^{-4}$)	-5.45	-1.57	**-2.83**	1.10	-6.12	-1.15	**-2.54**	1.37

All parameters derived from the heart rate profiles show good reproducibility on a test-to-test basis. However, only one level of physical work load was considered here. It is somewhere between light and moderate. It is suspected that this work load level may not be strenuous enough to discriminate among physically fit and less fit subjects. Green et al [10] reported that smokers showed a significantly greater heart rate response (the ratio of the difference between working and normal heart rates to normal heart rate) than non-smokers. This study also found similar results. Moreover, ambient temperature and certain types of apparel can also increase the level of the working heart rate as reported by Smith et al [7], Gertner et al [6], and Riley et al [8]. To avoid their possible effects, the room temperature was maintained at 75°F and all subjects were instructed to remove their shirts while performing the assigned work. On one occasion that the room temperature rose to 80°F, but an electric fan was then activated in order to dissipate the heat.

During the recovery period, subjects' heart rates decreased progressively from the time of cessation of work. This result is different from that reported by Das [4], but agrees with that mentioned in Grandjean [2]. Das [4] found no recovery and negative recovery. However, in his study only three sets of heart rate data were taken during the recovery period at one minute intervals. Significant fluctuation of heart rate during the recovery period was found in this study which could partly explain such adverse results, Das [4].

A continuation of this research study is being planned. We intend to investigate heart rate profiles under various physical work loads. More subjects with different physical conditions will be tested. It is anticipated that the results will be more quantitative and informative, thus leading to a more accurate determination of an index of physical fitness.

ACKNOWLEDGEMENTS

The authors wish to express their gratitude to Mr. Conrad A. Grant and Mr. Bang-Chiang Sung, both graduate students at NJIT, for their assistance in data collection and analysis. This study was funded by a separately budgeted research grant provided by the Department of Sponsored Programs, New Jersey Institute of Technology.

REFERENCES

[1] Astrand, P.-O. and Rodahl, K. (1986), Textbook of Work Physiology: Physiological Bases of Exercise, 3rd Edition, McGraw-Hill
[2] Grandjean, E. (1981), Fitting the Task to the Man: An Ergonomic Approach, 3rd Edition, Taylor & Francis
[3] Salvendy, G. and Knight, J. L. (1979), Physiological Basis of Machine-Paced and Self-Paced Work, Proceedings of the Human Factors Society, 23rd Annual Meeting: 158-162
[4] Das, B. (1982), Relationship of Work Pace to Worker Physiological Cost, Proceedings of the Human Factors Society, 26th Annual Meeting: 556-560
[5] Ilmarinen, J., Louhevaara, V. and Oja, P. (1984), Oxygen Consumption and Heart Rate in Different Modes of Manual Postal Delivery, Ergonomics, 27: 331-339
[6] Gertner, A., Israeli, R. and Cassuto, Y. (1984), Effects of Work and Motivation on the Heart Rates of Chronic Heat-Exposed Workers during Their Regular Work Shifts, Ergonomics, 27: 135-146
[7] Smith, L.A., Wilson, G.D. and Sirois, D.L. (1985), Heart-Rate Response to Forest Harvesting Work in the South-Eastern United States during Summer, Ergonomics, 28: 655-664
[8] Riley, M.W., Cochran, D.J. and Armstrong, J. W. (1980), Core Temperature, Heart Rate and Sweatsuits, Proceedings of the Human Factors Society, 24th Annual Meeting: 382-385
[9] Genaidy, A.M., Duyos, J.R. and Asfour, S.S. (1987), Physiological Stresses Associated with Manual Handling of Containers of Varied Sizes and Weights: A Case Study, Proceedings of the Human Factors Society, 31st Annual Meeting: 1326-1330
[10] Green, M.S., Luz, Y., Jucha, E., Cocos, M. and Rosenberg, N. (1986), Factors Affecting Ambulatory Heart Rate in Industrial Workers, Ergonomics, 29: 1017-1027

Trends in Ergonomics/Human Factors V
F. Aghazadeh (Editor)

ISOMETRIC-ISOTONIC EFFORT AND TRANSCUTANEOUS O2, CO2, HEAT

Charles A. Cacha

Center for Safety
OIOC, Hospital for Joint Diseases
New York University, NY, USA

Transcutaneous O2, CO2 and skin temperature sensors were placed on both biceps of young males. To determine local vs systemic effects, one arm remained at rest while the other did isometric or isotonic effort unto exhaustion. Sensor reading generally changed in opposite directions for working arm vs resting arm and isotonic vs isometric.

1. HISTORY, OBJECTIVES AND HYPOTHESES

There are available to the Ergonomist various noninvasive physiological measures which are capable of detecting, within the organism, work stresses either at a localized or systemic level (Geddes and Baker, 1968). Some of these measures are well established and frequently used and some of these measures are less frequently used. These techniques include heart rate, blood pressure, respiration rate/volume for measuring stresses upon the entire system and electromyograms, thermographs and blood circulaton rate for measuring local stresses upon a muscle or muscle group. The objective of this study is to potentially add two techniques to this repertory: transcutaneous gas movements and transcutaneous heat movements. Each will be briefly discussed.

1.1. Transcutaneous Gas Movements

As reported by Huch, Huch and Lubbers (1981) transcutaneous gas movements were first described by Verlacht a 19th century veterinarian. Verlacht adhered an inflated horse's bladder to his chest, allowing it to remain for 24 hours. After this period the gasses in the bladder were analyzed and it was discovered that the proportions of gasses within this bladder were not identical to the proportions of gasses found in ordinary ambient atmosphere. It was concluded that gasses such as oxygen (O2) and carbon dioxide (CO2) within the tissues of the organism had moved upward, towards the outside of the organism, by way of the porous structure of the skin. There has been recently devised commercially available equipment whose noninvasive skin mounted sensors will measure in millimeters of Mercury (mmHg) the quantity of these gasses at the surface of the skin (Huch et al., 1981). By use of an integrated heating unit which provides a maximum perfusion of blood to tissues in the immediate area, and by other adjustments in design, the sensor provides readings that reflect the amount of O2 and CO2 concentrations in the organism's bloodstream. These readings reach an accuracy equal to a +.9 or more correlation with actual arterial blood samples taken from the same organism (Lucey 1981).

1.1.1. Medicine. In Medicine, this equipment is most frequently used in pediatric clinics to monitor the vital signs of premature infants, thus generally obviating the need for traumatizing blood samples (Bahman 1981, Cassidy 1983). The readings provided are of a systemic nature in that they indicate the state of 02, C02 blood saturation throughout the circulatory system of the organism.
1.1.2. Other Uses. This equipment is also used medically to detect localized conditions (Dowd, Linge and Bentley 1983, Kram and Shoemaker 1983). The equipment, since it relies on a normal highly perfused circulation beneath the skin, may be placed on the extremities to detect circulatory abnormalities such as those of diabetics. These local abnormalities are detectable by low readings at the extremity as compared to higher readings which would occur at other normal parts of the anatomy.

1.2. Transcutaneous Heat

The measurement of transcutaneous heat is basically the process of taking skin temperatures. Techniques in thermometry are old and well established (Benedict 1972, Geddes and Baker 1978) and may include the use of standard noninvasive skin mounted thermometers such as gas or liquid in tubes, thermistor, thermocouples or radiometer. Transcutaneous passage of heat occurs when heat leaves a working or resting muscle by convection or conduction and increases or maintains the temperature of the skin directly above the muscle (Cooper, Randall and Hetzman 1959, Grant and Pearson 1937). In the case of a working muscle, the type and extent of exercise will determine the amount of heat produced and thus the magnitude of local skin temperature increase. Normal ambient atmospheric temperatures will minimally affect these readings. During muscle rest, local skin temperature maintenance is, in addition to resting heat produced by the muscle, substantially affected by ambient atmospheric temperature (Saltin, Gagge and Stalwijk, 1968).

1.3. Hypotheses

In order to determine the feasibility of using both of these pieces of equipment as Ergonomic tools, the concept of systemic vs local had to be tested. Identical sensors were used simultaneously above resting and above working muscles in order to determine systemic readings and localized readings. The major hypotheses were: 1) the sensors above resting muscles will maintain stable systemic readings while 2) the sensors above working muscles will indicate changed localized readings 3) whose magnitude and direction in movement depended on the type and extent of exercise.

2. PROCEDURE

2.1. Subjects

Forty one healthy male subjects were chosen from a group of science professionals and college students working at a hospital. Ages ranged from 18 to 32 with a mean age of 23.

2.2. Equipment

There were four basic units of equipment calibrated and used according to manufacturer's instructions.

2.2.1. O2, CO2. Transcutaneous oxygen (tcO2) and transcutaneous carbon dioxide (tcCO2) monitoring equipment (Novamatrix Corp., P.O. Box 690 Wallingford Conn. 06492, USA) included two combination tcO2 and tcCO2 sensors and appurtenant equipment. The sensors were secured to the skin by adhesive rings and additional surgical tape. The sensors had the necessary heating unit to bring about local skin blood perfusion which permits detection of O2,CO2 at skin surface. Digital readouts provided values at a resolution of 1 mmHg and an analog output provided values registered at .01 mmHg resolutions. The equipment also provided relative blood circulation values beneath the sensors stated in milliwatts (mW) of electricity need to heat the sensor.

2.2.2. Heat. Thermocouple temperature sensors of a flat design (Bailey Instruments, 154 Huron Ave., Clifton N.J. 07013 USA) were adhered to the skin with two thicknesses of surgical tape for security and insulation. Digital readings were at .1 degrees Centigrade (C) and analog output provided .01 degree C values.

2.2.3. Torque. A dynamometer (Cybex Corp., 2100 Smithtown Ave., Lake Ronkonkoma N.Y. 11779 USA) measured isometric torque amplitude against a padded bar. Torque values were in 1 foot pound (ft lb) resolutions with an analog output of .01 ft lbs.

2.2.4. Data. An IBM PC XT computer registered all analog outputs from the above equipment into memory on a floppy disk at .9 second intervals. Correction factors were included in the program to compensate for the nonlinear readings in the equipment.

2.3. Methods

All testing was done at normal ambient room temperatures and humidities which were monitored daily. Subjects remained dressed but exposed their arms. All testing was done by the same experimenter and assistant. Written instructions were used.

2.3.1. Isometric. The subject was seated in a straight backed, armless, casterless office chair. Using a steel tape, the tcO2, tcCO2 and temperature sensors were consistently placed in the same juxtaposition and same location on the center of the belly of the short head of each biceps. The required equipment equilibraton time of 20 minutes was allowed to lapse. The subject and chair were then positioned at the dynamometer in such a way that the subject's right supinated forearm, while at an extension of 90 degrees to the upper arm, was able to place pressure against the bar with the supinated wrist. The bar of the dynamometer was constantly set at 22 inches regardless of the subject's forearm length. The subject was then required to place maximal isometric pressure against the unmoveable bar for as long as possible and rest five seconds (first cycle), again place maximal pressure for as long as possible and rest five seconds (second cycle), and finally place maximal pressure for as long as possible (third cycle). During exercise the left arm was required to remain at rest in the lap. Instrument readings for both arms were continually recorded during the cycles and for 10 minutes thereafter (recovery).

2.3.2. Isotonic. After the 10 minute recovery period, a three pound dumbell was placed into the subject's left hand. While the right arm was at rest in the lap, the supinated left arm was required to fully flex and fully extend for at least 100 repetitions or until fatigue caused effort to cease. Each flexion and extension were done in time to a metronome beating at 3/4 second intervals. Instrument readings were continually recorded for both arms during the repetitions and for 10 minutes thereafter (recovery).

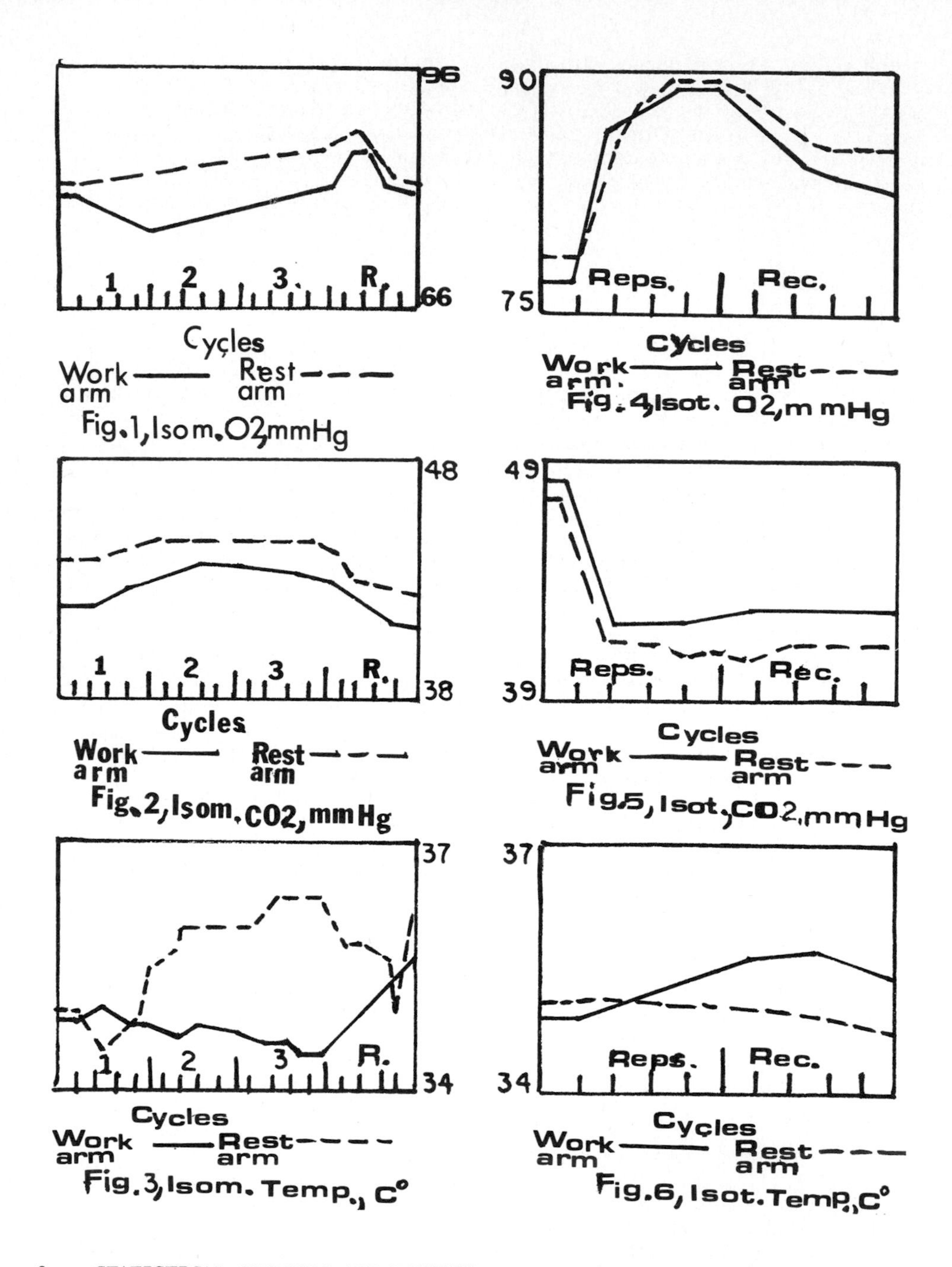

3. STATISTICAL ANALYSIS AND RESULTS

3.1. Statistical Analysis

Computer readings from all equipment for both arms and for both forms of exercise, as well as their recovery periods, were analyzed according to a before and after design. In order to provide graphic representation, the following were divided into five equal parts: each of the three isometric cycles, the isotonic repetitions and the isometric and isotonic recoveries. These intervals were also analyzed by an Analysis of Variance of Repeated Measures available from the Courant Institute of New York University. This analysis was followed by a Tukey Test of Multiple Comparisons. In all cases the level of significance sought was .05.

3.2. Results, Isometric

Averaged graphic data in Figures 1,2,3 provide the results of the isometric portion of the experiment. Generally all statistical results were at least at the .05 level, these results supporting the major occurences and conclusions which might be noted on the graphs. In some cases statistical results did not reach the .05 level but these negative results also supported conclusions on the graphs.

3.2.1. Oxygen. The Oxygen Graph indicates for the first cycle a steep drop in values for the working arm and a steady gradual increase of values for the resting arm. During the second and third cycles a steady gradual increase begins for the working arm and the same steady gradual increase continues for the resting arm. During the recovery there is an abrupt increase of values then an abrupt decline in values for both arms.

3.2.2. Carbon Dioxide. The Carbon Dioxide Graph indicates a great similarity of patterns for both working arm and resting arm in which both arms display an incline during the first cycle, a plateau during the second and third cycle, and then a drop and levelling off during the recovery.

3.2.3. Temperature. The Temperature Graph indicates a spike of abrupt rise and fall for the working arm and a spike of abrupt fall and rise of the resting arm. During the second and third cycles there was a gentle decline in values, with small spikes, of the working arm and a steep incline, with plateaus, of the resting arm. During recovery there was a steep incline of values of the working arm and a steep decline, with plateau, of the resting arm.

3.3. Results, Isotonic

Graphic trends supported by statistical tests similar to the those of the isometric experiment were executed (Figures 4,5,6,).

3.3.1. Oxygen. The Oxygen Graph shows a great similarity of pattern and magnitude for working arm and resting arm, indicating a steep incline of values during effort and gradual decline during recovery.

3.3.2. Carbon Dioxide. The Carbon Dioxide Graph indicates a similarity of change in pattern but not magnitude for working arm and resting arm, showing a steep decline during effort and a gradual incline during recovery.

3.3.3. Temperature. The Temperature Graph indicates a divergence in pattern and difference in magnitude for both arms, showing a steep incline during effort and a plateau during recovery for the working arm, and a gradual continuous decline over effort and recovery for the resting arm.

4. DISCUSSION

The major occurrences which were observed during both isometric and isotonic exercise (and their recoveries) generally coincide with established physiological occurrences related to muscle metabolism (Clarke 1975, Ganong 1981, Grandjean 1982, Jacob and Francone 1974 and Strauss, 1975).

4.1. Isometric

4.1.1. Oxygen. The lowering of oxygen in the working arm during the first cycle was due to partial occlusion of blood vessels which prevented oxygen from entering the muscle in order to replace the existing oxygen which was being metabolically consumed. The steep rise of working arm oxygen readings during recovery was due to repaying the "oxygen debt". The steady rise of resting arm oxygen values during work and their steep rise during recovery indicated an increase of systemic demands of oxygen in an attempt to supply oxygen to the working muscle.
4.1.2. Carbon Dioxide. The rise of carbon dioxide values of the working arm are due to the metabolic production of carbon dioxide during muscle effort. The rise of carbon dioxide values of the resting arm closely follow those of the working arm and are due to the solubility of carbon dioxide which is able to difuse the entire system despite the occlusion of blood vessels of the working muscle. The drop below normal of the working arm values during recovery may be the need of carbon dioxide as "building blocks" during the Krebs Cycle. The Krebs Cycle is the mechanism that reconstitutes ADP to ATP, the ultimate source of muscle energy.
4.1.3. Temperature. During the first cycle the working arm had a steep rise and immediate steep fall (spike) during effort due to metabolic heat production during exercise. During the first cycle the resting arm experienced a simultaneous reverse spike in which there was an abrupt fall due to loss of circulation in favor of the working arm. After the first cycle the working arm, during the second and third cycles, continued at a constant level with possible additional spikes due to effort. The resting arm values continually rose with plateaus which possibly coincided with the working arm spikes of effort. The literature has no ready explanation of these phenomona. During recovery the values of the working arm inclined steeply as a result of heat production during the "oxygen debt" and the resting arm values declined towards a stable condition.

4.2. Isotonic

4.2.1. Oxygen. The oxygen readings of both working arm and resting arm almost coincide in a steep increment of values. This circumstance is due to an increase of circulation within the working muscle due to light continuous isotonic exercise. This exercise carries oxygen to the muscle without any circulatory interference and thus the values of working arm and resting arm coincide.
4.2.2. Carbon Dioxide. As in the case of oxygen, there is a coincidence of working and resting arm values in that there is, during exercise, a steep drop of carbon dioxide readings for both arms. The reading for the resting arm, however, is more extreme. The reason for these phenomena may also be, as in the case of isometric exercise, the need of carbon dioxide as "building blocks" during the Krebs Cycle.

4.2.3. Temperature. In temperature readings there was a sharp rise in readings for the working arm and a gradual decline for the resting arm. This agrees with Saltin et al. (1968) who indicate that skin temperature above working muscle rises while skin temperatures in other areas fall and then move up or down towards ambient atmospheric temperatures.

4.3. Conclusions

Several conclusions result from the above experiments.
4.3.1. The oxygen, carbon dioxide and temperature sensors disclosed changes in values during both isometric and isotonic efforts as well as the recoveries that followed.
4.3.2. The magnitude and direction of change of all three sensors depended on whether the effort was isometric or isotonic.
4.3.3. In one half of the measurements, the working arm and resting arm nearly coincided in direction and magnitude, indicating that the reading over the working arm was not a localized reading. In one half of the measurements, the working arm and resting arm showed opposite directions and different magnitudes, indicating that the reading over the working arm was a localized reading.

4.4. Limitations

4.4.1. Minor. Some experimental limitations judged to be of a minor nature are briefly described. These limitations did not detract from the end result of the experiment which was to give extreme fatigue to the biceps muscles. These limitations are: 1)three left handed subjects, 2)bar of the dynamometer always the same length, 3)muscle size of subjects not controlled, 4)extent of honest effort by subjects indeterminable, 5)subject's effort sometimes augmented by shoulder shrugging, 6)one subject was a smoker, 7)actual arm flexion was by the brachialis not the biceps, 8)perspiration of subjects not fully controlled, 9)there were some experimenter errors and equipment failures, 10)a few subjects hyperventilated before exercise, 11)intra and interreliability of sensors was not perfect.
4.4.2. Major. Some major limitations are reported. Though of a greater concern, they are not believed to have a highly significant effect on the results of the experiments.
4.4.2.1. Circulation. Circulation readings were not continually recorded because of equipment limitations. Prior research, however, has substantiated the relationships of blood flow and the major forms of exercise (Clarke 1975, Ganong 1982, Strauss 1975).
4.4.2.2. Heat Contamination. The temperature sensors were placed on the skin in close proximity (1 cm.) to the tc02 and tcC02 sensors. The heating element of these sensors artificially increased the readings of the temperature sensors. This increase, however, was a constant which could be determined by applying the sensors in correct sequence and waiting adequate periods of time. This differential averaged 1.6 degrees C over all subjects. To determine absolute values from the temperature graphs 1.6 degrees C should be subtracted from all readings.
4.4.2.3. Isotonic Readings. It must be remembered that, 15 minutes prior, the isotonic working arm had been the isometric resting arm and the isotonic resting arm had been the isometric working arm. It is therefore possible that the isotonic readings may have been physiologically influenced by the prior isometric experiment. If this ocurred, it would be a question of magnitude rather than direction. As can be seen from the graphs and the conclusion section, isotonic trends were in opposite directions to isometric trends.

4.5. Practical Applications

It is expected that these results may be used for future clinical and industrial purposes. Oxygen, carbon dioxide and temperature readings on the same muscle may compare varying degrees of stress upon the muscle during various tasks assigned to the worker. Ultimately, comparisons may be made between well designed and poorly designed work stations and work procedures. It is expected from future research with this model that some of the three major variables may be eliminated and considered unnecessary. Alternatively, all three variables may be kept and greater weight may be placed on one or two variables which show the strongest relationship to trends during exercise or muscular effort. In addition, other noninvasive work variables such as pH, EMG may be added to the present list of O2, CO2 and temperature in order to make more powerful tests which will benefit the cause of designing healthy, comfortable workplaces for the worker.

REFERENCES

Bahman, V. (1981). Transcutaneous PO2 monitoring during pediatric surgery. Critical Care Medicine 9 (10), 714-716.

Benedict, R.P. (1972). Fundamentals of temperature and flow measurement. New York: John Wiley.

Cassidy, G. (1983). Transcutaneous monitoring in newborn infants. Journal of Pediatrics. 103 (6), 837-848.

Clarke, D.H. (1975). Exercise Physiology. Englewood Cliffs, N.J.: Prentice Hall.

Cooper, T., Randall, W.C. and Hertzman, A.B. (1959). Vascular convection of heat from active muscle to overlying skin. Journal of Applied Physiology. 14 (2), 202-211.

Dowd, G.S.E., Linge, K. and Bently, G. (1983). Measurement of transcutaneous oxygen pressure in normal and ischaemic skin. Journal of Bone and Surgery. 63 (1), 79-83.

Ganong, W.F. (1981). Review of medical physiology. 10th ed. Los Altos Cal.: Lange.

Geddes, L.A. and Baker, L.E. (1968). Applied biomedical instrumentation. (2nd ed.). New York: John Wiley and Sons.

Grandjean, E. (1982). Fitting the task to the man. London: Taylor and Francis.

Grant, R. and Pearson, S. (1937). Blood circulation in the human limb. Observations on the differences between the proximal and distal parts and remarks on the regulation of body temperature. Clinical Science. 3. 119-139.

Huch, R., Huch, A. and Lubbers, D.W. (1981). Transcutaneous PO2 New York: Thieme-Stratton.

Jacob, S.W. and Francone, C.A. (1974). Structure and function in man. (3rd ed.) Philadelphia: W.B. Saunders.

Kram, H.B. and Shoemaker, W.C. (1983). Use of transcutaneous monitoring in the intra management of severe peripheral vascular disease. Critical Care Medicine. 11 (6), 482-484.

Lucey, J.F. (1981). Clinical uses of transcutaneous oxygen monitoring. Advances in Pediatrics. 28. 29-55.

Saltin, P., Gagge, A.P. and Stolwijk, A.J. (1968). Muscle temperature during submaximal exercise in man. Journal of Applied Physiology. 25 679-687.

Strauss, R. (1979). Sports medicine and physiology. Philadelphia: W.B. Saunders.

Trends in Ergonomics/Human Factors V
F. Aghazadeh (Editor)
Elsevier Science Publishers B.V. (North-Holland), 1988

EFFECTS OF POSTURE ON THE METABOLIC EXPENDITURE REQUIRED TO LIFT A 50-POUND BOX

Sean GALLAGHER and Thomas G. BOBICK

Bureau of Mines, U.S. Department of the Interior, Pittsburgh Research Center, P.O. Box 18070, Pittsburgh, PA 15236

Eleven healthy, underground coal miners (36 yrs of age ± 8 SD) participated in an experiment examining the metabolic cost of lifting a 50-lb box asymmetrically in stooped and kneeling postures. Lifting periods were ten minutes in duration and the frequency was 10 lifts/min. Dependent measures included heart rate (HR), oxygen consumption (VO_2), ventilation volume (V_E), respiratory exchange ratio (R), and integrated electromyography (EMG) of eight trunk muscles. Results of an analysis of variance with repeated measures (ANOVR) showed that heart rate ($p < .05$), oxygen consumption ($p < .001$), ventilation volume ($p < .001$), and respiratory exchange ratio ($p < .05$) were all significantly elevated in the kneeling posture as opposed to stooped. A multivariate analysis of variance (MANOVA) indicated that the maximum integrated EMG of the trunk muscles was significantly different between the two postures ($p < .05$); univariate F-tests demonstrated that the right and left erectores spinae were the muscles accounting for the significant multivariate effects, being substantially higher in the kneeling posture ($p < .01$). These results indicate that workers may be more susceptible to muscular fatigue when handling materials in the kneeling posture, and that increased erectores spinae activity may cause higher compressive loads to be experienced by the spine when lifting in the kneeling posture. Results of this Bureau of Mines study suggest that decreasing the weight of loads handled in underground mines would be advisable, especially when lifting in the kneeling position.

1. INTRODUCTION

The metabolic and circulatory response of the human body to lifting loads has received a great deal of attention in the field of manual materials handling (MMH), particularly for high frequency lifting tasks [1-3]. Among the many factors known to affect the metabolic cost of performing lifting tasks are the weight of the load, frequency of lifting, vertical distance travelled in the lift, environmental conditions, and the technique or body posture utilized [4]. Posture may affect the metabolic cost of lifting by altering the load experienced by various muscle groups, and by the fact that different muscle groups may have varying metabolic efficiency (due in part to muscle fiber composition, moment arm and length of the muscle, and speed of contraction). It is therefore understandable that the position of the body can have a marked effect on the energy expenditure required to perform a lifting task. However, the literature on metabolic responses to body posture when lifting has primarily dealt with various techniques used during materials-handling tasks in unrestricted work postures [1,4].

Miners often lift in restricted work postures in underground coal mines. In low-seam mines (mines with a roof height of ≤ 48"), the two most prevalent postures are stooped and kneeling on two knees. Previous work reported by the Bureau of Mines demonstrated a significantly higher metabolic expenditure when lifting in the kneeling posture, even though significantly less weight was lifted in this posture [5]. The present study was performed to examine the metabolic cost of lifting a 50 lb weight (a typical supply item weight in underground mines) in both stooped and kneeling postures.

2. METHOD

2.1 Subjects

Eleven healthy male underground miners (Age: 36 years ± 8 SD) served as test subjects in this experiment. Subjects were paid volunteers and signed an informed consent form prior to participating in the experimental procedure. All participants were required to pass a thorough physical examination and an exercise tolerance test (at the expense of the Bureau of Mines) prior to taking part in the experiment. Each subject performed the lifting procedure once in both stooped and kneeling postures.

2.2 Apparatus

A 50.8- by 33.0- by 17.8-cm lifting box was used in this investigation. The box had two false bottoms under which weight was randomly varied, so the subject was not aware of the amount of weight he was lifting. The subject lifted the box across the front of his body from one end of a roller conveyor (starting on the subject's right) to the other, over a horizontal distance of 1.2 meters. The conveyor was slightly inclined so that the box would return automatically to the initial position. The frequency of lifting was controlled through the use of a computer-generated voice prompt. A Beckman* Dynograph Recorder was used to determine heart rate via surface electrocardiographic electrodes, and the metabolic demands of the lifting tasks were assessed using a Beckman Metabolic Measurement Cart I. Integrated electromyographic data was collected on-line from eight trunk muscles using surface electrodes, belt-wearable pre-amplifiers, and rack-mounted amplifiers/integrators (see Figure 1).

2.3 Procedure

Subjects were required to lift the box at a frequency of 10 lifts/min for a 10-minute period. A ten-second sample of heart rate was collected every minute during the test; the average of the last five values was taken as the mean HR for the condition. Metabolic data were collected approximately every thirty seconds during the last five minutes of the test, and average values were calculated for the rate of oxygen consumption, ventilation volume, and for the respiratory exchange ratio. Integrated EMG data was collected at minute 2 and minute 8 of the lifting periods from eight trunk muscles. The muscles studied in this investigation included left and right latissimus dorsi, erectores spinae, external oblique, and rectus abdominis. All EMG data was normalized to

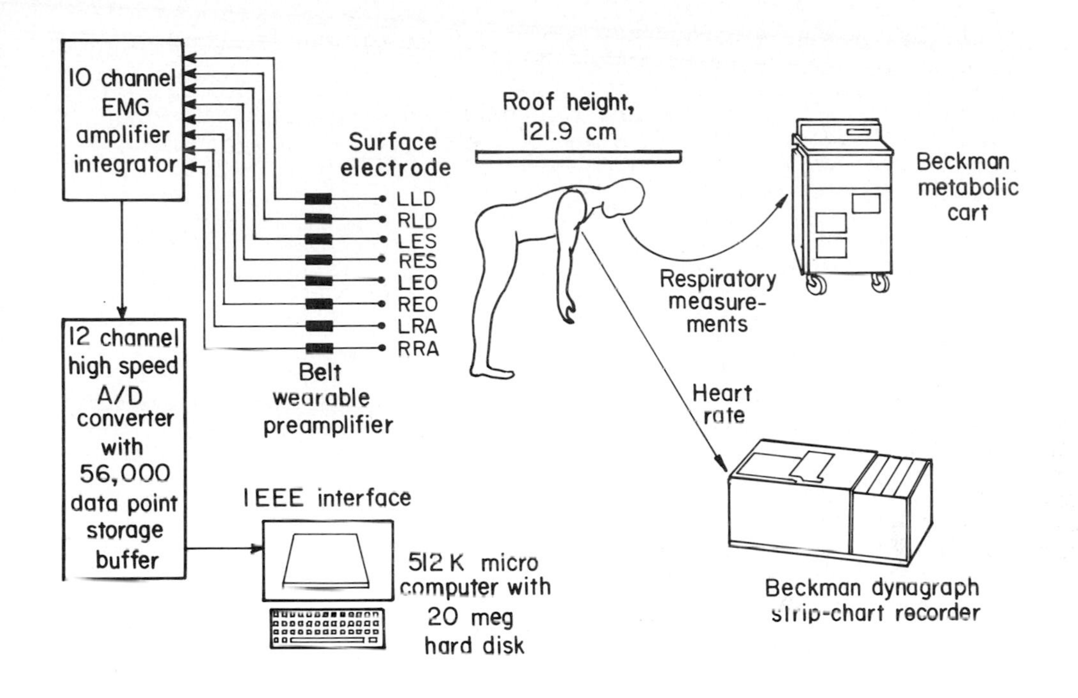

FIGURE 1
Schematic of apparatus used in the experiment.

maximum values obtained over all experimental conditions [6]. A twenty-minute break was provided between trials. Experimental conditions were randomized and presented in a counterbalanced order.

2.4 Data Treatment

Results of the heart rate, oxygen uptake, ventilation volume, and respiratory exchange ratio were analyzed using an analysis of variance with repeated measures (ANOVR) statistical package [7]. Maximum and mean integrated EMG data was analyzed using a multivariate analysis of variance (MANOVA). A significant MANOVA result was followed by a discriminant analysis and univariate F-tests to determine the individual muscles responsible for the significant MANOVA result [8]. The investigators were prepared to ignore results that did not reach a significance level of .05.

3. RESULTS

3.1 Physiological measures

Table 1 contains descriptive statistics for HR, VO_2, V_E, and R under the two experimental conditions. HR ($\underline{F}_{1,10}$ = 10.193, $\underline{p}$ < .05), VO_2 ($\underline{F}_{1,10}$ =

TABLE 1. - Physiological data for lifting a 50-lb box for all underground miners (N=11)

	Stooped	Kneeling	Significance
HR (beats/min)	116 (± 11)	129 (± 16)	$p < .01$
VO_2 (mL/kg/min)	12.6 (± 2.7)	14.9 (± 2.7)	$p < .001$
VE (L/min)	25.5 (± 5.5)	30.9 (± 5.9)	$p < .001$
R	.79 (± .08)	.82 (± .09)	$p < .05$

NOTE: Refer to text for abbreviations

26.234, $p < .001$), V_E ($F_{1,10} = 26.778$, $p < .001$), and R ($F_{1,10} = 6.444$, $p < .05$) were all significantly elevated in the kneeling posture as compared to stooping.

3.2 Electromyography

The MANOVA for maximum integrated EMG activity indicated a significant difference in muscular recruitment in the two postures ($F_{8,7} = 4.018$, $p < .05$), but not for time (minute 2 or minute 8 of the test) when the EMGs were taken ($F_{8,7} = 0.565$, $p = .78$). Results of the discriminant structure matrix for maximum EMG are given in table 2. Univariate ANOVAs demonstrate that the left and right erectores spinae were the significant individual variables accounting for the significant multivariate effects. The MANOVA for mean integrated EMG failed to achieve significance for posture ($F_{8,7} = 3.715$, $p = .050$) or time of EMG acquisition ($F_{8,7} = 0.649$, $p = .72$). No significant interactions were demonstrated for maximum or mean EMG ($p > .05$). The normalized values for maximum and

TABLE 2. - Pooled within-groups correlations between discriminating variable and canonical discriminant function for differences in maximum EMG activity due to posture assumed.

Trunk Muscle	Correlation to Discriminant Function
Left Erector Spinae	-0.51517
Right Erector Spinae	-0.43691
Right Latissimus Dorsi	-0.20979
Right Rectus Abdominis	0.20679
Right External Oblique	0.17527
Left Rectus Abdominis	0.10664
Left Latissimus Dorsi	-0.02211
Left External Oblique	0.00831

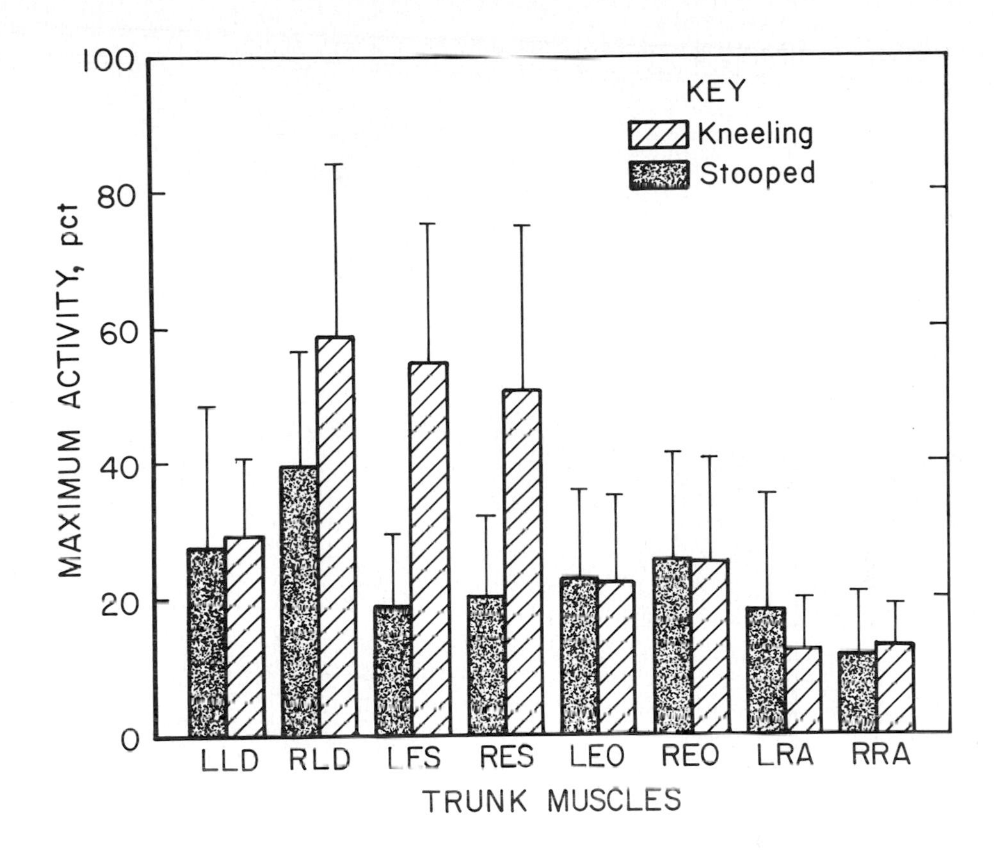

FIGURE 2.
Maximum normalized EMG activity during stooped and kneeling lifts (Note: LLD, RLD = Left and Right Latissimus Dorsi; LES, RES = Left and Right Erectores Spinae; LEO, REO = Left and Right External Obliques; LRA, RRA = Left and Right Rectus Abdominis).

mean integrated EMG activity for the two postures are shown in Figures 2 and 3.

4. DISCUSSION

Results of this investigation indicate that the metabolic cost of lifting is higher in the kneeling posture than in the stooped position during an asymmetric lifting task. This finding is in agreement with results previously reported by the Bureau of Mines [5]. In psychophysical studies performed by the Bureau, heart rate and oxygen consumption (but not V_E or R) were significantly higher in the kneeling posture, in spite of the fact that significantly less weight was lifted in this position. In this previous Bureau study, V_E and R probably did not achieve significance due to the lighter load lifted in the kneeling posture. The increased metabolic cost of lifting in the kneeling position suggests that fatigue is apt to occur more quickly in this restricted working posture.

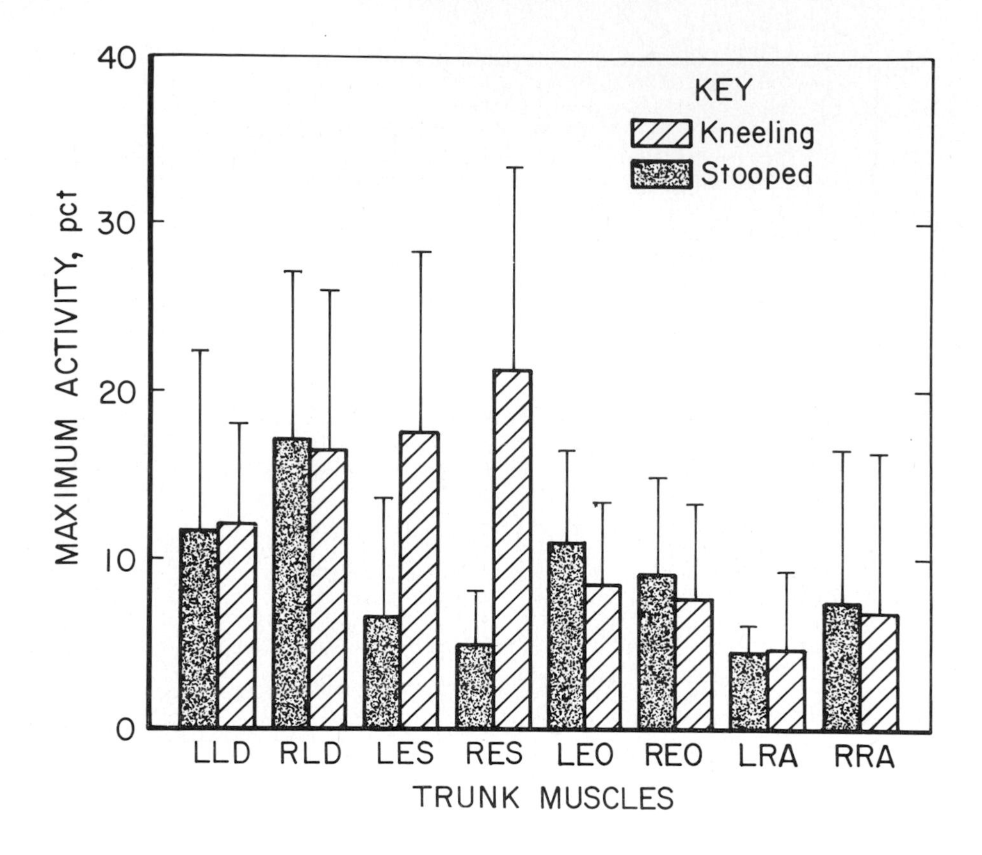

FIGURE 3.
Mean normalized EMG during stooped and kneeling lifts
(Note: Refer to Figure 2 for muscle abbreviations)

There would appear to be a couple of explanations for the increased metabolic rate of lifting when kneeling. It is important to note that it is not the work load itself that determines heart rate and oxygen uptake adjustments by the body, but the energy metabolism needed for the work load. Since the muscle mass that may be utilized to perform the lift would appear to be reduced in the kneeling posture, muscles available for the task may have to work substantially harder in order to accomplish a lift.

Studies in the field of exercise physiology have demonstrated that the smaller the muscle mass used to accomplish a given work load, the higher heart rate will rise and the sooner work will have to be interrupted due to exhaustion [9]. A second possible contributing factor for the increased metabolic rate may be that blood flow to the lower legs may be partially occluded in the kneeling position. Inhibition of blood flow to extremities has also been demonstrated to increase metabolic parameters, especially heart rate [9].

Results of the electromyographic data support the concept that certain muscles are having to bear an increased work load during lifting due to an overall decreased muscle mass in the kneeling posture. The erectores spinae are the predominant muscles affected by the posture assumed. Apparently, these muscles assume a greater responsibility for performing the lift when kneeling. It is plausible that these muscles develop the majority of the force required to counteract the moment due to the weight of the box being lifted when kneeling. In the stooped posture, it is likely that the paraspinal ligaments assume a larger responsibility for countering the moment due to lifting the load [10,11]. The increase in activity of the erectores spinae in the kneeling posture suggests an increased compressive load on the spine in this position compared to stooped. Current Bureau of Mines research is examining the biomechanical loading on the lumbar spine when lifting in these restricted working positions.

The results of this study imply that workers who must repeatedly lift a 50 lb weight in the kneeling position may experience fatigue more quickly than if they performed the same lift in the stooped posture. Therefore, it would appear to be advisable to allow more frequent rest breaks during extended materials-handling tasks in this posture. Previous psychophysical studies performed by the Bureau of Mines [4], using the criterion developed by Snook and Ciriello [12], indicate that 43.5 lbs is the recommended limit when lifting repetitively in the kneeling posture. Thus, it may also be advisable to reduce the weight of materials that must be lifted repetitively in the kneeling posture. This would not only reduce the metabolic expenditure necessary to perform the task, it would also reduce the compressive loading on the spine evidenced by the significantly increased activity of the erectores spinae muscles in this posture.

5. CONCLUSIONS

Heart rate, oxygen consumption, ventilation volume, and the respiratory exchange ratio are all significantly elevated when lifting a 50-lb box in the kneeling posture, compared to lifting the same weight in the stooped position.

The maximum EMG activity of the left and right erectores spinae muscles is significantly increased in the kneeling posture, compared to lifting the same weight stooped.

It may be advisable to allow more frequent rest breaks, and to reduce the weight of materials lifted, when performing lifts in the kneeling posture.

ACKNOWLEDGEMENTS

The authors would like to thank Dr. W. S. Marras, Mr. S. Lavender, and Mr. S. Rangarajulu of The Ohio State University Biodynamics Laboratory for their assistance in collection of data presented in this paper.

FOOTNOTE

* Reference to specific brands does not imply endorsement by the Bureau of Mines.

REFERENCES

[1] Basmajian, J. V., and De Luca, C. J., Muscles Alive. (Williams and Wilkins, Baltimore, 1985)
[2] Borgen, F. H., and Seling, M. J., Journal of Applied Psychology, (1978) 689.
[3] Floyd, W. F., and P. H. S. Silver, Journal of Physiology, (1955) 184.
[4] Gallagher, S., Back Strength and Lifting Capacity of Underground Miners, in: Proceedings of Bureau of Mines Technology Transfer Seminar, (Bureau of Mines Information Circular 9145, 1987) pp. 21-32.
[5] Games, P. A., Gray, G. S., Herron, W. L., and Pitz, G.F., Behavioral Research Methods and Instrumentation, (1980) 467.
[6] Garg, A., and Saxena, U., American Industrial Hygiene Association Journal, (1979) 894.
[7] Jorgensen, K., Ergonomics (1985) 365.
[8] Marras, W. S., Preparation, Recording and Analysis of the EMG Signal. in: S. S. Asfour (Ed.) Trends in Ergonomics/Human Factors IV, Part B. (Elsevier, Amsterdam, 1987) pp. 701-707.
[9] Mital, A., S. S. Asfour, and M. M. Ayoub., Journal of Human Ergology, (1982) 143.
[10] Stegemann, J., Exercise Physiology (Yearbook Medical Publishers, Chicago, 1981)
[11] U.S. Department of Health and Human Services. Work Practices Guide for Manual Lifting. NIOSH, Pub. 81-122, 1981, 183 pp.; NTIS PB 82-178-948.
[12] Snook, S. H., and Ciriello, V. M., Journal of Occupational Medicine (1974) 527.

Trends in Ergonomics/Human Factors V
F. Aghazadeh (Editor)

OXYGEN CONSUMPTION, HEART RATE, AND PERCEIVED EXERTION DURING WALKING IN SNOW WITH BOOTS OF DIFFERING WEIGHTS

Juhani SMOLANDER, Veikko LOUHEVAARA, Tarja HAKOLA, Esa AHONEN and Tapio KLEN

Institute of Occupational Health
Topeliuksenkatu 41 a A, SF-00250
Helsinki, Finland

In order to assess the physiological strain of different boot weights, seven male and three female subjects walked at a self-determined pace on a treadmill and a snowfield while wearing three types of boots: winter jogging shoes (WJS), rubber boots (RB), and rubber safety boots (RSB) weighing ($\bar{x} \pm$ SD) 0.9 ± 0.1, 1.9 ± 0.4, and 2.5 ± 0.2 kg, respectively. The mean (± SE) depths of footprint impression in snow while walking in the WJS, RB, and RSB were 26.1 ± 1.5, 25.6 ± 1.4, and 26.1 ± 1.5 cm (NS), respectively. During walking on the treadmill, the oxygen consumption was 0.79 ± 0.05, 0.81 ± 0.06, and 0.83 ± 0.04 $l \cdot min^{-1}$ (NS) and in snow 2.24 ± 0.18, 2.34 ± 0.17, and 2.34 ± 0.19 $l \cdot min^{-1}$ ($P < 0.01$) with the WJS, RB, and RSB, respectively. During the walking tests the corresponding mean heart rates were 106 ± 4, 93 ± 5, and 95 ± 5 beats $\cdot min^{-1}$ ($P<0.05$) on the treadmill, and 151 ± 11, 150 ± 11, and 151 ± 12 beats $\cdot min^{-1}$ (NS) in snow. No significant differences in ratings of perceived exertion were observed between the walking tests in snow with the three types of boots. In accordance with earlier studies, walking in snow was found to be strenuous work. In conclusion, the use of the RSB is recommended during logging work in snow, since they are known to provide greater protection while the increase in physiological strain was not appreciably different than that of boots of lighter weight.

1. INTRODUCTION

Safety boots are recommended to be worn in work tasks involving risks for lower limb injuries. Although safety boots provide greater protection, they often are considerably heavier than ordinary shoes. Loggers and other forestry workers comprise a large group of safety boot users. Even at present time, logging work is physically heavy [1] especially in wintertime when the work includes moving in snow and snow shoveling.

We questioned whether during walking in snow the increase in physiological strain due to the extra weight of the safety boots may become critically high and reduce working efficiency. This occurrence in turn could lead to the negligence of the use of the safety boots in logging work.

Thus, in the present study we compared the cardiorespiratory strain and perceived exertion during walking in snow between three types of boots of differing weights. To assess the effect of snow the three walking tests were done on a treadmill and on a snowfield.

2. SUBJECTS AND METHODS

2.1. Subjects

The study subjects were seven male and three female students from a forestry school. Their mean (±SD) age, body weight, and maximal oxygen consumption ($\dot{V}O_2max$) were 25 ± 6 years, 69 ± 14 kg and 51 ± 10 ml · min^{-1} kg^{-1}, respectively.

2.2. Procedures

The subjects walked in a random order at a self-determined pace on a level snowfield three times for ten minutes while wearing three types of boots: winter jogging shoes, rubber boots, and rubber safety boots weighing 0.9 ± 0.1 kg, 1.9 ± 0.4 kg, and 2.5 ± 0.2 kg, respectively. The walking routes were situated aside of a ice road on a lake. During the tests, the subjects used their own loose-fitting clothing with long underwear. During the test days, a cold weather dominated, but usually with a sunshine and negligible wind.

In addition to the tests in snow, on a separate day the subjects walked on a treadmill in a temperate environment three times for five minutes with the same clothing ensembles as during the snow walks. On the treadmill, the last walking test was done with the winter jogging shoes and continued as a maximum test for the determination of the VO_2max.

2.3. Measurements

In the snow-field, the oxygen consumption ($\dot{V}O_2$), was measured during the last 5 to 7 min of each 10 min walk with the portable Kofranyi-Michaelis gas meter. Gas analyses were performed with an infrared gas analyzer for carbon dioxide (Morgan 801d, P.K. Morgan, Ltd., England), and a paramagnetic gas analyzer for oxygen (Morgan 500d, P.K. Morgan, Ltd., England). In the laboratory, the gas exchange variables were measured with the Mijnhardt Oxycon-4 test system (Gebr. Mijnhardt B.V., Holland).

During the snow walking experiments, the heart rate (HR) was recorded every minute by a telemetric system (Sport tester, PolarElectro, Finland). Due to technical reasons, one subject's HR recording was not successful. For the analysis, the HR was averaged over the 5-7 min gas collection period. During the treadmill tests, electrocardiograms were monitored continuously (OLLI 208, Kone, Finland) and recorded by a Mingograf Minor 3

(Elema-Schönander, Sweden) during the last 15 s of each minute for the calculation of heart rate.

In the snowfield, the overall perceived exertion and the perceived exertion of the legs were assessed with a standard scale [2]. The depth of the footprint impression in snow from the surface to the point corresponding to the ball of the foot was calculated as an average of 40 measurements.

The effects of boot weight and snow on the physiological variables were evaluated with a two-way analysis of variance with repeated measures. Post-hoc comparisons were performed with the Newman-Keuls procedure. The differences were considered statistically significant when $P < 0.05$.

3. RESULTS

3.1 Walking Speed and Depth of Footprint Impression In Snow

The subject's self-determined walking speed in snow averaged 0.71 $\pm$ 0.18 $m \cdot s^{-1}$ (2.6 $\pm$ 0.6 $km \cdot hr^{-1}$). The mean ($\pm$ SE) depths of footprint depression while walking with the winter jogging shoes, rubber boots, and rubber safety boots were 26.1 $\pm$ 1.5 cm, 25.6 $\pm$ 1.4 cm, and 26.1 $\pm$ 1.5 cm (NS), respectively.

3.2 Oxygen Consumption and Heart Rate

Walking in snow substantially ($P<0.01$) increased $\dot{V}O_2$, and HR as compared to the walking tests on the treadmill (Table 1). During the tests on the treadmill and in snow, the VO_2 was slightly, but systematically higher with the heavier rubber boots and rubber safety boots than with the lighter winter jogging shoes. In snow, this difference was statistically significant ($P<0.01$).

The mean HR was about 50 beats $\cdot$ min^{-1} higher ($P<0.01$) during walking in snow than on the treadmill, but there were no significant differences between the three types of boots (Table 1). However, on the treadmill the HR was slightly ($P<0.05$) higher with the winter jogging shoes as compared to the rubber boots and rubber safety boots, which was probably caused by the psychological excitement due to the proximity of the maximum stress test (see Subjects and Methods).

Table 1. Mean cardiorespiratory data during walking on the treadmill and in snow with boots of differing weights.

	$\dot{V}O_2$ ($l \cdot min^{-1}$)	HR (beats . min^{-1})
Walking on treadmill		
Winter jogging shoes	0.79±0.05	106±4[a]
Rubber boots	0.81±0.06	93±5
Rubber safety boots	0.83±0.04	95±5
Walking in snow		
Winter jogging shoes	2.24±0.18	151±11
Rubber boots	2.34±0.17[b]	150±11
Rubber safety boots	2.34±0.19[b]	151±12

The values are the mean ± SE for 10 subjects except the heart rate (HR) in snow where N=9; $\dot{V}O_2$, oxygen consumption (STPD); a, $P<0.05$ as compared with the rubber boots and rubber safety boots; b, $P<0.01$ as compared to winter jogging shoes.

3.3 Ratings of Perceived Exertion.

The ratings of overall perceived exertion in snow varied from "very, very light" (0.5) to "heavy" (5). The mean values for the winter jogging shoes, rubber boots, and rubber safety boots were 2.3 ± 0.3, 2.2 ± 0.2, and 2.8 ± 0.5 (NS), respectively. The corresponding figures of perceived exertion for the legs were 2.0 ± 0.4, 1.6 ± 0.3, and 2.4 ± 0.4 (NS), respectively.

4. DISCUSSION

4.1 Effect of Boot Weight

In our study, walking on the treadmill or in snow in the heavier rubber safety boots (2.5 kg) did not increase the physiological strain appreciably as compared to the lighter rubber boots (1.9 kg) and winter jogging shoes (0.9 kg). Only during walking tests in snow there was a slight (100 $ml.min^{-1}$), significant increase in $\dot{V}O_2$ with the rubber safety boots and rubber boots as compared to the winter jogging shoes.

Expressed as a % per 100g of added total shoe weight the rise on $\dot{V}O_2$ averaged 0.3% both with the rubber safety boots and rubber boots, which is less than previously reported [3,4,5,6,7] values; the means ranging from 0.7% to 1.0%. This discrepancy may be related to the slow walking speed of our subjects, since at low walking speeds the relative increase in $\dot{V}O_2$ due to extra weight carried either on back [8] or on the foot [9] have been found to be smaller than at higher walking speeds.

In addition, the boots were not of a similar design. The rubber boots and rubber safety boots were stiffer than the winter jogging shoes, which perhaps could have improved balancing while walking in snow. Thus, there may have been a smaller increase in $\dot{V}O_2$ in these boots relative to the winter jogging shoes.

Besides safety boots, the loggers also have to wear other protective equipment, the chain saw, and tools. Three of the male subjects walked in snow with the total logging equipment including rubber safety boots (2.7 kg), safety clothing including the helmet (2.3 kg), tool belt (2.3 kg), a felling bar (1.5 kg), and a chain saw (6.3 kg). The mean HR increased from 114 ± 21 beats $\cdot$ min $^{-1}$ to 139 ± 20 beats $\cdot$ min $^{-1}$ and the $\dot{V}O_2$ rose from 2.08 ± 0.31 l $\cdot$ min $^{-1}$ to 2.31 ± 0.46 l $\cdot$ min $^{-1}$ while walking in the rubber safety boots and with the total equipment, respectively. Therefore, a clear increase in physiological strain was observed with the total equipment.

4.2 Effect of Snow

Walking in soft snow have been shown [10,11,12] to be strenuous work as seen also in the present study; the mean HR was about 50 beats $\cdot$ min^{-1} higher, and the mean $\dot{V}O_2$ was about three times higher during the snow walks with footprint depression of ~ 20cm than at nearly equivalent speeds on the treadmill. Greater amounts of lift work, balancing difficulties, and increased physical effort in pushing and pulling the foot into and out of the snow were the most likely causes for the increased physiological strain.

In conclusion, in accordance with earlier studies walking in snow was found to be strenous work. The use of the rubber safety boots is recommended during logging work in snow, since they are known to provide greater protection while the increase in physiological strain was not appreciably different than that of boots of lighter weight.

REFERENCES

[1] Kukkonen-Harjula, K., and Rauramaa, R., 1984, Oxygen consumption of lumberjacks in logging with a power-saw. Ergonomics, 27, 59-65.

[2] Borg, G., 1982, Psychological bases of perceived exertion. Medicine and Science in Sports And Exercise, 14, 377-381.

[3] Strydom, N.B., Van Graan, C.H., Morrison, J.F., Viljoen, J.H., and Heynes, A.J., 1968, The influence of boot weight on the energy expenditure of men walking on a treadmill and climbing stairs. Internationale Zeitschrift für angewandte Physiologie einschliesslich Arbeitsphysiologie, 25, 191-197.

[4] Soule, R.G., and Goldman, R.F., 1969, Energy cost of load carried on the head, hands, and feet. Journal of Applied Physiology, 27, 687-690.
[5] Jones, B.H., Toner, M.M., Daniels, W.L., and Knapik, J.J., 1984, The energy cost and heart rate response of trained and untrained subjects walking and running in shoes and boots. Ergonomics, 27, 895-902.
[6] Jones, B.H., Knapik, J.J., Daniels, W.L., and Toner, M.M., 1986, The energy cost of women walking and running in shoes and boots. Ergonomics, 29, 439-443.
[7] Miller, J.F., and Stamford, B.A., 1987, Intensity and energy cost of weighted walking vs running for men and women. Journal of Applied Physiology, 62, 1497-1501.
[8] Smolander, J., Louhevaara, V., Tuomi, T., Korhonen, O., and Jaakkola, J., 1984, Cardiorespiratory and thermal effects of wearing gas protective clothing. International Archives of Occupational and Environmental Health, 54, 261-270.
[9] Hettinger, T., and Müller, E.A., 1953, Der Einfluss der Schuhgewichtes auf den Energieumsatz beim Gehen und Lastentragen. Internationale Zeitschrift für angewandte Physiologie einschliesslich Arbeitsphysiologie, 15, 1015-1021.
[10] Heinonen, A.O., Karvonen, M.J., and Ruosteenoja, R., 1959, The energy expenditure of walking on snow at various depths. Ergonomics, 2, 389-393.
[11] Ramaswamy, S.S., Dua, G.L., Raizada, V.K., Dimri, G.P., Viswanathan, K.R., Madhaviah, J., and Srivastava, T.N., 1966, Effect of looseness of snow on energy expenditure in marching on snow-covered ground. Journal of Applied Physiology, 21, 1747-1749.
[12] Pandolf, K.B., Haisman, M.F., and Goldman, R.F., 1976, Metabolic energy expenditure and terrain coefficients for walking on snow. Ergonomics, 19, 683-690.

Trends in Ergonomics/Human Factors V
F. Aghazadeh (Editor)

EVALUATING THE CARDIOVASCULAR FITNESS OF DOWNS SYNDROME INDIVIDUALS

Kenneth H. PITETTI, Jeffrey E. FERNANDEZ, Jayne A. STAFFORD, and Nancy B. STUBBS

Departments of Health Science, Industrial Engineering, and Physical Education
The Wichita State University
Wichita, KS 67208-1595

Seven Down's syndrome individuals (mean IQ = 60), ages 14-28, were studied to determine whether a method which has been shown to accurately predict cardiovascular fitness (VO2 max) of non-Down's, moderately retarded individuals can also be used to predict VO2 max of moderately retarded, Down's syndrome individuals. A submaximal test on a Schwinn Air-Dyne ergometer (SAE) was used to predict VO2 max. The submaximal test consisted of two consecutive 3 minute bouts of exercise at a low (50 watts) and high (100-150 watts) intensity. VO2 max was predicted using Hellerstein's [1] formula (%VO2 max = (1.41 X %HR max) - 42) for the final heart rate at the high intensity level. A modified Balke test on the treadmill (TM) and a continuous bicycle exercise test on the SAE was used to determine actual VO2 max. Mean predicted VO2 max (31.0 ml/kg/min) was significantly ($p < 0.05$) higher than actual mean VO2 max for both TM (22.9 ml/kg/min) and SAE (21.6 ml/kg/min). No significant difference was observed for VO2 max between TM and maximal SAE tests. This difference between predicted vs actual cardiovascular fitness should merit attention by those responsible for assessing cardiovascular fitness levels of Down's syndrome individuals for job placement or physical education purposes.

1. INTRODUCTION

Down's syndrome is a condition usually caused by non-disjunction or translocation of the 21st chromosome resulting in an individual possessing three rather than two 21st chromosomes (trisomy 21). This condition is usually accompanied by moderate mental retardation and some obvious physical features that set Down's syndrome individuals apart within the population of the mentally retarded (MR). These physical features include a small, square head, upward slanting eyes, a large and protruding tongue, and short, stubby skeletal characteristics. Other differences between Down's and non-Down's MR individuals involve abnormalities in the Down's cardiovascular system which include septal defects in the heart accompanied by a small and narrow aorta and arteries [2]. These and other distinctive physical abnormalities in Down's syndrome (DS) individuals (e.g. small oral cavities which could restrict their ventalitory capacity) is thought to impose physical limitations not seen in non-Downs MR individuals [3]. Therefore there exist a need for a simple, yet accurate way to assess the cardiovascular fitness of DS individuals before they engage in stressful physical activity in job or recreational related endeavors.

In a study by Pitetti et al., [4], a simple, yet valid protocol was developed that could accurately predict the cardiovascular fitness (VO2 max) of the non-Down's mentally retarded individual. However, it is not known if this protocol is a valid prediction of cardiovascular fitness for DS individuals because of the distinctive physical differences of the Down's syndrome population. Therefore, the purpose of this study was to determine if the protocol used by Pitetti et al., [4] for predicting the cardiovascular fitness of moderately retarded individuals could also be applied to moderately retarded DS individuals.

2. METHODS AND PROCEDURES

2.1. Subjects

Four male and three female Down's syndrome subjects (IQs ranging from 50 to 67 with a mean of 60) participated in this study. The mean age was 25 years with a range of 14 to 28 years. All subjects had been medically cleared for the exercise tests, and all had been actively involved in the Special Olympic activities for at least one year at the time of the testing. An informed consent was obtained from either the subject (if legally capable) or guardian. The testing protocol in which the subjects participated was approved by the Institution Review Board of The Wichita State University, Wichita, Kansas.

2.2. Testing Protocol

The first test involved estimating VO2 max from heart rates during a submaximal exercise test on the Schwinn Air-Dyne (SAE) ergometer. The submaximal exercise test consisted of two continuous three minute bouts of exercise. The initial bout was at a low level of intensity (50 watts) and the second at a higher level of intensity (100 - 150 watts). The work level for the higher intensity was determined by the heart rates at the low level. If while exercising at the low level the subject's heart rate ranged below 65% of maximum heart rate, then 150 watts was selected for the second work level. If the subject's heart rate ranged at or above 65% of maximum, then 100 watts was selected for the second work level. Maximal heart rates were predicted by the formula, 220 - age, for females, and 205 -(age/2), for males [5]. This protocol produces heart rates at the second level in the vicinity of 80-85% of maximum, which has been shown to provide accurate estimates of VO2 max [6]. Heart rates were monitored continuously by telemetry (EST Sport Tester, Excotek) and recorded every 60 seconds. If the heart rate at three minutes was 5 beats/min greater than the two minute value, the exercise was continued for an extra minute and the fourth minute heart rate was considered to be the steady state value [7]. Blood pressure was measured at rest, at two minutes into each exercise level, and following exercise until it had returned to resting values. A six lead ECG was monitored throughout the test. The oxygen cost for exercise was estimated from the formula, VO2 (ml/min) = (Watts X 12) + 300. VO2 max was predicted from the formula, %VO2 max = (1.41 X %HR max) - 42 [1] (refer to Pitetti et al., [4], for further explanation of methodology). This protocol has been shown to be an accurate method to predict VO2 max for both the general population [6] and for the moderately mentally retarded [4].

The maximal exercise test on the SAE consisted of an initial exercise work load of 50 watts for two minutes, with incremental increases of 25 watts every two minutes until volitional exhaustion. The maximal exercise test on the treadmill (TM) consisted of walking at a speed ranging from 2.5 to 3.5 mph (whatever the subject could handle safely) at a grade of 2.5 degrees, with incremental increases in grade of 2.5 degrees every two minutes until volitional exhaustion. For both maximal tests, a Metabolic Measurement Cart (MMC) Horizon System or a Beckman Metabolic Measurement Cart was used to measure VO2 directly during the tests. Heart rate was monitored continuously and recorded at 30 second intervals.

Each subject performed all the tests on the same day. The submaximal test was given first to screen for any possible contraindications for exercise (i.e., any HR, blood pressure, or ECG abnormalities). Each of the two maximal tests were given following the submaximal test in a consecutive manner. The sequence of the maximal tests was chosen randomly, with three of the subjects performing the SAE max test first while four initially tested on the TM. Ample time was given between test (i.e., at least one hour) for recovery. No subject was forced to perform the maximal test if they did not feel totally recovered from the preceding test. An ECG was continuously monitored on the first maximal test.

2.3. Statistical Analysis

The means and standard deviations (SD) were calculated for all variables. A Pearson Correlation Coefficient was used to test for significant relationships between the following: 1) predicted VO2 max to observed VO2 max on the TM; and 2) predicted VO2 max to observed VO2 max on the SAE. A repeated measures one-way ANOVA followed by a Duncan multiple comparison was performed to detect specific differences among the following: 1) predicted VO2 max and observed VO2 max on the TM and SAE, respectively; and 2) predicted maximal HRs and observed maximal HRs on the TM and SAE, respectively. The results were judged to be significant at $p < 0.05$.

3. RESULTS

The means and SD for predicted VO2 max and observed VO2 max on the TM and SAE are listed in Table I. Predicted VO2 max was significantly higher than observed VO2 max on the TM and SAE, respectively. There was no significant difference between observed VO2 max on the TM or SAE. There was no significant correlation between predicted VO2 max and TM max (r = 0.74. t = 2.46), nor was there a significant correlation between predicted VO2 max and SAE max (r = .69, t = 2.15).

TABLE 1

Predicted and Observed VO2 Max (ml/kg/min)

SUBJECT	PREDICTED VO2 MAX	TM VO2 MAX	SAE VO2 MAX
1	32.7	19.4	26.0
2	39.2	25.8	25.3
3	34.7	24.5	25.1
4	31.4	25.7	19.2
5	34.4	26.3	17.8
6	22.3	19.7	17.8
7	22.3	19.7	18.9
Means	31.0 *	23.0	21.4
Standard Dev.	6.4	3.2	3.8

* significantly higher than TM and SAE

Means and SD for predicted and observed maximal HRs are listed in Table 2. The predicted values were significantly higher than observed for both the TM and SAE, respectively. There was no significant difference between observed max HR for the TM and SAE.

4. DISCUSSION

Direct measurements of maximum oxygen consumption (VO2 max) during high intensity exercise is the standard method of determining the functional capacity of an individual's cardiovascular system. However, this procedure requires expensive laboratory equipment and medical expertise usually not available in most workshop and/or activity settings for the MR. Therefore, the prediction of VO2 max has become the only alternative for the administrators at these settings. However many of the protocols used to predict the cardiovascular fitness of the general population are of questionable validity when administered to the MR [8]. Recently, Pitetti et al.,

[4] has demonstrated a field testing protocol that accurately assessed the cardiovascular fitness of non-Down's MR individuals. As already discussed, there are many physical characteristics that set Down's MR individuals apart within the population of the MR. It was the purpose of this study to determine if these differences would effect the accuracy of the protocol established by Pitetti et al., [4]. Although the number of subjects tested is small, the results suggest that this protocol [4] is not an accurate method of assessing the cardiovascular fitness of moderately retarded, Down's syndrome individuals.

TABLE 2

Predicted and Observed Maximal Heart Rates (Beats/Min)

SUBJECT	PREDICTED	OBSERVED TM	OBSERVED SAE
1	198	170	156
2	194	155	156
3	193	152	154
4	192	155	135
5	191	168	152
6	192	160	164
7	191	161	154
Means and SD	193.3 *	160.4	153.0
Standard Dev.	2.5	6.8	8.8

* Significantly higher than TM and SAE

The possibility exists that the initial submaximal test on the SAE fatigued the subjects to the extent that they were unable to adequately perform on first maximal test. This seems improbable in that the duration of the submaximal test is 6 to 7 minutes, with only a 3-4 minute period where they are above 75% of their estimated maximum HR. Furthermore, both their HRs and blood pressures returned to pretest levels within 3-5 minutes following the submaximal test. This short recovery time is another indicator that the submaximal test was not highly stressful. Additionally, another 60 to 80 minutes was allowed before the first maximal test. Therefore, we feel that the subjects had adequately recovered before the first maximal test enabling them to give their best effort for this test.

Similarly, the possibility exists that the combined submaximal and initial maximal test could have resulted in a poor final maximal test due to physical fatigue. However, of the three who took the SAE max test first (subjects 2,4,and 5), all showed lower VO2 max on the SAE than on the TM. Furthermore, Subject #1, who took the TM test first, showed a higher VO2 max on the SAE. Additionally, the average time for the maximal tests was only 11.5 minutes. Therefore, we feel that when considering both maximal tests, the factor of physical fatique had minimal effect on the results.

A consistent finding in this study is the lower than predicted maximum heart rates. It could be hypothesized that this is due to their low threshold of enduring the physical discomfort of high intensity exercise. This low threshold to high intensity exercise might also explain the lower than predicted VO2 maximums. Yet, mean respiratory quotients (RQ) for both the TM (1.06) and SAE (1.08) indicate that high intensities had been reached. Therefore, one is left to postulate that these low maximal heart rates are indicative of individuals who are either extremely unfit or who possess cardiovascular limitations characteristic of the condition of Down's Syndrome.

5. CONCLUSION

This study demonstrated that the submaximal exercise protocol previously shown to accurately assess the cardiovascular fitness of both the general population [6] and moderately MR individuals [4] is not as exact when administered to DS individuals. Furthermore, the results also show that these individuals are terminating exercise far short of predicted maximal heart rates. Therefore, it is essential that those concerned with the cardiovascular fitness of this special population of the MR continue their endeavors to develop a valid method for predicting the cardiovascular fitness of DS individuals.

REFERENCES

[1] Hellerstein, H., Hirsch, E.Z., Ades, R., Greenslott, N., and Segel, M. (1977). Principles of exercise prescription for normals and cardiac patients. In I.P. Naughton and H.K. Hellerstein (Eds). Exercise Testing and Exercise Training in Coronary Heart Disease. Academy Press, New York.

[2] Kock, R., and de la Cruz, F. (Eds). (1975). Down's Syndrome (Mongolism), Research, Prevention, and Management. New York: Brunner-Mazel.

[3] Cratty, B.J. (1980). Adapted Physical Education for Handicapped Children and Youth. Love Publishing Company, Denver, pgs 226 - 229.

[4] Pitetti, K.H., Fernandez, J.E., Pizarro, D.C., Stubbs, N.B., and Stafford, J.A. (In print, 1988). The cardiovascular fitness of non-Down's syndrome, moderately retarded individuals as an additional indice for job placement. Trends in Ergonomics/Human Factors V. Proceedings of the Annual International Industrial Ergonomics and Safety Conference.

[5] Hakki, A.H., Hare, T.W., Iskandrian, A.S., Lowenthal, D.T., and Segel, B.L. (1983). Prediction of maximal heart rates in men and women. Cardiovascular Review and Reports, 4(7), 887-999.

[6] Pitetti, K.H., Vaughan, R.H., and Snell, P.G. (1987). Estimation of VO2 max from heart rates during submax work on the Schwinn Air-Dyne ergometer. Medicine and Science in Sports and Exercise, 19(2): S64.

[7] Siconolfi, S.F., Culliname, E.M., Carleton, R.A., and Thompson, P.D. (1982). Assessing VO max in epidemiologic studies : modifications of the Astrand-Ryhming test. Med. Sc. Sports Exercise 14(5), 335-338.

[8] Seidl, C., Reid, S, and Montgomery, D.L. (1987). A Critique of cardio-vascular fitness testing with mentally retarded persons." Adapted Physical Activity Quarterly, 4, 106-116.

Trends in Ergonomics/Human Factors V
F. Aghazadeh (Editor)

PHYSIOLOGICAL RESPONSES WHILE PLAYING A VIDEO GAME

Jeffrey E. FERNANDEZ, Amy D. AKIN, Cathy L. COLLINS, and Jason F. VIRGILIO

Department of Industrial Engineering
The Wichita State University
Wichita, Kansas 67208

This study documents the measurement of physiological responses of video game players. Sixteen male subjects between the ages of 22 and 34 were divided into two player classes, novice and frequent, to play the GALAGA video game in a non-laboratory setting. Heart rate was measured before, during, and after each of three trials. Blood pressure was measured before and after each trial. The results of this study showed that heart rate increases significantly at $p < 0.05$ during each session at the video game. The experience level of each subject was also determined to be a significant factor in heart rate measurement. The frequent player's heart rate showed a rapid increase during play but leveled off at the end of each trial. However, the novice player's heart rate showed a slight increase during play but rose at the end of the trial. Level of experience also influenced the duration of play. The mean duration time of the frequent players (297 seconds) was approximately double the mean duration time of the novice players (157 seconds). There was no significant change in blood pressure for both the novice and frequent players. The conclusions drawn from this experiment were that the stress which resulted from playing the GALAGA video game caused increases in heart rate in males. The experience level of the player has a direct influence on their heart rate and duration time. It was also concluded that playing the GALAGA video game does not significantly change the blood pressure of the individual.

1. INTRODUCTION

Video games are played primarily for fun. Do they, however, subject the player to significant levels of stress? Does a frequent video game player exhibit higher levels of arousal than a novice player? Stress is assumed to act as an intermediate variable, causing physiological changes. Arousal, however, may be considered as a state of preparedness of the body [1]. Although much stress research has been performed in the work-related areas of man-machine interface, very little has been done in the recreational, self-subjected stress of video games. The studies that have been performed using heart rate and blood pressure to determine workload are inconsistent and even contradictory in their results. Ettema, et al. [2] found changes in both systolic and diastolic blood pressure as a function of difficulty of an auditory binary choice task. Mobbs [3] found no systematic relationship between mean heart rate and task difficulty. In a later study involving auditory choice reaction, Ettema and Zielhuis [4] found that blood pressure and heart rate increased with task load. Zwaga [5], in working with heart rate alone, found that it decreased with anticipation stress. He suggested that arousal is a more fundamental concept than load as far as heart rate is concerned. Boyce [6] found that heart rate increased with mental load. While Gaume and White [7] found no consistent pattern in heart rate, they did find a slight increase in blood pressure with increases in operator workload.

One of the researched methods of assessing operator mental workload is the use of physiological measures. The most popular method to measure workloads has to do with heart rate and heart rate variability. Similarly, but not as thoroughly researched, blood pressure has been analyzed as a potential indicator of mental workload [1]. Another method of assessing operator mental workload is the use of subjective measures, which deal with the quantification of the subject's perceived workload level [8]. Physiological measures in the form of heart rate and blood pressure were gathered, along with subjective measures, to determine the arousal of the subject. Heart rate variability has been suggested as a convenient and useful measure of mental workload [6]. Astrand and Rodahl [9] reported that workload may be assessed by the measurement of specific physiological responses. The following study attempts to determine the amount of physiological stress that is caused by playing a typical video game.

2. METHOD

2.1 Subjects

The subjects for this experiment consisted of sixteen male engineers between the ages of 22 and 34, with a mean age of 26.25 years and a standard deviation of 4.31 years. All subjects were tested on a voluntary bases. Eight of the subjects had played less than 10 games of GALAGA or had not played this game for at least 4 years. These subjects were considered novice players. The remaining eight subjects had played more than 10 games of GALAGA or had played this game within the last 4 years. These subjects were considered frequent players.

2.2 Apparatus

The apparatus used in gathering data included the GALAGA video game, an Exersentry heart monitor (by Respironics Inc.), a blood pressure gauge, and a digital stopwatch. The GALAGA video game was a seated version with ambidextrous controls. The GALAGA video game layout consisted of a small joystick located in the center of the control panel and a fire button located on each side of the joystick (providing equal ease of operation for both right and left handed players). The joystick controlled lateral movement of the player's spaceship across the playing boundary and the fire button activated missiles that were fired forward from the nose of the spaceships. The object of the game was to destroy as many arrays of enemy spaceships that randomly leaves a flight group to fire and/or ram your spaceship. Additional spaceships were awarded at predetermined point values.

A four legged non-adjustable wooden chair was used by all players. The video game used in this experiment was located in a relatively quiet, out-of-the-way, public location. The Exersentry heart monitor was attached to each subject with an elastic, adjustable shoulder harness. A PBC aneroid sphygmomanometer was used to determine the subject's blood pressure before and after each trial.

2.3 Instructions

Each subject was required to wear a short-sleeved shirt to facilitate the measuring of blood pressure. They were instructed that they would be monitored for heart rate, blood pressure, duration time of play, points scored, hit/miss ratio, and number of levels achieved. Subjects were also informed that they would complete a subjective survey after completing three games.

Subjects were allowed to watch visual instructions of the game for 1.5 minutes before starting the first game. After watching visual instructions, any questions on the game were answered.

2.4 Procedure

Upon entering the testing room, the subject was given verbal instructions on how to complete the subjective survey and play the GALAGA video game. The following data was obtained from each subject: name, age, perceived level of expertise, and dominant hand. After confirming that the subject understood the instructions, the experiment was conducted.

Each subject was fitted with the heart monitor harness. While seated, his resting heart rate and blood pressure were measured and recorded. The subject was not allowed to view any of his measured data until completion of the experiment. The subject then moved to the video game and watched the visual instructions and sample game for 1.5 minutes. He was instructed on the use of the controls for the video game and told to select the "one player" button when ready to begin playing. Any questions concerning operation of the game that the subject had were answered. The heart monitor was held behind the subject so as not to interfere with his playing ability.

Upon pushing the game's start button, the stopwatch was started and the first heart rate reading was recorded.

During the course of the game, heart rate was recorded at ten second intervals.

Upon completion of the game, the total points earned, hit/miss ratio, levels achieved, and duration time of play were logged. A post game blood pressure was then taken and recorded.

Each subject played three games and rested a minimum of ten minutes between games. Upon completion of three trials, subjects completed a subjective survey. The subjective survey covered such topics as how the game challenged the player, the stress level felt during play, effects of the controls and displays, and difficulty of operation of the game. The subjects were also not allowed to watch other players during the experiment to prevent their influencing the results.

3. RESULTS AND DISCUSSION

The data collected included age; perceived original experience level; preferred hand; blood pressure before and after; heart rate before, during, and after; hit/miss ratio; duration time of play; total points scored; and number of levels achieved. A statistical analysis was conducted using a SAS (version 5.16) package [10] on the mainframe IBM 3081 computer.

The means for each dependent variable were calculated with respect to different independent variables that could have influenced the results. The dependent variables that were studied included heart rate, duration of play, and diastolic and systolic blood pressure. The independent variables that may have influenced the results of the data were perceived experience level of the subject; trial being performed; and the pre-game, during, and post-game heart rate. Table 1 shows the means that were calculated for each of the independent variables that were applicable.

The results of the three-way ANOVA showed that the experience level, measurement, and experience level-measurement interaction had a significant effect at $p < 0.01$, but the trial had no significant effect at $p < 0.05$ on the heart rate as displayed in Table 2. The results of the Duncan multiple range test showed that the changes in heart rate were significant at $p < 0.05$ from pre-game to during game but were not significant at $p < 0.05$ from during game to post-game. Figure 1 graphically shows the mean heart rate achieved at pre-game, during, and post-game for each experience level.

Progression of trials and experience level were examined to determine their influence on the duration of play. The progression of trials had no significant effect on duration; however, the experience level was significant at $p < 0.01$. The duration time for frequent subjects was nearly double the duration time for the novice subjects.

TABLE 1

SUMMARY OF THE MEANS

Dependent Variable	Independent Variable	Level	Mean
Heart Rate	Experience Level	Novice	86.402
		Frequent	103.839
	Measurement	Pre-game	74.396
		During	102.362
		Post-game	108.604
	Trial	First	94.173
		Second	96.600
		Third	94.589
Duration	Experience Level	Novice	157.458
		Frequent	296.792
	Trial	First	196.060
		Second	238.810
		Third	246.500
Diastolic Blood Pressure	Experience Level	Novice	124.000
		Frequent	124.958
	Trial	First	126.906
		Second	122.219
		Third	124.313
Systolic Blood Pressure	Experience Level	Novice	77.188
		Frequent	78.354
	Trial	First	76.094
		Second	76.906
		Third	80.313

TABLE 2

ANALYSIS OF VARIANCE FOR HEART RATE

Source	Sum of Squares	df	Mean Squared	F	Probability
Experience	10944.82	1	10944.82	36.57	0.0001
Measurement	31860.41	2	15930.20	53.23	0.0001
Trial	161.73	2	80.86	0.27	0.7637
Experience* Measurement	3888.78	2	1944.39	6.50	0.0021
Error	37706.64	126	299.26	--	--

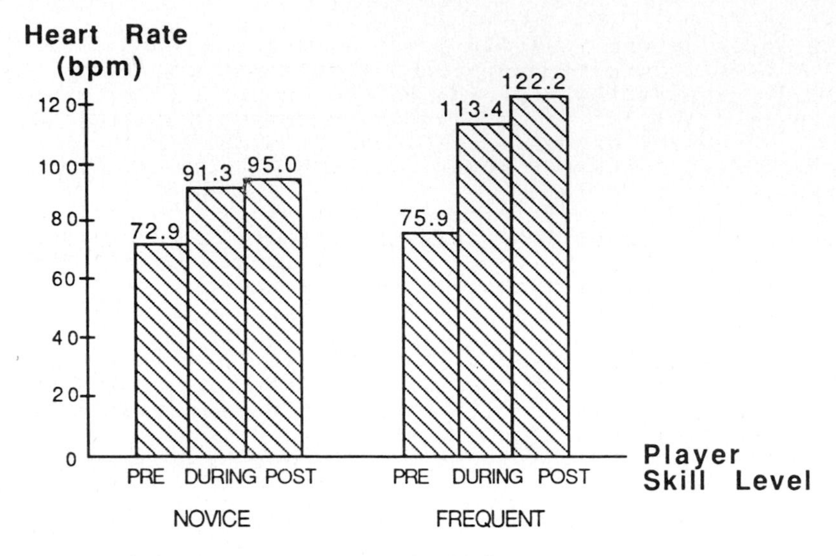

FIGURE 1
Mean Heart Rate Achieved at Pre-game, During, and Post-Game

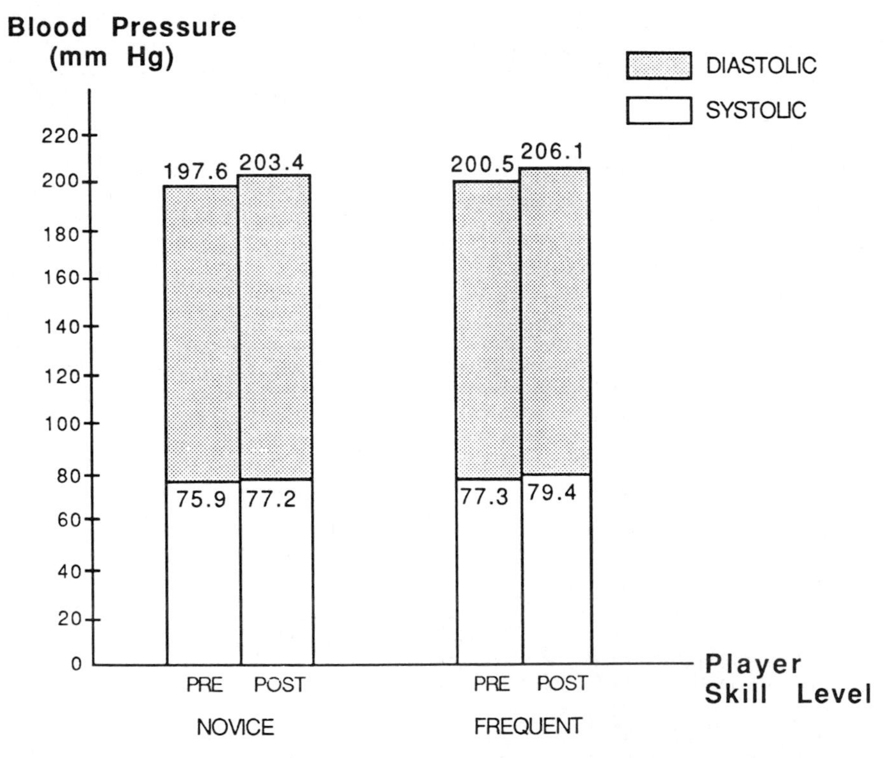

FIGURE 2
Mean Blood Pressure Achieved at Pre-game and Post-game

The blood pressure was analyzed to determine the influence of experience level, duration time, and a combination of the two. Blood pressure (both diastolic and systolic) did not change significantly at the $p < 0.05$ level from pre-game to post-game. Figure 2 graphically shows the mean blood pressure achieved for each experience level at pre-game and post-game.

The majority of players (93%) found the game to be of medium or greater difficulty; only 25% of the players reported above average stress in the subjective survey after their three trails while 31% of the subjects reported an average arousal level. It appears, however, that some level of arousal was expected, as 88% of the players found the experience quite stimulating. The current controls and displays on the game did not adversely affect the players, as all felt that the colors were suitable and the number of controls were not excessive.

4. CONCLUSIONS

The results of this experiment suggest that there are short-term physiological effects of playing the GALAGA video game. There was a significant increase in heart rate when measured before and during the game. This would lead one to believe that the subjects experienced mental stress to some extent while playing the video game. However, blood pressure, when measured before and after the game, showed insignificant increases.

The frequent users experienced a greater rise in heart rate from pre-game to during game measurements than the novice users. This is expected since the frequent user gets more involved and aroused in the game. The novice user could not anticipate the events of the game and, therefore, not experience as great an increase in the heart rate.

The experience level of each subject was significantly influenced by the duration of play. Those subjects that were considered frequent users were able to sustain play approximately twice as long as the novice user. This could be due to the knowledge of the game which might permit anticipation of actions and increase duration of play. To further study the physiological effects of the video game, the time for heart rate to return to the pre-game level might be examined in the future.

REFERENCES

[1] Wierwille, W.W. (1979). "Physiological Measures of AircrewMental Workload." Human Factors, 21(5), pp 575-593

[2] Ettema, J.H. (1969). "Blood Pressure Changes During Mental Load Experiments in Man." Phychotherapy and Psyhosomatics, vol. 17, pp 191-195.

[3] Mobbs, R.F., G.C. David, and J.M. Thomas. (1971). "An Evaluation of the Use of Heart Rate Irregularity as a Measure of Mental Workload in the Steel Industry." London, England: British Steel Corporation, BISRA, OR/HR/25/71, August.

[4] Ettema, J.H. and R.L. Zielhuis. (1971). "Physiological Parameters of Mental Load." Ergonomics, vol. 14, pp 137-144.

[5] Zwaga, H.J.G (1973). "Psychophysiological Reactions to Mental Tasks: Effort or Stress?" Ergonomics, vol. 16, pp 61-67.

[6] Boyce, P.R. (1974). "Sinus Arrhythmia as a Measure of Mental Load." Ergonomics, vol. 17, pp 177-183.

[7] Guame, J.G. and R.T. White. (1975). "Mental Workload Assessment, II. Physiological Correlates of Mental Workload: Report of Three Preliminary Laboratory Tests." St. Louis, Missouri: McDonnell Douglas Corporation, Report MDC J7023/01, December.

[8] Yung-Hui Terrence Lee, B.S. (1984). "Measurements of VDT Workload as a Function of Work/Rest Cycles." Unpublished Masters Thesis. Texas Tech University, Lubbock, TX.

[9] Astrand, Per-Olof and Kaare Rodahl. (1977). Textbook of Work Physiology. New York: McGraw-Hill Book Co.

[10] SAS User's Guide : Statistics. (1982). SAS Institute Inc. Box 8000, Cary, NC 27511.

XIV

ERGONOMICS IN REHABILITATION

Trends in Ergonomics/Human Factors V
F. Aghazadeh (Editor)

ERGONOMICS CONSIDERATIONS FOR THE REDUCTION OF PHYSICAL TASK DEMANDS OF LOW BACK PAIN PATIENTS

Elsayed Abdel-Moty, Tarek M. Khalil, Shihab S. Asfour, Renee S. Rosomoff and Hubert L. Rosomoff

Comprehensive Pain and Rehabilitation Center
Department of Industrial Engineering
and Department of Neurological Surgery
University of Miami
P.O.Box 248294
Coral Gables, FL 33124, USA

The implementation of ergonomics and biomechanical principles for the reduction of musculoskeletal stresses during work related activities has been repeatedly addressed in the literature. This paper presents three different examples of this type of application in a rehabilitation setting. The cases selected are low back pain patients who underwent a rehabilitation program at the University of Miami Comprehensive Pain and Rehabilitation Center. These patients sustained injuries to the low back due to job related factors. As a part of the Ergonomics and Bioengineering Division's involvement in the rehabilitation process, patients performed their job activities in a simulated environment. Principles of ergonomics, anthropometry, body mechanics, physical motion analysis, and biomechanics were applied to the analysis phase of this simulation. Based upon this analysis, necessary modifications and adjustments were implemented, and the tasks were re-evaluated. Procedures, methods, and results are presented. The value of applying these comprehensive methods to analyze job demands and as a strategy to aid rehabilitation and prevent further injury is also addressed.

1. INTRODUCTION

Injuries to the low back remain one of the most costly medical problems in the world today. Back pain can strike anyone, at any time, and due to many causes. It has been estimated that 80% of the people will suffer from some form of back ailment during their lifetime. The incidence of low back pain was often associated with industrial-type activities, such as driving heavy equipment and lifting. However, the available statistics have shown that people who are involved in sedentary-type and other non-industrial activities are equally prone to suffer low back pain [1]. It is also evident that task activities are not the major cause of low back pain. The most important and detrimental factor to the onset of back pain appears to be related to the way by which work activities are performed.

The rehabilitation of low back pain is a challenge to health care providers and to safety specialists. Current trends in low back pain rehabilitation show that the problem needs an interdisciplinary, holistic approach [2]. Experts have recognized that the physical, psychological, occupational, and other factors have to be treated as a part of a successful rehabilitation effort. The restoration of functional abilities of the low back pain sufferer used to be considered the sole domain of physicians and other health related professionals. The approach used at the University of Miami's Comprehensive Pain and Rehabilitation Center (UMCPRC) has proven that ergonomists working in clinical settings are a valuable addition to the rehabiliation team. Ergonomics, by its definition, studies the relationship between people and their environment, whether at work, at home, or elsewhere. Therefore, it is very beneficial to employ the knowledge of ergonomics in conjunction with biomechanical principles and proper body-mechanics, techniques of energy conservation, methods of physical motion analysis, and biofeedback to help the patient master critical job activities under supervision before re-entering real-life settings after rehabilitation. Generally the goal of this ergonomically-based approach is to determine optimal task activities (tools and personnel) to produce a man-machine system that is effective, productive, comfortable, and safe [3].

This paper presents one of the ergonomics and bioengineering contribution to low back pain rehabilitation. Three cases are used for illustrative purposes. The following general postural, engineering, and body-mechanical considerations provided the basis for task analysis and design [3,4,5,6,7,8,9].

1) Posture-Related Considerations:

The important element in this category is the alignment of body segments in order to avoid static stresses. The principles to be emphasized here are the following: to maintain the head, neck, and back as close to the same coronal plane as possible; to avoid static pressure on sensitive body parts; to maintain the anatomical axis of the joints in a neutral position; to keep the knees slightly higher than the hips level while sitting; and to provide a wide base of support at the feet while standing.

2) Engineering Considerations:

By re-engineering the environment, a more compatible and safer relationship between the people with low back pain and their working or living environment can be established [10]. Examples in this category would be: individualizing the design of appropriate seating devices to meet the needs of the "individual" in his/her work station [5]; keeping working objects close to the body through optimal layout of the work surface; designing tools to facilitate task performance; and balancing force vectors on the human body.

3) Body Mechanics Considerations:

In this category, the "do's" and "dont's" of body mechanics are taught and implemented. Among these are: avoiding repeated twisting and pivotal movements at the waist and ankles; avoiding localized stresses; minimizing over reaching; using ballistic movements at the joints; optimizing work-rest schedules; avoiding static postures; making use of stronger muscles (e.g. of the legs) instead of weaker muscles to perform the same task; and assuming pelvic tilt whenever possible.

Combining all these principles in the process of task analysis requires some form of scientific quantification. This is usually done through the use of biomechanical models in conjunction with accurate monitors of physiological responses of the body (such as electrical muscle activity, and heart rate).

3. CASE STUDIES

CASE STUDY #1:

<u>Case Description</u>: This is a 35 year old male. He weighs 165 lbs and stands 5' 10". He was admitted to the UMCPRC complaining of right lower extremity pain, especially at the right hip. He was given a diagnosis of multiple myofascial pain syndromes. He was then involved in an aggressive physical medicine program. The patient's treatment program was individualized and aimed at a simultaneous restoration of physical function together with vocational rehabilitation and behavioral modification [11].

<u>Task Description</u>: This patient works as a photo journalist. His job requires photographing news, events, locations, people, and other illustrative or educational material for use in publications or telecast. He uses still cameras, travels to assigned locations, and takes pictures. He may develop negatives and print films. He hangs a 15 lb bag and a 10 lb camera off the right shoulder, a 12 lb camera off the left shoulder, and two small cameras, 5 lbs each, around the neck (Figure 1a). Occasionally he carries a 40 lb suitcase in the left hand.

<u>Task Analysis Activities</u>: Job simulation activities included the possible physical task demands such as walking, standing, pulling, pushing, entering and exiting all types of vehicles, loading and unloading equipment, taking photographs from a variety of positions and obstacles, and desk-type work. A very important aspect of this simulation was the analysis of his photography task from an ergonomics and a biomechanical point

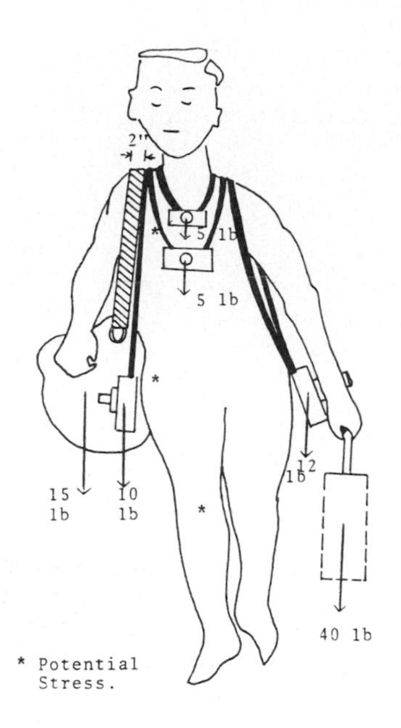

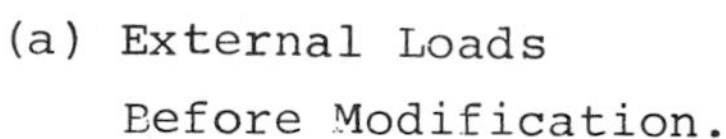
(a) External Loads Before Modification.

(b) New Vest Design.

Figure 2. Work Parameters of Case #1.

of view. Patient was asked to bring all tools for a realistic simulation. Ergonomics evaluation of functional abilities [5] showed the effect of the uneven external load distribution on the body. There was decreased muscle strength in the back and arms, and limited ranges of motion. Posture evaluation showed poor alignment of the right and left shoulders (right shoulder was 1.5" lower than the left one), and uneven hip level (left hip was 1" higher than right hip). There was decreased muscle mass in the right lower extremity. Biomechanical analysis was performed to determine the forces on the involved body parts [12]. Results (Table 1) showed that biomechanical forces were 23% to 60% higher when objects were included in the analysis. The goals were, then, to minimize the reaction forces on the hips, to distribute the weights evenly around the center of gravity of the body, and to reduce back muscular force by reducing the moment arms. In order to keep and maintain objects close to the center of gravity of the body, a specially made vest was designed. The different work tools were placed in the vest. Also, by attaching the side-cameras to the vest, friction with bony structures was prevented and cameras were kept close to the body during movements. Padding was also provided at the shoulder and the waist. By redistributing the weights (Figure 1b), the back muscular force was reduced by 7%, and the reaction forces on the hip joints where equalized to a value of 80 lbs (a decrease of 13%).

TABLE 1

Results of Bimechanical Analysis for Case #1
Before and After Task Modifications

Biomechanical Forces	No Tools*	With Tools	
		Before+	After+
Reaction Force on the Right Hip Joint	56.8	90.5	80
Reaction Force on the Left Hip Joint	56.8	70.1	80
Low Back Muscular Force	484.5	730.1	512.0
Reaction Force on the L5-S1 Joint	560.5	836.9	601.3

* Refers to Free Body in the Neutral Position.
+ Before and After Task Modifications.

CASE #2:

Case Description: This is a 69 year old man with several years history of intermittent low back and lower extremity pain, more on the right side. He was admitted to the UMCPRC for treatment and rehabilitation. As part of the conditioning program, job simulation and task analysis were performed.

Task Description: This patient works as a musical conductor. His job requires conducting symphony orchestras. He selects music to suite type of performance to be given. He directs group at performances up to 3 hours in length. He, also, schedules tours and performances and arranges for transportation and lodging.

Task Analysis Activities: A complete job simulation was performed with this subject. He was asked to bring musical notes and a stand. Patient simulated a live performance and was video taped throughout activity. A grid chart was placed behind him to monitor postural changes during simulation (Figure 2a). Muscle activity was continuously recorded from the right and left lumbar paraspinals muscles. Trunk motion pattern was monitored using an electronic goniometer. Pulse rate was also monitored. During a 90 minutes simulated performance, the following was observed (Figure 2 and Table 2):

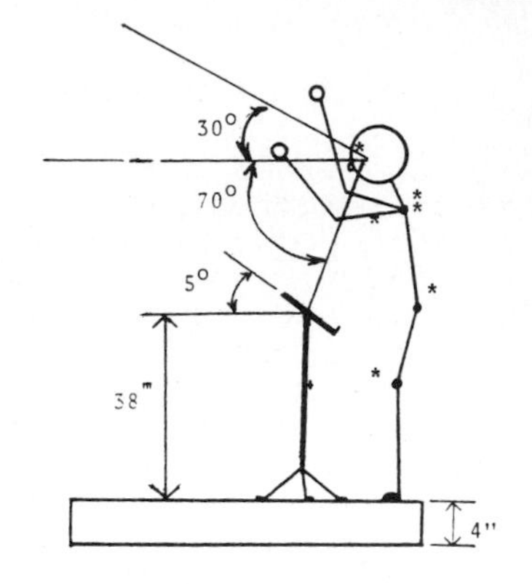

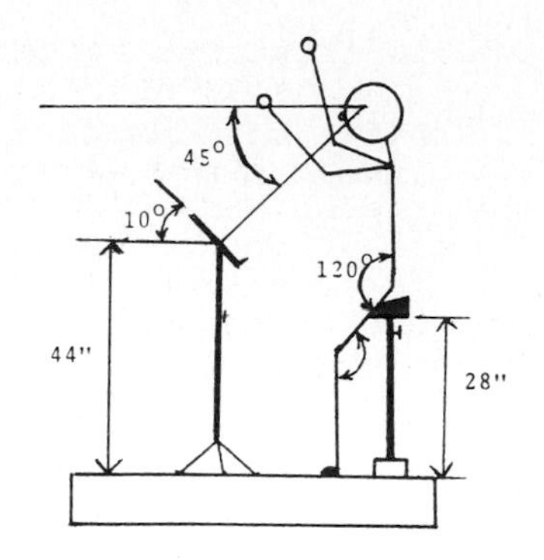

(a) Before Modification (b) After Modifications

Figure 2. Environment Parameters of Case #2.

TABLE 2

The Effects of Task Modifications on Pulse Rate and Muscle Activity Measures Obtained During a Simulated Task of 90 Minutes

	Before*		After*	
	Initial	Final	Initial	Final
Pulse Rate, BPM	60	113	59	86
EMG Value, uV*sec	33	36	20	15

* Refer to Before Modifications and After Modifications.

1. Angle of vision ranged from 30 degrees above horizontal to as low as 70 degrees below horizontal.
2. He was constantly shifting weight from the right to the left side.
3. Feet separation ranged from 10" to 32".
4. There was constant twisting and bending at the trunk and neck. The trunk moved between 28 and 43 degrees flexion during performance (pointing upward and turning pages of the musical notes).
5. Heart rate increased from 60 at rest to 113 beats per minute (BPM) while moving arms overhead. Recovery after performance was slow.

6. The level of the EMG activity of the lumbar paraspinals was high (about 33 uV*sec).

Video tape was reviewed with the patient and problems were identified. These had to do with leveling of the shoulders, weight distribution, constant twisting and forward flexion, and the angle of vision. It was interesting to notice that the poor alignment of the shoulder was, in part, due to the frequent use of the left arm to direct musicians at higher levels. Based on the analyses, recommendations for proper body mechanics to reduce muscle tension and cardiovascular stress were made. By assuming pelvic tilt, leveling shoulders, and aligning nech and back, the EMG activity decreased to below 20 uV*sec. He was instructed to widen base of support to balance weight on both legs. Proper turning of the body and avoidance of twisting were implemented. In order to further reduce static load on the spine during prolonged standing, he was provided with a high stool-type chair. This arrangement allowed him to assume a semi-sitting position whenever he fatigues or experiences increase in discomfort during long performances. This chair was designed to be adjustable, with a slight angle at the seat, and with soft cushioning (Figure 2b). With a height of 25", and by raising the notes stand to 44" (from 38") the EMG of the lumbar paraspinals decreased to below 15 uV*sec. With these modifications and training of proper body mechanics and efficient use of muscles, he performed for 90 minutes during which heart rate increased by 46% only, above resting level (as compared to the increase of 88% before modifications).

CASE #3:

Case Description: This is a 41 year old man who weighs 168 lb and stands 5′ 10" tall. He suffered low back pain which extended towards the buttock and the posterior thigh. Physical activity, especially weight bearing, was always associated with increased pain, numbness and cramping. After undergoing the multidisciplinary pain team evaluation, he was admitted to the 4-week treatment program with very limited standing tolerance (2 minutes), decreased muscle strength and decreased ranges of motion.

Task Description: As a radiologist, his job requires diagnosing diseases using X-rays and radioactive substances. He treats malignant growths by exposure to radiation. He wears a lead apron that weighs 15 lbs (Figure 3a) for as long as 8 hours every day. He ambulates carrying x-rays (up to 10 lbs), and he occasionally moves X-ray equipment.

Task Analysis Activities: Job simulation activities included wearing an apron, ambulating 464 ft while carrying 10 lbs, handling X-rays and X-ray machinery, moving persons on examination table from supine to prone position, and performing desk-type activities. Biomechanical analysis of the forces on the low back during activity showed that the use of the lead apron places an additional moment of 300 lb*inch on the low back region, not to mention the static loading on the shoulders, neck, upper back, and knees. The major modification in this case was the re-design of the lead apron. Figure 3b shows the new design where apron weight was reduced, distributed, and divided into two parts: top and bottom. With this new configuration, the moments on the low back region were balanced, static loading of the shoulders and neck was reduced by more than 80%, back muscle force was reduced dramatically, and posture was improved.

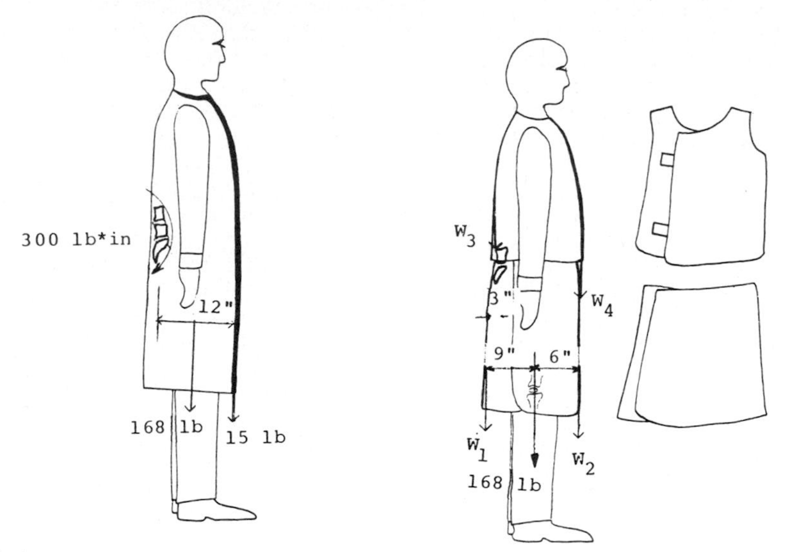

(a) Before Modifications. (b) After Modifications.

Figure 3. Apron Design and Biomechanical loads for Case #3.

DISCUSSION

The potential contribution of Ergonomics to rehabilitation at all its stages (pre-injury, post-injury rehabilitation, and post rehabilitation stages) is obvious [10]. In the post-injury rehabilitation stage particularly, the inclusion of task analysis and job simulations enables the patient to review work activities, receive suggestions from experts for modifications or adjustments, and implement recommendations in a simulated environment under supervision.

The cases presented in this paper are but few examples of the wide range of task activities awaiting the ergonomics insight to help prevent injury, pain and suffering. The minor, yet significant, adjustments that were made in the physical environment and in work behaviors have allowed these patients, and others, to comfortably and safely carry-out task activities, be subjected to less detrimental stresses while performing these and other tasks, and have contributed to reducing pain and improving function. These patients went on to successfully complete the 4-week rehabilitation program and have returned to full time employment immediately after. In a one year follow-up, they were with absolutely no restrictions, applying the new knowledge which they acquired during rehabilitation to their work and home environment, and functioning effectively and safely within the environment.

In conclusion, it is very critical, important, and cost effective to make use of the ergonomics knowledge available in the technical literature. When combined with biomechanical principles, this comprehensive approach can be used effectively in rehabilitation and "real life" situations.

REFERENCES

[1] Khalil, T.M., Asfour, S.S., Moty, E.A., Case Studies in Low Back Pain, Proceed Human Factors Society Annual Meeting (San Antonio: Texas 1984) pp. 465-470.

[2] Rosomoff, H.L., Rosomoff, R.S., Non Surgical Aggressive Treatment of Lumbar Spinal Stenosis, Spine: State of the Art Reviews Vol 1, No. 3 (Philadelphia: Hanley and Belfus, Inc., 1987)

[3] Khalil, T.M., Design Tools and Machines to Fit the Man, Indus Eng Jan (1972) 32-35.

[4] Abdel-Moty, E.A., Khalil, T.M., Computer Aided Design of the Sitting Workplace, Computers and Indust Eng 11 (1986) 23-26.

[5] Abdel-Moty, E., Khalil, T.M., A Computerized Expert System for Work Simplification and Workplace Design, Proceedings of the Annual International Meeting of the IIE (Washington, D.C.: 1987) pp. 17-21.

[6] Khalil, T.M., An Electromyographic Methodology for the Evaluation of Industrial Design, Human Factors 15 (1973) 257-264.

[7] Khalil, T.M., Ayoub, M.M., Work-rest Schedule Under Normal and Prolonged Vibration Envrionment. AIHA Journal, March, 1976.

[8] Khalil, T.M., Bell, B.H., Human Factors in Dental Health Care Delivery Systems. Technical Papers of the National Meeting of the Human Factors Society, New York, N.Y., pp. 14, 1971.

[9] Tichauer, E.R., The Biomechanical Basis of Ergonomics (New York: John Wiley & Sons, 1978)

[10] Khalil, T.M., Asfour, S.S., Moty, E.A., Rosomoff, R.S., Rosomoff, H.L., Ergonomics Contributions to Low Back Pain Rehabilitation, Pain Mang, In Press.

[11] Khalil, T.M., Asfour, S.S., Moty, E.A., Rosomoff, R.S., Rosomoff, H.L., The Management of Low Back Pain: A Comprehensive Approach. Proceedings of the 1983 Annual IE Conference (Louisvill: KY 1983) pp. 199-204.

[12] LeVeau, B., Biomechanics of Human Motion, 2nd ed. (W.B. Saunders Co.: Philadelphia, 1977)

Trends in Ergonomics/Human Factors V
F. Aghazadeh (Editor)
Elsevier Science Publishers B.V. (North-Holland), 1988

Psychological Indicators of Recovery from Back Pain

Michael J. Colligan*, Edward F. Krieg*
Steven E. Besing** & Thomas Bennett**

It has been estimated that nearly two percent of the total U.S. industrial workforce receives compensation for work-related back pain each year (Klein et al., 1984) and that as many as one- third of all sedentary workers and one-half of workers in heavy industry will suffer a significant episode of back pain during their working lives (Quinet, 1984). Contrary to popular belief, the vast majority of these cases are not victims of acute injury or strain. Rather, they describe their pain as having a vague and nonspecific origin, developing gradually over time, and resulting in a diffuse set of symptoms and complaints (Hadler, 1978; Quinet & Hadler, 1979).

The ambiguous nature of back pain, and the frequent absence of any clearly discernible organic or physical pathology make it difficult to predict recovery on an individual basis. This problem is further complicated by the fact that many back pain victims possess a psychological profile which may add to their overall distress and interfere with recovery. For example, Sternbach (Sternbach et al., 1973), based on a sample of 117 cases, described the chronic back pain patient as an individual who has developed a passive and dependent life-style characterized by depression, social withdrawal, and physical and psychological fatigue. Similarly, Levine (1971) identified depression as a significant component in the back pain syndrome. Still other researchers (Caldwell & Chase, 1977; Leavitt & Garron, 1982) present evidence that chronic back pain victims have elevated scores on the Hysteria, Depression, and Hypochondriasis scales of the Minnesota Multiphasic Personality Inventory (MMPI). This indicates, in addition to the depression, a preoccupation with physical health and pain and a tendency toward the dramatization and exaggeration of symptoms. These characteristics, by adversely affecting the individual's motivation and expectation of recovery, may play an important role in the rehabilitation process. To the extent that this

*Centers for Disease Control, National Institute for Occupational Safety & Health, Division of Biomedical & Behavioral Sciences, Cincinnati, Oh. 45226

**Indiana Center for Rehabilitation Medicine, Indianapolis, In. 46202

is the case, perceived ability to return to work following a debilitating back pain episode may be as much a function of psychological status as it is physical functional capacity. The purpose of the present paper is to examine the relative influence of select psychological and physical factors on judgments of an individual's ability to work following a compensable back pain episode.

Method

The records of 50 back pain patients (13 females and 37 males) were obtained from an occupational rehabilitation center. Upon admission, patients underwent an extensive physical and psychological evaluation. Information on age, sex, and months since injury (MSI) was also obtained. Psychological status was assessed with a battery of standardized self-report inventories including the Sickness Impact Profile (SIP), Beck Depression Inventory (BDI), State Anxiety Scale of the Spielberger State-Trait Anxiety Inventory (SAI), Lazarus Assertiveness Scale (LAS), Present Pain Index (PPI) and Pain Rating Intensity (PRI) scales of the McGill Pain Questionnaire, and the Minnesota Multiphasic Personality Inventory (MMPI). Insofar as previous research with the MMPI indicated that only the Depression (D), Hysteria (Hy), and Hypochondriasis (Hs) scales consistently related to reported back pain (Adams et al, 1981), these were the only scales of the MMPI considered in the present analysis.

Included in the physical status evaluation was a baseline EMG measure to be used in later biofeedback training, and a determination of functional capacity. The biofeedback (BFB) score consisted of the mean of lumbar paraspinal electromyographic (EMG) recordings sampled from each patient at 30-second intervals for a period of 4 minutes. Physical functional capacity (PFC) was evaluated via performance on seven standard exercises (e.g., straight leg raises, trunk flexion, abdominal strength) which were rated on a six-point scale ranging from "0" (maximum capacity) to "5" (unable to perform). Ratings on these seven exercises were averaged to produce a single value representing overall functional capacity.

Patients were also asked upon admission whether they expected to be able to return to work following therapy. This was an indication of their vocational expectancy (VE) prior to beginning treatment. VE was scored on a six-point scale ranging from "0" (able to return to former work without restrictions) to "6" (unable to work).

At the termination of the eight-week rehabilitation program, patients, their employers, and members of the rehabilitation staff met to evaluate the patient's ability to return to work using the same six-point scale described above for VE. For each case, this review produced a single value representing the unanimous agreement of the involved parties. No further follow-up of individual cases was performed. The determination of post-therapy vocational status (PTVS) served as the outcome measure in the present investigation.

For purposes of data analysis, both Ve and PTVS scales were recoded using a simpler 3-point format where "1" - "able to return to work", "0" = "uncertain of ability to work " and "-1" = "unable to return to work".

Results

Table 1 presents the variables comprising the data set. The Depression, Hysteria, and Hypochondriasis scales of the MMPI were combined into a single variable using principle components analysis. The first principle component accounted for 70% of the variance of the three scales (Lambda = 2.10). The Depression (.77), Hysteria (.92), and Hypochondriasis scales(.81) all correlated positively with this component.

Table 1

The Variables Used in the Analysis

Variable Name	Variable Description and Coding
SIP	The overall score from the Sickness Inventory Profile.
BDI	Score on the Beck Depression Inventory.
SAI	Score on the State Anxiety Index
LA	Score on the Lazarus Assertiveness Scale
BFB	Biofeedback measure.
PPI	Miller Pain Questionnaire Present Pain Index.
PRI	Miller Pain Questionnaire Pain Rating Intensity.
MMPI	The first principle component of D, Hy and Hs.
MSI	Months since injury.
Age	Age of the patient in years.
Sex	0 = Female, 1 = Male.
PFC	The mean of the physical therapy measures.
VE	Pre-therapy vocational status: 1 = Return to Work, 0 = Unsure of Status and -1 = Will not Return to Work.
PTVS	Post-therapy vocational status: 1 = Return to Work, 0 = Unsure of Status and -1 = Will not Return to Work.

Table 2 presents the Pearson product moment correlation matrix for all variables measured at admission. The number of cases varies across correlations because of missing values.

Table 2

A Correlation Matrix of the Variables Used in the Discriminant Analysis

	Age	MSI	SIP	BDI	SAI	LA	BFB	PPI	PRI	MMPI	PFC	VE	PTVS
Age	1.00	0.01	-0.06	0.02	-0.10	-0.35	-0.06	0.16	0.13	-0.06	0.14	-0.12	-0.11
MSI		1.00	0.22	0.15	0.11	-0.02	0.14	0.22	0.15	0.19	0.07	-0.07	-0.03
SIP			1.00	0.63*	0.75*	-0.34	0.24	0.27	0.46*	0.53*	0.51*	-0.16	-0.18
BDI				1.00	0.63*	-0.43*	0.22	0.43*	0.47*	0.52*	0.14	-0.25	-0.40*
SAI					1.00	-0.38*	0.11	0.35	0.62*	0.50	0.34	0.02	-0.28
LA						1.00	0.02	-0.32	-0.29	-0.36	-0.34	-0.36	0.16
BFB							1.00	0.02	0.04	0.10	0.13	-0.15	-0.33*
PPI								1.00	0.44*	0.15	0.27	-0.17	-0.13
PRI									1.00	0.33*	0.28	-0.11	-0.24
MMPI										1.00	0.20	0.12	-0.08
PFC											1.00	-0.18	-0.30*
VE												1.00	0.12
PTVS													1.00

*p < 0.05

At the univariate level, only the Beck Depression Inventory ($r = -.40$), BioFeedback Baseline ($r = -.33$) and Physical Functional Capacity ($r = -.30$) measures were significantly ($p < .05$) correlated with post-therapy vocational status (PTVS). Being rated as able to work following an eight-week treatment program was associated with lower depression, better performance on the functional capacity tests, and higher lumbar paraspinal EMG readings at the time of admission.

The variables listed in Table 1 were then submitted to a backward stepwise discriminant analysis using post-therapy vocational status as an outcome measure. This involved assigning cases to one of three PTVS groups: 1) those rated as able to work; 2) those whose ability to work was rated as uncertain; and 3) those rated as unable to work. All variables were then entered into the discriminant function to determine which combination of admission measures best predicted post-treatment ratings of ability to work. At each step, the variable with the smallest F-ratio was removed. This process was essentially an analysis of covariance with each variable examined individually and all other variables remaining in the discriminant functions treated as covariates. Variables were removed until the probabilities of the F-ratios of the remaining variables were .05 or less. This resulted in the identification of that combination of weighted admission measures which best predicted ability to work ratings following treatment. After the last non-significant variable was removed, the cases were classified based on the discriminant functions. Only 27 cases had complete data on all the variables and were used in this analysis. There were not enough cases for a cross-validation sample.

The combination of variables included in the discriminant functions which best predicted post-treatment ability-to-work ratings were the scores on the Beck Depression Inventory (BDI), the Spielberger State Anxiety Inventory (SAI), and the physical functional capacity (PFC) assessment. Table 3 presents the group means for these three variables.

Table 3

Group Means of the Variables in the Final Discriminant Functions

	Group		
Variable	Return	Unsure	Not Return
BDI	13.77	20.25	22.67
SAI	43.92	59.12	49.50
PFC	1.97	1.77	2.50

Individuals rated as unable to return to work following the eight-week treatment program had shown the highest level of depression at admission, while those rated as able to work had shown the lowest. Similarly, poorer performance on the physical functional capacity assessment upon admission was associated with lower post-treatment ability to work ratings. Finally, those individuals whose ability to work was rated as "uncertain" had shown the highest levels of anxiety at admission while those rated as able had shown the least.

The outcome of the classification of the cases based on the discriminant functions is presented in Table 4. Overall, the post-treatment vocational status of 78% of the patients was correctly identified from the weighted BDI, SAI, and PFC admission measures.

Table 4

The Classification of the Cases

Actual Group	Predicted Group: Return	Unsure	Not Return	% Correct
Return	9	2	2	69%
Unsure	1	7	0	88%
Not Return	0	1	5	83%

Discussion

As expected, the present findings indicate a relationship between pre-treatment physical status and vocational outcome. Those individuals evidencing the poorest physical functional capacity at the time of admission were judged as least able to return to work following the eight-week treatment program.

In addition, there was a significant relationship between pre-treatment psychological status and post-treatment judgments of ability to work. Individuals rated as unable to work had higher anxiety scores at admission than patients who were rated as able. The highest level of anxiety was reported by those individuals who received an "uncertain" ability to work rating. These individuals pose a particular problem for the rehabilitation therapist who must make a final recommendation as to whether or not the person should return to work. The excessive worry and apprehension expressed by the highly anxious patient may undermine the confidence of the evaluator in making an ability-to-work determination. As a consequence, resolution of the case may be delayed until the patient appears to reach a more stable psychological state. Without a clear understanding of this process and the appropriate psychological interventions, recovery may be slow and tenuous.

Finally there was an association between depression and PTVS. Those individuals expressing greater depression at the time of admission were more likely to be rated as unable to return to work following the eight-week treatment program. This is consistent with previous characterizations of the chronic back pain victim as suffering from learned helplessness, depression and social withdrawal (Sternbach et al, 1973; Levine, 1971; Marbach et al 1983). Furthermore, in the present case the depression experienced by the back pain patient appeared to be independent of their actual physical status. Table 2 reveals that scores on the Beck Depression Inventory were not significantly related to months since injury (MSI), suggesting that the depression was not the result of a lengthy disability. It should also be remembered that the stepwise discriminant analysis identified depression as a significant correlate of post-treatment vocational status independently of performance on the physical functional capacity measures (PFC).

Whether the depression existed prior to the reporting of back pain, or was a reaction to other stresses associated with the disability (e.g. financial pressures, marital strain, loss of esteem) cannot be determined in the present study. What these findings do indicate is that the depression experienced by some back pain sufferers is not simply a reaction to their reduced functional capacity. This being the case, it cannot be assumed that treatment programs which focus exclusively on physical rehabilitation will produce a corresponding improvement in psychological health, or that recovery of physical functional capacity alone insures that the individual may successfully return to work. As Hadler (1986) has argued, we should stop thinking of back pain as the simple result of acute injury or strain, and begin thinking of it as a complex syndrome having physical, psychological, and social components.

References

Adams, K.M., Heilbronn, M., Silk, S.D., Reider, E., & Blumer, D.P. (1981). Use of the MMPI with patients who report chronic back pain. Psychological Reports, 48, 855-866.

Caldwell, A.B., & Chase, C. (1977). Diagnosis and treatment of personality factors in chronic low back pain. Clinical Orthopaedics and Related Research, 129, 141-149.

Hadler, N.M. (1978). Legal ramifications of the medical definition of back disease. Annals of Internal Medicine, 89, 992-999.

Hadler, N.H. (1986). Is an aching back an injury? Occupational Problems in Medical Practice, 3 (1), 6-8.

Klein, B.P., Jensen, R.C. & Sanderson, L.M. (1984). Assessment of workers'compensation claims for back strains/sprains. Journal of Occupational Medicine, 26 (6), 443-448.

Levine, M.D. (1971). Depression, back pain, and disc protrusion: Relationships and proposed psychophysiological mechanisms. Diseases of the Nervous Systems, 32 (1), 41-45.

Marbach, J.J., Richlin, D.M., & Lipton, J.A. (1983). Illness behavior, depression and anhedonia in myofascial pain and back pain patients. Psychotherapy and Psychosomatics, 39, 47-54.

Quinet, R.J. (1986). Diagnosis of low back pain. Occupational problems in Medical Practice, 1 (3), 1-5.

Quinet, R.J., & Hadler, N.M. (1979). Diagnosis and treatment of backache. Seminars in Arthritis and Rheumatism, 8, 261-287.

Sternbach, R.A., Wolf, S.R., Murphy, R.W., & Akeson, W.H. (1973). Traits of pain patients: The low-back "loser". Psychosomatics, 14, 226-229.

Trends in Ergonomics/Human Factors V
F. Aghazadeh (Editor)

EFFECTIVENESS OF AGGRESSIVE TREATMENT OF BACK PAIN

T.M. Khalil, S.S. Asfour, S.M. Waly, R.S. Rosomoff, & H.L. Rosomoff

Comprehensive Pain and Rehabilitation Center
University of Miami
600 Alton Road
Miami Beach, Florida 33139

The main objective of the present investigation was to evaluate the role of the multidisciplinary rehabilitation program used at the University of Miami Comprehensive Pain and Rehabilitation Center (CPRC) in restoring the functional capabilities of low back pain (LBP) patients. Parameters monitored include both isometric strength and electromyographic signals of the lumbar paraspinal muscles. These are recorded before and after the administration of the treatment program. Forty LBP patients admitted to the CPRC were used in this study. The results obtained indicate that an intensive well coordinated rehabilitation program is a successful and effective approach in the restoration of trunk functional abilities of patients with chronic back pain. A detailed description of the conditioning program and the analysis of the results obtained are presented and discussed.

1. INTRODUCTION

Musculoskeletal injuries are among the most crucial and persistent problems in occupational medicine. These injuries have been listed as the second leading work–related illnesses [1]. The National Institute for Occupational Safety and Health, in 1981, [2] reported that over–exertion accounts for one–fourth of all reported occupational injuries in the United States. Approximately, two–thirds of these occupational injury claims involved lifting loads. In the United States, approximately 35% of all compensation claims are related to back injuries. Backaches are the second largest pain problem in our society, secondary only to headaches. The U.S. Public Health Service estimates that in 1978, between 12 and 15 million experienced chronic pain in the lower back, leading to impairment. Tabor [3] estimated the direct financial compensation of back injuries to be in the range of fourteen billion dollars. The indirect costs may be as much as four times this amount. These costs are usually associated with treatments, surgeries, lost productivity, claims, law suits, awards and settlements [4,5].

Back injuries have been recognized as a major occupational problem requiring control [1,4,5]. Health professionals have been confronted with a major challenge in developing effective approaches for dealing with the back pain problem. One approach that has been gaining popularity is the non–invasive aggressive treatment of back pain. Some rehabilitation programs which follow such an approach for the management of low back pain have shown a very high rate of success [6,7]. The goal of these rehabilitation programs is to return back injured individuals to their full working capabilities by restoring function and reducing pain. In order to achieve these objectives innovative and comprehensive treatment regimens are required.

2. MULTIDISPLINARY APPROACH TO TREATMENT

The treatment of back injuries requires a multidisciplinary approach since no one physician has the expertise or resources to manage this complex condition [6]. The Comprehensive Pain and Rehabilitation Center (CPRC) of the University of Miami follows an aggressive treatment program which makes use of several disciplines. The CPRC program is consituted of several components including aggressive, intense physical medicine, understanding and dedicated behavioral medicine, vocational and recreational rehabilitation, ergonomics and bioengineering. A model of the multidisplinary team concept of the CPRC is given in figure 1. This model illustrates the different disciplines involved in the diagnosis, evaluation, and treatment of low back pain (LBP) patients admitted to the rehabilitation program.

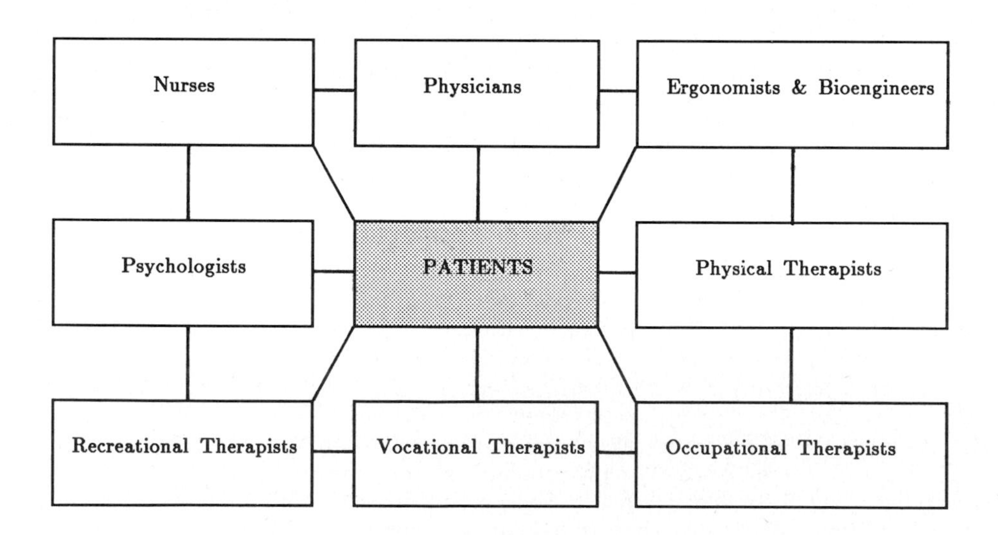

FIGURE 1: A model of the multidisplinary team of the CPRC.

The CPRC program is comprised of different components, as shown in figure 1, interacting together to create an integrated system. The main therapeutic objective of this system is to return LBP patients to their previous economic and social productive lifestyle upon completion of treatment. The goals of the CPRC team are as follows:

1. to increase functional capacities by freeing patients from pain, reducing suffering and discomfort, or having a better control of pain;

2. to eliminate or reduce drug intake and dependence which may interfere with normal function;

3. to increase both physical and mental capabilities through an individualized aggressive treatment program;

4. to conduct a quantitative evaluation of the individuals' physical, psychomotor, and mental capacities which assist in the selection of jobs that match the injured individuals upon completion of the treatment;

5. to provide job simulation, training, and workplace design to ensure more compatible and safer relationship between LBP individuals and their working or living environment.

6. to educate patient on proper body mechanics and methods of selfcare.

A detailed description of the CPRC treatment program can be found else–where [5, 7, 8]. This treatment program includes heavy emphasis on physical medicine, physical therapy and its modalities, beginning with reduction of trigger points by use of: physical modalities; judicious injection with either saline, steroids, or anasthetic agents; muscle stretching and strengthening to restore range of motion and endurance; gait training; and realignment of posture. Occupational therapy is used for the establishment of sitting, standing, walking, and driving tolerances; pacing with respect to daily and work activities; energy–saving techniques, occupational and recreational simulation; and use of proper body mechanics. Detoxification regimens, weight control, and nutritional education are introduced and enforced at an early stage of the treatment program. Biofeedback techniques are utilized to provide electromyographic control to assist muscle re–education and strengthening; monitor muscle tension in the active phase; maintain correct posture; and to utilize proper body mechanics during daily and work activities. The ergonomists and bioengineers establish a profile of performance capabilities of each patient admitted and progress is monitored throughout the program. Vocational counsellers in cooperation with ergonomists give major attention to job placement, job simulation, and work hardening, where applicable. The patient is taught to perform his/her assignment properly and ensure that he/she has been physically rehabilitated to handle the task demands of the job. Throughout the program, the behavioral medicine plays an integral role in the rehabilitation process.

3. EVALUATION OF THE TREATMENT PROGRAM

3.1. Subjects

The main objective of the present investigation was to evaluate the effect of the rehabilitation program used in restoring the functional capabilities of LBP patients. A random sample group of 40 patients suffering from chronic low back pain participated in this study. All subjects participating in this study received a diagnosis of myofascial syndrome as the etiology of their low back pain condition. Physical diagnosis related to the patients' pain condition were made independently by a neurosurgen and a physiatrist. Patients receiving one of the following diagnoses were not included in the sample used: (1) no evidence of low back pain; (2) evidence of radiculopathy or other non–muscular related organic disorders; and (3) other interfering medical problems (e.g., cardiovascular disorders). A description of the subjects by age, height, weight, and duration of pain is provided in Table 1.

TABLE 1: Subjects' demographic characteristics

Variable	Minimum	Maximum	Mean	Std Dev
Age (years)	22.00	71.00	43.33	13.64
Height (inches)	61.00	75.00	67.03	4.02
Weight (lbs)	101.00	253.00	163.18	39.83
Duration of pain (months)	6.00	480.00	71.68	107.15

3.2. Outcome Measures

In the present study, the outcome measures used to determine the effectiveness of the treatment program were: (1) electromyographic (EMG) activity of the lumbar paraspinals; (2) strength of low back extensors; (3) strength of low back flexors; (4) back extension range of motion; (5) back flexion range of motion; and (6) pain level at the low back area.

A Cyborg P303 EMG biofeedback apparatus with a Cyborg Q700 RMS data accumulator was used to monitor the EMG activity of the lumbar paraspinals. After the skin was prepared, surface electrodes were then placed on the right side of the lumbar paraspinal muscles. Both auditory and visual feedback were available for the subject during the evaluation session. Each session consisted of six trials; subjects rested 2 minutes between trials. For each trial, the subject was required to extend his/her low back by contracting the lumbar paraspinal muscles and raising up the head with chin tucked; during each trial, ankle weights were used to ensure that subjects' legs remained flat on the treatment table. Subjects were instructed to continually exert effort until their "acceptable maximum effort" (AME) level was reached and to maintain this effort for ten seconds. AME is a concept which was introduced by Khalil et. al. [9] and has been defined as a point of effort beyond which a person believes that pain would become intolerable (unacceptable). The average value of the six trials was considered the EMG score.

Measurement of the strength of low back extensors was carried out statically in the upright position. A special frame, which required subjects to stand erect facing it with their pelvis stabilized, was used. A belt was placed around the torso of the subject and was attached to a chain linked to a force transducer. Subjects were asked to apply pressure to the belt by performing low back extension. Subjects exerted pressure on the belt until their AME was achieved. The meaurement was repeated three times and the average value was considered the back extension strength score. The strength of the back flexors was measured with the same method, except that the subjects attempted back flexion.

The pain level in the low back area was recorded on a scale from 0 to 10, where 0 represents pain free sensation and 10 represents excruciating pain. Also, ranges of motion of the back extensors and back flexors were measured using a hand held gravity goniometer.

All the subjects participated in this study were evaluated at the beginning and after two weeks of the four–week CPRC program of treatment.

4. RESULTS AND DISCUSSION

Means and standard deviations for initial and final values on each outcome measure are presented in Table 2. A sereies of paired t–tests were used to compare pre to post treatment in the outcome measures. The results of these tests showed that there was a significant increase ($P < 0.01$) in EMG, extension strength, flexion strength, and trunk ranges of motion (both flexion and extension) and a significant decrease in the pain level.

Results of this study indicate that the multidisciplinary approach to treatment of chronic low back pain patients was an effective means for increasing functional abilities and reducing pain and discomfort. Static back extension strength after two weeks of the rehabilitation program increased from 48.9 lbs of force to 95.38 lbs (95.05%). The change in static back flexion strength was from 54.08 lbs to 96.48 lbs (78.4%). The EMG values obtained from the lumbar paraspinal muscles increased on the average from 59.75 micro volt to 97.37 micro volt (62.96%). The back extension range of motion increased 11.5 degrees (61.73%); while the increase in back flexion range of motion was 44.38 degrees (54.45%). Pain level was also significantly reduced.

TABLE 2: Mean and standard deviation of the intial and final outcome measures

Outcome measure	Initial Mean	Initial Std Dev	Final Mean	Final Std Dev
Pain Level*	6.63	2.53	5.4	2.27
EMG (Micro Volt)*	59.75	22.87	97.37	43.38
Extension Strength (lbs)*	48.90	35.34	95.38	51.35
Flexion Strength (lbs)*	54.08	34.22	96.48	47.28
Flexion (degrees)*	81.50	29.44	125.88	22.60
Extension (degrees)*	18.63	9.47	30.13	11.41

* Significant at the 0.01 level

5. CONCLUSIONS

The results obtained indicate that patients admitted to the CPRC of the University of Miami School of Medicine showed significant improvement in low back extension and flexion strength, low back flexion and extension ranges of motion and EMG values recorded from the lumbar paraspinals. Also, a significant decrease in the pain level was observed between the initial and the final measurements.

Based on the experience gained at the CPRC over the past ten years, during which over 6,000patients where evaluated and treated [8], and the results obtained from the present study, the following conclusions can be made:

1. Patients undergoing a multidisciplinary comprehensive rehabilitation program showed a significant improvement in their functional abilities as seen from the significant increase in the ranges of motion of the spine, trunk strengths and EMG recorded from the lumbar paraspinal muscles.

2. The use of rigorous and aggressive programs of rehabilitation is effective in the restoration of patients' functional abilities within a relatively short period of time (two weeks only).

3. A patient must have the desire and proper attitude in order to benefit most of the program components.

4. Back pain is a chronic problem that could reoccur. Patients should learn the proper techniques for energy savings, use of body mechanics, pain control, relaxation, and stress management and be able to apply them to improve their condition.

5. Additional controlled experimental investigations are needed to explain the effect(s) of each treatment modality and study the relationships between restoration of strength, ranges of motion and pain level in low back pain patients as well as the time needed to obtain significant improvements.

REFERENCES

[1] National Institute for Occupational Safety and Health (NIOSH), Proposed National Strategies for the Prevention of Leading Work-Related Diseases and Injuries - part 1 (NIOSH, 1986)

[2]. National Institute for Occupational Safety and Health (NIOSH), Work Practices Guide for Manual Lifting (NIOSH, Cincinnati, Ohio, 1981)

[3] Tabor, M., Occup Health Saf 51 (1982) 16.

[4] Asfour, S.S., Khalil T.M., Moty E.A., Steel R. and Rosomoff H.L., Proceedings of 7th International Conference on Production Research (Windsor, Canada, 1983)

[5] Khalil T.M., Asfour S.S., Moty E.A., Steel R. and Rosomoff H.L., Proceedings of the 1983 Annual Industrial Engineering Conference (Louisville, Kentucky, 1983)

[6] Rosomoff, H.L., Clin. Orthop 154 (1980) 83.

[7] Rosomoff, H.L., Green C., Gilbert M. and Steel R., Pain and Low Back Rehabilitaion Program at the University of Miami School of Medicine, in: Lorenz, K.Y. (ed.), New approaches to treatment of chronic pain: A review of multidisciplinary pain clinics and pain centers (NIDA Research Monograph 36, Department of Health and Rehabilitative Services, 1981) pp92-111

[8] Rosomoff, H.L. and Rosomoff, R.S., Spine: State of the Art Review 1 (1987) 383.

[9] Khalil T.M., Goldberg M.A., Asfour S.S., Moty E.A., Rosomoff R.S. and Rosomoff H.L., Spine 12 (1987) 372.

Trends in Ergonomics/Human Factors V
F. Aghazadeh (Editor)

RETURNING TO WORK: THE NEED TO MEASURE PHYSICAL ABILITIES

W.K. STOEFFLER, A.M. GENAIDY, and D.M. LYTH

Department of Industrial Engineering
Western Michigan University
Kalamazoo, Michigan 49008, U.S.A.

Traditionally, work measurement and ergonomics guidelines have been applied to adapt work environments to normal people. These guidelines are well documented in the literature. There is a lack, however, for such guidelines with respect to physically handicapped workers. The main objective of this paper is to review the techniques reported in the work measurement and ergonomics literature in order to develop these guidelines.

1. INTRODUCTION

Today's job market is no longer limited to healthy, strong able bodies. The population of the physically and mentally handicapped occuring during career years, is growing rapidly. This increase causes a greater financial burden to society. Some are not capable to work, however, the majority are able to return to or begin work and enter the job with more determination and enthusiasm than the non-disabled employee. Measuring work capabilities and fatigue limits is a critical part of job selection to assure placement in meaningful jobs. Physically disabled individuals may require special work accommodations to limit fatigue and increase output. Occasionally, retraining is necessary for individuals who cannot physically return to the same employment.

Based on a comprehensive review of the literature, it is felt that research on how to properly measure and evaluate abilities for the physically disabled is warranted to assist job placement for this population. Therefore, it is imperative to review and evaluate the existing literature on work measurement and ergonomic techniques utilized in assessing the physically disabled in order to develop a strategic model for job placement of this population.

2. REVIEW OF TECHNIQUES

There are few techniques reported in the literature which evaluate the functional abilities of physically handicapped workers. The techniques reviewed in this paper are disability index, manual abilities and scanning test, available motions inventory, and job selection and accommodation.

2.1. Disability Index

Birdsong [1], Birdsong and Chyatte [2], and Chyatte and Birdsong [3] developed an index, called the disability index (DI), to evaluate the anatomical impairment of the handicapped person. DI is a numerical deficit relative to the average person. It weighs the individual's handicapped body part against that of a non-handicapped person's body part using the procedure described below. The individual is first tested by performing MTM elements which correspond to a particular job or area of tasks he/she intends to perform. The job is filmed at 16 frames/sec. The film is then analyzed and the resulting times are determined. For each task, a performance index, which compares the individual's MTM score to the MTM norm, is developed. Next, a computer matches the individual's MTM abilities to the MTM requirements of the job. If an individual can perform the job, the performance index equals the disability index.

Birdsong proposed the use of DI for insurance claims following injury on a job. In this sense, DI could be used as a standard to which to base the claims to ensure a fairer system. DI can also be used to redesign a job or part of a job in which the worker has experience difficulty. It should be noted, however, that DI is task specific and needs to be developed separately for each job.

2.2. Manual Abilities and Scanning Test

The manual abilities scanning tests (MAST) are the outcome of the International MTM Directorate Rehabilitation Project [4]. The MAST system is classified into three divisions, namely, MAST-1, MAST-2, and MAST-3. The description of each division is given below.

MAST-1 tests basic manual tasks, quality or work performance, and work attitudes. A series of tests are administered which measure the individual's abilities to perform the basic manual motions, assembly abilities, and coordination abilities (Table 1). The results of these tests are on a percentage basis where 100% is average. Information from the tests is plotted on a manual abilities profile. By reading the profile, the evaluator can identify problem/weak areas of the individual. Each section has its own training program so that the program can be tailored to the individual's needs in the problem areas. MAST-1 includes a learning curve to predict the time and number of cycles of training necessary for the individual to reach a desired level of performance. The validity of each training method can be determined since the same tests are used for each of the subtests. Testers are also able to rate the individual's attitudes towards

Table 1. Summary of MAST-1 Tests

G	Industrial Action	Test Number	Test Description	Ability Measure
1	Basic motions	1-4	performing reach, grasp, move and release motions	simple dexterity, multi-limb, coordination kinesthetics
2	Selection and handling	5-7	introducing various conditions on the objects to be handled	discrimination, dexterity
3	Assemble	8	positioning of pegs to holes on the principle used by purdue pegboard	dexterity, coordination, use of tools
		9	positioning a number of different shapes of blanks to correspondingly shaped aperatures	
		10	assembling a series of rubber plugs to their mountings	
		11	assembling a hinged metal plate to a chassis using a variety of thredded fasteners (e.g., nuts, bolts)	
4	Disengage	12	disengaging a series of objects from their mountings	arm power
5	Surface	13-14	testing a number of circircular blanks are aligned to targets without the aid of guides or stops	tactual control
		15	examining the ability to align objects where the handicapped individual is given a pencil and preprinted paper on which a number of target crosses crosses are required to be accurately jointed together	
6	Turn	16-17	using a torque screwdriver to determine person when accomplishing turn motions	wrist power/ speed

Table 1. Cont'd

G	Industrial Action	Test Number	Test Description	Ability Measure
7	Left and right hand coordination	18	assembling a number of plugs and sockets with bayonet type fittings	coordination
8	Decide	19	using an electrical circuitry test to examine the person's ability on simple pass or fail situations from a signal light, and to take action appropriate to the signal	response orientation
9	Eye/hand/foot coordination	20	a foot-pedal equipment activates a plunger which pierces one of a series of targets printed on a sheet of paper	aiming and multi-limb coordination
10	Body motions	21	examining body motions such as walking, sitting, stooping, turning, and other important ambulatory features demanded by work situations	gross body coordinations

** G - group.

authority, behavior in the work environment and overall job motivation. This is called the work behavior and motivation index.

MAST-2 is a series of 11 tests for the more experienced individual and is designed to measure the individual's dexterity, precision, coordination, aptitude and initiative (Table 2). Some of the tests administered measure ability to couple pressure with hand control, to do electrical wiring from a manual, or the ability to position by feel and not by sight. Scoring is done on a percentage basis similar to MAST-1. The skills profile rates the tests given in different areas. MAST-2 also allows for a tailored training program.

MAST-3 provides a personal profile of the handicapped person over 28 different aspects in the areas of physical, educational, social, and physiological ability (Table 3). The aspects are specified at five levels in order that MAST-3 can specify the nature and degree of improvement that needs

Table 2. MAST-2 Tests

Test #	Features	Skill Measures	Test description
1	one handed control	speed/control	perform sustained pressure coupled with a high degree of control using only one hand
2	one handed control (no visual assistance)	speed/tactual control	effect position elements when vision is hindered and reliance is placed on the feel of the object being located
3	left hand speed/ right hand control simultaneous	multi-limb speed/control (in compatible actions)	perform speed actions with one hand and control simultaneous with the opposite without the assistance of vision
4	measure & adjust	precision measure (decide)	examine accuracy of mechanical adjustment that can be effected using simple tools
5	mechanical adjustment	precision, control judgement	examine the level of judgement and control that can be exercised by effecting precision on moving parts
6	assembly of flexi-part by remote handling	non tactual coordination and advanced kinesthetics	examine the proficiency of the client in assembling a flexible object a series of a very small aperatures, without being able to handle the objects except by means of tools
E	electrical assembly	initiative, aptitude, quality	assemble a number of electronic components and wiring from an instructional manual
S	threaded fastener assembly	finger/speed stamina	finger speed/stamina, dexterity test

Table 2. Cont'D

Test #	Features	Skill Measures	Test description
B1	fine manipulation assembly	finger speed dexterity	fine manipulation using fingers tweezers, and a balancing tool to effect the location of a number of very small steel balls
B2	tweezer manipulation test	tool dexterity	
B3	balance/manipulation test	tool/control dexterity	

Table 3. Summary of MAST-3 Tests

Functional Aspect \ Grade	A	B	C	D	E
1. Reading					
2. Speaking					
3. Writing					
4. Administration					
5. Arithmetic					
6. Measuring					
7. Gauging					
8. Reading signs					
9. Drawing					
10. Knowledge of materials					
11. Feeling for materials					
12. Knowledge of materials					
13. Aptitude for machinery					
14. Memory					
15. Eyesight					
16. Color discrimination					
17. Tactual control					
18. Hand/arm power					
19. Coordination					
20. Classification of movement					
21. Stamina					
22. Body posture					
23. Size discrimination					
24. Space discrimination					
25. Concentration					
26. Independence					

Table 3. Cont'd

\Grade Functional\ Aspect \	A	B	C	D	E
27. Social aptitude at work					
28. Social aptitude outside work					

NOTE: A - individual is unable to measure; B - individual is able to measure in cm/mm with a rule; C - can measure inside and outside diameters of round objects using calipers; D - can measure using a micrometer; E - can measure by gauges (e.g., shadowgraph) to very fine limits.

to be undertaken over these 28 functional aspects. This prepares the handicapped person to meet the demands of specific types of jobs or it can specify the types of jobs that are within the current range of the individual's capacity. The procedure starts with admitting the handicapped individual into the center, where an initial test will determine for him a provisional profile related to the 28 functional aspects. From this point, a training program is determined from the ability levels indicated by the individual's provisional profile, and the program is 'tailored' to suit the individual, for it must commence at an appropriate level.

Wilcock and Mink [4] cited the following benefits of MAST-1 and MAST-2: (1) they determine manual abilities and skills, (2) they identify manual limitations, (3) they predict type and amount of necessary training, (4) they have tailored training programs, (5) they can find benefits of particular training methods, (6) they can be used to provide equipment more compatible with the needs of individual clients and consequently improve their working environment, (7) they provide vocational guidance information, (8) they provide a method of matching a client with job requirements, (9) they prove the client's abilities to potential employers.

2.3. Available Motions Inventory

The available motions inventory (AMI) was developed by the Wichita State University Rehabilitation Engineering Center to evaluate the physical ability of individuals with neuromuscular impairments so that modifications could be made to tasks in an industrial setting involving light bench work [5-8]. AMI is designed to measure the fundamental functional output and the results can be described as functional abilities. It is a series of 71 tests to evaluate a person's ability to perform tasks in five basic areas: strength, rate, settings, assembly, and switches. An adjustable test frame is used to mount the subtest modules in several orientations to the subject. All tests are administered with the subject seated, and the test frame is adjusted to a standard position by referencing popliteal and maximal functional reach. The modular mounting systems allows for the location of subtests in eight vertical and two horizontal positions for each hand. Only five of these are used in a standard format. It should be noted that the scores of the AMI tests are calculated in terms of pounds for the strength tests, and time per correct actuation for all others.

2.4. Job Selection and Accommodation

Hasselquist [9] developed a system termed job selection and accommodation. This system consists of three steps: functional test, job selection, and job accommodation. The functional test is performed first where handicapped individuals are tested using a series of tasks for which the MTM analyses are already known. This enables the tester to determine the inabilities and abilities of the handicapped person. No details were given by Hasselquist on the extent and type of testing that was used. Job selection is the next step in this sytem. The MTM elements that the individuals are capable of doing are matched with the elements required of a job. A perfect match of the job requirements and the abilities of the client is sought. If a perfect match cannot be made, job accommodation is implemented where the individual's inabilities are alleviated by changing the layout of a work station or modifying equipment.

3. CONCLUSIONS

Four work measurement and ergonomic techniques, which are used to evaluate the functional abilities of handicapped workers, are reviewed. Extension of this research will involve testing and comparison of these and other techniques.

4. REFERENCES

[1] Birdsong, J.H., J. of MTM, XVII(4), 3-8.
[2] Birdsong, J.H. and Chyatte, S.B., J. of MTM, XV(2), 19.
[3] Chyatte, S.B. and Birdsong, J.H., J. MTM, XIII(2), 15.
[4] Wilcock, R. and Mink, J.A., J. of MTM, IX(2), 2.
[5] Malzahn, D., Proceedings of the Annual IE Conference, 44, 1979.
[6] Malzahn, D., Proceedings of the Annual IE Conference, 179, 1982.
[7] Malzahn, D. and Kapur, R., Proceedings of the Human Factors Society, 114, 1980.
[8] Rahimi, M. and Malzahn, D.E., Human Factors, 26(6), 715, 1984.
[9] Hasselquist, O., J. of MTM XVI(4), 6-12.

Trends in Ergonomics/Human Factors V
F. Aghazadeh (Editor)

Industrial Safety Programs in Sheltered Workshops

Paul C. Witbeck and Gene R. Simons, Ph.D.
Decision Sciences and Engineering Systems
Rensselaer Polytechnic Institute
Troy, New York 12180-3590

Abstract

A sheltered workshop engaged in fabrication, assembly and packaging operations presents a major problem in implementing a safety program due to the mix of handicapped and non-handicapped workers. This paper describes a two phase approach that first implemented a normal industrial safety and hygiene program and then modified and extended that program to meet the special needs of the clients.

1. Sheltered Workshops

A sheltered workshop is a facility where training and rehabilitation are provided for the physically and mentally handicapped. At the Workshops, Inc. in Menands, New York, over 400 handicapped people under the guidance of counselors and production supervisors perform a variety of tasks for which they are paid a nominal wage.

These tasks, which range from packaging and small parts assembly to cafeteria and janitorial services, are used to provide the clients with a sense of self-worth and in many cases to train them to perform similar jobs outside of the sheltered environment and thus become productive, self supporting members of the community.

To provide a steady supply of work for these clients as well as operating income for the Workshop, it is necessary to hire physically capable workers to perform some of the same tasks as the clients and to perform certain of the more complex tasks - especially those tasks that involve the operation of machinery. One such project is the fabrication of wooden products such as gun racks and office accessories which requires the use of wood working machinery.

Whereas the clients come under the oversight of various social agencies, the non-handicapped employees are covered by the same labor laws as anyone in general industry. This later group, however, is generally of lower caliber than most industrial organizations and is characterized by low skill and high turnover.

2. Safety Needs

The total workforce, both clients and employees, is engaged in manufacturing and is therefore subject to all the standards of OSHA as well as Workman's Compensation. It was necessary to provide a total safety program to comply with all legal requirements, to keep Workman's Compensation costs under control and above all to discharge the moral obligation to protect workers and clients from injury.

Safety is largely a matter of attitude and the single most important element of any safety program is **management commitment**. Other elements of a total safety program include safety rules, communications, awareness programs, training, personal protective equipment, record keeping, accident investigation, regular safety audits and medical surveillance.

Implementing a safety program at the Workshop presented several unique problems. Among the clients, recognized safety hazards are magnified by the client's handicaps, especially hearing and visual impairments. Others have varying degrees of speech impairment which makes communication difficult. Still others are emotionally disturbed which leads to fights and the possibility of injury. Some are subject to seizures and must be protected from collapsing. And a large group have below normal learning ability which makes safety training difficult.

The cadre of physically normal workers compounds the difficulties because they are an unstable workforce with low skills and pay rates. The injury rate among these workers was four times the rate of the clients.

Up to two years ago, safety had received little attention. A safety committee existed and met regularly but problems discovered during inspections were seldom followed up on or resolved. The committee had received no safety training and did not even have a copy of the pertinent OSHA standards. The plant nurse maintained an OSHA log but there was little understanding of the criteria for recording injuries. The nurse's primary responsibilities were "hand holding" for the clients and providing basic first aid. She was not involved in the employment or placement process and, as a result, the medical records of the employees were very sketchy. There was no physician on contract and anyone that required more than a Band Aid was sent to the local hospital emergency room.

In summary, there was no one specifically delegated the responsibility for safety and safety was not a topic of major emphasis.

3. Phase I - The Industrial Safety Program

In 1985, several events came together that lead to the implementation of a comprehensive safety program. First, a project was started in the conversion of toilet paper, i.e. slitting and cutting jumbo rolls into saleable units. The paper dust created by the slitting operation was perceived to be a potential health hazard and the Workshop asked for help from a local university in investigating the problem. A retired industrial safety professional, who was teaching at the university and who later joined the Board of Directors of the Workshop, set up an air monitoring program. The same individual recognized the need for a safety program and obtained approval from the Board to start one. Initial activities included:

a. Writing a statement of safety policy that was adopted by the Board that made the Executive Director responsible for the safe operation of the facility.

b. Setting up a student project with the university to identify and document all materials used in the Workshop to comply with OSHA's new Hazard Communication Standard.

c. Calling in the New York Consultative Service to conduct a complete safety and hygiene audit of the facilities. These audits were used to prioritize the remaining action items in the implementation program.

d. Having a retired Occupational Physician, who also joined the Board of Directors, work with the plant nurse to set up health records for each employee and to screen each new hire.

e. Setting up a monthly audit of the entire facility that is conducted by the Executive Director, the Safety Specialist (on the Board), the Nurse, and the two production supervisors. Minutes are kept and a followup procedure was adopted.

f. Adopting a safety glass program for all maintenance and certain operating personnel. A Board member, who is an Optometrist, conducts the examinations, prescribes and fits the glasses.

g. Bringing in a State Industrial Hygienist to address noise problems and to relocate noisy operations, prescribe ear protection and define followup procedures. When it was noted that disposable ear plugs were not being properly used, the effected employees were issued ear muffs.

The results of phase one were very encouraging. As problems were identified and corrected, employee confidence grew and the attitude toward safety improved. The OSHA incidence injury rate has dropped from 22 to 9 injuries per 200,000 manhours but there is still room for improvement.

4. Phase II - Special Needs

As Phase I progressed, the special needs of the mixed population of clients and workers began to be recognized and a second phase was implemented that specifically dealt with these differences. Specific areas addressed in Phase II were:

a. Increasing the visibility of the nurse by having her wear distinctive garb (white shoes, white pants and a pink blouse) without resorting to an anteseptic appearing uniform. She made daily tours of the facility with responsibility (and authority) to deal with safety problems. The nurse received additional training in safety and was considered to be the most knowledgable person on safety in the facility. Clients and employees were encouraged to talk to the nurse on the floor where a problem could be quickly resolved.

b. Making use of the talents on the Board of Directors as part of the safety program. This included the Safety Specialist, Occupational Physician and Optometrist previously mentioned as well as an Industrial Engineer and an insurance specialist. In addition, the Board has been kept informed as to the progress of the program and strongly supported the Executive Director's efforts both with their interest and with the required funds.

c. Instituting a program for the removal of hazards that recognized the limited ability of many of the clients to perceive danger. The priority was given to removing the hazard rather than providing personal protection to the worker who may not have the capacity to use it properly.

d. Taking advantage of the low ratio of employees to supervisors to institute intensive training for the supervisors. Unlike an industrial facility that may have a 10 or 15 to one employee to supervisor ratio, a workshop typically has a 5 to one ratio because the supervisor also trains and provides counseling. As a "parent" figure, it is essential that the supervisor understand the safety needs of his/her work group and take the program seriously.

e. Developing means of warning clients with sensory deprovision of hazards and emergencies including visual cues for the deaf and a buddy system for the blind. Clients with learning disabilities are also provided limited instruction.

5. The Future

With Phase I completed and Phase II well underway, the next phase is being designed. This includes more training, a hearing conservation program and implementation of a hazard communication (Hazcom) program. Investigation is underway to upgrade medical services to include access to a physician's group rather than depend on the emergency room.

Specific safety targets are now included in the goal oriented performance system for supervision for reduction in injury. The ultimate objective is to provide a safe work environment for both clients and workers. It is also expected that the knowledge gained from this program will assist the clients who leave the workshop for regular employment in dealing with the job related hazards that may exist in their new work environment.

Bibliography

1. Konikow, R. B. and McElroy, F. E., Communications for the Safety Professional, National Safety Council, Chicago, IL, 1975

2. DeReamer, R., Modern Safety and Health Technology, John Wiley and Sons, 1980

3. Hannaford, E. S., Supervisor's Guide to Human Relations, National Safety Council, Chicago, IL, 1967

4. McElroy, F. E. (Editor), Accident Prevention Manual for Industrial Operations, Sixth Edition, National Safety Council, Chicago, IL, 1981

5. U.S. Government Printing Office, Code of Federal Regulations 29, Parts 1900-1910, Juul 1, 1985

6. Supervisor's Safety Manual, Sixth Edition, National Safety Council, Chicago, IL, 1985

Trends in Ergonomics/Human Factors V
F. Aghazadeh (Editor)

THE CARDIOVASCULAR FITNESS OF NON-DOWNS SYNDROME, MODERATELY MENTALLY RETARDED INDIVIDUALS AS AN ADDITIONAL INDICE FOR JOB PLACEMENT

Kenneth H. PITETTI, Jeffrey E. FERNANDEZ, David C. PIZARRO, Nancy B. STUBBS, and Jayne A. STAFFORD

Departments of Health Science, Industrial Engineering, and Physical Education
The Wichita State University
Wichita, KS 67208-1595

The Schwinn Air-Dyne ergometer (SAE), which utilizes both arms and legs during exercise, was used to compare predicted and directly measured cardiovascular fitness (VO2 max) in 12 male and 4 female (mean age = 24.9 ± 4yrs) non-Downs, mentally retarded subjects (mean IQ = 64 ± 4). The submaximal test used to predict VO2 max consisted of 2 consecutive 3-min bouts of exercise at low (50 watts) and high (100 or 150 watts) intensity to produce heart rates (HR) in a range of 150-170 b/min during the high intensity level. VO2 max was predicted using Hellerstein's [1] formula (% VO2 max = (1.41 X %HR max) - 42) for the final HR at the high intensity level. A metabolic cart was used to directly measure VO2 max during a maximal exercise test on the SAE. There was no significant ($p < 0.05$) difference between predicted (36.1 ± 10 ml/kg/min) and actual (35.9 ± 0 ml/kg/min) VO2 max. Furthermore, the correlation coefficient between predicted and actual VO2 max was 0.90. Therefore, the protocol used in this study to estimate cardiovascular fitness merits consideration by those involved in determining job placement for mentally retarded individuals when cardiovascular fitness is of importance.

1. INTRODUCTION

Job placement for the mentally retarded (MR) is usually determined by thorough psychological and vocational evaluations and a comprehensive medical examination. However, the MR have been restricted from some physically demanding jobs because previous reports indicate that MR individuals fall below their non-MR peers on tests of physical fitness [2,3,4]. Yet, it would be unfair to categorize all MR individuals as physically unfit when some of the MR subjects involved in the above studies were proven to be in average physical condition. Furthermore, with the recent establishment of organizations such as Special Olympics, thousands of MR adults and children are now involved in year round programs of training and athletic competition which no doubt has improved the overall fitness of the participants. Therefore, there exists a need for a simple, yet accurate way to assess the physical fitness of MR individuals for job placement when cardiovascular fitness is of importance.

The standard method of determining the functional capacity of an individual's cardiovascular system is by direct measurements of maximum oxygen consumption (VO2 max) during high intensity exercise. However, this procedure requires expensive laboratory equipment and medical expertise that is usually not available in most MR evaluating settings. Therefore, estimating VO2 max is the only alternative for such settings. Unfortunately, many of the protocols used to estimate the cardiovascular fitness of the general population (e.g. Astrand -Rhyming bicycle test, Canadian Step Test, Cooper 12 minute Run/Walk Test) are of questionable validity when administered to the MR [5]. The difficulties encountered when applying these protocols to the MR for estimating VO2 max usually involve the inability of the MR to follow proper testing procedures which result in inaccurate assessments of cardiovascular fitness. Therefore, the purpose of this study was to develop a simple, yet valid field testing protocol for evaluating the cardiovascular fitness of MR individuals for job placement purposes.

2. METHODS AND PROCEDURES

2.1. Subjects

Twelve male and four female subjects, classified as moderately, mentally retarded (IQs ranging from 60 to 76), participated in this study. All the subjects were participants of the Kansas Special Olympics and were actively engaged in sports activities at the time of the testing. Sex, age, IQ, and anthropometric characteristics are listed in Table I. An informed consent was obtained from either the subject (if legally capable) or guardian. The testing protocols in which the subjects participated were approved by the Institutional Review Board of The Wichita State University, Wichita, Kansas.

2.2. Testing Protocol

The first test involved estimating VO2 max from heart rates during a submaximal exercise test on the Schwinn Air-Dyne ergometer. The submaximal exercise test consisted of two continuous three minute bouts of exercise initially at low (50 watts) and high (100-150 watts) levels of intensity. The work load chosen for the higher intensity was determined by the heart rate at the first level. If while exercising at the first level the subject's heart rate was below 65% of maximum, then 150 watts was selected for the second work level. If the subjects heart rate was at or above 65% of maximum, then 100 watts was selected for the second work level. Maximal heart rates were predicted by the formula 220 - age for females, and 205 - (age/2) for males [6]. This protocol usually produced

TABLE 1

Sex, Age, IQ, and Anthropometric Characteristics of 16 MR Subjects Involved in Testing Protocol

SUBJECT	SEX	AGE	HT(CM)	WT(KG)	IQ
1	F	28	157.5	65.8	60
2	M	21	166.4	76.2	62
3	M	27	165.4	61.7	60
4	M	24	179.7	60.8	61
5	F	22	151.8	48.5	71
6	M	22	195.2	71.0	62
7	M	21	161.3	41.1	61
8	M	27	165.2	77.3	65
9	F	36	160.0	67.4	65
10	F	28	173.5	77.3	60
11	M	28	177.8	69.9	61
12	F	24	168.9	54.9	63
13	M	22	182.8	78.0	67
14	M	26	182.3	96.6	68
15	M	25	182.9	89.4	66
16	M	18	169.5	55.8	76
Mean		24.9	171.3	68.2	64.2
Standard Dev.		4.2	11.5	14.5	4.5

heart rates in the vicinity of 80-85% of maximum for the high level of intensity. This protocol has been shown to provide accurate estimates of VO2 max when applied to the general population [7]. Heart rates were monitored continuously by telemetry (EST Sport Tester, Excotek) and recorded every 30 seconds. If the heart rate at three minutes was 5 beats/min greater than the two minute value, the exercise was continued for an extra minute and the fourth minute heart rate was considered to be the steady state value [8]. Blood pressure was taken before exercise, during the last minute of each exercise level, and following exercise until it returned to resting values. The oxygen cost for exercise was estimated from the formula, VO2 (ml/min) = (Watts X 12) + 300. VO2 max was predicted from the formula, %VO2 = (1.41 X % HR max) - 42 [1] (See Table 2 for methodology).

This protocol is accurate only if the subject maintains a constant work level during the three minute work intervals. Therefore, the protocol was initially demonstrated to the subject, then practiced by the subject to ensure constant exercise work levels. Furthermore, the staff assisted in ensuring that subjects focused on maintaining the "needle" on the proper work level or "number" (i.e., 1 = 50 watts, 2 = 100 watts).

TABLE 2

Methodology in Estimating VO2max Using Hellerstein's Formula

In an exercise test using a Schwinn Air-Dyne ergometer, a 24 year old male weighing 78 kgs attains a steady state heart rate of 155 beats/min on his second work level of 100 watts. Using the formula 205 - age/2, we estimate his maximal heart rate to be 193 beats/min. Therefore, his final heart rate of 155 on the exercise test represents 80% of maximum heart rate. Using Hellersteins's formula:

$$\%VO2\ MAX = (1.41 \times 80) - 42$$
$$\%VO2\ MAX = 70.8$$

Therefore, at this work level, it is estimated that he is working at 70.8% of his maximal capacity. Using the formula, VO2 (ml/kg) = (watts X 12) + 300, the oxygen cost at 100 watts is estimated to be 1500 ml/min. Again, this represents 70.8% of his estimated VO2 max, so by dividing 1500 by .708, we estimate his absolute VO2 max to be 2118.6 ml/min. Finally, dividing 2118.6 by his weight, 78kg, we arrive at his relative estimated V02 max to be 27.2 ml/kg/min.

Following the submaximal test, each subject performed a maximal exercise test on the Schwinn Air-Dyne ergometer. A Metabolic Measurement Cart (MMC) Horizon System or Beckman Metabolic Measurement Cart was used to measure VO2 directly during a maximal exercise test. Heart rates, blood pressure, and an ECG was monitored continuously during the maximal test. The test consisted of the subjects exercising at an initial work load of 50 watts for two minutes, then increasing the work load 25 watts every two minutes until volitional exhaustion.

2.3. Statistical Analysis

Means and standard deviations (SD) were calculated for all variables. A Pearson Correlation Coefficient was used to test for significant relationships between predicted and directly measured V02 max. A student t-test was used to determine if a significant difference existed between predicted and directly measured VO2 max. The results were judged to be significant at $p < 0.05$.

3. RESULTS

Means and standard deviations for age, height, weight and IQ are found in Table 1. Means and standard deviations for predicted and directly measured VO2 max are found in Table 3.

A significant, positive correlation coefficient ($r = 0.90$) was observed between predicted VO2 max and directly measured VO2 max (Table 3). There was no significant difference between predicted and directly measured VO2 max.

TABLE 3

Means and Standard Deviations for Predicted and Directly Measured VO2 max

SUBJECT	ESTIMATED VO2 max ML/KG/MIN	MEASURED VO2 max ML/KG/MIN
1	16.2	21.1
2	34.2	38.7
3	44.3	43.1
4	39.2	44.1
5	37.3	36.1
6	41.9	38.7
7	41.9	38.7
8	26.7	30.4
9	21.7	21.5
10	30.0	34.1
11	36.2	34.6
12	40.6	34.4
13	47.7	37.3
14	31.1	36.3
15	29.1	31.1
16	57.4	54.1
Means	36.0	35.9
Standard Dev.	10.2	8.0 *

* (r = 0.90)

4. DISCUSSION

As previously mentioned, many protocols used to estimate the cardiovascular fitness of the general population have proven to be difficult to administer to the MR. A study that used the Astrand-Rhyming bicycle test reported problems with maintaining the prescribed cadence/rhythm in pedaling which is essential in determing work level [9]. Although few problems have been observed with a Balke/Ware type treadmill test [1,10], the cost of such a machine might be prohibitive in most job evaluation settings. The proper pacing to sustain an "all out effort" with the Cooper 12 minute Run/Walk test has proven to be a significant barrier in producing valid and reliable results [9]. The Canadian Step Test has problems similar to the bicycle test in that proper cadence must be maintained [4]. Therefore, the Schwinn Air-Dyne ergometer was chosen because 1) it is relatively less expensive than a treadmill ($600.00 vs $2000.00), and 2) it was believed that the test procedure on the Schwinn Air-Dyne would be much easier to administer than either a standard bicycle ergometer or the Canadian Step Test. With the Schwinn ergometer the work load and pedal speed (cadence) are synonymous. Therefore, the subjects only had to concentrate on pushing (arms) and cycling "hard" enough to keep the needle of the work load indicator on the proper number. The high correlation coefficient (r = 0.90) observed when comparing predicted VO2 max to actual measured VO2 max demonstrates the validity of this exercise test. To further strengthen these findings , a student t-test determined no significant difference between these measurements.

Interestingly, when compared to the fitness levels of the general population established by the American College of Sports Medicine [11], the subjects tested in this study ranged from poor (14.0-24.9 ml/kg/min) to good (39.0-48.9 ml/kg/min), with one individual rating high (54.1 ml/kg/min). These findings strongly suggest that it could be unfair for employers to restrict the job opportunities of MR individuals based on assumed limited physical fitness.

5. CONCLUSION

Increasing the independence and productivity of the MR is important not only to these individuals and their families, but also to the society as a whole. It is essential that the mentally retarded are not restricted in their job opportunities because of previous findings that have questioned the physical capacities of these individuals. The fact that the mean, measured VO2 max of these subjects is considered "average" when compared to the general population strongly suggests that employers should properly evaluate the cardiovascular profile of each MR individual in the same objective manner that the MR's intellectual and psychological capacities are determined.

The results of this study have established a simple, yet valid method for assessing the cardiovascular fitness of MR individuals. It is believed that this test protocol would be useful for those involved in evaluating the cardiovascular fitness of the mild to moderately mentally retarded.

REFERENCES

[1] Hellerstein, H., Hirsch, E.Z., Ades, R., Greenslott, N., & Segel, M. (1973). Principles of exercise prescription for normals and cardiac patients. In I.P. Naughton and H.K. Hellerstein (Eds). Exercise Testing and Exercise Training in Coronary Heart Disease, Academy Press, New York.

[2] Bar-Or, O., Skinner, J.S., Bergsteinova, V., Shearburn, C., Royer, D., Bell, W., Haas, J., & Buskirk, E.R. (1971). Maximal aerobic capacity of 6-15 year old girls and boys with subnormal intelligent quotients. Acta Paediatrics Scandinavia, 60(Suppl. 217), 108-113.

[3] Rarick, G.L., Widdop, J.H., & Broadhead, G.D. (1970). The physical fitness and motor performance of educable mentally retarded children. Exceptional Children, 35, 508-519.

[4] Reid, G., Montgomery, D.L., & Seidl, C. (1985). Performance of mentally retarded adults on Canadian Standardized test of fitness. Canadian Journal of Public Health, 79, 187-190.

[5] Seidl, C., Reid, S., & Montgomery, D.L. (1987). A Critique of cardiovascular fitness testing with mentally retarded persons. Adapted Physical Activity Quarterly, 4, 106-116.

[6] Hakki, A.H., Hare, T.W., Iskandrian, A.S., Lowenthal, D.T., Segel, B.L. (1983). Prediction of maximal heart rates in men and women. Cardiovascular Review and Reports, 4(7), 888-999.

[7] Pitetti, K.H., Vaughan, R.H., & Snell, P.G. (1987). Estimation of the VO2 max from heart rates during submax work on the Schwinn Air-Dyne ergometer. Medicine and Science in Sports and Exercise, 19(2): S64.

[8] Siconolfi, S.F., Culliname, E.M., Carleton, R.A., and Thompson, P.D. (1982). Assessing VO2 max in Epidemiologic Studies: Modification of the Astrand-Ryhming test. Med. Sci Sports Exercise, 14(5), 335-338.

[9] Lavay, B., Giese, M., Bussen, M., Dart, S. (1987). Comparison of three measures of predictor VO2 maximum test protocols with mentally retarded adults: A pilot study. Mental Retardation, 25, 39-42.

[10] Andrew, G.M., Reid, J.G., Beck, S., & McDonald, W. (1979). Training of the developmentally handicapped young adult. Canadian Journal of Applied Sport Sciences, 4, 289-293.

[11] American College of Sports Medicine (1986): Guidelines for Exercise Testing and Prescription. Ed. 3. Philadelphia, Lea and Febiger, pg. 43.

XV

INDUSTRIAL APPLICATIONS

Trends in Ergonomics/Human Factors V
F. Aghazadeh (Editor)

COMMITTEE APPROACH TO ERGONOMIC CASES

Floyd T. Doxie

Senior Ergonomics Engineer
Reading Works
AT&T-Microelectronics

ABSTRACT

Background information is provided on the Committee approach to Ergonomics as used at the Reading Works of AT&T. Included is the structure of the Committee, case study procedures, and the application of Ergonomic concepts to selected case studies.

BACKGROUND

The Reading Works is a high technology, research, development and electronic component manufacturing plant that is part of AT&T Microelectronics. The plant is situated in the southeast part of Pennsylvania and offers more than 1.2 million square feet of manufacturing floor space. Over 3,000 people are employed at the facility of which approximately 2,000 are directs and 1,000 are engineers, other professionals, clerical and administrative personnel.

Some of the products produced at Reading include advanced integrated circuits, lightwave lasers, gallium arsenide integrated circuits and optoelectronics. A unique factory atmosphere is crucial to the quality and reliability of much of Reading's sensitive products. At critical points of manufacture, one speck of dust can render a device inoperative. With this in mind, much of the manufacturing at Reading is done in specially designed clean rooms. These clean rooms are capable of reducing the number of particles greater than .5 microns in size from more than one million per cubic foot of air to less than ten per cubic foot of air.

INTRODUCTION

Ergonomic principles and empirical data are only abstract thoughts or printed words until they are put to a useful purpose. How to develop and employ an effective ongoing Ergonomics program is a problem facing many companies today. Often the need to improve the man-task relationship is recognized by upper management but how to develop a program that effectively fits into the organizational structure becomes a question that may be difficult to answer.

No single approach is a panacea for the establishment of an Ergonomics program. The program must be tailored to the temperament of the organizational structure it is to serve. Factors such as the type of operations being performed, level of technical expertise available, economic considerations, individual employee/management acceptance and even corporate politics may be involved in the structure of any successful program.

Over the past twenty-five years, the Committee approach to processing Ergonomic cases has evolved into what is felt to be an effective program for the Reading Works of AT&T. The program makes use of the high level of engineering talent available and is structured to attack current ergonomic problems and to prevent future problems at the initial concept and design phase.

PURPOSE AND OBJECTIVES OF THE READING WORKS ERGONOMICS COMMITTEE

1. To guard the best interest of the worker in the man-task relationship. Promote the health and safety of the worker and prevent work situations that impede efficient and effective performance.

2. To review existing manufacturing and office areas and make recommendations for the implementation of basic principles of Ergonomics.

3. To provide Ergonomic data and information for expansion or modernization of existing or new areas and facilities with regard to operator, machine or process relationships.

4. To compile and disseminate Ergonomic information and provide engineers, designers, section chiefs and key operating personnel with an awareness of Ergonomic principles.

5. To review physical soreness cases and employee complaints concerning the worker-task relationship and recommend ways to alleviate potential problems.

MAKEUP OF THE COMMITTEE

The Committee consists of one full-time engineer trained in Human Engineering and Biomechanics, and six engineers on a part-time basis representing Factory and Plant Engineering, Design Engineering, Manufacturing Control Engineering and Environment/Safety/Health Engineering. What makes this Committee unique is the absence of representatives from Operating and Product Engineering as permanent members. It is not that we wish to exclude these groups, we just use them in a somewhat different way.

When a case before the Committee involves a specific Operating or Product group, the supervisors and engineers are invited to serve as ad hoc members of the Committee, their primary concern being a specific case rather than the more general concerns of the entire Committee agenda. This approach allows the Product Engineer and Operating Supervisor to work where they have the greatest expertise and impact. The Committee is chaired by the full-time Ergonomic Engineer and an attempt is made to rotate the other members every two or three years.

OBTAINING CASES

The following represent the principle routes by which cases are obtained by the Committee:

Review of Process Formats

All commitments of money and drafting requests are reviewed to alert the Committee as to what future action is planned. This provides an excellent opportunity to become involved at the outset or concept phase of design.

Medical Department

In addition to a review of work related injuries, the Committee chairman meets with the Medical Director each week to exchange information and review open cases.

3 x 5 Cards

At intervals of six to eight months, a cover letter with three cards attached is sent to all supervisors. The letter requests the supervisor to report problems, such as poor work positions, discomfort and risk to the worker, soreness cases and excess operator fatigue. The card is the vehicle to report the problem and serves as a memory jogger for problems that may occur in the future. Extra cards are retained by the supervisor.

TV Camera Review

The TV camera has been found to be a valuable data collection tool for engineering analysis. It instantly provides a permanent record of an operation or condition that can be reviewed by one or more people as many times as desired away from the conflicting activity of the shop. This technique has revealed Operating conditions to both the Product Engineer and the Operating Supervisor that went unnoticed during the routine activity of the day.

ASSIGNMENT OF CASES

Committee meetings are held each week in the Ergonomics Laboratory. At this time, the status of each open case is reviewed and new cases are assigned a case number and progress is documented as to times, dates and action taken. We feel that weekly meetings are necessary to provide good response to new cases and maintain a high level of activity for corrective action on open cases.

REPORTING RESULTS

Each quarter, a written and oral report on the Committee activities is given to the Environment/Safety/Health Engineering Advisory Committee for the Reading Works. A copy of the written report is sent to all supervisors giving a brief description of each case and its status/resolution. This keeps supervision aware of the Committee activities and provides examples of the kinds of problems the Committee investigates.

CASE STUDIES

The following case studies were selected to show the scope of the Committee's activities while at the same time being general enough to have possible application for the reader.

CASE I - FURNACE COMPLEX LOAD STATION

To save clean room floor space, two horizontal furnaces were stacked one on top of the other with a sealed ambient enclosure at one end for loading and unloading. The enclosure had two sets of armport openings with rubber gloves so that the operator could work at each furnace. Having two sets of gloves created problems for the operator:

1. The positive pressure inside the enclosure made the unused pair of gloves stand out and become a hindrance to the operator.

2. Operators had difficulty working through the upper armport opening and reaching the upper furnace without the use of a platform on which to stand.

By brainstorming the aforementioned problems, the Committee was able to make recommendations to Design Engineering that resulted in significant changes being made to the facility.

1. By careful study of the manual motion patterns required, it was possible to relocate the armport openings. A new viewing panel was made for the enclosure using two gloves instead of four. This allowed adequate arm and hand movement to work at both the upper and lower stations and eliminated the problem of the unused gloves being in the way of the operator.

2. When the front viewing panel was replaced, a 10 degree slope was provided to allow a more comfortable working position for the operator.

3. The design of a self-contained, fold-up platform for the operator to stand on was a major feature of this case. The problems were the lack of space under the enclosure to store a platform and the fact that the Safety Coordinator would not permit a free-standing platform. With traffic in the area and the operator restricted by the gloves of the enclosure, a free-standing platform was considered a potential safety hazard. As a solution to this problem, the Committee developed a unique fold-up platform unit 22 inches wide, 19 inches deep and 7 inches high that folds to a depth of 2 inches and can be inserted in any facility of bench top height. (See Figure 1)

PLATFORM IN UP POSITION

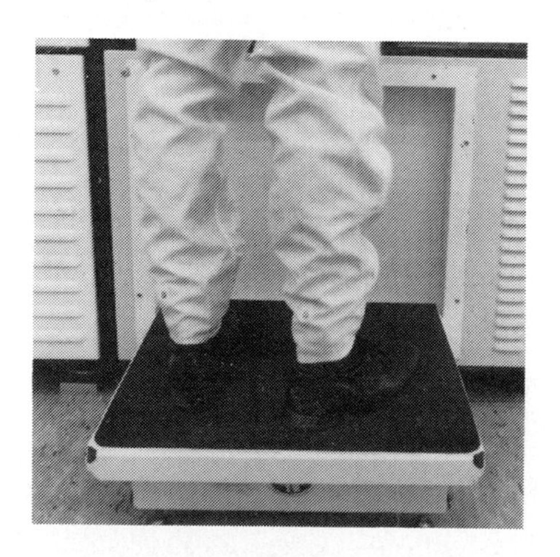

PLATFORM LOWERED FOR USE

FIGURE 1

Since the fold-up platform is a self-contained unit, the platform can be removed with the use of a screwdriver and reused at another facility at some later date.

CASE II - BACK INJURIES FROM LIFTING TOTE TRAYS LOADED WITH MAGNETS

At the request of the Operating Supervisor, the Committee investigated a series of minor back injuries and soreness problems to operators handling tote trays loaded with magnets. The investigation revealed that totes loaded with 3 inch and 4 inch magnets were on occasion over or under loaded causing a risk of injury to the employee handling the tray. It was also found that the lack of storage space forced operators to stack loaded totes on the floor and above shoulder height on the back of work benches. (See Figure 2)

Additionally, the method used to load the magnets in the totes caused weight shifting during handling and physical damage to the magnets. Some of the loaded tote trays were found to weigh in excess of 40 pounds. The Standard that the Committee established at Reading is: no push/pull or lifting force exceeding 21 pounds may be required of an employee performing routine shop operations without the approval of the Ergonomics Committee.

The Committee in cooperation with the Product Engineer and Operating Supervisor developed the following solutions to the problem:

1. Cardboard Separators
   Cardboard separators (inserts) are now used to solve the problems of overloading, weight shifting and magnet chipping. A separator is a piece of cardboard cut to fit the inside edge of the tote tray. Holes are cut in the cardboard for placing and retaining the magnets. In the case of 3 inch magnets, each piece of cardboard has 10 holes and there are 6 holes in each cardboard for 4 inch magnets. (See Figure 3)

   Limiting each tote to four layers of magnets (4 cardboards) provides a guarantee that the total weight of the loaded tote will not exceed 21 pounds. Since the cardboard is reusable, the cost to supply the line with inserts was 800 dollars, including tooling costs. The savings that resulted from the reduction of chipped and damaged magnets more than made up the tooling costs and cost of the inserts.

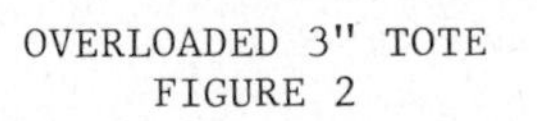

OVERLOADED 3" TOTE
FIGURE 2

CARDBOARD SEPARATORS
FIGURE 3

2. Self-Leveling Cart

   At the time of the investigation, two different carts were used to transport totes. A large cart suitable for use in a company truck was used to move the totes between buildings. The totes were then transferred to a smaller cart at Receiving because the large cart was too heavy to move easily and would not fit in most of the shop aisles.

   By designing a single cart that could be used in both the company truck and the shop aisles, it was possible to eliminate the cost of transferring the totes from one cart to another. To assist the operator in lifting and handling the loaded trays, the Committee explored the use of a self-leveling cart design. A self-leveling cart is both loaded and unloaded at waist height by use of springs inside the cart to maintain the tote on the top of the stack at waist height. (See Figure 4)

   The Committee worked with an outside manufacturer to provide a cart that would store and transport 10 tote trays weighing 21 pounds each and require no more than 21 pounds of force to push. The cart is enclosed and can be used for transport in the shop area or between buildings. A clear polycarbonate viewing panel was provided for inventory purposes. The cart is loaded and unloaded at bench top height. Since the totes can be stored in the cart until they are needed, material handling is reduced and the product is better protected.

SELF-LEVELING CART

TOTES HANDLED AT BENCHTOP HEIGHT

FIGURE 4

This case highlights a major advantage of the Committee concept. Since all functional parties are directly involved at the outset, the entire problem can be studied and decisions made that result in corrective action being implemented in the least amount of time.

CASE III - DIFFICULT LIFT OF REACTOR BASEPLATE

The baseplate used in the Gemini II reactors is approximately 3/4 of an inch thick, 27 inches in diameter and weighs about 15 pounds. To remove the baseplate, the operator reaches 24 inches to the center of the reactor, wearing heat-resistant gloves. Gloves are required because the plate is generally about 300 degrees F. when it is removed. (See Figure 5)

LIFTING BASEPLATE WITH FINGERS
FIGURE 5

Since the space between the base-plate and the quartz susceptor where it rests is less than 1/4 of an inch, the only way to re-move the hot plate is to press in with the arms and hands against the outside edge and lift. This puts a great deal of stress on the operator's back and arms and the plate is in constant danger of slipping. Not only is there a potential for back injury and burns to the operator, but the plate could fall and break an expensive quartz susceptor.

CORRECTIVE ACTION

A fixture was developed to assist the operator with the lift and reduce the potential danger to both the operator and the equip-ment. (See Figure 6)

FIXTURE DESIGNED TO LIFT BASEPLATE
FIGURE 6

LIFTING BASEPLATE WITH FIXTURE
FIGURE 7

The concept consists of a triangular aluminum frame with three specially coated contact points and two handles. The handles were located at the balance point and working position of the hands and arms during the lift. The frame slides onto the plate while it is in the reactor and the handles provide a safe means for lifting the plate from the quartz susceptor. (See Figure 7) This fixture not only has application at Reading, but at any location where Gemini II reactors are used. We have applied for a patent.

SUMMARY

An attempt has been made to share some insight into the Ergonomics Program at the Reading Works of AT&T. We feel that our Committee approach is somewhat unique and has evolved over the years as the best way we know of to accomplish our objectives within the framework of our organizational structure.

Trends in Ergonomics/Human Factors V
F. Aghazadeh (Editor)

IMPLEMENTING ERGONOMICS PROJECTS

John L. WICK, President

Occupational Ergonomics Associates
P.O. Box 151, Clinton, CT 06413

The implementation of an ergonomics project is quite often the most difficult step. Lack of acceptance and cooperation by all of the people involved in the project can negatively effect implementation. Such things as management lacking commitment, peer professionals not being involved, and the end user resisting change can deter successful implementation. This paper suggests ways which the ergonomics practitioner can facilitate the successful implementation by preparing management, by dealing with peer professionals and by working with the end user.

1. INTRODUCTION

Very often an ergonomics project will proceed smoothly through all of the steps of problem identification, analysis, design and testing only to fail at implementation. The reason for an unsuccessful or incomplete implementation frequently is lack of acceptance. This can come from management, peer professionals or the end users.

Upper and middle level management must understand the reasons for applying ergonomics to work and workplaces and, more importantly, understand how it will benefit the company as a whole, i.e. improve profitability. The task of educating the management team needs to be done at the very beginning to gain their commitment and support. More will be said here about this group because they are the most important. Without management support implementation is impossible. Without peer professionals or end users cooperation implementation is merely difficult.

Peer professionals such as industrial engineers, safety specialists and machinery engineers need to be made a part of the project to avoid offending them. This means involving them in each step of the project.

The end user - the worker whose workstation is being redesigned - must be a part of the process. It is "their" workplace and they feel they should be involved in any changes. It is necessary to provide them with an appreciation for the reasons for change and make them feel as though they are a part of the project. They need to know how the change will effect them.

2. PREPARING MANAGEMENT

2.1. ADDRESS THE NEEDS

Many managers have not heard of ergonomics or human factors. They have no idea what the benefits are or how one goes about applying ergonomics. Therefore, it is necessary to develop an understanding of and an appreciation for occupational ergonomics. The presentation of the material needs to be succinct so as not to waste valuable management time, but provide sufficient information for them to make sensible decisions regarding the application of ergonomics. The presentation meeting should take no longer than one hour and a half and the material should be well prepared and of high quality.

It is important to understand the needs of the group when preparing the material. If injuries are the concern, then how ergonomics will improve safety should be emphasized. If productivity improvement is the concern, then how ergonomics effects human efficiency should be the focus. In any case, cost reduction underlies all of the concerns and the major thrust of the presentation must be how ergonomics will improve profitability. There is a paradox here because many companies become interested in ergonomics to reduce the number of injuries. Yet, the accountants have difficulty with financing ergonomics with injury cost avoidance. It begs the question, "How long do you wait for an injury to not happen before you claim the savings?" Therefore, injury cost avoidance should be discussed, but savings of labor costs through productivity improvement is a more important benefit to management.

2.2. EXPLAIN ERGONOMICS

An effective way to present ergonomics to management is to use twenty minutes to explain all aspects of industrial ergonomics in language understandable to them. An ergonomist must use many terms in practice which are not in everyone's vocabulary. Avoid words which might be unfamiliar to the audience. For instance, use "hearing" instead of "audition", "wrist injury" instead of "carpal tunnel syndrome", etc. Photographs present much information in little time. Slides, especially of situations within the company, are very effective in explaining ergonomics in a short time. The slide-tape presentation on ergonomics from the University of Guelph [1] presents industrial ergonomics succinctly and understandably. It, or something like it, can be a used productively to provide a basic understanding of ergonomics.

2.3. DEFINE THE BENEFITS

Once management has an understanding of occupational ergonomics, they need to appreciate the benefits of application. An explanation of "cause and effect" is easily understood and presented. As an example, describe the epidemiology of back injuries and how ergonomics can be used to solve a back injury problem thereby saving the high costs of such injuries. Or, cite a low productivity situation and describe how the application of ergonomics will improve worker efficiency which will save labor costs. In all cases, the savings resulting from the application of ergonomics should be stressed. Improving profits is the goal.

2.4. GAIN COMMITMENT

The ergonomics effort must have the commitment of the management team so that it will have management support. Management support is requisite for the acquisition of funds and the assistance of others.

Discussion should be encouraged to ensure complete understanding of the material presented. Verbal agreement should be acquired during the discussion period. Often, the management team will decide to apply ergonomics to a situation they have identified as a result of having learned about ergonomics. This, of course, should be the next project and it must be a success.

It is desirable to have management demonstrate their commitment in some visible way. This can be accomplished through memoranda or newsletters, or through statements in various management meetings. The more visible management support is, the more others will cooperate and become involved.

3. WORKING WITH PEER PROFESSIONALS

3.1. GET THEM INVOLVED

Ergonomics does not stand alone. The ergonomics practitioner crosses into many different disciplines such as industrial engineering, safety, machinery design and facilities utilization. Very often professional jealousies can arise. They may even refuse to cooperate. This can be avoided by being sure to involve these professionals in the project.

It may be necessary to educate them in ergonomics much as was done for management. It would be advantagious to ensure they learn about ergonomics and how their various disciplines impact on ergonomics projects. Their understanding of ergonomics will guide them in future projects of their own and allow them to better assist in ergonomics projects.

Having the various other disciplines involved in the beginning of the project will bring out many useful ideas valuable in the problem solving phase. Seek their advice without being demanding. If a machine is required to relieve the stress on the worker or improve efficiency, bring it to the attention of the mechanical engineer who is the specialist in this discipline. Many additional improvements are sure to result. A team approach insures that all opportunities for improvement are realized [2].

3.2. RESPECT THEIR EXPERTISE

It is important to always remember that the other professionals are just that. They have specialized training and skills which, when applied to the ergonomics project, will enhance its success. Frequently, the ergonomics practitioner will only identify problems and varify the appropriateness of the solutions while providing guidance and advice during the design. In this way the ergonomist allows the other professionals to apply their special skills in the design and development.

If peer professionals are involved in the ergonomics project, they will work for its success. They can be counted on to assist in successful implementation.

4. OVERCOMING RESISTANCE OF THE USER

Again, education and involvement are key. The users need to know why change is required. This means they need to know something about ergonomics and how it works [3]. They need to understand how it will effect them. The more they understand the process, the more they develop a sense of ownership. Developing that sense of ownership is most important in overcoming resistance to change.

The end user is an important source of valuable data. Only they can say how the workplace effects them. The use of scales such as the Postural Discomfort/Pain scale by Corlett and Bishop [4] are useful data collection tools. They have the added value of involving the end user in the project.

The sense of ownership can be further developed by asking for their suggestions. No one knows the workplace better than the one using it daily. Quite often they have many of the answers to questions regarding what is wrong and what should be done for improvement. Active participation throughout the project by the end users builds ownership.

Use their suggestions. If they are not appropriate, demonstrate why and solicit more suggestions. Sometimes there is no "best" of more than one solution. In that case, the idea of the end user should apply.

Give them time. Their job may not normally require the sort of thought processes required for problem solving. They will need time to think. When testing a prototype, they will need time to become accustomed to the changes and become proficient with the new design.

The goal is to develop ownership in the end user of the result of the project. There is less likelihood of resistance to change if the end user has a role in bringing about that change.

5. FOLLOW-UP

Nothing is "bug-free". Therefore, follow-up begins on the first day of implementation. All who have had a hand in the design and development should be available to make any adjustments or changes as needed. Be sure adequate training has been provided. This includes the training of the supervisor.

A return one week later is warranted. The user has had an opportunity to become accustomed to the changes. More adjustments might be required.

6. CONCLUSION

The implementation of a project is the reason for all of the efforts beforehand. Without implementation the project has no meaning or worth.

The ergonomics practitioner's reputation rests on successful implementation. It is for those reasons every effort must be made to insure a successful implementation.

That success depends on the interest and cooperation of many people. The interest must be developed through education before the project begins and the cooperation must be developed through involvement during the project. This applies to management, peer groups and the end users. Their interest and cooperation is the foundation of successful implementation.

All of the suggestions offered in this paper are founded in practice and well tried. By applying these techniques, the ergonomics practitioner will be sure to enhance the success of implementation of ergonomics projects.

REFERENCES

[1] Webb, R.D.G., Man: The Fallible Machine, Slide-tape Presentation, University of Guelph, Guelph, Ontario, (1982).
[2] Wick, J.L., Productivity and ergonomic improvement of a packaging line: a case study, in: Asfour, S.S., (ed.), Trends in Ergonomics/Human Factors IV (North-Holland, Amsterdam, 1987) pp. 97-102.
[3] Drury, C.G. and Wick, J.L., Ergonomic applications in the shoe industry, Proceedings of the 1984 International Conference on Occupational Ergonomics, 1984, p 489.
[4] Corlett, C.N. and Bishop, R.P., A technique for assessing postural discomfort, Ergonomics (1976) 19, pp 175-182.

Trends in Ergonomics/Human Factors V
F. Aghazadeh (Editor)

Practical Ergonomic Application

Frank L. Mc Atee, Jr.

Ergonomics Coordinator
Buick-Oldsmobile-Cadillac Group
General Motors Corporation
P.O. Box 444
1500 E. Route A.
Wentzville, Mo. 63385

The purpose of this paper is to present a practical approach to the application of Ergonomics in an automobile assembly plant to reduce and/or eliminate strains, sprains and injuries.

INTRODUCTION

Good design of new, or modification of existing, tools and equipment is conducive to improved quality, increased productivity, health and safety, and worker satisfaction. In a new, highly automated assembly facility, one would think that only the best tools and equipment would be purchased with the human operator in mind. Tradition causes us to purchase the best. However, considering the human operator (our greatest asset) has not yet become traditional. The human operator must be considered prior to any design, construction, installation or purchase. Until that era arrives, we must react to poor design with reactive Ergonomics and continue to influence change with pro-active Ergonomics.

REDUCING REPETITIVE MOTIONS INJURIES IN OVERHEAD ASSEMBLY [1]

Frame to body secure using an articulating arm.

The Job

Operators assemble body bolts on automobiles in an overhead assembly line using a pistol grip air wrench. They assemble 225 bolts per hour and work an 8 hour shift with two 23 minute breaks and a 30 minute lunch break.

Repetitive Motions Injuries Observed

Approximately 10 people had done this job regularly over a one year period and 8 had experienced discomfort or injury.
Neck and shoulder strain and forearm tendonitis were the most common complaints.

Risk Factors

The following risk factors for repetitive motions injuries were present in this job:

- High forces exerted
- High impact forces in torquing
- Rotation while exerting high force
- High repetition rate
- Sustained holding (static loading of shoulder muscles)
- Shoulder and neck tension
- Elevated arms for extended periods

The hand holding the air wrench during bolt assembly repeatedly exerted high forces, especially at impact when torque is reached. The wrench was heavy and was not balanced properly for driving the bolts vertically upward. The operator's shoulder, elbow, and wrist were constantly under tension during bolt assembly.

Approach to Reduce the Risk Factors

To improve this job, a solution that would reduce the forces required, lower the shoulder and neck tension, and reduce continuous holding time was desired. An articulating arm was fabricated with an in-line (straight) air wrench attached. This permits the operator to load bolts to the wrench socket, position it for assembly, and then initiate the wrench action with his/her hands at chest level rather than overhead. The arm is spring-loaded; this reduces the amount of upward force the operator has to exert to drive the bolts.

Effectiveness of the Solution

In the two-year period since the job was modified using the articulating arm, there have been no complaints of sore arms, necks, shoulders, hands, or wrists and no diagnosed repetitive motions injuries. In addition, people who previously could not perform the job are now able to do it, making it a more desirable work situation for everyone.

Gas tank secure using semiautomatic fixture.

The Job

Two operators assemble gas tanks to the automobile in an overhead assembly line, lifting by hand and securing with pistol grip air wrenches.

Repetitive Motions Injuries Observed

Approximately 16 people had done this job regularly over a two year period and all had experienced discomfort or injury. Neck and shoulder strain, back pain and eye injuries were the most common complaints.

Risk Factors

The following risk factors for repetitive motions injuries were present:

- High forces exerted
- High impact forces in torquing
- Rotation while exerting high force
- High repetition rate
- Sustained holding (static loading of shoulder muscles)
- Awkward lifting positions
- Shoulder and neck tension
- Elevated arms for extended periods
- Foreign materials falling from tank (water, metal shavings)

The hand holding of the gas tank and air wrench during bolt assembly repeatedly exerted high forces. The operator's neck, shoulders, back and arms were constantly under tension during assembly of the tank to car.

Approach to Reduce the Risk Factors

To improve this job, a solution that would reduce the forces required, lower the shoulder and neck tension, eliminate foreign bodies falling in operators face, and reduce continuous holding time was desired. A gas tank loading fixture was designed and fabricated with two in-line air wrenches (straight). The gas tank is delivered by conveyor, operator pulls the tank from the conveyor onto the loader. The unit lifts the tank into position to assemble to the underbody of the car. The operator positions the straps and the wrenches to secure the tank (wrenches need only to be activated, they drive bolt to the prescribed torque specification, release and retract automatically). The operator controls the power assist to position and remove the tank loader.

Effectiveness of the Solution

In the nine-month period since the job was modified using the gas tank loading fixture, there have been no complaints of sore arms, necks, backs, shoulders, hands or wrists, foreign materials in eyes and no diagnosed repetitive motions injuries. In addition, female operators who previously could not perform this operation are now able to do so with minimal effort and only one (1) operator is required to perform the entire operation in place of two (2).

SIMPLE ERGONOMIC APPLICATIONS

Use of tilting mechanisms to reduce back strain.

Tilt-turn tables were purchased and installed to reduce back strain while lifting heavy parts such as alternators, air-conditioning compressors, starter motors, hubs, etc.

Engineering change to eliminate cumulative trauma.

The connection of the transmission neutral safety switch wire connector was made after the engine was installed in the car resulting in numerous hand and arm injuries. An engineering change was made to lengthen the wire to facilitate the connection while engine was being installed. The change resulted in elimination of all injuries and a reduction of two (2) operators per shift.

The Auto Jack was installed under the package shelf in the luggage compartment of the car, resulting in numerous cases of low back strain. An engineering change was made to change the jack style and location. The jack is now installed in the right rear quarter panel of the job resulting in elimination of back injuries and two (2) operators per shift.

CONCLUSION

Application of sound Ergonomic principles and the use of simple lifts, tables, tools, etc., can reduce and/or eliminate many potential injuries. When injuries are absent from the workplace, the human operator can perform his/her operation with minimum delay and maximum quality. It is our responsibility to provide a safe and healthy environment for each and every employe.

References

[1] Mc Atee, F. L., Reducing Repetitive Motions Injuries in Overhead Assembly, Seminars in Occupational Medicine, 1987, 2(1), 73-74.

Trends in Ergonomics/Human Factors V
F. Aghazadeh (Editor)

THE VALIDITY OF USING EXISTING ERGONOMIC NORMS IN INDUSTRIAL WORK DESIGN - SOME CASE STUDIES

S. P. DUTTA, Ph.D.

Department of Industrial Engineering
University of Windsor
Windsor, Ontario, Canada

The paper reports on the use of existing ergonomic norms related to static workloads in the automotive industry. Two specific workstations have been examined. In the first instance, the problem of static stress on finger, elbow and wrist joints has been investigated during the assembly of covers to seat cushions. Apparently no problems have been observed through standard work measurement techniques. Biomechanical studies, however, have revealed unacceptable levels of stress, particularly at vulnerable anatomical joints. In the second case, the problem of tool design for simple finishing tasks has been examined. Here again, work study has suggested certain levels of relaxation allowances to overcome pain at the wrist and shoulder as a result of continuous use of the tool. Simple biomechanical modelling has revealed the true extent of the problem and has led to a total redesign of both tool and workstation. In all the instances cited earlier, attempts have been made to use existing norms for static work available in handbooks and published literature. It has, however, been noted in many instances that proposed limits for stress at different body joints do not provide the clues necessary to solve such problems. The paper attempts to highlight the limitations of such published information in the context of their utilization in the real world situation.

INTRODUCTION

The effect of automated manufacturing systems has nowhere been felt as much as in the automotive assembly plants of North America. The changes brought about by the introduction of new plant and equipment, especially flexible handling and robotics, have affected work place, work load and the work environment in general. In spite of these modifications, human intervention in the process of assembling automotive systems is still significant and will probably continue to be so for at least the next decade. It is recognized, however, that the nature of human activity on the assembly line has shifted from largely dynamic or whole body activity to primarily static work, involving specific muscle groups of the forearm, upper torso or the lower back. Recent studies (Stetson et al., 1986) have reported significant incidence of occupationally related upper extremity cumulative trauma disorders (CTD's) in a variety of work stations in the automotive industry. These have been associated with high force and highly repetitive work, often regardless of posture.

Numerous studies have been conducted to evaluate the extent of these disorders and suggest methods of modifying work stations to ameliorate the effects of such conditions (Armstrong, 1986, Punnett and Keyserling, 1987). These have involved intensive on-site diagnosis and measurement of various biomechanical, psychophysical and physiological parameters, including electromyographic measurements of selected muscles, rated perceived exertion, heart rate and so on. It is often rather difficult to conduct investigations of this nature at a production site, for a variety of well-known reasons. Recently epidemiological studies have been carried out to establish guidelines related to risk factors at the workplace. However, because of the qualitative character of most current epidemiological studies, results cannot be readily implemented into quantitative ergonomic guidelines (Dul & Hildebrandt, 1987). A survey by the same authors of sixteen recent ergonomic, biomechanical and epidemiological books to obtain common ergonomic guidelines for both static and dynamic work shows that, in quantitative terms, a great variety exists amongst guidelines for similar types of tasks, possibly due to differences in evaluating and/or measurement criteria. In fact, the authors conclude that "there is great need for reliable and valid guidelines."

Similar observations are noted in the work of Punnett and Keyserling (1987) who conclude from their studies that there is great need for data on ergonomic exposures that can be generalized both within and across industries. In specific instances, such as jobs involving repetitive finger pinch grip and flat press (which are found in significant numbers in automotive assembly) the problem is rather acute, because differences in strengths between fingers vary widely for adults and the elderly (Imrhan, 1987). Similarly, Kumar (1987) has concluded that a strength norm obtained for population in an optimum posture can clearly not be considered as an appropriate design criterion for a task to be performed in an awkward posture with considerable mechanical disadvantage. However, "a silence in literature on this matter may have supported the idea of universal applicability of the strength values reported." Many jobs also involve asymmetric lifting. As indicated by Garg and Badger (1986), NIOSH guidelines (AL & MPL) normally used for setting work standards in industry are questionable in such instances, as the guidelines are intended to apply to two handed symmetric lifting in the sagittal plane with no axial rotation.

Again, in work situations involving static postural force exertions, parameters such as the maximum endurance time or the necessary recovery time after an exertion are of importance. As Rohmert et al. (1986) have indicated, the application of models to derive such values (Monod and Scherrer, 1965, Caldwell, 1964) in practical work situations may be difficult for several reasons, particularly in bent and twisted body postures where the trunk forms a complex structure including many weak links.

For these reasons, many job specifications in the automotive industry are still based on classical work study techniques, even when significant portions of the work cycle involve static muscular activity with postural constraints. The limitations of traditional work analysis methods are well known and have been emphasized by Punnett & Keyserling (1987), amongst others, because they provide limited information on inter-job and intra-worker variability. Mital, et al. (1987) have also reported that when weights or forces are involved, MTM (Methods Time Measurement) may severely overestimate work performance standards; moreover fatigue allowances given are subjective and almost impossible to estimate accurately.

The paper reports on the use of existing ergonomic norms related to static workloads in a large automotive assembly plant in Southwestern Ontario. It illustrates, through the use of two specific case studies, the problems encountered in applying available ergonomic norms to work related incidents where detailed biomechanical measurements are difficult or extremely time consuming. The limitations of published information in the context of their utilization in the real world situation are highlighted.

CASE STUDY - 1

Covers are assembled to padding to prepare seat cushions. The process involves "hogringing" the cover to listing wires embedded in the pad. A standard hogringing gun with a pistol grip is used for this purpose, and is held in the hand of the operator without any support. The jaws at the pistol head are used to pry open the folds of the pad within which the wires are embedded. Between 3 and 4 insertions are made for each seat cushion.

A number of problems were observed during the study of this process. For example, during the hogringing to the listing wire, the configuration of the upper torso changed such that an angle of 120°-130° was subtended at the hip joint at the end of every cycle. Significant shoulder abduction was observed, as was rotation at the lumbo-sacral disc. Again, the movement of the wrist during the insertion and hogging operation was significant, with a wrist angle of less than 90° near the end of the work cycle. This movement was caused by the pressure required to simultaneously push both inwards and sideways in order to widen the lips of the padding and obtain a proper grip on the wire prior to hogringing. Further, force exertions were recorded for some trials and it was observed that effort of up to 12 kilopond (1 kp≃1 kg) was called for to clamp the gun properly onto the wire. This did not include the effort required to press the trigger.

It may be noted that job standards set by time study and MTM did not consider these inappropriate postures. Rather, time standards for grasp, hold, move and so on, were employed to determine cycle time for each operation and the number of cushions to be prepared per shift were determined accordingly. Obviously, this was quite inadequate and led to a "work refusal" situation after numerous workers complained of wrist disorders, finger injuries and general inability to carry out the job.

The problem arose with the external load forces distributed over the palmar aspects of the fingers. Further, wrist deviations and exertions of the hand with a flexed wrist cause acute compression of the median nerve and result in acute symptom of carpal tunnel syndrome. Figure 1 indicates the type of forces developed under these conditions.

Assuming that the operator is holding the hogging tool, F_L represents the downward force. In this case, let F_L = weight of tool = 2.3 kg. The radius of the anatomical pulley, r, is the radius at the wrist, which may be assumed to be 3/2" = 1.5" = 38.4 mm. From literature, contact forces as defined in Fig. 1 can be obtained from the following equations:

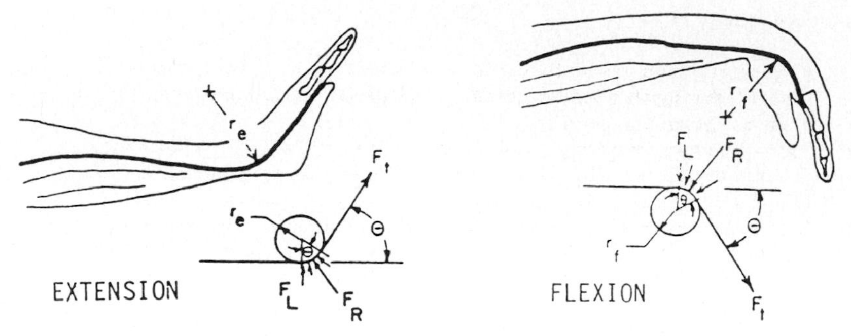

Fig. 1. The extrinsic finger flexor tendons are supported by anatomical pulleys with radii r_f and r_e during flexion and extension of the wrist. Intrawrist forces, F_L and F_R are described by Equations (1) and (2). From Armstrong and Chaffin, 1979.

$$F_L = F_t/r \quad (1)$$

$$F_R = 2F_t \sin\theta/2 \quad (2)$$

$$\therefore \quad F_t = F_L \times r = 2.3 \times 38.4 = 88.32 \text{ kg-mm.}$$

$$\therefore \quad \text{Resultant moment force } F_R = 2 \times 88.32 \times \sin \frac{90^\circ}{2}$$

$$= 2 \times 88.32 \times \sin 45^\circ$$

$$= 2 \times 88.32 \times 0.71 = 125.4 \text{ kg-mm.}$$

It may be noted that θ, total wrist deviation, has been observed to be 90° as indicated earlier. This force moment is counteracted in the vertical direction by the finger muscles as shown below:

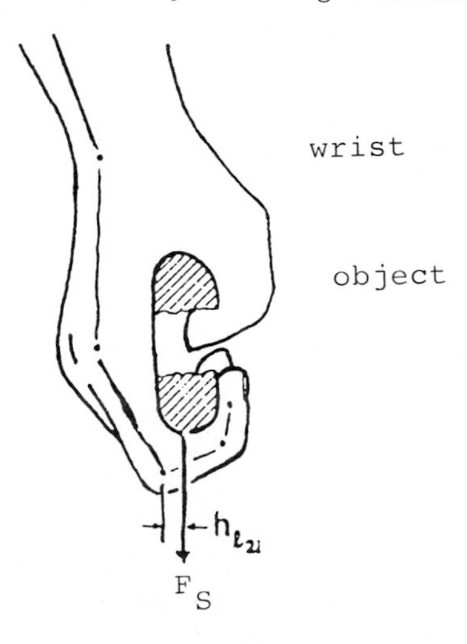

The major force-producing muscles in power-grip are the flexor digitorium profundis (FDP) and superficialis (FDS) located in the forearm. Table I indicates the maximum strength for gripping under static equilibrium with the loads assumed equally distributed between FDP and FDS. FDS represents the resultant force on each finger.

Noting that the moment arm at the fingers ($h_{\ell_{2i}}$) varies from 10 mm to 20 mm and that the moment resulting from using the hogringing tool is 125.4 kg-mm, the force on each digit of the fingers of the holding arm equals

$\frac{125.4}{10}$ = 12.54 kiloponds at the highest and $\frac{125.4}{20}$ = 6.27 kiloponds at the lowest. These values are within the limits shown in Table I.

Table I. Computation of FDP and FDS muscle moment arms (mm) and strength (kp) for each finger from grip strength measurement

Digit (i)	$h_{p_{3i}}$	$h_{p_{2i}}$	$h_{e_{2i}}$	F_{ℓ_i}	$h_{\ell_{2i}}$	F_{p_i}	F_{s_i}
2	4.2	9.1	8.1	20.1	14.4	16.8	16.8
3	4.3	9.4	8.5	26.7	10.0	15.5	15.5
4	4.0	9.2	8.2	20.4	15.5	18.2	18.2
5	3.6	8.8	7.8	12.4	19.5	14.6	14.6
Total	-	-	-	79.3	-	65.1	65.1

However, Table II describes the maximum press strength with each finger.

Table II. Prediction of maximum press strength (kp) with distal phalanx of each finger based on measurement of grip strength

Digit (i)	F_{p_i}	$h_{\ell_{3i}}$	$F_{\ell_{3i}}$	$F_{\ell_{3i}}$	Δ%	$h_{\ell_{2i}}$	F_s
2	16.8	10.7	6.6	6.6	0	38.5	12.5
3	15.5	11.8	5.5	8.9	-38	46.6	13.0
4	18.2	11.8	6.2	6.8	-6	44.4	13.2
5	14.6	13.0	4.0	4.1	0	36.7	2.3
Total	65.1	-	22.3	26.4	-15	-	41.0

As can be noted, the downward force does create a problem in the case of the 5th digit, where the maximum strength is 2.3 kiloponds.

All the previous calculations and observations have been based on the assumption that the weight of the tool is the critical factor in the operation. The basic problem seems to arise from the fact that the tool is used as a plier, a job for which it is not designed. It is important to consider changing the jaws of the tool to widen them in order to allow better insertion. It is also necessary to support the tool and make the handle vertical. Finally, the workbench should be tilted by 15-20° to bring the arm into a neutral position during the hogringing operation; thereby, it will also increase the angle at the hip joint and reduce bending as is the present situation.

CASE STUDY - 2

The problem relates to the use of a hand tool for finishing operations on the door of a particular type of automobile. Each worker is expected to complete 180 doors/shift. Observations indicate that about 60 hand movements are required to finish each door. The objective of using the tool (shown in Fig. 2) is to produce smooth and continuously distributed

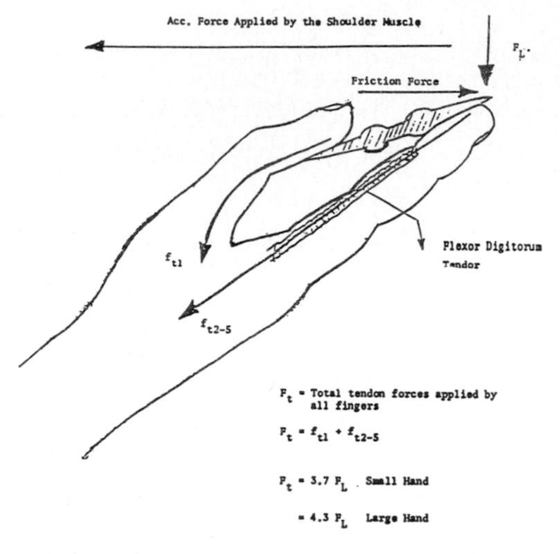

Fig. 2. Hand-tool force analyses of the present tool

forces over the door. The duration of this operating accounts for approximately 40% of the total cycle time.

The problems with this tool were fairly obvious. Tool and thumb contact produced a very high blood circulation due to the sharp edges of the tool (Fig. 2). Similarly, for the other four fingers, the reaction force F_L produced a very high stress on the tip of these fingers. Again, the small tool size required force exertion to a level which can cause problems with different operators and reduce the degree of flexibility of hand movement, specially with workers who have larger hands. Moreover, frequent use of this tool would be sufficient to cause the tendon muscles to gradually stretch, and if the load was applied too often or too quickly, recovery was not complete (adding as much as 1% to the tendon's length). With rest, of course, the tendon slowly recovers.

Based on simple biomechanical analysis, a modified tool was designed. The shape and dimensions of the modified tool are shown in Fig. 3. This tool eliminates many of the problems associated with the existing one. It allows greater flexibility of hand movements, drastically reduces the stress employed on the contacted surface of each finger involved, is easy to grasp and has no sharp edges that might contact any part of the hand during operation. The modified tool can also be used to produce greater forces on the contacted surface (between tool and part) if the operator performs the job by exerting forces from the top to the bottom using the front edge of the tool and pushes to lift horizontally using the side edge of the tool. This can be related to the abilities of people to produce greater forces in these directions as shown in the isometric strength data given in Table III and Figure 4.

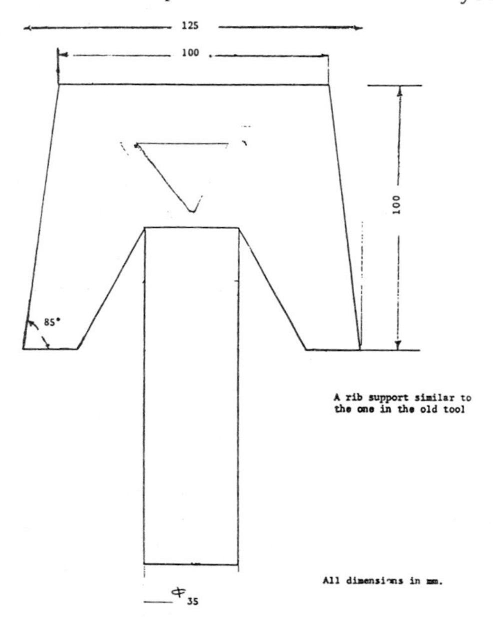

Fig. 3. Modified design of the suggested tool

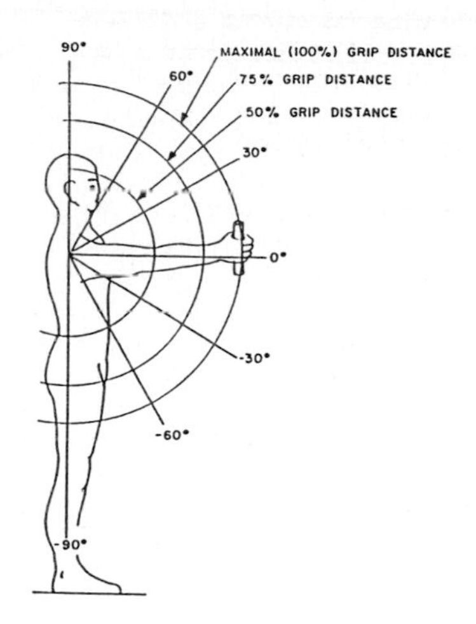

Fig. 4. Arm posture used in one arm strength study. The results are given in Table III.

Table III

Maximal Right-Handed Static Forces Exerted on a Vertical Hand-Grip by Standing Young Male Subjects

	Location of the Handgrip (see Figure 8.5)			
		At Percentages of Maximal Grip Distance		
Type of Exertion	Arm Angle (deg)	50%	75%	100%
		Force (N)		
Push outward, horizontal	30	71	108	142
	0	133	156	178
	−30	125	135	142
	−60	125	142	160
Pull inward, horizontal	30	85	98	116
	0	102	116	129
	−30	125	134	138
	−60	102	125	151
Push to the left, horizontal	30	156	136	107
	0	187	147	107
	−30	187	151	116
	−60	147	136	116
Push to the right, horizontal	30	107	98	93
	0	136	111	89
	−30	147	120	98
	−60	111	102	93
Lift, vertical	30	125	107	85
	0	151	116	80
	−30	222	178	125
	−60	280	227	182
Press down, vertical	30	338	258	182
	0	249	178	147
	−30	156	147	136
	−60	173	160	142

DISCUSSION AND CONCLUDING REMARKS

In Case Study-1, the results of the calculations have been compared with Tables I and II (Armstrong, 1982). Similar tables for pinch and hand grips by Imrhan (1987) provide figures which are significantly different. In fact, Armstrong (1982) has indicated that there is poor agreement between predicted and measured strengths

$$[F_{\ell_i} \text{ VS } F_{p_i} \text{ in Table I and } F_{\ell_{3i}} \text{ VS } F_{p_i} \text{ and } F_S \text{ in Table II}]$$

even for the data that he has published. One possible reason is the manner in which the load is characterized. As such, conclusions drawn from the calculations and outlined in Case Study-1 will change depending on whether the tables from Imrhan, predictive values from Armstrong or calculated values from Armstrong are used. Moreover, the variation of these strengths with change in shoulder abduction is not fully established. Neither do these figures address the question of repeated motions or how the posture affects maximum endurance time in such static force exertions. As indicated by Rohmert et al. (1986), attempts to use the traditional models proposed by Monod and Scherrer (1965) and others to establish this linkage may not provide the right answers because many postures (particularly bent and twisted) lack mechanical stability and do not allow full application of the potential maximal muscular strength. The exerted maximal forces, therefore, do not indicate the true maximal values as required for the model. Similarly for Case Study-2, the results of biomechanical analysis have been compared with Table III and

Figure 4. Grip strengths and force applications in the actual work situation involved asymmetric arm movement with the elbows bent and the trunk rotated outwards to align the tool to the job. Once again, the validity of the norms suggested in Table III are open to question. Kumar (1987) has shown that strength capability is inversely related to the reach distance of the task and is significantly reduced in an awkward posture. Moreover standard posture strength, which is a frequently used design criterion, does not match with any of the isometric or isokinetic strength studies.

It can therefore be concluded that adaptation of information in published literature with respect to static loading may not necessarily provide appropriate solutions to real life problems. Traditional biomechanical motion analysis, which is often the basis of such data, is restrictive in terms of the resources required to obtain fairly small data samples. It is therefore necessary, in order to estimate the variability of responses amongst individuals to exposure to diverse work stresses of a largely static nature, that epidemiological studies be carried out in conjunction with precise ergonomic analysis. Whilst there may be some trade-off between precision and ease of application of work norms, the results would provide work designers with more universal tools which are less restrictive in their utilization.

REFERENCES

1. Armstrong, T. J., Ergonomics and Cumulative Trauma Disorders, Hand Clinics, 2 (1986) 553-565.
2. Armstrong, T. J., Development of a Biomechanical Hand Model for Study of Manual Activities, in: Easterby, R., Kroemer, K.H.E., Chaffin, D.B. (eds.), Anthropometry & Biomechanics - Theory and Applications (Plenum Press, 1982) p. 183.
3. Armstrong, T.J., and Chaffin, D.B., Carpal Tunnel Syndrome and Selected Personal Attributes, Journal of Occupational Medicine, 21 (1979) 481-486.
4. Caldwell, L. S., Measurement of Static Muscular Endurance, Journal of Engineering Psychology, 3 (1964) 16-22.
5. Dul, J. and Hildebrandt, V.H., Ergonomic Guidelines for the Prevention of Low Back Pain at the Workplace, Ergonomics, 30 (1987) 419-429.
6. Garg, A., and Badger, D., Maximal Acceptable Weights and Maximal Voluntary Isometric Strengths for Asymmetric Lifting, Ergonomics, 29 (1986) 879-892.
7. Imrhan, S. N., An Analysis of Finger Pinch Strength in the Elderly, in: Asfour, S. S. (ed.), Trends in Ergonomics/Human Factors IV, (Elsevier, 1987) pp. 611.
8. Kumar, S., Arm Strength at Different Reach Distances, in: Asfour S. S. (ed.), Trends in Ergonomics/Human Factors IV, (Elsevier, 1987) pp. 623.
9. Mital, A., Asfour, S.S., and Aghazadeh, F., Limitations of MTM in Accurate Determination of Work Standards for Physically Demanding Jobs, in: Asfour, S.S. (ed.), Trends in Ergonomics/Human Factors IV, (Elsevier, 1987) pp. 979.
10. Monod, H., and Scherrer, J., The Work Capacity of Synergic Muscular Groups, Ergonomics, 8 (1965) 329-338.

11. Punnett, L., and Keyserling, W. M., Exposure to Ergonomic Stressors in the Garment Industry: Application and Critique of Job-site Work Analysis Methods, Ergonomics, 30 (1987) 1099-1116.

12. Rohmert, W., Wangenheim, M., Mainzier, J., Zipp, P., and Lesser, W., A Study Stressing the Need for a Static Postural Force Model for Work Analysis, Ergonomics, 20 (1986) 1235-1249.

13. Stetson, D. S., Armstrong, T. J., Fine, L.J., Silverstein, B.A., and Taynen, K., A Survey of Chronic Upper Extremity Disorders in an Automobile Upholstery Plant, in: Karwowski, W. (ed.), Trends in Ergonomics/Human Factors III, (Elsevier, 1986) pp. 623.

Trends in Ergonomics/Human Factors V
F. Aghazadeh (Editor)

PRACTICAL APPLICATIONS IN ERGONOMICS

Michael P. KLYM, Industrial Engineer

Ford Motor Company
Windsor/Essex Engine Plants
Windsor, Ontario
Canada N9A 6X3

When an individual in the Industrial World is working in the field of Ergonomics, odds are that along with this function, he is also expected to perform other Engineering duties. Extensive research and development from statistical data obtained from a controlled representative group are usually beyond realistic conditions. The "Ergonomist" in the Industrial environment relies primarily on R&D from the academia and rather than becoming a statistician, the Industrial related person must perform "Practical Applications in Ergonomics."

INTRODUCTION

This paper will discuss (2) practical applications in Ergonomics. The first topic will deal with (2) assemblers working on a power and free line doing an inspection procedure and installing drain plugs respectively while sitting on caster chairs. The second topic will take a historical look at three conveyor systems in three separate engine plants delivering the same component. These plants were built over approximately a period of (2) decades.

TOPIC 1

The power and free line is designed so the majority of labour completed by the operators are in a predominately neutral standing posture. (1) Two operators on this line perform a black-lite inspection function and install water drain plugs respectively. (Fig. 1 & 2).

The black-lite operator (Fig. 1) must inspect the lower half of each engine for potential oil leaks. In order to accomplish this, the operator must sit on a roller chair with the seat approximately 38 centimeters from the floor. The operator is equipped with an ultra-violet light that is suspended from a balancer and trolley above the conveyor. The major inspection area is approximately 74-91 centimeters from floor level. In order to complete the entire inspection, the operator must travel between the engines and periodically position himself/herself in an awkward posture, especially when inspecting the rear portion of the engine.

FIGURE 1

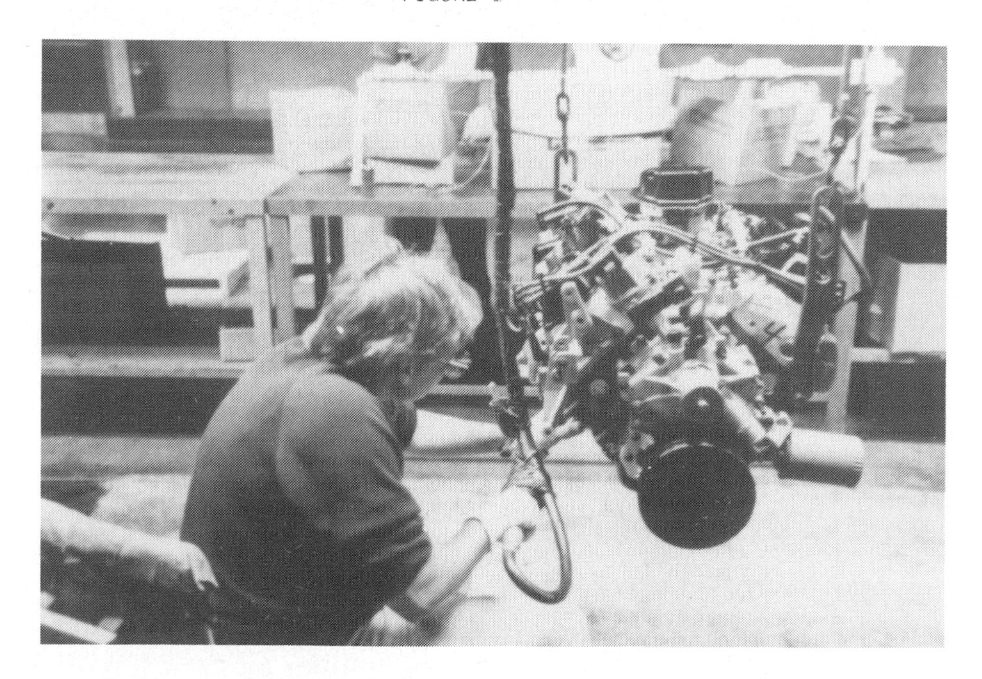

FIGURE 2

Similarly, the drain plug operator (Fig. 2) must sit on a roller chair and install (2) drain plugs-one on each side of the engine. As previously mentioned, the operator must travel between engines in an awkward posture but now there is the added torquing force of 25-40 newton metres from the air gun that installs the plugs. Linespeed in this section is set up at a rate of (180) jobs per hour.

In the analysis of these jobs, work posture proved to be the area of main concern. Job related injuries ranged from lower lumbar pains to shoulder and neck pains. Neck and shoulder pains can be attribued to the position of the operator during working conditions. Research has recommended that while in a sitting position the head/neck regions tilt should range from 17-29 degrees. (2) This was not the case as the operators were forced to position themselves beyond recommended limits in order to perform their jobs. Lower lumbar pains can be attributed to the static load being placed on the individual with the trunk position curved forward. (3)

FIGURE 3

After analysis, it was determined that the two jobs violated the basic principles of work station design (4) and action had to be taken. These two jobs had to be completed on the power and free line, therefore, job re-location was not feasible. It was mandatory to change the posture of the workers and it was felt that if the engines could be presented to the operators in a working standing position, that would be the ideal situation. Of several proposals considered, the most feasible (through considerations of cost and safety) was to raise a portion of the power and free conveyor approximately 61 centimeters, which would make the new work height approximately 137 centimeters. (Fig. 3 & 4) Modification to the design of the line not only reduced lumbar, shoulder and neck injury to the workers, it also reduced manpower turnovers thus increasing quality and productivity.

FIGURE 4

TOPIC 2

With the rapid advancement of technological knowledge, it is becoming more and more evident that the Plant Engineering functions, along with the remaining Engineering services, become more cognizant of the effects of proper Ergonomic applications.

Three Engine Plants on the Windsor/Essex site were constructed in 1964, 1967 and 1981 respectively. The Cylinder Head Assembly Conveyor System delivers cylinder heads from the feeder machining department to the Engine Assembly Line. At this point, the cylinder heads are manually unloaded and assembled to the engine. (Fig. 1, 2 & 3) The basic design of the conveyor system in all three plants is similar. Carriers of the 1967 (Fig. 2) plant to the 1981 (Fig. 3) plant are atypical.

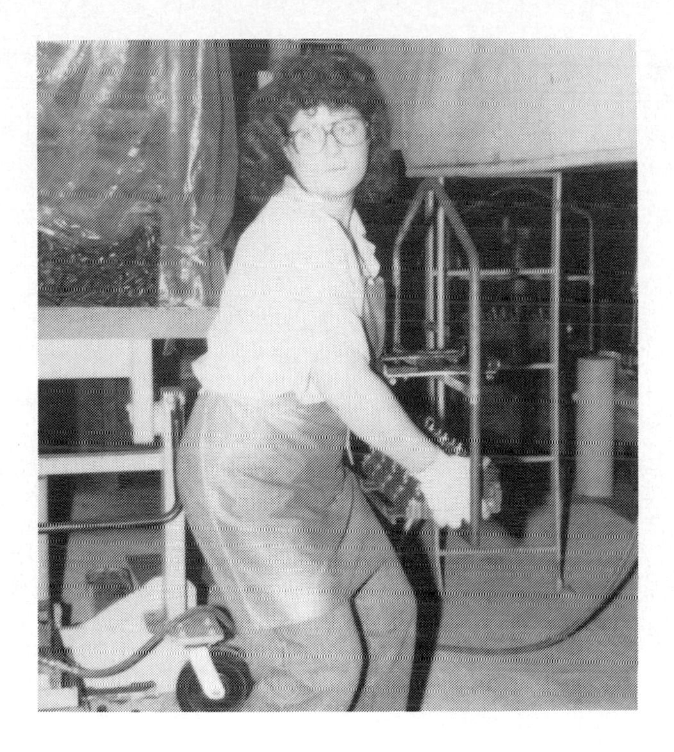

FIGURE 1

Operators must extend their arms over the line, pick the component from the carrier and place the Cylinder Head Assembly on the engine. Problems encountered by the operators were lower back, shoulder and wrist problems due to manual material handling especially in the later developed plants. Line rates varied in the plants from 150 to 180 pieces handled per hour.

In 1985 the latest developed plant investigated the feasibility of a Robotic application for installing the cylinder heads to the engine. In order to achieve this, both the conveyor and carriers had to be re-designed and installed. (Fig. 4) Re-design of the system to avoid manual material handling was initially considered an efficiency move rather than an Ergonomic consideration, but in the final analysis, injuries causing lost time, compensation and re-habilitation of workers was also a major consideration.

FIGURE 2

FIGURE 3

FIGURE 4

CONCLUSIONS

When projects are in the conceptual stages, along with engineering and design phases, Ergonomics should be a major consideration in order to eliminate potential interface problems with the working force. Additional capital expended initially to accommodate Ergonomic designs can potentially result in enlarged dividends.

Case in point, had Ergonomics been a consideration in the final development of the plant built in 1981, the probability of a robotic application for cylinder head installation would have been considered and the initial cost would have been far less than the cost to re-design the entire system (4) years later.

REFERENCES

(1) Clark, T.S., and Corlett, E.N., The Ergonomics of Workspaces and Machines, Taylor & Francis Ltd., London, 1984 P. 13-39.
(2) Grandjean, E., Fitting The Task To The Man - An Ergonomic Approach, Taylor & Francis Ltd., London, 1981, P. 46.
(3) Grandjean, E., Fitting The Task To The Man - An Ergonomic Approach, Taylor & Francis Ltd., London, 1981, P. 13.
(4) Alexander, David C., Pulat, B. Mustafa, Industrial Ergonomics A Practitioner's Guide, Industrial Engineering & Management Press, Atlanta, 1985, P. 58-60

Trends in Ergonomics/Human Factors V
F. Aghazadeh (Editor)

An Ergonomic Evaluation of a Steel Tube Furniture Manufacturer: Assembly Area Assessment and Recommendations

Lawrence J. H. Schulze and Jerome J. Congleton

Department of Industrial Engineering
Texas A&M University
College Station, Texas

A national manufacturer of classroom and office furniture, of the type remembered from our school days, processes almost all of its product components in-house. As with many industrial manufacturers, materials handling of component parts and finished products is associated with a disproportionately large share of the related worker compensation injuries. A comprehensive approach to operations evaluation was developed and implemented. The goal of the program to identify ergonomic problem areas within the manufacturing facility and to make recommendations to reduce or eliminate the identified problem areas.

1. INTRODUCTION

Inherent in many furniture manufacturing facilities is a hierarchy of many complex and integrated job-shop sub-organizations. The subject facility of this evaluation is no exception. Within this facility component parts for over fifty-two (52) separate end products (furniture types) are manufactured. Each component part is then transported to a designated area for assembly. Within and between each of the assembly areas, a disproportionate amount of manual materials handling and manipulation are required.

In order to determine and then examine the root cause(s) of the associated worker compensation injuries related to material handling, a phased effort was employed

2. METHOD

2.1 Injury Statistics

Analyses of injury statistics (both in-house and OSHA 200) were conducted to identify the areas within the manufacturing facility that were associated with the largest number of injuries. A summary of these data are presented in Figure 1. As a result of the injury data analyses, the assembly area of the plant was identified as being associated with the largest number of compensable injuries. Therefore, evaluation began with this area.

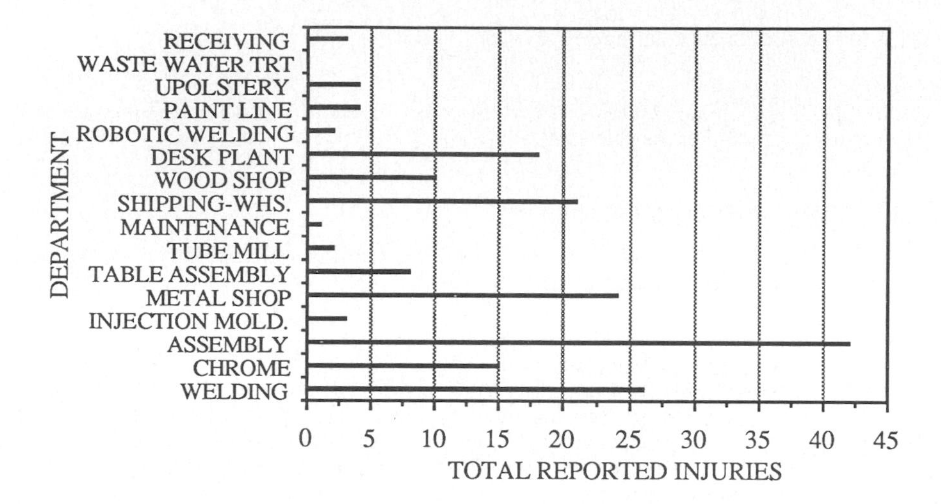

FIGURE 1. Number of OSHA 200 reported injuries by department.

2.2 Operations Process Analysis

The assembly operations were documented with both video tape and still photography. Operations process charts were developed to determine the basic operations required to manufacture (assemble) component parts into finished products. The results of the operations process analyses determined operations sources associated with worker injury probability.

2.3 Physiological Work Environment Analysis

2.3.1 Environmental Heat Stress

The ambient environment was analyzed regarding environmental heat stress. The *Wet Bulb Globe Temperature Index* (WBGT) was used to measure the environmental factors. The time-weighted average WBGT was determined for the measurements taken within the assembly area and were charted using the *Permissible Heat Exposure Threshold Limit Value* chart developed by the American Conference of Governmental Industrial Hygienists [1]. The average WBGT was determined by the following formula:

$$\text{Av. WBGT} = \frac{\text{WBGT}_1 \times t_1 + \text{WBGT}_2 \times t_2 + \ldots + \text{WBGT}_n \times t_n}{t_1 + t_2 + \ldots t_n}$$

2.3.2 Body Part Discomfort

Body part discomfort surveys were distributed in an attempt to document subjective assessments of the physiological consequences of required tasks. Past applications of this technique have indicated a direct relationship to both injury data and evaluation team observations and reviews of manufacturing procedures.

2.3.3 Manual Materials Handling Evaluation

The NIOSH Lifting Guide [2] was used as a guideline for evaluating material handling tasks in regard to Maximum Permissible Limits (MPL) and Action Limits (AL) for load lifting.

3. RESULTS and RECOMMENDATIONS

3.1 Operations Process Analysis

The results of the operations process analyses for the assembly area operation are presented in Tables 1 and 2, respectively. From Table 1 (chair assembly) it can be seen that 40% of the operations time is devoted to product manufacturing and packaging. Handling of component parts before assembly and product inspection accounted for 15% of the total time to manufacture a chair . Transportation of components and finished product plus delays inherent in the manufacturing area layout accounted for the remaining 45% of total time to manufacture a given chair component. Table 2 (student desk assembly) indicates similar results. For student desk products, 50% of the total manufacturing time was devoted to assembly of a given desk product. Handling of component parts and finished product accounted for 13% of total manufacturing time per unit, while transportation of parts and finished product accounted for 37% of the manufacturing time, identical to chair assembly.

TABLE 1. Operations process chart summary for chair assembly.

Symbol	Activity	%	Combined
○	ASSEMBLY	13.2%	39.8%
○	BANDING	26.6%	
⮚	HANDLING	6.7%	
⇨	TRANSPORT	37.4%	
□	INSPECTION	7.8%	
D	DELAY	8.3%	

TABLE 2. Operation process chart summary for student desk assembly.

Symbol	Activity	%	Combined
○	ASSEMBLY	45.0%	50.0%
○	PACKAGING	5.0%	
⮚	HANDLING	13.0%	
⇨	TRANSPORT	37.0%	

The time required to transport and handle both component parts and finished product were a result of the assembly area layout. Recommendations made included the application of industrial engineering principles to the layout of the assembly area to provide immediate access to component parts and elimination of the staging of product lotting before transport to the warehouse storage facility.

3.2 Physiological Work Environment

3.2.1 Environmental Heat Stress

The heat stress measurements were taken in August to represent a worst-case scenario. The result of the heat stress analyses conducted in both the chair and student desk assembly areas are presented in Figures 2 and 3, respectively. As can be seen from a review of Figure 3, both the clamping of seat pans and backs to the chair frames and preparation of seat pan and frame (DRILLING) as well as the quality control and packaging station were within the Permissible Heat Exposure Threshold Limit Value (PHE-TLV)for continuous work. However, the riveting operation (were seat pan is affixed to the chair frame) with a TWA WBGT of 81.75 and 390 kcal/Hr. work expenditure indicates that 75% work and 25% rest cycle is required based on the PHE-TLV.

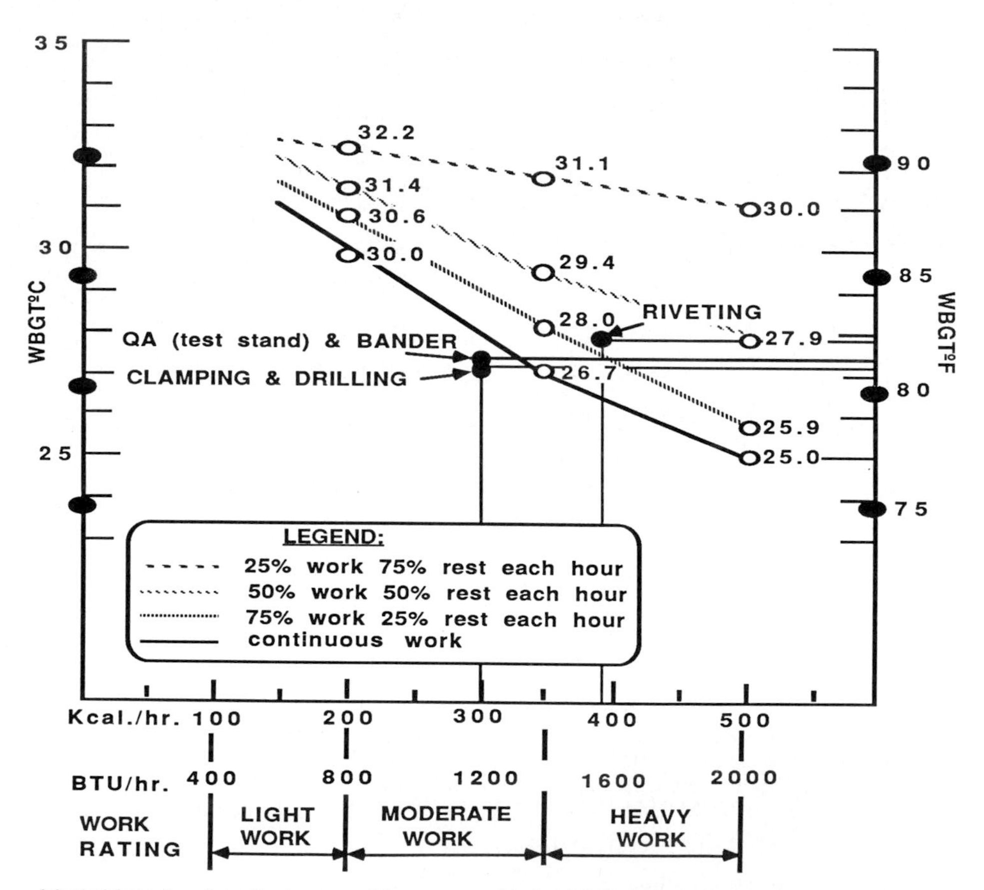

FIGURE 2. Environmental heat stress analysis evaluation results for the chair assembly area.

3.2.2 Body Part Discomfort

The result of the body part discomfort survey distributed in the assembly area are summarized in Figure 4. The primary body areas of worker discomfort were the shoulders, wrists, and legs. These results are consistent with the lifting, reaching, hammering, drilling, and transport operations prevalent in the assembly area operations. Secondary areas of discomfort were indicated as being the upper arm, lower arm and lower back for those workers involved in lifting and staging lots of packaged products.

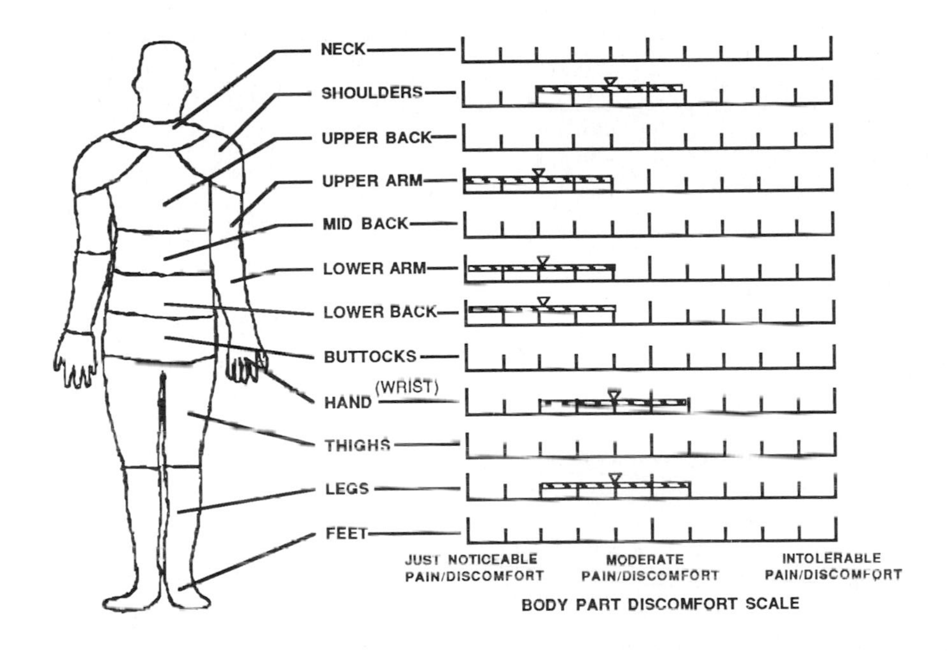

FIGURE 4. Body part discomfort survey summary for the assembly area.

Preliminary recommendations presented to alleviate body part discomfort included the design and implementation of an automated process to attach glides to chairs, re-orientation of the assembly work surfaces to eliminate extended reaches and multiple transportations, and provision of anti-fatigue mats where prolonged operator standing is required. Additional in-depth analyses are needed to facilitate redesign and upgrading of assembly workstations.

3.2.3 Manual Materials Handling

A representative sample of packaged products were evaluated following the guidelines presented in the NIOSH lifting guide for both *Action Limit* (AL) and *Maximum Permissible Limit* (MPL). Five products were chosen based on their weight and vertical location form start of lift. All products required a 24 inch horizontal location at the initiating point of lift.

Figure 3 presents the results of the heat stress analyses conducted in the desk (student) assembly area. As can be seen, the desk top preparation (drilling) for assembly operation results indicate a close proximity to the continuous work PHE-TLV. However, data points above a given PHE-TLV indicate the requirement to invoke the next highest work-rest cycle each hour. Therefore, a 75% work and 25% rest cycle per hour is indicated for table top preparation. the actual assembly operations (TWA WBGT 81.325, 360 kcal/Hr.) indicates that a 50% work and 50% rest cycle is required, following the guidance of the PHE-TLV chart.

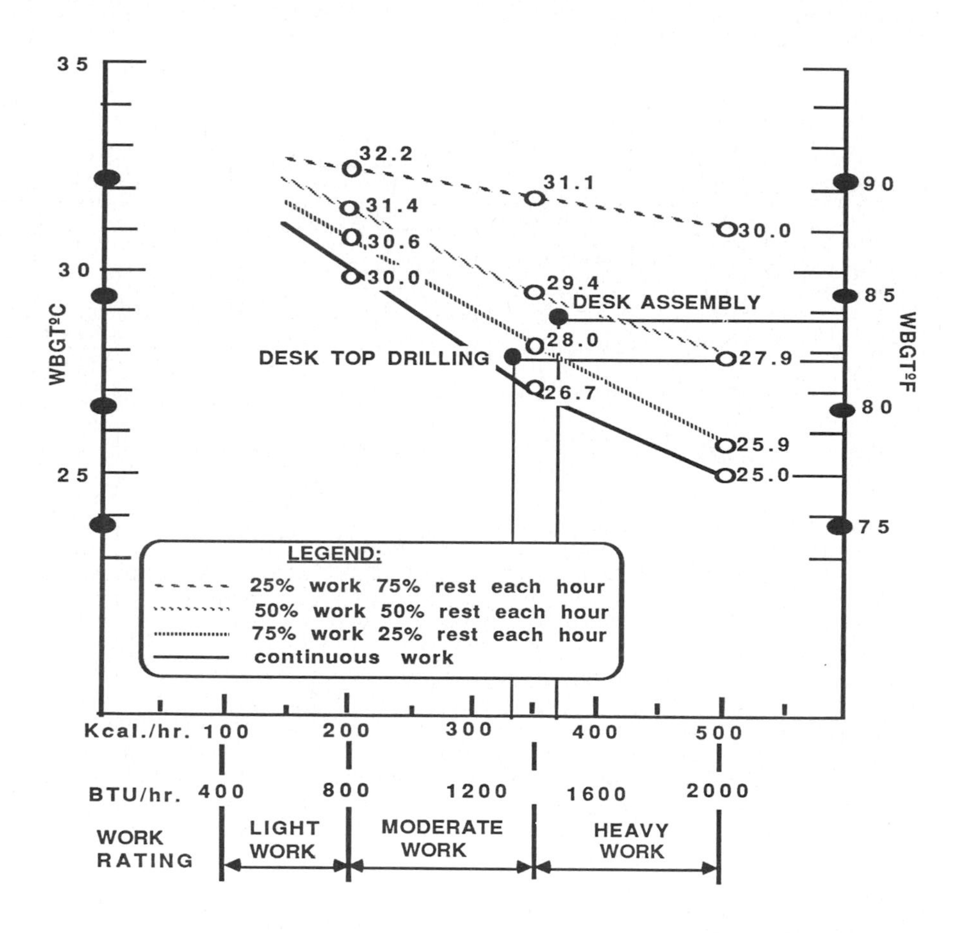

FIGURE 3. Environmental heat stress evaluation results for the student desk assembly area.

As a consequence of the findings of the environmental heat stress analyses, recommendations were made to adjust the work-rest cycles of the workers at risk as indicated by the PHE-TLV chart. This recommendation was presented as a temporary solution until engineering controls for reducing the TWA WBGT could be implemented and other engineering design solutions to reduce the work requirements of those workers most at risk investigated.

Results of the analyses conducted are presented in Table 3. All products evaluated either exceeded the AL and/or the MPL for the given combination of product weight, vertical and horizontal location , and nominal vertical distance traveled. In fact, most manual handling of final products involved lifts greater than the MPL. Recommendations were made to raise loads from the floor to knuckle height. Therefore, eliminating floor to conveyor height lifts. The installation of additional conveyors was also recommended to reduce the effort required for transportation of finished products. Thereby, eliminating potential compensable injuries.

TABLE 3. Summary of NIOSH lifting guide based analyses for the assembly area.

PRODUCT	WT lbs.	H LOC	V LOC	V DIST TRAV	FREQ OF LIFT	BAND & STACK	LIFT TO CONVEYOR
SERIES 0800	35	24	13.5	13.5	3/min	>AL (13) <MPL (39)	>MPL (34)
SERIES 9600	55	24	16.5	16.5	3/min	>MPL (36)	>MPL (36)
SERIES 7400	75	24	34	34	3/min	>MPL (34)	>MPL (36)
SERIES M457	95	24	33.5	33.5	3/min	>MPL (41)	>MPL (39)
CHAIRS	57	24	24	24	3/min	>MPL (35)	>MPL (34)

3.2.4 Employee Interviews and Evaluation Team Observations

Employee interviews were conducted to assess the practical validity of the assumptions made from the evaluation results. Interviews indicated that ambient temperature, equipment maintenance, facilities housekeeping, and assembly component logistics were important factors in getting the job done. Additional concern was expressedthat workers previously injured, returned to do work that was associated with their injury.

Evaluation team observations reiterated that equipment maintenance, facilities housekeeping, and assembly component logistics were operative factors in the time required to manufacture products and also related to documented compensable injuries. Additional observations noted that multiple transportations were required to move manufactured product to a lotting area and then to the conveyor for transport to the warehouse throughout an eight hour shift (first shift). Multiple transportations as described above were also required for the first 2-3 hours of the first shift to eliminate the queue created by the second shift. Warehouse operations did not support a second shift.

Equipment maintenance and facilities housekeeping programs were evaluated, revamped, and initiated to promote greater productivity, worker comfort, and safety. Recommendations regarding employee placement and moving second shift to first shift were presented and instituted.

4. CONCLUSIONS

Based on the analyses conducted by the evaluation team, a change in assembly area layout, the addition of conveyors to eliminate worker-at-risk lifts, and worker shift changes were warranted. Five assembly area reconfigurations (layouts) were developed based on peak production data and the flexibility requirements of the manufacturing facility. The first layout **(LAYOUT 1)** only added additional conveyors and associated package strappers to the current configuration to eliminate worker-at-risk lifts. The product flow was unchanged.The next two layouts (**LAYOUT 2** and **LAYOUT 3**) re-oriented the assembly lines to flow parallel with the center line of the assembly area, to facilitate ease of assembly component logistics and production flow. Layout 2 required additional conveyors and package strappers as in Layout 1. Layout 3 included additional conveyors for product line groupings. **LAYOUT 4** was an elaboration of Layout 1 by adding additional conveyors to eliminate worker-at-risk injuries due to manipulating the product queue created by the second shift. As with Layout 4, **LAYOUT 5** provided an elaboration of Layout 2 by adding additional conveyors to eliminate worker-at-risk lifts.

The feasibility of our preliminary recommendations were evaluated by costs/benefits analysis. The results of these analyses are presented in Table 4. The **COST** data was generated from the actual cost of the additional conveyors and support equipment; including installation. The **SAVINGS** data were generated from an average cost of a lift-associated injury and the elimination of time required to manipulate finished products before transport to the warehouse. **PAYBACK**was determined by the cost/benefits ratio. Shift change data was generated from the employee cost related to the creation and elimination of the second shift product queue and elimination of worker-at-risk lifts. The cost of adding a second shift to the warehouse was based on cost of additional employees, alone. Recommendations to implement **LAYOUT 2** and move second shift to first shift were presented. These recommendations are in the implementation phase at the time of this writing.

TABLE 4. Costs/benefits analysis results for the recommendations presented for the assembly area.

LAYOUT OPTION	COST $	SAVINGS $	PAYBACK
LAYOUT 1	34,950.00	76,000.00	(.46 YR)
LAYOUT 2	38,450.00	76,000.00	(.46 YR)
LAYOUT 3	35,000.00	86,800.00	(.44 YR)
LAYOUT 4	62,950.00	118,400.00	(.53 YR)
LAYOUT 5	79,450.00	118,400.00	(.67 YR)
2nd SHIFT TO 1st	---------	160,000.00	IMMEDIATE
ADD SHIFT (WHS)	81,600.00	---------	--------

This evaluation can be viewed as preliminary. In depth evaluations of each assembly workstation, toward optimizing and increasing productivity, as well as worker comfort and safety are currently in progress.

REFERENCES

[1] American Conference of Governmental Industrial Hygienists. (1983). *Threshold limit values for chemical substances and physical agents in the work environment*. (ISBN: 0-936712-45-7). Cincinnati, Ohio: Author.

[2] National Institute for Occupational Safety and Health. (1983). *A work practice guide for manual lifting*. (Technical Report No. 81-122). Cincinnati, Ohio: Author.

[3] Congleton, J. J., Ayoub, M. M. and Smith, J. L.. (1985). The design and evaluation of the neutral posture chair for surgeons. *Human Factors*. 27, 589-600.

Trends in Ergonomics/Human Factors V
F. Aghazadeh (Editor)

An Ergonomic Evaluation of a Steel Tube Furniture Manufacturer: Warehouse Assessment and Recommendations

Lawrence J. H. Schulze and Jerome J. Congleton

Department of Industrial Engineering
Texas A&M University
College Station, Texas

The warehouse operations of a national manufacturer of classroom and office furniture currently require intensive manual materials handling of packaged products during storage and order picking for shipping. As with industrial manufacturers that store and ship finished products from an on-site warehouse, materials handling is associated with a disproportionately large share of related worker compensation injuries. A comprehensive approach similar to that taken in the manufacturing facility was developed and implemented. The goal of the program was to identify ergonomic and operations problem areas within the warehouse and make recommendations to reduce or eliminate the identified problem areas.

1. INTRODUCTION

Finished products in the manufacturing facility are currently organized into lots before transport to the warehouse via a conveyor belt. As products arrive in the warehouse, they must be organized onto pallets and transported to storage locations. Transport of palletized products is conducted by means of fork-lift truck. Upon arrival at their destination storage area, the products are then manually placed in that storage location. Both boxes of variable weight and dimension as well as stacks of stacking chairs are stored in this manner.

When products are picked from storage for shipment, the majority of product manipulation and transport is conducted manually by workers using hand-trucks. large lots of a given product are transported by use of a fork-lift truck. However, the product is manually removed from storage and organized on a pallet for transport to the shipping dock.

In order to determine and then examine the source cause(s) of the associated worker compensation injuries related to the manual handling of stored product, a comprehensive approach similar to that taken in the manufacturing facility was developed and implemented.

2. METHOD

2.1 Injury Statistics

An analysis of injury data (both in-house and OSHA 200) identified the warehouse as being associated with the largest number of back related compensable injuries (Figure 1). In addition, the product storage effort in the warehouse was directly related to the output of the manufacturing facility. Therefore, evaluation of lifting and product manipulation operations within the warehouse were the main focus of this phase of our facility evaluation.

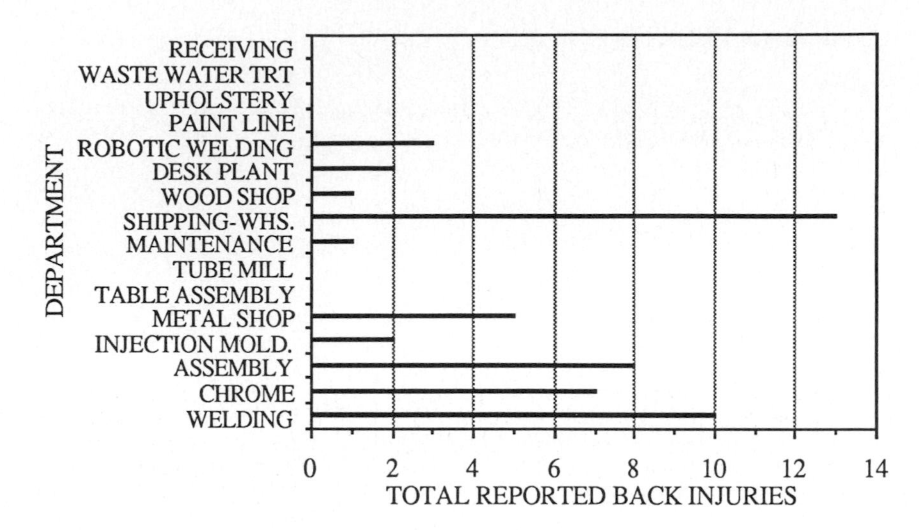

FIGURE 1. Total number of back injuries reported by department.

2.2 Operations Process Analysis

The product storage, order picking, and truck loading operations were documented with both video tape and still photography. Data was collected while following products through the storage, order picking, and truck loading operations. The video tapes were later evaluated and operations process charts were developed to determine the basic component operations required to store and retrieve products as well as load products for shipment. The results of the operations process analyses were also used to determine operations procedures associated with worker injury risk.

2.3 Physiological Work Environment Analysis

2.3.1 Environmental Heat Stress

The ambient environment was analyzed regarding environmental heat stress. The *Wet Bulb Globe Temperature Index* (WBGT) was used to measure the environmental factors. The *Time Weighted Average* (TWA) WBGT was determined for the measurements taken within the warehouse and shipping dock locations (including truck trailers). The results were charted using the *Permissible Heat Exposure Threshold Limit Value* chart developed by the American Conference of Governmental Industrial Hygienists [1].

2.3.2 Body Part Discomfort

Body part discomfort surveys [3] were distributed in an attempt to document subjective assessments of the physiological consequences of lifting and loading tasks. Past applications of this technique have indicated a direct relationship to both injury data and evaluation team observations and reviews of manual materials handling operations.

2.3.3 Manual Materials Handling Evaluation

The NIOSH Lifting Guide [2] was used as a guideline for evaluating material handling tasks in regard to *Maximum Permissible Limits* (MPL) and *Action Limits* (AL) for load lifting in the three operations.

3. RESULTS and RECOMMENDATIONS

3.1 Operations Process Analysis

The results of the operations process analyses conducted for product palletization, storage and retrieval operations are summarized in Tables 1, 2, and 3, respectively. As can be seen from Table 1, actual product manipulation accounted for only 28% of the time to load a complete pallet, delay (waiting for product to be in position to manipulate) and transportation of the product from the conveyor belt to the pallet accounts for 72% of the product palletization time.

The results presented in Tables 2, and 3 indicate that manual handling of products for the storage and retrieval operations required a disproportionately large number of transportations (and associated time) over mechanized methods of storage and retrieval. Consequently, mechanization (forklift trucks) of the storage and retrieval operations was recommended to reduce the work force engaged in these multiple transport operations. Thus, eliminating potential compensable injuries.

TABLE 1. Operations process evaluation summary for product palletization.

	HANDLING	16.8%
	LIFTING	11.5%
	DELAY	38.4%
	TRANSPORTATION	33.3%

TABLE 2. Operations process summary for product storage (MECH = Fork truck, EMP = Employee/operator).

	CHAIR STORAGE		BOX STORAGE	
	MECH	EMP	MECH	EMP
OPERATION HANDLING	28.6%	13.0%	25.0%	14.3%
TRANSPORT	50.0%	35.2%	62.5%	57.1%
STORAGE	21.4%	27.8%	12.5%	15.3%
DELAY	—	24.0%	—	13.3%
AVG TIME (min)	3:32	13:02	6:05	24:22

TABLE 3. Operations process summary for product retrieval for shipping (MECH = Fork truck, EMP = Employee/operator).

	CHAIRS		BOXED PROD.	
	MECH	EMP	MECH	EMP
TO STORAGE	33.3%	28.0%	41.3%	32.0%
OPERATION HANDLING	22.2%	32.0%	11.8%	22.0%
TO SHPNG DOCK	44.5%	40.0%	47.0%	46.0%
AVG TIME (min)	2:18	6:05	4:29	12:34

3.2 Physiological Work Environment

3.2.1 Environmental Heat Stress

The heat stress measurements were collected during August to represent a worst-case scenario. The results of the heat stress analyses conducted in both the warehouse and shipping dock area (including shipping trailers) is presented in Figure 2. As can be seen from a review of Figure 2, the palletization operation is associated with the highest work rate due to its continuous labor-intensive nature. The combination of TWA WBGT (81.75) and heavy work load (540 kcal/Hr.) indicates not only a 50% work and 50% rest cycle per hour but also that the combination is beyond the bounds of the PHE-TLV. Therefore, immediate rectification of current labor intensive operations by workers was required. A job rotation schedule was implemented as a temporary control until engineering controls could be designed and implemented to eliminate heat stress exposure.

Although not as labor intensive, both product storage (TWA WBGT 83.5, 390 kcal/Hr.) and order picking (TWA WBGT, 360 kcal/Hr.) operations are conducted in environments that indicate a 50% work and 50% rest cycle per hour is necessary for workers to safely perform required tasks. In conjunction with the Operations Process Analyses results, these data indicated that mechanization of product storage and retrieval operations would eliminate worker exposure to injury hazard, and has been implemented.

Truck loading operations (TWA WBGT 85.9, 500 kcal/Hr.), as can be seen from Figure 2, are conducted in environments exposing workers to both high temperature and heavy work. At the time of this evaluation, air circulation was minimal in truck trailers during loading operations due to the nature of the shipping trailers. Application of methods for mechanical circulation of air during loading operations is planned, as well as worker rotation to eliminate exposure to these heat stress conditions.

3.2.2 Body Part Discomfort

The results of the body part discomfort surveys distributed to workers in the warehouse are summarized in Figure 3. As can be seen, the primary areas of worker discomfort for the four warehouse operations (product palletization, storage, retrieval, and truck loading) were the mid and lower back. these results are consistent with the intensive lifting requirements of these operations. Secondary areas of discomfort included the upper back and lower arm as well as both the neck and shoulders. Again these results are consistent with intensive lifting

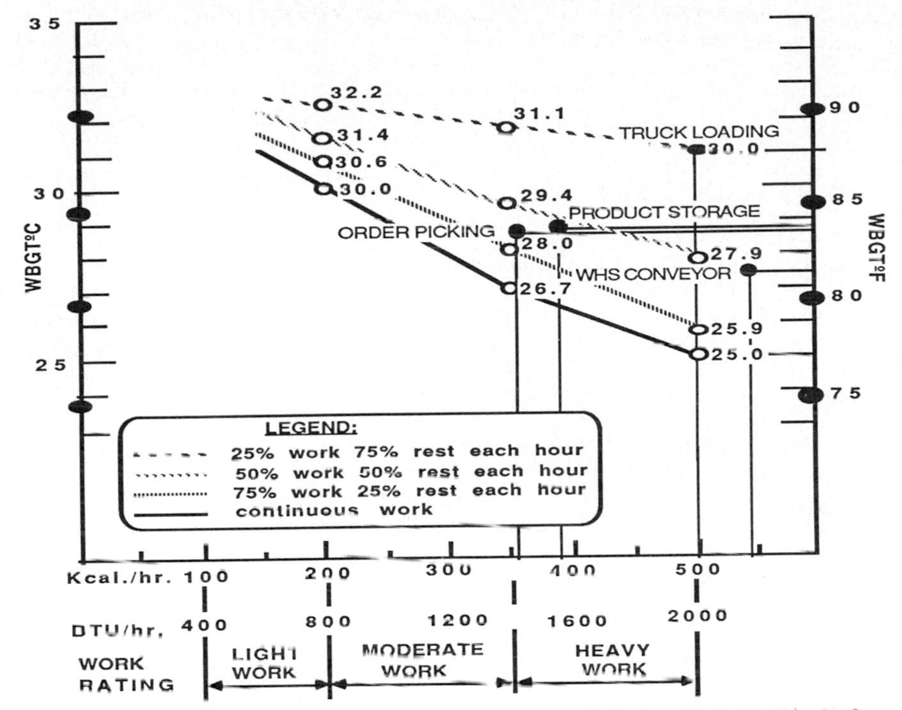

FIGURE 2. Environmental heat stress evaluation results for the warehouse operations.

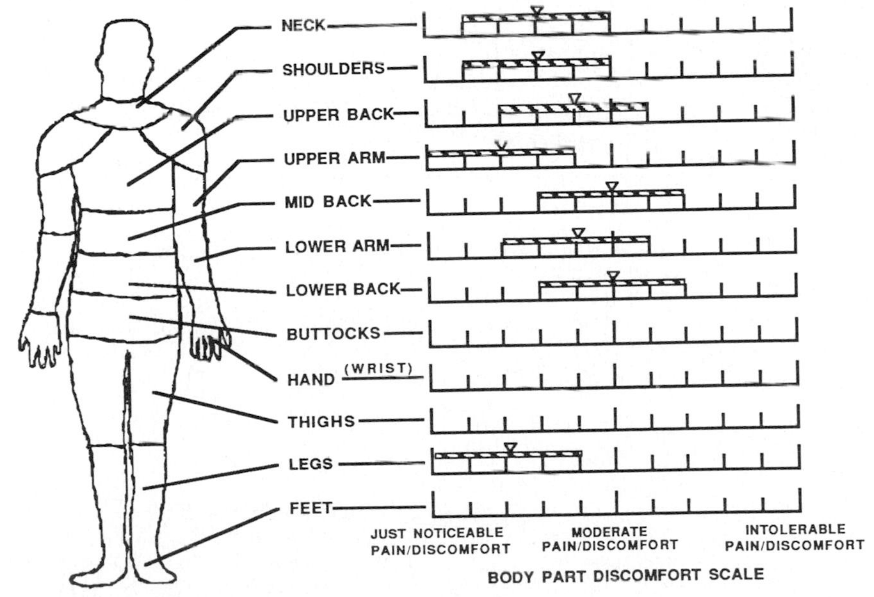

FIGURE 3. Summary of the body part discomfort survey results for the four warehouse operations.

activities. More tertiary discomfort was indicated for both the upper arm and legs. These results are consistent with the manual transport of product through the warehouse (1/7th mile long) by workers using hand trucks. In lieu of these findings, mechanization of product storage and retrieval and engineering assists for product palletization and truck loading were proposed.

3.2.3 Manual Materials Handling

A representative sample of packaged products were evaluated following the guidelines in the NIOSH lifting guide for both *Action Limit* (AL) and *Maximum Permissible Limit* (MPL). Five (5) products were chosen based on their weight, largest quantity of product type stored, and vertical location from start of lift. The results of the lifting analyses for product palletization and both product storage and retrieval (including truck loading) are presented in Tables 4 and 5, respectively.

As can bee seen from Table 4, all lifts engaged in by workers during product palletization were made from a vertical location of 32 inches; the height of the conveyor belt landing. The result of the analyses indicate that lifts of product weighing less than 60 pounds were greater than the AL but less than the MPL for and 8 hour shift. The two products weighing in excess of 60 lbs. (series 7400, 75 lbs., and series 9600, 95 lbs.) exposed workers to lifts greater than the MPL (64 and 65 lbs., respectively). Engineering assists to eliminate theses exposures are, therefore, needed.

TABLE 4. NIOSH lifting guide analysis results for the palletization operation.

PRODUCT	WT lbs.	H LOC	V LOC	V DIST TRAV	FREQ OF LIFT	PALLETTIZING
SERIES 0800	35	24	32	13.5	3/min	>AL (21) <MPL (64)
SERIES 9600	55	24	32	16.5	3/min	>AL (21) <MPL (64)
SERIES 7400	75	24	32	34	3/min	>MPL (64)
SERIES M457	95	24	32	33.5	3/min	>MPL (63)
CHAIRS	57	24	32	24	3/min	>AL (23) <MPL (69)

The results of Table 5, for product storage and retrieval operations (including truck loading), indicate that all lifts conducted in these operations exceed the MPL. These results further amplify the need for the mechanization of product storage and retrieval operations as well as the application of lifting aids to truck loading. Implementation of mechanized product storage and retrieval and application of lifting aids to both product palletization and truck loading are seen as direct means of eliminating exposures to potential compensable injuries.

TABLE 5. NIOSH lifting guide analysis results for the product storage and retrieval operations.

PRODUCT	WT lbs.	H LOC	V LOC	V DIST TRAV	FREQ OF LIFT	STORAGE AND RETRIEVAL
SERIES 0800	35	24	13.5	13.5	3/min	>MPL (34)
SERIES 9600	55	24	16.5	16.5	3/min	>MPL (36)
SERIES 7400	75	24	34	34	3/min	>MPL (36)
SERIES M457	95	24	33.5	33.5	3/min	>MPL (39)
CHAIRS	57	24	24	24	3/min	>MPL (37)

3.2.4 Employee Interviews and Evaluation Team Observations

Employee interviews were conducted to assess the practical validity of the assumptions made from the evaluation results. Employee interviews indicated that the ambient temperature (too hot in summer, too cold in winter, equipment maintenance, and facilities housekeeping were generally important factors in the ease of accomplishing assigned tasks. More important expressions of concern focused on the instability of current box and chair storage, excessive lifting of heavy products, transferring products from the shipping dock to trailers for shipping imposed slipping and falling hazards.

Evaluation team observations were made throughout the study. The observations reiterated the concern for ambient temperature, equipment maintenance, and facilities housekeeping. Additional observations also supported concern for the product storage procedures that were currently being followed. Excessive transportations during both product storage and retrieval operations were seen as associated with the same product type being stored in different locations within the warehouse and current storage procedure and storage space allocation. The inventory control procedures were noted as factors relating procedures and time to store, retrieve, and load products to documented compensable injuries.

Equipment maintenance and facilities housekeeping programs were evaluated and new procedures implemented. Additional fork-lifts were purchased to eliminate the worker-at-risk lifts directly associated with the disproportionately large number of compensable back injuries. The addition of the fork lift trucks has also substantially decreased the time to transport large numbers of product from storage to shipping. Alternate product storage and retrieval procedures are being evaluated for their application to the warehouse facility.

4. CONCLUSIONS

Based on the analyses conducted by the evaluation team, a recommended package of measures was presented for consideration and implementation. First, engineering assists, either in the form of pallet lifts or load lifting aids, were recommended for the **PALLETIZATION** area. Procurement of additional adjustable **HAND TRUCKS** to accommodate different types of products were recommended. Procurement of additional **FORK TRUCKS** was also recommended to eliminate worker-at-risk lifts and to substantially decrease time to store and retrieve products. The initiation of both **SAFETY PROCEDURES** and **OPERATIONS PROCEDURES** were recommended to eliminate hazard potential and streamline the warehouse operations. **HIGH DENSITY STORAGE**

RACKS were seen as a solution to the inventory control problems encountered during peak production and shipping times. Such storage racks would allow the dedication of product type location for ease of storage and retrieval and promote efficient use of available storage volume. Permanent installation of height adjustable ramps at the **SHIPPING DOCKS** was recommended to accommodate shipping trailers of different heights and eliminate slipping and falling hazards associated with portable non-height adjustable ramps.

The feasibility of the above preliminary recommendations were evaluated by costs/benefits analysis. The results of the analyses are presented in Table 6. The **COST** data was generated from the actual cost of the hardware and any associated installation costs. The **SAVINGS** data were generated from the average cost of injury , elimination of time required to perform excessive transportations, and man hours associated with storage and retrieval operations. **PAYBACK** was determined by the costs/benefits ratio.

TABLE 6. Costs/benefits analysis results for the recommendations presented for the warehouse.

OPTION	COST $	SAVINGS $	PAYBACK
PALLETIZATION	5,200.00	26,300.00	(.20 YR)
HAND TRUCKS	2,500.00	32,500.00	(.08 YR)
FORK TRUCKS	30,000.00	102,000.00	(.29 YR)
SAFETY PROCEDURES	----------	46,000.00	IMMEDIATE
OPERATIONS PROCEDURES	----------	82,800.00	IMMEDIATE
HIGH DENSITY STORAGE RACKS (INITIAL)	266,750.00	315,600.00	(.85 YR)
SHIPPING DOCKS	12,000.00	39,000.00	(.30 YR)
TOTAL	316,450.00	643,400.00	(.49 YR)

As with the assembly area evaluation, this evaluation can be viewed as preliminary. In depth evaluations of each of the warehouse operations, toward optimizing product storage, retrieval, and shipping, were recommended and are planned. At the time of this writing, the majority of the recommendations presented here are in the implementation phase.

REFERENCES

[1] American conference of Governmental Industrial Hygienists. (1983). *Threshold limit values for chemical substances and physical agents in the work environment.* (ISBN: 0-936712-45-7). Cincinnati, Ohio: Author.

[2] National Institute for Occupation Safety and Health. (1981). *A work practice guide for manual lifting.* (technical Report No. 81-122). Cincinnati, Ohio: Author.

[3] Congleton, J. J., Ayoub, M. M., and Smith, J. L.. (1985). The design and evaluation of the neutral posture chair for surgeons. *Human Factors*, 27, 589-600.

Trends in Ergonomics/Human Factors V
F. Aghazadeh (Editor)

An Ergonomic Investigation of Railroad Yard Worker Tasks

S.R. Kuciemba, G.B. Page, and C.J. Kerk*

Research and Test Department Safety Division
Association of American Railroads, Washington, D.C. 20001

The railroad industry in the United States incurs about 12,000 lost time injuries per year (Federal Railroad Administration,1987) costing $553 million in injury claims & suits (Lutz, 1987). Railroad yard workers suffer about 16.1% of the industry's lost time injuries. But they only account for 8% of the total man-hours worked (McMahan, Page, & Kuciemba 1987). We performed an ergonomic investigation of railroad yard worker activities to help identify which tasks are most stressful, estimate how often they are performed, identify the types of injuries that occur, and identify conditions contributing to injury. The investigation included: reviewing the local railroads' injury reports, interviewing eight yard crews (including a subjective difficulty survey), and observing the yard workers performing their tasks. Our findings suggest that adjusting railroad car drawbars (a coupling task) and throwing under-maintained hand switches are the most problematic tasks. Further research is warranted on the strength required and biomechanical stresses produced while aligning drawbars and throwing switches under various conditions.

1.0 INTRODUCTION

Lost-day injuries in the United States railroad industry have decreased from 38,470 in 1979 to 12,717 in 1986, a 67% decline (Federal Railroad Administration, 1987). Concurrently, the lost-day injury rate fell 41% while the severity rate decreased 30% over this period (McMahan, Page, & Kuciemba 1987). But injury claims and suits increased almost 28% (Lutz, 1987). Figure 1 shows these two trends.

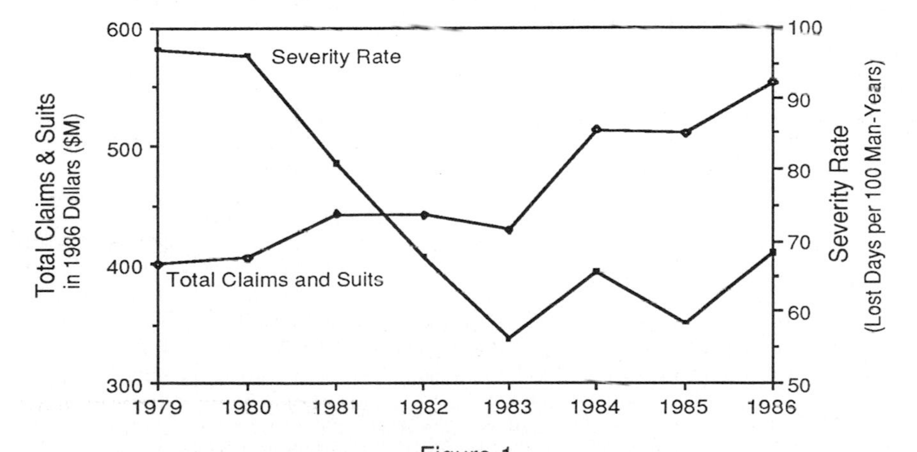

Figure 1
Industry trends: Annual claims & suits and severlty rates from 1979 to 1986

* Center for Ergonomics, The University of Michigan, Ann Arbor, MI 48109

Individual jobs with similar tasks are grouped into craft categories to simplify the analysis of injury statistics. Four of the twenty four craft categories accounted for over 57% of the lost day injuries. They also accounted for over 60% of the lost days, and over 60% of the total estimated claims & suits from 1979 to 1986. Conversely, the four crafts accounted for only 35% of the total man-hours worked. Table 1 describes the individual job titles associated with each of the four crafts. It also summarizes the relevant injury statistics for the four crafts and compares them with the total industry statistics.

Of the four crafts, yard workers have the highest lost day injury and severity rates. And they are also a sizable part of the industry's work force (over 8% of man-hours worked). This led us to our study of railroad yardworkers (herein called yardmen).

Table 1

Percent lost day injuries & man-hours, lost day injury & severity rate, and claims & suits by Craft for 1979 - 1986 (injury and severity rates describe injuries and lost days per 100 man-yrs, respectively)

Craft	Job Titles	Percent of Total Lost Day Injuries	Percent of Total Man Hours	Lost Day Injury Rate	Severity Rate	Estimated Claims & Suits ($M)
YARDMEN	Switchtenders, car retarder operators, ground service employees yard conductors, foremen, brakemen,& helpers	16.1%	8.1%	10.7	183.7	80.9
TRACKMEN	Track gang foremen, gang & section foremen, extra gang laborers, and section laborers	17.8%	9.3%	9.6	134.4	73.2
ROAD TRAINMEN	Road freight conductors, flagmen, and brakemen	15.5%	10.7%	7.2	131.3	82.7
CARMEN	Carmen	8.3%	7.3%	5.8	101.8	42.1
TOTAL INDUSTRY	All crafts including those above	100%	100%	5.0	79.1	553.0

Yardmen perform a variety of tasks. But for this study, we looked at the six tasks associated with the highest number of injuries and lost days. In addition to walking in the yard and other tasks, they perform the following tasks:

- *mounting and dismounting cars* — these two tasks describe getting on and off still or moving rail cars. Ladders are located on the sides and ends of most cars.

- *coupling and uncoupling cars* — these two tasks are unfortunately grouped in the injury statistics, but are separate tasks. Cars are coupled together using knuckles on the end of a bar called the drawbar. To couple two rail cars, one is sent rolling into the other, and the knuckles lock into each other. Often the knuckles do not interlock, and the drawbars must be manually realigned (called adjusting the drawbar). Uncoupling two cars is performed when both cars are moving. A release lever is pulled, lifting the pin that holds the knuckles together. This is also called pin pulling.

- *throwing hand switches* — this refers to the task of manually moving a hand switch lever to alter track alignment.
- *setting and releasing hand brakes* — again, these two different tasks are grouped together in injury statistics. Setting a hand brake can slow down a moving car and prevent a stationary car from moving. The hand brakes are mounted on the cars, and generally require a yardman to mount the car to operate them. Once on the car, the worker turns a wheel or pumps a ratchet-lever to set the brake. To release, the worker either turns a wheel or releases a lever.

Figure 2 shows the percent of lost day injuries and percent of lost days for yardmen performing these tasks. As the figure shows, each of the tasks similarly account for over 15 % of all lost days and lost day injuries to yardmen, except setting and releasing hand brakes (below 8%).

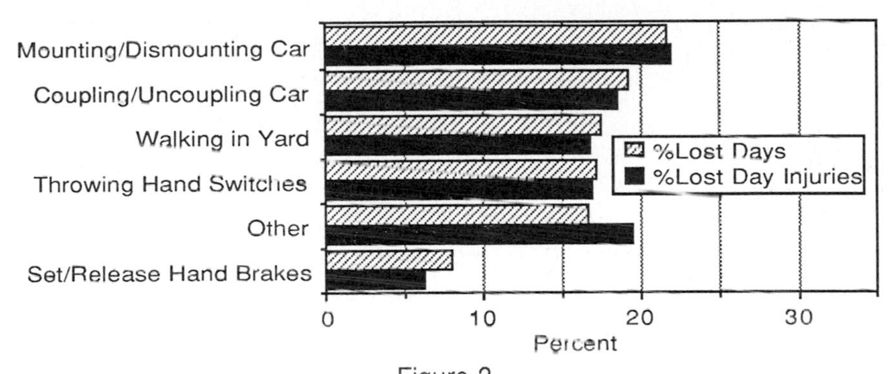

Figure 2
Percent of lost day injuries and lost days by task between 1979 and 1986 for Yardmen (AAR, 1987)

Given the large number of yardmen and their high injury and severity rates, we focussed our investigation upon the tasks they perform. This investigation intends to define where more detailed research efforts should be applied to develop remedies for reducing yardmens' risk of injury.

2.0 METHODS

We performed our investigation at four different railroad yards in the midwest—two yards for each of two different railroads. The investigation included:

(1) reviewing the injury reports of yardmen for injuries that occurred over the last 5 months or more if available,
(2) interviewing the yardmen which included using a subjective difficulty and frequency survey to rate their tasks, and
(3) observing and documenting (using a video camera) the yardmen performing their tasks.

Injury reports were requested from the safety officers of each yard. This data was collected to help provide further detail as to the cause of injury and injury type common to these yardmen. The reports were also collected to help verify national injury trends or identify discrepancies.

We interviewed eight yard crews during their work breaks. Each crew consisted of three or four yard workers. Railroad management suggested that we interview the workers in groups rather than individually to minimize distractions from the normal work routine. Safety officers and members of management were present during the 45 minute interviews. But they did not participate.

In each interview, we asked workers to:

- estimate how frequently they perform the following tasks: switch throwing, mounting a car, dismounting a car, coupling cars, uncoupling cars, setting and releasing hand brakes, and any other task they often perform,
- rate the difficulty or stressfulness of performing each task,
- identify the types of injuries associated with each task, and
- define factors contributing to injury.

The subjective difficulty rating system was adapted from Corlett and Bishops' (1976) subjective discomfort survey and Wiker's (1986) subjective physical discomfort and muscle fatigue survey. We could not perform a subjective discomfort survey itself for we could not interview workers individually. We also could not control:

- the time from task performance to the time we would ask the workers to rate the task (allowing recovery time from muscle fatigue to vary),
- when we surveyed workers within their shift (allowing the effects of cumulative task performance upon muscle fatigue to vary),
- the yardmens' work content before they responded to the survey (allowing the discomfort associated with each task to confound one another).

So we asked the workers to rate each task by the level of difficulty in performing the task once. This was done to help alleviate the confounding effects of shift duration, time within shift when surveyed, pace of work, and content of work upon the difficulty rating. The difficulty scale ranged from 1 (very easy) to 5 (very difficult). If a difficulty rating varied depending on the equipment type used for a specific task, we asked the yardmen to provide separate difficulty ratings for each equipment design.

We followed this by asking the yardmen how often they perform each task within an average 8-hour day. If frequency varied by job classification among the yardmen, workers from each job classification provided separate frequency estimates.

3.0 RESULTS

3.1 Local injury statistics

Only five months of data were available from the two railroads. During this period, 41 injuries occurred. Local records grouped coupling and uncoupling cars, as did national records. Table 2 summarizes the local injury statistics by the task being performed at the time of injury. Coupling operations (including uncoupling) and walking in the yard were associated with the highest number of injuries to yardmen—together they were responsible for 41%. Throwing hand switches and mounting & dismounting cars together accounted for 10% of the injuries.

Low-back pain accounted for 61% of all lost work days for both railroads. Leg/Foot injuries were responsible for another 20% of the lost work days.

Table 2
Summary of Local Injury Statistics by Task for Yardmen

Task	% Total Injuries	% Total Lost Days	Typical Injury Types	Contributing Factors
Throwing Switches	4.8	5.0	Back Sprain	Poor maintenance
Mount/Dismount Cars	4.8	4.0	Lower Ext Injuries Back Sprain	Slips,Falls
Couple (aligning Drawbar) Uncouple (pin pulling)	24.6	29.0	Back Sprain Lower Ext Injuries Back Sprain Lower Ext Injuries	Stubborn drawbar Pin fell out of knuckle Stubborn Pins Slips,Falls
Set/Release Brake	0.0	0.0		
Walking in Yard	17.0	34.6	Lower Ext Injuries Back Sprain	Poor yard maintenance
Other	48.8	27.4	Eye Irritation Lower Ext Injuries	Debris/Dust in yard Slips,Falls away from yard

3.2 Interviews

Figure 3 shows the estimated frequency of tasks performed in an average day. These values are for a yard crew of three workers (i.e., the whole crew performs this amount of work for an average day - individual crew members may perform these tasks at different frequencies).

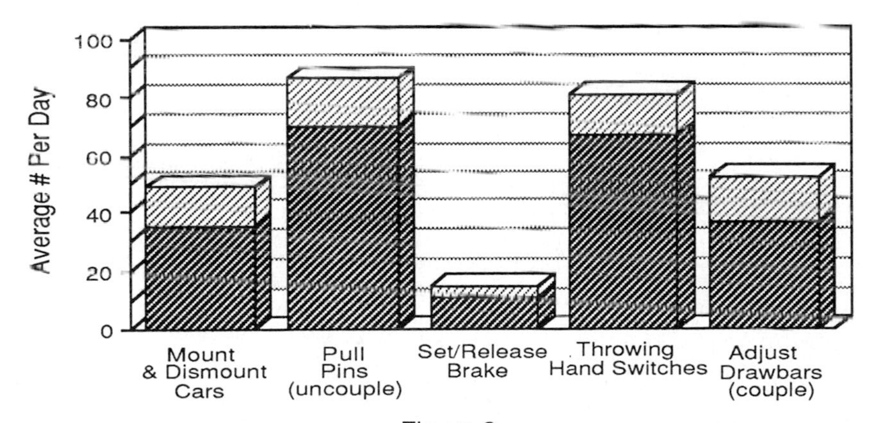

Figure 3
Average frequency of tasks performed on an average day (note: the top region represents the 95th percent confidence interval)

The estimated frequency was highest for uncoupling (pulling pins) with an average of 70 pins pulled per day. Throwing switches followed with an average of 67 switches thrown per day. Workers estimated that switch throwing and uncoupling are performed twice as often as coupling (adjusting drawbars) and mounting/dismounting cars. Setting and releasing hand brakes is performed the least often with an average of 11 brakes set/released per day.

Figure 4 shows the mean difficulty rating for the tasks performed. The difficulty scale ran from 1 (very easy) to 5 (very difficult). Adjusting Drawbars (a coupling task) was split by the workers into two groups by the length of the drawbar, *long* and *short*. The longer the car, the longer the drawbar, and the longer drawbars have a higher chance of mis-aligning (frequency estimates for alignment of either short or long drawbars were not available). Also split into two groups by the workers was throwing switches. Apparently there is a large difference in throwing switches dependant upon level of maintenance. Frequency estimates for throwing either poorly maintained or well maintained switches were not available.

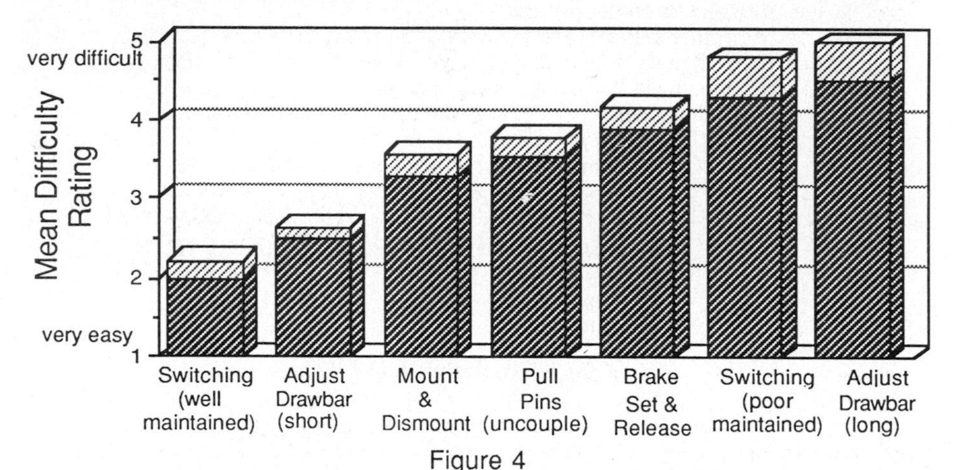

Figure 4
Mean perceived difficulty rating for tasks performed by yardmen (note: the top region represents the 95th percent confidence interval)

Adjusting long drawbars had the highest mean difficulty rating, with a mean of 4.5. Poorly maintained switches follow in difficulty at 4.2. Setting/releasing hand brakes, pulling pins, and mounting/dismounting cars follow in decreasing difficulty (or increasing ease). Adjusting short drawbars and throwing well maintained switches had the lowest mean difficulty rating (means of 2.5, and 2.0 respectively).

Typical injury types associated with tasks are listed in Table 3 along with their contributing factors. Low back pain and lower extremity injuries were the most frequently mentioned injury types. Poor yard and equipment maintenance were the most commonly listed contributing factors to injury. Inclement weather and poor equipment design were also mentioned.

Table 3
Summary of Typical Injury Types and Contributing Factors obtained through Yard Crew Interviews

Task	Typical Injury Types	Contributing Factors
Throwing Switches	Back Sprains, Upper Ext Strains, Knee/Foot Injuries, Hernia	Poor Switch Maintenance, Poor Weather, Age, Defective Equip
Mount/Dismount Cars	Low Ext Injuries, Bruised Knees, Upper Ext Strains, Back Sprains	Poor Yard Maintenance, Speed of Train, Slips/Falls
Coupling (adjusting Drawbar)	Back Sprain, Upper Ext & Low Ext Strain/Sprain, Groin/Stomach Pain	Stuck/Stubborn Drawbars, Pins Missing
Uncoupling (Pulling Pins)	Low Ext Injuries, Upper Ext Sprains, Finger/Hand Injuries	Slips/Falls, Poor Yard Maintenance, Stuck Pins
Set/Release Brake	Back Sprain, Upper Ext Strain	Stuck Brake Wheels

3.3 Site Visits

By observing yardmen performing their tasks and reviewing video documentation of their activities, we found:

- Factors affecting the throwing of switches - sand in switch points and large rock ballast hinders switch throwing; latched ground throw hand switches restrict throwing posture.
- Factors affecting mounting/dismounting cars - inability to see the bottom step and limited grab-irons on some cars increase difficulty; current dismounting methods are prone to ankle sprains.
- Simultaneous demands of pin pulling are numerous - pin pulling itself; timing of slack action; holding a radio and switch list while pulling release lever; walking between cars, over timbers, on ballast, and avoiding switches; and poor visibility during night or poor weather conditions.
- Adjusting drawbars requires excessive lifting, pushing, and pulling - the drawbar weighs up to 400 lbs; several different adjustment methods are used; adjustment often requires 3 to 5 workers; workers sometimes use ad hoc assists.

4.0 DISCUSSION

The subjective difficulty ratings and frequency estimates suggest that adjusting drawbars and throwing under-maintained hand switches are the most problematic yard tasks. Both tasks were rated as very difficult to perform. And yard workers appear to perform them often. Although, of those hand switches that yardmen throw, it is not known how many are under-maintained. The same is true with long drawbars. Yet the yard workers claim that they have to adjust long drawbars more often than short ones.

Both national and local injury statistics show that coupling and uncoupling tasks (including drawbar adjustment) account for a large portion of the lost day injuries experienced by yardmen, 19% and 25% respectively. This suggests that the drawbars workers adjust are indeed the more difficult, long drawbars. But coupling and uncoupling cars consists of several tasks. Thus, the portion of injuries attributed to adjusting drawbars or pulling pins (uncoupling) is not known.

Interestingly, there was a large discrepancy between the national and local injury statistics associated the throwing hand switches. The four yards we investigated only attributed 4.8% of their lost day injuries to this task, while national statistics suggest they account for 17% of lost day injuries. Therefore, hand switches at these yards may be better maintained than those in the industry as a whole. The local injury statistics also conflict with the reported perceived difficulty of under-maintained switches and the frequency at which yard workers throw hand switches. This discrepancy may suggest that the switches at these yards were well maintained and easy to throw too (the lowest difficulty rating reported). In addition, both discrepancies may be due in part to the short period of time under which injury reports were collected and the small workforce size analyzed.

Torso over-exertion injuries were reported (in local injury reports and the interviews) as the most common type of injury for both throwing hand switches and adjusting drawbars. Long draw bars weigh about 400 lbs. Their adjustment often requires pushing, pulling or lifting it horizontally a foot or more—sometimes requiring 3 to 5 workers to perform. Throwing a hand switch requires extreme flexion of the torso. But it can also require forceful twisting and lateral bending of the back which can significantly increase the risk of low-back pain (Punett et al., 1986).

Setting and releasing hand brakes were reported as difficult to perform, but they are rarely performed. National (Figure 2) and local (Table 2) injury statistics verify this. In fact, no injuries associated with operating hand brakes were reported at any of the yards for the 5 month period.

National injury statistics state that mounting and dismounting cars account for the most lost day injuries among yard workers (22%). But local injury statistics suggest otherwise, reporting that they account for only 4.8% of all lost day injuries. The local injury statistics correspond better to the moderate difficulty rating yard workers assigned to mounting and dismounting. Yard workers often stated that they slipped and fell while mounting or dismounting, suffering strains and sprains of the torso and lower extremities as a result.

Unfortunately the significance of the difficulty ratings and frequency estimates was limited by the small sample size of interviews performed. The large confidence intervals for both of these measures illustrate this. Furthermore, group bias and management's presence may have reduced the validity of our interview findings. And the perceived difficulty measure may have affected perceived frequency. Workers may overestimate the frequency they perform difficult tasks because of the tasks high level of difficulty. Estimated perceived frequencies could be verified through work sampling.

The subjective difficulty survey can be a useful method for task evaluation. It may solicite the worker's response to several types of stress, such as, perceived strength required, discomfort, localized muscle fatigue, whole-body fatigue, and required balance. But this versatility can also be a drawback to its use because of the difficulty in controlling for a specific stress response. Thus, one must determine what makes the task "difficult."

Our findings warrant further research on the causes of injury associated with hand switch throwing and drawbar adjustment. Since sprains and strains of the torso are the most common type of injury experienced while performing these tasks, it may be useful to investigate the stress imparted upon the low-back while performing these two tasks. The strength required to perform these tasks could also be investigated. These two measures could evaluate the effects of method, maintenance, and adjustment—leading to applicable remedies for reducing yard workers' risk of injury.

REFERENCES

Corlett, E.N. & Bishop, R.P. (1976). A technique for assessing postural discomfort. Ergonomics, 19(2), 175-182.

Federal Railroad Administration. (1987). Accident/incident bulletin: No. 155 calendar year 1986. Washington, DC: U.S. Department of Labor.

Lutz, L. (1987). Report of claim & litigation experience (personal injury & death) for 1986. Washington, DC: Association of American Railroads, Claims Research and Analysis.

McMahan, P.B., Page, G.P., & Kuciemba, S.R. (1987). Analysis of FRA casualty data: 1979-1986. Washington, DC: Association of American Railroads, Department of Safety Research.

Punnet, L., Fine, L.J., Keyserling, W.M., Herrin, G.D., & Chaffin, D.B. (1986). The health effects of non-neutral trunk and shoulder postures (Technical report). Ann Arbor: The University of Michigan, Center for Ergonomics.

Wiker, S.F. (1986). Effects of relative hand location upon human movement time and fatigue (Unpublished Doctoral Dissertation). The University of Michigan, Ann Arbor.

Trends in Ergonomics/Human Factors V
F. Aghazadeh (Editor)

WORKLOAD REDUCTION FOR DRIVERS OF HEAVY SHOVEL LOADERS IN UNDERGROUND MINING

D. LORENZ, W.F. MUNTZINGER, M. DANGELMAIER

Fraunhofer-Institut für Arbeitswirtschaft und Organisation
Stuttgart, F.R.G.

A field study reveals that the drivers of heavy shovel loaders in salt mines are subjected to high-level stresses by the stressors noise, vibration, climate and forced posture. The measured results are used for rating the three main characteristics of an ergonomically improved cab. A closed cab and a vibration-reducing cab suspension prove to be promising design features to reduce driver´s workload.

INTRODUCTION

In underground salt and ore mining, large-size shovel loaders with a payload of up to 20 t are used to pick up the mined material, to transport it over short distances and to make it available for further transport.

On the one hand, the working conditions of the driver are characterized by high stresses and strains resulting from the environmental factors - noise, vibration, climate, dust and exhaust gases - and, on the other hand, the driver is sitting in this vehicle in a transverse position to the vehicle´s longitudinal axis, as the vehicle moves along the same distances in forward and backward directions. This leads to a forced posture, because of the required rotation of the head.

The main characteristics of an improved cab, reducing stress and strain induced by these factors are :

- a closed cab (noise, climate, dust, exhaust gases),
- a swivelling system for turning the cab around its vertical axis (postural stress),
- a vibration-reducing cab suspension (vibration).

To permit the potential ergonomic usefulness of these measures to be estimated, it is necessary to measure the stresses. A field study concerned with noise, climate and whole-body vibration will be described in the following. The study was conducted in a salt mine approx. 700m below ground under realistic working conditions. The total weight of the shovel loader was 48 t at a payload of 12 t.

METHODS

The production cycle is used as a rating unit for the vibration and noise measurements. As there are no significant changes in the test route during the measuring period, the cycles are regarded as equivalent and the arithmetic mean of the stress characteristics are used to increase the confidence level. The average duration of the production cycles is 4 min., a single driving distance is approx. 190 m.

To obtain the vibrational stress, vibration accelerations are measured in three axes by the means of three piezoresitive accerelation sensors mounted on a cube. The axes can be seen in Figure 1. Because of the driver´s sitting position in transverse direction to the vehicle´s longitudinal axis, "y" indicates the forward driving direction. The measured signals pass through the measuring brigde and the pre-amplifier to the analog input of the recorder, in which the data is stored in pulse code modulated form on a video tape. The pre-amplifier and the recorder are accommodated in a protective box mounted on the vehicle. The test set is supplied with voltage from the vehicle electrical system. The individual phases of the production cycle (no-load run, loading, load run, unloading) are marked by the accompanying test leader on a fourth channel of the recorder with a d.c. voltage, manually switchable in steps.

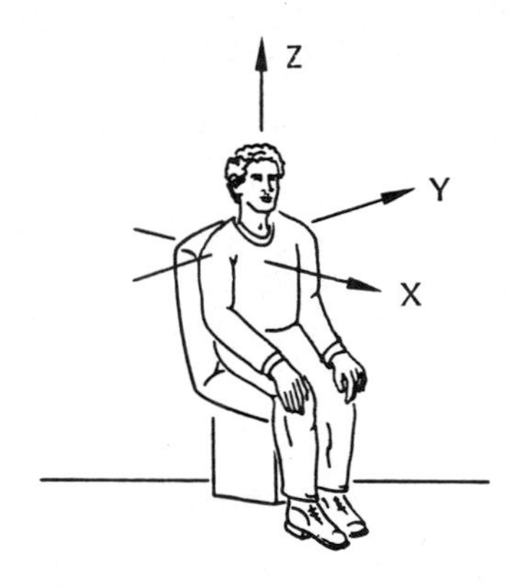

FIGURE 1
Coordinates for Vibration Measurements

In the evaluation above ground, the signals are reproduced in a real-time manner. They are supplied to the evaluating computer via the weighting filter and evaluated with a vibration-analysis program.

The K-value according to the VDI specification 2057 [1] commonly used in Germany is used as a stress characteristic. It is determined as a frequency-weighted and non-dimensional root mean square, with the averaging period extending over a complete phase or the total production cycle. The K-value is assigned the respective maximum permissible exposure time, which corresponds to the exposure limit (health and safety) in using the weighting mothod according to ISO 2631 [2].

The average stress characteristic K of six production cycles is used as a rating parameter of the vibrational stress.

The vibration measurements are immediately followed by the noise measurements. The noise to which the driver is subjected is picked up by a microphone fitted to the driver´s helmet. This microphone picks up the sound pressure at a distance of 10 cm from the driver´s right ear. A battery-operated special tape recorder for noise measurements is used for the A-weighted recording of the signals. Four cycles are recorded without interruption.

The data obtained is evaluated above ground with the same computer by means of standard software. The production cycle is divided into phases (no-load run backward, no-load run forward, loading, load run backward, load run forward, unloading) during playback of the tape, based on the driver´s announcements during the measuring period and on defined sound events in the cycle.

Energy-equivalent averaging levels are used as a stress characteristic, i.e., the levels in dB of the A-weighted, mean squared sound pressure (EEL), with the averaging period again extending over an entire phase or production cycle. The arithmetic mean of the EEL is used as a rating parameter.

For the assessment of the climatic stress, orientation measurements of the micro-climate on the loader and of the environmental climate were made in the same mine at an earlier test date with a "Hygrophil" (dry bulb temperature, wet bulb temperature, relative humidity) and a windmill-type anemomotor (air velocity); the contact temperatures which exist at the face were also measured. Both the Base Effective Temperature (BET) [3] and the Heat Stress Index (HSI) [4] for micro-climate and mine climate were estimated from that data.

RESULTS

The results obtained in the vibration measurements for the evaluation according to phases are summarized in Figure 2. A relatively low stress is present in the vertical direction (z), as a vibration-reducing seat is built in. The highest vibration to which the driver is subjected occurs during the rapid no-load run over the rough surface. Quieter riding is experienced during the load run. As expected, the highest value in the vehicle´s longitudinal axis (y) is obtained during loading, due to the excitation of the shovel. In all three coordinate directions, the lowest stress is present during the short unloading phase.

The mean rating parameters for the total cycle are reproduced in Table 1. It is interesting to see that the highest stress is clearly present in the vehicle´s transverse direction (x).

If this characteristic is used as a rating variable, the maximum permissible exposure time will be 3 hours; it will be as short as 1.5 hours, if K´ ("vector length

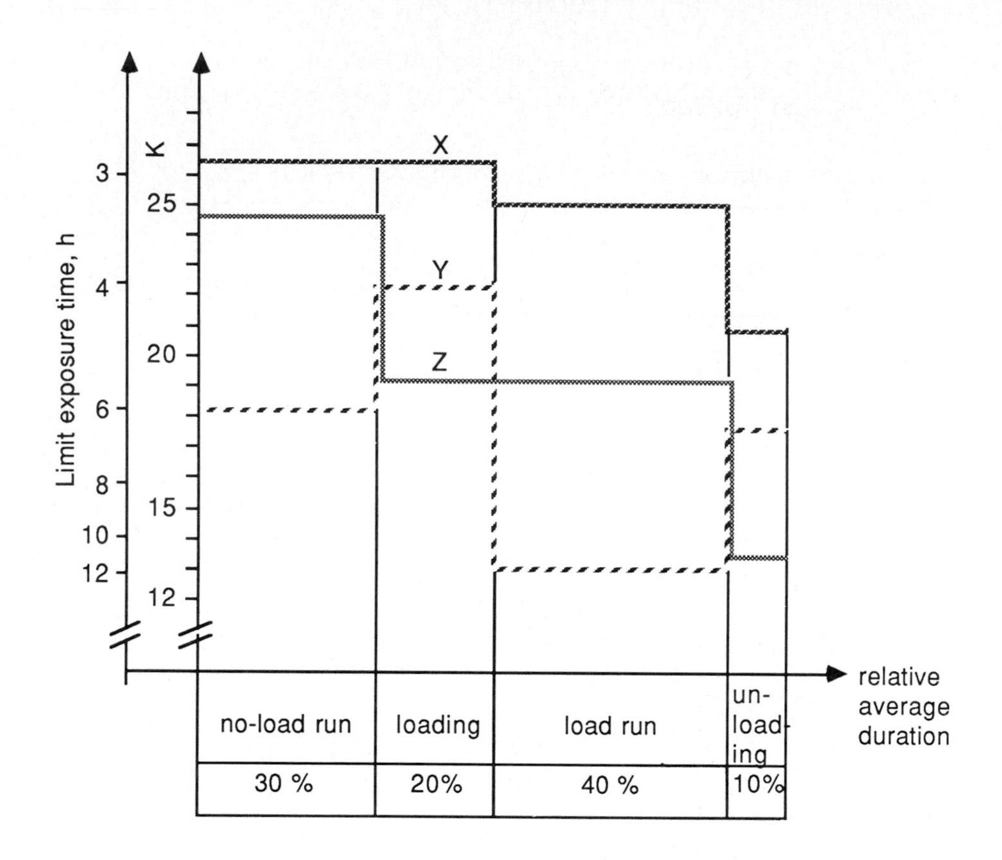

FIGURE 2
Average K-Values of the Phases of 6 Production Cycles

characteristic", see Table 1) is used as a criterion. Six hours (K=18) per shift, however, are to be demanded.

Analogous to Figure 2, the shape of the energy-equivalent sound pressure level is represented in Figure 3. Typically, it reaches its maximum during the load run backward; its minimum occurs during unloading. The shape is dependent on the test route involved, as a comparision with another test series has shown. It can be seen that the rating value for the total cycle of 93.7 dB(A) is subject to only minor fluctuations. The limit value, from which the wearing of ear plugs is indispensable, is 90 dB(A). The rating level at the workplace, however, should be less than 85 dB(A) [5], if possible.

Using the climatic variables measured, thermal indices can be specified for the climates as shown in Table 2.

The HSI indicates that a considerable additional stress is placed on the driver by the vehicle waste heat, apart from the uncomfortable mine climate. For the healthy, acclimatized driver, however, a HSI of 100 has to be regarded as a limit value, from which injury to the driver´s health can be expected.

TABLE 1
K-Values and Tolerable Exposition Times of 6 Entire Cycles

K-Value		Tolerable Exposition Time
K_x	25.8	3.0 h
K_y	16.9	6.9 h
K_z	20.5	4.7 h
$K' = \sqrt{K_x^2 + K_y^2 + K_z^2}$	37.0	1.5 h

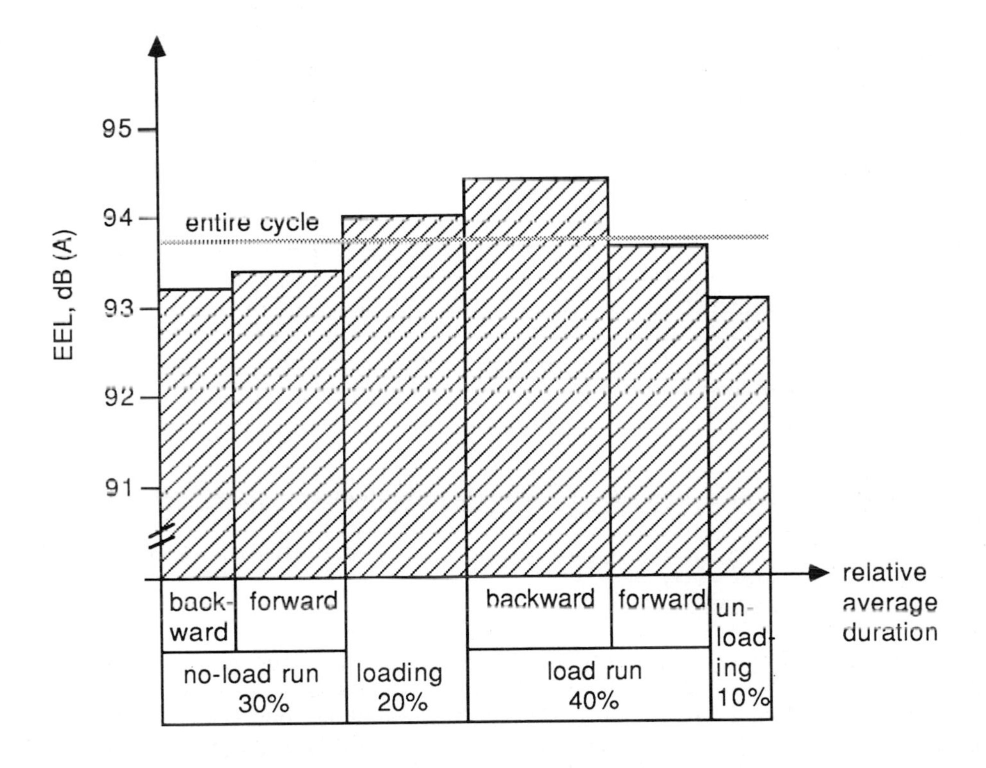

FIGURE 3
Average EELs of the Phases of 4 Production Cycles

POSTURAL STRESS

The musculotendious systems of the neck and the back are strained by the torsion of the vertebral column. The required turning of the head is used as a stress characteristic. For looking in driving direction, this angle is 90°. Actually, smaller angles (approx. 75°) are mostly observed, however, as the driver, in the close

TABLE 2
Thermal Indices

Index	mine climate	micro-climate
Air temperature	32.0°C	37.5°C
HSI	5	55
BET	25.8°C	26.8°C

range, orientates himself at the faces. The limit value is set to 52° [6], the maximal angle of twisting of the cervical spine.

CONCLUSION

Table 3 shows the attempt to integrate the foregoing results in a rating of the initially mentioned measures. From the rating characteristics measured or estimated and from an estimate of the technical success of the reduction measures to be expected, which cannot be discussed in this contribution, the relative stress reduction related to the limit values will be determined. To permit these values to be compared with each other, they will first have to be multiplied by a weighting factor, which is computed as the product of three factors, taking into account :

- the relative exceeding of the limit value,
- the relevance regarding injury to health,
- the impairment of driver´s comfort.

The low value for the swivelling cab must be ascribed to a conservative estimate, due to the lack of secured scientific findings. The high value for vibration-reducing measures can be explained by the high values of the individual weighting factors. All in all, the weighting method is characterized - due to its arithmetic structure - in that it upvalues over-proportionally in the upper range and downvalues overproportionally in the lower range.

Permissible ranking proves that the vibration-reducing measures have a higher priority in comparision to the noise-reducing measures. It remains doubtful, whether vibration-reducing measures should be preferred to the closed cab. Linear rating attempts, which permit the values for noise and climate to be added, however, indicate a greater usefulness of a closed cab. Further research work is required with respect to postural stress, to be able to better assess the usefulness of a swivelling cab.

The results obtained in this study apply to a special type of loader only and to a specific test route. Therefore, a generalization will only be possible under certain reservations.

TABLE 3
Rating of Design Features

1 Feature	swivelling cab	closed cab		vibration reduction system
2 Stress factor	postural stress	climate	noise	whole-body vibr.
3 Stress characteristic	angle of head rot.	HSI	EEL	K-Value
4 Limit value	52°	100	90 dB(A)	18
5 Actual value	75°	55	94 dB(A)	26
6 Attainable value	52°	0	80 dB(A)	18
7 Attainable reduction related to limit value	44%	55%	16%	44%
8 Weighting factor	16	24	90	112
9 Ratings	704	1320	1440	4928
10 Rank	4	3	2	1

REFERENCES :

[1] VDI 2057, Entwurf, Einwirkung mechanischer Schwingungen auf den Menschen (Verein Deutscher Ingenieure (ed.), Beuth-Verlag, Berlin, Köln, 1986).
[2] ISO 2631, Guide for the Evaluation of Human Exposure to Whole-Body Vibration, 1st Edition (International Standard, 1974).
[3] Yaglou, C.P., Temperature, Humidity and Air Movements in Industries : The Effective Temperature Index, J. Industr. Hyg. 9 (1927) 297.
[4] Belding, H.S. and Hatch,T.F., Index for Evaluating Heat Stress in Terms of Resulting Physiological Strains, ASHAE Trans. 62 (1956) 213.
[5] VDI 2058, Blatt 3, Beurteilung von Lärm am Arbeitsplatz unter Berücksichtigung unterschiedlicher Tätigkeiten (Verein Deutscher Ingenieure (ed.), Beuth-Verlag, Berlin, Köln, 1981).
[6] Schmidtke, H. (ed.), Lehrbuch der Ergonomie (Hanser, München, 1981).

Trends in Ergonomics/Human Factors V
F. Aghazadeh (Editor)

FINANCIAL TRADER WORKSTATION

J O H. H O L E N S T E I N
ARCHITECTURES, MINERVASTRASSE 117, 8032 Zurich, Switzerland,
Telephone 01 / 69 17 17

ABSTRACT: Big bang or the international stock exchange as brainwave for new office furniture systems with regards to information systems. Like SOFFEX (Swiss Options and Financial Futures Exchange), SIC (Swiss Interbank Clearing) and SWIFT (Society for worldwide Interbank Financial Telecommunications) there are other information systems in the back office which call for a furniture tool which enables a rapid change of hardware systems and dimensions without having to change the furniture itself. Information and Telecommunication Technology in the trading and financial world can lead to independent workstations in the back or front office, as well as new possibilities for homeworkers.

MOD.INT.DEP.

1. OUTLINE OF THE ISSUE

On the basis of my office furniture system ELAN SCALA (see proceedings of Industrial Ergonomics and Safety Conference 86, Trends in Ergonomics / Human Factors 3: Design for an office consisting of multi adaptable work-table and office environment) which integrates the ergonomic findings for office furnitures, offered the base for the further study of financial trader workstations.

OBJECTIVE

1.1. To design a trader workstation with maximal flexibility in monitoring reception and related equipment.

1.2. To create a modular furniture system with standard elements offering a wide range of combination varieties.

1.3. Open design, no built-in hardware. To include adaption possibility for heat extractors, where needed, in case of absence of air-conditioning.

1.4. To minimize dimensions of the furniture in order to use not more than 5 m2 of nett circulation and storage area (Fig. 1).

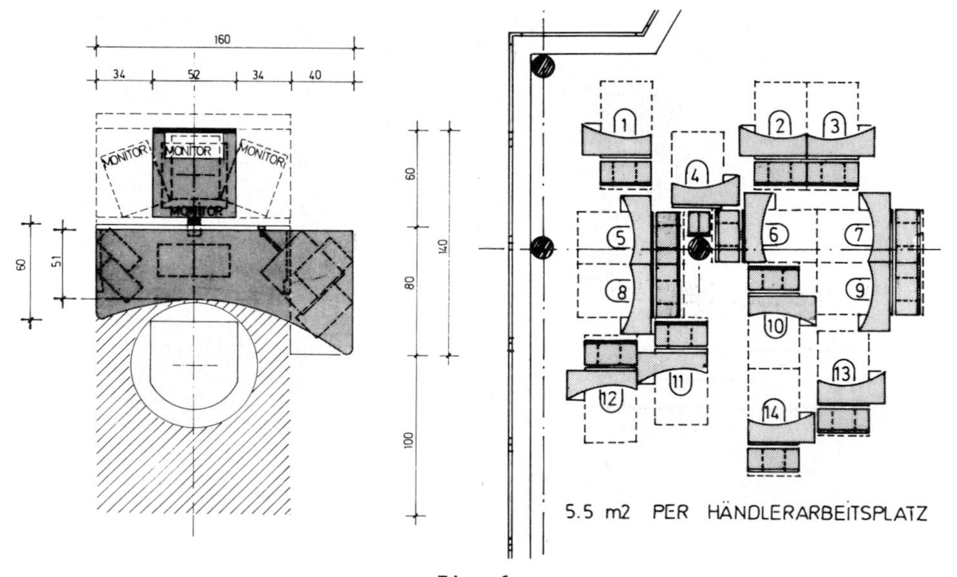

Fig. 1

1.5. Workstations to accommodate either one large 20 inch monitor or three 12 to 14 inch monitors.

1.6. To enable eye-contact between traders, the workstation's highest structure must not exceed 110 mm.

1.7. Traders telephone, intercom systems, keyboard, timing equipment and floppy disc units to be placed loose on the working table.

1.8. To enable complete leg movement (minimum 60 cm).

1.9. Viewing distance between eye to monitor corresponding to size of signs between 50 to 70 cm.

1.10. To facilitate manipulation of all hardware by trader in seated position (Fig. 2).

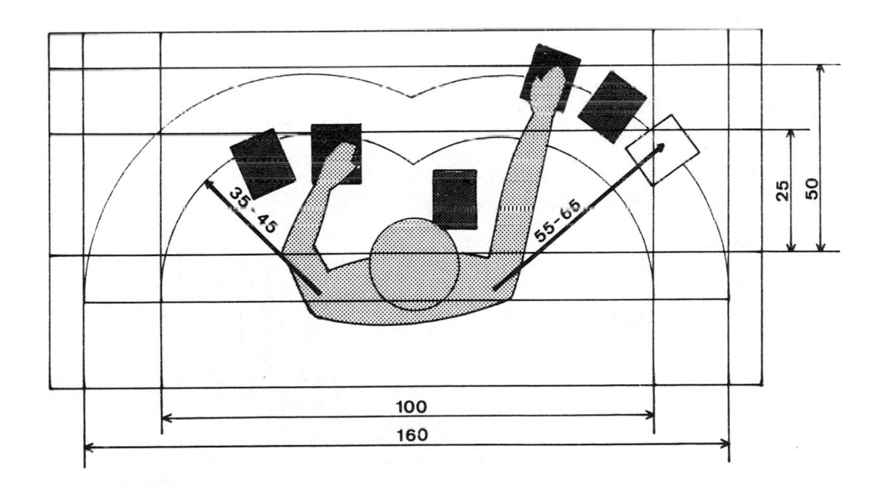

Fig. 2

1.11. To enable technical servicing and cable ducting while workstation is in use.

1.12. To radiate colours of calming effect on traders workstation.

1.13. To use sound absorbing materials where possible.

2. ANALYSIS

The result of the analysis showed that the flexibility of the working unit, which is to say the height and inclination adjustability of ELAN SCALA as the main work unit, should be extended to a further unit which independently offers the same height and inclination features. This also confirmed one of my earlier findings that electronic equipment should be placed off the main working surface (Fig. 3).

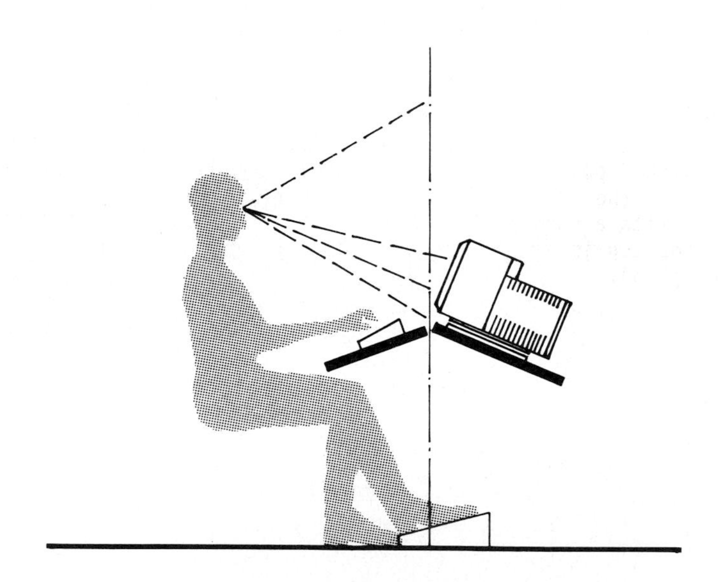

JOH. HOLENSTEIN **ARCHITECTURES**® MINERVASTR. 117 8032 ZÜRICH

Fig. 3

3. REPRESENTATION OF A MODEL

Via the combination of a Work-Modul with a Monitor-Modul, both offering the same but independent flexibility in position variations, the ergonomic findings as represented in Fig. 3 are ideally satisfied. (Fig. 4 - a combination of Work-Modul and Monitor-Modul).

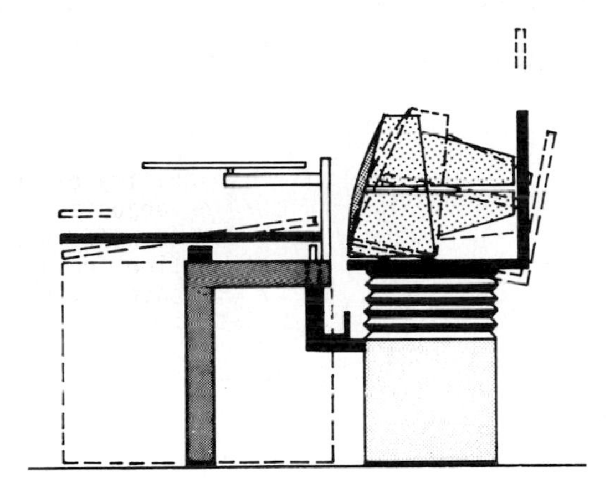

Fig. 4

3.1. THE WORK-MODUL

This modul with 2 points only touching the floor, offers complete leg freedom for the user. This modul is fitted with a top measuring 160/80-60 cm with a curved front, satisfying the ergonomic findings of Fig. 2. The top is inclinable and adjustable in the range of 65 to 80 cm (Fig. 5).

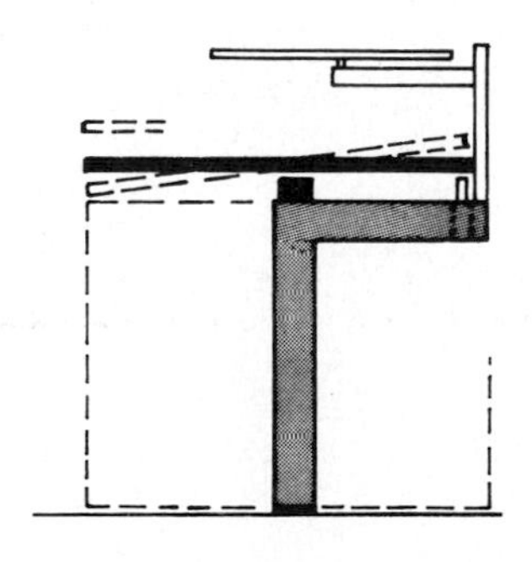

Fig. 5

3.2. THE MONITOR-MODUL

This modul consists of a cable duct base (see finding by same author registered in the proceedings of the Industrial Ergonomics and Safety Conference 1987, Trends in Ergonomics/Human Factors 4: A Meditation to Unconventional VDT Office Workplaces. Page 142, Fig. 4) with height and inclination adjustability, adjusting mechanism covered with a decorative flexible bellow. The monitor support shelve with a protective back comes in various sizes in order to satisfy various needs by different sized equipment. The cable duct base can be opened from all sides independently in order to assure servicing also in compact lay-out positions of work stations (Fig. 6).

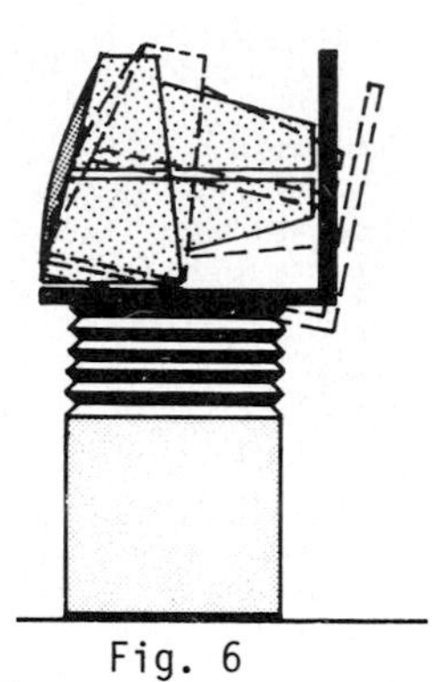

Fig. 6

Fig. 7

This workstation is internationally protected via WIPO, the International Bureau of the World Intellectual Property Organisation and won the competion for the financial traders workstation for the Union Bank of Switzerland UBS.

XVI

APPENDIX

Trends in Ergonomics/Human Factors V
F. Aghazadeh (Editor)

INTERNATIONAL FOUNDATION FOR INDUSTRIAL ERGONOMICS AND SAFETY RESEARCH

C O N S T I T U T I O N

BE IT KNOWN TO ALL THAT WE, THE UNDERSIGNED, HAVE VOLUNTARILY ASSOCIATED OURSELVES TOGETHER FOR THE PURPOSE OF FORMING A CORPORATION UNDER THE GENERAL NON-PROFIT CORPORATION LAW OF THE STATE OF OHIO, AND WE DO HEREBY CERTIFY:

ARTICLE 1 - Name and Incorporation

1. The organization is the International Foundation For Industrial Ergonomics And Safety Research (IFIESR), hereafter referred to as the FOUNDATION.
2. The FOUNDATION is a non-profit organization incorporated in the State of Ohio, United States of America.

ARTICLE 2 - Goals

The FOUNDATION goals are the advancement of industrial ergonomics and safety research by:

1. Organizing Annual International Industrial Ergonomics And Safety Conferences (the FOUNDATION shall be the primary sponsor),
2. Supporting Trends In Ergonomics/Human Factors Volumes,
3. Supporting research and application projects, world wide, either in cooperation with other government or private organizations or on its own initiative,
4. Cooperating with other organizations operating in this field,
5. Encouraging education in the field of industrial ergonomics and safety,
6. Exchanging information with other individuals and organizations,
7. Acting as information resource center in the field of industrial ergonomics and safety.
8. Publishing newsletters or other material to promote communication among members on matters of common interest and importance.
9. Establishing research priorities in the field on the basis of current and future needs.

ARTICLE 3 - Duration

The FOUNDATION is established for an indefinite period (perpetual existence).

ARTICLE 4

1. The FOUNDATION is a non-profit entity and is not authorized to issue shares of stock.
2. The FOUNDATION shall not have or exercise any power or authority, nor shall it directly or indirectly engage in any activity, that would prevent the FOUNDATION from qualifying and continuing to qualify as a corporation described in Section 501(c) (3) of the Internal Revenue Code of 1954, as now in force or afterwards amended, contributions to which are deductible for Federal Income Tax purposes.
3. The FOUNDATION shall be non-political and shall not engage in any transaction defined as prohibited under Section 503 of the Internal Revenue Code of 1954 as now in force or hereafter amended.
4. This FOUNDATION does not contemplate pecuniary gain or profit to the members thereof, and that the funds of the FOUNDATION, whether received by gift or income from membership dues or conferences, and regardless or the source thereof, shall be used exclusively for charitable, educational, and scientific purposes, objectives and activities of the FOUNDATION as the EXECUTIVE COUNCIL, hereafter referred to as the EC, may from time to time define and determine.
5. No compensation or payment shall ever be made to any member except as a reimbursement for actual expenditures made for this FOUNDATION.
6. Upon the termination, dissolution or winding up this FOUNDATION in any manner or for any reason whatsoever, any assets remaining after paying or adequately providing for the debts and obligations of the FOUNDATION shall be donated to recognized educational institutions that have established tax exempt status as defined in Section 501(c) (3) of the Internal Revenue Code of 1954 as now in force or hereafter amended.

ARTICLE 5 - Members

The FOUNDATION has members. The FOUNDATION can admit both individual members and institutional members, such as associations, research institutes, universities, industries, and the like, who/which are considered by the FOUNDATION to be representative in their country in the field of industrial ergonomics and safety. More than one member from each country can be admitted, provided they satisfy the admission criteria. The FOUNDATION decides on the admission of a member.

The FOUNDATION shall not invite individuals or institutions to join the FOUNDATION. Individuals or institutions shall, however, be admitted if they express an interest in the goals of the FOUNDATION and want to become a member.

ARTICLE 6 - Member Responsibilities

The FOUNDATION members shall undertake the responsibility to promote, support, and advocate the goals of the FOUNDATION (ARTICLE 2) in every reasonable and feasible way.

ARTICLE 7 - Termination of Membership

1. Membership can be terminated by a member by giving a written notice to the Chairman or Secretary of the FOUNDATION.
2. The FOUNDATION shall terminate membership if the member is no longer considered supporting the goals of the FOUNDATION or no longer actively represents the field of industrial ergonomics and safety.
3. Membership shall also be terminated by February 15th of the year if the Treasurer does not receive the annual membership dues for the year by January 15th of that year.

ARTICLE 8 - Membership Dues

1. All individual members shall be charged membership dues annually. The minimum annual contribution shall be determined by the FOUNDATION.
2. All institutional members shall be charged membership dues annually. The minimum annual contribution shall be determined by the FOUNDATION.
3. Institutional members, as well as individual members, can, and are encouraged to, contribute more than the minimum amount set by the FOUNDATION.

ARTICLE 9 - FOUNDATION Authority

1. The FOUNDATION members shall meet at least once every year. The regular meeting shall be held during the Annual International Industrial Ergonomics And Safety Conference in the country where the conference is held. The EC of the FOUNDATION can call additional FOUNDATION meetings at appropriate times and locations. The FOUNDATION members shall be convoked in writing.
2. The FOUNDATION can also make decisions without meeting, under the condition that relative proposals have been sent to all FOUNDATION members, that all members have given their vote in writing, and that no member has objected to this mode of decision making.
3. The FOUNDATION is entitled to enter into agreements that help it meet its goals. The Chairman and Secretary legally and otherwise, represent the FOUNDATION in such matters. The general principles of the agreement are approved by the FOUNDATION members.

4. Each member of the FOUNDATION shall have one

vote. The decision to admit an individual or an institution and the like shall be taken by simple majority or the registered members.

ARTICLE 10 - Executive Council (EC)

1. The FOUNDATION shall have an EC, headed by its Chairman. The EC shall consist of the following:
   a. The Founder and the five founding members,
   b. The General Editor (Founder) of the Trends In Ergonomics/Human Factors Volumes,
   c. Two persons assigned by institutional members representing all institutional members,
   d. Chairman, Secretary, and Treasurer of the FOUNDATION,
   e. Immediate past, present, and immediate future Chairmen of Annual International Industrial Ergonomics And Safety Conferences (these members will assume responsibility at the end of the regular annual FOUNDATION meeting; each member shall enjoy a three year term).
2. Each EC member shall have one vote. Decisions shall be taken by simple majority of the members present.
3. EC members shall keep office in accordance with this and section 5 of this Article. That does not apply to members who come under Section 1c and 1e of this article. Institutional representatives (para 1c) shall enjoy two year terms at the end of which institutional members shall ask the FOUNDATION to either reappoint them or furnish the FOUNDATION with the names of new representatives. The term of these representatives shall be from the end of a regular FOUNDATION meeting until the term end FOUNDATION meeting.
4. EC members described in Section 1a and 1b of this article shall be permanent members and shall not be removed without the individual's approval.
5. The FOUNDATION shall elect from its members, by simple majority, a Chairman, a Secretary, and a Treasurer, and such other persons the FOUNDATION thinks suitable. The term of these elected officials shall be three years. No officer shall serve two consecutive terms. He/she may be reelected to office after three years.
6. The Chairman and the Secretary are authorized to represent the FOUNDATION by law and otherwise.
7. The EC shall meet when so desired by a majority of its members.
8. The EC shall authorize expenditures only according to a written budget, approved by the FOUNDATION. Budgets must be approved every year at the regular FOUNDATION meeting. The Treasurer shall present the annual budget and the balance sheet to the members at the regular FOUNDATION meeting, in writing. Expenses not anticipated, but necessary, shall be approved by the FOUNDATION members either by a mail vote or at

other scheduled meetings. The statement of need and amount shall be prepared by the Treasurer.

ARTICLE 11 - Conference Locations

During each annual conference, the FOUNDATION members shall review the plans proposed by the EC and decide on the locations of future conferences. Locations of at least the next five conferences shall be finalized. The EC shall be entrusted with the task of appointing Conference Chairs. Any proposals submitted to the FOUNDATION, with the intent of hosting a future conference, shall also be reviewed at this time.

ARTICLE 12

The FOUNDATION shall neither engage in organizing exhibits at its annual conferences nor shall it permit the conference chairman to do so. Individual exhibitors shall be permitted to exhibit provided they independently make all the necessary arrangements and sign a contract with the conference chairman relieving him and the FOUNDATION of all liabilities, financial, legal, or other. The exhibitors shall pay the FOUNDATION a nominal fee (to be determined by the EC) for the privilege of exhibiting during the duration of the conference.

ARTICLE 13 - Conference Chair Authority

1. The Conference Chair shall formulate the Program and Organizing Committees and get the submissions reviewed.
2. He/she shall serve as Editor of Trends In Ergonomics/Human Factors Volume for that year and work in close association with the General Editor of the Trends Series.
3. The Conference Chair shall determine the registration fee.

ARTICLE 14 - Finances

The financial support of the FOUNDATION shall consist of:

1. The membership dues described in Article 7,
2. Contributions from conferences (to be determined by the FOUNDATION)
3. Any legal acquisitions by the FOUNDATION, such as legacies, gifts, and interest income from investment of reserves.

ARTICLE 15

The FOUNDATION can decide to modify this constitution or to dissolve itself. Seventy-five percent of the registered votes are required to do so. Article 10(4) of this constitution, however, shall neither be modified nor changed.

ARTICLE 16 - Proxy

In all FOUNDATION meetings, but not in those of EC, a member can replace himself/herself by a written proxy.

ARTICLE 17

The FOUNDATION, from time to time, shall honor individuals who have done their utmost for its development or for the advancement of industrial ergonomics and safety.

ARTICLE 18 - Bylaws

The FOUNDATION is entitled to lay down Bylaws with a view to the carrying into effect this constitution, and can modify the Bylaws. The Bylaws may not be in contravention of the constitution.

Finally, the founders declared that for the first time the FOUNDATION consists of the founder and the five founding members; this FOUNDATION shall expand in accordance with this constitution.

This instrument was executed in the month of March of the year Nineteen Hundred and Eighty-Six.

SIGNATURES:

Anil Mital (Founder & Chairman)

Biman Das (Founding Member & Secretary)

Waldemar Karwowski (Founding Member & Treasurer)

Shihab S. Asfour (Founding Member & Conference Chair)

Fereydoun Aghazadeh (Founding Member & Immediate Future Conference Chairman)

Bernard C. Jiang (Founding Member)

XVII

AUTHOR INDEX

AUTHOR INDEX

Abdel-Moty, E., 513, 959
Adams, Jr., C.C., 267
Aghazadeh, F., vii, 143, 249, 357, 549
Ahonen, E., 935
Aird, J.W., 705
Akin, A.D., 949
Alexander, D.C., 471
Allen, J.S., 297
Antin, J.F., 241
Archer, F.W., 699
Asfour, S.S., 959, 977
Asoudegi, E., 811
Ayoub, M.M., 837, 879
Azimullah, R., 249

Banaag, J., 871
Banhidi, L., 485
Behbehani, K., 219
Bennett, T., 969
Berg, V.J., 191
Besing, S.E., 969
Bishop, P.A., 433
Bishu, R.R., 417
Bittner, Jr., A.C., 15, 21, 529
Bobick, T.G., 521, 541, 927
Bullinger, H.J., 201
Burkhardt, A., 865
Butani, S.J., 741
Byun, S.N., 847

Cacha, C.A., 919
Carter, N., 567
Chaffin, D.B., 401, 847
Cheatwood, S.D., 479
Chen, H.C., 879
Chen, J.-G., 3, 101
Chlebicka, E., 351
Choi, S.Y., 63
Chung, J.H., 63
Cihangirli, M., 83
Clarkson, H., 227
Clay, D.J., 191
Cochran, D.J., 417
Colligan, M.J., 969
Collins, C.L., 949
Combs, R.B., 357
Congleton, J.J., 1045, 1053
Constable, S.H., 433
Cronin, S.A., 297
Czaja, S.J., 185

Dahalan, J., 759
Damos, D.L., 21
Dangelmaier, M., 1069
Das, B., 127
Dawood, R., 613
Deal, D., 101
Degreve, T.B., 71
Deivanayagam, S., 827
Dingus, T.A., 241
Doolittle, T.L., 767
Doxie, F.T., 1009
Dubrovsky, V., 135
Duchon, J.C., 151
Dutta, S.P., 1027

Eckert, R., 201
El-Nawawi, M.A., 407
Endrusick, T.L., 441, 449
Etherton, J.R., 629, 639
Eyada, O.K., 71

Fabó, L., 485
Fathallah, F.A., 191
Fernandez, J.E., 71, 83, 395, 759, 837, 941, 949, 999
Fey, G.E., 115
Field, W.E., 675
Freivalds, A., 161, 457

Gage, H., 911
Gallagher, S., 521, 779, 927
Garg, A., 871
Garza, J.R., 433
Genaidy, A.M., 613, 887, 985
George, L.J., 373
Gilad, I., 847
Goldberg, J.H., 161, 457
Gonzalez, A.J., 45
Gressel, M.G., 333
Grobelny, J., 77
Gros, E., 505
Guignard, J.C., 495, 529

Hacisalihzade, S.S., 297
Häkkinen, K., 573
Hakola, T., 935
Harter, Jr., B.T., 699
Harter, K.C., 699
Heitbrink, W.A., 333
Herranen, S., 211
Higginbotham, V.L., 191
Hobday, S.W., 321
Holenstein, J., 1077
Hommertzheim, D.L., 83
Hull, R.L., 621
Hulse, M.C., 241

Imrhan, S.N., 219, 235, 683, 775
Imrhan, V., 775

Jaraiedi, M., 811
Jensen, P.A., 333
Jensen, R.C., 639
Jeyakumar, V., 101

Kaiyala, K., 767
Karnes, E.W., 647, 669
Karwowski, W., 283, 803, 865
Kengskool, K., 365
Kerk, C.J., 1061
Khalil, T.M., 513, 959, 977
Kim, C.-H., 395
Kim, P.K.H., 689
Klein, C.M., 379
Klen, T., 935
Klym, M.P., 1037
Kondraske, G.V., 219
Krantz, J.H., 275
Krieg, E.F., 969
Kroemer, K.H.E., 313
Krupa, D., 647
Kuciemba, S.R., 1061
Kumar, S., 227
Kwon, Y.G., 345

Laurig, W., 53
Lazor, A., 821
Lee, C.H., 45
Lee, K.B., 63
Lee, K.D., 63
Lee, K.S., 143, 305, 549
Lee, S.D., 63
Leonard, S.D., 647, 669
Lorenz, D., 1069
Louhevaara, V., 895, 935
Lyth, D.M., 985

Madigan, E., 647
Malzahn, D.E., 395, 759

Marek, T., 283
Marley, R.J., 71, 759
Marras, W.S., 857
Martinez, S.E., 365
Mattila, M., 559
Mc Atee, Jr., F.L., 1023
McCright, P.R., 177
McGlothlin, J.D., 333
McMahan, P.B., 787
Menckel, E., 567
Miller, M.D., 257
Mital, A., 373, 425, 613, 887, 903
Miyao, M., 297
Morrissey, S.J., 15, 291
Moshref, S.B., 407
Muntzinger, W.F., 201, 1069

Nanthavanij, S., 911
Nestor, D., 729
Nielsen, R., 449
Notbohm, G., 505
Noworol, C., 283
Nunneley, S.A., 433
Nyran, P., 705

Obeidat, M., 887
Oh, S.H., 63
Ostberg, O., 305
Owen, B.D., 721

Page, G.B., 1061
Paluch, R., 821
Peacock, J.B., 3
Pedigo, W., 647
Pesonen, J., 573
Pieczonka-Osikowska, W., 283
Pin, R., 775
Pitetti, K.H., 941, 999
Pizarro, D.C., 999
Plummer, R.W., 713
Przetacznik, J., 283
Pulat, B.M., 37
Puppala, S., 887
Putz-Anderson, V., 601

Ramsey, J.D., 345
Rauko, M., 211
Redfern, M., 847
Redfern, M.S., 401
Riley, M.W., 417
Roberts, G., 705
Robinson-Edwin, L.J., 227
Rombach, V., 53
Roncini, G., 613
Rosomoff, H.L., 959, 977
Rosomoff, R.S., 959, 977

Sabuncuoglu, I., 83
Santee, W.R., 441
Schlegel, R.E., 3, 479
Schmidt, B., 479
Schneider, T., 647, 669
Schulze, L.J.H., 1045, 1053
Seppälä, A., 657
Sharit, J., 185
Shikdar, A.A., 127
Shuman, D., 169
Simons, G.R., 993
Smith, J.L., 345
Smith, L.A., 471
Smolander, J., 935
Stafford, J.A., 941, 999
Stark, L.W., 297
Stobbe, T.J., 713, 811
Stoeffler, W.K., 985
Stubbs, N.B., 941, 999
Szerdahelyi, J., 485

Tack, D.W., 751
Tayyari, F., 465
Teräslinna, P., 895
Tiitta, P., 663
Turner, M.L., 93

Unger, R.L., 521
Usher, J.M., 249

Valdes, J.C., 365
Väyrynen, S., 573
Ventura, J.A., 379
Virgilio, J.F., 949
Vuori, M., 211

Wagner, J.A., 591
Waikar, A.M., 143, 305, 549
Waly, S.M., 977
Warg, L.-E., 109
Way, T.R., 257
Weaver, T., 827
Webb, R.D.G., 751
Wells, L., 441
Wherry, Jr., R.J., 29
White, K.D., 169, 275
Wick, J.L., 1017
Wierwille, W.W., 241
Wilkinson, T.L., 675
Witbeck, P.C., 993
Woodcock Webb, K., 387, 583
Woods, C.B., 275
Wright, M.E., 257

Yates, J.W., 803

Zeinelabidien, A.A., 407
Zwahlen, H.T., 267